# MONOGRAPHS ON THE PHYSICS AND CHEMISTRY OF MATERIALS

# MONOGRAPHS ON THE PHYSICS AND CHEMISTRY OF MATERIALS

# Interfaces in Crystalline Materials

A. P. SUTTON

University Lecturer, Department of Materials, University of Oxford
and Fellow of Linacre College, Oxford

and

R. W. BALLUFFI

Professor, Department of Materials Science and Engineering,
Massachusetts Institute of Technology

CLARENDON PRESS · OXFORD

Oxford University Press, Great Clarendon Street, Oxford OX2 6DP
Oxford New York
Athens Auckland Bangkok Bogota Bombay Buenos Aires
Calcutta Cape Town Dar es Salaam Delhi Florence Hong Kong
Istanbul Karachi Kuala Lumpur Madras Madrid Melbourne
Mexico City Nairobi Paris Singapore Taipei Tokyo Toronto
and associated companies in
Berlin Ibadan

Oxford is a trade mark of Oxford University Press

Published in the United States by
Oxford University Press Inc., New York

First published 1995
First published in this binding (with corrections) 1996

A catalogue record for this book is available from the British Library

Library of Congress Cataloging in Publication Data
Sutton, A. P.
Interfaces in crystalline materials / A. P. Sutton and R. W.
Balluffi.
(Monographs on the physics and chemistry of
materials ; 51)
Includes bibliographical references.
1. Crystalline interfaces. 2. Dislocations in crystals.
3. Surface chemistry. I. Balluffi, R. W. II. Title. III. Series.
QC173.458.C78S88 1995 548'.8—dc20 94–24731 CIP
ISBN 0 19 850061 0

Printed in Great Britain by
Bookcraft (Bath) Ltd
Midsomer Norton, Avon

## To Pat and Ruth

*With love, and also admiration for seeing the difficulties in writing this book more clearly than we.*

# Preface

It is well recognized that many of the properties of polycrystalline materials are determined to a large extent by the properties of their internal interfaces. In fact, it has been commonly remarked that the study of the behaviour of polycrystalline materials is often reduced to the study of the behaviour of their interfaces. Since the properties of interfaces between crystals must inevitably depend upon their structure, the study of the structure and properties of interfaces has emerged as a central area in the larger field of the science and engineering of materials.

The study of interfaces in crystalline materials has undergone explosive growth during the last few decades due to: (1) the development of new high-resolution techniques for studying their atomic structures and behaviour and (2) increased modelling capability resulting from better theory and the rapid rise of powerful computer simulation methods. Major publications documenting progress in the field have consisted of the proceedings of frequently held international conferences and symposia and also edited collections of papers by different authors devoted to various aspects of interfaces. However, no single integrated text, spanning the field and written in a pedagogical manner, has yet appeared.

The present book is designed to fill this gap. In writing it we have attempted to develop the subject systematically and to concentrate on basic ideas and principles. Because of space limitations, no effort has been made to include detailed descriptions of experimental techniques. Many of these are covered in other texts. Attention is given to all major types of interfaces (e.g. homophase and heterophase, diffuse and sharp) between the major types of materials (e.g. metals, ceramics, semiconductors) to the greatest extent possible within the covers of a single book of reasonable length. Part I is concerned with structure and begins with the basic geometry and crystallography of interfaces (Chapter 1) and then proceeds to dislocation models (Chapter 2). The subject of interatomic forces is then taken up (Chapter 3) in order to set the stage for the description of the actual detailed atomic structures of interfaces (Chapter 4). Part II is concerned with the thermodynamics of interfaces. The basic equilibrium thermodynamic formalism is first laid down (Chapter 5) and serves as a basis for discussion of interfacial phases and phase transitions in Chapter 6. Finally, the equilibrium segregation of solute atoms to interfaces is treated in Chapter 7. Part III is concerned with basic kinetic phenomena. These include atomic diffusion at interfaces (Chapter 8) and both the conservative and non-conservative motion of interfaces (Chapters 9 and 10, respectively). Non-conservative motion includes the important notion of interfaces as sources or sinks for diffusional fluxes of atoms. Part IV is devoted to properties and focuses on the electrical and mechanical properties of interfaces (Chapters 11 and 12, respectively).

A systematic nomenclature is developed for describing the interfaces and interfacial quantities which are involved, and the terms which we have employed are summarized in a separate glossary. Wherever possible, we have used terms consistent with those used in the closely related field of surface science. Also, a set of standard symbols is adopted which we adhere to as closely as practicable (see List of symbols).

Despite the ambitious scope of the book it has proven impossible to be all-inclusive. The field is enormous and is still in a rapid state of development. Moreover, the study of many of its aspects is still in its infancy. In response to this we have added a considerable amount of new and unpublished material. In some particularly difficult areas we make the best of it and present simple, but plausible, models and discussion. Regrettably, several important topics have been omitted because of both limited space and sheer exhaustion on our part! Examples of such topics are corrosion and heterogeneous precipitation at interfaces, recrystallization, grain growth and texture development, and interfacial fatigue. We also decided to omit interfaces involving non-crystalline solids for similar reasons. Nevertheless, it is hoped that most of the fundamentals have been covered, and that the book will provide a springboard for the study of any omitted topics and further research into all areas of the science of interfaces in crystalline materials.

With respect to units, we have tried to be consistent by adopting SI units throughout the book. However, at the atomic scale we find the electron-Volt (eV) and the Angstrom (Å) more convenient and natural units of energy and length respectively. The reader may find it useful to note that $1\,eV/atom \equiv 96.4866\,kJ/mole$, and $1\,kJ/mole \equiv 0.0104\,eV/atom$.

We are indebted to many friends and colleagues for useful advice and helpful critiques. These include S. M. Allen, A. Argon, A. Atkinson, H. K. Birnbaum, P. D. Bristowe, J. W. Cahn, Y.-M. Chiang, J. W. Christian, M. Cohen, L. C. DeJonghe, M. W. Finnis, J. P. Hirth, S. Hofmann, R. Kirchheim, C. J. McMahon, Jr., G. B. Olson, D. G. Pettifor, R. C. Pond, D. N. Seidman, L. S. Shvindlerman, J. Tersoff, and R. Thomson.

We are also grateful to our departments for being sympathetic to our requests for time to write this book.

*Oxford*                                                                                   A. P. S.
*Cambridge, Massachusetts*                                                 R. W. B.
July 1994

# Acknowledgements

We are grateful to the following authors for permisssion to reproduce figures or tables from their papers or books: Prof. R. C. Pond (Figs 1.10, 1.12, 1.13,1.14, 1.16, 1.18, 1.21, 1.23, and 9.24b), Prof. J. W. Christian (Fig. 2.13), Prof. J. P. Hirth (Fig. 2.19), Prof. S. L. Sass (Fig. 2.20), Dr D. Wolf (Figs 2.25, 4.9, 4.22, 5.21a, 5.21b, and 5.22), Prof. P. D. Bristowe (Figs 2.31 and 9.13), Prof. A. H. King (Figs 2.36 and 10.8), Prof. J. Hafner (Fig. 3.4), Prof. D. G. Pettifor (Fig. 3.6), Dr M. J. Puska (Fig. 3.7), Dr D. J. Srolovitz (Figs 3.25 and 3.26), Dr M. W. Finnis (Figs 3.38, 3.40, 3.41, 4.40, and 4.41), Prof. S. M. Allen (Fig. 4.4a), Prof. M. Cohen (Figs 4.4b and 12.48b), Dr G. R. Barsch (Figs 4.5 and 4.6), Dr W. Krakow (Fig. 4.28), Prof. D. A. Smith (Figs 4.43, 6.5b, and 9.25), Dr P. W. Tasker (Figs 4.46 and 4.50), Dr. K. L. Merkle (Figs 4.47, 4.48, 4.51, 5.8, and 11.25), Dr D. Chatain (Fig. 4.54), AEA Technology Ltd (Fig. 4.55), Prof. W. Mader (Fig. 4.56), Prof. J. C. H. Spence (Figs 4.57 and 4.58), Dr D. R. Hamann (Fig. 4.59), Dr C. Herring (Fig. 5.6a), Dr A. Otsuki (Figs 5.20a and 5.23a), Dr T. Mori (Fig. 5.20b), Dr A. Revcolevschi, Dr G. Dhalenne, and Dr M. Dechamps (Fig. 5.23b), Prof. C. Rottman (Fig. 6.2), Dr W. Gust (Figs 6.9, 8.5, 8.7, 8.28, and 8.29), Dr M. P. Seah (Figs 7.1, 7.2, and 7.3), Prof. D. N. Seidman (Figs 7.4, 7.5, 7.6, 7.10, 7.18, and 7.19), Prof. C. J. McMahon, Jr. (Fig. 7.7), Prof. D. E. Luzzi (Fig. 7.8), Dr S. Hofmann (Figs 7.9 and 7.11), Dr C. L. Briant (Fig. 7.12), Prof. M. Doyama (Fig. 7.13), Dr R. Najafabadi (Figs 7.16, 7.17, and 12.47), Dr K. L. Kliewer (Figs 7.20 and 7.21), Prof W. Gust and Dr I. Kaur (Figs 8.5 and 8.7), Dr M. P. Seah (Fig. 8.7), Dr. A. Atkinson (Figs 8.9 and 8.23a), Dr J. Sommer (Fig. 8.11), Prof. M. Biscondi (Fig. 8.12), Prof. V. S. Stubicon (Figs 8.13 and 8.23b), Prof. D. R. Kirchheim (Fig. 8.24), Prof. L. C. De Jonghe (Fig. 8.26), Prof J. Washburn (Fig. 9.2), Prof. A. W. Sleeswyk (Fig. 9.5), Dr H. Fukutomi (Fig. 9.6), Dr U. Dahmen (Figs 9.7 and 9.9), Prof. I-Wei Chen (Figs 9.8 and 12.56), Dr Y. Ishida (Fig. 9.14), Dr J. W. Cahn (Figs 9.21a, 9.21b, and 9.31), Prof. K. T. Aust (Figs 9.21c, 9.21d, 9.28a, 9.28b, and 9.28c), Prof. L. S. Shvindlerman (Figs 9.22 and 9.29), Dr H. Gleiter (Fig. 9.24a), Prof. C. L. Bauer (Fig. 9.26a–f), Dr C. Gastaldi (Fig. 9.26g), Prof. K. Lücke (Figs 9.27 and 9.28d), Dr E. Nes (Fig. 9.34), Prof. W. W. Mullins (Fig. 9.37), Prof. R. D. Doherty (Figs 10.10a and 10.21), Prof. G. C. Weatherly (Fig. 10.10b), Prof. J. M. Howe (Fig. 10.11), Prof. G. W. Greenwood (Fig. 10.14), Prof. C. Elbaum (Fig. 10.17), Prof. R. Raj (Figs 10.18a, 10.18b, 10.18c, and 10.18d), Dr. R. F. Sekerka (Figs 10.19 and 10.20), Prof. S. G. Louie (Figs 11.3, 11.4, 11.5, and 11.6), Prof. W. Monch (Fig. 11.7), Prof. P. N. First (Fig. 11.8), Prof. R. H. Williams (Fig. 11.9), Dr J. Werner (Fig. 11.10), Dr J. Tersoff (Fig. 11.13), Dr I. Nakamichi (Figs 11.14, 11.15, and 11.16), Dr J. Y. W. Seto (Fig. 11.19), Prof. D. G. Ast (Fig. 11.20), Dr P. Chaudhari (Fig. 11.22), Dr J. M. Phillips (Fig. 11.24), Dr F. W. Young Jr. (Fig. 12.5), Dr J. Thibault (Fig. 12.9), Dr. W. Bollmann (Fig. 12.10), Prof H. Gleiter (Fig. 12.12), Dr C. A. P. Horton (Fig. 12.15), Dr. H. K. Birnbaum (Figs 12.17, 12.21, and 12.46), Prof M. J. Whelan (Fig. 12.18), Dr E. Smith (Fig. 12.19), Prof. L. E. Murr (Fig. 12.20), Prof. M. A. Myers (Fig. 12.24), Dr J. H. Schneibel (Fig. 12.25a), Prof. N. J. Grant (Fig. 12.25b), Prof. M. F. Ashby (Figs 12.27, 12.34, 12.38, 12.39, 12.53, 12.54, and 12.55), Dr W. Beere (Figs 12.29, 12.33, 12.35, and 12.36), Dr T. Mori (Fig. 12.30), Prof. R. Valiev (Fig. 12.31), Prof. T. Watanabe (Figs 12.32 and 12.37), Dr R. Thomson (Figs 12.40,

12.41, 12.42, and 12.43), Prof. J. F. Knott (Fig. 12.45), Dr J. J. Gilman (Fig. 12.48a), Prof. W. D. Nix (Fig. 12.57), Prof. A. G. Evans (Figs 12.59 and 12.60).

Figs 1.10, 1.12, and 1.13 are reprinted with permission from The Royal Society from Pond, R. C. and Bollman, W., The symmetry and interfacial structure of bicrystals, *Phil. Trans. R. Soc.*, **A292**, (1979) 449–72, figs 1 and 2. Fig. 1.14 is reprinted with permission from The Royal Society from Pond, R. C. and Vlachavas, D. S., Bicrystallography, *Proc. R. Soc* **A386** (1983), 95–143, fig 1. Fig. 1.16 is reproduced with permission from Elsevier Science Publishers BV, North Holland Imprint from Pond R. C. *et al.* (1984) *Mats. Res. Soc. Symp. Proc.* **25**, 273. Fig. 1.18 is reproduced with permission from the Institute of Materials (formerly the Institute of Metals) from Pond, R. C. (1985) in *Dislocations and properties of real materials* (ed. M. Lorretto), p. 71. Fig. 1.21 is reproduced with permission from Elsevier Science Publishers BV, North Holland Imprint from Pond, R. C. (1989) *Dislocations in Solids* **8** (ed. F. R. N. Nabarro). Fig. 1.23 is reproduced with permission from the Institute of Materials (formerly the Institute of Metals) from Pond, R. C. (1985) in *Dislocations and properties of real materials* (ed. M. Lorretto), p. 71. Figs 2.1, 2.3, and 2.4 are reproduced with permission from McGraw Hill Inc, N.Y. from Read W. T. (1953) *Dislocations in Crystals*. Fig. 2.10(a–e) is reproduced from *Acta Metall.* **30**, Balluffi R. W., Brokman, A., and King, A. H., CSL/DSC lattice model for general crystal–crystal boundaries and their line defects, 1453 ©(1982) with kind permission from Elsevier Science Ltd, The Boulevard, Langford Lane, Kidlington OX5 1GB. Fig. 2.19 is reproduced with permission from John Wiley & Sons, Inc from Hirth, J. P. and Lothe, J. (1982) *Theory of Dislocations*, © J. P. Hirth (1982). Fig. 2.20 is reproduced from Shieu. F.-S. and Sass. S. L.© (1990), Experimental and theoretical studies of the dislocation structure of NiO-Pt interfaces, *Acta Metall. Mater* **38**, 1653 with kind permission from Elsevier Science Ltd, The Boulevard, Langford Lane, Kidlington OX5 1GB. Fig. 2.25 is reproduced from Wolf, D. ©(1989), A Read–Shockley model for high-angle grain boundaries, *Scripta Metall.* **23**, 1713, with kind permission from Elsevier Science Ltd, The Boulevard, Langford Lane, Kidlington OX5 1GB. Fig. 2.31 is reproduced with permission fron The Japan Institute of Metals from Bristowe P. D. (1986) in *Grain Boundary Structure and Related Phenomena*, p89, Japan Institute of Metals, Sendai (Suppl. to *Trans. Jap. Inst. of Metals* **27**). Fig. 2.32 is reproduced with permission from Taylor and Francis from Schober, T. and Balluffi, R. W. (1970) **21**,109 *Phil. Mag.* Fig. 2.33 is reproduced with permission from Taylor and Francis from Babcock, S. E. and Balluffi, R. W. (1987) **55**, 643 *Phil Mag A*. Figs 2.34 and 2.35 are reproduced with permission of Taylor and Francis from Kvam E. P. and Balluffi R. W. (1987) **56**, 137, *Phil. Mag. A*. Fig. 2.36 is reproduced with permission from Taylor and Francis from Shin K. and King A. H. (1991) *Phil Mag A*. **63**,1023. Fig. 2.37 is reproduced from Hsieh, T. E. and Balluffi R. W. ©(1989) *Acta Metall.* **37**, Experimental study of grain boundary melting in aluminum, 1637, with kind permission from Elsevier Science Ltd, The Boulevard, Langford Lane, Kidlington OX5 1GB. Fig. 3.4 is reproduced with permission from Elsevier Science Publishers BV, North Holland imprint from Hafner, J. (1989) in *The Structures of Binary Compounds* (ed F. R. de Boer and D. G. Pettifor), p147. Fig. 3.6 is reproduced from Pettifor, D. G. and Ward, M. S. (1984) *Solid State Commun.* **49**, 291, An analytic pair potential for simple metals, with kind permission from Elsevier Science Ltd, The Boulevard, Langford Lane, Kidlington OX5 1GB. Fig. 3.7 is reproduced with permission from Puska M. J. (1990) in *Many-Atom Interactions in Solids* (ed R. M. Nieminen, M. J. Puska and M. J. Manninen) **48**, 134, Atoms embedded in electron gas, (Fig. 4), ©Springer-Verlag GmBH & Co. Figs 3.25 and 3.26 are reproduced from Lesar, R., Najafabadi, R. and Srolovitz, D. J. (1989), Finite-temperature defect properties from free-energy minimization, *Phys. Rev. Letters* **63**, 624, ©(1989) The American Physical Society. Figs 3.38 and 3.40 are reproduced from Finnis M. W. *Acta Metall. Mater.* **40**, Metal–ceramic cohesion and the image interaction, S25, ©(1992) with kind permission from Elsevier Science Ltd, The Boulevard, Langford Lane, Kidlington OX5 1GB. Fig. 3.41 is reproduced from M. W. Finnis *Surf. Sci.* **241**, 61, The interaction of a point charge with an aluminium (111) surface, ©(1991) with kind permission from Elsevier Science Ltd, The Boulevard, Langford Lane, Kidlington OX5 1GB. Fig. 4.4a is reproduced from Allen, J. M. and Cahn, J. W. *Acta Metall* **27**, 1085, A microscopic theory for antiphase boundary motion and its application to antiphase domain coarsening, ©(1979) with kind permission from Elsevier Science Ltd, The Boulevard, Langford Lane, Kidlington OX5 1GB. Fig. 4.4b is reproduced with permission of Les Editions de Physique from Olson, G. B. and Cohen M. (1982) *J. Physique* **43**, C4-75. Fig. 4.5 is reproduced from Barsch, G. R. and Krumhansl, J. A. (1984), Twin boundaries in ferroelastic media without interface dislocations, *Phys. Rev. Letters* **53**, 1069 ©(1989) The American Physical Society. Fig. 4.6 is reproduced from Cao, W. and Barsch, G. R. (1990), Landau–Ginzburg model of interphase boundaries in improper ferroelastic perovskites of $D_{4h}^{18}$ symmetry, *Phys. Rev. B* **41**, 4334 ©(1990) The American Physical Society. Fig. 4.9 is reproduced with permission from the Materials Research Society from Merkle, K. L. and Wolf, D. (1990) *MRS Bulletin* **15** No. 9, p. 42. Fig. 4.22 is reproduced with permission of Chapman and Hall, London from Wolf, D. and Merkle, K. L. (1992) in *Materials Interfaces* (ed D. Wolf and S. Yip), p87, Chapman and Hall, London. Fig. 4.27 is reproduced with permission from ASM International from Vitek, V., Sutton, A. P., Smith, D. A., and Pond, R. C. (1980) in *Grain boundary structure and kinetics* (ed. R. W. Balluffi), p. 115. Fig. 4.28 is reproduced with permission from Taylor and Francis from Krakow W. (1991) *Phil Mag A* **63**, 233. Figs 4.29, 4.30, 4.31, 4.32, 4.33, 4.34, 4.35 are reproduced with permission from The Royal Society from Sutton A. P. and Vitek, V., *Phil. Trans. R. Soc.*, On the structure of tilt grain boundaries in cubic metals I. Symmetrical tilt boundaries, **A309**, 1–35. Fig. 4.39 is reproduced with permission from Taylor and Francis from Schwartz, D., Vitek, V. and Sutton, A. P. (1985) *Phil. Mag A* **51**, 399. Fig. 4.40b is reproduced with permission from IOP Publishing Ltd from Finnis M. W. (1993) *Physics World* **6**, No. 7, 37. Fig. 4.41 is reproduced with permission from Taylor and Francis from Wolf, U., Ernst, F., Muschik, T., Finnis, M. W. and Fischmeister, H. F. (1992) *Phil. Mag. A* **66**, 991. Fig. 4.43 is reproduced with permission of The Japan Institute of Metals from Krakow, W. and Smith, D. A. (1986) In *Proc. JIMIS-4*, (Suppl to *Japan. Inst. Mets.* **27**) p. 277. Fig. 4.46 is reproduced with permission from Taylor and Francis from Duffy D. M. and Tasker P. W. (1983) *Phil. Mag. A* **47**, 817. Figs 4.47 and 4.48 are reproduced with permission from Merkle, K. L. and Smith, D. J. *Phys. Rev. Lett.*, Atomic structure of symmetric tilt grain boundaries in NiO, **59**, 2887 ©(1987) The American Physical Society. Fig. 4.50 is reproduced with permission from The American Institute of Physics from Duffy, D. M. and Tasker, P. W. (1984) *J. Appl. Phys.* **56**, 971. Fig. 4.51 is reproduced with permission of The Materials Research Society from Gao, Y. and Merkle, K. L. (1990) *J. Mater. Res.* **5**, 1995. Fig. 4.54 is reproduced by permission of the American Ceramic Society from Chatain, D., Chabert, F., Ghetta, V. and Fouletier, J. (1993) *J. Amer. Ceram. Soc.*, **76**, 1568, New experimental setup for wettability characterization under monitored oxygen activity. I. Role of oxygen-state and defect characterization on oxide wettability by gold, ©(1993). Fig. 4.55 is reproduced with permission of Kluwer Academic Publishers from Nicholas, M. G. (1989) in *Surfaces and Interfaces of Ceramic Materials* (ed. L. C. Dufour *et al.*) *NATO ASI SerE: Applied Sciences* **173**, 393, ©AEA Technology. Fig. 4.56 is reproduced from Mader, W. and Knauss, D. *Acta Metall. Mater.* **40**, S207, Equilibrium position of misfit dislocations at planar interfaces, ©(1992) with kind permission from Elsevier Science Ltd, The Boulevard, Langford Lane, Kidlington OX5 1GB. Figs 4.57 and 4.58 are reproduced with permission from Taylor and Francis from Cherns, D., Anstis, G. R., Hutchison, J. L. and Spence, J. C. H. (1982) *Phil. Mag. A.* **46**, 849. Fig. 4.59 is reproduced with permission from Hamann, D. R. *Phys. Rev. Lett.*, New silicide interface model from structural energy considerations, **60**, 313 ©(1988) The American Physical Society. Fig. 5.6a is reproduced with permission from Herring, C. *Phys. Rev.* **82**, 87, Some theorems on the free energies of crystal surfaces, ©(1951) The American Physical Society. Fig. 5.6b is reproduced with permission from The Materials Research Society from Omar, R. and Mykura, H. (1988) *Mats. Res. Soc. Symp. Proc.* **122**, 61. Fig. 5.8 is reproduced with permission from The Materials Research Society from Merkle, K. L. and Shao, B. (1988) *Mats. Res. Soc. Symp. Proc.* **122**, 69. Figs 5.11b and 5.11c are reproduced from Hsieh, T. E. and Balluffi, R. W. *Acta Metall.* **37**, 2133, Observations of roughening/de-faceting phase transitions in grain boundaries, ©(1989) with kind permission from Elsevier Science Ltd, The Boulevard, Langford Lane, Kidlington OX5 1GB. Figs 5.11d and 5.11e are reproduced from Ference, T. G. and Balluffi, R. W. *Scripta. Metall.* **22**, 1929, Observation of a reversible grain boundary faceting transition induced by changes in composition, ©(1988) with kind permission from Elsevier Science Ltd, The Boulevard, Langford Lane, Kidlington OX5 1GB. Figs 5.20a and 5.23a are reproduced with permission of The Japan Institute of Metals from Otsuki, A. and Mizuno, M. (1986) *Proc. JIMS* **4**, 789. Fig. 5.20b is reproduced with permission from Taylor and Francis from Mori, T., Miura, H., Tokita, T., Haji, J. and Kato, M. (1988) *Phil. Mag. Lett.* **58**, 11. Figs 5.21a and 5.21b are reproduced with permission from The Materials Research Society from Wolf, D. (1990) *J. Mater. Res.* **5**, 1708. Fig. 5.22 is reproduced from Wolf, D., *Acta Metall. Mater.* **38**, 791, Structure-energy correlation for grain boundaries in f.c.c. metals IV. Asymmetrical twist (general) boundaries, © (1990) with kind permission from Elsevier Science Ltd., The Boulevard, Langford Lane, Kidlington OX5 1GB. Fig. 5.23b is reproduced with permission of The American Ceramic Society from Dhalenne, G., Dechamps, M. and Revcolevschi, A. *Advances in Ceramics* **6** (ed M. Yan and A. Heuer), p139, fig 3, Energy of tilt boundaries and mass transport mechanisms in nickel oxide, ©(1983). Fig. 6.2 is reproduced with permission from Elsevier Science Publishers BV, North Holland Imprint from Rottman, C. and Wortis, M. (1984) *Phys. Reports* **103**, 59. Fig. 6.5a is reproduced from Goodhew, P. J., Tan, T. Y. and Balluffi, R. W. *Acta Metall.* **26**, 557, Low energy planes for tilt grain boundaries in gold, ©(1978) with kind permission from Elsevier Science Ltd, The Boulevard, Langford Lane, Kidlington OX5 1GB. Fig. 6.5b is reproduced with permission from the Japan Institute of Metals from Krakow, W. and Smith, D. A. (1986) in *Grain boundary structure and related phenomena*, Proceeedings of JIMIS 4, Supplement to translation, p. 277. Fig. 6.9 is reproduced from Straumal, B., Muschik, T., Gust, W. and Predel, B. *Acta Metall. Mater.* **40**, 939, The wetting transition in high and low energy grain boundaries in the Cu(In) system, ©(1992) with kind permission from Elsevier Science Ltd, The Boulevard, Langford Lane, Kidlington OX5 1GB. Figs 7.1, 7.2, and 7.3 are reproduced with permission from Elsevier Science Publishers BV, North Holland Imprint from Hondros, E. D. and Seah, M. P. (1983) in *Physical Metallurgy* (ed R. W. Cahn and P. Hassen) p. 855. Figs 7.4 and 7.5 are reproduced with permission of Les Editions de Physique from Seidman, D. N. *et al.* (1990) *J. de Phys. Colloque C1* **51** C1-147. Fig. 7.6, 7.18, and 7.19 are reproduced from Seki, A., Seidman, D. N., Oh, Y. and Foiles, S. M. *Acta Metall. Mater.* **39**, 3167, Monte Carlo simulations of segregation at [001] twist boundaries in a Pt(Au) alloy – I. Results, ©(1991) with kind permission from Elsevier Science Ltd, The

Boulevard, Langford Lane, Kidlington OX5 1GB. Fig. 7.7 is reproduced from Ogura, T., McMahon, C. J., Feng, H. C. and Vitek, V. *Acta Metall.* **26**, 1317, Structure-dependent intergranular segregation of phosphorus in austenite in a Ni–Cr steel, ©(1978) with kind permission from Elsevier Science Ltd, The Boulevard, Langford Lane, Kidlington OX5 1GB. Fig. 7.8 is reproduced with permission from Yan, M., Sob, M., Luzzi, D. E., Vitek, V., Ackland, G. J., Methfessel, M. and Rodriguez, C. O. (1993) *Phys. Rev. B* **47**, 5571 ©(1993) The American Physical Society. Fig. 7.9 is reproduced from Hofmann, S. and Lejcek, P. (1991) *Scripta. Metall. Mater.* **25**, 2259, Correlation between segregation enthalpy, solid solubility and interplanar spacing of $\Sigma$ = tilt grain boundaries in $\alpha$-iron, ©(1991) with kind permission from Elsevier Science Ltd, The Boulevard, Langford Lane, Kidlington OX5 1GB. Fig. 7.10 is reproduced with permission from Elsevier Science Publishers BV, Academic Publishing Division from Seidman, D. N. (1991) *Mat. Sci. and Eng. A* **137**, 57. Fig. 7.11 is reproduced with permission from La Societe Française de Chimie from Hoffman, S. (1987) *J. de Chimie Physique* **84**, 141. Fig. 7.12 is reproduced with permission from The Materials Research Society from Briant, C. L. (1990) *Mats. Res. Soc. Bulletin* **15**, No. 10, 26. Fig. 7.13 is reproduced from Hashimoto, M., Ishida, Y., Yamamoto, R. and Doyama, M. (1984) *Acta Metall.* **32**, 1, Atomistic studies of grain boundary segregation in Fe–P and Fe–B alloys – I. Atomic structure and stress distrubution, ©(1984) with kind permission from Elsevier Science Ltd, The Boulevard, Langford Lane, Kidlington OX5 1GB. Figs 7.16 and 7.17 are reproduced from Najafabadi, R., Wang, H. Y., Srolovitz, D. J. and LeSar R. (1991) *Acta Metall. Mater.* **39**, 3071, A new method for the simulation of alloys: application to interfacial segregation, ©(1991) with kind permission from Elsevier Science Ltd, The Boulevard, Langford Lane, Kidlington OX5 1GB. Figs 7.20 and 7.21 are reproduced with permission from Kliewer, K. L. and Koehler, J. S. (1965) *Phys. Rev. A* **140**, 1226 ©(1993) The American Physical Society. Fig. 8.2 is reproduced from Cahn, J. W. and Balluffi, R. W. (1979) *Scripta Metall.*, **13**, 499, On diffusional mass transport in polycrystals containing stationary or migrating grain boundaries, ©1979 with kind permission from Elsevier Science Ltd, The Boulevard, Langford Lane, Kidlington OX5 1GB. Fig. 8.6 is reproduced from Ma, Q. and Balluffi, R. W. (1993) *Acta Metall. Mater.* **41**,133, Diffusion along [001] tilt boundaries in the Au/Ag system – I. Experimental results, ©(1993) with kind permission from Elsevier Science Ltd, The Boulevard, Langford Lane, Kidlington OX5 1GB. Fig. 8.7 is reproduced with permission of The Royal Society from Bernardini *et al.* (1982), *Proc. R. Soc.* **A379**, 159, (figs 9 and 10). Fig. 8.9 is reproduced by permission of Kluwer Academic Publishers from Atkinson, A. (1989) in Surfaces and Interfaces of Ceramic Mats. (ed L.-C. Dufour, C. Monty and G. Petot-Ervas) p.273. Fig. 8.10 is reproduced from Balluffi, R. W. and Brokman, A. (1983) *Scripta. Metall.* **17**, 1027, Simple structural unit model for core-dependent properties of symmetrical tilt boundaries, ©(1983) with kind permission from Elsevier Science Ltd, The Boulevard, Langford Lane, Kidlington OX5 1GB. Fig. 8.11 is reproduced with permission of Scietec Publications Ltd. from Sommer, J., Herzig, C., Mayer, S. and Gust W. (1989) *Defect and Diff. Forum* **66–69**, 843. Fig. 8.12 is reproduced from Herbeuval, I. and Biscondi, M. (1974) *Canadian Metall. Quaterly* **13**, 171, Diffusion du zinc dans les joints de flexion syms triques de l'aluminium, ©(1974) with kind permission from Elsevier Science Ltd, The Boulevard, Langford Lane, Kidlington OX5 1GB. Figs 8.13 and 8.23b are reproduced with permission of Plenum Publishing Corporation, Plenum Press imprint from Stubicon, V. S. (1985), in *Transport in non-stoichiometric compounds* (ed G. Simkovich and V.S. Stubicon) p345. Figs 8.14 and 8.15 are reproduced with permission of the American Institue of Physics from Brokman, A., Bristowe, P. D. and Balluffi, R. W. (1981) *J. Appl. Phys.* **52**, 6116. Figs 8.16 and 8.17

are reproduced with permission of Academic Press Inc. from Balluffi, R. W. (1984) in *Diffusion in Crystalline Solids* (ed G. E. Murch and A. S. Nowick) p320. Fig. 8.23a is reproduced with permission of Les Editions de Physique from Atkinson, A. (1985) *J. Physique* **46** Colloque C4–379. Fig. 8.24 is reproduced from Mütschele, T. and Kirchheim, R. (1987) *Scripta. Metall.* **21**, 135, Segregation and diffusion of hydrogen in grain boundaries of palladium, ©(1983) with kind permission from Elsevier Science Ltd, The Boulevard, Langford Lane, Kidlington OX5 1GB. Fig. 8.26 is reproduced with permission from Chapman and Hall Ltd. from De Jonghe, L. C. (1979) *J. Mats. Sci.* **14**, 33 (Fig. 14), Fig. 8.27 is reproduced from Balluffi, R. W. and Cahn, J. W. (1981) *Acta Metall.* **29**, 493, Mechanism for diffusion induced grain bundary motion, ©(1981) with kind permission from Elsevier Science Ltd, The Boulevard, Langford Lane, Kidlington OX5 1GB. Figs 8.28 and 8.29 are reproduced from Schmelzle, R., Giakupian, B., Muschik, T., Gust, W. and Fournelle, R. A. (1992) *Acta Metall. Mater.* **40**, 997, Diffusion induced grain boundary migration of symmetric and asymmetric ⟨011⟩{011} tilt boundaries during the diffusion of Zn into Cu, ©(1992) with kind permission from Elsevier Science Ltd, The Boulevard, Langford Lane, Kidlington OX5 1GB. Fig. 9.2 is reproduced from Li, C. H., Edwards, E. H., Washburn,, J. and Parker, E. R. (1953) *Acta Metall.*, **1**, 223, Stress-induced movement of crystal boundaries, ©(1953) with kind permission from Elsevier Science Ltd, The Boulevard, Langford Lane, Kidlington OX5 1GB. Fig. 9.5 is reproduced with permission from Taylor and Francis from Sleeswyk, A. W. (1974) *Phil. Mag.* **29**, 407. Fig. 9.6 is reproduced with permission from Dr Y. Ishida from Horiuchi, R., Fuktomi, H., and Takahashi, T. (1987) in *Fundamentals of diffusion bonding* (ed. Y. Ishida), p. 347. Fig. 9.7 is reproduced from Dahmen, U. (1987) *Scripta. Metall.* **21**, 1029, Surface relief and the mechanism of a phase transformation, ©(1987) with kind permission from Elsevier Science Ltd, The Boulevard, Langford Lane, Kidlington OX5 1GB. Fig. 9.8 is reproduced from Chiao, Y.-H. and Chen, I.-W. (1990) *Acta Metall. Mater.* **38**, 1163, Martensitic growth in $ZrO_2$ – an *in situ*, small particle, TEM study of a single interface transformation, ©(1990) with kind permission from Elsevier Science Ltd, The Boulevard, Langford Lane, Kidlington OX5 1GB. Fig. 9.9 is reproduced from Dahmen, U. (1987) *Scripta. Metall.* **21**, 1029, Surface relief and the mechanism of a phase transformation, ©(1987) with kind permission from Elsevier Science Ltd, The Boulevard, Langford Lane, Kidlington OX5 1GB. Fig. 9.13 is reproduced from Jhan, R.-J. and Bristowe P. D. (1990) *Scripta. Metall. Mater.* **24**, 1313, A molecular dynamics study of grain boundary migration without the participation of secondary grain boundary dislocations, ©(1990) with kind permission from Elsevier Science Ltd, The Boulevard, Langford Lane, Kidlington OX5 1GB. Fig. 9.14 is reproduced with permission of Les Editions de Physique from Ichinose, H., and Ishida, Y. (1990) *J. de Phys.* **51**, Colloque, C1–185. Fig. 9.15 is reproduced from Babcock, S. E. and Balluffi, R. W. (1989) *Acta Metall. Mater.* **37**, 2367, Grain boundary kinetics – II. *In situ* observations of the role of grain boundary dislocations in high-angle boundary migration, ©(1989) with kind permission from Elsevier Science Ltd, The Boulevard, Langford Lane, Kidlington OX5 1GB. Figs 9.21a and 9.21b are reproduced from Cahn, J. W. (1962) *Acta Metall.* **10**, 789, The impurity-drag effect in grain boundary motion, ©(1962) with kind permission from Elsevier Science Ltd, The Boulevard, Langford Lane, Kidlington OX5 1GB. Figs 9.21c and 9.21d are reproduced with permision of Academic Press Inc. from Simpson, C. J., Winegard, W. C. and Aust, K. T. (1976) in *Grain Boundary Structure and Properties* (ed G. A. Chadwick and D. A. Smith) p201. Fig. 9.22 is reproduced with permission from Zeitschrift für Metallkunde from Fridman, E. M., Kopezsky, C. V. and Shvindlerman, L. S. (1975) *Z. Metallkd.* **66**, 533. Fig. 9.24a is reproduced from Gleiter, H. (1969) *Acta Metall.* **17**, 565, The mechanism of grain boundary migration, (1969) with kind permission from Elsevier Science Ltd, The Boulevard, Langford Lane, Kidlington OX5 1GB. Fig. 9.24b is reproduced from Dingley, D. J. and Pond, R. C. (1979) *Acta Metall.* **27**, 667, On the interaction of crystal dislocations with grain boundaries, ©(1979) with kind permission from Elsevier Science Ltd, The Boulevard, Langford Lane, Kidlington OX5 1GB. Fig. 9.25 is reproduced with permission from ASM International from Smith, D. A., Rae, C. M. F., and Grovenor, C. R. M. (1980) in Grain boundary structure and kinetics (ed. R. W. Balluffi), p. 35. Figs 9.26a–f are reproduced with permission of the Materials Research Society from Bauer, C. L., Gastaldi, J., Jourdan, C. and Grange, G. (1988) *Mats. Res. Soc. Symp. Proc.* **122** (ed. M. H. Yoo, W. A. T. Clark and C. L. Briant) p199. Fig. 9.26g is reproduced with permission of Les Editions de Physique from Gastaldi, C., Jourdan, C. and Grange, G. (1990) *J. de Phys.* **51**, Colloque C1–405. Fig. 9.27 is reproduced from Lücke, K. (1974) *Canad. Metall. Quart.* **13**, 261, The orientation dependence of grain boundary motion and the formation of recrystallization textures, ©(1974) with kind permission from Elsevier Science Ltd, The Boulevard, Langford Lane, Kidlington OX5 1GB. Fig. 9.28a is reproduced from Rutter, J. W. and Aust, K. T. (1965) *Acta Metall.* **13**, 181, Migration of ⟨100⟩ tilt grain boundaries in high purity lead, ©(1965) with kind permission from Elsevier Science Ltd, The Boulevard, Langford Lane, Kidlington OX5 1GB. Figs 9.28b and 9.28c are reproduced with permission of Academic Press Inc. from Simpson, C. J., Winegard, W. C. and Aust, K. T. (1976) in *Grain Boundary Structure and Properties* (ed G. A. Chadwick and D. A. Smith) p.201. Fig. 9.28d is reproduced with permission of Plenum Publishing Corporation, Plenum Press imprint from Lücke, K., Rixen, R. and Rosenbaum, F. W. (1972) in *The Nature and Behaviour of Grain Boundaries* (ed H. Hu) p. 245. Fig. 9.29 is reproduced with permission of the American Institute of Physics from Molodov, D. A., Kopetskii, C. V. and Shvindlerman, L. S. (1981) *Sov. Phys. Solid State* **23**, 1718. Fig. 9.31 is reproduced with permission of the Institue of Physics Publishing, Ltd. from Cahn, J. W., Taylor, J. E. and Handwerker, C. A. (1991) in *Sir Charles Frank, an 80th Birthday Tribute* (ed R. G. Chambers, J. E. Enderby, A. Keller, A. R. Lang and J. W. Steeds) p. 88. Fig. 9.34 is reproduced from Nes, E., Ryum, N. and Hunderi, O. (1985) *Acta Metall.* **33**, 11, On the Zener drag, ©(1985) with kind permission from Elsevier Science Ltd, The Boulevard, Langford Lane, Kidlington OX5 1GB. Fig. 9.37 is reproduced from Mullins, W. W. (1958) *Acta Metall.* **6**, 414, The effect of thermal grooving on grain boundary motion, ©(1958) with kind permission from Elsevier Science Ltd, The Boulevard, Langford Lane, Kidlington OX5 1GB. Fig. 10.6 is reproduced from Siegel, R. W., Chang, S. M. and Balluffi, R. W. (1980) *Acta Metall.* **28**, 249, Vacancy loss at grain boundaries in quenched polycrystalline gold, ©(1980) with kind permission from Elsevier Science Ltd, The Boulevard, Langford Lane, Kidlington OX5 1GB. Fig. 10.7 is reproduced with permission from Taylor and Francis from Komen, Y., Petroff, P. and Balluffi, R. W. (1972) *Phil. Mag A* **26**, 239. Fig. 10.8 reproduced with permission from Taylor and Francis from King, A. H. and Smith, D. A. (1980) *Phil. Mag. A.*, **42**, 595. Fig. 10.9 is reproduced from Balluffi, R. W., Brokman, A. and King, A. H. (1982) *Acta Metall.* **30**, 1453, CSL/DSL lattice model for general crystal–crystal boundaries and their line defects, ©(1989) with kind permission from Elsevier Science Ltd, The Boulevard, Langford Lane, Kidlington OX5 1GB. Fig. 10.10a is reproduced from Rajab, K. E. and Doherty, R. D. (1989) Acta Metall. **37**, 2709, Kinetics of growth and coarsening of faceted hexagonal precipitates in an F.C.C. matrix – I. Experimental observations, ©(1989) with kind permission from Elsevier Science Ltd, The Boulevard, Langford Lane, Kidlington OX5 1GB. Fig. 10.10b is reproduced from Weatherly,

G. C. (1971) *Acta Metall.* **19**, 181, The structures of ledges at plate-shaped precipitates, ©(1971) with kind permission from Elsevier Science Ltd, The Boulevard, Langford Lane, Kidlington OX5 1GB. Fig. 10.16 is reproduced with permission of Metallurgical and Materials Transactions from Bastin, G. F. and Rieck, G. D. (1974) *Met. Trans.* **5**, 1817. Fig. 10.17 is reproduced from Basu, B. K. and Elbaum, C. (1965) *Acta Metall.* **13**, 1117, Surface vacancy pits and vacancy diffusion in aluminum, (1965) with kind permission from Elsevier Science Ltd, The Boulevard, Langford Lane, Kidlington OX5 1GB. Figs 10.18a and 10.18c are reproduced with permission of Metallurgical and Materials Transactions from Raj, R. and Ashby, M. F. (1971) *Met. Trans.* **2**, 1113. Figs 10.18b and 10.18d are reproduced with permission of Metallurgical and Materials Transactions from Raj, R. and Ashby, M. F. (1972) *Met. Trans.* **3**, 1937. Fig. 10.19 is from Langer, J. S. and Sekerka, R. F. (1975) *Acta Metall.* **23**, 1225, Theory of departure from local equilibrium at the interface of a two-phase diffusion couple, ©(1975) with kind permission from Elsevier Science Ltd, The Boulevard, Langford Lane, Kidlington OX5 1GB. Fig. 10.20 is reproduced with permission of the American Institue of Physics from Mullins, W. W. and Sekerka, R. F. (1963) *J. Appl. Phys.* **34**, 324. Fig. 10.21 is reproduced with permission of Keter Publishing House Ltd. from Bainbridge, B. G. and Doherty, R. D. (1969) in *Quantitative Relation Between Properties and Microstructure* (ed D. G. Brandon and A. Rosen) p. 427. Fig. 11.3 is reproduced with permission from Louie, S. G., Chelikowsky, J. and Cohen, M. L. (1977) *Phys. Rev.* **B15**, 2154 ©(1977) the American Physical Society. Figs 11.4, 11.5, and 11.6 are reproduced with permission from Louie, S. G. and Cohen, M. L. (1976) *Phys. Rev.*, **B13**, 2461. Fig. 11.7 is reproduced with permission of IOP Publishing Ltd. from Mönch, W. (1990) *Rep. Prog. Phys.* **53**, 221, On the physics of metal–semiconductor interfaces. Fig. 11.10 is reproduced with the permission of the American Institute of Physics from Werner, J. H. and Güttler, H. H. (1991) *J. Appl. Phys.* **69**, 1522. Fig. 11.13 is reproduced with permission from Elsevier Science Publications BV, North Holland Imprint from Tersoff, J. (1987) in *Heterojunction Band Discontinuities : Physics and Device Applications* (ed. F. Capasso and G. Margaritondo) p. 3. Figs 11.14, 11.15, and 11.16 are reproduced from the Publishing Committee of the Journal of Science of Hiroshima Univesity from Nakamichi, I. (1990) *J. Sci. Hiroshima Univ. Ser A* **54**, 49. Fig. 11.19 is reproduced with the permission of the American Institute of Physics from Seto, J. Y. W. (1975) *J. Appl. Phys.* **46**, 5247. Fig. 11.22 is reproduced with permission from Dimos, D., Chaudhari, P. and Mannhart, J. (1990) *Phys. Rev.* **B41**, 4038 ©(1990) the American Physical Society. Fig. 11.24 is reproduced with permission from Eom, C. B., Marshall, A. F., Suzuki, Y., Geballe, T. H., Boyer, B., Pease, R. F. W., van Dover, R. B. and Phillips, J. M. (1992) *Phys. Rev.* **B46**, 11902 ©(1990) the American Physical Society. Fig. 11.25 is reproduced with permission from Elsevier Science Publications BV, North Holland Imprint from Gao, Y., Bai, G., Lam, D. J. and Merkle, K. L. (1991), *Physica. C* **173**, 487. Fig. 12.3 is reproduced with permission from Elsevier Science Publications BV, North Holland Imprint from Balluffi, R. W., Komem, Y. and Schober, T. (1972) *Surf. Sci.* **31**, 68. Fig. 12.5 is reproduced with the permission of the American Institute of Physics from Young, F. W. (1958) *J. Appl. Phys.* **29**, 760. Fig. 12.6 is reproduced with permission of Taylor and Francis from Sun, C. P. and Balluffi, R. W. (1982) *Phil. Mag. A.* **46**, 63. Fig. 12.9 is reproduced with permission of Taylor and Francis from Elkajbaji, M. and Thibault-Desseaux, J. (1988) *Phil. Mag. A*, **58**, 325. Fig. 12.10 is reproduced with permision of Akademie Verlag Gmblt from Bollmann, W., Michaut, B. and Sainfort, G. (1972) *Phys. Stat. Sol. (a)*, **13**, 637. Fig. 12.11 is reproduced with permission of Taylor & Francis from Sun, C. P. and Balluffi, R. W. (1982) *Phil. Mag. A*, **46**, 49. Fig. 12.12

is reproduced with permission of Taylor and Francis from Pumphrey, P. H. and Gleiter, H. (1974) *Phil. Mag.*, **30**, 593. Fig. 12.15 is reproduced courtesy of National Power PLC and with the permission of Taylor & Francis from Kegg, G. R., Horton, C. A. P. and Silcock, J. M. (1973), *Phil. Mag.*, **27**, 1041. Figs 12.17 and 12.21 are reproduced with the permission of Metallurgical Transactions from Lee, T. C., Robertson, I. M. and Birnbaum, H. K. (1990) *Met. Trans.*, **21A**, 2437. Fig. 12.18 is reproduced with the permission of The Royal Society from Whelan, M. J., Hirsch, P. B. and Horne R. W. (1957) *Proc. R. Soc. London A* **240**, 524, (figs 13 and 19). Fig. 12.19 is reproduced from Worthington, P. J. and Smith, E. (1964) *Acta Metall.* **12**, 1277, The formation of slip bands in polycrystalline 3% silicon iron in the pre-yield microstrain region, ©(1964) with kind permission from Elsevier Science Ltd, The Boulevard, Langford Lane, Kidlington OX5 1GB. Fig. 12.20 is reproduced with the permission of Metallurgical Transactions from Murr, L. E. (1975) *Met. Trans.* **6A**, 505. Fig. 12.23 is reproduced from Hauser, J. J. and Chalmers, B. (1961) *Acta Metall* **9**, 802. Fig. 12.24 is reproduced with the permission of Taylor and Francis from Myers, M. A. and Ashworth, E. (1982) *Phil. Mag. A*, **46**, 737. Fig. 12.25a is reproduced with permission of the Japan Institute of Metals from Schneibel, J. H. and Peterson, G. F. (1986) JIMIS 4, *Grain Boundary Structure and Related Phenomena, Suppl. to Trans. Jap. Inst. Mets.* **27**, 859. Fig. 12.25b is reproduced from Mullendore, A. W. and Grant, N. J. (1963) *Trans. A.I.M.E.* **227**, 319 with the permission from Transactions of the Metallurgical Society, a publication of The Metals and Materials Society, Warrendale , Pennsylvania 15086. Fig. 12.27 is reproduced with permission from Elsevier Science Publications BV, North Holland Imprint from Ashby, M. F. *Surf. Sci.* **31**, 498. Figs 12.29, 12.33, 12.35, and 12.36 are reproduced with the permission of The Royal Society from Beere, W. (1978) *Phil. Trans. R. Soc. Lond.* **A288**, 177 (figs 5, 10, 15,16, 20, and 21). Fig. 12.30 is reproduced with the permission of Taylor and Francis from Kato, M. and Mori, T. (1993) *Phil. Mag.* **A68**, 939. Fig. 12.31 is reproduced with permision of Akademie Verlag Gmblt from Valiev, R. Z., Kaibyshev, O. A., Astanin, V. V. and Emaletdinov, A. K. (1983) *Phys. Stat. Sol. (a)* **78**, 439. Fig. 12.32 is reproduced with the permission of Taylor and Francis from Kokawa, H., Watanabe, T. and Karashima, S. (1981) *Phil. Mag.*, **A44**, 1239. Fig. 12.34 is reproduced from Ashby, M. F. (1972) *Acta Metall.* **20**, 887, A first report on deformation-mechanism maps, ©(1972) with kind permission from Elsevier Science Ltd, The Boulevard, Langford Lane, Kidlington OX5 1GB. Fig. 12.37 is reproduced with permission from Trans. Tech. Publi. Ltd. from Watanbe, T. (1989) *Mats. Sci. Forum* **46**, 25. Fig. 12.38 is reproduced from Ashby, M. F., Gandhi, C. and Taplin, D. M. R. (1979) *Acta Metall.* **27**, 699, Fracture-mechanism maps and their construction for F.C.C. metals and alloys, ©(1979) with kind permission from Elsevier Science Ltd, The Boulevard, Langford Lane, Kidlington OX5 1GB. Fig. 12.39 is reproduced from Gandhi, C. and Ashby, M. F. (1979) *Acta Metall.* **27**, 1565, Fracture-mechanism maps for materials which cleave: F.C.C. and B.C.C. and H.C.P. metals and ceramics, ©(1979) with kind permission from Elsevier Science Ltd, The Boulevard, Langford Lane, Kidlington OX5 1GB. Figs 12.40, 12.41, 12.42, and 12.43 are reproduced with permission from Elsevier Science Publications BV, North Holland Imprint from Thomson, R. M. (1983) in *Physical Metallurgy* (ed R. W. Cahn and P. Haasen) p. 1487. Fig. 12.45 is reproduced with permission from Butterworth Heinemann Ltd. from Knott, J. F. (1973) *Fundamentals of Fracture Mechanics*. Fig. 12.46 is reproduced from Lee, T. C. , Robertson, I. M., and Birnbaum, H. K. (1989) *Acta Metall.* **37**, 407, An HVEM *in situ* deformation study of nickel doped with sulfur, ©(1989) with kind permission from Elsevier Science Ltd, The Boulevard, Langford Lane, Kidlington OX5 1GB. Fig. 12.47 is reproduced with permission from Chapman and Hall,

London from Srolovitz, D. J., Yang, W. H., Najafabadi, R., Wang, H. Y. and Lesar, R. (1992) in *Materials Interfaces* (ed D. Wolf and S. Yip) p691. Fig. 12.48a is reproduced from Gilman, J. J. (1958) *Trans. A.I.M.E.* **212**, 783 with the permission from Transactions of the Metallurgical Society, a publication of The Metals and Materials Society, Warrendale , Pennsylvania 15086. Fig. 12.48b is reproduced with permission from MIT Press from Hahn, G. T., Averbach, B. L., Owen, W. S. and Cohen, M. (1959) in *Fracture* (ed B. L. Averbach, D. K. Felbeck, G. T. Hahn and D. L. Thomas) p. 91. Figs 12.52, 12.53, 12.54 and 12.55 are reproduced from Cocks, A. C. F. and Ashby, M. F. (1982) *Prog. in Mats. Sci.* **27**, 189, On creep fracture by void growth, ©(1982) with kind permission from Elsevier Science Ltd, The Boulevard, Langford Lane, Kidlington OX5 1GB. Fig. 12.56 is reproduced with permission from Metallurgical and Materials Transactions from Chen, I-W. (1983) *Met. Trans.* **14A**, 2289. Fig. 12.57 is reproduced from Nix, W. D., Matlock D. K. and Dimelfi, R. J. (1977) *Acta Metall.* **25**, 495, A model for creep fracture based on the plastic growth of cavities at the tips of grain boundary wedge cracks, ©(1977) with kind permission from Elsevier Science Ltd, The Boulevard, Langford Lane, Kidlington OX5 1GB. Fig. 12.58 is reproduced with permission from the Institute of Materials from Soderberg, R. (1975) *Met. Sci.* **9**, 275. Fig. 12.59 is reproduced with permission of Elsevier Science Publishers B.V., North Holland imprint from Evans, A. G., Rùhle, M., Dalgleish, B. J. and Charalambides, P. G. (1990) *Mats. Sci. Eng.* **A126**, 53. Fig. 12.60 is reproduced with permission from Evans, A. G. and Dalgleish, B. J. (1992) *Acta Metall. Mater.* **40**, S295, The fracture resistance of metal–ceramic interfaces, ©(1992) with kind permission from Elsevier Science Ltd, The Boulevard, Langford Lane, Kidlington OX5 1GB.

# Contents

## PART III   INTERFACIAL KINETICS

# Symbols

The following common symbols are used throughout to the greatest extent possible. Exceptions to this usage are indicated locally in the text.

| | |
|---|---|
| $A$ | area |
| $a$ | basis vector of a lattice |
| $B$ | Net Burgers vector of interfacial dislocations intersecting arbitrary line in interface |
| $b$ | Burgers vector of a dislocation |
| $C$ | number of components in thermodynamic system |
| $c$ | time-averaged site occupancy |
| $D$ | diffusivity |
| $d$ | grain size in a polycrystal |
| $d_{\mathrm{F}}$ | degrees of freedom of a thermodynamic system |
| $\delta$ | effective thickness of an interface core |
| $E; e$ | internal energy; internal energy per unit volume |
| $\mathbf{E}$ | identity matrix |
| $\varepsilon_{ij}$ | elastic strain |
| $F$ | Helmholtz free energy |
| $f$ | correlation factor for diffusion |
| $\phi$ | number of phases in thermodynamic system |
| $f_{ij}$ | interface stress |
| $G; g$ | Gibbs free energy; Gibbs free energy per unit volume |
| $H$ | enthalpy |
| $h$ | height of step in interface |
| $K$ | compressibility |
| $k$ | Boltzmann constant |
| $M_i$ | diffusion potential of component $i$ |
| $\mu$ | elastic shear modulus |
| $\mu_i$ | chemical potential of component $i$ |
| $N_0$ | Avogadro constant |
| $N_i; n_i$ | number of entities of type $i$; number of entities of type $i$ per unit volume |
| $\hat{n}$ | unit vector normal to interface |
| $\nu$ | Poisson ratio |
| $\nu_0$ | atomic vibrational frequency |
| $\Omega$ | atomic volume |
| $P$ | hydrostatic pressure |
| $p$ | pressure on interface |
| $Q$ | activation energy |
| $R$ | radius of curvature |
| $\mathbf{R}$ | lattice vector |
| $\mathbf{R}$ | rotation matrix |
| $r$ | position vector in space |

*Symbols*

| | |
|---|---|
| $\hat{\rho}$ | unit vector along rotation axis in axis/angle pair description of crystal misoriention across an interface, i.e., $\hat{\rho}/\theta$. |
| $\rho^{R}$ | Rodrigues vector |
| $S$; $s$ | entropy; entropy per unit volume |
| $s$ | interface step vector |
| $\sigma$ | interfacial free energy per unit area |
| $T$ | absolute temperature |
| $T_{m}$ | absolute melting temperature |
| $\mathbf{T}$ | transformation matrix relating the two crystal lattices adjoining an interface |
| $t$ | time |
| $t$ | rigid body translation between lattices adjoining an interface |
| $\tau_{ij}$ | elastic stress |
| $u$ | displacement |
| $\theta$ | rotation angle around rotation axis in axis/angle pair description of crystal misorientation across an interface, i.e. $\hat{\rho}/\theta$. |
| $\upsilon$ | velocity |
| $V$ | volume |
| $X$ | atom fraction |
| $\hat{\xi}$ | unit vector tangent to dislocation along positive direction |

# Glossary

**Anticoherency dislocation**
An interfacial dislocation which destroys locally the continuity of the reference lattice, and hence the coherency, across the interface.

**Boundary**
A term used synonymously with 'interface'.

**Coherency dislocation**
An interfacial dislocation which maintains locally the continuity of the reference lattice, and hence the coherency, across the interface.

**Coherent interface**
An interface across which continuity of the reference lattice is maintained throughout.

**Commensurate interface**
An interface in which some translation vectors of the two adjoining crystal lattices are equal. This gives rise to one- or two-dimensional periodicity in the interface plane. The lattice translation vectors are not necessarily primitive crystal lattice vectors.

**CSL boundary**
A boundary having a reference structure in which the two adjoining lattices in their unstressed states form an exact CSL. Compare with *Near CSL boundary*.

**Diffuse interface**
An interface, the width of which is large compared to a nearest neighbour atomic separation. Diffuse interfaces are normally associated with incipient chemical or mechanical instabilities in the system.

**DSC dislocation**
An interfacial dislocation with a Burgers vector that is a translation vector of a reference DSC lattice. The set of DSC dislocations includes primary and secondary dislocations.

**Extrinsic interfacial dislocation**
A stress generator dislocation which entered an interface that was originally free of long-range stresses, while the orientational/deformational relationship between the crystal lattices was constrained. It becomes an intrinsic dislocation when the orientational/deformational relationship between the crystal lattices is relaxed to accommodate the added dislocation.

**General interface**
An interface, the free energy of which is at or near a local maximum with respect to one or more macroscopic geometrical degrees of freedom. It is contrasted with singular and vicinal interfaces, which are at or near a local minimum of the free energy with respect to one or more macroscopic geometrical degrees of freedom. Because there are five macroscopic degrees of freedom it is possible for an interface to be 'general' with respect to some and 'singular' or 'vicinal' with respect to other macroscopic degrees of freedom.

**Grain boundary**
A homophase interface involving a misorientation between the adjoining crystal lattices.

**Heterophase interface**
An interface between two crystals of different phase.

**Homophase interface**
An interface between two crystals of the same phase. It includes grain boundaries, stacking faults, inversion domain boundaries, and anti-phase domain boundaries.

**Incoherent interface**
An interface in which the continuity of the reference lattice across the interface, and hence the coherency, is destroyed throughout. The density of anticoherency dislocations within the interface is so great that their cores are indistinguishable.

**Incommensurate interface**
An interface in which the translation vectors of the adjoining crystal lattices, in each direction within the interface, are irrational multiples of each other.

**Intrinsic interfacial dislocation**
Either a stress generator or stress annihilator dislocation in an interface that is free of long-range stresses.

**Large-angle grain boundary**
A grain boundary for which the disorientation angle is in the 'large angle regime', i.e. larger than about 15°. With increasing disorientation the cores of primary dislocations interact increasingly as their separation decreases. The interactions lead to distortions of the structure of the cores compared with the core structure of an isolated crystal lattice dislocation of the same line direction and Burgers vector. In the large-angle regime the distortions are sufficiently large that the core structures of intrinsic dislocations are topologically distinct from those of isolated dislocations.

**Macroscopic geometrical degrees of freedom**
The five geometric variables characterizing the inclination of the average interfacial plane (2), and the orientational relationship (3) between the adjoining crystal lattices, measured far from the interface. In an enantiomorphic crystal the handedness (left or right) is a sixth macroscopic degree of freedom.

**Microscopic geometrical degrees of freedom**
The four geometric variables characterizing the relative rigid body translation of the two adjoining crystal lattices (3) and, in the case where the atomic basis associated with a lattice site in either crystal comprises two or more atoms, the location of the boundary plane along the boundary normal (1).

**Near CSL boundary**
A boundary having a reference structure in which the two adjoining lattices in their unstressed states almost form an exact CSL. The reference bicrystal structure is constructed by deforming homogeneously either or both adjoining crystal lattices so that they form an exact CSL. Compare with *CSL boundary*.

**Partial interfacial dislocation**
An interfacial dislocation with a Burgers vector that is not a translation vector of a reference DSC lattice.

**Partial step**
An interfacial line defect with step character in which at least one of the step vectors is not a translation vector of its respective crystal lattice. Compare *Partial interfacial dislocation*.

**Perfect interfacial dislocation**

An interfacial dislocation with a Burgers vector that is a translation vector of a reference DSC lattice. Synonymous with a *DSC dislocation*.

**Perfect step**

An interfacial line defect with step character in which the step vectors are translation vectors of their respective crystal lattices. Compare *Perfect interfacial dislocation*.

**Primary interfacial dislocation**

An interfacial dislocation with a Burgers vector that is a translation vector of a reference crystal lattice.

**Pure step**

An interfacial line defect with step character but no dislocation or disclination character, for which the step vectors in the black and white crystals are equal.

**Reference lattice**

A lattice that is used to define the degree of coherency of an interface and the Burgers vector of an interfacial dislocation. It is either a single crystal lattice or a DSC lattice. See *Reference structure*.

**Reference structure**

Either a single crystal or a bicrystal to which the dislocation content of a given bicrystal is related. The reference lattice is a crystal lattice or a DSC lattice if the reference structure is a single crystal or bicrystal respectively.

**Secondary interfacial dislocation**

An interfacial dislocation with a Burgers vector that is a translation vector of a reference DSC lattice, excluding those vectors that are also crystal lattice translation vectors.

**Semicoherent interface**

An interface in which there are regions where the continuity of the reference lattice across the interface is maintained elastically, separated by distinct regions (i.e. the cores of anticoherency dislocations) where the continuity of the reference lattice is destroyed.

**Sharp interface**

An interface, the width of which is of the order of a nearest neighbour atomic separation.

**Singular interface**

An interface, the free energy of which is a local minimum with respect to at least one macroscopic geometrical degree of freedom.

**Small-angle grain boundary**

A grain boundary for which the disorientation angle is in the 'small-angle regime', i.e. smaller than about 15°. This is the disorientation range where mutual interactions between cores of primary dislocations are sufficiently small that the core structures are not topologically distinct from those of an isolated crystal lattice dislocation of the same Burgers vector and line direction. It is contrasted with a large-angle boundary.

**Stress-annihilator interfacial dislocation**

An interfacial dislocation whose long-range stress field tends to cancel that of stress-generator dislocations.

**Stress-generator interfacial dislocation**

An interfacial dislocation that effects the orientational/deformational relation between the adjoining crystal lattices from some reference state. If the Burgers vector density of stress generator dislocations is not cancelled by stress-annihilator dislocations the interface is associated with a long-range stress field.

**Vicinal interface**

An interface, the free energy of which is near to a local minimum with respect to at least one of the macroscopic geometrical degrees of freedom. The structure of such an interface consists of the singular interface at the local minimum with a superimposed array of discrete line defects which may be dislocations (that may or may not be associated with steps), or pure steps. The line defects are defined with the bicrystal containing the singular interface as the reference structure.

# PART I

# INTERFACIAL STRUCTURE

# 1

# The geometry of interfaces

## 1.1 INTRODUCTION

Interfaces often exist in polycrystalline, multiphase materials in a large number of configurations, including non-planar ones. However, the simplest interface is a single isolated planar interface separating two otherwise perfect crystals of the same, or different, crystalline phases, that is, a planar interface in a bicrystal. Just as the concept of a perfect single crystal is the basis for all discussion of point and line defects and elementary excitations in crystalline matter, so the concept of an ideal bicrystal containing a planar interface is an essential experimental and theoretical tool for understanding the properties of interfaces. The first part of this chapter is concerned with the geometrical description of such a bicrystal. This involves the relationship between the crystals on either side of the interface as well as the interface plane. The geometrical specification of a bicrystal forms a part of the thermodynamic degrees of freedom associated with an interface, and it is therefore of central importance to much that follows in this book.

In the second part of the chapter the discussion moves onto bicrystallography—the crystallography of bicrystals. The translational and orientational order that exists in the adjoining crystals of a bicrystal is the source of all structural order that may exist in the interface. Bicrystallography is concerned with the symmetries characterizing the structural order that exists in a bicrystal. First, we wish to enumerate those symmetries and how they may be determined from the space groups of the adjoining crystals and the transformation relating the two crystals. Secondly, many of the symmetry elements of the adjoining crystals will not, in general, be present in the bicrystal. Those symmetry elements that have been 'lost' relate equivalent bicrystal variants. The existence of bicrystal variants is a consequence of a fundamental principle known as the principle of symmetry compensation (Shubnikov and Koptsik 1977): *if symmetry is reduced at one structural level it arises and is preserved at another.* Perhaps the most fundamental application of symmetry in physics is the establishment of conservation laws. For example, translational invariance of the Hamiltonian leads to conservation of linear momentum and time invariance of the Hamiltonian leads to conservation of energy. The principle of symmetry compensation expresses another kind of conservation law. Thus, when two crystals join to form a bicrystal there is a definite sense (i.e. the existence of variants) in which none of the symmetry elements of either crystal is lost even though a particular bicrystal is generally an object with lower symmetry than that of the original crystals. Rather than speaking of symmetry being 'lost' we should therefore speak of symmetry being 'suppressed', since it arises again through the existence of variants. The use of symmetry arguments in this chapter is entirely based on the principle of symmetry compensation. Perhaps the most useful consequence of the enumeration of the variants of an interface is that it immediately leads to a classification of interfacial line defects that separate coexisting variants. This classification has many powerful applications, some of which we illustrate in further sections of this chapter with specific examples for homophase and heterophase interfaces. The same ideas are applied to enumerating the possible forms of crystals embedded within other crystals, such as embedded precipitates. Most recently, ideas from the fields of incommensurate structures and quasicrystals have been imported into the study of irrational interfaces, i.e. interfaces in which there is at most one direction of translational symmetry. As a final topic, we discuss these ideas and show the relationship between them and other concepts developed in this chapter.

Some of the results we shall derive in the second part of this chapter were first obtained by Van Tendeloo and Amelinckx (1974). Three groups of workers have rediscovered these results and extended them significantly: Pond and Bollmann (1979), Gratias and Portier (1982), Kalonji and Cahn (1982), Cahn and Kalonji (1982), Pond and Vlachavas (1983), Pond (1985), Pond (1989) and Pond and Dibley (1990). The approach we follow here follows closely that of Pond and coworkers, whose work is based entirely on the principle of symmetry compensation.

## 1.2 ALL THE GROUP THEORY WE NEED

The following sections of this chapter use some concepts and simple results from group theory (see Cracknell (1968) for an introduction to group theory). For clarity and completeness we explain them here, beginning with the *definition of a group*: A set of elements $g_1, g_2, g_3, \ldots$ forms a group $G = \{g_1, g_2, g_3, \ldots\}$ if we can define a 'product' of two elements $g_i g_j$ in some sense and if

(1) the product of any two elements is also an element of the group: $g_i g_j = g_k \in G$ (the symbol $\in$ means 'is an element of');

(2) one of the elements of the group is an identity element, $e$: $g_i e = e g_i = g_i$;

(3) the elements of the group obey the associative law of 'multiplication': $(g_i g_j) g_k = g_i (g_j g_k)$;

(4) for every element of the group there is an inverse: $g_i g_i^{-1} = e$ where $g_i^{-1} \in G$.

The number of elements of the group is defined as the *order of the group*. A simple example of a group is the point group $2/m$, which is a group of order 4: $\{1, 2, \bar{1}, m\}$. This means that the environment around the point leads to the existence of certain symmetry elements at the point: the identity, a two-fold rotation axis, a centre of inversion and a mirror plane. Figure 1.1(a) illustrates a body in which the centre has $2/m$ symmetry. In Fig. 1.1(b) we show the stereographic projection that is generated by the symmetry elements of $2/m$. Using Fig. 1.1(b) we can construct the 'group multiplication table' very easily:

|       | 1         | 2         | $\bar{1}$ | $m$       |
|-------|-----------|-----------|-----------|-----------|
| 1     | 1         | 2         | $\bar{1}$ | $m$       |
| 2     | 2         | 1         | $m$       | $\bar{1}$ |
| $\bar{1}$ | $\bar{1}$ | $m$   | 1         | 2         |
| $m$   | $m$       | $\bar{1}$ | 2         | 1         |

Using this multiplication table it is easy to verify that rules (1)–(4) above for a group are satisfied.

If a set $H = \{h_1, h_2, h_3, \ldots\}$ is a subset of a group $G = \{g_1, g_2, g_3, \ldots\}$, which is written as $H \subset G$, and if $H$ possesses the properties of a group, then $H$ is called a *subgroup* of $G$. For example, the group $2/m$ has the following subgroups: $\{1\}$, $\{1, 2\}$, $\{1, \bar{1}\}$, $\{1, m\}$. All subgroups of a finite group (i.e. a group with a finite number of elements) may be found with the aid of *Lagrange's theorem*: the order of a subgroup $H$ of a finite group $G$ is a divisor of the order of the $G$. The group $2/m$ is of order 4.

**Fig. 1.1** (a) An object displaying $2/m$ point group symmetry at its centre, which is marked by a cross. The two-fold rotation axis is shown and the mirror plane (not shown) passes through the cross perpendicular to the two-fold axis. (b) A stereographic projection showing $2/m$ symmetry.

Therefore its subgroups are of order 1 or 2. If $G$ is of order 12 the subgroups are of order 1, 2, 3, 4, and 6.

For any subgroup $H$ of the group $G$ we may define *left and right cosets*:

$$g_i H = \{g_i h_1, g_i h_2, \dots\}; Hg_i = \{h_1 g_i, h_2 g_i, \dots\}$$

where $g_i \notin H$. Demanding that $g_i \notin H$ ensures that the cosets are *not* groups because they cannot contain the identity; for if $g_i h_m = e$ then $h_m = g_i^{-1}$, but $H$ must contain $h_m^{-1}$ which is $g_i$ and therefore $g_i \in H$ in contradiction to $g_i \notin H$. It can be shown that the elements of two cosets are either disjoint (i.e. they have no elements in common) or they are identical. Using this fact we can carry out a *decomposition of the group $G$ with respect to the subgroup $H$*. This means that we can enumerate all the elements of $G$ according to the distinct cosets to which they belong:

$$G = H \cup g_1 H \cup g_2 H \dots \cup g_j H. \tag{1.1}$$

The symbol $\cup$ means union: $A \cup B$ denotes the set of elements contained in both $A$ and $B$. Using our example of the point group $2/m$ eqn (1.1) becomes

$$2/m = \{1, 2\} \cup \bar{1}\{1, 2\} = \{1, 2\} \cup \{\bar{1}, m\} = \{1, 2, \bar{1}, m\}.$$

The number of cosets $j$ in the decomposition of a group with respect to a subgroup is called the *index of the subgroup*. Lagrange's theorem indicates that the index of a subgroup is always a divisor of the order of the group.

An *invariant subgroup* is a particularly important type of subgroup. This is characterized by the fact that the left and right cosets are the same. Thus if $T = \{t_1, t_2, \dots\}$ is an invariant subgroup of $G = \{g_1, g_2, \dots\}$ then for any $g_i$ the equality of the left and right cosets of $T$ requires $g_i t_j = t_k g_i$, that is, $t_j = g_i^{-1} t_k g_i$. Thus if $t_k$ is an element of an invariant subgroup then $g_i^{-1} t_k g_i$ is also, where $g_i$ is any element of the group $G$.

It sometimes happens that two groups may, for algebraic purposes, be regarded as the same group. We say that two groups $G$ and $H$ are *isomorphic* if a one-to-one correspondence $G_i \leftrightarrow H_i$ may be set up between the elements of $G$ and the elements of $H$ in such a way that if $G_i G_j = G_k$ then $H_i H_j = H_k$. Two isomorphic groups therefore have the same multiplication table.

## 1.3 THE RELATIONSHIP BETWEEN TWO CRYSTALS

### 1.3.1 Crystals and lattices

A lattice is a set of vectors that is closed under addition and subtraction: the sum and difference of any two vectors in the set are also in the set. When there exists a shortest length vector in the lattice the lattice is described as periodic; otherwise, it is described as quasiperiodic. We shall not say any more about quasiperiodicity until Section 1.9. A crystal is obtained by decorating each lattice site of a periodic lattice with a basis comprising one or more atoms arranged in an identical fashion. The symmetry of the crystal can, therefore, be lower than that of its lattice, e.g. each lattice site is an inversion centre in the lattice but not necessarily in the crystal. If the crystals on either side of an interface have the same chemical composition and structure the interface is of the homophase type. Heterophase interfaces separate crystals of differing composition and/ or structure. Heterophase interfaces appear, for example, in first-order phase transformations where a new phase nucleates and grows within an existing phase. Common examples of homophase interfaces are grain boundaries separating identical crystals with differing orientations, inversion boundaries, and stacking faults.

If a crystal space group does not contain mirror glide planes or screw rotation axes it is described as symmorphic. Non-symmorphic operations are characterized by the requirement of a supplementary displacement, that is not a lattice vector, in addition to a point group operation such as a rotation or reflection. If a crystal contains a centre of inversion it is described as centrosymmetric. Some crystals display rotational symmetries that are combined with inversion operations; such operations are called improper rotations. Crystals that do not contain any mirror planes or an inversion centre may have a 'handedness' and exist in enantiomorphic pairs. When we describe the relationship between two crystals we may use *any* linear transformation that is not a symmetry operation of either crystal, including inversions, mirror reflections, improper rotations, proper rotations, and homogeneous deformations. Exceptionally, the supplementary displacement, associated with a non-symmorphic operation relating two crystals, may be ignored because it could be argued that the crystals meeting at the interface will translate with respect to each other so as to minimize the interfacial energy. The supplementary displacement then becomes incorporated into this relative translation.

### 1.3.2  Vector and coordinate transformations

Consider now the relationship between the crystal lattices that meet at an interface. This relationship describes how one crystal lattice may be deformed and/or rotated into the other. Mathematically the relationship is a homogeneous linear transformation in which points, lines, and planes of one lattice are mapped onto points, lines, and planes respectively of the other lattice. Such a transformation is called affine. The transformation is independent of the position and orientation of the interface. Therefore, we can dispense with the interface for the time being and imagine that the two crystal lattices interpenetrate and fill all space. If the sites of one lattice are coloured black and those of the other lattice are coloured white we obtain a *dichromatic pattern*. Let $a_1^w$, $a_2^w$, $a_3^w$ be *any* three non-coplanar, primitive vectors of the white lattice and let $a_1^b$, $a_2^b$, and $a_3^b$ be a similar set of the black lattice. The choice of coordinate system in which to express these vectors is arbitrary and we choose the coordinate frame defined by the vectors $a_1^w$, $a_2^w$, and $a_3^w$, i.e. $a_1^w = [1, 0, 0]$, $a_2^w = [0, 1, 0]$, and $a_3^w = [0, 0, 1]$. The vectors $a_1^w$, $a_2^w$, $a_3^w$ form a 'basis set' for the white lattice which means that the white lattice is generated by forming linear combinations of them. (The term 'basis set' is not to be confused with the 'atomic basis', which is the atomic motif in the crystal at each lattice site.) Each of the $a_i^b$'s may be expressed as a linear combination of the $a_i^w$'s:

$$a_i^b = \sum_{j=1}^{3} T_{ji} a_j^w. \tag{1.2}$$

The reason for writing the components of **T** in this form will become clear shortly. The meaning of eqn (1.2) is that the components of $a_i^b$ in the white basis are $[T_{1i}, T_{2i}, T_{3i}]$. Equation (1.2) is described as a *vector transformation* because the coordinate system is fixed and vectors are rotated or deformed or both. If $\mathbf{a}^w$ denotes the column matrix of basis vectors $a_i^w$ and $\mathbf{a}^b$ the column matrix of basis vectors $a_i^b$, then eqn (1.2) may be expressed as

$$\mathbf{a}^b = \mathbf{T}^t \mathbf{a}^w, \qquad (1.3)$$

where $\mathbf{T}^t$ denotes the transpose of the matrix **T**. We may wish to relate the coordinates of a given point in space with respect to two distinct coordinate systems, one in the white lattice and the other in the black. Let the coordinate system of the black lattice be aligned along the three basis vectors $a_i^b$. If the coordinates of the fixed point are $(r_1^w, r_2^w, r_3^w)$ and $(r_1^b, r_2^b, r_3^b)$ with respect to the white and black coordinate systems respectively, then

$$\sum_{j=1}^{3} r_j^w a_j^w = \sum_{i=1}^{3} r_i^b a_i^b$$

and using eqn (1.2) we obtain

$$\sum_{j=1}^{3} r_j^w a_j^w = \sum_{i=1}^{3} \sum_{j=1}^{3} r_i^b T_{ji} a_j^w$$

which implies

$$r_j^w = \sum_{i=1}^{3} T_{ji} r_i^b. \qquad (1.4)$$

Letting $r^w$ and $r^b$ denote $(r_1^w, r_2^w, r_3^w)$ and $(r_1^b, r_2^b, r_3^b)$, eqn. (1.4) may be rewritten as

$$r^w = \mathbf{T} r^b. \qquad (1.5)$$

The interpretation of eqn (1.5) is that **T** represents the transformation of the components of a given vector expressed in the coordinate system of the black lattice into the components expressed in the coordinate system of the white lattice. Since the vector remains fixed in space, and only its components change, eqn (1.5) represents a *coordinate transformation*.

Up to now we have assumed that the lattices have at least one site in common. This is not necessarily the case, and if $t$ denotes the requisite rigid body displacement of the white lattice to bring at least one of its sites into coincidence with a black lattice site, then eqn (1.2) becomes:

$$a_i^b = \sum_{j=1}^{3} T_{ji} a_j^w + t, \qquad (1.6)$$

whereas eqn (1.5) becomes

$$r^w = \mathbf{T} r^b + t. \qquad (1.7)$$

The meaning of the 'vector addition' of $t$ in eqn (1.6) is that the vectors $\sum_{j=1}^{3} T_{ji} a_j^w$ are rigidly shifted by $t$; it describes a change of origin and not a vector addition in the normal sense. In this chapter we will interpret the transformation **T** as a vector transformation relating black and white lattice vectors through eqn (1.6). We shall use the coordinate system of the white lattice. It is sometimes necessary to re-express an operator **M** referred to the black lattice coordinate frame in the coordinate system of the white lattice. The new matrix is $\mathbf{TMT}^{-1}$ and can be used to act on vectors expressed in the white frame. The sequence of operations (from right to left) contained in $\mathbf{TMT}^{-1}$ can be regarded as first converting the coordinates of a vector expressed in the white frame into those of the black frame, while leaving the vector invariant, secondly transforming the vector according to the matrix $\mathbf{M}$ in the black frame, and finally transforming the coordinates of this resultant vector back into the white frame.

The matrix $\mathbf{T}$ can be factored into the product of a proper rotation $\mathbf{R}$ and a pure deformation $\mathbf{P}$:

$$\mathbf{T} = \mathbf{PR}. \tag{1.8}$$

Note that the order of this factorization is important mathematically (but not physically) because the matrices $\mathbf{P}$ and $\mathbf{R}$ do not commute in general. The specification of the transformation matrix $\mathbf{T}$ in eqn (1.8) is not unique. The lack of uniqueness stems from the point group and translational symmetries present in both crystal lattices. Different choices of primitive cells in the two lattices to be related in eqn (1.2) may be effected by replacing $\mathbf{T}$ by $(\mathbf{TU}^b\mathbf{T}^{-1})\mathbf{TU}^w = \mathbf{TU}^b\mathbf{U}^w$ where $\mathbf{U}^w$ and $\mathbf{U}^b$ are unimodular matrices with integer elements. (A unimodular matrix has a determinant of $\pm 1$.) Similarly $\mathbf{T}$ may be replaced by $(\mathbf{TG}^b\mathbf{T}^{-1})\mathbf{TG}^w = \mathbf{TG}^b\mathbf{G}^w$ where $\mathbf{G}^w$ and $\mathbf{G}^b$ represent point group operations of the white and black lattices expressed in their own coordinate systems. We shall prove in Section 1.5 that for fixed structures and orientation of the two lattices the dichromatic pattern is completely unaffected by the specification of $\mathbf{T}$. However, we shall see in Chapter 2 that the choice of $\mathbf{T}$ is significant in the dislocation model of interfaces and at martensitic interfaces, because there it is assumed that the transformation $\mathbf{T}$ is actually carried out by the interface when it migrates.

For a given pair of crystal lattices the pure deformation matrix $\mathbf{P}$ may be fixed in some convenient form, for example the form that minimizes the principal strains. In a homophase boundary $\mathbf{P}$ is most conveniently taken to be the identity or the inversion. The totality of dichromatic patterns will then be generated by varying the proper rotation matrix $\mathbf{R}$ in eqn (1.8).

### 1.3.3  Descriptions of lattice rotations

The rotation between two misoriented lattices may be described mathematically in several ways. These include the rotation matrix, the axis/angle pair, the Rodrigues vector, and the quaternion representations. Each may be used to advantage in certain situations and is described in the following.

### 1.3.3.1  *Vector and matrix representations*

Rotations in 3D space are often represented by orthogonal $3 \times 3$ matrices. There are three degrees of freedom associated with a rotation, two for the axis of rotation and one for the angle. Therefore, it must be possible to represent all rotations by vectors in 3D space. As will be seen in the next subsection this is indeed the case. But first we derive the matrix representation of an arbitrary rotation.

Consider the rotation of a vector $r$ in a clockwise sense about an axis $\hat{\rho}$ through an angle $\theta$ to a new position $r'$. In Fig. 1.2, $r$ is denoted by $OP$, $r'$ by $OQ$, and $\hat{\rho}$ lies along $ON$. The vector $ON$ is equal to $(\hat{\rho} \cdot r)\hat{\rho}$. Figure 1.3 shows a plan of the vectors in the plane normal to the axis of rotation. $NP = OP - ON = r - (r \cdot \hat{\rho})\hat{\rho} = \hat{\rho} \times (r \times \hat{\rho})$ and since, $NS$ is perpendicular to both $NQ$ and $ON$ we see that $NS = \hat{\rho} \times r$. Thus,

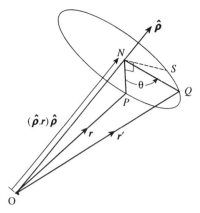

**Fig. 1.2** Diagram illustrating the rotation of the vector *r* to *r'* about the rotation axis $\hat{\rho}$ by the angle $\theta$. See eqn (1.9).

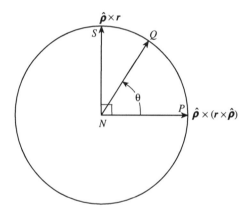

**Fig. 1.3** Plan view of the circle normal to the rotation axis in Fig. 1.2. The vectors *r* and *r'* touch the circle at *P* and *Q* respectively, and $\theta$ is the angle of rotation.

$$NQ = \cos\theta[r - (\hat{\rho}\cdot r)\hat{\rho}] + \sin\theta(\hat{\rho} \times r)$$

and

$$r' = OQ = ON + NQ = (\hat{\rho}\cdot r)\hat{\rho} + \cos\theta[r - (\hat{\rho}\cdot r)\hat{\rho}] + \sin\theta(\hat{\rho} \times r)$$

$$= r\cos\theta + \hat{\rho}(\hat{\rho}\cdot r)(1 - \cos\theta) + \sin\theta(\hat{\rho} \times r). \tag{1.9}$$

Equation (1.9) is sometimes called the 'rotation formula'. When it is expressed in component form we may recover the matrix **R** representing the rotation. That is,

$$r_i' = \sum_{j=1}^{3} R_{ij} r_j$$

where

$$R_{ij} = \cos\theta\delta_{ij} + \rho_i\rho_j(1 - \cos\theta) - \sum_{k=1}^{3} \varepsilon_{ijk}\rho_k \sin\theta, \tag{1.10}$$

$\delta_{ij}$ is the Kronecker delta:

$$\delta_{ij} = \begin{cases} 1 \text{ if } i = j \\ 0 \text{ if } i \neq j \end{cases} \tag{1.11}$$

and $\varepsilon_{ijk}$ is the permutation tensor:

$$\varepsilon_{ijk} = \begin{cases} 1 \text{ if } i, j, k \text{ are an even permutation of } 1, 2, 3 \\ -1 \text{ if } i, j, k \text{ are an odd permutation of } 1, 2, 3 \\ 0 \text{ otherwise.} \end{cases} \tag{1.12}$$

Using eqn (1.10) it may be deduced that the trace of the rotation matrix $= R_{11} + R_{22} + R_{33} = 1 + 2\cos\theta$ and the rotation axis is parallel to $[R_{32} - R_{23}, R_{13} - R_{31}, R_{21} - R_{12}]$.

### 1.3.3.2   *The Frank–Rodrigues map*

Following Frank (1988a, 1988b) we define the *Rodrigues vector* $\rho^R$ as follows:

$$\rho^R = \hat{\rho} \tan \theta/2. \tag{1.13}$$

Using eqn (1.13) the set of all possible rotations may be mapped onto 3D space forming the Frank–Rodrigues map. Any rotation may be expressed as modulo $2\pi$, and we choose to represent all rotations between $-\pi$ and $+\pi$. Since a rotation of $\pi$ is equivalent to a rotation of $-\pi$ the map is connected at infinity. This means that if we increase the angle fractionally above $\pi$ the Rodrigues vector disappears at plus infinity and reappears on the same axis at minus infinity. Using eqns (1.10) and (1.13) we may recover the matrix representation of a rotation from the Rodrigues vector as follows:

$$(1 + \rho^{R^2})R_{ij} = (1 - \rho^{R^2})\delta_{ij} + 2\left[\rho_i^R \rho_j^R - \sum_{k=1}^{3} \varepsilon_{ijk} \rho_k^R\right] \tag{1.14}$$

where $\rho^{R^2} = |\rho^R|^2 = \tan^2 \theta/2$ and $\rho_i^R$ is the $i$th component of $\rho^R$.

With matrix representations of rotations we combine rotations by matrix multiplication. Thus a rotation $\mathbf{R}_A$ followed by a rotation $\mathbf{R}_B$ is represented by $\mathbf{R}_B \mathbf{R}_A$. The corresponding rule for the Rodrigues vectors is

$$(\rho_A^R, \rho_B^R) = \frac{\rho_A^R + \rho_B^R - \rho_A^R \times \rho_B^R}{1 - \rho_A^R \cdot \rho_B^R} \tag{1.15}$$

where $(\rho_A^R, \rho_B^R)$ denotes the Rodrigues vector obtained by carrying out the rotation corresponding to $\rho_A^R$ first and following it with the rotation corresponding to $\rho_B^R$. If we wish to change the orientation of a body from that represented by the matrix $\mathbf{R}_A$ to that represented by $\mathbf{R}_B$ we carry out the rotation $\mathbf{R}_B \mathbf{R}_A^{-1}$. $\mathbf{R}_B \mathbf{R}_A^{-1}$ corresponds to a single rotation about some axis starting from the orientation $\mathbf{R}_A$. Similarly, the Rodrigues vector representing this change in orientation is $\rho_{AB}^R = (-\rho_A^R, \rho_B^R)$. This is readily confirmed by using eqn (1.15) to show that $(\rho_A^R, \rho_{AB}^R) = \rho_B^R$. Equation (1.15) indicates that the axis of the rotation $\rho_{AB}^R$ is *not* parallel to $\rho_B^R - \rho_A^R$, except in the trivial case that $\rho_A^R$ and $\rho_B^R$ share the same axis.

Equation (1.15) leads to certain rectilinearity properties of the Rodrigues map: a continuing rotation about a fixed axis $\hat{\rho}_B$, starting from an arbitrary rotation $\rho_A^R$, is represented by a straight-line trajectory in the Rodrigues map. Let $\rho_B^R = \lambda\hat{\rho}_B$ denote the varying rotation about the axis $\hat{\rho}_B$. Then $\rho^R = (\rho_A^R, \rho_B^R)$ is given by eqn (1.15) as follows:

$$\rho^R = \frac{\rho_A^R + \lambda\hat{\rho}_B^R - \rho_A^R \times \lambda\hat{\rho}_B^R}{1 - \rho_A^R \cdot \lambda\hat{\rho}_B^R}$$

$$= \rho_A^R + \frac{\lambda}{1 - \lambda\rho_A^R \cdot \hat{\rho}_B} \left[ (\rho_A^R \cdot \hat{\rho}_B)\rho_A^R + \hat{\rho}_B - \rho_A^R \times \hat{\rho}_B \right] \qquad (1.16)$$

$$= \rho_A^R + (\text{scalar function of } \lambda) \times (\text{vector independent of } \lambda)$$

which is the equation of a straight line passing through $\rho_A^R$. Note that the direction of the line depends on both $\hat{\rho}_B$ and $\rho_A^R$.

We may wish to change the origin of the map to a new reference misorientation $\rho_r^R$. $\rho_r^R$ is brought to the origin of the map by applying the rotation $-\rho_r^R$ to the whole map. Any other point $\rho^R$ in the map becomes $\rho^{R'} = (\rho^R, -\rho_r^R)$. It may be shown that the straight line $\rho^R = \rho_0^R + \lambda\hat{\rho}$ is transformed into another straight line given by

$$\rho^{R'} = \rho_0^{R'} + \frac{\lambda}{1 + (\rho_0^R + \lambda\hat{\rho}) \cdot \rho_r^R} \left[ \hat{\rho} + \hat{\rho} \times \rho_r^R - (\hat{\rho} \cdot \rho_r^R)\rho_0^{R'} \right] . \qquad (1.17)$$

Applying this result to all straight lines lying in a plane it is seen that a plane remains a plane after the change of origin.

The concept of an orientationally equidistant boundary between two given orientations A and B may now be defined. It is the locus of map points C such that the angle of rotation from A to C is equal to the angle of rotation from B to C. It comprises two planes, one near to A and B and the other at infinity. The plane at infinity exists because the Rodrigues map is self-connected between diametrically opposite points at infinity. In practice, unless at least one of the two points A and B is very far from the origin, one plane of the orientationally equidistant boundary will be so remote that we can ignore it; the orientationally equidistant boundary is then just one plane. The important point is that in the Rodrigues map the boundary is a plane, as we now demonstrate. First we bring $\rho_B^R$ to the origin by applying $-\rho_B^R$ to $\rho_A^R$ and $\rho_B^R$; $\rho_A^R$ becomes $\rho_A^{R'} = (\rho_A^R, -\rho_B^R) = \hat{\rho}_A' \tan\theta_A'/2$ and $\rho_B^{R'} = 0$, see Fig. 1.4(b). We then apply the rotation $\theta_A'/2$ along $-\hat{\rho}_A'$ to bring $\rho_A^{R'}$ to $\rho_A^{R''} = (\rho_A^{R'}, -\hat{\rho}_A' \tan\theta_A'/4) = \hat{\rho}_A' \tan\theta_A'/4$ and $\rho_B^{R'}$ to $\rho_B^{R''} = (0, -\hat{\rho}_A' \tan\theta_A'/4) = -\hat{\rho}_A' \tan\theta_A'/4$; see Fig. 1.4(c). A″ and B″ are now equidistant from the origin. Consider any point C on the perpendicularly bisecting plane, which is shown

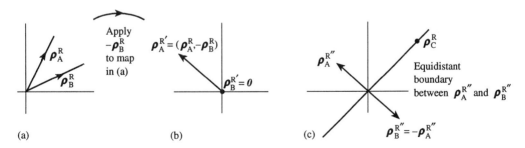

**Fig. 1.4** Diagram showing that the orientationally equidistant boundary between two points (defined by $\rho_A^R$ and $\rho_B^R$ in (a)) in the Frank–Rodrigues map is a plane. In (b) the rotation $-\rho_B^R$ is applied to the whole Rodrigues map which brings the rotation $\rho_B^R$ to the origin and the rotation $\rho_A^R$ to $\rho_A^{R'}$. In (c) a rotation of minus half the angle corresponding to $\rho_A^{R'}$ about the axis $\hat{\rho}_A^{R'}$ is then applied to the whole map which brings $\rho_A^{R'}$ and $\rho_B^{R'}$ to lie at equal orientational distances from the origin. The plane passing through the origin at right angles to $\rho_A^{R''}$ and $\rho_B^{R''}$ contains the rotations, such as $\rho_C^R$, which are equidistant from them.

in Fig. 1.4(c). Since $\rho_A^{R''} = -\rho_B^{R''}$ and $\rho_A^{R''} \cdot \rho_C^{R} = 0$ it follows from eqn (1.15) that the angle between A″ and C is equal to the angle between B″ and C. Thus the orientationally equidistant boundary between A″ and B″ is the perpendicularly bisecting plane. Since it has already been shown that under a change of reference orientation planes in the map remain planes it follows that the orientationally equidistant boundaries between A′ and B′ and between A and B are also planes. Applying the rotation, $\theta_A'/2$ along $\hat{\rho}_A'$ to bring $\rho_A^{R''}$ to $\rho_A^{R'}$ and $\rho_B^{R''}$ to the origin it is found, using eqn (1.15), that the orientationally equidistant boundary between A″ and B″, given by $\rho^{R''} \cdot \hat{\rho}_A' = 0$, becomes

$$\rho^{R'} = \hat{\rho}_A' \tan \theta_A'/4 + (\rho^{R''} - \rho^{R''} \times \hat{\rho}_A' \tan \theta_A'/4). \tag{1.18}$$

Since the vector in brackets is perpendicular to $\hat{\rho}_A'$ we conclude that the orientationally equidistant boundary between A′ and the origin is still perpendicular to $\hat{\rho}_A'$ but at $\tan \theta_A'/4$ from the origin, rather than $(\tan \theta_A'/2)/2$ as might have been thought. This is of crucial importance for the development of the next concept, namely the fundamental zone of rotations in a crystal.

### 1.3.3.3  *Fundamental zones*

When a crystal possesses proper rotation symmetries there is a multiplicity of rotations from the reference orientation, all of which are physically indistinguishable. However, these rotations are represented by different points in the map. The idea of the fundamental zone is to limit the map to a region in which physically indistinguishable rotations are represented once only. It is analogous to a Brillouin zone in reciprocal space which arises from the indistinguishability of wave vectors differing by reciprocal lattice vectors. In order for the fundamental zone to display (at least) the rotational symmetry of the crystal we select the map point which lies closest to the origin and reject the more remote equivalent map points. Similarly, we align the axes of the map with the principal symmetry directions in the crystal. Any orientation appearing in the zone is the smallest possible of all the equivalent orientations and it is called the *disorientation*. The accepted points fall within a region that may or may not be bounded, depending on the point group symmetry. If the region is bounded it is a polyhedron that is defined by planes which are orientationally equidistant between the origin and neighbouring points that are equivalent by a symmetry rotation to the origin. Any point lying outside one of these planes has an equivalent point inside the fundamental zone, which is accepted instead of it. In this way the presence of $N$ proper rotational symmetries in the point group of the crystal is seen to result in the division of the infinite Frank–Rodrigues map into $N$ zones.

So far we have limited the relationship between the crystals to being a proper rotation. But it is possible that the crystals may be related by improper operations, namely a centre of inversion, a mirror reflection, or an improper rotation, all of which involve an inversion. (A mirror reflection can be described as a two-fold improper rotation.) All point group operations are either proper or improper. Since the determinants of proper and improper operations are $+1$ and $-1$ respectively, the product of two improper operations is a proper operation, while the product of a proper and an improper operation is improper. It follows that any point group $G$ can be decomposed as follows:

$$G = P \cup K_i P \tag{1.19}$$

where $P$ is the group of proper rotations contained in $G$ and $K_i$ is any improper operation of the point group $G$. Elements of the group $P$ define the fundamental zone which we considered above, and which we now call the proper fundamental zone to emphasize the nature of the point group operations used to define it. The coset $K_iP$ is the set of all improper operations contained in $G$, which we have so far ignored in the definition of the fundamental zone. There are eleven point groups that contain only proper operations and they are listed in the top row of Table 1.1. They are the proper point groups, and they have great significance for the definition of fundamental zones as we shall now show.

If the crystal is centrosymmetric then we may choose the inversion operation, $\bar{1}$, for $K_i$ so that $G = P \cup \bar{1}P$. The proper components of the improper operations $\bar{1}P$, i.e. elements of $P$, may be used to define an improper fundamental zone which will obviously be identical to the proper zone. A bicrystal represented by a point in an improper fundamental zone differs from a bicrystal represented by the same point in a proper fundamental zone by the action of an inversion operation applied to one crystal. In this case, because the crystal is centrosymmetric, there is no distinction between the two bicrystals. All bicrystals in centrosymmetric crystals may therefore be represented in the proper fundamental zone and we may dispense with the improper zone. We emphasize that only the proper operations are used to define the fundamental zone for centrosymmetric crystals. The centrosymmetric point groups appear in the second row of Table 1.1.

For noncentrosymmetric crystals the coset of improper operations $K_iP$ does not contain the inversion. Nevertheless, all operations in $K_iP$ can be expressed as $\bar{1}R_j$, where $R_j$ is a proper rotation excluding the identity. The set of proper rotations $\{R_j\}$ defines the improper fundamental zone. There can be no elements common to $\{R_j\}$ and $P$ because $\bar{1}$ is not contained in $G$. Therefore, the proper and improper fundamental zones are defined by disjoint sets of operations, in contrast to the centrosymmetric case where they are identical. Let points within the proper and improper fundamental zones be coloured white and black respectively, and let the zones share the same origin. We obtain a grey zone in the region where the zones overlap. A point outside the grey zone is equivalent to a point inside by either a proper or an improper symmetry operation. The grey zone is therefore the fundamental zone in the absence of a centre of symmetry and it is defined by the union of the proper rotations in $P$ and $\{R_j\}$. The group $G = P \cup \bar{1}\{R_j\}$ is isomorphic to the proper point group $G' = P \cup \{R_j\}$. Therefore the fundamental zone is defined by the elements of the proper point group $G'$. The noncentrosymmetric point groups appear in the third row of Table 1.1 according to the proper point groups to which they are isomorphic. We see that for the noncentrosymmetric case, as well as the centrosymmetric case, the fundamental zone is defined by the elements of a proper group.

We have already remarked that the proper operations $P$ divide the whole Frank–Rodrigues map, so that it is possible to access any point in the map by a proper rotation alone. The same remark does not hold for improper rotations because the combination of two improper rotations yields a proper rotation. In the case of noncentrosymmetric crystals the presence of improper operations in the point group can be thought of as giving rise to a further division of the Frank–Rodrigues map, in addition to the division achieved by the proper symmetry rotations. Exceptionally, there is no division of the map by the proper operation (i.e. 1) in the point group $m = \{1, m\}$ and the division into two zones is provided by the improper operation $m$.

At each point inside the grey zone there are two bicrystal structures with the same orientation but differing by an inversion operation. But the inversion is equivalent to a proper rotation in those groups containing improper rotational symmetries, which includes mirror reflections. Therefore, in noncentrosymmetric groups containing improper

**Table 1.1**  The 11 fundamental zones for the 32 point groups

| System | Triclinic | Monoclinic | Orthorhombic | Trigonal | | Tetragonal | | Hexagonal | | Cubic | |
|---|---|---|---|---|---|---|---|---|---|---|---|
| Proper point group $G_p$ | 1 | 2 | 222 | 3 | 32 | 4 | 422 | 6 | 622 | 23 | 432 |
| Centrosymmetric group | $\bar{1}$ | $2/m$ | $mmm$ | $\bar{3}$ | $\bar{3}m$ | $4/m$ | $4/mmm$ | $6/m$ | $6/mmm$ | $m\bar{3}$ | $m\bar{3}m$ |
| Improper groups isomorphic with $G_p$ | | $\bar{2}$ | $2mm$ | | $3m$ | $\bar{4}$ | $4mm$ $\bar{4}2m$ | $\bar{6}$ | $6mm$ $\bar{6}m2$ | | $\bar{4}3m$ |
| Fundamental zone | all space $t = 2\tan\pi/4$ | | cube | slab $t = 2\tan\pi/6$ hexagonal prism | | slab $t = 2\tan\pi/8$ octagonal prism | | slab $t = 2\tan\pi/12$ dodecagonal prism | | octahedron | semiregular truncated cube |

operations all nonequivalent bicrystal structures may be generated by proper rotations alone, but not all of those proper rotations fall within the fundamental zone. In order to represent all nonequivalent bicrystal structures within the fundamental zone it is necessary to associate two bicrystal structures with each point inside the zone, related by an inversion operation applied to one crystal. With enantiomorphic structures there are no improper operations in the point group. Therefore the two structures that exist at each point within the fundamental zone may only be related by an inversion applied to one crystal: there is no proper rotation that will relate the two structures. One of the two structures at each orientation contains crystals with the same handedness and the other crystals of opposite handedness. The 11 proper groups are the only groups that may display enantiomorphism.

To recapitulate, there are only 11 fundamental zones and they are defined by the 11 proper point groups. With the exception of crystals belonging to one of the 11 proper point groups, all bicrystal structures may be generated by proper rotations alone. Crystals belonging to any of the 11 proper point groups may display enantiomorphism, and it is an additional degree of freedom possessed by an interface whenever it arises. For noncentrosymmetric crystals the fundamental zone contains all possible bicrystal structures only if each point within the zone is associated with two bicrystal structures that are related by an inversion of one crystal. All bicrystal structures between centrosymmetric crystals are represented once only within the fundamental zone.

The fundamental zones are listed in the last row of Table 1.1. Since the triclinic point groups contain no rotational symmetries the fundamental zone consists of the entire Frank–Rodrigues map. For the point groups 2, 3, 4, and 6 rotations along the symmetry axis are limited to $\pm\pi/n$ where $n = 2, 3, 4$, and 6 respectively. In those cases the fundamental zone is a slab of infinite area normal to the symmetry axis. The faces of the slab are at a distance of $\tan(\pi/2n)$ from the origin. The other fundamental zones are polyhedra, and they are illustrated in Fig. 1.5. In the orthorhombic system the fundamental zone is a cube of length $2\tan(\pi/4) = 2$; see Fig. 1.5(a). The fundamental zone for the point groups 32, $\bar{3}m$, and $3m$ of the trigonal system is a hexagonal prism with hexagonal faces normal to the triad axis, and square prism faces (with edge length $2\tan\pi/6$) normal to the diad axes at unit distance from the origin; see Fig. 1.5(b). The fundamental zone for the point groups 422, $4/mmm$, $4mm$, and $\bar{4}2m$ of the tetragonal system is an octagonal prism with octagonal faces normal to the tetrad axis at $\tan(\pi/8)$ from the origin and square prism faces normal to the diad axes at unit distance from the origin; see Fig. 1.5(c). In the hexagonal system for the point groups 622, $6/mmm$, $6mm$, and $\bar{6}m2$ the fundamental zone is a dodecagonal prism with the dodecagonal faces normal to the hexad axis at $\tan(\pi/12)$ from the origin. The 12 prism faces are squares with edge lengths equal to $2\tan(\pi/12)$, and they are normal to diad axes at unit distance from the origin; see Fig. 1.5(d). There are two fundamental zones in the cubic system. For the point groups 23 and $m\bar{3}$ the fundamental zone is an octahedron with faces at $\tan(\pi/6)$ from the origin; see Fig. 1.5(e). For the point groups 432, $m\bar{3}m$, and $\bar{4}3m$ the fundamental zone is a semiregular truncated cube, with six octagonal faces normal to the tetrad axes at a distance from the origin of $\tan(\pi/8)$, and eight triangular faces normal to triad axes at a distance from the origin of $\tan(\pi/6)$, see Fig. 1.5(f).

We shall now show that when a point exits the zone through a face it re-enters the zone at the opposite face but with a rotation about the zone face centre. This is the analogue of the Umklapp process in a Brillouin zone. Consider again the fundamental zone in an orthorhombic crystal. The point $\rho_A^R = (-1, y, z)$, where $|y| \leqslant 1$ and $|z| \leqslant 1$,

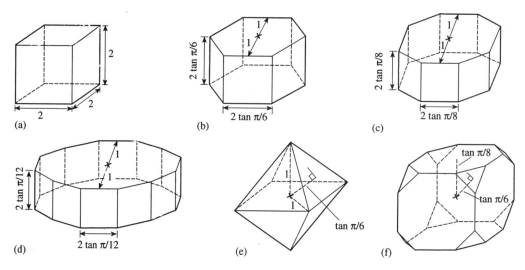

**Fig. 1.5** The fundamental zones of the forms of closed polyhedra: (a) cube for the point groups 222, 2*mm*, *mmm* of the orthorhombic system; (b) hexagonal prism for 32, $\bar{3}m$, 3*m* of the trigonal system; (c) octagonal prism for 422, 4/*mmm*, 4*mm*, and $\bar{4}2m$ of the tetragonal system; (d) dodecagonal prism for 622, 6/*mmm*, 6*mm*, and $\bar{6}m2$ of the hexagonal system; (e) octahedron for 23 and $m\bar{3}$ of the cubic system; (f) semiregular truncated cube for 432, $m\bar{3}m$, and $\bar{4}3m$ of the cubic system.

lies on the face $x = -1$. $\rho_A^R$ is equivalent to a point on the $x = 1$ face which is equal to $(\rho_A^R, [1, 0, 0] \tan \omega/2)$ where we shall take the limit $\omega \to \pi$. Using eqn (1.15) we find

$$(\rho_A^R, [1, 0, 0] \tan \omega/2) = \frac{[-1, y, z] + [1, 0, 0] \tan \omega/2 - [0, z, -y] \tan \omega/2}{1 + \tan \omega/2}$$

and taking the limit $\omega \to \pi$ we obtain $(\rho_A^R, [1, 0, 0] \tan \omega/2) = [1, -z, y]$. Thus the point $\rho_A^R$ is equivalent to a point on the opposite face that has been rotated by $\pi/2$ about the face centre. Similarly, points that re-enter at opposite faces related by triad, tetrad, or hexad axes are rotated by $\pi/3$, $\pi/4$, and $\pi/6$ respectively about the centres of those faces. Thus the self-connectedness of the zone has a twisted character, rather like a Möbius strip.

The point group symmetry of the crystal may be used further to deduce equivalent crystal directions. For example, in a cubic crystal there are up to 48 equivalent directions $\langle U, V, W \rangle$. Thus in the cubic fundamental zone there are up to 48 Rodrigues vectors $\langle U, V, W \rangle \tan \theta/2$ that give rise to equivalent dichromatic patterns. An irreducible wedge in a cubic fundamental zone may be defined, by analogy with the irreducible wedge of a Brillouin zone in a cubic crystal, to be the region bounded by the [100], [110], and [111] crystal axes and the faces of the zone. This is illustrated in Fig. 1.6 for the octahedral fundamental zone. Similar irreducible zones may be defined for the other fundamental zones of Table 1.1.

At a heterophase interface the two crystals generally have differing rotational symmetries, for example one crystal may have tetragonal point group symmetry 4/*mmm* and the other may have 222 orthorhombic point group symmetry. In that case it is necessary first to choose a suitable reference orientation. The reference orientation is arbitrary but a convenient choice is to align the principal rotational axes of the two crystals. In our

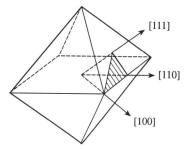

**Fig. 1.6** Irreducible wedge (shaded) in the octahedral fundamental zone for the point groups 23 and $m\bar{3}$ of the cubic
system. Salient crystal directions are shown.

example the tetrad axis of the tetragonal crystal is aligned with one of the diad axes of
the orthorhombic crystal, and the perpendicular diad axes of the tetragonal crystal are
aligned with the remaining diad axes of the orthorhombic crystal. The fundamental
zone is then determined by using all the rotational symmetries of the proper groups to
which the crystals are related in Table 1.1. In our example this will lead to the octagonal
fundamental zone of the tetragonal crystal. But new fundamental zones are generated
by some combinations of crystals. For example, if one crystal has point symmetry 6
and the other point symmetry 32, the reference orientation is defined by orienting the
hexad and triad axes parallel to each other. The fundamental zone is then a hexagonal
prism of height $2 \tan(\pi/12)$, rather than $2 \tan(\pi/6)$ for the trigonal crystal alone. To
find the fundamental zone in the general case first construct the fundamental zones
for the two crystals separately. Orient them so that the principal axes are aligned and
superimpose them concentrically. The fundamental zone is then the region that is
common to both zones. Mathematically, the fundamental zone is determined by the
union of the point group operations of the proper groups defining the fundamental zones
of the separate crystals.

### 1.3.3.4 *Quaternions*

Handscomb (1958) introduced the quaternion representation to address the problem of
finding the range of disorientations between two cubes possessing $m\bar{3}m$ point symmetry.
He showed that the quaternion representation has many of the advantages of the
Rodrigues representation, although it has the disadvantage of being four dimensional.
A rotation of $\theta$ about the axis $\hat{\rho}$ is represented by the unit quaternion

$$Q = (e_1, e_2, e_3, e_4) = (\rho_1 \sin \theta/2, \ \rho_2 \sin \theta/2, \ \rho_3 \sin \theta/2, \ \cos \theta/2). \qquad (1.20)$$

The four components $e_1, e_2, e_3, e_4$ are related by

$$e_1^2 + e_2^2 + e_3^2 + e_4^2 = 1 \qquad (1.21)$$

and thus there are only three independent components in the quaternion. Using eqns
(1.10), (1.20), and (1.21) the matrix representation may be deduced from the quaternion
representation as follows:

$$R_{ij} = (e_4^2 - e_1^2 - e_2^2 - e_3^2)\delta_{ij} + 2e_ie_j - 2 \sum_{k=1}^{3} \varepsilon_{ijk}e_ke_4 \qquad (1.22)$$

The Rodrigues vector is easily found to be $\rho^R = (e_1, e_2, e_3)/e_4$. If $Q_A$ and $Q_B$ denote
$(e_1^A, e_2^A, e_3^A, e_4^A)$ and $(e_1^B, e_2^B, e_3^B, e_4^B)$ then the components of the quaternion $Q = (e_1, e_2,$

$e_3, e_4$) obtained by performing the rotation represented by $Q_A$ followed by the rotation represented by $Q_B$, which we denote by $(Q_A, Q_B)$, are as follows:

$$\left.\begin{aligned}
e_1 &= e_1^A e_4^B + e_4^A e_1^B - e_2^A e_3^B + e_3^A e_2^B \\
e_2 &= e_2^A e_4^B + e_4^A e_2^B - e_3^A e_1^B + e_1^A e_3^B \\
e_3 &= e_3^A e_4^B + e_4^A e_3^B - e_1^A e_2^B + e_2^A e_1^B \\
e_4 &= e_4^A e_4^B - e_1^A e_1^B - e_2^A e_2^B - e_3^A e_3^B
\end{aligned}\right\} . \qquad (1.23)$$

It may be shown that this multiplication law is consistent with the corresponding rule for Rodrigues vectors expressed in eqn (1.15). The formula for $e_4$ in eqn (1.23) yields a useful expression for the angle, $\theta$, of the resultant rotation $(Q_A, Q_B)$ in terms of the component rotation angles $\theta_A$, $\theta_B$ and axes $\hat{\rho}_A, \hat{\rho}_B$:

$$\cos(\theta/2) = \cos(\theta_A/2)\cos(\theta_B/2) - \hat{\rho}_A \cdot \hat{\rho}_B \sin(\theta_A/2)\sin(\theta_B/2). \qquad (1.24)$$

If $\theta$ in eqn (1.20) is replaced by $\theta + 2\pi$ it is found that the quaternion changes sign. Thus, in the quaternion representation every rotation is represented by a pair of antipodal points on the unit sphere in 4D space, given by eqn (1.21). The identity is represented by $\pm(0, 0, 0, 1)$ and the inverse to $Q = (e_1, e_2, e_3, e_4)$ is represented by $\bar{Q} = \pm(e_1, e_2, e_3, -e_4)$. The change in orientation from B to A is obtained by forming $(Q_A, \bar{Q}_B)$ and using eqn (1.24):

$$\cos(\theta/2) = \cos(\theta_A/2)\cos(\theta_B/2) + \hat{\rho}_A \cdot \hat{\rho}_B \sin(\theta_A/2)\sin(\theta_B/2). \qquad (1.25)$$

From the expression for $e_4$ in eqn (1.23) we see that $\cos(\theta/2)$ in eqn (1.25) is just the 'dot product' of $Q_A$ and $\bar{Q}_B$ regarded as vectors in 4D space. Therefore, we may represent eqn (1.25) as $\cos(\theta/2) = Q_A \cdot \bar{Q}_B$. It follows that the condition for $Q$ to be orientationally equidistant from $Q_A$ and $Q_B$ is that

$$\bar{Q} \cdot (Q_A - Q_B) = 0 \qquad (1.26)$$

which is the equation of a plane midway between $Q_A$ and $Q_B$ on the 4D sphere. Since $-Q_A$ and $-Q_B$ also represent orientations A and B there is another orientationally equidistant plane orthogonal to that described by eqn (1.26), given by $\bar{Q} \cdot (Q_A + Q_B) = 0$. The fundamental zone of orientations may now be derived using eqn (1.26). Taking the reference orientation as $Q_r = (0, 0, 0, 1)$, which corresponds to $\theta = 0$, the fundamental zone is the polyhedron formed from planes that are orientationally equidistant between the reference orientation and symmetry rotations of the crystal. For example, consider an orthorhombic crystal with symmetry 222. The diads about the $\langle 100 \rangle$ directions are represented by $\pm(1, 0, 0, 0)$, $\pm(0, 1, 0, 0)$, $\pm(0, 0, 1, 0)$. Using eqn (1.26) to obtain the planes that are orientationally equidistant between the reference orientation and the diad rotations we obtain $\pm e_1 = e_4$, $\pm e_2 = e_4$, and $\pm e_3 = e_4$, which are the six planes of a 3D cube. The cube length is found by choosing say $\pm e_1 = e_4$ and setting $e_2 = e_3 = 0$. Then eqn (1.21) leads to the result that $\pm e_1 = e_4 = 1/\sqrt{2}$, and therefore the extremal values of the orientation along the $\langle 100 \rangle$ axes are $\pm 90°$. The corners of the cube are found by setting $\pm e_1 = \pm e_2 = \pm e_3 = e_4$ and using eqn (1.21) again. The result is $e_4 = 1/2$ and therefore the maximum disorientation is $120°$ about a $\langle 111 \rangle$ axis. The fundamental zone is therefore identical, apart from a length scaling, to that shown in Fig. 1.5(a). Indeed, the 11 fundamental zones obtained with the quaternion representation are identical to those obtained with the Rodrigues vector representation, apart from a trivial scaling of length. Thus, Handscomb (1958) was the first to obtain the fundamental zone for the $432$, $m\bar{3}m$, and $\bar{4}3m$ point groups.

Quaternions are particularly useful for finding the 24 equivalent representations of a rotation in a cubic crystal with point group symmetry 432, $m\bar{3}m$, or $\bar{4}3m$. The 23 proper rotations which define the fundamental zone are 3 of $\pi$ about $\langle 1, 0, 0 \rangle$, 8 of $\pm 2\pi/3$ about $\langle 1, 1, 1 \rangle$, 6 of $\pi$ about $\langle 1, 1, 0 \rangle$, and 6 of $\pi/2$ about $\langle 1, 0, 0 \rangle$. Consider a general rotation represented by the unit quaternion $(e_1, e_2, e_3, e_4)$. By using the 23 quaternions representing the proper rotational symmetries it is easy to show that the 23 equivalent rotations are represented by the following quaternions:

$$\pm (e_4, e_3, -e_2, -e_1)$$
$$\pm (-e_3, e_4, e_1, -e_2)$$
$$\pm (e_2, -e_1, e_4, -e_3)$$
$$\cdot \; \pm (e_1 + e_4 + e_2 - e_3, e_2 + e_4 + e_3 - e_1, e_3 + e_4 + e_1 - e_2, e_4 - e_1 - e_2 - e_3)/2$$
$$\pm (e_1 - e_4 - e_2 + e_3, e_2 - e_4 - e_3 + e_1, e_3 - e_4 - e_1 + e_2, e_4 + e_1 + e_2 + e_3)/2$$
$$\pm (e_1 - e_4 + e_2 - e_3, e_2 + e_4 - e_3 - e_1, e_3 + e_4 + e_1 + e_2, e_4 + e_1 - e_2 - e_3)/2$$
$$\pm (e_1 + e_4 - e_2 + e_3, e_2 - e_4 + e_3 + e_1, e_3 - e_4 - e_1 - e_2, e_4 - e_1 + e_2 + e_3)/2$$
$$\pm (e_1 + e_4 + e_2 + e_3, e_2 - e_4 + e_3 - e_1, e_3 + e_4 - e_1 - e_2, e_4 - e_1 + e_2 - e_3)/2$$
$$\pm (e_1 - e_4 - e_2 - e_3, e_2 + e_4 - e_3 + e_1, e_3 - e_4 + e_1 + e_2, e_4 + e_1 - e_2 + e_3)/2$$
$$\pm (e_1 + e_4 - e_2 - e_3, e_2 + e_4 + e_3 + e_1, e_3 - e_4 + e_1 - e_2, e_4 - e_1 - e_2 + e_3)/2$$
$$\pm (e_1 - e_4 + e_2 + e_3, e_2 - e_4 - e_3 - e_1, e_3 + e_4 - e_1 + e_2, e_4 + e_1 + e_2 - e_3)/2$$
$$\pm (e_2 - e_3, e_4 - e_1, e_4 + e_1, -e_2 - e_3)/\sqrt{2}$$
$$\pm (e_4 + e_2, e_3 - e_1, e_4 - e_2, -e_1 - e_3)/\sqrt{2}$$
$$\pm (e_4 - e_3, e_4 + e_3, e_1 - e_2, -e_1 - e_2)/\sqrt{2}$$
$$\pm (-e_2 - e_3, e_4 + e_1, -e_4 + e_1, -e_2 + e_3)/\sqrt{2}$$
$$\pm (e_4 - e_2, e_3 + e_1, -e_4 - e_2, -e_1 + e_3)/\sqrt{2}$$
$$\pm (e_4 + e_3, -e_4 + e_3, -e_1 - e_2, -e_1 + e_2)/\sqrt{2}$$
$$\pm (e_1 + e_4, e_2 + e_3, e_3 - e_2, e_4 - e_1)/\sqrt{2}$$
$$\pm (e_1 - e_3, e_2 + e_4, e_3 + e_1, e_4 - e_2)/\sqrt{2}$$
$$\pm (e_1 + e_2, e_2 - e_1, e_3 + e_4, e_4 - e_3)/\sqrt{2}$$
$$\pm (e_1 - e_4, e_2 - e_3, e_3 + e_2, e_4 + e_1)/\sqrt{2}$$
$$\pm (e_1 + e_3, e_2 - e_4, e_3 - e_1, e_4 + e_2)/\sqrt{2}$$
$$\pm (e_1 - e_2, e_2 + e_1, e_3 - e_4, e_4 + e_3)/\sqrt{2} \tag{1.27}$$

These quaternions represent the 23 equivalent axis/angle descriptions of any given rotation. The disorientation corresponds to the quaternion with the largest fourth component. For example, consider the rotation represented by $(4, 3, 1, 5)/\sqrt{51}$. The disorientation is obtained with the quaternion

$$(e_1 - e_4 - e_2 + e_3, e_2 - e_4 - e_3 + e_1, e_3 - e_4 - e_1 + e_2, e_4 + e_1 + e_2 + e_3)/2$$
$$= (-3, 1, -5, 13)/2\sqrt{51},$$

which represents a rotation of $2 \cos^{-1} 13/2\sqrt{51} = 48.94°$ about $[-3, 1, -5]$. In a cubic crystal we may change the sign of any component of a quaternion because all $\langle U, V, W \rangle$ crystal directions are equivalent, and changing the sign of $\cos(\theta/2)$ merely corresponds to replacing $\theta$ by $\theta + 2\pi$. Moreover, since $(e_1, e_2, e_3, e_4)$ is equivalent to $\pm(e_4, e_3, -e_2, -e_1)$, $\pm(-e_3, e_4, e_1, -e_2)$ and $\pm(e_2, -e_1, e_4, -e_3)$ we may interchange the order of the components of the quaternion in any way we choose. Thus we may choose to make all the components of the quaternion positive and order them such that $e_4 \geqslant e_3 \geqslant e_2 \geqslant e_1 \geqslant 0$. Then the disorientation is found by selecting the quaternion from only the following *three* that has the largest fourth component:

$$\left.\begin{array}{l} (e_1, e_2, e_3, e_4) \\ (e_1 - e_2, e_2 + e_1, e_3 - e_4, e_4 + e_3)/\sqrt{2} \\ (e_1 - e_4 - e_2 + e_3, e_2 - e_4 - e_3 + e_1, e_3 - e_4 - e_1 + e_2, e_4 + e_1 + e_2 + e_3)/2 \end{array}\right\} \quad (1.28)$$

In our example of the quaternion $(4, 3, 1, 5)/\sqrt{51}$ we first rewrite it as $(1, 3, 4, 5)/\sqrt{51}$ and compare $5/\sqrt{51}$ with the fourth components of $(-2, 4, -2, 9)/\sqrt{102}$ and $(-3, -5, 1, 13)/2\sqrt{51}$. Thus, we conclude, as before, that the disorientation is $2 \cos^{-1}(13/2\sqrt{51})$ about a $\langle 5, 3, 1 \rangle$ axis.

## 1.4  GEOMETRICAL SPECIFICATION OF AN INTERFACE

### 1.4.1  Macroscopic and microscopic geometrical degrees of freedom

The minimum number of geometric variables required to specify a complete geometrical characterization of the interface is called the number of geometrical degrees of freedom. We distinguish between macroscopic and microscopic degrees of freedom of an interface. The macroscopic degrees of freedom can be thought of as the information required to manufacture a bicrystal from given crystals, with a particular orientation relation between the crystals and a particular interfacial plane. As we shall see in the next section there are five such degrees of freedom. The microscopic degrees of freedom are a summary description of the atomic structure of the interface, and they are determined by relaxation processes at the interface. If the atomic structure of the interface is periodic the rigid body displacement, $t$, of one crystal relative to the other defines three microscopic degrees of freedom. In the absence of periodicity in the interface these three degrees of freedom reduce to one: the component of $t$ normal to the interface, which is called the expansion of the interface. The location of the interface plane is, in general, a further microscopic degree of freedom if either crystal atomic basis is greater than monatomic. Since the four microscopic degrees of freedom are determined by relaxation processes they may not be varied independently of the macroscopic degrees of freedom. Thus the macroscopic degrees of freedom specify boundary conditions far from the interface and the microscopic degrees of freedom adjust in such a way as to minimize the free energy of the system subject to those boundary conditions. We shall see in Section 5.5 that the macroscopic degrees of freedom constitute geometric thermodynamic variables which are required in a full thermodynamic description of the interface. Similarly, any chemical composition variations at the interface will also be driven by the minimization of the free energy of the system subject to the same boundary conditions. The interface normal may change locally through faceting (see Section 5.6.3), and in that case the specification of the interface normal as a macroscopic degree of freedom is meaningful only in the sense that it defines the average interface normal (see Section 6.2).

In Section 1.3.3.3 it was pointed out that if either crystal displays enantiomorphism then there is an additional degree of freedom associated with the handedness of either crystal. At a given orientation between the crystals and interface normal there are two bicrystals that are related by an inversion of an enantiomorphic crystal. It is not possible to transform one bicrystal into the other except by inverting one crystal structure. We classify this as a sixth macroscopic degree of freedom because whether an enantiomorphic crystal displays the left- or right-handed form is independent of the relaxation processes at the interface. It forms part of the boundary conditions imposed on the relaxation at the interface. We shall not comment further on enantiomorphism because it is independent of the other macroscopic degrees of freedom.

In summary, there are, in general, ten degrees of geometrical freedom associated with an interface: six macroscopic and four microscopic.

### 1.4.2  Macroscopic geometrical degrees of freedom of an arbitrary interface

In this section we discuss the macroscopic geometrical degrees of freedom associated with the misorientation relation betwee the crystals and the inclination of the interface normal. It was pointed out in Section 1.3.3.3 that for bicrystals containing nonenantiomorphic crystals it is always possible to express the relationship between two given crystals as a proper rotation. In such cases bicrystals obtained by applying inversions, improper rotations, or mirror reflections may always be generated alternately by proper rotations. There are three degrees of freedom associated with the specification of a proper rotation: two for the unit vector along the rotation axis, $\hat{\rho}$, and one for the rotation angle, $\theta$. Two further degrees of freedom are then required to specify the interface unit normal, $\hat{n}$. Therefore, there are always five macroscopic degrees of freedom associated with the orientation relation between the crystals and the interface normal.

In deformation twinning and martensitic transformation the relationship between the lattices is an invariant plane strain. This is an affine transformation which has the property that the interface is an undistorted and unrotated plane of the transformation. Again, five macroscopic degrees of freedom are associated with such an interface and all five are required to specify the invariant plane strain transformation. We shall return to invariant plane strain deformations in the next chapter. In this section we confine our attention to crystals related by proper rotations.

It is sometimes useful to specify the five macroscopic degrees of freedom in a way which focuses attention more on the interface plane normal in both crystals rather than the relationship between the crystals. After all, we can regard the creation of an interface as the sum of two operations: first bring together two crystal surfaces with normals $\hat{n}$ and $\hat{n}'$, and then rotate (twist) one crystal with respect to the other about $\hat{n}$ (or $\hat{n}'$) by the angle $\theta_{\text{twist}}$. Since $\hat{n}$ and $\hat{n}'$ are unit vectors only four degrees of freedom are consumed in the first operation. The fifth is then the angle, $\theta_{\text{twist}}$. The relationship between the crystals can be deduced from $\hat{n}$, $\hat{n}'$, and $\theta_{\text{twist}}$ once certain conventions are established for the senses of the normals and the coordinate systems in which they are expressed. In noncubic crystals these relationships are very cumbersome and there seems little point in pursuing them. In the next section we therefore confine the discussion to grain boundaries in cubic materials.

### 1.4.3  Grain boundaries in cubic materials

#### 1.4.3.1  *The median lattice and the mean boundary plane*

In order to manipulate vectors belonging to both crystals we must first choose a coordinate system. An obvious choice is the coordinate system of either crystal, but this may not be the most convenient choice. A particularly useful choice is the coordinate system of the median lattice, which is used throughout Section 1.4.3. The two crystal lattices are obtained from the median lattice by equal and opposite rotations, as illustrated in

## The geometry of interfaces

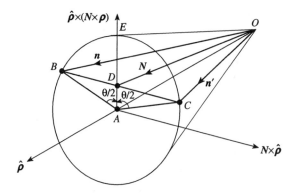

**Fig. 1.7** Diagram illustrating the derivation of eqn (1.29). Prior to the creation of the grain boundary two surface normals $n$ and $n'$ of a single crystal, which is called the median lattice, are selected. The mean boundary plane is defined as $N = (n + n')/2$. The components of $n$, $n'$, and $N$ are referred to the median lattice. The surface normals $n$ and $n'$ are brought into parallel alignment, along **OE**, at the grain boundary by rotations of equal and opposite amounts of $\theta/2$ about the axis $\hat{\rho}$. During this operation the vectors $n$ and $n'$ are rotated but their components remain unchanged and referred to the median lattice.

Fig. 1.7. A grain boundary is created by rotating surface normals $n$ and $n'$ by $\theta/2$ in opposite senses about the rotation axis $\hat{\rho}$. Note that $n$ and $n'$ are not necessarily unit vectors any more but they must have the same length. Prior to the rotations the mean boundary plane is along $N = (n + n')/2$. After the rotation the vectors $n$ and $n'$ are parallel but their components remain as they were in the unrotated state. By considering the triangle OBC bounded by $n$ and $n'$ it may be deduced that

$$n = N - \mu(N \times \hat{\rho})/|N \times \hat{\rho}|$$

and

$$n' = N + \mu(N \times \hat{\rho})/|N \times \hat{\rho}|,$$

where $\hat{\rho}$ is the rotation axis and $\mu$ is equal to $BD = DC$ and $\tan \theta/2 = \mu/AD$. But $AD$ is equal to $|N - (N \cdot \hat{\rho})\hat{\rho}| = |N \times \hat{\rho}|$. Thus $\mu = |N \times \hat{\rho}| \tan \theta/2 = |N \times \hat{\rho}^R|$, and therefore

$$n = N - N \times \rho^R$$

and

$$n' = N + N \times \rho^R, \tag{1.29}$$

where $\rho^R$ is the Rodrigues vector $\hat{\rho} \tan \theta/2$. Therefore, $n$ and $n'$ are rational provided $N$ and $\rho^R$ are rational. $N$ is rational provided the mean boundary plane is rational. As will be shown in Section 1.5.4 the Rodrigues vector is rational in a cubic system at any coincidence site lattice orientation. It follows that all grain boundaries at coincidence site lattice orientations with rational mean boundary planes are themselves rational with respect to both grains. Conversely, all rational grain boundaries in a coincidence system may be generated from eqn (1.29) by allowing the mean boundary plane to range over all rational crystal lattice planes.

### 1.4.3.2  *Tilt and twist components*

It is often useful to express the total misorientation of a boundary in terms of tilt and twist components. We imagine that the boundary is created by two successive rotations about perpendicular axes. First a tilt rotation is performed, which is defined by the condition that the rotation axis is in the boundary plane. Secondly a twist rotation is performed about the boundary plane normal, as described above. The order in which these rotations are performed could be reversed although, since they do not commute, different rotation angles would be involved. The decomposition into tilt and twist components is dependent on the boundary plane as well as the total rotation. Similarly, the decomposition is dependent on which of the 24 equivalent descriptions of the rotation is selected since the angles between the boundary plane normals, $\hat{n}$ and $\hat{n}'$, and the rotation axis differ from one description to the next. There is, therefore, a lack of uniqueness in the specification of the tilt and twist components of a boundary. In the following we assume a particular choice of rotation description has been made, which is consistent with a boundary having normals $\hat{n}$ and $\hat{n}'$ in either crystal.

The tilt axis must be along $\hat{n} \times \hat{n}'$, since this is perpendicular to both $\hat{n}$ and $\hat{n}'$. The tilt angle, $\theta_{\text{tilt}}$, is the angle between $\hat{n}$ and $\hat{n}'$, $\hat{BOC}$ in Fig. 1.7, and is given by

$$\tan^2(\theta_{\text{tilt}}/2) = (\hat{N} \times \rho^R)^2. \tag{1.30}$$

Therefore the Rodrigues vector representing the tilt rotation is

$$\rho^R_{\text{tilt}} = |\hat{N} \times \rho^R| (\hat{n} \times \hat{n}')/|\hat{n} \times \hat{n}'|. \tag{1.31}$$

Since $\rho^R = (\rho^R_{\text{tilt}}, \rho^R_{\text{twist}})$ where $\rho^R_{\text{twist}}$ is the Rodrigues vector for the twist rotation it follows from eqn (1.15) that

$$\rho^R = \rho^R_{\text{tilt}} + \rho^R_{\text{twist}} - \rho^R_{\text{tilt}} \times \rho^R_{\text{twist}}, \tag{1.32}$$

and therefore

$$\tan^2(\theta/2) = \tan^2(\theta_{\text{tilt}}/2) + \tan^2(\theta_{\text{twist}}/2) + \tan^2(\theta_{\text{tilt}}/2)\tan^2(\theta_{\text{twist}}/2). \tag{1.33}$$

Using eqns (1.33) and (1.30) we obtain

$$\tan^2 \theta_{\text{twist}}/2 = \frac{(\rho^R \cdot N)^2}{N^2 + (N \times \rho^R)^2}. \tag{1.34}$$

Equations (1.30) and (1.34) enable the tilt and twist components of any grain boundary to be written down in terms of the the mean boundary plane and the Rodrigues vector for the total misorientation. We note that the relationship (eqn 1.33)) between $\theta$, $\theta_{\text{tilt}}$, and $\theta_{\text{twist}}$ may be expressed more compactly using eqn (1.24):

$$\cos(\theta/2) = \cos(\theta_{\text{tilt}}/2)\cos(\theta_{\text{twist}}/2). \tag{1.35}$$

Sometimes the grain boundary is specified by the Rodrigues vector $\rho^R$ and the boundary plane normal $\hat{n}$ in one grain. The boundary plane in the other grain $\hat{n}'$ is then given by

$$(1 + \rho^{R^2})\hat{n}' = 2(\rho^R \cdot \hat{n})\rho^R + (1 - \rho^{R^2})\hat{n} - 2\hat{n} \times \rho^R, \tag{1.36}$$

The mean boundary plane is parallel to

$$N = (\rho^R \cdot \hat{n})\rho^R + \hat{n} - \hat{n} \times \rho^R. \tag{1.37}$$

The tilt component is a rotation of $\theta_{\text{tilt}}$ about an axis parallel to

$$\text{tilt axis} = (\rho^R \cdot \hat{n})(\hat{n} \times \rho^R - \hat{n}) + \rho^R, \tag{1.38}$$

where

$$\tan^2(\theta_{tilt}/2) = \frac{\rho^{R^2} - (\rho^R \cdot \hat{n})^2}{1 + (\rho^R \cdot \hat{n})^2}. \tag{1.39}$$

Using eqn (1.33) it follows that the twist angle $\theta_{twist}$ is given by:

$$\tan(\theta_{twist}/2) = \pm \rho^R \cdot \hat{n}. \tag{1.40}$$

It is sometimes desirable to know the Rodrigues vectors that can generate a boundary plane with normals $n$ and $n'$. This is readily solved using eqn (1.29), from which we deduce that

$$n' - n = (n + n') \times \rho^R. \tag{1.41}$$

The Rodrigues vectors that satisfy this equation are of the following form:

$$\rho^R = \lambda n' \times n + \mu N. \tag{1.42}$$

$\mu$ is an arbitrary number but $\lambda$ is determined by the condition that the lengths of the vectors on either side of eqn (1.41) are equal:

$$\rho^R = \frac{n' \times n}{n^2 + n \cdot n'} + \mu N. \tag{1.43}$$

As an example of the use of eqn (1.43) suppose we required the set of Rodrigues vectors that would generate a boundary with $n = [22\bar{1}]$ and $n' = [001]$. We first multiply $n'$ by 3 to give it the same length as $n$: $n' = [003]$. Then eqn (1.43) becomes

$$\rho^R = [1\bar{1}0] + \mu[111]. \tag{1.44}$$

Thus for $\mu = 0$ we have $\rho^R = [1\bar{1}0]$, which represents a rotation of $2\tan^{-1}\sqrt{2}$ about $[1\bar{1}0]$, while for $\mu = \frac{1}{2}$ we have $\rho^R = \frac{1}{2}[3\bar{1}1]$ which represents a rotation of $2\tan^{-1}(\sqrt{11}/2)$ about $[3\bar{1}1]$, and so on. There is an infinite set of $n = [22\bar{1}]$, $n' = [003]$ boundaries corresponding to the infinite number of choices for $\mu$, but only rational values of $\mu$ will generate periodic boundary structures. Equation (1.44) is the equation of a straight line in the Frank–Rodrigues map. The point $\rho^R = [1\bar{1}0]$ is the rotation required to generate the pure asymmetric tilt $(22\bar{1})/(001)$ boundary. A subsequent, arbitrary twist rotation about the boundary normal sends the Rodrigues vector along the line through $[1\bar{1}0]$ parallel to $[111]$.

### 1.4.3.3 *Symmetric and asymmetric tilt boundaries*

Pure tilt boundaries are defined by the condition that $\hat{\rho}$ lies in the boundary plane. In that case $\theta = \theta_{tilt}$ and $\theta_{twist} = 0$. Two types of tilt boundary are normally distinguished: symmetric and asymmetric. The plane of a symmetric tilt boundary has the same Miller index form in both crystals, e.g. (571) in one crystal and $(75\bar{1})$ in the other. Symmetric tilt boundaries are also frequently called type I twin boundaries or simply twins. The relationship between the crystal structures across a type I twin may be described as a mirror reflection in the boundary plane, which is equivalent to a two-fold rotation about the boundary plane normal in a centrosymmetric crystal. Thus, a symmetric tilt boundary in a centrosymmetric crystal may also be described as a 180° twist boundary (Christian 1975, p. 52). An asymmetric tilt boundary plane has different Miller index forms in either crystal, e.g. (430) in one crystal and (010) in the other.

The distinction between symmetric and asymmetric tilt boundaries is particularly clear in terms of the mean boundary plane. The mean boundary plane of any symmetric tilt boundary must be a mirror plane of the perfect crystal, whereas the mean boundary plane of an asymmetric tilt boundary must not be a mirror plane of the perfect crystal. The above example of a symmetric tilt boundary has a mean boundary plane normal of $[571] + [75\bar{1}] = [12, 12, 0]$ which is normal to the (110) mirror plane. To find the mean boundary plane of an asymmetric tilt boundary we first ensure that $n$ and $n'$ have the same length and add them. In the above example we add [430] and [050] to get [480] which is normal to the (120) mean boundary plane. We stress that these relations follow only if the coordinate system of the median lattice is used.

Symmetric tilt boundaries exist on all planes. To find the symmetric tilt boundary on $(hkl)$ we first set $n = [hkl]$, and $n' = [h\overline{kl}]$ so that the mean boundary plane normal is along a mirror plane normal, in this case [100]. The tilt angle is then given by $\cos\theta = n \cdot n' / |n|^2 = (h^2 - k^2 - l^2)/(h^2 + k^2 + l^2)$ and the tilt axis by $n \times n'$ which in this case is parallel to $[0\overline{l}k]$. Since we could have chosen either of the other two $\{100\}$ mirror planes or one of the six $\{110\}$ mirror planes for the mean boundary plane there are up to nine possible descriptions of the $(hkl)$ symmetric tilt boundary. The macroscopic degrees of freedom of a symmetric tilt boundary are defined by the plane on which it lies, and obviously there are only two degrees of freedom associated with it.

Since the mean boundary plane of a symmetric tilt boundary must always be either $\{110\}$ or $\{100\}$, the tilt axis must always lie in the zone of either $\langle 110 \rangle$ or $\langle 100 \rangle$. Conversely, there are no symmetric tilt boundaries with tilt axes that do not lie in one of these zones. For example, there are no symmetric $\langle 123 \rangle$ tilt boundaries. Asymmetric tilt boundaries may be generated only from mean boundary planes that are not mirror planes. The lowest index mean boundary plane satisfying this condition is $\{111\}$. There are four macroscopic degrees of freedom associated with an asymmetric tilt boundary, since the requirement that the cross product of the plane normals is along the tilt axis consumes one degree of freedom.

In general a boundary has tilt and twist components and it is then often described as a mixed tilt and twist boundary. Nevertheless, any boundary may be described as symmetric or asymmetric according to whether the components of $n$ and $n'$ have the same Miller index form (Wolf 1989). Thus a symmetric twist boundary is the same as what is normally called a pure twist boundary. An asymmetric twist boundary is an asymmetric tilt boundary that has been subjected to a further twist about the boundary normal. As we have already remarked, a 180° symmetric twist boundary $(hkl)$ is the symmetric tilt boundary $(hkl)$. Asymmetric twist boundaries are the most general boundaries and are associated with five macroscopic degrees of freedom.

## 1.5  BICRYSTALLOGRAPHY

### 1.5.1  Introduction

In this section we will describe the method introduced by Pond and Vlachavas (1983) to enumerate and classify the symmetries of a planar, homophase, or heterophase interface between two crystals. There are both geometrical and physical applications of this theory. Firstly, a systematic classification enables the grouping of bicrystals into classes and permits any generic relations between different types of interfaces to be established. Secondly, the work reveals variants of an interface that are related by symmetry. This is arguably the most significant application of the theory to date because it immediately leads to a classification of all interfacial line defects separating energetically degenerate regions. There is a direct analogy here with perfect line defects in single crystals because they are defined by symmetry operations of the crystal. For example, a perfect dislocation in a single crystal is classified by its Burgers vector, which must be a lattice translation vector. Similarly, a perfect disclination in a single crystal is defined by one of the rotational symmetries of the crystal. Thus, in both single crystals and bicrystals symmetry enables perfect line defects to be defined. The physical application of the theory, which has not been developed extensively, is to use irreducible representations of the symmetry groups to label eigenfunctions of the Hamiltonian for a bicrystal. This labelling describes the way that the eigenfunctions transform under all symmetry operations of the bicrystal and gives a considerable amount of information about the degeneracies. In this chapter we shall confine our attention to the geometrical aspects of the theory.

### 1.5.2  Outline of crystallographic methodology

A bicrystal is a composite object of two crystals. The symmetry of a heterophase bicrystal, i.e. one in which the component crystals do not share the same structure, is equal to the intersection of the symmetries of the component crystals. That is to say, the symmetry of the bicrystal is determined by the symmetry elements of the component crystals that survive when the bicrystal is created. In general the symmetry of the bicrystal will be lower than the symmetries of the component crystals and this lowering of the symmetry is referred to as *dissymmetrization*. However, there is an important difference when a homophase bicrystal is created, i.e. one in which the component crystals share the same structure. New symmetry elements may arise that relate the two component crystals, which do not exist in either component crystal in isolation. These new symmetry operations always relate one crystal to the other. They are called *colour reversing* or *antisymmetry* operations in the literature and we shall use the expression antisymmetry operation. The notion of colour is introduced to distinguish between operations relating sites in the same crystal from those relating sites in different crystals. If we label all the sites in one crystal black and those in the other white then *ordinary* operations relate black sites to black and white sites to white, while antisymmetry operations relate black sites to white and white sites to black. Antisymmetry operations may exist only at a homophase bicrystal, such as a grain boundary. Symbolically, antisymmetry operations are distinguished by a prime. Thus 2 and 2′ denote an ordinary and antisymmetry two-fold rotation axes respectively. As an example consider Fig. 1.8, which shows a composite object formed from two identical but misoriented rectangles. There is a diad axis at the centre of the two rectangles, which is all that survives of the ordinary operations of the isolated rectangles. But there are also two *m′* planes, at right angles to each other that relate one rectangle to the other. Clearly the existence of the *m′* planes is determined by the presence of *both* rectangles. Moreover, the orientation of the *m′* planes to the sides of the rectangles changes as the misorientation between the rectangles changes. An example of an interface displaying colour symmetry is a twin boundary. In a cubic crystal a twin boundary may be generated by reflecting one crystal across a plane that is *not* an ordinary mirror plane of the crystal. This is illustrated in Fig. 1.9 for the (111) twin in an f.c.c. crystal, which is shown in projection along $[1\bar{1}0]$. The twin boundary is an antimirror plane and there is also an antidiad along $[11\bar{2}]$. Since $[1\bar{1}0]$ is common to both crystals there is an ordinary mirror along this direction. Thus the point group symmetry of this interface is $2′mm′$. There is a convention that the symmetry elements in the point group specification are associated with a right handed cartesian frame with the $z$-axis

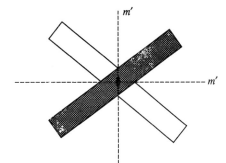

**Fig. 1.8** Antisymmetry mirror planes, $m'$, in a composite object of two identical but misoriented rectangles, one of which is black and the other white. There is also an ordinary diad axis normal to the page through the centres of the rectangles.

**Fig. 1.9** The atomic structure of the (111) twin in f.c.c. crystals seen in projection along $[1\bar{1}0]$. Atoms in the two $(2\bar{2}0)$ planes in each $\frac{1}{2}[1\bar{1}0]$ period are distinguished by crosses and triangles. Atoms in the lower crystal are coloured black and those in the upper crystal white. The boundary plane coincides with an antisymmetry mirror plane and an antisymmetry diad axis. The plane of the drawing (i.e. $(1\bar{1}0)$) is an ordinary mirror plane.

along the boundary normal. Thus, in $2'mm'$ there is a $2'$ axis along $x$, an ordinary mirror plane along $y$, and the boundary plane is an antimirror plane.

Pond has introduced a systematic method for deriving the symmetry of an interface consisting of four stages. At each stage the symmetry of the object is either the same or lower than the symmetry at the previous stage. This four-stage process of dissymmetrization enables variants to be identified at each stage. Thus, four classes of variants are identified and they each have distinct geometrical significance.

Since we are considering a process of dissymmetrization it is necessary to start with an object that has maximal symmetry. The choice of this object is not trivial and the treatment of Pond and coworkers differs from that of others on this point. There is no fundamental principle to help us choose an object with maximal symmetry. Rather, the choice we make will lead to certain classes of variants and the best choice will be the one which can account for all the experimentally observed variants.

Pond starts with the *lattices* of the separate black and white crystals. The decoration of each lattice with atomic bases to produce the black and white crystal structures is

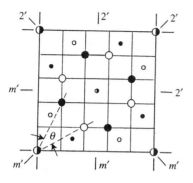

**Fig. 1.10**  [001] projection of the dichromatic pattern formed by rotating two f.c.c. lattices (one coloured black and the other white) by $\theta = 2\tan^{-1}\frac{1}{3}$ about [001]. The (projected) CSL unit cell coincides with the bounding square. The fine square mesh inside the CSL cell is the (projected) DSC lattice. (From Pond and Bollman (1979).)

deferred to a later stage. The space groups of the black and white lattices are denoted by $\Phi^{b^*}$ and $\Phi^{w^*}$. Since these are Fedorov space groups of ordinary symmetry operations the Russian 'f', $\Phi$, is used. The asterisk indicates that the groups are holosymmetric. The structures of the black and white crystals may have lower symmetries than these maximal symmetry groups, depending on the atomic bases.

The first stage of dissymetrization is to allow the black and white lattices to inter-penetrate to form the *dichromatic pattern*. In the creation of the dichromatic pattern one lattice is rotated to introduce the relative orientation of the two crystals that will exist in the final interface. Figure 1.10 illustrates a dichromatic pattern formed by two f.c.c. lattices misoriented by a rotation of $\theta = 2\tan^{-1}\frac{1}{3}$ about $\hat{\rho} = [001]$. Further examples are shown in Fig. 1.12. Both ordinary and antisymmetry operations may exist in the dichromatic pattern, as seen in Fig. 1.10. The space group of the dichromatic pattern may therefore contain both types of operation and such a group is called a colour group or antisymmetry group. The symbol used to denote a colour group is a Russian 'sh', Ш, after the Russian crystallographer Shubnikov, who pioneered the development of colour groups. The symmetry of the dichromatic pattern depends, in general, on the relative translation of the two lattices. We may adjust the relative translation of the two lattices to obtain the colour group with maximal symmetry, which we denote by Ш*(p). The 'p' indicates that it is the colour group of the dichromatic pattern. The ordinary symmetry elements of $\Phi^{b^*}$ and $\Phi^{w^*}$ that are not present in Ш*(p) relate equivalent dichromatic patterns. This is the first example of dissymetrization and both point group and translational operations may be suppressed in Ш*(p). The variants of a particular interface that exist in the equivalent dichromatic patterns are called *orientational* or *translational variants* depending on whether they arise from the suppression of point group or translational symmetry operations in Ш*(p). The translational variants are generated by translating one crystal lattice relative to the other by any combination of black and white crystal lattice vectors that is not a translation vector of the dichromatic pattern. The effect of such translations is to recreate the identical dichromatic pattern but with a shift of origin. An example of orientational variants is the formation of self-accommodating groups of variants of a martensitic phase within a parent phase. Each variant produces a certain shape change of the parent phase. By forming orientational variants the resultant shape change is reduced, resulting in a lower accommodation energy in the parent phase.

In the second stage of dissymetrization we allow for the fact that both crystals meeting at the interface may not have holosymmetric or symmorphic space groups. A lattice complex is the set of points obtained by carrying out on a chosen point all the symmetry operations of the *crystal's* space group, as opposed to the crystal lattice's space group. Each point will have an identical environment except possibly for orientation. The space groups of the black and white lattice complexes are denoted by $\Phi^b$ and $\Phi^w$, and they are not necessarily the same as $\Phi^{b^*}$ and $\Phi^{w^*}$. If we allow the two lattice complexes to interpenetrate we obtain the *dichromatic complex*. The holosymmetric colour space group of the dichromatic complex is called Щ*(c). For holosymmetric and symmorphic crystals $\Phi^b = \Phi^{b^*}$, $\Phi^w = \Phi^{w^*}$, and Щ*(c) = Щ*(p). But for non-holosymmetric and/or non-symmorphic crystals Щ*(c) is a subgroup of Щ*(p) and then there exist *complex variants* of Щ*(c) related by the symmetry elements Щ*(p) that are not present in Щ*(c). Complex variants give rise to domain structures on a given interfacial plane.

Up to now we have still not introduced an interface between the crystals. The third stage of dissymmetrization consists of sectioning the dichromatic complex on the interface plane and discarding one lattice complex on one side of the section and the other lattice complex on the other side. Thus we create an unrelaxed bicrystal. The symmetry of the bicrystal depends, in general, on the relative translation of the two complexes and on the location of the interface plane. The holosymmetric space group of the unrelaxed bicrystal is denoted Щ*(i). In general Щ*(i) is a subgroup of Щ*(c) and the symmetry elements that are present in Щ*(c) but not present in Щ*(i) relate *morphological variants*. Morphological variants that arise from the suppression of point symmetry elements in Щ*(i) may determine the symmetry of a crystal embedded within another crystal, although other shapes are possible as will be discussed in Section 1.8.

The final stage of dissymmetrization consists of 'switching on' the interatomic forces to enable the bicrystal to minimize its free energy at a particular temperature and pressure. (We assume that the boundary conditions applied to the bicrystal are such as to ensure that the interface does not migrate right out of it!) During the relaxation the interface plane may migrate into either crystal, cooperative relaxation involving one crystal translating rigidly with repect to the other may occur, each atom may adjust its position to relax any resultant force acting on it, and material may be added or removed at the interface. At the end of this process the symmetry of the bicrystal is denoted by Щ(i). If Щ*(i) contains symmetry elements that are not present in Щ(i) then *relaxational variants* of the relaxed bicrystal will exist that are related by the suppressed elements of Щ*(i). The existence of relaxational variants cannot be predicted by symmetry arguments, because they depend on physical and chemical interactions in the bicrystal.

We may summarize the four-stage procedure of dissymmetrization as follows:

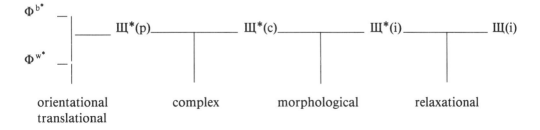

### 1.5.3  Introduction to Seitz symbols

Before we proceed to discuss the symmetries of dichromatic patterns, complexes, and bicrystals we introduce some notation that will be useful. A space group operation generally consists of a point group operation combined with a lattice translation operation. In a non-symmorphic space group there are operations, i.e. screw rotation axes and mirror glide planes, that consist of a point group operation combined with a supplementary displacement. The supplementary displacement is never a lattice translation operation. In the Seitz notation an arbitrary space group operation of the black crystal is denoted by $[\mathbf{D}_j^b | R_i^b + \alpha_j^b]$. Here $\mathbf{D}_j^b$ denotes the $(3 \times 3)$ matrix representing the $j$th point group operation of the black crystal; in the case of a non-symmorphic operation $\alpha_j^b$ denotes the supplementary displacement associated with $\mathbf{D}_j^b$. $R_i^b$ denotes the $i$th lattice translation vector of the black crystal. An arbitrary vector $r$, when acted upon by this space group operation, transforms as follows:

$$[\mathbf{D}_j^b | R_i^b + \alpha_j^b] r = \mathbf{D}_j^b r + R_i^b + \alpha_j^b. \tag{1.45}$$

The 'product' of two Seitz symbols is obtained by applying eqn (1.45) twice:

$$[\mathbf{D}_k^b | R_n^b + \alpha_k^b] [\mathbf{D}_j^b | R_i^b + \alpha_j^b] r = [\mathbf{D}_k^b | R_n^b + \alpha_k^b] (\mathbf{D}_j^b r + R_i^b + \alpha_j^b)$$

$$= \mathbf{D}_k^b \mathbf{D}_j^b r + \mathbf{D}_k^b (R_i^b + \alpha_j^b) + R_n^b + \alpha_k^b$$

$$= [\mathbf{D}_k^b \mathbf{D}_j^b | \mathbf{D}_k^b (R_i^b + \alpha_j^b) + R_n^b + \alpha_k^b] r.$$

Therefore,

$$[\mathbf{D}_k^b | R_n^b + \alpha_k^b] [\mathbf{D}_j^b | R_i^b + \alpha_j^b] = [\mathbf{D}_k^b \mathbf{D}_j^b | \mathbf{D}_k^b (R_i^b + \alpha_j^b) + R_n^b + \alpha_k^b]. \tag{1.46}$$

Note that in general the product of two Seitz symbols does not commute. It follows from the 'multiplication law', eqn (1.46), that the inverse of the Seitz symbol $[\mathbf{D}_j^b | R_i^b + \alpha_j^b]$ is $[\mathbf{D}_j^{b-1} | -\mathbf{D}_j^{b-1} (R_i^b + \alpha_j^b)]$:

$$[\mathbf{D}_j^b | R_i^b + \alpha_j^b] [\mathbf{D}_j^{b-1} | -\mathbf{D}_j^{b-1} (R_i^b + \alpha_j^b)] = [\mathbf{E} | \mathbf{0}]. \tag{1.47}$$

The identity operation $\mathbf{E}$ is the $3 \times 3$ identity matrix.

### 1.5.4  Symmetry of dichromatic patterns

In order to discuss the symmetry of a dichromatic pattern we first have to specify the relationship between the black and white lattices. Each basis vector in the black lattice $a_i^b$ is related to the basis set of the white lattice by the vector transformation described by eqn (1.6):

$$a_i^b = \sum_{j=1}^{3} T_{ji}' a_j^w + t, \tag{1.48}$$

where

$$\mathbf{T} = \mathbf{PR}. \tag{1.49}$$

The prime on the $T_{ji}$ in eqn (1.48) denotes colour reversal because the operation is relating lattice vectors of different colours. It will be recalled that $t$ denotes the displacement of the black lattice that is required to bring at least one of its sites into coincidence with a white lattice site. We have already remarked in Section 1.3.2 that the specification of $\mathbf{T}$ is not unique. In this section we will show that despite this the symmetry of the dichromatic pattern is invariant with respect to alternative equivalent specifications of **T**.

The problem at hand is the following: given the space groups $\Phi^{b^*}$ and $\Phi^{w^*}$ of the black and white lattices and the relationship between the lattices $[\mathbf{T}'|t]$, what are the space group operations comprising $Ш^*(p)$?

The group of ordinary operations $\Phi(p)$ of a dichromatic pattern is equal to the intersection between the black and white lattice space group operations:

$$\Phi(p) = \Phi^{w^*} \cap [\mathbf{T}'|t]\Phi^{b^*}[\mathbf{T}'|t]^{-1}. \tag{1.50}$$

This is a rather symbolic way of saying that we seek the intersection between the infinite sets of operations comprising $\Phi^{b^*}$ and $\Phi^{w^*}$ when they are both expressed in the coordinate system of the white lattice. We shall consider the translation and point group operations of $\Phi(p)$ separately.

The ordinary translation group $T(p)$ of the dichromatic pattern comprises the lattice translation operations that satisfy eqn (1.50). That is,

$$T(p) = [\mathbf{E}|R_i^w] \cap [\mathbf{T}'|t][\mathbf{E}|R_j^b][\mathbf{T}'|t]^{-1}. \tag{1.51}$$

Therefore the elements of $T(p)$ satisfy the following equation:

$$[\mathbf{E}|R_i^w] = [\mathbf{T}'|t][\mathbf{E}|R_j^b][\mathbf{T}'|t]^{-1}. \tag{1.52}$$

Using eqns (1.46) and (1.47) to multiply out the right-hand side we find that

$$[\mathbf{E}|R_i^w] = [\mathbf{E}|\mathbf{T}'R_j^b],$$

or

$$R_j^b = \mathbf{T}'^{-1}R_i^w. \tag{1.53}$$

In eqn (1.53) $R_j^b$ and $R_i^w$ are translation vectors of the black and white lattices expressed in their own coordinate systems. Therefore eqn (1.53) expresses the condition that $R_j^b$ and $R_i^w$ are coincident vectors when they are expressed in the same coordinate system. If $T(p)$ is not an empty set its elements constitute a *coincidence site lattice* (CSL), which may be one, two, or three dimensional. Examples are shown in Figs 1.10 and 1.12. As seen in eqn (1.53) the existence of a coincidence site lattice is independent of the relative translation $t$ of the two lattices. Grimmer (1976) has proved that in order for a 3D CSL to exist all nine elements of the matrix **T** must be rational. If the black and white lattices are identical apart from a difference of orientation then **P** in eqn (1.8) may be taken as the identity and **T** reduces to the rotation matrix **R**. Thus Grimmer's theorem states that all nine elements of the rotation matrix **R** must be rational. Using eqn (1.14) we may express the rotation matrix **R** in terms of the Rodrigues vector $\rho^R$:

$$(1 + \rho^{R^2})R_{ij} = (1 - \rho^{R^2})\delta_{ij} + 2\left(\rho_i^R \rho_j^R - \sum_{k=1}^{3} \varepsilon_{ijk} \rho_k^R\right). \tag{1.14}$$

It is important to note that in this equation the components, $\rho_i^R$, of the Rodrigues vector are expressed in an orthogonal coordinate system. In order for each $R_{ij}$ to be rational *in a cubic lattice* it is necessary and sufficient that each $\rho_i^R$ is rational. Thus rational Rodrigues vectors correspond to coincidence site lattice orientations. An arbitrary rational Rodrigues vector is

$$\rho^R = \frac{m}{n}[HLK] \tag{1.54}$$

where $m, n, H, K,$ and $L$ are integers. This represents a rotation $\theta$ given by

$$\tan \theta/2 = m(H^2 + K^2 + L^2)^{\frac{1}{2}}/n \tag{1.55}$$

about an axis $[HKL]$. In a non-cubic material the occurrence of a 3D CSL is much more restricted, except in special cases where the axial ratios of the crystal lattices assume particular values. For example, let the $c/a$ ratio in a tetragonal lattice be $\lambda$. To obtain a rotation matrix with rational elements we again require a Rodrigues vector of the form of eqn (1.54). However, simply demanding that the axis of rotation is a rational direction no longer guarantees that $\rho^{R^2}$ is rational because for $\rho^R = [HKL]m/n$ we have

$$\rho^{R^2} = \tan^2 \theta/2 = \frac{m^2(H^2 + K^2 + L^2\lambda^2)}{n^2}. \tag{1.56}$$

Therefore $\lambda^2$ must be rational also. The same conclusion applies to a hexagonal lattice: in general the existence of a CSL requires that $(c/a)^2$ is rational. The obvious exceptions to this rule are rotations about the $c$-axis in the hexagonal and tetragonal lattices for which the existence of a CSL is clearly independent of the $c/a$ ratio.

When a 3D CSL exists the ratio of the volume of a CSL primitive cell to a crystal lattice primitive cell is denoted by $\Sigma$. Grimmer (1976) proved that $\Sigma$ is the least positive integer such that if $\mathbf{R}$ is expressed in a primitive crystal basis then $\Sigma\mathbf{R}$ and $\Sigma\mathbf{R}^{-1}$ are matrices with integer elements. For primitive cubic crystal lattices we see from eqns (1.14) and (1.55) that this leads to the result that

$$\Sigma = n^2 + m^2(H^2 + K^2 + L^2), \tag{1.57}$$

which is originally due to Ranganathan (1966). For cubic lattices $\Sigma$ is always odd, so that any factors of two in eqn (1.57) must be cancelled out.

Returning to eqn (1.50) we now examine the condition for the existence of ordinary point group operations in the dichromatic pattern. Those operations are given by solutions of

$$[\mathbf{D}_i^w|\mathbf{0}] = [\mathbf{T}'|t][\mathbf{D}_j^b|\mathbf{0}][\mathbf{T}'|t]^{-1}.$$
$$= [\mathbf{T}'\mathbf{D}_j^b\mathbf{T}'^{-1}| -\mathbf{T}'\mathbf{D}_j^b\mathbf{T}'^{-1}t + t] \tag{1.58}$$

Therefore,

$$\mathbf{D}_i^w = \mathbf{T}'\mathbf{D}_j^b\mathbf{T}'^{-1} \tag{1.59}$$

and

$$\mathbf{T}'\mathbf{D}_j^b\mathbf{T}'^{-1}t = t. \tag{1.60}$$

Equation (1.59) expresses the requirement of bringing point symmetry operations of the same type into orientational alignment. Otherwise those symmetry operations are not elements of the ordinary point group of the dichromatic pattern. Equation (1.60) expresses the fact that a general relative displacement of the lattices takes the point symmetry operations out of translational alignment. Only those displacements that are unaffected by the point group operation, $\mathbf{D}_i^w = \mathbf{T}'\mathbf{D}_j^b\mathbf{T}'^{-1}$, enable the point operation to survive. Thus an ordinary rotation axis or mirror plane is conserved only by displacements *parallel* to the rotation axis or plane of symmetry respectively. The location of the ordinary symmetry elements is unaffected by such displacements.

Let us now consider the effect of alternative but equivalent specifications of $[\mathbf{T}'|t]$. In eqn (1.48) we may replace $[\mathbf{T}'|t]$ by $[\mathbf{D}_k^b|R_n^b][\mathbf{T}'|t][\mathbf{D}_j^w|R_i^w]$ because this amounts to nothing more than acting on the black and white lattices with symmetry operations of those lattices. In this expression $[\mathbf{D}_k^b | R_n^b]$ is expressed in the coordinate frame of the black lattice. Evidently there is an infinite number of equivalent specifications of $[\mathbf{T}' | t]$. If these alternative specifications are substituted into eqn. (1.50) we find that the condition for the existence of an ordinary point group operation in the dichromatic pattern becomes

$$[\mathbf{D}_l^w | R_m^w] = [\mathbf{D}_j^w | R_i^w]^{-1} [\mathbf{T}' | t]^{-1} [\mathbf{D}_k^b | R_n^b]^{-1} [\mathbf{D}_p^b | R_q^b] [\mathbf{D}_k^b | R_n^b] [\mathbf{T}' | t] [\mathbf{D}_j^w | R_i^w]$$

Therefore,

$$[\mathbf{D}_j^w | R_i^w]^{-1} [\mathbf{D}_l^w | R_m^w] [\mathbf{D}_j^w | R_i^w] = [\mathbf{T}' | t]^{-1} [\mathbf{D}_k^b | R_n^b]^{-1} [\mathbf{D}_p^b | R_q^b] [\mathbf{D}_k^b | R_n^b] [\mathbf{T}' | t]. \tag{1.61}$$

The term on the left-hand side of this equation is a space group operation of the white lattice. The right-hand side of the equation is of the form $[\mathbf{T}' | t]^{-1} [\mathbf{D}_f^b | R_q^b] [\mathbf{T}' | t]$. Therefore eqn (1.61) is exactly the same form as before (see eqn (1.50)) and $\Phi^*(p)$ is unaffected by alternative specifications of $[\mathbf{T}' | t]$.

Antisymmetry operations may exist in the dichromatic pattern only when the black and white lattices are identical except for a difference of orientation. There can be no antitranslation operations owing to the difference in orientation of the two lattices. Therefore we have only to consider antisymmetry point group operations. Because an antisymmetry operation relates the two lattices it must be equivalent to one of the specifications of $[\mathbf{T}' | t]$ that has the form of a symmetry operation. When the black and white lattices are identical there is an isomorphism between the symmetry operations of both lattices. In that case the most general specification of $[\mathbf{T}' | t]$, that does not involve a lattice translation vector, is $[\mathbf{T}' | t][\mathbf{D}_j^w | 0]$. This follows because it is always possible to express $[\mathbf{D}_p^b | 0]$ as $[\mathbf{T}' | t][\mathbf{D}_p^w | 0] \mathbf{T}' | t]^{-1}$, owing to the isomorphism between $\Phi^{b*}$ and $\Phi^{w*}$. The antisymmetry point group operations are therefore those members of the set $\{\mathbf{T}' \mathbf{D}_j^w\}$ which have the form of a symmetry operation. Alternatively, if antisymmetry is present then there exists a description of the transformation $[\mathbf{T}' | t][\mathbf{D}_j^w | 0]$ as a vector transformation relating the white lattice to the black, and simultaneously there must exist a vector transformation relating the black lattice to the white. The latter is $[\mathbf{T}' | t]^{-1}[\mathbf{D}_k^b | 0]$, in the coordinate frame of the black lattice. Re-expressing this in the white frame, i.e. $[\mathbf{T}' | t][\mathbf{T}' | t]^{-1}[\mathbf{D}_k^b | 0][\mathbf{T}' | t]^{-1} = [\mathbf{D}_k^b | 0] [\mathbf{T}' | t]^{-1}$ it follows that

$$[\mathbf{T}' | t] [\mathbf{D}_j^w | 0] = [\mathbf{D}_k^b | 0] [\mathbf{T}' | t]^{-1}. \tag{1.62}$$

The antisymmetry operations are given by solutions of eqn (1.62). Expanding eqn (1.62) we obtain

$$[\mathbf{T}' \mathbf{D}_j^w | t] = [\mathbf{D}_k^b \mathbf{T}'^{-1} | -\mathbf{D}_k^b \mathbf{T}'^{-1} t]. \tag{1.63}$$

Therefore an antisymmetry point group operation satisfies

$$\mathbf{T}' \mathbf{D}_j^w = \mathbf{D}_k^b \mathbf{T}'^{-1}. \tag{1.64}$$

The presence of the translation $t$ will destroy the antisymmetry point group operation unless

$$\mathbf{D}_k^b \mathbf{T}'^{-1} t = -t. \tag{1.65}$$

This equation has to be interpreted carefully. The convention we have adopted is that $t$ is the translation of the white lattice required to bring at least one white site into coincidence with a black site. This convention still applies after the action of the antisymmetry operation $\mathbf{D}_k^b \mathbf{T}'^{-1}$. But in eqn (1.65) we have not yet taken into account the change of colour which is part of the antisymmetry operation. When we do take account of the change of colour the minus sign on the right of eqn (1.65) disappears because of the definition of $t$. In this sense $t$ is not an ordinary vector since colour is included in its definition, and eqn (1.65) is not correct because it does not take account of this subtlety. We may express an antisymmetry operation, $\mathbf{A}_n'$, as $\mathbf{A}_n$ followed (or preceded) by $1'$, where $\mathbf{A}_n = \mathbf{D}_k^b \mathbf{T}^{-1} = \mathbf{T}\mathbf{D}_j^w$ and $1'$ is the colour reversing operation. In this notation eqn (1.65) becomes

$$A_n t = -t, \qquad (1.66)$$

whereas

$$A_n' t = t. \qquad (1.67)$$

Alternatively, $\mathbf{A}_n'$ may also be expressed as

$$A_n' = \bar{1}A_n. \qquad (1.68)$$

This means that we may always regard an antisymmetry operation as an ordinary operation followed (or preceded) by an inversion of both lattices. The inversion in eqn (1.68) changes the sign of $t$ in eqn (1.65), so that eqns (1.67) and (1.68) are consistent with each other. Note that eqn (1.67) has the same form as eqn (1.60) which expresses the condition for survival of an ordinary point group operation in the presence of a displacement $t$. Antisymmetry rotation axes of order higher than $2'$ are destroyed by any non-zero displacement $t$. However, $2'$ axes and $m'$ planes are conserved by displacements $t$ that are *perpendicular* to these axes and planes respectively. But the locations of the $2'$ axes and $m'$ planes are altered by $t/2$. This is illustrated in Fig. 1.11. We see that ordinary and antisymmetry point group operations are affected by $t$ in complementary ways.

The product of two antisymmetry operations must be an ordinary operation of the dichromatic pattern. Therefore $\mathbf{T}'\mathbf{D}_j^w \mathbf{T}'\mathbf{D}_k^w = \mathbf{D}(p)_m$ where $\mathbf{D}(p)_m$ is an ordinary point group operation of the dichromatic pattern. Redefining $\mathbf{T}'$ to be the antisymmetry operation $\mathbf{T}'\mathbf{D}_k^w$ this equation becomes $\mathbf{T}'\mathbf{D}_k^{w-1}\mathbf{D}_j^w \mathbf{T}' = \mathbf{T}'\mathbf{D}_q^w \mathbf{T}' = \mathbf{D}(p)_m$ and therefore the antisymmetry operation $\mathbf{T}'\mathbf{D}_q^w$ must be expressible as

$$\mathbf{T}'\mathbf{D}_q^w = \mathbf{D}(p)_m \mathbf{T}'^{-1}. \qquad (1.69)$$

Since the left-hand side of eqn (1.69) is an antisymmetry operation so is its inverse and therefore all the antisymmetry operations may be generated from one of them by multiplying it by the group of ordinary point group operations of the dichromatic pattern. Thus the set of antisymmetry operations is the coset $\{\mathbf{T}'\mathbf{D}(p)_j\}$ where $\mathbf{T}'$ is any antisymmetry operation.

In conclusion, the space group $Щ^*(p)$ is given by the union of the ordinary group $\Phi^*(p)$ and the coset of antisymmetry operations $\mathbf{T}'P^*(p)$ where $P^*(p)$ is the holosymmetric point group of ordinary symmetry operations of the dichromatic pattern, and $\mathbf{T}'$ is an antisymmetry operation:

$$Щ^*(p) = \Phi^*(p) \cup \mathbf{T}'P^*(p). \qquad (1.70)$$

When the two crystal lattices are not identical there are no antisymmetry operations and $Щ^*(p) = \Phi^*(p)$. $\Phi^*(p)$ is given by eqn (1.50) in terms of the space groups, $\Phi^{b*}$ and $\Phi^{w*}$, of the black and white lattices. Further examples of dichromatic patterns in f.c.c. lattices are given in Fig. 1.12.

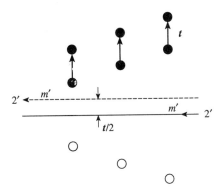

**Fig. 1.11** The black lattice is displaced by $t$ perpendicular to the 2′ and $m'$ antisymmetry elements with the effect that these symmetry elements are displaced by $t/2$.

The symmetry elements that are present in $\Phi^{b^*}$ and $\Phi^{w^*}$ and absent in Щ$^*$(p) relate equivalent dichromatic patterns. Both point group and translational operations in $\Phi^{b^*}$ and $\Phi^{w^*}$ may be suppressed in Щ$^*$(p). By decomposing $\Phi^{b^*}$ and $\Phi^{w^*}$ with respect to $\Phi^*$(p) we may enumerate the orientational and translational variants of the dichromatic pattern. Lagrange's theorem tells us that, since $\Phi^*$(p) is a subgroup of $\Phi^{b^*}$ and $\Phi^{w^*}$, the number of orientational variants is equal to the order of the point group of $\Phi^{b^*}$ plus the order of the point group of $\Phi^{w^*}$ both divided by twice the order of the point group of $\Phi^*$(p). For example, consider, the $\Sigma = 3$ dichromatic pattern formed from twin related f.c.c. lattices shown in Fig. 1.12(a). One description of the relationship between these lattices is 60° about the [111] axis of the white crystal. In this case $\Phi^{b^*}$ and $\Phi^{w^*}$ are both $Fm\bar{3}m$ and the holosymmetric ordinary point group of the dichromatic pattern is $\bar{3}m$, which is a group of order 12. Thus of the 96 point group operations present initially in $\Phi^{b^*}$ and $\Phi^{w^*}$, i.e. 48 black and 48 white, only 12 black and 12 white operations survive as ordinary operations of the dichromatic pattern. It follows from Lagrange's theorem that $96/24 = 4$ variants arise as a consequence of the dissymmetrization. These four variants may be found by decomposing the point group of $\Phi^{w^*}$, i.e. $m\bar{3}m$, with respect to the ordinary point group of $\Phi^*$(p), i.e. $\bar{3}m$. The four variants correspond to 60° rotations about the four body diagonals of the white unit cell. Note that the existence or otherwise of antisymmetry operations in the dichromatic pattern does not influence the number of orientational variants because those operations arise as a consequence of symmetrization rather than dissymmetrization.

Finally in this section we discuss limits on the relative translation $t$ between the black and white lattices. We saw in eqn (1.53) that varying $t$ had no influence on the translational symmetry group $T$(p) of the dichromatic pattern. On the other hand eqns (1.60) and (1.67) showed that ordinary and antisymmetry point group operations are affected by $t$ in complementary ways.

Consider a dichromatic pattern in which a CSL exists. Let the $t = 0$ reference state correspond to the situation in which black and white lattice sites coincide at the origin. The CSL is then a lattice of grey sites, formed by a superposition of black and white sites. If we translate the black lattice by any vector, $t$, separating a black lattice site from a white lattice site, i.e. $R_i^w - \mathbf{T}'R_j^b$ expressed in the white lattice frame, we will recreate

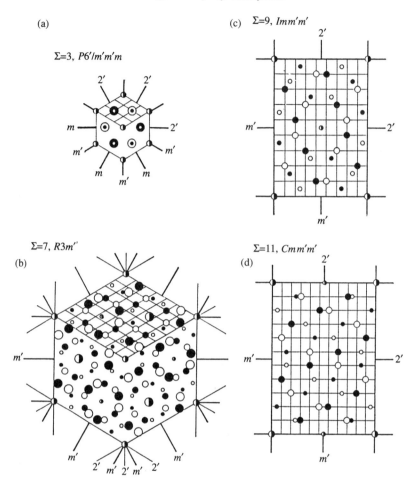

**Fig. 1.12** Dichromatic patterns formed by misoriented f.c.c. lattices, in each case viewed in projection along the rotation axis. In each case a unit cell of the CSL is shown and the fine mesh is the DSC lattice. The different circle sizes refer to lattice sites at different positions along the direction of projection. (a) $\Sigma = 3$: 60° about [111]; (b) $\Sigma = 7$: 38.21° about [111]; (c) $\Sigma = 9$: 38.94° about [110]; (d) $\Sigma = 11$: 50.48° about [110]. (From Pond and Bollmann (1979).)

the identical CSL. But unless $t$ is one of the translation vectors of the dichromatic pattern the origin will change after the displacement. The set of vectors $\{R_i^w - T'R_j^b\}$ forms a lattice known as the *DSC* lattice. It was originally found by Bollmann (1970). It may be described as the coarsest lattice that contains the black and white crystal lattices as sublattices. Examples are shown in Figs 1.10, 1.12, and 1.14. Translation vectors of the DSC lattice that are not CSL vectors relate the translational variants of the dichromatic pattern. Following Pond and Bollmann (1979) we interpret DSC as standing for *dis*placements which are *s*ymmetry *c*onserving, although this is not the original meaning intended by Bollmann (1970).

It follows that dichromatic pattern symmetry varies periodically with $t$ when a CSL exists. Therefore $t$ may always be expressed uniquely as $t$ modulo a DSC vector. Thus when we study the variation of dichromatic point group symmetry with $t$ it is necessary and sufficient to consider only those displacements which fall within the Wigner–Seitz

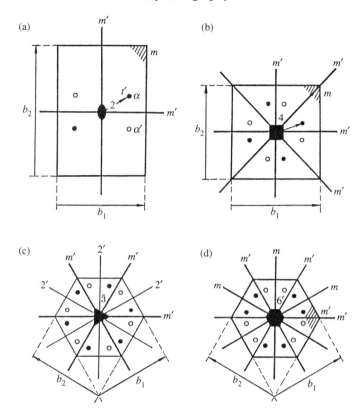

**Fig. 1.13** Projection along [*hkl*] of the Wigner–Seitz cell of the DSC lattices associated with the dichromatic patterns shown in Figs 1.10 and 1.12. (a) $\Sigma = 9$ [*hkl*] = [110], $b_1 = \frac{1}{18}[11\bar{4}]$, $b_2 = \frac{1}{9}[2\bar{2}1]$ or $\Sigma = 11[hkl] = [110]$, $b_1 = \frac{1}{22}[3\bar{3}2]$, $b_2 = \frac{1}{11}[11\bar{3}]$; (b) $\Sigma = 5$ [*hkl*] = [001], $b_1 = \frac{1}{10}[310]$, $b_2 = \frac{1}{10}[1\bar{3}0]$; (c) $\Sigma = 7$ [*hkl*] = [111], $b_1 = \frac{1}{14}[3\bar{2}1]$, $b_2 = \frac{1}{14}[23\bar{1}]$; (d) $\Sigma = 3$ [*hkl*] = [111], $b_1 = \frac{1}{6}[2\bar{1}1]$, $b_2 = \frac{1}{6}[1\bar{2}1]$. The black and white circles illustrate the symmetry within each Wigner–Seitz cell. (From Pond and Bollman (1979).)

cell of the DSC lattice. Examples of Wigner–Seitz cells of DSC lattices associated with dichromatic patterns containing CSLs are shown in Fig. 1.13. As the size of the primitive cell of the CSL increases there tends to be more non-equivalent members of the set of $R_i^w - T'R_j^b$ inside the CSL cell, with the result that the DSC lattice becomes finer and its Wigner–Seitz cell smaller. In the limit of no periodicity in the dichromatic pattern the CSL unit cell tends to an infinite size and the DSC lattice shrinks to a point. In that case an arbitrary relative displacement of the two crystal lattices recreates any finite patch of the dichromatic pattern, but the shift of origin of the pattern may be very large. These remarks are related to the concept of local isomorphism in quasicrystals, which we shall describe in Section 1.9.

The reciprocal relation between the sizes of the CSL and DSC lattices has been stated precisely by Grimmer (1974): the DSC lattice formed from the black and white crystal lattices is the reciprocal lattice of the CSL formed from the reciprocal lattices of the black and white crystal lattices. Conversely, the CSL formed from the black and white crystal lattices is reciprocal to the DSC lattice formed by the reciprocal lattices of the black and white crystal lattices. These results apply to crystal lattices of any symmetry. In the special case where the black and white crystal lattices have the simple cubic structure (and hence their reciprocal lattices are identical to their direct lattices) Grimmer's result becomes: the DSC lattice is the reciprocal of the CSL lattice.

### 1.5.5  Symmetry of dichromatic complexes

To extend the treatment to interfaces between non-symmorphic and/or non-holosymmetric crystals it is necessary to introduce the idea of lattice complexes, as described in Section 1.5.2. The space groups of the black and white lattice complexes are denoted by $\Phi^b$ and $\Phi^w$ and will be isomorphic to subgroups of $\Phi^{b^*}$ and $\Phi^{w^*}$. They may contain mirror glide planes and screw rotation axes. If $\Phi^w$ is a subgroup of lower order than $\Phi^{w^*}$ then a multiplicity of white crystals exists given by decomposing $\Phi^{w^*}$ with respect to $\Phi^w$. Similarly, if the space group, Щ(c), of the dichromatic complex, formed by interpenetrating the black and white lattice complexes, is a subgroup of Щ$^*$(p) then a multiplicity of equivalent dichromatic complexes exists given by decomposing Щ$^*$(p) with respect to Щ(c). The variants that arise as a consequence of this dissymmetrization are called complex variants. Clearly, if in this process of dissymmetrization we had omitted the stage of constructing the dichromatic pattern, and immediately formed the dichromatic complex from the lattice complexes, we would have overlooked the existence of complex variants.

The group $\Phi(c)$ of ordinary operations of the dichromatic complex is given by an analogous equation to eqn (1.50):

$$\Phi(c) = \Phi^w \cap [T'|t]^{-1}\Phi^b[T'|t]. \tag{1.71}$$

Since the the black and white lattice translation vectors are the same in the lattice complexes the translation vectors of the dichromatic complex are identical to those of the dichromatic pattern. Thus eqn (1.53) also expresses the condition for a translation vector of the dichromatic complex.

The supplementary displacements associated with non-symmorphic operations introduce an additional requirement to those embodied in eqns (1.59) and (1.60) for the existence of a non-symmorphic point group operation in the dichromatic complex. For non-symmorphic operations eqn (1.58) becomes:

$$[D_i^w|\alpha_i^w] = [T'|t][D_j^b|\alpha_j^b][T'|t]^{-1}. \tag{1.72}$$

The new condition, in addition to eqns (1.59), and (1.60), is

$$\alpha_j^b = T'^{-1}\alpha_i^w. \tag{1.73}$$

In this equation $\alpha_j^b$ is expressed in the coordinate frame of the black lattice and $\alpha_i^w$ in the frame of the white lattice. Therefore, this equation expresses the condition that the supplementary displacements associated with the coincident non-symmorphic operations must be identical and collinear, as a coordinate transformation. This additional requirement implies that in general ordinary operations will be rarer in a dichromatic complex created from non-symmorphic lattice complexes than symmorphic lattice complexes.

The generalization of eqn (1.62) for the existence of antisymmetry operations for non-symmorphic operations $[D_j^w|\alpha_j^w]$ and $[D_k^b|\alpha_k^b]$ leads, in addition to eqns (1.64) and (1.67), to

$$\alpha_k^b = T'\alpha_j^w. \tag{1.74}$$

In this equation both $\alpha_k^b$ and $\alpha_j^w$ are expressed in the frame of the white lattice. Therefore, it expresses the condition, as a vector transformation, that the supplementary displacements $\alpha_k^b$ and $\alpha_j^w$ are brought into coincidence by the transformation $T'$.

Having established rules for the variation of symmorphic and non-symmorphic symmetry elements in a dichromatic complex with $t$ we may find the holosymmetric dichromatic complex with space group symmetry Щ*(c). Cystallographically equivalent variants of the holosymmetric complex arise if the order of the point symmetry of Щ*(c) is less than that of the holosymmetric dichromatic pattern Щ*(p). The interrelation of the complex variants may be found by decomposing Щ*(p) with respect to Щ*(c). The dissymmetrization of Щ*(p) can arise as a consequence of the fact that the crystal space groups are either non-symmorphic and/or non-holosymmetric. This is illustrated in Fig 1.14. The space group of the dicromatic pattern in Fig. 1.14(a) is Щ*(p) = $I4/mm'm'$, and the order of the point symmetry is 16. On the other hand, the space group symmetry of the complex in Fig. 1.14(b) is Щ(c) = $\bar{I}42'm'$, and the order of the point symmetry in this case is 8. (In fact there are at least two dichromatic complexes which have symmetry Щ*(c) that is isomorphic with Щ*(p) (see Pond (1983)), but that does not affect the present argument.) There are two complex variants related by an operation present in Щ*(p) but absent in Щ(c): for example the ordinary mirror plane parallel to the plane of Fig. 1.14(a). Since the diamond cubic crystal has a space group $Fd\bar{3}m$, which is isomorphic to the f.c.c. space group $Fm\bar{3}m$, it is holosymmetric. Dissymmetrization of the dichromatic pattern occurs because the diamond cubic structure is non-symmorphic. In such cases where dissymmetrization arises because of non-symmorphic but holosymmetric groups the relation between the set of complex variants can also be expressed as rigid body displacements $t$. In Fig. 1.14(b) the displacement relating the two variants is 1/10/[210] using the coordinate system of either crystal. This displacement brings large and small squares into coincidence rather than large and small circles. If, in Fig. 1.14(b), circles represent zinc atoms and squares represent sulphur atoms the figure can also be regarded as the dichromatic complex for two crystals having the zincblende structure. The appropriate dichromatic pattern is still Fig. 1.14(a). But the space groups of the crystals have been reduced to $F\bar{4}3m$, i.e. they are non-holosymmetric but symmorphic. The two complex variants are still related by the ordinary mirror parallel to the plane of the figure. But there is no displacement relating the two complex variants anymore. Instead the two complex variants are composed of pairs of black and white zincblende lattice complexes in which the Zn and S sites have been interchanged.

For non-holosymmetric symmorphic crystals there will in general be fewer solutions of eqns. (1.59) and (1.60) than for holosymmetric symmorphic crystals, leading to fewer

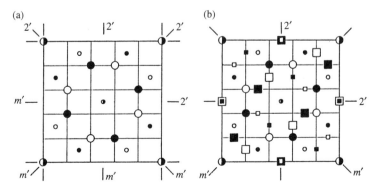

**Fig. 1.14** Projection along [001] of one CSL unit cell of: (a) the $\Sigma = 5$ dichromatic pattern formed by interpentrating two f.c.c. lattices; (b) the $\Sigma = 5$ dichromatic complex formed by interpenetrating two diamond cubic crystals. The fine mesh is the DSC lattice in (a) and (b). (From Pond and Vlachavas (1983).)

ordinary operations in Щ*(c) than in Щ*(p). Since the set of antisymmetry operations in Щ*(c) is the coset $\{[\mathbf{T}'\,|\,t]\,[\mathbf{D}(p)_i\,|\,\alpha(p)_i]\}$ it follows that there will also be fewer anti-symmetry operations in Щ*(c). Consequently particular holosymmetric interfacial structures created from Щ*(c) by sectioning will generally have lower symmetries than the corresponding interfaces created from Щ*(p). But, in accordance with the principle of symmetry compensation, additional interfacial structures can be created from the variant dichromatic complexes. For grain boundary type interfaces, for example, it is possible for non-holosymmetric domains to arise such as polarity reversed domains in sphalerite.

### 1.5.6 Symmetry of ideal bicrystals

We create an ideal bicrystal from a dichromatic complex by first choosing an interface plane with a specified normal. The dichromatic complex is then sectioned on the interface plane and the white lattice complex is discarded on one side of the interface and the black on the other side. The symmetry operations of the dichromatic complex that survive the sectioning process are defined by the condition that they leave the interface normal invariant. They form a group which we call Щ(i). This group can only be an ordinary group $\Phi(\mathrm{i})$ if the adjoining crystals do not have identical structures.

Щ(i) may contain zero, one, or two non-collinear translation axes depending on the translational symmetry present in the dichromatic pattern. It is always possible to choose sectioning planes that contain no translation vectors of the dichromatic pattern even when a 3D CSL exists. An ordinary operation of the dichromatic complex $[\mathbf{D}(c)_j\,|\,\alpha(c)_j]$ will survive the sectioning process provided

$$\mathbf{D}(c)_j\hat{n} = \hat{n} \tag{1.75}$$

and provided $\alpha(c)_j$ is one of a set of particular fractional vectors of the set of translation operations of the bicrystal. More explicitly, $\alpha(c)_j$ must satisfy $(\alpha(c)_j + \mathbf{R}_k)\cdot\hat{n} = 0$, where $\mathbf{R}_k$ is any lattice translation vector of the black or white lattice. When we consider the survival of an antisymmetry operation we have to be more careful and take into account the definition of the boundary normal sense. We shall adopt the convention that the boundary normal always points into the white crystal. Therefore the boundary normal, like the translation $t$, has colour included in its definition and this must be taken into account. An antisymmetry operation of the dichromatic complex $[\mathbf{D}'(c)\,|\,\alpha(c)_i]$ will survive the sectioning process provided

$$\mathbf{D}'(c)_i\hat{n} = \hat{n} \tag{1.76}$$

and provided $\alpha(c)_i$ satisfies $(\alpha(c)_i + \mathbf{R}_k)\cdot\hat{n} = 0$. In eqn (1.76) the interface normal changes sign twice: once when $\mathbf{D}(c)_i$ transforms the black crystal into the white and vice versa and once when the colour reversal is carried out.

In the case of heterophase bicrystals the group $\Phi(\mathrm{i})$ is invariant with location of the interface plane along its normal because only ordinary operations are present and eqn (1.75) is obeyed regardless of the location of the interface. Thus the holosymmetric bicrystal space group $\Phi^*(\mathrm{i})$ is readily obtained. For homophase bicrystals, however, the presence of certain antisymmetry operations in Щ(i) is dependent on the location of the interface. For example, anti-inversion centres, antidiads, and antimirrors are present in Щ(i) only when the particular interface location contains them. In such cases different bicrystals, obtained by relocating the interface plane, occur in symmetry related sets, which can be obtained by decomposing the holosymmetric group Щ*(i) with respect to the ordinary group $\Phi(\mathrm{i})$.

By decomposing Щ*(c) with respect to Щ*̇(i) we may enumerate the morphological variants of an interface. If Щ*(i) is an invariant subgroup of Щ*(c) an ideal bicrystal can exist only with the particular inclination. But for other interface normals Щ*(i) need not be an invariant subgroup of Щ*(c) and it is then useful to visualize the morphological variants in terms of the interfaces of a black crystal embedded in a white crystal (or white embedded in black). These equivalent interfaces define a regular polyhedral form for the embedded crystal, which may or may not be closed. It is clear that if the embedded crystal is bounded by interfacial structures, some of which only have point symmetry 1, the symmetry of the polyhedron will be consistent with Щ*(c). The precise relationship between the polyhedral form of the embedded crystal and a real precipitate morphology hinges on several assumptions about the existence of defects in the interfaces, as will be discussed in Section 1.8.

### 1.5.7 Symmetry of real bicrystals

This is the stage at which we switch on the interatomic forces. When this happens relaxation will take place possibly resulting in rigid body translation of one crystal relative to the other, migration of the interface plane along its normal, local atomic relaxation, and insertion or removal of additional material at the interface. It is therefore quite possible that the symmetry Щ(i) of the relaxed bicrystal is lower than that of the holosymmetric bicrystal Щ*(i). The variants obtained by decomposing Щ*(i) with respect to Щ(i) are relaxational variants.

A rigid body translation $t$ of one crystal relative to the other is known to be an important mode of relaxation at interfaces possessing one- or two-dimensional translational symmetry. In general $t$ has components both parallel and perpendicular to the interface, the perpendicular component often being called the interfacial expansion. The component of the translation parallel to the interface plane is known as the in-plane translation. If there is no translational symmetry in the interface an arbitrary relative in-plane translation produces bicrystal structures that are locally isomorphic. We shall return to this point in Section 1.9. But if the interface possesses translational symmetry then two in-plane translations that differ by a translation vector of the interface produce equivalent structures. There is thus a 'cell of non-identical displacements' (c.n.i.d.) which contains all possible in-plane translations that are not equivalent by adding a bicrystal translation vector (Vitek *et al.* 1980). In Chapter 4 we will prove that the c.n.i.d. is defined as the Wigner–Seitz cell of the lattice that is reciprocal to the coincidence lattice of reciprocal crystal lattice vectors in the interface plane. For bicrystals that contain only one direction of translational symmetry the c.n.i.d. is of infinite size in the perpendicular direction.

Since the translational symmetry of the dichromatic pattern is invariant with rigid body displacements it follows that any translational symmetry present in the bicrystal cannot be destroyed by a rigid body translation. Whether a bicrystal point symmetry operation survives a rigid body translation $t$ depends on the rules that we have derived before (eqns (1.60) and (1.67):

$$\text{ordinary operations: } \mathbf{D}(i)_j t = t \tag{1.77}$$

$$\text{anti-operations: } \mathbf{D}'(i)_j t = t. \tag{1.78}$$

We see that if $t$ is a pure expansion, i.e. it is normal to the plane of the interface, both ordinary and anti-operations survive. Equation (1.78) indicates also that if an anti-inversion centre exists in the interface it cannot be destroyed by a translation $t$. Anti-inversion centres cannot exist in heterophase or grain boundary type bicrystals. They can exist only in stacking fault bicrystals in centrosymmetric crystals, or in inversion boundaries where the black and white crystal lattices are identical in structure and orientation but the black and white crystal bases are related by inversion symmetry.

As discussed in Section 1.5.6 it is possible to modify the symmetry of a bicrystal by migration of the interface plane along its normal when antisymmetry is present in the holosymmetric bicrystal. For grain boundary type bicrystals, where the basis is a single atom, this mode of relaxation is not distinct from rigid body translation. In that case the various structures that are generated by relocating the boundary plane, while maintaining a constant in-plane translation, may all be obtained at a fixed boundary location by varying the in-plane translation within the c.n.i.d.. However, if the basis is not monatomic this is not generally true, and the boundary plane location is then an independent degree of freedom. If the numbers of atoms in the black and white crystal bases are $n^b$ and $n^w$ then at a fixed relative translation there will be a maximum of $n^b n^w$ distinct interfacial structures (Pond *et al.* 1991). These structures may be obtained by relocating the interface between all possible pairs of atomic planes parallel to the interface. $n^b n^w$ is an upper limit because two or more basis atoms may lie in the interface plane. There are thus up to $n^b n^w$ distinct c.n.i.d.s for a given interface.

Local atomic relaxation is another important mode of relaxation in which atoms are displaced from the positions they occupy in their parent crystals. By contrast, in the rigid body translation mode of relaxation atoms do not move from their parent crystal sites. Thus, the final relaxation displacement of an atom in a bicrystal may always be decomposed into a rigid body displacement, which is shared by all atoms on one side of the interface, and a local atomic relaxation displacement that will vary from atom to atom. This decomposition is achieved in practice by examination of the rigid body displacement far from the interface, where local atomic relaxations may be neglected. Local atomic relaxations may destroy both ordinary and antisymmetry point group operations. They may also modify the translational symmetry group T(i) by, for example, doubling the primitive translation vector in one or more directions in the interface.

Removal or insertion of material comprising less than a complete number of whole bases is another possible mode of reducing the free energy of a bicrystal, and may also modify the holosymmetric symmetry. Insertion of additional material at the interface, but not at sites belonging to either the white or the black crystal, as in an extrinsic stacking fault in f.c.c. materials, is another possibility.


## 1.6  TWO EXAMPLES

### 1.6.1  Lattice matched polar–non-polar epitaxial interfaces

Our first application is to interfaces between closely lattice-matched polar and non-polar semiconductors such as GaAs–Ge and GaP–Si (Pond *et al.* 1984; 1985). Such interfaces may be grown by molecular beam epitaxy (MBE) and it has been found experimentally that some inclinations of the interface normal can induce antisite disorder in the polar compound. The antisite disorder consists of domains in the GaAs separated by inversion boundaries that emanate from the interface with the Ge and thread through the GaAs layer: see Fig. 1.15. Across the inversion domain boundaries the sublattices occupied by the Ga and As sites are interchanged. (The reason why we are not calling these domain

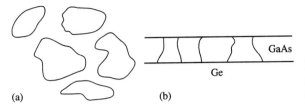

(a)                    (b)

**Fig. 1.15** Schematic illustration of antisite disorder in a layer of GaAs grown on Ge: (a) plan view; (b) side view.

boundaries antiphase domain boundaries will become clear below.) In this example we shall first explain the origin of the antisite disorder for GaAs films grown on Ge (001) substrates. We will then go on to show how the crystallographic analysis can be used to predict other substrate normals on which the disorder will arise and those where it will not. The basis of the analysis is that the inversion boundaries are nucleated at the interface with the substrate and separate energetically degenerate domains of that interface.

For GaAs–Ge and GaP–Si epitaxial specimens the lattices of the substrate and epitaxial layers are both f.c.c., have parallel alignment and can be taken to have the same lattice constant. The dichromatic pattern is therefore a single f.c.c. lattice of grey sites, formed from superimposed black and white sites; hence $Ш^*(p) = Fm\bar{3}m1'$, where $m\bar{3}m1'$ denotes a grey group of 96 elements: $m\bar{3}m1' = m\bar{3}m \cup (m\bar{3}m)'$. The Ge crystal structure has space group $\Phi^w = Fd\bar{3}m$ while the crystal structure of the polar compound has a space group of lower symmetry $\Phi^b = F\bar{4}3m$. The dichromatic complex therefore has the symmetry $Ш^*(c) = F\bar{4}3m1'$, where $\bar{4}3m1'$, is a grey group of order 48. Since the order of the point group of the dichromatic complex is 48, and is therefore only half the order of the point group of the dichromatic pattern, it follows that there are two complex variants of the dichromatic complex. This may be seen by decomposing the group $m\bar{3}m$ with respect to $\bar{4}3m$:

$$(m\bar{3}m) = (\bar{4}3m) \cup \bar{1} \cdot (\bar{4}3m). \tag{1.79}$$

The 48 operations in the cosets $\bar{1} \cdot (\bar{4}3m)$ and $\bar{1}' \cdot (\bar{4}3m)$ relate the two complex variants and they are called complex exchange operations. The 24 operations in $\bar{1} \cdot (\bar{4}3m)$ comprise the inversion, three rotations of 90° about $\langle 001 \rangle$, three rotations of 270° about $\langle 001 \rangle$, three mirror reflections in $\{001\}$, six rotations of 180° about $\langle 110 \rangle$, four improper rotations of 120° about $\langle 111 \rangle$, and four improper rotations of 240° about $\langle 111 \rangle$. The two complex variants are called $\alpha$ and $\beta$ and they are shown in Fig. 1.16. It may be verified that they are related by the complex exchange operations listed above. It is seen that the $\alpha$ variant can be transformed into the $\beta$ variant by simply interchanging the Ga and As atoms, although, after interchanging, the superimposed Ge and As atoms at the origin in the $\alpha$ structure are identical to the Ge and As atoms at the point $\frac{1}{4}, \frac{1}{4}, -\frac{1}{4}$ in the $\alpha$ structure.

The next stage is to create a bicrystal from the dichromatic complex by sectioning. Let us consider the $\alpha$ dichromatic complex variant. Any operations in this dichromatic complex which do not appear in the bicrystal after sectioning relate equivalent bicrystals. They are the morphological variants of the bicrystal. They form a set of bicrystals which may all be created from the $\alpha$ complex variant. It is also possible to create another equivalent set of bicrystals, called the $\beta$ set, by sectioning the $\beta$ complex variant. The normals of the two sets of bicrystals are related by the set of 48 complex exchange

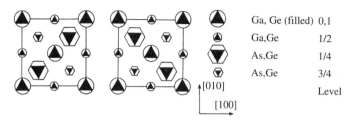

| | | |
|---|---|---|
| Ga, Ge (filled) | 0,1 |
| Ga,Ge | 1/2 |
| As,Ge | 1/4 |
| As,Ge | 3/4 |
| | Level |

**Fig. 1.16**  Two complex variants of the dichromatic complex formed between Ge and GaAs. (From Pond *et al.* (1984).)

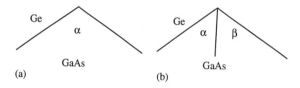

(a)                                    (b)

**Fig. 1.17**  Schematic illustration of two distinct kinds of facet junction formed between Ge and GaAs crystals: (a) both facets are formed from the same complex variant $\alpha$ and there is no defect emanating from the facet junction. (b) The facets are formed from different complex variants $\alpha$ and $\beta$, and a defect emanating from the facet junction inside the GaAs is necessary where the complex variants meet.

operations. The energies of all interfaces in both sets of variants are the same because all the interfaces are related by symmetry operations.

Consider now a bicrystal of Ge and GaAs with a faceted interface, as shown in Fig. 1.17. If adjacent facets have been created by sectioning either the $\alpha$ or the $\beta$ dichromatic complex, as in Fig. 1.17(a), no defect is required to exist along their intersection (unless additional relaxation processes take place in the two interfacial structures such as the development of rigid body translations). However, it is also possible for adjacent facets, one created from the $\alpha$ complex and one from the $\beta$ complex, to coexist in a bicrystal provided a defect is present along their intersection, as shown in Fig. 1.17(b). Such defects arise because of the incompatibility between the $\alpha$ and $\beta$ variants owing to the reversal in the occupancies of the two f.c.c. sublattices in the GaAs crystal. Thus the defect that is necessary along the $\alpha/\beta$ facet intersection is an inversion domain boundary extending into the GaAs crystal from the facet intersection. We do not call this boundary an antiphase domain boundary because none of the 48 complex exchange operations relating the $\alpha$ and $\beta$ complex variants involves a translation.

Certain interface normals $\hat{n}^*$ exist which allow both $\alpha$ and $\beta$ variants to coexist on the same interface plane. The normals to such interfaces are those which are left invariant by at least one of the set of 48 complex exchange operations. (Note that any improper operation always inverts the interface normal.) It is readily confirmed that normals of the type $\hat{n}^* = \langle hk0 \rangle$ satisfy this condition. It follows that on such planes energetically degenerate domains of the $\alpha$ and $\beta$ variants of the interface normal can coexist with arbitrary shapes. That is, the defects separating the $\alpha$ and $\beta$ domains can have any position and orientation on such planes because they are not restricted to lie along facet junctions. The set of planes $\{hk0\}$ includes (100) and (110), but not, for example, $\{111\}$ and $\{112\}$, which is entirely consistent with the experimental observations of the antisite disorder. Thus, the crystallographic analysis has revealed the planes on which antisite disorder *can* exist and those on which it *cannot*. This is all that we are entitled to expect of an analysis based on symmetry considerations alone.

Finally, we return to the shift, $s = \frac{1}{4}[111]$, of origin required to bring the $\alpha$ and $\beta$ complexes into registry. Its significance is that it can give rise to an interfacial step on special facets of height $s \cdot \hat{n}^*$. (See Sections 1.7.1 and 1.7.2 for a full discussion of interfacial steps.) Thus on (001) the step height is $\frac{1}{4}$, which corresponds to one atomic (004) plane. But on the $(1\bar{1}0)$ plane, the step height is zero.

### 1.6.2 Lattice matched metal-silicide silicon interfaces

In this example we consider epitaxial layers of $CoSi_2$, $NiSi_2$, and $Pd_2Si$ grown on Si substrates and the analysis follows Cherns and Pond (1984).

Let the Si substrate be designated the white crystal. Then $\Phi^{w*} = Fm\bar{3}m$ and $\Phi^w = Fd\bar{3}m$. $CoSi_2$ and $NiSi_2$ both have the flourite structure, which is a symmorphic and holosymmetric space group: $\Phi^{b*} = \Phi^b = Fm\bar{3}m$. In the case of $Pd_2Si$ $\Phi^{b*} = P6/mmm$ and $\Phi^b = P\bar{6}m2$. We assume that the small difference in lattice parameters between $CoSi_2$ and Si and between $NiSi_2$ and Si can be accommodated by anticoherency dislocations (see Chapter 2) which perturb only the underlying periodic interfacial structures. The analysis addresses only those coherent patches of underlying periodic structure.

In the cases of $NiSi_2$ and $CoSi_2$ oriented parallel to the underlying Si substrate the dichromatic pattern has ordinary point group symmetry $\Phi^*(p) = Fm\bar{3}m1'$ and the dichromatic complex has ordinary point group symmetry $\Phi^*(c) = F\bar{4}3m1'$. This situation is analogous to the Ge-GaAs dichromatic pattern and complex and there are two complex variants, which we again label $\alpha$ and $\beta$, and may be seen in Fig. 1.16. The $\alpha$ complex can be regarded as an array of black sites with symmetry $Fm\bar{3}m$, superimposed on an array of white points with the diamond structure (and having identical lattice parameters) with basis sites at $0, 0, 0$ and $\frac{1}{4}, \frac{1}{4}, \frac{1}{4}$. The $\beta$ complex corresponds to the identical array of black sites with symmetry $Fm\bar{3}m$ superimposed on an array of white sites with the diamond structure with basis sites at $0, 0, 0$ and $\frac{1}{4}, \frac{1}{4}, -\frac{1}{4}$. Thus the $\beta$ complex can be transformed into the $\alpha$ complex by shifting the diamond array by $\frac{1}{4}[\bar{1}\bar{1}1]$.

We have already seen in Section 1.6.1 that an interface may consist of equivalent facets from $\alpha$ and $\beta$ complexes provided line defects are present along their intersections. In the present case the defects are dislocations with Burgers vectors equal to $\frac{1}{4}[\bar{1}\bar{1}1]$, which effect the transition from the $\beta$ to the $\alpha$ structures. It was also shown that if there are interfaces with normals $\hat{n}^*$ which are left invariant by any of the complex exchange operations in the coset $\bar{1} \cdot (\bar{4}3m)$ then degenerate interfacial domains from $\alpha$ and $\beta$ complexes may coexist on those interfaces separated by $\frac{1}{4}\langle 111 \rangle$ dislocations. The normals $\hat{n}^*$ are again of the form $\{hk0\}$. Experimentally it has been found (Cherns and Pond 1984) that $\frac{1}{4}\langle 111 \rangle$ dislocations occur on $NiSi_2/(001)$Si facets and along $\{001\}/\{111\}$ and $\{111\}/\{111\}$ facet junctions. These findings are entirely consistent with the crystallographic theory. On the (001) facets the $\frac{1}{4}\langle 111 \rangle$ dislocations have a component, $\frac{1}{4}\langle 110 \rangle$, parallel to the interface. Thus the $\frac{1}{4}\langle 111 \rangle$ dislocations may relieve the $\approx 0.4$ per cent misfit between the natural Si and $NiSi_2$ lattice parameters. Since $CoSi_2$ is isomorphous with $NiSi_2$ we expect similar defects in $CoSi_2/S$.

When $NiSi_2$ is deposited on Si(111) another orientation relationship is found experimentally. This is sometimes referred to as the 'twinned' orientation, and corresponds to a 180° rotation of the $NiSi_2$ crystal about the [111] normal. The dichromatic pattern and complex are then found to have the same ordinary point group symmetries, $P\bar{3}m1$, and hence there are no complex variants. Therefore, complex variants separated by $\frac{1}{4}\langle 111\rangle$ dislocations are not predicted to occur on any interfaces with the twinned orientation relationship.

For $Pd_2Si$–Si (111) interfaces lattice matching exists only for the parallel (0001) and (111) planes, so that $\Phi^*(p)$ and $\Phi^*(c)$ contain only 2D translational symmetry: $\Phi^*(p) = p\bar{3}m1$ and $\Phi^*(c) = p3m1$. Since $\bar{3}m$ is a group of order 12, while $3m$ is a group of order 6, it follows that there are 2 complex variants. The complex exchange operations include $\bar{1}$ and a 60° rotation about [0001]. Thus domains from the two complex variants may coexist on $(111)Si/(0001)Pd_2Si$ with line defects separating them that have no dislocation character or steps associated with them. The interfacial domains correspond to the intersection of a domain boundary in the $Pd_2Si$ crystal with the interface. These predictions are in accord with experimental observations of this interface by Cherns *et al.* (1982).


## 1.7  CLASSIFICATION OF ISOLATED INTERFACIAL LINE DEFECTS

In Section 1.6 we saw how the enumeration and interrelation of the variants of each interface provided information about the possible line defects in those interfaces. This is one of the most important applications of bicrystallography and it is directly analogous to the classification of line defects in single crystals. In this section we will develop the classification following closely the work of Pond (1985, 1989). *It is restricted to those line defects which separate regions of an interface that are related by symmetry, and which are therefore energetically degenerate.* This is only a subset of the totality of possible interfacial line defects but it is a very important subset all the same. We will see how the geometrical character of all the possible line defects (in this subset) in a given interface may be predicted. The characterization uses the full space group symmetries of the adjoining crystals and the transformation relationship between them.

There are some subtle points about the characterization of line defects in crystalline materials that we describe first. The most widely used method for characterizing line defects is circuit mapping. For example, dislocations in single crystals are normally characterized by a Burgers circuit procedure (Hirth and Lothe 1982). We first construct a closed circuit around the defect in the distorted medium. This circuit is then mapped onto a reference lattice, which is the perfect crystal lattice, and the closure failure in this reference lattice then defines the Burgers vector. Notice that the Burgers vector is a property of the reference lattice, and not the distorted medium.

To implement this procedure we must first define a suitable reference lattice. As discussed in Chapter 2 the choice of a suitable reference lattice for an interface is often not unambiguous, especially when it contains arrays of dislocations. But even for an isolated dislocation in an interface the choice may be far from obvious when there is no translational symmetry in the dichromatic pattern. Faced with this problem many authors follow a procedure in which they seek the 'nearest, reasonable' reference lattice, in which there is translational symmetry in the dichromatic pattern, and justify their choice on the grounds that they can account for the observed diffraction contrast in the electron microscope. Clearly one cannot attempt a predictive classification of the possible defects following such an approach, since it relies ultimately on experimental verification.

A second fundamental problem arises from the fact that we have to be able to map the closed circuit in the distorted medium onto a circuit in the reference lattice. But this assumes a knowledge in advance of the distortion field around a defect in order that an initial circuit can be constructed. This is clearly a problem if we wish to be able to predict the character of previously unknown defects.

A new school of thought has emerged in recent years which proposes an alternative characterization of *isolated* interfacial line defects. A new procedure based on the Somigliana dislocation has been developed (Pond 1985; Bonnet *et al*. 1985) that circumvents the above problems. A Somigliana dislocation in an elastic continuum is made by introducing a cut C on a surface bounded by a curve c. Each pair of points on opposite sides of the cut is given a relative displacement *d*, which can *vary* with the position of the points on C. After scraping away material where there would be interpenetration, and filling in any gaps, the sides of the cut are rebonded. The elastic field that results in the body is completely determined by the displacement *d* as a function of position on the cut C, the boundary conditions on the surface of the body, and the elastic constants of the body. Volterra dislocations are a special case of Somigliana dislocations where *d* has a constant value. A Volterra disclination is produced when *d* corresponds to a rigid body rotation of angle $\omega$.

The rigid body displacement *d*, or rotation angle $\omega$, is a convenient means of characterizing a Volterra defect. Note that this parameter is the operation involved in the creation of the defect and at no stage is it necessary to map the distortion field of the defect onto a reference lattice. Thus the defect is characterized by a vector or angle that is defined in the medium in which the defect resides. In an elastic continuum there is no restriction on the admissible values of *d* and $\theta$. But in a crystal, unless these operations correspond to symmetry operations of the crystal then the line defect will introduce a planar defect, such as a fault. Perfect line defects in single crystals are therefore characterized by symmetry operations of the crystal. There are three principal categories:

(1) dislocations, which are characterized by translation operations of the perfect crystal (Burgers vectors);

(2) disclinations, which are characterized by symmetry rotation operations of the perfect crystal;

(3) dispirations, which are characterized by screw-rotation operations of a non-symmorphic crystal.

A dispiration has both disclination and dislocation character. The angle of the disclination component is the angle of rotation associated with the screw rotation operation. The translation of the dislocation component is the supplementary displacement associated with the screw rotation operation.

Using the concept of Volterra dislocations and disclinations we have shown that perfect line defects in a single crystal may be characterized by symmetry operations of the crystal. We now extend this approach to the characterization of isolated perfect line defects in interfaces. A perfect interfacial line defect is defined by the condition that the regions of interface it separates must be related by some symmetry operation. Only when this condition is satisfied can we be certain that the regions of interface, on either side of the defect, are energetically degenerate. The key difference in the case of an interfacial line defect is that the symmetry operation by which it is characterized is a combination of symmetry operations from the adjoining crystals. The mathematical formulation of the general expression defining such compound operations is remarkably simple. Yet despite this simplicity the diversity of potential defects predicted is great. The origin of the diversity is that the general expression involves not only the symmetry operations from either crystal but also the relative orientation and position of the adjoining crystals.

It is emphasized that the method of characterizing line defects described in this section is limited to *isolated* defects and cannot be used to characterize *arrays* of line defects with the same degree of rigour or accuracy. This is because the method characterizes the defect in terms of an operation that is a property of the medium in which the defect is embedded, and which is assumed to be unaffected by the presence of the defect. If the presence of the defect altered this operation in some way then the method would become ambiguous: should we take the operation in the presence or absence of the defect? This is precisely the difficulty presented by an array of line defects. An array of line defects alters the geometric relationship between the crystals on either side of the interface. For example, the array may contribute to the misorientation relation, or a change in the crystal structures on either side of the interface, or a change of the average interface normal. Thus, for an array of line defects it is not possible to characterize each defect by an operation that is independent of the array because the array is an inseparable part of the medium in which the operation is defined. Arrays of line defects have been characterized only by the more traditional circuit mapping approach, and this is developed in full in the next chapter.

### 1.7.1   General formulation

Consider a bicrystal comprising an interface between a black crystal and a white crystal, arranged as shown in Fig. 1.18(a). We imagine that it is manufactured by bonding together the flat surfaces of two semi-infinite black and white crystals. Since the crystals exhibit symmetry it is possible to expose in both of them new surfaces which are equivalent, through symmetry operations, to the initial surfaces.

For example, in Fig. 1.18(b) we separate the two crystals and remove, or add, material from, or to, the two surfaces in the region to the right of A so that steps are formed as shown. The new surfaces that are formed to the right of the steps are equivalent to the original surfaces. In reality there will be some relaxation in both surfaces associated with the introduction of the steps. The existence of the relaxation causes no difficulties, even though we cannot say anything about it without taking into account interatomic forces. The point is that provided each step separates regions of the surface that are separated by a lattice translation vector in the *unrelaxed* state then, at sufficiently large distances from the step that the relaxation caused by the step is negligible, the structures of the surfaces on either side of the step will be equivalent and energetically degenerate. If the two crystals are now bonded back together so that the points indicated by the A's and B's are forced into registry an interfacial line defect will be produced in the AB region. Furthermore, the line defect will be bounded by equivalent interfaces on both sides. The line defect is characterized by the operation which brings the new black surface onto the new white surface. Thus the characterizing operation is a combination of black and white crystal symmetry operations. In the example shown in Fig. 1.18(b) it is a pure translation operation and hence the defect is a dislocation, which also has step character.

In Fig. 1.18(c) we show the two separated crystals prior to the creation of an interfacial disclination. In this case each surface contains two equivalent facets prior to bonding together. In order for each pair of facets to be equivalent they must be related by a symmetry rotation of their respective crystals. On bonding the surfaces together we will create an interfacial disclination separating regions of an interface that are equivalent sufficiently far from the disclination that relaxation may be ignored. The disclination is also characterized by the operation which brings the new surfaces into contact and in this case it is combination of symmetry rotations from both crystals. In the general case an interfacial line defect is characterized by a combination of symmetry operations from both crystals involving point symmetry operations and translations. Such a defect has both dislocation and disclination character.

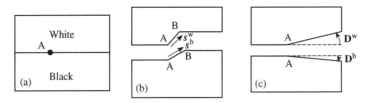

**Fig. 1.18** The formation of an interface by bonding black and white crystals together with (a) perfect surfaces; (b) surfaces in which steps with vectors $s^w$ and $s^b$ exist prior to bonding; (c) surfaces in which equivalent facets exist related by rotational symmetries $\mathbf{D}^w$ and $\mathbf{D}^b$ prior to bonding. (From Pond (1985).)

It may be noticed that the operation characterizing an interfacial line defect is independent of the interface normal. Whether the interface has a periodic structure or not is therefore irrelevant to the existence and classification of the defect. This is in sharp contrast with the circuit mapping approach where the reference structure is frequently taken to be some periodic configuration. As we shall see in Section 1.9, at an aperiodic interface the regions of interface on either side of the defect are locally isomorphic, and therefore they are also energetically degenerate, as they must be if they are related by symmetry.

Let the space group operation of the black crystal relating the equivalent surfaces of the black crystal prior to bonding be $S^b = [D^b_k | \alpha^b_k + R^b_l]$. We have allowed for the possibility that the crystal may be non-symmorphic through the inclusion of the supplementary displacement $\alpha^b_k$. Similarly let the space group operation of the white crystal relating its equivalent surfaces be $S^w = [D^w_i | \alpha^w_i + R^w_j]$. These space group operations are currently expressed in the coordinate systems of the respective crystals. In order to express them both in the coordinate system of the white crystal we re-express $S^b$ as $[T' | t][D^b_k | \alpha^b_k + R^b_l][T' | t]^{-1}$. Now in order to bring the new surface of the black crystal onto the new surface of the white crystal we have to perform the operation $S^w S^{b-1}$ on the new surface of the black crystal. Thus the interfacial defect is characterized by the operation

$$S^w S^{b-1} = [\mathbf{D}^w_i | \alpha^w_i + R^w_j][\mathbf{T'} | t][\mathbf{D}^b_k | \alpha^b_k + R^b_l]^{-1}[\mathbf{T'} | t]^{-1} \qquad (1.80)$$

in the coordinate system of the white crystal. If $S^w S^{b-1}$ has the form $[\mathbf{E} | \tau]$, where $\tau$ is a displacement, it describes an interfacial dislocation of Burgers vector $\tau$. If $S^w S^{b-1}$ has the form $[\mathbf{D} | 0]$, where $\mathbf{D}$ is a proper rotation, it describes an interfacial disclination with a rotation angle described by $\mathbf{D}$. Finally, if $S^w S^{b-1}$ has the form $[\mathbf{D} | \tau]$, where neither $\mathbf{D}$ nor $\tau$ lead to admissible operations, it describes an interfacial dispiration.

We see that the set of all possible perfect interfacial line defects is characterized by the set of proper operations $\{S^w S^{b-1}\}$ which is simply the set $\{S^w S^b\}$ since $S^{b-1}$ is another $S^b$. So far we have restricted $S^w$ and $S^b$ to be space group operations of the white and black *crystals*. If either crystal is non-holosymmetric then there are space group operations of the crystal lattice that are not shared by the crystal itself. If one of those operations is used in $S^w S^{b-1}$ then a fault or domain boundary will emanate from the interfacial defect into the non-holosymmetric crystal. However, the interfacial defect will still separate symmetry related regions of interface, although those regions will belong to different dichromatic complexes. It follows that provided we accept the existence of faults or domain boundaries emanating from an interfacial line defect the operations $S^w$

and $S^b$ may be any space group operations of the white and black *crystal lattices* even for non-holosymmetric crystals.

If the dichromatic pattern contains coincident symmetry elements there will be some members of the set $\{S^w S^b\}$ which correspond to the identity. In those cases the interfacial defect will have no dislocation or disclination character although there may be a step or facet junction associated with the defect. In the extreme case where the black crystal is identical to the white crystal, and they share the same orientation, the set $\{S^w S^b\}$ reduces to the set $\{S^w\}$. On the other hand, if the space group of the dichromatic pattern is simply 1 then all the operations of the set $\{S^w S^b\}$ will correspond to line defects with dislocation and/or disclination character. Here we have another example of the principle of symmetry compensation. The absence of any symmetry in the dichromatic pattern (and hence in the interface) is exactly compensated by the existence of equivalent dichromatic patterns (and hence interfaces) related by the sets $\{S^b\}$ and $\{S^w\}$. Thus the absence of any symmetry in an interface is exactly compensated by the existence of equivalent interfaces and hence interfacial line defects, with dislocation and/or disclination character, that may separate them. Put another way, the greater the degree of dissymmetrization in the dichromatic pattern the greater the number and variety of possible interfacial line defects.

An important aspect of interfacial line defects is that there will generally be a morphological feature associated with the defect core such as a step or a facet junction. This may be understood by reference to Fig. 1.18(b). The new white surface is related to the initial one by a translation operation $[\mathbf{E}\,|\,R_j^w]$. Thus a surface step of height

$$h^w = \hat{n} \cdot R_j^w \tag{1.81a}$$
$$= \hat{n} \cdot s^w \tag{1.81b}$$

is present on the surface, where $\hat{n}$ is the unit normal to the surface and hence to the eventual interface. $\hat{n}$ is expressed in the coordinate system of the white crystal and we remind the reader that our convention for the sense of $\hat{n}$ is that it points into the white crystal. In writing eqn (1.81b) we have adopted a simpler, more widespread, notation by replacing $R_j^w$ by the vector $s^w$ which may be regarded as the 'step vector' of the line defect. Similarly, a step separates the new and initial surfaces of the black crystal, with height

$$h^b = \hat{n} \cdot T' R_l^b = \hat{n} \cdot s^b, \tag{1.82}$$

where $s^b$ is the step vector in the black crystal. We note that the senses of these step vectors are fully specified since they always point in the direction from A to B. The height of the step associated with the interfacial defect is then approximately equal to the average height of the two surface steps, taking their sense into account, i.e.

$$h \simeq \tfrac{1}{2}\left(h^w + h^b\right) = \tfrac{1}{2}\hat{n}\cdot\left(R_j^w + T' R_l^b\right) = \tfrac{1}{2}\hat{n}\cdot\left(s^w + s^b\right). \tag{1.83}$$

This is an approximate estimate of the interfacial step height because it does not take into account the elastic distortion of the crystals. We have assumed so far that $s^w$ and $s^b$ are crystal lattice vectors $R_j^w$ and $R_l^b$. If non-symmorphic operations exist in either crystal then $s^w = \alpha_i^w + R_j^w$ and $s^b = \alpha_k^b + R_l^b$ in general.

If the line defect has no dislocation or disclination character, but there is a step, then the defect is called a pure step. Pure steps arise when $S^b$ and $S^w$ are pure translation operations and where the step vectors, $s^b$ and $s^w$, are equal. In order for the step vectors to be equal they must be equal to a translation vector of the dichromatic pattern. Thus, pure steps can exist only when there is translational symmetry in the dichromatic pattern.

Steps, whether pure or not, may be classified as perfect, when the step vectors are crystal lattice translation vectors, or partial, when the step vectors include supplementary displacements. Thus, the step vectors of partial steps separate sites of different sublattices within a crystal. Our use of the adjectives 'perfect' and 'partial' is analogous to their use in perfect and partial crystal lattice dislocations, where the Burgers vectors are whole and fractional crystal lattice translation vectors respectively. An example of a partial step is a 'demi-step' on a (001) surface of a diamond cubic crystal, which is a step of a/4 height and the step vector includes the supplementary displacement $\frac{1}{4}\langle 111 \rangle$.

Finally in this section we point out that perfect line defects of either adjoining crystal are always admissible perfect, interfacial line defects. This immediately follows from eqn (1.80), where either $S^w$ or $S^b$ is set equal to the identity and it holds regardless of the interfacial normal. Conversely, a perfect interfacial line defect characterized by $S^w S^{b-1}$, where neither $S^w$ nor $S^b$ is the identity, is not, in general, an admissible perfect line defect of either crystal. (The exceptions arise when $S^w$ or $S^b$ is a symmetry operation of the dichromatic pattern.) Such defects are confined to the interface. There are many experimental observations of crystal lattice dislocations in interfaces. Small-angle grain boundaries are normally formed from arrays of crystal lattice dislocations, although in many instances the dislocations are dissociated into complex networks (see Section 2.8). Crystal lattice dislocations are also found in epitaxial systems where they act as anti-coherency dislocations to relieve the coherency strains caused by the change in lattice parameters across the interface (see Section 2.7.4).

### 1.7.2 Interfacial dislocations

Interfacial dislocations are characterized by operations $S^w S^{b-1}$ which are of the form $[E|\tau]$, where $\tau$ is a displacement. Expanding eqn (1.80) we obtain:

$$S^w S^{b-1} = [\mathbf{D}_i^w \mathbf{T}' \mathbf{D}_k^{b-1} \mathbf{T}'^{-1} | -\mathbf{D}_i^w \mathbf{T}' \mathbf{D}_k^{b-1} (\mathbf{T}'^{-1} t + \alpha_k^b + R_l^b) + \mathbf{D}_i^w t + \alpha_i^w + R_j^w].$$
(1.84)

#### 1.7.2.1 *DSC dislocations*

The first class of dislocations is obtained by choosing $S^w = [E|R_j^w]$ and $S^b = [E|R_l^b]$. Then eqn (1.81) reduces to

$$S^w S^{b-1} = [E|R_j^w - \mathbf{T}' R_l^b]$$
(1.85)

We recognize the vectors $R_j^w - \mathbf{T}' R_l^b$ as vectors of the DSC lattice described in Section 1.5.4. For this reason we call these dislocations DSC dislocations. The Burgers vector of each dislocation is given by

$$b = R_j^w - \mathbf{T}' R_l^b = s^w - s^b$$
(1.86)

and is dependent on $\mathbf{T}'$, but independent of the relative translation $t$ of the two crystals and of the interface normal. The Burgers vectors of these dislocations are unaffected by the presence of non-symmorphic operations in a crystal since they do not alter the lattice translation vectors. The crystallographic reason for the existence of DSC dislocations is that the translational symmetries of the adjoining crystal lattices have been broken in the formation of the bicrystal.

A DSC dislocation may move along the interface in general by a combination of glide and climb. The climb component is the component, $b_c$, of the Burgers vector normal to the interface, which is simply the difference in the step heights on the two crystal surfaces:

$$b_c = \hat{n} \cdot b = \hat{n} \cdot (s^w - s^b) = h^w - h^b. \tag{1.87}$$

It is clear that the step height is not simply related to the climb (or glide) component of the Burgers vector of the dislocation.

It is noted that the Burgers vectors defined by eqn (1.86) change continuously with continuous changes in the transformation $\mathbf{T}'$. By contrast, in the circuit mapping approach the Burgers vectors are determined by the reference lattice. In that case a small change in the transformation $\mathbf{T}'$ results in a small change in the average spacing of the defects but their Burgers vectors remain constant. The differences between these Burgers vectors are related to the errors in assigning Burgers vectors to dislocations in arrays, and are probably too small to be resolved experimentally. Another way to see the distinction between them is that a small change in $\mathbf{T}'$ causes discrete perturbations to the spacing of the dislocations derived from the reference lattice in the circuit mapping approach. But, as we shall see in the next chapter, a perturbation in the spacing of an array of dislocations is equivalent to introducing another dislocation into the array and retaining their unperturbed spacings. The Burgers vector of the new dislocation is the one that would be predicted by the present treatment, whereas the Burgers vector of the dislocations in the array is the one that would be predicted by the circuit mapping treatment. This amounts to saying that the Burgers vectors defined by eqn (1.85) are the same as those obtained by the circuit mapping approach if the reference structure is simply the ideal bicrystal containing the defect free interface, shown in Fig. 1.18(a). In that way the reference structure changes continuously with $\mathbf{T}'$, in the present approach, whereas in the circuit mapping approach the reference structure changes discontinuously at some $\mathbf{T}'$ from one bicrystal to another.

There are many experimental observations of DSC dislocations in homophase and heterophase interfaces. For example, Figs 2.32, 2.34, 2.37 (a–d), 10.7, and 12.11 show DSC dislocations in twist and tilt grain boundaries in Au, Al, and MgO. Figure 1.19 shows a computer simulated structure of a $\frac{1}{10}[310]$ DSC dislocation in a $(1\bar{3}0)$ twin boundary in aluminium. Note the step vectors and the step at the core of the dislocation of two $(2\bar{6}0)$ planes height. In Fig. 1.20 we show the computer-simulated structure of a $\frac{1}{6}[11\bar{2}]$ DSC dislocation in an interface between f.c.c. and h.c.p. crystals. The interface is on (111) in the f.c.c. phase and (0001) in the h.c.p. phase. The step vectors are shown and the dislocation is associated with a step of two (111) planes height.

### 1.7.2.2  Supplementary displacement dislocations

These dislocations arise from supplementary displacements associated with non-symmorphic space group operations. Therefore, they arise only when at least one of the crystals is non-symmorphic. Let $S^w = [\mathbf{D}_i^w | \alpha_i^w]$ and $S^b = [\mathbf{D}_k^b | \alpha_k^b]$ in eqn (1.84). Equation then simplifies to

$$S^w S^{b-1} = [\mathbf{D}_i^w \mathbf{T}' \mathbf{D}_k^{b-1} \mathbf{T}'^{-1} | -\mathbf{D}_i^w \mathbf{T}' \mathbf{D}_k^{b-1} (\mathbf{T}'^{-1} t + \alpha_k^b) + \mathbf{D}_i^w t + \alpha_i^w]. \tag{1.88}$$

In order for this to be of the form $[\mathbf{E} | \tau]$ we require

$$\mathbf{D}_i^w = \mathbf{T}' \mathbf{D}_k^b \mathbf{T}'^{-1}. \tag{1.89}$$

This expresses the condition for $\mathbf{D}_i^w$ and $\mathbf{D}_k^b$ to be coincident operations in the dichromatic complex. When this condition is satisfied the term $\mathbf{D}_i^w \mathbf{T}' \mathbf{D}_k^{b-1}$ is equal to $\mathbf{T}'$ and eqn (1.88) becomes:

$$S^w S^{b-1} = [\mathbf{E} | \alpha_i^w - \mathbf{T}' \alpha_k^b + (\mathbf{D}_i^w - \mathbf{E})t]. \tag{1.90}$$

Thus the Burgers vector of the dislocation is given by $b = \alpha_i^w - \mathbf{T}' \alpha_k^b + (\mathbf{D}_i^w - \mathbf{E})t$. If the rigid body translation $t$ is zero, or such that it does not misalign the coincident point

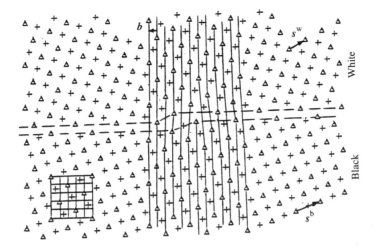

**Fig. 1.19** Computer simulated structure of a $b = \frac{1}{10}[310]$ DSC edge dislocation in a ($\bar{1}$30) symmetric tilt boundary in aluminium, viewed in projection along [001]. The triangles and crosses denote atoms on the two (002) planes in each [001] period along the direction of projection. The last planes of the black and white crystals adjacent to the boundary are shown by broken lines. A step of two ($\bar{2}$60) planes height is seen. (310) planes are shown by solid lines, and the dislocation centre is seen at the step by the terminating (310) plane. The DSC lattice is shown at the lower left. Also, the step vectors, $s^w$ and $s^b$, and the Burgers vector, $b$, are shown, where $b = s^w - s^b$.

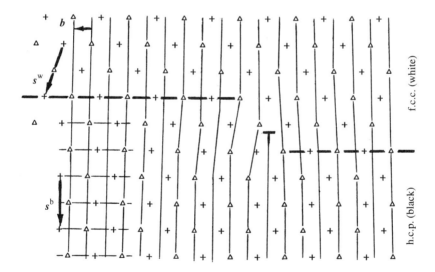

**Fig. 1.20** Computer-simulated structure of a $\frac{1}{6}[11\bar{2}]$ dislocation in an interface between f.c.c. and h.c.p. crystals. At the interface (111) of the f.c.c. crystal is parallel to (0001) of the h.c.p. crystal and close packed directions are parallel. The nearest neighbour spacings in the two crystals are equal. The direction of projection is [1$\bar{1}$0] in the f.c.c. crystal, and the triangles and crosses show atoms on the two (2$\bar{2}$0) planes in each crystal period. The interface (broken lines) has a step of two close packed planes height. The step vectors $s^b$ and $s^w$ are shown, as is the Burgers vector and the DSC lattice (at the lower left). The dislocation centre may be identified at the terminating (11$\bar{2}$) plane.

group operations (in which case $\mathbf{D}_i^w t = t$), then $b$ is determined entirely by the supplementary displacements and the transformation $\mathbf{T}'$. Otherwise, the rigid body translation $t$ contributes to the Burgers vector of the dislocation.

Let us assume $\mathbf{D}_i^w t = t$ for the present. There are two principal cases of interest corresponding to whether or not the point symmetry operations leave the orientation of a particular interface normal invariant. Consider first the case where $\hat{n}$ is left invariant by $\mathbf{D}_i^w$. In this case the dislocation can exist on the interface with normal $\hat{n}$, and the dislocation has been introduced by bonding white and black surfaces exhibiting steps with heights $h^w = \hat{n} \cdot \alpha_i^w$ and $h^b = \hat{n} \cdot \mathbf{T}' \alpha_k^b$. The resulting dislocation has a climb component $\hat{n} \cdot b = \hat{n} \cdot (\alpha_i^w - \mathbf{T}' \alpha_j^b)$ and an associated step of height $h = \frac{1}{2} \hat{n} \cdot (\alpha_i^w + \mathbf{T}' \alpha_j^b)$. The second case of interest is where $\hat{n}$ is not left invariant by $\mathbf{D}_i^w$. In this case the dislocation, with the same Burgers vector as in the first case, must exist at the junction between facets with normals $\hat{n}$ and $[\mathbf{D}_i^w | \alpha_i^w] \hat{n}$.

The NiSi$_2$–Si interfaces discussed in Section 1.6.2 provide examples of supplementary displacement dislocations. Since $\mathbf{T} = \mathbf{E}$ and $t = 0$ in that case, and the NiSi$_2$ phase is symmorphic, it follows that the Burgers vectors are expected to be $[\mathbf{E} | \alpha_i^w - \mathbf{T}' \alpha_k^b] = [\mathbf{E} | -\alpha_k^b]$. For Si the supplementary displacement is $\frac{1}{4}[111]$ and hence the Burgers vectors are expected to be $\frac{1}{4}[\bar{1}\bar{1}\bar{1}]$ modulo $\mathbf{R}_j^b$. Since all the point symmetry operations are coincident the dislocations can exist on any interfacial plane that is left invariant by any of the nonsymmorphic operations of the Si crystal. As we saw in Section 1.6.2 this set of interfaces has normals parallel to $\langle hk0 \rangle$. Thus we predict that dislocations with Burgers vectors equal to $\frac{1}{4}[\bar{1}\bar{1}\bar{1}]$ (modulo $\mathbf{R}_j^b$) can exist on $\{hk0\}$ interfaces only, or at junctions between equivalent $\{hkl\}$ facets, where $h, k,$ and $l$ are all non-zero. This is in agreement with the analysis described in Section 1.6.2 and the experimental observations of supplementary dislocations reported there.

Figure 1.21(a) is a schematic representation of NiSi$_2$ and Si (001) crystal surfaces before bonding. The Si crystal surface has a partial step with height $a/4$, whereas the NiSi$_2$ surface is perfect. The surfaces on either side of the partial step expose atoms from the two f.c.c. sublattices in the Si crystal. Those surfaces are related by nonsymmorphic operations of the Si, such as $4_1$ screw rotation and diamond glide operations, whereas the corresponding operations in the NiSi$_2$ crystal, namely 4 and $m$ leave its surface invariant. On bonding the partial step becomes the core of a supplementary

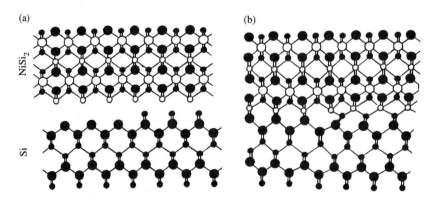

**Fig. 1.21** Schematic illustration of the formation of a supplementary displacement dislocation between Si and NiSi$_2$ crystals: (a) NiSi$_2$ (upper crystal) and Si surfaces prior to bonding: the NiSi$_2$ surface is perfect but there is a step on the Si surface; (b) after bonding the step on the Si surface gives rise to a dislocation. (From Pond (1989).)

displacement dislocation, as shown in Fig. 1.21(b). Note that the interfacial step height in this case is the same as the climb component of the dislocation and equal to $-\frac{1}{2}\hat{n}\cdot\alpha_j^b$. It follows that the dislocations may be glissile (i.e. have no climb component) only on {110} interface planes.

When the rigid body translation $t$ is non-zero we have seen that the Burgers vectors of the supplementary displacement dislocations are altered. This is an obvious difference between this class of dislocations and DSC dislocations. We shall consider the effect of $t$ next.

### 1.7.2.3 *Relaxation displacement dislocations*

These dislocations have their origins in the rigid body displacement $t$ in the operation $[\mathbf{T}'|t]$ relating the crystals. The magnitude and direction of $t$ is determined by relaxation processes at the interface and cannot be predicted by a crystallographic analysis. But its consequences, whenever it should arise, can be predicted by such an analysis, as we describe in this section. It will be recalled from the final part of Section 1.5.4 that $t$ may always be expressed modulo a vector of the DSC lattice. Therefore, we are concerned about the effect of $t$ only in those cases where there are coincident translation vectors in the dichromatic pattern.

Using eqn (1.84) the effect of $t$ is to introduce an additional translational component to $S^wS^{b-1}$ of $[\mathbf{E}|\mathbf{D}_i^w(\mathbf{E} - \mathbf{T}'\mathbf{D}_k^{b-1}\mathbf{T}'^{-1})t]$. This means that an additional displacement is required to bring the new black and white surfaces together. For both crystal lattice dislocations and DSC dislocations $\mathbf{D}_i^w$ and $\mathbf{D}_k^b$ were set equal to $\mathbf{E}$ and hence this additional displacement was zero. This result simply means that there is no additional Burgers vector because the rigid body displacement $t$ is the same on either side of the defects. But in the case of supplementary displacement dislocations $\mathbf{D}_i^w$ and $\mathbf{D}_k^b$ are not $\mathbf{E}$ but coincident point symmetry operations, i.e. $\mathbf{D}_i^w = \mathbf{T}'\mathbf{D}_k^b\mathbf{T}'^{-1}$ (eqn 1.89). Substituting this equality into the expression for the additional Burgers vector we find that it simplifies to $(\mathbf{D}_i^w - \mathbf{E})t$, which is seen to be part of the Burgers vector of a supplementary displacement dislocation in eqn (1.90).

Relaxation displacement dislocations arise whenever there are coincident point group operations that are destroyed by a displacement $t$. To prove this we set $S^w = [\mathbf{D}_i^w|\mathbf{0}]$, $S^b = [\mathbf{D}_k^b|\mathbf{0}]$, and $\mathbf{D}_i^w = \mathbf{T}'\mathbf{D}_k^b\mathbf{T}'^{-1}$. Then $S^wS^{b-1}$ in eqn (1.81) simplifies to $S^wS^{b-1} = [\mathbf{E}|(\mathbf{D}_i^w - \mathbf{E})t]$. The Burgers vector will be zero for special values of $t$ for which $\mathbf{D}_i^wt = t$.

We conclude that interfacial dislocations can arise whenever a rigid body displacement breaks an initially coincident symmetry operation, irrespective of whether the operation was symmorphic or not. These are the dislocations we refer to as relaxation displacement dislocations. Such dislocations separate degenerate interfacial domains whose structures are related by the initially coincident operation. They exist either on a particular interface plane or at a facet junction depending on whether or not the coincident operation leaves the interface normal invariant.

In the case of grain boundaries it is possible that a rigid displacement $t$ may break an antisymmetry operation. In that case a defect will separate degenerate structures on interfaces with normals related by the antisymmetry operation, and the defects can arise on planar interfaces only in those cases where the interface normal is left invariant by the antisymmetry operation. Antisymmetry operations do not belong to the space group of either crystal and therefore cannot be substituted into eqn (1.80). Nevertheless, a general expression for characterization of defects in this class can be obtained by forming the compound operation which first transforms the black crystal surface into the white, i.e. $S^{b'-1} = [\mathbf{T}'|t][\mathbf{D}_i^{b'}|\boldsymbol{\alpha}_i^b]^{-1}[\mathbf{T}'|t]^{-1}$, where $[\mathbf{D}_i^{b'}|\boldsymbol{\alpha}_i^b]$, is the antioperation expressed in the coordinate frame of the black crystal, followed by $S^{w'} = [\mathbf{D}_i^{w'}|\boldsymbol{\alpha}_i^w]$, which transforms the white surface back to the black and is the anti-operation expressed in the white frame. Thus the compound operation is given by

$$S^{w'}S^{b'-1} = [\mathbf{D}_i^{w'}|\boldsymbol{\alpha}_i^w][\mathbf{T}'|t][\mathbf{D}_i^{b'}|\boldsymbol{\alpha}_i^b]^{-1}[\mathbf{T}'|t]^{-1}$$
$$= [\mathbf{E}|(\mathbf{D}_i^{w'} - \mathbf{E})t] \qquad (1.91)$$

provided $\mathbf{D}_i^{w'} = \mathbf{T}'\mathbf{D}_i^{b'}\mathbf{T}'^{-1}$ and $\boldsymbol{\alpha}_i^w = \mathbf{T}'\boldsymbol{\alpha}_i^b$. Thus the Burgers vector of a dislocation separating the two variants related by a destroyed antisymmetry operation is given by $b = (\mathbf{D}_i^{w'} - \mathbf{E})t$. The Burgers vector is the displacement operation required to move the black crystal from the relative position characteristic of one variant to that of another related to the former by the broken anti-operation. In practice the anti-operation is first carried out as if it were an ordinary operation and then an inversion is applied, as in eqn (1.68).

The first experimental observation of relaxation displacement dislocations was made by Pond (1977), who saw a $\frac{1}{3}[111]$ DSC dislocation in a $(11\bar{2})$ twin in aluminium dissociate into two relaxation displacement dislocations with approximate Burgers vectors $\frac{1}{9}[111]$ and $\frac{2}{9}[111]$. Figure 1.22 shows computer-simulated relaxation displacement dislocations in a $(1\bar{3}0)$ twin in aluminium. In Fig. 1.22(a) the dislocation has no step and its Burgers vector is approximately $\frac{1}{15}[310]$, while in Fig. 1.22(b) there is a step of two $(2\bar{6}0)$ planes height and the Burgers vector is approximately $\frac{1}{30}[310]$. Both dislocations separate energetically degenerate domains of the $(1\bar{3}0)$ interface that are related by the $2'$ or the $m'$ symmetries that have been suppressed by the relative translation $t$ of approximately $\frac{1}{30}[310]$. These relaxation displacement dislocations are formed by the dissociation of the $\frac{1}{10}[310]$ DSC dislocation shown in Fig. 1.19: $\frac{1}{10}[310] = \frac{1}{15}[310] + \frac{1}{30}[310]$.

### 1.7.2.4   *Non-holosymmetric crystals and interfacial defects*

When we discussed interfaces between Ge and GaAs in Section 1.6.1 we found that the reduction of symmetry in the GaAs crystal from $Fm\bar{3}m$ to $F\bar{4}3m$ resulted in the existence of two complex variants. Whenever they coexisted at an interface a domain boundary was required, emanating from an interfacial line defect. We re-examine this example in terms of the operations characterizing possible line defects.

Recall that each point symmetry operation in the Ge and GaAs crystals is coincident. The supplementary displacements are also identical in the Ge and GaAs crystals, but in the GaAs case they correspond to interchanging the Ga and As sites and hence to complex exchange operations. If we select the same operations for $S^w$ and $S^b$, and since $t = 0$, then $S^wS^{b-1} = [\mathbf{E}|\boldsymbol{0}]$. Moreover, if $S^w$ and $S^b$ are non-symmorphic we will have a line defect in the interface with no dislocation character separating the two complex variants. An antisite domain boundary will emanate from this defect into the GaAs crystal. The defect can exist only on those interfaces that are left invariant by at least one of the non-symmorphic operations. As in the NiSi$_2$–Si example this set of interfaces is $\{hk0\}$. Thus, in agreement with our conclusions in Section 1.6.1, we find that antisite domain boundaries can intersect $\{hk0\}$ interfaces with Ge and separate energetically degenerate interfacial regions. However, unlike the closely related example of the NiSi$_2$–Si interfaces, we find that the defects in the Ge:GaAs case have no dislocation character, provided there is no relaxation displacement at the inversion domain boundary.

Figure 1.23 is a schematic illustration of the overgrowth of GaAs on (001) Ge. The

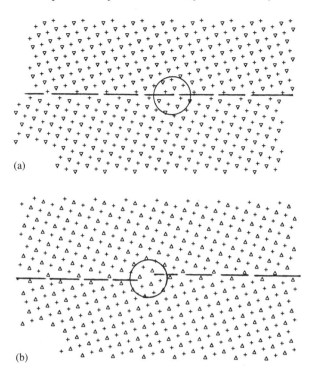

**Fig. 1.22** Relaxation displacement dislocations in a $(1\bar{3}0)$ symmetric tilt boundary in aluminium obtained by computer simulation. The structure is seen in projection along [001], and the triangles and crosses refer to atoms on successive (002) planes. (a) The core of the dislocation ($b \simeq \frac{1}{15}[310]$) is encircled. The dislocation is not associated with a step, as may be seen by the planarity of the boundary shown by the broken line. (b) In this case $b \simeq \frac{1}{30}[310]$ and a step of two $(2\bar{6}0)$ planes height is associated with the core (encircled).

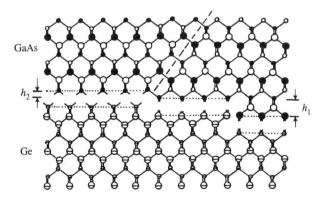

**Fig. 1.23** Schematic illustration of a GaAs crystal grown on a Ge (001) substrate with two defects in the interface. The step on the right, of height $h_1$, is a pure step which is also perfect because the step vector is a crystal lattice vector. Consequently, this step not associated with an antisite domain boundary in the GaAs. The step on the left, of height $h_2$, is also a pure step but it is a partial step because the step vector is not a crystal lattice vector since it connects a Ga site and an As site. Consequently, there is an antisite domain boundary in the GaAs, shown by the dashed line, emanating from this step. (From Pond (1985).)

large step, on the right, is a perfect step and it does not require an antisite domain boundary in the GaAs. However, the partial step, on the left, requires an antisite domain boundary in the GaAs.

### 1.7.2.5 *Interfacial disclinations and dispirations*

Interfacial disclinations are described by eqn (1.84) when $S^w = [D_i^w | \mathbf{0}]$ and $S^b = [D_k^b | \mathbf{0}]$, for then

$$S^w S^{b-1} = [D_i^w T' D_k^{b-1} T'^{-1} | -D_i^w T' D_k^{b-1} T'^{-1} t + D_i^w t].$$

This compound operation will be a proper operation provided $S^w$ and $S^b$ are either both proper or both improper. $S^w S^{b-1}$ must be a proper operation for otherwise the defect will invert the black and white crystals, which is topologically impossible. Note that the resulting rotation, $D_i^w T' D_k^{b-1} T'^{-1}$, will be quite small if $D_i^w$ and $D_k^b$ represent the same type of point symmetry operation and $T'$ represents a small misorientation. The disclination acquires some dislocation character, with $b = D_i^w (E - T' D_k^{b-1} T'^{-1}) t$, in the presence of a rigid body displacement $t$. The topological feature at a disclination is a discontinuous change in the interfacial inclination, depending on the compliances of the adjoining crystals. If the compliances of the adjoining crystals are the same, the inclination of an interface initially having a normal $\hat{n}$ changes discontinuously to a value intermediate between $[D_i^w | \mathbf{0}]\hat{n}$ and $T'[D_k^b | \mathbf{0}]T'^{-1}\hat{n}$.

Interfacial dispirations form when at least one of the adjoining crystals is non-symmorphic. Setting $R_j^w = R_l^b = \mathbf{0}$ in eqn (1.84), $S^w S^{b-1}$ becomes $[D_i^w T' D_k^{b-1} T'^{-1} | - D_i^w T' D_k^{b-1} (T'^{-1} t + \alpha_k^b) + D_i^w t + \alpha_i^w]$. It is clear that a dispiration has both disclination and dislocation character even when $t = \mathbf{0}$.

Since the elastic strain field energy of a disclination varies quadratically with the angle of rotation and the the size of the specimen, both disclinations and dispirations are unlikely to occur at interfaces except when the angle of rotation is small and/or the specimen is small.


## 1.8  THE MORPHOLOGIES OF EMBEDDED CRYSTALS

In this section we consider the morphology of one crystal entirely embedded within another, such as a precipitate or a stacking fault tetrahedron (Hirth and Lothe 1982) or an 'island grain'. At equilibrium the morphology of the embedded crystal is that which minimizes the free energy of the system (see Section 5.6.2). The free energy depends on (i) the interfacial free energies, (ii) the elastic strain field energy in the system, and (iii) the energies of any defects present in the interfaces or within the embedded crystal. This is a considerably more complex problem than determining the morphology of a crystal growing in a fluid. Our approach follows Pond and Dibley (1990) which generalizes the work of Cahn and Kalonji (1982) and Kalonji and Cahn (1982).

Kalonji and Cahn (1982) introduced the concept of the group of the Wulff Form, $P(c)$. For a heterophase system this is the point group of the ordinary space group, $\Phi(c)$, of the dichromatic complex. It is equal to the intersection of the point groups of the two crystals, taking into account their relative orientation and translation (see eqn (1.71)). Two interfaces with normals $\hat{n}$ and $D(c)_j \hat{n}$, where $D(c)_j$ is an element of $P(c)$, are equivalent and hence energetically degenerate. They are morphological variants of the same interface. These equivalent interfaces define a shape for the embedded crystal, which may or may not be a closed polyhedron. If the embedded crystal is bounded by interfacial structures, some of which have only point symmetry 1, the symmetry of the polyhedron will be consistent with $P(c)$. According to this analysis the symmetry group of the morphological form of the embedded crystal is $P(c)$.

When antisymmetry is present, as in the case of an embedded island grain, there are as many antisymmetry point group operations as there are ordinary point group operations in the dichromatic complex. The group of the Wulff Form must remain a mono-chromatic group since it describes the morphological symmetry of a single crystal. This is achieved by multiplying all the antisymmetry operations present in the dichromatic complex by $\bar{1}$ and ignoring the colour reversing character of the resulting operations (see eqn (1.68)). The multiplication by the inversion takes into account the fact that an embedded crystal can have only outward pointing normals.

The group of the Wulff Form for a crystal embedded in a fluid is simply the point group of the crystal. This is a familiar result which follows from the Wulff–Herring construction (see Section 5.6.2). It is based on the minimization of the surface free energy with the constraint that the volume of the crystal is conserved. But for non-holosymmetric crystals the principle of symmetry compensation indicates that it is possible for the morphological symmetry of the crystal to be that of the lattice rather than the crystal structure, provided we accept the existence of defects separating complex variants. For example, consider the morphology of a single crystal of sphalerite in a fluid. The point group is $\bar{4}3m$. Let us suppose that the energy of the {111} surfaces is sufficiently lower than the other surface normals that the crystal is bounded only by {111} surfaces. A perfect sphalerite crystal with {111} faces would have a tetrahedral morphology, which would be consistent with the point group. But the morphological form of the crystal could have $m\bar{3}m$ symmetry, the symmetry of the crystal lattice, if antisite domain boundaries within the crystal were allowed. Thus an octahedral morphology would be attained by introducing antisite domain boundaries inside the octahedron. This is illustrated in Fig. 1.24, where the antisite domain boundaries are shown on {100} planes for simplicity. In this example the surfaces in both morrhological forms are of the same type and there is no energy gain by introducing antisite domains. But the example illustrates the principle that symmetry compensation can lead to morphologies with symmetries of the crystal lattice rather than the crystal structure. For a crystal wholly embedded within another

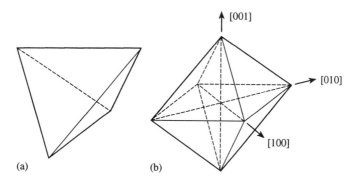

**Fig. 1.24** Illustration showing that a sphalerite crystal in a fluid could, in principle, adopt a morphology with a symmetry higher than its point group symmetry $\bar{4}3m$. (a) A tetrahedral form bounded by {111} planes, (with symmetry $\bar{4}3m$) which would be adopted if {111} surfaces had a much lower energy than any other surface normal. (b) An octahedral form, again bounded by {111} surfaces, with higher symmetry $m\bar{3}m$. In order for the sphalerite to adopt this fully compensated form there would have to be antisite domain boundaries shown on the {100} planes inside the octahedron.

the energy cost associated with the introduction of defects may be offset by the energy gain associated with the introduction of new interface normals. In that case the Wulf–Herring construction is not applicable, since the total energy of the system is determined by a balance between interfacial energies and energies associated with defects separating equivalent interfaces (aside from any elastic strain field energies).

Pond and Dibley have pursued the logical consequences of the principle of symmetry compensation further. If a cavity is prepared in the host crystal, bounded by degenerate sets of surfaces, then it can be filled by the embedded crystal in such a way that the sets of initially degenerate surfaces become sets of degenerate interfaces, provided interfacial defects are introduced. In this case the symmetry of the embedded crystal is identical to the symmetry of the host crystal. This remarkable conclusion is independent of the symmetries and structures of the two crystals concerned and their relative orientation, although those factors affect the nature of the geometrically necessary defects. We shall derive this result in detail below. Pond and Dibley refer to this extreme case where the morphological symmetry of the embedded crystal is identical to that of the host crystal as 'fully compensated' symmetry since all the symmetry of the host crystal has been retained. Lower symmetry morphologies are possible, requiring fewer defects. The case considered by Cahn and Kalonji is where no defects are required and the morphological symmetry is simply the point group of the dichromatic complex. Even lower symmetries may be envisaged where other factors, such as the elastic strain field energy, prevent the symmetry of the Wulff plot from being attained. We see that symmetry arguments alone cannot predict morphologies for embedded crystals, but they enable us to see possibilities that we may not have otherwise thought of. In particular, they reveal the important role that interfacial defects may play in affecting the embedded crystal morphology.

Let the embedded crystal be black and the surrounding crystal white. We assume that the crystals have different structures and/or compositions so that there are no anti-symmetry operations relating them and there are no coincident translation vectors. Since there are no coincident translation vectors the space group of the dichromatic complex, $\Phi(c)$, reduces to a point group. $\Phi(c)$ depends on the relative orientation and position of the crystals and it is the intersection of the black and white crystal space groups: $\Phi^b \cap \Phi^w$. We are interested in the orientational variants of the primary dichromatic complex that are related by white crystal point group symmetry operations that do not appear in $\Phi(c)$. They are obtained by decomposing $\Phi^w$ with respect to $\Phi(c)$ into cosets:

$$\Phi^w = \Phi(c) \cup \Phi(c)\mathbf{W}_1 \cup \Phi(c)\mathbf{W}_2 \cup \Phi(c)\mathbf{W}_3 \cup \ldots \qquad (1.92)$$

where $\mathbf{W}_1, \mathbf{W}_2, \mathbf{W}_3$ are coset representatives, and correspond to point group operations of the white crystal that do not appear in $\Phi(c)$.

Consider the creation of a bicrystal from the primary complex and let the normal to the interface be represented by $\hat{n}$. The symmetry of the resulting bicrystal is represented by $\Phi(i)$. The operations of the dichromatic complex that are absent in $\Phi(i)$ relate equivalent bicrystals which we called morphological variants in section 1.5.6. The morphological variants are found by decomposing $\Phi(c)$ with respect to $\Phi(i)$ into cosets:

$$\Phi(c) = \Phi(i) \cup \Phi(i)\mathbf{C}_1 \cup \Phi(i)\mathbf{C}_2 \cup \Phi(i)\mathbf{C}_3 \cup \ldots \qquad (1.93)$$

where $\mathbf{C}_1, \mathbf{C}_2, \mathbf{C}_3$ are coset representatives and correspond to suppressed operations of the dichromatic complex.

The total set of equivalent bicrystals obtained from all the equivalent dichromatic complexes can now be enumerated. Consider first the set of equivalent bicrystals that are generated from the primary complex. They are morphological variants with normals $\hat{n}$, $\mathbf{C}_1\hat{n}, \mathbf{C}_2\hat{n}, \mathbf{C}_3\hat{n}$, etc., and let there be $x$ of them. $x$ is given by Lagrange's theorem:

$$x = \frac{O(\Phi(c))}{O(\Phi(i))} \tag{1.94}$$

where $O(\Phi)$ denotes the order of the group $\Phi$, i.e. the number of elements contained in the group $\Phi$. There are $y$ equivalent dichromatic complexes, where $y$ is also given by Lagrange's theorem:

$$y = \frac{O(\Phi^w)}{O(\Phi(c))}. \tag{1.95}$$

The total number of equivalent bicrystals is therefore

$$xy = \frac{O(\Phi^w)}{O(\Phi(i))}. \tag{1.96}$$

If we choose $\hat{n}$ such that $\Phi(i) = 1$, and hence $O(\Phi(i)) = 1$, then $xy = O(\Phi^w)$. That is, there are as many equivalent interfaces as there are point group symmetry elements in $\Phi^w$. They are interrelated by the point group symmetry operations of the white crystal. It is possible, therefore, to join the white components of this set into a contiguous assembly without distortion. Indeed, we can arrange these components so they create a closed cavity within the white crystal. If the white crystal exhibits low symmetry the cavity may not necessarily be a closed volume. In that case further, non-equivalent, sets of bicrystals must be introduced with interfacial normals $\hat{n}_1, \hat{n}_2$ from the primary dichromatic complex. It is then always possible to create a closed cavity. Moreover, the cavity will always display the symmetry of the white crystal provided one of the interface normals is associated with point symmetry 1. On the other hand if all of the interface normals are associated with symmetry higher than 1 then the resulting cavity may exhibit higher symmetry than the white crystal.

Although the white components of the bicrystals can be joined contiguously the black components can not, in general, without the introduction of defects. Consider two adjacent facets that are related by a symmetry operation, $\mathbf{W}_i$, of the white crystal. The defect separating the two facets is characterized by the operation that brings the two black components into coincindence. This is $\mathbf{W}_i$, but we are at liberty to precede this operation with any symmetry operation of the black crystal, say $\mathbf{B}_j^{-1}$. Thus appropriate defects are characterized by $\mathbf{Q}_{ij} = \mathbf{W}_i \mathbf{T}' \mathbf{B}_j \mathbf{T}'^{-1}$, in the coordinate system of the white crystal. We are at liberty to select any $\mathbf{B}_j$ in $\mathbf{Q}_{ij}$ and the appropriate choice is the one that minimizes the free energy of the system. Thus, we might choose the $\mathbf{B}_j$ that led to the smallest Burgers vector and/or angular closure failure in $\mathbf{Q}_{ij}$. Alternatively, we could select for $\mathbf{B}_j$ a symmetry operation relating complex variants (one of the $\mathbf{C}_i$'s) and the interfacial defect would then be associated with an extended defect in the black crystal. It is also possible to replace $\mathbf{B}_j$ by a twinning operation of the black crystal, although the interfacial defect will then be associated with a twin boundary in the black crystal.

The total disclination and dislocation content of the embedded crystal must sum to zero because we have assumed that the crystal in which it is embedded is perfect. Thus the configuration of line defects must be in the form of closed loops around the particle. Defects are required at junctions between facets created from different dichromatic complexes. Since the number of dichromatic complexes is fixed, and does not depend on the number of interfacial orientations $\hat{n}_1, \hat{n}_2$, etc., used from the primary complex, the geometrically necessary defect content does not depend on the number of normals invoked from the primary complex.

It may be thought that the presence of defects in an induded crystal would increase the free energy of the system. But the energy cost associated with the defects may be offset by the concomitant increase in area of low energy interfaces bounding the particle. Since the elastic energy associated with disclinations is rather high they are seen extremely rarely in precipitates. Pond and Dibley reviewed experimental observations of the morphological symmetries of precipitates and found examples ranging between the fully compensated form to forms with symmetry lower than the intersection symmetry.

Two interesting examples are provided by $NiSi_2$ precipitation in Si (Augustus 1983). As we saw in Section 1.6.2 $NiSi_2$ has the fluorite structure and its spacegroup is $\Phi^b = Fm\bar{3}m$. The spacegroup of Si is $\Phi^w = Fd\bar{3}m$. As before we neglect the very small difference in lattice parameters. Silicide precipitation occurs in two distinct orientations in Si. In the type A orientation the two crystals have parallel orientations, whereas in type B the silicide is rotated with respect to the Si by 180° about [111]. The intersection symmetry, $\Phi(c)$, for the type A and B orientations is $F\bar{4}3m$ and $P\bar{3}m$ respectively, whereas the fully compensated symmetry in both cases is $\Phi^w = Fd\bar{3}m$. For the type A orientation the intersection symmetry would be consistent with a tetrahedral morphology while the fully compensated symmetry would be consistent with an octahedral form. In fact the octahedral morphology is observed. The geometrically necessary defects in this case are $\frac{1}{4}\langle 111 \rangle$ dislocations, as we saw in Section 1.6.2, which delineate all the edges of the octahedral precipitate. On the other hand, for the type B orientation, trigonal platelets are observed consistent with intersection rather than fully compensated symmetry. In this case the fully compensated morphology would require large-angle disclinations, which are presumably too costly energetically.

## 1.9  QUASIPERIODICITY AND INCOMMENSURATE INTERFACES

The general formulation of operations characterizing perfect interfacial line defects was discussed in Section 1.7.1. The space group operations of the black and white crystals were used explicitly to characterize an interfacial line defect separating equivalent interfacial regions. New, equivalent surfaces of the black and white crystals are created, prior to bonding the two crystals together, by operating with space group operations $S^w$ and $S^b$ on the original surfaces. The defect is characterized by the operation, $S^w S^{b-1}$, which brings the new surface of the black crystal onto the new surface of the white crystal. The equivalence of the new interface to the old is also apparent by considering the dichromatic pattern. The dichromatic pattern remains invariant after the operations $S^w$ and $S^b$ are applied to their respective crystal lattices. That does not mean $S^w$ and $S^b$ are symmetry operations of the dichromatic pattern: in general they are not. The point is that the operations $S^w$ and $S^b$ are applied to the two crystal lattices *separately*. To make this more explicit we could express the operation characterizing the line defect as $(S^w, S^b)$ where we understand by this notation that $S^w$ acts only on the white crystal lattice and $S^b$ acts only on the black. The relationship between the crystals appears when we wish to express $S^w$ and $S^b$ in the same coordinate system.

We have seen that although a bicrystal may display no symmetry itself the symmetry of the adjoining crystals is manifested through the variants of the bicrystal and the line defects that separate them. This is a direct consequence of the principle of symmetry compensation. The absence of symmetry operations of the adjoining crystals in the bicrystal is compensated by the existence of the variants, which are related by the absent symmetry operations. The variants are more numerous, and hence the variety of interfacial defects is greater, the lower the symmetry of the bicrystal. No special crystallographic significance is attached to bicrystals displaying translational or point symmetry operations in this scheme. In such cases the number of variants of the bicrystal is simply reduced appropriately.

For non-holosymmetric or non-symmorphic crystals we distinguish between the dichromatic pattern and the dichromatic complex. The full set of all possible variants of an interface is obtained by considering operations $(S^w, S^b)$ where $S^w$ and $S^b$ are space group operations of the crystal lattices. The set $\{(S^w, S^b)\}$ includes the orientational, translational, complex, morphological, and relaxational variants of Section 1.5. The set is independent of the interface normal. Moreover, the number of operations in the set is independent of the relationship between the crystals, although when coincident symmetry is present not all the operations $(S^w, S^b)$ will be distinct.

When antisymmetry is present there are operations relating the two crystal lattices and for every ordinary operation $(S^w, S^b)$ there is an antisymmetry operation $(S^b, S^w)'$. $(S^b, S^w)'$ means that first $S^b$ and $S^w$ are applied to the black and white lattices separately and then the colours of the lattices are changed. The reason for reversing the order of the operations will be explained below.

Gratias and Thalal (1988) arrived at the same conclusions by embedding the black and white lattices in a 6D lattice. If the first three dimensions are filled with the white crystal lattice and the second three with the black then $(S^w, S^b)$ is a space group operation of the 6D lattice. Let $E_1, E_2 \ldots E_6$ be the six basis vectors of the 6D crystal lattice. Then $E_1 = [a_1^w, 0]$, $E_2 = [a_2^w, 0]$, $E_3 = [a_3^w, 0]$, $E_4 = [0, a_1^b]$, $E_5 = [0, a_2^b]$, $E_6 = [0, a_3^b]$, where the $a_i^w$'s and $a_i^b$'s are the basis vectors of the white and black crystals, as described in Section 1.3.2. At this stage there is no misorientation between the white and black crystal lattices. The misorientation between the crystals will be introduced later. The components of the $a_i^w$'s and $a_i^b$'s are expressed in the same coordinate system and reflect any differences in the lattice structures as may exist. From eqns (1.2) and (1.8)

$$a_i^b = \sum_{j=1}^{3} P_{ji} a_j^w, \tag{1.97}$$

where $\mathbf{P}$ is a pure deformation matrix.

An arbitrary lattice vector in the 6D lattice may be expressed as $r^{(6)} = (r^w, r^b)$ where $r^w$ and $r^b$ are lattice vectors of the white and black crystals. The space group operation $(S^w, S^b)$ may be represented as follows:

$$(S^w, S^b) = \begin{bmatrix} [S^w] & [0] \\ [0] & [S^b] \end{bmatrix}. \tag{1.98}$$

The quantities in square brackets are $3 \times 3$ matrices. When $(S^w, S^b)$ acts on $(r^w, r^b)$ it transforms $(r^w, r^b)$ into $([S^w]r^w, [S^b]r^b) = (D_i^w r^w + R_j^w, D_k^b r^b + R_n^b)$ where $S^w = [D_i^w | R_j^w]$ and $S^b = [D_k^b | R_n^b]$.

If the black and white crystal lattices are identical there is a mirror plane m$'$ in the 6D lattice relating the two 3D subspaces:

$$m' = \begin{bmatrix} [0] & [1'] \\ [1'] & [0] \end{bmatrix}, \tag{1.99}$$

where the primes on the $3 \times 3$ identity matrices denote colour reversal. The set of all antisymmetry operations displayed by the 6D lattice is thus $\{m'(S^w, S^b)\} = \{(S^b, S^w)'\}$.

Note the reversal in the order of the operations. The symmetry operations displayed by the 6D lattice comprise $G^{6D} = \{(S^w, S^b)\} \cup \{(S^b, S^w)'\}$. If the black and white crystal lattices are not identical then $G^{6D} = \{(S^w, S^b)\}$, because $m'$ is not a symmetry operation of the 6D lattice.

The white and black crystal lattices are recovered in their final misoriented state from the 6D lattice by applying the $6 \times 3$ projection matrix ([1], $[\mathbf{R}^t]$) to the 3D 'faces' $(r^w, \mathbf{0})$ and $(\mathbf{0}, r^b)$ of the 6D crystal lattice. $[\mathbf{R}^t]$ is transpose of the $3 \times 3$ rotation matrix in eqn (1.8). It follows that the dichromatic pattern may be viewed as a 3D projection of two orthogonal 3D faces of a 6D crystal lattice. The 6D lattice is independent of the relative orientation between the crystal lattices: that is introduced in the projection operation.

One of the key points of this analysis is that an equivalent dichromatic pattern is created if a symmetry operation of the 6D crystal lattice is performed prior to the projection. This is equivalent to the above remark that the set of operations $\{(S^w, S^b)\}$ generates identical dichromatic patterns provided we ensure that $S^w$ acts only on the white lattice and $S^b$ only on the black. But in that case $S^w$ and $S^b$ are generally not symmetry operations of the dichromatic pattern. By expressing $(S^w, S^b)$ in the form of eqn (1.98) we have not only ensured that $S^w$ and $S^b$ act only on their respective lattices, but we realize that the object for which $(S^w, S^b)$ is a symmetry operation is a 6D crystal lattice.

It is clear that the two approaches we have described in this section are entirely equivalent. Both approaches reveal the symmetries $\{(S^w, S^b)\}$ and $\{(S^b, S^w)'\}$ that lay 'hidden' from view in a bicrystal possessing no symmetry itself. In both approaches these hidden symmetries are manifested by the existence of an exactly compensating set of variants of the bicrystal, which are related by the symmetry elements $\{(S^w, S^b)\}$ and $\{(S^b, S^w)'\}$ that are not displayed by the bicrystal. The underlying principle uniting both approaches is the principle of symmetry compensation.

At this point we can make contact with the fields of quasiperiodicity and incommensurate structures. Consider the electron diffraction pattern of a bicrystal. The incident beam undergoes multiple scattering events in both the black and white crystals and the diffracted amplitudes consist of a convolution of the diffracted amplitudes from the two separate crystals. For example, a reflection, $g^b$, in the black crystal subsequently enters the white crystal and is elastically scattered again so that the reflection becomes $g^b + g^w$. The set of reflections is therefore $\{g^b + g^w\}$, where $g^b$ and $g^w$ range over all reciprocal lattice vectors of the two crystals. The diffracted amplitude from the bicrystal is therefore described as follows:

$$\rho(k) = \sum_{g^b, g^w} A(g^b + g^w)\delta(k - g^b - g^w) \qquad (1.100)$$

where $A(g^b + g^w)$ is the amplitude of the reflection $g^b + g^w$, arising from all the multiple diffraction events. We are ignoring here any extra reflections that may arise from relaxation at the interface. They are irrelevant to the present argument.

At a commensurate interface there is a 2D repeat cell of finite area in the interface plane. If there is a matching periodicity in only one direction the interface is said to be commensurate along that direction and incommensurate in all other directions. If there is no matching periodicity in any direction in the interface plane the interface is described as incommensurate. In the case of commensurate grain boundaries with CSL misorientations it is possible for the adjoining crystals to be commensurate with each other. In that case not only is the interface commensurate but there is a matching periodicity normal to the interface.

The set of reflections $\{g^w + g^b\}$ from the bicrystal may be indexed with at most six independent sets of *integers*, three for the white crystal reflections and three for the black. This would be the case where there is no direction of matching periodicity in the bicrystal; this is the completely incommensurate case. For a rational grain boundary at a CSL misorientation all the reflections may be indexed with only three integers; this is the completely commensurate case. The intermediate cases require four or five integers to index all the reflections. A function in 3D space whose Fourier spectrum requires more than three independent sets of integers to index all the reflections is *defined* as a quasiperiodic function. As we shall see a quasiperiodic function is a projection of a strictly periodic function in a higher dimensional space.

The question now is what is the quasiperiodic function in 3D space that gives rise to the Fourier spectrum described by eqn (1.100)? Since the spectrum is a convolution of the diffracted amplitudes from the adjoining crystals the simplest function that will give rise to the set of reflections $\{g^w + g^b\}$ is

$$\rho(r) = \rho^w(r)\rho^b(r), \tag{1.101}$$

where $\rho^w(r)$ is the mass density distribution function in the white cyrstal, and similarly for $\rho^b(r)$. Evidently, $\rho(r)$ is *not* the mass density distribution of the bicrystal, or even of the dichromatic complex or pattern. The mass density of the dichromatic complex is $\rho_{dc}(r) = \rho^w(r) + \rho^b(r)$. But it is clear that if $\rho^w(r)$ and $\rho^b(r)$ are commensurate with each other along a particular direction then $\rho(r)$ will be a periodic function in that direction. If $r$ is confined to the plane $r \cdot \hat{n} = 0$ then $\rho(r)$ will reflect whatever periodicities exist in the interface created by sectioning the dichromatic pattern on the same plane. For example, consider the mass distributions along the $x$-direction of an interface plane with normal $\hat{n}$. We may express $\rho^w(x)$ and $\rho^b(x)$ as follows:

$$\left.\begin{array}{l} \rho^w(x) = \displaystyle\sum_{n=-\infty}^{\infty} W_n \exp\left(2\pi inx\right) \\[1.5em] \rho^b(x) = \displaystyle\sum_{n=-\infty}^{\infty} B_n \exp\left(2\pi inx/\lambda\right). \end{array}\right\} \tag{1.102}$$

The wavelength of the periodic density distribution in the white crystal has been set equal to 1 and in the black crystal to $\lambda$. We require $\lambda \leq 1$ without loss of generality. The priodicity of the interface in the $x$-direction is determined by the priodicity of $\rho^w(x) + \rho^b(x)$. If $\lambda$ is an irrational number $\rho^w(x) + \rho^b(x)$ is not periodic. If $\lambda$ is rational, say $\lambda = p/q$, where $p$ and $q$ are coprime integers, then $\rho^w(x) + \rho^b(x)$ is periodic with a wavelength of $p$. Now consider $\rho(x) = \rho^w(x)\rho^b(x)$. If $\lambda$ is irrational then $\rho(x)$ is also not periodic, whereas if $\lambda = p/q$ then the periodicity of $\rho(x)$ is also $p$. Thus the periodicity of $\rho^w(x) + \rho^b(x)$ is the same as the periodicity of $\rho^w(x)\rho^b(x)$. Since $x$ can be any direction in the interface it follows that the periodicity $\rho^w(r) + \rho^b(r)$ is the same as the periodicity of the function $\rho^w(r)\rho^b(r)$, where $r$ is any vector in the interface plane.

The Fourier spectrum of $\rho_{dc}(r) = \rho^w(r) + \rho^b(r)$ is a subset of the spectrum of $\rho(r) = \rho^w(r)\rho^b(r)$. This is seen immediately by expressing $\rho^w(r)$ and $\rho^b(r)$ as Fourier series:

$$\rho^w(r) = \sum_{g^w} W(g^w) \exp\left(ig^w \cdot r\right) \tag{1.103}$$

$$\rho^b(r) = \sum_{g^b} B(g^b) \exp\left(ig^b \cdot r\right), \tag{1.104}$$

where $W(g^w)$ and $B(g^b)$ are appropriate structure factors. Whereas the Fourier spectrum of $\rho_{dc}(r)$ comprises the sets $\{g^w\}$ and $\{g^b\}$, the Fourier spectrum of $\rho(r)$ also includes $\{g^w + g^b\}$.

Since the periodicities of $\rho(r)$ and $\rho_{dc}(r)$ are the same $\rho(r)$ can be used to determine the long range order in the dichromatic pattern, or in an interface if $r$ is confined to lie in a plane. Long-range order exists if the autocorrelation function, $C(r)$, otherwise known as the Patterson function, continues to take on finite values as $|r|$ tends to infinity. It is defined by

$$C(r) = \int \rho(r')\rho(r + r')dr'. \tag{1.105}$$

We shall see that regardless of whether the interface is periodic or not, long-range order exists always. This is one of the main insights gained by making contact with the field of quasiperiodicity.

Let us use $\rho(r) = \rho^w(r)\rho^b(r)$, eqn (1.101), together with eqns (1.103) and (1.104), to explore the properties of $\rho(r)$ further. It has the following Fourier series representation:

$$\rho(r) = \sum_{g^w} \sum_{g^b} W(g^w)B(g^b) \exp\left(ig^w{\cdot}r + ig^b{\cdot}r\right). \tag{1.106}$$

Consider the most incommensurate case, where there is no matching periodicity in any direction. Six independent sets of integers will be required to index all the reflections $\{g^w + g^b\}$. Although $\rho(r)$ is not periodic it is closely related to a periodic function in 6D space as may be seen by expressing $\rho(r)$ as follows:

$$\rho(\mathbf{r}) = \mathbb{R}(r, r), \tag{1.107}$$

where

$$\mathbb{R}(r^w, r^b) = \sum_{g^w} \sum_{g^b} W(g^w)B(g^b) \exp\left(ig^w{\cdot}r^w + ig^b{\cdot}r^b\right) = \rho^w(r^w)\rho^b(r^b). \tag{1.108}$$

We recognize the density distribution $\mathbb{R}(r^w, r^b)$ as that of the 6D crystal, which is obviously periodic in 6D space. It is clear from eqn (1.107) that $\rho(r)$ is the 3D density distribution in the hyperplane $r^b = r^w$ of the 6D crystal. On the other hand, we recall that $\rho_{dc}(r)$ is the linear superposition of the mass densities on the $(r^w, 0)$ and $(0, r^b)$ 'faces' of the 6D crystal.

If $\rho(r)$ and $\rho_{dc}(r)$ are not periodic functions they are quasiperiodic functions, which means that one can define 'translation vectors' for them (for the mathematics of quasi-periodicity see Besicovitch (1932)). $\tau$ is said to be a translation vector, belonging to $\varepsilon \geq 0$, of a function $f(r)$ if $|f(r + \tau) - f(r)| \leq \varepsilon$ for all $r$. Obviously if $f(r)$ is periodic and $\tau$ is a repeat vector then $\varepsilon = 0$. But for a quasiperiodic function $\varepsilon$ must be greater than zero and the inequality expresses the approximate nature of the replication of the function. A translation vector belonging to $\varepsilon$ belongs also to any $\varepsilon' > \varepsilon$. If $\tau$ is a translation vector belonging to $\varepsilon$, then so is $-\tau$. If $\tau_1, \tau_2$ are translation vectors belonging respectively to $\varepsilon_1, \varepsilon_2$ then $\tau_1 \pm \tau_2$ is a translation vector belonging to $\varepsilon_1 + \varepsilon_2$. Let the set of all translation vectors of $f(r)$ belonging to $\varepsilon$ be denoted by $E\{\varepsilon, f(r)\}$.

It is convenient to rewrite eqn (1.106) for $\rho(r)$ as follows:

$$\rho(r) = \sum_{n} A_n \exp\left(ig_n{\cdot}r\right), \tag{1.109}$$

where $g_n$ is a combination of $g^w$ and $g^b$, $A_n$ is the corresponding product of structure factors, and the $g_n$'s are numbered in order of increasing magnitude. In order for $\tau$ to be a translation vector of $\rho(r)$ belonging to $\varepsilon$ it must satisfy some number $N$ of Diophantine inequalities $|g_n \cdot \tau - 2\pi v| < \delta$ for $n = 1, 2, \ldots N$, where $v$ is an integer and $\delta$ is a positive number that is less than $\pi$. The same set of $N$ diophantine inequalities has to be satisfied by all translation vectors of $E\{\varepsilon, \rho(r)\}$. The proof of this statement is very informative about the properties of translation vectors of quasiperiodic functions.

In order to determine the translation vector of $\rho(r)$ belonging to $\varepsilon$ we first note that the series representation in eqn (1.109) is convergent. This follows from the Parseval identity:

$$\text{the mean value of } |\rho(r)|^2 = \sum_{n=0}^{\infty} |A_n|^2, \tag{1.110}$$

and the fact that the left-hand side is finite. Therefore we can approximate $\rho(r)$ by a finite expansion $s(r)$:

$$s(r) = \sum_{n=0}^{N} H_n \exp(ig_n \cdot r), \tag{1.111}$$

where $N$ and $H_n$ are undetermined. For reasons that will be clear shortly we require $|\rho(r) - s(r)| < \varepsilon/3$ for all $r$. An obvious means of ensuring this inequality is satisfied is to set $H_n = A_n$ and to increase $N$ until it is satisfied. The point is that it must be satisfiable for some finite value of $N$ provided $\varepsilon > 0$. Having found a suitable $s(r)$ let $C = |H_1| + |H_2| + \ldots + |H_N|$. Now,

$$|s(r + \tau) - s(r)| = \left| \sum_{n=0}^{N} H_n \exp(ig_n \cdot r)(\exp(ig_n \cdot \tau) - 1) \right| \tag{1.112}$$

and again we require $|s(r + \tau) - s(r)| < \varepsilon/3$ for all $r$. To achieve this we require that

$$\left| \exp(ig_n \cdot \tau) - 1 \right| = \delta \tag{1.113}$$

for $n = 1, 2, \ldots N$. Then $\exp(ig_n \cdot \tau) - 1 = \delta \exp(i\varphi_n)$ and

$$|s(r + \tau) - s(r)| = \left| \sum_{n=0}^{N} H_n \exp(ig_n \cdot r) \exp(i\varphi_n)\delta \right| \leq \delta \sum_{n=0}^{N} |H_n| = \delta C. \tag{1.114}$$

Therefore for $|s(r + \tau) - s(r)| < \varepsilon/3$ we must require $\delta < \varepsilon/3C$ in eqn (1.113). Since we have required $|\rho(r) - s(r)| < \varepsilon/3$ for all $r$ we may now write

$$|\rho(r + \tau) - \rho(r)| \leq |\rho(r + \tau) - s(r + \tau)| + |s(r + \tau) - s(r)| + |s(r) - \rho(r)|$$
$$< \varepsilon/3 + \varepsilon/3 + \varepsilon/3 = \varepsilon \tag{1.115}$$

Thus $\tau$ is a translation vector of $\rho(r)$ belonging to $\varepsilon$. To find $\tau$ we have to solve $N$ diophantine inequalities (eqn 1.113):

$$\left| \exp(ig_n \cdot \tau) - 1 \right| < \varepsilon/3C, \quad n = 1, 2, \ldots N, \tag{1.116}$$

i.e. $|g_n \cdot \tau - 2\pi k| < 2\sin^{-1}(\varepsilon/6C)$, where $k$ is an integer.

$N$ is determined by the following condition:

$$\left| \rho(r) - \sum_{n=0}^{N} H_n \exp(ig_n \cdot r) \right| < \varepsilon/3. \tag{1.117}$$

As $\varepsilon$ is reduced $N$ increases in order to satisfy eqn (1.117), and the increased number of diophantine inequalities to be satisfied in eqn (1.116) results in fewer translation vectors. For any choice of $\varepsilon$ greater than zero the set of translation vectors forms a relatively dense set. This means that there is a cubic lattice, with a lattice constant $L$, such that every cell contains the head of at least one translation vector. As $\varepsilon$ is reduced, $L$ increases in size because there are fewer translation vectors. We see that although $\rho(r)$ is not a periodic function nevertheless it does repeat, to an accuracy specified by $\varepsilon$, at least once somewhere inside every cell of a cubic lattice. As the accuracy parameter $\varepsilon$ is reduced so the cubic lattice becomes coarser and the repetitions less frequent. It should be appreciated that not only is the value of the function $\rho(r_o)$ repeated (to within $\varepsilon$) at $r_o + \tau$ but the value of the function $\rho(r_o + \delta r)$ is repeated (to within $\varepsilon$) at $r_o + \tau + \delta r$, where $\delta r$ is *any* vector. Therefore a Maxwell demon sitting at $r_o$ could not distinguish the *environment* around him from that at $r_o + \tau$ to within an accuracy of $\varepsilon$.

Consider now the autocorrelation function $C(r)$, defined by eqn (1.105), for the quasi-periodic function $\rho(r)$. Substituting eqn (1.109) for $\rho(r)$ into eqn (1.105) we obtain:

$$C(r) = \sum_n |A_n|^2 \exp(ig_n \cdot r). \tag{1.118}$$

Therefore, $C(r)$ is itself a quasiperiodic function and therefore it repeats, within an accuracy of $\varepsilon$, *ad infinitum*. We conclude that $\rho(r)$ displays long-range order regardless of whether or not the crystals are commensurate.

Since $\rho_{dc}(r)$ can also be expressed in the form of eqn (1.109), where the $g_n$'s are either $g^w$'s or $g^b$'s, it follows that the mass density of the dichromatic complex is also a quasiperiodic function with its own set of translation numbers and an autocorrelation function that always displays long-range order. Moreover, since the long-range order displayed by an interface depends on the long-range order of the dichromatic complex from which it was sectioned, it follows that a planar interface always displays long-range order.

The final point we shall make in this section concerns local isomorphism. Two structures are said to be locally isomorphic if, and only if, given any point $P$ in either structure and any finite distance $d$, there exists a translation of the other structure such that the structures coincide exactly from $P$ out to at least distance $d$. Thus every finite region of one structure appears in the other and vice versa. The diffraction patterns and free energies of two locally isomorphic structures are indistinguishable. If the black and white crystals are incommensurate with each other then two dichromatic complexes, which differ only by a relative displacement of the crystal lattices, are locally isomorphic. The relative displacement of the crystal lattices is the 'phason' degree of freedom associated with quasiperiodic structures. The significance of this is that two dichromatic complexes that are related by symmetry operations of either crystal are locally isomorphic. The symmetries of the adjoining crystals, which are symmetry operations of the 6D crystal, become pseudosymmetry operations after the irrational projection to form an incommensurate dichromatic complex. They are pseudosymmetry operations because after they are applied, unlike a normal symmetry operation, not all of the structure is superimposable on the original structure. Instead, an infinite number of points in the two structures are superimposable, but not the whole structures. We shall return to the concept of local isomorphism in the next chapter.

## REFERENCES

Augustus, P. D. (1983). *Inst. Phys. Conf. Ser. No.* 67, 229.
Besicovitch, A. S. (1932). *Almost periodic functions.* Cambridge University Press, Cambridge.

Bollman, W. (1970). *Crystal defects and crystalline interfaces*, Springer, Berlin.

Bonnet, R., Marcon, G., and Ati, A. (1985). *Phil. Mag. A*, **51**, 429.

Cahn, J. W. and Kalonji, G. (1982). In *Solid state phase transformations* (ed. H. I. Aaronson). American Society for Metals, Metals Park, Ohio.

Cherns, D. and Pond, R. C. (1984). *Mat. Res. Soc. Symp. Proc.*, **25**, 423.

Cherns, D., Smith, D. A., Krakow, W., and Batson, P. E. (1982). *Phil. Mag. A*, **45**, 107.

Christian, J. W. (1975). *The theory of transformations in metals and alloys*, Part I. Pergamon, Oxford.

Cracknell, A. P. (1968). *Applied group theory*. Pergamon, Oxford.

Frank, F. C. (1965). *Acta Cryst.*, **18**, 862.

Frank, F. C. (1988*a*). *MRS Bulletin*, March issue, 24.

Frank, F. C. (1988*b*). *Met. Trans.*, **19A**, 403.

Gratias, D. and Portier, R. (1982). *J. Physique*, **43**, C6-15.

Gratias, D. and Thalal, A. (1988). *Phil. Mag. Letts.*, **57**, 63.

Grimmer, H. (1974). *Scr. Metall.*, **8**, 1221.

Grimmer, H. (1976). *Acta Cryst.*, **A32**, 783.

Handscomb, D. C. (1958). *Canad. J. Math.*, **10**, 85.

Hirth, J. P. and Lothe, J. (1982). *Theory of dislocations*. Wiley, New York.

Kalonji, G. and Cahn, J. W. (1982). *J. Physique*, **43**, C6-25.

Pond, R. C. (1977). *Proc. R. Soc. Lond.*, **A357**, 471.

Pond, R. C. (1983). *Inst. Phys. Conf. Ser. No. 67*, 59.

Pond, R. C. (1985). In *Dislocations and Properties of Real Materials*, p. 71. (ed. M. Lorretto), Institute of Metals, London.

Pond, R. C. (1989). In *Dislocations in solids*, Vol. 8 (ed. F. R. N. Nabarro). North-Holland: Amsterdam.

Pond, R. C. and Bollman, W. (1979). *Phil. Trans. R. Soc. Lond.*, **A292**, 449.

Pond, R. C. and Dibley, P. E. (1990). *J. Physique*, **51**, C1-25.

Pond, R. C. and Vlachavas, D. S. (1983). *Proc. R. Soc. Lond.*, **A386**, 95.

Pond, R. C., Gowers, J. P., Holt, D. B., Joyce, B. A., Neave, J. H., and Larsen, P. K. (1984). *Mat. Res. Soc. Symp. Proc.*, **25**, 273.

Pond, R. C., Gowers, J. P., and Joyce, B. A. (1985). *Surf. Sci.*, **152/3**, 1191.

Pond, R. C., Bacon, D. J., Serra, A. and Sutton, A. P. (1991). *Metall. Trans. A*, **22**, 1185.

Ranganathan, S. (1966). *Acta Cryst.*, **21**, 197.

Shubnikov, A. V. and Koptsik, V. A. (1977). *Symmetry in science and art*. Plenum, New York.

Van Tendeloo, G. and Amelinckx, S. (1974). *Acta Cryst.*, **A30**, 431.

Vitek, V., Sutton, A. P., Smith, D. A., and Pond, R. C. (1980). In *Grain boundary structure and kinetics* (ed. R. W. Balluffi), p. 115. American Society for Metals, Metals Park, Ohio.

Wolf, D. (1989). *Acta Metall.*, **37**, 1983.

# 2

# Dislocation models for interfaces

## 2.1 INTRODUCTION

In this chapter we shall consider the modelling of interfaces by continuously or discretely distributed arrays of dislocations. Many models use the known properties of discrete dislocations to deduce properties of an interface. In this way energies, elastic stress and strain fields, diffusion coefficients, point defect source and sink efficiencies, segregation kinetics, roughening transitions, migration, sliding, and internal friction of interfaces have been modelled. This extensive list of properties makes the dislocation model of prime importance in the field of interfaces.

The history of the model spans more than half a century and its conception dates from G. I. Taylor's pioneering work on crystal plasticity. Taylor (1934) calculated the elastic displacement and stress fields of a wall of crystal lattice edge dislocations forming what he called a 'surface of misfit', illustrated schematically in Fig. 2.1. Today we would call Taylor's surface of misfit a small-angle symmetrical tilt boundary. J. M. Burgers (1939) calculated the elastic displacement and stress fields of a small-angle symmetrical tilt boundary and also an epitaxial heterophase interface containing a single array of crystal lattice edge dislocations with Burgers vectors parallel to the interface, illustrated in Fig. 2.2. J. M. Burgers (1940) recognized that an asymmetric tilt boundary could be constructed from two sets of edge dislocations (see Fig. 2.3). Similarly, twist grain

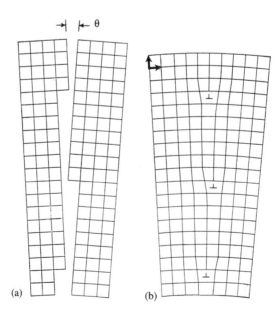

**Fig. 2.1** Schematic illustration of the formation of symmetric tilt boundary of misorientation $\theta$ in (b) by bonding together two crystals in (a) with high index free surfaces. The steps on the free surfaces in (a) become edge dislocations in the boundary in (b). (From Read (1953)).

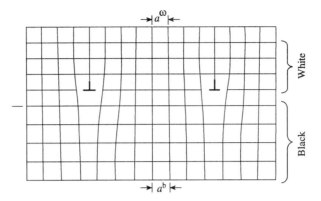

**Fig. 2.2** Schematic illustration of an epitaxial interface between black and white crystals with lattice parameters $a^b$ and $a^w$, where $a^b > a^w$. The difference in lattice parameters is accommodated by edge dislocations.

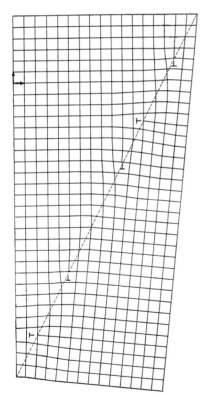

**Fig. 2.3** A schematic illustration of an asymmetric tilt boundary along the dashed line. Note that there are two sets of edge dislocations. (From Read (1953)).

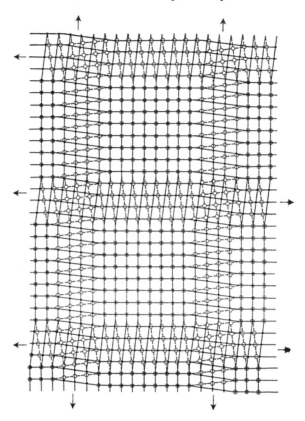

**Fig. 2.4**   A schematic illustration of a twist boundary seen in plan view. Sites of the two crystal lattices are represented by dots and circles. The boundary contains two sets of screw dislocations along the arrows which tend to localize the twist misorientation leaving relatively large patches of almost perfect crystal between them. (From Read (1953)).

boundaries could be constructed from grids of screw dislocations, as shown in Fig. 2.4. These early authors were primarily concerned with small-angle grain boundaries, such as sub-grain boundaries comprising the mosaic structure formed during cold work. It was not until 1947 that W. G. Burgers (the brother of J. M. Burgers) gave a description of large-angle grain boundaries in terms of dislocations. This proposal was consistent with the growing concensus that large-angle grain boundaries were not amorphous layers some 100 atoms in thickness as had been argued much earlier by Rosenhain and coworkers, e.g. Rosenhain and Humphrey (1913). However, it was recognized that the description of a large-angle boundary in terms of crystal lattice dislocations was limited in its usefulness owing to the close separation of the dislocations.

    Read and Shockley (1950) considered the energy of a small-angle grain boundary composed of an array of crystal lattice dislocations. The classic Read-Shockley formula for the energy-misorientation relation, $\sigma(\theta) = \sigma_0\theta(A - \ln\theta)$, appeared in this paper. This formula predicts that $\sigma(\theta)$ increases with increasing $\theta$ and possesses a sharp cusp as $\theta \to 0$. Although this formula was derived for small-angle boundaries, $\theta \leqslant 15°$, the paper also addressed the structures and energies of large-angle grain boundaries. Read and Shockley recognized that their formula applied only when the dislocations in the boundary were uniformly spaced. But this is possible only in special cases when the spacing of the dislocations is an integer multiple of some crystal lattice spacing. The non-uniformities in the spacings of the dislocations, which exist in more general cases, may be thought of as perturbations superimposed on an array of uniformly spaced dislocations. Read and Shockley showed that the perturbations may be thought of as another array of dislocations with smaller (weaker) Burgers vectors. Thus, there are additional weaker cusps in the energy-misorientation relation at periodic boundaries. These ideas are equivalent to modern dislocation models of large-angle grain boundaries involving DSC dislocations (see Section 1.7.2.1). We shall derive the Read–Shockley formula in Section 2.10.3.

To understand current dislocation models it is essential to distinguish two types of interface. The first has no stress field in either crystal far from the interface. The stress field associated with the interface decays exponentially over a distance comparable to a characteristic wavelength in the interface, such as the average spacing of an array of dislocations. This is the type of interface that is produced when two stress-free crystals are bonded together with no constraints applied externally to the bicrystal. The second type of interface is associated with a long-range stress field. Far from the interface the stress tensor approaches a constant non-zero value. This type of interface occurs naturally, for example in thin-film specimens containing interfaces between thin epitaxial layers and thick substrates. Once the epilayer reaches a certain critical thickness it becomes energetically favourable for the interface to relieve the long-range stress field by the introduction of an array of dislocations at the interface. This transition is discussed in Section 2.10.6.

In this chapter we shall be concerned mainly with interfaces that are free of long-range stresses. For such an interface Frank (1950) and Bilby (1955) showed that the dislocation content is determined by a purely geometrical condition. They viewed the interface as a transformation front; as it moves it transforms one crystal lattice adjoining the interface into the other. The transformation may be a rotation, as in the case of a grain boundary, or a general affine transformation. In general, the transformation would open gaps or produce overlapping material at the interface. The interface is then said to contain incompatibilities. Dislocations are required geometrically to eliminate the incompatibilities. It is only when the interface is fully compatible that it is free of stress at long range. Once a description of the relationship between the crystal lattices has been selected, the dislocation content of the interface is obtained in a straightforward way from the Frank–Bilby equation, which is derived in Section 2.3. One of the difficulties with the Frank–Bilby theory is that the relationship between the crystal lattices for a given interface may be specified in an infinite number of ways. It follows that there is an infinite number of Frank–Bilby descriptions of the dislocation content of an interface, all satisfying the condition of no long-range stresses. Problems associated with this feature of dislocation models of interfaces and the question of whether there is a 'best' description are discussed in Section 2.4.

There is an important distinction between the interfacial dislocations that were considered in Section 1.7 and the interfacial dislocations that are used in dislocation models of interfaces. The interfacial dislocations of Section 1.7 are *isolated* Volterra dislocations. Their Burgers vectors are defined without reference to any other structure. By contrast, a dislocation model of an interface involves one or more *arrays* of dislocations. The existence of an array of dislocations alters the orientational and/or deformational relationship between the adjoining crystal lattices from that of some reference structure, where the corresponding array of dislocations is absent. It is therefore more appropriate for the Burgers vectors of dislocations in an array to be defined by a Burgers circuit construction in a reference lattice. Such a procedure enables the systematic variations of the spacings and line directions of the dislocations in the interface to be modelled as the orientational and/or deformational relationship between the adjoining crystal lattices varies. As we shall see in Section 2.2 the choice of reference lattice is not unique. Provided the closure failure of the Burgers circuit is always measured in the reference lattice the value obtained is independent of the spacing of the dislocations in the array. On the other hand the 'local' Burgers vector (Hirth and Lothe 1982), which is obtained by measuring the closure failure in the interface where the dislocations are located, varies as the dislocation spacing varies.

## 2.2  CLASSIFICATION OF INTERFACIAL DISLOCATIONS

In Section 1.7 we classified interfacial line defects according to bicrystal symmetry. The classification was based on the condition that the line defect separated regions of interface that were related by symmetry and hence energetically degenerate. An interfacial dislocation was created by a Volterra process involving a uniform displacement across a cut and subsequent rebonding. The Burgers vector of the dislocation was equal to the uniform displacement across the cut. An isolated perfect dislocation in a crystal lattice may be created by a similar Volterra process in which the displacement across the cut is a translation vector of the crystal lattice. Similarly, the Burgers vector of a *perfect* interfacial dislocation is a translation vector of the reference lattice. However, it was noted in Section 1.7 that the classification was not complete because other interfacial dislocations could be envisaged that did not separate energetically degenerate regions of an interface. For example, a dislocation with a Burgers vector that is not a translation vector of the reference lattice will not, in general, separate energetically degenerate regions of the interface. Such dislocations are called *partial* interfacial dislocations, by analogy with partial crystal lattice dislocations such as Frank and Shockley partial dislocations in an f.c.c. lattice (Read 1953). However, we note that the analogy with the classification of crystal lattice dislocations is not exact. Whereas partial interfacial dislocations can separate energetically degenerate interfacial domains (e.g. see Section 1.7.2), a partial crystal lattice dislocation introduces a stacking fault, which has a different energy from the rest of the crystal (Read 1953).

Over the past 30 years there has been a proliferation in the types of interfacial dislocations which have been identified. In addition, dislocations of a given type have often been called by different names, e.g. see Balluffi and Olson (1985). Thus, intrinsic (Hirth and Balluffi 1973), extrinsic (Hirth and Balluffi 1973), coherency (Olson and Cohen 1979), anticoherency (Olson and Cohen 1979), twinning (Christian 1981), transformation (Christian 1981), primary or crystal lattice interfacial dislocations (Bollmann 1970), secondary (Bollmann 1970), tertiary (Bollmann 1970), virtual (Hirth and Balluffi 1973), surface (Bilby 1955), misfit (Frank and van der Merwe 1949), perfect (Pond 1977), partial (Pond 1977), Somigliana (Bonnet *et al.* 1985), Volterra (Pond and Vlachavas 1983), and DSC dislocations (Bollmann 1970) have appeared in the literature. One reason for this proliferation is the different roles that interfacial dislocations are perceived to play in determining the stress field of an interface. Another reason is the concept of a reference structure and the different choices of reference structure that may be adopted, as discussed later in this section and in Section 2.4.

In recent years Olson and Cohen (1979) and Bonnet (1981*a,b,c*, 1982, 1985) have developed a general approach to describe the elastic fields of interfaces in terms of dislocations. A useful discussion of the method has been given by Dupeux (1987). The basic idea is to use two arrays of dislocations with Burgers vector densities of opposite sign and different distributions. The short-range elastic field of an interface that has no long-range stress field is modelled by two cancelling arrays of dislocations. Thus the net dislocation content of such an interface is zero. One array may be regarded as an array of *stress-generator dislocations* and the other as an array of *stress-annihilator dislocations*. The elastic field of an interface that is associated with a long-range stress field is modelled by an incomplete cancellation of the two arrays, or, in the simplest case, by only one array. In such cases the net dislocation content of the interface is not zero. In our view this approach has many advantages and we shall therefore adopt it throughout this book.

To illustrate the use of stress generator and annihilator arrays consider the formation of a grain boundary that is free of long-range stresses by the following imaginary four-step process (see Fig. 2.5). We begin with a reference structure consisting of a stress-free, single crystal, Fig. 2.5(a). The crystal is gripped and bent elastically through an angle $\theta$ between the grips, Fig. 2.5(b). The long-range stress field that is set up by this elastic distortion is modelled by a continuous distribution of dislocations acting as stress generators in the distorted region. We emphasize that the bending of the crystal is elastic and that the dislocations that are continuously distributed are not 'real' dislocations, in the conventional sense, but they serve as a device to model the elastic stress field of the bent crystal. They may be compared with the use of continuous distributions of dislocations to model the elastic fields of loaded cracks (Hirth and Lothe 1982). The bent crystal is in mechanical equilibrium only because the crystal halves are being clamped in position far from the boundary plane. If the clamps were removed the bicrystal would return to Fig. 2.5(a). The long-range stress field is eliminated by introducing a second distribution of crystal lattice dislocations acting as stress annihilators, with exactly the opposite Burgers vector density, see Fig. 2.5(c). These dislocations are 'real' dislocations. It is emphasized that in this representation the stress field produced by the stress annihilator dislocations is that produced under the constraint that no bending (rotation) of the crystal is allowed during their introduction. Finally, the elastic energy of the system is minimized by localizing all dislocations in the boundary plane, as shown in Fig. 2.5(d). The details of the short-range stress field of the boundary then depend on the distribution of the Burgers vector density that is assumed for the stress annihilators.

In Fig. 2.5 one half of the bicrystal is produced from the other by the transformation **T** (see Section 1.3.2), which in this case is a rotation, where the half-crystals are rotated by equal and opposite amounts.

Prior to the introduction of the stress annihilators (Fig. 2.5(b)) the continuity of the reference lattice is maintained across the interface, although it is elastically deformed; in this state the interface is said to be in a state of forced elastic coherence or simply 'coherent'.

To illustrate the same approach for a heterophase interface, that is free of long-range stresses, we consider an analogous four step process, illustrated in Fig. 2.6. We begin with a reference structure consisting of a single crystal of the $\alpha$ phase, Fig. 2.6a. One half of the reference crystal lattice is elastically transformed into the $\beta$ phase, by the transformation **T**, which produces a change of shape of the reference crystal as shown in Fig. 2.6(b). Again the crystal halves are clamped in position far from the interface to prevent the $\beta$ half transforming back to $\alpha$ or the $\alpha$ half transforming to $\beta$. Continuity of the reference lattice is maintained across the interface by an elastic distortion of the adjoining $\alpha$ and $\beta$ crystals, and the interface is coherent. The interface is associated with a long range stress field, which is modelled by the continuous distribution of stress

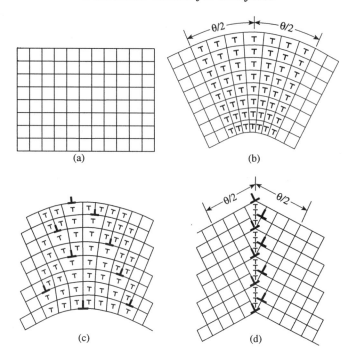

**Fig. 2.5** The formation of a grain boundary that is free of long-range stresses by a four-step process. Starting from a single crystal reference lattice in (a) the crystal is gripped at the left and right sides and bent elastically to introduce the misorientation $\theta$ in (b). The long-range elastic field is modelled by a continuous distribution of stress generator dislocations. In (c) stress annihilator dislocations are introduced, whose Burgers vector density cancels that of the stress generator dislocations. There is now no long-range elastic field associated with the boundary. The elastic energy is minimized by rearranging the dislocations into the boundary plane in (d).

generators, shown in Fig. 2.6(b). The stress annihilators, which may be crystal lattice dislocations, are introduced in the third step and cancel the longe-range field, Fig. 2.6(c). In the final step the stress generators and annihilators rearrange their positions to minimize the interfacial energy, Fig. 2.6(d), which need not restrict them to lying in the chemical interface depending on the relative elastic constants of the two phases (see Section 4.4.3).

For interfaces that are free of long-range stresses the cancellation of the Burgers vector densities of the stress generator and annihilator arrays ensures that no macroscopic change in the relative orientation or structures of the adjoining crystals is effected by the two arrays. It is because of this that the method models the elastic field that would be produced by bonding together parallel surfaces of two crystals that are *already* misoriented or transformed in the desired way. We note that when the two sets of dislocations are in balance, neither set can be uniquely identified as stress generators or stress annihilators, since it is clear that the order in which the above set of operations is carried out can be altered without affecting the final state of the system.

This approach may be contrasted with the more commonly used Read–Shockley approach for grain boundaries. Dislocations in a single crystal are run into a plane, which becomes the boundary plane. No constraints are applied externally to the crystal and one crystal half rotates with respect to the other by $\theta$ about an axis $\hat{\rho}$. If the rotation were

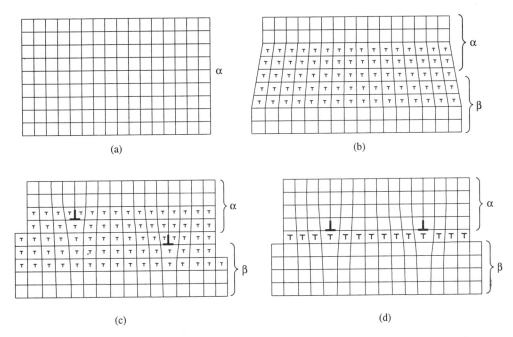

**Fig. 2.6** The formation of a heterophase interface that is free of long-range stresses by a four-step process. One half of the reference single crystal lattice of the $\alpha$ phase in (a) is transformed elastically in (b) into the the lattice of the $\beta$ phase. The long-range stress field of the resulting bicrystal is modelled by a continuous distribution of stress generator dislocations. Stress annihilator dislocations are introduced in (c) with an equal and opposite Burgers vector density to cancel that of the stress generators. The interface is now free of long-range stresses. In (d) the stress generator and annihilator dislocations rearrange so as to minimize the elastic strain energy.

suppressed by external constraints a long-range stress field would be set up and the dislocations would be an array of stress generators. But the coexistence of the rotation and the dislocations in the boundary produces a bicrystal that is free of long-range stresses. Thus, in the Read–Shockley approach the dislocation array serves two purposes: (i) its distortion field effects a change in the orientational relationship between the crystal lattices, and (ii) it introduces a Burgers vector density to relieve incompatibilities at the interface that are produced by (i). Provided the change in the relationship between the crystal lattices in (i), and the net Burgers vector density of the dislocations in (ii), are related by the Frank–Bilby equation (see Section 2.3) the interface is free of long-range stresses. The stress field in the bicrystal is the same as that obtained by bonding parallel surfaces of two crystals that are already misoriented by $\theta$ about $\hat{\rho}$. Thus the Read–Shockley and Olson–Cohen–Bonnet approaches describe the same final state: *in the absence of long-range stresses the two approaches are equivalent.* As discussed in Section 2.10.4, great care with the boundary conditions far from the interface has to be taken with the use of elasticity theory in the Read–Shockley approach to ensure that the absence of external constraints on the bicrystal is implicit in the elastic distortion field. Otherwise, the Frank–Bilby equation is not satisfied by the distortion field and spurious long-range stresses are found. One of the advantages of the Olson–Cohen–Bonnet approach is that an interface *is guaranteed* to have no long-range stress field by using cancelling arrays of dislocations. But more significantly, it has the added advantage that it can be readily applied to interfaces that do have long-range stress fields, because of particular circumstances in which those interfaces were produced, so that the Burgers vector densities of the stress generator and annihilator arrays do not cancel.

An example of an interface with a long-range stress field that looks like that produced by several closely spaced dislocation loops is a lenticular mechanical twin within a crystal, as illustrated in Fig. 2.7. In this case the longe-range field is caused by the coexistence of a sheared volume inside an unsheared crystal. An enlarged view of the interface near one of the dislocations is shown in Fig. 2.8(d) where it is seen that the dislocation has step character. The dislocation is a stress generator. Continuity of the reference lattice, which in this case is the untwinned crystal lattice, is maintained everywhere and therefore the stepped interface is coherent. No stress annihilators are present and therefore the stress field of the stress generators is not cancelled: the interface has a long-range stress field.

Similar, constrained, changes of shape occur when small precipitates of new phases are formed within a material. Continuity of some reference lattice is maintained across the interface for precipitate sizes below some critical value. The interface is coherent and the stress fields of these inclusions may be modelled by continuously or discretely distributed arrays of stress generators at the interface. In both the twin and the precipitate cases, however, the long-range stress fields can be eliminated by running crystal lattice dislocations into the interfaces. These then are the stress annihilators.

In the previous examples the choice of reference lattice was the lattice of a single crystal. With the help of stress-generator dislocations continuity of the reference lattice across the interface was maintained and the interface was then described as coherent. The coherency of the interface was disrupted by the introduction of stress-annihilator crystal lattice dislocations. But we may also have situations where it is useful to start with a reference structure consisting of a bicrystal. By applying a suitable transformation, $T$, to one of the crystals of such a reference bicrystal a new interface can be produced having geometrical parameters which deviate only slightly from those of the original reference interface. Provided the reference interface has a relatively low energy the deviated interface may be expected to relax to a configuration consisting of patches of the reference interface separated by appropriate interfacial dislocations. In this way the deviated interface is described as the reference interface with a superimposed array of interfacial dislocations. Such situations can also be described in terms of cancelling arrays of stress generator and stress annihilator dislocations.

Suppose that we choose a reference interface that has a quasiperiodic structure. This is conceivable because relatively low-energy quasiperiodic interfaces can exist, e.g. (100)/(110) grain boundaries in Al have been observed to be of relatively low energy (Dahmen and Westmacott 1988). Choosing a quasiperiodic reference structure is equivalent to choosing a periodic reference structure in the limit that the period tends to infinity. A possible drawback of choosing a periodic reference structure with a long period is that there may be local relaxation patterns *within* the period of the reference structure which will not be described by this choice. To account for such local relaxations it is necessary to choose the appropriate shorter period reference structure that is being preserved by the local relaxations. Similarly, there may be local relaxations at a quasiperiodic reference structure, which would be arranged quasiperiodically. Between these local relaxations there is some periodic approximant to the quasiperiodic interface. To account for them one should choose the periodic approximant that is being preserved. This is known as the 'near CSL model', where the existence of a nearby periodic interfacial structure implies the existence of a nearby CSL (and periodic DSC lattice) in one, two, or three dimensions. However, it is conceivable that there are no local relaxations at a

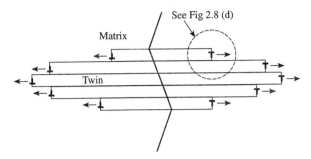

**Fig. 2.7** Schematic illustration of a lenticular mechanical twin within a crystal. The twin is 5 layers thick. Each layer is bounded by a dislocation loop and an enlarged view of the step associated with each dislocation is shown in Fig. 2.8(d). The twin elongates by the lateral motion of the dislocations as indicated by the arrows.

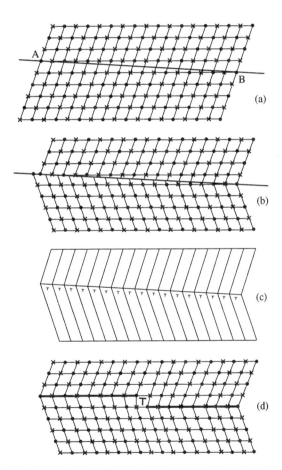

**Fig. 2.8** The formation of a step in a (111) mechanical twin boundary in an f.c.c. lattice. Starting from a single f.c.c. lattice in (a), seen in projection along $[1\bar{1}0]$, where the dots and crosses refer to lattice sites on successive $(2\bar{2}0)$ planes, the crystal lattice is cut along AB. The lower half undergoes the mechanical twinning shear in (b), and it is seen that the two halves no longer fit together along AB. In (c) a continuous distribution of stress generator dislocations (i.e. coherency dislocations) is introduced along AB in order to make the two crystal halves commensurate. These continuously distributed dislocations localize into a $\frac{1}{6}[11\bar{2}]$ coherency dislocation associated with a step of one (111) plane height in (d).

quasiperiodic reference structure (though on physical grounds it would seem unlikely) and in that case the quasiperiodic interface itself can serve as a reference structure to describe local relaxations in deviated interfaces.

To illustrate a reference structure consisting of a bicrystal consider a short period symmetric [001] tilt boundary between two crystals misoriented to produce the $\Sigma = 5$ CSL and DSC lattice as shown in Fig. 2.9(a). The deviated boundary, obtained by a transformation, **T**, corresponding to a small tilt rotation, $\Delta\theta$, around [001], is shown schematically in Fig. 2.9(b). By analogy with Fig. 2.5 we may imagine producing the deviated boundary by first gripping the reference bicrystal and bending it elastically so that the misorientation is increased by $\Delta\theta$. The elastic stress field of the bicrystal may then be modelled by a continuous distribution of stress generators along the boundary, as shown in Fig. 2.9(b). The stress generators maintain the continuity of the DSC lattice across the interface, as opposed to the crystal lattice in Fig. 2.5(b). The reference lattice is thus the DSC lattice in the present case. The long-range stress field of the bicrystal is then eliminated by introducing a cancelling distribution of stress annihilators, as shown in Fig. 2.9(c). The Burgers vectors of the stress annihilators are DSC lattice vectors: these stress annihilators are, therefore, often called DSC dislocations. The final state, Fig. 2.9(c), being free of long-range stresses, is analogous to Fig. 2.5(d).

The procedure may be readily generalized (Bonnet and Durand 1975, Balluffi *et al.* 1982) to include the near CSL model for a quasiperiodic interface. Two crystals which meet this near CSL criterion are shown in Fig. 2.10, where the black and white crystal lattices are almost commensurate. They almost form a two-dimensional CSL with a corresponding DSC lattice that is periodic in two dimensions. Here, $CSL^w$ and $CSL^b$ are sublattices of the two crystal lattices which are almost identical. If $T^{(1)}$ is the transformation generating the $CSL^b$ sublattice from the $CSl^w$ sublattice we may apply $T^{(1)}$ to the white crystal lattice to generate the white′ lattice shown in Fig. 2.10(d). Then it is seen (Figs. 2.10(b,d)) that the black sublattice $CSL^b$ and the sublattice, $CSL^{w'}$, of the white′ lattice form a CSL identical to $CSL^b$. A periodic DSC lattice is also formed which we label the $DSC^b$ lattice. In a similar manner we may operate on the black lattice with the inverse of $T^{(1)}$ to produce a black′ lattice which, in concert with the white crystal lattice, produces a CSL identical to $CSL^w$ and a periodic DSC lattice which we label the $DSC^w$ lattice (Figs 2.10(a,c)). Note that the $CSL^b$ may be generated from the $CSL^w$, or the $DSC^b$ lattice may be generated from the $DSC^w$ lattice, by applying $T^{(1)}$. We may now choose as a reference structure a specially constructed bicrystal consisting of the white crystal and the black crystal meeting along a periodic interface. The $DSC^w$ lattice will now serve as a common framework across the interface. Using this reference structure we may produce the original quasiperiodic bicrystal by a procedure exactly analogous to those used previously. First we apply the transformation $T^{(1)}$ to the black crystal elastically to return the black crystal to its original structure. During this process the continuity of the $DSC^w$ lattice is maintained elastically and the interface is coherent. The bicrystal will be associated with a long-range stress field which may be modelled by a continuous distribution of stress generator dislocations in the interface, as shown in Fig. 2.10(e). The long-range elastic field of the interface is eliminated by introducing stress annihilators which have Burgers vectors of the $DSC^w$ lattice (see Fig. 2.10(e)). But because the original interface is quasiperiodic the spacing of the stress annihilators will also be quasiperiodic.

Having related the quasiperiodic reference structure to a periodic reference structure we may now consider the effect of an additional misorientation, $T^{(2)}$, applied to the black crystal of the quasiperiodic reference structure. Instead of applying only the

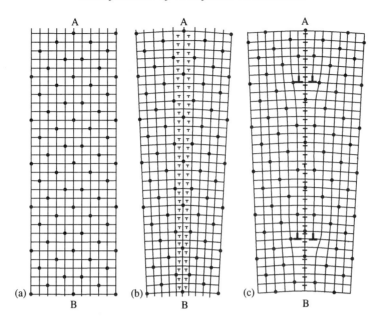

**Fig. 2.9** The use of a bicrystal as a reference structure. (a) The reference structure, consisting of a bicrystal containing a $\Sigma = 5$ (210) symmetric tilt boundary along AB; the reference lattice is the $\Sigma = 5$ DSC lattice, also shown. In (b) an additional misorientation $\Delta\theta$ is introduced elastically about the [001] tilt axis. This gives rise to a continuous distribution of stress generator dislocations, which maintain coherency (continuity) of the DSC reference lattice. In (c) an array of discrete stress annihilator dislocations, with Burgers vectors of the DSC lattice, has been introduced which cancels the long-range stresses and which destroys locally the continuity of the reference lattice.

transformation $\mathbf{T}^{(1)}$ to the black crystal of the periodic reference structure, as we did before, we now apply $\mathbf{T}^{(2)}\mathbf{T}^{(1)}$ to it instead. In this way we see that deviations from a quasiperiodic reference structure may always be considered as slightly different deviations from a nearby periodic reference structure.

In all cases considered thus far we began with a reference structure and an associated reference lattice. When the reference structure was a single crystal the reference lattice was the corresponding crystal lattice. When the reference structure was a bicrystal the reference lattice was the corresponding DSC lattice which served as the common framework for the bicrystal. It is obvious that a single crystal may also be regarded as a bicrystal in which the two crystal halves are the same. Similarly a single crystal lattice may be regarded as a DSC lattice of a bicrystal in which there is no change of the crystal lattice across the interface. In this sense a single crystal reference structure and a single crystal reference lattice are special cases of a bicrystal reference structure and a DSC reference lattice. It follows that the reference structure is always a bicrystal, which may happen to be a single crystal, and that the reference lattice is always a DSC lattice, which may happen to be a single crystal lattice.

The array of stress generator dislocations maintained the continuity of the reference lattice across the interface, that is, they maintained the coherency of the interface. In the language of Olson and Cohen (1979), the stress generators are *coherency dislocations*. On the other hand, the stress annihilator dislocations destroyed the continuity of the reference lattice across the interface locally, that is, they destroyed the coherency of the

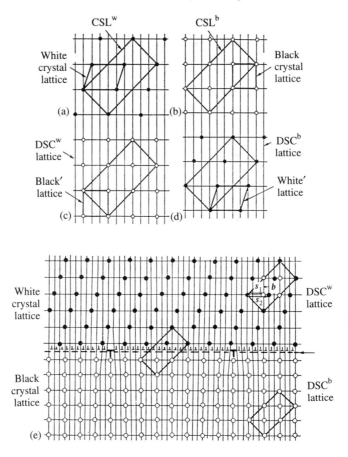

**Fig. 2.10** The near CSL model for a quasiperiodic interface. The incommensurate white and black crystal lattices are shown in (a) and (b). The cells labelled CSL$^\mathrm{w}$ and CSL$^\mathrm{b}$ are nearly commensurate with each other. In (c) the black crystal lattice is deformed slightly to produce the black' lattice, with the result that the CSL$^{\mathrm{b}\prime}$ cell is commensurate with the CSL$^\mathrm{w}$ cell and a periodic DSC lattice is produced, labelled DSC$^\mathrm{w}$. Alternatively, in (d), the white lattice is deformed slightly so that the CSL$^{\mathrm{w}\prime}$ cell is commensurate with the CSL$^\mathrm{b}$ cell and a periodic DSC lattice is produced, labelled DSC$^\mathrm{b}$. Choosing the reference structure to be a bicrystal comprising the white and black' lattices meeting along a periodic interface, the DSC$^\mathrm{w}$ lattice becomes the reference lattice. Allowing the black' lattice to return to the undeformed black lattice in (e) then introduces a continuous distribution of stress generator dislocations, maintaining the continuity of the DSC$^\mathrm{w}$ lattice, with an array of discrete stress annihilator dislocations with Burgers vectors of the DSC$^\mathrm{w}$ lattice. (From Balluffi *et al.* (1982).)

interface locally. Olson and Cohen (1979) called the stress annihilators *anticoherency dislocations*. This will serve as a useful further basis for classifying interfacial dislocations throughout the remainder of this book. It obviates the need for introducing a number of the dislocation types listed at the beginning of this section such as, for example, 'virtual' interfacial dislocations. We emphasize that the systematic use of such a method of classification depends upon the clear identification of the assumed reference lattice. Similarly a 'coherent' interface refers to the continuity of a particular reference lattice across the interface. It is noted that in all previous examples the stress generators were coherency dislocations. However, this need not be the case as will be seen below in other situations.

The above concepts and terminology also allow us to classify interfaces as coherent, semicoherent, or incoherent. At a coherent interface the continuity of the reference lattice, whatever that is chosen to be, is maintained, and there are no anticoherency dislocations. In the limit of an incoherent interface the continuity of the reference lattice is destroyed everywhere along the interface. This limit is reached when the spacing of the anticoherency dislocations is comparable to the width of their cores. In a semicoherent interface the spacing of the anticoherency dislocations is greater than their core width so that significantly large regions of forced elastic coherence exist between successive anticoherency dislocations.

In later chapters we shall see that many of the physical properties of these different types of interfaces differ considerably. For example, in Chapter 9 we shall see that the motion of an interface possessing some degree of coherency often causes a change in the macroscopic shape of the bicrystal. This arises from the maintenance of continuity of the reference lattice across the coherent regions of the interface. On the other hand, no shape change is induced by the motion of an incoherent interface since there is no continuity of the reference lattice to be maintained.

The concepts of 'intrinsic' and 'extrinsic' dislocations in interfaces are related to the concepts of stress generators and annihilators. In the absence of external constraints we do not expect a grain boundary to be associated with a long-range stress field. This is achieved by obtaining an exact balance between the stress annihilator and stress generator Burgers vector densities. These dislocations constitute then the 'intrinsic' dislocation content of the interface. As shown below in Sections 2.3 and 2.5 the Burgers vector density of the intrinsic stress annihilator dislocations (or, alternatively, the negative of the Burgers vector density of the stress generators) of an interface may be deduced from the Frank–Bilby equation.

The concept of an extrinsic dislocation is often associated with a crystal lattice dislocation that enters a grain boundary, which is initially free of long-range stress, as discussed by Hirth and Balluffi (1973) and Dupeux (1987). The dislocation is called extrinsic because it is not part of the intrinsic boundary structure, but is an extra dislocation which entered the boundary from the bulk. More precisely, it is extrinsic because its Burgers vector is not part of the Burgers vector density of the stress anni-hilators required by the Frank–Bilby equation with the boundary misorientation fixed at its value before the dislocation entered. It is therefore associated with a long-range stress field. However, it becomes part of the intrinsic array once the boundary misorientation, as measured far from the boundary plane, changes slightly. The small change in ·mis-orientation is equivalent to a small change in the Burgers vector density of the stress generator array, the integral of which just cancels the change in Burgers vector density of the stress annihilators due to the added dislocation. Once this happens the extrinsic dislocation becomes an intrinsic dislocation and the long-range field associated with it is removed. The distinction between intrinsic and extrinsic dislocations may be difficult to establish experimentally, because it relies on a very accurate measurement of the misorientation far from the boundary plane (see Dupeux 1987). Experimental observa-tions of the conversion of extrinsic dislocations to intrinsic dislocations are described in Section 12.4.

We may illustrate many of the concepts described in this section with the example of the interfaces associated with mechanical twins in f.c.c. crystals. It is well known (Christian 1981) that such twinning takes place on (111) planes with a shear of $2^{-\frac{1}{2}}$ in a $[11\bar{2}]$ direction. Also, this is achieved by gliding interfacial dislocations, with $b = \frac{1}{6}[11\bar{2}]$, on successive (111) planes, as shown schematically in Fig. 2.7. Each of these dislocations is associated with a boundary step of height equal to one (111) spacing as seen in Fig. 2.8(d). Consider first the nature of these dislocations. We begin by taking the reference structure to be the single crystal (Fig. 2.8(a)) and the transformation, **T**, to be the above shear. We then cut it along AB and transform the lower half by **T**. As seen in Fig. 2.8(b) the two crystal halves are no longer commensurate along AB. (In the language of deformation twinning the plane AB is not an invariant plane of the shear transformation.) In order to join the crystal halves together commensurately, so that the single crystal reference lattice is continuous across the interface, and the interface is coherent, it is necessary to introduce a distribution of stress generators (coherency dislocations) along AB, as in Fig. 2.8(c). The interface is now associated with a long-range stress field. Because the coherent interface has a particularly low energy when it lies on the (111) plane the coherency dislocations localize as shown in Fig. 2.8(d). The localized step seen in Fig. 2.8(d) is thus associated with a discrete $\frac{1}{6}[11\bar{2}]$ coherency dislocation. If a train of such dislocations were present, as in Fig. 2.7 and 2.11(a), the longe-range stress field of the interface could be cancelled by introducing crystal lattice dislocations (which would be classified as stress annihilator or anticoherency dislocations) of total Burgers vector $-\frac{1}{2}[11\bar{2}]$ for every three coherency dislocations. These could be $\frac{1}{2}[\bar{1}01]$ and $\frac{1}{2}[0\bar{1}1]$ crystal lattice dislocations, for example, as shown schematically in Fig. 2.11(a).

If these dislocations group together so that for every three coherency dislocations there are two anticoherency dislocations, line defects of the type shown in Fig. 2.11(b) will be formed. These line defects have no net Burgers vector content and, hence, are pure steps of height three (111) spacings. However, the reference lattice, i.e. the crystal lattice, is discontinuous at the pure step owing to its anticoherency dislocation content, and the interface is, therefore, semicoherent. If a large number of such steps accumulate, a giant step will be formed which would then more appropriately be regarded as a vertical segment of a symmetric tilt boundary running parallel to $(11\bar{2})$ with the intrinsic stress generator and stress annihilator content shown in Fig. 2.11(c).

Let us now consider the same boundary but with the $\Sigma = 3$ DSC lattice as the reference lattice, rather than either crystal lattice. The boundary belongs to the $\Sigma = 3$ coincidence system because it may be created by a rotation of $\cos^{-1}\frac{1}{3}$ about the $[1\bar{1}0]$ axis. Both the CSL and DSC lattice are shown in Fig. 2.12(a). The dislocation shown previously in Fig. 2.8(d) can now be introduced by a Volterra cut and displacement procedure in the DSC lattice to produce the configuration shown in Fig. 2.12(b) which is exactly the same physical object as that shown in Fig. 2.8(d). The dislocation shown in Fig. 2.12(b) has destroyed locally the continuity of the reference DSC lattice across the interface. Therefore, in the framework of the DSC lattice the dislocation is of the anticoherency type, although it is still a stress generator. Furthermore, the interface must now be regarded as semicoherent.

Consider next the configuration shown in Fig. 2.11 where stress annihilator lattice dislocations have been added in order to eliminate the long-range stresses produced by the stress generators. These dislocations will also disrupt the continuity of the reference lattice (i.e. the DSC lattice) and, hence, they must be classified as anticoherency, stress annihilator dislocations. But, if we gather these dislocations up to form a step of the type shown in Fig. 2.11(b), we obtain the step configuration shown in Fig. 2.12(c). The atomic configuration at the step in Fig. 2.12(c) is the same as that shown in Fig. 2.11(b) but it is shown in the framework of the DSC lattice as the reference lattice, rather than the

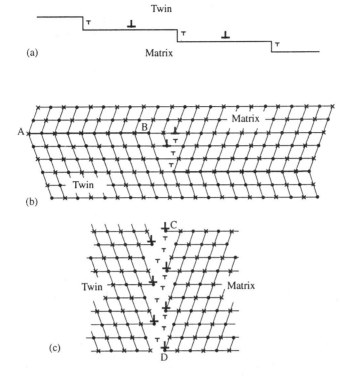

**Fig. 2.11** (a) An interface between a deformation twin and matrix in an f.c.c. lattice, inclined to the (111) invariant plane, in which the stress fields of the stress generator (or coherency) dislocations associated with the steps in the interface are cancelled by crystal lattice (or anticoherency) dislocations. (b) A pure step of 3 (111) planes height in the matrix–twin interface formed by grouping together 3 coherency dislocations and two cancelling crystal lattice anticoherency dislocations. (c) A segment of the $(11\bar{2})$ symmetric tilt boundary formed by running together pure steps of the type shown in (b). With respect to the matrix (or twin) reference crystal lattice the $(11\bar{2})$ boundary is classified as incoherent.

crystal lattice as reference. Again, the total Burgers vector content of the step is zero, and furthermore the anticoherency effects of the stress generators and stress annihilators just cancel so that the continuity of the DSC lattice at the step is maintained. Therefore, with the DSC lattice as the reference lattice, both the step and the interface are coherent. Again, an accumulation of a large number of such steps would form a symmetric tilt boundary segment parallel to $(11\bar{2})$ as shown in Fig. 2.12(d). The boundary in Fig. 2.12(d) is the same physical object as that shown in Fig. 2.11(c). However, in the DSC reference lattice framework of Fig. 2.12(d) we must regard it as a coherent interface devoid of any net dislocation content. It could be generated from the reference bicrystal (Fig. 2.12(a)) by simply rotating the boundary plane from (111) to $(11\bar{2})$ by a process of shuffling atoms within the framework of the CSL without the introduction of any dislocations. Even more simply, it could be adopted as a reference structure itself, that is, a coherent bicrystal in the $\Sigma = 3$ DSC lattice with its interface parallel to $(11\bar{2})$.

Our example of a (111) twin boundary demonstrates that the same physical dislocation may be classified as either coherency or anticoherency, depending on the choice of reference lattice. It is therefore necessary to specify the choice of reference lattice when

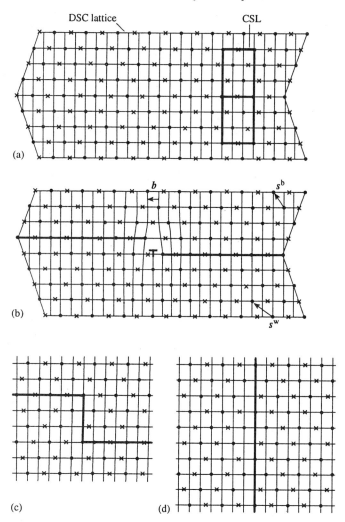

**Fig. 2.12** (a) The (111) interface between matrix and twin in an f.c.c. lattice showing the $\Sigma = 3$ DSC lattice and two cells of the $\Sigma = 3$ CSL. (b) The same $\frac{1}{6}[11\bar{2}]$ stress generator dislocation as seen in Fig. 2.8(d), and the step vectors $s^b$ and $s^w$. This dislocation destroys the continuity of the $\Sigma = 3$ DSC reference lattice, and it is now therefore classified as anticoherency. (c) The same pure step as seen in Fig. 2.11(b), but now in the framework of the $\Sigma = 3$ DSC reference lattice: note the continuity of the DSC lattice. (d) The same $(11\bar{2})$ boundary as shown in Fig. 2.11(c) showing the continuity of the $\Sigma = 3$ DSC lattice across the interface. With respect to the $\Sigma = 3$ DSC reference lattice the $(11\bar{2})$ boundary is coherent.

classifying dislocations as coherency or anticoherency, in the same way as it is when classifying interfaces as coherent, semicoherent, or incoherent.

## 2.3 THE FRANK–BILBY EQUATION

Frank (1950) and Bilby (1955) addressed the following question: given an affine transformation relating two lattices, what is the Burgers vector density that is required to make the two lattices fit together compatibly at the interface? In general, when one lattice is acted on by the transformation, overlapping regions and/or gaps are produced at the interface, as already seen in Figs 2.5 and 2.6. In other words, the interface is not, in general, an invariant plane of the transformation. To remove the incompatibilities between the lattices, dislocations are required in the interface. We may take these to be stress annihilators for reasons that are described below. Once the incompatibilities are removed the bicrystal is free of stress far from the interface. Frank–Bilby theory provides the net Burgers vector, $B$, of the stress annihilators crossing a vector $p$ lying in the interface. Since the total Burgers vector content of the stress generators is just the negative of that of the stress annihilators, we may also recover $B$ for the stress generators. The resolution of $B$ into Burgers vectors of individual dislocations is not unique geometrically and may be without physical significance if $|B|$ is large. In any event the interface may always be described as a 'surface dislocation', which is a single entity that is characterized by a second rank tensor. There are two methods of deriving the main result: one using a Burgers circuit construction and the other is based on the theory of continuous distributions of dislocations. Both methods are instructive in that they illustrate different aspects of the theory, and we shall develop them both. Our treatment follows Christian (1981).

Consider two lattices which meet along a flat interface, with normal $\hat{n}$. Let the lattices on the positive and negative sides of the normal $\hat{n}$ be coloured black and white. In general the lattices are not identical and there is a misorientation between them. The two lattices are considered to be generated from a reference lattice by affine transformations $S^b$ and $S^w$. In accordance with the discussion of Section 2.2 the two lattices may be black and white crystal lattices or DSC$^b$ and DSC$^w$ lattices (see Fig. 2.10), depending on whether we are working with a single crystal or bicrystal reference structure. The corresponding reference lattice would then be either the lattice of the white crystal or the DSC$^w$ lattice. In cases where the two lattices may be related by a pure rotation, as in the case of grain boundaries in cubic materials, the reference lattice could be the same lattice at a median orientation, i.e. a median lattice. In the language of the crystallographic theory of martensite, $S^b$ and $S^w$ are the lattice deformations. They are represented by $3 \times 3$ matrices, and for a grain boundary in a cubic material they are usually rotation matrices. When $S^b$ acts on the components of a vector of the reference lattice it transforms the components into those of a vector in the black lattice, and similarly for $S^w$. Let the reference and black and white lattices share a common origin at O, see Fig. 2.13. Here we are disregarding any relative displacement of the black and white lattices, since it does not affect the result. Let $OP = p$ be a *large* vector in the boundary, and consider a right-handed Burgers circuit $PA_1OA_2P$, as shown in Fig. 2.13. The vector $p$ must be large compared with any substructure within the interface, e.g. vicinal steps or twins emanating from the interface into the adjoining crystals. The corresponding path $Q_1B_1OB_2Q_2$ in the reference lattice is obtained by applying the inverse transformations $S^{b-1}$ and $S^{w-1}$ to the parts $PA_1O$ and $OA_2P$ of the circuit respectively. The closure failure in the reference lattice, using the FS/RH convention, is $Q_2Q_1 = OQ_1 - OQ_2 = (S^{b-1} - S^{w-1})p$. This is the net Burgers vector of stress annihilators crossing the interfacial vector $p$:

$$B = (S^{b-1} - S^{w-1})p. \tag{2.1}$$

Equation (2.1) was first derived by Frank (1950) for grain boundaries and it was generalized to heterophase boundaries by Bilby (1955). It is known as the Frank–Bilby equation. Note that the Burgers vector $B$ is expressed in the coordinate system of the

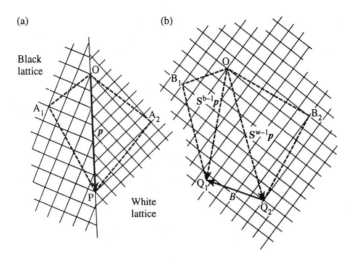

**Fig. 2.13** The derivation of the Frank–Bilby equation. (a) A closed right-handed circuit $PA_1OA_2P$ enclosing the interface vector $p$ is mapped in (b) onto a reference lattice where it becomes the path $Q_1B_1OB_2Q_2$ with closure failure $Q_2Q_1$. (From Christian (1981).)

reference lattice. For example, if the black lattice is selected as the reference lattice, eqn (2.1) becomes

$$B = (E - S^{w-1})p, \tag{2.2}$$

where $E$ is the identity matrix. We shall derive the more popular 'Frank formula' for grain boundaries from eqn (2.1) in Section 2.5.

If the net dislocation content $B$ did not exist in the interface, the vectors $S^{b-1}p$ and $S^{w-1}p$ would be brought into parallel alignment in the interface, along $p$, only by straining the black and white lattices elastically. The interface would then be in a state of forced elastic coherence, and it would be associated with a long-range stress field. This state is modelled, as previously, by a uniform and continuous distribution of stress generator dislocations at the interface. The Burgers vector density of this array is the negative of $B$, i.e. $-(S^{b-1} - S^{w-1})p$. This is the reason why we identify the dislocations, whose Burgers vector density is predicted by the Frank–Bilby equation, as stress annihilators. The notion of compatibility is defined more precisely in the second derivation of the Frank–Bilby equation which we move on to now.

We begin with the concept of the continuously dislocated state of a lattice. Consider an element of a lattice containing several dislocation lines. The dislocations are discrete objects but we obtain a convenient mathematical description by imagining that the number of dislocation lines of each type tends to infinity while the Burgers vectors of each tends to zero in such a way that the product remains constant. The continuous dislocation distribution that we obtain is characterized by a tensor $\alpha_{ij}(r)$, which defines the net Burgers vector in the $x_i$ direction at $r$ of dislocations threading through a surface element of unit area perpendicular to the $x_j$ axis. Let $C$ be a closed curve bounding an area $A$ and let $S$ be any cap ending on $C$. Conservation of Burgers vector requires that the resultant Burgers vector of dislocations threading through $A$ must equal the resultant Burgers vector of dislocations threading through $S$. Thus the resultant Burgers vector threading through $A$ is given by

$$b_i = \iint_S \alpha_{ij}\, dS_j, \tag{2.3}$$

and $b_i$ is independent of the choice of $S$. Summation over $j = 1, 2, 3$ is implied by the repeated suffix convention in eqn (2.3) and elsewhere in this chapter, except where explicitly stated. When we have a continuous distribution of dislocations there is no longer 'good lattice' in which to make a Burgers circuit in order to define the resultant Burgers vector. Bilby *et al.* (1955) showed how this difficulty may be overcome by making use of a local correspondence between the real lattice and a reference lattice. As we have already discussed in Section 1.7 the same difficulty arises in defining the Burgers vectors of interfacial dislocations in densely spaced arrays.

At any point of the real lattice choose three independent basis vectors, $a_i^c$, which correspond to a set of basis vectors, $a_i^r$, of the reference lattice. The local vectors $a_i^c$ may be regarded as being generated from the reference vectors $a_i^r$ at each point of the lattice by a local deformation:

$$a_i^c = D_{ij}\, a_j^r. \tag{2.4}$$

The components $D_{ij}$ vary from point to point in the lattice. Clearly it is necessary that throughout the dislocated lattice we consistently relate the same crystallographic vectors $a_i^c$ to the reference vectors $a_i^r$. The vectors $a_i^c$ define local variations in the lattice, although to an observer moving in the real lattice the local $a_i^c$ vectors are everywhere parallel, and any two parallel vectors, defined by reference to the local lattice, have the same $a_i^c$ components.

Let $\hat{e}_i$ be a set of fixed orthonormal vectors. With respect to a Cartesian frame aligned along the $\hat{e}_i$ a displacement in the real lattice from $x_i$ to $x_i + dx_i$ can be written as the vector $dx_i \hat{e}_i$. If $C$ is a small closed circuit in the real lattice we have

$$\int_C dx_i \hat{e}_i = 0. \tag{2.5}$$

Let the vectors $d_i$ be obtained from the vectors $\hat{e}_i$ by the local deformation $\mathbf{D}$, i.e. $d_i = \mathbf{D}_{ij}\hat{e}_j$. Then eqn (2.5) becomes

$$\int_C dx_i D_{ij}^{-1}\, d_j = 0. \tag{2.6}$$

We now require the closure failure of the corresponding circuit in the reference lattice. Each vector $d_k$ of the real lattice, is replaced by its corresponding vector, $\hat{e}_k$, in the reference lattice, and the closure failure giving the net Burgers vector of the distribution encircled by $C$ is the vector sum of the reference lattice displacements, i.e.

$$b = -\int_C dx_i D_{ij}^{-1} \hat{e}_j. \tag{2.7}$$

The minus sign is inserted in eqn (2.7) to be consistent with the FS/RH convention (Hirth and Lothe 1982). In order to compare directly with eqn (2.3) we use Stokes' theorem to transform the line integral in eqn (2.7) into a surface integral over any cap $S$ having $C$ as its limit. Stokes' theorem may be stated as follows:

$$\int_C f_i dx_i = \iint_S \varepsilon_{ijk} \frac{\partial f_k}{\partial x_j}\, dS_i, \tag{2.8}$$

where $f_i$ is a differentiable vector function, $\varepsilon_{ijk}$ is the permutation tensor of eqn (1.12), and $\mathrm{d}S_i$ denotes the $i$th component of the unit normal to the surface multiplied by the element of area $\mathrm{d}S$. Applying Stokes' theorem to eqn (2.7), by identifying $f_i$ with $D_{ij}^{-1}\hat{e}_j$, we obtain

$$b = -\iint_S \varepsilon_{imn} \frac{\partial D_{nj}^{-1}}{\partial x_m} \hat{e}_j \, \mathrm{d}S_i$$

$$= -\iint_S \varepsilon_{jmn} \frac{\partial D_{ni}^{-1}}{\partial x_m} \hat{e}_i \, \mathrm{d}S_j. \tag{2.9}$$

Comparing eqns (2.3) and (2.9) we see that

$$\alpha_{ij} = -\varepsilon_{jmn} \frac{\partial D_{ni}^{-1}}{\partial x_m}. \tag{2.10}$$

This equation shows us how to evaluate the dislocation tensor at each point in the continuously dislocated lattice even though there is no good material to construct an ordinary Burgers circuit. Imagine the reference lattice is cut into small volume elements, each of which is then given the local deformation $D_{ij}$. If the separate elements can be glued back together contiguously, without any holes or overlapping regions, so at to form a continuous lattice in ordinary space, the deformations are said to be *compatible*, and a continuous deformation field exists. Small lattice vectors about a point in the real lattice may be written as $\mathrm{d}y_i d_i$, where $\mathrm{d}y_i$ is a system of local coordinates based on the vectors $d_i$. Since $d_i = D_{ij}\hat{e}_j$, the relation between the local and reference coordinates may be obtained from $\mathrm{d}y_i d_i = \mathrm{d}y_i D_{ij}\hat{e}_j = \mathrm{d}x_j\hat{e}_j$, implying that

$$\mathrm{d}x_i = D_{ij}\mathrm{d}y_j, \quad \mathrm{d}y_i = D_{ji}^{-1}\mathrm{d}x_j. \tag{2.11}$$

In order for the deformation field $D_{ij}$ to be compatible it is necessary and sufficient that the values of $y_i$ at any point $Q$ may be found from their values at any other point $P$ by integrating $\mathrm{d}y_i = D_{ji}^{-1}\mathrm{d}x_j$ along *any* path from $P$ to $Q$. For the integral to be independent of the path from $P$ to $Q$ it is necessary and sufficient that

$$\frac{\partial D_{ni}^{-1}}{\partial x_k} = \frac{\partial D_{ki}^{-1}}{\partial x_n}. \tag{2.12}$$

We see from eqn (2.10) that this corresponds to zero dislocation density. Thus *the condition for compatibilility is the absence of dislocations.* Conversely, when the local deformations of the separate volume elements are not compatible, dislocations are required to fit these elements together.

Now suppose that we have a continuous distribution of dislocations specified by the tensor $\alpha_{ij}$ and concentrate all the dislocations into a shell of thickness $t$ so that $\alpha_{ij}$ vanishes outside the shell. Let $t$ tend to zero and $\alpha_{ij}$ tend to infinity in such a way that the product $t\alpha_{ij}$ remains finite and tends to $\beta_{ij}$. $\beta_{ij}$ is called the *surface dislocation tensor*. This limiting process means that the deformation field $D_{ij}$ is constant outside the shell and equal to $(D^b)_{ij}$ and $(D^w)_{ij}$ above and below the shell. The deformation field changes from $(D^b)_{ij}$ to $(D^w)_{ij}$ through the shell. Consider a Burgers circuit that is intersected by the shell as shown in Fig. 2.14. The Burgers vector of the circuit can be written as $b_i = -\int_c (\mathrm{d}y_i - \mathrm{d}x_i)$ since this is just the negative of the difference in the circuits in the real and reference lattices. Using eqn (2.11) this may be rewritten as

$$b_i = -\int_C (\delta_{ij} - D_{ji}) \, \mathrm{d}y_j \tag{2.13}$$

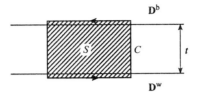

**Fig. 2.14** The circuit $C$ enclosing the shaded area $S$ which intersects a shell of thickness $t$ containing, and parallel to, the final interface. Across the shell the deformation tensor changes from $\mathbf{D}^b$ above to $\mathbf{D}^w$ below.

where the the circuit $C$ is carried out in the real lattice, as in Fig. 2.14. Using Stokes' theorem this line integral may be converted into a surface integral:

$$b_i = \iint_S \varepsilon_{jkn} \frac{\partial D_{ni}}{\partial y_k}\, dS_j, \tag{2.14}$$

where the surface $S$ is any surface bounded by the circuit $C$. Choosing the shaded area in Fig. 2.14 as the surface $S$ and noting that, for small $t$,

$$\frac{\partial D_{ni}}{\partial y_k} = \frac{(D^b)_{ni} - (D^w)_{ni}}{t} n_k, \tag{2.15}$$

we find that

$$b_i = \iint_S \varepsilon_{jkn} \frac{(D^b)_{ni} - (D^w)_{ni}}{t} n_k\, dS_j, \tag{2.16}$$

and therefore

$$\beta_{ij} = \lim_{t \to 0}(t\alpha_{ij}) = \varepsilon_{jkn}((D^b)_{ni} - (D^w)_{ni})n_k. \tag{2.17}$$

It is important to note that $\mathbf{D}^b$ and $\mathbf{D}^w$ gave the relations between the basis vectors of the black and white lattices and the reference lattice, whereas $\mathbf{S}^b$ and $\mathbf{S}^w$ transformed the components of a reference lattice vector into those of a vector in the black and white lattices. Therefore $\mathbf{S}^{b-1} = (\mathbf{D}^b)^t$ and $\mathbf{S}^{w-1} = (\mathbf{D}^w)^t$ (see Section 1.3.2) and eqn (2.17) may be rewritten as follows:

$$\beta_{ij} = \varepsilon_{jkn}((S^{b-1})_{in} - (S^{w-1})_{in})n_k. \tag{2.18}$$

Now consider the resultant Burgers vector of dislocations which cross a small area of the shell defined by the vector $p$ and the vector thickness $t\hat{n}$. The normal to this area has components $dS_j = t\varepsilon_{jab}p_a n_b$ and substituting this in eqn (2.16) we obtain

$$
\begin{aligned}
b_i &= \varepsilon_{jkn} \frac{(D^b)_{ni} - (D^w)_{ni}}{t} n_k t\varepsilon_{jab} p_a n_b \\
&= \varepsilon_{jkn}\varepsilon_{jab}((D^b)_{ni} - (D^w)_{ni})p_a n_b n_k \\
&= (\delta_{ka}\delta_{nb} - \delta_{kb}\delta_{na})((D^b)_{ni} - (D^w)_{ni})p_a n_b n_k \\
&= ((D^b)_{ni} - (D^w)_{ni})p_n \\
&= ((S^{b-1})_{in} - (S^{w-1})_{in})p_n, \tag{2.19}
\end{aligned}
$$

where we have used $p \cdot \hat{n} = 0$, which follows when $t \to 0$. Thus we have rederived the Frank–Bibly equation, eqn (2.1). The derivation shows explicitly that the Burgers vector density given by eqn (2.1) is such as to make the black and white lattices compatible in the interface.

From eqns (2.16) and (2.17) we see that the surface dislocation tensor $\beta_{ij}$ gives the $i$th component of the resultant Burgers vector of dislocation lines cutting unit length in the surface perpendicular to the $j$th direction. The interface may be thought of as a single entity called a *surface dislocation*, characterized by the tensor $\beta_{ij}$. The tensor $\beta_{ij}$ is the analogue for a surface dislocation of the Burgers vector for a line dislocation. Moreover, whereas a line dislocation may be defined as a line discontinuity separating areas in the slip plane where the amount of slip is different, a surface dislocation is a surface discontinuity separating volumes of material where the lattice deformation is different. This concept, which was introduced by Bilby (1955), is central to the crystallographic theory of martensitic transformations and deformation twinning developed by Bullough and Bilby (1956).

## 2.4 COMMENTS ON THE FRANK–BILBY EQUATION AND THE DISLOCATION CONTENT OF AN INTERFACE

The Frank–Bilby equation is the cornerstone of many attempts to account for the line defects that are observed in the electron microscope at grain boundaries and especially at heterophase boundaries. In this section we shall discuss some complicating aspects of the theory in more detail.

The first point to note is that it is a continuum theory. The nature of the adjoining black and white crystal or DSC lattices is taken into account only in the choices of unit cells that are assumed to be related to some unit cell of the reference lattice. These choices establish $S^b$ and $S^w$, and together with the interface normal they determine the surface dislocation tensor in eqn (2.18). The point group and translational symmetries of the adjoining crystals are not built into the theory, and the surface dislocation tensor is not invariant with respect to these symmetry operations. For example, if $S^b$ is replaced by where $U^b S^b$, where $U^b$ is a point group operation of the black lattice, or a unimodular matrix which effects a change in the unit cell of the black lattice to be related to the reference lattice, the surface dislocation tensor changes in general. Yet the dichromatic pattern is completely unaffected by this replacement (see Section 1.5.4), and therefore the atomic structure of the interface is unaffected also.

The reason why the theory is not invariant with respect to different choices of $U^b$ (or $U^w$) is clear from an examination of the Burgers circuit construction in Fig. 2.13. Each $U^b$ produces a different circuit $Q_1 B_1 O$ in the reference lattice, and hence a different closure failure $Q_2 Q_1$. The fact that the infinite lattices generated with and without $U^b$ are equivalent, in type and orientation, is not reflected automatically by the theory. As an example, consider a 'grain boundary' obtained by a 120° rotation about a [111] axis in a cubic lattice. Since the [111] axis has three-fold symmetry this large-angle boundary is nothing more than a perfect single crystal, yet it would be described by a high-density of crystal lattice dislocations on the 'boundary plane'. The origin of the ambiguity in the dislocation description of an interface can be traced back to eqn (2.4) which defines the local deformation $D_{ij}$. This is the device that Bilby *et al.* (1955) introduced to define a Burgers vector in a continuously dislocated lattice. In eqn (2.4) $D_{ij}$ may be replaced by $U_{ik}^{-1} D_{kj}$ where $U$ is a unimodular matrix that describes local point symmetry operations of the dislocated crystal, or effects different choices of unit cell vectors $c_i$. To a local observer in the dislocated crystal $\mathbf{U}$ is a constant matrix, but to an external observer $\mathbf{U}$ is a function of position. The dislocation tensor $\alpha_{ij}$ in eqn (2.10) is thus also affected by different choices of $\mathbf{U}$.

There is an infinite number of unimodular matrices and therefore *there is an infinite number of dislocation descriptions of a particular interface*. However, *all* dislocation descriptions of a particular interface are *descriptions of a unique physical object* with a particular atomic structure and energy in its fully relaxed state, and in an *exact* theory, involving full atomic relaxation, all these descriptions would yield identical atomic structures, energies, and elastic fields!

The classic example of this unsatisfactory aspect of dislocation models of interfaces is the symmetrical tilt grain boundary. The relationship between the crystal lattices may be described as (i) a rotation about an axis in the boundary plane as in the usual tilt boundary description, (ii) a simple shear on the boundary plane, as in a deformation twin, or (iii) as a 180° rotation about the boundary normal, corresponding to a 180° twist boundary. This is illustrated in Fig. 2.15. There is an infinity of other descriptions. According to (i) the resultant crystal lattice dislocation content of the boundary consists of a single array of edge dislocations, with Burgers vectors normal to the boundary plane. Since the crystal lattices are fully compatible with each other on the shear plane, i.e. the boundary plane, there is no resultant dislocation content in the boundary for the second description. In the third description the boundary would be described by a dense network of screw dislocations forming a 180° twist boundary. Again, the same atomic structure would be produced after full relaxation of the boundary, for all these boundary models, that is, the final boundary structure would be independent of the route taken to generate it geometrically.

As a specific example, consider again the (111) $\Sigma = 3$ twin boundary illustrated previously in Figs 2.8 and 2.11. There we assumed that the transformation, $\mathbf{T}$, relating the crystal lattices was a shear parallel to (111). Hence the section of boundary along AB in Fig. 2.11(b) was devoid of any dislocation content. On the other hand, the segment parallel to $(11\bar{2})$ along CD in Fig. 2.11(c) possessed the crystal lattice dislocation content shown. If instead we now assume that the transformation, $\mathbf{T}$, relating the crystal lattices is a rotation of $\cos^{-1}\frac{1}{3} \simeq 70.5°$ around $[1\bar{1}0]$, and follow the procedure shown in Fig. 2.5, we obtain the crystal lattice dislocation structures shown along AB and CD in Figs 2.16(a,b) for these two boundary segments. Very different dislocation structures are therefore obtained for these identical physical objects.

We are therefore faced with the rather unsatisfactory question of whether any of the possible dislocation descriptions is in some sense preferable to all the others. It is tempting to adopt as the 'best' description the one which corresponds to the smallest dislocation content, since this would presumably be the simplest. But, according to this criterion, the best description of a symmetrical tilt boundary would correspond to zero dislocation content regardless of the tilt angle. However, at small tilt angles experimental observations reveal an array of discrete line singularities which can be identified as crystal lattice edge dislocations corresponding to description (i) above. The presence of these line singularities cannot be accounted for by description (ii), and their origin would have to be attributed to some form of local relaxation in the boundary, which does not relieve any incompatibility. We emphasize that both descriptions apply to the same physical object consisting of atoms arranged in a particular configuration.

Cases where there is justification for choosing a particular $\mathbf{S}^b$ and $\mathbf{S}^w$ occur when migration of the interface leads to a change of shape of the bicrystal as a result of the conservative motion of interfacial dislocations, as discussed in Chapter 9 (see Table 9.2).

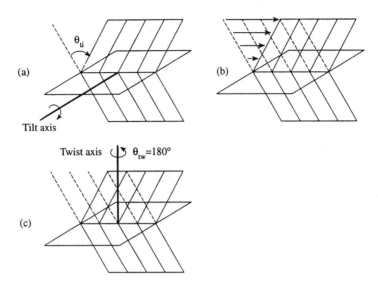

**Fig. 2.15** Diagram showing that a symmetric tilt boundary may be regarded as either (a) a tilt boundary of misorientation $\theta_{ti}$, or (b) a deformation twin boundary created by a simple shear on the boundary plane, or (c) a 180° twist boundary.

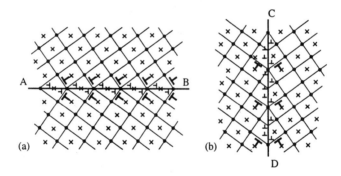

**Fig. 2.16** (a) The (111) twin boundary shown in Fig. 2.11(b) and 2.12(a) regarded as a symmetric tilt boundary of misorientation $\approx 70.5°$. The $(11\bar{2})$ boundary shown in Figs 2.11(c) and 2.12(d) is now regarded as a symmetric tilt boundary of misorientation $\approx 70.5°$. Note the differences in the dislocation descriptions for the same interfaces.

In that case $S^b$ and $S^w$ describe how planes and directions of one lattice are transformed (at least locally) at the interface into planes and directions of the other lattice. This is the case in martensitic transformations and deformation twinning, where there is a local correspondence between unit cells of either crystal that is determined by local atomic movements as the interface migrates. It is also the case in the glissile motion of a small-angle tilt grain boundary as discussed in Section 9.2.1.1. The success of the crystallographic theory of martensite (Bullough and Bilby 1956) is to determine $S^b$ and $S^w$, the macroscopic shape change, the Burgers vector content of the interface, and the interface plane self-consistently (see Section 9.2.1.3).

We have described one source of ambiguity in the dislocation description of an interface arising from the infinity of descriptions of $\mathbf{S}^b$ and $\mathbf{S}^w$ for a given choice of reference lattice. Another source of ambiguity is the choice of reference structure. An example has already been given in Section 2.2 where a comparison was made of the dislocation descriptions that are derived for a grain boundary using either a single crystal or a bicrystal as the reference structure. Useful reference structures are often interfaces in which one or two directions are nearly commensurate in both crystal lattices enabling near-CSLs to be generated. In many cases, especially for heterophase interfaces, a variety of possible choices of a near-CSL exists (Balluffi *et al.* 1982). However, the appropriate choice of reference structure is the one for which the experimentally observed dislocations in the interface have Burgers vectors of the corresponding reference lattice. But since Frank–Bilby theory does not address atomic relaxation processes it cannot predict the appropriate choice.

Finally, we emphasize that the Burgers vector content $B$, eqn (2.1), is the sum of the Burgers vectors of all dislocations crossing a vector $p$ lying in the interface. Consistent with the continuum nature of the theory the vector $B$ is not discretized into Burgers vectors of crystal lattice dislocations or DSC dislocations. Such a discretization is dedendent on the relaxation processes within the interface which are beyond Frank–Bilby theory. Some further aspects of this problem are discussed in Section 2.8. Any such discretization is of little practical interest if the density of dislocations is so large that their cores overlap. In this situation it may be more useful to think of the whole interface as a surface dislocation characterized by a surface dislocation tensor, eqn (2.18).

## 2.5 FRANK'S FORMULA

For a grain boundary, where the two adjoining lattices are related by a rotation, $\mathbf{S}^b$ and $\mathbf{S}^w$ in the Frank–Bilby equation may be replaced by rotation matrices $\mathbf{R}^b$ and $\mathbf{R}^w$:

$$B = (\mathbf{R}^{b-1} - \mathbf{R}^{w-1})p. \qquad (2.20)$$

Using the rotation formula, eqn (1.9), where $\mathbf{R}^b$ corresponds to a rotation of $\theta^b$ about $\hat{\rho}^b$ and $\mathbf{R}^w$ to a rotation of $\theta^w$ about $\hat{\rho}^w$, we obtain

$$B = p(\cos\theta^b - \cos\theta^w) - \sin\theta^w(p \times \hat{\rho}^w) + \sin\theta^b(p \times \hat{\rho}^b)$$
$$+ (1 - \cos\theta^b)(\hat{\rho}^b{\cdot}p)\hat{\rho}^b - (1 - \cos\theta^w)(\hat{\rho}^w{\cdot}p)\hat{\rho}^w. \qquad (2.21)$$

Frank showed that this formula is simplified considerably if the reference lattice is taken to be the median lattice. The black and white lattices are obtained from the median lattice by equal and opposite rotations of $\theta/2$ about a common axis $\hat{\rho}$. Setting $\hat{\rho}^b = \hat{\rho}^w = \hat{\rho}$ and $\theta^w = -\theta^b = \theta/2$ in eqn (2.21) we obtain

$$B = 2\sin(\theta/2)(p \times \hat{\rho}), \qquad (2.22)$$

which is known as Frank's formula. The modulus of $|B|$ varies, in general, as $p$ varies in orientation in the boundary plane. If $\gamma$ is the angle between $\hat{\rho}$ and $p$ then

$$|B| = 2\sin(\theta/2)|p|\sin\gamma, \qquad (2.23)$$

where

$$\sin\gamma = \{\sin^2\mu\sin^2\varphi + \cos^2\varphi\}^{\frac{1}{2}}, \qquad (2.24)$$

and $\varphi$ is the angle between the rotation axis $\hat{\rho}$ and the boundary normal $\hat{n}$ and $\mu$ is the angle between the projection of $\hat{\rho}$ onto the boundary plane and $p$ (see Fig. 2.17). The maximum value of $|B|$, $|B|_{max}$, is thus obtained when $\sin\mu = \pm 1$:

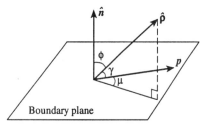

Boundary plane

**Fig. 2.17** The terms used in eqns (2.23) and (2.24).

$$|B|_{max} = 2 \sin (\theta/2) |p|. \qquad (2.25)$$

The minimum value, $|B|_{min}$, is obtained when $\sin \mu = 0$:

$$|B|_{min} = 2 \sin (\theta/2) |p| \cos \varphi. \qquad (2.26)$$

From eqn (2.25) we see that $|B|_{max}$ increases monotonically with misorientation $\theta$. For a pure tilt boundary $|B|$ varies between zero for $p$ along the tilt axis and $|B|_{max}$ for $p$ perpendicular to the tilt axis. For a pure twist boundary $|B|$ is equal to $|B|_{max}$ for all orientations of $p$.

## 2.6 THE O-LATTICE

The O-lattice (Bollmann 1970) is a further construction which has been widely used in analysing the dislocation structure of interfaces. Although the O-lattice was developed independently of the Frank–Bilby theory it turns out that they are very closely related. Indeed, the equation that defines the O-lattice may be regarded as a quantized form of the Frank–Bilby equation, in which the vectors $B$ and $p$ are discrete rather than continuous. The principle underlying the O-lattice is the belief that optimal structures are produced in interfaces in regions centred on points where the two lattices adjoining the interface 'match'. The definition of 'matching' that has been developed in the context of the O-lattice will be given shortly. We have already defined the CSL (Section 1.5.4) as the one-, two-, or three-dimensional lattice of common lattice sites which may be produced by two interpenetrating crystal lattices at particular relative orientations. If the two lattices adjoining the interface produce a CSL then the lattice points of the CSL in the interface are one example of points where the two adjoining lattices are said to 'match' each other. However, a CSL exists only in rather special circumstances, e.g. at certain misorientations between identical lattices, the values of which depend on the lattice symmetry, or when there are equal lattice parameters in the case of differing crystal lattices. If we take two arbitrary crystal lattices, with an arbitrary orientation between them, the chances are there will be no CSL. However, Bollmann (1970) showed that other more general points of lattice matching may be defined. If it is imagined that the two adjoining lattices interpenetrate, there are points in space which occupy equivalent positions in the unit cells of the two lattices. More precisely, if for any cell of one lattice the internal coordinates of a point, expressed as fractions of the cells edges, are identical with the fractional coordinates of the same point measured relative to a cell of the other lattice, the point is a point of 'lattice matching' and it is called an O-point. The set of all O-points constitutes the O-lattice which, as seen below, may be a point, line or planar lattice. Also, as seen below, such a lattice will always exist regardless of the misorientation between the two interpenetrating lattices. The O-lattice may therefore be thought of as a generalization of the concept of the CSL. In the particular case where a CSL exists, the CSL is a sublattice of the O-lattice.

Consider the case where we are working in the framework of a single crystal reference system, and the two lattices adjoining the interface are black and white crystal lattices, not necessarily of the same type. To find the O-lattice we imagine that the two crystal lattices are interpenetrating. The sites of the white lattice are denoted by the infinite set of vectors $\{R^w\}$. We take the white lattice as the reference lattice and consider an arbitrary point with coordinates $r^w$ expressed in the frame of the white lattice. The coordinates of this point are transformed into the coordinates with respect to the black lattice by the transformation $r^b = S^b r^w$. Thus, if $r^w$ and $r^b$ have the same internal cell coordinates then the point is an O-point. But if that is true then $r^b$ and $r^w + R^w$ have the same internal cell coordinates. Thus the general condition for an O-point is the following:

$$r^b = r^w + R^w$$

where

$$r^b = S^b r^w. \tag{2.27}$$

Then $r^b$ is an O-point, $R^o$. Equation (2.27) is illustrated in Fig. 2.18. Eliminating $r^w$ from Eqn. (2.27) we obtain an explicit equation for the O-points:

$$(E - S^{b-1})R^o = R^w \tag{2.28}$$

or

$$R^o = (E - S^{b-1})^{-1} R^w. \tag{2.29}$$

Equation (2.29) is the defining equation of the O-lattice. By substituting for $R^w$ the three base vectors of the (reference) white lattice in turn, the columns of the matrix $(E - S^{b-1})^{-1}$ define the corresponding base vectors of the O-lattice. The matrix $(E - S^{b-1})$ may be of rank 3,2 or 1. When it is of rank 3 the solutions $R^o$ represent a lattice of points in three dimensions. This is the case when all three eigenvalues of $S^b$ are not unity, as occurs at most heterophase boundaries, where there is not a

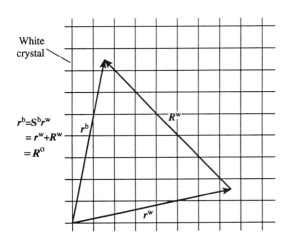

White crystal

$r^b = S^b r^w$
$= r^w + R^w$
$= R^o$

$r^b$

$R^w$

$r^w$

**Fig. 2.18** The derivation of the O-lattice equation, eqns (2.27)–(2.29).

coincidental relationship between the lattice parameters of the two crystal lattices. When the matrix is of rank 2 the O-lattice becomes a lattice of lines, called O-lines. The lines are parallel to the eigenvector of $S^b$ corresponding to the eigenvalue equal to unity. For example, if the black and white lattices are identical, apart from a rotation $\mathbf{R}^b = \mathbf{S}^b$, where $\mathbf{R}^b$ represents a rotation of $\theta^b$ about $\hat{\rho}^b$, then the O-lines are parallel to $\hat{\rho}^b$. Thus, for grain boundaries the O-lattice may always be described as a lattice of O-lines by choosing to describe the relationship between the lattices as a rotation. In a plane normal to $\hat{\rho}^b$ the rotation matrix may be represented by a $2 \times 2$ matrix, which is of rank 2:

$$\mathbf{R}^b = \begin{pmatrix} \cos\theta & -\sin\theta \\ \sin\theta & \cos\theta \end{pmatrix} \tag{2.30}$$

and

$$(\mathbf{E} - \mathbf{R}^{b-1})^{-1} = \begin{pmatrix} \tfrac{1}{2} & \tfrac{1}{2}\cot\tfrac{1}{2}\theta \\ -\tfrac{1}{2}\cot\tfrac{1}{2}\theta & \tfrac{1}{2} \end{pmatrix}. \tag{2.31}$$

It follows from eqns (2.29) and (2.31) that the basis vectors of the O-lattice vary continuously with the misorientation $\theta$. Since CSL sites are coincidences of lattice sites with internal cell coordinates $(0,0,0)$ it follows that CSL sites are also O-points but not all O-points are CSL sites. At irrational orientations where there is no CSL the O-lattice continues to exist. If one of the eigenvalues of $S^b$ is equal to unity the transformation $S^b$ is described as an invariant line strain for obvious reasons. If $S^b$ is an invariant plane strain there are two eigenvalues of $S^b$ that are equal to unity and then the matrix $(\mathbf{E} - \mathbf{S}^{b-1})$ becomes of rank 1. The O-lattice then becomes a lattice of parallel planes, called O-planes. A general invariant plane strain $S^b$ is represented by

$$(S^b)_{ij} = \delta_{ij} + e_i n_j, \tag{2.32}$$

where $e$ is a constant vector and $\hat{n}$ is normal to the invariant plane. Thus, if $p$ is any vector lying in the invariant plane then $(S^b)_{ij} p_j = p_i$ confirming that $p$ is invariant under the transformation. The inverse of $S^b$ is represented by

$$(S^{b-1})_{ij} = \delta_{ij} - \frac{e_i n_j}{1 + e_k n_k}, \tag{2.33}$$

as may be confirmed directly by forming $(S^{b-1})_{ij}(S^b)_{jk} = \delta_{ik}$. Then $(\mathbf{E} - \mathbf{S}^{b-1})$ becomes

$$\delta_{ij} - (S^{b-1})_{ij} = \frac{e_i n_j}{1 + e_k n_k}, \tag{2.34}$$

which is a matrix of rank 1, as can be seen by writing out its determinant, $\Delta$:

$$\Delta = e_1 e_2 e_3 n_1 n_2 n_3 \begin{vmatrix} 1 & 1 & 1 \\ 1 & 1 & 1 \\ 1 & 1 & 1 \end{vmatrix}. \tag{2.35}$$

Another, and equally important, interpretation of the O-lattice is that it is the lattice of possible origins for the relation $r^b = S^b r^w$. To show this consider a shift of origin to the O-lattice point $R^o$. The relation $r^b = S^b r^w$ becomes

$$r^b - R^o = S^b r^w - R^o$$

$$= S^b(r^w - S^{b-1}R^o)$$

$$= S^b(r^w + R^w - R^o), \tag{2.36}$$

where we have used eqn (2.29). This is of the original form $r^b = S^b r^w$ but corresponding points of the two crystals are changed. Thus point $r^b$ was derived from point $r^w$ before the shift of origin but is now derived from the point $r^w + R^w$, measured from the first origin.

So far we have not attributed any physical significance to the O-points other than to say that they are points of generalized crystal lattice coincidence. It is evident from eqn (2.29) that the O-lattice depends on the choice of the description $S^b$ of the relationship between the crystal lattices. For a given pair of crystal lattices, at a particular relative orientation, alternative specifications of $S^b$ correspond to choosing alternative unit cells from either lattice. Mathematically, this is achieved by pre- and post-multiplying $S^b$ by unimodular matrices. It is, therefore, not surprising that the O-lattice changes with different selections of $S^b$ because the cells, with respect to which the internal cell coordinates are measured, change when an alternative $S^b$ is selected. Only the coincident crystal lattice sites, when they exist, are invariant with respect to alternative specifications of $S^b$.

In order to make further progress it is necessary to introduce a *physical assumption*. We create an interface by defining a cut plane through the O-lattice and discarding black crystal lattice sites on one side of the cut and white crystal lattice sites on the other. It is then assumed that if an O-element (point, line, or plane) exists in the interface then it is a region of 'good fit' meaning that it is energetically favourable. The O-points are described as points of geometric registry between the crystal lattices. There is no rigorous justification for these assertions, or even a clear idea of what they mean in terms of atomic arrangements, but some intuitive reasoning may be offered.

If we take the trivial case where the black and white crystal lattices are identical with no misorientation forming a single (grey) crystal lattice then the 'interfacial energy' associated with any plane in the lattice is obviously zero. The O-lattice for this trivial case is a continuum extending throughout the crystal lattice since every point has the same internal cell coordinates in the 'black and white lattices'. As soon as we introduce a misorientation between the crystal lattices, or we deform one crystal lattice, the continuum of O-points is destroyed and we have an O-lattice. At small misorientations or deformations the O-lattice is coarse and it becomes finer as the misorientation or the lattice deformation increase in magnitude. This seems paradoxical, because it suggests that we discontinuously change from a state of perfect matching (the single crystal) to a state of widely spaced regions of matching (small misorientations or deformations) to a state of improved matching at higher misorientations or deformations. The resolution to the paradox is that we have ignored relaxation within the interface. At small misorientations or deformations we can reasonably expect to see discrete lattice dislocations in the interface. Since the O-points are assumed to be associated with regions of good fit within the interface we expect to see one O-point between each successive pair of dislocations. Therefore the average spacing of dislocations along a particular direction within the interface should correspond to the O-lattice spacing in that direction. Thus, if we assume that the relaxation within the interface introduces *localized* dislocations then at small misorientations or deformations there are large regions of relatively good fit separated by small regions of bad fit at the dislocation cores. Although the spacing of the O-points is large at small misorientations or deformations it is assumed that the energy of the interface derives mainly from the relatively small regions of misfit between the O-points. If we assume this argument may be extrapolated to higher angles or deformations we conclude that the energy of the interface increases as the average spacing of the dislocations decreases, which is equivalent to the O-lattice spacing. We see that it is not essential to understand precisely what is meant by saying that an O-point represents a point of geometric registry or good fit, because all that it implies is that *between the O-points* there is an accumulating disregistry which, it is assumed, is concentrated into dislocations approximately midway between them. Put another way, it is the dislocation content of the interface that is physically significant and the O-lattice is merely a geometrical construction to obtain it.

Following Christian (1981), we shall now make the relationship between the O-lattice and the Frank–Bilby equation for the dislocation content of an interface explicit. We separate the different O-elements (points, lines, or planes) by Wigner–Seitz cell walls. On constructing the interface by sectioning the O-lattice, as described above, the intersections of the Wigner–Seitz cell walls are identified with line discontinuities within the interface which are called 'mathematical dislocations'. Each O-lattice point may be considered as the origin for the relation $r^b = \mathbf{S}^b r^w$ within its own cell. When a cell wall is crossed the relation between corresponding lattice points changes by the vector $R^w$ which is the difference vector between two reference lattice points. We saw in eqn (2.29) that $(\mathbf{E} - \mathbf{S}^{b-1})R^\circ = R^w$. By identifying $R^\circ$ with $p$, and $B$ with $-R^w$, this is identical to the Frank–Bilby equation, eqn (2.1), where $\mathbf{S}^w = \mathbf{E}$. The only difference between eqn (2.29) and the Frank–Bilby equation is that the continuous variables $p$ and $B$ are quantized in the O-lattice equation equation because the discreteness of the black and white crystal lattices has been taken into account in the O-lattice treatment but not in the Frank–Bilby treatment. Bollmann calls the discrete dislocations that are defined by eqn (2.29) *primary* dislocations. They are distinguished by having Burgers vectors that are lattice vectors of either crystal.

The physical assumption that the dislocation content of the interface is minimized is equivalent to maximizing the spacing of O-elements within the interface. This suggests a criterion for selecting a description of the relationship $\mathbf{S}^b$ from among the infinity of possible description of $\mathbf{S}^b$ which maximizes the size of the one-, two-, or three-dimensional unit cell of the O-lattice. We have already remarked (Section 2.4), however, that this criterion is unsatisfactory when it is applied to symmetrical tilt boundaries. In that case the simple shear description gives an O-plane lattice with the boundary plane coinciding with one of the O-planes. As before, there are no dislocations required by such a description of the interface, and therefore at small angles of misorientation, where line defects possessing dislocation character are seen experimentally, the criterion does not lead to the most useful description.

Unlike the mathematical dislocations, whose positions are defined by the O-lattice cell walls, the physical dislocations are not, in general, uniformly spaced along the interface. The physical spacings are determined by the discreteness of the crystal lattice. For example, if the average spacing of the dislocations is $2\frac{1}{2}a$ this will be the spacing of the mathematical dislocations but the physical dislocations will be spaced alternately by $2a$ and $3a$.

Consider next the case where we are using a bicrystal reference structure and the two relevant lattices are the $DSC^w$ and $DSC^b$ lattices. These two DSC lattices form an O-lattice, which is sometimes called an O2-lattice, given by

$$(\mathbf{E} - \mathbf{D}^{b-1})R^\circ = \mathbf{R}^{DSC^w}. \tag{2.37}$$

Here we take the $DSC^w$ lattice as the reference lattice, and $\mathbf{D}^b$ is the transformation which generates the $DSC^b$ lattice from the $DSC^w$ lattice. The dislocations between successive O-elements of the O2-lattice have Burgers vectors of the $DSC^w$ lattice and they are called secondary dislocations. The Wigner–Seitz construction may be applied in the same way to the O2-lattice to determine the patches of good DSC lattice matching, i.e. the patches where continuity of the DSC lattice across the interface is maintained.

There have been numerous applications of the O-lattice theory to interpreting observed line defect contrast at grain and interphase boundaries in the electron microscope. Both primary and secondary dislocation arrays have been identified and characterized, and it is remarkable how successful the theory has been considering its simplicity and the rather unsatisfactory physical assumptions it contains.

One weak point of the theory is of course the physical assumption that determines the choice of the description of $S^b$. Precisely the same problem arose in the Frank–Bilby theory, and there we concluded that the only rigorous way to decide the most appropriate description was to observe the motion of atoms as the interface migrated. In the O-lattice case the most appropriate description of $S^b$ (or $D^b$) has another interpretation. It is the description which most accurately describes the pattern of local relaxation responsible for the network of dislocations with Burgers vectors $R^w$ (or $R^{DSC^w}$). Thus, at a small-angle grain boundary we would choose the disorientation description for $S^b$ because this most simply describes the local changes in misorientation that occur between successive dislocations in the interface. As Bollmann has pointed out the O-lattice is effectively a first approximation at describing the relaxation pattern within the interface. If a different $S^b$ is selected then a different relaxation pattern will be derived. The trick is to *guess* the description of $S^b$ (or $D^b$) which most closely approximates the local relaxations within the interface. This is not always straightforward as demonstrated by the experimental observations of Goodhew *et al.* (1976). Those authors found that the dislocation structure of small-angle (1–5°) (110) twist boundaries in gold could be explained by an O-lattice construction only if the transformation relating the two crystal lattices varied with position in the boundary plane: in some regions it was a simple rotation whereas in others it was a rotation plus a translation. This simply amounts to saying that the actual dislocation structure is dictated by minimization of the boundary energy, and since the O-lattice construction is geometrical in nature it may or may not predict the correct structure.

## 2.7 THE GEOMETRY OF DISCRETE DISLOCATION ARRAYS IN INTERFACES

### 2.7.1 The general interface

In this section we assume that a description of the relationship between the lattices which adjoin the interface has been selected, and that the interface is free of stresses at long range. The Burgers vector density of stress annihilator dislocations in the interface is assumed to be represented by $i$ independent sets of discrete dislocations, the $j$th set with Burgers vector $b_j$, line vector $\hat{\xi}_j$, and spacing $d_j$. The question we address is the following: given the set of Burgers vectors $b_j$ what are the line directions $\hat{\xi}_j$ and spacings $d_j$? This problem was first considered by Frank (1950), Read (1953), and Hirth and Lothe (1982), who considered grain boundaries, and more recently the heterophase interface problem was addressed by Sargent and Purdy (1975) and Knowles (1982). We shall follow the approach of Knowles (1982) for heterophase interfaces. Our analysis includes grain boundaries as the special case where the relationship between the crystal lattices is often a pure rotation.

Our starting point is the Frank–Bilby equation, eqn (2.1):

$$B = (S^{b-1} - S^{w-1})p, \qquad (2.1)$$

Recall that $S^b$ and $S^w$ are the affine transformations that generate the black and white crystal lattices from some reference lattice, and $B$ is the sum of the Burgers vector of stress annihilator dislocations crossing the interfacial vector $p$. Since $B$ is expressed in the reference lattice we assume that the discrete Burgers vectors of the dislocations, into which $B$ is assumed to decompose, are lattice vectors of the reference lattice. That is, we assume that $B$ may be written as

$$B(p) = \sum_{j=1}^{i} c_j(p)b_j, \qquad (2.38)$$

where the sum is taken over the $i$ sets of discrete dislocations that are assumed to exist in the interface, and the vectors $b_j$ are assumed to be lattice vectors of the reference lattice, although they are not necessarily primitive vectors. However, it is quite possible for the dislocations to dissociate into partial dislocations forming complex networks in the interface involving stacking faults. Such an eventuality cannot be predicted by this geometrical theory which can treat only the average Burgers vector content crossing a macroscopic vector $p$ in the interface and its decomposition into assumed discrete Burgers vectors. To account for this eventuality we must consider local interactions among the $i$ predicted sets of dislocations, and this will be discussed briefly in Section 2.8.

Let $W = S^{b-1} - S^{w-1}$ and consider the decomposition of $B$ described by eqn (2.38). Let $\hat{n}$ be the interface normal, pointing from the white crystal into the black. Let $N_j$ be a vector lying in the interface, normal to the sense vector $\hat{\xi}_j$ of the dislocations with Burgers vector $b_j$, and of length equal to the reciprocal of the spacing $d_j$:

$$N_j = \frac{\hat{n} \times \hat{\xi}_j}{d_j}. \qquad (2.39)$$

The number of dislocations intersected by a vector $p$ is $N_j \cdot p$, and a dislocation cut by $p$ is counted as a positive contribution to $B$ if $p \times \hat{n}$ has a positive component along $\hat{\xi}_j$. It follows that $c_j(p)$ in eqn (2.38) is equal to $N_j \cdot p$, and therefore

$$B = \sum_{j=1}^{i} (N_j \cdot p)b_j = Wp. \qquad (2.40)$$

Our task is to find the line directions $\hat{\xi}_j$ and spacings $d_j$ given $W$ and the Burgers vectors $b_j$.

The first point to observe is that if there are more than three sets of dislocations in eqn (2.40), i.e. $i > 3$, then the $i$ Burgers vectors $b_1, b_2, b_3, \ldots b_i$ are linearly dependent. In that case there is no unique solution. In the general case three independent Burgers vectors are required to satisfy eqn (2.40), and if there are less than three then only certain special types of interface may be described. In this section we consider the general case of $i = 3$ and solve eqn (2.40) for the $N_j$, $j = 1, 2, 3$. In general we cannot assume that $|W| \neq 0$, since $W$ may be of rank 2 or 1. Writing out enq (2.40) in full we have

$$(N_1 \cdot p)b_1 + (N_2 \cdot p)b_2 + (N_3 \cdot p)b_3 = Wp \qquad (2.41)$$

and multiplying both sides by $b_2 \times b_3$ we obtain

$$(N_1 \cdot p)(b_1 \cdot b_2 \times b_3) = Wp \cdot (b_2 \times b_3). \qquad (2.42)$$

In particular if $p$ is parallel to $\hat{\xi}_1$ we have $N_1 \cdot p = 0$ and therefore

$$W\hat{\xi}_1 \cdot b_1^* = 0 \qquad (2.43)$$

where $b_1^*$ is a reciprocal Burgers vector defined by

$$b_1^* = \frac{b_2 \times b_3}{b_1 \cdot b_2 \times b_3}. \tag{2.44}$$

Similar formulae hold for $\hat{\xi}_2$ and $\hat{\xi}_3$ and $b_2^*$ and $b_3^*$. Now $\mathbf{W}\hat{\xi}_1 \cdot b_1^* = \hat{\xi}_1 \cdot \mathbf{W}^t b_1^*$, where $\mathbf{W}^t$ is the transpose of $\mathbf{W}$, and therefore $\hat{\xi}_1$ is perpendicular to both $\mathbf{W}^t b_1^*$ and $\hat{n}$. Therefore $\hat{\xi}_1$ is parallel to $(\mathbf{W}^t b_1^*) \times \hat{n}$ and

$$N_1 = \frac{\hat{n} \times \hat{\xi}_1}{d_1} = \frac{\hat{n} \times ((\mathbf{W}^t b_1^*) \times \hat{n})}{d_1 |(\mathbf{W}^t b_1^*) \times \hat{n}|}. \tag{2.45}$$

Multiplying both sides of eqn (2.41) by $b_1^*$, for any interfacial vector $p$, we obtain

$$(N_1 \cdot p) = \mathbf{W} p \cdot b_1^* \tag{2.46}$$

and substituting eqn (2.45) into eqn (2.46) we find that

$$d_1 = 1/|(\mathbf{W}^t b_1^*) \times \hat{n}| \tag{2.47}$$

and

$$N_1 = \mathbf{W}^t b_1^* - (\hat{n} \cdot \mathbf{W}^t b_1^*)\hat{n} \tag{2.48}$$

with

$$\hat{\xi}_1 = \frac{((\mathbf{W}^t b_1^*) \times \hat{n})}{|(\mathbf{W}^t b_1^*) \times \hat{n}|}. \tag{2.49}$$

Equations (2.47)–(2.49) represent the solution to the task we set ourselves. Equation (2.48) states that $N_1$ is the projection of $\mathbf{W}^t b_1^*$ onto the interface plane. These formulae hold regardless of the rank of $\mathbf{W}$. But in the particular case where the rank of $\mathbf{W}$ is equal to three the inverse of $\mathbf{W}$ exists and eqn (2.41) may be rewritten as follows:

$$(N_1 \cdot p)\mathbf{W}^{-1} b_1 + (N_2 \cdot p)\mathbf{W}^{-1} b_2 + (N_3 \cdot p)\mathbf{W}^{-1} b_3 = p. \tag{2.50}$$

The vectors $\mathbf{W}^{-1} b_j$, are recognized as O-lattice vectors (see eqn (2.29)) and the lattice formed from them is a point O-lattice. Setting $a_1^o = \mathbf{W}^{-1} b_1$, $a_2^o = \mathbf{W}^{-1} b_2$, $a_3^o = \mathbf{W}^{-1} b_3$ as the basis vectors of the O-lattice, eqn (2.50) becomes

$$(N_1 \cdot p)a_1^o + (N_2 \cdot p)a_2^o + (N_3 \cdot p)a_3^o = p. \tag{2.51}$$

Setting $p$ parallel to $\hat{\xi}_1$, so that $N_1 \cdot p = 0$, and multiplying by $a_2^o \times a_3^o$ we obtain

$$\hat{\xi}_1 \cdot (a_2^o \times a_3^o) = 0. \tag{2.52}$$

Thus $\hat{\xi}_1$ is perpendicular to $a_1^{o*} = (a_2^o \times a_3^o)/(a_1^o \cdot a_2^o \times a_3^o)$ and $\hat{n}$:

$$\hat{\xi}_1 = \frac{a_1^{o*} \times \hat{n}}{|a_1^{o*} \times \hat{n}|}. \tag{2.53}$$

Substituting this expression for $\hat{\xi}_1$ into eqn (2.39) and using $(N_1 \cdot p) = p \cdot a_1^{o*}$, which follows from eqn (2.51) by multiplying both sides by $a_1^{o*}$, we deduce that

$$N_1 = a_1^{o*} - (a_1^{o*} \cdot \hat{n})\hat{n} \tag{2.54}$$

and

$$d_1 = 1/|a_1^{o*} \times \hat{n}|. \tag{2.55}$$

Equation (2.54) states that $N_1$ is the projection of $a_1^{o*}$ onto the interface plane. Equations (2.53), (2.54), and (2.55) are equivalent to eqns (2.49), (2.48), and (2.47) if

$$\mathbf{W}^t b_1^* = a_1^{o^*},\qquad(2.56)$$

for $|\mathbf{W}| \neq 0$, and similar relations hold for $a_2^{o^*}$ and $a_3^{o^*}$. We shall prove that this relation holds by showing that it is consistent with $a_1^o \cdot a_1^{o^*} = 1$ and $a_2^o \cdot a_1^{o^*} = a_3^o \cdot a_1^{o^*} = 0$. Using the definition $a_1^o$ we obtain:

$$a_1^o \cdot a_1^{o^*} = (\mathbf{W}^{-1} b_1) \cdot (\mathbf{W}^t b_1^*),$$

which in component form becomes

$$(a_1^o)_i (a_1^{o^*})_i = W_{ij}^{-1} (b_1)_j W_{ki} (b_1^*)_k.\qquad(2.57)$$

Performing the sum on $i$, and using $W_{ki} W_{ij}^{-1} = \delta_{kj}$, we obtain

$$a_1^o \cdot a_1^{o^*} = b_1 \cdot b_1^* = 1.\qquad(2.58)$$

Similarly, $a_2^o \cdot a_1^{o^*}$ becomes

$$a_2^o \cdot a_1^{o^*} = (\mathbf{W}^{-1} b_2) \cdot (\mathbf{W}^t b_1^*),$$

which in component form reads

$$(a_2^o)_i (a_1^{o^*})_i = W_{ij}^{-1} (b_2)_j W_{ki} (b_1^*)_k.\qquad(2.59)$$

Performing the sum on $i$ again we deduce that

$$a_2^o \cdot a_1^{o^*} = b_2 \cdot b_1^* = 0.\qquad(2.60)$$

Similarly $a_3^o \cdot a_1^{o^*}$ may also be shown to be zero. This proves that eqn (2.56) is correct.

If we now add a fourth set of dislocations, with specified Burgers vector $b_4$, line sense $\hat{\xi}_4$, and spacing $d_4$, we find that the $N_i$ change, where $i = 1, 2, 3$, by $-(b_4 \cdot b_i^*)N_4$. As before, $N_4 = \hat{n} \times \hat{\xi}_4 / d_4$. A fourth set of dislocations may be produced by local interactions between the first three sets of dislocations.

To *summarize*, eqns (2.47) and (2.49) give the spacings and line directions of the three independent sets of dislocations in an interface with normal $\hat{n}$ for an arbitrary choice of the matrix $\mathbf{W}$. In the particular case where the rank of $\mathbf{W}$ equals 3 these equations are equivalent to eqns (2.55) and (2.53), which are expressed in terms of reciprocal lattice vectors of the corresponding point O-lattice.

### 2.7.2 Application to a grain boundary with arbitrary geometrical parameters

Consider a grain boundary requiring three independent sets of dislocations to describe the stress annihilator dislocation content. We have already seen that if the adjoining lattices are related by a rotation, and if the median lattice is selected as the reference lattice, the Burgers vector content has a particularly simple form:

$$B = 2(p \times \hat{\rho}) \sin(\theta/2).\qquad(2.22)$$

Here the black and white lattices are created from a median lattice by equal and opposite rotations of $\theta/2$ about the axis $\hat{\rho}$. At small angles one would generally choose the median lattice to be a crystal lattice with an orientation half way between the two crystal lattices. In that case the Burgers vectors $b_i$ are lattice vectors of the crystal lattice and the discrete dislocations are primary dislocations. The angle $\theta$ is interpreted as the total boundary misorientation. At large angles one could choose another large-angle boundary as the reference structure, and regard the black and white crystal lattices as being generated by equal and opposite rotations from their orientations in this reference structure. In that case the dislocations are of the secondary type, and the Burgers vectors are lattice vectors of the DSC lattice of the reference boundary. The angle $\theta$ is interpreted as the mis-orientation from the large-angle boundary reference structure. Thus,

$$2 \sin (\theta/2) (p \times \hat{\rho}) = \sum_{i=1}^{3} (N_i \cdot p) b_i, \qquad (2.61)$$

and after multiplying both sides with respect to $b_j^*$ we deduce

$$2 \sin (\theta/2) b_j^* \cdot (p \times \hat{\rho}) = N_j \cdot p. \qquad (2.62)$$

In particular, setting $p = \hat{\xi}_j$ we find that $\hat{\xi}_j$ is perpendicular to $\hat{\rho} \times b_j^*$. Therefore,

$$\hat{\xi}_j = \frac{(\hat{\rho} \times b_j^*) \times \hat{n}}{|(\hat{\rho} \times b_j^*) \times \hat{n}|}, \qquad (2.63)$$

and

$$N_j = 2(\hat{\rho} \times b_j^* - [\hat{n} \cdot (\hat{\rho} \times b_j^*)] \hat{n}) \sin (\theta/2), \qquad (2.64)$$

$$d_j = \frac{1}{2 \sin (\theta/2) |(\hat{\rho} \times b_j^*) \times \hat{n}|}. \qquad (2.65)$$

Equations (2.63) and (2.65) give the line directions and spacings of the three sets of dislocations with Burgers vectors $b_1$, $b_2$, and $b_3$ required to represent the Burgers vector content of a grain boundary with arbitrary geometrical parameters. Note that all vectors are expressed in the coordinate frame of the median lattice.

### 2.7.3  Grain boundaries containing one and two sets of dislocations

We have seen that in general we need three non-coplanar Burgers vectors of discrete dislocations to account for the Burgers vector content of a grain boundary with arbitrary geometrical parameters. In this section we ask what kinds of boundaries may be formed from just one or two independent sets of dislocations and still satisfy Frank's formula.

If there is *only one set of dislocations*, with Burgers vector $b$, then Frank's formula becomes

$$2(p \times \hat{\rho}) \sin (\theta/2) = (N \cdot p) b. \qquad (2.66)$$

Therefore $b$ is perpendicular to $p$ and $\hat{\rho}$ regardless of the orientation of $p$ within the boundary plane. Thus, $b$ must be perpendicular to the boundary plane, and $\hat{\rho}$ must lie in the boundary plane. Thus the boundary is a tilt boundary. It is emphasized that these vectors are expressed in the median lattice. If the normal to the boundary plane, $\hat{n}$, is a mirror plane of the median lattice the boundary is of the symmetric tilt type, otherwise it is an asymmetric tilt boundary. Setting $p$ parallel to $\hat{\xi}$ we further deduce that $\hat{\xi} \times \hat{\rho} = 0$ and therefore the dislocation lines are parallel to the rotation axis. The spacing, $d$, of the dislocation lines is given by

$$d = |b|/2 \sin (\theta/2) \qquad (2.67)$$

If there are *two independent sets of dislocations* there are just two types of boundary that may be formed, and we shall consider them in turn. Both types of boundary satisfy Frank's formula:

$$2(p \times \hat{\rho}) \sin (\theta/2) = (N_1 \cdot p) b_1 + (N_2 \cdot p) b_2. \qquad (2.68)$$

Multiplying both sides by $b_1 \times b_2$ we deduce

$$p \cdot \hat{\rho} \times (b_1 \times b_2) = 0. \tag{2.69}$$

The first way of satisfying this equation is for $\hat{\rho} \times (b_1 \times b_2)$ to be perpendicular to $p$. Since this has to be true for any $p$ lying in the boundary plane it can be satisfied only if $\hat{\rho} \times (b_1 \times b_2)$ is parallel to the boundary normal. Thus $\hat{\rho}$ and $(b_1 \times b_2)$ lie in the boundary plane and the boundary is a tilt boundary. Setting $p$ parallel to $\hat{\rho}$ in eqn (2.68) we find

$$(N_1 \cdot \hat{\rho}) b_1 + (N_2 \cdot \hat{\rho}) b_2 = 0. \tag{2.70}$$

Since $b_1$ and $b_2$ are not parallel, by assumption, it follows that $N_1 \cdot \hat{\rho}$ and $N_2 \cdot \hat{\rho}$ are zero and hence $\hat{\xi}_1$ and $\hat{\xi}_2$ are parallel to $\hat{\rho}$. The asymmetric tilt boundary to which this case corresponds is sketched schematically in Fig. 2.3. To find the spacings, $d_1$ and $d_2$, of the dislocations we set $\hat{\xi}_1 = \hat{\xi}_2 = \hat{\rho}$ and $p = \hat{\rho} \times \hat{n}$ in eqn (2.68):

$$-2 \sin(\theta/2)\hat{n} = \frac{b_1}{d_1} + \frac{b_2}{d_2}. \tag{2.71}$$

Therefore,

$$\frac{d_1}{d_2} = -\frac{(b_1 \times \hat{n}) \cdot (b_1 \times b_2)}{(b_2 \times \hat{n}) \cdot (b_1 \times b_2)}. \tag{2.72}$$

Substituting this into eqn (2.71) and multiplying both sides by $\hat{n}$ we obtain

$$d_1 = \frac{(b_2 \cdot \hat{n})(b_1 \times \hat{n}) \cdot (b_1 \times b_2) - (b_1 \cdot \hat{n})(b_2 \times \hat{n}) \cdot (b_1 \times b_2)}{2 \sin(\theta/2)(b_2 \times \hat{n}) \cdot (b_1 \times b_2)}$$

and

$$d_2 = \frac{(b_1 \cdot \hat{n})(b_2 \times \hat{n}) \cdot (b_1 \times b_2) - (b_2 \cdot \hat{n})(b_1 \times \hat{n}) \cdot (b_1 \times b_2)}{2 \sin(\theta/2)(b_1 \times \hat{n}) \cdot (b_1 \times b_2)}. \tag{2.73}$$

The second type of boundary that satisfies eqn (2.69) is that where $\hat{\rho}$ is parallel to $b_1 \times b_2$. Setting $\hat{\rho} = b_1 \times b_2/|b_1 \times b_2|$ in eqn (2.68) we obtain

$$\frac{2 \sin(\theta/2)}{|b_1 \times b_2|} [(p \cdot b_2)b_1 - (p \cdot b_1)b_2] = (N_1 \cdot p)b_1 + (N_2 \cdot p)b_2. \tag{2.74}$$

Since this must hold for any vector $p$ in the boundary it follows that

$$N_1 \cdot p = \frac{2 \sin(\theta/2)}{|b_1 \times b_2|}(p \cdot b_2),$$

and

$$N_2 \cdot p = -\frac{2 \sin(\theta/2)}{|b_1 \times b_2|}(p \cdot b_1). \tag{2.75}$$

Setting $p$ parallel to $\hat{\xi}_1$ and $\hat{\xi}_2$ in turn we deduce that $\hat{\xi}_1 \cdot b_2 = \hat{\xi}_2 \cdot b_1 = 0$. Thus,

$$\hat{\xi}_1 = \frac{b_2 \times \hat{n}}{|b_2 \times \hat{n}|},$$

and

$$\hat{\xi}_2 = \frac{b_1 \times \hat{n}}{|b_1 \times \hat{n}|}. \tag{2.76}$$

Therefore $N_1$ becomes

$$N_1 = \frac{b_2 - (\hat{n} \cdot b_2)\hat{n}}{|b_2 \times \hat{n}| d_1},$$

whereas

$$N_2 = -\frac{b_1 - (\hat{n} \cdot b_1)\hat{n}}{|b_1 \times \hat{n}| d_2}. \tag{2.77}$$

Substituting $p = N_1$ and $p = N_2$ in turn into eqn (2.75) and using eqn (2.77) for $N_1$ and $N_2$ we find that

$$d_1 = \frac{|b_1 \times b_2|}{2 \sin (\theta/2) |b_2 \times \hat{n}|}$$

and

$$d_2 = \frac{|b_1 \times b_2|}{2 \sin (\theta/2) |b_1 \times \hat{n}|}. \tag{2.78}$$

Thus,

$$N_1 = \frac{b_2 - (\hat{n} \cdot b_2)\hat{n}}{|b_1 \times b_2|} 2 \sin (\theta/2)$$

and

$$N_2 = -\frac{b_1 - (\hat{n} \cdot b_1)\hat{n}}{|b_1 \times b_2|} 2 \sin (\theta/2). \tag{2.79}$$

When $\hat{n}$ is parallel to $b_1 \times b_2$ we have a pure twist boundary because $\hat{p}$ is then parallel to $b_1 \times b_2$. Then,

$$N_1 = \frac{b_2}{|b_1 \times b_2|} 2 \sin (\theta/2)$$

and

$$N_2 = -\frac{b_1}{|b_1 \times b_2|} 2 \sin (\theta/2). \tag{2.80}$$

Therefore, we see that the vectors $N_1$ and $N_2$ for boundaries of arbitrary $\hat{n}$ in eqn (2.79) are the projections onto that boundary of the $N_1$ and $N_2$ for the pure twist boundary. Hence, it may be shown (Frank 1950, Hirth and Lothe 1982) that the dislocation grid for a boundary of arbitrary $\hat{n}$ is just the projection onto it of the dislocation grid of the pure twist boundary. The dislocation structures of all boundaries of this type therefore consist simply of two families of intersecting dislocations.

### 2.7.4  Epitaxial interfaces

An epitaxial interface is formed when one material (the epilayer) is grown on a thick substrate. If the epilayer has a thickness below some critical value it is energetically favourable for the epilayer to be elastically strained so that a commensurate interface is developed between the epilayer and the substrate. At larger thicknesses a transition occurs in which stress annihilator dislocations are introduced into the interface to relieve the long range stress field in the epilayer. The transition is discussed in Section 2.10.6. In this section we shall confine our attention to the kinds of misfit that may be relieved by the existence of one or two independent sets of stress annihilator dislocations in the interface. We follow the procedure described by Sargent and Purdy (1975), although some of our conclusions are quite different. The purpose is to present an analysis for epitaxial interfaces that is analogous to the analysis of Section 2.7.3 of the types of grain boundaries that may be formed from one or two independent sets of dislocations.

Since the misfit strain is confined largely to the thin epilayer we assume that the reference lattice is the crystal lattice of the substrate. The Frank–Bilby equation is then

$$B = (S^{-1} - E)p \tag{2.81}$$

where $S$ transforms the components of a vector expressed in the substrate crystal lattice into those of a vector expressed in the crystal lattice of the epilayer. Since the interface is epitaxial we assume that $S$ satisfies $S\hat{n} = \hat{n}$. This simply ensures that the interface normal is the same direction in both the substrate and deposit lattices.

Let there be only *one set* of crystal lattice dislocations in the interface with Burgers vector $b$ and line sense $\hat{\xi}$, expressed in the substrate lattice. Consider the form of the transformation $S$ that can satisfy the Frank–Bilby equation, eqn (2.1), given this one set of dislocations. We have

$$(N \cdot p)b = (S^{-1} - E)p \tag{2.82}$$

and setting $p = \hat{\xi}$ we deduce that $S^{-1}\hat{\xi} = \hat{\xi}$. If $\hat{\xi} = [1, 0, 0]$, $\hat{N} = [0, 1, 0]$, and $\hat{n} = [0, 0, 1]$ then $S^{-1}$ has the form

$$S^{-1} = \begin{pmatrix} 1 & S_{12}^{-1} & 0 \\ 0 & S_{22}^{-1} & 0 \\ 0 & S_{32}^{-1} & 1 \end{pmatrix}. \tag{2.83}$$

Setting $p = N$ in eqn (2.82) we deduce that $N^2 b = (S^{-1} - E)N$ and therefore $S_{12}^{-1} = b_1/d$, $S_{22}^{-1} = b_2/d + 1$, $S_{32}^{-1} = b_3/d$.

For pure edge dislocations with $b \cdot \hat{n} = 0$ then $b_1 = 0$, $b_2 = b$, $b_3 = 0$, and $S^{-1}$ describes a tetragonal distortion:

$$S^{-1} = \begin{pmatrix} 1 & 0 & 0 \\ 0 & (b/d + 1) & 0 \\ 0 & 0 & 1 \end{pmatrix}. \tag{2.84}$$

For pure edge dislocations with $b \cdot N = 0$ then $b_1 = 0$, $b_2 = 0$, $b_3 = b$, and $S^{-1}$ describes a simple shear normal to the interface plane:

$$S^{-1} = \begin{pmatrix} 1 & 0 & 0 \\ 0 & 1 & 0 \\ 0 & b/d & 1 \end{pmatrix}. \tag{2.85}$$

For pure screw dislocations $b_1 = b$, $b_2 = 0$, $b_3 = 0$, and $\mathbf{S}^{-1}$ describes a simple shear in the interface plane:

$$\mathbf{S}^{-1} = \begin{pmatrix} 1 & b/d & 0 \\ 0 & 1 & 0 \\ 0 & 0 & 1 \end{pmatrix}. \tag{2.86}$$

Consider now the form of the transformation $\mathbf{S}$ that is possible when *two independent sets of dislocations* are present in the interface. Let the Burgers vectors of the two sets be $b_1$ and $b_2$ and let the line senses be $\hat{\xi}_1$ and $\hat{\xi}_2$. The Frank–Bilby equation becomes:

$$(N_1 \cdot p)b_1 + (N_2 \cdot p)b_2 = (\mathbf{S}^{-1} - \mathbf{E})p. \tag{2.87}$$

Setting $p$ equal to $\hat{\xi}_1$ and $\hat{\xi}$ in turn we deduce that

$$(N_2 \cdot \hat{\xi}_1)b_2 = (\mathbf{S}^{-1} - \mathbf{E})\hat{\xi}_1 \tag{2.88}$$

and

$$(N_1 \cdot \hat{\xi}_2)b_1 = (\mathbf{S}^{-1} - \mathbf{E})\hat{\xi}_2 \tag{2.89}$$

Setting $\hat{\xi}_1 = [1, 0, 0]$, $N_1 = [0, 1/d_1, 0]$, $\hat{n} = [0, 0, 1]$, $\hat{\xi}_2 = [\hat{\xi}_{21}, \hat{\xi}_{22}, 0]$, $N_2 = 1/d_2[N_{21}, N_{22}, 0]$, $b_1 = [b_{11}, b_{12}, b_{13}]$ and $b_2 = [b_{21}, b_{22}, b_{23}]$ and substituting these vectors into eqns (2.88) and (2.89) we find that

$$\mathbf{S}^{-1} = \begin{pmatrix} -\xi_{22}b_{21}/d_2 + 1 & b_{11}/d_1 + \xi_{21}b_{21}/d_2 & 0 \\ -\xi_{22}b_{22}/d_2 & b_{12}/d_1 + \xi_{21}b_{22}/d_2 + 1 & 0 \\ -\xi_{22}b_{23}/d_2 & b_{13}/d_1 + \xi_{21}b_{23}/d_2 & 1 \end{pmatrix}. \tag{2.90}$$

If the two sets of dislocations are pure edge dislocations with Burgers vectors in the plane of the interface, then $\mathbf{S}^{-1}$ has the form:

$$\mathbf{S}^{-1} = \begin{pmatrix} (\sin^2\theta)b_2/d_2 + 1 & -(\sin\theta\cos\theta)b_2/d_2 & 0 \\ -(\sin\theta\cos\theta)b_2/d_2 & 1 + b_1/d_1 + (\cos^2\theta)b_2/d_2 & 0 \\ 0 & 0 & 1 \end{pmatrix}, \tag{2.91}$$

where $b_1 = |b_1|$, $b_2 = |b_2|$, and $\hat{\xi}_2 = [\cos\theta, \sin\theta, 0]$. When the angle $\theta$ between the dislocation lines is $90°$, $\mathbf{S}^{-1}$ reduces to a diagonal matrix describing two orthogonal tensile strains, of magnitudes $b_1/d_1$ and $b_2/d_2$, in the plane of the interface. For $\theta$ between 0 and $90°$, $\mathbf{S}^{-1}$ is no longer diagonal but it is symmetric. When it is diagonalized it is found to represent two orthogonal principal strains in the plane of the interface of magnitude $(x \pm (x^2 - 4(b_1 b_2/d_1 d_2)\sin^2\theta)^{\frac{1}{2}})/2$, where $x = b_1/d_1 + b_2/d_2$. However, the directions of these tensile strains no longer coincide with the Burgers vectors $b_1$ and $b_2$. Instead, the two angles between the principal strain axes and $b_1$ are given by

$$\tan\varphi = \tan\theta - \frac{x}{(b_2/d_2)\sin(2\theta)}\left(1 \mp \left[1 - \frac{4b_1 b_2 \sin^2\theta}{d_1 d_2 x^2}\right]^{\frac{1}{2}}\right). \tag{2.92}$$

If the two sets of dislocations are pure screw dislocations then $\mathbf{S}^{-1}$ has the form:

$$\mathbf{S}^{-1} = \begin{pmatrix} 1 - \sin\theta\cos\theta\, b_2/d_2 & b_1/d_1 + \cos^2\theta\, b_2/d_2 & 0 \\ -\sin^2\theta\, b_2/d_2 & 1 + \sin\theta\cos\theta\, b_2/d_2 & 0 \\ 0 & 0 & 1 \end{pmatrix}. \tag{2.93}$$

As before $\hat{\xi}_2 = [\cos\theta, \sin\theta, 0]$. $\mathbf{S}^{-1}$ describes a shear transformation in the interface plane, although it is neither pure nor simple in general. The principal strains are $\pm\sqrt{b_1 b_2/d_1 d_2}\sin\theta$ and the angles between the principal directions and $b_1$ are given by

$$\tan\varphi = \frac{\left[\dfrac{b_2}{d_2}\right]^{\frac{1}{2}}\sin\theta}{\left[\dfrac{b_2}{d_2}\right]^{\frac{1}{2}}\cos\theta \mp \left[\dfrac{b_1}{d_1}\right]^{\frac{1}{2}}}. \tag{2.94}$$

## 2.8 LOCAL DISLOCATION INTERACTIONS

The analysis of the stress annihilator dislocation content of an interface presented thus far takes no account of interactions between the dislocations. The dislocations were all straight, even when they crossed each other, and the spacing of parallel dislocations was constant. We have already remarked that the spacing of parallel dislocations is quantized by the crystal lattice, so that we expect variations in the spacing such that the average spacing is the same as that predicted by the theory. When dislocation lines cross they interact and may form new networks. It is obvious that the decomposition of the net Burgers vector content $B$ into Burgers vectors of discrete dislocations is not unique, and the correct description is the one corresponding to the lowest boundary energy.

The local interactions that arise from dislocations crossing each other depend on their self-energies and interaction energies and thus they depend on the crystal structure and the character of the interface. This has been considered in detail by Amelinckx and Dekeyser (1959) for small-angle grain boundaries in cubic crystals and by Amelinckx (1979). Here we confine ourselves to a few general remarks and refer the reader to Amelinckx and Dekeyser (1959) and Amelinckx (1979) for a more specific treatment. Provided the final dislocation network in the interface still satisfies the Frank–Bilby equation the interface will remain free of long-range stresses.

One possible way of satisfying this condition is for an array of parallel dislocations with Burgers vector $b$, line direction $\hat{\xi}$, and $N = (\hat{n} \times \hat{\xi})/d$, to split into two arrays of Burgers vector $b$, line directions $\hat{\xi}_1$ and $\hat{\xi}_2$, and $N_1 = (\hat{n} \times \hat{\xi}_1)/d_1$ and $N_2 = (\hat{n} \times \hat{\xi}_2)/d_2$), such that $N = N_1 + N_2$. This follows from $(N \cdot p)b = (N_1 \cdot p)b + (N_2 \cdot p)b$ for any interfacial vector $p$, and hence the Frank–Bilby equation is unaffected.

Another way of not affecting the Frank–Bilby equation is by repeating local changes regularly throughout the network, such that the average directions of the dislocations are not altered. These local rearrangements do not change the number of dislocations cut by a large vector $p$ in the interface and thus the Frank–Bilby equation remains satisfied. An example is shown in Fig. 2.19 where a lozenge-shaped network has interacted locally to produce a hexagonal network. The interaction can be visualized as occurring in two steps. In the first, each straight dislocation in the sets 1 and 2 in Fig. 2.19(a) is transformed into a zig-zag configuration conforming to the geometry in Fig. 2.19(b). In the second, segments of dislocations 1 and 2 which overlap react to form segments of new dislocations of type 3 according to the Burgers vector reaction $b^{(1)} + b^{(2)} = b^{(3)}$. If a reaction does not occur the network has a lozenge-shaped mesh, as shown in Fig. 2.19(b). In general, twist grain boundaries are expected to contain hexagonal networks of dislocations except when the lines are within a few degrees of being orthogonal. Whether or not reactions take place is determined by a balance of elastic energies that is considered in detail by Hirth and Lothe (1982).

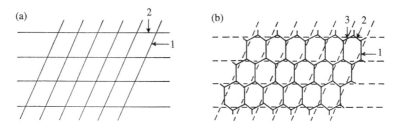

**Fig. 2.19** Local interactions in a lozenge‒shaped network of two types of dislocation in (a) to produce a hexagonal network of three types of dislocation in (b). (From Hirth and Lothe (1982).)

## 2.9  TWO EXAMPLES

In this section we shall illustrate the application of the theory of dislocation arrays at interfaces to two heterophase interfaces. In the first example we consider interfaces, produced by diffusion bonding at 1200 °C, between single crystals or polycrystals of Pt and a (001) substrate of NiO, following Shieu and Sass (1990). The interesting feature about some of these interfaces is that there is a misorientation of the f.c.c. crystal axes in addition to the misfit arising from the difference in lattice parameters $a^{Pt}$ and $a^{NiO}$. In the second example we consider Al-Al$_3$Ni interfaces in a directionally soldified eutectic alloy, following Knowles and Goodhew (1983$a$, $b$). Those authors determined the line directions of three sets of stress annihilator dislocations within interfaces bounding the Al$_3$Ni rods and then used the theory in a rather novel way to deduce the transformation relating the Al and Al$_3$Ni phases and the Burgers vectors of the dislocations.

### 2.9.1  Pt-NiO interfaces

Shieu and Sass (1990) observed three types of orientation relation between single crystals of Pt, or polycrystalline films of Pt, and a single crystal (001) substrate of NiO. Here we shall confine our attention to the perfect epitaxial orientation ((001)$^{Pt}$ on (001)$^{NiO}$ and [110]$^{Pt}$ parallel to [110]$^{NiO}$) and the 'twist misfit' interface in which the [110]$^{Pt}$ and [110]$^{NiO}$ axes are misoriented by an angle $\theta$. When $\theta = 0$ we recover the perfect epitaxial orientation. For the perfect epitaxial orientation a square network of dislocations was observed with line directions along [110] and [1$\bar{1}$0] and a spacing of 5.2 ±0.5 nm. At the twist misfit interfaces square arrays of dislocations were also observed but the spacings of the dislocations decreased as $\theta$ increased and the line directions changed rapidly from ⟨110⟩. For example, for $\theta = 1.5°$ the measured spacing of the dislocations was 3.8 ± 0.5 nm and the line directions were approximately 25 ° away from ⟨110⟩. An example of the dislocation structure observed in a misoriented interface is shown in Fig. 2.20. Our task is to explain the observed spacings and line directions of the dislocations as a function of $\theta$. Our analysis is similar to that of Shieu and Sass (1990) except that we take the Pt crystal lattice, rather than the NiO lattice, as our reference lattice and we use the Frank–Bilby theory rather than the O-lattice formulation. Our reason for choosing the Pt crystal as the reference is that we believe that the dislocations could lie slightly on the Pt side because NiO is elastically harder. Of course, whether we use the Frank–Bilby theory or the O-lattice theory makes no difference to the predicted line directions and spacings.

We define the misfit parameter $\alpha$ as follows:

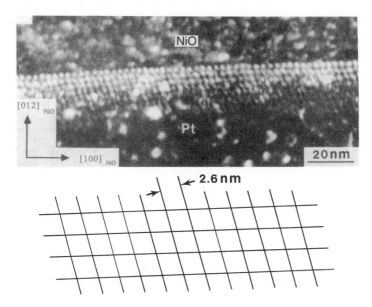

**Fig. 2.20** Dark-field electron micrograph of a NiO–Pt interface with a misorientation comprising ≃3.2° about the axis normal to the interface and ≃1° about the [1$\bar{1}$0] axis of NiO in the plane of the interface. A network of two sets of dislocations is seen as shown in the schematic diagram. From Shieu and Sass (1990).

$$\alpha = \frac{a^{\text{NiO}} - a^{\text{Pt}}}{a^{\text{Pt}}}. \tag{2.95}$$

The appropriate values of $a^{\text{NiO}}$ and $a^{\text{Pt}}$ are those at the temperature at which the interfaces were formed, i.e. 1200 °C. Using the value of the lattice parameter for Pt at 20 °C (0.392 39 nm) and the linear thermal expansion coefficient for Pt ($9 \times 10^{-6}\,°\text{C}^{-1}$) we obtain $a^{\text{Pt}} = 0.396\,56$ nm at 1200 °C. The lattice parameter of NiO at 275 °C is 0.419 46 nm, but the linear thermal expansion coefficient is not available. If we assume the thermal expansion coefficient is the same as that of MgO (i.e. $12.8 \times 10^{-6}\,°\text{C}^{-1}$) we obtain $a^{\text{NiO}} = 0.424\,43$ nm at 1200 °C. Therefore at 1200 °C the misfit parameter $\alpha = 0.0703$. In view of the uncertainty about the NiO lattice parameter at 1200 °C it is unreasonable to specify $\alpha$ so precisely and we shall assume $\alpha = 0.07$. If the $[110]^{\text{NiO}}$ axis is misoriented from the $[110]^{\text{Pt}}$ axis by $\theta$ then the transformation **S** relating the NiO lattice to the Pt lattice is

$$\mathbf{S} = (1 + \alpha) \begin{pmatrix} \cos\theta & -\sin\theta & 0 \\ \sin\theta & \cos\theta & 0 \\ 0 & 0 & 0 \end{pmatrix}, \tag{2.96}$$

and therefore

$$\mathbf{W} = \mathbf{S}^{-1} - \mathbf{E} = \frac{1}{1+\alpha} \begin{pmatrix} 1 + \alpha - \cos\theta & -\sin\theta & 0 \\ \sin\theta & 1 + \alpha - \cos\theta & 0 \\ 0 & 0 & \alpha \end{pmatrix}. \tag{2.97}$$

We *assume* that the Burgers vectors of the observed dislocations are $b_1 = \frac{1}{2}[110]$ and $b_2 = \frac{1}{2}[1\bar{1}0]$ lattice vectors of the Pt crystal. The Frank–Bilby equation becomes:

$$(N_1 \cdot p)b_1 + (N_2 \cdot p)b_2 = Wp, \qquad (2.98)$$

and multiplying both sides of this equation with $b_1$ and setting $p = \hat{\xi}_1$ we deduce that $W^t b_1 \cdot \hat{\xi}_1 = 0$. Therefore $\hat{\xi}$ is perpendicular to $W^t b_1$ and [001], i.e.

$$\hat{\xi}_1 = \frac{[(1 + \alpha - \cos\theta - \sin\theta),\, -(1 + \alpha - \cos\theta + \sin\theta),\, 0]}{\sqrt{2(1 + \alpha - \cos\theta)^2 + 2(\sin\theta)^2}}. \qquad (2.99)$$

The angle $\eta_1$ between $\hat{\xi}_1$ and $b_1$ indicates the edge vs. screw character of the dislocations:

$$\tan\eta_1 = \frac{(1 + \alpha - \cos\theta)}{\sin\theta}. \qquad (2.100)$$

We observe that when $\theta = 0$, i.e. at the exact epitaxial interface, the dislocations are of pure edge type, and that the angle $\eta_1$ decreases rapidly from $90\,°$ at small values of $\theta$. For $\theta = 1.5\,°$ we obtain $\eta = 70\,°$. Similar results are obtained for $\hat{\xi}_2$ and $\eta_2$:

$$\hat{\xi}_2 = \frac{[(-1 - \alpha + \cos\theta - \sin\theta),\, -(1 + \alpha - \cos\theta - \sin\theta),\, 0]}{\sqrt{2(1 + \alpha - \cos\theta)^2 + 2(\sin\theta)^2}}, \qquad (2.101)$$

and $\eta_2 = \eta_1$, so that the dislocations remain orthogonal regardless of $\theta$. Thus, we have shown that the network of $\frac{1}{2}\langle 110 \rangle$ dislocations remains orthogonal as the twist angle is varied and that the network is rotated, initially very rapidly, as the twist angle is increased. The estimation that the network rotates by $20\,°$ when $\theta = 1.5\,°$ compares very well with the experimentally observed rotation of $25\,°$ in view of the extreme sensitivity of $\eta_1$ to both $\alpha$ and $\theta$.

The dislocation spacing $d_1$ is readily found from

$$\frac{b_1}{d_1} = W\hat{n} \times \hat{\xi}_1, \qquad (2.102)$$

which is obtained by setting $p = N_1$ in eqn (2.98). From this we deduce that

$$d_1 = \frac{(1 + \alpha)a^{Pt}}{\sqrt{2(1 + \alpha + \cos\theta)^2 + 2(\sin\theta)^2}}$$

$$= \frac{a^{NiO}a^{Pt}}{2[(a^{Pt})^2 + (a^{NiO})^2 - 2a^{Pt}a^{NiO}\cos\theta)]^{\frac{1}{2}}}. \qquad (2.103)$$

When $\theta = 1.5\,°$ we obtain $d_1 = 4.0\,\text{nm}$, which compares very well with the experimentally observed spacing of $3.8 \pm 0.5\,\text{nm}$. At $\theta = 0$ we obtain $4.3\,\text{nm}$, which is somewhat smaller than the experimentally observed spacing of $5.2 \pm 0.5\,\text{nm}$.

## 2.9.2 Al–Al$_3$Ni eutectic interfaces

Knowles and Goodhew (1983a,b) studied the dislocation structures of interfaces between Al$_3$Ni fibres in an Al matrix by transmission electron microscopy. The specimen was produced by directional solidification. Al$_3$Ni has a complicated orthorhombic structure and the Al matrix has an f.c.c. structure. The [010] axis of the fibres was parallel to the [1$\bar{1}$0] axis of the matrix and the (111) plane in the matrix was parallel to the (102) plane in the fibres. There was pronounced faceting of the fibres and three predominant interface planes, which were high index in both fibre and matrix, were observed. It was found that three sets of dislocations appeared in the interfaces, with spacings and directions that varied with the interface normal $\hat{n}$. It was not possible to determine the Burgers vectors of the dislocations experimentally but it was possible to find three vectors $r_1$, $r_2$, and $r_3$ such that for each interface normal the three sets of observed line directions and spacings satisfied

$$\hat{\xi}_i \text{ parallel to } r_i \times \hat{n}$$

and

$$d_i = \frac{1}{|r_i \times \hat{n}|}. \tag{2.104}$$

The significance of this is realized by comparing these equations with eqns (2.47) and (2.49): $r_i$ is to be identified with $\mathbf{W}^t b_i^*$, which is a basis vector of the reciprocal lattice of the O-lattice (see eqn (2.56)). The three $r_i$ were found to be:

$$r_1 = 0.131 [\cos 63.1°, \cos 31.6°, \cos 74.7°]$$

$$r_2 = 0.625 [\cos 138.7°, \cos 130.8°, \cos 84.8°]$$

$$r_3 = 0.151 [\cos 44.7°, \cos 53.4°, \cos 68.0°] \tag{2.105}$$

in units of $nm^{-1}$, using an axis system parallel to the crystal axes in the Al matrix. The problem now is to determine the transformation matrix $\mathbf{W}$ and the set of three $b_i$ which satisfy the three equations

$$r_i = \mathbf{W}^t b_i^*. \tag{2.106}$$

The transformation matrix $\mathbf{W}$ in these equations describes the deviation in the actual relationship between the crystal lattices from some bicrystal reference structure. The Burgers vectors $b_i$ are defined with respect to a corresponding reference lattice. There are two related questions to be answered: what is the appropriate reference structure that defines $\mathbf{W}$ and what are the appropriate Burgers vectors? In principle there is an infinite number of possible reference structures but in practice there are restrictions that may be placed on the solutions in order for them to make physical sense. Knowles and Goodhew (1983b) made the common assumption of the near CSL model for their selection of possible reference structures as described in Section 2.2. According to this model we must seek $CSL^w$ and $CSL^b$ cells (Fig. 2.10(a)) in the matrix and fibre lattices that are almost coincident at the observed orientation relationship between the matrix and fibre lattices. The strain that is required to make them coincident is then the source of the misfit in the interface that gives rise to the observed dislocations, as was the case in Fig. 2.10(b). It seems reasonable to demand that there is at least one near CSL period between successive dislocations. If this were not the case then it would not make sense to describe the dislocations as localizing the misfit from the near CSL reference structure and, therefore, the reason for the existence of the dislocations would disappear. The Burgers vectors are vectors of the reference DSC lattice and it is possible that they are non-primitive vectors, although we would not expect them to be much larger than primitive DSC lattice vectors for reasons of energetics. This restriction was used by Knowles and Goodhew (1983b), together with the restriction described earlier for the choice of reference state, to eliminate many possible solutions of eqn (2.106).

Eight possible choices of reference structure were considered by Knowles and Goodhew

(1983b) and all of them involved non-primitive DSC lattice vectors for the $b_i$. The most successful choice of reference state was judged by comparing the predicted and observed spacings of dislocations on the three interface facets. The Burgers vectors of the interfacial dislocations were thus found to be $b_1 = \frac{1}{2}[1\bar{2}1]$, $b_2 = \frac{1}{4}[11\bar{2}]$, and $b_3 = [010]$ with respect to the matrix coordinate system. However, it was necessary to use the lattice parameters of Al and $Al_3Ni$ at an elevated temperature in order to obtain satisfactory agreement. It was assumed that the thermal expansion of $Al_3Ni$ was negligible in comparison with that of Al and the room temperature lattice parameters of $Al_3Ni$ were used. The near coincident unit cell was of a relatively small volume, i.e. $12.5(a^m)^3$ and $4abc$ where $a^m$ is the lattice parameter in the matrix and $a,b,c$ are the lattice parameters in the fibre.

The analysis of Knowles and Goodhew raises the general question of whether the matrix **W** can be specified uniquely if we know the Burgers vectors, line directions, and spacings of all dislocations in a particular interface. The answer is *no* because $B = Wp$ may always be replaced by $B = WIp$, where **I** is an invariant plane strain (see eqn (2.32)), which, by definition, leaves all interface vectors $p$ invariant. Knowles and Goodhew overcame this difficulty by analysing the line directions and spacings of dislocations in non-parallel interfaces.

## 2.10 ELASTIC FIELDS OF INTERFACES

### 2.10.1 Introduction

In Section 2.2 we discussed two formulations of the dislocation content of a flat interface that is free of long-range stresses. In Bonnet's formulation the elastic field is modelled by cancelling arrays of stress generator and annihilator dislocations. There is no net dislocation content of the interface. In the Read–Shockley formulation there is a net dislocation content of the interface, which alters the relationship between the crystal lattices from that of some reference structure. We have maintained that the two approaches are equivalent provided there is no stress field far from the interface. In this section we consider the elastic field of such an interface. Each crystal is approximated by an elastic continuum and continuity of displacements and tractions are assumed to be maintained at the interface. The dislocations in the interface are regarded mathematically as source functions which generate the stress field of the interface. In linear elastic theory the field of the interface is given by a linear superposition of the elastic fields of all the dislocations in the interface.

There is no inconsistency in applying the Read–Shockley formulation (see Section 2.2) to the elastic field of a grain boundary. In the absence of externally applied constraints to the bicrystal the distortion field of the dislocations reduces to a pure rotation far from the boundary plane and the strain tensor tends to zero. Since the stress is proportional to the strain it follows that the stress field of the dislocation array should also tend to zero far from the boundary plane. We expect this remark to apply irrespective of whether the elastic constants of the two crystals are the same, or whether isotropic or anisotropic elasticity is assumed. The only condition is that the relationship between the crystal lattices is a pure rotation. In this section we examine whether our expectation is satisfied when linear elasticity is used to model the elastic fields of dislocation arrays in the Read–Shockley formulation. Because we are restricting ourselves to linear elasticity the misorientation must be small. We shall see that our expectation is always satisfied, but in some cases care has to be taken with the boundary conditions far from the interface.

In particular, simply summing the elastic fields of dislocations accommodating a misorientation between crystals of the same structure in anisotropic elasticity, or between crystals with different structures in isotropic or anisotropic elasticity, does not give zero long-range stresses unless care is taken to ensure that no tractions are applied externally to the bicrystal.

The same cannot be said of applying the Read–Shockley formulation to a heterophase interface. The distortion field of the dislocations effects a change in crystal structure, which is not merely a rotation. Since the strain tensor is finite far from the interface it follows that there is a long-range stress field associated with the dislocations. The absence of a long-range stress field in reality can be modelled correctly with interfacial dislocations by using Bonnet's formulation. For example, at an epitaxial interface the uniform state of stress in a thin epilayer (i.e. below the critical thickness) can be modelled by a uniform, continuous distribution of dislocations at the interface. These dislocations are stress generators. Their long-range field is cancelled at large thicknesses by the introduction of discrete stress annihilator dislocations. Thus, both arrays are necessary to describe the stress field of the interface in the absence of long-range stresses.

### 2.10.2 Stress and distortion fields of grain boundaries in isotropic elasticity

In Section 2.7.2 we saw how a general grain boundary may be modelled by three sets of parallel dislocations, with line directions and spacings given by eqns (2.63) and (2.65). In Section 2.7.3 we discussed the types of grain boundary that may be modelled with just one or two independent sets of dislocations. In this section we obtain expressions for the linear isotropic elastic stress fields of these arrays of dislocations. We assume that the dislocation lines are straight. Local dislocation interactions may violate this assumption and it is then necessary to sum the stress fields of dislocation segments. Although Frank's formula prescribes the net Burgers vector density required to produce a particular relative rotation, it is perhaps not obvious that the sum of the elastic distortions produced by an array of dislocations, with the required net Burgers vector density, effects the required rotation. We demonstrate that this is satisfied in the approximation that $2 \sin (\theta/2) \cong \theta$, i.e. in the small-angle regime. Moreover, we demonstrate that long-range stresses when they exist, are consistent with the presence of long-range strains and Hooke's law. The existence of long-range strains implies a change of crystal structure across the interface. To model correctly the absence of long-range stresses at such an interface requires cancelling arrays of stress generator and annihilator dislocations.

Following Hirth and Lothe (1982) we consider an array of straight dislocations parallel to the $z$-axis, and let the boundary normal be along the $x$-axis, as shown in Fig. 2.21. Let the Burgers vector have components $(b_x, b_y, b_z)$. We may obtain the stress field of this array of dislocations by considering the stress fields of three components separately and summing them, as shown schematically in Fig. 2.21.

The first array has $b = (b_x, 0, 0)$ and is shown in Fig. 2.21(a). This is the array expected in a symmetric tilt boundary. For an isolated edge dislocation at the origin with Burgers vector $(b_x, 0, 0)$ the displacement, $(u_x, u_y, 0)$, and stress field, $\tau$, at the point $(x, y, z)$ are

$$u_x = \frac{b_x}{2\pi} \left[ \tan^{-1}\left(\frac{y}{x}\right) + \frac{xy}{2(1-\nu)(x^2+y^2)} \right] \qquad (2.107)$$

$$u_y = -\frac{b_x}{2\pi} \left[ \frac{(1-2\nu)}{4(1-\nu)} \ln(x^2+y^2) + \frac{x^2-y^2}{4(1-\nu)(x^2+y^2)} \right] \qquad (2.108)$$

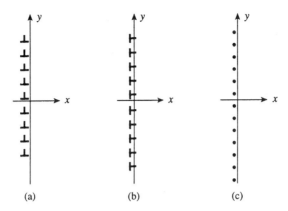

**Fig. 2.21** Schematic illustration of the manner in which the linear elastic stress field of an array of dislocations along $z$ with Burgers vector $(b_x, b_y, b_z)$ is broken down into component arrays with Burgers vector (a) $(b_x, 0, 0)$, (b) $(0, b_y, 0)$ and (c) $(0, 0, b_z)$.

$$\tau_{xx} = -\frac{\mu b_x}{2\pi(1-\nu)}\frac{y(3x^2 + y^2)}{(x^2 + y^2)^2} \tag{2.109}$$

$$\tau_{yy} = \frac{\mu b_x}{2\pi(1-\nu)}\frac{y(x^2 - y^2)}{(x^2 + y^2)^2} \tag{2.110}$$

$$\tau_{xy} = \frac{\mu b_x}{2\pi(1-\nu)}\frac{x(x^2 - y^2)}{(x^2 + y^2)^2} \tag{2.111}$$

$$\tau_{zz} = \nu(\tau_{xx} + \tau_{yy}) = -\frac{\mu b_x \nu}{\pi(1-\nu)}\frac{y}{(x^2 + y^2)} \tag{2.112}$$

$$\tau_{xz} = \tau_{yz} = 0 \tag{2.113}$$

Here $\nu$ is Poisson's ratio. Using eqn (2.109) we deduce that the stress $\tau_{xx}$ due to the array of dislocations shown in Fig. 2.21(a) is given by

$$\tau_{xx} = -\frac{\mu b_x}{2\pi(1-\nu)D}\sum_{n=-\infty}^{\infty}\frac{(3X^2 + (Y-n)^2)(Y-n)}{(X^2 + (Y-n)^2)^2} \tag{2.114}$$

where $X = x/D$, $Y = y/D$, and $D$ is the dislocation spacing. Performing the sum we obtain:

$$\tau_{xx} = -\tau_0 \sin(2\pi Y)(\cosh(2\pi X) - \cos(2\pi Y) + 2\pi X \sinh(2\pi X)) \tag{2.115}$$

where

$$\tau_0 = \frac{\mu b_x}{2D(1-\nu)(\cosh(2\pi X) - \cos 2\pi Y)^2}. \tag{2.116}$$

The other stress components are similarly found using eqns (2.110) and (2.111):

$$\tau_{yy} = -\tau_0 \sin(2\pi Y)(\cosh(2\pi X) - \cos(2\pi Y) - 2\pi X \sinh(2\pi X)) \tag{2.117}$$

$$\tau_{xy} = \tau_0 2\pi X(\cosh(2\pi X)\cos(2\pi Y) - 1). \tag{2.118}$$

At large distances form the boundary, where $x \gg D/2\pi$, the stresses decay exponentially. For example, $\tau_{xx}$ behaves as follows:

$$\tau_{xx} \cong - \frac{\mu b_x \sin(2\pi y/D)}{(1-\nu)D} (2\pi x/D) \exp(-2\pi x/D). \qquad (2.119)$$

Thus, there are no long-range stresses. Indeed, at $x = 2D$ the the stresses have decayed to approximately 1 per cent of the stress from a single dislocation, in agreement with St Venant's principle. In the boundary plane, where $x = 0$, the ratio of the stress, $\tau_{xx}$, at $(0, y)$ to the stress due to a single dislocation at the origin is equal to $(\pi y/D) \cot(\pi y/D)$. Thus at separations of up to $D/2\pi$ from a given dislocation in the array, more than 90 per cent of the stress $\tau_{xx}$ of the array is attributable to the given dislocation.

The second array, Fig. 2.21(b), has $b = (0, b_y, 0)$. The stress fields of this array are given by

$$\tau_{xy} = \tau_0 \sin(2\pi Y)(\cosh(2\pi X) - \cos(2\pi Y) - 2\pi X \sinh(2\pi X)) \qquad (2.120)$$

$$\tau_{xx} = -\tau_0 2\pi X(\cosh(2\pi X)\cos(2\pi Y) - 1) \qquad (2.121)$$

$$\tau_{yy} = \tau_0[2\sinh(2\pi X)(\cosh(2\pi X) - \cos(2\pi Y)) - 2\pi X(\cosh(2\pi X)\cos(2\pi Y) - 1)] \qquad (2.122)$$

As $x \to \pm\infty$ it is seen that $\tau_{xx}$ and $\tau_{xy}$ tend to zero exponentially, but

$$\tau_{yy} \to \frac{\mu b_y}{D(1-\nu)} \operatorname{sgn}(x), \qquad (2.123)$$

where $\operatorname{sgn}(x) = 1$ if $x > 0$ and $\operatorname{sgn}(x) = -1$ if $x < 0$. Therefore, there is a long-range stress field, $\tau_{yy}$, associated with this array of dislocations. This simply means that for a grain boundary to satisfy Frank's formula there must be no net Burgers vector parallel to the boundary plane associated with edge dislocations. For example, for the asymmetric tilt boundary, which can be modelled by two independent arrays of edge dislocations, the components of the Burgers vector densities parallel to the boundary plane cancel, as seen in eqn (2.71). An array of dislocations of the type shown in Fig. 2.21(b) can appear in an epitaxial interface. In that case the stress field far from the interface is cancelled by the stress field of the stress generator array.

The final array, shown in Fig. 2.21(c), is associated with the following stress fields:

$$\tau_{xz} = - \frac{\mu b_z}{2D} \frac{\sin(2\pi Y)}{\cosh(2\pi X) - \cos(2\pi Y)} \qquad (2.124)$$

$$\tau_{yz} = \frac{\mu b_z}{2D} \frac{\sinh(2\pi X)}{\cosh(2\pi X) - \cos(2\pi Y)}. \qquad (2.125)$$

As $x \to \pm\infty$ it is seen that $\tau_{xz}$ decays to zero exponentially but

$$\tau_{yz} \to \frac{\mu b_z}{2D} \operatorname{sgn}(x). \qquad (2.126)$$

To derive these formulae we have used the stress field of a single screw dislocation with Burgers vector $(0, 0, b_z)$:

$$\tau_{xz} = - \frac{\mu b_z}{2\pi} \frac{y}{x^2 + y^2}, \qquad (2.127)$$

$$\tau_{yz} = \frac{\mu b_z}{2\pi} \frac{x}{x^2 + y^2}, \qquad (2.128)$$

$$\tau_{xy} = \tau_{xx} = \tau_{yy} = \tau_{zz} = 0. \qquad (2.129)$$

The displacement field for the isolated screw dislocation is:

$$u_z = -\frac{b_z}{2\pi}\tan^{-1}(x/y); u_x = u_y = 0. \tag{2.130}$$

Equation (2.126) shows that a single array of screw dislocations in a grain boundary is associated with a long-range shear stress $\tau_{yz}$. In a pure twist boundary this is cancelled by another array of screw dislocations along $y$. The combined effect of the two arrays of screw dislocations is to produce a rotation about the boundary normal $x$, and all stress components decay exponentially at large separations from the boundary plane.

The distortion tensor $\mathbf{u}$ is defined by the nine derivatives of the displacements $u_x, u_y, u_z$:

$$\mathbf{u} = \begin{bmatrix} u_{x,x} & u_{x,y} & u_{x,z} \\ u_{y,x} & u_{y,y} & u_{y,z} \\ u_{z,x} & u_{z,y} & u_{z,z} \end{bmatrix}. \tag{2.131}$$

where $u_{x,y} = \partial u_x/\partial y$, etc. The strain tensor components are symmetric combinations of the distortion tensor, e.g.

$$\left.\begin{aligned} \varepsilon_{xy} &= (u_{x,y} + u_{y,x})/2 \\ \varepsilon_{xx} &= u_{x,x} \end{aligned}\right\} \tag{2.132}$$

whereas the antisymmetric components of $\mathbf{u}$ define the rotation vector, $\boldsymbol{\theta}$, at a point:

$$\left.\begin{aligned} \theta_x &= (u_{y,z} - u_{z,y})/2 \\ \theta_y &= (u_{z,x} - u_{x,z})/2 \\ \theta_z &= (u_{x,y} - u_{y,x})/2. \end{aligned}\right\} \tag{2.133}$$

Consider the long-range distortion field of the array of edge dislocations shown in Fig. 2.21(a). Using eqns (2.107) and (2.108) for the displacement fields $u_x$ and $u_y$ we obtain the distortion tensor far from the interface as follows:

$$\mathbf{u}_a = b_x \operatorname{sgn}(x)/2D \begin{bmatrix} 0 & 1 & 0 \\ -1 & 0 & 0 \\ 0 & 0 & 0 \end{bmatrix}. \tag{2.134}$$

Thus the long-range distortion is a relative rotation of the two grains about the $z$-axis of $b_x/D$. This is consistent with Frank's formula because $2\sin(\theta/2) \cong \theta = b_x/D$ in the small-angle limit where linear elastic theory is valid. It is also consistent with the absence of any strains and hence any stresses far from the interface.

The long-range distortion field of the dislocation array shown in Fig. 2.21(b) is given by:

$$\mathbf{u}_b = b_y \operatorname{sgn}(x)/2D \begin{bmatrix} f & 0 & 0 \\ 0 & 1 & 0 \\ 0 & 0 & 0 \end{bmatrix} \tag{2.135}$$

where

$$f = -\nu/(1-\nu) \tag{2.136}$$

This represents a relative tetragonal distortion of $b_y/D$ along the $y$-axis, and a Poisson contraction of $fb_y/D$ along $x$. Hooke's law for an isotropic medium may be stated as follows:

$$\left.\begin{array}{l} \tau_{xx} = 2\mu\varepsilon_{xx} + \lambda e \\ \tau_{yy} = 2\mu\varepsilon_{yy} + \lambda e \\ \tau_{zz} = 2\mu\varepsilon_{zz} + \lambda e \\ \tau_{xy} = 2\mu\varepsilon_{xy} \\ \tau_{yz} = 2\mu\varepsilon_{yz} \\ \tau_{zx} = 2\mu\varepsilon_{zx}. \end{array}\right\} \tag{2.137}$$

where

$$\lambda = 2\nu\mu/(1-2\nu) \tag{2.138}$$

and

$$e = \varepsilon_{xx} + \varepsilon_{yy} + \varepsilon_{zz} \tag{2.139}$$

When the long-range disortions, eqn (2.135), are substituted into Hooke's law, using eqn (2.132), we recover the long-range stress $\tau_{yy}$, given by eqn (2.123), and all other stress components are zero.

The long-range distortion field of the array shown in Fig. 2.21(c) is given by:

$$\mathbf{u}_c = b_z \,\mathrm{sgn}\,(x)/2D \begin{bmatrix} 0 & 0 & 0 \\ 0 & 0 & 1 \\ 0 & 0 & 0 \end{bmatrix}. \tag{2.140}$$

The presence of a long-range shear distortion is consistent with the long-range stress $\tau_{yz}$, given by eqn (2.126).

Any grain boundary composed of sets of straight dislocations can be analysed using the equations we have given for the stress and distortion fields for the three components of the Burgers vector. Although a given set of dislocations may be associated with a long-range stress field the combination of all sets must be free of long-range stresses if the boundary satisfies Frank's formula.

### 2.10.3  Grain boundary energies

Read and Shockley (1950) used dislocation theory to analyse grain boundary energies in the isotropic elastic approximation. In this section we shall derive their celebrated formula for small-angle grain boundaries.

Consider a small-angle symmetric tilt boundary, which is composed of a wall of edge dislocations of spacing $D$. We saw in the previous section that the stress field of each dislocation extends a distance of roughly $D$. Thus, the energy per unit length of each dislocation is approximately given by

$$E_d \cong \frac{\mu b^2}{4\pi(1-\nu)}\ln(D/r_o) + E_c. \tag{2.141}$$

The core radius of the dislocation is $r_o$ and $E_c$ is the energy per unit length of the dislocation core. Expressing $r_o$ as $\alpha b$, where $\alpha$ is a constant of the order of unity for a crystal lattice dislocation, and using $\theta \cong b/D$, the energy per unit area of the boundary becomes

$$\sigma \cong \sigma_o \theta(A - \ln\theta) \tag{2.142}$$

where

$$\sigma_o = \frac{\mu b}{4\pi(1-\nu)} \tag{2.143}$$

$$A = \frac{4\pi(1-\nu)E_c}{\mu b^2} - \ln\alpha. \tag{2.144}$$

The parameter $\sigma_o$ involves only the elastic constants while the parameter $A$ involves unknown quantities about the dislocation cores, namely the core energy and radius.

Although this derivation is rather intuitive it contains the essential physics. We shall follow a derivation presented by Read (1953) which can be readily extended to show that the same form of $\sigma$ can be expected even when the boundary contains several arrays of dislocations. Consider the symmetric tilt boundary again. The bicrystal is divided into parallel strips centred on each dislocation as shown in Fig. 2.22. Let the energy associated with each strip be $E_{\text{strip}}$. This energy can be divided into an elastic strain energy, $E_s$, and a core energy, $E_c$:

$$E_{\text{strip}} = E_s + E_c. \tag{2.145}$$

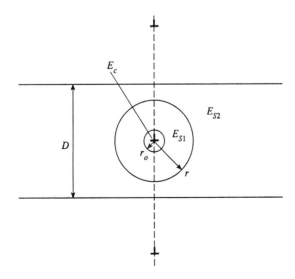

Fig. 2.22 The calculation of grain boundary energy. The array of dislocations, with spacing $D$, is divided into strips of width $D$. The elastic strain energy is divided into a contribution, $E_{s1}$, in a cylindrical region of radius $r$ around the dislocation line, and the energy, $E_{s2}$, in the volume of the strip outside this region. The core energy, $E_c$, is confined to the cylindrical region of radius $r_o$.

The strain energy $E_s$ is further divided into the elastic strain energy, $E_{s1}$, in a cylindrical region around the dislocation line and the energy in the volume of the strip outside this region, $E_{s2}$ (see Fig. 2.22). The radius of the cylindrical region is $r = cD$, where the constant $c < 1$ must be large enough to ensure $cD \gg r_o$ and small enough for the stress field inside the cylinder to be approximately equal to that of the enclosed dislocation alone.

Suppose the boundary misorientation decreases by $d\theta$, leading to increases in $D$ and $r$ given by:

$$-d\theta/\theta = dD/D = dr/r. \qquad (2.146)$$

We now consider the changes in the three contributions to the energy of the strip, $dE_c$, $dE_{s1}$, and $dE_{s2}$. Provided the dislocation spacing $D$ is large enough we can reasonably expect the change in the dislocation core energy, $dE_c$, to be zero. Consider $dE_{s2}$. The volume of the strip increases but the elastic energy density decreases because the dislocations are further apart. We now show that these two effects cancel so that $dE_{s2} = 0$. The elastic energy density varies as the square of the elastic strain. The elastic strain in this region varies as $b/D$. Thus the elastic strain energy in this region is proportional to $1/D^2$. On the other hand, the area of a cylindrical element of area varies as $D^2$. Thus, the elastic strain energy $E_{s2}$ is invariant when $D$ changes: $dE_{s2} = 0$.

Consider $dE_{s1}$. The stress in this region depends only on the included dislocation and therefore the elastic energy density does not change. But the area of the region increases from that of a cylinder of radius $cD$ to a cylinder of radius $c(D + dD)$. The elastic stress field in the increased area is that of the included dislocation. The self-energy of a dislocation is equal to the work done on the slip plane in a virtual process in which the dislocation is created by introducing a relative displacement equal to $b$ across a cut on the slip plane. The work is done against the shear stress acting on the slip plane and in the slip direction. The increase in the radius of the cylinder by $dr = cdD$ increases the area of the slip plane by $dr$ per unit length of dislocation. During the virtual process the stress at any point rises from its initial value of zero to its final value of $\tau_o b/r$, so that the work done on the part of the slip plane $dr$ is $\frac{1}{2}\tau_o b^2 dr/r$ of creating the dislocation. $\tau_o$ is uniquely determined by the elastic constants of the material and this expression for the stress is valid provided $r \gg b$ and it remains valid even in anisotropic elasticity. Therefore,

$$dE_{\text{strip}} = dE_{s1} = \tfrac{1}{2}\tau_o b^2 \, dr/r = -\tfrac{1}{2}\tau_o b^2 \, d\theta/\theta. \qquad (2.147)$$

Integration of this expression results in the following:

$$E_{\text{strip}} = \tfrac{1}{2}\tau_o b^2 (A - \ln\theta). \qquad (2.148)$$

$A$ is an integration constant. The energy per unit area of the boundary is $E_{\text{strip}}/D$, or

$$\sigma = \tfrac{1}{2}\tau_o b\theta (A - \ln\theta), \qquad (2.149)$$

which is identical to eqn (2.142).

If the boundary contains more than one array of dislocations we have to take account of the interactions between the arrays. But the above analysis shows that the interaction energy can contribute only to the elastic strain energy terms $E_{s1}$ for each array. The above argument may thus be repeated for each array separately, and the interaction energy absorbed into the constants of integration. The result is that $\sigma$ has the form of

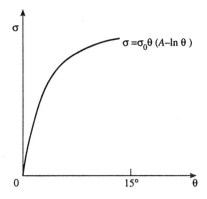

$\sigma = \sigma_0 \theta (A - \ln \theta)$

**Fig. 2.23** A sketch of the energy, $\sigma$, of a small–angle grain boundary as a function of the misorientation, $\theta$, according to eqn (2.142).

eqn (2.142) for each array and therefore the resultant boundary energy still has the form of eqn (2.142). We conclude, therefore, that the predicted variation of $\sigma$ with $\theta$ has the same form for all small-angle grain boundaries. This form is sketched in Fig. 2.23. It is seen that as $\theta \to 0$ the boundary energy shows a cusp.

The derivation assumes that the dislocations are uniformly spaced, but this is possible only when the dislocation spacing corresponds to some crystal repeat distance. Thus, uniformly spaced dislocations occur only at particular misorientations. For intermediate angles the irregularities in the dislocation spacings introduce additional terms in the energy. For a small deviation $\delta\theta$ from the nearest rational angle $\theta = 1/m$ (for rotations about $\langle 001 \rangle$ axes in cubic crystals), the extra energy is of order

$$\frac{-\sigma_0 \delta\theta}{m} \ln \delta\theta. \tag{2.150}$$

As $\delta\theta \to 0$ the slope of the $\sigma$ versus $\theta$ curve becomes infinite and the rational angles $\theta = 1/m$ correspond to cusps in the true $\sigma$ versus $\theta$ curve. Thus we obtain the $\sigma$ versus $\theta$ curve that is sketched in Fig. 2.24. The broken line represents eqn (2.142), and at each rational orientation there is small cusp corresponding to the term described by (2.150). The 'strength' of these minor cusps is proportional to $1/m$, and thus we expect deeper cusps at boundaries with smaller dislocation spacings. However, as the dislocation spacing decreases the whole theory becomes doubtful.

The energy term (2.150) may be interpreted as the energy arising from an additional array of dislocations with Burgers vector $b/m$, superimposed on an array of uniformly spaced dislocations with Burgers vector $b$. In this picture the non-uniformities in the spacings of the dislocations near $\theta = 1/m$ are modelled by new dislocations with Burgers vector $b/m$ superimposed on the boundary of orientation $\theta = 1/m$. We recognize the new dislocations as secondary dislocations of the $\theta = 1/m$ reference state. Thus, secondary dislocations are equivalent to non-uniformities in the spacings of primary dislocations. Tertiary dislocations are equivalent to non-uniformities in the spacings of secondary dislocations, and so on. We call the cusps caused by primary and secondary dislocations primary and secondary cusps.

The agreement between the predicted $\sigma$ versus $\theta$ curve, eqn (2.142), for primary cusps, i.e. small-angle grain boundaries, and experimental results is very good, as seen in Fig. 5.20. The agreement is in some cases too good since it extends to quite large angles where the linear elastic theory must break down. The existence of secondary cusps at

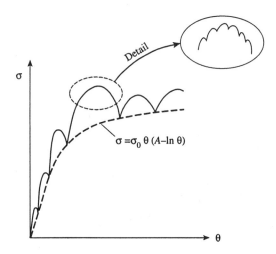

**Fig. 2.24** A sketch of the energy, $\sigma$, of a small—angle grain boundary as a function of the misorientation, $\theta$, according to the Read–Shockley analysis. Primary cusps occur at those orientations where dislocations are uniformly spaced, where the energy is given by the Read–Shockley formula, eqn (2.142), shown by the dashed line. Further, secondary, cusps exist on the primary cusps and so on.

rational grain boundary orientations has been much more controversial, and was reviewed by Sutton and Balluffi (1987). These cusps are often very shallow and may disappear altogether at finite temperatures. Entropic contributions to the boundary free energy are ignored in the Read–Shockley analysis. Their chief effect is to delocalize secondary dislocation cores and thereby weaken the elastic fields of the dislocations. We shall return to this subject in Section 2.13.3.

An interesting *empirical* observation has been made by Wolf (1989) concerning a Read–Shockley type formula for large-angle grain boundaries. Wolf carried out computer simulations of relaxed grain boundary energies at 0 K for several series of boundaries. In each series the mean boundary plane was constant and either the tilt or the twist angle was varied systematically. Thus, in each series only one of the five macroscopic degrees of freedom was varied. Certain features of the $\sigma$ versus $\theta$ curves were found to be common to all the curves: there were no major cusps except at the endpoints of the $\sigma(\theta)$ curve, and $\sigma(\theta)$ was found to be a smooth function of $\theta$, with vanishing slope in the middle of the misorientation range between the cusps at either end (see Fig. 2.25). Wolf found that the Read–Shockley formula with $\theta$ replaced by $\sin\theta$ could be fitted to all the calculated $\sigma$ versus $\theta$ curves very well throughout the entire misorientation range:

$$\sigma(\theta) = \sigma_o \sin\theta (A - \ln(\sin\theta)). \tag{2.151}$$

The parameters $\sigma_0$ and $A$ were varied to give the best fit to the calculated values of $\sigma(\theta)$. However, it was also necessary to take account of crystal symmetry, if the rotation axis were a symmetry axis, by suitable scaling of $\theta$, and also to adjust the endpoints of the fitted $\sigma(\theta)$ curve if they corresponded to grain boundaries rather than the perfect crystal.

There does not appear to be any theoretical justification that can be given for eqn (2.151). Frank's formula indicates that at large angles $\theta$ should be replaced by $2\sin(\theta/2)$ and not $\sin\theta$, but then $\sigma(\theta)$ would not have the required property of vanishing

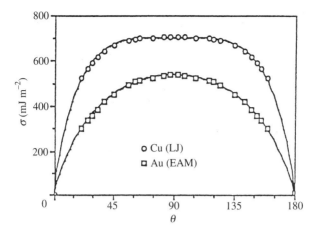

**Fig. 2.25** Computed energies, $\sigma$, of (001) twist grain boundaries in Cu and Au using Lennard–Jones and embedded atom potentials respectively. The misorientation, $\theta$, has been multiplied by 2. (From Wolf (1989).)

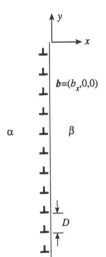

**Fig. 2.26** Schematic illustration of a symmetric tilt boundary comprising an array of edge dislocations with Burgers vector $b = (b_x, 0, 0)$, and spacing $D$, between two phases $\alpha$ and $\beta$.

slope at $\theta = 90°$. The usefulness of eqn (2.151) is that it suggests that large-angle grain boundary energies may be determined from the energies of small-angle boundaries by an *empirical* extrapolation. Further discussion of grain boundary energies in the framework of the structural unit model is given in Section 4.3.1.8.

### 2.10.4 Stress fields of heterophase interfaces in isotropic elasticity

Consider an array of edge dislocations forming a flat, symmetric tilt boundary between two phases $\alpha$ and $\beta$, as shown in Fig. 2.26. The array of dislocations produces an elastic field which we shall consider in this section in the isotropic elastic approximation. This field is in addition to any other strain field that existed in the bicrystal prior to the introduction of the dislocation array. As far as the Frank–Bilby equation is concerned the interface is effectively a grain boundary because the dislocations accommodate a misorientation between the two crystals. Indeed, the application of Frank's formula to the array of edge dislocations is exactly the same as for a symmetric tilt grain boundary. The misorientation between the crystals should be independent of their elastic constants because it depends only on the magnitude, $b_x$, of the Burgers vector of the dislocations and on their spacing, $D$. Thus, the misorientation produced by the dislocation array is given by $\theta_z \cong 2 \sin(\theta_z/2) = b_x/D$. This is the misorientation we expect to obtain from the elastic field of the dislocation array, and it is perhaps not obvious how this result is independent of the elastic constants.

In order for the elastic distortion field of the dislocation array to satisfy Frank's formula it is necessary that there are no net surface tractions far from the interface. Otherwise the net surface tractions will have to be balanced by externally applied forces to maintain equilibrium. Such applied loads will lead to distortions, in addition to those prescribed by Frank's formula. It is perhaps surprising that this rather obvious condition has sometimes been overlooked and consequently a great deal of confusion has resulted. For example, Chou and Lin (1975) found that simply summing the stress fields of the dislocations comprising the array produced finite stresses infinitely far from the interface. They concluded (erroneously) that such tilt walls could not be mechanically stable in heterophase interfaces. Hirth *et al.* (1979) showed that their result was consistent with Frank's formula being violated by the elastic field and showed how an elastic field consistent with Frank's formula could be obtained. It follows that symmetric tilt walls of dislocations are indeed mechanically stable in heterophase interfaces. We shall follow the analysis of Hirth *et al.* (1979).

We begin with the formulae derived by Nakahara *et al.* (1972) for the displacement field of a single edge dislocation with Burgers vector $[b_x, 0, 0]$ in an $\alpha/\beta$ interface. These formulae are the heterophase analogues of eqns (2.107) and (2.108). They are:

$$u_x^\alpha = \frac{\mu^\beta K^{\beta\alpha} b_x}{\pi \mu^\alpha} \frac{xy}{x^2 + y^2} + \frac{b_x(1 + K^{\alpha\beta} - K^{\beta\alpha})}{2\pi} \tan^{-1}(y/x) \qquad (2.152)$$

$$u_y^\alpha = \frac{(K^{\alpha\beta} + K^{\beta\alpha} - 1)b_x}{4\pi} \ln(x^2 + y^2) + \frac{\mu^\beta K^{\beta\alpha} b_x}{2\pi\mu^\alpha} \frac{y^2 - x^2}{x^2 + y^2} \qquad (2.153)$$

where

$$K^{\alpha\beta} = \mu^\beta [\mu^\beta + \kappa^\beta \mu^\alpha]^{-1}, \qquad (2.154)$$

and

$$\kappa^\alpha = 3 - 4\nu^\alpha. \qquad (2.155)$$

Here Poisson's ratio has been denoted by $\nu$. The corresponding formulae in the $\beta$ phase are obtained by interchanging the $\alpha$ and $\beta$ superscripts. Using these formulae we can calculate the rotations far from the interface of the $\alpha$ and $\beta$ phases about the tilt axis, $z$. We obtain the following rotation of the $\alpha$ phase relative to the $\beta$ phase:

$$\theta_z = \frac{b_x}{D} \left[ 1 + \tfrac{1}{2}(\mu^\beta - \mu^\alpha)\{K^{\beta\alpha}/\mu^\alpha - K^{\alpha\beta}/\mu^\beta\} \right]. \qquad (2.156)$$

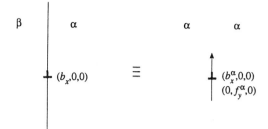

Fig. 2.27 A dislocation with Burgers vector $(b_x, 0, 0)$ between two phases $\alpha$ and $\beta$ is equivalent to a dislocation with Burgers vector $(b_x^\alpha, 0, 0)$ and a line force $(0, f_y^\alpha, 0)$ in an infinite medium of the $\alpha$ phase.

It is clear that Frank's formula is not satisfied unless $\mu^\beta = \mu^\alpha$. Using eqns (2.152)–(2.153) it can be shown that the long-range stresses $\tau_{xx}$ and $\tau_{yy}$ arising from the array of edge dislocations are zero in both phases. However, the long-range stress $\tau_{xy}$ tends to the following value at long range:

$$\tau_{xy}^\alpha = \tau_{xy}^\beta = \frac{b_x}{D}\left[\mu^\alpha K^{\alpha\beta} - \mu^\beta K^{\beta\alpha}\right]. \tag{2.157}$$

The key to understanding the origin of this long-range stress field is a result due to Dundurs and Sendeckyj (1965). They noted the existence of net tractions in each half-crystal associated with an interfacial dislocation. The tractions are equal in magnitude and opposite in sign in the two half crystals, giving no net force on the dislocation. This means that in each half-space the field of an interfacial dislocation may be represented by that of an infinite medium dislocation (i.e. as though the other half-space had the same elastic constants) that is coincident with a line of force, as shown schematically in Fig. 2.27. More precisely, the elastic field in the $\alpha$ phase of an isolated interfacial dislocation with Burgers vector $(b_x, 0, 0)$ is equivalent to that of an infinite medium dislocation with Burgers vector $(b_x^\alpha, 0, 0)$ and an infinite medium line force $(0, f_y^\alpha, 0)$ given by

$$b_x^\alpha = b_x[1 - K^{\beta\alpha} + K^{\alpha\beta}] \tag{2.158}$$

$$f_y^\alpha = 2b_x[\mu^\beta K^{\beta\alpha} - \mu^\alpha K^{\alpha\beta}]. \tag{2.159}$$

In the $\beta$ phase the Burgers vector $(b_x^\beta, 0, 0)$ and line force $(0, f_y^\beta, 0)$ describing the elastic field are given by

$$b_x^\beta = b_x[1 - K^{\alpha\beta} + K^{\beta\alpha}] \tag{2.160}$$

$$f_y^\beta = -f_y^\alpha. \tag{2.161}$$

In the $\alpha$-phase the displacements arising from the line of force, given by eqn (2.159), are as follows:

$$u_x^\alpha = \frac{f_y^\alpha}{2\pi\mu^\alpha(\kappa^\alpha + 1)}\frac{xy}{x^2 + y^2} \tag{2.162}$$

$$u_y^\alpha = \frac{-f_y^\alpha}{4\pi\mu^\alpha(\kappa^\alpha + 1)}\left[\kappa^\alpha \ln(x^2 + y^2) + \frac{x^2 - y^2}{x^2 + y^2}\right]. \tag{2.163}$$

When these displacements are added to those of the infinite medium dislocation, given by eqns (2.107)–(2.108) with Burgers vector given by eqn (2.158), we obtain the displacements of the interfacial dislocation given by eqns (2.152)–(2.153). Thus, we have proved the equivalence of the field of an interfacial dislocation to the field of an infinite medium dislocation and an infinite medium line of force.

It has already been shown in eqns (2.114)–(2.119) that an array of infinite medium dislocations, forming a symmetric tilt boundary, has no long-range stress field. Therefore the source of the long-range stress field found in eqn (2.157) must be the lines of force at the interface. The net tractions on a cut normal to the $x$-axis are readily evaluated using eqns (2.162)–(2.163) and found to be $-\frac{1}{2}f_y^\alpha$ in the $\alpha$ phase ($x > 0$) and $\frac{1}{2}f_y^\alpha$ in the $\beta$ phase ($x < 0$). These tractions are independent of the position of the cut plane along $x$ in each phase. The shear stress, $\tau_{xy}^\alpha$, they generate is equal to $-\frac{1}{2}f_y^\alpha/D$ which is exactly equal to the long range shear stress of the array of interfacial dislocations given in eqn (2.157).

In a finite bicrystal mechanical equilibrium requires that the long-range shear stress $\tau_{xy}^\alpha$ is balanced by externally applied forces. In other words the bicrystal is *not* in equilibrium unless forces are applied to it externally. This is to be expected because Frank's formula is not satisfied, as seen in eqn (2.156). With no constraints applied to the surface of a finite bicrystal the long-range shear stress $\tau_{xy}^\alpha$ is cancelled by an image stress $-\tau_{xy}^\alpha$. The image stress produces uniform simple shears, parallel to the interface, of $-\tau_{xy}^\alpha/\mu^\alpha$ and $-\tau_{xy}^\alpha/\mu^\beta$, in the $\alpha$ and $\beta$ phases respectively. These simple shears introduce a change of orientation of the $\alpha$ phase relative to the $\beta$ phase equal to:

$$\tau_{xy}^\alpha \left[ \frac{1}{2\mu^\alpha} - \frac{1}{2\mu^\beta} \right] = \frac{b_x}{2D} (\mu^\beta - \mu^\alpha) \left[ \frac{K^{\alpha\beta}}{\mu^\beta} - \frac{K^{\beta\alpha}}{\mu^\alpha} \right] \tag{2.164}$$

Adding this to $\theta_z$ in eqn (2.156) we see that elimination of the long-range shear stress field results in Frank's formula being satisfied. In this final relaxed state the elastic field of the dislocation array in the $\alpha$ phase may be thought of as a superposition of three fields: (i) the field of an array of infinite medium dislocations with effective Burgers vector $(b_x^\alpha, 0, 0)$, (ii) the field of an array of infinite medium line forces with forces per unit length $(0, f_y^\alpha, 0)$, (iii) a uniform simple shear parallel to the interface of $-\tau_{xy}^\alpha/\mu^\alpha$ to cancel the long-range shear stress generated by the array of line forces. The field in the $\beta$ phase is the same except that quantities associated with the $\alpha$ phase are replaced by those associated with the $\beta$ phase. It is interesting to note that the effective Burgers vectors $b_x^\alpha$ and $b_x^\beta$ depend on the elastic constants of the two phases. Therefore the rotations of the two phases are not equal in magnitude, in contrast to the grain boundary case, although their relative rotation is the same as in the grain boundary case.

To *summarize*, the elastic field of the heterophase interface does satisfy Frank's formula when the adjoining crystals are free of net surface tractions. This is physically sensible because we can always imagine that the $\alpha/\beta$ interface is created by bonding together the surfaces of two stress free $\alpha$ and $\beta$ phases with a relative misorientation of $\theta = b_x/D$. Then, according to St Venant's principle the stress field of the interface should decay over a distance comparable to $D$ from the interface. The relaxation of the long-range stresses does not introduce any further incompatibilities into the interface. This may be seen in two ways. First, the average effective Burgers vector, $(b_x^\alpha + b_x^\beta)/2$, is equal to $b_x$. Secondly, the long-range shear stresses are relieved by uniform simple shear strains parallel to the interface, leaving the interface invariant. These two statements hold also in the anisotropic elastic case.

Finally, we note that the difficulty we encountered with long-range tractions far from the interface does not arise if cancelling arrays of stress generator and annihilator dislocations are used to model the elastic field of the interface.

### 2.10.5 Dislocation arrays at interfaces in anisotropic elasticity

Although the analysis is more involved it can be shown (Hirth *et al.* 1979) that the conclusions of the previous section apply in the anisotropic elastic case as well. The anisotropic elastic field of an interface dislocation in an infinite bicrystal may be represented by a sum of two fields in each half-space. The first is that of a dislocation with an effective Burgers vector in an infinite medium of the half-space. There are no net surface tractions on any plane associated with an infinite medium dislocation. The second is a line of force coinciding with the dislocation. Barnett and Lothe (1974) generalized the result of Dundurs and Sendeckyj (1965) to give an expression for the line force in anisotropic elasticity. There are net surface tractions acting on planes parallel to the interface due to these lines of force at the interface. Furthermore, Frank's formula is not satisfied by the interface so long as the net surface tractions are not relaxed. In a finite bicrystal with no surface constraints the surface tractions of the dislocation array are cancelled by image stresses. The strains associated with these image stresses leave the interface invariant and do not, therefore, introduce any further incompatibilities into the interface. In the final relaxed state of the finite unconstrained bicrystal Frank's formula is satisfied and there are no net tractions acting any plane parallel to the interface.

Alternatively, and arguably more conveniently, the anisotropic elastic field can always be modelled by cancelling arrays of stress generator and annihilator dislocations. The net tractions produced by the two arrays cancel automatically and the condition of no stresses far from the interface is guaranteed.

### 2.10.6 Isotropic elastic analysis of epitaxial interfaces

Frank and van der Merwe (1949) presented an elastic stress analysis of an epilayer grown on a flat substrate with a slightly different lattice parameter. Their one-dimensional analysis was extended to two dimensions by Jesser and Kuhlmann-Wilsdorf (1967). The essence of the analysis is to demonstrate that once the thickness of the epilayer exceeds a certain critical value it becomes energetically favourable to introduce crystal lattice stress annihilator dislocations at the interface, and thereby relieve the elastic strain energy.

For simplicity we consider cubic lattices for the substrate and epilayer, with lattice parameters $a^s$ and $a^e$. The interface is assumed to be on a (001) plane in each crystal. The 'misfit' is defined as

$$f = (a^e - a^s)/a^s \tag{2.165}$$

and represents the elastic strain in the [100] and [010] directions in the interface if the two lattices are in perfect registry. In this elastically strained state the interface contains an array of coherency dislocations and is fully coherent, as in Fig. 2.6b. This condition is sometimes described as 'commensurate' or a 'state of forced elastic coherence'. The epilayer has a free surface and the surface tractions must be zero. Since the epilayer is much thinner than the substrate we assume that the strain is all taken up in the epilayer. The epilayer is thus in a state of biaxial stress, which in isotropic elasticity is given by

$$\sigma = 2\mu \frac{(1 + \nu)}{(1 - \nu)} f \tag{2.166}$$

where $\mu$ and $\nu$ are the shear modulus and Poisson ratio in the epilayer. The biaxial stress state may therefore be modelled by two uniform, continuous distributions of stress generator, coherency dislocations with Burgers vector densities equal to [$f$00] and [0$f$0]. The elastic strain energy of the epilayer, per unit area, is

$$E_{el} = 2\mu \frac{(1 + \nu)}{(1 - \nu)} f^2 h,$$    (2.167)

where $h$ is the film thickness. Since this increases linearly with $h$ there must be a critical thickness at which it becomes energetically favourable to introduce anticoherency stress annihilator dislocations. To find this critical thickness we first calculate the energy of the interface when a fraction of the misfit is relieved by stress annihilator dislocations. We assume the stress annihilator dislocations form a square array, with Burgers vector $b$ and spacing $D$. The remaining elastic strain $\varepsilon$ is given by

$$\varepsilon = f - b/D.$$    (2.168)

The energy of the dislocation array, per unit area, is

$$E_d = 2 \times \frac{\mu b^2}{4\pi (1 - \nu)D} \ln \left[ \frac{eh}{r_o} \right]$$    (2.169)

where $r_o$ is the core radius of the stress annihilator dislocations, and the factor of 2 arises from the existence of two orthogonal sets of edge dislocations. By putting $h$ inside the logarithm of eqn (2.169) we have assumed that the film thickness is less than the spacing of the stress annihilators. The total energy, per unit area, of the film is obtained by adding $E_d$ to $E_{el}$ with $\varepsilon$ in place of $f$ in eqn (2.167):

$$E = 2\mu \frac{(1 + \nu)}{(1 - \nu)} \varepsilon^2 h + \frac{\mu b (f - \varepsilon)}{2\pi (1 - \nu)} \ln \left[ \frac{eh}{r_o} \right].$$    (2.170)

For a given film thickness the total energy is minimized for a value of $\varepsilon = \varepsilon_o$, given by

$$\varepsilon_o = \frac{b}{8\pi (1 + \nu)h} \ln \left[ \frac{eh}{r_o} \right].$$    (2.171)

If $\varepsilon_o$ is larger than $f$ then all the misfit is accommodated elastically with no stress annihilators at the interface. If $\varepsilon_o$ is less than $f$ then some of the misfit is relaxed by stress annihilators, the spacing of which is given by

$$D = b/(f - \varepsilon_o).$$    (2.172)

The critical film thickness, $h_c$, at which it is energetically favourable for the first stress annihilator to be introduced is obtained by setting $\varepsilon_o$ equal to $f$:

$$h_c = \frac{b}{8\pi (1 + \nu)f} \ln \left[ \frac{eh_c}{r_o} \right].$$    (2.173)

This is an implicit relation for $h_c$. We note that it depends on the value assumed for the core radius, $r_o$, and the Burgers vector, $b$, of the stress annihilators. There is thus some uncertainty about the value of $h_c$. But perhaps the most important limitation of this analysis is that it has ignored the question of where the dislocations come from. Matthews and Blakeslee (1974) pointed out that in order to introduce dislocations there has to be a mechanism for doing so. Thus, the Frank–van der Merwe criterion is a necessary but not sufficient condition for stress annihilators to be introduced. For all these reasons comparison of eqn (2.173) with experimental measurements is problematic. Observed values of $h_c$ are generally larger than those of eqn (2.173), partly because of the insensitivity of the experimental techniques employed and partly for kinetic reasons.

Matthews and Blakeslee (1974) developed a somewhat different criterion for the introduction of stress annihilator dislocations at the interface. They suggested that the dislocations are generated by the glide of existing dislocations which 'thread' through the substrate and epilayer. Other mechanisms have been proposed in recent years: see Hirsch (1991) for a review. In order to move a threading dislocation the force on it has to be sufficient to overcome the line tension of the interfacial dislocations. This criterion again leads to a critical epilayer thickness at which stress annihilators appear, and Willis *et al.* (1990) have shown that, at equilibrium, it is identical to the critical thickness predicted by Frank and van der Merwe. Willis *et al.* argue that apparent differences between the two treatments are due entirely to the use of different approximations to the energy of a single dislocation. Furthermore, Jain *et al.* (1992) criticize both the Frank–van der Merwe and Matthews–Blakeslee analyses because they consider $h_c$ to be determined by the stress at which a *single* stress annihilator dislocation is introduced into the coherent interface. Jain *et al.* argue instead that if an *array* of stress annihilator dislocations is introduced at the transition from the coherent to the semicoherent states not only is a lower energy equilibrium state achieved but also the criterion for $h_c$ is altered qualitatively. In practice, however, the new criterion does not predict a value of $h_c$ which is very different from the values predicted by the Frank–van der Merwe and Matthew–Blakeslee analyses. The validity of the criticism from a theoretical point of view depends on whether stress annihilator dislocations enter the coherent interface one at a time or in unison in the form of an array. This is very difficult to establish experimentally. But as theory of the *equilibrium state* of the interface the treatment by Jain *et al.* is a significant improvement on the earlier models.

### 2.10.7 Stress fields of precipitates and non-planar interfaces

Eshelby (1957) considered the elastic fields of ellipsoidal precipitates. He noted that the elastic field could be modelled by thinking of the interface surrounding the precipitate as a Somigliana dislocation (see Section 1.7). Although this observation was made in 1957 it is only in recent years that it has been exploited. Bonnet *et al.* (1985) have given elastic solutions for Somigliana dislocations that are particularly useful in modelling the elastic fields of faceted precipitates. As noted by those authors it allows morphologies of precipitates other than ellipsoidal to be treated with relative ease, including precipitates with corners such as cubes. For example, for the cubic morphology the elastic field is calculated by summing the separate contributions of the six faces of the cube, each of which is treated as a Somigliana dislocation defined by a square cut. If the relaxation at each face of the cube gives rise to a rigid body translation of the matrix with respect to the precipitate then there will be Volterra dislocations at the twelve edges of the cube to account for the changes in translation vector between adjoining facets. The elastic fields of these dislocations can then be added to those of the Somigliana dislocations to get the total elastic field of the precipitate. As Bonnet (1988) has demonstrated the elastic fields of quite complex, non-planar heterophase interfaces may be modelled by using appropriate distributions of Somigliana and Volterra dislocations. However, such an analysis is dependent on detailed information from high-resolution electron microscopy of the strain at the interface (see Bonnet (1988)). Specifically, the distribution of dislocations and their degree of localization is essential to formulate the elastic problem.

## 2.11  DEGREE OF LOCALIZATION OF THE CORES OF INTERFACIAL DISLOCATIONS

### 2.11.1  Introduction

Until now we have assumed that discrete dislocations in an interface are line defects with Burgers vector densities that are localized within the plane of the interface. But any dislocation has a tendency to delocalize the distribution of its Burgers vector density in order to reduce the energy of its elastic strain field. On the other hand this spreading of the dislocation core will generally disrupt the structure of the surrounding material and so tend to raise the energy of the system. A dislocation will therefore delocalize to the point where these two opposing tendencies are in balance. In addition, the analyses of the elastic fields of interfacial dislocations discussed in Section 2.10 were based on the assumption of a linear elastic continuum for both the interface and the adjoining crystals. In this approximation the stress and strain become singular at the core of each dislocation. But in reality these singularities are relieved by relaxations within the core because at smaller length scales the discrete atomic structure of the medium cannot be ignored, and the approximation of linear elasticity breaks down. Important questions to raise at this point are therefore the following: (i) to what extent are the cores of interfacial dislocations localized? and (ii) how are their stress–strain fields affected by the degree of localization?

If a perfect interfacial dislocation has a Burgers vector that is a non-primitive lattice vector of the reference lattice it may be able to reduce its elastic energy by *dissociating* into two, or more, interfacial dislocations with smaller (primitive) Burgers vectors. This phenomenon is described in Sections 9.2.2.2 and 12.4.3.1. Of course it may be regarded as a form of core delocalization. However, we are not concerned with this phenomenon here: instead, we are concerned with the degree of localization of an interfacial dislocation which is unable to dissociate into discrete dislocations with finite Burgers vectors of the reference crystal or DSC lattices.

For such a dislocation embedded in an interface it is easily seen that the Burgers vector density parallel to the interface may delocalize preferentially in the plane of the interface, if the interface has a relatively low resistance to a change in the translational disposition, parallel to the interface, of the two adjoining crystals. In such a case this portion of the Burgers vector density will have a tendency to spread out in the interface, thereby shearing the interface locally. The resistance of the interface to this disruptive shearing process, averaged over the whole interfacial area, is measured by the slope of the $\gamma$-surface, which is the surface representing the ground state energy of the interface as a function of an imposed, rigid-body translation, $t$, parallel to the interface (Vitek 1968). As discussed in Section 4.3.1.1, when the interface has a periodic structure the $\gamma$-surface is a periodic function with the periodicity of the c.n.i.d. (defined in Section 1.5.7). Therefore, the $\gamma$-surface of a periodic interface has maxima and minima in general, and the interface will then encounter some resistance to a change in its translational state from the ground state configuration. It follows that the cores of dislocations lying in such an interface are localized in directions parallel to the interface to some extent at least. The degree of localization depends on the slope of the $\gamma$-surface.

As the c.n.i.d. decreases in size the $\gamma$-surface tends to become flatter, as shown in Section 4.3.1.1. In the limit when the area of the c.n.i.d. is reduced to zero, as happens in the case of a quasiperiodic interface, the $\gamma$-surface becomes flat. In that case there is no resistance on average to changes in the translational state of the interface, in the ground state. Therefore, in the ground state, a dislocation inserted into such an interface will tend to undergo complete delocalization of its Burgers vector density parallel to the interface plane. In that case, the stress and strain fields due to the component of the Burgers vector density parallel to the interface will vanish. However, it is emphasized that the γ-surface measures the *average* resistance to a change of the rigid-body translation of one crystal relative to the other parallel to the interface. The *local* resistance may vary considerably from one region of the interface to another, such that the average resistance is zero. Therefore there may be an activation barrier for the core delocalization to take place, with the result that localized dislocation cores may persist until sufficient thermal activation has been supplied. A discussion of such delocalization processes at elevated temperatures is given in section 12.4.3.2.

The degree of localization of the Burgers vector density normal to the boundary plane is determined by the cohesive forces acting across the interface plane. Provided the interface is stable with respect to cleavage, as is generally the case, there will always be a force tending to localize this Burgers vector density irrespective of the form of the γ-surface.

On the basis of the above, we expect wide variations in the degree to which the cores of interfacial dislocations are delocalized in the ground state. For widely spaced crystal lattice dislocations in small-angle grain boundaries the situation should be similar to that for isolated lattice dislocations. However, as the spacing is reduced, the situation will change as discussed below in Section 2.11.2. For dislocations in large-angle boundaries large variations will occur, depending on the presence, or absence, of periodicity and the form of the γ-surface, and kinetic factors which control the extent to which the interface attains the ground state. These variations will be associated with corresponding variations in the surrounding stress and strain fields.

We have already argued that the localization of the Burgers vector density parallel to the interface may decrease as the temperature is raised at a quasiperiodic interface. Even at a periodic interface the degree of core localization may vary with temperature because the γ-surface is expected to vary with temperature. At a finite temperature it is the free energy of the interface, rather than the internal energy, which determines the γ-surface. As discussed in Section 3.9 the vibrational entropy of the boundary leads, effectively, to a softening of the interatomic forces acting in the material. In addition the enhanced anharmonicity of the atomic environment of a grain boundary in metals, compared with the bulk, leads to an increased thermal expansion which further weakens the cohesive forces across the boundary plane (see Section 4.3.1.10). In that case the ground state of the interface may change with temperature from one which supports localized Burgers vector densities parallel to the interface to one that does not. However, experimental evidence, cited in Section 2.12.2, indicates that for at least a range of grain boundaries in Al, the change in the γ-surface with temperature is not sufficient to lead to significant delocalization.

In the following sections we describe the results of a number of calculations of the detailed core structure of interfacial dislocations at 0 K. They range from results obtained by analytical methods using simple lattice theories (Section 2.11.2) to those obtained by computer simulation using interatomic force models (Section 2.11.3).

## 2.11.2 Lattice theories of dislocation arrays

### 2.11.2.1 *Introduction*

In this section we describe analytic models in which the dislocation core region is treated discretely using simple assumed force laws. These include the Peierls–Nabarro (Peierls

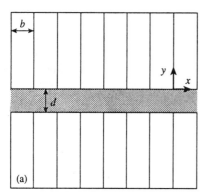 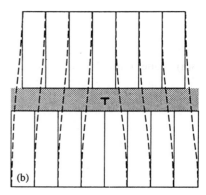

**Fig. 2.28** To illustrate the Peierls–Nabarro model for an edge dislocation. In (a) we see two semi-infinite crystals, modelled as elastic continua, separated by an inelastic slab of width $d$. A dislocation with Burgers vector $b$, equal to the spacing of the lattice planes in (a), is introduced into the inelastic slab in (b).

1940; Nabarro 1947; Hirth and Lothe 1982) and van der Merwe (1950) models. These more realistic models remove the elastic singularity at the core and provide information about the degree of localization of the core. In addition, they yield expressions for the surrounding stress and strain fields which differ somewhat from those obtained using the elastic continuum approximation.

The idea underlying the Peierls–Nabarro and van der Merwe models is that the core is allowed to spread within an inelastic slab, as sketched in Fig. 2.28 for an edge dislocation. The inelastic slab is bonded to elastic continua on either side of it. Far from the dislocation core the two faces of the slab are displaced relative to one another by the Burgers vector of the dislocation. The object of the model is to determine how rapidly this relative displacement is accumulated through the core. A force law is defined within the inelastic slab which reflects the crystal lattice periodicity and is matched to the elastic constants at small strains. By constructing a force law with this periodicity we are restricting the models to small-angle interfaces containing crystal lattice dislocations. If we wanted to develop analogous models or secondary dislocation arrays at large-angle grain boundaries we would have to construct force laws with appropriate periodicities. In that case it is much less obvious what elastic constants should be fitted in the boundary core region. A balance of forces is established at the boundaries of the inelastic and elastic regions. The inelastic region tries to localize the core as much as possible while the elastic strain field energy in the adjacent continua favours an infinitely wide core. At the force balance the core has a finite width and the elastic singularity is removed.

Van der Merwe (1950) considered models for epitaxial interfaces and grain boundaries of the twist and symmetric tilt types. Bullough and Tewary (1979) presented a Peierls–Nabarro-type model of a symmetric tilt boundary. In this section we introduce the principles of such models by briefly describing the Peierls–Nabarro model for an isolated edge dislocation. We then discuss a symmetric tilt boundary treated by a Peierls–Nabarro model and then by a van der Merwe model. There seems little point in a thorough exposition of these models since their deficiencies are well known. In particular the close proximity of the elastic continua to the dislocation cores and the uncertainty in the inelastic force laws make these models of qualitative interest only.

### 2.11.2.2 *Peierls–Nabarro model for an isolated edge dislocation*

An edge dislocation is introduced into the crystal by the following imaginary process, which is illustrated in Fig. 2.28. First, the crystal is cleaved along the slip plane, Fig. 2.28(a). One lattice plane, spacing $b$, is removed from the upper half-crystal, Fig. 2.28(b). The lattice planes are no longer in registry across the cleavage plane. The disregistry of the bottom half-plane with respect to the top one is $\varphi(x) = b/2$ $(x > 0)$ and $\varphi(x) = -b/2$ $(x < 0)$, Fig. 2.28(b). The two half crystals are rebonded and displacements $u(x, y)$ are introduced by relaxation. The displacements are assumed to be antisymmetric about the plane $y = 0$: $u(x, -\varepsilon) = -u(x, \varepsilon)$. The disregistry is now

$$\left.\begin{array}{l} \varphi(x) = 2u(x, -\varepsilon) + b/2 \quad (x > 0) \\ \varphi(x) = 2u(x, -\varepsilon) - b/2 \quad (x < 0) \end{array}\right\} \tag{2.174}$$

with the boundary conditions $u(\infty, -\varepsilon) = -u(-\infty, -\varepsilon) = -b/4$. These boundary conditions ensure that the disregistry is zero far from the core. For brevity we shall write $u(x)$ for $u(x, -\varepsilon)$, so that $u(x)$ refers to the displacement of the bottom surface. The two surfaces above and below $y = 0$ in Fig. 2.28(b) are now assumed to interact by an inelastic force law. In a local disregistry approximation the force law is given by the negative of the gradient of the $\gamma$-surface. That is, the restoring force at $x$, where the disregistry is $\varphi(x)$, is given by the negative of the slope of the $\gamma$-surface at a constant disregistry equal to $\varphi(x)$. The approximation consists of ignoring additional terms arising from the coexistence of a continuous set of disregistries in the dislocation core.

The restoring force has the periodicity of the crystal lattice, which is equal to $b$ in the direction of the Burgers vector. Thus the simplest form of the restoring force is

$$\tau_{xy} = A \sin\left[\frac{2\pi\varphi}{b}\right] \tag{2.175}$$

where $A$ is a constant. Substituting eqn (2.174) for $\varphi(x)$ we obtain

$$\tau_{xy} = -A \sin\left[\frac{4\pi u}{b}\right]. \tag{2.176}$$

We may think of this as the first term of an infinite Fourier series representation of $\tau_{xy}$. The constant $A$ is determined, as in the Frenkel model of the theoretical shear strength, by requiring that Hooke's law is satisfied, for small displacements $u$. Hence in the isotropic elastic approximation we have

$$\tau_{xy} = 2\mu\varepsilon_{xy} = \frac{\mu\varphi}{d} = A\frac{2\pi\varphi}{b} \tag{2.177}$$

where $d$ is the interplanar spacing. It is noted that the shear is inhomogeneous here since it is applied only across one pair of lattice planes. Thus, the equalities in eqn (2.177) are not exact, but approximate, since they apply only in the case of a homogeneous shear $\varepsilon_{xy}$. In this approximation we obtain

$$\tau_{xy} = -\frac{\mu b}{2\pi d} \sin\left[\frac{4\pi u}{b}\right]. \tag{2.178}$$

We now consider the dislocation to be represented by a continuous distribution of dislocations along $x$, with a Burgers vector density $\rho(x)$. The meaning of $\rho(x)$ is that $\rho(x)dx$ is the Burgers vector lying between $x$ and $x + dx$. From eqn (2.174) we have

$$\rho(x) = -2\frac{du}{dx},$$

(2.179)

and

$$b = \int_{-\infty}^{\infty} \rho(x)\,dx = -2\int_{-\infty}^{\infty}\frac{du}{dx}\,dx.$$

(2.180)

The shear stress produced on the plane $y = 0$ by this continuous distribution of dislocations at $(x, 0)$ is given by

$$\tau_{xy}(x, 0) = -\frac{\mu}{2\pi(1-\nu)}\,\mathbb{P}\int_{-\infty}^{\infty}\frac{\rho(x')}{(x-x')}\,dx' = \frac{\mu}{\pi(1-\nu)}\,\mathbb{P}\int_{-\infty}^{\infty}\frac{(du/dx)_{x=x'}}{(x-x')}\,dx'$$

(2.181)

where $\mathbb{P}$ denotes the principal value of the integral. Combining eqns (2.178) and (2.181) we obtain the following integral equation for the displacements $u(x)$:

$$\mathbb{P}\int_{-\infty}^{\infty}\frac{(du/dx)_{x=x'}}{(x-x')}\,dx = \frac{b(1-\nu)}{2d}\sin\left(\frac{4\pi u}{b}\right)$$

(2.182)

for which the solution is

$$u(x) = -\frac{b}{2\pi}\tan^{-1}\left(\frac{x}{\zeta}\right)$$

(2.183)

where

$$\zeta = \frac{d}{2(1-\nu)}.$$

(2.184)

Equation (2.183) satisfies the boundary conditions that $u(\infty) = -u(-\infty) = -b/4$. In addition $u(\zeta) = -b/8 = \frac{1}{2}u(\infty)$. We may define $2\zeta$ as the width of the dislocation since in the region $-\zeta < x < \zeta$ the disregistry is greater than one half of its maximum value.

The Burgers vector density has the form of a Lorentzian:

$$\rho(x) = \frac{b}{\pi}\frac{\zeta}{x^2 + \zeta^2}$$

(2.185)

Using this Burgers vector density and the standard expressions for the stress field of an isolated edge dislocation (eqns (2.109)–(2.113)) expressions may be derived for the Peierls–Nabarro edge dislocation by integration. It is found that the stress field of the dislocation is no longer singular, the singularity being removed by the finite width of the dislocation core.

### 2.11.2.3  *Peierls–Nabarro model for a symmetrical tilt boundary*

Consider a symmetrical tilt boundary as sketched in Fig. 2.29. The Peierls–Nabarro model for this boundary was considered by Bullough (1955) (see Bullough and Tewary 1979). The model consists of a vertical stack of infinite elastic plates, each of thickness $D$ and separated by inelastic slabs of thickness $d$. The dislocations are assumed to delocalize within the inelastic slabs. The aim is to derive the variation of the dislocation core widths with the boundary misorientation. The problem reduces to the equilibrium of a single elastic plate, with its centre at the origin, whose surfaces are subjected to the displace-

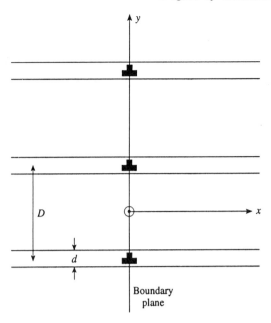

**Fig. 2.29** Peierls–Nabarro model of a symmetric tilt boundary. The dislocation spacing is $D$ and the thickness of each inelastic slab is $d$. (From Bullough and Tewary (1979).)

ments $u(x)$ and $-u(x)$. The same restoring force law, eqn (2.178) is assumed to act across each inelastic slab. The elastic shear stress is obtained from eqn (2.118) by setting $Y = 0$:

$$\tau_{xy}(x, D/2) = -\frac{\mu}{D(1 - \nu)} \int_{-\infty}^{\infty} \frac{\pi(t - x)/D}{\sinh^2[\pi(t - x)/D]} \frac{du}{dt} \, dt. \tag{2.186}$$

At equilibrium this stress balances the stress from eqn (2.178):

$$\sin \frac{4\pi u(x)}{b} = -\frac{2\pi d/b}{D(1 - \nu)} \int_{-\infty}^{\infty} \frac{\pi(t - x)/D}{\sinh^2[\pi(t - x)/D]} \frac{du}{dt} \, dt. \tag{2.187}$$

Note that when $D \to \infty$ this equation reverts to eqn (2.182) for an isolated edge dislocation. Although an exact solution to eqn (2.187) has proved elusive an approximate solution, that is quite adequate for misorientations up to about 5°, has been given by Bullough. Bullough's solution is the following:

$$u(x) = \frac{b}{2\pi} \tan^{-1} \left[ \frac{\sinh(\pi x/\delta)}{\sin(\pi \varepsilon/\delta)} \right] \tag{2.188}$$

where $\varepsilon$ and $\delta$ are defined by

$$D = \delta - 2\varepsilon$$

$$\tan(\pi \varepsilon/\delta) = b\pi/2\delta(1 - \nu). \tag{2.189}$$

The dislocation core width is obtained from eqn (2.189) as follows:

$$2\zeta = \frac{2\delta}{\pi} \sinh^{-1} \left[ \sin\left(\frac{\pi \varepsilon}{\delta}\right) \right]. \tag{2.190}$$

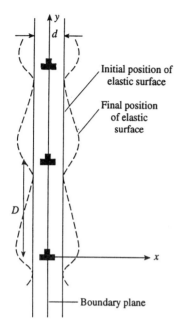

Initial position of elastic surface

Final position of elastic surface

Boundary plane

**Fig. 2.30**   Van der Merwe model of a symmetric tilt boundary. The dislocation spacing is $D$ and the inelastic slab, of thickness $d$, is now in the boundary plane. The dashed line shows the final shape of the surfaces of the adjoining elastic continua, once the elastic relaxation has taken place. (From Bullough and Tewary (1979).)

It follows from eqn (2.190) that the core width decreases as the dislocation spacing $D$ decreases, i.e. as the tilt angle increases. This is the main result of the model. One of the consequences is that it predicts that the stress required to move the boundary by dislocation glide increases as the misorientation increases.

### 2.11.2.4   *The van der Merwe model for a symmetrical tilt boundary*

Once the boundary misorientation exceeds a few degrees there is no longer sufficient good material between the dislocation cores for the Peierls–Nabarro model to be applicable. The van der Merwe model (van der Merwe 1950) may be used to model the boundary when the non-Hookean relaxations are confined largely to the boundary plane. Thus the model consists of two elastic half-spaces separated by an inelastic slab of thickness $d$ in the boundary plane (see Fig. 2.30). The object of the model is to study the core relaxation in the boundary plane as a function of the tilt angle. A key difference between this model and the Peierls–Nabarro model is that the restoring force acting between the elastic slabs is tensile. It is again approximated by a sinusoidal force, although in reality it is not a periodic function of the relative separation of the elastic half-spaces. We shall not discuss the model in detail here but state its main results. Figure 2.30 shows the periodic displacement relaxations for the symmetrical tilt boundary. It is seen that each dislocation has a wider core on the compressive side of the dislocation than on the tensile side, as would be expected from the anharmonicity of atomic interactions. The displacement field $u(y)$ is given by

$$u(y) = b/4 + \frac{b}{2\pi}\tan^{-1}\left[A \sin Y/(1 - A \cos Y)\right] \qquad (2.191)$$

where

$$
\left.
\begin{aligned}
A &= \sqrt{1 + \beta^2} - \beta \\
Y &= 2\pi y/D \\
\beta &= \frac{\pi b(1 - 2\nu)}{2D(1 - \nu)} \cong \frac{\pi(1 - 2\nu)}{2(1 - \nu)}\theta.
\end{aligned}
\right\}
\tag{2.192}
$$

As the boundary misorientation increases and the dislocation cores approach each other the compressive field of one dislocation becomes closer to the tensile field of the dislocation above it. The width of the dislocation cores in the boundary plane therefore increases as the boundary misorientation increases.

### 2.11.3 Atomistic models using computer simulation and interatomic forces

Numerous calculations of the atomic core structures of interfacial dislocations (especially secondary dislocations) have been made using computer simulation techniques and interatomic force laws. We cite here a few examples that illustrate some of the points that were made in Section 2.11.2.1. The cores of secondary edge dislocations with $b$ normal to the boundary plane are found to be highly localized (see Section 4.3.1.8). This result may be expected because of the large cohesive forces acting across the boundary. The cores of some relaxed secondary edge dislocations with $b$ parallel to the interface have already been shown in Fig. 1.20 and 1.21. These cores are seen to be localized as indicated by the pattern of distortion of the DSC lattices, which is easily detected. The localization of these cores is expected because the c.n.i.d.s for these interfaces are quite large and therefore significant variations of the interfacial energy can occur as a function of the relative translation of the adjoining crystals parallel to the interface. On the other hand, Bristowe (1986) has found considerable variation of the degree of localization of a number of secondary screw dislocations in [001] twist boundaries in f.c.c. metals represented by different interatomic force models. Some results are shown in Fig. 2.31, where the calculated width of an isolated secondary screw dislocation in a $\Sigma = 5$ boundary is shown using four different interatomic force laws. It was found that the degree of delocalization tended to increase for a force law which gave a $\gamma$-surface with a smaller slope, as expected. Similarly, it was found that the degree of delocalization tended to increase as the size of the c.n.i.d. decreased, which is also expected because the range of values of $\sigma$ decreases as the c.n.i.d. decreases in size.

## 2.12 EXPERIMENTAL OBSERVATIONS OF ARRAYS OF INTERFACIAL DISLOCATIONS

### 2.12.1 Mainly room-temperature observations

There have been many observations, by transmission electron microscopy, of arrays of dislocations at homophase and heterophase interfaces at room temperature. In Section 2.9 we discussed the application of the Frank–Bilby analysis to two experimental observations of dislocation arrays at heterophase interfaces. Despite the large number of experimental studies there have been only a few instances in which the Burgers vectors of the dislocations have been positively identified from the image contrast. As pointed out by Forwood and Clarebrough (1985) the conditions under which the usual $g \cdot b = 0$

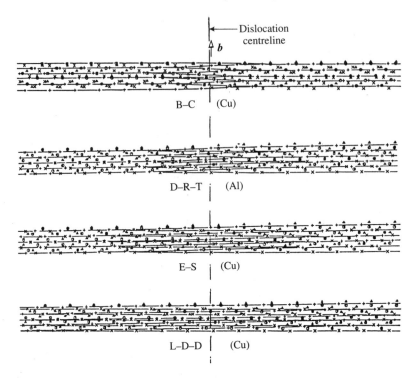

**Fig. 2.31** Relaxed structures of an isolated screw dislocation in a $\Sigma = 5(001)$ twist boundary computed at 0 K with 4 different potentials for f.c.c. metals. The horizontal lines show the (310) planes in the plane of the boundary. The width of the dislocation increases from the top of the series to the bottom. (From Bristowe (1986).)

criteria for determining the Burgers vector of a dislocation are more restrictive for interfacial dislocations than they are for a dislocation in the bulk crystal. Frequently the Burgers vectors have been inferred from the measured misorientation from an assumed reference state and the spacings of the dislocations, e.g. Schober and Balluffi (1970), Bollmann *et al.* (1972), Clark and Smith (1978), Knowles and Goodhew (1983a), and Babcock and Balluffi (1987). In this section we shall not present a catalogue of all such observations, which would fill a book in itself. Instead we focus on those observations which have a direct bearing on the theory that we have described in this chapter.

Schober and Balluffi (1970) and Tan *et al.* (1975) observed square grids of line contrast in certain (001) twist grain boundaries in gold (see Fig. 2.32), by transmission electron microscopy. The grain boundaries were manufactured by bonding together two (001) films of gold at selected misorientations. Misorientations close to 0, $2\tan^{-1}\frac{1}{7}$, $2\tan^{-1}\frac{1}{5}$, $2\tan^{-1}\frac{1}{4}$, $2\tan^{-1}\frac{1}{3}$ were studied, corresponding to boundaries vicinal to $\Sigma = 1$, 25, 13, 17, and 5 respectively. The lines of contrast were assumed to correspond to dislocations with Burgers vectors of the appropriate DSC lattices. The observed average dislocations spacings were found to satisfy Frank's formula:

$$d = |b|/2\sin(\Delta\theta/2) \qquad (2.193)$$

where $\Delta\theta$ is the measured misorientation from the nearby exact coincidence site lattice orientation. This is illustrated in Fig. 2.33. As the spacing of the dislocations decreased,

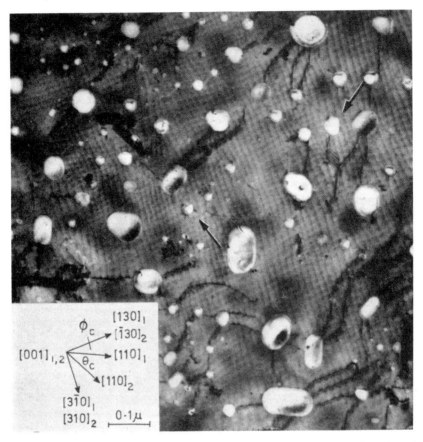

**Fig. 2.32** Electron micrograph of a manufactured (001) twist grain boundary in gold with a misorientation vicinal to $\Sigma = 5$, showing a square array of $a/10\langle 310 \rangle$ DSC screw dislocations. (From Schober and Balluffi (1970).)

and as the magnitude of the Burgers vectors decreased, the line contrast faded and the dislocations became increasingly difficult to detect. Thus, in Fig. 2.33 it is seen (black dots only) that the arrays of screw dislocations were detectable only within 9° of the $\theta = 0°$ orientation, 2° of the $\Sigma = 5$ orientation, and about 0.6° of the $\Sigma = 13$ and $\Sigma = 17$ orientations. However, no dislocations were seen in the vicinity of the $\Sigma = 25$ orientation. This raised the questions of whether the dislocations were present near the $\Sigma \geq 25$ orientations and at greater deviations from the exact coincidence orientations, or whether they were simply indetectable. From the viewpoint of Frank–Bilby theory the dislocations are always present for purely geometrical reasons. Physically, the question concerns the localization of the Burgers vector density into discrete dislocations and whether forces to bring about this localization exist in the boundary plane. It was also not clear from these experiments whether the line contrast was indeed due to dislocations or Moiré fringes. As discussed by Babcock and Balluffi (1987), if Moiré fringes were present they could have the same directions and spacings as the reported dislocation networks.

Babcock and Balluffi (1987) clarified these questions by repeating the experiments for manufactured (001) twist boundaries in gold and silver. In addition to repeating the earlier

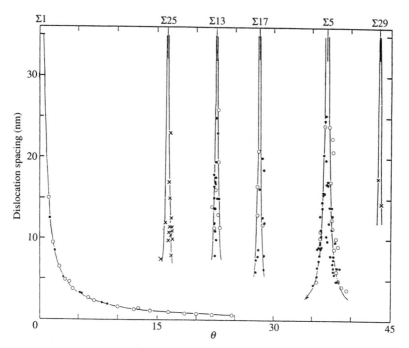

**Fig. 2.33** Observed dislocation spacings in manufactured (001) twist grain boundaries in gold as a function of the twist angle. The lines show the spacings expected by Frank's formula, eqn (2.193), for DSC dislocations of the $\Sigma 1$, 5, 13, 17, 25, and 29 orientations. (From Babcock and Balluffi (1987).)

observations of arrays of line contrast near $\Sigma = 1, 5, 13$, and 17, they also observed arrays of line contrast near $\Sigma = 25$ and $\Sigma = 29$ (see Fig. 2.33). They showed, by a series of tests, that in each case the line contrast was indeed due to dislocations and not Moiré fringes. The fact that dislocations were found in boundaries vicinal to $\Sigma \geq 25$ reflects the increased resolution of the more modern instruments and the improvements in specimen preparation.

Similar observations of arrays of screw dislocations in twist boundaries in NiO and MgO have been made by Liou and Peterson (1981) and Sun and Balluffi (1982). One of the interesting features about these observations is that there is some evidence that the Burgers vectors of the dislocations are not always primitive DSC vectors. Obviously it is not a requirement of the Frank–Bilby theory that the Burgers vectors of discrete dislocations, which make up the Burgers vector content of the boundary, are primitive DSC vectors.

The contrast from dislocation arrays depends on the Burgers vectors of the dislocations, the spacing of the dislocations, the foil thickness, and the angle at which the array is viewed. It also depends on the resolution of the microscope and the magnification at which it is operated. Under certain conditions arrays of coarsely spaced dislocations are detected. But if the imaging conditions are changed, for the same specimen, additional, finer spaced networks of dislocations may be detected. This was first demonstrated by Cosandey and Bauer (1981) and it was studied systematically by Kvam and Balluffi (1987) who examined a large number of manufactured symmetric [001] tilt boundaries in gold. Dislocation-like strain contrast was detected effectively throughout the 90° misorientation

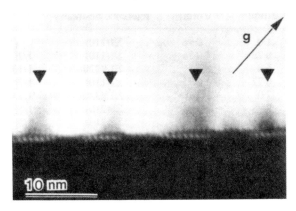

**Fig. 2.34** Electron micrograph of a boundary vicinal to a $\Sigma = 13$ (320) symmetric [001] tilt boundary in gold seen almost edge on, showing two patterns of strain contrast. The fine scale contrast is due to $\frac{1}{2}$ [110] crystal lattice dislocations, while the coarser scale contrast (arrowed) is due to $\frac{1}{13}$ [320] DSC dislocations. (From Kvam and Balluffi (1987).)

range. This indicates that at all misorientations there are forces acting in the boundary plane tending to localize the Burgers vector density of the Frank–Bilby equation into discrete dislocations.

In Fig. 2.34 we show a boundary with a misorientation of 21.7°, which is less than 1° from that of the $\Sigma = 13$ (320) symmetric tilt boundary. The boundary is viewed at a highly oblique angle and therefore it appears very narrow. The strain contrast associated with it appears on two length scales. The fine contrast is due to $\frac{1}{2}$[110] edge dislocations and their measured spacing is 0.77 nm, as compared with the predicted spacing from Frank's formula of 0.75 nm. The strain contrast of this array decays very rapidly into the adjoining crystals, as expected from the elastic field analysis of Section 2.10. The coarser scale contrast, arrowed in Fig. 2.34, is associated with an array of $\frac{1}{13}$[320] dislocations, which are DSC dislocations of the $\Sigma = 13$ coincidence site lattice. Their measured spacing is 7 nm, as compared with the predicted spacing of 7.1 nm. In Fig. 2.35 we show a plot of the observed dislocation spacings in symmetric [001] tilt boundaries as a function of the tilt angle. Taken together with Fig. 2.34 this provides strong experimental confirmation of the Frank–Bilby analysis of the Burgers vector content of interfaces and of the localization of the Burgers vector density into discrete dislocations. The localization is a result of relaxation processes within the interface, whose origin is beyond the scope of the Frank–Bilby theory.

For grain boundaries between h.c.p. crystals the selection of a suitable reference structure is even more complex. In general, CSLs are very rare in h.c.p. crystal lattices where $(c/a)^2$ is not a rational fraction. In that case the near CSL model is applied in which the actual value of $(c/a)^2$ is approximated by a rational approximant (Chen and King 1988, Antonopoulos *et al.* 1990, Shin and King 1991). Thus the selection of a reference structure entails first the choice of a rational approximant to $(c/a)^2$ to produce a 'constrained' CSL. Secondly with a chosen rational approximant to $(c/a)^2$ there is an infinity of possible constrained CSLs to choose from, just as there is for cubic crystals. Stress annihilator dislocations are required to accommodate the deviation of the actual value of $(c/a)^2$ from the chosen rational approximant as well as the misorientation of the boundary from the orientation relation for the constrained CSL. The selection of the

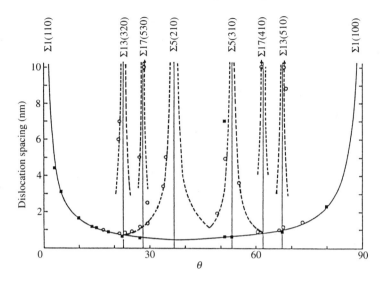

**Fig. 2.35** A plot of the observed dislocation spacings (shown as points) in [001] symmetric tilt boundaries in gold as a function of tilt angle, $\theta$. The solid line shows the spacings expected by Frank's formula for crystal lattice dislocations, and the dashed lines show the expected spacings of DSC dislocations. (From Kvam and Balluffi (1987).)

rational approximant to $(c/a)^2$ and the constrained CSL for that rational approximant is again made (Shin and King 1991) on the grounds that the predicted secondary dislocation configuration is consistent with the strain contrast in the electron microscope, at the operating level of spatial resolution.

An interesting example of this approach was presented by Shin and King (1991) who analysed the linear strain contrast features in grain boundaries in zinc at temperatures between 140 and 300 K. In zinc the $(c/a)$ ratio varies markedly with temperature. Therefore the most appropriate choice of reference structure may change from one rational approximant to another as $(c/a)^2$ changes with temperature. Correspondingly the predicted secondary dislocation configuration may change discontinuously. Since this would involve a discontinuous change in the reference periodic boundary structure and the Burgers vectors of the secondary dislocations it amounts to a first-order phase transformation of the boundary structure. Possible indications of a reversible transformation of this kind at grain boundaries in zinc were presented by Shin and King and are shown in Fig. 2.36. In Fig. 2.36(a) we see a room-temperature observation of a grain boundary in which the misorientation is 86.2° [2.0, −0.983, −1.017, 0.0] and the observed dislocation configuration is consistent with that predicted by choosing a $\Sigma = 15$ reference structure. Figure 2.36(b) shows the same boundary after cooling to 77 K. The dislocation configuration is heavily disrupted, which Shin and King interpret as evidence that the dislocation structure is transforming to that of another reference structure. But because the temperature is so low the transformation cannot be completed in the time available in the experiment. Estimates of the elastic strain field energy of the boundary indicate that the lower energy dislocation configuration does indeed change to that of a $\Sigma = 32$ reference structure at the lower temperature. The disruption to the dislocation array of Fig. 2.36(b) is reversed upon returning to room temperature, and the structure shown in Fig. 2.36(a) is again obtained. This example illustrates once again the difficulty in

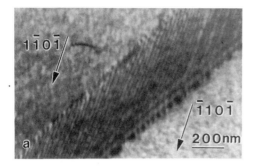

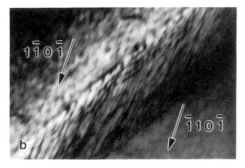

**Fig. 2.36** Electron micrographs of a grain boundary in zinc with a misorientation of 86.2° [2.0, −0.983, −1.017, 0.0] imaged (a) at room temperature and (b) at 77 K. Note the disruption of the dislocation configuration in (b). (From Shin and King (1991).)

selecting a suitable reference structure to account for observed contrast features in the electron microscope at interfaces.

Positive identification of Burgers vectors of dislocations detected in boundaries vicinal to $\Sigma = 3, 9, 27$ and 81 coincidence site lattices has been made by Clarebrough and Forwood (1980a,b), and Forwood and Clarebrough (1985, 1986) using the image matching technique of Head *et al.* (1973). Transmission electron microscope images of a dislocation were compared with calculated images computed for different Burgers vectors. Boundaries within 0.1° of the exact coincidence orientations were selected from a Cu + 6 at % Si alloy. In all cases the Burgers vectors of the dislocations were identified as DSC vectors, and in most cases, though not all, the vectors were the smallest possible DSC vectors in their respective lattices.

It is not always the case that the Burgers vectors of dislocations accommodating a misorientation from a coincidence boundary are DSC vectors. It is sometimes possible for a DSC dislocation to dissociate into partial dislocations separating regions of a boundary with different structures. If the regions are related by symmetry then they are energetically degenerate (see Section 1.7.2.3). Periodic arrays of such partial dislocations have been identified in a boundary vicinal to $\Sigma = 5$ in germanium (Bacmann *et al.* 1981 and in boundaries vicinal to $\Sigma = 9$ and $\Sigma = 27$ boundaries in a copper silicon alloy by Forwood and Clarebrough (1982, 1983). Forwood and Clarebrough (1986) showed that the partial dislocations could also separate grain boundary structures that are not related by symmetry. Two of the three independent arrays of dislocations, in a $\Sigma = 3$ $(2\bar{4}1)/(\bar{5}\ 10\ 8)$ $\cos^{-1}(-2/3)/[210]$ asymmetric tilt boundary facet, were dissociated into partial DSC dislocations that separated non-equivalent regions of the boundary plane. They estimated that the difference in energy of the two regions was 2.5 mJ m$^{-2}$.

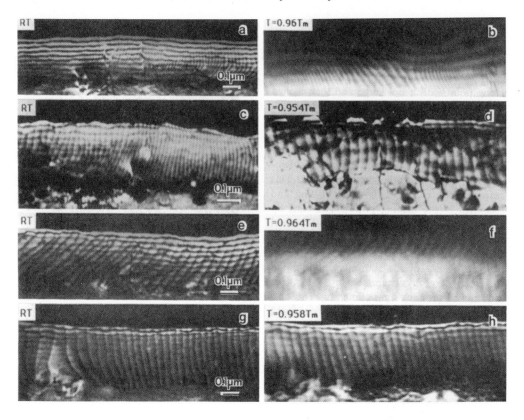

**Fig. 2.37** Dark-field electron micrographs of localized grain boundary dislocation arrays in various boundaries in aluminium at room temperature (left) and at the highest temperature of observation (right). (a) and (b) $\Sigma = 13$ (510) symmetric tilt boundary showing grain boundary DSC edge dislocations with $b = \frac{1}{26}\langle 510 \rangle$. (c) and (d) $\Sigma = 17$ (410) symmetric tilt boundary showing grain boundary DSC edge dislocations with $b = \frac{1}{17}\langle 410 \rangle$. (e) and (f) incommensurate (100) twist boundary with $\theta = 45°$ showing dislocations with $b = \frac{1}{2}\langle 001 \rangle$. (g) and (h) incommensurate symmetric [001] tilt boundary with $\theta = 45°$ showing dislocations with $b = \frac{1}{2}\langle 100 \rangle$. (From Hsieh and Balluffi (1989).)

### 2.12.2 High-temperature observations

Experiments designed to reveal any extensive delocalization of the cores of primitive secondary interfacial dislocations at elevated temperatures have been performed by Hsieh and Balluffi (1989). The experiments consist of following the change in the strain contrast of dislocation arrays as the specimen is heated in the electron microscope. The results for aluminium indicate that sufficient localization of the dislocation strain field to give line contrast features in the electron microscope exists up to 96 per cent of the melting point of the material. These authors manufactured bicrystals close to (310), (410), and (510) symmetric tilt orientations and also a $\theta = 45°$ $\langle 100 \rangle$ tilt boundary and a {100} twist boundary with $\theta = 45°$. The first three boundaries vicinal to misorientations at which there is two-dimensional periodicity in the boundary plane. There is no periodicity in the boundary plane of the 45° twist boundary. The 45° $\langle 100 \rangle$ tilt boundary is periodic only along the tilt axis. In all cases dislocation arrays were seen in these boundaries at temperatures up to 0.96 of the melting point. This must mean that the cores remained localized to a sufficient extent that the elastic strain field associated with them produced the contrast in the microscope. This is shown in Fig. 2.37.

# REFERENCES

Amelinckx, S. (1979). In *Dislocations in solids*, Vol. 2 (ed. F. R. N. Nabarro), p. 67. North-Holland, Amsterdam.

Amelinckx, S. and Dekeyser, W. (1959). *Solid State Physics*, **8**, 325.

Antonopoulos, J. G., Delavignette, P., Karakostas, Th., Komninou, Ph., Laurent-Pinson, E., Lay, S., Nouet, G., and Vincens, J. (1990). *J. Phys.* (Paris), **51**, C1-61.

Babcock, S. E. and Balluffi, R. W. (1987). *Phil. Mag. A*, **55**, 643.

Bacmann, J. J., Silvestre, G., Petit, M., and Bollmann, W. (1981). *Phil. Mag. A*, **43**, 189.

Balluffi, R. W. and Olson, G. B. (1985). *Met. Trans. A*, **16**, 529.

Balluffi, R. W., Brokman, A., and King, A. H. (1982). *Acta Metall.*, **30**, 1453.

Barnett, D. M. and Lothe, J. (1974). *J. Phys. F: Metal Phys.*, **4**, 1618.

Bilby, B. A. (1955). *Report on the conference on defects in crystalline solids*, p. 123. The Physical Society, London.

Bilby, B. A., Bullough, R., and Smith, E. (1955). *Proc. Roy. Soc.*, **231A**, 263.

Bollmann, W. (1970). *Crystal defects and crystalline interfaces*. Springer-Verlag, Berlin.

Bollmann, W., Michaut, B., and Sainfort, G. (1972). *Phys. Stat. Sol.* (*a*), **13**, 637.

Bonnet, R. (1981*a*). *Acta Metall.*, **29**, 437.

Bonnet, R. (1981*b*). *Phil. Mag. A*, **43**, 1165.

Bonnet, R. (1981*c*). *Phil. Mag. A*, **44**, 625.

Bonnet, R. (1982). *J. Physique*, **43**, C6-215.

Bonnet, R. (1985). *Phil. Mag. A*, **51**, 51.

Bonnet, R. (1988). *Mat. Res. Soc. Symp. Proc.*, **122**, 281.

Bonnet, R., Marcon, G., and Ati, A. (1985). *Phil. Mag. A*, **51**, 429.

Bristowe, P. D. (1986). In *Grain boundary structure and related phenomena*, p. 89. Japan Institute of Metals, Sendai (Suppl. to *Trans. Jap. Inst. Mets.* **27**.)

Bullough, R. (1955). PhD thesis, University of Sheffield.

Bullough, R. and Bilby, B. A. (1956). *Proc. Phys. Soc. B*, **69**, 1276.

Bullough, R. and Tewary, V. K. (1979). In *Dislocations in solids*, Vol. 2. (ed. F. R. N. Nabarro), p. 1. North-Holland, Amsterdam.

Burgers, J. M. (1939). *Proc. Kon. Ned. Akad. V. Wet. Amsterdam*, **42**, 293.

Burgers, J. M. (1940). *Proc. Phys. Soc.*, **52**, 23.

Burgers, W. G. (1947). *Proc. Kon. Ned. Akad. V. Wet. Amsterdam*, **50**, 595.

Chen, F.-R. and King, A. H. (1988). *Phil. Mag. A*, **57** 431.

Chou, Y. T. and Lin, L. S. (1975). *Mat. Sci. Eng.*, **20**, 19.

Christian, J. W. (1981). *The theory of transformations in metals and alloys*, Part I. Pergamon, Oxford.

Clarebrough, L. M. and Forwood, C. T. (1980*a*). *Phys. Stat. Sol.* (*a*), **58**, 597.

Clarebrough, L. M. and Forwood, C. T. (1980*b*). *Phys. Stat. Sol.* (*a*), **59**, 263.

Clark, W. A. T. and Smith, D. A. (1978). *Phil. Mag. A*, **38**, 367.

Cosandey, F. and Bauer, C. L. (1981). *Phil. Mag. A*, **44**, 391.

Dahmen, U. and Westmacott, K. H. (1988). *Scr. Metall.*, **22**, 1673.

Dundurs, J. and Sendeckyj, G. P. (1965). *J. Appl. Phys.*, **36**, 3353.

Dundurs, J. (1969). *Mathematical theory of dislocations*, p. 70. American Society of Mechanical Engineers, New York.

Dupeux, M. (1987). *Phil. Mag. Letts.*, **55**, 7.

Eshelby, J. D. (1957). *Proc. Roy. Soc.*, *A*, **241**, 376.

Forwood, C. T. and Clarebrough, L. M. (1982). *Acta Metall.*, **30**, 1443.

Forwood, C. T. and Clarebrough, L. M. (1983). *Phil. Mag.*, **47**, L35.

Forwood, C. T. and Clarebrough, L. M. (1985). *Aust. J. Phys.*, **38**, 449.

Forwood, C. T. and Clarebrough, L. M. (1986). *Phil. Mag. A*, **53**, 863.

Frank, F. C. (1950). *Symposium on the plastic deformation of crystalline solids*, p. 150. Office of Naval Research, Pittsburgh, Pennsylvania.

Frank, F. C. and van der Merwe, J. H. (1949). *Proc. R. Soc. A*, **198**, 205.

Goodhew, P. J., Darby, T. P., and Balluffi, R. W. (1976). *Scr. Metall.*, **10**, 495.

Head, A. K., Humble, P., Clarebrough, L. M., Morton, A. J., and Forwood, C. T. (1973). *Computed electron micrographs and defect identification*. North Holland, Amsterdam.

Hirsch, P. B. (1991). *Polycrystalline semiconductors II*, Springer Proc. Phys. Vol. 54, (eds. J. H. Werner and H. P. Strunk), p. 54. Springer-Verlag, Berlin.

Hirth, J. P. and Balluffi, R. W. (1973). *Acta Metall.*, **21**, 973.

Hirth, J. P. and Lothe, J. (1982). *Theory of dislocations*. Wiley, New York.

Hirth, J. P., Barnett, D. M., and Lothe, J. (1979). *Phil. Mag. A*, **40**, 39.

Hsieh, T. E. and Balluffi, R. W. (1989). *Acta Metall.*, **37**, 1637.

Jain, S. C., Gosling, T. J., Willis, J. R., Totterdell, D. H. J., and Bullough, R. (1992). *Phil. Mag. A*, **65**, 1151.

Jesser, W. A. and Kuhlmann-Wilsdorf, D. (1967). *Phys. Stat. Sol.* **19**, 95.

Knowles, K. M. (1982). *Phil. Mag. A*, **46**, 951.

Knowles, K. M. and Goodhew, P. J. (1983*a*). *Phil. Mag. A*, **48**, 527.

Knowles, K. M. and Goodhew, P. J. (1983*b*). *Phil. Mag. A*, **48**, 555.

Kvam, E. P. and Balluffi, R. W. (1987). *Phil. Mag. A*, **56**, 137.

Liou, K.-Y., and Peterson, N. L. (1981). In *Surfaces and interfaces in ceramics and ceramic—metal systems* (eds. J. A. Pask and A. G. Evans). Plenum, New York.

Matthews, J. W. and Blakeslee, A. E. (1974). *J. Cryst. Growth*, **27**, 118.

Nabarro, F. R. N. (1947). *Proc. Phys. Soc.*, **59**, 256.

Nakahara, S., Wu, J. B. C., and Li, J. C. M. (1972). *Mat. Sci. Eng.*, **10**, 291.

Olson, G. B., and Cohen, M. (1979). *Acta Metall.*, **27**, 1907.

Peierls, R. E. (1940). *Proc. Phys. Soc.*, **52**, 23.

Pond, R. C. (1977). *Proc. Roy. Soc.*, **A357**, 471.

Pond, R. C. and Vlachavas, D. S. (1983). *Proc. Roy. Soc. Lond., A*, **386**, 95.

Read, W. T. (1953). *Dislocations in crystals*. McGraw-Hill, New York.

Read, W. T. and Shockley, W. (1950). *Phys. Rev.*, **78**, 275.

Rosenhain, W. and Humphrey, J. C. W. (1913). *J. Iron and Steel Inst.*, **87**, 219.

Sargent, C. M. and Purdy, G. R. (1975). *Phil. Mag.*, **32**, 27.

Schober, T. and Balluffi, R. W. (1970). *Phil. Mag.*, **21**, 109.

Shieu, F.-S., and Sass, S. L. (1990). *Acta Metall. Mater.*, **38**, 1653.

Shin, K. and King, A. H. (1991). *Phil. Mag. A*, **63**, 1023.

Sun, C. P. and Balluffi, R. W. (1982). *Phil. Mag. A*, **46**, 49.

Sutton, A. P. and Balluffi, R. W. (1987). *Acta Metall.*, **35**, 2177.

Tan, T. Y., Sass, S. L., and Balluffi, R. W. (1975). *Phil. Mag.*, **31**, 575.

Taylor, G. I. (1934). *Proc. Roy. Soc.*, **145A**, 388.

van der Merwe, J. H. (1950). *Proc. Phys. Soc.*, **A63**, 616.

Vitek, V. (1968). *Phil. Mag.*, **18**, 773.

Willis, J. R., Jain, S. C., and Bullough, R. (1990). *Phil. Mag. A*, **62**, 115.

Wolf, D. (1989). *Scr. Metall.*, **23**, 1713.

# 3

# Models of interatomic forces at interfaces

## 3.1 INTRODUCTION

At the most fundamental level, the properties of interfaces are determined by the chemical and physical aspects of interatomic forces which act in their vicinities. In subsequent chapters we will describe models of the atomic and electronic structures of interfaces that are based on various descriptions of these atomic interactions. There is a very wide range of descriptions of interatomic forces, ranging from empirical pair potentials to those derived from self-consistent density functional theory. In this chapter we introduce the more important theories, and we provide some critical insight into their strengths and limitations.

A major theme of this chapter is the relationship between interatomic forces and the local atomic environment. At an interface there is a variety of atomic environments, which is one of the main reasons why the properties of an interface may be distinct from those of the adjoining crystals. It is desirable, therefore, to relate the properties of atoms at various sites in the interface to their atomic environments. Physicists and chemists have traditionally viewed the electronic structure of solids in rather different ways. Physicists tend to think in terms of Bloch's theorem, eigenstates extending over the whole crystal, energy bands, and Brillouin zones, with the analysis being formulated in reciprocal space. Chemists, on the other hand, tend to think in terms of the local atomic environment, and chemical bonding and antibonding states in real space. Over the last twenty years there has been a convergence of these two viewpoints, to the extent that one can write down formal links between them, and understand the role of the local atomic environment quantitatively. This has been achieved largely through a particular ansatz for the wave function of an electron in the solid, namely a linear combination of atomic orbitals (LCAO), which is discussed in Section 3.8.

At an interface there may or may not be periodicity parallel to the interfacial plane, and in a bicrystal there is no periodicity normal to the interface. At first sight it seems that band theory, which relies on translational symmetry, cannot be applied to interfaces. Periodicity parallel to the interface may be introduced by appropriate strains of the adjoining crystals to force them to be commensurate with each other. The strains may be made as small as desired but at the expense of increasing the size of the two-dimensional unit cell (see Sections 2.2 and 2.9.2). Periodicity normal to the interface may be restored by considering a unit cell containing two bicrystals, as shown in Fig. 3.1. (There is a complication here concerning the compatibility of the translation states of the two interfaces in each unit cell, but we defer this point until Section 4.3.1.6) Thus we have recovered three-dimensional translational symmetry and Bloch's theorem but the supercell contains a large number of atoms, typically 100–10 000. The Brillouin zone is correspondingly small and contains as many bands as there are orbitals in the supercell, forming what has been described as a 'can of spaghetti' (Heine 1980). It is not obvious what the band structure tells us about bonding in the interface. Indeed, if we doubled the size of the supercell parallel to the interface the Brillouin zone would shrink by a factor of two and the bands would be folded to form a narrower can of spaghetti. Yet there would be *no change* in the local chemistry in the interface. The framework that

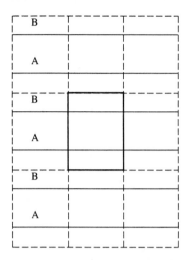

**Fig. 3.1** To illustrate the use of periodic boundary conditions to model interfaces. In each cell, outlined by dashed lines, there are two interfaces, represented by continuous lines, separating regions A and B, which may be two different phases or two misoriented regions of the same phase. The emboldened cell is the computational cell and the surrounding cells are periodic replicas. Periodic boundary conditions are also applied normal to the page.

is introduced in Section 3.8 enables us to extract a local picture of bonding in the interface from the band structure.

At homophase interfaces, such as grain boundaries, stacking faults, inversion domain bundaries, etc., obviously the same description of interatomic forces applies on both sides of the interface as well as at the interface itself. This simplicity is the reason why most atomistic modelling of interfaces has been applied to homophase interfaces. At hetero-phase interfaces, not only is the interface a region where the bonding changes from that of one crystal to the other in some way which is often not understood, but there may be additional cohesive terms that exist only at the interface. For example, at an interface between a metal and an ionic crystal the ionic lattice induces a lattice of image charges in the metal which gives an important contribution to the binding of the bicrystal even if there is no chemical interaction at the interface. Thus, heterophase interfaces are intrinsically more complex to model, as discussed further in Section 3.11. One approach to dealing with this complexity is to start from a fundamental description of cohesion, in which all types of bonding may be described on an equal footing. This is the approach of density functional theory, which is described in Section 3.2. Not only does density functional theory provide an exact solution of the full many-body problem of bonding, but, at least in principle, it may also be used to analyse and bolster simpler models of bonding in the solid state such as effective medium theory and tight binding models.

Models of interatomic forces traditionally fall into three broad classes: metallic, covalent, and ionic. In metallic and covalent bonding atoms are charge neutral and bonding is a consequence of electron delocalization from atomic states into itinerant states forming giant molecular orbitals. The electron delocalization is a result of quantum mechanical tunnelling (called hopping) of each electron from the attractive potential well of one atom into the potential well on a neighbouring atom. From a chemical point of view the main qualitative difference between metallic and covalent bonding is that they involve unsaturated and saturated bonds respectively. A saturated bond is one in which there are two electrons occupying bonding states, and no electrons occupying antibonding states. If there is an energy gap between bonding and antibonding states then the material is a semiconductor or insulator depending on whether the gap is small or large compared with $kT$. In a metallic bond there are always less than two electrons occupying bonding states, and there is no energy gap between the highest occupied state and the lowest unoccupied state so that the material has a finite electronic conductivity at absolute zero. We note that in both covalent and metallic bonding electrons are 'free', in the sense that they are itinerant, and bonding may be directional in the metallic case as well as the covalent case.

In an ionic crystal the tendency for charge transfer to take place between atoms is much greater than the tendency for electrons to delocalize and form wide bands. This balance is determined by the difference in the electronic energy levels on the anion and cation versus the probability of an electron tunnelling per unit time from one to the other. Thus, ionic crystals are characterized by wide band gaps and relatively small band widths. The charge transfer between anions and cations results in electrostatic fields that give rise to most of the cohesion of the solid. This is the basis of the rigid ion model of ionic crystals. In a low symmetry environment, such as in the core of a defect, the polarizability of the ions is an important additional contribution to the binding energy. The shell model is designed to account for this contribution.

Most simple models of interatomic forces use the concept of an interatomic potential. An interatomic potential is a function of the relative positions of two or more atoms (or ions), and it may also be a function of other parameters such as the charges and polarizabilities of the ions or the local atomic density. The total potential energy of an assembly of atoms/ions is then the sum of all such potentials. For example, in a model based on pair potentials the total potential energy of the system is given by the sum of pair potentials between all pairs of atoms. All interatomic potentials rely on the validity of the Born–Oppenheimer approximation in which it is assumed that the electrons in the system are always in their ground state as the nuclear positions change. This is usually an excellent approximation since the relaxation time for the electrons is typically three orders of magnitude smaller than that of the nuclear coordinates. However, the approximation obviously breaks down in some very important dynamical phenomena such as a chemical reaction where an electron jumps (tunnels) from one energy surface to another. With this approximation the total potential energy can certainly be expressed as a function of *all* the nuclear coordinates. However, this is not much use in practice unless we can express the total potential energy as a sum of functions of much smaller numbers of nuclear coordinates. This is achieved by making the physically reasonable assumption that the local environment around each atomic site is more important than the positions of more remote atoms in determining the local bonding. There are often many more assumptions underlying the use of a potential, but perhaps the most severe is the assumption that it is transferable from one atomic environment to another, where the local atomic density and coordination may differ significantly. In such cases it is preferable to return to a more fundamental description of bonding to attempt to derive the variation in the functional form of the potentials with the local atomic environment from first principles. That is the approach of Pettifor's bond order potentials, which are briefly described in Section 3.8.7.

We use Dirac notation to denote integrals. The meaning of this notation is as follows:

$$\left.\begin{aligned}
\langle f | g \rangle &= \int f^*(r) g(r) \, \mathrm{d}^3 r, \\
\langle f | O | g \rangle &= \int f^*(r) O(r) g(r) \, \mathrm{d}^3 r,
\end{aligned}\right\} \tag{3.1}$$

where $f$ and $g$ are integrable functions of $r$ and $O(r)$ is an operator. The asterisk denotes complex conjugation. The integral $\langle f | O | g \rangle$ is sometimes called a matrix element of the operator $O$.

In this chapter, except where stated otherwise, we use atomic units: $h/2\pi = 1$, $e^2 = 2$, $m = \frac{1}{2}$. The unit of length is the Bohr, which is 0.5292 Å, and the unit of energy is the Rydberg, which is 13.6058 eV.

## 3.2  DENSITY FUNCTIONAL THEORY

### 3.2.1  The variational principle and the Kohn–Sham equations

Each electron in an atom, molecule, or solid interacts with all the other electrons through electrostatic Coulomb forces. In addition, electrons interact with atomic nuclei through similar Coulomb forces. In the Born–Oppenheimer approximation the electronic and nuclear coordinates are decoupled. The much lighter electrons respond so quickly to the electrostatic field of the heavier nuclei that the nuclei may be regarded as stationary. The nuclear coordinates $\{R_i\}$ are then parameters which appear in the quantum state of the electrons. This quantum state is described as a many-body state because the mutual interactions between the electrons do not allow us to discuss the motion of any one electron independently of the others. If the electronic coordinates are denoted by $r_1, r_2, \ldots r_N$ then the quantum state of the system is described by the many electron wave-function $\psi(r_1, r_2, \ldots r_N)$, which depends parametrically on the nuclear coordinates $\{R_i\}$.

In a solid with $10^{23}$ electrons per cubic centimetre the determination of the many electron wavefunction is a hopeless task. Such was the state of the theory prior to 1964. The breakthrough came when Hohenberg and Kohn (1964) proved a fundamental theorem: *all aspects of the electronic structure of a system of interacting electrons, in an external potential $v(r)$ and a non-degenerate ground state, are completely determined by the electronic charge density $\rho(r)$.* The 'external potential' in the present context is the potential due to the atomic nuclei. This theorem immediately leads to a major simplification: rather than working with the many electron wavefunction we work with a function of just three variables, $\rho(r)$. Hohenberg and Kohn also proved that the ground state energy of the interacting electron gas is a *unique functional* of $\rho(r)$, for a given external potential, and that this functional is minimized by the correct function $\rho(r)$. (A functional is a function of a function.) The only drawback is that the functional is not known since it amounts to a solution of the full many electron problem. However, several useful approximations for the functional have been proposed, and they have led to remarkable success in predicting the equilibrium structures and ground state properties of materials.

Kohn and Sham (1965) made the second vital breakthrough. Starting from the energy variational principle of Hohenberg and Kohn they derived a system of self-consistent *one-particle* equations for the description of electronic ground states. The interacting $N$ electron problem is thus mapped rigorously onto $N$ single electron equations, in which each electron is moving independently of the other electrons, but it experiences an effective potential which emulates all the interactions with the other particles. In other words, the electrostatic interactions between the electrons have effectively been switched off so that the motion of each electron is independent of the other electrons. In place of these electrostatic interactions each independent electron is now moving in an effective single particle potential which simulates exactly the interactions with all the other electrons. Moreover, the effective single-particle potential is a unique functional of the electron density. Again, the only difficulty is that the exact form of the single-particle potential is not known. Almost invariably a local density approximation is invoked to overcome this difficulty, as described below. The one-particle equations are known as the Kohn–Sham equations.

All electronic structure calculations for systems larger than simple molecules make the 'one-electron approximation'. In essence this approximation consists of treating each particle as moving independently of the other particles by replacing the electron–electron Coulomb interactions by an effective single-particle potential. Until the work of Kohn and Sham (1965) there was no rigorous justification for doing this. An example of this practice is in Hartree theory, where each electron moves in the mean field created by all electrons in the system, called the Hartree potential, and the nuclear potential. The Hartree potential $V_H(r)$ is defined by the classical electrostatic formula:

$$V_H(r) = \int \frac{\rho(r')}{|r - r'|} \, dr'. \tag{3.2}$$

This is an approximation to the true electron–electron interaction because in this formula each electron is also incorrectly interacting with itself. The effective single particle potential in which each electron moves is then

$$V_{eff}(r) = V_H(r) + V_N(r) \tag{3.3}$$

where the potential due to the nuclei, of nuclear charge $Z_i$, is given by

$$V_N(r) = \sum_i \frac{Z_i}{|r - R_i|}. \tag{3.4}$$

The effective single particle equations in the Hartree approximation are thus:

$$(-\nabla^2 + V_{eff}(r))\varphi_i(r) = \varepsilon_i\varphi_i(r). \tag{3.5}$$

These single particle equations have to be solved self-consistently with the condition that

$$\rho(r) = \sum_i n_i|\varphi_i(r)|^2, \tag{3.6}$$

which specifies the relationship between the single-particle wavefunctions and the charge density that is used to generate the Hartree potential in eqn (3.2). $n_i$ is the occupation of state $i$, as determined by Fermi–Dirac statistics. Equations (3.3), (3.5), (3.6) are a set of self-consistent equations which when satisfied minimize the total energy of the system in the Hartree approximation:

$$E[\rho(r)] = \int V_N(r)\rho(r) \, dr + \tfrac{1}{2} \int \frac{\rho(r)\rho(r')}{|r - r'|} \, dr \, dr' + T_s[\rho(r)], \tag{3.7}$$

where $T_s[\rho(r)]$ is the kinetic energy of a non-interacting electron gas of density $\rho(r)$. The first term describes the electron–nuclear electrostatic interaction energy, and the second term describes the electron–electron electrostatic energy, including unphysical electron self-interactions.

The problem with Hartree theory is that the electron self-interactions are not negligible. An electron repels other electrons for two reasons. The net effect is to lower the energy of the system compared with the Hartree estimate. The first reason is simply the Coulomb repulsion, and the energy lowering that this leads to is called the correlation energy. Secondly, electrons sharing the same spin will repel each other because of the exclusion principle, and the concomitant energy lowering is called the exchange energy. Thus, in a solid each electron is surrounded by a 'hole' in the electron density, which is called the exchange-correlation hole. Since each electron and its exchange-correlation hole cannot be separated the two together form a quasiparticle. We can visualize the exchange-correlation hole by thinking of the image charge that an external electron induces inside a metal. As the external electron approaches the surface of the metal the image charge also approaches the surface. As the electron enters the metal the classical part of the image charge becomes the correlation hole and the quantum mechanical part becomes the exchange hole. The exchange-correlation energy, $E_{xc}$, is simply the electrostatic interaction energy between the electron density $\rho(r)$ and the exchange-correlation hole surrounding each electron. When this is added to the classical electrostatic interaction energy, $E_H$, where

$$E_H(r) = \tfrac{1}{2}\int \rho(r) V_H(r)\,dr, \qquad (3.8)$$

the total interaction energy is between each electron and $N - 1$ electrons, as it should be. An estimate of the size, $r_s$, of the exchange-correlation hole is

$$4\pi r_s^3/3 = 1/\langle\rho(r)\rangle, \qquad (3.9)$$

which is typically of the order of an angstrom. Thus the quasiparticles in a solid are much larger than the bare electrons.

Hohenberg and Kohn (1964) *proved* that the total ground state energy of a system of interacting electrons may be written exactly as follows:

$$E_G[\rho(r)] = \int V_N(r)\rho(r)\,dr + \tfrac{1}{2}\int \frac{\rho(r)\rho(r')}{|r - r'|}\,dr\,dr' + G[\rho(r)]. \qquad (3.10)$$

The first and second terms are the same as in Hartree theory, eqn (3.7). The third term accounts for the kinetic energy of the interacting electron gas and the corrections to the Hartree energy for the exchange-correlation hole surrounding each electron. They also proved that this functional is minimized by the correct electron density.

Since the kinetic energy of the interacting electron gas is not known the functional $G[\rho(r)]$ is written as follows:

$$G[\rho(r)] = T_s[\rho(r)] + E_{xc}[\rho(r)] \qquad (3.11)$$

where $T_s[\rho(r)]$ is the kinetic energy of a non-interacting electron gas of the same density. $E_{xc}[\rho(r)]$ is now the exchange-correlation energy of the non-interacting electron gas together with a correction to account for the difference between the kinetic energies of the interacting and non-interacting electron gases. The ground state energy functional is thus

$$E_G[\rho(r)] = \int V_N(r)\rho(r)\,dr + \tfrac{1}{2}\int \frac{\rho(r)\rho(r')}{|r - r'|}\,dr\,dr' + T_s[\rho(r)] + E_{xc}[\rho(r)]. \qquad (3.12)$$

If the last term were omitted the problem of minimizing $E_G[\rho(r)]$ with respect to $\rho(r)$ would be identical to the minimization of the total energy in the Hartree approximation, as may be seen by comparing eqn (3.7). Demanding that $E_G[\rho(r)]$ is a minimum with respect to $\rho(r)$, subject to the condition that the total number of electrons is conserved:

$$\int \rho(r)\,dr = N, \qquad (3.13)$$

yields the condition that

$$\frac{\delta T_s[\rho]}{\delta \rho(r)} + V_N(r) + V_H(r) + V_{xc}(r) = \mu, \tag{3.14}$$

where $V_{xc}(r)$ is called the exchange-correlation potential:

$$V_{xc}(r) = \frac{\delta E_{xc}[\rho]}{\delta \rho(r)} \tag{3.15}$$

and $\mu$ is the chemical potential of the electrons. If the exchange-correlation potential were absent in eqn (3.14) the functional minimization of $E_G[\rho(r)]$ would be achieved by the self-consistent Hartree equations, eqn (3.3), (3.5), (3.6). Kohn and Sham (1965) proved that $E_G[\rho(r)]$ may be minimized by solving self-consistently the equations:

$$(-\nabla^2 + V_{eff}(r))\psi_i(r) = \varepsilon_i^{KS}\psi_i(r) \tag{3.16}$$

where

$$V_{eff}(r) = V_N(r) + V_H(r) + V_{xc}(r) \tag{3.17}$$

and

$$\rho(r) = \sum_i n_i |\psi_i(r)|^2. \tag{3.18}$$

Equations (3.16)–(3.18) are the Kohn–Sham equations. The only difference from the self-consistent Hartree equations is the inclusion of the very important exchange-correlation potential, $V_{xc}$. No approximation has been made to the many-electron problem. We have obtained an exact mapping of the fully interacting many-electron problem onto a set of self-consistent field equations for non-interacting electrons. Strictly speaking, the single particle eigenvlaues $\varepsilon_i^{KS}$ and eigenfunctions $\psi_i(r)$ have no meaning other than that they are devices that are used to construct the charge density $\rho(r)$, which is the central quantity of density functional theory. In practice the occupied Kohn–Sham eigenvalues are often identified with the physical energies required to excite electrons to the free continuum, and the unoccupied eigenvalues are identified with excited single particle states. There is no rigorous justification for doing this because the quasiparticle spectrum is determined by the solutions of the Dyson equation:

$$-\nabla^2 \tilde{\psi}_i(r) + \int \Sigma(r,r',\tilde{\varepsilon}_i)\tilde{\psi}_i(r')\,dr' = \tilde{\varepsilon}_i\tilde{\psi}_i(r), \tag{3.19}$$

where $\Sigma(r,r',\tilde{\varepsilon}_i)$ is a non-local, energy-dependent quantity called the mass operator. The mass operator is not identifiable with the local, energy-independent Kohn–Sham effective potential in eqn (3.17). The energy dependence of the mass operator renders the operator acting on $\tilde{\psi}_i$ on the left of eqn (3.19) non-Hermitian, and hence the eigenvalues $\tilde{\varepsilon}_i$ of this equation are complex in general. The imaginary part of $\tilde{\varepsilon}_i$ reflects the finite lifetime of excitations in the system. There is a special case, however. For an infinite system, whose highest occupied Kohn–Sham states are extended throughout the system, the highest occupied eigenvalue, $\tilde{\varepsilon}_N^{KS}$, equals the chemical potential $\mu$. We note, however, that this does not mean that the Fermi surface generated by the Kohn–Sham equations is necessarily identical with the real Fermi surface of a metal. Indeed, Mearns (1988) has shown that they are not equivalent, although the differences are small for a wide range of metals.

Once the Kohn–Sham equations have been solved self-consistently the total energy, $E^{HKS}$, may be expressed as

$$E^{HKS}[\rho^{sc}(r)] = T_s[\rho^{sc}(r)] + \int V_N(r)\rho^{sc}(r)\,dr + \tfrac{1}{2}\int \frac{\rho^{sc}(r)\rho^{sc}(r')}{|r-r'|}\,dr\,dr'$$

$$+ E_{xc}[\rho^{sc}(r)] + E_{NN}, \tag{3.20}$$

where $\rho^{sc}(r)$ is the self-consistent charge density and $E_{NN}$ is the nuclear–nuclear Coulomb interaction energy:

$$E_{NN} = \tfrac{1}{2}\sum_{i\neq j}\frac{Z_i Z_j}{|R_i - R_j|}. \tag{3.21}$$

In practice this total energy is evaluated by using the Kohn–Sham eigenvalues to obtain $T_s[\rho^{sc}]$, which is an unknown functional:

$$\sum_i n_i \varepsilon_i = \sum_i n_i \langle \psi_i| -\nabla^2 + V_{eff}(r) |\psi_i(r)\rangle$$

$$= T_s[\rho^{sc}] + \int \rho^{sc}(r)(V_N(r) + V_H(r) + V_{xc}(r))\,dr. \tag{3.22}$$

Thus, the total energy may be rewritten in the following useful way:

$$E^{HKS}[\rho^{sc}(r)] = \sum_i n_i \varepsilon_i - \int \rho^{sc}(r)\left(\frac{V_H(r)}{2} + V_{xc}(r)\right)dr + E_{xc}[\rho^{sc}(r)] + E_{NN}. \tag{3.23}$$

The integral on the right of this equation corrects for the fact that the electron–electron Coulomb interactions are double counted in the eigenvalue sum and also that the eigenvalue sum includes the exchange-correlation potential. Equation (3.23) is the form of the total energy that is minimized in an all-electron calculation, such as an augmented plane wave (APW) or muffin tin orbital (MTO) method.

Although the Kohn–Sham equations are exact, their solution requires approximation because the exact forms of the exchange-correlation energy and potential are unknown. Indeed their determination is equivalent to solving the full many-electron problem! The most widely used approximation is the *local density approximation (LDA)*. In this approximation the exchange-correlation energy is expressed as:

$$E_{xc}[\rho] = \int \varepsilon_{xc}(\rho(r))\rho(r)\,dr, \tag{3.24}$$

where $\varepsilon_{xc}(\rho)$ is the exchange-correlation energy per electron of a uniform electron gas of density $\rho$. This works in the following way. In a particular volume element $dr$ at $r$ the electron density is $\rho_o = \rho(r)$. The exchange-correlation energy per electron of the electron gas in this volume element is approximated by the exchange-correlation energy per electron of a uniform electron gas of density $\rho_o$. The point is that reliable expressions for the latter have been derived over a wide range of electron densities. Because only the electron density at $r$ is taken into account, regardless of the variation of the electron density in the vicinity of $r$, the approximation is 'local'. We would expect this approximation to work only if $\rho(r)$ varies slowly in space so that various gradient corrections are negligible. But in fact it works much better than we might have thought. The reason for the success of the local density approximation is that it satisfies the condition that the exchange-correlation hole surrounding each electron amounts to a depletion of the electron density by exactly one electron, as it should. Thus, failures of the local density approximation may be attributed to the incorrect description of the shape and size of the exchange-correlation hole, but not its content.

**Table 3.1** Comparison of calculated and experimental band gaps in electron volts from Louie (1989). LDA: local density functional theory. GW: GW approximation for the mass operator in quasiparticle theory. Note that germanium is predicted to be metallic in density functional theory.

|  | LDA | GW | Experiment |
|---|---|---|---|
| Diamond | 3.9 | 5.6 | 5.48 |
| Silicon | 0.52 | 1.29 | 1.17 |
| Germanium | <0 | 0.75 | 0.744 |
| LiCl | 6.0 | 9.1 | 9.4 |

In spite of the reservations that have been expressed about the meaning of the Kohn–Sham eigenvalues, and the local density approximation, it is undoubtedly true that local density functional calculations have transformed the predictive power of modern theories of cohesion and structure in ionic, covalent, and metallic crystals. Atoms, molecules, and magnetic materials are treated by local spin density functional theory, where electron densities with up and down spins are distinguished. For heavy elements it is necessary to solve the Dirac equation rather than the Schrödinger-like Kohn–Sham equations. In this way all the elements of the periodic table may be studied, essentially from first principles. The calculations involve intensive computing, but it is now possible to calculate, using the atomic number and some local density functional as the only input, the following properties more or less routinely: the stability of elements and alloys in various observed or hypothetical crystal structures (as a function of atomic volume or some affine transformation such as a shear), equilibrium lattice constants, stacking fault and surface energies, elastic constants, phonon dispersion relations, band structures, and Fermi surfaces (based on Kohn–Sham eigenvalues), and the relaxed atomic and electronic structures and energies of periodic interfaces with small planar unit cells. One of the principal failures of density functional theory is the underestimation of band gaps in semiconductors and insulators, by between 50 and 100 per cent. This is due to a discontinuous change in the exchange-correlation potential between the valence and conduction bands (Perdew and Levy 1983; Sham and Schlüter 1983; Kohn 1986). The underestimation of the band gap is largely a consequence of the discontinuity in the exchange-correlation potential, and much less a failure of the local density approximation. Another questionable aspect of density functional theory (Heine 1991) is its failure, for example in a transition metal, to distinguish between the exchange-correlation hole seen by a d-electron as compared with the hole seen by an sp-electron. Thus the relative positions of the sp and d bands will be incorrect, which in turn affects the amount of sp-d hybridization and hence the chemical behaviour of the metal.

More recently, efforts have been made to go beyond density functional theory to improve the prediction of ground state and excited state properties and to solve the quasiparticle equation, eqn (3.19), rather than the Kohn–Sham equations (see Hybertsen and Louie 1986, and references therein). The idea is to calculate the mass operator $\Sigma$, in eqn (3.19), using an approach known as the GW approximation. It has been applied successfully to a wide variety of solids including semiconductors, ionic insulators, and metals, as well as surfaces and interfaces (for a review see Louie (1989)). Table 3.1 shows a comparison of calculated and experimental band gaps for semiconductors and insulators. It is seen that the local density functional results consistently underestimate the band gaps, but the quasiparticle approach is much more successful.

Accurate band gaps, dispersion relations, optical transition energies, and photoemission spectra have been obtained for both homopolar and heteropolar semiconductors with the quasiparticle approach. This is significant for the prediction of band offsets at heterojunction interfaces, and interface states (see Section 11.3.2). Interestingly, it is found that the wave functions of the quasiparticle equation are almost the same as the Kohn–Sham eigenfunctions in the local density approximation, even though the eigenvalue spectra of the two equations are so different. The charge density is therefore well described in local density functional theory.

### 3.2.2 The Harris–Foulkes energy functional

The variational principle on which the Hohenberg–Kohn–Sham scheme rests allows us to make rather crude estimates of the charge density and still obtain acceptable total energies. Indeed, the principle implies that if $\Delta\rho$ is the error in the charge density then the error in the total energy is of order $(\Delta\rho)^2$. Thus a 10 per cent error in the charge density gives an error in the total energy of order 1 per cent. A typical self-consistent density functional calculation begins by making a first guess at the charge density, $\rho^{in}(r)$, from which an input Kohn–Sham potential is constructed using eqn (3.17). Solving the Kohn–Sham equations produces a set of eigenvalues $\varepsilon_i^{out}$ and eigenfunctions $\psi_i^{out}$, from which an output charge density can be constructed:

$$\rho^{out}(r) = \sum_i n_i |\psi_i^{out}(r)|^2. \tag{3.25}$$

The output charge density is used to construct a new effective potential in the Kohn–Sham equations and the process is repeated until self-consistency between the input potential and the output eigenfunctions is achieved. But let us suppose we stop after the first cycle and ask for the best estimate of the total energy, given our trial input charge density and the output charge density it produces. The usual procedure is to write down the total energy in a form which we can be sure is either above or equal to the ground state energy. This is done by ensuring that the functional is expressed in terms of only the output charge density, by subtracting from the eigenvalue sum all the contributions that explicitly contain the input charge density:

$$E^{HKS}[\rho^{out}(r)] = \sum_i n_i^{out}\varepsilon_i^{out} - \int \frac{\rho^{out}(r)\rho^{in}(r')}{|r-r'|}\, dr\, dr' + \tfrac{1}{2}\int \frac{\rho^{out}(r)\rho^{out}(r')}{|r-r'|}\, dr\, dr'$$

$$- \int \rho^{out}(r) V_{xc}(\rho^{in}(r))\, dr + E_{xc}[\rho^{out}(r)] + E_{NN}. \tag{3.26}$$

This expression is guaranteed to lie above the ground state energy unless $\rho^{in} = \rho^{out}$, which would mean that self-consistency has been attained and we are therefore at the ground state energy. We note that this expression involves the explicit evaluation of the output charge density. To second order, the error in this estimate of the total energy is

$$E^{HKS}[\rho^{out}] - E^{HKS}[\rho^{sc}] = O([\rho^{out}(r) - \rho^{in}(r)][\rho^{out}(r') - \rho^{sc}(r')]), \tag{3.27}$$

where $O(x)$ means of the order of $x$.

Harris (1985) and Foulkes (1987) (see also Foulkes and Haydock (1989)) introduced independently another energy functional that is also stationary at the self-consistent charge density, although it is not variational. Again, the intention is to avoid self-consistency cycling and yet still obtain a reasonable estimate of the total energy by solving the Kohn–Sham equations only once. Denoting their functional by $E^{HF}$, it is

$$E^{HF}[\rho^{in}, \rho^{out}] = \sum_i n_i^{out} \varepsilon_i^{out} - \tfrac{1}{2} \int \frac{\rho^{in}(r)\rho^{in}(r')}{|r - r'|} \, dr \, dr' - \int \rho^{in}(r) V_{xc}(\rho^{in}(r)) \, dr$$

$$+ E_{xc}[\rho^{in}(r)] + E_{NN}. \tag{3.28}$$

The attraction of this functional is four-fold. First, it is equal to the Kohn–Sham total energy when $\rho^{in} = \rho^{out} = \rho^{sc}$. Secondly, in contrast to the previous functional, $E^{HKS}[\rho^{out}(r)]$, it does not involve the explicit evaluation of $\rho^{out}$. Thirdly, by choosing $\rho^{in}$ judiciously it is possible to make the evaluation of the integrals very simple. Finally, experience indicates that it is usually a more accurate estimate than $E^{HKS}[\rho^{out}(r)]$ (Polatoglou and Methfessel, 1988, Read and Needs 1989, Finnis 1990), although there does not appear to be a fundamental reason for this behaviour. To second order, the error in the Harris–Foulkes estimate of the total energy is

$$E^{HF}[\rho^{in}, \rho^{out}] - E^{HKS}[\rho^{sc}] = O([\rho^{out}(r) - \rho^{in}(r)][\rho^{in}(r') - \rho^{sc}(r')]). \tag{3.29}$$

Because the Harris–Foulkes functional is a functional of two charge densities it is no longer variational, so it is quite possible for the Harris–Foulkes estimate of the total energy to lie above or below the true ground state energy. Robertson and Farid (1991) have shown that the Harris–Foulkes functional possesses either a saddle point or a local minimum at the self-consistent ground state charge density. The Harris–Foulkes scheme has provided good estimates of bond lengths and vibrational frequencies of a number of homopolar dimer molecules (Harris 1985, Foulkes and Haydock 1989, Averill and Painter 1990), and cohesive properties (i.e. cohesive energies, lattice constants, and bulk moduli) of metallic, covalent, and ionic solids (Polatoglou and Methfessel 1988). In these calculations the input charge density was taken as a superposition of free atomic charge densities. It is particularly interesting that the scheme works for NaCl, which is an ionic material. In that case it was found (Polatoglou and Methfessel 1988) that the exact density functional results for the binding energy, lattice constant, and bulk modulus were estimated to within 10 per cent, 1 per cent, and 11 per cent respectively, by taking a superposition of neutral atomic charge densities as the input. Finnis (1990) tested the scheme for the (111) surface energy of Al and the vacancy formation energy in Al. Using a superposition of free atomic charge densities, significant errors were found in the surface and vacancy formation energies compared with the fully self-consistent results. However, dramatic improvements were found if the atomic charge densities were 'renormalized' to shrink the range of their tails. The renormalization would be achieved if further Kohn–Sham self-consistency cycles were allowed beyond the one cycle that was permitted. It is equivalent to allowing the free atomic charge densities to respond to their new environment inside the metal. In general, we would expect free atomic charge densities to contract in this way when bonding occurs because the piling up of bond charge between the atoms depletes the charge on the atoms, which in turn reduces the screening of the atomic nuclei. The remaining charge density on each atom is thus more strongly attracted to the nucleus and the charge density contracts. This picture is consistent with the virial theorem as pointed out by Pettifor (1971).

The Harris–Foulkes scheme underpins semi-empirical approaches to electronic structure, such as the tight binding bond model discussed in Section 3.8.5.

## 3.3  VALENCE AND CORE ELECTRONS: PSEUDOPOTENTIALS

It is convenient to classify electrons in an atom as core electrons or valence electrons. Valence electrons are in the outermost shells and are affected by the potentials of neighbouring atoms. They are responsible for the formation of bonds between atoms. Core electrons are in deeper shells, closer to the nucleus, and do not participate in bond formation. Their states are virtually unaffected by placing the atom in the environment of a solid. The distinction between states according to their energy is equivalent to distinguishing between their spatial extent: the deeper the energy of the state, the more spatially localized it becomes. Core states are thus more closely bound in space to the nucleus. In the frozen core approximation the core electrons are grouped with the nucleus to form an ion core that remains the same in whatever environment the atom is placed. In this approximation only the valence electron states respond to changes in the atomic environment. It is then necessary to solve the Kohn–Sham equations only for valence electron states, which is a considerable simplification. For example, in aluminium the outermost three electrons are the valence electrons and the remainder are core electrons. In fact the assumption that the core charge densities are frozen is not a very good one (von Barth and Gelatt 1980) but the core state eigenvalues are well approximated by assuming they remain frozen. As noted by Foulkes and Haydock (1989) this is because the variational principle ensures that the error in the core state eigenvalues is second order in the error in the core state charge density used to generate them. Anyway, the frozen core approximation underlies the pseudopotential approximation which we move on to now.

The concept of a pseudopotential stems from the observation that a valence electron experiences not only the potential of the ions and the other valence electrons but an effective repulsive potential at the ion cores arising from the increase in kinetic energy there. This is easy to understand in a classical picture, where the gain in kinetic energy as the valence electron enters the core cancels exactly the drop in potential energy. This cancellation survives fairly well quantum mechanically. The exclusion principle demands that the valence electron wavefunction be orthogonal to the core states. For this reason the valence electron wavefunction oscillates rapidly in the core, and hence its kinetic energy increases dramatically, which largely cancels the drop in potential energy. The cancellation of the kinetic and potential energies can be translated into a cancellation of an effective repulsive potential and the real attractive potential. The result is that the valence electron experiences only a relatively weak 'pseudopotential' from the ion core.

The cancellation between the real attractive and the effective repulsive potential is virtually complete provided the atomic core contains states of the required angular momentum. For example, in germanium the core contains s,p, and d states. The orthogonality requirement can, therefore, cancel s,p, and d components of the attraction of the valence state to the core. Thus in germanium the pseudopotential for s,p, and d valence state components is weak. But in carbon the core contains only s electrons. The valence states have s and p components. Thus, while there is good cancellation for the s component of the valence state there is not for the p component. The p component of the carbon pseudopotential is therefore strong, which is why the 2p wave function has a maximum at about the same position as the 2s state, which in turn is why carbon chemistry is so different (fortunately for us!) from the chemistry of other group IV elements. Similarly, in the 3d series of the transition metals the ionic cores do not contain d electrons and therefore the d component of their pseudopotentials is strong, leading to narrow d-bands and the possibility of ferromagnetism.

It has long been recognized that the electronic structure of sp bonded solids is very well described by nearly free electron theory. Even in diamond the bands are well described as free electron bands with gaps at the zone boundaries due to Bragg reflection (Mott and Jones 1936). The important point is that the deviations from the free electron behaviour, i.e. the band gaps, are small compared with the band width or the free electron Fermi energy. The small magnitudes of the band gaps imply weak scattering by any one atom. But this does not mean that the actual atomic potentials are weak, as envisaged in the original nearly free electron approximation, but only that the scattering from them is weak. It is convenient, for computational purposes, to replace the strong actual potential by a weaker effective potential which gives the same scattering power. This is an alternative way of picturing the pseudopotential as simply a mathematical trick to guarantee the same scattering behaviour of the valence states by the ion cores, and hence the same band structures of the valence states. At the same time the weakness of the pseudopotential gives rise to a pseudo wave function for the valence states that is nodeless in the ion core. Such a smooth wave function is described by far fewer plane waves than the true valence wave function, which makes the calculation of the band structure using a plane wave basis set possible.

Let us now express the concept of a pseudopotential mathematically, following Phillips and Kleinman (1959). Let $|\psi\rangle$ denote a real valence state and $|\varphi\rangle$ denote a smooth, nodeless pseudo wave function. The exclusion principle requires that the valence state is orthogonal to all the core states, any one of which we represent by $|c\rangle$. $|\psi\rangle$ is therefore expressed as follows:

$$|\psi\rangle = |\varphi\rangle - \sum_c |c\rangle\langle c|\varphi\rangle. \tag{3.30}$$

It follows that $\langle\psi|c\rangle = 0$ by construction, as required. (Note that the core states form an orthonormal set: $\langle c|c'\rangle = \delta_{cc'}$). We substitute this form for the valence state wave function into the Schrödinger equation, $H|\psi\rangle = E|\psi\rangle$, and obtain:

$$H|\psi\rangle = H|\varphi\rangle - \sum_c H|c\rangle\langle c|\varphi\rangle = H|\varphi\rangle - \sum_c E_c|c\rangle\langle c|\varphi\rangle = E|\psi\rangle$$

$$= E|\varphi\rangle - \sum_c E|c\rangle\langle c|\varphi\rangle, \tag{3.31}$$

where we have used $H|c\rangle = E_c|c\rangle$, and $E_c$ is the eigenvalue of the core state $|c\rangle$. The trick is to rewrite this as an equation for defining the pseudo wave function $|\varphi\rangle$:

$$H|\varphi\rangle + \sum_c (E - E_c)|c\rangle\langle c|\varphi\rangle = E|\varphi\rangle, \tag{3.32}$$

and writing $H$ as the sum of a kinetic, $T$, and potential energy, $V$, we have

$$T|\varphi\rangle + \left[V + \sum_c (E - E_c)|c\rangle\langle c|\right]|\varphi\rangle = E|\varphi\rangle. \tag{3.33}$$

The quantity in square brackets is the pseudopotential. The attractive, negative real potential $V$ is reduced by the positive potential $\sum_c (E-E_c)|c\rangle\langle c|$ (it is positive because the valence eigenvalue, $E$, is more positive than the core eigenvalues, $E_c$).

One of the remarkable features about this equation for the pseudo wave function is that its eigenvalues are precisely those of the true valence states $|\psi\rangle$. This is readily demonstrated by multiplying both sides of eqn (3.32) by $\langle\psi|$:

$$\langle\psi|H|\varphi\rangle + \sum_c (E - E_c)\langle\psi|c\rangle\langle c|\varphi\rangle = E\langle\psi|\varphi\rangle, \tag{3.34}$$

and since $\langle\psi|H|\varphi\rangle = E\langle\psi|\varphi\rangle$, and $\langle\psi|c\rangle = 0$, the equality follows provided $\langle\psi|\varphi\rangle \neq 0$. The surprising feature about this result is that it would have remained true if we had replaced the terms $(E - E_c)$ by anything else. Thus the pseudopotential is far from unique. The pseudo wave functions are also not unique because if we add a linear combination of core states to any pseudo wave function we still satisfy eqn (3.32).

Modern pseudopotentials are not based on the Phillips–Kleinman prescription, but nevertheless it illustrates important features of all pseudopotentials. First, the pseudo-potential is generally an energy dependent, non-local operator. It is non-local owing to the projection operator, $|c\rangle\langle c|$, onto the core states, whose action on a state $|\varphi\rangle$ involves a spatial integration: $|c\rangle\langle c|\varphi\rangle$. Secondly, because the summation in eqn (3.30) is over core states on all atoms, and $V$ is the true potential of all atoms, the pseudopotential is in general dependent on the atomic environment. Modern pseudopotentials seek to minimize the energy and environment dependencies, and we shall discuss them shortly. But first we note that the Phillips–Kleinman pseudo wave functions are not the same as the true valence wave functions. This may be seen directly from eqn (3.30):

$$\langle\psi|\psi\rangle = \langle\varphi|\varphi\rangle - \sum_c \langle\varphi|c\rangle\langle c|\varphi\rangle. \tag{3.35}$$

It is clear that the normalizations of the true and pseudo wave functions are not equal. Physically, the true charge density in the ion cores is lower than the pseudo charge density owing to the the exclusion principle. The rapid oscillations of the true wave function in the ion core result in a decrease of the charge density (crudely because the average of $\sin^2 x$ is $\frac{1}{2}$ and not 1). The depletion in the charge density at the ion core is called the orthogonality hole. By contrast, the pseudo charge density does not have an orthogonality hole, and consequently the pseudo charge density outside the core is lower than the true charge density. This is a serious deficiency in a self-consistent calculation which uses such pseudo wave functions instead of true wave functions, because the pseudo charge distribution and Hartree and exchange-correlation potentials will be wrong. In principle the pseudo wave functions could be orthogonalized to the cores before the charge density is evaluated but this is not really practicable.

The problem of the orthogonality hole is a feature of the Phillips–Kleinman construction, but as we have noted the pseudopotential is not unique. Modern pseudo-potentials overcome the problem and require just two features of the pseudo wave function: (i) it should be nodeless, (ii) it should, after it is normalized, become identical to the true valence wave function beyond some core radius, $R_c$. There are many ways of constructing such pseudo wave functions. The second requirement ensures that the pseudo charge density and potential are identical to the true charge density and potential beyond $R_c$. The pseudopotential inside $R_c$ mimics the scattering of the true ionic potential inside $R_c$ for a particular energy and angular momentum component of the valence state. Such 'norm-conserving' pseudopotentials can be constructed for any element of the periodic table (Bachelet *et al.* 1981, 1982). One of the consequences of conserving the normalization of the pseudo wave function is that the pseudopotential for the ionic core is more transferable from one environment to another, e.g. in a molecule, liquid, or solid. More precisely, the scattering properties of the pseudopotential and the true potential have the same energy variation to first order when transferred to other environments (Bachelet *et al.* 1982). This optimizes transferability to leading order in energy. This is important because a pseudopotential constructed for an isolated ionic core is of no use if it cannot be used in the solid state environment. Fortunately, the range of energies of valence states in atoms, molecules, and solids is quite small (compared with core state energies) and therefore norm-conserving pseudopotentials should be transferable.

Applying the frozen core approximation to eqn (3.23) and describing the potential due to the frozen, spherically symmetric, non-overlapping cores by norm conserving pseudo-potentials we arrive at the central equations of self-consistent, local density functional, pseudopotential methods. The Kohn–Sham equations become

$$(-\nabla^2 + W_{ps}(r) + V_H(\rho_v) + V_{xc}(\rho_v))\psi^{ps}_{iv}(r) = \varepsilon^{KS}_{iv}\psi^{ps}_{iv}(r), \qquad (3.36)$$

where $W_{ps}(r)$ is the sum of ionic pseudopotentials and $\psi^{ps}_{iv}$ is the $i$th pseudo wave function with eigenvalue $\varepsilon^{KS}_{iv}$. The pseudo charge density is identical to the true valence charge density owing to the the norm conservation of the pseudo wave functions. The Hartree and exchange-correlation potentials are defined in terms of only the valence charge density, $\rho_v$, which is constructed from the eigenfunctions $\psi^{ps}_{iv}$. The total energy is given by

$$E[\rho_v + \rho_c] = \sum_i n_i \varepsilon^{KS}_{iv} - \int \rho_v(r) \left[\frac{V_H(\rho_v)}{2} + V_{xc}(\rho_v)\right] dr + E_{xc}[\rho_v(r)] + E_{ii} + E_c.$$

$$(3.37)$$

The total ion–ion electrostatic energy is denoted by $E_{ii}$, and $E_c$ denotes the energies of the frozen cores whose frozen charge density is $\rho_c$. The simplification achieved in eqns (3.36) and (3.37) is that the problem is now expressed entirely in the valence orbital subspace of the whole Hilbert space. In fact, there are some non-trivial, additional approximations, involving the exchange-correlation energy, that have been made to achieve this simplicity, which are discussed in Louie *et al.* (1982).

So far we have discussed ionic pseudopotentials. We shall also come upon atomic pseudopotentials, particularly in the context of tight binding theories. An atomic pseudo-potential describes the scattering by the ion core and its screening cloud of valence electrons. Thus it is an effective total potential, representing the ion core, the Hartree potential, and the exchange-correlation potential of the valence electrons. Clearly, these potentials are not transferable since the screening of the ion core by the valence electrons is certainly environment dependent. The screening will be considered explicitly in Section 3.5 where we discuss the pseudopotential theory of sp-bonded metals ('simple' metals) and effective pair interactions between neutral pseudo-atoms.

## 3.4 THE FORCE THEOREM AND HELLMANN–FEYNMAN FORCES

We have seen how the total energy of a system in its ground state may be expressed in self-consistent density functional theory, using either the bare nuclear potential in an all-electron calculation, eqn (3.23), or an ionic pseudopotential, eqn (3.37). For the former, core and valence electrons are treated on an equal footing, whereas for the latter only the valence electrons are considered explicitly. In this section we consider the change in the total energy when an atom is moved in order to calculate the force on the atom. The expression for the force is remarkable for its simplicity. It was derived by Pettifor (1976, 1978) and Mackintosh and Andersen (1980), and a clear account of its lengthy derivation is given in Heine (1980). Here we shall simply state the result, and discuss its implications.

The starting point is a self-consistent solution of the Kohn–Sham equations for some atomic configuration. Let us define what we mean by the force on an 'atom' in a solid as the change, $\delta E$, in the total energy when the entire contents of a cell enclosing the atom (usually the Voronoi cell, but in principle it can be any cell) is displaced by a small amount $\delta R$, while keeping the rest of the system fixed. The force on the atom is $-\delta E/\delta R$, where $E$ is given by eqn (3.23) or eqn (3.37). When the cell is displaced there will be overlapping regions and holes in the valence charge density at the cell boundary.

After the cell is displaced the charge density is no longer self-consistent. Therefore we might expect a contribution to the force to come from the fact that the self-consistent potential in the displaced system is not the same as it is in the undisplaced system. If this were true then the calculation of the force would be very arduous indeed. It turns out that this contribution is zero to first order in the displacement $\delta R$, and thus does not contribute to the force. The reason is simply the variational principle. Since we started from a self-consistent charge density, $\delta E/\delta \rho$ is zero, and therefore there is no change in the total energy to first order when the charge density changes by $\delta \rho$. The implication is that we *should not* allow the charge density and the potential to become self-consistent again after the displacement to calculate the force. (If we were interested in force constants, which correspond to second-order changes in the energy, then the self-consistent redistribution of charge should be taken into account.) Thus the one electron potential outside the cell is taken to be unchanged, while the potential inside is simply rigidly displaced along with the cell. Apart from the virtual rigid displacement the effective one-electron potential is thus frozen.

The result is the 'force theorem':

$$\delta E = \delta \left( \sum_i n_i \varepsilon_i \right) + \delta \mathrm{E_{es}}. \tag{3.38}$$

The second term $\delta \mathrm{E_{es}}$ is the change in the *total* electrostatic interaction energy between the cell and its surroundings. If the content of the cell were a neutral spherically symmetric atom then $\delta \mathrm{E_{es}}$ would be zero. In an ionic crystal $\delta \mathrm{E_{es}}$ corresponds to the change in the Madelung energy. The first term is the change in the occupied one-electron eigenvalues, which is calculated using the rigidly displaced, otherwise frozen, one-electron potential. This term is important in metallic and covalent systems. The force theorem explains the success of quantum chemists over many years at calculating total energy changes of electrically neutral atoms by considering only the change in the eigenvalue sum. While the total energy itself is obviously not given by just the eigenvalue sum, first-order changes are given by just changes in this term, so long as the change in the total electrostatic energy is negligible.

The most important aspect of the force theorem is that the potential used to calculate the change in the sum of occupied eigenvalues is frozen. If it were not frozen then we would have to take into account the changes in other terms in the total energy, such as the correction for the double counting of the Hartree energy, which do not appear in eqn (3.38). We would then find that all the additional terms would cancel, to first order, owing to the variational principle.

We have seen that the force theorem is very significant from a conceptual point of view. We shall see in Section 3.8.5 that it underlies the calculation of forces in the tight binding bond model as well. But in local density functional calculations interatomic forces are computed more directly using the Hellmann–Feynman theorem. For example, consider the total energy, eqn (3.37), computed with a pseudopotential. According to the Hellmann–Feynman theorem the force on ion $i$ is given by

$$-\partial E/\partial \mathbf{R}_i = -\int \rho_v(\mathbf{r}) \partial V_{ps}(\mathbf{r})/\partial \mathbf{R}_i \, d\mathbf{r} - \partial E_{ii}/\partial \mathbf{R}_i. \qquad (3.39)$$

The Hellmann–Feynman force is thus the electrostatic force that the ion experiences due to the valence charge density and the other ions. In an all-electron calculation the same expression applies, except that the valence charge density is replaced by the total charge density (i.e. core + valence) and the pseudopotential and interionic Coulomb energy are replaced by the nuclear potential and the internuclear Coulomb energy.

In practice there are several problems with implementing the Hellmann–Feynman forces in a relaxation calculation, as discussed by Scheffler *et al.* (1985). There are additional contributions to the force if self-consistency in the charge density has not been achieved exactly. If the charge density is not self-consistent then $\delta E/\delta \rho$ is not zero and the true energy gradient becomes dependent on the redistribution of electronic charge. This contribution does not correspond to a physical force, because in a real material the charge density is always self-consistent. But in a calculation of the energy gradient it can be a major contribution because the degree of self-consistency attained in the calculation may not be sufficiently high. In addition, there is a further complication if the basis functions, in which the eigenstates of the system are expanded, are anchored to atomic sites, e.g. an atomic orbital basis set consisting of gaussians or Slater orbitals. In that case when an ion is moved the orbitals associated with it move also. This gives rise to a contribution to the energy gradient, often called the Pulay force, which is essentially a consequence of the fact that the basis set is not complete. Provided the Hilbert space spanned by the basis set is independent of the atomic distortions the Pulay force is zero. Finally we note that the errors in the charge density in an all-electron calculation must be much smaller than in a pseudopotential calculation to obtain Hellmann–Feynman forces of comparable accuracy. That is because the nuclear potential is so much stronger than the ionic potential.

## 3.5 COHESION AND PAIR POTENTIALS IN sp-BONDED METALS

The metals of the periodic table that are normally described as nearly free electron metals are listed in Table 3.2. Cohesion in these metals is provided by the outermost s and p electrons forming free electron bands with small gaps at the Brillouin zone boundaries due to weak ionic pseudopotentials. Because the pseudopotentials are weak we may apply perturbation theory to them. The zeroth and first-order terms of the perturbation expansion yield 90 per cent or more of the cohesive energies of the metals. These terms are independent of the crystal structure since they depend only on the density of the material. The dependence on the crystal structure appears in the second- and higher-order terms. The perturbation expansion is carried out in reciprocal space, but when the second-order term is transformed into real space it is seen to be equivalent to a sum of pair interactions. The pair interactions are also dependent on the density of the material. This section provides the only justification for the widespread use of pair potentials to describe atomic interactions in metals. We shall see that the conditions for the validity of this description are quite strict, and that the density dependence of the potentials raises questions about their application to interfaces. In transition metals there is no theoretical justification for a pair potential description, because the pseudopotential is too strong to apply perturbation theory. This section is based largely on Heine and Weaire (1970), Pettifor (1983), and Hafner (1989).

**Table 3.2** Equilibrium bulk properties of the simple and noble metals. au denotes atomic units. (From Pettifor (1983).)

| Metal | $Z$ | Quantity | | | | | |
|---|---|---|---|---|---|---|---|
| | | $E_{coh}/Z$ (eV/electron) eqn (3.49) | $r_{ws}$ (au) eqn (3.40) | $r_s$ (au) eqn (3.9) | $r_c$ (au) eqn (3.47) | $B/B_{fe}$ eqn (3.51) | $B/B_{fe}$ experiment |
| Li | 1 | 1.7 | 3.27 | 3.27 | 1.32 | 0.63 | 0.50 |
| Na | 1 | 1.1 | 3.99 | 3.99 | 1.75 | 0.83 | 0.80 |
| K | 1 | 0.9 | 4.86 | 4.86 | 2.22 | 1.03 | 1.10 |
| Rb | 1 | 0.9 | 5.31 | 5.31 | 2.47 | 1.14 | 1.55 |
| Cs | 1 | 0.8 | 5.70 | 5.70 | 2.76 | 1.29 | 1.43 |
| Be | 2 | 1.7 | 2.36 | 1.87 | 0.76 | 0.45 | 0.27 |
| Mg | 2 | 0.8 | 3.35 | 2.66 | 1.31 | 0.73 | 0.54 |
| Ca | 2 | 0.9 | 4.12 | 3.27 | 1.73 | 0.95 | 0.66 |
| Sr | 2 | 0.9 | 4.49 | 3.57 | 1.93 | 1.05 | 0.78 |
| Ba | 2 | 0.9 | 4.67 | 3.71 | 2.03 | 1.11 | 0.84 |
| Zn | 2 | 0.7 | 2.91 | 2.31 | 1.07 | 0.60 | 0.45 |
| Cd | 2 | 0.6 | 3.26 | 2.59 | 1.27 | 0.71 | 0.63 |
| Hg | 2 | 0.3 | 3.35 | 2.66 | 1.31 | 0.73 | 0.59 |
| Al | 3 | 1.1 | 2.99 | 2.07 | 1.11 | 0.69 | 0.32 |
| Ga | 3 | 0.9 | 3.16 | 2.19 | 1.20 | 0.74 | 0.33 |
| In | 3 | 0.9 | 3.48 | 2.41 | 1.37 | 0.83 | 0.39 |
| Tl | 3 | 0.6 | 3.58 | 2.49 | 1.43 | 0.87 | 0.39 |
| Cu | 1 | 3.5 | 2.67 | 2.67 | 0.91 | 0.45 | 2.16 |
| Ag | 1 | 3.0 | 3.02 | 3.02 | 1.37 | 0.71 | 2.94 |
| Au | 1 | 3.8 | 3.01 | 3.01 | 1.35 | 0.69 | 4.96 |

**Fig. 3.2**  To illustrate jellium. The shaded region represents the uniform positive background charge density. The thick solid line represents the electron density, which spills beyond the positive background at the edges.

The lowest level of approximation to an sp-bonded metal is a free electron gas, of the appropriate density, with a compensating uniform background positive charge. This is called jellium, a metallic jelly, and it is sketched in Fig. 3.2. The background positive charge represents the charges on the ion cores, which have been smeared out into a continuous, uniform distribution. Thus, every volume element is electrically neutral, except at the surfaces where the electrons spill out into the vacuum beyond the edge of the positive charge distribution, as shown in Fig. 3.2. We ignore these surface effects.

Let the valence of the metal be $Z$, i.e. this is the number of valence electrons per atom and the charge on the ion core. In jellium the only variable is the electron density. The Wigner–Seitz radius, $r_{WS}$, is defined as the radius of a sphere with volume equal to the atomic volume. The number of electrons per unit volume is thus

$$\rho = \frac{3Z}{4\pi r_{WS}^3}. \tag{3.40}$$

Comparing this with eqn (3.9) we see that

$$r_{WS} = Z^{\frac{1}{3}} r_s. \tag{3.41}$$

$r_s$ is related to the Fermi wave vector $k_F$ by:

$$k_F = \left(\frac{9\pi}{4}\right)^{\frac{1}{3}} \frac{1}{r_s}. \tag{3.42}$$

The total energy per electron of the electron gas is (Nozieres and Pines 1958):

$$E_{eg} = 2.21/r_s^2 - 0.916/r_s - (0.115 - 0.031\,3 \ln r_s). \tag{3.43}$$

The first term in eqn (3.43) represents the average kinetic energy, i.e. $3E_F/5$, where $E_F$ is the Fermi energy, and represents a repulsive contribution to the total energy. The second term is the exchange energy which is attractive because the Pauli exclusion principle keeps electrons with parallel spin apart and thus reduces the electron Coulombic repulsion. The final term represents the correlation energy and is also attractive because it represents the tendency for electrons to be kept apart by Coulomb repulsions. Minimizing $E_{eg}$ with respect to $r_s$ we find that the electron gas is in equilibrium at $r_s = 4.2$ au, or 2.3 Å, with a binding energy per electron of 0.16 Ry or 2.2 eV. To distinguish between the metals, or alternatively to account for why they have different values of $r_s$, we must therefore proceed to the next level of approximation.

In the next level of approximation we imagine that the uniform background positive charge is condensed into point ions with charge $+Z$. The electrostatic interaction between the point ions and the electron gas introduces an attractive term, $E_{epi}$. To estimate the magnitude of this term we employ the atomic sphere approximation, which turns out to be excellent for f.c.c., h.c.p., and b.c.c. crystal structures. A sphere is centred on each point ion with radius equal to $R_{WS}$. Each sphere is electrically neutral because it contains an ionic charge of $+Z$ and $Z$ electrons. The spheres may overlap a little, but that is ignored. The electrostatic energy per atom is thus the electrostatic energy contained within each sphere, which is as follows:

$$E_{epi} = - \frac{3Z}{4\pi r_{WS}^3} \int_0^{r_{WS}} dr \, 4\pi r^2 \, 4\pi r^2 \frac{Ze^2}{r} = - \frac{3Z^2 e^2}{2r_{WS}}. \tag{3.44}$$

Since $e^2 = 2$ in atomic units this becomes

$$E_{epi} = - \frac{3Z^2}{r_{WS}}. \tag{3.45}$$

The electrons within each atomic sphere repel each other and this gives rise to a positive contribution $E_{ee}$, which is as follows:

$$E_{ee} = \frac{1.2Z^2}{r_{WS}}. \tag{3.46}$$

The point ion model leads to an over estimate of the cohesive energy because it does not account for the repulsion between the valence and core electrons, which gives rise to the orthogonality hole. The point ion model is therefore replaced by a pseudopotential which describes this effect, and the simplest is the empty core pseudopotential, due to Ashcroft (1966). According to this model there is an exact cancellation between the attractive Coulomb potential and the repulsive core orthogonality contribution within some core radius $r_c$, for all angular momentum components of the valence wave function. Thus the ionic pseudopotential is local and has the following form in atomic units:

$$V_{ps}(r) = \begin{cases} 0 & \text{for } r < r_c \\ -2Z/r & \text{for } r > r_c \end{cases} \tag{3.47}$$

and it is sketched in Fig. 3.3. Core radii are given in Table 3.2. The electron–ion attractive interaction is now reduced from $E_{epi}$, eqn (3.45), to $E_{ei}$ given by

$$E_{ei} = - \frac{3Z^2}{r_{WS}} \left(1 - (r_c/r_{WS})^2\right). \tag{3.48}$$

The cohesive energy per atom is obtained by adding $Z$ times the electron gas contribution per electron, eqn (3.43), to the electron–electron and electron–ion electrostatic contributions per atom, eqns (3.49) and (3.48) respectively:

$$E_{coh} = ZE_{eg} + E_{ee} + E_{ei}. \tag{3.49}$$

The equilibrium atomic radius, $r_{WS}$, is determined by minimizing the cohesive energy with respect to $r_{WS}$, which yields an equation relating $r_{WS}$ and the core radius $r_c$:

$$(r_c/r_{WS})^2 = \frac{1}{5} + \frac{0.102}{Z^{\frac{2}{3}}} + \frac{0.0035 r_{WS}}{Z} - \frac{0.491}{Z^{\frac{1}{3}} r_{WS}}. \tag{3.50}$$

Girifalco (1976) used the experimental values of the Wigner–Sitz radius to determine an effective empty core radius, $r_c$, from eqn (3.50). The resultant values appear in Table 3.2, where as expected the core size increases with atomic number in a given group of

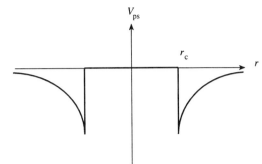

**Fig. 3.3** Plot of the Ashcroft empty core pseudopotential. Within the core radius $r_c$ the pseudopotential, $V_{ps}$, is zero. Outside the core radius the pseudopotential is the usual Coulomb potential of a positive ion.

the periodic table. It is interesting that only sodium has a value of $r_s$ that is close to the free electron gas value of 4.2 au. This indicates that although the pseudopotential is weak in these metals its effect on the cohesive energy is crucial.

The bulk modulus, $B = V(\mathrm{d}^2 E_{coh}/\mathrm{d}V^2)$, may be expressed in the form

$$B/B_{fe} = 0.2 + 0.815 r_c^2/r_s \tag{3.51}$$

at equilibrium, where $B_{fe}$ is the bulk modulus of the non-interacting free electron gas, i.e.

$$B_{fe} = 0.586/r_s^5. \tag{3.52}$$

In eqn (3.51) we have ignored the very small correlation contribution. The bulk moduli are compared with experimental values in Table 3.2. It is seen that the presence of the ion core is also crucial for obtaining reasonable values of the bulk modulus. However, the first-order expression, eqn (3.51), leads to poor estimates of the bulk modulus in polyvalent metals because the second-order terms are not negligible (Ashcroft and Langreth 1967). It is also clear that noble metals are not well described by the nearly free electron approximation, and this is because of hybridization with d electrons.

The first-order treatment is still independent of the structure of the metal, since it depends only on the density through $r_{WS}$. To account for differences in energy at a given density we must consider the terms that are second order in the pseudopotential. The resulting total energy per atom (Finnis 1974) may be expressed in the following physically transparent form:

$$E_{coh} = Z(E_{eg} - \Omega B_{eg}/2) + \tfrac{1}{2}\varphi(R = 0, r_s) + \tfrac{1}{2}\sum_{R \neq 0} \varphi(R, r_s) \tag{3.53}$$

where $B_{eg}$ represents the bulk modulus of the electron gas, $B_{eg} = \Omega(\mathrm{d}^2 E_{eg}/\mathrm{d}\Omega^2)$ and $\Omega$ is the atomic volume. The first two terms are structure-independent, density-dependent terms and account for more than 90 per cent of the total energy. Only the third term, which consists of a sum of density dependent pair potentials, depends on the crystal structure. It contributes only a small fraction of the total binding energy.

In this picture the solid is described as an assembly of neutral pseudo-atoms. Each ion core is associated with its own screening cloud which neutralizes the core. The second term in eqn (3.53) is one half of the electrostatic interaction energy between an ion core

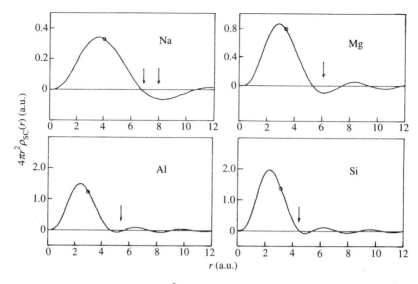

**Fig. 3.4** Screening charge densities, $4\pi r^2 \rho_{sc}(r)$, for Na, Mg. Al, and Si. The vertical arrows show the nearest neighbour in f.c.c. Al and h.c.p. Mg, diamond cubic Si, and the nearest and next nearest neighbours in b.c.c. Na. The open circle denotes the Wigner–Seitz radius. (From Hafner (1989).)

and its own screening cloud. The factor of one half is consistent with the virial theorem because the equilibrium binding energy of all electrons in a free atom is half of the total potential energy. If an ion is displaced then the resulting charge redistribution is simply obtained by a rigid displacement of the whole pseudo-atom. The energy change accompanying such a displacement is simply the change in the electrostatic energy between the displaced ion and all the other fixed pseudoatoms. Since this energy change is described by the change in the last term in eqn (3.53), we identify $\varphi(R \neq 0)$ as being the potential of the ion in the electrostatic field of a pseudo-atom at $R$. It consists of the Coulomb repulsion between the ions, plus the attraction of the first ion to the screening cloud of the second.

The efficiency of the screening of each ion core by the electron gas is quite remarkable. In Fig. 3.4 we show the screening charges associated with ion cores in Na, Mg, Al, and Si (from Hafner 1987). It is seen that the vast majority of the screening charge is located at the ionic core so that the first neighbours see an essentially neutral pseudo-atom. This is the basis of the pseudo-atom picture. Only about 25 per cent of the electronic charge lies beyond the Wigner–Seitz radius, compared to 40 per cent for free-atom charge densities. There are small Friedel oscillations in the screening charge density at longer range, which are quantum mechanical in origin. They arise from the sharp cutoff in the occupation of states in reciprocal space at the Fermi wave vector.

The effective pair potential $\varphi(R, r_s)$ is dependent on the density because the screening by the electron gas is density dependent. For a local pseudopotential it may be expressed as (Hafner and Heine 1986)

$$\varphi(R \neq 0, r_s) = \frac{2Z^2}{R} \left[ 1 - 16 \int_0^\infty \frac{\chi(q, r_s)}{\varepsilon(q, r_s)} \frac{|\hat{V}_{ps}(q)|^2}{q^2} \frac{\sin(qR)}{q} \, dq \right]. \tag{3.54}$$

The first term in eqn (3.54) gives the direct ion–ion interaction and the second term describes the attraction between the ion core and the electronic screening cloud on the

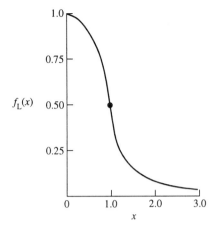

**Fig. 3.5** The Lindhard function, given by eqn (3.56). The circle shows the weak singularity at $x = 1$.

other ion. $\hat{V}_{ps}(q)$ is the normalized Fourier component of the local ionic pseudopotential $V_{ps}(r)$, i.e. $\hat{V}_{ps}(q) = (\Omega q^2/8\pi)V_{ps}(q)$. For an Ashcroft empty core pseudopotential $\hat{V}_{ps}(q) = \cos(qr_c)$. $\chi(q, r_s)$ is the response function of a free electron gas, which relates the first-order change in the electron density to a perturbing potential. For a non-interacting electron gas it is given by

$$\chi(q, r_s) = - 1.5(\rho/E_F)f_L(q/2k_F), \tag{3.55}$$

where $f_L(x)$ is the Lindhard function (Lindhard 1954):

$$f_L(x) = \tfrac{1}{2} + \frac{1 - x^2}{4x} \ln \left| \frac{1 + x}{1 - x} \right|, \tag{3.56}$$

which is shown in Fig. (3.5). For small $q/2k_F$, or a high electron density, the Lindhard function tends to unity. The response function of the free electron gas then tends to $-k_F/2\pi^2$.

The dielectric function $\varepsilon(q, r_s)$ is defined by

$$\varepsilon = 1 - 8\pi\chi/q^2, \tag{3.57}$$

and for small $q/2k_F$ it becomes

$$\varepsilon = 1 + k_{TF}^2/q^2 \tag{3.58}$$

where $k_{TF}$ is the Thomas–Fermi wave vector:

$$k_{TF} = (4k_F/\pi)^{\frac{1}{2}}. \tag{3.59}$$

If the exchange and correlation interactions are taken into account the electron Coulomb interaction, $8\pi/q^2$, in eqns (3.57) and (3.58), must be replaced by an effective electron-electron interaction $(1 - G(q))(8\pi/q^2)$, where $G(q)$ is a local field factor. In the local density functional approximation

$$G(q) = - \frac{q^2}{8\pi} \frac{d^2(\rho E_{xc})}{d\rho^2} = \alpha(q/k_F)^2, \tag{3.60}$$

where $\alpha$ is determined by a sum rule for the compressibility of the electron gas (Taylor 1978).

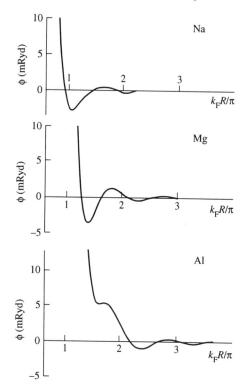

**Fig. 3.6**   Pair potentials for Na, Mg and Al, plotted as a function of $k_F R/\pi$. (From Pettifor and Ward (1984).)

The (weak) logarithmic singularity in the Lindhard function gives rise to long-range Friedel oscillations in the effective pair interaction $\varphi(R, r_s)$:

$$\varphi(R) \sim |V_{ps}(2k_F)|^2 \cos(2k_F R)/(2k_F R)^3. \tag{3.61}$$

These oscillations are very inconvenient in a real space atomistic simulation because they are so long ranged. But Pettifor and Ward (1984) pointed out that since most elemental metals do not have reciprocal lattice vectors spanning the Fermi surface these long-range oscillations must interfere destructively in the lattice sums. Therefore they removed the logarithmic singularity by replacing the Lindhard function by a rational function, which is correct to third order in $q^2$ and $1/q^2$. Thus they were able to evaluate the integral in eqn (3.54) analytically and for $r_s > 1.66$ au they showed the effective pair potential is a sum of three damped oscillatory terms:

$$\varphi(R, r_s) = (2Z^2/R) \sum_{n=1}^{3} A_n \cos(k_n R + \alpha_n) e^{-\kappa_n R}. \tag{3.62}$$

Figure 3.6 shows the analytic pair potentials computed by Pettifor and Ward (1984) for Na, Mg, and Al.

Hafner and Heine (1983, 1986) have pointed out that the variation of $\varphi(R, r_s)$ across the periodic table is determined by just two parameters: the electron density, which determines $r_s$, and the effective radius of the ion core $r_c$. This is apparent in eqn (3.54). The valence $Z$ scales the amplitude of the potential but does not affect its shape. At high electron densities the near-neighbour region of the effective pair potential assumes a Thomas–Fermi form (Taylor 1978) and the effective pair potential can become repulsive at the first neighbour separation. This is seen in Fig. 3.6 where, for Al, the repulsive core moves over the first minimum of the oscillatory part of the potential. Varying the core radius $r_c$ at constant electron density results in changes in the amplitude and phase of the oscillations. The oscillations become weaker as $V_{ps}(2k_F)$ decreases in magnitude (see eqn (3.61)). Hafner and Heine (1983,1986) and Hafner (1987) have discussed the variations in the crystal structures of the sp-bonded elements in terms of $r_s$ and $r_c$.

In closing this section let us make some remarks about the density dependence of the effective pair interactions $\varphi(R, r_s)$. The $r_s$ that appears in $\varphi$ is an average computed for the whole system. However, if a vacant site or surface is created there is clearly a need for a more local definition of the electron density. At defects such as these it seems that linear response and second-order perturbation theory are no longer adequate. However, the theory that we have outlined in this section has been applied to liquids of elemental metals and alloys with considerable success (see Hafner (1987)). This indicates that the theory remains valid if there are variations of the local density such as one would find in a liquid. At a large-angle grain boundary in an elemental metal there is almost always an expansion, which is largely confined to the boundary plane (see Section 4.3.1). Thus the ions adjacent to the geometrical boundary plane see a locally reduced electron density, which leads to a change in the effective size of the atoms. We can understand this in terms of competing electronic rearrangements. First, the smaller electron density leads to a smaller orthogonality hole at the ionic core, leading to more electronic charge at the core. Secondly, screening ensures that the atom remains electrically neutral. Therefore, the screening length increases to disperse the screening charge over a greater radius, although the vast majority of the screening charge remains on the atom. The end result is that the atom remains neutral but its effective size has increased because the screening charge has moved outwards from the nucleus. Therefore the local effective pair potential should become more repulsive in the first neighbour region, corresponding to an increase in the effective atomic size in the boundary region. This is entirely analogous to the 'chemical compression' observed in metallic alloys (Hafner 1987), where the effective atomic size of elements with lower valence decreases on alloying with elements of higher valence. Our picture of the metal is therefore the following (Finnis 1974): it consists of neutral pseudo-atoms, with screening charges that are dynamically contracting and expanding in response to local variations in the electron density caused by neighbouring atoms moving around, to ensure that each pseudo-atom remains electrically neutral at all times. This picture carries over into tightly bound systems (Sections 3.8.5 and 3.8.6), where nearly free electron theory breaks down.

We have noted that the effective pair potential describes energy changes following an ionic rearrangement at constant density. More precisely, the condition is that the screening clouds associated with each atom retain the same shape and size. In a compression wave this condition is not satisfied. If we compute the phonon dispersion relations with a fixed potential function, which does not take into account the (virtually instantaneous) local changes in the screening clouds, we find $\sim 50$ per cent errors in the bulk modulus of polyvalent metals like Al and Pb, using the long wavelength limit (Finnis 1974, Rosenfeld and Stott 1987). This amounts to a recognition that there are important $N$-body interactions because the effective pair interaction between two pseudo-atoms is influenced by the disposition of all the neighbours to those two atoms. Such $N$-body interactions are much easier to handle in a tight binding framework (Section 3.8).

In view of the above remarks it is difficult to assess the reliability of using effective pair potentials, eqns (3.54) and (3.62), to describe interactions between pseudo-atoms with frozen screening charge distributions, at interfaces in metals. Clearly, a self-consistent treatment is needed in which the screening clouds can respond to the local density variations in such a way as to maintain each pseudo-atom electrically neutral. This need is recognized and fulfilled by the tight binding bond model, Section 3.8.5.

## 3.6 EFFECTIVE MEDIUM THEORY

In effective medium theory (Norskov and Lang 1980, Stott and Zaremba 1980) the idea is to relate the energy of an atom in a solid to the energy of embedding an atom in jellium. The jellium calculation can be performed self-consistently, within local density functional theory, once and for all over a range of electron densities. The physical reasoning underlying the approach is that in a metal screening is so efficient that the chief role of the local atomic environment is to define a local electron density, and the atomic structure of the local environment (at a given electron density) is of less importance. The method was applied initially to chemisorption of light gas atoms (hydrogen, oxygen, carbon) on metals (e.g. Chakraborty *et al.* 1985). Effective medium theory underlies the widely used empirical approaches such as the embedded atom method (Daw and Baskes 1984). Daw and Baskes (1984) generalized effective medium theory to a description of metallic cohesion by recognizing that each atom in a solid may be regarded as being embedded in an electron gas provided by neighbouring atoms. Jacobsen *et al.* (1987) carried out a detailed theoretical study of the concept, and identified the energy contributions in the solid that are not considered by embedding the atom in jellium. Much of this section is based on Jacobsen (1990) and Puska (1990).

Consider an atom embedded within jellium of density $\rho_h$, which is called the embedding density. The atom induces changes in the electron density and the electrostatic potential; indeed, there is no electrostatic potential in the jellium before the atom is embedded. Let $\Delta\rho(r)$ denote the charge density distribution that is induced by the atom: it is the difference between the electron density of the jellium plus embedded atom system and the original jellium. The integral of $\Delta\rho(r)$ equals the negative of the nuclear charge of the atom. The induced density varies as the embedding density changes. Similarly, let $\Delta\varphi(r)$ denote the total electrostatic potential of the atom plus jellium system.

The total electron density of the solid, $\rho(r)$, is expressed as a superposition of induced densities $\Delta\rho_i(r)$ centred at sites $i = 1, 2, \ldots$:

$$\rho(r) = \sum_i \Delta\rho_i(r). \tag{3.63}$$

This is the basic ansatz of the theory. The variational principle of density functional theory has been invoked by Jacobsen *et al.* (1987) to argue that the errors in this charge density ansatz give rise to only second-order errors in the total energy. Equation (3.63) does not define $\rho(r)$ until we specify the embedding density for each atom. If the atomic sites are not equivalent the embedding densities should differ. This is achieved by defining the embedding density, $\bar{\rho}_i$, at site $i$ as the average density, inside a neutral sphere centred on atom $i$, due to all the other atoms in the system:

$$\bar{\rho}_i = \left\langle \sum_{j \neq i} \Delta\rho_j \right\rangle. \tag{3.64}$$

The radius of the neutral sphere is arbitrary, but a sensible choice is to require that the total electronic charge within the sphere cancels the nuclear charge on the atom. The total electronic charge within the sphere consists of the tails of the electron charge densities from the neighbouring atoms, and the electron charge density of the central atom which falls within the sphere. Thus we have defined a self-consistent problem: the embedding densities are defined in terms of the $\Delta\rho_i$'s, which are in turn dependent on the embedding densities. Thus, it is argued that the electronic self-consistency problem of the atom in a real solid is equivalent to the problem of the atom in a homogeneous electron gas of a self-consistently chosen density. The arbitrary choice of a sphere for defining the embedding density is reasonable in a close packed metal, but less so in a more open crystal structure.

Let $s$ be the radius of the neutral sphere. To a good approximation (Jacobsen 1990), for usual metallic densities, the embedding density $\bar\rho$ is related to $s$ exponentially:

$$\bar\rho(s) = \rho_0 \exp[-\eta(s - s_0)]. \tag{3.65}$$

As the embedding density increases the neutral sphere decreases in size. In the language of the previous section, the decreasing sphere radius reflects the contraction of the screening charge density to counteract the increasing orthogonality hole and maintain local charge neutrality. In effect eqn (3.65) describes the 'chemical compression' of the size of an atom as the density of the electron gas in which it is embedded increases. The parameter $\eta$ depends only on the atomic number of the embedded atom. The parameters $\rho_0$ and $s_0$ are discussed below. Values of $\eta$, $\rho_0$, and $s_0$ have been tabulated by Puska (1990) for 30 elements.

Jacobsen *et al.* (1987) discussed the derivation of the binding energy in effective medium theory based on the ansatz for the charge density, eqn (3.63), and the variational principle of density functional theory. The result is that the binding energy is given by

$$E_B = \sum_i E_i(\bar\rho_i) + E_{ov} + E_{1e}. \tag{3.66}$$

$E_i(\bar\rho_i)$ is the energy of embedding the $i$th atom in a neutral sphere in which the average electron density is $\bar\rho_i$. This is expressed as the energy of embedding the $i$th atom in jellium of density $\bar\rho_i$, with a correction, $-\bar\rho_i\varphi_i(\bar\rho_i)$, to account for the electrostatic interaction between the uniform positive background density, $-\bar\rho_i$, of the jellium and the Coulomb potential of the central atom inside the neutral sphere. But the embedding density $\bar\rho_i$ is defined by the charge density tails of the neighbouring atoms, see eqn (3.64). If we ignore their non-spherically symmetric components, which is reasonable in a close packed environment, we may intrepret the correction as follows: it describes, within the neutral sphere, the electrostatic attraction between the charge density tails of the neighbouring atoms and the central atom. Figure (3.7) shows the total energy, $E_i$, of embedding He, Ne, H, and O atoms as a function $\bar\rho_i$. The inert gas atoms are repelled by all electron densities but there is an optimium embedding density for H and O. This suggests that H and O atoms will seek atomic sites in a solid where they can find these optimum embedding densities, but He and Ne will always seek those sites of lowest electron density. Optimum sites may be found at defects such as interfaces. At high embedding densities the energy of an atom in jellium increases rapidly, owing to the rise in kinetic energy of the conduction electrons to satisfy the orthogonality constraint with the bound atomic states. On the other hand the electrostatic interaction energy, $-\bar\rho_i\varphi_i(\bar\rho_i)$, is attractive and becomes increasingly so with increasing electron density. At high electron densities the repulsion due to the kinetic energy always dominates. This explains the general shape of the H and O curves in Fig. 3.7. Puska (1990) has parametrized and tabulated the function $E_i(\bar\rho_i)$ for 30 elements.

The second term is called the overlap energy. It describes the electrostatic repulsion

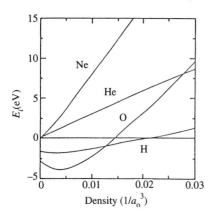

**Fig. 3.7** A plot of the energies (in eV) of embedding Ne, He, H, and O atoms in a homogeneous electron gas as a function of the electron density (in inverse cubic Bohr radii). (From Puska (1990).)

that occurs when neutral spheres are forced to overlap. For a close packed metal crystal the neutral spheres are the Wigner–Seitz spheres, and the overlap between neighbouring spheres is small. But in more open structures the overlap can become important. The overlap is important even in a close packed metal if we want to calculate the shear elastic constants, as first shown by Fuchs (1935).

The third term is called the one-electron energy and is important in cases where the atomic environment into which the atom is embedded deviates markedly from a free electron gas. More precisely, if the local density of one-electron states in the metal is not free-electron-like then the third term is important. Thus it is reasonable to neglect this term for sp-bonded metals but not for transition metals, which are dominated by d–d bonding with prominent, narrow d bands in the density of states. In those cases where the one-electron term is important it is more sensible to start with a tight binding model of the solid, and to abandon the concept of embedding the atom in a homogeneous electron gas.

The effective medium theory seems well suited to close-packed, sp-bonded metals, where the one electron term, $E_{1e}$, can be ignored, and the spherical averaging should not introduce too many errors. For example, a simple picture can be given to explain the inward relaxation observed at many simple metal surfaces. The reduced electron density at the surface layer reduces the kinetic energy repulsion due to the formation of the orthogonality hole. The electrostatic attraction between the atoms of the surface layer and the layer below can then draw the surface atoms inwards. This is essentially the same explanation given by Finnis and Heine (1974). The inward relaxation of a clean metal surface can be removed by absorption of atoms on the surface, because they replenish the surface electron density. The outward relaxation is expected to be dependent on the surface coverage, as found experimentally for H on Cu (110) (Baddorf *et al.* 1987).

Let us compare the effective medium theory with the second-order perturbation theory of the previous section for sp-bonded close-packed metals. The large, purely density-dependent energy of the perturbation theory is similar in some respects to the energy of embedding the atom in jellium in effective medium theory. They are both density-dependent and they both dominate the binding energy of the metal. Equation (3.65) defining the effective size of an atom in terms of a local electron density is reminiscent of the fact that the effective pair potentials in perturbation theory are density-dependent. Effective medium theory has the advantage of a self-consistent algorithm for determining the embedding charge density and hence the embedding energy. In second-order perturbation theory the local density variations are ignored partly because there is no equivalent, self-consistent algorithm for defining the local density. Thus the validity of effective medium theory as a description of atomic interactions at surfaces should be greater than second-order perturbation theory. On the other hand, the long-range oscillatory form of the pair potential in second-order perturbation theory is known to be important for stacking fault energies and structural stability at constant density. But in the empirical implementations of effective medium theory the induced charge densities $\Delta\rho_i(r)$ are often taken to be short-ranged, monotonically decreasing functions of $r$. The ability of such empirical treatments to predict structural energy differences is therefore called into question. It is possible that a better approach would be to use the induced charge densities that are computed by local density functional theory for an atom embedded in jellium. These induced charge densities exhibit long-range Frieder oscillations, and may give the required structural stability.

## 3.7 THE EMBEDDED ATOM METHOD

The embedded atom method (Daw and Baskes 1983, 1984) is an empirical implementation of eqn (3.66) for the binding energy of the metal as a sum of two terms. The first term is the attractive energy of embedding each atom in the electron density provided by neighbouring atoms. The second term describes the electrostatic repulsive overlap inter-action of neutral atoms. The 'one-electron' term of eqn (3.66) is omitted. The cohesive energy of an elemental metal is expressed as follows:

$$E_{\text{coh}} = \sum_i f(\rho_i) + \tfrac{1}{2} \sum_{i \neq j} \sum U(r_{ij}),$$

$$\rho_i = \sum_{j \neq i} \rho^{\text{a}}(r_{ij}). \qquad (3.67)$$

$\rho_i$ is the charge at the $i$th nucleus due to the spherically symmetric atomic charge densities, $\rho^{\text{a}}(r_{ij})$, of neighbouring atoms, and $f(\rho_i)$ is the embedding function. The over-lap interaction is represented by a pair potential $U(r_{ij})$. The embedding function $f$ and the pair potential $U$ are found empirically by fitting bulk properties such as the lattice constant, cohesive energy, elastic constants, and vacancy formation energy. For dilute binary alloys the heats of formation were also fitted (Foiles *et al.* 1986). The embedded atom method is closely related to the Finnis–Sinclair model (Finnis and Sinclair 1984), which we discuss in Section 3.8.6, and to the glue model of Ercolessi *et al.* (1988).

The embedded atom method (EAM) has been widely used to model interfaces in f.c.c. elemental metals and alloys. It is an improvement over models in which *all* the cohesion is expressed solely by pair potentials for two principal reasons. Firstly, the EAM correctly describes the tendency in metallic bonding for the bond strength to increase as the coordination number decreases. This is arguably the most significant aspect of the EAM and related schemes. We shall discuss this in more detail below. Secondly, the Cauchy relation between the elastic constants in cubic crystals, $c_{12} = c_{44}$, is inescapable with pair potentials but not in the EAM. The computational cost of using the EAM is no more than twice that of using a simple pair potential model of comparable range, and is often slightly less.

In a nearest neighbour, pair potential model, where *all* the cohesion is represented by a sum of pair interactions, the energy of any particular atom is proportional to its coordination number. But in a metal the energy of each atom contains a term that is proportional to a non-linear function of the coordination number that is often well approximated by the square root. In the EAM the function of the coordination number is identified with the embedding function $f(\rho)$. We shall explain the mathematical origin of the square root dependence in Section 3.8.6. Here, we shall give a qualitative discussion of the metallic bond, to give some insight into the origin of the non-linear function.

In a molecule like $H_2$ the bond is saturated because the bonding state is occupied by two electrons and the antibonding state is empty. In a solid the bonding and antibonding states become bands, but one can still talk about saturated bonds containing two electrons. For example, in diamond each C–C bond is saturated and contains two electrons. The C–C bond in diamond has very similar properties to the C–C bond in ethane. Thus, the properties of this C–C bond are transferable from one situation to another, provided the two carbon atoms remain tetrahedrally coordinated. It is as if the C–C bond has its own pair of electrons which it does not share with surrounding atoms.

An unsatuarated bond has fewer than two electrons and the properties of the bond are not transferable to other environments. In a metal the coordination number is too large to allow each bond to contain two electrons. The valence electrons lower their kinetic energy, owing to the uncertainty principle, by distributing themselves in molecular orbitals extending over as many atoms as possible. The electrons 'resonate' in Pauling's language (Pauling 1950) between the available bonds. At any given instant there may be two electrons in a particular bond, but at a later time there will be fewer. The delocalization or resonance of each electron among all the neighbours leads to an intrinsically multi-atom force which cannot be expressed as a sum of two-body, three-body, ... forces. This is the origin of the non-linear dependence of the energy on the coordination number. The concept of multi-atom forces in metals is consistent with the discussion of interatomic forces in second-order perturbation theory in Section 3.5. There the cohesive energy of the metal is dominated by a density-dependent term. This is also an intrinsically multi-atom force since the definition of a local volume, through a Voronoi cell construction for example, entails all the neighbours of an atom.

The non-linear nature of the function leads to the conclusion that increasing the coordination number of an atom decreases the energy associated with each of its bonds. If the bond energies were represented by a pair potential it would have to be dependent on the coordination number to simulate this effect. In the EAM and Finnis–Sinclair models it is achieved through the function $f(\rho)$ having a positive second derivative. To see this let us expand $f(\rho)$ about some reference density $\bar{\rho}$, which may be the value of $\rho$ in a perfect crystal or at a defect such as a surface or interface:

$$f(\rho) = f(\bar{\rho}) + (\rho - \bar{\rho})f'(\bar{\rho}) + (\rho - \bar{\rho})^2 f''(\bar{\rho})/2 + \ldots . \tag{3.68}$$

$\rho$ is a sum of spherically symmetric charge densities, which is simply a sum of pair potentials. Thus the term $(\rho - \bar{\rho})f'(\bar{\rho})$ is a sum of pair potentials, and the next term is a sum of three-body interactions and so on. Provided $|\rho - \bar{\rho}|$ is small this series is convergent. Note that the expansion does not invalidate our earlier remarks about the difficulty in describing metallic bonding by a convergent series of two-body, three-body, ... interactions! The two-body, three-body, ... terms in eqn (3.68) are in addition to the term $f(\bar{\rho})$, where the greatest contribution and the intrinsically multi-atom interactions reside. The coefficients in the expansion are dependent on the reference density $\bar{\rho}$. If $f''(\rho)$ is positive then $|f'(\rho_1)|$ is less than $|f'(\rho_2)|$ for $\rho_1$ larger than $\rho_2$, where we recall that $f(\rho)$ is negative. This is illustrated in Fig. 3.8. For a given bond length $\rho_1/\rho_2$ implies that the coordination number for $\rho_1$ is larger. Thus the pair potentials obtained from the series expansion about the larger coordination number environment are weaker. As this

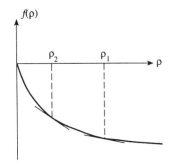

**Fig. 3.8** To illustrate that the magnitude of the slope of the embedding function $f(\rho)$ for a larger embedding density $\rho_1$ is smaller than the magnitude of the slope for a smaller embedding density $\rho_2$.

is a fundamental property of metallic bonding the second derivative of the embedding function is always positive in the EAM and related schemes.

The positive second derivative of the embedding function is also crucial for the violation of the Cauchy relation $c_{12} = c_{44}$ in the EAM. For a cubic material one can show (Daw and Baskes 1984) that

$$c_{12} - c_{44} = \text{positive constant} \times f''(\bar\rho) \tag{3.69}$$

where $\bar\rho$ is the value of $\rho$ in the perfect crystal. Thus in the EAM and related models the Cauchy discrepancy, $c_{12} - c_{44}$, is always positive. In fact, the Cauchy discrepancy is positive for most metals, but for rhodium and iridium, both of which have f.c.c. structures, it is negative. This indicates that there must be important additional terms in the energy to account for the negative Cauchy discrepancy, and their origin is mentioned in Section 3.8.7.

The EAM is best suited to sp-bonded f.c.c. metals where the primary consideration is dense packing and the influence of the one electron term in eqn (3.66) is negligible. Thus the EAM should not be expected to be a good description of the transition metals, even if they are f.c.c., or other materials where there is directional bonding such as Si. But the EAM has often been regarded as a purely empirical approach, and therefore several groups have tried fitting b.c.c. and h.c.p. metals with varying degrees of success (see papers in Vitek and Srolovitz (1989), Johnson (1990)). As we shall see in Section 3.8.4, whether a metal adopts a b.c.c., f.c.c., or h.c.p. crystal structure is determined by higher-order terms, which, *in principle*, are beyond the capability of eqn (3.67).

For small local changes in the function $\rho$ we have seen in eqn (3.68) that we may derive two-body, three-body, etc., interactions from the embedding function $f(\rho)$. If the series is truncated after the two-body interactions the cohesive energy, eqn (3.67), becomes:

$$E_{\text{coh}} = N(f(\bar\rho) - \bar\rho f'(\bar\rho)) + \tfrac{1}{2}\sum\sum_{i\neq j} U(r_{ij}) + 2f'(\bar\rho)\rho^{a}(r_{ij}), \tag{3.70}$$

where $N$ is the total number of atoms. Thus, we can define an effective, pair potential as follows:

$$U_{\text{eff}}(r_{ij}) = U(r_{ij}) + 2f'(\bar\rho)\rho^{a}(r_{ij}). \tag{3.71}$$

We stress two points about the effective pair potential. First, it is dependent on the choice of $\bar\rho$. Second, even for small departures of $\rho$ from the reference value the cohesive energy consists not only of the sum of effective pair potentials, but also a sum of one-body terms, $f(\bar\rho) - \bar\rho f'(\bar\rho)$, which obviously depend on the choice of $\bar\rho$. However, the one-body terms are cancelled when we consider the *energy differences* between two configurations, and the EAM reduces to a pair potential model, *provided* the local variations of $\rho$ from the reference $\bar{\rho}$ are small.

Some workers have argued that the pair potential should always be positive and the embedding function always negative, while others prefer to use an embedding function derived from immersing the atom in an electron gas where it can be positive or negative. Thus, it appears that the embedding function and the pair potential are not unique. This is indeed the case (Ercolessi *et al.* 1988) because if we replace $f(\rho)$ by $g(\rho)$ and $U(r)$ by $V(r)$ where

$$g(\rho) = f(\rho) + k\rho$$

and

$$V(r) = U(r) - 2k\rho^a(r), \qquad (3.72)$$

where $k$ is a constant, then the cohesive energy is invariant:

$$E_{\text{coh}} = \sum_i f(\rho_i) + \tfrac{1}{2}\sum_{i \neq j}\sum U(r_{ij}) = \sum_i g(\rho_i) + \tfrac{1}{2}\sum_{i \neq j}\sum V(r_{ij}). \qquad (3.73)$$

Thus we can add any multiple of $\rho$ to the embedding function and compensate for it in the pair potential. Johnson (1990) pointed out that if we choose $k = -f'(\bar{\rho})$ then the pair potential $V(r)$ becomes the effective pair potential $U_{\text{eff}}(r)$ given in eqn (3.71). This is a convenient way to 'normalize' apparently different EAM functional forms so that they may be compared meaningfully. The embedding function becomes

$$g(\rho) = f(\rho) - f'(\bar{\rho})\rho, \qquad (3.74)$$

and thus its first derivative at $\rho = \bar{\rho}$ is zero. The significance of this result is that to calculate *energy differences*, such as a vacancy formation energy or a grain boundary energy, the EAM functional form, eqn (3.67), may be replaced by the effective pair potential $U_{\text{eff}}(r)$ because the errors incurred by ignoring the changes in $g(\rho)$ are *second order* in $\rho - \bar{\rho}$. But to calculate the cohesive energy we are obliged to take both $U_{\text{eff}}(r)$ and $g(\rho)$ into account. The empirical pair potentials, which were used for many years before the advent of the EAM and related models, should therefore be identified with the effective pair potentials $U_{\text{eff}}$ of the EAM. As Johnson (1990) has noted, this implies that while it is sensible to fit empirical pair potentials to the vacancy formation energy they should not be fitted to the cohesive energy. Indeed, they should not be expected to reproduce the cohesive energy. On the other hand, the equilibrium lattice constant is determined by only $U_{\text{eff}}$ and hence it is correct to fit the lattice constant when fitting an empirical pair potential. Obviously, the second-order changes in $g(\rho)$ are *not* negligible when one is interested in second-order changes in the energy, such as in phonon dispersion relations and elastic constants, or when $(\rho - \bar{\rho})$ is large as at a surface. Therefore, empirical pair potentials should not be expected to reproduce elastic constants, phonon dispersion relations, or surface energies and reconstructions.

At a grain boundary in a close-packed metal the second-order changes in the embedding energy are small because the changes of the coordination number and the boundary expansion are usually small (see Section 4.3.1). The EAM functional form may therefore be replaced by the effective pair potential without incurring significant errors in modelling the structure and energy of a grain boundary in a close packed metal.

We may conclude this section by noting that the structures of a number of [001] and [111] twist boundaries in Au, as calculated with the EAM, have been found to be in substantial agreement with the strutcures found experimentally by means of X-ray diffraction (Majid *et al.* 1989, Majid and Bristowe 1992). For both types of boundaries the general forms of the atomic displacement fields and the magnitudes of the displacements were well described by the simulations. Moreover, simpler pair potential calculations have been found (Bristowe 1993) to yield similar results, as we have just argued above.

## 3.8 TIGHT BINDING MODELS

### 3.8.1 Introduction

Tight binding is a semi-empirical, LCAO theory in which we interpret the basis functions as *atomic-like* orbitals, displaying the same angular momentum as real atomic orbitals (s,p,d, etc., corresponding to angular momentum $l = 0,1,2$, etc.) but with radial dependencies that can be quite different from free atomic orbitals. Tight binding is often regarded as the opposite of the nearly free electron approximation. The usual argument (e.g. Ziman (1972)) is that for the tight binding approximation to be valid we require the overlap between atomic basis functions on neighbouring atoms to be small. With a small overlap the atomic states are weakly coupled and therefore the probability of an electron tunnelling from one atom to the next is small. Since the electron spends most of the time in the vicinity of some atom or other the approximation that its wave function is a linear combination of atomic orbitals should be valid. Indeed, this is true for the d-bands of the transition metals where the overlap between atomic d states on neighbouring atoms is small, and tight binding describes the bonding very well. But the tight binding and nearly free electron treatments converge to the same physical picture in sp-bonded materials where there is *large overlap* between atomic basis functions. For example, in diamond we gain a perfectly acceptable description of the valence bands if we use one s and three p type atomic-like basis functions. And yet the resultant bands look very free-electron-like, as we remarked in Section 3.5. This is completely consistent with the fact that the valence electrons are not spread out uniformly in the manner of a free electron gas but are located almost entirely in the bonds between atoms, with almost no electron density in the tetrahedral holes. We can understand this paradox by using an argument due to J. C. Phillips (quoted by Heine *et al.* (1991)). The charge density becomes non-uniform through standing waves being set up by Bragg reflection of free electron travelling waves at zone boundaries. Since the kinetic energy of the standing waves is exactly the same as the kinetic energy of the travelling waves, the dispersion relation remains free-electron-like even though the charge distribution is no longer uniform. The non-uniformity in the charge distribution is simply a result of quantum interference. Whether we choose an LCAO or plane wave basis set is largely a matter of convenience and taste. Mathematically, it does not matter which basis set one chooses for the expansion so long as the basis set is complete.

### 3.8.2 The diatomic molecule

Consider a diatomic molecule comprising an A atom and a B atom. We make the simplifying assumption that the effective potential felt by a valence electron is simply the linear superposition of the individual atomic pseudopotentials $v_A(r)$ and $v_B(r)$. The assumption is equivalent to a Harris–Foulkes approximation for the input charge density as a superposition of free atomic charge densities (see Section 3.2.2). Let a solution of the Schrödinger equation for a valence electron state of an isolated A atom be $\psi_A$, with energy $E_A$:

$$H_A \psi_A = E_A \psi_A \tag{3.75}$$

where

$$H_A = -\nabla^2 + v_A(r). \tag{3.76}$$

$v_A(r)$ is the effective potential due to atom A. Similar equations are satisfied by $\psi_A$ and $E_B$. Let the wave function for the AB molecule be $\psi_{AB}$. We assume that $\psi_{AB}$ may be expressed as a linear superposition of $\psi_A$ and $\psi_B$, with variable coefficients $c_A$ and $c_B$:

$$\psi_{AB} = c_A \psi_A + c_B \psi_B. \tag{3.77}$$

Physically, this assumption amounts to saying that the molecular orbital $\psi_{AB}$ retains, at least partially, the character of an occupied valence state of a free atom when the electron is near to either atomic core. We can improve the approximation by including higher-energy atomic states $\psi_A$ and $\psi_B$ in the set of states in which $\psi_{AB}$ is expanded. Clearly, if $\psi_{AB}$ were expanded in a complete set of states the representation of $\psi_{AB}$ would be exact.

The Schrödinger equation for the molecular orbital $\psi_{AB}$ is as follows:

$$(-\nabla^2 + v_A + v_B)(c_A \psi_A + c_B \psi_B) = E_{AB}(c_A \psi_A + c_B \psi_B). \tag{3.78}$$

Multiplying from the left by $\psi_A^*$ and $\psi_B^*$ in turn and integrating over all space we obtain:

$$\begin{pmatrix} (E_A' - E_{AB}) & (h - E_{AB}S) \\ (h^* - E_{AB}S^*) & (E_B' - E_{AB}) \end{pmatrix} \begin{pmatrix} c_A \\ c_B \end{pmatrix} = \begin{pmatrix} 0 \\ 0 \end{pmatrix} \tag{3.79}$$

where $E_A'$ and $E_B'$ are the free atomic eigenvalues corrected by the electrostatic field due to the other atom:

$$\left. \begin{array}{l} E_A' = E_A + \langle \psi_A | v_B | \psi_A \rangle \\ E_B' = E_B + \langle \psi_B | v_A | \psi_B \rangle \end{array} \right\}. \tag{3.80}$$

$h$ is a negative quantity called the hopping integral, and describes the coupling between the atomic states $\psi_A$ and $\psi_B$ through the Hamiltonian:

$$h = \langle \psi_A | -\nabla^2 + v_A + v_B | \psi_B \rangle = E_A S + \langle \psi_A | v_B | \psi_B \rangle = E_B S + \langle \psi_A | v_A | \psi_B \rangle, \tag{3.81}$$

where $S$ is a positive quantity called the overlap integral, measuring the extent to which the atomic basis functions $\psi_A$ and $\psi_B$ overlap in space:

$$S = \langle \psi_A | \psi_B \rangle. \tag{3.82}$$

The asterisks in eqn (3.79) denote complex conjugation. We may always choose real basis functions for $\psi_A$ and $\psi_B$, and therefore $h^* = h$ and $S^* = S$. Equation (3.79) is called a secular equation and non-trivial solutions are obtained by setting the determinant of the $2 \times 2$ matrix to zero. The exact solutions are

$$E_{AB} = \frac{\bar{E}' - hS}{(1 - S^2)} \pm \frac{1}{(1 - S^2)} \left( (\Delta E)^2 (1 - S^2) + (h - S\bar{E}')^2 \right)^{\frac{1}{2}} \tag{3.83}$$

where

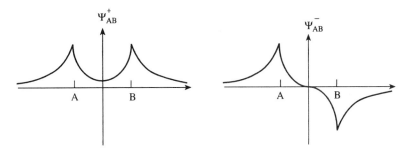

**Fig. 3.9** The bonding ($\psi_{AB}^+$) and antibonding ($\psi_{AB}^-$) molecular orbitals of two hydrogen atoms at A and B.

$$\left. \begin{array}{l} \bar{E}' = (E_A' + E_B')/2 \\ \Delta E = (E_A' - E_B')/2 \end{array} \right\}. \tag{3.84}$$

These results become more physically transparent if we set the overlap, $S$, to zero. This is obviously a simplification but it does not alter the qualitative picture we shall now derive. The eigenvectors corresponding to the two eigenvalues in Eqn (3.84) are now given by

$$\frac{c_A}{c_B} = \frac{h}{\bar{E}' \pm Q - E_A'}, \tag{3.85}$$

where

$$Q = ((\Delta E)^2 + h^2)^{\frac{1}{2}}. \tag{3.86}$$

If the A and B atoms are of the same type then $\bar{E}' = E_A' = E_B' = E'$ and $\Delta E = 0$, and the two solutions are

$$E_{AB}^{\pm} = E' \pm h \tag{3.87}$$

corresponding to the normalized eigenvectors $(c_A, c_B) = (1, 1)/\sqrt{2}$ and $(1, -1)/\sqrt{2}$ respectively. The state with the lower eigenvalue, $E' + h$, is called a bonding state because its energy is lower than the free atomic eigenvalues $E_A = E_B$. Similarly the state with the higher eigenvalue is called an antibonding state because its energy is higher than $E_A$ and $E_B$. As shown in Fig. 3.9 the bonding state corresponds to a symmetric combination of the atomic states $\psi_A$ and $\psi_B$ with charge accumulating in the bond between the atoms. The formation of the bond may be seen explicitly by evaluating the electron density $\rho_{AB}^b = 2|\psi_{AB}^b|^2$ of two electrons occupying the bonding state (one from each atom) as follows:

$$\rho_{AB}^b = \rho_A + \rho_B + \rho_{bond}, \tag{3.88}$$

where $\rho_{bond}$ represents the redistribution of the electron charge density due to the formation of the covalent bond:

$$\rho_{bond} = 2\psi_A\psi_B \tag{3.89}$$

This equation shows that the formation of the bond is a quantum interference effect, the bond charge density being due to the interference term $\psi_A\psi_B$. In the antibonding state the charge density falls to zero midway between the atoms, as shown in Fig. 3.9.

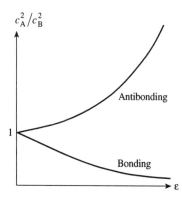

**Fig. 3.10** The ratio of the charges on the A and B atoms for the bonding and antibonding states, given by eqns (3.93) and (3.94) respectively, of the diatomic molecule as a function of $\varepsilon$, given by eqn (3.92).

For A and B atoms that are not of the same type $\Delta E$ is not zero and the eigenvalues of the molecule become

$$E^{\pm}_{AB} = \bar{E}' \pm ((\Delta E)^2 + h^2)^{\frac{1}{2}}. \tag{3.90}$$

The positive root is now the antibonding state since its energy is higher than $\bar{E}'$. The corresponding eigenvectors are:

$$(c_A, c_B) = (1, -\varepsilon \pm \sqrt{1 + \varepsilon^2})/(2(1 + \varepsilon^2 \mp \varepsilon\sqrt{1 + \varepsilon^2}))^{\frac{1}{2}} \tag{3.91}$$

where

$$\varepsilon = \frac{\Delta E}{h}. \tag{3.92}$$

The difference in the atomic levels $E'_A$ and $E'_B$ introduces an asymmetry into the charge distribution of the bonding and antibonding molecular orbitals. The magnitude of the asymmetry is determined by the parameter $\varepsilon$. If there is one electron in the bonding state with energy level $E^{-}_{AB}$ then the ratio of the contributions from the A and B atoms to this charge is given by

$$c^2_A/c^2_B = 1/(1 + 2\varepsilon^2 + 2\varepsilon\sqrt{1 + \varepsilon^2}). \tag{3.93}$$

This is sketched in Fig. 3.10. As $\varepsilon$ increases more weight of the bonding charge distribution transfers to the B atom. This is simply a consequence of the fact that as $\varepsilon$ increases charge flows from the higher energy level, $E'_A$, to the lower $E'_B$. For the antibonding state the reverse situation applies:

$$c^2_A/c^2_B = 1/(1 + 2\varepsilon^2 - 2\varepsilon\sqrt{1 + \varepsilon^2}), \tag{3.94}$$

which is also sketched in Fig. 3.10. Thus, when $\Delta E$ is not zero the bond becomes partially ionic.

Harrison (1980) defines the polarity of a bond (i.e. the extent to which it is polar or ionic) by the quantity

$$\alpha_p = \frac{\varepsilon}{\sqrt{1 + \varepsilon^2}}, \tag{3.95}$$

and the covalency of a bond (i.e. the extent to which it is covalent) by

$$\alpha_c = \frac{1}{\sqrt{1 + \varepsilon^2}} \tag{3.96}$$

so that

$$\alpha_c^2 + \alpha_p^2 = 1. \tag{3.97}$$

The completely ionic limit is attained when $\varepsilon \to \infty$, and the completely covalent limit when $\varepsilon = 0$. As we shall see in the next section similar concepts apply in crystals.

### 3.8.3 Bands, bonds, and Green functions

Consider an assembly of $N$ atoms, which may or may not be in a crystalline arrangement. Let $|\alpha\rangle$ denote an atomic-like state, where $\alpha$ is a composite index for the position of the atom to which it is anchored and the angular momentum of the orbital (i.e., s,p,d etc.). The set of all $|\alpha\rangle$'s forms a basis set, in which an eigenstate $|n\rangle$ of the whole assembly is expanded:

$$|n\rangle = \sum_\alpha c_{n\alpha} |\alpha\rangle. \tag{3.98}$$

The eigenstate $|n\rangle$ is a giant molecular orbital. Inserting this trial wave function into the Schrödinger equation we obtain

$$\sum_\alpha H_{\beta\alpha} c_{n\alpha} = E_n \sum_\alpha S_{\beta\alpha} c_{n\alpha}. \tag{3.99}$$

The Hamiltonian matrix element $H_{\beta\alpha}$ represents $\langle\beta|H|\alpha\rangle$. Similarly $S_{\beta\alpha}$ is an element of the overlap matrix and represents $\langle\beta|\alpha\rangle$. Equation (3.99) is the secular equation, and non-trivial solutions (i.e. $c_{n\alpha} \neq 0$ for all $\alpha$) are found by solving the secular determinant,

$$\text{Det}\,(H_{\alpha\beta} - ES_{\alpha\beta}) = 0. \tag{3.100}$$

Suppose we have two eigenstates $|n\rangle = \sum_\alpha c_{n\alpha}|\alpha\rangle$, and $|m\rangle = \sum_\alpha c_{m\alpha}|\alpha\rangle$. The eigenstates are orthonormal, which means that $\langle m|n\rangle = \delta_{mn}$. Therefore,

$$\sum_{\alpha,\beta} c_{n\alpha}^* c_{m\beta} S_{\alpha\beta} = \delta_{mn}. \tag{3.101}$$

Using this relation and the matrix form of the Schrödinger equation, eqn (3.99), it follows that

$$E_n = \sum_{\alpha,\beta} c_{n\beta}^* c_{n\alpha} H_{\beta\alpha}. \tag{3.102}$$

This is a relation between the energy of an eigenstate of the entire system and the Hamiltonian matrix elements, which are relatively short ranged, local quantitites. However, this relation is not quite what we are looking for because the individual eigenenergies are not well-behaved quantities. To see this we only have to recall that the eigenstates of the system are strongly affected by the boundary conditions, i.e. whether the external shape of the crystal is cubic or spherical for example. A quantity that is much more stable is the sum of the occupied eigenstates, which is known as the band energy:

$$E_{\text{band}} = 2\sum_n{}' E_n = \sum_{\alpha,\beta} \left( 2\sum_n{}' c_{n\alpha} c_{n\beta}^* \right) H_{\beta\alpha}, \tag{3.103}$$

where the prime on the sum means that only occupied states are included at $0\,\text{K}$, and the factor of 2 is for spin degeneracy. The quantity in brackets in eqn (3.103) is called the *density matrix*, $\rho_{\alpha\beta}$, and it is of central importance:

$$\rho_{\alpha\beta} = 2\sum_n{}' c_{n\alpha} c_{n\beta}^*. \tag{3.104}$$

The number of electrons in the system is given by

$$N = \mathrm{Tr}\,\rho S, \tag{3.105}$$

where Tr denotes trace. Similarly, the band energy can be expressed in terms of the density matrix:

$$E_{\mathrm{band}} = \mathrm{Tr}\,\rho H. \tag{3.106}$$

Thus we have established the following relation:

$$2\sum_n{}' E_n = \sum_{\alpha,\beta} \rho_{\alpha\beta} H_{\beta\alpha}. \tag{3.107}$$

This is the link we have been seeking between the energies of eigenstates extending throughout the solid and a real space breakdown of the energies in terms of 'on-site' energies, $\rho_{\alpha\alpha} H_{\alpha\alpha}$, and bond energies ($\rho_{\alpha\beta} H_{\beta\alpha} + \rho_{\beta\alpha} H_{\alpha\beta}$). The off-diagonal elements of the density matrix are known as *partial bond orders* (Coulson 1939). If the solid is periodic then the eigenstates, $|nk\rangle$, are labelled by a wave vector $k$ in the first Brillouin zone and a band index $n$. For each eigenstate $|nk\rangle$ there is a band energy $E_n(k)$ and an eigenvector $c_{n\alpha}(k)$. The density matrix is given by an integration over the first Brillouin zone:

$$\rho_{\alpha\beta} = 2\int_{\mathrm{1st\,BZ}} \mathrm{d}k \sum_n{}' c_{n\alpha}(k) c_{n\beta}^*(k). \tag{3.108}$$

It is also useful to define the *spectral density matrix* as follows:

$$\rho_{\alpha\beta}(E) = \sum_n \delta(E - E_n) c_{n\alpha} c_{n\beta}^*, \tag{3.109}$$

where the sum is again over all the eigenstates of the system. The density matrix is related to the spectral density matrix by an integration up to the Fermi energy $E_F$:

$$\rho_{\alpha\beta} = 2\int_{-\infty}^{E_F} \rho_{\alpha\beta}(E)\,\mathrm{d}E. \tag{3.110}$$

The *global density of states* for the system is given by:

$$N(E) = \sum_n \delta(E - E_n) = \sum_{\alpha,\beta} \rho_{\alpha\beta}(E) S_{\beta\alpha} = \mathrm{Tr}\,\rho(E)S. \tag{3.111}$$

It is also useful to define a *local density of states* (LDOS) for each basis state $|\alpha\rangle$:

$$n_\alpha(E) = \sum_\beta \rho_{\alpha\beta}(E) S_{\beta\alpha} = \sum_n \sum_\beta \delta(E - E_n) c_{n\alpha} c_{n\beta}^* S_{\beta\alpha}. \tag{3.112}$$

The local density of states differs from the global density of states by a local weighting factor, $\sum_\beta c_{n\alpha} c_{n\beta}^* S_{\beta\alpha}$, for each eigenstate. The sum of all local densities of states equals

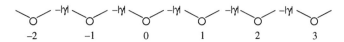

**Fig. 3.11** Schematic illustration of part of an infinite linear chain, with hopping integral $-|\gamma|$ between neighbouring atoms.

the global density of states. The band energy may also be represented in terms of the local densities of states as follows:

$$E_{\text{band}} = 2\sum_{\alpha} \int_{-\infty}^{E_F} n_\alpha(E)E\,\mathrm{d}E \tag{3.113}$$

Thus the band energy has an 'on-site' representation involving local densities of states and an intersite representation, eqn (3.106), involving bond orders.

Off-diagonal elements of the spectral density matrix are known as COOP curves in chemistry (Hoffmann 1988). Energies at which $\rho_{\alpha\beta}(E)$ is positive (negative) correspond to bonding (antibonding) states. The *local chemistry and environment dependence is contained in the spectral density matrix.*

The local density of states takes on a particulary transparent form when the basis set $\{|\alpha\rangle\}$ is orthonormal. The overlap matrix $S_{\alpha\beta}$ reduces to the identity $\delta_{\alpha\beta}$ matrix and the LDOS becomes

$$n_\alpha(E) = \sum_n \delta(E - E_n)|c_{n\alpha}|^2. \tag{3.114}$$

Each eigenstate is weighted by a factor $|c_{n\alpha}|^2$, which equals $|\langle\alpha|n\rangle|^2$. This factor is a measure of the extent to which the basis state $|\alpha\rangle$ participates in the eigenstate $|n\rangle$.

Let us illustrate the concepts that have been introduced in this section with the example of an infinite linear chain, shown in Fig. 3.11. Our purpose is to extract a picture of bonding in the chain from its band structure. Each atom is assumed to be associated with one basis state and the basis state at the $n$th atom is denoted by $|n\rangle$. The set of all basis states is assumed to be orthonormal: $\langle n|m\rangle = \delta_{mn}$. The Hamiltonian matrix elements are all set to zero except the hopping integrals between nearest neighbours, which are set to $-|\gamma|$. Owing to translational symmetry along the chain we can use Bloch's theorem and immediately write down the normalized eigenstates $|k\rangle$:

$$|k\rangle = \frac{1}{\sqrt{N}}\sum_n e^{ikn}|n\rangle. \tag{3.115}$$

$N$ is the total number of atoms in the chain (assumed infinite). The eigenstate expansion coefficients are identified as $e^{ikn}/\sqrt{N}$.

Substituting this eigenstate into the Schrödinger equation we deduce that the corresponding eigenvalue is

$$E(k) = -2|\gamma|\cos k. \tag{3.116}$$

The wave vector $k$ is uniquely defined in the first Brillouin zone, which we choose to lie between $-\pi$ and $+\pi$. The band structure $E(k)$ is shown in Fig. 3.12. At the bottom of the band, where $k = 0$, the eigenstate has the greatest bonding character for neighbouring atoms with all expansion coefficients having the same magnitude and sign, as shown in Fig. 3.13. At the top of the band, where $|k| = \pi$, the eigenstate has the greatest antibonding character for neighbouring atoms with successive expansion coefficients

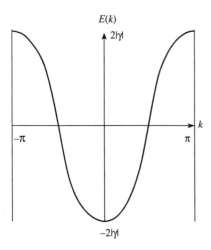

**Fig. 3.12** The band structure, $E(k)$, for the one-dimensional infinite linear chain, given by eqn (3.116).

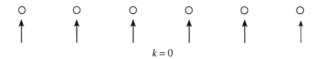

$$k = 0$$

**Fig. 3.13** Schematic representation of the $k = 0$ state for the infinite linear chain. All atoms are in phase and this is the most bonding state for the chain.

alternating in sign as shown in Fig. 3.14. These two extreme cases are analogous to the bonding and antibonding states of the diatomic molecule discussed in the previous section. Between the two extremes there is a continuous change from the most bonding to the most antibonding state. For example, in the middle of the band, at $|k| = \pi/2$, the phase of the eigenstate has a period of four atoms as shown in Fig. 3.15, and it is neither bonding nor antibonding for neighbouring atoms.

As in eqn (3.108) the spectral density matrix may be expressed as an integral over the first Brillouin zone:

$$\rho_{mn}(E_o) = \frac{1}{2\pi}\int_{-\pi}^{+\pi} \delta(E_o - E(k))e^{ik(m-n)}\,dk. \tag{3.117}$$

There are two values of $k$ where $E(k) = E_o$, which we call $k_o$ and $-k_o$. The integral may be evaluated as follows:

$$\rho_{mn}(E_o) = \frac{1}{2\pi}\int_{-\pi}^{+\pi} dk \left[\frac{\delta(k - k_o)}{|E'(k_o)|} + \frac{\delta(k + k_o)}{|E'(-k_o)|}\right] e^{ik(m-n)}$$

$$= \frac{\cos\,[k_o(m-n)]}{2\pi|\gamma|\sin k_o}, \tag{3.118}$$

where $E'(k_o) = dE/dk$ at $k = k_o$, and $E_o = -2|\gamma|\cos k_o$.

The local density of states is obtained by setting $m = n$. Owing to the translational symmetry it is the same on all sites:

$$n(E) = \frac{1}{2\pi|\gamma|\sin k} = \frac{1}{\pi\sqrt{4\gamma^2 - E^2}}. \tag{3.119}$$

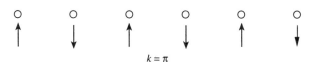

**Fig. 3.14** Schematic representation of the $k = \pi$ state for the infinite linear chain. This is the most antibonding state, but notice that next nearest neighbouring pairs of atoms are exactly in phase.

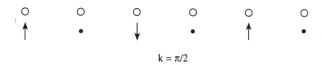

**Fig. 3.15** Schematic representation of the $k = \pi/2$ state for the infinite linear chain. This state is neither bonding nor antibonding but notice next nearest neighbouring pairs of atoms are $\pi$ out of phase.

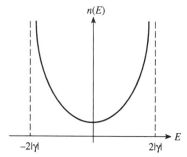

**Fig. 3.16** The local density of states for an atom in an infinite perfect linear chain, as given by eqn (3.119).

As seen in Fig. 3.16 the local density of states is singular at the band edges and has a minimum in the centre of the band. The number of electrons per atom is related to the Fermi energy (at $0\,\mathrm{K}$) by

$$N_e(E_F) = 1 + \frac{2}{\pi} \sin^{-1} \frac{E_F}{2|\gamma|} \tag{3.120}$$

where $E_F$ lies between $-2|\gamma|$ and $+2|\gamma|$, and a factor of 2 has been included for spin degeneracy.

Consider the spectral density matrix element between neighbouring atoms, for which $m - n = \pm 1$. From eqn (3.118) it is given by

$$\rho_{01}(E_o) = \frac{\cos k_o}{2\pi |\gamma| \sin k_o}. \tag{3.121}$$

As shown in Fig. 3.17 $\rho_{01}(E)$ is positive in the bottom half of the band and negative in the top half. Thus the bottom half of the band contains bonding states for neighbouring atoms and the top half of the band contains antibonding states. The middle of the band is neither bonding nor antibonding. However, the same conclusions do not apply to next nearest neighbour bonds. For these, $m - n = \pm 2$ and the spectral density matrix element is

$$\rho_{02}(E_o) = \frac{\cos 2k_o}{2\pi |\gamma| \sin k_o}. \tag{3.122}$$

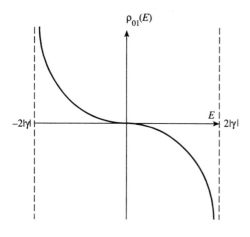

**Fig. 3.17** The spectral density matrix element $\rho_{01}(E)$ for nearest neighbours in the infinite linear chain.

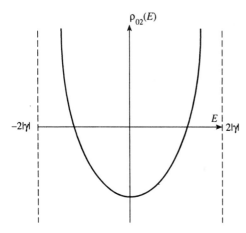

**Fig. 3.18** The spectral density matrix element $\rho_{02}(E)$ for next nearest neighbours in the infinite linear chain.

As seen in Fig. 3.18 the middle of the band is now antibonding for next nearest neighbours, in agreement with the alternating phases of nearest neighbour atoms at $k=0$. The top and bottom of the band correspond to bonding states for next nearest neighbours.

If it were always necessary to obtain the spectral density matrix from the eigenstates of the system it would have little computational value. It acquires much greater significance from the fact that it is related directly to the electronic Green function, which has powerful analytic properties and can be calculated independently of the eigenstates of the system. The Green function operator is defined by

$$G(z) = (z - H)^{-1}, \tag{3.123}$$

where $z$ is an arbitrary complex number. It may be expressed in terms of the eigenvectors and eigenvalues of the Hamiltonian as follows:

$$G(z) = \sum_n \frac{|n\rangle\langle n|}{z - E_n}, \tag{3.124}$$

as may be verified by substituting this expression in the definition, eqn (3.123), and using

$$H = \sum_n |n\rangle E_n \langle n|. \tag{3.125}$$

Matrix elements of the Green function are thus given by

$$G_{\alpha\beta}(z) = \sum_n \frac{c_{n\alpha} c_{n\beta}^*}{z - E_n}. \tag{3.126}$$

We shall call the Green function operator and its matrix elements simply Green functions, since it is clear whether we are discussing the operator or a matrix element from the context. The Green function is singular at the eigenvalues of the Hamiltonian, which lie on the real axis. As the complex number $z$ approaches the real axis it tends to some energy $E$, and if $E$ coincides with one of the eigenvalues the Green function is not well defined. Whether we approach an eigenvalue from the upper or lower half of the complex plane determines the sign of the imaginary component of the Green function. To understand this better consider the Fourier transform of the Green function:

$$G_{\alpha\beta}(t) = \frac{1}{2\pi i} \int_{-\infty}^{+\infty} G_{\alpha\beta}(E) \exp\left(-iEt/\hbar\right) \, dE. \tag{3.127}$$

Here $t$ is time. We may replace this by a contour integral in the complex plane, which for $t \geqslant 0$ we close by a semicircle in the lower half of the plane. Does the contour include or exclude the poles at the eigenvalues on the real axis? If it excludes the poles then $G_{\alpha\beta}(t)$ is zero. If $z$ approaches the real axis from the upper half plane, so that $z = E + i\varepsilon$, where $\varepsilon$ is a positive, real, infinitesimal number, then the contour runs along just above the eigenvalues and they are within the semicircle. If, however, $z$ approaches from the lower half plane, so that $z = E - i\varepsilon$, then the contour runs along just below the eigenvalues and they are excluded from the semicircle. Thus for $t \geqslant 0$ we must take $z = E + i\varepsilon$ and we obtain:

$$G_{\alpha\beta}(t) = \sum_n \langle \alpha | n \rangle \exp\left(-iE_n t/\hbar\right) \langle n | \beta \rangle = \langle \alpha | \exp\left(-iHt/\hbar\right) | \beta \rangle. \tag{3.128}$$

The operator $\exp(-iHt/\hbar)$ is recognized as the time development operator. $G_{\alpha\beta}(t)$ describes the probability amplitude of an electron propagating from the state $|\beta\rangle$ at time $t = 0$ to the state $|\alpha\rangle$ at time $t \geqslant 0$. This is why Green functions are sometimes called propagators. If the direction of time were reversed we would have to close the contour in the upper half of the complex plane and then we would take $z = E - i\varepsilon$ to get a non-zero $G_{\alpha\beta}(t)$. Our conclusion is that we take $z = E + i\varepsilon$ as $z$ approaches the real axis in order to describe the evolution of states in the usual forward direction of time. Setting $z = E + i\varepsilon$ in the Green function we have

$$G_{\alpha\beta}(E + i\varepsilon) = G_{\alpha\beta}(E^+) = \sum_n \frac{c_{n\alpha} c_{n\beta}^*}{E + i\varepsilon - E_n}. \tag{3.129}$$

Using the mathematical identity

$$\frac{1}{E + i\varepsilon - a} = \mathbb{P}\frac{1}{E - a} - \pi i \delta(E - a), \tag{3.130}$$

where $a$ lies on the real axis and $\mathbb{P}$ denotes Cauchy principal value, we obtain

$$G_{\alpha\beta}(E^+) = \mathbb{P}\sum_n \frac{c_{n\alpha} c_{n\beta}^*}{E - E_n} - \pi i \sum_n c_{n\alpha} c_{n\beta}^* \delta(E - E_n). \tag{3.131}$$

We recognize the second sum as being the spectral density matrix element $\rho_{\alpha\beta}(E)$. Thus we have proved the following very useful relation

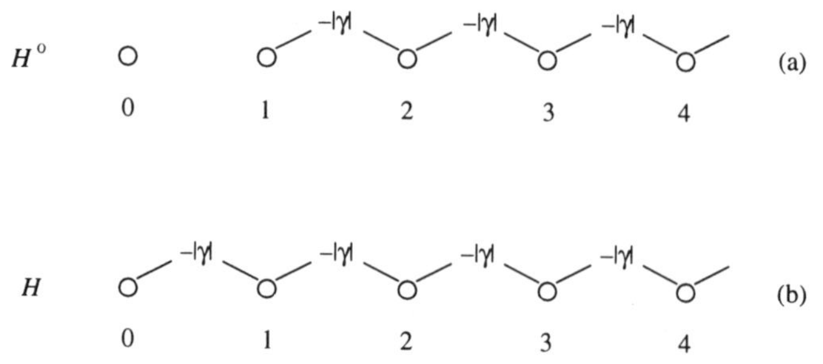

**Fig. 3.19** To illustrate the evaluation of the Green function matrix element for the end atom of a semi-infinite linear chain. In (a) the end atom of the chain is decoupled from the rest in the 'unperturbed' Hamiltonian, $H^\circ$. Coupling this end atom in (b) in the 'perturbed' Hamiltonian, $H$.

$$\rho_{\alpha\beta}(E) = -\frac{Im}{\pi} G_{\alpha\beta}(E^+), \qquad (3.132)$$

where Im denotes imaginary part. In particular the local density of states $n_\alpha(E)$ is given by

$$n_\alpha(E) = -\frac{Im}{\pi} G_{\alpha\alpha}(E^+). \qquad (3.133)$$

In the following we drop the + superscript on the energy argument of the Green functions, with the understanding that the energy $E$ is interpreted as $E + i\varepsilon$.

Having related the spectral density matrix to the Green function we shall now outline how the Green functions may be determined without solving the Hamiltonian for the eigenvectors and eigenvalues. There are essentially three ways of doing this.

The first is to use a Green function, $G^\circ$, that is already determined for some unperturbed Hamiltonian $H^\circ$ to obtain the Green function $G$ for a perturbed Hamiltonian $H = H^\circ + V$. The relationship between $G$, $G^\circ$, and V is the Dyson equation:

$$G = G^\circ + G^\circ V G, \qquad (3.134)$$

which is readily derived from the definitions $G = (E - H^\circ - V)^{-1}$ and $G^\circ = (E - H^\circ)^{-1}$. The Dyson equation is solved exactly by

$$G = (1 - G^\circ V)^{-1} G^\circ \qquad (3.135)$$

or approximately by the perturbation expansion

$$G = G^\circ + G^\circ V G^\circ + G^\circ V G^\circ V G^\circ + G^\circ V G^\circ V G^\circ V G^\circ + \dots \qquad (3.136)$$

where successive terms are one higher order in the perturbation $V$.

As an example consider the evaluation of the Green function $G_{00}(E)$ for the end atom of a semi-infinite constant linear chain (see Fig. 3.19). As in the infinite linear chain example above all the Hamiltonian matrix elements are zero except between nearest neighbours, where the hopping integrals are $-|\gamma|$. We define our 'unperturbed' Hamiltonian to be the system shown in Fig. 3.19(a). Atom 0 is decoupled from the remainder of the chain. The perturbation $V$ consists of coupling atom 0 to the chain by setting the matrix $V_{01} = V_{10} = -|\gamma|$; all other matrix elements of $V$ are zero. Since atom 0 is isolated in the unperturbed Hamiltonian we have $G_{00}^0 = 1/E$. Furthermore we

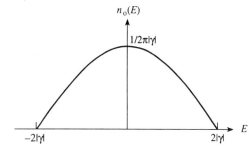

**Fig. 3.20** The local density of states, $n_o(E)$, for the end atom of a semi-infinite linear chain, given by eqn (3.141).

also have $G_{11}^\circ = G_{00}$, which is the matrix element we want. Using the Dyson equation for $G_{00}$ we have:

$$G_{00} = G_{00}^\circ + G_{00}^\circ V_{01} G_{10}. \tag{3.137}$$

The Green function $G_{10}$ is not known but we can use the Dyson equation again:

$$G_{10} = G_{11}^\circ V_{10} G_{00} = G_{00} V_{10} G_{00}, \tag{3.138}$$

where we have used $G_{10}^\circ = 0$, because atoms 0 and 1 are not coupled in the unperturbed Hamiltonian, and $G_{11}^\circ = G_{00}$. Substituting this expression for $G_{10}$ into eqn (3.17) for $G_{00}$ we have

$$G_{00} = G_{00}^\circ + G_{00}^\circ V_{01} G_{00} V_{10} G_{00} = 1/E + (\gamma^2/E)(G_{00})^2. \tag{3.139}$$

The solution of this quadratric equation is

$$G_{00} = \frac{E \pm (E^2 - 4\gamma^2)^{\frac{1}{2}}}{2\gamma^2}. \tag{3.140}$$

The two roots correspond to the two possibilities $E \mp i\varepsilon$, and therefore we choose the negative square root. The density of states on the terminating atom is thus

$$n_o(E) = -\frac{Im}{\pi} \frac{E - (E^2 - 4\gamma^2)^{\frac{1}{2}}}{2\gamma^2} = \frac{(4\gamma^2 - E^2)^{\frac{1}{2}}}{2\pi\gamma^2}, \tag{3.141}$$

for $|E| \leqslant 2|\gamma|$ and zero otherwise, which is recognized as the extent of the band in the infinite linear chain. The density of states is sketched in Fig. 3.20. Comparing it with the density of states on an atom in the infinite linear chain (Fig. 3.16) we see that it is higher in the middle of the band and lower at the band edges. All Green function matrix elements for the constant semi-infinite linear chain may be determined by defining suitable unperturbed and perturbed Hamiltonians, as shown by Sutton *et al.* (1988). The Green function $G_{jk}(E)$, where $j \leqslant k$, is given by

$$G_{jk}(E) = -\frac{\sin(j\theta)e^{ik\theta}}{|\gamma| \sin\theta}, \tag{3.142}$$

where $E = -2|\gamma|\cos\theta$ and $G_{jk} = G_{kj}$. The semi-infinite linear chains may be coupled together by another Dyson equation to produce an infinite linear chain. Thus, we may obtain the Green functions for the infinite linear chain without using its eigenvectors and eigenvalues:

$$G_{jk}(E) = -\frac{e^{i|j-k|\theta}}{2|\gamma| \sin\theta}. \tag{3.143}$$

Returning to the quadratic equation for the Green function matrix element on the terminating atom of the chain, eqn (3.139), we observe that it may be expressed as a continued fraction:

$$G_{00} = \frac{1}{E - \gamma^2 G_{00}} = \cfrac{1}{E - \cfrac{\gamma^2}{E - \cfrac{\gamma^2}{E - \dots}}}. \tag{3.144}$$

The continued fraction representation of $G_{00}$ allows us to make contact with the second method of obtaining Green functions, namely the recursion method. The mathematical origin of the continued fraction may be understood when it is recalled that $G_{00}$ is $\langle 0|(E - H)^{-1}|0\rangle$, which is the leading diagonal element of the inverse of the infinite matrix

$$(E - H) = \begin{pmatrix} E & |\gamma| & 0 & 0 & 0 & . \\ |\gamma| & E & |\gamma| & 0 & 0 & . \\ 0 & |\gamma| & E & |\gamma| & 0 & . \\ 0 & 0 & |\gamma| & E & |\gamma| & . \\ 0 & 0 & 0 & |\gamma| & E & . \\ . & . & . & . & . & . \end{pmatrix}. \tag{3.145}$$

Thus, $G_{00}$ is the cofactor of the leading diagonal element divided by the determinant of the whole matrix. Using the Cauchy expansion of a determinant we obtain the continued fraction given in eqn (3.144).

If $H$ were the following infinite tridiagonal matrix

$$H = \begin{pmatrix} a_0 & b_1 & 0 & 0 & 0 & . \\ b_1 & a_1 & b_2 & 0 & 0 & . \\ 0 & b_2 & a_2 & b_3 & 0 & . \\ 0 & 0 & b_3 & a_3 & b_4 & . \\ 0 & 0 & 0 & b_4 & a_4 & . \\ . & . & . & . & . & . \end{pmatrix} \tag{3.146}$$

the Green function $G_{00}$ could similarly be obtained by a Cauchy expansion as the following continued fraction:

$$G_{00} = \cfrac{1}{E - a_0 - \cfrac{b_1^2}{E - a_1 - \cfrac{b_2^2}{E - a_2 - \dots}}}. \tag{3.147}$$

The matrix $G_{00}$ element is the Green function for the terminating atom of a semi-infinite linear chain in which the Hamiltonian matrix elements are as shown in Fig. 3.21.

If we could transform a Hamiltonian for a 3D problem into the tridiagonal form of eqn (3.146) we could immediately write down the Green function matrix element $G_{00}$ as the continued fraction given in eqn (3.147). The transformation of the Hamiltonian to tridiagonal form is precisely what the recursion method (or Lanczos algorithm) does. The

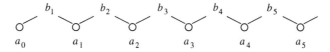

**Fig. 3.21** A semi-infinite linear chain with on-site Hamiltonian matrix elements $a_0$, $a_1$, $a_2$, etc., and intersite matrix elements $b_1$, $b_2$, $b_3$, etc., for which the Green function matrix element of the end atom atom is given by the continued fraction of eqn (3.147).

method involves setting up a new basis $u_n$, $n = 0, 1, 2, \ldots$, the first of which, $u_0$, we choose as the particular orbital $|\alpha\rangle$ or combination of orbitals we want the Green function matrix element for. For example, if we choose $u_0^{\pm} = (|\alpha\rangle \pm |\beta\rangle)/\sqrt{2}$ then $G_{00}^{\pm}$ will equal $(G_{\alpha\alpha} + G_{\beta\beta})/2 \pm G_{\alpha\beta} \pm G_{\beta\alpha}$, and $G_{\alpha\beta} + G_{\beta\alpha} = (G_{00}^+ - G_{00}^-)/2$. In this way we can determine off-diagonal elements of the Green function, as well as diagonal elements by choosing $u_0 = |\alpha\rangle$. Having chosen $u_0$ we define $u_1$ by

$$b_1 u_1 = H u_0 - a_0 u_0, \qquad (3.148)$$

where

$$a_0 = \langle u_0 | H | u_0 \rangle. \qquad (3.149)$$

The constant $b_1$ is determined by the normalization condition $\langle u_1 | u_1 \rangle = 1$. The remaining basis functions $u_n$ are determined by the recurrence relation

$$b_{n+1} u_{n+1} = H u_n - a_n u_n - b_n u_{n-1}. \qquad (3.150)$$

The coefficients $a_n$ and $b_n$ are such as to orthogonalize $H u_n$ to the preceding vectors $u_n$, $u_{n-1}$, and $b_{n+1}$ is the coefficient to normalize $u_{n+1}$ to unity. Thus the new basis set is generated by repeated action with the Hamiltonian. The matrix elements of the Hamiltonian in the new basis are

$$\langle u_n | H | u_n \rangle = a_n ; \langle u_{n-1} | H | u_n \rangle = \langle u_{n-1} | H | n_n \rangle = b_n \qquad (3.151)$$

and all other matrix elements are zero. Thus the Hamiltonian is transformed into tridiagonal form. Note that the method does not rely on translational symmetry and no use of Bloch's theorem or the eigenstates of the system has been made. The Green function $G_{00}$ is obtained by repeated operations of the Hamiltonian on the chosen starting orbital. Each successive operation with the Hamiltonian brings in information from sites that are one more electron hop remote from the chosen starting orbital. As we go down the continued fraction we are therefore including information about sites more remote from our starting orbital. This feature is closely mirrored in the third and final method of obtaining Green's functions, namely the equation of motion method.

In the equation of motion method (MacKinnon 1985) we solve the time-dependent Schrödinger equation:

$$H|\psi\rangle = i\hbar \frac{\partial |\psi\rangle}{\partial t} \qquad (3.152)$$

for which the formal solution is

$$|\psi(t)\rangle = e^{-iHt/\hbar} |\psi(0)\rangle. \qquad (3.153)$$

We recognize $e^{-iHt/\hbar}$ as the time-development operator, eqn (3.128). We can obtain the energy-dependent Green function from the time evolution operator by Laplace transformation:

$$\langle \psi(0) | G(E + i\varepsilon) | \psi(0) \rangle = -\frac{i}{\hbar} \langle \psi(0) | \int_0^\infty \exp\left(i(E + i\varepsilon)t/\hbar\right) e^{-iHt/\hbar} \, dt \, | \psi(0) \rangle$$

$$= \langle \psi(0) | (E + i\varepsilon - H)^{-1} | \psi(0) \rangle. \tag{3.154}$$

This is the Green function matrix element, as a function of energy, for the initial state $|\psi(0)\rangle$. For example, let us assume an orthonormal tight binding basis $\{|\alpha\rangle\}$, and expand $|\psi(t)\rangle$ in this basis:

$$|\psi(t)\rangle = \sum_a c_\alpha(t) |\alpha\rangle. \tag{3.155}$$

At time $t = 0$ let us assume that all $c_\alpha(0)$ are zero except $c_\beta(0)$ which we set equal to 1. This initial state describes an electron which is localized on state $|\beta\rangle$ and subsequently propagates through the surrounding system. Thus

$$c_\alpha(0) = \delta_{\alpha\beta} \quad \text{and} \quad |\psi(0)\rangle = |\beta\rangle. \tag{3.156}$$

Then $G_{\beta\beta}(E + i\varepsilon)$ is given by eqn (3.154) as the Laplace transform of $c_\beta(t)$. This is evaluated by solving the differential equation, eqn (3.152), numerically to get $c_\beta(t)$ and then evaluating the Laplace transform numerically. See MacKinnon (1985) for technical details. At each time-step in the numerical solution of the time-dependent Schrödinger equation the Hamiltonian $H$ acts on the state $|\psi(t)\rangle$. Therefore, as time proceeds the electron is able to hop to more remote environments from its initial position at $|\beta\rangle$. There is a striking comparison here with the inclusion of further levels in the recursion method, each of which allows the electron to hop to one more remote shell of neighbours. Indeed, the information content obtained in both methods after $N$ operations with the Hamiltonian is the same. The important physical picture to emerge from the equation of motion method, however, is that the local density of states $n_\beta(E)$ is related directly to the evolution in time of the probability amplitude, $c_\beta(t)$, of finding the electron in the state $|\beta\rangle$, with the initial condition that this probability amplitude is unity, and the initial probability amplitudes on all other states are zero. This is an extraordinary and useful insight.

The recursion method and equation of motion method are not only practical ways of computing the Green functions for disordered solids. They provide a direct way of quantifying the effect of variations in the local atomic environment on the bonding in the solid through its close relation (eqn (3.132)) with the spectral density matrix. In the next section we show how moments of the spectral density matrix may be obtained directly from the local atomic environment without evaluating the Green function or the spectral density matrix.

### 3.8.4 Moments of the spectral density matrix

There is a remarkable theorem (Cyrot-Lackmann 1968) which allows us to calculate the moments of the spectral density matrix exactly from the topology of the local atomic environment without knowing the functional form of the spectral density matrix in advance! Before we prove the theorem let us consider what information the moments contain. Throughout this section we assume the basis set $\{|\alpha\rangle\}$ is orthonormal. The $p$th moment of the spectral density matrix is defined by

$$\mu_{\alpha\beta}^{(p)} = \int_{-\infty}^{+\infty} \rho_{\alpha\beta}(E)(E - H_{\alpha\alpha})^p \, dE. \tag{3.157}$$

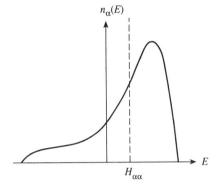

**Fig. 3.22** A local density of states, $n_\alpha(E)$, with a large negative third moment. The centre of gravity, $H_{\alpha\alpha}$, is shown by the dashed line.

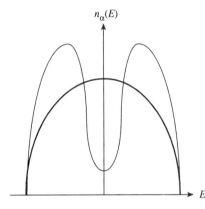

**Fig. 3.23** The bold curve shows a unimodal local density of states, and the other curve shows a bimodal local density of states. The unimodal local density of states has a larger value of $\mu^{(4)}/(\mu^{(2)})^2$.

The site diagonal elements of the spectral density matrix are the local densities of states $n_\alpha(E)$. In that case the $p$th moment is simply

$$\mu_\alpha^{(p)} = \int_{-\infty}^{+\infty} n_\alpha(E)\,(E - H_{\alpha\alpha})^p \, \mathrm{d}E. \tag{3.158}$$

The zeroth moment of the LDOS is one owing to the normalization of the basis states. The first moment is the centre of gravity of the LDOS relative to $H_{\alpha\alpha}$. We shall see shortly that this is always zero: the centre of gravity of the LDOS is $H_{\alpha\alpha}$. The second moment is the moment of inertia of the LDOS relative to the centre of gravity. The square root of the second moment is a measure of the *width* of the LDOS in the root mean square sense. The third moment measures the *skewness* of the LDOS about the centre of gravity. Indeed, if the LDOS is symmetrical about the centre of gravity then all the odd moments are zero. A large negative value of $\mu^{(3)}$ corresponds to a long tail in the LDOS below the centre of gravity of the band and a more compressed peak above the centre of gravity, as shown in Fig. 3.22. The fourth moment measures the tendency for a *gap* to form in the middle of the band (Gaspard and Lambin 1985, Carlsson 1990), as shown in Fig. 3.23. A low value of the normalized fourth moment, $\mu^{(4)}/(\mu^{(2)})^2$, corresponds to two well separated peaks, or bimodal behaviour, in the density of states whereas a large value corresponds to a central peak or unimodal behaviour. Higher moments are important in questions of stability of different crystal structures and defects, but the shape of the LDOS is most easily interpreted in terms of the second, third, and fourth moments (Carlsson 1990).

The $p$th moment of an off-diagonal element of the spectral density matrix is given by

$$\mu_{\alpha\beta}^{(p)} = \int_{-\infty}^{+\infty} \rho_{\alpha\beta}(E)(E - H_{\alpha\alpha})^p \, \mathrm{d}E. \tag{3.159}$$

The zeroth moment is zero because the basis states $|\alpha\rangle$ and $|\beta\rangle$ are orthogonal. We shall see below that the first moment is equal to the Hamiltonian matrix element $H_{\alpha\beta}$, linking the two orbitals. It is therefore a measure of the extent to which the two orbitals are coupled through the Hamiltonian. For example, in a nearest neighbour tight binding model $H_{\alpha\beta}$ is zero except for neighbouring atoms. Since the average of $\rho_{\alpha\beta}(E)$ is always zero there must be nodes in the function between the band limits. If all the odd moments, including the first, are zero then $\rho_{\alpha\beta}(E)$ is an even function and the number of nodes is even. On the other hand, if all the even moments are zero then $\rho_{\alpha\beta}(E)$ is an odd function and the number of nodes is odd. In general the function is neither odd nor even, but the following result always applies: if the first $n$ moments, including $\mu^{(0)}$, of $\rho_{\alpha\beta}(E) + \rho_{\beta\alpha}(E)$ are zero then $\rho_{\alpha\beta}(E) + \rho_{\beta\alpha}(E)$ has at least $n$ nodes between the band limits. This observation follows from a second theorem concerning moments of the spectral density matrix, due to Ducastelle and Cyrot-Lackmann (1971), which is discussed below.

The derivation of the theorem of Cyrot-Lackmann (1968) is very instructive. To begin, we recall from eqn (3.125) that the Hamiltonian operator may be expressed in terms of its eigenstates $|n\rangle$ and eigenvalues $E_n$ as follows:

$$H = \sum_n |n\rangle E_n \langle n|. \tag{3.125}$$

Since the eigenstates of the Hamiltonian form an orthonormal set then

$$H^p = \sum_n |n\rangle E_n^p \langle n|. \tag{3.160}$$

It follows that for any analytic function $g(E)$ we have

$$g(H) = \sum_n |n\rangle g(E_n)\langle n|. \tag{3.161}$$

Substituting eqn (3.114) for the LDOS in eqn (3.158) for the $p$th moment we obtain

$$\mu_\alpha^{(p)} = \sum_n (E_n - H_{\alpha\alpha})^p c_{n\alpha} c_{n\alpha}^* \tag{3.162}$$

and since the basis is orthonormal $c_{n\alpha}$ equals $\langle \alpha | n \rangle$ and we have

$$\mu_\alpha^{(p)} = \sum_n \langle \alpha | n \rangle (E_n - H_{\alpha\alpha})^p \langle n | \alpha \rangle. \tag{3.163}$$

Using eqn (3.161) we identify this as the $\alpha\alpha$ matrix element of $(H - H_{\alpha\alpha})^p$. Using the closure relation of the basis set

$$\sum_\alpha |\alpha\rangle\langle\alpha| = 1, \tag{3.164}$$

we obtain Cyrot-Lackmann's result:

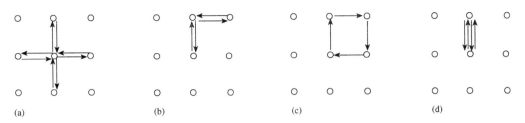

**Fig. 3.24** To illustrate the evaluation of the second and fourth moments of the local density of states in a square lattice. In (a) the second moment consists of the sum of four closed paths of two electron hops from the central atom to each of the four neighbours. There are three types of path that contribute to the fourth moment shown in (b), (c), and (d).

$$\mu_\alpha^{(p)} = \langle\alpha|\, (H - H_{\alpha\alpha})^p\, |\alpha\rangle$$

$$= \sum_\gamma \sum_\delta \sum_\varepsilon \cdots \sum_\kappa \langle\alpha|H - H_{\alpha\alpha}|\gamma\rangle\langle\gamma|H - H_{\alpha\alpha}|\delta\rangle\langle\delta|H - H_{\alpha\alpha}|\varepsilon\rangle \cdots \langle\kappa|H - H_{\alpha\alpha}|\alpha\rangle.$$

$$(3.165)$$

Each product of Hamiltonian matrix elements contains $p$ terms. If all the on-site Hamiltonian matrix elements are the same, each product of terms in the summation has a simple interpretation: it describes the path of an electron starting in orbital $|\alpha\rangle$ and hopping through the local environment, arriving after $p$ hops, back at the orbital $|\alpha\rangle$. The $p$th moment is given by the sum of all such paths of length $p$ hops. This is illustrated in Fig. 3.24 for a square lattice. If the on-site Hamiltonian matrix elements vary from site to site, or there is more than one orbital type per site, then there are contributions to the $p$th moment from terms involving on-site Hamiltonian matrix element differences $H_{\gamma\gamma} - H_{\alpha\alpha}$. The first moment of the LDOS is always zero.

To illustrate the power of this theorem consider the second moments of the local densities of states for the terminating atom of the semi-infinite chain and the infinite chain discussed in the previous section. There we obtained explicit expressions, eqns (3.119) and (3.141), for the local densities of states. The second moment of the local density of states of the terminating atom of the semi-infinite chain is $\gamma^2$ because there is only one path of length two hops from this atom. On the other hand, the second moment for an atom in the infinite linear chain is $2\gamma^2$, because the electron can hop in both directions. These results may be confirmed by direct integrations using the formulae for the local densities of states. The difference in the second moments is reflected in the shapes of the densities of states shown in Figs 3.16 and 3.20. We may immediately generalize this important result: if the local coordination number is reduced then the local density of states will have a reduced root mean square width.

The obvious conclusion that follows from this theorem is that the low-order moments of the local density of states are determined entirely by the local atomic environment. Suppose we cut out and removed a cluster of atoms from a solid, in which the smallest number of nearest neighbour hops from the central atom to an atom on the surface of the cluster, is $m$. An electron hopping from the central atom along any closed path of length less than $2m + 2$ would not know it is in a cluster. Thus, the first $2m + 2$ moments (i.e. $\mu^{(0)}$ to $\mu^{(2m + 1)}$) of the LDOS on the central atom in the cluster are identical to the first $2m + 2$ moments of the LDOS of the atom when it is in the solid. Thus, we have a rather graphic picture of how features in the LDOS depend on the atomic environment.

The picture is consistent with the recursion method described in the previous section because the low-order moments of the density of states are related directly to the low order $a$'s and $b$'s in the tri-diagonalized Hamiltonian. For example, $\mu^{(2)} = b_1^2$, $\mu^{(3)} = b_1^2(a_1 - a_0)$. The picture is also consistent with the equation of motion method because, using eqns (3.153) and (3.155), it may be shown that:

$$\langle \alpha | H^r | \alpha \rangle = \left( -\frac{\hbar}{i} \right)^r \left[ \frac{d^r c_\alpha(t)}{dt^r} \right]_{t=0} \tag{3.166}$$

with the initial condition that $c_\beta(0) = \delta_{\alpha\beta}$. Therefore, the $p$th moment of the local density of states $n_\alpha(E)$ is determined by a linear combination of the first $p$ time derivatives of the probability amplitude $c_\alpha(t)$, evaluated at $t = 0$, with the initial condition $c_\beta(0) = \delta_{\alpha\beta}$. The first $p$ derivatives determine $c_\alpha(t)$ after $p$ actions with the Hamiltonian in the numerical solution of the time dependent Schrödinger equation for $c_\alpha(t)$.

Turning to the off-diagonal spectral density matrix elements the proof of the theorem is identical and the result is the same as eqn (3.165) except that the final ket is $|\beta\rangle$:

$$\mu_{\alpha\beta}^{(p)} = \langle \alpha | (H - H_{\alpha\alpha})^p | \beta \rangle. \tag{3.167}$$

Once again if all the on-site Hamiltonian matrix elements are the same then the $p$th moment represents the sum of all possible paths for an electron to hop from orbital $|\alpha\rangle$ to orbital $|\beta\rangle$, with no contributions from on-site Hamiltonian matrix elements. The first moment is always given by

$$\mu_{\alpha\beta}^{(1)} = \langle \alpha | (H - H_{\alpha\alpha}) | \beta \rangle = H_{\alpha\beta}. \tag{3.168}$$

Since we now have a simple algorithm for determining the moments of the spectral density matrix it is tempting to ask whether we can determine the spectral density matrix elements from the moments. This is the classic 'moments problem', which has been discussed by Haydock (1980) and Gaspard and Lambin (1985).

One major application of Cyrot-Lackmann's theorem has been to understand the origin of the differences in energies of rival crystal structures of the elements and compounds. Each crystal structure presents its own set of closed paths, called *rings*, of length 2, 3, . . . $n$ hops, and the theorem relates these paths to the moments of the density of states. For example, Cressoni and Pettifor (1991) have analyzed the structural trends within the sp-bonded elements from this point of view. They observed that close-packed structures (f.c.c. and h.c.p.) have densities of states skewed to lower energies due to the presence of three membered rings leading to large, negative third moments. By contrast, the more open structures, such as the cubic and hexagonal diamond structures, contain only even membered rings. Thus the electronic contribution to the cohesive energy favours close-packed structures when the band is less than half full, to take advantage of the long tail in the density of states below the centre of gravity. Another example of the utility of the theorem is provided by the following observation of Burdett and Lee (1985). Consider an s-band model, with nearest neighbour hopping only, of an f.c.c. and an h.c.p. crystal with an ideal $c/a$ ratio. Burdett and Lee proved that the densities of states have identical moments, and therefore, the densities of states are identical. In fact their result is even more general: any polytypic sequence of close packed layers has the same density of states. Therefore, if we wish to account for a finite stacking fault energy in a nearest neighbour tight binding model of an f.c.c. crystal we must include the angular dependence of valence orbitals, i.e. p, d, etc., states. This result is *not* obvious. In a simple pair potential model the difference between f.c.c. and h.c.p. structures does not appear until the third neighbour interactions. In the nearest neighbour tight binding model these interactions appear in the fourth moment. Burdett and Lee's result also applies to cubic and hexagonal diamond crystals.

The second theorem concerning moments, due to Ducastelle and Cyrot-Lackmann (1971), is particularly useful for comparing the electronic contributions to the energy of two crystal structures, perhaps one containing a defect. It also has other applications as discussed by Heine and Samson (1980, 1983). Consider a spectral density $\rho(E)$ which is defined for some energy band $a \leqslant E \leqslant b$. The spectral density is zero outside the band and it must be real inside the band. An off-diagonal spectral density matrix element, $\rho_{\alpha\beta}(E)$, may be complex, in which case we consider $\rho(E) = \rho_{\alpha\beta}(E) + \rho_{\beta\alpha}(E)$. Let the first $n$ moments (i.e. $\mu^{(0)}$, $\mu^{(1)}$, $\mu^{(2)}$, ..., $\mu^{(n-1)}$) of $\rho(E)$ be zero. The theorem states that $\rho(E)$ crosses the energy axis at least $n$ times within the band, i.e. in $a < E < b$. We have already used this theorem to deduce a minimum number of zeros for the symmetrized off-diagonal elements of the spectral density matrix. As another example let $\rho(E)$ be the difference in the densities of states for an atom near an interface and an atom in the bulk crystal. We assume no relaxation at the interface so that all bond lengths and angles in the two crystals are the same, right up to the interfacial plane, as those in the bulk crystal. This is a reasonable description of a lattice-matched epitaxial interface. Let the shortest path from the atom of interest to the interface involve $n$ hops. All closed paths from this atom of length $2n + 1$ or less do not see the interface and they are identical to those in the bulk. It follows that $\mu^{(0)}, \mu^{(1)}, \ldots, \mu^{(2n+1)}$ are identical and therefore the difference in the densities of states has at least $2n + 2$ zeros. The further the atom from the interface the greater the number of oscillations in the difference of densities of states. Heine and Samson (1980) have considered the origin of Friedel oscillations from a similar point of view.

Ducastelle and Cyrot-Lackmann's theorem is readily generalized to integrated quantities of the spectral density matrix. If $\delta n_\alpha(E)$ is a difference between two densities of states, and $\mu_\alpha^{(0)}, \mu_\alpha^{(1)}, \ldots, \mu_\alpha^{(n-1)}$ of $\delta n_\alpha(E)$ are zero, then $\delta N_\alpha(E_F)$ has $n - 1$ zeros and $\delta U_\alpha(E_F)$ has $n - 2$ zeros, as $E_F$ varies through the band, where

$$\delta N_\alpha(E_F) = \int_{-\infty}^{E_F} \delta n_\alpha(E) \, \mathrm{d}E \tag{3.169}$$

and

$$\delta U_\alpha(E_F) = \int_{-\infty}^{E_F} (E - E_F) \delta n_\alpha(E) \, \mathrm{d}E. \tag{3.170}$$

$\delta N_\alpha(E_F)$ is the difference in the occupations of the state $|\alpha\rangle$ and $\delta U(E_F)$ is the difference in the local contributions to the band energy.

Turchi and Ducastelle (1985) found that the fifth moment of the density of states is crucial for determining the relative stabilities of the b.c.c., f.c.c., and h.c.p. crystal structures in the transition metal series. The significance of this result is the following: any empirical model which does not capture the information contained in the fifth moment of the density of states of a nearest neighbour, tight binding d-orbital Hamiltonian does not contain the physics that discriminates between the crystal structures of the transition metals. This comment certainly applies to the embedded atom and related schemes (Sections 3.7, 3.8.6), although it is sometimes possible to fit the empirical models so that the desired structural energy differences are reproduced. Cressoni and Pettifor (1991) found that the sixth moment is also important for determining the relative stabilities of the cubic or hexagonal diamond lattices in sp-bonded elements.

### 3.8.5 The tight binding bond (TBB) model

Having seen how the local atomic environment controls the local electronic structure we now wish to construct a tight binding description of the cohesive energy suitable for atomistic simulations. This is the tight binding bond (TBB) model of Pettifor and Podloucky (1986), Pettifor (1987), and Sutton *et al.* (1988). In the TBB model the input charge density, $\rho^{\text{in}}$, is expressed as a superposition of atomic like charge densities, $\rho_i$:

$$\rho^{\text{in}} = \sum_i \rho_i. \tag{3.171}$$

The one electron Hamiltonian is thus given by

$$H = T + V^{\text{in}} + V^{\text{in}}_{\text{xc}} + v, \tag{3.172}$$

where $T$ is the kinetic energy, $V^{\text{in}}$ is an input Hartree potential, which is a linear super-position of atomic Hartree potentials:

$$V^{\text{in}} = \sum_i V^{\text{in}}_i. \tag{3.173}$$

$V^{\text{in}}_{\text{xc}}$ is the exchange-correlation potential for the input charge density and $v$ is the total ionic pseudopotential. The total energy is given in the Harris-Foulkes approximation by eqn (3.28), which may be rewritten as follows:

$$E_{\text{total}} = \text{Tr}\,\rho^{\text{out}} H - \text{Tr}\,\rho^{\text{in}}(V^{\text{in}}/2 + V^{\text{in}}_{\text{xc}}) + E_{\text{xc}}[\rho^{\text{in}}] + E_{ii}. \tag{3.174}$$

The output density operator, $\rho^{\text{out}}$, is constructed from the eigenstates of $H$:

$$\rho^{\text{out}} = 2 \sum_n{}' |n\rangle\langle n|. \tag{3.175}$$

We recognize $\text{Tr}\,\rho^{\text{out}} H$ as the band energy. Note that eqn (3.174) is independent of the basis set.

Many empirical tight binding schemes are based on eqn (3.174). e.g. Chadi (1984). The total energy is expressed as the band energy plus a repulsive pair potential representing the remaining terms on the right of eqn (3.174). There is no real justification for a pair potential representation of the remaining terms. But that is not the main objection to such schemes. The scheme is in no sense self-consistent, and as will be shown below some form of self-consistency is essential, particularly when the band is not half full, for reproducing correctly second-order changes in the energy, such as the elastic properties of the solid.

A much more physically transparent form (Harris 1985, Sutton *et al.* 1988) for the energy may be obtained by adding and subtracting $\text{Tr}\,\rho^{\text{in}} H$ on the right hand side of eqn (3.174) and using eqn (3.172) for $H$:

$$\begin{aligned}
E_{\text{total}} = &\; \text{Tr}(\rho^{\text{out}} - \rho^{\text{in}})H \\
&+ \sum_i \text{Tr}\,\rho_i\left(\sum_{j \neq i} V^{\text{in}}_j/2 + v_j\right) + E_{ii} \\
&+ E_{\text{xc}}[\rho^{\text{in}}] - \sum_i E_{\text{xc}}[\rho_i] \\
&+ \sum_i\left[\text{Tr}\,\rho_i(T + V^{\text{in}}_i/2 + v_i) + E_{\text{xc}}[\rho_i]\right].
\end{aligned} \tag{3.176}$$

The ionic pseudopotential at site $i$ has been represented by $v_i$. The last line in eqn (3.176) is the energy of the isolated atoms in their non-interacting states, i.e. before they are condensed to form the solid. When this term is taken over to the left-hand side of eqn (3.176) the remaining terms on the right-hand side are equal to the binding energy, $E_B$. We have assumed here that the free atom is not spin-polarized. In fact spin polarization can be an important contribution to the energy of an atom and it has to be added separately to the last line in eqn (3.176) in order to make comparisons with full local spin density functional calculations or experimental measurements of binding energies of solids. Consider the terms on the second line. They are equal to the change in the *total* electrostatic energy, $\Delta E_{es}$, of all valence electrons and ion cores when the atoms are condensed from infinity to make the solid. It is emphasized that $\Delta E_{es}$ includes electron–electron, electron–ion, and ion–ion electrostatic interactions. Similarly the terms in the third line are the change in the exchange and correlation energy, $\delta E_{xc}$, in forming the solid from isolated atoms. Thus, the following transparent expression is obtained for the binding energy of the solid:

$$E_B = \text{Tr}\left(\rho^{\text{out}} - \rho^{\text{in}}\right)H + \Delta E_{es} + \Delta E_{xc}. \tag{3.177}$$

The errors in this expression for the binding energy are second order in the error in the input charge density, compared with the self-consistent charge density, as discussed in Section 3.2.2. We note that $\Delta E_{es}$ and $\Delta E_{xc}$ are functionals of $\rho^{\text{in}}$ and not $\rho^{\text{out}}$. Sutton *et al.* (1988) showed that $\Delta E_{es} + \Delta E_{xc}$ may be approximated by a sum of pair potentials, an approximation that has been discussed by Foulkes and Haydock (1989) and Skinner and Pettifor (1991). In the following we shall make the pair potential approximation for $\Delta E_{es} + \Delta E_{xc}$.

The term $\text{Tr}\left(\rho^{\text{out}} - \rho^{\text{in}}\right)H$ describes the bonding and rehybridization that occurs when the solid is condensed from free atoms. Choosing an LCAO basis set we immediately find that $\rho^{\text{in}}$ is diagonal, with diagonal elements equal to the number of electrons in each state of the free atom. For example, in carbon there are two electrons in the 2s shell and two electrons in the 2p shell. The output density matrix has diagonal and off-diagonal elements. The diagonal terms of $\text{Tr}\left(\rho^{\text{out}} - \rho^{\text{in}}\right)H$ are called the *promotion energy*, $E_{\text{prom}}$, because they describe the energy of promoting electrons from free atomic configurations into those of the solid. For example, in diamond formed from atoms in $2s^2 2p^2$ states the promotion energy is $\varepsilon_p - \varepsilon_s$ if the state of hybridization in diamond is $2s^1 2p^3$, where $\varepsilon_s$ and $\varepsilon_p$ are the s and p diagonal elements of the crystal Hamiltonian $H$. The off-diagonal elements of $\text{Tr}\left(\rho^{\text{out}} - \rho^{\text{in}}\right)H$ are entirely due to the non-zero bond orders of $\rho^{\text{out}}$, and they describe the *covalent bond energy*, $E_{\text{cov}}$:

$$E_{\text{cov}} = \sum_{\alpha \neq \beta} \sum \rho_{\alpha\beta}^{\text{out}} H_{\beta\alpha}. \tag{3.178}$$

This is literally the sum of energies of individual covalent bonds. For example, consider the bond between orbitals $|\alpha\rangle$ and $|\beta\rangle$ which are coupled through the Hamiltonian with matrix elements $H_{\alpha\beta}$. The energy of the bond is $\rho_{\alpha\beta}^{\text{out}} H_{\beta\alpha} + \rho_{\beta\alpha}^{\text{out}} H_{\alpha\beta}$. Although each bond energy has a two-centre form it is a many-body energy because each density matrix element is determined by the environment around the orbitals $|\alpha\rangle$ and $|\beta\rangle$. The covalent bond energy may also be represented in a useful site-diagonal representation using the densities of states:

$$E_{\text{cov}} = 2\sum_{i\alpha} \int_{-\infty}^{E_F} n_{i\alpha}(E)\,(E - \varepsilon_{i\alpha})\,\text{d}E. \tag{3.179}$$

Here we have used separate site and orbital indices for clarity. The diagonal Hamiltonian matrix element $H_{i\alpha i\alpha}$ is denoted by $\varepsilon_{i\alpha}$.

Thus, the binding energy is expressed in the TBB model as follows:

$$E_{\mathrm{B}} = E_{\mathrm{cov}} + E_{\mathrm{prom}} + E_{\mathrm{pair}}, \tag{3.180}$$

where $E_{\mathrm{pair}}$ represents $\Delta E_{\mathrm{es}} + \Delta E_{\mathrm{xc}}$.

There is considerable freedom in the choice of atomic like basis set and we can exploit this in a particularly useful way due to Anderson (1969). Anderson's idea is to use the valence orbitals on neighbouring atoms to screen their potentials. In that way each electron in the solid feels the full potential of the atom it is sitting on, together with weak, screened potentials due to neighbouring atoms. Anderson defines localized orbitals $|\alpha\rangle$, where $\alpha$ is a composite index denoting site and angular momentum component, by

$$\left\{ -\nabla^2 + V_\alpha + \sum_{\beta \neq \alpha} (1 - |\beta\rangle\langle\beta|) V_\beta \right\} |\alpha\rangle = E_\alpha |\alpha\rangle \tag{3.181}$$

where $V_\beta$ is the potential due to atom $\beta$. The potential $(1 - |\beta\rangle\langle\beta|) V_\beta$ is called a chemical pseudopotential. If the set of orbitals $|\beta\rangle$ on each atom were complete the chemical pseudopotential would be zero. Expanding the eigenstates of the system $|n\rangle$ in this basis set:

$$|n\rangle = \sum_\alpha c_\alpha |\alpha\rangle, \tag{3.182}$$

and inserting $|n\rangle$ into the Schrödinger equation

$$\left( -\nabla^2 + \sum_\alpha V_\alpha \right) |n\rangle = E_n |n\rangle, \tag{3.183}$$

we obtain the secular equation

$$\sum_\beta [(E_\beta - E_n)\delta_{\alpha\beta} + \langle\alpha| V_\alpha |\beta\rangle (1 - \delta_{\alpha\beta})] c_\beta = 0. \tag{3.184}$$

For non-trivial solutions we require

$$\det [(E_\beta - E_n)\delta_{\alpha\beta} + \langle\alpha| V_\alpha |\beta\rangle (1 - \delta_{\alpha\beta})] = 0. \tag{3.185}$$

Let us examine what has been achieved here. The construction of the local orbitals $\{|\alpha\rangle\}$ in eqn (3.181) is such as to satisfy the Schrödinger equation, eqn (3.183), *exactly*. This may be contrasted with earlier LCAO theories where there was always an uncertainty about the selected, *fixed* orbitals being a good basis for molecular functions or Bloch functions of a solid. In those earlier treatments one took fixed atomic orbitals and appealed to the variational principle to bolster the hope that they were adequate. Many basis functions were needed by these earlier treatments, often having long-range tails that necessitated the evaluation of many two- and three-centre integrals in the Hamiltonian matrix. For example, Jansen and Sankey (1987) found that up to sixth nearest neighbour interactions need to be included to obtain a reasonable band structure in silicon if a basis of fixed atomic orbitals is used. By contrast, the orbitals in eqn (3.181) are more localized and the Hamiltonian matrix is of two-centre form by construction. The penalty that is paid is two-fold: (i) the orbitals are now *environment dependent* and have to be determined self-consistently via eqn (3.181); (ii) the Hamiltonian matrix is no longer Hermitian (see Anderson (1984), Heine (1980), and Bullett (1980) for discussions of this point). A complementary analysis has been carried out by Andersen and Jepsen (1984) in the framework of linear muffin tin orbital theory. The orbitals appear to be orthogonal in the secular equation, eqn (3.184). Their non-orthogonality is hidden in the diagonal matrix elements, $E_\alpha$, which may be obtained directly from eqn (3.181):

$$E_\alpha = \langle \alpha | -\nabla^2 + V_\alpha | \alpha \rangle + \sum_{\beta \neq \alpha} \langle \alpha | V_\beta | \alpha \rangle - S_{\alpha\beta} \langle \beta | V_\beta | \alpha \rangle. \tag{3.186}$$

Thus $E_\alpha$ is the term value, $\langle \alpha | -\nabla^2 + V_\alpha | \alpha \rangle$, of the free atom, corrected by the crystal field terms, $\Sigma_{\beta \neq \alpha} \langle \alpha | V_\beta | \alpha \rangle$, and another term, $-\Sigma_{\beta \neq \alpha} S_{\alpha\beta} \langle \beta | V_\beta | \alpha \rangle$, due to the non-orthogonality of the basis set. The latter is a positive quantity (because the overlap matrix elements are postive and $\langle \beta | V_\beta | \alpha \rangle$ is negative). The larger the overlap the larger this positive correction to $E_\alpha$. This simply reflects the increase in kinetic energy to maintain eigenstate orthogonality when there is large overlap of the basis functions on neighbouring sites. In the tetrahedrally bonded semiconductors this positive contribution is the main source of the repulsion under a compression, because the overlap is so large (Harrison 1980). Since the overlap terms arise only in the diagonal Hamiltonian matrix elements, they are often extracted from the promotion energy and added to the energies $\Delta E_{es} + \Delta E_{xc}$. In empirical tight binding treatments of tetrahedrally bonded semiconductors the pair potential is identified with the non-orthogonality term.

*To recapitulate*, in the TBB model the binding energy is expressed in a two-centre, tight binding Hamiltonian by eqn (3.180), where $E_{cov}$ and $E_{prom}$ are the covalent bond and promotion energies computed as though the basis set were orthogonal and $E_{pair}$ is a sum of pair potentials representing (i) the change in the total electrostatic and exchange-correlation energies in forming the solid from free atoms and (ii) the repulsive energy arising from the non-orthogonality of the basis set. In tetrahedrally bonded semiconductors the pair potential term is dominated by the non-orthogonality contribution. It is implicit that the basis set changes with the atomic environment, and this is reflected in the variations of the diagonal and off-diagonal Hamiltonian matrix elements and the pair potential.

So far we have used the Harris–Foulkes approximation to derive an expression for the binding energy that is correct to first order in the charge density. In an atomistic calculation we require a reasonable description of elastic properties by whatever scheme we are using to evaluate atomic interactions. But elastic constants are determined by changes in the energy that are *second* order in the atomic displacements. Physically this amounts to saying that the charge redistribution that takes place in a deformation of the system makes a significant contribution to the energy change to second order (Sutton *et al.* 1988). Therefore, we must go beyond the Harris–Foulkes approximation if we are to reproduce elastic constants. On the other hand, the force theorem (Section 3.4) tells us that the self-consistent redistribution of charge should not contribute to the force on an atom. Both of these requirements are satisfied in the TBB model by introducing some form of self-consistency. The simplest form of self-consistency is to assume that each atom remains charge neutral. This is an excellent approximation in metals where the screening length is normally less than a nearest neighbour separation. It is also an excellent approximation in semiconductors where the band gap is small compared with the band width (Heine and Weaire 1970). Local charge neutrality is achieved in the TBB model by adjusting all the on-site Hamiltonian matrix elements on a given atom by the same amount. If one wants to study charged defect centres in semiconductors the requirement of local charge neutrality must be relaxed, for example by setting the site-diagonal Hamiltonian matrix elements on a given atom to be proportional to the charge excess or deficit on the atom.

To be consistent with the force theorem the shifts in the site diagonal Hamiltonian matrix elements, to maintain local charge neutrality, should *not* contribute to the force acting on an atom. The gradient of the binding energy with respect to atom $i$ is given by:

$$\nabla_i E_B = \text{Tr}(\rho^{\text{out}} - \rho^{\text{in}})\nabla_i H + \nabla_i E_{\text{pair}}. \qquad (3.187)$$

Because of local charge neutrality there are no diagonal contributions to the trace provided all the Hamiltonian matrix elements on a given site are adjusted by the same amount, regardless of their angular momentum components. Thus, we are left with only off-diagonal components in the trace, which describe covalent bond contributions to the gradient. The TBB model therefore satisfies the force theorem explicitly. On the other hand if local charge neutrality were introduced into the band model, where the total energy is the band energy plus a sum of pair potentials, the force theorem would be satisfied only if appropriate terms in the band energy and the pair energy were *allowed* to cancel. In an empirical parametrization of the pair potential such a cancellation can never be achieved and hence empirical band models do not satisfy the force theorem. This is the main reason for working with the covalent bond energy in the TBB model rather than the band energy (Pettifor 1990). The point is that in an empirical model the cancellation of terms arising from the self-consistent redistribution of charge, in order to satisfy the force theorem, must be explicit and this is done in the TBB model by the regrouping of terms in eqn (3.176). Thus, eqn (3.176) is not merely a re-expression of the total energy in a much more transparent form, it is also an *essential* regrouping of terms in an empirical parametrization to satisfy the force theorem.

The TBB model has been used by Paxton and Sutton (1989) to model grain boundaries in silicon, and this work is discussed in Section 4.3.3. Wilson *et al.* (1990) compared the application of the TBB model, with and without local charge neutrality, to surfaces of silicon. It was found that local charge neutrality made almost no difference to the relaxed structures and energies of the surfaces. On the other hand the bulk modulus for b.c.c. transition metals, as a function of d-band filling, has been computed by Sutton *et al.* (1988) and found to be strongly dependent on whether local charge neutrality is imposed. Up to 50 per cent errors were found if no form of self-consistency is imposed in the evaluation of the bulk modulus. However, the errors were zero when the band was half full. This is consistent with the result of Wilson *et al.* for silicon, which may be regarded as a half full band since the band gap is only about 5 per cent of the combined valence and conduction band widths. Wilson *et al.* also compared the relaxed structures obtained by the TBB model with those obtained by the classical potential of Stillinger and Weber (1985). The Stillinger–Weber potential failed to describe the surface reconstructions, which were well accounted for by the TBB model. For example the Stillinger–Weber potential failed to describe the 30° tilting of $[1\bar{1}0]$ chains of atoms on the (110) surface, and it did not yield the $\pi$-bonded chain $2 \times 1$ reconstruction on the (111) surface. The calculations of Wilson *et al.* indicated that some surface reconstructions are driven by Jahn–Teller distortions, and while a classical potential cannot be expected to describe these effects the TBB model can and does.

A major difficulty with the TBB model is that it is computationally quite intensive, even though a number of approximations are made. For example, it is typically two to three orders of magnitude slower to carry out a tight binding calculation for silicon than it is to use a classical potential such as that of Stillinger and Weber (1985). This is a considerable limitation. There have been two significant attempts to derive interatomic potentials from the TBB model, and they are described in the next two sections. The idea is that a potential allows many more atoms to be treated than the TBB model, and the aim is to achieve this speed while not sacrificing too much of the physics of the TBB model.

### 3.8.6 The second moment appproximation

Finnis and Sinclair (1984) simplified the TBB expression for the binding energy into the form of an embedded atom potential (Section 3.7) by expressing the covalent bond energy in a form that utilizes information about the second moment of the local density of states. Information about higher moments of the local density of states is not included. The promotion energy is neglected and the binding energy is simply the sum of the attractive covalent bond energy and a repulsive pair potential:

$$E_{\mathrm{B}} = E_{\mathrm{cov}} + E_{\mathrm{pair}}. \tag{3.188}$$

The neglect of the promotion energy is consistent with the application of the method to just one type of orbital, such as the d-band in the transition metals.

The Finnis–Sinclair model expresses the covalent bond energy as the *square root* of a sum of pair potentials which, together with the repulsive pair potentials comprising $E_{\mathrm{pair}}$, are fitted empirically. Thus the form of the binding energy is exactly the same as eqn (3.67) in the embedded atom method, with the embedding function being the square root. The physical origin of the square root functional form is most easily understood by considering an explicit functional form for the local density of states. With the transition metals in mind we consider a d-band which can contain up to 10 electrons (including spin degeneracy). The zeroth moment of the local density of states is thus 5. Let the centre of gravity of the local density of states at site $i$ be $\varepsilon_i$, and let the second moment be $\mu_i^{(2)}$. Thus, if the local density of states at site $i$ is $n_i(E)$ then

$$\int_{-\infty}^{+\infty} n_i(E)\,\mathrm{d}E = 5$$

$$\int_{-\infty}^{+\infty} n_i(E)\,(E - \varepsilon_i)\,\mathrm{d}E = 0 \tag{3.189}$$

$$\int_{-\infty}^{+\infty} n_i(E)\,(E - \varepsilon_i)^2\,\mathrm{d}E = \mu_i^{(2)}.$$

The second moment is given by the sum of all paths of length two from site $i$, as discussed in Section 3.8.4. This is given by

$$\mu_i^{(2)} = 5 \sum_{j \neq i} h^2(r_{ij}), \tag{3.190}$$

where $h(r_{ij})$ is the root mean square hopping integral given by:

$$h^2(r_{ij}) = (\mathrm{dd}\,\sigma^2(r_{ij}) + 2\mathrm{dd}\,\pi^2(r_{ij}) + 2\mathrm{dd}\,\delta^2(r_{ij}))/5. \tag{3.191}$$

We now write down a functional form for the local density of states that satisfies eqn (3.189), and a Gaussian form is chosen for convenience:

$$n_i(E) = \frac{5}{\sqrt{(2\pi\mu_i^{(2)})}}\exp-\frac{(E - \varepsilon_i)^2}{2\mu_i^{(2)}}. \tag{3.192}$$

As shown by Ackland *et al.* (1988) the final result, that the covalent bond energy associated with site $i$ is proportional to the square root of is $\mu_i^{(2)}$, independent of the choice
of the functional form of $n_i(E)$.

If $E_F$ is the Fermi energy then the covalent bond energy associated with site $i$ is
given by

$$E_{cov}^{(i)} = 2 \int_{-\infty}^{+E_F} n_i(E)\,(E - \varepsilon_i)\,dE$$

$$= -\frac{10}{\sqrt{2\pi}} \left[\mu_i^{(2)}\right]^{\frac{1}{2}} \exp - \frac{(E_F - \varepsilon_i)^2}{2\mu_i^{(2)}}. \tag{3.193}$$

In early versions of the second moment approximation (Allan and Lannoo 1976, Heine
1980, Masuda and Sato 1981, Finnis and Sinclair 1984) no variation in $\varepsilon_i$ from site to
site was envisaged. For a half-filled band $E_F = \varepsilon_i$ and $E_{cov}^{(i)}$ is proportional to the square
root of $\mu_i^{(2)}$. But for a band that is not half full the exponential factor in eqn (3.193) is
dependent on the second moment. Indeed, Masuda and Sato (1981) included the exponential factor in their study of the Peierls stress of a screw dislocation. Ackland *et al.*
(1988) pointed out that if the $\varepsilon_i$'s are adjusted to achieve local charge neutrality then the
exponential factor becomes a constant which is independent of the site. The assumption
of local charge neutrality is consistent with the TBB model. Local charge neutrality
holds if $x_i = (E_F - \varepsilon_i)/(\mu_i^{(2)})^{\frac{1}{2}}$ is constant at all sites. Thus, for a given Fermi energy,
$\varepsilon_i$ is adjusted to ensure that $x_i$ is constant, and $E_{cov}^{(i)}$ is proportional to $(\mu_i^{(2)})^{\frac{1}{2}}$, *for all
band fillings.*

To illustrate Finnis–Sinclair potentials consider the potentials for f.c.c. metals constructed by Sutton and Chen (1990). Their form is as follows:

$$E_B = \varepsilon \left( \sum_{i \neq j} \sum \left(\frac{a}{r_{ij}}\right)^n - c \sum_i \sqrt{\rho_i} \right), \tag{3.194}$$

where

$$\rho_i = \sum_{j \neq i} \left(\frac{a}{r_{ij}}\right)^m. \tag{3.195}$$

$\varepsilon$ is a parameter with the dimensions of energy, $a$ is a parameter with the dimensions of
length, $c$ is a positive dimensionless parameter, and $m$ and $n$ are positive integers. For
a given $n$ and $m$ the parameter $c$ is determined by the equilibrium condition, i.e. the energy
of the perfect crystal does not change to first order when the lattice parameter is varied
about its equilibrium value. The parameters $\varepsilon$ and $a$ were found for each metal by fitting
the cohesive energy and lattice parameter exactly. The integers $m$ and $n$ were found from
the best fit to the elastic constants $c_{11}$, $c_{12}$, and $c_{44}$, with the requirement that $n > m$.
The fitting for Ir and Rh was only partly successful because for those metals $c_{44} > c_{12}$,
which cannot be achieved with Finnis–Sinclair potentials because of eqn (3.69).

By comparing the Finnis–Sinclair potential of eqn (3.194) with eqn (3.67) of the
embedded atom method it is seen that the two are equivalent if the embedding func-
tion $f(\rho)$ is identified with $-\varepsilon c \sqrt{\rho}$, the pair potential $U(r)$ is identified with $\varepsilon (a/r)^n$
and the pair potential $\rho^a(r)$ is identified with $(a/r)^m$. Applying the transformation given
in eqn (3.72) to put the above Finnis–Sinclair potentials in 'normalized form' we obtain:

$$g(\rho) = -\varepsilon c \rho^{\frac{1}{2}} [1 - \rho^{\frac{1}{2}}/(2\bar{\rho}^{\frac{1}{2}})],$$

and

$$V(r) = \varepsilon \left( \left( \frac{a}{r} \right)^n - c(\bar{\rho})^{-\frac{1}{2}} \left( \frac{a}{r} \right)^m \right).$$ (3.196)

The value of $\rho_i$, given by eqn (3.195), in the perfect crystal is denoted by $\bar{\rho}$. It is seen that the effective pair potential $V(r)$ is a Lennard–Jones n–m potential. All the comments that were made in Section 3.7 in connection with the invariance of embedded atom potentials under the transformation of eqn (3.72) apply equally to Finnis–Sinclair potentials.

Pettifor (1987, 1990) has discussed the physics of the second moment approximation using a simple *first nearest neighbour model*. While the model correctly describes the trends of the band width, equilibrium bond length, cohesive energy and bulk modulus across the 4d transition series it also has some spectacular failures. The most important failure of the model is that, for a given transition metal series and band filling, the cohesive energy is independent of the crystal structure. Thus, the second moment approximation is unable to predict the relative stabilities of rival crystal structures. This is consistent with our discussion in Section 3.8.4 where higher moments of the density of states, particularly $\mu^{(5)}$ and $\mu^{(6)}$, were identified as being important in discriminating between rival crystal structures in a nearest neighbour tight binding model. Empirical fittings of Finnis–Sinclair potentials often build in the constraint that the desired crystal structure has the lowest energy, invariably by extending the range of the potentials to beyond first neighbours. However, it should be borne in mind that the Finnis–Sinclair model (and the embedded atom method) do not embody the real factors controlling structural stability. Pettifor has also noted that $c_{44}$ is zero in his nearest neighbour Finnis–Sinclair model. This is consistent with the fact that $c_{44}$ oscillates in sign for a given crystal structure as the d band is filled (Pettifor 1987). It is negative at those band fillings where the crystal structure is unstable. Again, the variation of $c_{44}$ with band filling is determined by higher moments of the density of states than the second. Pettifor (1990) has shown that the nearest neighbour Finnis–Sinclair model also predicts that the unrelaxed vacancy formation energy is zero, whereas pair potential models predict that it equals the cohesive energy. The experimental vacancy formation energy (which includes the relatively small relaxation energy) is about half of the cohesive energy. The example of the vacancy illustrates another failure of Finnis–Sinclair potentials: they predict an *inward* relaxation of the first neighbour shell of atoms around a vacancy in metals such as Nb or Mo. In contrast more accurate TBB model calculations by Ohta *et al.* (1987) show a slight *outward* relaxation.

The two most significant improvements introduced by the Finnis–Sinclair and embedded atom models of metallic cohesion, over a model based entirely on pair potentials (i.e. with no density dependent energy), are (i) the strengthening of surviving bonds when the coordination number is reduced; and (ii) $c_{12}$ is always predicted to be greater than $c_{44}$. By contrast, in a description based entirely on pair potentials, there is no environment dependence of the energy of a given bond, and $c_{12} = c_{44}$ always. Comparing the practical use of Finnis–Sinclair potentials (or embedded atom potentials) with pair potentials derived from second order perturbation theory (Section 3.5), the Finnis–Sinclair potentials remove the uncertainty about how to treat the local variations of the density dependent energy and the density dependence of the pair potential. The discussion in the previous paragraph highlights the shortcomings of the Finnis–Sinclair and embedded atom methods. Clearly, structural stability and the resistance to shear are more subtle quantities that are controlled by higher moments of the density of states. It is worth bearing in mind that in the Finnis–Sinclair and embedded atom models there is no directional bonding and atoms are still treated as fuzzy spheres, in much the same way as they are in a pair potential model. It is often argued that this is all that is required to describe an f.c.c. metal, but the negative Cauchy discrepancies $(c_{12}-c_{44})$ in Ir and Rh counter the claim.

### 3.8.7 Beyond the second moment approximation

Pettifor (1989, 1990) and Pettifor and Aoki (1991) have introduced the idea of a potential for the bond order. The idea has its roots in the TBB model where the binding energy of the solid is expressed as the sum of the three terms shown in eqn (3.180). The covalent bond energy may be broken down into a sum of energies of individual covalent bonds, as in eqn (3.178). The bond order $\rho_{i\alpha j\beta}$ is simply one half the difference between the number of electrons in the bonding $(|i\alpha\rangle + |j\beta\rangle)/\sqrt{2}$ and antibonding $(|i\alpha\rangle - |j\beta\rangle)/\sqrt{2}$ states. The analytic dependence of the bond order on its surroundings is derived by using the recursion method to obtain the bonding $(\langle i\alpha| + \langle j\beta|)G(|i\alpha\rangle + |j\beta\rangle)/2$ and anti-bonding $(\langle i\alpha| - \langle j\beta|)G(|i\alpha\rangle - |j\beta\rangle)/2$ Green functions. The bond order is the integral of the difference of the imaginary part of these Green functions up to the Fermi energy. The recursion method yields the Green functions as continued fractions with coefficients $a_i$ and $b_i$ as given in eqn (3.147). The coefficients are related directly, through the moments theorem (Section 3.8.4), to ring paths in the local atomic environment starting on atom $i$ and ending on atom $j$. The further we go down the continued fraction the more information that is incorporated about the local atomic environrunent. For example, at the first level of the continued fraction the bonding and antibonding Green functions are those of an isolated diatomic molecule. The second level of the continued fraction begins to reflect the influence of the first shell of neighbours about the bond on the bonding and antibonding Green functions. Further levels of the continued fraction sample further shells of neighbours about the bond. Interestingly, it is the three, four, and higher membered rings that are responsible for the differences between the bonding and antibonding recursion coefficients. We note that in the second moment approxima-tion to the local density of states of the previous section only two membered rings are taken into account. Thus, in the second moment approximation to the local density of states it is not possible to describe the bond order. Bond order potentials are currently being developed for semiconductors, transition metals, and intermetallic compounds (see references cited at the beginning of this section). They hold out the promise of incorporating the physics and chemistry of the TBB model, but at a fraction of the computational time needed by the TBB model.

We conclude this section by mentioning briefly the work of Moriarty (1972, 1988, 1990) and Moriarty and Phillips (1991), in which an expansion of the total energy for transition metals in terms of one, two, three, and four body potentials has been obtained from first principles. In a perfect elemental metal, with atomic volume $\Omega$, the total energy functional has the form:

$$E_{\text{tot}}(r_1, r_2, \ldots r_N) = NE_{\text{vol}}(\Omega) + \tfrac{1}{2}\sum_{i \neq j} v_2(r_{ij}, \Omega) + \tfrac{1}{6}\sum_{i \neq j \neq k} v_3(r_{ij}, r_{jk}, r_{ki}, \Omega)$$

$$+ \tfrac{1}{24}\sum_{i \neq j \neq k \neq l} v_4(r_{ij}, r_{jk}, r_{kl}, r_{li}, r_{ki}, r_{lj}, \Omega). \tag{3.197}$$

This expansion has been obtained within the framework of 'generalized pseudopotential theory' (Moriarty 1972) and local density functional theory. In transition metals the pseudopotential is weak for the nearly free electron s and p electrons but is strong for the more localized d electrons. A mixed basis set, consisting of plane wave states for the s and p electrons and localized d states centred on each site, is chosen. The interatomic potentials $v_2, v_3$, and $v_4$ are expressed in terms of weak pseudopotential and d-state tight-binding and hybridization matrix elements coupling different sites, and the series is rapidly convergent beyond $v_3$. The leading term $E_{vol}$ includes all intra-atomic contributions to the total energy and leads to a good description of transition metal cohesion in lowest order. The expansion of eqn (3.197) is a generalization of the expansion in eqn (3.53) for simple metals. But in contrast to the expansion for simple metals the three and four body potentials are essential in the central transition metals, reflecting the importance of directional bonding in those metals. The density dependence of the potentials $v_2$, $v_3$ and $v_4$ raises the familiar problem of defining a local volume in a defective environment such as a surface or an interface. Moriarty and Phillips (1991) addressed this problem by transforming the total energy functional, eqn (3.197), into the form of an embedded atom potential or a Finnis–Sinclair potential. In this theory the bulk volume term and interatomic potentials are modulated by the environment either through the local electron density, as in the embedded atom and effective medium schemes, or through the local density of states as in the Finnis–Sinclair scheme. $E_{vol}$, of eqn (3.197), is identified with the embedding function of the embedded atom method and effective medium theory, or with the square root of the second moment of the local d-electron density of states in the Finnis–Sinclair approach. The scheme of Moriarty and Phillips (1991) goes beyond the embedded atom and Finnis–Sinclair approaches by explicitly including the three and four body potentials, whose density dependence is now expressed in terms of a locally defined electron density or second moment. Although these potentials are very new, it is clear that they are a more accurate description of transition metal bonding than is achieved by second moment and embedded atom models.

## 3.9 TEMPERATURE DEPENDENCE OF ATOMIC INTERACTIONS

The issue we turn to now is the determination of the atomic structure and excess thermodynamic quantities of an interface at an elevated temperature. Since the atoms are vibrating we have to define what we mean by 'the atomic structure'. This is true even at 0 K owing to the zero point energy. Of course, we mean the time-averaged positions of the atoms, assuming that the system is in thermodynamic equilibrium. The time-averaged structure is what is measured in an X-ray diffraction experiment for example. As the temperature changes the time-averaged positions of the atoms change until the time-averaged forces on them are zero. The time averaged forces vary with temperature because of the anharmonicity of the atomic interactions. The time-averaged structure may be determined, therefore, by requiring that the time-averaged force on each atom is zero. The ergodic hypothesis of statistical mechanics asserts that the time-averaged position of a quantity is identical to the ensemble average. The ensemble average of the force on an atom is given by the negative of the gradient of the ensemble free energy with respect its position. Thus, if we can express the free energy of the system as a function of the average atomic coordinates, we can obtain the time-averaged structure directly by simply minimizing the free energy with respect to the position of each atom. This idea was introduced independently by LeSar *et al.* (1989) and Sutton (1989). The key point is that direct minimization of the free energy involves little more than the minimization of any other function of the atomic coordinates. By contrast, evaluation of the free energy by Monte Carlo or molecular dynamics techniques is a much more computationally intensive task. Once we have obtained the structure at some elevated temperature we may evaluate other thermodynamic functions such as the excess specific heat, entropy and internal energy.

Henceforth we shall refer to the time-averaged atomic positions simply as the atomic positions. In general, it is necessary to make an approximation in order to write down the free energy of the system as a function of the atomic positions. In the harmonic approximation the potential energy is expanded to second order in the displacements of the atoms from their mean positions. Let $u_{i\alpha}$ be the displacement of atom $i$ in the $\alpha$ direction ($\alpha = x, y,$ or $z$) from the mean position $r_i$. The potential energy is given by

$$E_p = E_p(r_1, \ldots, r_N) - \sum_{i\alpha} f_{i\alpha} u_{i\alpha} + \tfrac{1}{2} \sum_{i\alpha} \sum_{j\beta} D_{i\alpha j\beta} u_{i\alpha} u_{j\beta}, \qquad (3.198)$$

where $f_{i\alpha} = -\partial E_P/\partial r_{i\alpha}$ and $D_{i\alpha j\beta} = \partial^2 E_P/\partial r_{i\alpha} \partial r_{j\beta}$ are evaluated at the equilibrium positions. It follows that $f_{i\alpha}$ are all zero. The potential energy $E_P$ is expressed either in terms of some interatomic potential, such as a sum of pair potentials or embedded atom potentials, or as the potential energy surface defined by an electronic structure calculation. The equations of motion are as follows:

$$m_i \ddot{u}_{i\alpha} = -\sum_{j\beta} D_{i\alpha j\beta} u_{j\beta}, \qquad (3.199)$$

where $m_i$ is the mass of atom $i$. Since we are assuming that each atom is performing harmonic vibrations we write $u_{i\alpha}(t) = u_{i\alpha} e^{i\omega t}$ where $u_{i\alpha}$ is now time independent and a complex number in general. We obtain the following time-independent equations of motion:

$$m_i \omega^2 u_{i\alpha} = \sum_{j\beta} D_{i\alpha j\beta} u_{j\beta}. \qquad (3.200)$$

If we set $\tilde{u}_{i\alpha}$ equal to $\sqrt{m_i} u_{i\alpha}$ and $\tilde{D}_{i\alpha j\beta}$ equal to $D_{i\alpha j\beta}/\sqrt{m_i m_j}$ we can eliminate the masses from the equations of motion:

$$\omega^2 \tilde{u}_{i\alpha} = \sum_{j\beta} \tilde{D}_{i\alpha j\beta} \tilde{u}_{j\beta}. \qquad (3.201)$$

These equations of motion are similar to the tight binding equations, eqn (3.99), with three 'basis functions' per atomic site and the identity for the overlap matrix. The eigenvalues $\omega_n^2$ may be determined by solving the secular determinant $\det(\omega^2 I - \tilde{D}) = 0$ and the normalized eigenvectors $\tilde{u}_{i\alpha}^{(n)}$ may then be obtained from eqn (3.201) by setting $\omega^2$ equal to $\omega_n^2$. The total density of states is given by

$$Y(\omega^2) = \sum_n \delta(\omega^2 - \omega_n^2). \qquad (3.202)$$

There are standard expressions for all the thermodynamic functions in the harmonic approximation in terms of the total density of states as a function of $\omega$, rather than $\omega^2$. The Helmholtz free energy is given by

$$F = E_P + kT \int_0^\infty N(\omega) \ln\left[2 \sinh\left(\frac{\hbar\omega}{4\pi kT}\right)\right] d\omega, \qquad (3.203)$$

where the density of states $N(\omega)$ is related to the density of states $Y(\omega^2)$ by

$$Y(\omega^2) = \frac{N(\omega)}{2\omega}. \tag{3.204}$$

The equilibrium position of atom $i$ is given by the condition that $-\nabla_i F = 0$. From eqn (3.203) for the free energy we see that there are two contributions to $-\nabla_i F$. The first is simply the temperature independent force due to the potential energy $-\nabla_i F$. This is the force that we would consider in the absence of any vibratory motion of the atoms. The second contribution, $\nabla_i(F - E_P)$, is temperature dependent and arises from the fact that the density of states changes as atom $i$ undergoes a virtual displacement because elements of the matrix $\tilde{D}$ change. We note that this contribution is not zero even when $T = 0$ owing to the zero point motion of the atoms. The changes in the matrix $\tilde{D}$ as an atom is displaced are due to the anharmonicity of the potential energy $E_P$. That is, the potential energy has non-zero third and higher derivatives. Thus, although the expansion of the potential energy is carried out to only second order in the atomic displacements, the matrix elements $D_{i\alpha j\beta}$ vary as the equilibrium atomic positions change because of higher-order derivatives of the potential energy.

In order to evaluate the temperature dependent contribution, $-\nabla_k(F - E_P)$, let us first rewrite the integral in eqn (3.203) as a discrete sum over the normal modes:

$$F - E_P = kT \sum_n \ln\left[2 \sinh\left(\frac{h\omega_n}{4\pi kT}\right)\right]. \tag{3.205}$$

Using eqn (3.201) and the normalization of the eigenvectors $u^{(n)}$, with components $u_{j\beta}^{(n)}$, it is easily shown that

$$\nabla_k \omega_n^2 = \sum_{i\alpha, j\beta} \tilde{u}_{i\alpha}^{(n)} \tilde{u}_{j\beta}^{(n)} \nabla_k \tilde{D}_{j\beta i\alpha}. \tag{3.206}$$

Therefore, using eqns (3.205) and (3.206) we obtain

$$\nabla_k(F - E_P) = \sum_n \frac{h}{4\pi} \coth\left(\frac{h\omega_n}{4\pi kT}\right) \nabla_k \omega_n$$

$$= \frac{1}{2} \sum_{i\alpha, j\beta} \left[\frac{h}{4\pi\sqrt{m_i m_j}} \sum_n \coth\left(\frac{h\omega_n}{4\pi kT}\right) \frac{\tilde{u}_{j\beta}^{(n)} \tilde{u}_{i\alpha}^{(n)}}{\omega_n}\right] \nabla_k D_{i\alpha j\beta}. \tag{3.207}$$

Note the absence of the tilda over the $D_{i\alpha j\beta}$. The term in square brackets has a simple physical interpretation. It is in fact the equal time, displacement–displacement correlation function $\langle u_{j\beta} u_{i\alpha} \rangle$, where the $\langle \rangle$ brackets denote a thermal average. For example, the diagonal element, $\langle u_{i\alpha} u_{i\alpha} \rangle = \langle u_{i\alpha}^2 \rangle$, is simply the mean square displacement of atom $i$ in the direction $\alpha$. Therefore, eqn (3.207) can be expressed in the following physically transparent form, which may be compared with the Hellmann–Feynman theorem, eqn (3.187), for a tight binding Hamiltonian:

$$\nabla_k(F - E_P) = \frac{1}{2} \sum_{i\alpha, j\beta} \langle u_{j\beta} u_{i\alpha} \rangle \nabla_k D_{i\alpha j\beta}. \tag{3.208}$$

This formula is *exact* within the quasiharmonic approximation. Its main advantage over eqn (3.207) is that it does not require a knowledge of the normal modes of the system, provided the correlation functions $\langle u_{j\beta} u_{i\alpha} \rangle$ can be determined by some other means.

The condition that the energy of the atomic assembly is invariant with respect to a rigid translation leads to the following relation:

$$D_{i\alpha i\beta} = - \sum_{j \neq i} D_{i\alpha j\beta}. \tag{3.209}$$

This may be used to rewrite eqn (3.208) in terms of intersite matrix elements of **D** only:

$$\nabla_k(F - E_P) = \tfrac{1}{2} \sum_{\substack{i\alpha, j\beta \\ i \neq j}} (\langle u_{j\beta} u_{i\alpha} \rangle - \langle u_{i\beta} u_{i\alpha} \rangle) \nabla_k D_{i\alpha j\beta}. \tag{3.210}$$

In a pair potential model for the atomic interactions $\nabla_k D_{i\alpha j\beta}$ is zero unless $k$ coincides with $i$ or $j$. This is the analogue of a 'two-centre' approximation in tight binding. In that case eqn (3.210) becomes:

$$\nabla_k(F - E_P) = \sum_{\substack{j\alpha\beta \\ j \neq k}} (\langle u_{j\beta} u_{k\alpha} \rangle - \langle u_{k\beta} u_{k\alpha} \rangle) \nabla_k D_{k\alpha j\beta} \tag{3.211}$$

and we see that $\nabla_k(F - E_P)$ is given by a sum of contributions from the neighbours to atom $k$, where the range is determined by the range of the interatomic pair potential. However, it should be kept in mind that although the sum in eqn (3.211) has the appearance of a sum of pairwise contributions the correlation functions are determined by all the neighbours to atoms $k$ and $j$.

Consider the classical limit where the thermal energy $kT$ is much greater than any phonon energy $\hbar\omega$. In that case it is not difficult to show that the term in square brackets in eqn (3.207) becomes

$$\langle u_{j\beta} u_{i\alpha} \rangle = kT(D^{-1})_{j\beta i\alpha}. \tag{3.212}$$

In the classical limit we see that the displacement–displacement correlation functions are proportional to matrix elements of the inverse of the matrix of second derivatives of the potential energy. Substituting these correlation functions in eqn (3.208) for $\nabla_k(F - E_P)$ we obtain:

$$\nabla_k(F - E_P) = \frac{kT}{2} \sum_{i\alpha, j\beta} (D^{-1})_{j\beta i\alpha} \nabla_k D_{i\alpha j\beta}$$

$$= \frac{kT}{2} \operatorname{Tr} \mathbf{D}^{-1} \nabla_k \mathbf{D}$$

$$= \frac{kT}{2} \nabla_k \ln |\mathbf{D}|. \tag{3.213}$$

Owing to eqn (3.209) the determinant of the matrix **D**, $|\mathbf{D}|$, is zero! This simply means that three of the normal modes of the system correspond to rigid body translations along $x, y$, and $z$, and they have zero frequency. These modes must be eliminated by reducing the dimension of **D** by three before eqn (3.213) can be evaluated.

LeSar *et al.* (1989) introduced an Einstein approximation to evaluate the determinant $|\tilde{\mathbf{D}}|$. In the Einstein approximation each atom vibrates independently of the other atoms and the matrix $\tilde{\mathbf{D}}$ reduces to block diagonal form where each block is a 3 by 3 matrix associated with a particular atom. The elements of the $i$th 3 by 3 block are $\tilde{D}_{i\alpha i\beta}$ and its determinant is $|\tilde{\mathbf{D}}_i| = \omega_{i1}^2 \omega_{i2}^2 \omega_{i3}^2$, where $\omega_{i1}$, $\omega_{i2}$, and $\omega_{i3}$ are local Einstein frequencies of the $i$th atom obtained by diagonalizing the matrix $\tilde{\mathbf{D}}_i$. In this approximation the determinant $|\tilde{\mathbf{D}}|$ simplifies to a product of the $N$ determinants $|\tilde{\mathbf{D}}_i|$, and the classical free energy becomes:

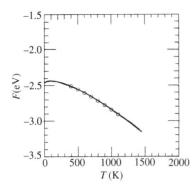

**Fig. 3.25** Calculated free energy vs. temperature at zero pressure for a perfect Cu crystal modelled by a truncated Morse pair potential. The solid curve is obtained with the Einstein model, eqn (3.214), and the circles from a Monte Carlo thermodynamic integration procedure, which is essentially exact. The estimated errors in the Monte Carlo calculations are smaller than the symbol size. (After LeSar *et al.* (1989))

$$F_{\text{clas}} = E_{\text{p}} + 3kT \sum_{j=1}^{N} \ln \left[ \frac{h|\tilde{\mathbf{D}}_j|^{\frac{1}{6}}}{2\pi kT} \right]. \tag{3.214}$$

The chief justification for introducing this approximation to the full quasiharmonic expression in the classical limit is the observation, in eqn (3.209), that the diagonal elements of the matrix **D** are equal to sums of off-diagonal elements. Therefore, there is some reason to expect that the contribution from the diagonal elements will be dominant. The equilibrium position of atom $k$ is determined by $\nabla_k F_{\text{clas}} = 0$, which is now straightforward to evaluate:

$$\nabla_k F_{\text{clas}} = \nabla_k E_{\text{p}} + \frac{kT}{2} \sum_j \nabla_k \ln |\mathbf{D}_j|. \tag{3.215}$$

Comparing this approximation with the exact result, eqn (3.210), in the classical limit, eqn (3.212), we see that in the Einstein approximation the intersite correlation functions, $\langle u_{j\beta} u_{i\alpha} \rangle$, are set to zero. No further approximations are made. If the off-diagonal elements of **D** are an order of magnitude smaller than the diagonal elements then the intersite correlation functions, $\langle u_{j\beta} u_{i\alpha} \rangle$, will be an order of magnitude smaller than the on-site correlation functions $\langle u_{i\beta} u_{i\alpha} \rangle$. Therefore, the contribution from intersite correlations in eqn (3.210) would be an order of magnitude smaller than those included in the Einstein approximation in eqn (3.215). It follows that provided the diagonal elements of **D** are dominant then the Einstein approximation, in the classical limit, should be a good approximation. LeSar *et al.* (1989) compared the estimation of the Helmholtz free energy of a perfect Cu crystal using the above Einstein model with a more exact Monte Carlo procedure for the classical limit. The Cu crystal was modelled by a pairwise Morse potential truncated between the second and third neighbours. The two sets of results are indistinguishable, as seen in Fig. 3.25. They also compared the vacancy formation free energy in the two methods, shown in Fig. 3.26. The errors range from 0 to 1.2 per cent as the temperature is increased from zero up to about 75 per cent of the melting point of the model for Cu. This agreement is perhaps surprisingly good in view of the approximations that are made in the Einstein model. Both the Einstein model and the Monte Carlo procedure yield incorrect free energies at temperatures below the Debye temperature of the model because they neglect the quantum freezing out of modes.

The fact that the Helmholtz free energy in eqn (3.203) involves the global density of phonon states $N(\omega)$ suggests an alternative strategy (Sutton 1989) which is similar in spirit to the development of Finnis–Sinclair potentials for tight binding Hamiltonians,

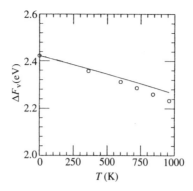

**Fig. 3.26** The free energy of formation, $\Delta F_{\mathrm{v}}$, of a vacancy in a 108-atom system, with periodic boundary conditions, as a function of temperature. $\Delta F_{\mathrm{v}} = (107/108)F_{\mathrm{p}} - F_{\mathrm{v}}$ where $F_{\mathrm{p}}$ is the free energy of a perfect 108-particle system and $F_{\mathrm{v}}$ is the free energy of the 107-particle system with a vacancy at the same volume. The solid curve is obtained with the Einstein model, and the circles by a Monte Carlo thermodynamic integration procedure. The estimated errors in the Monte Carlo procedure are smaller than the symbol size. (From Le Sar *et al.* (1989).)

described in Section 3.8.6. We can always write the total density of states $N(\omega)$ exactly as a sum of local densities of states:

$$N(\omega) = \sum_{j=1}^{N} n_j(\omega) = 2\omega \sum_{j=1}^{N} y_j(\omega^2), \qquad (3.216)$$

where the local density of states $y_j(\omega^2)$ is defined by (compare eqn (3.114) in the tight binding case):

$$y_j(\omega^2) = \sum_n |\tilde{u}_j^{(n)}|^2 \delta(\omega^2 - \omega_n^2), \qquad (3.217)$$

and the sum is over all $3N$ normal modes. In the local density of states for atom $j$ each normal mode contribution is weighted by a factor representing the squared amplitude of vibration of the atom in that mode. The Einstein model approximates $y_j(\omega^2)$ by a sum of three delta functions, one for each of the three normal modes obtained by diagonalizing the matrix $\tilde{D}_j$. The alternative strategy is to characterize the local density of states through its moments, which may be related directly to the local atomic environment through Cyrot–Lackmann's moments theorem of Section 3.8.4.

Let $M_j^{(p)}$ denote the $p$th moment of the local density of states $y_j(\omega^2)$:

$$M_j^{(p)} = \int_0^\infty y_j(\omega^2)\omega^{2p} \, \mathrm{d}\omega^2. \qquad (3.218)$$

It is easily verified that $M_j^{(p)}$ is equal to the $2p$th moment, $\mu_j^{(2p)}$, of the local density of states $n_j(\omega)$:

$$M_j^{(p)} = \mu_j^{(2p)} = \int_0^\infty n_j(\omega)\omega^{2p} \, \mathrm{d}\omega. \qquad (3.219)$$

Using the moments theorem, eqn (3.165), we can express this as a sum of closed paths starting and ending on atom $j$:

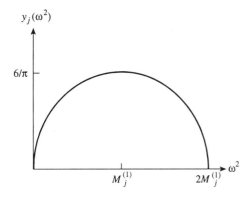

Fig. 3.27 The phonon local density of states $y_j(\omega^2)$, given by eqn (3.223). The centre of gravity is $M_j^{(1)}$, which is related to the local stiffness through eqn (3.221).

$$M_j^{(p)} = \sum_{\alpha=1}^{3} \sum_{k\beta} \sum_{m\theta} \sum_{rr} \sum_{sx} \tilde{D}_{j\alpha k\beta} \tilde{D}_{k\beta m\theta} \cdots \tilde{D}_{rrsx} \tilde{D}_{sxj\alpha}, \tag{3.220}$$

where each term in this sum consists of a product of $p$ elements of the $\tilde{D}$ matrix. Thus the second and fourth moments of $n_j(\omega)$ are given by

$$\mu_j^{(2)} = M_j^{(1)} = \sum_{\alpha=1}^{3} \tilde{D}_{j\alpha j\alpha} = \nabla_j^2 E_p / m_j \tag{3.221}$$

$$\mu_j^{(4)} = M_j^{(2)} = \sum_{\alpha=1}^{3} \sum_{k\beta} \tilde{D}_{j\alpha k\beta} \tilde{D}_{k\beta j\alpha}. \tag{3.222}$$

Sutton (1989) fitted the first moment $M_j^{(1)}$ to an assumed functional form for the local density of states $y_j(\omega^2)$. In a three-dimensional crystal the density of states must vary like the square root of $\omega^2$ at the band edges. The lower band edge is always at $\omega^2 = 0$, and the integral of $y_j(\omega^2)$ over the whole band must equal 3. The simplest choice of functional form for $y_j(\omega^2)$, satsifying these constraints, is the following:

$$y_j(\omega^2) = (6/\pi M_j^{(1)}) [ (M_j^{(1)})^2 - (\omega^2 - M_j^{(1)})^2 ]^{\frac{1}{2}}. \tag{3.223}$$

which is sketched in Fig. 3.27. This form is a semi-elliptic density of states, which is non-zero between $\omega^2 = 0$ and $2M_j^{(1)}$, with the centre of gravity at $M_j^{(1)}$. The corresponding local density of states $n_j(\omega)$ is given by

$$n_j(\omega) = \frac{12\omega^2}{\pi(\mu_j^{(2)})^2} (2\mu_j^{(2)} - \omega^2)^{\frac{1}{2}}, \tag{3.224}$$

which is proportional to $\omega^2$ at low frequencies, and hence it leads to the $T^3$ dependence of the low-temperature specific heat. The approximation of the exact local density of states $n_j(\omega)$ by the form given in eqn (3.224) is a second moment approximation, similar to the Finnis–Sinclair approach described in Section 3.8.6. The functional form of $n_j(\omega)$ is sketched in Fig. 3.28. The information content of the second moment, $\mu_j^{(2)}$, is less than the determinant, $|\tilde{D}_j|$, used in the Einstein model, eqn (3.214), because only the trace of $\tilde{D}_j$ is captured by $\mu_j^{(2)}$. The off-diagonal elements of $\tilde{D}_j$ describe the resistance to local shears, while the diagonal elements of $\tilde{D}_j$ describe the resistance to local volume changes. The normalized first moment of $n_j(\omega)$ is given by

$$\mu_j^{(1)} / \mu_j^{(0)} = \frac{32\sqrt{2}}{15\pi} [\mu_j^{(2)}]^{\frac{1}{2}} \simeq 0.96 [\mu_j^{(2)}]^{\frac{1}{2}}. \tag{3.225}$$

$n_j(\omega^2)$

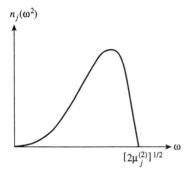

$[2\mu_j^{(2)}]^{1/2}$

**Fig. 3.28** The phonon local density of states $n_j(\omega)$, given by eqn (3.224).

At $T = 0\,\mathrm{K}$ the Helmholtz free energy differs from the potential energy due to zero point motion. Taking the limit that $T = 0$ in eqn (3.203) we obtain

$$F_{\text{zero}} = E_{\mathrm{p}} + \frac{h}{4\pi} \sum_j \mu_j^{(1)} \simeq E_{\mathrm{p}} + 1.44 \frac{h}{2\pi} \sum_j [\mu_j^{(2)}]^{\frac{1}{2}}. \tag{3.226}$$

Thus, at $T = 0\,\mathrm{K}$ the Helmholtz free energy has the form of a Finnis–Sinclair potential, with the square root term arising from zero point motion. The contribution of zero point motion to the free energy of the system increases as the atomic mass decreases and as the Laplacian of the local potential energy increases. Equation (3.226) can be used to define an average 'Einstein frequency' $\omega_j^{\mathrm{E}}$, since in an Einstein model the zero point energy of atom $j$ is $1.5 h\omega_j^{\mathrm{E}}/2\pi$. Thus $\omega_j^{\mathrm{E}} = 0.96[\mu_j^{(2)}]^{\frac{1}{2}}$.

At intermediate temperatures, where quantum freezing out of modes is still important, we must use the full form of the Helmholtz free energy:

$$F = E_{\mathrm{P}} + kT \sum_{i=1}^{N} \int_0^\infty n_i(\omega) \ln\left[2 \sinh\left(\frac{h\omega}{4\pi kT}\right)\right] \mathrm{d}\omega. \tag{3.227}$$

In our second moment approximation this becomes

$$F = E_{\mathrm{P}} + \sum_{i=1}^{N} F_i \tag{3.228}$$

where

$$F_i = \frac{48kT}{\pi} \int_0^1 x^2 (1 - x^2)^{\frac{1}{2}} \ln\left[2 \sinh\left(\frac{c_i x}{2}\right)\right] \mathrm{d}x, \tag{3.229}$$

and

$$c_i = \frac{h[2\mu_i^{(2)}]^{\frac{1}{2}}}{2\pi kT}. \tag{3.230}$$

The free energy in eqn (3.228) has the form of the potential energy $E_{\mathrm{P}}$ plus a sum of projections $F_i$ of the vibrational free energy of the whole system onto individual sites. The projection is effected by the local densities of states, which are projections of the global density of states onto individual atomic sites. Loosely speaking, we may think of the projection $F_i$ as the local vibrational free energy associated with atom $i$, although it is clear that each $F_i$ is determined by the local atomic environment through $\mu_i^{(2)}$. When $c_i$ is infinite the temperature is zero and $F$ reduces to $F_{\text{zero}}$ given in eqn (3.226). At the

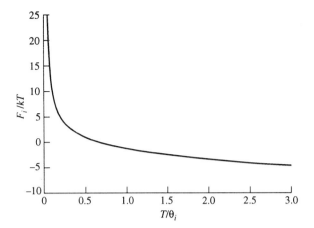

**Fig. 3.29** The local contribution, $F_i$, to the Helmholtz vibrational free energy, normalized with respect to $kT$, plotted as function of the normalized temperature $T/\theta_i = 1/c_i$, where $F_i$, $\theta_i$, and $c_i$ are given by eqns (3.229), (3.232), and (3.230) respectively.

other extreme limit where $c_i$ tends to zero we obtain the classical limit, which in our second moment model is given by

$$F_i^{\text{clas}} = 3kT[\ln(c_i/2) + \tfrac{1}{4}].\tag{3.231}$$

Quantum effects begin to be important when $c_i \geqslant 1$. At temperatures greater than

$$\theta_i = \frac{h[2\mu_i^{(2)}]^{\frac{1}{2}}}{2\pi k}\tag{3.232}$$

the thermal energy $kT$ is greater than the maximum local phonon energy, $h[2\mu_i^{(2)}]^{\frac{1}{2}}/2\pi$, of the local density of states and all the local modes are excited. Thus $\theta_i$ has the same meaning as a local Debye temperature. In Fig. 3.29 we plot $F_i/kT$ as a function of the normalized temperature $T/\theta_i$.

Other standard thermodynamic functions projected onto site $i$ may be derived from the theory of lattice dynamics. The entropy projected onto site $i$ is given by

$$S_i = \frac{48k}{\pi}\int_0^1 x^2(1-x^2)^{\frac{1}{2}}\left(\left(\frac{c_ix}{2}\right)\coth\left(\frac{c_ix}{2}\right) - \ln\left[2\sinh\left(\frac{c_ix}{2}\right)\right]\right)dx.\tag{3.233}$$

In Fig. 3.30 we plot $S_i/k$ as a function of the normalized temperature $T/\theta_i$. $S_i$ tends to zero as the temperature tends to zero as $T^3$, while at high temperatures, where $T \gg \theta_i$, the entropy becomes

$$S_i = 3k[\tfrac{3}{4} - \ln(c_i/2)].\tag{3.234}$$

The vibrational contribution to the internal energy can similarly be projected onto atomic sites. At site $i$ it is given by

$$\frac{U_i}{kT} = \frac{24}{\pi}c_i\int_0^1 x^3(1-x^2)^{\frac{1}{2}}\coth\left(\frac{c_ix}{2}\right)dx,\tag{3.235}$$

which is plotted in Fig. 3.31 as a function of the normalized temperature $T/\theta_i$. At high temperatures $U_i$ tends to $3kT$ as required by the law of Dulong and Petit, and at zero kelvin $U_i$ becomes the zero point energy at site $i$.

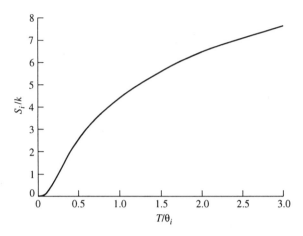

**Fig. 3.30** The local contribution, $S_i$, to the vibrational entropy, normalized with respect to $k$, plotted as function of the normalized temperature $T/\theta_i$, where $S_i$ is given by eqn (3.233).

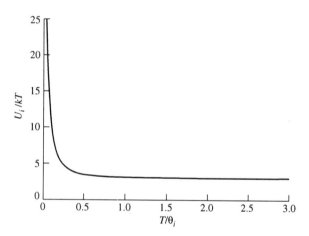

**Fig. 3.31** The local contribution, $U_i$, to the vibrational internal energy, normalized with respect to $kT$, plotted as a function of the normalized temperature $T/\theta_i$, where $U_i$ is given by eqn (3.235).

The projected specific heat at constant volume at site $i$ is given by

$$\frac{C_\mathrm{v}^i}{k} = \frac{12c_i^2}{\pi} \int_0^1 \frac{x^4(1-x^2)^{\frac{1}{2}}}{\sinh^2(c_i x/2)} \, dx, \tag{3.236}$$

which is plotted in Fig. 3.32 as a function of the normalized temperature $T/\theta_i$. At high temperatures $C_\mathrm{v}^i$ tends to $3k$, while as $T$ tends to zero $C_\mathrm{v}^i$ tends to

$$C_\mathrm{v}^i \to \frac{1152k}{\pi} c_i^{-3}, \tag{3.237}$$

which varies as $T^3$.

Two other quantities that characterize the vibrational freedom of an atom are the mean square displacement $\langle u_i^2 \rangle$ and the local Grüneisen parameter $\gamma_i$. The mean square displacement is given by

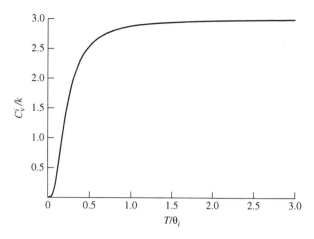

**Fig. 3.32** The local contribution, $C_v^i$, to the vibrational specific heat, normalized with respect to $k$, plotted as a function of the normalized temperature $T/\theta_i$, where $C_v^i$ is given by eqn (3.236).

$$\langle u_i^2 \rangle = \frac{6h^2}{\pi^3 m_i k T c_i} \int_0^1 x(1-x^2)^{\frac{1}{2}} \coth\left(\frac{c_i x}{2}\right) dx. \qquad (3.238)$$

At high temperatures, $\langle u_i^2 \rangle$ tends to $6kT/\nabla_i^2 E_P$, and at zero Kelvin $\langle u_i^2 \rangle$, due to zero point motion, is $4h/(\pi^2 m_i [2\mu_i^{(2)}]^{\frac{1}{2}})$. The Grüneisen constant is the central quantity in determining the thermal expansion coefficient and other anharmonic properties of a crystal. We can define a local Grüneisen parameter, $\gamma_i$, which is a measure of the anharmonicity of the local atomic environment around site $i$. In our second moment model $\gamma_i$ is independent of temperature because it is the same for all modes. It is given by

$$\gamma_i = -\frac{1}{2} \frac{\partial \left( \ln \nabla_i^2 E_p \right)}{\partial \left( \ln V \right)}, \qquad (3.239)$$

where $V$ is the volume of the whole system. The total pressure in the system is given by

$$P = -\left(\frac{\partial F}{\partial V}\right)_T = -\frac{\partial E_P}{\partial V} + \sum_j \frac{U_j \gamma_j}{V}. \qquad (3.240)$$

It is the balance between these two terms in the pressure that determines the thermal expansion of the solid. In the case of a bicrystal the thermal expansion parallel and perpendicular to the interface will in general be different, and we should treat the Grüneisen parameter and thermal expansion coefficient as second-rank tensors. At low temperatures, below $\theta_i$, only the lower-frequency modes are excited and, in reality, their Grüneisen parameters may be quite different from the average over the whole band. For example, in open crystal structures, such as those of the zincblende type, the lowest-frequency modes are often associated with relative shears, as sketched in Fig. 3.33 (Barron *et al.* 1980). Since the average separation of the atoms along the bond is smaller such modes lead to a thermal contraction. At low temperatures, therefore, many zincblende structures have negative thermal expansion coefficients. Such behaviour is beyond our second moment model and to describe it we would have to take into account higher moments of the local densities of states. In particular $\mu_i^{(4)}$ contains information, in matrix elements like $\tilde{D}_{i\alpha j\beta}$, of the local resistance to shear displacements. Therefore, we

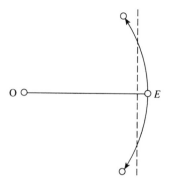

**Fig. 3.33** A low-frequency bond bending mode, in which an atom with an equilibrium position at $E$, vibrates in the manner indicated by the arrows. The dashed line shows that the time-averaged distance of the vibrating atom from the atom at the origin O, projected along the equilibrium bond vector $OE$, is smaller.

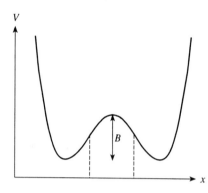

**Fig. 3.34** Schematic illustration of a double potential well as a function of some configurational coordinate, $x$, where the configurations between the dashed lines are inaccessible in quasiharmonic theory even though the barrier $B$ may be small compared to $kT$. The region between the dashed lines is defined by the condition that $d^2 V/dx^2 \leqslant 0$.

should think of the local Grüneisen parameter defined in eqn (3.239) as an average over the whole band, which is applicable only in the classical regime where $T > \theta_i$.

We shall now derive the temperature-dependent contribution to the 'force' on an atom using our second moment model. The 'force' acting on atom $k$ is given by the negative gradient of the total free energy, eqn (3.227), with respect to the position of atom $k$. We therefore obtain:

$$-\nabla_k F = -\nabla_k E_P - \sum_j \frac{U_j}{2} \nabla_k \left( \ln \nabla_j^2 E_p \right), \qquad (3.241)$$

where $U_j$ is the internal energy projected onto site $j$, eqn (3.235), and is plotted in Fig. 3.31. Comparing this with the force in the local Einstein model, eqn (3.215), we see that the two are very similar in that they both ignore intersite displacement–displacement correlations. Both models suffer from a feature that is due, ultimately, to the expansion of the potential energy to second order in the displacements. To illustrate the problem consider an atom in the one-dimensional potential double well shown in Fig. 3.34. If the height of the barrier between the adjacent wells is of order $kT$ we expect the atom to jump over the barrier from time to time. But from eqn (3.241) we see that the atom will be confined to whichever well it first finds itself in. As the atom approaches the inflection point of the double well the temperature dependent force becomes infinite and keeps the atom within a region of the potential where the second derivative is positive. Such double wells are thought to be important in glassy metals where they lead to a low-temperature specific heat that is proportional to $T$ rather than $T^3$. They may also be important in interfaces.

$y_i(\omega^2)$

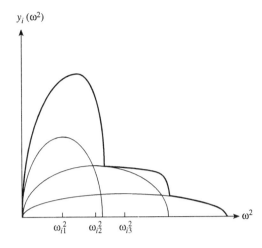

$\omega_{i1}^2 \quad \omega_{i2}^2 \quad \omega_{i3}^2$

$\omega^2$

**Fig. 3.35** Plot of the local phonon density of states $y_i(\omega^2)$ (bold curve) as a sum of three semi-elliptic local densities of states, one centred on each local Einstein frequency, $\omega_{i1}$, $\omega_{i2}$, and $\omega_{i3}$.

We have already indicated the possible importance of matrix elements such as $\tilde{D}_{j\alpha j\beta}$ in controlling the low-temperature excited modes. Such matrix elements contribute to $|\tilde{D}_j|$ but not to $\mathrm{Tr}\,\tilde{D}_j$. The Einstein model displays more information about the local environment than the second moment model because in the former the local density of states, $y_j(\omega^2)$, is represented by *three* delta functions. By contrast, in the second moment model the centre of gravity and the band width are determined by just *one* piece of local information, namely $\mathrm{Tr}\,\tilde{D}_j$. On the other hand the second moment model uses a continuous density of states, with the correct singularities at the band edges. A model that combines all the information contained in each matrix $\tilde{D}_j$ with a continuous density of states $y_j(\omega^2)$, displaying square root singularities at the band edges, would be preferable to both the Einstein model and the second moment model. We now propose such a model.

The idea is to replace the three delta functions representing the local density of states in the Einstein model by three semi-elliptic bands, one for each Einstein mode, as sketched in Fig. 3.35. First we diagonalize the three by three matrix $\tilde{D}_i$ and obtain the three eigenvalues $\omega_{i\nu}^2$ ($\nu = 1, 2, 3$) and the corresponding orthonormal eigenvectors $\tilde{x}_{i\alpha}^{(\nu)}$:

$$\sum_\beta \tilde{D}_{i\alpha i\beta}\tilde{x}_{i\beta}^{(\nu)} = \omega_{i\nu}^2 \tilde{x}_{i\alpha}^{(\nu)}. \tag{3.242}$$

It follows that

$$\omega_{i\nu}^2 = \sum_\alpha \sum_\beta \tilde{x}_{i\beta}^{(\nu)} \tilde{x}_{i\alpha}^{(\nu)*} \tilde{D}_{i\alpha i\beta}, \tag{3.243}$$

and the gradient of each eigenvalue with respect to the position of atom $j$($j$ may equal $i$) is

$$\nabla_j \omega_{i\nu}^2 = \sum_\alpha \sum_\beta \tilde{x}_{i\beta}^{(\nu)} \tilde{x}_{i\alpha}^{(\nu)*} \nabla_j \tilde{D}_{i\alpha i\beta}. \tag{3.244}$$

The local density of states, $y_i(\omega^2)$, for atom $i$ is now replaced by three semi-elliptic local densities of states, one for each of the three eigenvalues $\omega_{i\nu}^2$:

$$y_i(\omega^2) = \sum_{\nu=1}^{3} y_{i\nu}(\omega^2), \tag{3.245}$$

where

$$y_{i\nu}(\omega^2) = \begin{cases} \dfrac{2}{\pi\omega^2_{i\nu}}[\omega^4_{i\nu} - (\omega^2 - \omega^2_{i\nu})^2]^{\frac{1}{2}}, & 0 \leqslant \omega^2 \leqslant 2\omega^2_{i\nu} \\ 0 & \text{otherwise} \end{cases} \tag{3.246}$$

As before we can define projections of the vibrational free energy and internal energy onto site $i$ and mode $\nu$:

$$F_{i\nu} = \frac{16kT}{\pi} \int_0^1 x^2 (1 - x^2)^{\frac{1}{2}} \ln\left[2 \sinh\left(\frac{c_{i\nu}x}{2}\right)\right] dx \tag{3.247}$$

$$U_{i\nu} = \frac{8kT}{\pi} c_{i\nu} \int_0^1 x^3 (1 - x^2)^{\frac{1}{2}} \coth\left(\frac{c_{i\nu}x}{2}\right) dx \tag{3.248}$$

where

$$c_{i\nu} = \frac{h[2\omega^2_{i\nu}]^{\frac{1}{2}}}{2\pi kT}. \tag{3.249}$$

Thus the total free energy is given by

$$F = E_{\mathrm{p}} + \sum_{i=1}^{N} \sum_{\nu=1}^{3} F_{i\nu}. \tag{3.250}$$

The effective 'force' on atom $j$ is now readily evaluated:

$$\begin{aligned} -\nabla_j F &= -\nabla_j E_{\mathrm{P}} - \sum_{i=1}^{N} \sum_{\nu=1}^{3} \nabla_j F_{i\nu} \\ &= -\nabla_j E_{\mathrm{P}} - \sum_{i=1}^{N} \sum_{\nu=1}^{3} \frac{U_{i\nu}}{2\omega^2_{i\nu}} \nabla_j \omega^2_{i\nu} \\ &= -\nabla_j E_{\mathrm{P}} - \sum_{i=1}^{N} \sum_{\nu=1}^{3} \frac{U_{i\nu}}{2\omega^2_{i\nu}} \sum_{\alpha=1}^{3} \sum_{\beta=1}^{3} \tilde{x}^{(\nu)}_{i\beta} \tilde{x}^{(\nu)*}_{i\alpha} \nabla_j \tilde{D}_{i\alpha i\beta} \end{aligned} \tag{3.251}$$

where we have used eqn (3.244) in the last step. In the high-temperature limit, where $U_{i\nu}$ becomes $kT$, we recover eqn (3.215) of the Einstein model. But at lower temperatures, where quantum effects reduce $U_{i\nu}$, we obtain a more accurate description of the forces and thermodynamic functions than in either the Einstein model or the second moment model.

Finally we remark that the thermodynamic functions and the temperature-dependent contributions to the interatomic forces are integrals over the whole density of states times various functions. It is at low temperatures, where quantum effects freeze out the higher frequency modes, that the shape of the density of states can make the most significant difference. Otherwise, the detailed shape of the density of states tends to be averaged out by the integrations. We shall discuss the application of the Einstein model and the second moment model to grain boundaries in metals in Section 4.3.1.10.

## 3.10 IONIC BONDING

The first principles methods that were described in Sections 3.2–3.4 are just as applicable to ionic solids as they are to any other solid. But, as in the rest of this chapter, we are most concerned with simple models that allow us to understand bonding and cohesion

Electron
energy

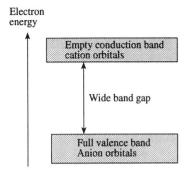

**Fig. 3.36** Schematic illustration of the energy bands of an ionic solid. The band gap is wide compared to the widths of the valence and conduction bands.

in a class of solids, both qualitatively and quantitatively for atomistic modelling of interfaces. The ionic model of polar solids is a particularly successful example. Cohesion is modelled through the transfer of electronic charge from cations to anions and the electrostatic attraction of the ensemble of ions that results. The charge transfer leads to closed shells of electrons on the anions and cations, and thus to Harrison's picture of an ionic solid (Harrison 1980) as comprising inert gas atoms interacting through a predominantly repulsive interaction, with modified *nuclear* charges, which drive the formation of close-packed structures of alternate positive and negative ions. As pointed out by Finnis (1992) this picture is applicable to a wider range of polar solids than might have been thought, and it circumvents some difficulties with the classical Born model that are described below.

The applicability of the ionic model to a solid may be judged by examining its band structure. In Section 3.8.3 we defined the polarity (eqn (3.95)) and the covalency (eqn (3.96)) of a diatomic bond in terms of the parameter $\varepsilon$ defined in eqn (3.92). In a similar way the ionic model is applicable to an insulator with a band gap that is comparable to, or greater than, the widths of the bands. In Fig. 3.36 we illustrate this schematically, with a narrow, unoccupied, conduction band, associated with the cation orbitals, separated by a relatively large gap from a narrow, full valence band associated with the anion orbitals. This picture applies to NaCl for example. The ionic model applies rigorously in the limit that the ratio of the band gap to the band widths becomes infinite, which corresponds to $\varepsilon$ becoming infinite for the diatomic molecule. The band width is an indication of the ease with which an electron can hop from site to site and thereby form a significant bond order. Thus, a large band width, compared with the band gap, indicates that covalent bonding dominates and that the ionic model is not appropriate. For example, even though GaAs and SiC are polar materials their large band widths compared with their band gaps indicate that the ionic model is not appropriate to describe the bonding in these materials (see also Majewski and Vogel (1989)). At charged defects in these materials the electrostatic forces on the atoms must obviously be taken into account, but they are in addition to the forces arising from the underlying covalent bonding.

In the Born model we imagine the creation of an ionic solid from separated neutral atoms by the following two steps. We first remove all the electrons in the outer shell of the cation species, and add those electrons to the anion species so that their outer shells become complete. This step costs a certain amount of energy, which we call the preparation energy, and is independent of the final positions of the ions in the crystal. Each ion has the same number of electrons as one of the inert gases, but not the same number of protons. In the second step the ions are condensed onto their final positions of the crystal. There are two contributions to the energy change accompanying this process. The most important is the Madelung energy, which is the net attractive, electrostatic interaction between each ion pair and the remainder of the crystal. It is calculated by assuming that the ions remain spherically symmetric throughout the condensation process, which allows them to be treated as point charges. Consider a binary solid with composition $C_c A_a$, where $C$ and $A$ are the cation and anion species and $c$ and $a$ are the numbers of each species per formula unit. Charge neutrality requires that $cZ_c = aZ_a$, where $Z_c$ and $Z_a$ are the valences of the cation and anion. The Madelung energy (in Rydbergs), per molecular unit, is given by

$$E_{\text{Mad}} = -\frac{\alpha(c + a)Z_c Z_a}{d} \tag{3.252}$$

where $\alpha$ is the Madelung constant (of order unity), which has been calculated for many common crystal structures by Johnson and Templeton (1961), and $d$ is the nearest neighbour spacing. At equilibrium this attractive interaction is balanced by a short-range interaction arising primarily from the overlap of the inert gas electron charge densities. This short-range interaction is often described empirically by a pair potential of the Buckingham form

$$\varphi(r) = A_{\text{BM}} e^{-r/\rho} - C_{\text{D}}/r^6. \tag{3.253}$$

The parameters $A_{\text{BM}}$, $\rho$, and $C_{\text{D}}$ are fitted empirically. Attempts have also been made to calculate the short-range interactions form first principles; see Harding (1990) and references therein. The first term of $\varphi(r)$ is a Born–Mayer repulsion representing the overlap of the inert gas charge densities and the second describes the relatively weak dispersion forces. The Born model thus boils down to a sum of pair potentials between all ion pairs given by the sum of $\varphi(r_{ij})$ and the Coulomb interaction $(\pm)Z_i Z_j/r_{ij}$, where $Z_i$ is the valence of ion $i$ and the negative sign is chosen if ions $i$ and $j$ are of opposite sign.

The Born model regards each ion as a rigid object that cannot be polarized. It has long been appreciated that the ions are indeed polarizable, and at a defect the ionic polarizations can make very significant contributions to the energy and structure. The shell model of Dick and Overhauser (1958) is a simple mechanical model that couples the ionic polarization to the short-range pair potential, $\varphi(r)$. It is illustrated in Fig. 3.37. Each ion consists of a massless shell of charge $Y|e|$ tied to an undeformable core of charge $X|e|$ by a harmonic spring of constant $K$, where $(X + Y)|e|$ is the ionic charge. The polarizability of the ion is $Y^2/2K$ in atomic units. The parameters $Y$ and $K$ are generally obtained by fitting to dielectric data, although both elastic and phonon properties are sometimes included. It is normally assumed that the short-range forces act only between the shells, while the long-range Coulomb forces act between all cores and shells excluding the core and shell on the same ion. It is tempting to think that the shell charge represents the valence electrons, but in some fitted shell models the shell charges turn out to be positive. No physical significance can be attributed to the positions of the shells, and it seems more sensible to think of the shell model as simply a (successful) way of coupling the electronic polarization on each ion to the structural distortion.

The Born model, combined with the shell model, has been used extensively to model point defects, dislocations, surfaces, and grain boundaries in ionic crystals with considerable success (see Catlow and Mackrodt (1982)). We shall review the applications to grain boundaries in ionic materials in Section 4.3.2. Here we shall discuss the model more

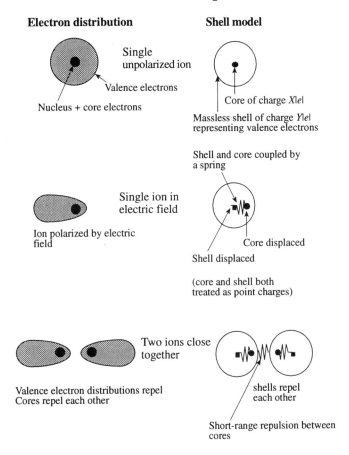

**Fig. 3.37** To illustrate the shell model of interactions between polarizable ions. The left-hand side shows the electron distribution and the right-hand side shows how the shell model represents this distribution.

critically. Since the Born model is a pair potential model it satisfies the Cauchy relation $c_{12} = c_{44}$. This relation is only approximately satisfied in the alkali halides, and in many other compounds it is not satisfied at all, e.g. in MgO $c_{12}/c_{44} = 0.59$. One solution to this problem is the breathing shell model of Schroder (1966) which allows spherically symmetric distortions of the shell, that are determined by the local environment. But the breathing shell model can only describe those Cauchy violations where $c_{12} < c_{44}$, and violations with $c_{12} > c_{44}$ are frequently found, e.g. AgCl. For this reason Sangster (1974) extended the model by allowing ellipsoidal deformations of the shell as well, from which it is possible to account for $c_{12} > c_{44}$. As emphasized by Harding (1990) the problem with these refinements to the shell model is the rapid increase in the number of fitted parameters, with no insight available as to reasonable bounds for those parameters. Perhaps more fundamentally the experimentally observed Cauchy violations indicate the importance of covalent contributions to the bonding, which are completely ignored by the Born model and the shell model. In that case the refinements to the shell model that we have mentioned may cure the symptoms but not the disease.

Even in a material like NaCl the success of the ionic model is paradoxical in view of the assumption that the ionic charges are spherically symmetric and non-overlapping.

This is an extremely bad approximation to the true, self-consistent charge density in NaCl because the ionic charges overlap very considerably. Therefore the assumption that the ions may be treated as point charges would seem to break down completely. Finnis (1992) has shown how Harrison's picture of an ionic solid as comprising inert gas atoms with differing *nuclear* charges resolves this paradox very neatly. Consider the specific case of corundum, $Al_2O_3$, as an illustrative example. The electronic configurations of the ionized states, $Al^{3+}$ and $O^{2-}$, correspond to the rare gas neon. In the first step, therefore, we imagine that neutral neon gas atoms are condensed onto the final positions of the $Al^{3+}$ and $O^{2-}$ ions of the corundum crystal. We can calculate the energy of the neon gas solid rather well by summing Lennard–Jones pair interactions:

$$V(r) = 4\varepsilon [(\sigma/r)^{12} - (\sigma/r)^6] \tag{3.254}$$

with parameters $\varepsilon$ and $\sigma$ appropriate for neon given by Gordon and Kim (1972). The equilibrium nearest neighbour separation in solid neon is considerably larger than in corundum. Therefore, the Lennard–Jones pair interactions are dominated by the repulsive interactions between nearest neighbours in the corundum crystal, and it is necessary to sum the pair interactions over nearest neighbours only. Thus, for each formula unit we have an energy of the rare gas solid given by:

$$E_{rgs} = 2 \times z_{Al} \times 4\varepsilon [(\sigma/d)^{12} - (\sigma/d)^6] \tag{3.255}$$

where the factor of 2 enters because there are 2 Al ions in each formula unit, $z_{Al}$ is the coordination of Al, and $d$ is the average Al–O bond length, which we are aiming to calculate.

In the second step we convert the neutral neon gas atoms into $Al^{3+}$ and $O^{2-}$ ions by removing two protons from those neon atoms that will become $O^{2-}$ ions and adding the protons to other neon atoms that will become $Al^{3+}$ ions. This step is a redistribution of the nuclear charge, while the electronic configurations remain unaltered. Strictly speaking, we create different isotopes of $Al^{3+}$ and $O^{2-}$ because the numbers of neutrons are not the same, but that is an irrelevant complication. The most important change in the energy of the system is the electrostatic attraction between the $3+$ and $2-$ charges that we have created, which is precisely the Madelung energy, $E_{Mad}$, with a Madelung constant for the corundum structure of 1.68. At this point we have recovered the repulsive and attractive terms of the standard ionic model, but without the misleading picture of non-overlapping electronic charge densities. In the revised picture the Madelung energy arises from nuclear charges, which may certainly be treated as non-overlapping point charges. The total energy of the system is thus:

$$E = E_{rgs} + E_{Mad}. \tag{3.256}$$

The electronic charge densities on the $Al^{3+}$ ions and $O^{2-}$ are contracted and expanded respectively compared with neon atoms. We assume that the repulsive interactions between ions with inert gas configurations are similar to those between inert gas atoms, and we therefore ignore changes in $E_{rgs}$. This seems a reasonable assumption since one type of ion contracts while the other expands. Equation (3.256) for the total energy is the final result, and Finnis (1992) has applied it to a number of ceramics. For each ceramic the equilibrium bond length is calculated by minimizing the total energy. For corundum Finnis obtains an average bond length of 1.84 Å, in good agreement with the experimentally observed average bond length of 1.91 Å. Figure 3.38 summarizes Finnis's results. The agreement between calculated and observed bond lengths in such diverse materials as $ZrO_2$ and KCl, with no fitting of parameters to any experimental data, is encouraging

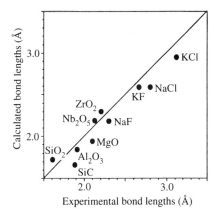

**Fig. 3.38** Bond lengths for ionic and partially ionic materials calculated by Finnis (1992) using eqn (3.256) for the total energy, compared with experimental values.

for the ionic model. The worst case tested is SiC, which is normally regarded as a covalent compound, but even here the ionic model does better than we might have expected. A more demanding test of the ionic model would be to compute the phonon dispersion relations. Here we would not expect the ionic model to fare so well since second-order changes in the energy are much more sensitive to the redistribution of electronic charge accompanying a distortion. In general we expect the ionic model to fare better in close packed structures where the charge density is more spherically symmetric, in keeping with the assumptions of the model.

As noted by Finnis (1992), it is because of the proximity of adjacent ions in NaCl that the difference between overlapping ionic or *atomic* charge densities and the self-consistent charge density is only about 10 per cent. This suggests that we should be able to use the Harris–Foulkes functional (see Section 3.2.2) to model the properties of NaCl very well if we took as our input charge density a superposition of free Na and Cl *atomic* charge densities. Polatoglou and Methfessel (1988) have performed this calculation and shown that the Harris–Foulkes functional does indeed obtain excellent results with a superposition of free atomic charge densities as input. For example, they obtain the lattice constant and bulk modulus to within 1 per cent of the experimental values and the cohesive energy to within 10 per cent of the experimental value. The larger error in the cohesive energy is more a consequence of the local density approximation, than the assumed input charge density. However, we would not expect to obtain good agreement with experiment if we used the Harris–Foulkes functional to compute the splitting between the transverse and longitudinal optic phonon frequencies since this depends more critically on the charge density. Indeed, the error in the energy computed by the Harris–Foulkes scheme is second order in the error in the assumed input charge density, and the phonon frequencies are determined by second-order changes in the energy.

## 3.11 INTERATOMIC FORCES AT HETEROPHASE INTERFACES

At some heterophase interfaces the interatomic forces are no more difficult to describe than they are in an alloy. For example, at a metal–metal interface the interatomic forces may be modelled in the same way as for an alloy of the two metals. Provided we have interatomic potentials for such an alloy we may proceed to model the metal–metal interface. One possible complication is that an electrostatic dipole will be formed in general at the interface owing to the difference in the work functions of the two metals. To

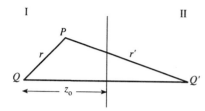

I　　　　　　　　　　　II

**Fig. 3.39** To illustrate the image charge construction for a real charge $Q$ a distance $z_0$ from an interface between two dielectric media, I and II. The potential at a point $P$ in region I is given by eqn (3.257).

describe the formation of this dipole, and its effect on the interatomic forces, requires a self-consistent method.

At other heterophase interfaces there may be a qualitative change in the bonding across the interface. The best example is an interface between a metal and an ionic crystal. While we may have appropriate models for describing interatomic forces in the metal and in the ionic crystal separately, it is not obvious how we should describe the forces at the interface. One solution to the problem is to use a first-principles density functional approach, where it is not necessary to think in terms of an interatomic potential. A few such calculations have been performed, as described in Section 4.4.3, and they have provided some important insights into the nature of the bonding at the interface. The disadvantages of this approach are that the calculations demand massive computational resources and that the number of atoms that may be treated is very limited, although this number will increase with improved computing power. We need simpler models to understand trends in cohesion at interfaces, and in this section we focus on some of the recent work that has been done to develop such models. It is fair to say that this work is still in a very developmental stage, but in view of the importance of metal/non-metal interfaces we feel the work should be discussed. Let us state at the outset that we are addressing interfaces in which no reaction layer occurs. At many important metal–ceramic interfaces a thin layer of a new phase develops that is formed by inter-mixing and possibly some chemical reaction between the metal and the ceramic (e.g. see de Bruin (1988)). The new phase may be crystalline or amorphous. We are not aware of any simple models to describe the interatomic forces at such interfaces, and the thermodynamic stability of the interface layer is often analysed simply in terms of the phase diagram of the metal–ceramic system. In this section we confine our attention to much simpler heterophase interfaces, in particluar those between a noble metal and an ionic crystal, where the composition profile at the interface is abrupt on an atomic scale, and where the interface is flat.

There are several types of interaction that we may think of at an interface between a noble metal and an ionic crystal. There are short-range repulsive forces, due to the overlap of atomic charge densities, there are attractive dispersion (van der Waals) forces, and the spill-over of the conduction electrons of the metal into the ionic crystal may affect the interatomic forces at the interface. Stoneham and Tasker (1985) have argued that the most important attractive interaction between a non-reactive metal and an ionic crystal is the polarization of the electronic charge density at the surface of the metal, which is equivalent to the formation of a lattice of image charges in the metal.

Following Landau and Lifshitz (1984, p. 37), consider a plane interface between two semi-infinite dielectric continua with dielectric constants $\varepsilon^{\mathrm{I}}$ and $\varepsilon^{\mathrm{II}}$, as shown in Fig. 3.39. A point charge $Q$ in region I, at a distance $z_0$ from the interface, experiences a repulsive or attractive force due to the change in polarizability at the interface. If region

II is free space then the interface is a free surface and $\varepsilon^{II} = 1$; if it is a metal $\varepsilon^{II}$ is infinite. The electrostatic potential, $\varphi^I$, in region I is obtained by imagining region I extended to fill all space and summing the potentials of the real charge $Q$ at $z_o$ and a fictitious image charge $Q'$ at the mirror image position $-z_o$:

$$\varphi^I = \frac{Q}{\varepsilon^I r} + \frac{Q'}{\varepsilon^I r'}, \tag{3.257}$$

where $r$ and $r'$ are the distances from $Q$ and $Q'$ (see Fig. 3.39). The potential, $\varphi^{II}$, in region II is given by that of another fictitious charge $Q''$ at the site of the real charge $Q$:

$$\varphi^{II} = \frac{Q''}{\varepsilon^{II} r}. \tag{3.258}$$

The boundary conditions of (i) continuity of the potential $\varphi$, and (ii) continuity of the normal component of the electric displacement vector at the interface lead to the following relations for $Q'$ and $Q''$:

$$\left. \begin{array}{c} Q' = Q\dfrac{\varepsilon^I - \varepsilon^{II}}{\varepsilon^I + \varepsilon^{II}}, \\[3mm] Q'' = \dfrac{2\varepsilon^{II} Q}{\varepsilon^I + \varepsilon^{II}} \end{array} \right\}. \tag{3.259}$$

As $\varepsilon^{II} \to \infty$ we have $Q' = -Q$ and $\varphi^{II} = 0$, as it should be for a conductor. The energy of interaction between the charge $Q$ at $z_o$ and its image charge $Q'$ at $-z_o$ is given by

$$U(z) = \frac{Q^2}{4z_o \varepsilon^I} \frac{\varepsilon^I - \varepsilon^{II}}{\varepsilon^I + \varepsilon^{II}}. \tag{3.260}$$

Thus the interaction is attractive if $\varepsilon^{II} > \varepsilon^I$ and repulsive if, $\varepsilon^{II} < \varepsilon^I$. Physically the reason for this is that if $\varepsilon^{II} > \varepsilon^I$ then the higher polarizability of region II reduces the electrostatic energy of the charge compared with region I. In the metallic limit, $\varepsilon^{II} \to \infty$, the interaction energy becomes

$$U(z) = -\frac{Q^2}{4z_o \varepsilon^I}. \tag{3.261}$$

Note that this is *half* the electrostatic energy of a charge $Q$ interacting with a charge $-Q$, separated by $2z_o$, in region I. The other half of this electrostatic energy is cancelled by the electrostatic self-interaction energy of the induced charge density on the metal surface. At the interface between the two dielectrics there is an induced charge density. $\sigma(\rho)$, given by

$$\sigma(\rho) = \frac{Q}{2\pi} \frac{(\varepsilon^I - \varepsilon^{II})}{\varepsilon^I(\varepsilon^{II} + \varepsilon^I)} \frac{z_o}{(z_o^2 + \rho^2)^{\frac{3}{2}}}, \tag{3.262}$$

where $\rho$ is the distance of a point in the interface from the $z$ axis.

If a semi-infinite lattice of point charges is brought into contact with a semi-infinite metal then a lattice of image charges will be generated in the metal, as shown in Fig. 3.40. It is clear that the interaction between the real charges of the ionic crystal and the image charges is very significant because in some cases, as in Fig. 3.40(a), the image lattice is

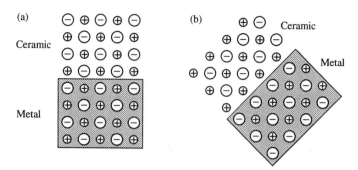

**Fig. 3.40** Illustrating the lattices of image charges created by a real ionic crystal near a metal surface: (a) with a non-polar surface next to the metal; (b) with a polar surface next to the metal. (From Finnis (1992).)

simply a continuation of the real lattice, if the image plane is appropriately positioned. However, the image lattice is in general not simply a continuation of the real lattice, as illustrated in Fig. 3.40(b). Using this attractively simple idea of image interactions Stoneham and Tasker (1985) rationalized a number of experimental observations of wetting and non-wetting behaviour of oxides by liquid metals. Stoneham (1982/83) noticed that the oxides that were wetted by non-reactive metals, notably Au, had dielectric constants greater than a critical value. Such oxides disorder easily through the formation of ions in a variety of ionized states, e.g. NiO may contain $Ni^{2+}$ and $Ni^{3+}$ ions with compensating Ni vacancies. The key point is that the ions with the higher charges will give a stronger image interaction with the metal because of the $Q^2$ dependence of the image interaction energy in eqn (3.260). This suggests that the most stable interfaces between noble metals and ionic crystals will be those in which there are defects with large charges in the ionic crystal as close as possible to the interface. This has been demonstrated by Stoneham and Tasker (1988) who modelled an (001) interface between NiO and BaO. BaO has a high dielectric constant, and the most stable interface had two layers of $Ni^{3+}$ ions and compensating $Ni^{2+}$ vacancies adjacent to the first layer of BaO.

Although the simplicity of the image charge idea is attractive there is a number of difficulties associated with it in a simulation of an interface between a noble metal and ionic crystal. From a practical point of view the precise position of the image plane for the metal is crucial, but not easily defined at an interface. From a fundamental point of view the idea is rooted in classical electrostatics, and the analysis in eqns (3.260)–(3.261) must break down right at the image plane where the image potential diverges (i.e. $z_0 \to 0$). Consider an electron approaching a metal surface very slowly. Far from the metal, classical electrostatics describes the interaction very well. But as the electron approaches the surface its energy does not diverge. Instead, its energy approaches the mean inner potential of the metal, which is approximately given by the Fermi energy plus the work function. The separation at which the classical picture breaks down corresponds to the point at which there is significant overlap between the wave function of the approaching electron and the wave functions of the electrons in the metal. Appelbaum and Hamann (1972) showed that this separation is about 2 Å in a simple metal, and more recently Jennings and Jones (1988) showed that the critical separation is about 5 Å for a transition metal. In addition, we would expect exactly the same classical interaction for an electron approaching a metal surface as we would a proton. But the electron is indistinguishable from the other electrons in the metal, and thus there is an exchange

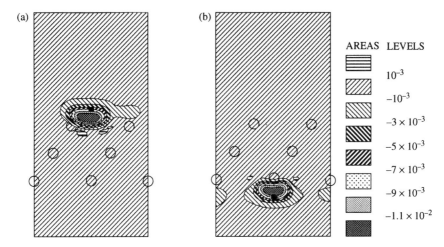

| AREAS | LEVELS |
|-------|--------|
| | $10^{-3}$ |
| | $-10^{-3}$ |
| | $-3 \times 10^{-3}$ |
| | $-5 \times 10^{-3}$ |
| | $-7 \times 10^{-3}$ |
| | $-9 \times 10^{-3}$ |
| | $-1.1 \times 10^{-2}$ |

**Fig. 3.41** The self-consistent induced charge density in a slab of three layers of aluminium for an external charge of one electron (solid circle). The atomic sites are indicated by open circles. (a) External charge above hollow site; (b) external charge above atom site. (From Finnis (1991).)

interaction with an electron but not for a proton. Thus, the quantum mechanical interactions are different.

The physical reason for the failure of the classical picture at small separations was stated very clearly by Appelbaum and Hamann (1972). In the classical picture the induced polarization surface charge is located in a mathematical plane with no thickness. In a real metal the induced charge distribution has a finite thickness in the direction of the surface normal. This is because an infinitesimally thin induced charge distribution would correspond to an infinite electronic kinetic energy (which depends on the second derivative of the wave function). Another way of saying this is that the Fermi wavelength is the smallest possible wavelength of charge fluctuations that can be sustained in the metal. Since the Fermi wavelength is of the order of an angstrom this is the expected thickness of the induced charge density. This is indeed the thickness of the induced charge distribution that is found in self-consistent local density functional calculations (Finnis 1991, Inglesfield 1991). In Fig. 3.41(a) and (b) we show the charge induced by one external electron above a slab of three layers of Al, computed fully self-consistently in the local density approximation by Finnis (1991). Note that the induced charge density is slightly different when the external electron is above a hollow site (a) and an atom (b). It is striking that the induced charge density is confined to the tails of the metallic charge density, in the vacuum above the first layer of atoms. The screening is extremely efficient since the charge density below the first layer of Al atoms is practically unaffected. The fact that the induced charge density has a finite thickness means that there is a dipole potential associated with it, and higher multipole potentials, which remove the divergence of the classical image potential, as pointed out by Appelbaum and Hamann (1972).

J. R. Willis, quoted in Duffy *et al.* (1992), has calculated the minimum energy charge distribution induced by a point charge outside a metal when the wave vectors of the induced charge fluctuations in the metal are less than or equal to the Fermi wave vector. The resulting potential in the vacuum outside the metal, caused by the induced charge distribution, has the form

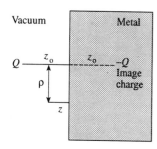

Fig. 3.42 To illustrate the quantities used in eqn (3.263).

$$\varphi(\rho, z) = \frac{-Q}{(\rho^2 + Z^2)^{\frac{1}{2}}} + \frac{Q}{2\pi} \int_0^{2\pi} d\theta \, \frac{\exp[-k_F(Z + i\rho \cos \theta)]}{Z + i\rho \cos \theta}, \qquad (3.263)$$

where $Z = z_o + z$ (see Fig. 3.42) and $k_F$ is the Fermi wave vector. The integral in this expression is the correction to the classical result. It decreases in magnitude exponentially with distance from the image plane $z_o$, and oscillates with varying $\rho$. When $k_F Z \gg 1$ the potential is well described by the classical result. The removal of the classical divergence as $Z \to 0$ is easily seen by considering $\rho = 0$ when the potential becomes

$$\varphi(0, Z) = -\frac{Q}{Z}[1 - \exp(-k_F Z)]. \qquad (3.264)$$

Using eqn (3.263) Duffy et al. (1992) were able to compute the electrostatic interaction between the charge of an ion in an ionic crystal and the potential produced by the image charge distribution in the metal of all the ions. Repulsive interactions between the metal (Ag) and the ions of the ionic crystal (MgO) were determined from Hartree–Fock calculations. Thus, they were able to use the machinery of a standard package for relaxing surfaces of ionic crystals to model the (001) Ag–MgO interface. The weakness of this approach is that it is necessary to define the image plane, $z_o$, for the metal surface. Lang and Kohn (1973) showed how it may be defined rigorously for jellium models of the metal. Smith et al. (1989) have determined the position of the image plane for the (100), (110), and (111) planes of six f.c.c. metals, including the noble metals, from experimental data. However, higher-index planes introduce steps into the surface of the metal and the position of the image plane is not so easily defined.

Finnis (1992) has provided additional insight into the image interaction, and suggested a different approach to model an interface between a metal and an ionic crystal that overcomes the image plane problem. To begin with Finnis considered the important question of whether the Harris–Foulkes functional (see Section 3.2.2) could describe the image interaction between metals and ionic crystals. The answer is no because it turns out that the image interaction corresponds to a second-order change in the energy of the electron system of the metal caused by the external potential of the ionic lattice. If $\Delta n(r)$ is the induced charge density in the metal caused by the potential $\Delta V_{\text{ext}}(r)$ of the ionic lattice, then the change in the energy of the bicrystal as a result of 'switching on' the ionic charges, is given by

$$\Delta E = \Delta E_1 + \Delta E_2 + \Delta E_{\text{ion}}, \qquad (3.265)$$

where $\Delta E_1$ is the first-order change

$$\Delta E_1 = \int dr \, n_0(r) \Delta V_{ext}(r), \tag{3.266}$$

and $\Delta E_2$ is the second-order change

$$\Delta E_2 = \tfrac{1}{2} \int dr \, \Delta n(r) \Delta V_{ext}(r). \tag{3.267}$$

$n_0(r)$ is the unperturbed charge density of the metal, and $\Delta E_{ion}$ is the interaction energy between the ion cores of the metal and the ions of the ionic lattice. Thus, $\Delta E_{ion}$ and $\Delta E_1$ may be combined to give the electrostatic energy of interaction between the ionic lattice and the unperturbed neutral metal, which is a short-range interaction. We are left with longer range with $\Delta E_2$, which must, therefore, represent the image interaction. Indeed, $\Delta E_2$ has the form of an image interaction since it is *half* the electrostatic energy of interaction between the induced charge density $\Delta n(r)$ and the inducing potential $\Delta V_{ext}(r)$. Finnis has shown that the other half of the electrostatic interaction energy is cancelled not only by the electrostatic self-interaction of the induced charge density in the metal, as it is in the classical model, but also by the changes in the kinetic energy of the electrons and the exchange-correlation energy. This very surprising result was also derived by Lang (1983). It shows that not only is the image interaction the dominant attractive interaction, but it can be justified fully from a quantum mechanical point of view. The conclusion is that the only significant quantum correction to the simple classical picture is that the induced charge density has a finite thickness. It is worth stressing, however, that the analysis is based on second-order perturbation theory, or linear response of the electron gas in the metal. There are metals for which the distribution of the induced charge is not well described by linear response theory, as discussed by Inglesfield (1991).

To model an interface between a noble metal and an ionic crystal, Finnis (1992) first considers each ion to be a neutral inert gas atom, as discussed in Section 3.10. The interaction between a metal and an inert gas solid is rather well understood, using effective medium theory (see Section 3.6). At short range it is repulsive, and the energy is linear in the local electron density, as seen in Fig. 3.7. If the metal charge density is represented as a sum of atomic charge densities the interaction of the inert gas atoms with the metal atoms can be adequately represented by a pair potential, such as a suitably parametrized Lennard–Jones potential. The inert gas solid is converted back into an ionic crystal by switching on the protons at the cation sites and switching them off at the anion sites. The interactions within the ionic crystal may be described by the ionic model, combined with the shell model, as described in Section 3.10. The interaction, $\Delta E_1 + \Delta E_{ion}$, between the unperturbed metal and the ions of the ionic crystal is a short-range repulsive interaction, which could be added to the Lennard–Jones potentials. The dominant attractive term is expected to be $\Delta E_2$, representing the image interaction. To model the image interaction without the uncertainty in the definition of the image plane Finnis (1991) introduced the 'discrete classical model' (DCM). In this model each atomic site of the metal is required to be at the same electrostatic potential. This condition is met by allowing charge to transfer between sites and by associating a dipole of varying strength with each site. The dipoles are essential to mimic the finite thickness of the induced charge density in the metal. The magnitude of the induced charge and the dipole moment associated with each site are determined self-consistently, with the constraint that the total induced charge is the negative of the total charge in the ionic crystal. See Finnis (1992) for details. Once the induced charges and dipoles at each site of the metal are known their electrostatic interaction with the ionic charges of the ionic crystal can be found by standard Ewald summation techniques.

So far we have concentrated on the image interaction at an interface between a metal and an ionic crystal. If the ionic crystal is electrically neutral overall there is no net induced charge in the metal. But in general there is a net transfer of charge at the interface in order to equalize the Fermi energies in the metal and the ionic crystal. This charge transfer produces a net dipole at the interface which rigidly shifts electrostatic potentials of all atoms on one side of the interface relative to those on the other. For many purposes, such as calculating the image interaction, this rigid shift is unimportant. But it becomes important when an ion moves through the dipole layer at the interface, for example during segregation to the interface or intermixing across the interface. To calculate the dipole requires a self-consistent approach.

Finally, we mention the work of Gubanov and Dunaevskii (1977) who used a variational approach to model interfaces between simple metals and alkali halide crystals. Although this work has been superceded by modern density functional calculations it is remarkable that it considers, explicitly and quantum mechanically, the induced charge density in the metal, its self-interaction and the interaction between it and the ionic lattice. Thus, all the ingredients of the metal–ionic crystal interaction that we have focussed on in this section were considered in this work. There is also another interesting paper by the same authors on the interaction between two ionic crystals (Gubanov and Dunaevskii 1976). We should also mention the work of Mauritz *et al.* (1973), who modelled the interaction between crystalline polyethylene and alkali halides, where the attractive interactions are the induced dipoles in the polyethylene and dispersion forces.

## REFERENCES

Ackland G. J., Finnis, M. W., and Vitek V. (1988). *J. Phys. F: Met. Phys.*, **18**, L153.
Allan, G. and Lannoo M., (1976). *J. Phys. Chem. Solids*, **37**, 699.
Anderson, P. W. (1969). *Phys. Rev.*, **181**, 25.
Anderson, P. W. (1984). *Physics Reports*, **110**, 311.
Andersen O. K. and Jepsen, O. (1984). *Phys. Rev. Letts.*, **53**, 2571.
Appelbaum, J. A. and Hamann, D. R. (1972). *Phys. Rev. B*, **6**, 1122.
Ashcroft, N. W. (1966). *Phys. Lett.*, **23**, 48.
Ashcroft, N. W. and Langreth, D. C. (1967). *Phys. Rev.*, **155**, 682.
Averill, F. W. and Painter, G. S. (1990). *Phys. Rev. B*, **41**, 10344.
Bachelet, G. B., Greenside, H. S., Baraff, G. A., and Schlüter, M. (1981). *Phys. Rev. B*, **24**, 4745.
Bachelet, G. B., Hamann, D. R., and Schlüter, M. (1982). *Phys. Rev. B*, **26**, 4199.
Baddorf, A. P., Lyo, I.-W., Plummer, E. W., and Davis, H. L. (1987). *J. Vac. Sci. Tech. A*, **5**, 782.
Barron, T. H. K., Collins, J. G., and White, G. K. (1980). *Adv. Phys.*, **29**, 609.
Bristowe, P. D. (1993). Private communication.
Bullet, D. W. (1980). *Solid State Physics*, **35**, 129.
Burdett, J. K. and Lee, S. (1985). *J. Am. Chem. Soc.*, **107**, 3063.
Carlsson, A. E. (1990). *Solid State Physics*, **43**, 1.
Catlow, C. R. A. and Mackrodt, W. C. (eds) (1982). *Computer simulation of solids*, Springer lecture notes in physics Vol. 166. Springer-Verlag, Berlin.
Chadi, D. J. (1984). *Phys. Rev. B*, **29**, 785.
Chakraborty, B., Holloway, S., and Norskov, J. K. (1985). *Surf. Sci.* **152/153**, 660.
Coulson, C. A. (1939). *Proc. R. Soc. A*, **169**, 413.
Cressoni, J. C. and Pettifor, D. G. (1991). *J. Phys.: Condens. Malt.*, **3**, 495.
Cyrot-Lackmann, F. (1968). *J. Phys. Chem. Solids*, **29**, 1235.
Daw, M. S. and Baskes, M. I. (1983). *Phys. Rev. Letts.*, **50**, 1285.

Daw, M. S. and Baskes, M. I. (1984). *Phys. Rev. B*, **29**, 6443.

de Bruin, H. J. (1988). In *Surface and near-surface chemistry of oxide materials* (eds. J. Nowotny and L.-C. Dufour), p. 507. Elsevier, Amsterdam.

Dick, B. G. and Overhauser, A. W. (1958). *Phys. Rev.*, **112**, 90.

Ducastelle, F. and Cyrot-Lackmann, F. (1971). *J. Phys. Chem. Solids*, **32**, 285.

Duffy, D. M., Harding, J. H and Stoneham, A. M. (1992). *Acta Metall.*, **40**, Suppl. S11.

Ercolessi, F., Parrinello, M., and Tossatti, E. (1988). *Phil. Mag. A*, **58**, 213.

Finnis, M. W. (1974). *J. Phys. F.: Met. Phys.*, **4**, 1645.

Finnis, M. W. (1990). *J. Phys.: Condens. Matt.*, **2**, 331.

Finnis, M. W. (1991). *Surf. Sci.*, **241**, 61.

Finnis, M. W. (1992). *Acta Metall.*, **40**, Suppl. S25.

Finnis, M. W. and Heine, V. (1974). *J. Phys. F: Met. Phys.*, **4**, L37.

Finnis, M. W. and Sinclair, J. E. (1984). *Phil. Mag. A*, **50**, 45.

Foiles, S. M., Baskes, M. I., and Daw, M. S. (1986). *Phys. Rev. B*, **33**, 7983.

Foulkes, W. M. C. (1987). PhD thesis, University of Cambridge.

Foulkes, W. M. C. and Haydock, R. (1989). *Phys. Rev. B*, **39**, 12520.

Fuchs, K. (1935). *Proc. Roy. Soc. A*, **151**, 585.

Gaspard, J. P. and Lambin, P. (1985). In *The recurston method and its applications* (eds. D. G. Pettifor and D. L. Weaire), Springer Series in Solid State Sciences Vol. 58, p. 72. Springer-Verlag, Berlin.

Girifalco, L. A. (1976). *Acta Metall.*, **24**, 759.

Gordon, R. G. and Kim, Y. S. (1972). *J. Chem. Phys.*, **56** 3122.

Gubanov, A. I. and Dunaevskii, S. M. (1976). *Sov. Phys. Sol. Stat.*, **18**, 1309.

Gubanov, A. I. and Dunaevskii, S. M. (1977). *Sov. Phys. Sol. Stat.*, **19**, 795.

Hafner, J. (1989). In *The structure of binary compounds*, Cohesion and structure Vol. 2 (eds. F. R. de Boer and D. G. Pettifor), p. 147. North-Holland, Amsterdam.

Hafner, J. and Heine, V. (1986). *J. Phys. F: Met. Phys.*, **16**, 1429.

Hafner, J. and Heine, V. (1983). *J. Phys. F: Met. Phys.*, **13**, 2479.

Harding, J. H. (1990). *Rep. Prog. Phys.*, **53**, 1403.

Harris, J. (1985). *Phys. Rev. B*, **31**, 1770.

Harrison, W. A. (1980). *Electronic structure*. Freeman, San Francisco, California.

Haydock, R. (1980). *Solid State Physics*, **35**, 216.

Heine, V. (1980). *Solid State Physics*, **35**, 1.

Heine, V. (1991). In discussion to L. J. Sham, *Phil. Trans. R. Soc. Lond. A*, **334**, 481.

Heine, V. and Samson, J. H. (1980). *J. Phys. F: Met. Phys.*, **10**, 2609.

Heine, V. and Samson, J. H. (1983). *J. Phys. F: Met. Phys.*, **13**, 2155.

Heine, V. and Weaire, D. L. (1970). *Solid State Physics*, **24**, 250.

Heine, V., Robertson, I. J., and Payne, M. C. (1991). *Phil. Trans. R. Soc. Lond. A*, **334**, 393.

Hoffmann, R. (1988). *Solids and surfaces*. VCH Publishers, New York.

Hohenberg, P. and Kohn, W. (1964). *Phys. Rev.*, **136**, B864.

Hybertsen, M. S. and Louie, S. G. (1986). *Phys. Rev. B*, **34**, 5390.

Inglesfield, J. E. (1991). *Phil. Trans. R. Soc. Lond. A*, **334**, 527.

Jacobsen, K. W. (1990). In *Many-atom interactions in solids* (eds. R. M. Nieminen, M. J. Puska, and M. J. Manninen), Springer Proc. Phys. Vol. 48, p. 34. Springer-Verlag, Berlin.

Jacobsen, K. W., Norskov, J. K., and Puska, M. J. (1987). *Phys. Rev. B*, **35**, 7423.

Jansen, R. W. and Sankey, O. F. (1987). *Phys. Rev. B*, **36**, 6520.

Jennings, R. W. and Jones, R. O. (1988). *Adv. Phys.*, **37**, 341.

Johnson, R. A. (1990). In *Many-atom interactions in solids* (eds. R. M. Nieminen, M. J. Puska, and M. J. Manninen), Springer Proc. Phys. Vol. 48, p. 85. Springer-Verlag, Berlin.

Johnson, Q. C. and Templeton, D. H. (1961). *J. Phys. Chem.*, **34**, 2004.

Kohn, W. (1986). *Phys. Rev. B*, **33**, 4331.

Kohn, W. and Sham, L. J. (1965). *Phys. Rev. A*, **140**, 1133.

Landau, L. D. and Lifshitz, E. M. (1984). *Electrodynamics of continuous media*. Pergamon, Oxford, 2nd edn.

Lang, N. D. (1983). In *Theory of the inhomogeneous electron gas* (eds. S. Lundqvist and N. H. March), p. 309. Plenum, New York.

Lang, N. D. and Kohn, W. (1973). *Phys. Rev. B*, **7**, 3541.

LeSar, R., Najafabadi, R., and Srolovitz, D. J. (1989). *Phys. Rev. Letts.*, **63**, 624.

Lindhard, J. (1954). *K. Dan. Vidensk. Selsk. Mat. Fys. Medd.*, **28**, 28.

Louie, S. G. (1989). In *Atomistic simulation of materials* (eds. V. Vitek and D. J. Srolovitz), p. 125. Plenum, New York.

Louie, S. G., Froyen, S. and Cohen, M. (1982). *Phys. Rev. B*, **26**, 1738.

MacKinnon, A. (1985). In *The recursion method and its applications* (eds. D. C. Pettifor and D. L. Weaire), p. 84. Springer-Verlag, New York.

Mackintosh, A. R. and Andersen, O. K. (1980). In *Electrons at the Fermi surface* (ed. M. Springford), Chap. 5.3. Cambridge University Press, Cambridge.

Majewski, J. A. and Vogl, P. (1989). In *The structure of binary compounds*, Cohesion and structure Vol. 2 (eds. F. R. de Boer and D. G. Pettifor), North-Holland, Amsterdam.

Majid, I. and Bristowe, P. D. (1992). *Phil. Mag. A.*, **66**, 73.

Majid, I., Bristowe, P. D., and Balluffi, R. W. (1989). *Phys. Rev. B*, **40**, 2779.

Masuda, K. and Sato, A. (1981). *Phil. Mag. A*, **44**, 799.

Mauritz, K. A., Baer, E., and Hopfinger, A. J. (1973). *J. Polym. Sci: Polym. Phys.*, **11**, 2185.

Mearns, D. (1988). *Phys. Rev. B*, **38**, 5906.

Moriarty, J. A. (1972). *Phys. Rev. B*, **5**, 2066.

Moriarty, J. A. (1988). *Phys. Rev. B*, **38**, 3199.

Moriarty, J. A. (1990). *Phys. Rev. B*, **42**, 1609.

Moriarty, J. R. and Phillips, R. (1991). *Phys. Rev. Letts.*, **66**, 3036.

Mott, N. F. and Jones, H. (1936). *The theory of the properties of metals and alloys*. Clarendon Press, Oxford. Republished by Dover, New York (1958).

Norskov, J. K. and Lang, N. D. (1980). *Phys. Rev. B*, **21**, 2131.

Nozieres, P. and Pines, D. (1958). *Phys. Rev.*, **111**, 442.

Ohta, Y., Finnis, M. W., Pettifor, D. G., and Sutton, A. P. (1987). *J. Phys. F: Met. Phys.*, **17**, L273.

Pauling, L. (1950). *The nature of the chemical bond*. Oxford University Press, London, 2nd edn.

Paxton, A. T. and Sutton, A. P. (1989). *Acta Metall.*, **37**, 1693.

Perdew, J. P. and Levy, M. (1983). *Phys. Rev. Letts.*, **51**, 1884.

Pettifor, D. G. (1971). *J. Phys. F: Met. Phys.*, **1**, L62.

Pettifor, D. G. (1976). *Commun. Phys.*, **1**, 141.

Pettifor, D. G. (1978). *J. Chem. Phys.*, **69**, 2930.

Pettifor, D. G. (1983). In *Physical Metallurgy* (eds. R. W. Cahn and P. Haasen), Chap. 3. North-Holland, Amsterdam.

Pettifor, D. G. (1987). *Solid State Physics*, **40**, 43.

Pettifor, D. G. (1989). *Phys. Rev. Letts.*, **63**, 2480.

Pettifor, D. G. (1990). In *Many-atom interactions in solids* (eds. R. M. Nieminen, M. J. Puska, and M. J. Manninen), Springer Proc. Phys. Vol. 48, Springer-Verlag, Berlin.

Pettifor, D. G. and Aoki, M. (1991). *Phil. Trans. R. Soc. Lond. A*, **334**, 439.

Pettifor, D. G. and Podloucky, R. (1986). *J. Phys. C: Sol. Stat. Phys.*, **19**, 285.

Pettifor, D. G. and Ward, M. S. (1984). *Sol. Stat. Commun.*, **49**, 291.

Phillips, J. C. and Kleinman, L. (1959). *Phys. Rev.*, **116**, 287.

Polatoglou, H. M. and Methfessel, M. (1988). *Phys. Rev. B*, **37**, 10403.

Puska, M. J. (1990), In *Many-atom interactions in solids* (eds. R. M. Nieminen, M. J. Puska, and M. J. Manninen), Springer Proc. Phys. Vol. 48, p. 134. Springer-Verlag, Berlin.

Read, A. J. and Needs, R. J. (1989). *J. Phys.: Condens. Matt.*, **1**, 7565.

Robertson, I. J. and Farid, B. (1991). *Phys. Rev. Letts.*, **66**, 3265.

Rosenfeld, A. and Stott, M. J. (1987). *J. Phys. F: Met. Phys.*, **17**, 605.

Sangster, M. J. (1974). *J. Phys. Chem. Solids*, **35**, 195.

Scheffler, M., Vigneron, J.-P., and Bachelet, G. B. (1985). *Phys. Rev. B*, **31**, 6541.

Schroder, U. (1966). *Sol. Stat. Commun.*, **4**, 347.

Sham, L. J. and Schlüter, M. (1983). *Phys. Rev. Letts.*, **51**, 1888.

Skinner, A. J. and Pettifor, D. G. (1991). *J. Phys.: Condens. Matt.*, **3**, 2029.

Smith, N. V., Chen, C. T., and Weinert, M. (1989). *Phys. Rev. B*, **40**, 7565.

Stillinger, F. H. and Weber, T. A. (1985). *Phys. Rev. B*, **31**, 5262.

Stoneham, A. M. (1982/83). *Applic. Surf. Sci.*, **14**, 249.

Stoneham, A. M. and Tasker, P. W. (1985). *J. Phys. C: Sol. Stat. Phys.*, **18**, L543.

Stoneham, A. M. and Tasker, P. W. (1988). In *Surface and near-surface chemistry of oxide materials* (eds. J. Nowotny and L.-C. Dufour), p. 1. Elsevier, Amsterdam.

Stott, M. J. and Zaremba, E. (1980). *Phys. Rev. B*, **22**, 1564.

Sutton, A. P. (1989). *Phil. Mag.*, *A*, **60**, 147.

Sutton, A. P. and Chen, J. (1990). *Phil. Mag. Letts.*, **61**, 139.

Sutton, A. P., Finnis, M. W., Pettifor, D. G and Ohta, Y. (1988). *J. Phys. C: Sol. Stat. Phys.*, **21**, 35.

Taylor, R. (1978). *J. Phys. F: Met. Phys.*, **8**, 1699.

Turchi, P. and Ducastelle, F. (1985). In *The recursion method and its applications* (eds. D. G. Pettifor and D. L. Weaire), Springer Series in Solid State Sciences Vol. 58, p. 104. Springer-Verlag, Berlin.

Vitek, V. and Srolovitz, D. J. (eds) (1989). *Atomistic simulation of materials*. Plenum, New York.

von Barth, U. and Gelatt, C. D. (1980). *Phys. Rev. B*, **21**, 2222.

Wilson, J. H., Todd, J. D. and Sutton, A. P. (1990). *J. Phys.: Condens. Matt.*, **2**, 10259.

Ziman, J. M. (1972). *Principles of the theory of solids*. Cambridge University Press, Cambridge, 2nd edn.

# Models and experimental observations of atomic structure

## 4.1 INTRODUCTION: CLASSIFICATION OF INTERFACES

The structure of an interface may be defined in various ways depending on the resolution limit of the instrument being used to examine it. At large length scales we observe that the interface is flat or curved, and that it separates two crystals with a particular orientation relationship. At smaller length scales, but still larger than the atomic scale (often called the 'mesoscopic', scale), we might observe arrays of dislocations and/or steps in the interface. At the atomic scale we see the various atomic configurations and chemical makeup of the interface. Properties of interfaces may be influenced by structural features at all these length scales. In Chapter 1 we described the characterization of an interface at large length scales in terms of its macroscopic degrees of freedom. In Chapter 2 we discussed the dislocation model of interfaces, which has been used widely to characterize structure at the mesoscopic scale. In this chapter we shall discuss structure at the atomic scale.

Much insight into the atomic structure of interfaces, and grain boundaries in particular, has been gained from computer simulation. That is why a reasonably complete description of models of interatomic forces at interfaces was given in the previous chapter. High-resolution electron microscopy (HREM) (Gronsky 1980) and X-ray diffraction (Sass and Bristowe 1980) have been the major experimental tools for examining the structure of interfaces. The HREM results have not only been able to check many of the predictions of the simulations for tilt grain boundaries, but they have also revealed new relaxation mechanisms that had not been envisaged by the simulations. Some of these are discussed in section 4.3.1.9. The X-ray diffraction results (see Balluffi et al. 1989, Majid et al. 1989) have been useful for determining the structures of twist grain boundaries, and checking the results of computer simulations, as discussed in Section 3.7.

One shortcoming of both computer simulation and experimental techniques is that although they provide very detailed information about the particular interface under investigation, it is often difficult to deduce systematic *trends* in structure and properties from one interface to another. In this chapter we shall expose the underlying principles governing the relationship between structure and energy of large-angle grain boundaries in metals by making extensive use of simple analytic models. These models enable much more of the 5D space of the macroscopic geometrical degrees of freedom to be explored than could ever be achieved with computer simulation or experiment. Moreover, these principles are so fundamental they are at least a starting point for analysis of other kinds of interfaces. However, it is necessary to examine critically the simplifying assumptions of an analytic model by comparison with experimental observations and computer simulations in specific cases.

Compared with grain boundaries our knowledge of the structures of heterophase interfaces is quite scant. Undoubtedly this is because of the difficulty in modelling interatomic forces at such interfaces. Most of the information we have has come from HREM, and this is discussed in Section 4.4. There we also present some of the computer simulations of heterophase interfaces that have been carried out, and what we have learnt from them.

In Table 4.1 we present a classification of interfaces between crystals that we adopt throughout the remainder of this book.

At a homophase interface the same crystal phase exists on either side of the interface. This includes grain boundaries, and interfaces which do not involve a misorientation between the crystal lattices, namely stacking faults, antiphase boundaries, and inversion domain boundaries. At a heterophase interface there are different crystal phases on either side of the interface.

It is convenient to divide all interfaces into those which are structurally abrupt or sharp on an atomic scale, and those which are wide or diffuse on an atomic scale. In making this distinction we disregard elastic displacements which may exist in the near vicinity of the interface, due, for example, to an array of discretely spaced dislocations in the interface. Therefore, a small-angle grain boundary, consisting of an array of crystal lattice dislocations with localized cores, is taken to be a sharp interface. Solid-state chemical reactions sometimes take place at heterophase interfaces where two phases are in intimate contact. Phases $\alpha$ and $\beta$ meeting at an interface may react to form a third phase, $\gamma$. The original $\alpha$–$\beta$ interface is then replaced by two new interfaces $\alpha$–$\gamma$ and $\gamma$–$\beta$, each of which may be sharp or diffuse. These 'chemical reactions' include many bulk phase transformations, such as eutectoid recombination ($\alpha + \beta \rightarrow \gamma$), heterogeneous precipitation at grain boundaries, and the formation of ordered compounds. Discussion of some of these reactions and others is given in Section 4.4 and in Chapters 9 and 10. In many polyphase ceramic materials glassy layers are frequently found at the interfaces. We shall exclude interfaces containing glassy layers from the present discussion and refer the reader to Clarke (1987) for further discussion.

As seen in Table 4.1 there are examples of sharp and diffuse interfaces among both homophase and heterophase interfaces. Sharp interfaces have been subdivided in Table 4.1 as singular, vicinal, or general. The meanings of these terms, and the reasons for their introduction, will become clear in section 4.3.1.5. A glossary giving brief definitions of the different types of interfaces that are described in this book (and the different types of interfacial line defects) is given on p. xxix. In the next section we begin with diffuse interfaces, which can be treated by analytic methods, before going on to the more prevalent sharp interfaces in Sections 4.3 and 4.4.

## 4.2 DIFFUSE INTERFACES

The cores of certain types of crystalline interfaces may be relatively wide and straddle an appreciable number of interplanar spacings, i.e. they may be 'diffuse'. Such interfaces (Table 4.1) are most likely to appear in systems exhibiting incipient chemical or mechanical instabilities of various kinds. These include systems with phase diagrams possessing miscibility gaps which are near their critical temperatures and systems with long-range order which are near their critical temperatures (chemical instabilities). Also included are interfaces in systems which can undergo displacive type phase transitions due, for example, to phonon mode softening (mechanical instabilities). Since the interface cores are relatively wide, their structures and properties can be treated by means of continuum as well as atomistic methods. The most widely used continuum methods have been variants of the so-called Landau–Ginsburg model (Landau and Lifshitz 1935, Cao and Barsch 1990), originally formulated to treat the structure of interfaces between domains with different directions of magnetization in ferromagnetic materials (i.e. 'Bloch walls'). In the following we describe models for representative examples of these different types of interfaces.

**Table 4.1**  Classification of interfaces

|  | Sharp | Diffuse |
|---|---|---|
| Homophase | Stacking faults<br>Inversion domain boundaries<br>Antiphase domain boundaries<br>Grain boundaries:<br>  Small-angle<br>  Large-angle { singular<br>  vicinal<br>  general | Anti-phase domain boundaries in<br>  systems with long-range order<br>  near the critical temperature |
| Heterophase | Singular<br>Vicinal<br>General | Interfaces in systems with a<br>  miscibility gap<br>Interfaces involved in displacive<br>  transformations near a<br>  mechanical instability |

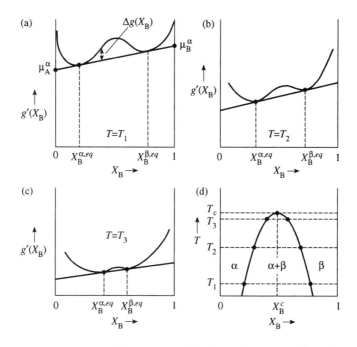

**Fig. 4.1**  Free energy versus composition curves, (a)–(c), and corresponding phases diagram, (d), for a binary solid solution system possessing a miscibility gap. $g'(X_B)$ = free energy per atom. The common tangent construction in (a)–(c) yields the compositions of the $\alpha$ and $\beta$ phases in equilibrium (i.e., $X_B^{\alpha,eq}$ and $X_B^{\beta,eq}$) and also the chemical potentials of A and B in the two phases. Note that $\mu_A^\alpha = \mu_A^\beta$ and $\mu_B^\alpha = \mu_B^\beta$.

### 4.2.1 Heterophase interfaces in systems with a miscibility gap

The free energy versus composition curves and corresponding phase diagram for a typical binary system composed of A and B atoms with a miscibility gap (Christian 1981) are shown in Fig. 4.1. Here, $X_B$ = atom fraction of B, and $g'(X_B)$ = free energy per atom. Below $T_c$ in the miscibility gap region the free energy curves have the characteristic 'double minima' shapes shown. As is well known (Christian 1981), the compositions of the coexisting $\alpha$ and $\beta$ phases at equilibrium, $X_B^{\alpha,\,eq}$ and $X_B^{\beta,\,eq}$, are given by the common tangent construction illustrated in Fig. 4.1, and the chemical potentials of components A and B throughout the $\alpha$ and $\beta$ phases are given by the intercepts of the common tangent lines. Consider now a flat interface between the coexisting $\alpha$ and $\beta$ phases at a temperature below $T_c$. We assume initially that the atomic volumes of A and B are constant and equal throughout. It is anticipated that the interface will be diffuse, and that the composition profile across it can be represented by a smooth function of the form shown schematically in Fig. 4.2(a). Both the form of this profile and the energy of the interface may be obtained by a continuum approach which involves the excess 'gradient energy' which is present because of the non-uniformity of the material in the diffuse boundary core. The standard model for such a problem is the general Landau–Ginsburg model (Landau and Lifshitz 1935, Barsch and Krumhansl 1984, Cao and Barsch 1990) in which the local free energy of the material is expressed as a function of a suitable independent variable in the form of an expansion in powers of this variable and its spatial gradients. For present purposes we follow Cahn and Hilliard (1958) and adopt a modified version of this model in which the independent variable is the concentration, $X_B$, and the local free energy per atom, $g$, is assumed to be of the form $g = g(X_B, \nabla X_B, \nabla^2 X_B, \ldots)$. Expanding $g$ around its value for a uniform concentration, and retaining only the important leading terms, we then have

$$g(X_B, \nabla X_B, \nabla^2 X_B, \ldots) = g'(X_B) + \sum_i L_i \partial X_B/\partial x_i + \sum_{ij} K_{ij}^{(1)} \partial^2 X_B/\partial x_i \partial x_j$$

$$+ \frac{1}{2} \sum_{ij} K_{ij}^{(2)} (\partial X_B/\partial x_i)(\partial X_B/\partial x_j) + \ldots \qquad (4.1)$$

where

$$L_i = \frac{\partial g}{\partial (\partial X_B/\partial x_i)}$$

$$K_{ij}^{(1)} = \frac{\partial g}{\partial (\partial^2 X_B/\partial x_i \partial x_j)} \qquad (4.2)$$

$$K_{ij}^{(2)} = \frac{\partial^2 g}{\partial (\partial X_B/\partial x_i)\, \partial (\partial X_B/\partial x_j)}.$$

Here $x_i$ is a spatial variable, $x_1$, $x_2$, or $x_3$ and the sums over $i$ and $j$ are from 1 to 3. The quantity $g'(X_B)$ is the free energy per atom of the uniform solution of composition $X_B$, and the remaining terms represent the 'gradient energy'. Now, $g$ must be invariant to the symmetry operations of the crystal; using standard methods (e.g. Nye 1957), eqns (4.2) for a cubic crystal (and also an isotropic material) then reduce to

$$g = g'(X_B) + K^{(1)} \nabla^2 X_B + K^{(2)} |\nabla X_B|^2 + \ldots, \qquad (4.3)$$

where

$$K^{(1)} = \frac{\partial g}{\partial (\nabla^2 X_B)}, \tag{4.4}$$

$$K^{(2)} = \frac{\partial^2 g}{\partial (|\nabla X_B|)^2}. \tag{4.5}$$

The total free energy of a non-uniform region is then

$$G = (1/\Omega) \int_V \{ g'(X_B) + K^{(1)} \nabla^2 X_B + K^{(2)} |\nabla X_B|^2 \} \, dV. \tag{4.6}$$

The second term in the integrand may now be expressed in another form. In general $\nabla \cdot K^{(1)} \nabla X_B = \nabla X_B \cdot \nabla K^{(1)} + K^{(1)} \nabla \cdot \nabla X_B$. Therefore,

$$\int (K^{(1)} \nabla^2 X_B) \, dV = \int (\nabla \cdot K^{(1)} \nabla X_B) \, dV - \int (\nabla X_B \cdot \nabla K^{(1)}) \, dV. \tag{4.7}$$

Invoking the divergence theorem,

$$\int_V (\nabla \cdot K^{(1)} \nabla X_B) \, dV = \int_S K^{(1)} \nabla X_B \cdot n \, dS, \tag{4.8}$$

where the right-hand side is an integral over the surface, $S$, which bounds the volume. However, $S$ can always be chosen so that the surface integral vanishes. Therefore,

$$\int (K^{(1)} \nabla^2 X_B) \, dV = - \int (\nabla X_B \cdot \nabla K^{(1)}) \, dV. \tag{4.9}$$

We may also write

$$\nabla X_B \cdot \nabla K^{(1)} = \frac{dK^{(1)}}{dX_B} |\nabla X_B|^2. \tag{4.10}$$

Therefore, from eqns (4.6–4.10) we obtain

$$G = (1/\Omega) \int \{ g'(X_B) + K |\nabla X_B|^2 \} \, dV, \tag{4.11}$$

where

$$K = K^{(2)} - \frac{dK^{(1)}}{dX_B} = \frac{\partial^2 g}{\partial (|\nabla X_B|)^2} - \frac{\partial^2 g}{\partial X_B \partial (\nabla^2 X_B)}. \tag{4.12}$$

Equation (4.11) expresses the result that the energy of a small volume of non-uniform solution can be written as the sum of two contributions; the energy that the volume would have if it were of uniform composition, plus an extra 'gradient energy' due to the non-uniformity.

Having this result, we may now derive the energy and form of the interface in Fig. 4.2(a). Using eqn (4.11) with $g'(X_B)$ given by the standard expression for a uniform solution, $g'(X_B) = [X_B \, \mu_B(X_B) + (1 - X_B) \, \mu_A(X_B)]$, we have for the system containing the interface

$$G = (1/\Omega) \int_{-\infty}^{\infty} \{ [X_B \mu_B(X_B) + (1 - X_B) \mu_A(X_B)] + K \, (dX_B/dx)^2 \} \, dx. \tag{4.13}$$

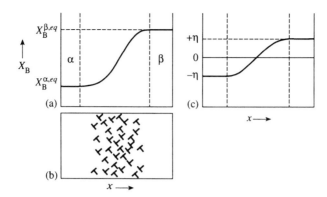

**Fig. 4.2** Diffuse heterophase interface between $\alpha$ and $\beta$ phases in binary solid solution system possessing a miscibility gap as in Fig. 4.1. The temperature is slightly below the critical temperature $T_c$ (Fig. 4.1(d)). (b) Accommodation of a lattice parameter mismatch in (a) by the introduction of anticoherency lattice edge dislocations in the diffuse interface region. (c) Diffuse antiphase domain boundary in binary system possessing long range order. $\eta$ = long-range order parameter.

The interfacial energy, $\sigma$, is, by definition (Section 5.2), the difference between the energy of the non-uniform system containing the interface and the energy that it would have if both phases were uniform right up to the interface. In order to evaluate the latter energy we may imagine that each slab of thickness, $dx$, and composition, $X_B$, in the interface is first decomposed into a two-phase mixture of the equilibrium $\alpha$ and $\beta$ phases. If the total amounts of $\alpha$ and $\beta$ phases obtained in this way are then consolidated, we can produce the desired system in which the two phases are uniform right up to a sharp interface at the dividing surface. The free energy per atom of the two-phase mixture obtained by each decomposition is given by $[X_B \mu_B^\alpha + (1 - X_B)\mu_A^\alpha]$ and corresponds to a point along the common tangent line in Fig. 4.1(a). Using this result, the interfacial energy is then

$$\sigma = (1/\Omega) \int_{-\infty}^{\infty} \{ [X_B \mu_B(X_B) + (1 - X_B) \mu_A(X_B)] - [X_B \mu_B^\alpha + (1 - X_B)\mu_A^\alpha] \quad (4.14)$$

$$+ K(dX_B/dx)^2] \} \, dx$$

$$= (1/\Omega) \int_{-\infty}^{\infty} \{ \Delta g(X_B) + K(dX_B/dx)^2 \} \, dx \quad (4.15)$$

where

$$\Delta g(X_B) = X_B[\mu_B(X_B) - \mu_B^\alpha] + (1 - X_B)[\mu_A(X_B) - \mu_A^\alpha]. \quad (4.16)$$

and has the simple representation shown in Fig. 4.1(a).

Our task now is to find the profile which minimizes the integral in eqn (4.15), and this can be accomplished by a variational method. The integrand is of the form $I = I(X_B, dX_B/dx)$, and a standard treatment (Margenau and Murphy 1943) shows that the variation of $\sigma$, i.e. $\delta\sigma$, due to a variation in $X_B$ throughout the volume, i.e. $\delta X_B(x)$, is given by

$$\delta\sigma = (1/\Omega) \int_{-\infty}^{\infty} \left[ \frac{\partial I}{\partial X_B} - \frac{d}{dx} \frac{\partial I}{\partial (dX_B/dx)} \right] \delta X_B(x) \, dx. \quad (4.17)$$

The condition for a minimum is then

$$\frac{\partial I}{\partial X_B} - \frac{d}{dx}\frac{\partial I}{\partial(dX_B/dx)} = 0. \tag{4.18}$$

However (Margenau and Murphy 1943), eqn (4.18) can be put in the alternative form

$$\frac{\partial I}{\partial x} - \frac{d}{dx}\left[I - (dX_B/dx)\frac{\partial I}{\partial(dX_B/dx)}\right] = 0. \tag{4.19}$$

Since $I \neq I(x)$, this implies

$$I - (dX_B/dx)\frac{\partial I}{\partial(dX_B/dx)} = \text{constant}. \tag{4.20}$$

Substituting for the integrand from eqn (4.15), we then have

$$\Delta g(X_B) - K[dX_B/dx]^2 = \text{constant}. \tag{4.21}$$

The constant in eqn (4.21) must be equal to zero, since both $\Delta g(X_B)$ and $dX_B/dx$ tend to zero as $x \to \pm\infty$. Therefore,

$$dX_B/dx = [\Delta g(X_B)/K]^{\frac{1}{2}}, \tag{4.22}$$

and eqn (4.15) takes the form

$$\sigma = (2/\Omega)\int_{-\infty}^{\infty} \Delta g(X_B)\, dx. \tag{4.23}$$

It is seen from Fig. 4.1 that $\Delta g(X_B)$ tends to zero as $T$ approaches $T_c$. Therefore, according to eqn (4.22), $dX_B/dx$ must approach zero, and we have the important result that the interface width tends to increase with increasing temperature and will approach infinity as the critical temperature is approached. At the same time, according to eqn (4.23), the interface energy must decrease and approach zero at $T_c$. Cahn and Hilliard (1958) studied the behaviour of the interface near $T_c$ in more detail by means of a Taylor expansion of $\Delta g(X_B, T)$ near $X_B^c$ and $T_c$ and found that the interface width varies with temperature approximately as $(T_c - T)^{-\frac{1}{2}}$ and that the interface energy, $\sigma$, varies as $(T_c - T)^{\frac{3}{2}}$.

The compositional profile across the interface can, in principle, be evaluated by integrating eqn (4.22) when $\Delta g(X_B)$ and $K$ are known. The function $\Delta g(X_B)$ consists of a hump between the points of common tangency at $X_B^{\alpha,\mathrm{eq}}$ and $X_B^{\beta,\mathrm{eq}}$ (Fig. 4.1), and when this shape is introduced into eqn (4.22) it can be seen immediately that that the profile must be sigmoidal as anticipated in Fig. 4.2(a).

Further insight into the characteristics of such interfaces has been obtained by using atomistic models. Cahn and Hilliard (1958) employed a regular solution model for a system composed of atoms interacting via pairwise potentials; they found a gradient energy coefficient of the form

$$K = \sum_n z_n r_n^2 E_n/6 \tag{4.24}$$

where $z_n$ = number of atoms in the $n$th coordination shell of radius $r_n$ and

$$E_n = E_n^{12} - \tfrac{1}{2}[E_n^{11} + E_n^{22}], \tag{4.25}$$

where the $E_n^{ij}$ are the interaction potentials between the like $(i = j)$ and unlike $(i \neq j)$ atoms at the distance $r_n$. (We note that $E_n$ is the well-known energy parameter which can be expected (Christian 1981) to appear in such pairwise interaction models for binary systems). As shown by Cahn and Hilliard (1958), specific expressions for the energy and profile of the interface can then be obtained through the use of these results. Atomistic models for such interfaces using the cluster variation method have also been studied (Kikuchi 1972).

The continuum model described above neglected any effects due to possible variations in the atomic volumes of the A and B components. Such variations will generally be present, and they can be accommodated in the diffuse interface region either in a coherent fashion by elastic strains or by the introduction of anticoherency misfit lattice disloca-tions. If they are accommodated elastically, an additional elastic strain must be added to the integrand in eqn (4.11) as described by Cahn (1961). If they are accommodated by dislocations, the interface will require a distribution of anticoherency, lattice edge dislocations as illustrated schematically in Fig. 4.2(b). We note that if all long-range stress is to be eliminated, the sum of the Burgers vectors of these distributed dislocations must equal the sum of the Burgers vectors of the more localized dislocations which would be present if the interface were sharp and free of long-range stress as illustrated in Fig. 2.6. The distribution of misfit dislocations will be a function of the variation of lattice parameter with composition and the form of the composition profile, and will contribute an additional energy density which must be added to the integrand in eqn (4.11) as discussed further in Section 10.5 in the development leading to eqn (10.92). In both the elastic accommodation and dislocation accommodation cases the solution of the variational problem for the interface profile will therefore differ somewhat from our previous simpler solution. However, the general sigmoidal shape will be retained along with the divergent behaviour at the critical temperature.

### 4.2.2 Antiphase domain boundaries in systems with long-range order

As is well known (Krivoglaz and Smirnov 1965, Barrett and Massalski 1980), many binary systems in which A type atoms find it favourable to surround themselves with B type atoms develop long-range order. A classic example is the so-called B2 type structure illustrated in Fig. 4.3. This structure may be regarded as two interpenetrating simple cubic sublattices translated with respect to each other by the vector $\frac{1}{2}[111]$ as shown. A atoms can then preferentially surround themselves with B atoms if they preferentially occupy one of the sublattices and hence generate long-range order. In the limit when all of the A atoms are on one sublattice we have the maximum possible degree of long-range order. The degree to which this is achieved is generally specified by the long-range order parameter, $\eta$, defined (Kittel 1967) so that

$$(1 + \eta)/2 = \text{probability that an A atom is on sublattice 1,}$$

$$(1 - \eta)/2 = \text{probability that an A atom is on sublattice 2.}$$

Therefore, when there is no long-range order, the probability that an A atom is on sublattice 1 is $\frac{1}{2}$ and $\eta = 0$. When there is complete long-range order, the probability that an A atom is on sublattice 1 is either 1, or 0, and $\eta$ is then either 1, or $-1$. At low temperatures where entropy effects are generally small, the system will generally be completely ordered with $\eta = \pm 1$. As the temperature is raised, the system will tend to disorder cooperatively by undergoing a second-order phase transition culminating sharply

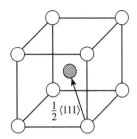

**Fig. 4.3** B2 type structure consisting of two interpenetrating simple cubic lattices translated with respect to each other by $\frac{1}{2}\langle 111\rangle$.

at the critical temperature, $T_c$. An approximate nearest-neighbour pairwise interaction model (Kittel 1967) shows that $\eta$ decreases with increasing temperature during such a disordering transition according to

$$\frac{1-\eta}{1+\eta} = \exp\left[\frac{8E_1\eta}{kT}\right], \tag{4.26}$$

where $E_1$ is given by eqn (4.25).

Antiphase boundaries (APBs) can appear in a single crystal of such a system if the sublattice occupied by A atoms shifts from one sublattice to the other in different regions. The interfaces bounding these shifted regions (i.e. domains) are then antiphase boundaries. Since $\eta$ simply changes sign when going from an unshifted to a shifted region, we expect $\eta$ to vary across such an interface in the gradual manner illustrated schematically in Fig. 4.2(c).

The diffuse antiphase boundary in Fig 4.2(c) can be analysed (Allen and Cahn 1979) using the same gradient energy approach described previously in Section 4.2.1 for diffuse interfaces in miscibility gap systems. However, in this case the independent variable is the order parameter, $\eta$, rather than the concentration, $X_B$. In addition, the free energy per atom of the uniformly ordered system, i.e. $g'(\eta)$, must be an even function of $\eta$, since states of order $\eta$ are physically identical to states of of order $-\eta$. This function will again possess a double minimum (when $T < T_c$) and will appear as indicated in Fig. 4.4(a) (Allen and Cahn 1979). The equilibrium degree of order corresponds to the points of common tangency as illustrated. By again carrying out a Taylor expansion to obtain the free energy of a non-uniform system, the free energy of the non-uniform system containing unit APB area may be written in the form

$$G = (1/\Omega)\int_{-\infty}^{\infty} [g'(\eta^+) + \Delta g(\eta) + K(d\eta/dx)^2]\,dx. \tag{4.27}$$

The quantity $\Delta g(\eta)$ is illustrated in Fig. 4.4(a), and eqn (4.27) is analogous to our previous eqn (4.13). The energy of the system if both domains were uniform right up to the interface is

$$G^\circ = (1/\Omega)\int_{-\infty}^{\infty} g'(\eta^+)\,dx, \tag{4.28}$$

and therefore

$$\sigma = G - G^\circ = (1/\Omega)\int_{-\infty}^{\infty} [\Delta g(\eta) + K(d\eta/dx)^2]\,dx, \tag{4.29}$$

which is exactly analogous to eqn (4.15). The form of the equilibrium interface profile and the corresponding energy can be found by employing the same variational procedures as used previously with the results

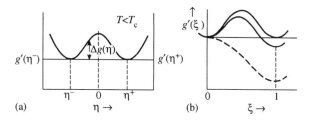

**Fig. 4.4** Bulk free energy (per atom) of ordered binary alloy, $g'(\eta)$, as a function of the long-range order parameter, $\eta$. $g'(\eta)$ is an even function of $\eta$. Common tangent construction yields the equilibrium order parameter. $\Delta g(\eta)$ is the free energy difference between a state with order $\eta$ and that with $\eta = \eta^+$ or $\eta^-$. (From Allen and Cahn (1979).) (b) Series of curves showing bulk free energy (per atom) of parent bulk phase, $g'(\xi)$, as it is deformed continuously along a deformation path to produce a new phase. $\xi$ = parameter measuring the degree of deformation. When $\xi = 0$, white crystal exists at a minimum in $g'(\xi)$. When $\xi = 1$, black crystal exists at a second minimum. Series of curves in the direction of the arrow shows the effect of an increasing thermodynamic driving force (due, e.g., to increasing stress or decreasing temperature) to form the black crystal. The dashed curve corresponds to a sufficiently large driving force to make the white crystal mechanically unstable. (From Olson and Cohen 1982.)

$$d\eta/dx = [\Delta g(\eta)/K]^{\frac{1}{2}}, \tag{4.30}$$

and

$$\sigma = (2/\Omega) \int_{-\infty}^{\infty} \Delta g(\eta) \, dx, \tag{4.31}$$

which may be compared with eqns (4.22) and (4.23). As $T \rightarrow T_c$, $\eta^+ = -\eta^- \rightarrow 0$, and $\Delta g(\eta) \rightarrow 0$. Therefore, as in the previous miscibility gap system, the energy of the diffuse interface will tend to decrease, and the width will tend to increase, as the temperature is increased until at the critical temperature the energy approaches zero and the width approaches infinity.

Further analysis of diffuse antiphase boundaries has been carried out using various atomistic models including the pair approximation and Bragg–Williams approximation and also the cluster variation method with multiatom interactions (Kikuchi and Cahn 1962, 1979; Cahn and Kikuchi 1966).

### 4.2.3 Displacive transformation interfaces in systems near a mechanical instability

We are now concerned with diffuse interfaces which may appear in systems which are near a mechanical instability. Such interfaces can be analysed by imagining that the structure of the white crystal adjoining the interface can be transformed continuously into that of the black crystal along a deformation path. If $\xi$ is a 'deformation parameter' which measures progress along this path, we would then expect the free energy of the crystal as a function of $\xi$ to appear as indicated schematically in Fig. 4.4(b). When $\xi = 0$, we have the white crystal and a corresponding minimum in the free energy versus $\xi$ curve: when $\xi = 1$, we have the black crystal and a second local minimum in the free energy. Between these limits a continuous series of intermediate structures is present. Many of these intermediate structures are of relatively higher energy, and an energy barrier to the continuous transformation from the white crystal to the black crystal exists in one form or another. When the thermodynamic conditions are changed, the energies of all of the structures will change relative to one another. A series of illustrative curves is drawn in Fig. 4.4(b), showing how these energies might change as a function of varying, for example, the temperature or stress. The dashed curve corresponds to the critical condition where the barrier is reduced to zero, and the white crystal becomes mechanically unstable with respect to the formation of the black crystal. The free energy curves in Fig. 4.4(b) are seen to be generally similar to those in Fig. 4.1(a) and 4.4(a). In all cases they refer to two structures which can be graded into each other continuously and which are separated by a hump in free energy of one form or another. When the system at some point along the deformation path is near a condition of mechanical instability, the hump will tend to be relatively low. The interface between the white and black crystals can then become diffuse with all structures along the deformation path appearing in sequence along the interface profile.

The above situation may occur for interfaces between a wide variety of structures which are related by displacive type transformations. As discussed by Cohen *et al.* (1979), displacive transformations can be broadly classified into two major types, i.e. 'lattice distortive' and 'shuffle-type'. In the lattice distortive type the major distortion (deformation) associated with the transformation involves a distortion of the unit cell of the lattice. In the shuffle-type, the major distortion corresponds to the shuffling of atoms (sometimes in cooperative groups) within the unit cell acting as a relatively fixed framework. (We note the the displacive transformation relating the lattices in Figs 4.5 and 9.7 are obviously of the lattice distortive type.)

In view of the similarity between Fig. 4.4(b) and Figs 4.1(a) and 4.4(a), it is evident that diffuse interfaces between crystals related by a lattice distortive transformation may also be treated within the framework of the general Landau–Ginsburg continuum model. As shown by Olson (1982, 1986, 1989), this can be accomplished by employing a version of the model in which: (i) the deformation parameter, $\xi$, is used as the independent variable; (ii) the free energy term corresponding to the uniform material (see eqn (4.11)) is expressed as a power series in $\xi$, i.e. a Landau-type expansion (Landau and Lifshitz 1935), with coefficients which are related to the second-, third-, and fourth-order elastic constants of the white phase; and (iii) the gradient energy is represented by a term of the form $K|\nabla\xi|^2$ (see eqn (4.11)). Expressions for the interface energy and width can then be obtained by the usual variational procedure, and various aspects of interfaces in systems undergoing lattice-displacive transformations can then be analysed (Olson 1982, 1986, 1989).

A more elaborate version of the general model may also be used. Barsch and Krumhansl (1984) obtained the diffuse {110} twin interface in a tetragonal ferroelastic material shown in Fig. 4.5 (a) by taking strain as the independent variable and expressing the free energy of the distorted material as a function of strain (including terms up to fourth order) and strain gradient (including terms up to second order). In this case the white and black tetragonal phases are unstable with respect to a cubic phase whose structure is close to various structures formed along the deformation path (as may be seen directly in Fig. 4.5(a) by examining the structures near the centre of the diffuse interface). The intermediate energy hump is therefore relatively small, and the resulting interface is then diffuse.

Diffuse interfaces between structures related by a displacive transformation which is primarily of the shuffle-type may also be analysed in a generally similar way. As an example, consider the cubic $ABX_3$ perovskite structure shown in Fig. 4.6(a). This structure is stable at elevated temperatures but possesses soft phonon modes (Nakanishi

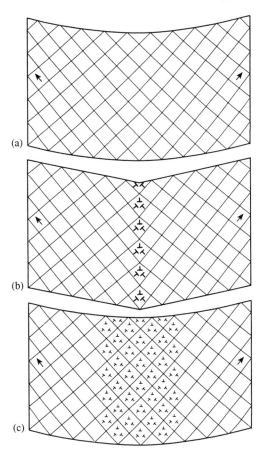

(a)

(b)

(c)

**Fig. 4.5** (a) Diffuse [110] twin boundary in a tetragonal ferroelastic material. (From Barsch and Krumhansl (1984).) (b) Same interface as in (a) but in the configuration it would adopt if it were sharp. This interface may be regarded as containing three sets of closely spaced coherency dislocations whose Burgers vectors sum to zero. (c) Same diffuse interface as in (a). Its diffuse character may be attributed to the partial delocalization of the closely spaced coherency dislocations shown in the sharp interface structure in (b). Strengths of the delocalization products are not indicated quantitatively.

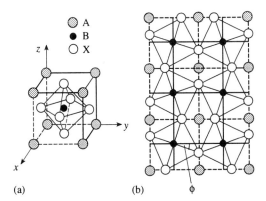

(a)                                    (b)          φ

**Fig. 4.6** (a) Cubic $ABX_3$ perovskite structure which is stable at elevated temperatures but possesses soft phonon modes corresponding to antiphase rotations of the $BX_3$ octahedra around any of the $\langle 100 \rangle$ axes. (b) Stable low-temperature tetragonal phase formed from the structure in (a) by condensing out of the soft phonon modes. Unit cell corresponds to 4 unit cells of the parent structure shown in (a) viewed along $z$. (From Cao and Barsch (1990).)

*et al.* 1982, Burns 1985) corresponding to antiphase rotations of the $BX_3$ octahedra in adjacent unit cells around a common cubic axis. As the temperature is lowered, these phonons tend to condense out (i.e. their frequency tends towards zero), and a tetragonal phase with the larger unit cell shown in Fig. 4.6(b) eventually becomes stable. Three variants are possible corresponding to rotational shuffles around the three $\langle 100 \rangle$ axes of the original cubic phase. Interfaces between these variants in the form of twins can then be formed (Cao and Barsch 1990) which tend to be diffuse because of the incipient instability of the system to the soft mode rotations. Taking $\phi$ as the $BX_3$ rotation angle (see Fig. 4.6(b)), Cao and Barsch (1990) introduced the parameter $Q = (a/2) \tan \phi$ as the primary independent variable used to describe the continuous change in structure (i.e. change in $Q$) across the diffuse interface. They then developed a Landau-type expansion for the free energy of the intermediate structures across the diffuse interface which included gradient energy terms and additional strain energy terms which had to be included because of significant distortions of the unit cell which accompany the $BX_3$ rotational shuffles. Further details are given by Cao and Barsch (1990), including a calculated profile of the diffuse interface in the perovskite $SrTiO_3$, which has a width close to that suggested by experiment.

## 4.3  SHARP HOMOPHASE INTERFACES: LARGE-ANGLE GRAIN BOUNDARIES

In the absence of impurities, grain boundaries in metallic, ionic, and covalent crystals, are generally sharp interfaces, with the change in crystal lattice orientation occurring within a few atomic planes. However, as discussed in Section 4.3.1.9, at a somewhat larger length scale the occurrence of microfaceting and boundary dissociation can lead to an apparent width which is considerably larger.

Grain boundaries are normally classified as small-angle or large-angle, with the division appearing somewhat arbitrarily at a disorientation of 15°. Small-angle boundaries are usually well described by dislocation models, in which the reference lattice is a crystal lattice and the reference structure is a single crystal. As discussed in Chapter 2, dislocation models provide a description of the elastic stress and strain fields and the elastic energy of the boundary, but they are unable to describe the atomic structure or energy of the dislocation cores. Experimental observations (e.g. Bourret and Desseaux 1979*a,b*, Krakow and Smith 1986*a*) have indicated that dislocation configurations in small-angle boundaries are often much more complex than might have been imagined. For example, there may be a considerable number of redundant (stress annihilator) dislocations, in the sense that the sum of their Burgers vectors is zero, or the crystal lattice dislocations may be dissociated into partial dislocations forming complex arrays with stacking faults. However, the observed *net* Burgers vector density is always consistent with the Frank-Bilby equation of Section 2.3, provided the boundary is free of stress at long range.

The transition to the large-angle regime is usually defined by the disorientation at which the dislocation cores start to overlap, whereupon the dislocation core energy dominates the boundary energy. In that case, elasticity theory is no longer applicable, and it is necessary to consider atomic interactions explicitly. Such large-angle grain boundaries are discussed in this section.

As we have seen in the last chapter, atomic interactions differ widely in metallic, ionic, and covalent crystals. By far the easiest to model are grain boundaries in metals. If it is assumed that directional bonding is negligible then a short-range pair potential model (short range compared with Coulomb interactions in ionic systems) may be used. At a grain boundary the more sophisticated embedded atom and Finnis–Sinclair potentials (see Sections 3.7 and 3.8.6) may be replaced quite legitimately by their effective pair potentials because the change in density at the interface is small compared with, for example, a free surface. But it is clear that pair potentials and embedded atom and Finnis–Sinclair potentials are inadequate when directional bonding is significant. It is quite likely that in b.c.c. transition metals, and even certain f.c.c. metals, such as Ir, directional bonding should be taken into account. For example, Campbell *et al.* (1993) carried out a detailed comparison of simulated atomic structures of the (310) twin in Nb with an HREM image, and concluded that only generalized pseudopotential theory gave agremment with experiment. As discussed in Section 3.8.7, generalized pseudopotential theory describes the directional bonding in Nb. On the other hand, the structures derived from embedded atom and Finnis–Sinclair potentials did not agree with the HREM image.

The simplicity of the pair potential model is such that it enables an analytic model of grain boundary energy to be developed. This model reveals the fundamental principles on which much of the rest of the chapter is based. Using this model we shall explore the relationship between the structure and energies of grain boundaries *throughout* the 5D space of the macroscopic degrees of freedom. The relative importance of the microscopic degrees of freedom, namely translations parallel to the interface and the boundary expansion are revealed clearly. We shall also discuss the various forms of relaxation away from the interface into the adjoining crystals analytically. Although these insights are derived on the basis of a pair potential model for metals we shall see that they are also relevant to other kinds of interfaces where a simple pair potential model could not be justified.

### 4.3.1  Large-angle grain boundaries in metals

#### 4.3.1.1  *The significance of the rigid body displacement parallel to the boundary plane*

The rigid body displacement, $t$, was introduced in Section 1.4.1. It is a vector with two components parallel to the boundary, which we call $\tau$, and a third component, $e\hat{n}$, normal to the boundary, which is the boundary expansion. In this section we shall determine the geometrical limits on $\tau$ for a fixed position of the boundary plane. Limits on $\tau$ exist in the sense that if $\tau$ lies outside these limits then there is an equivalent value of $\tau$ within the limits. These limits define the cell of non-identical displacements (c.n.i.d.) which was introduced in Section 1.5.7. The physical significance of the c.n.i.d. is that as its size shrinks to zero the variations in the interfacial energy as a function of $\tau$ must also tend to zero, i.e. the *ground state* interfacial energy becomes independent of $\tau$. Thus the size of the c.n.i.d. is related to the extent to which the interfacial energy may be lowered by a rigid body translation $\tau$. The size and shape of the c.n.i.d. will be shown to be directly related to whether there is 1D or 2D periodicity in the boundary plane. The possible existence of a 3D CSL relationship between the crystals is irrelevant to the argument, which focuses entirely on the two dimensions of the boundary plane.

If we assume that the periodicity of the boundary does not change on relaxation we may, for present purposes, consider the unrelaxed state. Furthermore, it is necessary to consider only one crystal lattice plane on either side of the boundary plane, because the periodicity, parallel to the boundary plane, of the remaining crystal halves is the same as these planes. Thus, we have parallel black and white crystal lattice planes $(hkl)^b$ and $(hkl)^w$, at a distance $z$ apart, as shown in Fig. 4.7. Let a lattice site in each plane coincide at the origin O. One plane may be rotated relative to the other by an arbitrary angle, $\theta_{twist}$, about an axis normal to the parallel planes through O. Let $\tau$ denote an

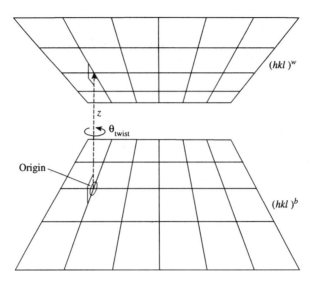

**Fig. 4.7** Two parallel 2D lattices separated by a distance $z$. $(hkl)^b$ and $(hkl)^w$ denote the Miller indices of the planes of the 3D black and white crystal lattices which these 2D lattices correspond to. They are shown in the reference state where a lattice site of each 2D lattice is coincident at the origin O. The white lattice may be rotated by $\theta_{twist}$ about the common normal through O, and it may be translated relative to the black lattice by $\tau$ in the plane parallel to the two 2D lattices.

arbitrary translation of the white lattice relative to the black lattice parallel to the interface from this reference state. In the following we assume that each lattice site is associated with an atom and that there are no other atoms in the crystal basis. This assumption does not affect the conclusion, and can be easily relaxed. We assume a simple pair potential model, although, being geometrical in nature, the final result is independent of the description of interatomic forces assumed.

Let the lattice sites in the plane of the black crystal be denoted by the set $\{X^b\}$ and those in the white by $\{X^w\}$. Consider the potential $V^b$ at a position $r = (x, z)$ arising from all atoms in the black lattice plane, as sketched in Fig. 4.8. Here $x$ is an arbitrary vector parallel to the boundary plane, and $|r|^2 = |x|^2 + z^2$. The potential $V^b$ may be expressed as follows:

$$V^b(x, z) = \sum_{X^b} v(x - X^b, z) \tag{4.32}$$

where $v(r) = v(|r|)$ is the pair potential at a separation $|r|$. Since the potential $V^b(x, z)$ is periodic parallel to the lattice it may be expanded in a 2D Fourier series:

$$V^b(x, z) = \sum_{G^b} \tilde{V}^b(G^b, z) \exp(iG^b \cdot x) \tag{4.33}$$

where

$$\tilde{V}^b(G^b, z) = \frac{1}{N_c^b A_c^b} \int V^b(x, z) \exp(-iG^b \cdot x) \, dx, \tag{4.34}$$

$A_c^b$ is the area of a primitive unit cell of the black lattice plane and $N_c^b$ is the number of such unit cells, which is infinite. If $a_1^b$ and $a_2^b$ are primitive, non-parallel translation

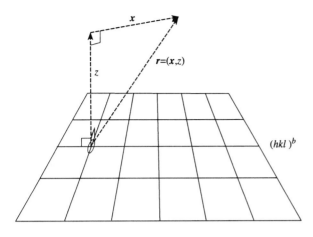

**Fig. 4.8** To illustrate the coordinate system for a point $r = (x, z)$ above the black 2D lattice plane $(hkl)^b$.

vectors of the black lattice plane and $a_3^b$ is any vector normal to the plane, then $a_1^{b*} = 2\pi (a_2^b \times a_3^b)/(a_1^b \cdot a_2^b \times a_3^b)$ and $a_2^{b*} = 2\pi (a_3^b \times a_1^b)/(a_1^b \cdot a_2^b \times a_3^b)$ are basis vectors of the *layer* reciprocal lattice comprising the vectors $\{G^b\}$. In general, the reciprocal lattice vectors $\{G^b\}$ are *not* reciprocal lattice vectors of the black crystal lattice. Substituting eqn (4.32) and eqn (4.34) into eqn (4.33) we find

$$V^b(x, z) = \frac{1}{A_c^b} \sum_{G^b} \tilde{v}(G^b, z) \exp(iG^b \cdot x) \tag{4.35}$$

where $\tilde{v}(G^b, z)$ is the 2D Fourier transform of the pair potential:

$$\tilde{v}(G^b, z) = \int v(x, z) \exp(-iG^b \cdot x)\, dx. \tag{4.36}$$

The integral in eqn (4.36) extends over all 2D space.

The energy of interaction of the two lattices per unit area is given by:

$$E(\tau, z) = \frac{1}{N_c^w A_c^w} \sum_{X^w} V^b(X^w + \tau, z) \tag{4.37}$$

Substituting eqn (4.35) for $V^b$ into eqn (4.37) we obtain

$$E(\tau, z) = \frac{1}{N_c^w A_c^w A_c^b} \sum_{G^b} \sum_{X^w} \exp(iG^b \cdot X^w) \exp(iG^b \cdot \tau) \tilde{v}(G^b, z). \tag{4.38}$$

Using the relation

$$\sum_{X^w} \exp(iG^b \cdot X^w) = N_c^w \delta_{G^b, G^w}, \tag{4.39}$$

where $\delta_{G^b, G^w} = 1$ if $G^b = G^w$ and zero otherwise, it is finally deduced that

$$E(\tau, z) = \frac{1}{A_c^b A_c^w} \sum_{G^c} \tilde{v}(G^c, z) \exp(iG^c \cdot \tau) \tag{4.40}$$

where $G^c$ is a *common* reciprocal lattice vector of the sets $\{G^b\}$ and $\{G^w\}$. If the two lattices are incommensurate with each other the only $G^c$ entering the sum in eqn (4.40) is $G^c = 0$, and the interaction energy is independent of the relative translation of the two lattices. At commensurate orientations of the black and white lattices the energy of interaction, $E(\tau, z)$, is a periodic function of $\tau$. The periodicity of $E(\tau, z)$ is given by the reciprocal lattice of the one or 2D lattice of $G^c$ vectors. Thus, $\tau$ may be stated uniquely if it is expressed modulo a reciprocal lattice vector of the one or 2D lattice of $G^c$ vectors. This would place $\tau$ within the Wigner–Seitz cell of the lattice that is reciprocal to the lattice of $G^c$ vectors. The Wigner–Sietz cell defines the c.n.i.d. that was introduced in Section 1.5.7.

In general, the lattice that is reciprocal to the lattice of $G^c$ vectors is a sublattice of the DSC lattice plane parallel to the boundary plane. That is because the DSC lattice is defined (see Section 1.5.4) in *three* dimensions, whereas our analysis is confined to translations in *two* dimensions which create equivalent boundary structures at the same location of the boundary plane. The excluded DSC vectors are those that generate equivalent boundary structures only if the position of the boundary plane is allowed to move into one of the adjoining crystals. Thus, a dislocation with a Burgers vector equal to one of the excluded DSC vectors would be associated with a step in the boundary plane.

As a first example consider $\theta = \cos^{-1}\frac{1}{5}$ (001) twist boundary in f.c.c. crystals, which lies in the $\Sigma = 13$ CSL. We consider black and white (002) f.c.c. lattice planes, separated by a distance $z$, and misoriented by $\theta = \cos^{-1}\frac{1}{5}$. The basis vectors of the real space lattices $\{X^b\}$ and $\{X^w\}$ are $\frac{1}{2}\langle 110 \rangle$, in their respective coordinate systems. Similarly, the basis vectors of the layer reciprocal lattices $\{G^b\}$ and $\{G^w\}$ are $2\pi\langle 110 \rangle$. The lattice of common reciprocal lattice vectors $\{G^c\}$ is square with basis vectors $2\pi\langle 510 \rangle$. The reciprocal lattice of the $\{G^c\}$ lattice is also square and has a basis $\frac{1}{26}\langle 510 \rangle$. Therefore the energy of the (001) twist boundary is a periodic function of $t$, with periodicity equal to $\frac{1}{26}[510]$ by $\frac{1}{26}[\bar{1}50]$. On the other hand, the periodicity of the boundary structure is $\frac{1}{2}[\bar{1}50]$. In this example the periodicity of the boundary energy is the same as the periodicity of the DSC lattice in the plane of the boundary.

For the second example consider the (510) symmetric tilt boundary in an f.c.c. lattice, which also lies in the $\Sigma = 13$ CSL. In the coordinate system of the median lattice the boundary plane is (510) and (5$\bar{1}$0) in the black and white crystals respectively. We consider parallel black and white $(10, 2, 0)^b$ and $(10, \bar{2}, 0)^w$ f.c.c. lattice planes, separated by a distance $z$. The basis vectors of the real space lattices $\{X^b\}$ and $\{X^w\}$ are $\frac{1}{2}[\bar{1}50]^b$ and $[001]^b$ and $\frac{1}{2}[150]^w$ and $[001]^w$ respectively, where $\frac{1}{2}[\bar{1}50]^b$ is coincident with $\frac{1}{2}[150]^w$ and $[001]^b$ is coincident with $[001]^w$. The basis vectors of the layer reciprocal lattices $\{G^b\}$ and $\{G^w\}$ are $(2\pi/13)[\bar{1}50]^b$, $2\pi[001]^b$, and $2\pi/13[150]^w$, $2\pi[001]^w$, which also define the lattice of common reciprocal lattice vectors $\{G^c\}$. The reciprocal lattice of the $\{G^c\}$ lattice has the same basis as the black and white $(10, 2, 0)^b$ and $(10, \bar{2}, 0)^w$ lattice planes. It is a sublattice of the corresponding plane of the $\Sigma = 13$ DSC lattice. In this case, as in all symmetric tilt boundaries, the periodicity of the boundary energy as a function of $\tau$ is the same as the periodicity of the boundary structure, which is the same as the periodicity of the corresponding plane of the crystal lattice.

The magnitude of the contribution to the interaction energy from each matching $G^c$ vector is a maximum of $|\tilde{v}(G^c, z)|$ and it is modulated by the phase factor $\exp(iG^c \cdot \tau)$. For a Lennard–Jones pair potential of the form

$$v(r) = \varepsilon \left[ \left( \frac{r_0}{r} \right)^{12} - 2 \left( \frac{r_0}{r} \right)^6 \right], \tag{4.41}$$

which has a minimum of depth $\varepsilon$ at $r = r_0$, the 2D Fourier transform is given by (Sutton 1991*a*):

$$\tilde{v}(q,z) = \frac{\pi \varepsilon r_{\mathrm{o}}^2}{4} \left[ \left( \frac{q r_{\mathrm{o}}}{z/r_{\mathrm{o}}} \right)^5 \frac{K_5(qz)}{480} - 2 \left( \frac{q r_{\mathrm{o}}}{z/r_{\mathrm{o}}} \right)^2 K_2(qz) \right] \qquad (4.42)$$

where $K_5$ and $K_2$ are modified Bessel functions. At large values of $qz$ the asymptotic form is

$$\tilde{v}(q,z) = \frac{\pi \varepsilon r_{\mathrm{o}}^2}{4} \left[ \frac{1}{480} \left( \frac{q r_{\mathrm{o}}}{z/r_{\mathrm{o}}} \right)^5 - 2 \left( \frac{q r_{\mathrm{o}}}{z/r_{\mathrm{o}}} \right)^2 \right] \left( \frac{\pi}{2qz} \right)^{\frac{1}{2}} \exp\left( -qz \right). \qquad (4.43)$$

As discussed by Sutton (1991a), the sign of the Fourier transform $\tilde{v}(q,z)$ is a complicated function of the wave vector $q$ and the separation $z$, although for $z/r_{\mathrm{o}} \ll 1$ it is always positive due to the dominance of the repulsive part of the potential at short range. But the important point to note is that at large values of $qz$ the coefficient decays exponentially. This indicates that for an unrelaxed boundary the significance of periodicity within the boundary plane decays exponentially as the size of the c.n.i.d. approaches zero. From a physical point of view this is a necessary result because in the incommenurate limit, when the c.n.i.d. shrinks to a point, the boundary energy is independent of $\tau$ (see also Section 1.9). When the boundary structure is allowed to relax the c.n.i.d. does not change, assuming that whatever periodicity is present in the unrelaxed state is maintained in the relaxed state. It would be unphysical for the energy of an interface to change by a large amount when $\tau$ is changed by a small amount. Therefore, if the c.n.i.d. is very small the variation in the *relaxed* boundary energy with $\tau$ must also be small. This reasoning, which is primarily geometrical in nature, is expected to apply to *all* kinds of interfaces between crystals.

There is considerable evidence in support of the above arguments for relaxed grain boundaries. For example, Bristowe and Crocker (1978) found that the energies of relaxed large-angle (001) twist boundaries in f.c.c. metals were relatively insensitive to $\tau$. For these boundaries the c.n.i.d. is always quite small. There is also compelling evidence from experiment. Sutton and Balluffi (1987) reviewed the results of rotating ball experiments where small single crystal balls are placed on a single crystal substrate and allowed to rotate during high-temperature annealing to low-energy misorientations. While the grain boundary plane on the substrate side is fixed it is completely free on the ball side. For any given plane on the ball side there is an infinite number of possible misorientations because the ball may rotate about an axis normal to the substrate without changing the boundary plane. It was found that for each plane selected on the ball side the observed misorientations always corresponded to those with the largest c.n.i.d. In some cases not only the largest c.n.i.d. misorientation was seen but also the two or three next largest. Our rationalization for this observation is that the largest c.n.i.d. (hence smallest $|G^c|$) on a given boundary plane is associated with the largest variations in the boundary energy as a function of $\tau$, and thus the boundary can attain the lowest energy state on a given plane at these orientations by adopting a suitable value of $\tau$. In a few cases balls also became trapped at smaller local minima where the c.n.i.d. was relatively large, but not the largest available on a given plane.

The rotating ball experiments are consistent with computer simulations of the boundary energy as a function of the twist angle on a given plane. In Fig. 4.9 we show the calculated boundary energy for four boundary planes in Cu represented by a Lennard–Jones pair potential, as a function of twist angle (Merkle and Wolf 1990). For the (111) pure twist boundaries shown in Fig. 4.9(a), $\theta_{\mathrm{twist}} = 0$ corresponds to the perfect crystal and $\theta_{\mathrm{twist}} = 60°$ corresponds to the (111) symmetric tilt boundary. At either end of the series of (115)/(111) mixed tilt and twist boundaries, also shown in Fig. 4.9(a), we have the

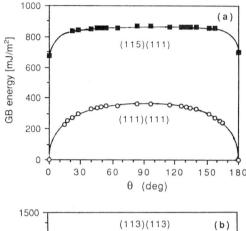

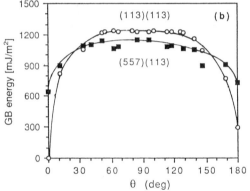

**Fig. 4.9** Energies of grain boundaries as a function of twist misorientation, as calculated by Merkle and Wolf (1990) using a Lennard–Jones potential constructed for Cu. (a) (111) pure twist boundaries and (115)/(111) mixed tilt and twist boundaries. Note that the twist misorientation has been scaled by a factor of three for the (111) pure twist boundaries. (b) (113) pure twist boundaries and (557)/(113) mixed tilt and twist boundaries.

asymmetric tilt boundaries $\Sigma = 3$ $(\bar{1}\bar{1}5)/(111)$ and $\Sigma = 9$ $(115)/(111)$. For the (113) pure twist boundaries, shown in Fig. 4.9(b), $\theta_{\text{twist}} = 0$ corresponds to the perfect twist crystal and $\theta_{\text{twist}} = 180°$ to the (113) symmetric tilt boundary. At either end of the (557)/(113) series of mixed tilt and twist boundaries, shown in Fig. 4.9(b), we have the $\Sigma = 3$ (557)/ $(\bar{1}\bar{1}3)$ and $\Sigma = 11$ (557)/(113) asymmetric tilt boundaries. In each series the c.n.i.d.s of the boundaries at the ends have the same size and they are also the largest c.n.i.d.s that occur for each series. For each series it is seen that the ends coincide with the lowest energy, and hence in the rotating ball experiments the balls would tend to rotate to these misorientations on (111) and (113) substrates. There are also some subsidiary minima at relatively large sizes of the c.n.i.d. where rotating balls may get trapped.

The criterion for low interfacial energy *on a given boundary plane* of selecting the twist angle which maximizes the size of the c.n.i.d is equivalent to the restricted form of the maximum planar coincidence site density criterion first proposed by Wolf (1985). This criterion was shown to be highly successful for grain boundaries in metallic and ionic systems by Sutton and Balluffi (1987), who considered the body of experimental data on energies of homophase and heterophase interfaces. In the unrestricted form of this criterion (Brandon *et al.* 1964) one simply selects the boundary which maximizes the density of coincidence sites in the boundary plane regardless of whether the planes are the same or not. More generally, one selects the boundary whose 2D repeat cell in the boundary plane has the smallest area. Sutton and Balluffi (1987) showed that this criterion fails as often as it succeeds. In the highly successful restricted form of the criterion it is applied only to boundaries sharing the same plane. The reason for the failure of the unrestricted form of the criterion will be given in the next section.

During relaxation of an interface with a twist angle close to that of a relatively low-energy interface containing non-parallel $G^c$ vectors, two or more sets of screw dislocations may be introduced to create localized regions where the twist angle is almost the same as that of the low energy interface. Experimental observations of such arrays of screw dislocations were mentioned in Section 2.12. Boundaries in which $G^c$ vectors exist along only one direction in the interface have also been found experimentally to have relatively low energy (Pumphrey 1972, Schindler *et al.* 1979). These boundaries are among those that have been called 'plane-matching' or 'coincident axial direction' boundaries. The driving force for these relaxations is the energy gain from introducing patches of boundary where relatively small common $G$ vectors exist along one direction in the boundary plane. Against this energy gain must be set the energy cost of the strain introduced by the relaxation. We have here the physical basis for many of the observed modes of relaxation at interfaces, such as the localization of particular arrays of dislocations, and the near coincidence model for incommensurate interfaces.

### 4.3.1.2 *The significance of the expansion normal to the boundary plane*

In the previous section we have seen that the relative rigid body displacement, $\tau$, parallel to the boundary, may lead to a lowering of the boundary energy whenever there is 1D or 2D periodicity in the boundary plane. But regardless of whether there is periodicity, the boundary may always undergo an expansion, $e$, (or contraction) normal to the plane. (An expansion or contraction parallel to the boundary plane is ruled out in an infinite specimen because it would give rise to a long-range elastic field, the energy of which would diverge with the specimen size.) In this section we shall separate the roles of $\tau$ and $e$ by considering the 'incommensurate limit' of all boundaries that may exist on a given plane (Mooser and Schlüter 1971, Brokman and Balluffi 1981, Wolf 1984, Wolf 1990a, Sutton 1991a, Sutton 1992). The incommensurate limit is attained at the maximum in the energy as a function of $\theta_{twist}$. These boundaries are the least affected by any commom reciprocal lattice vectors, which may occur at other values of $\theta_{twist}$. There is no twist periodicity in these boundaries and therefore the boundary energy is independent of $\tau$. One of the aims of this section is to estimate the energies of all boundaries in the incommensurate limit. To do this we shall consider *all* possible boundary plane normals in the black and white crystals. Thus we shall consider the entire parameter space associated with 4 of the 5 macroscopic degrees of freedom. The expansion will be identified as the most important relaxation parameter for these boundaries. The relationship between the energies of these boundaries and the plane normals in both crystals will be made explicit, as will the relationship to the free surface energies. The conceptual importance of the incommensurate limit as a means of separating out the role of the boundary plane was first recognized by Wolf and coworkers, e.g. Wolf and Phillpot (1989). In the following we shall follow Sutton (1991a, 1992).

Consider a flat grain boundary between two f.c.c. crystals, in which the spacings of the crystal lattice planes parallel to the boundary in the black and white crystals are $d^b$ and $d^w$. The boundary is imagined to be created by the following four steps:

1. Flat surfaces of two well separated, semi-infinite perfect crystals are aligned parallel to each other. These surfaces will become the boundary plane in each crystal.

2. One crystal is rotated about the common normal of both surfaces to the incommensurate limit for the grain boundary that is formed in the third step.

3. An ideal, unrelaxed grain boundary is formed by bringing the semi-infinite perfect crystals together, until the separation of the lattice planes at the boundary is $(d^b + d^w)/2$. At this separation the density of the bicrystal is equal to that of either perfect crystal, and the expansion is defined to be zero.

4. The two semi-infinite crystal halves are now allowed to float normal to the boundary plane until the energy of the bicrystal is minimized. No relaxation is allowed of the atoms individually within the semi-infinite crystals, so that the crystal halves remain perfect. The nature and significance of this relaxation will be discussed in the next section. For the present the only relaxation variable is the boundary expansion, $e$.

We shall now calculate the expansion which minimizes the energy of the boundary for any pair of $d^b$ and $d^w$. To do this we shall again use the Lennard–Jones potential of eqn (4.41). The energy of the boundary in the incommensurate limit, $\sigma_{IL}$, may be written down immediately using eqn (4.40), for the energy per unit area of interaction between two parallel layers:

$$\sigma_{IL} = \sigma_s^b + \sigma_s^w + \frac{1}{A_c^b A_c^w} \sum_{l^b = 1}^{\infty} \sum_{l^w = 1}^{\infty} \tilde{v}(0, (l^b d^b + l^w d^w - (d^b + d^w)/2 + e)). \quad (4.44)$$

The energies of the non-interacting free surfaces are denoted by $\sigma_s^b$ and $\sigma_s^w$. The energy of interaction between the crystal halves is expressed as a sum of interactions between layers in the black and white crystal halves. The layers in both crystals have been numbered from 1 to $\infty$, and the geometrical boundary plane is between layer 1 of both crystals. The separation between layer $l^w$ in the white crystal and layer $l^b$ in the black crystal is $l^b d^b + l^w d^w - (d^b + d^w)/2 + e$, where $e$ is the boundary expansion. Since we are considering the incommensurate limit the only term in the sum in eqn (4.40) over the common reciprocal lattice vectors that appears in eqn (4 .44) is $G^c = 0$. The Fourier component, $\tilde{v}(0, z)$, of the potential is given by

$$\tilde{v}(0, z) = \pi \varepsilon r_o^2 \left( \frac{1}{5} \left( \frac{r_o}{z} \right)^{10} - \left( \frac{r_o}{z} \right)^4 \right). \quad (4.45)$$

Substituting this expression into eqn (4.44) for $\sigma_{IL}$ we obtain

$$\sigma_{IL} = \sigma_s^b + \sigma_s^w + \frac{\pi \varepsilon r_o^4}{5\Omega^2} g(d^b/r_o, d^w/r_o, e/r_o) \quad (4.46)$$

where $\Omega$ is the atomic volume and $g$ is a dimensionless function, which we call the binding function:

$$g = S_{10} - 5S_4 \quad (4.47)$$

and

$$S_n = \sum_{l^b = 1}^{\infty} \sum_{l^w = 1}^{\infty} \frac{(d^b/r_o)(d^w/r_o)}{\left[ \dfrac{l^b d^b + l^w d^w - (d^b + d^w)/2 + e}{r_o} \right]^n}. \quad (4.48)$$

To evaluate $S_n$ it is convenient to rewrite it as an integral:

$$S_n = \frac{(d^b/r_o)(d^w/r_o)}{(n-1)!} \int_0^{\infty} dy \frac{y^{(n-1)} \exp\left[ -(e + \langle d \rangle) y/r_o \right]}{(1 - \exp[-(d^b y)/r_o])(1 - \exp[-(d^w y)/r_o])} \quad (4.49)$$

where $\langle d \rangle = (d^b + d^w)/2$ is the average interplanar spacing. From eqn (4.46) the ideal cleavage energy, $\sigma_s^b + \sigma_s^w - \sigma_{IL}$, of the boundary is given by:

$$\sigma_{cl} = -\pi \varepsilon r_o^4 g / 5\Omega^2. \qquad (4.50)$$

Thus, the quantity that is calculated directly is the ideal cleavage energy, and to evaluate the grain boundary energy we must also know the energies of the two free surfaces.

We now minimize the boundary energy in eqn (4.46) with respect to the boundary expansion to obtain the following implicit relationship between e and the interplanar spacings $d^b$ and $d^w$:

$$S_{11} = 2S_5. \qquad (4.51)$$

Once eqn (4.51) is solved for $e$, for a given $d^b$ and $d^w$, we substitute $e$ back into eqn (4.47) to get the binding function.

Let us consider pure twist boundaries first, for which $d^b = d^w = \langle d \rangle$. In Fig. 4.10 we show the binding function, $g$, as a function of the expansion, $e$, for three boundaries corresponding to (111), (113) and irrational planes, for which the interplanar spacings are $0.5774a$, $0.3015a$, and 0, where $a$ is the lattice parameter of the f.c.c. crystal. The expansion is expressed in units of $r_o$. (An f.c.c. crystal in which atomic interactions are described by the Lennard–Jones potential of eqn (4.41) is in equilibrium, at 0 K, when $a = 1.3735r_o$; therefore $r_o = 0.7281a$.) As the expansion tends to infinity so the binding function $g$ tends to zero, and the 'grain boundary' energy, eqn (4.46), tends to the sum of the energies of the two free surfaces. The binding function rises steeply at small (or negative) expansions, which is a consequence of the short range repulsion in the atomic interactions. Although at the state of zero expansion the spacing of planes at the boundary is the same as the equilibrium spacing in the bulk, the relative rotation by $\theta_{twist}$ has introduced overlap between atoms at the boundary, and hence strong repulsion. In the irrational limit, where the interplanar spacing is zero the repulsion becomes infinite, and the unrelaxed boundary energy is also infinite. The equilibrium values of the expansion and binding function occur at the minimum of each curve. As the interplanar spacing decreases the equilibrium expansion increases and the depth of the binding function decreases. Thus, the ideal cleavage energy decreases and the equilibrium expansion increases as $\langle d \rangle$ decreases.

In Figs 4.11 and 4.12 we see the equilibrium expansion and binding function plotted as a function of the interplanar spacing (the latter in units of $r_o$). These curves confirm the trend seen in Fig. 4.10 of increasing expansion and decreasing binding with decreasing interplanar spacing. The equilibrium expansion is predicted to be zero only when the interplanar spacing is $0.8794r_o = 0.6420a$, which is greater than the maximum spacing ($0.5774a$) in f.c.c. crystals. Thus the model predicts that there is always an expansion at twist boundaries in f.c.c. crystals in the incommensurate limit. The model also predicts that as the interplanar spacing tends to zero the expansion tends to a maximum value of $15^{-\frac{1}{6}}r_o = 0.4636a$ and the binding function tends to $-15^{\frac{3}{4}}/24 = -1.5414$.

For mixed tilt and twist boundaries where $d^b \neq d^w$ it is convenient to work with the average interplanar spacing $\langle d \rangle$ and $\delta = (d^b - d^w)/(d^b + d^w)$. Figures 4.13 and 4.14 show the equilibrium expansion and binding function for two extreme sets of boundaries as a function of $\langle d \rangle$. The set represented by circles corresponds to the case of $\delta = 0$ and $\langle d \rangle = d^b = d^w$, which is also shown in Figs 4.11 and 4.12. The set represented by squares corresponds to the case of $d^w = 0$, $d^b = 2\langle d \rangle$, $\delta = 1$ (or $d^b = 0$, $d^w = 2\langle d \rangle$, $\delta = -1$). It is seen that for a given average interplanar spacing $\langle d \rangle$ the boundary with $d^b = d^w$ always has a greater expansion and smaller cleavage energy than the boundary

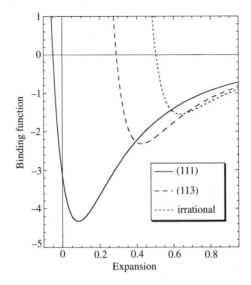

**Fig. 4.10** The dimensionless binding function, $g$, given by eqn (4.47), for (111), (113), and irrational twist boundaries in the incommensurate limit, as a function of the boundary expansion in units of $r_0$ ($r_0 = 0.7218a$). As the interplanar spacing decreases from $0.5774a$ for (111), to $0.3015a$ for (113), to 0 for the irrational twist boundary, the equilibrium expansion increases and the binding energy, which is proportional to the depth of the well, decreases.

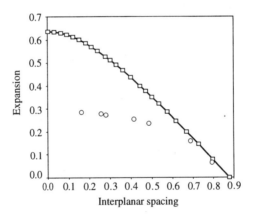

**Fig. 4.11** The predicted equilibrium expansion of twist boundaries in the incommensurate limit as a function of the interplanar spacing (both in units of $r_0$). The squares show the expansions which minimize the boundary energy in eqn (4.46). The circles show the expansions found for the twist boundaries, listed in Table 4.2, by full atomistic relaxation. Note that the predicted monotonic increase of expansion with decreasing interplanar spacing is confirmed by the simulations.

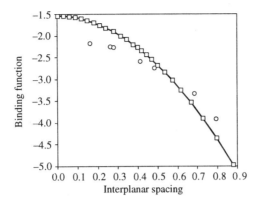

**Fig. 4.12** The equilibrium dimensionless binding function, $g$, in eqn (4.46), as a function of the interplanar spacing (in units of $r_o$), for twist boundaries in the incommensurate limit is shown by the squares. The circles show the same dimensionless function obtained from the ideal cleavage energy (last column of Table 4.2) for fully relaxed twist boundaries near the incommensurate limit. Note that the prediction, of a monotonic decrease in the binding with decreasing interplanar spacing, by the analytic model, is confirmed by the simulations.

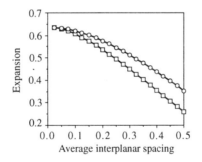

**Fig. 4.13** The equilibrium expansions (in units of $r_o$), predicted by minimizing the energy in eqn (4.46) for mixed tilt and twist boundaries in the incommensurate limit, as a function of the average interplanar spacing (in units of $r_o$). The circles correspond to $\delta = 0 \Rightarrow d^b = d^w$. The squares correspond to $\delta = 1 \Rightarrow d^w = 0$ and $d^b = 2\langle d \rangle$ (or $\delta = -1 \Rightarrow d^b = 0$ and $d^w = 2\langle d \rangle$). At a given average interplanar spacing, $\langle d \rangle$, the predicted expansions of all boundaries, in the incommensurate limit, lie between the two curves.

with $d^b = 2\langle d \rangle$, $d^w = 0$. Moreover, the larger the average interplanar spacing the greater the difference in expansions and cleavage energies for the two extreme cases. *All other boundaries, in the incommensurate limit, lie between the curves in Figs 4.13 and 4.14.* For example, the intermediate cases where $d^b$ varies between 0 and $\langle d \rangle$ for $\langle d \rangle = 0.5r_o$ are illustrated in Figs 4.15 and 4.16. Similar curves are obtained for other values of $\langle d \rangle$. It is seen that the boundary expansion increases monotonically, and the cleavage energy decreases monotonically as $d^b$ increases from 0 to 0.5, i.e. as $\delta$ decreases from 1 to 0. Thus the model predicts that the equilibrium expansion of an incommensurate mixed tilt and twist boundary is less than that of an incommensurate twist boundary with the same value of $\langle d \rangle$. Furthermore, the cleavage energy of the former is predicted always to be greater than that of the latter.

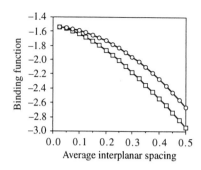

**Fig. 4.14** The same as Fig. 4.13, but showing the equilibrium dimensionless binding function, $g$, of eqn (4.46).

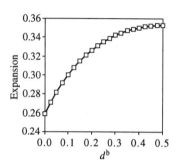

**Fig. 4.15** The equilibrium expansions for mixed tilt and twist boundaries with an average interplanar spacing of $0.5r_0$, as a function of the interplanar spacing, $d^b$, in the black crystal (in units of $r_0$). The boundaries represented in this graph correspond to $\langle d \rangle = 0.5r_0$ in Fig. 4.13.

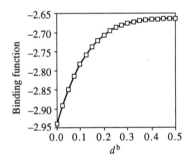

**Fig. 4.16** The dimensionless binding function for the boundaries shown in Fig. 4.15. The boundaries represented in this graph correspond to $\langle d \rangle = 0.5r_0$ in Fig. 4.14.

In the next section we shall see that the trends predicted by our analytic theory are confirmed by full atomistic relaxations of twist boundaries near the incommensurate limit.

### 4.3.1.3 *Testing the analytic model*

The analytic model described in the previous section ignored all individual atomic relaxation. In a computer simulation the forces on all atoms are relaxed to zero, and in this section we shall describe some simulations that were done to test the model. The methods of computer simulation for interfaces will be described in Section 4.3.1.6. For the present it is sufficient to say that the energy of each grain boundary was minimized with respect to all atomic coordinates, the relative rigid body displacement, $\tau$, and the boundary expansion, $e$. The same Lennard–Jones potential, eqn (4.41), was used as in the analytic model. Because periodic boundary conditions are used parallel to the boundary plane it is necessary to consider boundaries that are periodic in two dimensions. To approach the incommensurate limit values of $\theta_{twist}$ of periodic boundary structures close to the maximum in the energy vs. $\theta_{twist}$ relation were selected. Seven pure twist boundaries were relaxed, and the results are tabulated in Table 4.2. In each case the smallest $G^c$ vectors are also shown, and it is seen that they are all large, and therefore only very small contributions to the boundary energy are likely to arise from them (see eqn (4.43)). The results for the equilibrium expansion and binding function are shown by circles on Figs 4.11 and 4.12. The trends shown by the analytic model are seen to be well reproduced by the computer simulations. However, as the interplanar spacing tends to zero the analytic model over-estimates the expansion by about a factor of two. This increasing discrepancy simply reflects the increasing importance of individual atomic relaxation as the interplanar spacing decreases. We shall discuss individual atomic relaxation in Section 4.3.1.4.

The success of the analytic model is that it correctly predicts the trends for boundaries in the incommensurate limit, that as the average interplanar spacing decreases the expansion increases and the cleavage energy decreases. Let us consider the physical reasons for these trends. Clearly the expansion occurs to relieve the repulsion arising from the atomic overlap that exists in the ideal unrelaxed state, where the separation of planes at the boundary is $(d^b + d^w)/2$. As the boundary expands the repulsive forces do work against the attractive forces across the boundary, and both the repulsive and attractive forces across the boundary are weakened. The equilibrium expansion is determined by the balance of these weakened repulsive and attractive forces. Therefore, although it is the repulsive forces which cause the expansion in the first place, the attractive forces are just as important in determining the equilibrium expansion and cleavage energy. The greater the initial overlap the greater the expansion that is required to relieve it, and the weaker the residual attractive forces across the boundary, leading to a smaller cleavage energy.

Note that the analytic model derives the cleavage energy and not the boundary energy, $\sigma_{IL}$. Thus although the cleavage energy shows a clear dependence on the boundary expansion and the interplanar spacings the corresponding boundary energy variations are complicated by the values of the free surface energies which enter the expression for $\sigma_{IL}$ in eqn (4.46).

### 4.3.1.4 *The significance of individual atomic relaxation*

Individual atomic relaxation displacements may be uncorrelated over distances greater than a few nearest neighbour distances. In such cases their role is merely to optimize local atomic environments in the sense of relieving atomic overlap and maximizing the local coordination number. Examples of such relaxations will given in Section 4.3.1.7. But individual atomic relaxations may also be correlated over larger distances. Correlated relaxations are associated, for example, with the introduction of dislocations which effect local changes of boundary misorientation. We can distinguish two kinds of variation in local misorientation: (i) variations in local twist angle, $\theta_{twist}$ and (ii) variations in local tilt angle, $\theta_{tilt}$. The former are effected by screw dislocations resulting in correlated displacements parallel to the boundary plane. The latter are effected by edge dislocations with correlated displacements normal to the boundary plane.

Consider local variations in $\theta_{twist}$. These variations are driven by the energy gain of introducing patches of a lower energy reference boundary, against which must be set the energy cost of the elastic strain field associated with the concomitant screw dislocations. The energy gain is approximately proportional to the appropriate Fourier coefficient $\tilde{v}(G^c, z)$, where $G^c$ is the smallest common reciprocal lattice vector of the reference

**Table 4.2** Results of atomistic relaxations for seven pure twist boundaries near the incommensurate limit, as computed with the Lennard–Jones potential of eqn (4.41). The twist angle is $\theta_{twist}$. The smallest common layer reciprocal lattice vectors are shown in third column. $d$ is the interplanar spacing and $e$ is the boundary expansion. The relaxed boundary energy is denoted by $\sigma$. The energy of the corresponding relaxed free surface is denoted by $\sigma_s$. The cleavage energy, $\sigma_{cl} = 2\sigma_s - \sigma$, is expressed in the last column in terms of the dimensionless parameter, $g$, defined by eqn (4.50). Note that as $d$ decreases, $e$ increases and the cleavage energy decreases monotonically.

| Boundary plane | $\cos\theta_{twist}$ | Smallest $G^c \times a/2\pi$ | $d(r_o)$ | $e(r_o)$ | $\sigma(\varepsilon/r_o^2)$ | $\sigma_s(\varepsilon/r_o^2)$ | $-g = \dfrac{5\sigma_{cl}\Omega^2}{\pi\varepsilon r_o^4}$ |
|---|---|---|---|---|---|---|---|
| (111) | $-\frac{1}{7}$ | $\frac{2}{3}[-1,5,-4]$ $\frac{2}{3}[1,4,-5]$ | 0.7930 | 0.0647 | 0.9927 | 3.4222 | 3.9085 |
| (100) | $\frac{20}{29}$ | $[0,3,7]$ $[0,7,-3]$ | 0.6868 | 0.1589 | 2.1240 | 3.5536 | 3.3284 |
| (110) | $-\frac{1}{17}$ | $[2,-2,3]$ $[3,-3,-4]$ | 0.4856 | 0.2369 | 3.4040 | 3.7535 | 2.7406 |
| (113) | $-\frac{1}{10}$ | $[0,-6,2]$ $-\frac{1}{11}[20,-2,-6]$ | 0.4141 | 0.2541 | 3.6630 | 3.7667 | 2.5851 |
| (112) | $-\frac{1}{49}$ | $\frac{1}{3}[22,-2,-10]$ $[-1,9,-4]$ | 0.2804 | 0.2718 | 4.1033 | 3.7475 | 2.2654 |
| (115) | $-\frac{1}{7}$ | $\frac{1}{3}[1,19,-4]$ $\frac{1}{3}[22,-2,-4]$ | 0.2643 | 0.2759 | 4.1624 | 3.7677 | 2.2529 |
| (114) | $-\frac{1}{17}$ | $[4,0,-1]$ $[1,-17,4]$ | 0.1619 | 0.2847 | 4.3067 | 3.7802 | 2.1733 |

boundary, and $z$ is the separation of layers on either side of the boundary (see eqn (4.40)). We have seen in eqn (4.43) that at large values of $|G^c|z$ the Fourier coefficient decays exponentially as $\exp(-|G^c|z)$. On the other hand, at a given misorientation from the reference boundary, the elastic distortion energy varies linearly with the Burgers vector, $b$, of the relevant screw dislocations. Now $b$ scales with the size of the c.n.i.d. of the reference boundary, which we shall call $c$. Thus, the elastic energy is proportional to $c$. On the other hand $|G^c|$ scales as the inverse of $c$, and therefore the energy gain is proportional to $-\exp(-z/c)$. The total energy of introducing screw dislocations is thus proportional to $c - a \exp(-z/c)$, where $a$ is a positive constant. This function is sketched in Fig. 4.17, where it is seen that the energy of the boundary is lowered only if $c$ is larger than a critical value, $c_0$. It follows that relaxations resulting in the introduction of screw dislocations into a boundary are favourable energetically only when the size of the c.n.i.d. of the reference boundary exceeds some critical value, i.e. when the period of the reference boundary is less than some upper limit. Our conclusion amounts to a physical restriction on the boundaries that can sustain arrays of localized screw dislocations.

Let us now consider correlated displacements normal to the boundary. Figures 4.18 and 4.19 show the displacements of layers normal to the boundary in the fully relaxed (100) and (114) twist boundaries appearing in Table 4.2. Each layer displacement is obtained by averaging the displacements of all non-equivalent atoms within a layer. In our analytic model these curves would be step functions between the first layers of either crystal, i.e. layers $-1$ and $0$. In the case of the (100) twist boundary the vast majority of the boundary expansion is indeed taken up right at the boundary plane, and there are no oscillations in the interplanar spacings as we move away from the boundary plane. But although most of the expansion is again taken up at the boundary plane in the case of the (114) boundary there are oscillations in the interplanar spacing that decay exponentially away from the boundary plane.

The oscillations in the interplanar spacing seen in Fig. 4.19 are typical of most grain boundaries not only in metals, but also in covalent and ionic solids. Indeed, oscillations do not occur only at very low index grain boundaries, such as (111), (100), and (110) in f.c.c. crystals. We shall now discuss the origin of these oscillations both geometrically and physically.

The geometrical origin of the oscillations is the local bending of planes associated with the formation of edge dislocations in the boundary, as illustrated schematically in Fig. 4.20. The local boundary plane becomes lower index, with a larger interplanar spacing, between successive edge dislocations. The local value of $\theta_{tilt}$ becomes close to that of the low-index boundary plane between successive edge dislocations. This kind of relaxation has also been called 'plane coalescence' (Crocker and Faridi 1980) because the high-index planes between successive edge dislocations coalesce and make a segment of a lower-index plane. As discussed in Section 2.10.2 the elastic strain field of the dislocations decays exponentially in the bulk, and hence the oscillations in the interplanar spacing also decay exponentially. We may conclude that the exponentially decaying oscillations are a signature of local relaxations to introduce edge dislocations into the boundary plane. Exponentially decaying oscillations are also seen frequently beneath high-index surfaces, where they are again caused by bending of high-index planes due to relaxation of steps at the surface.

The forces promoting the formation of localized edge dislocations are the cohesive forces acting across the boundary plane. These forces always exist, provided the boundary is stable against spontaneous cleavage, but they decrease as the boundary expansion increases.

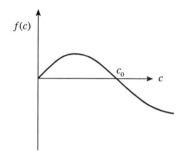

**Fig. 4.17** Schematic plot of $f(c) = c - a\exp(-z/c)$, where $a$ and $z$ are positive constants. Here, $c$ is the size of the c.n.i.d., which is inversely proportional to the period of the twist boundary structure. The function $f(c)$ is proportional to the energy of localizing screw dislocations from a reference boundary with a c.n.i.d. of size $c$, which is valid in the limit of small $c$. Note that $f(c)$ is negative only for $c > c_0$, indicating that it is energetically favourable to localize screw dislocations only from reference twist boundaries with c.n.i.d.s larger than some critical value, or, equivalently, with boundary periods smaller than some critical value.

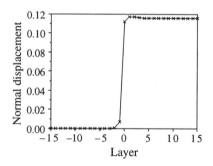

**Fig. 4.18** Normal displacements of layers (averaged over non-equivalent atoms within a layer) obtained by atomistic relaxation for the (100) twist boundary listed in Table 4.2. The geometrical boundary plane is between layers $-1$ and 0. Notice that most of the boundary expansion is taken up between layers $-1$ and 0, in agreement with the assumptions of the analytic model. Note also that there are no oscillatory normal displacements.

The form of the relaxation normal to the boundary plane may be derived analytically using the harmonic approximation. Let the expansion be taken up entirely between the first layers of either crystal, as in the analytic model of Section 4.3.1.2, but let it be such that there are residual repulsive forces, per unit area, equal to $-\partial\sigma_{\mathrm{IL}}/\partial e$ and $+\partial\sigma_{\mathrm{IL}}/\partial e$ acting on the first layers. These are the forces that generate the relaxation displacements in other layers, and in the theory of lattice statics they are sometimes called 'defect forces'. At equilibrium the relaxation displacements in other layers produce forces on the first layers which balance the defect forces. In this way some of the repulsive force between the crystal halves is balanced by forces arising from relaxation displacements within the crystal halves. Thus the expansion of the bicrystal is reduced compared with the value deduced from eqn (4.46). Following Allan (1987) and Houchmandzadeh *et al.* (1992) we use the harmonic approximation to express the energy of the bicrystal as follows:

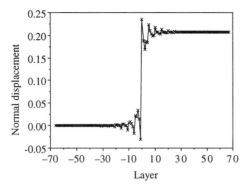

**Fig. 4.19** Normal displacements of layers (averaged over non-equivalent atoms within a layer) obtained by atomistic relaxation for the (114) twist boundary listed in Table 4.2. The geometrical boundary plane is between layers −1 and 0. Although most of the boundary expansion is taken up between layers −1 and 0 there are significant additional displacements of layers normal to the boundary plane, which are oscillatory and decay exponentially away from the boundary plane.

$$\sigma = \sigma_{\mathrm{IL}} - \sum_i f_i u_i + \frac{1}{2}\sum_i\sum_j D_{ij}u_i u_j, \tag{4.52}$$

where $\sigma_{\mathrm{IL}}$ is given by eqn (4.46). Here $u_i$ is the relaxation displacement of layer $i$ (averaged over all non-equivalent atoms in the layer) normal to the boundary. The sums are taken over all layers parallel to the boundary. The defect forces, $f_i$, are assumed to be non-zero only for the first layers of either crystal. The matrix elements $D_{ij}$ are the second derivatives $\partial^2\sigma/\partial u_i\,\partial u_j$, and for present purposes it is acceptable to assume they are the same as the corresponding layer force constants of the perfect crystal.

Minimization of $\sigma$ with respect to $u_m$ leads to

$$f_m = \sum_j D_{mj}u_j. \tag{4.53}$$

Except at the first layers $f_m$ is zero and therefore $\Sigma_j D_{mj}u_j = 0$. On the other hand, the equation for the bulk phonon spectrum is:

$$\omega^2 u_m = \sum_j D_{mj}u_j, \tag{4.54}$$

where $\omega$ is the angular frequency. We see that the solution for the relaxation displacements is a linear superposition of bulk phonons with zero frequency (i.e. $\omega^2 = 0$). Thus we seek wave vectors, $k$, normal to the boundary plane, for which the bulk phonon spectrum is zero. With the trivial exception of phonons with zero wave vector, the zeros in $\omega^2(k)$ occur at complex wave vectors $k = k_1 + ik_2$. Since we are interested only in solutions that decay away from the boundary they will be of the form:

$$u_m = \sum_{\omega^2(k_1 + k_2) = 0} C_\omega \cos\left(mk_1 d + \varphi_\omega\right)\exp\left(-mk_2 d\right), \tag{4.55}$$

where $d$ is the interplanar spacing and $mk_2 > 0$. The coefficients $C_\omega$ and the phase factors $\varphi_\omega$ are determined by fitting $u_1$ to half the boundary expansion, $e/2$. We see in

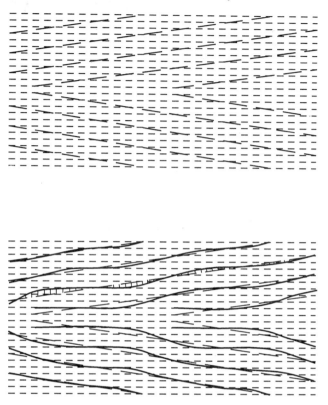

**Fig. 4.20**   To illustrate the relationship between the oscillatory relaxations in the spacing of high index planes parallel to a high index boundary and the localization of edge dislocations. The upper diagram depicts the unrelaxed boundary, which runs from left to right in the middle of both diagrams. The horizontal dashed lines represent the high index planes parallel to the boundary. The sloping dashed lines represent low index planes which enter the boundary on both sides. The lower diagram shows the relaxed configuration. The low index planes have become bent through the localization of edge dislocations. The bending of one low index plane is shown by vertical lines to reveal the elastic displacement field. It is seen that the displacements oscillate in sign as we move away from the boundary and their magnitude decays exponentially. The displacements of the high index planes are not shown for clarity, but their spacings must oscillate and decay with distance from the boundary in order to bring about the bending of the low index planes.

eqn (4.55) that the wavelength and damping of the oscillations are determined by the bulk complex phonon spectrum.

   If the interplanar spacing $d$ is sufficiently large that the force constants extend only to nearest neighbour planes then the only zero frequency phonon travelling normal to the boundary is $k = 0$. In that case there are no oscillations in the interplanar spacing and the relaxation is confined to those layers that experience defect forces. This is the case for the (100) twist boundary shown in Fig. 4.18, and other low-index boundary planes.

   The energy associated with the local relaxation displacements is readily evaluated. Substituting the equilibrium values of the displacements given by eqn (4.53) into the expression for the boundary energy eqn (4.52), we obtain:

$$\sigma = \sigma_{\mathrm{IL}} - \tfrac{1}{2}\left(f_1 u_1 + f_{-1} u_{-1}\right), \tag{4.56}$$

where $f_1$ and $f_{-1}$ are the defect forces, $-\partial\sigma_{\mathrm{IL}}/\partial e$ and $+\partial\sigma_{\mathrm{IL}}/\partial e$ respectively, and $u_1$ and $u_{-1}$ are the displacements of the first layers of either crystal, which we have labelled 1 and $-1$. The factor of $\tfrac{1}{2}$ in eqn (4.56) arises from the energy cost associated with the relaxation displacements of other layers. Using $u_1 - u_{-1} = e$, we can rewrite eqn (4.56) as follows:

$$\sigma = \sigma_{\mathrm{IL}} + \frac{e}{2}\,\partial\sigma_{\mathrm{IL}}/\partial e. \tag{4.57}$$

Comparing this equation with eqn (4.46) for $\sigma_{\mathrm{IL}}$ we see that the effect of including the relaxation displacements normal to the grain boundary in the harmonic approximation is to introduce the term $(e/2)\,\partial\sigma_{\mathrm{IL}}/\partial e$. This new term is negative provided $\partial\sigma_{\mathrm{IL}}/\partial e < 0$ (because $e$ is always greater than zero in the incommensurate limit), which is true provided the expansion is smaller than the equilibrium value predicted by $\sigma_{\mathrm{IL}}$ alone.

In Fig. 4.21 we plot the dimensionless binding functions $g(e)$ and $\{g(e) + (e/2)\,\mathrm{d}g/\mathrm{d}e\}$ for the (111) twist boundary in the incommensurate limit, using eqn (4.47). From eqn (4.50) we have seen that the binding function $g(e)$ is equal to $[\sigma_{\mathrm{IL}}(e) - \sigma_{\mathrm{s}}^{\mathrm{b}} - \sigma_{\mathrm{s}}^{\mathrm{w}}]5\Omega^2/\pi\varepsilon r_{\mathrm{o}}^4$. From eqn (4.57) we see that the binding function $\{g(e) + (e/2)\,\mathrm{d}g/\mathrm{d}e\}$ is equal to $[\sigma(e) - \sigma_{\mathrm{s}}^{\mathrm{b}} - \sigma_{\mathrm{s}}^{\mathrm{w}}]5\Omega^2/\pi\varepsilon r_{\mathrm{o}}^4$. It is seen in Fig. 4.21 that the equilibrium expansion for $\sigma$ is indeed smaller than that for $\sigma_{\mathrm{IL}}$, and that the minimum in $\sigma$ is also lower. Moreover, the stiffness of the interface to changes in expansion, about its equilibrium value, is increased when relaxation is included. However, it is found that eqn (4.57) does not yield sensible functional forms for $\sigma(e)$ for small interplanar spacings, which is because the relaxation displacements are too large to be described well by the harmonic approxmation.

The new equilibrium expansion is determined by

$$\partial\sigma/\partial e = \frac{3}{2}\,\partial\sigma_{\mathrm{IL}}/\partial e + \frac{e}{2}\,\partial^2\sigma_{\mathrm{IL}}/\partial e^2 = 0. \tag{4.58}$$

If we expand $\sigma_{\mathrm{IL}}$ in a Taylor series to second order about its equilibrium expansion $e_{\mathrm{IL}}^{\mathrm{o}}$ (given by eqn (4.51)) we can write

$$\sigma_{\mathrm{IL}} = \sigma_{\mathrm{IL}}^{\mathrm{o}} + \frac{\kappa}{2}\,(e - e_{\mathrm{IL}}^{\mathrm{o}})^2, \tag{4.59}$$

where $\kappa$ is the stiffness $(\partial^2\sigma_{\mathrm{IL}}/\partial e^2)_{e=e_{\mathrm{IL}}^{\mathrm{o}}}$. The new equilibrium expansion is obtained by solving eqn (4.58):

$$e^{\mathrm{o}} = \tfrac{3}{4}\,e_{\mathrm{IL}}^{\mathrm{o}} \tag{4.60}$$

and the new boundary energy is given by

$$\sigma = \sigma_{\mathrm{IL}}^{\mathrm{o}} - \frac{\kappa}{16}\,(e_{\mathrm{IL}}^{\mathrm{o}})^2. \tag{4.61}$$

Thus the effect of including relaxations normal to the boundary plane is to reduce the boundary expansion by 25 per cent, at least for large interplanar spacings, where the harmonic approximation applies.

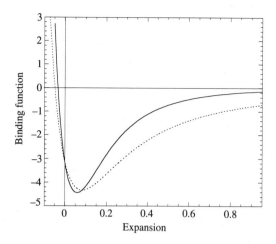

**Fig. 4.21**  The dimensionless binding function of (111) twist boundaries in the incommensurate limit, as a function of the boundary expansion in units of $r_o$. The broken line shows the binding function $g$, given by eqn (4.47), which takes into account only a rigid expansion confined between the first layers of either crystal. The solid line shows the binding function, $g + (e/2) \, dg/de$, which takes into account changes in the average interplanar spacings of all layers parallel to the boundary, within the harmonic approximation. Notice that the equilbrium expansion is smaller, and the depth of the binding function is greater, for the solid curve than the broken curve.

### 4.3.1.5  *Discussion: singular, vicinal, and general interfaces*

A complete description of the structure of a flat interface involves the specification of all atomic coordinates in the bicrystal up to some distance from the interface. Such detail is useful for testing models of specific structures, but it does not lead to an appreciation of the underlying factors controlling trends in structure and properties of interfaces. The approach we have taken in this section so far has been to identify the most significant, collective modes of relaxation using simple models that can be solved analytically as far as possible, and to test the predicted trends using computer simulations. Thus we have discussed the roles of the rigid body displacement parallel to the boundary plane, the expansion normal to the boundary plane, and correlated individual atomic relaxations parallel and perpendicular to the boundary plane. Having analysed these relaxation modes separately, we shall now bring them all together in order to classify large-angle grain boundaries and sharp heterophase interfaces as singular, vicinal or general.

Let us consider first boundaries in the incommensurate limit. We have defined an unrelaxed interface as one in which the adjoining crystals are perfect and the separation of planes on either side of the geometrical interface plane is $(d^b + d^w)/2$. In the incommensurate limit for large-angle grain boundaries we have seen that the energy of the unrelaxed state diverges to infinity as the interplanar spacings $d^w$ and $d^d$ tend to zero. By including *one* relaxation parameter, the boundary expansion, not only is the divergence in the energy eliminated but the predicted expansion and ideal cleavage energy show the same trends with the average interplanar spacing, $\langle d \rangle$, as fully relaxed computer simulations. In the incommensurate limit, it is not possible to obtain such success with any other relaxation parameters. It is for this reason that we have identified the boundary expansion as the most important relaxation parameter for the incommensurate limit.

In the incommensurate limit the analytic model described earlier predicts that the boundary expansion increases and the ideal cleavage energy decreases as $\langle d \rangle$ decreases. The physical reasons for these trends were also given. It is important to emphasize that the model does not predict a monotonic variation of the boundary energy with $\langle d \rangle$, but that it does for the ideal cleavage energy. The difference between the ideal cleavage energy and the boundary energy is, of course, the energy of the two free surfaces forming the boundary plane. For example, the spacings of $\{551\}$ and $\{171\}$ planes are equal in cubic crystals. Whereas the model predicts that (551) and (117) pure twist boundaries and (551)/(117) mixed tilt and twist boundaries in the incommensurate limit have the same ideal cleavage energy their boundary energies may all differ owing to the difference between the $\{551\}$ and $\{117\}$ surface energies.

Consider the (113) twist boundary in the incommensurate limit. The interplanar spacing (0.3015a) is just over half the maximum interplanar spacing found in f.c.c. crystals. The (10, 10, 31) plane is only 0.72° from the (113) plane, but its spacing is only 4.9 per cent of that of the (113) plane. The analytic model predicts that the expansion and cleavage energy of the (10, 10, 31) twist boundary in the incommensurate limit are radically different from those of the (113) boundary. The closer we get to the (113) plane, such as (100, 100, 301), (1000, 1000, 3001), the greater the discontinuous change in predicted expansion and cleavage energy. We see that the analytic model predicts discontinuous changes in the expansion and ideal cleavage energy at all rational boundary planes! What has gone wrong here is that in the analytic model the high index planes near (113) are only allowed to be displaced rigidly normal to the boundary plane. It does not take into account local relaxation which bends the slightly non-parallel (113) planes on either side of the boundary into parallel alignment, so that most of the local boundary plane is parallel to (113) with the deviation from the exact (113) orientation localized at edge dislocations. This local relaxation was discussed in detail in Section 4.3.1.4.

Local, individual atomic relaxation eliminates the discontinuities in the expansion and cleavage energy that are predicted by the analytic model at low index boundaries in the incommensurate limit. At the same time a lower limit on the average interplanar spacing, $\langle d \rangle$, must be imposed for the analytic model to remain valid. Boundaries with interplanar spacings smaller than this lower limit are then regarded as deviations from one (or more) of the boundaries above the limit. This limit, though somewhat arbitrary, is crucial because it acknowledges the importance of individual atomic relaxation, without which we have discontinuities. The trends that follow from the analytic model apply only to those boundaries with $\langle d \rangle$ greater than the lower limit.

The introduction of a lower limit on the average interplanar spacing separates boundaries in the incommensurate limit into three kinds. In the first are boundaries with $\langle d \rangle$ greater than the lower limit, which are local minima in the boundary energy as a function of changes of $\theta_{\text{tilt}}$. These boundaries are examples of 'singular' interfaces. The remainder are related to one or more of these local minima through local relaxations involving the introduction of arrays of edge dislocations. If it is close in misorientation space to just one singular boundary then it is called 'vicinal', which means literally that it is in the vicinity of a singular interface. Otherwise, if the relaxation is as much affected by the proximity of two or more singular interfaces then it is called 'general'.

Let us now consider boundaries that are not at the incommensurate limit, but at some value of $\theta_{\text{twist}}$ where there is 1D or 2D periodicity in the boundary plane. Using eqns (4.40) and (4.44) the boundary energy is given by:

$$\sigma = \sigma_{IL} + \sum_{G^c \neq 0} \tilde{V}(G^c, e) e^{i G^c \cdot \tau} \tag{4.62}$$

where

$$\tilde{V}(G^c, e) = \frac{1}{A_c^b A_c^w} \sum_{l^w = 1}^{\infty} \sum_{l^b = 1}^{\infty} \tilde{v}(G^c, (l^b d^b + l^w d^w - (d^b + d^w)/2 + e)). \tag{4.63}$$

At this stage the only relaxation that is allowed is the expansion, $e$, and the rigid body displacement, $\tau$, parallel to the boundary. In eqn (4.62) it is clear that the presence of 1D or 2D periodicity in the boundary plane introduces additional terms in the boundary energy. Because these terms decay exponentially with $|G^c|z$, where $z$ is the smallest separation of layers on either side of the boundary, only the smallest common reciprocal lattice vectors will make a significant contribution. It is then possible to choose $\tau$ in such a way that the significant new terms give an attractive contribution to the boundary energy. Thus we see that $\tau$ is an important mode of relaxation whenever there is 1D or 2D periodicity. Note also that periodicity will also modify the predicted expansion compared with the incommensurate limit because the new terms on the right of eqn (4.62) are also functions of $e$. Therefore the expansion does not depend only on $d^b$ and $d^w$ but also on periodicity in the boundary plane. Similarly, the ideal cleavage energy and the boundary energy itself do not depend on only $d^b$ and $d^w$ because they are also affected by new terms arising from the presence of periodicity on the right of eqn (4.62). It follows that the monotonic relation between the ideal cleavage energy and the boundary expansion obtained for the incommensurate limit breaks down when periodicity is present in the interface.

Equation (4.62) explains why the unrestricted form of the planar coincidence site density criterion (Brandon *et al.* 1964) fails (Sutton and Balluffi 1987). By not restricting the criterion to boundaries sharing the same plane we see that the term $\sigma_{IL}$ is not cancelled for boundaries on different planes. Differences in $\sigma_{IL}$ may easily outweigh any differences arising from the periodicities in the boundary planes. But by limiting the criterion to boundaries sharing the same plane (Wolf 1985) the term $\sigma_{IL}$ is cancelled and differences in $\sigma$ arise only from the Fourier coefficients $\tilde{V}(G^c, e)$.

Sutton and Balluffi (1987) found that the criterion for low boundary energy of maximizing $\langle d \rangle$ (Wolf 1985) failed as often as it succeeded when it was tested against the body of experimental data. There are three reasons. First, as we have repeatedly emphasized, it is the ideal cleavage energy, not the boundary energy itself, which is expected to correlate with $\langle d \rangle$. Secondly, any correlation with $\langle d \rangle$ may be obscured by the presence of periodicity which will lower $\sigma$ for a given $\langle d \rangle$; therefore, the correlation with $\langle d \rangle$ should be sought only among boundaries in the incommensurate limit. Thirdly, any criterion based on $\langle d \rangle$ should be restricted to singular, not vicinal, interfaces, which implies that a lower limit on $\langle d \rangle$ must be imposed before the criterion is applied either to the boundary energy or the ideal cleavage energy. The first two objections may also be raised to the proposal that low interfacial energy should be correlated with a small expansion. Indeed, results of computer simulations (Wolf and Merkle 1992) indicate some degree of correlation between grain boundary energy and expansion, but there is considerable scatter, as shown in Fig. 4.22.

Nevertheless, there is more recent experimental evidence in support of the criterion that the lowest possible interfacial energies in a system are associated with the highest possible values of $\langle d \rangle$. This evidence comes from observations of faceting of grain boundaries (Ichinose and Ishida 1985, Dahmen and Westmacott 1988, Krakow and Smith 1986b, Merkle 1991) and heterophase interfaces surrounding precipitates (Necker and Mader 1988, Merkle 1991, Mader and Maier 1990, Mader 1989, Ernst *et al.* 1989, Merkle 1990, Mader and Necker 1990). By restricting the criterion to no more than the highest available

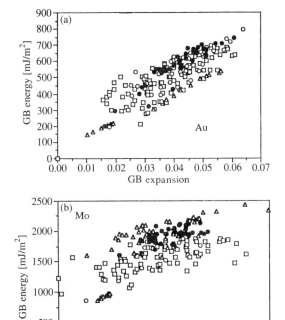

**Fig. 4.22** The energies of relaxed grain boundaries in (a) Au and (b) Mo as a function of their expansion, as computed by Wolf and Merkle (1992). Squares denote symmetrical tilt boundaries, triangles denote pure twist boundaries, open circles denote asymmetrical tilt boundaries and filled circles denote mixed tilt and twist boundaries. The boundary expansion is in units of the crystal lattice parameter. In (a) an embedded atom potential for Au was used (see Section 3.7). In (b) a Finnis–Sinclair potential for Mo was used (see Section 3.8.6). Note that as the boundary expansion increases the boundary energy tends to increase but there is considerable scatter.

values of $\langle d \rangle$ we do not make the mistake of applying it to vicinal interfaces. By 'highest available' the experimental results indicate that in some cases we should consider no more than the 3 or 4 highest possible values of $d^b$ or $d^w$. Moreover, these planes often coincide with the lowest available surface energies, and thus the difference between the ideal cleavage energy and the interfacial energy does not affect the correlation. Finally, planes associated with the highest available $\langle d \rangle$ will also be the densest packed and may therefore contain the smallest possible common reciprocal lattice vectors. Therefore, the reduction in the interfacial energy arising from any periodicity (or elastically forced periodicity) is likely to be greater in these interfaces than any others. Thus, not only can the above experimental observations be explained, but also the failure of the high $\langle d \rangle$ criterion for low interfacial energy in less restrictive applications can also be understood.

Returning to eqn (4.62), just as we found discontinuities in our analytic model for boundaries in the incommensurate limit so we find discontinuities of a different kind in the boundary energy as a function of $\theta_{\text{twist}}$. According to eqn (4.62) there are additional terms, $G^c \neq 0$, only if the boundary displays periodicity. But if the value of $\theta_{\text{twist}}$ deviates infinitesimally so that formally the periodicity is destroyed the boundary energy becomes equal to $\sigma_{\text{IL}}$. As we discussed previously this is another consequence of ignoring local, individual atomic relaxation. In this case correlated atomic relaxations introduce an array of screw dislocations which ensure that much of the boundary may continue to benefit from the nearby common reciprocal lattice vectors. We also gave a physical argument for why there is an upper limit on the period of a boundary that can sustain an array of screw dislocations. Therefore the effect of admitting individual atomic relaxations is again to divide boundaries into three classes. The first comprises boundaries that correspond to local minima in the energy as a function of $\theta_{twist}$. These are the boundaries displaying 1D or 2D periodicity, and they are also called singular. Sometimes exact periodicity can not exist at any value of $\theta_{twist}$, but there may still be local minima in $\sigma(\theta_{twist})$ where there are relatively small layer reciprocal lattice vectors on either side of the interface that are forced to be equal through elastic distortions (as in the near coincidence model discussed in Section 2.2). Such cases would also be classified as singular. Boundaries with values of $\theta_{twist}$ sufficiently close to one singular boundary that the relaxation is dominated by the proximity of the singular boundary are called vicinal. Finally, boundaries in which the relaxation of the local value of $\theta_{twist}$ is equally influenced by two or more (or no) singular boundaries are called general.

We have defined a singular interface as one that is at a local minimum in $\sigma$ with respect to variations in $\theta_{tilt}$, or $\theta_{twist}$. It is quite possible for a boundary to be singular with respect to $\theta_{tilt}$ but general with respect to $\theta_{twist}$. For example all twist boundaries on low index planes are singular with respect to $\theta_{tilt}$, but in the incommensurate limit they are general with respect to $\theta_{twist}$. Thus, when we classify a boundary as singular we must also be clear with respect to which variations.

In the literature on grain boundaries one frequently finds our 'singular' boundaries classified as 'special'. We have avoided the word 'special' because its usage has become so wide than we believe it has little meaning. For example, a boundary which is at a local extremum of *any* property such as energy, fracture toughness, diffusivity, propensity for segregation, migration rate, sliding rate, corrosion rate, etc., has been called special. Our usage of 'singular' is intended to convey that the boundary is at a local minimum in the energy with respect to variations in any of the five macroscopic geometrical degrees of freedom. We believe this usage is the same as in surface science where a singular surface corresponds to a local minimum in the energy as a function of the two macroscopic degrees of freedom characterizing the surface normal. Similarly, the usage of 'vicinal', which was first proposed by Wolf (1990*b*), follows the terminology of surface science.

Local relaxations at interfaces are not confined to local changes in the tilt and twist misorientations. In Section 4.3.1.9 faceting and dissociation will be discussed. Local relaxations to maximize the local coordination number in metals are described in Section 4.3.1.7.

The significance of a singular interface is that, being at a local minimum in the energy with respect to one (or more) of the macroscopic geometrical degrees of freedom, it will resist changes to that degree of freedom. This resistance will result in localized defects if such changes are imposed on the interface. For example, an interface with a large value of $\langle d \rangle$ will resist changes in $\theta_{tilt}$ and changes in the boundary inclination. The former will result in localized dislocations with $b$ normal to the boundary plane to localize changes in $\theta_{tilt}$. Changes in the boundary inclination will result in localized steps, each of which may be associated with a dislocation. However, the Frank–Bilby equation requires that the net Burgers vector be zero if there is no change in the orientation relation or structure of the adjoining crystals. Similarly, interfaces displaying 1D or 2D periodicity, or elastically forced periodicity, will resist changes in $\tau$ and $\theta_{twist}$ provided the smallest $|G^c|$ are less than some upper limit (see Section 4.3.1.4). Such interfaces will localize dislocations with Burgers vectors parallel to the interface. Conversely, if the interfacial energy is insensitive to changes in $\tau$ then, *in the ground state*, it will not support localized dislocations with $b$ parallel to the interface. It is emphasized that this remark applies to such an interface only if it is able to relax to the ground state, which may involve passing through thermally activated states. On the other hand, any interface, whether it be singular, vicinal, or general, will support localized dislocations with $b$

normal to the interface because there are always cohesive forces acting across the interface to localize the Burgers vector density. However, these forces are weaker the greater the boundary expansion. When the crystal lattice planes parallel to the boundary are high index in both crystals, the Burgers vectors of these dislocations will be very small when the boundary is in its ground state, and, therefore, they will be of little physical significance.

The conclusions and classification we have discussed in this section will be used frequently in later chapters. They are based on an analysis which has ignored the influence of temperature and the fact that we should be considering the boundary free energy rather than the internal energy at absolute zero. The effect of temperature will be considered in Section 4.3.1.10. Moreover, our conclusions and classification, which have been formulated for large-angle grain boundaries in 'Lennard–Jonesium', will be applied to all sharp interfaces in both homophase and heterophase systems! This may seem unjustified in view of the wide variety of atomic interactions that we have not considered. But the conclusions we have reached have been based on essentially two fundamental points, which hold for all interfaces. The boundary expansion in the incommensurate limit is caused by the *repulsion* generated by atomic overlap in the unrelaxed state: the greater the overlap, the greater the tendency to expand. The expansion is limited by the work done against the *attraction* across the interface. All interfaces (that are stable with respect to cleavage) will share these basic features of a short-range repulsion arising from atomic overlap and attractive, cohesive forces acting across the interface. Similarly, whatever the detailed nature of the atomic interactions, if an interface displays 1D or 2D periodicity an equation of the form of eqn (4.62) must apply, although the Fourier coefficients $\tilde{V}(G^c, e)$ will not have the simple form of eqn (4.63). Thus our conclusions about the significance of $\tau$, and the resistance to changes of $\tau$, apply to any interface that is periodic (or forced elastically to be periodic) in at least one direction.

We are aware that the distinctions we have made between singular, vicinal, and general interfaces are arbitrary to some extent, because we have not specified the lower limit on $\langle d \rangle$, or the upper limit on $|G^c|$, at which an interface ceases to be singular. These limits are determined by local relaxations and therefore they may be expected to differ from one system to another. However, as we shall see in later chapters, the arbitrariness does not diminish the usefulness of the classification for understanding a wide variety of properties of interfaces, in the same way as the distinction between singular and vicinal surfaces is so useful for understanding surface properties, and is equally arbitrary.

### 4.3.1.6 *Methods of computer simulation*

There are three principal methods of computer simulation that have been applied to interfaces: energy minimization, molecular dynamics, and Monte Carlo. The atoms of a bicrystal in some initial configuration are represented by their nuclear coordinates in a computational cell. The interface is usually flat, although it may be curved or faceted provided it satisfies the border conditions that are applied to the surface of the computational cell. We begin by describing these border conditions, which present quite unique problems for modelling interfaces owing to the rigid body displacement, $t$, of one crystal relative to the other.

A popular strategy, which is particularly suitable for energy minimization simulations, is to divide the cell into two regions as shown in Fig. 4.23. The interface is contained in region I and all atoms are allowed to move to any position within the region, according to the forces acting on them. The embedding of region I in an infinite bicrystal is simulated by region II, which comprises two slabs of perfect crystal above and below region I. The slab below region I is usually fixed in position as a reference. All atoms

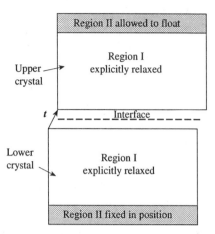

**Fig. 4.23**  To illustrate the region I–region II boundary conditions which have been used widely for computer simulations of interfaces. The atoms in region I are explicitly relaxed. The atoms in region II are held in perfect crystal positions. The rigid body displacement, $t$, of the upper crystal with respect to the lower crystal is established within region I. The upper region II is allowed to float so as not to restrict $t$.

in the slab above region I may be displaced rigidly parallel and perpendicular to the interface. The displacement of the upper slab is the rigid body translation, $t$, of the bicrystal. Atoms in region I interact with each other and with atoms in region II. It is assumed that region I is sufficiently large normal to the interface that holding the atoms of region II in perfect crystal positions does not affect the relaxation of region I. It is also assumed that the thickness of the region II slabs is greater than *twice* the range of interatomic forces. The reason why this is necessary is explained below. Born–von Karman periodic border conditions are applied to the four faces of the computational cell normal to the interface. Thus one (or more) repeat cell(s) of an interface that is periodic in two dimensions is (are) simulated. If there are $N$ atoms in region I then the energy is a function of $3N + 3$ variables, the last three being the components of $t$ for the upper region II slab.

The disadvantage with using the border conditions shown in Fig. 4.23 for a dynamical simulation of the bicrystal at a finite temperature is that the atoms in region II remain frozen, although the slabs may be displaced or even vibrate rigidly. Therefore, there are two further, artificial interfaces introduced between region I and the two region II slabs, where vibrating atoms in region I meet frozen atoms of region II. It is found that the lattice parameter assumed for the perfect crystal in region II must reflect the thermal expansion of the perfect crystal. Otherwise considerable misfit stresses may be set up at the interfaces between regions I and II. In view of these problems some authors prefer to omit region II altogether and let the upper and lower surfaces of region I be free. This second kind of border condition has been used for dynamical and energy minimization simulations. However, in both cases the size of region I normal to the interface has to be sufficiently large that either the frozen slabs or the free surfaces do not affect the dynamics at the interface.

The border conditions we have described so far are suitable for descriptions of inter-atomic forces that are based on interatomic potentials of some form or another. But in an electronic structure calculation the embedding of region I becomes much more critical.

The upper and lower free surfaces of region II in Fig. 4.23 may be associated with surface electronic states which can interact with electronic states we may wish to study at the interface. For example, in real-space electronic structure calculations, such as the recursion method, the border conditions shown in Fig. 4.23 may be used provided the thickness of the region II slabs is greater than the distance an electron can hop from region I. The rigorous way to deal with this problem is to embed the finite bicrystal of region I in two semi-infinite regions II by using Green function methods (Pollmann and Pantelides 1980). However, for the reasons we have already discussed, this is not suitable for a dynamical simulation.

The most commonly employed border conditions for dynamical simulations are Born–von Karman periodic border conditions both parallel and pendicular to the interface, as shown in Fig. 4.24. In general two interfaces are now required in the computational cell in order for the top and bottom surfaces to marry onto each other. The relaxations of the two interfaces are independent. Let $t_1$ and $t_2$ be their rigid body displacements. Since $t_1 \neq -t_2$ in general the shape and size of the computational cell will change. The Parrinello–Rahman method of molecular dynamics (Parrinello and Rahman 1981) has been used to deal with this problem, where the size and shape of the cell fluctuate dynamically according to the resultant stresses acting within it.

Our final remark about border conditions is that, for certain kinds of interfaces, modified periodic border conditions in 3D may be used with only *one* interface in the computational cell. Displaced periodic border conditions (Allen and Tildesley 1987) may be used to model stacking faults as shown in Fig. 4.25. In this case the rigid body displacement, $t$, is the fault vector and successive cells normal to the fault are displaced by $t$. The displacement of successive cells by $t$ is an alternative to changing the shape of the computational cell in the Parrinello–Rahman method. Parallel to the fault usual Born–von Karman periodic border conditions are applied. Helical periodic border conditions may be used to model twist grain boundaries (O. Hardouin Duparc, private communication 1993). Again, parallel to the interface usual Born–von Karman periodic border conditions are applied. Normal to the interface *all* atoms in successive cells undergo a rotation of $\theta_{\text{twist}}$ about the interface normal and a translation, $t$, equal to the rigid body displacement of the interface, as shown in Fig. 4.26. The rotation is in addition

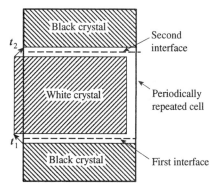

**Fig. 4.24** To illustrate the application of Born–von Karman periodic boundary conditions to modelling interfaces. Each computational cell, shown in bold, is repeated periodically in all three directions. There are two interfaces in the computational cell, for which the rigid body displacements, $t_1$ and $t_2$, are generally independent. The size and shape of the computational cell change as $t_1$ and $t_2$ vary during the simulation.

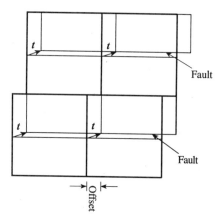

**Fig. 4.25** To illustrate the application of displaced periodic boundary conditions to model a stacking fault characterized by a rigid body displacement $t$. The diagram shows four computational cells in bold. In each cell there is one fault running from left to right in the middle of the cell. The computational cell is repeated periodically in all three directions, but in the direction normal to the fault successive cells are displaced horizontally by the offset indicated. In this way the shape of the cell does not change, in contrast to the periodic boundary conditions shown in Fig. 4.24. However, if the fault vector, $t$, has a component normal to the fault plane (i.e. an expansion) then the computational cell must be allowed to lengthen normal to the fault plane to eliminate any normal stresses in this direction. Notice that the crystal phases on either side of the fault must be identical.

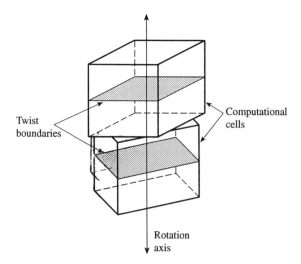

**Fig. 4.26** To illustrate the application of helical periodic boundary conditions to twist grain boundaries (O. Hardouin Duparc, private communication 1993). The diagram shows two computational cells (outlined in bold) each containing only *one* twist boundary (shaded). Within each computational cell the upper half is rotated with respect to the lower by $\theta_{\text{twist}}$ about the rotation axis. Successive computational cells along the rotation axis are also rotated rigidly by $\theta_{\text{twist}}$. This ensures that the upper crystal half of the lower computational cell has the same orientation as the lower crystal half of the upper computational cell. In addition, successive computational cells along the rotation axis may be rigidly displaced by $t$ to allow a rigid body displacement between the two crystal halves of each cell. Usual Born von Karman periodic boundary conditions are applied to the four faces of the computational cell normal to the rotation axis.

to the relative rotation by $\theta_{\text{twist}}$ about the interface normal of one crystal half with respect to the other within each bicrystal. Thus a helix of bicrystals is generated where succesive bicrystals are rotated by $\theta_{\text{twist}}$ and translated by $t$. If $x$ and $y$ are parallel to the interface and $z$ is normal to the interface then the coordinates $(x_i, y_i, z_i)$ of an atom in one cell become

$$\begin{bmatrix} x_i \\ y_i \\ z_i \end{bmatrix} \rightarrow R \begin{bmatrix} x_i \\ y_i \\ z_i \end{bmatrix} + \begin{bmatrix} t_x \\ t_y \\ t_z + c_z \end{bmatrix} \tag{4.64}$$

where

$$R = \begin{bmatrix} \cos \theta_{\text{twist}} & -\sin \theta_{\text{twist}} & 0 \\ \sin \theta_{\text{twist}} & \cos \theta_{\text{twist}} & 0 \\ 0 & 0 & 1 \end{bmatrix} \tag{4.65}$$

in the next cell along $z$. Here $c_z$ is the length of the cell along $z$. For example, for a symmetric tilt boundary, for which $\theta_{\text{twist}} = 180°$, the transformation relating atomic coordinates in successive cells along the interface normal is $(x_i, y_i, z_i) \rightarrow (-x_i + t_x, -y_i + t_y, z_i + c_z + t_z)$. If $\theta_{\text{twist}} = 0$ the helical border conditions become the displaced periodic border conditions shown in Fig. 4.25. Helical border conditions cannot be used if the boundary is not a pure twist boundary, such as an asymmetric tilt boundary or a mixed tilt and twist boundary, or an inversion domain boundary.

In an energy minimization simulation the internal energy or enthalpy of the bicrystal is minimized 0 K. If the total volume of the computational cell is kept constant then it is the internal energy that is being minimized. Any expansion at the interface, which must always be normal to the interface owing to the periodicity parallel to the interface, has to be compensated by an elastic compression in the remaining bicrystal. If a net expansion or contraction of the computational cell normal to the interface is allowed then it is the enthalpy that is being minimized, and there is no net normal stress on any plane parallel to the interface. We call such simulations zero normal stress simulations. However, as in the case of a free surface, the average stress components parallel to the interface, $f_{xx}$, $f_{yy}$ and $f_{xy}$ are not zero in general. They are the components of the interface stress tensor, discussed in Section 5.4. Energy minimization simulation techniques have also been applied to the direct minimization of the free energy of a grain boundary at a finite temperature, where the function that is being minimized is an approximate free energy functional of the atomic coordinates, as described in Section 3.9. Whether it is the internal energy or enthalpy at 0 K, or the free energy at a finite temperature that is minimized, each iteration of the simulation involves the evaluation of appropriate derivatives of the function with respect to the atomic coordinates. These derivaties are used to minimize the function by a range of standard numerical algorithms such as steepest descents, conjugate gradients, or variable metric methods (see Press *et al.* 1989).

Molecular dynamics techniques have been applied to grain boundaries to simulate phenomena such as diffusion and grain boundary sliding and migration, and to study structural phase transformations at interfaces. Each atom now has kinetic energy as well as potential energy, and its motion is determined by the numerical integration of Newtonian equations. In each iteration or 'time-step' the changes in the position and velocity of the atom are computed using the force acting on the particle. In addition, all the atomic velocities may be scaled in each iteration by some factor to simulate the immersion of the bicrystal in a heat bath. This is a means of controlling and varying the temperature in the system. The Parrinello–Rahman technique also allows an external stress to be applied to the system by changing the shape and size of the computational cell.

The Monte Carlo method is used to generate equilibrium atomic structures at a finite temperature, and it is particularly useful for modelling segregation of impurities to interfaces at constant chemical potential, as discussed in Section 7.5. In this section we shall discuss the application of the Monte Carlo method to elemental solids. In that case the Monte Carlo method simulates annealing of the specimen at a chosen temperature. In each Monte Carlo step an atom is picked at random and it is displaced by a random amount, up to some preset limit, in a random direction. If the move lowers the potential energy of the system the move is accepted. The new configuration then contributes to the averaging which is performed at the end of the simulation. If the move raises the potential energy by $\Delta V$ the move is still accepted with probability $\exp -(\Delta V/kT)$. This is achieved by comparing $\exp -(\Delta V/kT)$ with a random number generated between 0 and 1. If $\exp -(\Delta V/kT)$ is greater than the random number then the move is accepted, otherwise the atom is returned to its original position and the original configuration contributes again to the averaging. At the end of the simulation the average position of each atom is computed from all the accepted moves. At the same time other configurational averages may be computed, such as the average potential energy.

The final topic we shall consider in this section is the evaluation of the rigid body displacement, $t$, with the border conditions shown in Fig 4.23. We shall also indicate how the same method may be applied to the helical periodic border conditions shown in Fig. 4.26. Let the $L$ layers of region I parallel to the interface be numbered $1, 2, 3, \ldots L$ from the bottom to the top. In layer $l$ let there be $n_l$ atoms. Layer $L + 1$ is the first layer of the upper region II slab. The point at issue is this: the rigid body displacement is a *collective* mode of relaxation in the sense that a whole crystal is translated rigidly according to forces that act very locally at the interface. It is quite different, therefore, from the relaxation of individual atoms arising from forces acting on them alone. In the following we will see how the collective nature of the relaxation is made explicit by transforming the variables of the relaxation.

Let the coordinates of atom $i$ in layer $l$ be $(x_{li}, y_{li}, z_{li})$. We define variables $\bar{x}_l$ by

$$\bar{x}_l = \sum_{i=1}^{n_l} x_{li} \qquad (4.66)$$

and similar formulae for $\bar{y}_l$ and $\bar{z}_l$. Here, $1 \leqslant l \leqslant L$. For the region II slab above region I we define:

$$\bar{x}_{L+1} = \sum_{l>L} \sum_{i=1}^{n_l} x_{li} \qquad (4.67)$$

and similar formulae for $\bar{y}_{L+1}$ and $\bar{z}_{L+1}$. The sum over $l$ in eqn (4.67) is over all layers in region II. We also define 'differential displacement' variables for each layer as follows:

$$X_l = \bar{x}_l - \bar{x}_{l-1}$$

with

$$X_1 = \bar{x}_1 \qquad (4.68)$$

and similar formulae for $Y_l$ and $Z_l$, where $1 \leqslant l \leqslant L + 1$. It follows that

$$\bar{x}_l = \sum_{i=1}^{l} X_i. \tag{4.69}$$

The $x$-component of the rigid body translation, $t$, of the upper region II slab is given by

$$n_{\mathrm{II}} t_x = \Delta \bar{x}_{L+1} = \sum_{i=1}^{L+1} \Delta X_i, \tag{4.70}$$

where $n_{\mathrm{II}}$ denotes the number of atoms in the slab and $\Delta s$ denotes the change in $s$ after relaxation.

The $3n_l$ relaxation variables associated with layer $l$ of region I are now chosen to be $X_l$, $Y_l$, $Z_l$, $u_{l2}$, $v_{l2}$, $w_{l2}$, ..., $u_{ln_l}$, $v_{ln_l}$, $w_{ln_l}$, and for region II they are just the three variables $X_{L+1}$, $Y_{L+1}$, $Z_{L+1}$. Here, $u_{li}$, $v_{li}$, and $w_{li}$ are defined by

$$\left.\begin{aligned} u_{li} &= x_{li} - \bar{x}_l \\ v_{li} &= y_{li} - \bar{y}_l \\ w_{li} &= z_{li} - \bar{z}_l \end{aligned}\right\}. \tag{4.71}$$

Consider $\partial E / \partial X_l$, where $E$ is the energy. Using eqns (4.66) and (4.69) and the chain rule it may easily be shown that

$$\frac{\partial E}{\partial X_l} = \sum_{j \geqslant l} \sum_{i=1}^{n_j} \frac{\partial E}{\partial x_{ji}}. \tag{4.72}$$

Thus, the force that is conjugate to the differential displacement variable $X_l$ is equal to the sum of the forces acting on atoms in layer $l$ and all atoms in layers $j$ above layer $l$. To evaluate this force we draw an imaginary cut between the atoms of layers $l$ and $l-1$ and sum the forces acting on all atoms above the cut due to all atoms below it. In particular, $\partial E / \partial X_{L+1}$ is equal to the sum of the forces acting on the atoms of the upper region II slab due to atoms in region I. Therefore, it is essential that the thickness of the slab be greater than twice the range of the interatomic forces to ensure that those atoms in region II that are interacting with region I are unaware of the free surface at the top of the slab. From eqn (4.69) we see that the change in the layer variable $\bar{x}_l$ is equal to the sum of the changes of the differential displacements in all layers below it. Thus, the cumulative and collective nature of the rigid body displacement is made explicit through the use of differential displacement variables.

In order for non-equivalent atoms within a layer to undergo relaxations in addition to the rigid displacement of the whole layer, it is necessary to use the derivatives $\partial E / \partial u_{li}$, $\partial E / \partial v_{li}$, and $\partial E / \partial w_{li}$. Using eqn (4.66) and (4.71) and the chain rule it is straightforward to show that

$$\frac{\partial E}{\partial u_{li}} = -\sum_{\substack{j=1 \\ j \neq i}}^{n_l} \frac{\partial E}{\partial x_{lj}}. \tag{4.73}$$

Given the changes in the differential variables and $u_{li}$ the change in the atomic coordinate $x_{li}$ is given by

$$\Delta x_{li} = \Delta \bar{x}_l + \Delta u_{li}$$

$$= \sum_{i=1}^{l} \Delta X_i + \Delta u_{li} \tag{4.74}$$

and similar formulae hold for $\Delta y_{li}$ and $\Delta z_{li}$.

If the helical or displaced periodic border conditions are used the same transformed variables may be used. However, layers greater than $L$ are now in the next computational cell. Atomic coordinates within layer $L + l$ are related to those in layer $l$ (where $1 \leqslant l \leqslant L$) by eqn (4.64):

$$\begin{bmatrix} x_{L+l,i} \\ y_{L+l,i} \\ z_{L+l,i} - c_z \end{bmatrix} = \begin{bmatrix} \cos\theta_{\text{twist}} & -\sin\theta_{\text{twist}} & 0 \\ \sin\theta_{\text{twist}} & \cos\theta_{\text{twist}} & 0 \\ 0 & 0 & 1 \end{bmatrix} \begin{bmatrix} x_{li} \\ y_{li} \\ z_{li} \end{bmatrix} + \begin{bmatrix} t_x \\ t_y \\ t_z \end{bmatrix}. \tag{4.75}$$

Instead of eqn (4.67) we define variables $\bar{x}_{L+l}$ in the next computational cell in an analogous manner to $\bar{x}_l$ of eqn (4.66):

$$\bar{x}_{L+l} = \sum_{i=1}^{n_l} x_{L+l,i}. \tag{4.76}$$

The relaxation variables are $X_l$, $Y_l$, $Z_l$, $u_{l2}$, $v_{l2}$, $w_{l2}$, ..., $u_{ln_l}$, $v_{ln_l}$, $w_{ln_l}$ for each of the $L$ layers in the computational cell, plus $X_{L+1}$, $Y_{L+1}$, $Z_{L+1}$, making $3N + 3$ variables altogether, where $N$ is the number of atoms in the cell. As before, $X_{L+1} = \bar{x}_{L+1} - \bar{x}_L$, etc. The rigid body displacement is obtained

$$\Delta\bar{x}_{L+1} = \sum_{l=1}^{L+1} \Delta X_l$$

and

$$\begin{bmatrix} t_x \\ t_y \\ t_z \end{bmatrix} = \begin{bmatrix} \Delta\bar{x}_{L+1} \\ \Delta\bar{y}_{L+1} \\ \Delta\bar{z}_{L+1} \end{bmatrix} - \begin{bmatrix} \cos\theta_{\text{twist}} & -\sin\theta_{\text{twist}} & 0 \\ \sin\theta_{\text{twist}} & \cos\theta_{\text{twist}} & 0 \\ 0 & 0 & 1 \end{bmatrix} \begin{bmatrix} \Delta x_1 \\ \Delta y_1 \\ \Delta z_1 \end{bmatrix}. \tag{4.77}$$

We note that dynamics may be incorporated straightforwardly with the above relaxation variables and helical periodic border conditions. For example, the equations of motion for the variables $X_l$ and $u_{li}$ are obtained from eqns (4.66), (4.68), and (4.71) as follows:

$$\frac{\partial^2 X_l}{\partial t^2} = \frac{\partial^2 \bar{x}_l}{\partial t^2} - \frac{\partial^2 \bar{x}_{l-1}}{\partial t^2}$$

$$= \sum_{i=1}^{n_{l-1}} \frac{1}{m_{l-1,i}} \frac{\partial E}{\partial x_{l-1,i}} - \sum_{i=1}^{n_l} \frac{1}{m_{l,i}} \frac{\partial E}{\partial x_{li}} \tag{4.78}$$

and

$$\frac{\partial^2 u_{li}}{\partial t^2} = \sum_{\substack{j=1 \\ j \neq i}}^{n_l} \frac{1}{m_{lj}} \frac{\partial E}{\partial x_{lj}} \tag{4.79}$$

where $m_{li}$ is the mass of atom $i$ in layer $l$. The positions of the particles at any given instant are then obtained from eqn (4.74) and $t$ from eqn (4.77).

### 4.3.1.7 *The polyhedral unit model*

In this and the following section we consider the atomic structure of the core of grain boundaries in metals in greater detail. In the polyhedral unit model, the boundary core is described in terms of arrays of polyhedra consisting of closely packed clusters of atoms.

As we have discussed in Chapter 3, metallic bonding favours dense packing. In the second moment approximation (see Section 3.8.6) the cohesive bond energy varies as the square root of the local coordination number. Therefore, by maximizing the local coordination number, and avoiding repulsive atomic overlap, the energy of the system is minimized. This is the physical basis on which the polyhedral unit model rests.

The requirements of dense packing and small atomic overlap limit the number of possible polyhedra, the edges of which are formed by lines joining the centres of neighbouring atoms, to just eight (Ashby *et al.* 1978). Five of these were found by Bernal (1964) in hard-sphere models of liquid structures. The atom centres are the vertices of convex deltahedra: polyhedra whose faces are equilateral triangles. Ashby *et al.* argued that any grain boundary between metallic crystals could be described as a 2D packing of deltahedra. Independently, Pond *et al.* (1978) proposed that certain polyhedral clusters of atoms appeared in grain boundaries in metals. Figure 4.27 shows the polyhedra that have been found in computer simulations. Note that an f.c.c. perfect crystal is made up of octahedra and tetrahedra.

The difference between the formulations of the model by Ashby *et al.* and Pond *et al.* centres on the treatment of distortions of the polyhedra. The clusters are invariably distorted owing to the requirements of compatibility with surrounding atoms. The limit on the acceptable distortions imposed by Ashby *et al.* is given by the condition that another atom cannot be inserted into the hole at the centre of the polyhedron. This condition is generous and allows large distortions of bond lengths. If a grain boundary is described as a contiguous arrangement of highly distorted deltahedra we might question the information content of such a description. Pond *et al.* limited the maximum acceptable distortion to 15 per cent of an edge length from the ideal nearest neighbour spacing. But the consequence of this more stringent limit is that in general the require-

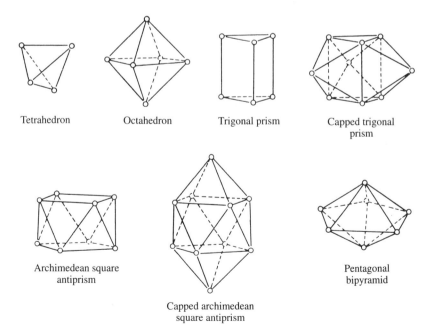

Tetrahedron     Octahedron     Trigonal prism     Capped trigonal prism

Archimedean square antiprism     Capped archimedean square antiprism     Pentagonal bipyramid

**Fig. 4.27** Polyhedral clusters of atoms that have been identified in computer simulations of grain boundaries in metals (Vitek *et al.* 1980).

ments of filling space and compatibility with the adjoining grains cannot be satisfied by configurations composed wholly of such polyhedra.

Figure 4.28 shows an HREM image of a $\Sigma = 5$ [001] curved tilt boundary in Au (Krakow 1991). It is seen that local relaxations have introduced capped trigonal prisms. As noted by Ashby *et al.*, capped trigonal prisms may be regarded as 'units of tilt'. Similarly, Archimedian square antiprisms may be regarded as 'units of twist', and have been identified at the $\theta = 22.6°$ (001) twist boundary in Au by X-ray diffraction (Fitzsimmons and Sass 1989) and computer simulations (Vitek *et al.* 1980). Polyhedral units may be identified in all HREM images of tilt boundaries in metals; see the papers by Krakow and Smith (1986*a,b*) for many examples.

Perhaps the most important physical aspect of the polyhedral unit model is the central role it gives to *local* relaxation. The model is not concerned with longe-range periodicity in the boundary plane, or even if the boundary is flat or curved. It is concerned entirely with the *local* packing forces on atoms and how those forces act to minimize the *local* free volume at the interface. In doing so it recognizes the in-built competition and frustration of local relaxation, where clusters of atoms compete for neighbours and distort each other to greater or lesser extents in the drive to maximize the local coordination number and minimize the local free volume. The result is that some patches of the boundary win over others and thus attain higher coordinated atoms and denser local packing. This picture is highly reminiscent of Mott's early model of grain boundaries as regions of 'good fit' separated by regions 'bad fit' (Mott 1948). It describes the local relaxation seen in Fig. 4.28 rather well. Further experimental evidence in support of this view of local relaxation may be found in the review by Merkle (1991).

The presence of polyhedral clusters at interfaces unlike those found in the bulk may also be significant for segregation of impurities. For example, as discussed in Section 7.3.3, P solute atoms may be attracted to trigonal prisms at grain boundaries in Fe because they are then in an environment similar to that found in $Fe_3P$.

The chief weakness of the polyhedral unit model is that it is a description that may be applied only when the atomic structure has been determined either experimentally or by simulation. It is not a predictive description. By contrast, the structural unit model, which we move onto next, does have a predictive capability.

### 4.3.1.8 *The structural unit model*

The structural unit model considers the systematic changes in the structure of grain boundaries as the macroscopic degrees of freedom, such as the misorientation or inclination, are varied. The basic idea, which was first proposed by Bishop and Chalmers (1968), is that certain relatively short period boundaries form the basic building blocks, or structural units, of longer period boundaries nearby in the misorientation or inclination range. Further developments appeared in the papers by Bishop and Chalmers (1971), and Weins *et al.* (1969, 1970, 1971). The model was formalized and extended by Sutton and Vitek (1983*a,b,c*) and our discussion is based on those and subsequent papers. Throughout this section the coordinate system is that of the median lattice (see Section 1.4.3.1).

We shall begin with a series of symmetric $[1\bar{1}0]$ tilt boundaries in Al as computed by Sutton and Vitek (1983*a*) by energy minimization and the boundary conditions shown in Fig. 4.23, with the total volume of the computational cell constant. Table 4.3 lists the periodic boundaries selected in the misorientation range between $\theta_{tilt} = 31.59°$ and 50.48°. The relaxed (115) and (113) boundaries, which delimit the range of misorientation, are shown in Fig. 4.29. In f.c.c. lattices periodic $[1\bar{1}0]$ tilt boundary planes, where

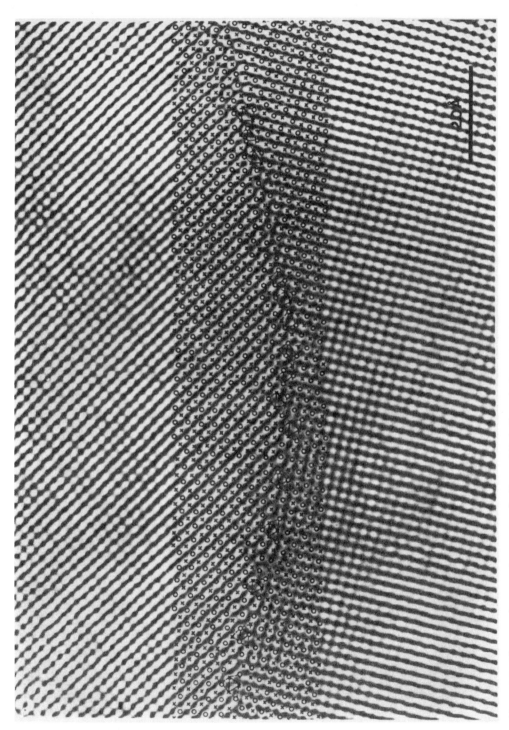

**Fig. 4.28** HREM image of a curved segment of a $\Sigma = 5$ [001] tilt grain boundary in Au, viewed along the tilt axis (Krakow 1991). The atomic columns are white. The circles and crosses distinguish atomic columns that are shifted by $\frac{1}{2}$ [001] (i.e. normal to the page) owing to the stacking sequence of (002) planes. Notice the capped trigonal prisms in the boundary.

**Table 4.3** Parameters of the symmetric [1$\bar{1}$0] tilt boundaries studied by Sutton and Vitek (1983*a*), and their structural unit representations.

| Plane | $\theta(°)$ | $\Sigma$ | Period vector | Structure |
|-------|-------------|----------|---------------|-----------|
| (115) | 31.59 | 27 | $\frac{1}{2}[55\bar{2}]$ | $\lvert A.A \rvert$ |
| (3, 3, 14) | 33.72 | 107 | $[77\bar{3}]$ | $\lvert AAAAAB \rvert$ |
| (229) | 34.89 | 89 | $\frac{1}{2}[99\bar{4}]$ | $\lvert AAAB \rvert$ |
| (5, 5, 21) | 37.22 | 491 | $\frac{1}{2}[21, 21, \bar{10}]$ | $\lvert ABABA.ABABA \rvert$ |
| (114) | 38.94 | 9 | $[22\bar{1}]$ | $\lvert AB \rvert$ |
| (5, 5, 19) | 40.83 | 411 | $\frac{1}{2}[19, 19, \bar{10}]$ | $\lvert BABAB.BABAB \rvert$ |
| (3, 3, 11) | 42.18 | 139 | $\frac{1}{2}[11, 11, \bar{6}]$ | $\lvert BBA.BBA \rvert$ |
| (227) | 44.00 | 57 | $\frac{1}{2}[77\bar{4}]$ | $\lvert BBBA \rvert$ |
| (113) | 50.48 | 11 | $\frac{1}{2}[33\bar{2}]$ | $\lvert B.B \rvert$ |

the Miller indices are all odd, belong to base-centred orthorhombic coincidence site lattices (CSLs) with the base-centred plane parallel to the boundary plane. Thus, there are two CSL sites in each period of the boundary viewed along the tilt axis as in Fig. 4.29. For this reason the structure of the boundaries in Fig. 4.29 repeats in each half-period except that atoms in one half are displaced by $\frac{1}{4}[1\bar{1}0]$ relative to the other. Periodic $\langle 110 \rangle$ tilt boundaries, where the Miller indices are all odd, are therefore called 'centred'. Since (115) is a centred boundary its basic building block, or structural unit, occupies just one half of the period of the boundary normal to the tilt axis, and it comprises one of the irregular pentagons labelled *p* and one tetrahedron labelled *t*. Let us call this structural unit *A*. In the boundary plane it is delineated by the vectors $\frac{1}{4}[55\bar{2}]$ and $\frac{1}{2}[1\bar{1}0]$. This is the only constraint on the size of the structural unit. The size normal to the boundary plane is arbitrary, and it is convenient to choose the smallest size that enables the structural unit to be identified. We may represent the structure of the (115) boundary by $\lvert A \cdot A \rvert$, where the vertical lines delineate one period of the boundary normal to the tilt axis, and the dot signifies that the contents of each half-period are identical except for a relative displacement of $\frac{1}{4}[1\bar{1}0]$. The (113) boundary shown in Fig. 4.29(b) is also centred and its structural unit, which we call *B*, is delineated by the vectors $\frac{1}{4}[33\bar{2}]$ and $\frac{1}{2}[1\bar{1}0]$. Each *B* unit comprises a capped trigonal prism, and the boundary structure is represented by $\lvert B \cdot B \rvert$.

It is emphasized that the *area* of the structural unit in the boundary plane is determined by the translational symmetry of the boundary. There is an infinite choice of primitive unit cells in a 2D lattice but, provided the area associated with each of them is equal to the area associated with one lattice site, any of them will generate the lattice through translation operations. The *shape* of the primitive cell is entirely a matter of convenience. In the present case we shall be looking at sequences of structural units normal to the tilt axis. It is therefore convenient to choose rectangular structural units with one side parallel to the tilt axis. Having made this choice the sizes of the *A* and *B* units in the boundary plane are completely determined by the translational symmetries of the (115) and (113) boundaries and they are the same in all materials with an f.c.c. lattice. This is analogous to the fact that there is a conventional choice of unit cell for all crystals which have an f.c.c. lattice. But the atomic motif which decorates a structural unit will differ from one material to another. This is analogous to the different atomic motifs which decorate the f.c.c. lattices of Al and Si for example. There are examples in the literature where

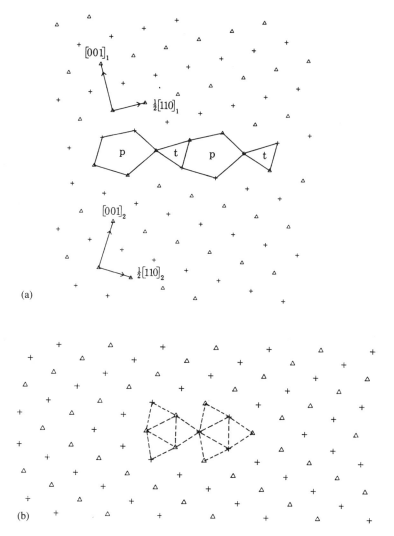

**Fig. 4.29** Relaxed structures of (a) (115) and (b) (113) symmetric tilt boundaries in Al. The structures are seen in projection along the [1$\bar{1}$0] tilt axis. The triangles and crosses distinguish atoms on successive (2$\bar{2}$0) planes. The boundaries run from left to right in the middle of each picture. Two '*A*' structural units are identified in (a), each comprising an irregular pentagon, labelled 'p', and a tetrahedron, labelled 't'. Two '*B*' structural units, each comprising a capped trigonal prism, are outlined in (b). (From Sutton and Vitek (1983*a*).)

'structural units' have been defined that occupy only a fraction of the area associated with one CSL site of the boundary from which they originate. A contiguous sequence of such units would not generate the boundary of which they are, therefore, an incomplete representation. They are not structural units as we have defined them, or as they were defined by Sutton and Vitek (1983*a*), but more like the polyhedral units of the previous section, whose sizes are not constrained by any crystallographic conditions.

Figure 4.30(a) shows the relaxed structure of the (3, 3, 14) boundary, which is 2.1° from the (115) boundary. This boundary is not centred and in each period we can see five

(a)

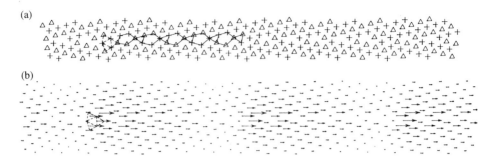

(b)

**Fig. 4.30** (a) Relaxed structure of (3, 3 , 14) symmetric tilt boundary in Al. Five $A$ units (solid lines) and one $B$ unit (broken lines) are shown in a boundary period. (b) Hydrostatic stress field map of the boundary. The hydrostatic pressure (eqn (4.85)) at each atomic site is represented by an arrow centred on the site. The length of the arrow is proportional to the magnitude of the pressure. Arrows pointing to the right denote compression and those to the left denote tension. Note that $B$ units coincide with an abrupt transition from maximum to minimum compression in the boundary. (From Sutton and Vitek (1983$a$).)

slightly distorted $A$ units, shown by solid lines, and one slightly distorted $B$ unit, shown by broken lines. Therefore, this boundary structure may be represented by $|AAAAAB|$. The period vector of the boundary is $[77\bar{3}]$ and the decomposition of the structure into five $A$ units and one $B$ unit is reflected by the vector decomposition:

$$[77\bar{3}] = \tfrac{5}{4}[55\bar{2}] + \tfrac{1}{4}[33\bar{2}]. \tag{4.80}$$

Figure 4.31(a) shows the relaxed structure of the (229) boundary. There are now three $A$ units to each $B$ unit (shown by broken lines) and the structure of the boundary is $|AAAB|$. The period vector, $\tfrac{1}{2}[99\bar{4}]$, undergoes the following decomposition:

$$\tfrac{1}{2}[99\bar{4}] = \tfrac{3}{4}[55\bar{2}] + \tfrac{1}{4}[33\bar{2}]. \tag{4.81}$$

The full lines show pairs of {115} planes of the black and white crystals entering the boundary and terminating at $B$ units: $B$ units coincide with the cores of edge dislocations. The Burgers vector of the dislocations is $\tfrac{2}{27}[115]$ because two {115} planes terminate. They are anticoherency dislocations because they destroy the parallel alignment of the {115} planes of the reference (115) boundary structure. In accordance with the discussion of Section 2.2, there is a continuous distribution of coherency dislocations between successive $B$ units, with net Burgers vector $-\tfrac{2}{27}[115]$. The coherency dislocations bring the slightly misaligned {115} planes on either side of the (229) boundary into (almost) parallel alignment. The total Burgers vector content of the coherency and anticoherency dislocations is zero, and there is no long-range stress field. We shall see that the stress field of the boundary is consistent with our dislocation description. The anticoherency dislocations are (non-primitive) DSC dislocations of the $\Sigma = 27$ coincidence system which the (115) reference boundary structure belongs to.

Figure 4.32(a) shows the relaxed structure of the (5, 5, 21) boundary. This centred boundary contains three $A$ units for every two $B$ units and its structure could be $|AAABB \cdot AAABB|$ or $|AABAB \cdot AABAB|$. Figure 4.32(a) shows the latter is correct. Pairs of {115} planes are again seen entering the boundary and terminating at $B$ units. Therefore $B$ units are again located at the cores of $\tfrac{2}{27}[115]$ anticoherency dislocations.

Figure 4.33(a) shows the relaxed structure of the (114) boundary. Each period of this

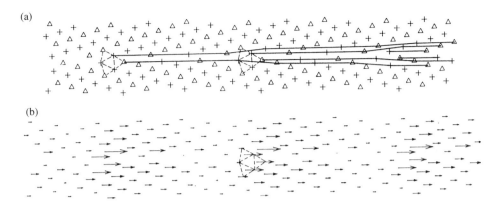

**Fig. 4.31** (a) Relaxed structure of (229) symmetric tilt boundary in Al. Three *A* units may be seen between successive *B* units (the latter shown by broken lines). The full lines show pairs of {115} planes which enter the boundary and terminate at *B* units. (b) Corresponding hydrostatic stress field map. (From Sutton and Vitek (1983*a*).)

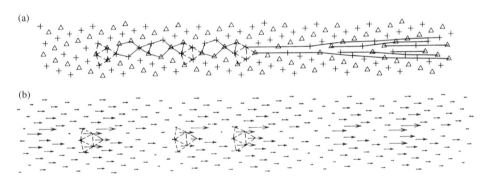

**Fig. 4.32** (a) Relaxed structure of (5, 5, 21) symmetric tilt boundary in Al. Each half-period of this centred boundary comprises three *A* units and two *B* units. Pairs of {115} planes are shown entering and terminating in the boundary at *B* units. (b) Corresponding hydrostatic stress field map. (From Sutton and Vitek (1983*a*).)

boundary contains one *A* and one *B* unit: $|AB|$. At this boundary pairs of {115} planes again enter the boundary and terminate at *B* units, but also pairs of {113} planes enter the boundary and terminate at *A* units. Therefore *B* units coincide with the cores of $\frac{2}{27}[115]$ dislocations preserving the *A* unit reference structure, or *A* units coincide with the cores of $-\frac{2}{11}[113]$ dislocations preserving the *B* unit reference structure. The Burgers vector is negative because the angular deviation from the (113) reference orientation is negative. Since there are just as many *A* units as there are *B* units the descriptions are equally valid.

Increasing the deviation from (115) further we reach the (5, 5, 19) boundary, whose relaxed structure is shown in Fig. 4.34(a). There are now three *B* units for every two *A* units, and the structure of the boundary is $|BBABA \cdot BBABA|$. The decomposition of each half of the period vector is:

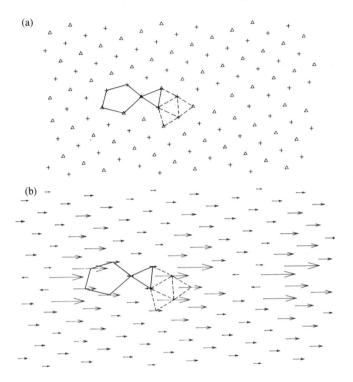

**Fig. 4.33** (a) Relaxed structure of (114) symmetric tilt boundary in Al. Each boundary period comprises one *A* unit and one *B* unit. (b) Corresponding hydrostatic stress field map. (From Sutton and Vitek (1983*a*).)

$$\tfrac{1}{4}[19, 19, \overline{10}] = \tfrac{3}{4}[33\overline{2}] + \tfrac{2}{4}[55\overline{2}]. \qquad (4.82)$$

The solid lines show pairs of {113} planes from the black and white crystals entering the boundary and terminating at A units. Since there are now more *B* units than *A* units the *A* units correspond to the cores of $-\tfrac{2}{11}[113]$ DSC dislocations preserving the B reference structure.

Finally, in Fig. 4.35(a) we show the relaxed structure of the (3, 3, 11) boundary. There are two *B* units and one *A* unit in each half-period of this centred boundary. Again, the *A* units correspond to the cores of $-\tfrac{2}{11}[113]$ dislocations preserving the *B* unit reference structure.

To summarize, the relaxed boundary structures between (115) and (113) are all composed of *A* and *B* structural units. Between the orientations of the (115) and (114) boundaries there are more *A* than *B* units, and *B* units then correspond to the cores of $\tfrac{2}{27}[115]$ anticoherency $\Sigma = 27$ DSC dislocations preserving the (115) reference boundary structure. Between the orientations of the (114) and (113) boundaries there are more *B* than *A* units and *A* units then correspond to the cores of $-\tfrac{2}{11}[113]$ anticoherency $\Sigma = 11$ DSC dislocations preserving the (113) reference boundary structure. At the (114) boundary there is a 1:1 mixing of *A* and *B* units and both descriptions in terms of dislocations are equally valid. The descriptions of the boundary structures in terms of structural units are entirely equivalent to the DSC dislocation model of grain boundaries depicted in Fig. 2.9.

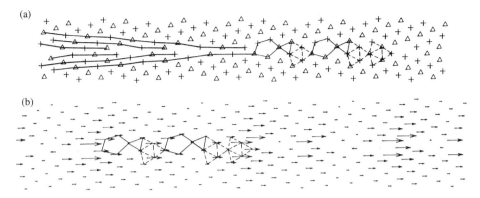

**Fig. 4.34** (a) Relaxed structure of (5, 5, 19) symmetric tilt boundary in Al. Each half-period of this centred boundary comprises two *A* units and three *B* units. Solid lines show pairs of {113} planes which enter the boundary and terminate at *A* units. (b) Corresponding hydrostatic stress field map. (From Sutton and Vitek (1983*a*).)

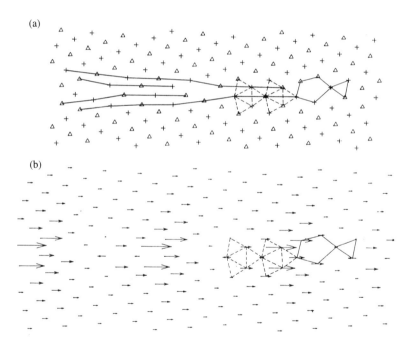

**Fig. 4.35** (a) Relaxed structure of (3, 3, 11) symmetric tilt boundary in Al. Each half-period of this centred boundary comprises two *B* units and one *A* unit. (b) Corresponding hydrostatic stress field map. (From Sutton and Vitek (1983*a*).)

The descriptions of the relaxed structures in terms of coherency and anticoherency dislocations also describe the stress field of the relaxed boundaries quite well. To see this we must first define the meaning of stress at the atomic scale.

Consider an arbitrary assembly of interacting atoms. Let the force that atom $l'$ exerts on atom $l$ be $f^{ll'}$. Let the position vector of atom $l'$ relative to atom $l$ be $r^{ll'}$. If an infinitesimal homogeneous strain $\delta\varepsilon_{\alpha\beta}(\alpha,\beta = x, y, \text{ or } z)$ is applied to the assembly then the change in the energy of the system to first order in $\delta\varepsilon$ is

$$\delta E = \sum_l V^l \sum_{\alpha\beta} \delta\varepsilon_{\alpha\beta}\tau^l_{\alpha\beta} \tag{4.83}$$

where $\tau^l_{\alpha\beta}$ is the atomic level stress tensor acting at site $l$, and $V^l$ is the atomic volume associated with site $l$. By equating this to the first-order change in the energy arising from the work done by the atomic interactions it may be shown that

$$\tau^l_{\alpha\beta} = \frac{1}{2V^l} \sum_{l'} f^{ll'}_\alpha r^{ll'}_\beta. \tag{4.84}$$

The hydrostatic pressure at site $l$, $p^l$, is given by

$$p^l = -\frac{1}{3} \sum_\alpha \tau^l_{\alpha\alpha} = -\frac{1}{6V^l} \sum_{l'} r^{ll'} \cdot f^{ll'}. \tag{4.85}$$

In the bulk the equilibrium condition demands that $p^l = 0$ at all lattice sites. At an edge dislocation the region of the core just above the termination of the extra half-plane(s) is in a state of compression, and the region just below is in a state of tension. In Fig. 4.30(b) we show a map of the atomic level hydrostatic stresses, eqn (4.85), of the (3, 3, 14) grain boundary. Arrows pointing to the right denote compression and those to the left denote tension. The magnitude of each arrow is proportional to the stress. (Owing to the expansion at the boundary the cell is in a state of overall compression, because the total volume of the computational cell was conserved during the relaxation.) We see that the hydrostatic compression increases gradually between $B$ units and decreases abruptly at each $B$ unit. The same is true also in Figs 4.31(b) and 4.32(b). On the other hand for the (3, 3, 11) boundary shown in Fig. 4.35(b) the hydrostatic compression increases gradually between successive $A$ units and decreases abruptly at each $A$ unit. The same is also seen in Fig. 4.34(b). The hydrostatic stress field maps are therefore consistent with the statement that minority units correspond to the cores of anticoherency dislocations preserving the reference structure of the majority units. At the (114) boundary the stress field, Fig. 4.33(b), is consistent with either dislocation description.

The hydrostatic stress field maps show that the DSC dislocation descriptions are physically meaningful throughout the 19° misorientation range between the (115) and (113) reference boundary structures. This simply reflects the bending of (115) and (113) planes in the black and white crystals of the intervening boundaries to localize the angular deviation from the reference structures at minority units. The bending is driven by cohesive forces acting across each intervening boundary plane, as discussed in Section 4.3.1.5.

It is also possible to choose reference structures other than pure $A$ or pure $B$. For example, suppose we choose the (114) boundary, whose structure is $|AB|$, as a reference structure. Let an '$AB$' unit be called $C$. The structure of the (5, 5, 21) boundary would then be written as $|CCA \cdot CCA|$. The 1:1 mixing boundary is (3, 3, 13) and its structure would be written as $|CA \cdot CA|$. Between the (115) and (3, 3, 13) orientations there are more $A$ than $C$ units and the $C$ units would then correspond to the cores of $\frac{2}{27}[115]$ dislocations preserving the $A$ reference structure. But between $(3, 3, 13)$ and $(114)$ there are more $C$ than $A$ units and the $A$ units would then correspond to the cores of $-\frac{1}{18}[114]$ $\Sigma = 9$ dislocations preserving the $\Sigma = 9$ $(114)$ reference structure. The $(114)$ boundary in this example is an illustration of what Sutton *et al.* (1981) called a multiple unit reference structure. We could continue and define $AC$ as a new structural unit $D$. In this way we can construct a hierarchy of structural unit/dislocation descriptions. The more structural units we combine in a multiple unit reference structure the smaller the angular range that can be spanned by dislocations preserving that reference structure. On the other hand, the description becomes more accurate because the distortions the units suffer are less. There is therefore a trade-off between the predictive capacity of the model, in the sense of the angular range of boundary structures spanned by the structural unit description, and the accuracy of the model, in the sense that the structural units undergo greater distortions the larger the angular range they span. Obviously, the dislocation model corresponding to a multiple unit reference structure does not account for the stress variations within the multiple units of the reference, and it is therefore not such a complete description of the stress field as that obtained with the fundamental structural units as the reference. As discussed in Section 2.12, the strain fields of dislocation arrays based on alternative reference structures for a given interface may be seen at different levels of spatial resolution in the electron microscope.

We have seen that the multiplicity of possible reference structures in dislocation models is mirrored exactly in a multiplicity of choices of structural units to delimit a range of misorientations. In doing this we have distinguished between reference structures comprising the fundamental structural units and those comprising mixtures of them. But in what sense are the fundamental units different?

Sutton and Vitek (1983a) defined a fundamental unit as a unit that could not be broken down into units from other boundaries. A boundary composed of a contiguous sequence of such units they called 'favoured'. Thus, the above $(115)$ and $(113)$ boundaries are favoured. But the distinction between a favoured boundary and a multiple unit reference structure may be difficult to maintain unless some limit on the distortion of a fundamental unit is introduced. For example, Krakow (1992) has argued that the $A$ unit in Fig. 4.29(a) is composed of the capped trigonal prism of the $B$ unit in Fig. 4.29(b), together with a rather distorted unit of the $(001)$ plane of the perfect crystal. The only reason for rejecting this description is that the distortion of the perfect crystal unit is 'unacceptably' large. But how large is acceptable? The answer to this question is arbitrary. Most authors deal with the problem by referring to the reference boundaries, whose units appear in the boundaries throughout the misorientation range between them, as 'delimiting' boundaries, because they delimit the misorientation range. Whether the delimiting boundaries are favoured boundaries or multiple unit reference structures is entirely a matter of choice, and one's criterion for deciding which are the fundamental structural units.

Perhaps the most sensible basis for the choice of delimiting boundaries is the modelling of physical properties. For example, if the range of misorientation between delimiting boundaries is too large then deducing properties of intervening boundaries by interpolating between properties of the delimiting boundaries will be inaccurate. But if the range is too small then the properties of some delimiting boundaries could have been deduced from those of other delimiting boundaries. It boils down, again, to a balance between accuracy of the interpolations and the range of misorientations over which the interpolations are done: how much information do we have to put into the model to get

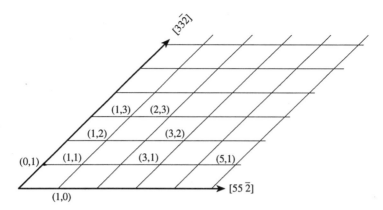

**Fig. 4.36** The decomposition lattice for the sequence of boundaries listed in Table 4.3. The axes are along the period vectors of the delimiting (115) and (113) boundaries, i.e. [55$\bar{2}$] and [33$\bar{2}$] respectively, and the basis vectors (1, 0) and (0, 1) are $\frac{1}{4}$[55$\bar{2}$] and $\frac{1}{4}$[33$\bar{2}$]. Nodes in this lattice correspond to period vectors (or half-period vectors in the case of centred boundaries) of boundaries in the misorientation range between the delimiting boundaries. For example, the node (3, 1) corresponds to the period vector $\frac{3}{4}$[55$\bar{2}$] + $\frac{1}{4}$[33$\bar{2}$] = $\frac{1}{2}$[99$\bar{4}$], which is the period vector of the (229) boundary. Therefore, this boundary period is composed of 3 *A* units and 1 *B* unit. The nodes corresponding to all boundaries listed in Table 4.3 are shown.

a certain amount of new information out of it? It is also quite possible that more delimiting boundaries may be needed to span a given misorientation range for some properties than others. We shall see that this is indeed the case at the end of this section.

For each of the boundaries between the (113) and (115) orientations the decomposition of the structure into *A* and *B* structural units is reflected by the decomposition of the period vector, or half of the period vector in the case of a centred boundary, into a linear combination of the vectors $\frac{1}{4}$[55$\bar{2}$] and $\frac{1}{4}$[33$\bar{2}$] belonging to the *A* and *B* units. The vectors $\frac{1}{4}$[55$\bar{2}$] and $\frac{1}{4}$[33$\bar{2}$] form the basis of a 2D *decomposition lattice*, as shown in Fig. 4.36. The period vectors of all boundaries lying between (113) and (115) are vectors of this lattice. For example, the period vector (3, 3, 14) boundary is [77$\bar{3}$] which, as we saw in eqn (4.80), decomposes into $\frac{5}{4}$[55$\bar{2}$] and $\frac{1}{4}$[33$\bar{2}$]. The (3, 3, 14) boundary is therefore represented by the node (5, 1) in the decomposition lattice. The point (13, 19) of the lattice corresponds to the vector $\frac{13}{4}$[55$\bar{2}$] + $\frac{19}{4}$[33$\bar{2}$] = $\frac{1}{2}$[61, 61, $\overline{32}$], which is the period vector of the $\Sigma$ = 4233 (16, 16, 61) boundary.

There is a selection rule on the choice of delimiting boundaries to ensure that the structures of all intervening boundaries *can* decompose into structural units of the delimiting boundaries. The rule arises from the existence of centred boundaries. At least one member of each pair of delimiting boundaries must be centred to enable each half-period of an intervening centred boundary to decompose into units of the delimiting boundaries. This would not be possible if both delimiting boundaries were not centred. The existence of a *geometrical* selection rule indicates that the choice of delimiting boundaries is not dictated only by the boundary *energy*. Delimiting boundaries are *not* necessarily singular. In fact, they may even be unstable, as we shall see below.

The numbers of *A* and *B* units in each intervening boundary satisfy a simple lever rule (Nazarov and Romanov 1989). Let *p* be the length of the repeat sequence of units in some intervening boundary, where the tilt angle is $\theta$. Let $l_A$ be the length of the undistorted

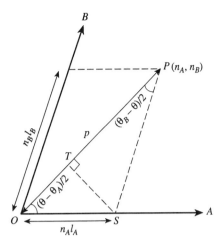

**Fig. 4.37** To illustrate the derivation of eqns (4.86) and (4.87). The diagram shows the decomposition lattice. The node $(n_A, n_B)$ corresponds to a boundary in which the period vector has length $OP = p$. The misorientation, $\theta$, of this boundary lies between the misorientations, $\theta_A$ and $\theta_B$, of the delimiting $A$ and $B$ boundaries.

A structural unit normal to the tilt axis, and let $\theta_A$ be the tilt angle of the $A$ reference structure. Let $l_B$ and $\theta_B$ have similar meanings for the $B$ reference structure. Assume $\theta_A < \theta < \theta_B$. In the decomposition lattice shown in Fig. 4.37 the node $(n_A, n_B)$ is represented by the point $P$, and the distance $OP$ is $p$. Since $OP = OT + TP$ we have

$$p = n_A l_A \cos((\theta - \theta_A)/2) + n_B l_B \cos((\theta_B - \theta)/2). \tag{4.86}$$

By considering $TS$ in Fig. 4.37 we deduce that

$$n_A l_A \sin((\theta - \theta_A)/2) = n_B l_B \sin((\theta_B - \theta)/2)$$

and therefore

$$\frac{n_A}{n_B} = \frac{l_B \sin((\theta_B - \theta)/2)}{l_A \sin((\theta - \theta_A)/2)}. \tag{4.87}$$

This is the lever rule.

If the minority units are $B$ then the reference structure is $A$. Let the Burgers vectors of each of the $n_B$ dislocations in each repeat $p$ be $b_A$. Frank's formula gives

$$n_B b_A = 2p \sin((\theta - \theta_A)/2). \tag{4.88}$$

Using eqns (4.86) and (4.87) we deduce that

$$p = n_B l_B \frac{\sin((\theta_B - \theta_A)/2)}{\sin((\theta - \theta_A)/2)} \tag{4.89}$$

and substituting this in eqn (4.88) we obtain

$$b_A = 2l_B \sin((\theta_B - \theta_A)/2). \tag{4.90}$$

Similarly the Burgers vector, $b_B$, of dislocations preserving the $B$ reference structure, when $A$ units are in the minority, is given by

$$b_B = -2l_A \sin((\theta_B - \theta_A)/2). \tag{4.91}$$

We shall return to the significance of these formulae later in the section.

An algorithm to determine the sequence of structural units is also necessary and Sutton and Vitek (1983*a*) were the first to recognize this and develop one. If the repeat sequence of units in some intervening boundary comprises $n_A$ $A$ units and $n_B$ $B$ units then the number of ways of arranging these units in a periodic fashion is

$$W = \frac{(n_A + n_B - 1)!}{n_A! n_B!}. \qquad (4.92)$$

For example, for the (16, 16, 61) boundary we have seen that $n_A = 13$ and $n_B = 19$ and therefore $W = 10,855,425$. Clearly the model has no predictive value unless we have an algorithm to select the correct sequence from this huge number of possibilities.

The algorithm is based on the assumption that as the misorientation varies the boundary structure changes in as smooth and continuous manner as possible, given the constraint that it is made up of $A$ and $B$ units. Continuity is assured by making longer sequences through interpolation of known shorter sequences. For example, the 3:2 sequence is deduced by interpolating between the known sequences 2:1 and 1:1. Thus, $3:2 = 2:1 + 1:1 = AAB + AB = AABAB$. As we noted above there is another 3:2 sequence, namely $AAABB$. But this violates the continuity principle because in the 2:1 and 1:1 structures there are no $BB$ sequences. The algorithm always results in the minority units being separated as much as they possibly can be in a discrete sequence, and two minority units never appear together.

To illustrate the algorithm let us derive the sequence for the above 13:19 boundary. Since 13:19 lies between 1:1 and 1:2 it will be made up of $AB$ and $ABB$ sequences. Moreover, $13:19 = 7(1:1) + 6(1:2)$; there are 7 $AB$ sequences and 6 $ABB$ sequences. Let us call $C = AB$ and $D = ABB$. Then our sequence consists of a 7:6 mixture of $C$ and $D$. But $7:6 = 5(1:1) + 2:1$. Let us call $E = CD$ and $F = CCD$. Then our sequence consists of a 5:1 mixture of $E$ and $F$, which can only be $EEEEEF$. But

$$EEEEEF = [5(CD)]CCD = CDCDCDCDCDCCD$$

$$= ABABBABABBABABBABABBABABBABABABB.$$

Since the boundary is periodic the sequence may be permuted cyclically. Note that the minority units, $A$, are always separated by at least one $B$ unit.

It is also possible to determine the sequence of units directly from the decomposition lattice using the 'strip' method of quasicrystallography (Sutton (1992) and references therein). This is significant because it enables direct contact to be made between irrational boundaries and quasiperiodicity. In the strip method a slice of a higher-dimensional lattice (in our case 2D) is projected onto a lower-dimensional plane (in our case a line). The 2D lattice is the decomposition lattice and the line is the repeat vector of the boundary. If the boundary is rational then the repeat vector is finite and there is a periodic sequence of units. If the boundary is irrational the sequence never repeats and it is quasiperiodic.

It is convenient to map the decomposition lattice onto a square lattice. The mapping does not alter the sequences we shall derive because it is an affine transformation, and it simply enables the construction to be seen more clearly. The basis vectors of the lattice outline the unit square, as shown in Fig. 4.38. The line $W$ passes through the point (13, 19) and it therefore represents the period vector of the (16, 16, 61) boundary. Consider the strip obtained by sliding the unit square, labelled $S$, along the line $W$. Inside the strip there is a unique broken line, shown bold in Fig. 4.38, which joins all the lattice points inside the strip. This line is now projected orthogonally onto $W$ and it is seen that two 'tiles', $A$ and $B$, are produced which are the projections of the horizontal and vertical edges of the unit square. These tiles are the structural units $A$ and $B$ and it is readily

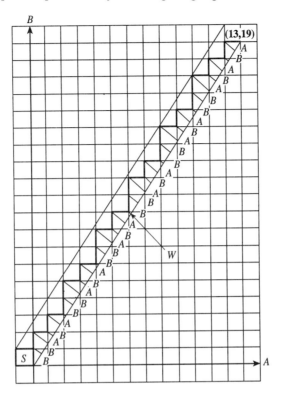

**Fig. 4.38** To illustrate the strip method of quasicrystallography to determine the sequence of 13 *A* and 19 *B* units in the (16, 16, 61) symmetric tilt boundary, which is represented by the node (13, 19) in the decomposition lattice.

confirmed that the sequence of tiles along *W* is equivalent to the sequence derived above by a cyclic permutation.

The unique broken line inside the strip is the sequence of steps on the free surface of a 2D crystal cut parallel to *W*. The (13, 19) symmetrical tilt boundary is constructed by bringing this free surface into contact with a mirror image of itself obtained by reflection in the plane containing *W* and the tilt axis. On bonding the two surfaces together the steps become the cores of anticoherency dislocations as cohesive forces acting across the boundary pull the two surfaces together and physically effect the projection operation of the strip method. Thus, the strip method is not just an abstract geometrical construction; the projection operation is realized by the local relaxation at a symmetrical tilt boundary.

If the ratio $n_A : n_B$ is irrational then the slope of the line *W* also becomes irrational. But the broken line inside the strip is still unique and therefore the sequence of structural units is also unique. Here we have a graphical illustration of the statement we proved analytically in Section 1.9 that irrational interfaces display long-range order. The origin of the long-range order is simply the ordered sequence of steps on an irrational, unrelaxed surface of a crystal.

Sutton (1988) showed that if the ratio $n_A : n_B$ is a quadratic irrational number then the sequence of structural units is self-similar. A quadratic irrational is an irrational number that satisfies a quadratic equation with integer coefficients. The existence of self-similarity means that there is an inflation algorithm for generating the sequence by recursively substituting a sequence of units for an *A* or *B* unit. The mathematics underlying this beautiful geometry is very elegant and involves the remarkable properties of continued fractions. See Sutton (1988, 1992) for details.

The methods we have described yield sequences for the ground state at absolute zero. The elementary excitations of the sequence involve interchanges of neighbouring $A$ and $B$ units. They correspond to the formation of a double kink on the anticoherency dislocation line threading the minority unit. If whole columns (along the tilt axis) of adjacent $A$ and $B$ units interchange the dislocation climbs. The energy cost associated with $A \leftrightarrow B$ column interchanges can be very small if no structural unit sequences are introduced which were not already present elsewhere in the sequence (Lu and Birman 1986). For example, between the boundaries composed of only $ABB$ and $AB$ sequences, continuity of boundary structure would forbid the occurrence of $AA$ or $BBB$ sequences. Thus $ABBAB$ and $ABABB$ are acceptable $A \leftrightarrow B$ column changes whereas $ABBAB$ and $ABBBA$ are not. The energy cost associated with the former interchange is likely to be less than that for the latter because only the former ensures that the anticoherency dislocations threading minority units (i.e. $A$) remain separated by at least one majority unit. These considerations affect the source/sink efficiency of the boundary for vacancies (see Section 10.3.3).

So far we have assumed that the atomic motifs decorating the structural units $A$ and $B$ are unique. If either boundary delimiting a range of misorientations can exist in more than one metastable state then a multiplicity of structures of intervening boundaries becomes possible (Wang *et al.* 1984). Suppose the two delimiting boundaries have $\nu_A$ and $\nu_B$ different structures, respectively. Thus, there are $\nu_A$ distinct structural units associated with the first delimiting boundary, all of them sharing the same area in the boundary plane but differing in the atomic motifs they contain. An intervening boundary in which each period contains $n_A$ units of the first boundary and $n_B$ of the second may exist in $(\nu_A^{n_A})(\nu_B^{n_B})$ possible structures. But we have assumed here that the intervening boundary remains periodic both parallel and perpendicular to the tilt axis. As the temperature is raised the configurational entropy will favour the destruction of both periodicities, rather like an ordered alloy will tend to become a random alloy at higher temperatures (for further discussion see Section 6.3.2.1). Two alternative structural units of the (210) symmetric tilt boundary in Au have been seen coexisting in the same patch of boundary by HREM (Krakow 1991).

There have been few systematic HREM studies to test the validity of the structural unit model throughout a range of misorientations. This is obviously a difficult task experimentally because it involves growing numerous bicrystals throughout the misorientation range. But there is a growing concensus that the model is useful for interpreting HREM images of grain boundaries in a wide variety of crystalline materials.

Penisson *et al.* (1988) analysed the structure of a $\theta = 14 \pm 0.5°$ [001] symmetric tilt boundary in Mo in terms of structural units. The boundary plane is close to (810), and according to the structural unit model it should be composed of a roughly equal mixture of sequences of structural units comprising the (710) and (910) boundaries. Penisson *et al.* found that this was the case but there were two complications. The first is that the there were steps of one (200) plane height in the boundary plane. For each step up there was a step down, so that the mean boundary plane was preserved. Secondly, the repeat sequences of units comprising the (710) and (910) boundaries should alternate in the (810) boundary, but this was found to be true only on average. The bicrystals were subjected to an anneal of 8 hours at 2300 K, and it is quite conceivable, therefore, that configurational entropy may have played a significant role in determining the boundary structure.

Probably the most detailed HREM study to date was carried out by Rouviere and Bourret (1990) for [001] for symmetric tilt boundaries in Si and Ge. They concluded that the structural unit model agreed with some of their images but not all. The breakdown of the structural unit model seemed to be associated with extraneous factors such as the tendency for faceting on asymmetrical boundary planes or the presence of stress in the bicrystal favouring the introduction of new structural units. The fact that the bicrystals were produced at high temperature also introduced a considerable structural multiplicity.

The applicability of the structural unit model to relaxed structures of (001) twist boundaries in metals, obtained by computer simulation, was demonstrated by Sutton (1982) and Schwartz *et al.* (1985). Figure 4.39(a) shows the relaxed structure of a $\theta_{\text{twist}} = 8.8°$ ($\Sigma = 85$) (001) twist boundary. The crosses and squares show the (002) planes on either side of the geometrical boundary plane. As this is a small-angle boundary most of the area associated with the 2D repeat motif, bounded by $\frac{1}{2}\langle 670 \rangle$ vectors, is occupied by structural units, $A$, of the perfect crystal. The $A$ units are bounded by $\frac{1}{2}\langle 110 \rangle$ vectors, and there is a $5 \times 5$ patch of them in the repeat cell. At the corner of the repeat cell there is a structural unit, $B$, of the $\theta_{\text{twist}} = 36.9°$ ($\Sigma = 5$) boundary. The $B$ unit is bounded by $\frac{1}{2}\langle 210 \rangle$ vectors. Thus the 2D decomposition of the unit cell may be represented by the following two vector decompositions:

$$\left. \begin{array}{l} \frac{1}{2}[760] = \frac{5}{2}[110] + \frac{1}{2}[210] \\ \frac{1}{2}[6\bar{7}0] = \frac{5}{2}[1\bar{1}0] + \frac{1}{2}[1\bar{2}0] \end{array} \right\}. \tag{4.93}$$

However, the $A$ and $B$ units do not fill the repeat cell, and additional 'filler' units are required. The filler units correspond to the cores of $\frac{1}{2}\langle 110 \rangle$ screw dislocations which intersect at the minority $B$ units and preserve the $A$ unit majority reference structure. Figure 4.39(b) is a map of the shear stress field of the boundary. The maximum shear stress in the (001) plane at each atomic site $i$ is evaluated as $\tau^i = [(\tau^i_{13})^2 + (\tau^i_{23})^2]^{\frac{1}{2}}$, where $\tau^i_{13}$ and $\tau^i_{23}$ are the atomic level shear stress components in the (001) plane at site $i$. The stress, $\tau^i$, is represented by an arrow centred at the atom site, with length proportional to its magnitude, and direction parallel to the maximum shear stress in the (001) plane. Figure 4.39(b) shows the stress field of a square array of screw dislocations which is entirely consistent with the positions of the screw dislocations given above.

The application of the structural unit model to asymmetric tilt boundaries was considered by Sutton and Vitek (1983b). The additional degree of freedom displayed by asymmetric tilt boundaries over symmetric tilt boundaries introduces another mode of decomposition, namely faceting. At a given misorientation a boundary may always facet onto other boundary planes. Faceting may also be regarded as decomposition into structural units of other boundaries, and a decomposition lattice to describe it can be constructed (Sutton and Vitek 1983b). If a segment of boundary is constrained at its ends and undergoes a faceting decomposition then the spatial extent of the facets must be such that the average boundary plane normal is conserved. Apart from this there are no geometrical constraints on the lengths of the individual facets, which is in contrast to the well-defined sequences of structural units that exist when decomposition involving changes of local boundary misorientation occurs. As discussed in Section 4.3.1.9, the length scales of the individual facets are determined by energetics not geometry. Faceting is discussed further in Sections 5.6 and 6.3.1.

During faceting the misorientation remains constant but the mean boundary plane varies from one facet to another. As in the case of symmetric tilt boundaries, asymmetric tilt boundaries may also undergo decompositions where the local misorientation varies.

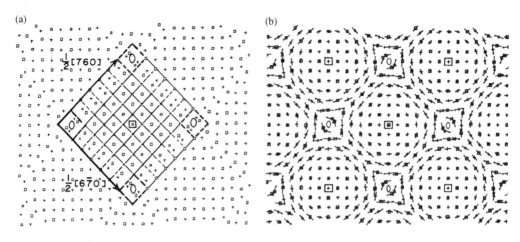

**Fig. 4.39** (a) Relaxed structure of a $\theta_{\text{twist}} = 8.8°$ ($\Sigma = 85$) (001) twist boundary in Cu (Schwartz *et al.* 1985). The first atomic planes on either side of the geometrical boundary plane are distinguished by crosses and squares. The CSL vectors $\frac{1}{2}\langle 760\rangle$ outline a repeat period of the boundary structure. The positions of ordinary diad and tetrad symmetry axes normal to the plane of the paper are shown. A $5 \times 5$ patch of perfect crystal units is visible, centred on the tetrad axis. Structural units, $B$, of the $\theta_{\text{twist}} = 36.9°$ boundary are seen (broken lines) centred on the diad axes. The filler units lie between the $B$ units parallel to the CSL vectors. (b) Shear stress map of the boundary. The arrows are centred on atomic sites. Their magnitudes and directions indicate the maximum shear stress acting in the boundary plane at each atomic site. The shear stress map is consistent with screw dislocations lying along filler units and intersecting at $B$ units.

In that case the mean boundary plane is conserved and the local misorientation varies. Thus, the two modes of decomposition are complementary. This second type of decomposition of asymmetric tilt boundaries has also been studied by computer simulation (Sutton and Vitek 1983*b*), and shown to be entirely analogous to the structural unit decompositions of symmetric tilt boundaries we have already discussed.

In order for different structural units to coexist in a boundary without generating a long-range stress field they must satisfy certain compatibility conditions. The first is that the translation states of the boundaries from which the units originate must be compatible. This condition is sufficiently strong that the minority units of a boundary may be forced by the majority units to adopt a metastable structure. As the boundary misorientation approaches that of the minority unit there is eventually a structural transformation where the minority units can no longer be stabilized and they transform to a stable configuration. An example of such a transition was found by Sutton and Vitek (1983*a*) for symmetric $\langle 111\rangle$ tilt boundaries in Al. At small angles of misorientation there is no rigid body displacement along the tilt axis because the majority units are those from the perfect crystal. The minority units belong to the (341) boundary. However, there is only one stable structure of the (341) boundary and it has a rigid body displacement along the tilt axis of 0.278*a*. Thus the minority units in the small-angle boundaries belong to a structure of the (341) boundary which is not even metastable. As the misorientation increases towards (341) it is found that boundaries can exist in two metastable states, one with and one without a rigid body displacement along the tilt axis. Only boundaries with a rigid body displacement along the tilt axis are found as the misorientation approaches (341).

Assuming that the rigid body displacements of two structural units are compatible, are there other restrictions to ensure that the units may coexist in some intervening boundary without generating long-range compatibility stresses? For example, it is obvious that structural units of (100) and (110) twist boundaries may not be mixed together to make some new boundary without generating compatibility stresses. Sutton and Balluffi (1990) proved that there are only two kinds of permissible structural unit decomposition: (i) faceting and (ii) those which conserve the mean boundary plane. All other conceivable forms of structural unit decomposition result in long-range compatibility stresses. The conclusion applies to tilt, twist, and mixed tilt and twist boundaries. If two boundaries share the same misorientation but different inclinations then boundaries with intermediate inclinations may facet into them. But there is *no* permissible way of interpolating between two boundaries that do not have the same misorientation if they do not share the same mean boundary plane. If they share the same mean boundary plane, but not the same boundary plane, they differ by tilt rotations about some axis or axes in the mean boundary plane. Interpolating between these two boundaries will then involve the introduction of anticoherency edge dislocations. If they share the same boundary plane then they are related by twist rotations about the boundary normal. Interpolating between them will then involve the introduction of anticoherency screw dislocations. We see that the structural unit model can be applied along only very specific trajectories in the 5D parameter space.

We have described the various ways in which the structural unit model may serve as an interpolation scheme for the structures of boundaries between two known delimiting boundaries. No other model of grain boundary structure has provided such a predictive capability or insight into the possible relationships between different boundary structures. The model is particularly useful for describing variations in structure along trajectories that involve changes in $\theta_{\text{twist}}$ or $\theta_{\text{tilt}}$, because it also provides the sequences of units into which intervening boundaries decompose.

As we have already discussed, the usefulness of the model hinges on the misorientation range between successive delimiting boundaries being sufficiently large. For example, the total misorientation range of (221) twist boundaries is 180°. If only, say, 6 delimiting boundaries are required to span this misorientation range then the model is a much more useful interpolation scheme than if 30 are required. With five times more delimiting boundaries the predictive capability of the model is reduced by a factor of five, even though the model may provide quite accurate interpolations between the small misorientation ranges. We have already discussed the selection of delimiting boundaries qualitatively. In the following we shall address the question more quantitatively, following Sutton (1989a). We shall examine the relationship between the Burgers vectors of anticoherency dislocations and the selection of delimiting boundaries, and the consequences of those Burgers vectors for the distortions of the delimiting boundary units.

The distortions of the units depend on whether sufficient local relaxation takes place to alter the local misorientations to those of the ideal delimiting boundary units. Another way of saying the same thing is that the distortions are less if the anticoherency dislocations associated with the minority units are more localized. If the Burgers vector is too large then it will not be localized and the distortions will be reduced to an acceptable level only by decreasing the misorientation range between delimiting boundaries. This follows directly from eqn (4.90) for the Burgers vector, $b_A$, of the anticoherency dislocation associated with the minority unit $B$. Mutliplying both sides of this equation by $l_A$ we obtain $b_A l_A = 2\mathcal{C}$, or

$$b_A = 2\mathcal{Q}/l_A \tag{4.94}$$

where $\mathcal{Q}$ is the area of a primitive cell in the decomposition lattice. Assuming that $\mathcal{Q}$ is the minimum value it may attain for a given rotation axis the only way to decrease $b_A$ is to increase $l_A$. If the periods of delimiting boundaries increase then the ranges of misorientation between them decreases, and the model has less predictive capability. In general $b_A$ will increase as the rotation axis becomes higher index, although this is not a smooth trend. The conclusion is that there is a tendency for more delimiting boundaries to be required in a given range of misorientations as the rotation axis becomes higher index, and the predictive capacity of the model decreases correspondingly.

Our discussion of the structural unit model has focussed so far on its ability to describe changes in boundary structure as the macroscopic degrees of freedom are varied. The changes in structure bring about changes in properties, and the model has been used to discuss grain boundary diffusivity and energy of tilt boundaries as a function of misorientation. The diffusivity is discussed in Section 8.2.3.2. We shall conclude this section with a discussion of grain boundary energy following Wang and Vitek (1986).

Consider a tilt boundary composed of $n_A$ A units and $n_B$ B units, such that $n_A > n_B$. Since B units are in the minority they correspond to the cores of anticoherency edge dislocations, with Burgers vector $b_A$, preserving the A unit reference structure. The energy of the boundary is divided into a core energy associated with a narrow strip where the grains meet, and an elastic field energy generated by the anticoherency dislocations. The elastic energy, in the isotropic approximation, for an array of edge dislocations of average spacing D, is given by (see Section 2.10.3)

$$\sigma_{el} = \frac{\mu b_A^2}{4\pi(1-\nu)D} \ln(eD/2\pi r_o). \tag{4.95}$$

The core energy is determined using the structural unit model. If there were no energy of interaction between A and B units then we could write the core energy as a weighted average of the energies of the delimiting A and B boundaries. But the interaction is not negligible, and since B units in the boundary are always separated by A units two kinds of interaction can be distinguished: A–A and A–B–A. The energy associated with an A–A interaction is equated with the energy per unit area of the A delimiting boundary, $\sigma_A$, multiplied by the length, $l_A \cos((\theta - \theta_A)/2)$, of one A unit *in the intervening boundary*. The energy associated with an A–B–A interaction is equated to the energy, $\sigma_{AB}$, per unit area of the AB boundary multiplied by the length, $l_{AB} = l_A \cos((\theta - \theta_A)/2) + l_B \cos((\theta_B - \theta)/2)$, of one AB repeat sequence in the intervening boundary. Since the boundary contains $n_A$ A units and $n_B$ B units there are $(n_A - n_B)$ A–A segments and $n_B$ A–B–A segments. Thus, the core energy is given by

$$\sigma_c = \frac{(n_A - n_B)l_A \cos((\theta - \theta_A)/2)\sigma_A + n_B[l_A \cos((\theta - \theta_A)/2) + l_B \cos((\theta_B - \theta)/2)]\sigma_{AB}}{n_A l_A \cos((\theta - \theta_A)/2) + n_B l_B \cos((\theta_B - \theta)/2)} \tag{4.96}$$

where we have used eqn (4.86) for the length of the repeat sequence of $n_A$ A units and $n_B$ B units. Using eqn (4.87) for the relationship between $n_A, n_B$, and $\theta$ and eqn (4.90) for $b_A$ it can be shown that

$$\sigma_c = \sigma_A + \frac{2\sin((\theta - \theta_A)/2)}{b_A} \frac{l_{AB}}{l_B}(\sigma_{AB} - \sigma_A). \tag{4.97}$$

It can also be shown that when $\theta = \theta_{AB}$ this equation gives $\sigma_c = \sigma_{AB}$, as it should do. If $\theta - \theta_A$ is small then this relationship may be well approximated by

$$\sigma_c = \sigma_A + \frac{\theta - \theta_A}{b_A} \frac{l_{AB}}{l_B} (\sigma_{AB} - \sigma_A) \tag{4.98}$$

and we see that the core energy is a linear interpolation between $\sigma_A$ and $\sigma_{AB}$. The same result applies to other properties that depend only on the core, such as the boundary diffusivity discussed in Section 8.2.3.2. The total energy of the boundary is given by the sum of the elastic energy, eqn (4.95), and the core energy:

$$\sigma = \sigma_A + (\theta - \theta_A) \left[ \frac{1}{b_A} \frac{l_{AB}}{l_B} (\sigma_{AB} - \sigma_A) + \frac{\mu b_A}{4\pi(1 - \nu)} \left[ \ln \left( \frac{eb_A}{2\pi r_o} \right) - \ln (\theta - \theta_A) \right] \right].$$
$$\tag{4.99}$$

The unlinearized formula is regained from this expression simply by replacing $(\theta - \theta_A)$ by $2\sin((\theta - \theta_A)/2)$. As noted by Wang and Vitek (1986) the elastic energy must be positive for obvious physical reasons and this imposes the following upper limit on $(\theta - \theta_A)$ for the Read–Shockley formula to be valid:

$$\sin((\theta - \theta_A)/2) \leqslant \frac{eb_A}{4\pi r_o}. \tag{4.100}$$

The only unknown quantity in eqn (4.99) for $\sigma$ is the core radius $r_o$. For the model to be internally consistent it should be roughly $l_{AB}/2$. Wang and Vitek showed that was the case for [001] and [111] symmetric tilt boundaries by fitting eqn (4.99) to a few computed boundary energies between delimiting boundaries. For the [001] symmetric tilt boundaries five delimiting boundaries between the (110) and (001) perfect crystal orientations were required to reproduce $\sigma(\theta)$ throughout the 90° misorientation range. However, the structures of the boundaries throughout the 90° misorientation could be reasonably well described with only two delimiting boundaries (i.e. (310) and (210)) between the perfect crystal orientations. The reason why more delimiting boundaries were required for $\sigma$ is that the maximum permissible deviation from a reference structure, given by eqn (4.100), is often small and therefore further delimiting boundaries are required. It should be noted that this limitation arises from the elastic energy, not the core energy. In section 8.2.3.2 we will see that the structural unit model describes properties that depend only on the boundary core using fewer delimiting boundaries. This example illustrates the point we made earlier that the selection of delimiting boundaries may vary with the property under study. But, in general we expect the choice of delimiting boundaries to model properties that depend only on the boundary core (such as boundary diffusivity) to be the same as the choice for describing the structure of the core.

### 4.3.1.9  *Three-dimensional grain boundary structures*

It is quite common in materials with a low stacking-fault energy for small-angle boundaries to contain crystal lattice dislocations that have dissociated into partial dislocations and stacking faults forming complex, 3D arrays. In these materials there are also HREM observations of faults emanating from some large-angle, sharp grain boundaries. In this way the relaxation extends occasionally into the adjoining crystals on inclined close-packed planes, and endows the boundary with a 3D form. Such interfaces cannot be described as diffuse, and since their core regions are sharply defined we are content to classify them as sharp.

One of the clearest examples of the formation of faults at a large-angle boundary is the discovery that certain tilt boundaries in Cu (Wolf *et al.* 1992) and Ag (Ernst *et al.* 1992) in the $\Sigma = 3$ coincidence system relax by forming a 1–2 nm layer of 9R phase at the boundary. In the $\Sigma = 3$ coincidence system in Cu, Omar (1987) observed that the (211) symmetric tilt boundary, which has a mean boundary plane of (100), occurred less frequently than the $(9\bar{6}\bar{6})/(11, 4, 4)$ asymmetric tilt boundary, which has a mean boundary plane of $(10, \bar{1}, \bar{1})$. The inclination of the asymmetric tilt boundary to the symmetric tilt is given by the angle between their mean boundary planes, which is just $\cos^{-1}(10/\sqrt{102}) = 8.05°$. Wolf *et al.* (1992) carried out thermal grooving experiments for a range of $\Sigma = 3$ $[01\bar{1}]$ tilt boundaries between the symmetric (111) and (211) inclinations and found that there is a local energy minimum at about 8° from the (211) inclination. This was also found by computer simulations using an embedded atom potential for Cu.

Figure 4.40(a) shows an HREM image of the $(9\bar{6}\bar{6})/(11, 4, 4)$ boundary. Figure 4.40(b) shows a computer simulation of the boundary which agrees very well with the HREM image. The shaded region at the boundary in Fig. 4.40(b) is a slab of 9R phase between the adjoining f.c.c. crystals. Inside the rectangle of the shaded area we can see the ... *ABCBCACAB* ... stacking sequence of the 9R phase. The 9R stacking sequence is obtained from the *ABC* stacking sequence of the f.c.c. phase by inserting an intrinsic stacking fault every third {111} plane. The stacking faults are bounded by Shockley partial dislocations at the edges of the slab. The arrays of Shockley partials produce sub-boundaries at the edges of the 9R phase of misorientation $2\sin^{-1}(\frac{1}{2}\sqrt{18}) = 13.54°$. Approximately half of the sub-boundary misorientation is taken up on either side of each sub-boundary. Therefore, the spacings of the close-packed planes in the 9R slab and in the adjoining f.c.c. crystals are very nearly equal.

The rotation of the close-packed planes inside the 9R slab is the key to understanding why the asymmetric tilt boundary has a lower energy than the (211) symmetric tilt boundary. The relaxed structure of the (211) boundary is shown in Fig. 4.41, which was also confirmed by HREM (Wolf *et al.* 1992). The boundary is also wide, as found originally by Crocker and Faridi (1980). The tendency to form the 9R phase inside a slab at the boundary is again evident, but the sub-boundary rotations must now be taken up entirely within the 9R phase because the close-packed planes in the adjoining f.c.c. crystals are constrained to be at 90° to the (211) boundary plane. The asymmetric distribution of the sub-boundary misorientation forces the close-packed planes in the 9R phase to have a smaller separation than those in the adjoining f.c.c. crystals, as illustrated schematically in Fig. 4.42. The increased atomic overlap generates hydrostatic stresses in the boundary plane and raises the boundary energy. The fault vectors in the 9R phase are slightly smaller than those of the asymmetric tilt boundary so as to reduce the sub-boundary rotation and the degree of atomic overlap.

The stacking sequence of close-packed planes in the 9R phase is a compromise between the reversed sequences that occur in the adjoining f.c.c. crystals on either side of a $\Sigma = 3$ grain boundary, and presumably this is why it is energetically favourable for it to occur. It is interesting to note that very small precipitates of Cu in an $\alpha$–Fe matrix have been found to adopt the 9R structure (Othen *et al.* 1991).

Although these boundaries are wide in comparison to other grain boundaries in metals, we still classify them as sharp interfaces because they are the products of a type of interface dissociation (see Fig. 6.4) involving the formation of an intermediate new phase, which is 9R in the present case. A full discussion of interface dissociation is deferred to section 6.3.1.3 because it is a type of interface phase transition. For the present we note the dissociation is energetically favourable if

$$\sigma^{1,2} > (\sigma^{1,3} + \sigma^{3,2} + wg^3), \tag{4.101}$$

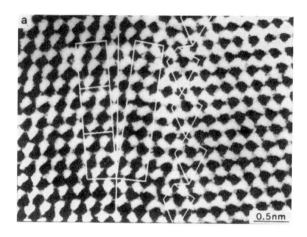

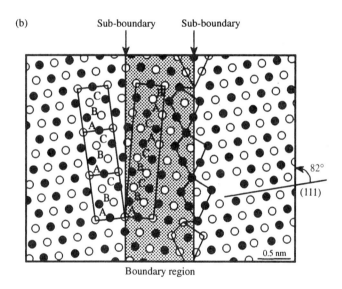

**Fig. 4.40** (a) HREM image of the $(9\bar{6}\bar{6})/(11, 4, 4)$ asymmetric $[01\bar{1}]$ tilt boundary in Cu (courtesy of F. Ernst), viewed in projection along the tilt axis. The grain boundary runs from top to bottom in the image. The atomic columns are black. The white lines indicate features which are seen in the computer relaxed structure (Finnis 1993) shown in (b). The match between (a) and (b) is almost perfect. The grey shaded stripe in (b) indicates a region which has transformed to the 9R phase.

where $\sigma^{1,2}$ is the energy per unit area of the undissociated grain boundary between phases 1 and 2, $\sigma^{1,3}$ and $\sigma^{2,3}$ are the energies per unit area of the interfaces bounding the new phase 3 (i.e. 9R), and $g^3$ is the excess free energy per unit volume of the new phase and $w$ its width. Similar grain boundary dissociations may be anticipated in other low stacking fault materials such as Co and SiC. In SiC we may speculate that other polytypic structures may be formed at a wide range of grain boundaries (not just $\Sigma = 3$), because $g^3$ may become negative for certain polytypes as the temperature is raised, and this may be a mechanism by which polytypic transformations are effected in the solid state. In this connection it is interesting to note that Merkle (1991) has observed stacking

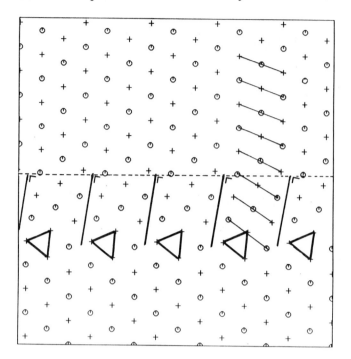

**Fig. 4.41** Relaxed structure of the (211) symmetric tilt boundary in Cu (Wolf *et al.* 1992). The boundary runs from left to right. The boundary region comprises a slab of 9R phase, which extends from the capped trigonal prisms (shown as triangles) to the sub-boundary (shown by the broken line). Stacking faults, bounded by Shockley partials, are indicated on every third (1̄11) plane inside the 9R phase.

faults emanating from several grain boundaries in Au, which are not in the $\Sigma = 3$ coincidence system.

   Another way in which grain boundaries acquire a 3D appearance is through faceting on a microscopic scale. Figure 4.43 shows an HREM image of a (114)/(110) asymmetric tilt boundary in the $\Sigma = 3$ coincidence system in Au. It has faceted into the (111) and (112) symmetric tilt inclinations and the individual facets are 30–40 Å in size. As discussed in Section 5.6.3, faceting is driven by an overall reduction in interfacial energy. The question we address here is what determines the length scale of the facets?

   The regularity of the faceting in Fig. 4.43 is remarkable, and other observations of such regularly faceted interfaces have appeared in the literature. As discussed in Sections 5.6.2 and 5.6.3, facet junctions are generally associated with dislocations (see Fig. 5.9) and lines of force. The Burgers vectors of the dislocations alternate in sign and they are equal to $\pm b$, where $b = (t_1 - t_2)$ and $t_1$ and $t_2$ are the rigid body displacements of the two facets. The lines of force arise from the different interface stress tensors in the two kinds of facet (interface stress is defined and discussed in Section 5.4). We shall now analyse the contributions to the total energy from the dislocations and lines of force at the facet junctions. We will see that a regular periodic array of the facets is stabilized by the lines of force, and that there is a characteristic length scale for the microfaceting. This is directly analogous to the stability of elastic stress domains at free surfaces (see Vanderbilt *et al.* (1989) and references therein).

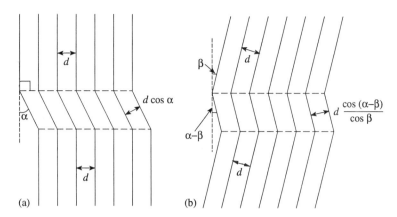

**Fig. 4.42** Schematic diagrams of the structures of the (a) (211) and (b) $(9\bar{6}\bar{6})/(11, 4, 4)$ $[01\bar{1}]$ tilt boundaries in Cu. Both boundaries contain slabs of 9R phase between the horizontal broken lines. The material inside each slab is rotated by $\alpha = 2 \sin^{-1} (\frac{1}{2}\sqrt{18}) = 13.54°$ by the sub-boundaries along the broken lines. (In the diagrams $\alpha$ has been enlarged for clarity.) The traces of close-packed planes are shown crossing each boundary. The spacing of the close-packed planes is marked as $d$. In the (211) symmetric tilt boundary, (a), the spacing of the close-packed planes in the 9R phase has been reduced to $d \cos \alpha$, i.e. about 3 per cent. By contrast, in the $(9\bar{6}\bar{6})/(11, 4, 4)$ asymmetric tilt boundary, (b), the spacing of the close-packed planes has been increased to $d_1 = d \cos (\alpha - \beta)/\cos \beta$. Since $\beta = 8.05°$, then $d_1$ is 0.5 per cent larger than $d$. The greater atomic overlap arising from the decreased spacing of the close-packed planes of the 9R phase in the (211) boundary is the reason for the greater energy of this boundary.

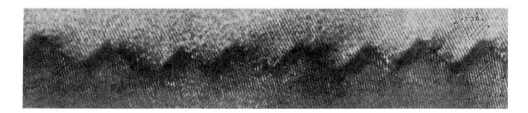

**Fig. 4.43** HREM image of a $\Sigma = 3$ (114)/(110) asymmetric $[1\bar{1}0]$ tilt boundary in Au (Krakow and Smith 1986*b*). Notice the regular faceting into the (111) and (112) symmetric tilt inclinations. Individual facets are 30–40 Å in size.

The elastic energy of the dislocations can be divided into three terms. The first is the energy per unit area, $\sigma_1$, of the dislocations with Burgers vector $\boldsymbol{b}$, in the absence of the dislocations with Burgers vector $-\boldsymbol{b}$. The second is the energy, $\sigma_2$, of the dislocations with Burgers vector $-\boldsymbol{b}$, in the absence of the dislocations with Burgers vector $\boldsymbol{b}$. Clearly, $\sigma_1 = \sigma_2 > 0$. The third is the energy of interaction, $\sigma_{12}$, between the dislocations in the first two sets, and $\sigma_{12} < 0$. The sum, $\sigma_d^{\text{el}} = \sigma_1 + \sigma_2 + \sigma_{12}$, will always be positive and it will have the form

$$\sigma_d^{\text{el}} = \frac{K_{ij}b_ib_j}{d} \{\ln(d/r_0) - C\} \tag{4.102}$$

where the facet length is $2d$, $C$ is a constant arising from the interaction energy $\sigma_{12}$, $K_{ij}$ depends on the elastic constants, and summation over $i$ and $j$ from 1 to 3 is implied. The core radius is $r_0$. The energy per unit area associated with the cores of the facet junctions is $\sigma^{\text{core}} = C'/d$, where $C'$ is another positive constant. Therefore the total energy of the dislocations, $\sigma_d^{\text{el}} + \sigma_d^c$, decreases monotonically as the facet length, $2d$, increases. The presence of dislocations favours the maximum possible facet length.

The physical origin of the line force is that each interface is in a state of 2D stress. Although each interface stress tensor has components only in the plane of the interface the line of force will, in general, have three non-zero components because the facets are inclined to each other. The force per unit length, $\varphi$, is equal to the discontinuities in the resolved components of the interface stresses parallel and perpendicular to the junction line. The lines of force alternate in sign. They induce an elastic relaxation, the energy of which is *negative*. To see this let $f(r)$ be some distribution of body force, and let $u(r)$ be the relaxation displacement field induced by $f(r)$. The relaxation energy, to second order in $u(r)$, is given by

$$E_{\text{rel}} = -\int f(r)u(r)dr + \frac{1}{2}\int\int u(r')D(r' - r)u(r)\,dr\,dr' \qquad (4.103)$$

where $D(r - r')$ couples the displacement fields at $r$ and $r'$, and mechanical stability requires that the second term on the right is positive definite. In a discrete lattice model $D(r - r')$ is the matrix of force constants. Minimizing the relaxation energy with respect to $u(r)$ gives

$$f(r) = \int D(r - r')u(r')\,dr'. \qquad (4.104)$$

Substituting this back into eqn (4.103) we obtain

$$E_{\text{rel}} = -\frac{1}{2}\int f(r)u(r)\,dr = -\frac{1}{2}\int\int u(r')D(r' - r)u(r)\,dr\,dr', \qquad (4.105)$$

which is always negative because the double integral is positive definite. Since the displacement field associated with a straight line of force has the same functional form as that of a straight dislocation, with $\varphi$ replacing $b$, and a different combination, $Q_{ij}$, of elastic constants replacing $K_{ij}$ (see Bacon *et al.* 1978) the relaxation energy per unit area has the form

$$\sigma_f^{\text{el}} = -\frac{Q_{ij}\varphi_i\varphi_j}{d}\ln(d/r_0). \qquad (4.106)$$

Use has been made of St Venant's principle here to limit the range of the strain field to cylinders of radius $d$ centred on each line of force. The same argument was used to derive the Read–Shockley formula in eqn (2.141). The total energy per unit area of the facet junctions is therefore given by

$$\sigma_{\text{junctions}} = \sigma_d^{\text{el}} + \sigma_f^{\text{el}} + \sigma^{\text{core}}$$

$$= \frac{K_{ij}b_ib_j}{d}\{\ln(d/r_0) - C\} - \frac{Q_{ij}\varphi_i\varphi_j}{d}\ln(d/r_0) + \frac{C'}{d}$$

$$= -\frac{C_{\text{el}}}{d}\ln(d/r_0) + B/d \qquad (4.107)$$

where

$$C_{el} = Q_{ij}\varphi_i\varphi_j - K_{ij}b_ib_j \qquad (4.108)$$

and

$$B = C' - K_{ij}b_ib_jC.$$

We see that $C_{el}$ and $B$ may each be positive or negative. Differentiating the total energy of the facet junctions we find that it is an extremum when

$$\ln(d/r_o) = \frac{C_{el} + B}{C_{el}} \qquad (4.109)$$

whereupon the total energy becomes

$$\sigma_{\text{junctions}} = -\frac{C_{el}}{d} = -\frac{C_{el}}{r_o\exp\left((C_{el} + B)/C_{el}\right)}. \qquad (4.110)$$

We conclude that if $C_{el} > 0$ then the presence of facet junctions lowers the energy of the system, and there is an equilibrium facet length of $2r_o\exp\left((C_{el} + B)/C_{el}\right)$. This facet length may be large compared with $r_o$.

The interface stress tensor represents a source of potential energy in the system. Some of the potential energy is released when faceting occurs because work is done, through relaxation displacements, by lines of force that are generated at facet junctions by discontinuities in the interface stresses. It is the release of potential energy which favours the fine scale of the faceting, against which must be balanced the energy cost of cores of the facet junctions. The presence of dislocations at the facet junctions alters the balance, but they do not by themselves favour faceting on a fine scale. *The fine scale of the faceting is therefore a form of interface reconstruction which is driven by interface stresses.* As we will discuss in Section 6.3.1.3, interface dissociation may also involve faceting (see Figs 6.4 and 6.5), and the length scale of the facets is again governed by the same considerations we have discussed here.

### 4.3.1.10  *The influence of temperature*

We may expect interfaces to disorder in various ways during heating to produce structures of higher entropy than those we have discussed so far in this chapter. In this section we shall confine ourselves to a discussion of the consequences of vibrational disorder for the atomic structure and thermodynamic properties of grain boundaries in metals. Other forms of disorder, such as roughening and melting transitions, are deferred to Sections 6.3.2.1 and 6.3.2.4. Other temperature-induced phase transitions at interfaces are also discussed in Chapter 6. A full discussion of the thermodynamics of interfaces is given in the next chapter.

The excess entropy of a grain boundary in a single component solid is given by $\Delta S = -(\partial\sigma/\partial T)_p$ (see eqn (5.27)). In Table 5.1, measured average values $\Delta S$ for general boundaries in Au and Ni are given as 0.1 and 0.2 mJ m$^{-2}$ K$^{-1}$ respectively. Murr (1970, 1972) measured the ratio of (111) twin boundary energy, $\sigma_t$, to general grain boundary energy, $\sigma_g$, in Cu at 700, 800, and 900 °C and obtained a mean value of $\sigma_t/\sigma_g = 0.034$ *at each temperature*. If it is assumed that the excess entropy and enthalpy of the boundary are independent of temperature, then the fact that $\sigma_t/\sigma_g$ is independent of temperature implies that the excess enthalpy is directly proportional to the excess entropy: $\Delta H = T_c\Delta S$. The constant of proportionality, $T_c$, has been called the 'compensation tempera-

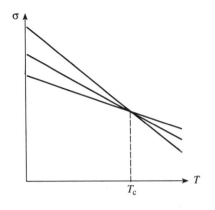

**Fig. 4.44** A schematic plot of the free energies, $\sigma$, of three grain boundaries as a function of temperature. $T_c$ is called the compensation temperature. As $T$ approaches $T_c$ the differences in the boundary free energies diminsh, until, at $T = T_c$, the free energies are equal. At $T > T_c$ the order of the free energies of the boundaries is reversed.

ture' by Shvindlerman and coworkers (Fridman *et al.* 1974, Kopetsky *et al.* 1978, Molodov *et al.* 1984, Straumal *et al.* 1984).

The significance of the compensation temperature is that the free energies of all interfaces in a given material are equal at the compensation temperature, as shown schematically in Fig. 4.44. The compensation temperature may be greater than the melting point. In the case of Cu, Murr (1972) quotes a value of the excess entropy for a general grain boundary in Cu of $0.5 \, \text{mJ m}^{-2} \text{K}^{-1}$, and the excess free energy of a general boundary at 700 °C of $730 \, \text{mJ m}^{-2}$. It follows that $T_c = 2433 \, \text{K}$ in Cu. As the temperature is increased towards $T_c$ the difference in free energies between any two interfaces in a material decreases. Experimental evidence for this has been provided by the rotating spheres on a plate experiments of Erb and Gleiter (1979). Small Cu spheres were sintered on a single crystal plate and rotated to low-energy misorientations under the driving force of the interfacial energy between the sphere and the plate. The number of misorientations into which the spheres rotated decreased as the temperature increased. Assuming that a sphere is trapped at a given misorientation only if the depth of the local energy well is greater than some critical minimum value, then as temperature increases the number of trapping wells decreases as the energy differences between boundaries become less than the critical value.

Klam *et al.* (1987) measured the average thermal expansion coefficient of grain boundaries in polycrystalline Cu specimens. They obtained values of $40$–$80 \times 10^{-6} \text{K}^{-1}$ at room temperature, which are between 2.5 and 5 times the thermal expansion coefficient of a single crystal of Cu. Fitzsimmons *et al.* (1988), using X-ray diffraction techniques, measured the thermal expansion coefficients of the $\Sigma = 13$ (001) twist boundary in Au both parallel and perpendicular to the boundary plane. At 255 K they obtained $12 \pm 2 \times 10^{-6}$ and $43 \pm 4 \times 10^{-6}$ respectively, which should be compared with the bulk thermal expansion coefficient at room temperature of $14 \times 10^{-6}$. They also measured the mean square displacements in the grain boundary and in the bulk at 298 K and obtained $0.012 \pm 0.002 \, \text{Å}^2$ and $0.008 \pm 0.0005 \, \text{Å}^2$ respectively.

The picture that emerges from these experimental measurements is that grain boundaries are more anharmonic environments than the adjoining crystals. Moreover, the anharmonicity is not isotropic, being larger normal than parallel to the boundary plane. It also appears that the excess entropy is proportional to the excess enthalpy, at least over a limited temperature range. However, it must be said that the experimental database is very small indeed. Computer simulations have been carried out using the temperature-dependent interatomic forces described in Section 3.9. Before we discuss these simulations

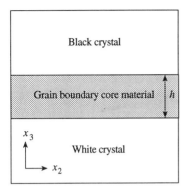

Black crystal

Grain boundary core material   $h$

$x_3$

White crystal

$x_2$

**Fig. 4.45**   To illustrate the model used in Section 4.3.1.10.

we shall first describe a phenomenological model that is suggested by the experimental data. We shall see that the model can explain some of the results of the computer simulations, in particular the relationship between the temperature dependence of the excess entropy and the excess thermal expansion. It also sheds further light on the possible existence of a compensation temperature. We also refer the reader to the analytic models of Ewing (1971), Ewing and Chalmers (1972), and Provan and Bamiro (1977), for insight into the atomistic origins of the excess entropy and thermal expansion.

Consider a bicrystal in which the grain boundary core material is a homogeneous slab of thickness $h$, as shown in Fig. 4.45. At some reference temperature, $T°$, the boundary expansion is $e°$ and excess internal energy and entropy are $\Delta U°$ and $\Delta S°$ per unit area respectively. (In the following we shall work with the Helmholtz rather then the Gibbs free energy, but the difference between them is negligible at atmospheric pressure. Consequently we consider the excess internal energy rather than the excess enthalpy.) For our purposes the reference temperature should be above the 'Debye temperature', $\theta$, given by eqn (3.232) (but lower than the melting point!). Outside the slab the Debye temperature is $\theta$, and inside it is $\theta_{gb}$. In general, we expect $\theta_{gb}$ to be less than $\theta$, and to become lower still for a larger boundary expansion, $e°$, and thermal expansion. We assume the material outside the slab has cubic point group symmetry, with a coefficient of linear thermal expansion $\alpha$, a Gruneisen constant $\gamma$, and a bulk modulus $B$. As the temperature is raised the slab must remain commensurate with the adjoining material and, therefore, the thermal strain parallel to the boundary must be the same inside and outside the slab. This boundary condition is consistent with the measurements of the thermal expansion coefficients by Fitzsimmons *et al.* (1988). But normal to the boundary plane the thermal strain in the slab is expected to be larger. Thus the slab has tetragonal symmetry. Let $x_1, x_2$ be parallel to the slab and $x_3$ be perpendicular. The thermal strains $\varepsilon_{11}, \varepsilon_{22}$, and $\varepsilon_{33}$ in the slab set up elastic stresses $\tau_{11}, \tau_{22}$, and $\tau_{33}$ given by

$$\left.\begin{aligned} \tau_{11} &= c_{11}\varepsilon_{11} + c_{12}\varepsilon_{22} + c_{13}\varepsilon_{33} \\ \tau_{22} &= c_{12}\varepsilon_{11} + c_{11}\varepsilon_{22} + c_{13}\varepsilon_{33} \\ \tau_{33} &= c_{13}\varepsilon_{11} + c_{13}\varepsilon_{22} + c_{33}\varepsilon_{33} \end{aligned}\right\} \tag{4.111}$$

which involve four independent elastic constants in the slab: $c_{11}, c_{12}, c_{13}$, and $c_{33}$. Setting $\varepsilon_{11} = \varepsilon_{22} = \varepsilon$ and $\varepsilon_{33} = \varepsilon_\perp$ we can express the elastic energy density in the slab as

$$W_{el} = c_{11}\varepsilon^2 + c_{12}\varepsilon^2 + 2c_{13}\varepsilon\varepsilon_\perp + \tfrac{1}{2}c_{33}\varepsilon_\perp^2. \tag{4.112}$$

The thermal free energy outside the slab is given by eqn (3.231):

$$F = 3kT[\tfrac{1}{4} - \ln(2T/\theta)] \quad \text{per atom} \tag{4.113}$$

and the entropy is given by eqn (3.234):

$$S = 3k[\ln(2T/\theta) - \tfrac{3}{4}] \quad \text{per atom.} \tag{4.114}$$

The same expressions apply for the free energy and entropy per atom within the slab if $\theta$ is replaced by $\theta_{gb}$. The thermal internal energy per atom is $3kT$, and it is the same inside and outside the slab.

We shall first calculate the linear thermal strain, $\varepsilon$, outside the slab. This is also the thermal strain inside the slab parallel to the boundary plane. The free energy per atom of the bulk in the presence of the thermal strain may be written as

$$F_{bulk} = \frac{9B\Omega\varepsilon^2}{2} + 3kT[\tfrac{1}{4} - \ln(2T/\theta)] \quad \text{per atom} \tag{4.115}$$

where the first term accounts for the elastic energy per atom, and $\Omega$ is the atomic volume. Minimizing $F_{bulk}$ with respect to $\varepsilon$, at constant temperature, yields

$$\varepsilon = \frac{k\gamma T}{B\Omega} \tag{4.116}$$

where we have used $d\ln\theta/d\ln V = -\gamma$ and $d\ln V/d\varepsilon = 3$. Thus, the linear coefficient of thermal expansion, $\alpha$, in the bulk is given by $k\gamma/B\Omega$, which is a standard result.

The free energy in the slab is given by

$$F_{gb} = \Omega_{gb} W_{el} + 3kT[\tfrac{1}{4} - \ln(2T/\theta_{gb})] \quad \text{per atom} \tag{4.117}$$

where $\Omega_{gb}$ is the atomic volume in the slab, and $W_{el}$ is the elastic energy density given by eqn (4.112). Minimizing this free energy with respect to $\varepsilon_{33} = \varepsilon_\perp$ we obtain

$$\varepsilon_\perp = \alpha_\perp T$$

where

$$\alpha_\perp = \frac{k}{c_{33}}\left[\frac{3\gamma_{gb}}{\Omega_{gb}} - \frac{2c_{13}\gamma}{B\Omega}\right] \tag{4.118}$$

where $\gamma_{gb}$ is the Grüneisen constant inside the slab. The excess thermal expansion coefficient of the boundary, which leads to the larger thermal strain normal to the boundary plane, is given by

$$\alpha_\perp - \alpha = \frac{3k}{c_{33}}\left[\frac{\gamma_{gb}}{\Omega_{gb}} - \frac{\gamma(2c_{13} + c_{33})}{\Omega(2c_{13}^{bulk} + c_{33}^{bulk})}\right] \tag{4.119}$$

where we have expressed the bulk modulus in the bulk as $(2c_{13}^{bulk} + c_{33}^{bulk})/3$. We see that the excess thermal expansion stems from the differences in the elastic constants, the atomic volumes, and the Grüneisen constants.

The total expansion of the boundary comprises the expansion, $e^\circ$, of the reference state plus the excess thermal expansion, $(\alpha_\perp - \alpha)(T - T^\circ)$. It is important to note that the change in excess free energy per atom of the slab, as a function of temperature, is *second* order in the excess thermal expansion. That is because the equilibrium condition for the slab requires that there is no first-order variation in the free energy with thermal expansion. By contrast, the contribution to the excess free energy arising from the expansion, $e^\circ$, of the reference state, may vary linearly with $e^\circ$. This distinction turns out to be crucial.

The excess entropy per atom is given by eqn (4.114):

$$\Delta S = 3k \ln(\theta/\theta_{gb}). \tag{4.120}$$

Expanding $\ln \theta$ about the reference temperature we obtain, to first order,

$$\ln \theta = \ln \theta^\circ + \frac{d \ln \theta}{d \ln V} \frac{d \ln V}{dT} (T - T^\circ) \tag{4.121}$$

and hence we can write

$$\Delta S = \Delta S^\circ + 3k \{2(\gamma_{gb} - \gamma)\alpha + \gamma_{gb}\alpha_\perp - \gamma\alpha\}(T - T^\circ) \text{ per atom} \tag{4.122}$$

where $\Delta S^\circ = 3k \ln(\theta^\circ/\theta_{gb}^\circ)$. If we make the simplifying assumption that $\gamma_{gb} = \gamma$ then this expression becomes

$$\Delta S = \Delta S^\circ + 3k\gamma(\alpha_\perp - \alpha)(T - T^\circ) \text{ per atom} \tag{4.123}$$

and we see that the temperature dependence of the excess entropy is directly proportional to the excess thermal expansion coefficient, $\alpha_\perp - \alpha$, given by eqn (4.119) with $\gamma_{gb} = \gamma$. We shall see below that this relationship is borne out by the computer simulations, and that $\gamma_{gb} \simeq \gamma$ is not unreasonable.

Does our simple model imply the existence of a compensation temperature? We will show that it does provided two conditions are satisfied: (i) the temperature dependence of the excess entropy is neglected and (ii) the excess internal energy, $\Delta U^\circ$, of the reference state is proportional to the boundary expansion, $e^\circ$. There is no excess thermal internal energy because the internal thermal energy per atom inside and outside the slab is $3kT$. Hence, the excess internal energy is $\Delta U^\circ$. Since the excess entropy per atom is $\Delta S^\circ$, a compensation temperature, $T_c$, exists if

$$\Delta U^\circ = T_c \Delta S^\circ. \tag{4.124}$$

Consider $\Delta S^\circ$. It is given, per atom, by

$$\Delta S^\circ = 3k \ln(\theta^\circ/\theta_{gb}^\circ). \tag{4.125}$$

Expanding $\ln(\theta_{gb}^\circ)$ about $\ln(\theta^\circ)$ to first order in the expansion $e^\circ$, we have

$$\ln(\theta_{gb}^\circ) = \ln(\theta^\circ) + \frac{d \ln \theta}{d \ln V} \frac{d \ln V}{de} e^\circ \tag{4.126}$$

where $d \ln \theta/d \ln V = -\gamma$ and $d \ln V/de = 1/h$. Therefore,

$$\Delta S^\circ = 3k\gamma e^\circ/h \text{ per atom.} \tag{4.127}$$

To express $\Delta S^\circ$ per unit area we multiply $\Delta S^\circ$ per atom by the volume of the slab per unit area, which is just $h$, and divide by the atomic volume in the slab, which is $\Omega_{gb}$:

$$\Delta S^\circ = 3k\gamma e^\circ/\Omega_{gb} \text{ per unit area.} \tag{4.128}$$

This expression was derived earlier by H. O. K. Kirchner (private communication via J Thibault). For $e^\circ = 0.5$ Å, $\Omega_{gb} = 2 \times 10^{-29}$ m$^3$, and $\gamma = 3$ we obtain $\Delta S^\circ \simeq 0.3$ mJ m$^{-2}$ K$^{-1}$, which is in reasonable agreement with the values given in Table 5.1.

From eqns (4.124) and (4.128) we conclude that for a compensation temperature to exist the excess internal energy, $\Delta U^\circ$, must also be proportional to $e^\circ$. We have seen in section 4.3.1.5 that this relationship holds approximately (see Fig. 4.22). If we write

$\Delta U^\circ = 2\sigma_s e^\circ / L$ per unit area, where $\sigma_s$ is the (average) free surface energy, and $L$ is a length of the order of the lattice parameter, then the compensation temperature is given by

$$T_c = \frac{2\sigma_s \Omega_b}{3k\gamma L}. \tag{4.129}$$

Let us insert some typical values of these variables for Cu. With $\sigma_s = 2 \, J \, m^{-2}$, $\Omega_b = 2 \times 10^{-29} \, m^3$, $\gamma = 2$, $k = 1.4 \times 10^{-23} \, J \, K^{-1}$, and $L = 3.7 \times 10^{-10} \, m$, we obtain $T_c = 2570 \, K$, which is very close to the value (2433 K) we deduced from Murr's experimental data!

We can estimate the severity of ignoring the temperature dependence of the excess entropy by calculating the temperature difference, $T - T^\circ$, at which the temperature-dependent contribution in eqn (4.123) is equal to $\Delta S^\circ$. Equating these two terms (per unit area) we have

$$\frac{3k\gamma e^\circ}{\Omega_{gb}} = 3k\gamma (\alpha_\perp - \alpha)(T - T^\circ) h / \Omega_{gb}$$

or

$$e^\circ / h = (\alpha_\perp - \alpha)(T - T^\circ). \tag{4.130}$$

The meaning of this equation is clear. The thermal contribution to the excess entropy is equal to excess entropy of the reference state when the excess thermal strain between $T^\circ$ and $T$ is equal to the strain $e^\circ / h$ of the reference state. Using the value of $\alpha_\perp - \alpha$ measured by Fitzsimmons *et al.* (1988) of approximately $30 \times 10^{-6} \, K^{-1}$, and assuming $e^\circ = 0.5 \, Å$ and $h = 6 \, Å$, we obtain $(T - T^\circ) \simeq 2780 \, K$. The justification for choosing $h \simeq 6 \, Å$ will be given by the simulations discussed below. We conclude that the concept of a compensation temperature is meaningful over temperature ranges of a few hundred kelvin.

Early atomistic simulations of the thermodynamic properties of interfaces, e.g. Hashimoto *et al.* (1981) calculated the local phonon density of states at each atomic site 0 K. In harmonic lattice theory there are standard results relating thermodynamic quantities, such as the entropy, free energy, and specific heat to integrals over the phonon density of states (see Section 3.9). The weakness of this approach is that it is not thermodynamically self-consistent. At $T \neq 0 \, K$ the average atomic positions are determined by minimization of the free energy, not the internal energy. As the thermal contribution to the free energy increases, the boundary structure changes, and it is this change which was ignored in the early treatments. In the treatments discussed below the structure evolves as the temperature is increased, such that the atomic positions are those which minimize the free energy at a given temperature. This is what we mean by thermodynamic self-consistency. Although the analytic models of Ewing (1971), Ewing and Chalmers (1972), and Provan and Bamiro (1977) were also not self-consistent, those authors attempted to take account of the change of structure with temperature by adding a term to account for the temperature-dependence of the entropy. That term was called the 'anharmonic entropy', because it originates from the anharmonic nature of atomic interactions.

A thermodynamically self-consistent simulation of the structure of an interface was carried out by Sutton (1989*b*). The simulation was based on the quasiharmonic lattice theory discussed in Section 3.9, and in particular on the temperature-dependent forces

**Table 4.4** Calculated boundary expansion ($e$) as a ratio of the equilibrium crystal lattice parameter, $a_T$, at temperature $T$, excess free energy ($\Delta F$), excess entropy ($\Delta S$) and excess specific heat ($\Delta C_v$) for the $\Sigma = 13$ (001) twist boundary in Au at three temperatures, as computed by Sutton (1989$b$).

| $T$ (K) | $e(a_T)$ | $\Delta F$ (mJ m$^{-2}$) | $\Delta S$ (mJ m$^{-2}$ K$^{-1}$) | $\Delta C_v$ (mJ m$^{-2}$ K$^{-1}$) |
|---|---|---|---|---|
| 0 | 0.1096 | 1091 | 0 | 0 |
| 600 | 0.1148 | 1028 | 0.0664 | $2.556 \times 10^{-3}$ |
| 1200 | 0.1231 | 948 | 0.1555 | $1.163 \times 10^{-3}$ |

given by eqn (3.241). Sutton considered the $\Sigma = 13$ (001) ($\theta = 22.6°$) twist boundary in Au that was studied experimentally by Fitzsimmons *et al.* (1988). The potential energy was described by a Lennard–Jones pair potential. The boundary was simulated at 0, 600, and 1200 K. The free energy, given by eqn (3.227), was minimized using the border conditions shown in Fig. 4.23. Each simulation was carried out at constant temperature, and the perfect crystal slabs of region II were expanded to the lattice parameter, $a_T$, which minimized the free energy of the perfect crystal at the temperature, $T$, of the simulation. The periodic border conditions parallel to the boundary plane ensured that the thermal expansion of the grain boundary core material parallel to the boundary plane equalled the thermal expansion of the adjoining crystals. But the thermal expansion normal to the boundary plane was not constrained.

The main results of the simulations are summarized in Table 4.4. The fact that the boundary expansion, $e$, is an increasing function of $a_T$ means that the thermal expansion of the boundary core material normal to the boundary plane is greater than the thermal expansion of the perfect crystal. The excess free energy decreases by about 13 per cent over the 1200 K temperature range. Notice that the excess entropy increases by a factor of about 2.3 between 600 K and 1200 K. This indicates a rather strong temperature dependence for the excess entropy, and would not be consistent with the existence of a compensation temperature. In the final column we see the excess specific heat. In the classical limit the specific heat per atom is $3k$, regardless of where the atom is located in the solid. Thus, at high temperatures the excess specific heat converges to zero. At very low temperatures, where quantum effects dominate, the excess specific heat is again zero, because the specific heat per atom converges to zero, regardless of where the atom is located. The excess specific heat must increase to a maximum at some temperature between 0 K and the Debye temperature and decrease to zero again at higher temperatures. It follows that quantum effects play a crucial role in the variation of the excess specific heat with temperature.

By far the most significant effect of temperature on the boundary structure was the increase in expansion. Atomic separations in the plane of the boundary changed by an order of magnitude less than separations normal to the boundary plane. The mean square displacements $\langle u_i^2 \rangle$ differed from the bulk values by as much as 110 per cent in the first atomic layer adjacent to the boundary plane. But by the third layer this difference had dropped to less than 3 per cent. Thus the width, $h$, of the boundary core material is about three (002) interplanar spacings or 6 Å. The local Grüneisen constant, given by eqn (3.239), differed from the bulk value by less than 10 per cent at all atomic sites except one, where it differed by 17 per cent. This justifies the simplifying assumption, made in eqn (4.123), that $\gamma_{gb} \simeq \gamma$. As had been noticed by Hashimoto *et al.* (1981) there is a strong correlation between the local hydrostatic stress state, eqn (4.85), and the local

Debye temperature $\theta_i$: tensile sites are associated with low values of $\theta_i$. This is physically obvious because tensile sites are softer. They are therefore more able to absorb heat at low temperatures. Because of the highly non-linear dependence of thermodynamic functions, such as the entropy, on the local stiffness it is quite possible for the thermodynamic properties of a grain boundary to be dominated by the one or two softest sites in each boundary period. These sites are readily identified because they have large bond lengths to neighbouring sites: the atom is rattling around in a rather large cage.

A systematic, thermodynamically self-consistent study of twelve (001) twist boundaries in Au for $0 \leqslant T \leqslant 700$ K was carried out by Najafabadi *et al.* (1991). The misorientations of the twist boundaries varied between 16° and 41°. They used the classical Einstein model of eqn (3.214) with temperature-dependent forces given by eqn (3.215), and an embedded atom potential for Au representing the potential energy, $E_P$. The border conditions of Fig. 4.23 were used. Because they used the classical limit of the Einstein approximation the values of thermodynamic functions they computed are valid only at temperatures greater than the Debye temperature, which is 170 K in Au. However, the average value of the excess entropy they obtained at 200 K was about 0.09 mJ m$^{-2}$ K$^{-1}$ and increased to about 0.12 mJ m$^{-2}$ K$^{-1}$ at 700 K. This variation of the excess entropy with temperature is in line with the estimates we made using eqn (4.130). The temperature variation of the excess entropy mirrored the temperature variation of the excess linear thermal expansion coefficient, in agreement with eqn (4.123). The excess internal energy was found to increase only very slightly with temperature, in agreement with the argument preceding eqn (4.124).

At room temperature, Najafabadi *et al.* found that mean square displacements are approximately 25 per cent larger than those of the bulk at the (002) planes immediately adjacent to the geometrical boundary plane, and decayed to less than 2 per cent of the bulk value by the second (002) plane. Once again we see that the width of the boundary core material, $h$, as measured by the excess mean square displacements, is about 6 Å.

### 4.3.2  Grain boundaries in ionic crystals

To a large extent the structures of grain boundaries in real ionic materials are strongly affected by the presence of impurities, which are extremely difficult to remove. In many technologically important ionic ceramics, additives are deliberately incorporated in order to improve the sintering of a powder compact. One of the ways in which these additives effect an improvement is by forming a thin, glassy layer at the grain boundaries. In this section we shall not address such interfaces, and concentrate instead on 'pure' ionic materials, where, according to available HREM evidence, intermediate glassy layers do not form. Since most of the available information from HREM and computer modelling has been obtained for NiO, this material will be considered in greatest detail.

Most of the concepts that have been developed for grain boundaries in metals are also applicable to ionic crystals. But there are three fundamental differences. The first is that the repulsive electrostatic interaction between ions of like sign is very strong, and the boundary will undergo relaxations and reconstructions to avoid ions of like sign being nearest neighbours. Therefore, the polyhedral unit model is not applicable, but the structural unit model still applies in the same way as for metals. Secondly, the presence of two of more ionic species introduces the possibility of a greater variety of possible structures for a particular boundary plane, for the reasons described in Section 1.5.7. Finally, the boundary core may acquire a net charge, which must be compensated by space charge layers that can extend considerable distances into the adjoining crystals. The net charge may arise either from inbalances in the formation energies of intrinsic defects (Duffy and Tasker 1984a) or from the segregation of impurities (see Section 7.6).

Grain boundaries in ionic crystals have not met with the same degree of agreement between experiment and computer simulations as grain boundaries in metals. At the time of writing there is no concensus about the structure of tilt boundaries, although twist boundaries are thought to be reasonably well understood. In this section we shall propose some general principles which we believe may be extracted from the experimental and theoretical work to date. However, we are conscious that the database from which we are drawing is very small, and therefore our analysis is speculative.

Early computer simulations of (001) twist boundaries in oxides with the NaCl structure (Wolf 1980, Wolf and Benedek 1981, Wolf 1982) suggested that the boundary expansion was so large that the boundary was barely stable with respect to cleavage into two free (001) surfaces. This result was at variance with experiments by Sun and Balluffi (1982) in which (001) twist boundaries in MgO were found to be stable and to display arrays of screw dislocations, in much the same way as (001) twist boundaries in Au.

The reason for the low stability of the computer generated structures was that, in the first layers on either side of the geometrical boundary plane, pairs of ions of like charge became nearest neighbours. Since the c.n.i.d. for these boundaries is small it is not possible to remove the repulsion between ions of like charge through the rigid body translation parallel to the boundary plane. The repulsion was relieved only by increasing the expansion normal to the boundary plane. In fact the expansion was so great ($0.33a$ at the $\Sigma = 5$ boundary in MgO) that the weak residual attraction between the two crystals resulted in a cleavage energy of only 50 mJ m$^{-2}$. Having identified this as the cause, Tasker and Duffy (1983) solved the problem by removing one ion in each pair of like charge. In the resulting structure the nearest neighbours of each ion had the opposite charge, and the remaining Coulomb interactions across the boundary were all attractive. For the $\Sigma = 5$ boundary the expansion fell to $0.033a$ and the ideal cleavage energy rose to 1530 mJ m$^{-2}$. Further removal of ions raised the energy of the boundary. Therefore, at 0 K the *ground state* of the boundary contains a well-defined concentration of vacant cation and anion sites. In single crystals vacant sites are sometimes called Schottky defects (pairs of anion and cation vacancies), but in the grain boundaries we are discussing this is entirely inappropriate because the 'defects' are an intrinsic part of the ground state; if the 'defects' are removed by filling the vacant sites the energy of the boundary increases. For (110) twist boundaries Duffy (1986) calculated that the (220) planes on either side of the boundary plane are also less dense than in the perfect crystal.

A similar conflict between early computer simulations and later HREM experiments has taken place with tilt boundaries. Computer simulations of [001] symmetric tilt boundaries in NiO (Duffy and Tasker 1983) were found to have very open structures, although the ideal cleavage energies were all greater than 1 J m$^{-2}$. In Fig. 4.46 we show the relaxed structure for the (310) boundary. It was obtained by minimizing the internal energy of the bicrystal at 0 K. Electrostatic interactions favour ions of the upper grain lying directly above ions of opposite charge in the lower grain. The resulting structure has large open channels along the tilt axis. Figure 4.47 shows an HREM micrograph of the same boundary in NiO, where the ions are black (Merkle and Smith 1987a). Two distinct structures, A and B, are seen, separated by a step of two (620) planes height. Judging from the absence of strain contrast at the step there is no dislocation associated with it. The step of two (620) planes introduces a change in the rigid body translation, $t$, parallel to the boundary, of $a/10[\bar{1}30]$ between structures A and B. This is apparent in the models shown in Fig. 4.48(a) and (b) which were deduced from the HREM image.

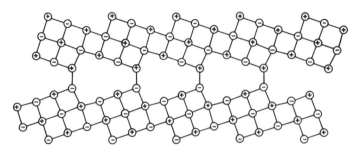

**Fig. 4.46**   Relaxed structure (at 0 K) of the (310) symmetric tilt boundary in NiO seen in projection along the [001] tilt axis (Duffy and Tasker 1983). Notice the large open channels along the tilt axis.

**Fig. 4.47**   HREM image of the (310) symmetric tilt boundary in NiO seen in projection along the [001] tilt axis (Merkle and Smith 1987a). Two structures, A and B, are seen, separated by a step in the middle of the micrograph. The atomic columns are dark.

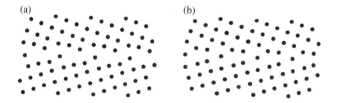

**Fig. 4.48**   Structural models of the (310) symmetric tilt boundary in NiO deduced from the HREM image shown in Fig. 4.46 (Merkle and Smith 1987a). (a) corresponds to structure A and (b) to structure B. The atomic columns are represented by solid circles because it is not known which columns are cations and which are anions, or if the columns are mixtures of cations and anions.

Neither of the structures shown in Fig. 4.48(a) and (b) is the same as Fig. 4.46. Whereas the expansion normal to the boundary plane in Fig. 4.46 is 1.1 Å, the expansions of structures A and B are 0.4 and 0.3 Å respectively. There is approximately one (620) plane missing from the structure in Fig. 4.46 compared with either structure derived from experiment.

One possible reason for the disagreement is that the assumption in the simulation of constant ionic charges may break down at the grain boundary. In a defective environment the tendency for charge transfer from cation to anion will differ from that in the perfect crystal. For example, if two cations are nearest neighbours the electrostatic potential will favour them having a charge of less than $+2$ each. In effect we are saying that the interactions may be more covalent in the boundary plane. There would then not be such a large energy penalty for ions of like charge being nearest neighbours. Only a self-consistent calculation can decide whether this is in fact the case.

Another possible reason for the discrepancy is that the structure of the boundary is expected to vary with variations of the bulk stoichiometry of the oxide. In particular, variations in the chemical potential of oxygen may result in a local excess or deficit of oxygen at the grain boundary. The bicrystals were grown from the melt by the Verneuil technique. They were annealed in a $CO/CO_2$ atmosphere at 1500 °C to equilibrate the structures under close to stoichiometric conditions.

Two related experimental observations indicate a different reason for the disagreement. The first is that for the (510) boundary in NiO, intensity scans across atomic columns in HREM images have revealed that two columns in each structural unit are reproducibly less intense than others (Merkle *et al.* 1988). The columns are along the tilt axis and the observation suggests that approximately 25 per cent of sites along the columns are vacant. The same observation has not been made at the (310) boundary, but if more columns are less vacant then it is possible that the reduction in intensity is not detectable. Further evidence of a reduction in density at large-angle grain boundaries in NiO comes from the strong Fresnel fringe contrast they produce (Merkle *et al.* 1985). Fresnel fringe contrast arises from the reduction in the average electrostatic potential at the grain boundary compared with the adjoining crystals. The difference in the average electrostatic potential is proportional to the change in density at the grain boundary. (Grain boundaries in metals do not usually produce strong Fresnel fringes.) These two observations suggest how the structure shown in Fig. 4.46 may be transformed into that of Fig. 4.48(b). If some ions in the columns bordering the open channels in Fig. 4.46 migrate into the open channels, we would generate a structure that looks very much like Fig. 4.48(b) when viewed along the tilt axis. However, the columns of ions would contain vacant sites.

It is important to keep in mind that the experimental images were obtained from bicrystals that were annealed at 1500 °C, and then quenched to room temperature from 1100 °C. It follows that the structures that are observed are likely to be the ground state *at high temperature*. On the other hand the simulations of Duffy and Tasker were based on minimization of the internal energy at 0 K. It is conceivable that the simulated structure shown in Fig. 4.46 is the correct ground state for the boundary at low temperatures, and that it turns into the structure shown in Fig. 4.48(b) at high temperatures. To simulate the transition molecular dynamics must be used to allow ion jumps to occur, and the size of the period imposed along the tilt axis must be increased to allow non-equivalent sites in that direction.

Such a simulation has been performed by Meyer and Waldburger (1993). At low temperatures the structure they obtained is virtually the same as that of Duffy and Tasker shown in Fig. 4.46. During a very long anneal (by molecular dynamics standards!) of 0.480 ns at 2500 K both anions and cations in the columns bordering the open channels were seen hopping into and out of the open channels. This suggests that the ions bordering the open channels are located in double potential wells, of the form shown in Fig. 4.49, where $\Delta E$ is the energy difference between the potential minima. The minima in the potential are located at the column and the channel. At low temperatures the ions are trapped in the potential well of the column, and we have the structure shown in Fig. 4.46. As the temperature is increased the probability of the ion jumping over the barrier into the potential well of the channel increases. During a long anneal at high temperature the site occupations in the channels and the bordering columns will equilibrate, and there will be a considerable amount of disorder along the columns. When the sample is quenched to room temperature this disordered structure is frozen in, and that is what is seen in Fig. 4.48(b). The origin of the structure shown in Fig. 4.48(a) has not yet been resolved.

The statistical mechanics of a two-level system such as that shown in Fig. 4.49 is easily derived. We can estimate the contribution to the free energy of the boundary from each two-level system as follows. The partition function for the two-level system is

$$Z = \exp\left(-E_1/kT\right) + \exp\left(-E_2/kT\right) \tag{4.131}$$

where $E_1$ and $E_2$ are the energies of the potential minima. Let $\Delta E = E_2 - E_1$. The free energy of the two-level system, relative to $E_1$, is then

$$F = -kT\ln\left[1 + \exp-\left(\Delta E/kT\right)\right]. \tag{4.132}$$

The internal energy, relative to $E_1$, is

$$U = \frac{\Delta E}{1 + \exp\left(\Delta E/kT\right)}, \tag{4.133}$$

from which we see that the probability of the higher potential well being occupied is the Fermi–Dirac factor $1/(1 + \exp[\Delta E/kT])$. Using the experimental result that the columns bordering the channels contain $\simeq 25$ per cent vacant sites, we equate this to the Fermi–Dirac factor and obtain $\Delta E/kT \simeq 1.1$. Taking $T$ to be the temperature of the anneal (1500 °C) we obtain $F = -0.044$ eV and $\Delta E \simeq 0.17$ eV. If each two-level system occupies an area of about 10 Å$^2$, then its contribution to the free energy of the boundary is $-70$ mJ m$^{-2}$. Clearly, this is a significant contribution, particularly since there may be more than one two-level system in each structural unit of the boundary.

We note in passing that the two-level system is associated with a peak in the specific heat as a function of temperature. This is known as a Schottky anomaly. From eqn (4.133) for the internal energy we can derive the following expression for the specific heat per two-level system:

$$C_V = k\frac{(\Delta E/2kT)^2}{\cosh^2(\Delta E/2kT)}, \tag{4.134}$$

which has a maximum of $0.44k$ when $\Delta E/2kT \simeq 1.2$. For $\Delta E = 0.17$ eV the maximum occurs at $T \simeq 550$ °C. This may give rise to an observable peak in the specific heat of nanocrystalline samples of NiO as a function of temperature.

Let us attempt a synthesis of these experimental and computer modelling results on tilt boundaries and twist boundaries in NiO. There appears to be a high density of vacant sites (both anion and cation) at all temperatures inside the boundary core, which has a

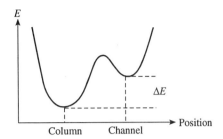

**Fig. 4.49** Schematic diagram of the potential energy, $E$, of an ion as a function of its position in the (310) symmetric tilt boundary in NiO. When it occupies one of the atomic columns in Fig. 4.46 its potential energy is minimized. However, if the ion migrates into one of the large open channels seen in Fig. 4.46 it may occupy another potential energy minimum, but $\Delta E$ higher in energy. At high temperatures ions can jump between both potential energy minima.

thickness of 5–10 Å. At low temperatures the vacant sites form ordered arrays. But at high temperatures the vacant sites migrate between well-defined sites in the core. Vacant sites are required to avoid the energetically unfavourable situation of nearest neighbours having like charge. They are not defects but an intrinsic part of the ground state of the boundary structure.

Since vacant sites exist to ensure that ions of like charge are not nearest neighbours, the remaining Coulomb interactions across the boundary are entirely attractive. The ideal cleavage energy of the boundary is then determined by the balance between core–core repulsive interactions and attractive Coulomb interactions. The situation is now very much like the model we described in section 4.3.1.2, especially at high temperatures where the distribution of vacant sites is disordered and allows the boundary expansion to decrease to its smallest possible value (neglecting thermal expansion). Therefore, we can expect grain boundaries in ionic crystals, at least in their high-temperature forms, to display the same variations of the expansion and ideal cleavage energy with the interplanar spacings $d^b$ and $d^w$ as we derived for metals. Moreover, if we ignore the anisotropy in the free surface energies, we expect the grain boundary energy itself (in the incommensurate limit) to decrease as the interplanar spacings $d^b$ and $d^w$ increase. However, the caveats that were discussed in section 4.3.1.5 apply here also.

Merkle and Smith (1987b) have reported HREM observations of [001] tilt boundaries in NiO which have faceted onto low energy planes. The samples were prepared by the same procedure as described above, and therefore the boundary structures are those of the high temperature state. Both symmetric and asymmetric tilt boundaries were found, and in each case $\langle d \rangle$ was one of the highest available for the given misorientation relation. It is particularly noteworthy that the asymmetric tilt boundary parallel to (100) in one grain always appeared because this boundary corresponds to the highest $\langle d \rangle$ at any tilt misorientation about [001]. For example at $\theta = 37°$ there are $(100)^b/(340)^w$ ($\langle d \rangle = 0.3a$) asymmetric and $(210)^b/(2\bar{1}0)^w$ ($\langle d \rangle = 0.22a$) symmetric facets. At $\theta = 26°$ there are symmetric tilt boundary facets close to $(410)^b/(4\bar{1}0)^w$ ($\langle d \rangle = 0.12a$) and asymmetric, incommensurate facets close to $(100)^b/(210)^w$ ($\langle d \rangle = 0.36a$) facets. At $\theta = 43°$ facets close to the asymmetric, incommensurate $(100)^b/(110)^w$ ($\langle d \rangle = 0.43a$) boundary planes are observed. An incommensurate symmetric tilt boundary at $\theta = 43°$ decomposed into $(100)^b/(110)^w$ and $(010)^b/(\bar{1}10)^w$ microfacets.

In conclusion, the above analysis suggests that, despite the apparent complication of

Coulomb interactions, the energies of grain boundaries in ionic crystals follow the same trends with $\langle d \rangle$ and the boundary expansion as grain boundaries in metals. There is some experimental evidence to support this view. The same caveats (see section 4.3.1.5) to these trends are expected to apply to boundaries in ionic crystals as in metals.

Finally, in this section, we mention a phenomenon which is of importance for certain interfaces in ionic crystals called 'dipolar' interfaces. Such interfaces may exist when the interface plane is parallel to atomic planes in the bulk that carry a net charge. An example is a (311) symmetric tilt boundary in NiO. To reveal the possible dipolar nature of such an interface, consider first a bulk crystal made of a stack of (311) planes as in Fig. 4.50. Each plane is composed entirely of either anions or cations, and the entire block is charge neutral. However, if the planes are of infinite extent, the uncompensated terminating planes are surfaces of infinite energy, and a potential difference is established across the crystal (Duffy and Tasker 1984b). This unrealistic situation can be alleviated by reconstructing the surfaces and transporting charge to produce the configuration shown schematically in Fig. 4.50(b), where the field has been removed. In this process half of the negative charge on the terminating plane on the left has been moved to create a new half-plane on the right. If the symmetric tilt boundary is now introduced by a 180° rotation of half the crystal around [311] so that the boundary plane is along AB, as in Fig. 4.50(c), calculations of the atomic structure (Duffy and Tasker 1984b) show that a dilation (expansion) occurs in the core which produces the situation shown schematically. Since the stack on the left has a net charge of $-\sigma/2$ and that on the right $+\sigma/2$, this dilation produces a potential difference between the right and left sides equivalent to that produced by a dipole layer of spacing $\delta$ and charge density $\sigma/2$, as shown in Fig. 4.50(d). Detailed calculations (Duffy and Tasker 1984b) indicate that the potential change across the (311) symmetric tilt boundary in NiO is relatively large, i.e. $\sim 2\,V$. However, if mobile charged defects are present they will tend to eliminate the potential difference by redistributing themselves to produce space charge regions in the form of a compensating dipole layer. If the defects are of the Schottky type, and for simplicity, any point defect source/sink action at the interface is suppressed, excess cation vacancies accumulate to produce a region of negative space charge on the right, excess anion vacancies accumulate on the left, and the final screened potential then appears as illustrated in Fig. 4.50(e). At large distances from the interface the potential is zero. However, there is a potential barrier at the interface (for reasons described in Section 11.1), since the charge cancellation is incomplete in that highly localized region. Asymmetric barriers that are characteristic of dipolar boundaries can be an order of magnitude larger than the symmetric barrier due to the thermal formation of point defects described in Section 7.6. If a dipolar boundary also acts as a point defect source/sink (see Section 7.6), the exact solution for the final potential can only be found by taking into account the coupling between the different effects producing the two barriers.

## 4.3.3 Grain boundaries in covalent crystals

The atomic structures of grain boundaries in covalently bonded materials are intimately related to their electronic structures, and we cannot discuss one without paying attention to the other. It is the formation of covalent bonds at the boundary that primarily determines its atomic structure. Conversely, the frustration that results from bringing two misoriented crystals into contact does not allow all atoms to satisfy their ideal complement of bonds simultaneously. In that case distorted or broken ('dangling') bonds are expected to be associated with electronic states somewhere in the fundamental energy gap.

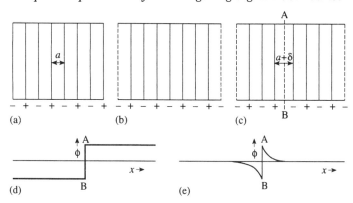

**Fig. 4.50** (a) Stack of (311) planes in NiO. Each successive plane is composed entirely of either anions or cations. (b) Same as (a) except that the surfaces have been reconstructed (to reduce the energy) by transferring half of the negative charge on the terminating plane on the left to create a new negative half-plane on the right. (c) (311) symmetric tilt boundary created along AB by rotation of half of the previous single crystal by 180° about [311]. Creation of the boundary causes a dilation, $\delta$, at the boundary core. (d) Potential difference caused by the dilation in (c). (e) The screened potential, $\varphi$, remaining after accumulation of excess cation vacancies on the right and excess anion vacancies on the left. (From Duffy and Tasker (1984$b$).)

Grain boundaries in covalent semiconductors, such as Si and Ge, are often associated with electronic states. The question that has long dominated the field has been whether these states are caused by the intrinsic structural frustration, or by impurities segregated to the boundaries.

Grain boundaries in Si and Ge were among the first to be studied by HREM because of their relatively large lattice parameters. However, only [110] and [001] tilt boundaries have been observed directly by HREM, and we know of only one experimental study of the structure of a twist boundary (see Sutton (1991$b$) and refs. therein.) In all cases for which a structural model has been determined, all atoms are four-fold coordinated. Furthermore, in the case of (001) twist boundaries, with $0 \leqslant \theta_{\text{twist}} \leqslant 40°$, measurements (Nelson 1991) of the potential barrier to electronic transport revealed an *occupied* electronic state density of the order of $3 \times 10^{10}\,\text{cm}^{-2}$, which was virtually independent of $\theta_{\text{twist}}$. This is too low to be associated with intrinsic structural features of the boundaries, and is more likely to be caused by impurities. It appears, therefore, that there is no experimental evidence for an intrinsic origin to the occupied electronic states at grain boundaries. On the other hand, there is considerable experimental evidence that electronic states are associated with colonies of metallic precipitates (10–100 nm in diameter) at grain boundaries in Si (see papers in Möller *et al.* (1989) and Werner and Strunk (1991)). Maurice (1990) found that even junctions between (111) and (11$\bar{2}$) twin boundary facets in Si are not associated with localized states until the sample was annealed at 900°C, followed by a 5°C s$^{-1}$ cool. After this heat treatment Cu and Ni rich precipitates were found at one facet junction, but not the other. There was a one-to-one correspondence between the facet junctions with precipitates and those displaying states in the gap at which minority carrier recombination was observed.

Computer simulations of tilt boundaries in Si and Ge, using a variety of descriptions of atomic interactions, indicate that there are no states at all in the fundamental gap (see Sutton (1991$b$) and references therein). In every case the boundary reconstructed in such a way that each atom was four-fold coordinated and there were no dangling bonds. That is not to say, however, that there were no intrinsic electronic states localized at the grain boundaries. Band structure calculations revealed states in the energy gap, between certain ranges of the wave vector parallel to the boundary, at energies below the valence band maximum and above the conduction band minimum. These states do not show up in the density of states because they are buried under the bulk conduction and valence band edges.

Computer simulations of twist boundaries have been less clear cut. As we discussed on general grounds in Section 4.3.1.1, the size of the c.n.i.d. limits the effectiveness of the rigid body translation, $\tau$, parallel to the interface as a means of lowering the boundary energy. As the c.n.i.d. decreases in size, the variations in the boundary energy as a function of $\tau$ must decrease, and vanish when the area of the c.n.i.d. vanishes. If the boundary does not facet or dissociate the only mode of relaxation available to it is individual atomic relaxation. This includes removing or inserting atoms at the boundary, while each atom attempts to optimize its own nearest neighbour environment, given the overall constraints of the adjoining misoriented crystals and the fixed boundary plane normal. The key question is how effective is this local constrained optimization at a twist boundary in a covalently bonded material, where the directionality of the bonds introduces additional constraints?

The $\Sigma = 9$ (115) ($\theta_{twist} = 120°$) twist boundary has been observed in Si by the $\alpha$-fringe technique and modelled using a Keating valence force field (Cheikh *et al.* 1991). It was found that the boundary reconstructs fully if one atom is removed from each unit cell. There are then no dangling bonds and the boundary energy is comparable to that obtained for tilt boundaries with the same potential. The same group (Ralantoson *et al.* 1993) has also modelled the $\Sigma = 5$ (001) ($\theta_{twist} = 36.9°$) twist boundary in Si and Ge with valence force fields. They found two relaxed configurations, both of which contain only four-fold coordinated atoms. In the diamond cubic crystal structure there are two atoms in the atomic basis, one at $(0, 0, 0)$ and one at $(\frac{1}{4}, \frac{1}{4}, \frac{1}{4})$. This means that there are two sets of non-equivalent structures of a (001) twist boundary which differ by the insertion or removal of a (004) atomic plane. Boundary structures belonging to the same set may be transformed into each other by translations within the c.n.i.d. But, for a given $\theta_{twist}$, boundaries belonging to different sets may be transformed into each other only by inserting or removing a (004) plane. Alternatively, at a fixed boundary location the two structures may be obtained by rotations of $\theta_{twist}$ and $90 - \theta_{twist}$. (See Section 1.5.7 for a general discussion of this point.) In the case of the $\Sigma = 5$ boundary these two sets of possible structures have become known as $\Sigma 5$ and $\Sigma 5^*$. Ralantoson *et al.* (1993) found one stable structure for $\Sigma 5$ and one for $\Sigma 5^*$. Interestingly, there is a *contraction* normal to the boundary plane of $\frac{1}{17}[001]$ in both structures.

Tarnow *et al.* (1990) present an altogether different picture of the $\Sigma = 5 (001)$ twist boundary in Ge, based on their simulations using local density functional theory. They found a multiplicity of local minima, differing in the local bonding arrangements, and the number of atoms in the boundary plane, and concluded that the bonding topology is complex in the sense that it is not uniquely defined. There was a tendency for dimer bonds to form between atoms in the same grain rather than forming bonds across the grain boundary, which suggests that these boundaries are less stable against cleavage. In one of the many configurations they generated, all atoms were four-fold coordinated, but the energy was not lower than many of the other configurations. The boundary energies varied within a range of about 15 per cent from one configuration to another. Presumably, these energy variations are sufficient to localize arrays of DSC screw dislocations if $\theta_{twist}$ deviates slightly from the $\Sigma = 5$ misorientation.

At a general boundary the c.n.i.d. vanishes, and the only mode of relaxation available is local atomic relaxation. The calculations of Tarnow *et al.* (1990) suggest that the structure of such a boundary cannot be defined uniquely, and that the local atomic structure varies from one region to another in the boundary plane. The boundary is also likely to be capable of absorbing or emitting vacant sites with virtually no energy cost.

One of the most significant results of the study by Tarnow *et al.* (1990) is that they find several electronic states distributed throughout the band gap in their relaxed configurations. But all these states are *unoccupied*, and, therefore, they do not give rise to a potential barrier to current flow because the boundary is not charged (see Section 11.5). This is consistent with the experimental observations of Nelson (1991). Similar results from simulations of twist boundaries have been reported by Kohyama *et al.* (1993).

There have been a few HREM studies of tilt boundaries in compound semiconductors with the sphalerite structure (see Sutton (1991*b*) and references therein). The two f.c.c. sublattices are now occupied by different atoms and this introduces a new degree of complexity. For example at the (111) twin boundary in the AC sphalerite compound, where A is the anion and C the cation, three distinct structures of the boundary can be envisaged involving only A–C bonds or only A–A or only C–C bonds across the interface. Experimental observations of the (111) twin in ZnSe and GaAs have revealed only C–A bonds across the interface (Sutton (1991*b*) and references therein). But in general the stoichiometry one finds at the interface depends on the chemical potential difference between C and A, as discussed in Section 7.4.1. Kohymama *et al.* (1991) have carried out a detailed study of the (221) symmetric tilt boundary in SiC where they find localized states at the band edges due to the presence of Si–Si and C–C bonds. They found that one or other of the polar interfaces is always more stable than the non-polar interface, and in a C-rich atmosphere the boundary contains C–C bonds and in a Si-rich atmosphere it contains Si–Si bonds, as we might expect intuitively.

## 4.4 SHARP HETEROPHASE INTERFACES

### 4.4.1 Introduction

Heterophase interfaces occur naturally in solid state phase transformations, such as precipitation and eutectoidal decomposition, and in the processing and manufacture of materials, for example through sintering of polyphase powder compactions, diffusion bonding, brazing, and epitaxial growth from the vapour phase. Before we discuss different types of heterophase interfaces in detail we shall make some comments of a general nature. Heterophase interfaces are inherently more complex than homophase interfaces for two principal reasons. First, there is a chemical dimension due to the coexistence of different atomic species across the interface. Secondly, the difference in the crystal unit cells on either side of the interface introduces a source of misfit in addition to any misfit arising from misorientation of the principal crystal axes. We shall now discuss these two general points in more detail.

A full discussion of the thermodynamics of interfaces appears in the next chapter. For the present purposes it is essential to note that whenever there is more than one atomic species present the free energy of an interface depends explicitly on the chemical potentials of all species in the system. This comment applies as much to homophase interfaces in alloys, such as ionic crystals, as it does to heterophase interfaces. Unless the system is in thermodynamic equilibrium at a given temperature, pressure, and set of chemical potentials (or chemical compositions of all phases) the free energy of the interface is *not defined*. As will be shown in Section 5.2, equilibrium between all phases present must be taken into account explicitly in the evaluation of the interfacial energy, regardless of whether those phases abut or are remote from the interface of interest. If there is a gradient in the chemical potential of some species across an interface there will be a tendency for diffusion to take place. For example, interdiffusion of atomic species by the vacancy mechanism across an interface may produce a Kirkendall effect in which the interface moves into one phase, as discussed in Section 10.4.2.2. An example is the case of an interface between initially pure Cu and pure Ag (Laffont and Bonnet 1982). The inequality of chemical potentials across the interface may also produce a chemical reaction in which a new intermediate phase is created. The spatial extent of the chemical reaction or interdiffusion depends on kinetic factors largely controlled by the temperature. In the case of the creation of a new phase there will be a nucleation barrier due to the formation of new interfaces (d'Heurle 1988). The reaction may be either diffusion or nucleation controlled, as discussed in Section 10.4.2.2. In some cases the free energy decrease accompanying a chemical reaction at an interface is so large that an amorphous phase is frozen in (solid-state amorphization) before crystallization can be effected (Johnson 1986).

To emphasize the chemical dimension of a heterophase interface consider an interface in an Fe–C steel between the ferrite and cementite phases of pearlite. Carbon has a small solubility in ferrite which increases as the temperature is raised towards the eutectoid temperature, $T_E$. Therefore, the change in the carbon content across a ferrite-cementite interface is sharp, and the interface is well defined. But at the eutectoid temperature a bulk phase transformation takes place in which the ferrite and cementite phases recombine and are replaced entirely by a new single phase, austenite. At equilibrium the interface does not even exist at temperatures greater than $T_E$! Alternatively, we could describe the phase transformation as a (solid-state) chemical reaction at the interface where the ferrite and cementite phases meet. Surprisingly little is known about the mechanisms of chemical reactions in the solid state, but see, for example, d'Heurle (1988), and the discussion of the motion of heterophase interfaces in Chapter 10.

The change in crystal structure and/or lattice parameters is particularly significant at interfaces surrounding very small precipitates, and in thin epitaxial films and super-lattices. The misfit at the interface produces a continuous distribution of coherency dislocations, which generates a long-range stress field, as we discussed in Section 2.2 (see Fig. 2.6). Whether or not these dislocations are cancelled (at least partially) by a distribution of anticoherency dislocations depends on an energy balance between the energy decrease of the bulk crystals, concomitant with the (partial or complete) removal of the long range elastic field, and the increase of the interfacial energy resulting from the loss of coherency. Thus, there is a critical thickness of an epilayer, or a critical dimension of a growing precipitate, at which it becomes energetically favourable to introduce anticoherency dislocations. It is clear, then, that the misfit at the interface may generate long-range elastic fields in the adjoining crystals.

A related phenomenon is that of topotaxy, in which the orientation of an embedded crystal precipitate is parallel to that of the surrounding crystal. In some cases the structure adopted by the embedded crystal is changed to that of the surrounding crystal. For example, solid rare gas inclusions often take on the crystal structure of the host lattice (Horsewell *et al.* 1993). Thus, solid Kr is f.c.c. in Al and Cu matrices, and b.c.c. in Mo. Solid rare gas inclusions in metals show strong misorientation relationships with the host and faceted surfaces. The gas is under very high pressures, typically 1–6 GPa, and the close-packed planes of the f.c.c. solid rare gas are parallel to the {111} planes of an f.c.c. host or {110} planes of a b.c.c. host. It is tempting to conclude that these interfaces are formed because they have particularly low energy. But, as noted by Finnis (1987), that cannot be true because rare gas atoms repel each other, as well as metal atoms, and a close-packed gas surface has more gas–gas bonds than an open, higher-index surface. The resolution (Finnis 1987) of the paradox is that a more efficient use is made of the available volume inside the host cavity when the gas surfaces are close packed and parallel to the close packed planes of the cavity surface. This lowers the bulk free energy of the gas by $P\Delta V$, where $P$ is the pressure of the gas and $\Delta V$ is the difference in volume occupied by the gas with more open surfaces and close-packed surfaces. Finnis (1987) showed that the gain in the bulk free energy of the gas more than compensates the energy cost of close packed interfaces.

Other examples of topotaxy may be found in small metallic inclusions in metals. For example, Horsewell *et al.* (1993) found that small ($\leqslant 10$ nm) particles of Na embedded in Al are generally f.c.c. and that larger Na particles occur with the b.c.c. structure of bulk Na. Similarly, in Al–In alloys small In inclusions do not occur with the equilibrium tetragonal structure but with the f.c.c structure. In the early stages of Cu precipitation in ferritic steels coherent b.c.c. Cu-rich clusters nucleate and grow and transform to the 9R phase (see Section 4.3.1.9) before eventually transforming to the bulk f.c.c. phase (Othen *et al.* 1994).

It is clear from these examples of topotactic growth that the interface surrounding a small precipitate is not necessarily that with the lowest energy. If the precipitation is accompanied by a volume change and/or a shape change the precipitate will be subjected to a significant stress by the constraint of the surrounding medium. This stress may be the dominating factor in determining not only the structure of the interface, but even the crystal structure of the precipitate. Similar considerations may also apply to interfaces in thin films and multilayer structures, where the elastic energy of the system may be the dominating factor in controlling the interfacial structure and even the crystal structure of one or more adjoining phases.

Once any long-range elastic field has been eliminated, the misfit between the crystal structures contributes to the Burgers vector density of anticoherency dislocations through the Frank–Bilby equation (see Section 2.3). This misfit is in addition to the misfit arising from a misorientation of the crystal axes, and the two sources of misfit may partially cancel each other. By reducing the Burgers vector density in this way, one can argue that the energy of the short-range elastic strain field of the interface is reduced and therefore the interfacial energy is lowered. This line of reasoning has been applied, with some success, to observed orientation relationships between b.c.c. metallic deposits on f.c.c. metallic substrates. For example, Smith *et al.* (1987) examined the orientation relationship between Cr vapour deposited on a Au (100) single crystal substrate. Gold is f.c.c. with a lattice parameter at room temperature of 4.0788 Å. Chromium is b.c.c. with a lattice parameter at room temperature of 2.8846 Å. At an interface where the $\langle 100 \rangle$ axes of the two crystals are parallel the misfit would be 29 per cent, and the dislocation spacing would be too small for the interface to be anything other than incoherent. Smith *et al.* observed that $(100)_{Cr}$ is parallel to the $(100)_{Au}$ substrate but $[011]_{Cr}$ is parallel to $[001]_{Au}$. The twist of 45° about the interface normal reduces the misfit to less than 0.02 per cent, and the interface is either coherent or semicoherent. Smith *et al.* reported other f.c.c.–b.c.c. systems where the observed orientation relationships resulted in overall misfits of less than 20 per cent. But equally, there are contrary observations where the misfit appears to play no role in the observed orientation relation. For example, McCafferty *et al.* (1992) observed that Ag spheres deposited on a (100) single crystal NaCl substrate rotate to orientations where a {100} plane of the Ag is parallel to the substrate, and either

$\langle 010 \rangle_{Ag}$ is parallel to $\langle 010 \rangle_{NaCl}$, or $\langle 010 \rangle_{Ag}$ is parallel to $\langle 011 \rangle_{NaCl}$. Despite the much greater misfit (30 per cent) at the former orientation, it was observed as frequently as the the latter orientation, where the misfit was only 2 per cent. Moreover, for Au spheres on the same substrate, only the orientation with the large misfit was observed, even though the misfits in the Ag/NaCl and Au/NaCl systems are nearly equal.

It seems that the strength of the interaction across the interface may play a role in determining whether the misfit is a significant factor. For most metal–metal interfaces the bonding across the interface is significantly stronger than it is between noble metals and ionic crystals, where the interaction is primarily an induced polarization charge density on the metal surface (see Section 3.11), and is relatively weak. But in our view the real significance of minimization of misfit at a metal–metal interface is not the corresponding reduction in elastic field energy, although that certainly plays a contributory role, but the reduction in the core energy of the interface when there are matching layer reciprocal lattice vectors of small length, as discussed in Section 4.3.1.1. After all, a small misfit is not considered significant if the size of the matching 1D or 2D cell in the interface is very large.

It is clear from the above discussion that there are often constraints placed on a heterophase interface during its formation, which complicate and can even dominate the factors that determine its structure and energy. But when the interface is free of long-range stresses, and chemical equilibrium has been achieved, the atomic structure is controlled by local interactions across the interface. In practice it may sometimes be difficult to establish whether this is true for any particular case. But when it is true, and if the strength of the interactions across the interface is comparable to those within each phase, considerations similar to those that we discussed for sharp homophase interfaces come to the fore. In particular, large interplanar spacings parallel to and on either side of the interface, and the existence of small common layer reciprocal lattice vectors, are likely to be associated with low interfacial energy. However, the same caveats that were discussed in section 4.3.1.5 apply here also.

### 4.4.2  Metal–metal interfaces

An excellent model system to study metallic heterophase interfaces is provided by the Ag–Ni system because Ag and Ni are immiscible, with no intermetallic compound formation. The same is true for the Au–Ni system. Therefore, compositionally sharp interfaces can be expected in these systems. Ag, Au, and Ni possess f.c.c. crystal structures and the misfit, $\delta$, defined by

$$\delta = 2|a_1 - a_2|/(a_1 + a_2),\qquad\qquad (4.135)$$

where $a_1$ and $a_2$ are the lattice parameters, is about 15 per cent in both the Ag–Ni and Au–Ni systems. Gao and Merkle (1990) manufactured thin foils of Ag containing holes, varying from 5 to 500 nm in diameter, into which Ni was deposited. The resulting embedded Ni particles grew with their $\langle 100 \rangle$ axes parallel to those of Ag, and they were surrounded by $\{111\}$ interfaces. As seen in Fig. 4.51 the large misfit at the interface is localized at edge dislocations with Burgers vector $\frac{a}{2}\langle 110 \rangle$. The spacing of the dislocations agrees with the spacing predicted by the Frank–Bilby equation for 15 per cent misfit. Curved regions of the interface were faceted on a microscopic scale on $\{111\}$ planes. Steps were also seen on some of the large $\{111\}$ facets, and the smallest step height was one $\{111\}$ interplanar spacing. $\{110\}$ interfaces were found to be faceted on $\{111\}$ planes on an atomic scale forming a saw-tooth morphology. Because the interfaces were created

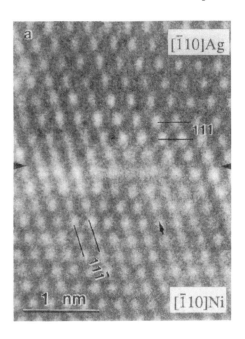

**Fig. 4.51** HREM image of (111) Ag–Ni interace (Gao and Merkle 1990). The ⟨100⟩ crystal axes are parallel in the Ag and Ni. White dots represent atomic ⟨110⟩ columns in both Ag and Ni. The interface runs from left to right in the middle of the image. Despite the 15 per cent atomic misfit, a localized crystal lattice edge dislocation is clearly visible, as indicated by the inclined arrow.

by Ni atoms filling existing holes in Ag foil, it is unlikely that the Ni particles were subjected to large constraint stresses. Therefore, we can be reasonably confident that the predominance of {111} facets indicates that the lowest energy interfaces are those on {111} planes. Interfaces between Au and Ni behave in a similar manner to those between Ag and Ni (Gao and Merkle 1990).

Gao *et al.* (1989) argued that {111} interfaces have the lowest energy because the number of Ag–Ni bonds is minimized on this plane. The number of Ag–Ni bonds per atom at {111}, {100}, and {110} interfaces is roughly the same as the number of broken bonds at the corresponding free surfaces, i.e. 3, 4, and 5 respectively. Gao *et al.* (1989) found that the average energies of simulated {111}, {100}, and {110} Ag–Ni twist interfaces increased in agreement with this simple rule. They also noted that the criterion for low interfacial energy of minimizing the number of Ag–Ni bonds is equivalent to the criterion of maximizing the average interplanar spacing $\langle d \rangle$. The atomic planes of an interface with a large value of $\langle d \rangle$ are more closely packed, and therefore there are more neighbours of the same species in the plane, and fewer neighbours in adjacent planes of the other species.

Zero creep experiments have been carried out for Ag–Ni multilayer structures (Josell and Spaepen 1993). The layers of Ag and Ni were polycrystalline, with a strong (111) texture, and therefore the Ag–Ni interfaces were parallel to {111} planes, but there was presumably a range of twist angles about the common ⟨111⟩ interface normal. The zero creep measurements gave an energy of $\sigma_{\text{Ag-Ni}} = 0.76 \pm 0.12 \text{ J m}^{-2}$ for the average Ag–Ni {111} interfacial energy. This energy is considerably higher than that calculated with embedded atom potentials for {111} Ag–Ni interfaces by Gao *et al.* (1989), who obtained 0.42 and 0.47 J m$^{-2}$ for $\theta_{\text{twist}} = 0$ and 90° respectively. Thermal grooving experiments for the grain boundaries in the Ni and Ag layers gave $\sigma_{\text{Ni-Ni}}/\sigma_{\text{Ag-Ni}} = 1.03 \pm 0.15$, and $\sigma_{\text{Ag-Ag}}/\sigma_{\text{Ag-Ni}} = 0.21 \pm 0.19$.

There is some experimental evidence that observed orientation relations and interface plane normals between metallic phases are such that there is 1D or 2D matching, or near matching, of layer reciprocal lattice vectors. But there is also considerable variability in these relations depending on many factors, including the conditions under which the interface was produced. In considering interfaces in eutectic metallic alloys Kraft (1962) proposed that a low-energy interface is formed when:

1. the density of atoms in the interface plane is high;

2. the density of atoms in the interface plane is similar in both phases.

The first criterion is equivalent to the high interplanar spacing criterion, and the second is related to the criterion of minimizing the misfit in the interface as defined by $2(\rho_\alpha - \rho_\beta)/(\rho_\alpha + \rho_\beta)$, where $\rho_\alpha$ and $\rho_\beta$ are the atomic densities in the interface plane.

Cantor (1981) has reviewed the experimental data on orientation relationships and plane normals in metallic eutectic systems. He concluded that the above criteria are unable to explain the high index interface planes in Al–Zn and Cu–Cr, the low index, high misfit interface in $Ni_3Al$–Mo, the wide variety of interface facets which are seen in Al–$Al_3Ni$ and Pb–Sn, or the variability of interface orientation in Al–$Al_2Cu$ where 22 different orientation relationships have been reported in eutectic and precipitation studies. He concluded that the suggestion that, as a general principle, each eutectic alloy has its own strongly preferred, or even unique, crystallography is incorrect. Eutectic crystallography comes from a balance between competing factors including anisotropic solid–liquid interfacial energies, anisotropic growth kinetics, and nucleation crystallography in addition to interfacial energy. Similar complexity may be expected for other heterophase interfaces produced by phase transformations.

### 4.4.3   Metal–insulator interfaces

It was noted in Section 4.4.1 that the strength of the interaction across an interface, relative to the interactions within each phase, is a significant factor in determining the structure and energy of a heterophase interface. This is particularly true for interfaces between metals and non-metals, which are frequently divided into those with strong interactions and those with weak. Those with weak interactions are characterized by a relatively small cleavage energy, an insensitivity of the interfacial energy to translations parallel to the interface, and a small tendency to localize misfit dislocation cores in the interface, resulting in (virtually) incoherent interfaces. By contrast, those with strong interactions have a large cleavage energy, strongly preferred translations parallel to the interface, and highly localized misfit dislocations.

A useful indication of the strength of the interaction between a metal and a non-metal is provided by the contact angle between a liquid drop of the metal in equilibrium with a solid surface of the non-metal. Consider the balance of capillarity forces (see Section 5.6.4) parallel to the surface of the non-metal shown in Fig. 4.52. At equilibrium (and neglecting torque terms (see Section 5.6.4)) we have

$$\cos\theta = \frac{\sigma_{cv} - \sigma_{cm}}{\sigma_{mv}} \tag{4.136}$$

where $\sigma_{cv}$, $\sigma_{cm}$, and $\sigma_{mv}$ are identified in Fig. 4.52. Equation (4.136) is known as the Young–Dupre equation. If $\theta < 90°$ the metal is said to wet the surface, because it has a tendency to spread over the surface. The reversible work required to separate the drop from the surface, called the work of adhesion, $W_{ad}$, is related to the contact angle through

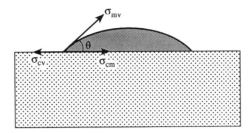

**Fig. 4.52** To illustrate the interfacial energies appearing in the Young–Dupre equation, eqn (4.136), for a liquid drop forming a contact angle $\theta$ on a solid substrate. $\sigma_{mv}$ is the surface energy of the liquid drop, in equilibrium with its vapour. $\sigma_{cv}$ is the surface energy of the substrate surface, in equilibrium with its vapour. $\sigma_{cm}$ is the interfacial energy of the interface between the liquid drop and the solid substrate. All interfacial energies are expressed per unit area.

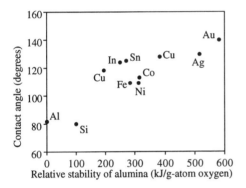

**Fig. 4.53** Correlation between wetability (contact angle) of an alumina substrate by a liquid metal, and the stability (i.e. standard free energy of formation per gram atom of oxygen) of the alumina relative to the stability of the oxide of the liquid metal. (Data from Chatain *et al.* (1986).)

$$W_{ad} = \sigma_{mv} + \sigma_{cv} - \sigma_{cm} = \sigma_{mv}(1 + \cos\theta). \qquad (4.137)$$

Thus, the smaller the contact angle the greater the wetability and the greater the work of adhesion.

Naidich (1981) has collected together available data on contact angles for a wide variety of systems. However, much of this data is not reliable owing to the inadequate control of the purity of the materials and the environment in which the measurements were made. Consequently, many of the results have not been reproduced. Nevertheless, using reproducible data for the wetting of $Al_2O_3$ by various liquid metals (Chatain *et al.* 1986), it is found that there is a correlation between the contact angle and the oxygen affinity of Al relative to that of the liquid metal (see Fig. 4.53). As the relative oxygen affinity of the Al increases the $Al_2O_3$ becomes increasingly stable and the wetability decreases. This suggests that the strength of the interaction between the liquid metal and the $Al_2O_3$, as measured by $W_{ad}$, decreases as the stability of the $Al_2O_3$ against reduction by the liquid metal increases.

However, this stability does not depend only on the free energy of formation of the oxides under standard conditions, but on the chemical potential of oxygen in the system. For instance, increasing the oxygen partial pressure, so that the O/U ratio of $UO_2$ changes from 2.001 to 2.084, causes the contact angle of liquid Cu at 1150 °C to change from 116° (non-wetting) to 84° (wetting) (Nicholas 1989). A particularly interesting experiment was carried out by Chatain *et al.* (1993), who measured the contact angle of

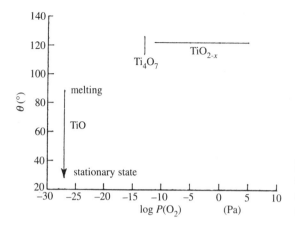

**Fig. 4.54** Contact angle of a liquid Au drop on titanium oxide as a function of the oxygen partial pressure. (From Chatain *et al.* (1993).)

a liquid Au drop on titanium oxide as a function of the oxygen partial pressure. $TiO_2$ is an insulator, $TiO_{2-x}$ is an ionic and electronic conductor, and TiO is an electronic conductor (i.e. it is metallic). In addition, $TiO_{2-x}$ is highly defective ($x_{max}$ is about 1 per cent at 1300 K) before it changes into a Magnelo phase $Ti_nO_{2n-1}$. Thus, these experiments enabled the dependence of the wetability on the conductivity of the oxide and its intrinsic defect population to be studied. The results are shown in Fig. 4.54. When the oxygen partial pressure is decreased from $10^5$ to $10^{-11}$ Pa the oxygen composition decreases from $TiO_2$ to $TiO_{2-x}$ and the electronic conductivity increases by a factor of 1000. However, the experimental contact angle remains constant and equal to $122° \pm 5°$. Therefore, intrinsic defects do not affect the wetability of the oxide. For $Ti_4O_7$ and TiO the contact angles, measured as soon as the Au melts, are equal to $120° \pm 6°$ and $88° \pm 10°$. Thus, the wetability of Au on the titanium oxide increases only when the stoichiometry of the oxide changes. At very low oxygen partial pressures, corresponding to TiO stoichiometry, the contact angle of the liquid Au drop decreases with time to $28° \pm 3°$. It was found that near the interface the Au was quickly saturated with Ti and O and the solid phase changed to $Au_4Ti$. This suggests that the good wetability of TiO by Au is not due to the high electronic conductivity of TiO, but to the dissolution of some of the oxide and the formation of an intermediate compound.

Similar variations of the wetability of more covalent non-metals with stoichiometry have been found. For example, in Fig. 4.55 we show the contact angle of liquid Cu on titanium carbide as a function of the Ti/C ratio.

It is clear from these observations that wetability depends on the chemical potentials of the active species present in the system. The chemical potential of oxygen in a metal may be lowered or increased by alloying with other metals that have a larger or smaller affinity for oxygen respectively. At very low oxygen activities a reaction may be induced in which an intermetallic phase is created, and again the wetability is increased. In both cases increased wetability, and hence increased work of adhesion, is attained with a compositionally wider interface (Chatain *et al.* 1993). However, in the solid state one may wish to avoid intermetallic reaction layers because of their brittleness.

A particularly striking illustration of the role of the oxygen chemical potential in determining the structure and composition of a solid metal–oxide interface was given by Wagner *et al.* (1992). These authors passed an electric current through a galvanic cell in which the solid electrolyte, yttria-stabilized cubic zirconia, was sandwiched between solid

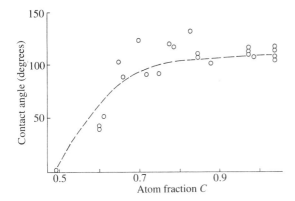

**Fig. 4.55** Contact angle of liquid Cu in titanium carbide as a function of the Ti/C ratio. (From Nicholas (1989).)

Ni electrodes. The Ni–zirconia interfaces were produced by diffusion bonding. The current was conducted through the zirconia by the motion of oxygen anions and electrons. Very large differences in the oxygen chemical potential at the two electrodes were effected by passing a current through the cell. When the electrochemical potential exceeded a critical value, about 1V, the intermetallic compound $Ni_5Zr$ formed at the interface with a low oxygen chemical potential, while NiO was formed at the interface with a high oxygen chemical potential.

Interfaces between metals and ceramics may be produced directly by diffusion bonding or by brazing a ceramic with a liquid metal, or by epitaxial growth. They may also be produced by internal oxidation of a metallic alloy or composite or by internal reduction of an oxide alloy or composite. Perhaps the most commonly found metal–ceramic interface arises in the external oxidation of a metal. Some of the most reproducible and best controlled interfaces are produced by internal oxidation and reduction. During internal oxidation of a metallic alloy the more reactive component is oxidized, resulting in a dispersion of oxide particles inside the more noble metallic host. During internal reduction a mixed oxide is placed in a reducing environment and the more noble metal component forms precipitates inside the oxide matrix. In both cases the chemical reaction takes place first in the outer skin of the specimen and spreads inwards as the new equilibrium concentrations of point defects, as determined by the external oxygen partial pressure, spread inwards by mass transport. Well-defined orientation relationships are frequently observed, with well-defined faceted precipitates. It is often assumed that this consistency in the observed orientation relationships and plane normals is evidence that the interfaces have low energy. But there are often very large changes in volume accompanying these internal reactions, and the surrounding matrix may provide a significant constraint on the growing precipitate if it is not relaxed. The degree to which these stresses are relaxed has not been settled conclusively. This is particularly pertinent to internal reduction in a hard oxide matrix, where the constraint of the oxide matrix may be sufficient to force a topotaxial relationship for the reasons we discussed earlier in connection with noble gas inclusions in metals. But for internal oxidation of a metallic alloy it may be argued that, at the high temperature of the oxidation, the rate at which vacancies in the surrounding metal arrive at a growing oxide particle is sufficient to accommodate the change in volume. Some evidence that the constraint of the surrounding metallic matrix is relaxed is provided by the observation that no difference in the lattice parameter of the oxide particle, compared with that of the bulk oxide, can normally be detected.

Both weakly interacting and strongly interacting interfaces have been produced by internal oxidation of metallic alloys. Irrespective of the strength of the interaction it is frequently observed that the oxide particles grow with their close packed planes and directions parallel to those of the metallic matrix. For oxides with the NaCl crystal structure growing in an f.c.c. metallic matrix this means that the oxide particles grow in parallel alignment ('cube-on-cube') with the metallic matrix, where the $\langle 100 \rangle$ axes in the particles are parallel to those of the matrix. Thorium oxide particles are then octahedral, bounded by $\{111\}$ planes. (See Mader (1989, 1992), Mader and Necker (1990), and Gao and Merkle (1990) for reviews.) But there are exceptions to this general observation. Precipitates of NiO appear (Merkle 1990) in Pd with a twin orientation relation (60° about $\langle 111 \rangle$) as frequently as those with parallel alignment, with interfaces on symmetric $\{211\}$ planes and asymmetric $\{111\}$ $\{511\}$ and $\{100\}$ $\{122\}$ planes. Two orientation relations in addition to the parallel alignment were found (Ernst 1990) for particles of MnO in Cu formed by internal oxidation of Cu–Mn alloys, namely the twin relationship and a 55° rotation about [110] yielding $(111)_{Cu}$ parallel to $(002)_{MnO}$. The latter orientation relation was also seen in the rotating ball experiments of Fecht and Gleiter (1985) involving Au and Cu spheres on ionic crystal substrates. Ernst (1990) also observed MnO particles misoriented randomly with respect to a Cu matrix.

In the NaCl crystal structure $\{111\}$ planes contain only anions or only cations. At a $\{111\}$ $\{111\}$ interface between an oxide with the NaCl structure and a metal it is possible, therefore, for the oxide to be terminated by a layer of anions or cations. Which terminating layer is found depends in part on the oxygen chemical potential; a relatively high oxygen chemical potential (as required for oxidizing conditions) will favour an oxygen terminating layer. Necker and Mader (1988) concluded, on the basis of HREM image simulations, that at Ag/CdO interfaces the terminating oxide layer is made up of oxygen. Jang *et al.* (1993) examined the composition of the $\{111\}$ Cu–MgO interface directly by atom-probe field ion microscopy and showed that the terminating oxide layer is 100 per cent oxygen with the penultimate layer 100 per cent magnesium.

At most metal–oxide interfaces that have been observed by HREM the misfit is found to be localized, to some degree at least, at dislocations. In general the degree of localization decreases as the misfit increases and as the strength of interaction across the interface decreases. Some degree of localization has been observed in systems with misfits as large as 22 per cent, e.g. Au–ZrO$_2$ (Gao and Merkle 1992). In some cases no misfit localization has been reported, e.g. at Ag/CdO interfaces (Necker and Mader 1988) where the misfit is 14 per cent. However, it is unclear whether this is simply a consequence of the fact that in some cases the expected misfit dislocations are inclined to the electron beam and are not, therefore, seen end-on.

Misfit dislocations at interfaces between Nb and Al$_2$O$_3$ produced by internal oxidation of Nb–Al alloys have been observed (Mader 1987, 1989) to be located 3–4 atomic spacings from the interface in the Nb (see Fig. 4.56). This 'stand-off' position of the dislocation is readily understood in terms of a balance of forces (Kamat *et al.* 1988). First there is a driving force for the misfit dislocations to be located in the interface, with no stand-off, to relieve the coherency stresses in as large a volume of the Nb as possible. However, each dislocation is repelled by the elastically hard oxide (this is the image interaction discussed in Section 12.3) and its final position is therefore some distance away from the interface. There is then a slab of Nb parallel to the interface which is elastically stretched into coherence with the larger atomic spacing of the Al$_2$O$_3$. The thickness of the slab is equal to the stand-off distance. A detailed linear elastic analysis has been given by Mader and Knauss (1992). It follows that if misfit dislocations are located in the

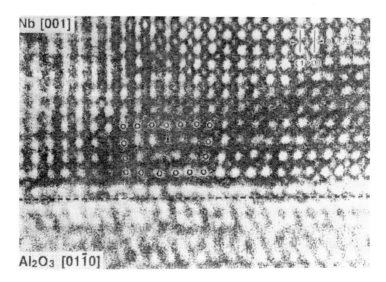

**Fig. 4.56** HREM image of a misfit dislocation close to the Nb–Al$_2$O$_3$ interface (the latter shown by the dashed line). The dislocation is located four (110)$^{\text{Nb}}$ planes from the interface. (From Mader and Knauss (1992).)

elastically harder material then no stand-off occurs because the image interaction is attractive.

However, a stand-off is not always observed at Nb–Al$_2$O$_3$ interfaces, depending on how the interface was manufactured. Mayer *et al.* (1990) grew single crystal Nb films by molecular beam epitaxy on (0001), (1$\bar{1}$00), and (1$\bar{2}$10) substrates of Al$_2$O$_3$. Misfit dislocations were observed with no stand-off in all three cases. Mayer *et al.* speculate that this may be because the growth temperature was too low to allow the dislocations to climb to their equilibrium stand-off positions, in contrast to Nb–Al$_2$O$_3$ interfaces produced by internal oxidation or diffusion bonding. Another interesting observation by Mayer *et al.* was that the Nb layers grew with the same orientation relationship on all three alumina substrates, namely (0001)$_{\text{Al}_2\text{O}_3}$ parallel to (111)$_{\text{Nb}}$ and [2$\bar{1}$$\bar{1}$0]$_{\text{Al}_2\text{O}_3}$ parallel to [1$\bar{1}$0]$_{\text{Nb}}$. In this orientation relation the three-fold axes of Nb and Al$_2$O$_3$ are always parallel. But this orientation relationship differs from that obtained by internal oxidation of Nb–Al alloys, where (0001)$_{\text{Al}_2\text{O}_3}$ is parallel to (110)$_{\text{Nb}}$, and [01$\bar{1}$0]$_{\text{Al}_2\text{O}_3}$ is parallel to [001]$_{\text{Nb}}$ (Mader 1989). It appears, therefore, that the observed orientation relationship depends on the growth conditions, which, for this system, are 2D during molecular beam epitaxy and 3D during internal oxidation.

Bonding at interfaces between an Al$_2$O$_3$ (0001) substrate and vapour deposited mono-layers of transition metals (Ti, Nb, Ni, and Cu) has been studied by Ohuchi and Kohyama (1991) using photoemission electron spectroscopy and semi-empirical tight binding band structure calculations. The calculations were neither self-consistent nor were the atoms relaxed, but they nevertheless provided useful insights into the experimental results. It was found experimentally that Ti and Nb interact strongly with the oxygen of the alumina, whereas for Cu and Ni the dominant interaction was with the aluminium. In the calculations it was assumed that the Nb atoms occupy sites that would have been occupied by Al if the alumina crystal were continued across the interface, although the bond lengths were adjusted to equal the sum of Slater's atomic radii for the atoms concerned. Strong bonds were found between Nb and oxygen, which were partially covalent and partially ionic. These bonds were formed between p-orbitals of the surface oxygen atoms and d-orbitals of the Nb. Electronic charge transfer from the Nb to the oxygen atoms was also found. In the experiments further deposition of the Nb beyond a monolayer did not alter these bonding features at the interface, which simply reflects the efficiency of metallic screening. Similar calculations were carried out for a series of 3d and 4d transition metals on the (0001) surface of alumina. In each case hybridization between p-states on the oxygen and d-states on the transition metal was seen. As we move across the transition metal series from left to right the energy level of the d-states decreases and approaches the oxygen 2p level. Thus, the bonding becomes more covalent and less ionic. But, just as the bond order in the pure transition metals decreases once the d-band is more than half-full (owing to the increasing occupation of antibonding states), so the bond order between transition metals and oxygen decreases as antibonding p–d hybrids are increasingly occupied. This suggests that in the second half of the transition metal series the interaction with oxygen becomes weaker. In that case the transition metal may form a stronger interaction with the aluminium than the oxygen, possibly resulting in an intermetallic layer. An intermetallic alloy between Ni and Al has been observed under high vacuum for Ni deposited on $Al_2O_3$ by Zhong and Ohuchi (1990).

Another metal–oxide interface that has been studied both experimentally and theoretically is the Ag–MgO (100) interface in the cube-on-cube orientation relationship. Trampert *et al.* (1992) deposited layers of 100 nm thickness of Ag on (100) MgO substrates by molecular beam epitaxy. There is a 3 per cent misfit which is accommodated by a square array of $a_{Ag}/2[010]$ and $a_{Ag}/2[001]$ partial edge dislocations. Between the dislocations the Ag atoms sit either on top of Mg atoms or oxygen atoms. This indicates that the energy difference between the Ag–O and Ag–Mg patches is not sufficient to offset the elastic energy gain from dissociating perfect $a/2\langle011\rangle_{Ag}$ dislocations into $a/2\langle010\rangle_{Ag}$ partial dislocations. Using a full-potential linear muffin tin orbital method in the local density approximation Schonberger *et al.* (1992) calculated that when the Ag is on top of the Mg the interfacial energy is a maximum, and 0.8 J m$^{-2}$ higher than the minimum energy of the interface which occurs when the Ag is above the O. They found that the bonding across the Ag–MgO interface is primarily electrostatic in nature, with the polarization of the oxygen and silver atoms playing an important role. The hybridization between the Ag d-orbitals and the O p-orbitals played no significant role, in agreement with the comments of the previous paragraph, and earlier local density functional calculations by Blochl *et al.* (1990). Freeman *et al.* (1990) have also applied a density functional approach to the Ag–MgO (100) interface and shown directly that the charge density on the Ag atoms is polarized by the oxygen atoms beneath them. The importance of the atomic polarizability of the metal at metal–oxide interfaces was emphasized by Finnis (1991), and it was discussed briefly in Section 3.11.

### 4.4.4  Metal–semiconductor interfaces

Interfaces between metals and semiconductors are key elements of semiconductor device technology. The Schottky barrier has been studied intensively, and various models for the Schottky barrier will be discussed in detail in Section 11.2. But despite this huge effort there is still no consensus about the origin of the effect. This extraordinary state of affairs has arisen partially because the vast majority of these studies have not considered metal–semiconductor interfaces that are fully characterized both structurally and compositionally. It has therefore not been possible to test models quantitatively for specific interfaces. The interfaces between transition metal silicides and silicon are exceptional in this regard, and in the remainder of this section we shall discuss the structures of interfaces between $NiSi_2$ or $CoSi_2$ and Si.

Silicide layers are grown epitaxially on silicon substrates under UHV conditions by three techniques. In the solid phase epitaxy technique a layer of the transition metal is deposited at room temperature and is then annealed under UHV conditions. During the anneal the metal reacts with the Si to produce a sequence of silicides. For, example, in the case of Ni on Si(111) the anneal results in the formation of $Ni_2Si$, NiSi, and $NiSi_2$ at progressively higher temperatures with $NiSi_2/Si(111)$ being formed at $T \geqslant 750\,°C$ (Tu *et al.* 1974). The second growth technique is molecular beam epitaxy, where metal and Si atoms in the required stoichiometric ratio are co-evapourated onto a clean Si substrate, which is held at $T \simeq 450-600\,°C$. Finally the 'template' technique (Tung *et al.* 1983*a,b*) is a two-step process involving first the growth of a very thin silicide 'template' layer (less than 20 Å) by deposition of a few monolayers of the metal at room temperature. The reaction to form the silicide is limited by the extent of intermixing between the metal and Si, but since the layer is so thin it proceeds quite quickly even at room temperature. Most significantly, it is found that slightly different thicknesses of the deposited metal layer (between 10 and 20 Å in the case of Ni) result in silicide layers with different orientations to the substrate: these may be controlled in a reproducible manner by controlling the thickness of the metal layer. Further growth on the template layer, by molecular beam epitaxy or solid phase epitaxy, simply thickens the silicide layer while keeping the misorientation constant. In this way the template technique may be used to grow thick silicide layers with a predetermined orientation relationship to the Si substrate.

$NiSi_2$ and $CoSi_2$ have the fluorite crystal structure, in which each metal atom is surrounded by eight Si atoms at the corners of a cube, and each Si is surrounded by four metal atoms at the corners of a tetrahedron. The lattice parameters of $NiSi_2$ and $CoSi_2$ are, respectively, 0.4 per cent and 1.2 per cent smaller than that of Si (5.428 Å). The metal–Si bond lengths in $NiSi_2$ and $CoSi_2$ and the Si–Si bond length in Si are all approximately 2.34 Å.

There are two observed orientations for the silicide layers on Si(111), which have become known as type A and type B. In the type A case the orientation of the silicide is the same as that of the Si substrate. In type B the silicide layer is rotated by 180° about the common [111] interface normal. Cherns *et al.* (1982) studied both interfaces in $NiSi_2$ by HREM. They were both observed to be atomically sharp. Assuming that each Si atom remains four-fold coordinated across the interface they proposed two structures for each interface, which are shown in Figs. 4.57 and 4.58. For the type A interface, the Ni atoms at the interface are either seven-fold coordinated, (Fig. 4.57(a)) or five-fold coordinated (Fig. 4.57(b)). Similarly in the type B interface the Ni atoms are either seven-fold coordinated (Fig. 4.58(a)), or five-fold coordinated (Fig. 4.58(b)). By careful matching between simulated images derived from these models and the experimental images, Cherns *et al.* concluded that the structures with seven-fold coordinated Ni atoms gave better agreement with the experimental images. Their conclusion was supported by total energy, local density functional calculations by Hamann (1988). He found that the energy of the five-fold coordinated type A structure for $NiSi_2$ was 4.34 times greater than that of the seven-fold coordinated structure, which was $0.44\,J\,m^{-2}$. Since the difference between the type A and type B structures appears only at the third neighbours of the interfacial atoms, a similar energy difference may be expected between the five-fold and seven-fold coordinated type B structures. The calculated energy (Hamann 1988) of the

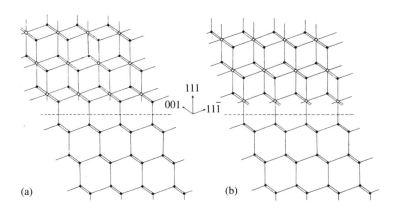

**Fig. 4.57** Two alternative structures for the type *A* (111) NiSi$_2$–Si interface, viewed along [1$\bar{1}$0]. The interface is along the broken line. Si atoms are black and Ni atoms are white. In (a) the Ni atoms are seven-fold coordinated. In (b) the Ni atoms are five-fold coordinated. (From Cherns *et al.* (1982).)

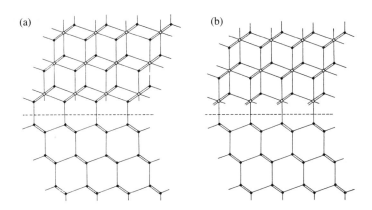

**Fig. 4.58** Two alternative structures for the type *B* (111) NiSi$_2$–Si interface, viewed along [1$\bar{1}$0]. See caption to Fig. 4.57. The Ni atoms are seven-fold coordinated in (a) and five-fold coordinated in (b). (From Cherns *et al.* (1982).)

type B seven-fold coordinated structure was only 50 mJ m$^{-2}$ greater than that of the type A, which is consistent with both structures being found experimentally.

Interestingly, the structure of the interface for CoSi$_2$ on Si(111) has been more controversial. Gibson *et al.* (1982) found only the type B orientation, and concluded that a better match between the experimental and simulated images was obtained with the five-fold coordinated model, shown in Fig. 4.58(b). However, Hamann (1988) found that the energy of the five-fold coordinated structure (2.76 J m$^{-2}$) was much greater than the seven-fold (1.08 J m$^{-2}$). Furthermore, the lowest energy configuration contained only eight-fold coordinated Co atoms with three-fold coordinated Si atoms, as shown in Fig. 4.59. The calculated energy of the eight-fold coordinated structure is 0.67 J m$^{-2}$. Hamann also calculated the energy of an eight-fold coordinated type A structure and found it was higher in energy by 0.18 J m$^{-2}$, which is more than three times the energy

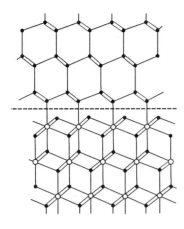

**Fig. 4.59** Structure of the type B (111) $CoSi_2$–Si interface, viewed along [1$\bar{1}$0]. The Si atoms are black and the Co atoms are white. The Co atoms at the interface are eight-fold coordinated and the Si atoms at the interface are three-fold coordinated. (After Hamann (1988).)

difference between type A and B structures in $NiSi_2$. Hamann pointed out that the eight-fold coordinated structure was also consistent with the experimental observations.

The question now arises as to why $NiSi_2$ and $CoSi_2$, which have such similar atomic and electronic structures, should form different stable interfaces on Si(111). In the eight-fold coordinated $CoSi_2$ structure (Fig. 4.59) Co atoms bond directly to the Si substrate, and the interfacial Si atoms are three-fold coordinated. By contrast, in the seven-fold coordinated $NiSi_2$ type A and B structures (Figs 4.57(b) and 4.58(b)) Si atoms bond directly to the Si substrate, and the interfacial Si atoms are four-fold coordinated. There is one more electron in Ni than in Co. A detailed 'frontier orbital' analysis of bonding at these interfaces has been presented by van den Hoek *et al.* (1988). They found that the local density of states in the seven-fold coordinated structure is enhanced for bonding states just below the Fermi energy. The extra electron in $NiSi_2$ can thus enter a bonding state in the seven-fold coordinated structure. In the eight-fold coordinated structure the splitting between bonding and antibonding states is greater, with more weight being shifted from bonding to antibonding states. $CoSi_2$ can take advantage of the lower lying bonding states without having to occupy any of the higher antibonding states. This explains why the eight-fold coordinated structure is more stable for $CoSi_2$. On the other hand, $NiSi_2$, with its extra electron, would have to occupy antibonding states in the eight-fold coordinated structure. Therefore the eight-fold coordinated structure is destabilized by the extra electron in $NiSi_2$. We see here the important roles played by the local electronic structure and the electron to atom ratio in determining the stability of these silicide interfaces.

## REFERENCES

Allan, G. (1987). *Prog. Surf. Sci.*, **25**, 43.
Allen, M. P. and Tildesley, D. J. (1987). *Computer simulation of liquids*, Oxford University Press, Oxford.
Allen, S. M. and Cahn, J. W. (1979). *Acta Metall.*, **27**, 1085.
Ashby, M. F., Spaepen, F., and Williams, S. (1978). *Acta Metall.*, **26**, 1647.
Bacon, D. J., Barnett, D. M., and Scattergood, R. O. (1978). *Prog. Mat. Sci.*, **23**, 51.
Balluffi, R. W., Majid, I., and Bristowe, P. D. (1989). *Mat. Res. Soc. Symp. Proc.*, **138**, 457.
Barrett, C. S. and Massalski, T. B. (1980). *Structure of metals*. Pergamon, Oxford.
Barsch, G. R. and Krumhansl, J. A. (1984). *Phys. Rev. Lett.*, **53**, 1069.

Bernal, J. D. (1964). *Proc. Roy. Soc. Lond.*, **280A**, 299.

Bishop, G. H., and Chalmers, B. (1968). *Scr. Metall.*, **2**, 133.

Bishop, G. H. and Chalmers, B. (1971). *Phil. Mag.*, **24**, 515.

Blochl, P., Das, G. P., Fischmeister, H.F., and Schonberger, U. (1990). In *Metal-Ceramic Interfaces* (eds M. Rühle, A. G. Evans, M. F. Ashby, and J. P. Hirth), p. 9. Pergamon, Oxford.

Bourret, A. and Desseaux, J. (1979*a*). *Phil. Mag A*, **39**, 405.

Bourret, A. and Desseaux, J. (1979*b*). *Phil. Mag A*, **39**, 419.

Brandon, D. G., Ralph, B., Ranganathan, S., and Wald, M. S. (1964). *Acta Metall.*, **12**, 813.

Bristowe, P. D. and Crocker, A. G. (1978). *Phil. Mag. A*, **38**, 487.

Brokman, A. and Balluffi, R. W. (1981). *Acta Metall.*, **29**, 1703.

Burns, G. (1985). *Solid state physics*. Academic Press, New York.

Cahn, J. W. (1961). *Acta Metall.*, **9**, 795.

Cahn, J. W. and Hilliard, J. E. (1958). *J. Chem. Phys.*, **28**, 258.

Cahn, J. W. and Kikuchi, R. (1966). *J. Phys. Chem. Solids*, **27**, 1305.

Campbell, G. H., Foiles, S. M., Gumbsch, P., Rühle, M., and King, W. E. (1993). *Phys. Rev. Lett.*, **70**, 449.

Cantor, B. (1981). Crystallography and interfaces in eutectic alloys (unpublished).

Cao, W. and Barsch, G. R. (1990). *Phys. Rev. B*, **41**, 4334.

Chatain, D., Rivollet, I., and Eustathopoulos, N. (1986). *J. Chimie Physique*, **83**, 561.

Chatain, D., Chabert, F., Ghetta, V., and Fouletier, J. (1993*a*). *J. Am. Ceram. Soc.*, **76**, 1568.

Chatain, D., Ghetta, V., Chabert, F., and Fouletier J. (1993*b*). *Mat. Sci. Forum*, **126–28**, 715.

Cheikh, M., Deyehe, M., Hairie, A., Hairie, F., Nouet, G., and Paumier, E. (1991). *Springer. Proc. Phys.*, **54**, 200.

Cherns, D., Anstis, G. R., Hutchison, J. L., and Spence, J. C. H. (1982). *Phil. Mag. A*, **46**, 849.

Christian, J. W. (1981). *The theory of transformations in metals and alloys*. Pergamon, Oxford.

Clarke, D. R. (1987). *J. Am. Ceram. Soc.*, **70**, 15.

Crocker, A. G. and Faridi, B. A. (1980). *Acta Metall.*, **28**, 549.

Dahmen, U., and Westmacott, K. H. (1988). *Scr. Metall.*, **22**, 1673.

d'Heurle, F. M. (1988). *J. Mater. Res.*, **3**, 167.

Duffy, D. M. and Tasker, P. W. (1983). *Phil. Mag. A*, **47**, 817.

Duffy, D. M. and Tasker, P. W. (1984*a*). *Phil. Mag. A*, **50**, 143.

Duffy, D. M. and Tasker, P. W. (1984*b*) *J. Appl. Phys.*, **56**, 971.

Duffy, D. M. (1986). *J. Phys. C: Sol. State Phys.*, **19**, 4393.

Erb, U., and Gleiter, H. (1979). *Scr. Metall.*, **13**, 61.

Ernst, F. (1990). *Mat. Res. Soc. Symp. Proc.*, **183**, 49.

Ernst, F., Pirouz, P., and Heuer, A. H. (1989). *Mat. Res. Soc. Symp. Proc.*, **138**, 557.

Ernst, F., Finnis, M. W., Hofmann, D., Muschik, T., Schonberger, U., Wolf, U., and Methfessel, M. (1992). *Phys. Rev. Lett.*, **69**, 620.

Ewing, R. H. (1971). *Acta Metall.*, **19**, 1359.

Ewing, R. H. and Chalmers, B. (1972). *Surf. Sci.*, **31**, 1961.

Fecht, H. J. and Gleiter, H. (1985). *Acta Metall.*, **33**, 557.

Finnis, M. W. (1991). *Surf. Sci.*, **241**, 61.

Finnis, M. W. (1993). *Physics World*, **6**(7), 37.

Fitzsimmons, M. R. and Sass, S. L. (1989). *Acta Metall.*, **37**, 1009.

Fitzsimmons, M. R., Burkel, E., and Sass, S. L. (1988). *Phys. Rev. Lett.*, **61**, 2237.

Freeman, A. J., Li, C., and Fu, C. L. (1990). In *Metal-ceramic interfaces* (eds M. Rühle, A. G., Evans, M. F., Ashby, and J. P. Hirth), p. 2. Pergamon, Oxford.

Fridman, E. M., Kopetskii, Ch. V., and Shvindlerman, L. S. (1974). *Sov. Phys. Solid State*, **16**, 1152.

Gao, Y. and Merkle, K. L. (1990). *J. Mater. Res.*, **5**, 1995.

Gao, Y. and Merkle, K. L. (1992). *Mat. Res. Soc. Symp. Proc.*, **238**, 775.

Gao, Y., Shewmon, P. G., and Dregia, S. A. (1989). *Acta Metall.*, **37**, 3165.

Gibson, J. M., Bean, J. C., Poate, J. M., and Tung, R. T. (1982). *Appl. Phys. Lett.*, **41**, 818.

Gronsky, R. (1980). In *Grain boundary structure and kinetics* (ed. R. W. Balluffi), p. 45. American Society for Metals, Metals Park, Ohio.

Hamann, D. R. (1988). *Phys. Rev. Lett.*, **60**, 313.

Hashimoto, M., Ishida, Y., Yamamoto, R., and Doyama, M. (1981). *Acta Metall.*, **29**, 617.

Horsewell, A., Johnson, E., and Bourdelle, K. K. (1993). *Mat. Sci. Forum*, **126-28**, 647.

Houchmandzadeh, B., Lajzerowicz, J., and Salje, E. (1992). *J. Phys.: Condens. Matter*, **4**, 9779.

Ichinose, H. and Ishida Y. (1985). *J. Physique*, **46**, C4-39.

Jang, H., Seidman, D. N., and Merkle, K. L. (1993). *Mat. Sci. Forum*, **126-28**, 639.

Johnson, W. L. (1986). *Prog. Mat. Sci.*, **30**, 81.

Josell, D. and Spaepen, F. (1993). *Acta Metall.*, **41**, 3017.

Kamat, S. V., Hirth, J. P., and Carnahan, B. (1988). *Mat. Res. Soc. Symp. Proc.*, **103**, 55.

Kikuchi, R. (1972). *J. Chem. Phys.*, **57**, 777, 783, 787.

Kikuchi, R. and Cahn, J. W. (1962). *J. Phys. Chem. Solids*, **23**, 137.

Kikuchi, R. and Cahn, J. W. (1979). *Acta Metall.*, **27**, 1337.

Kittel, C. (1966). *Introduction to solid state physics*, 3rd edn. Wiley, New York.

Klam, H. J., Hahn, H., and Gleiter, H. (1987). *Acta Metall.*, **35**, 2101.

Kohyama, M., Kose, S., and Yamamoto, R. (1991). *J. Phys.: Condens. Matter*, **3**, 7555.

Kohyama, M., Kose, S., and Yamamoto, R. (1993). *Mat Sci. Forum*, **126-28**, 213.

Kopetsky, Ch. V., Shvindlerman, L.S., and Sursaeva, V. G. (1978). *Scr. Metall.*, **12**, 953.

Kraft, R. W. (1962). *Trans. Met. Soc. AIME*, **224**, 65.

Krakow, W. (1991). *Phil. Mag. A*, **63**, 233.

Krakow, W. (1992). *Acta Metall. Mater.*, **40**, 977.

Krakow, W. and Smith, D. A. (1986*a*). *J. Mater. Res.*, **1**. 47.

Krakow, W. and Smith, D. A. (1986*b*). *Proc. JIMIS*-4, Suppl. to *Jap. Inst. Met.*, 277.

Krivoglaz, M. A. and Smirnov, A. A. (1965). *The theory of order-disorder in alloy*. Elsevier, New York.

Laffont, A. and Bonnet, R. (1982). *Acta Metall.*, **30**, 763.

Landau, L. and Lifshitz, E. (1935). *Phys. Z. Sowjetunion*, **8**, 153.

Lu, J. P. and Birman, J. L. (1986). *J. Physique*, **47**, C3-251.

Mader, W. (1987). *Mat. Res. Soc. Symp. Proc.*, **82**, 403.

Mader, W. (1989). *Z. Metallkde.*, **80**, 139.

Mader, W. (1992). *Mat. Res. Soc. Symp. Proc.*, **238**, 763.

Mader, W. and Knauss, D. (1992). *Acta Metall. Mater.*, **40**, S207.

Mader, W. and Maier, B. (1990). *J. Physique*, **51**, C1-867.

Mader, W. and Necker, G. (1990). In *Metal-ceramic interfaces* (eds M. Rühle, A. G. Evans, M. F. Ashby, and J. P. Hirth), p. 222, Pergamon, Oxford.

Majid, I., Bristowe, P. D., and Balluffi, R. W. (1989). *Phys. Rev. B*, **40**, 2779.

Margenau, H. and Murphy, G. M. (1943). *The mathematics of physics and chemistry*, Van Nostrand, New York.

Maurice, J. L. (1990). *J. Physique*, **51**, C1-581.

Mayer, J., Flynn, C. P., and Rühle, M. (1990). *Ultramicroscopy*, **33**, 51.

McCafferty, K., Soper, A., Shirokoff, J., and Erb, U. (1992). *Mat. Res. Soc. Symp. Proc.*, **238**, 47.

Merkle, K. L. (1990). In *Metal-ceramic interfaces* (eds M. Rühle, A. G. Evans, M. F. Ashby, and J. P. Hirth), p. 242. Pergamon, Oxford.

Merkle, K. L. (1991). *Ultramicroscopy*, **37**, 130.

Merkle, K. L. and Smith, D. J. (1987*a*). *Phys. Rev. Lett.*, **59**, 2887.

Merkle, K. L. and Smith, D. J. (1987*b*). *Ultramicroscopy*, **22**, 57.

Merkle, K. L. and Wolf, D. (1990). *MRS Bulletin*, September issue, p. 42.

Merkle, K. L., Reddy, J. F., and Wiley, C. L. (1985). *J. Physique*, **46**, C4-95.

Merkle, K. L., Reddy, J. F., Wiley, C. L., and Smith, D. J. (1988). *J. Physique*, **49**, C5-251.

Meyer, M. and Waldburger, C. (1993). *Mat. Sci. Forum*, **126-28**, 229.

Möller, H. J., Strunk, H. P., and Werner, J. H. (ed.) (1989). *Polycrystalline semiconductors*, Springer Proc. Phys. **35**. Springer-Verlag, Berlin.

Molodov, D. A., Straumal, B. B., and Shvindlerman, L. S. (1984). *Scr. Metall.*, **18**, 207.

Mooser, E. and Schluter M. (1971). *Phil. Mag.*, **23**, 811.

Mott, N. F. (1948). *Proc. Phys. Soc.*, **60**, 391.

Murr, L. E. (1970). *Phys. Stat. Sol.(a)*, **3**, 447.

Murr, L. E. (1972). *Scr. Metall.*, **6**, 203.

Naidich, Yu. (1981). *Prog. Surf. Memb. Sci.*, **14**, 353.

Najafabadi, R., Srolovitz, D. J., and LeSar R. (1991). *J. Mater. Res.*, **6**, 999.

Nakanishi, N., Nagasawa, A., and Murakami, Y. (1982). *J. Physique*, **43**, C4–35.

Nazarov, A. A. and Romanov, A. E. (1989). *Phil. Mag. Lett.*, **60**, 187.

Necker, G. and Mader, W. (1988). *Phil. Mag. Lett.*, **58**, 205.

Nelson, S. F. (1991). PhD thesis, Cornell University.

Nicholas, M. G. (1989). In *Surfaces and interfaces of ceramic materials* (eds L. C. Dufour, C. Monty, and G. Petot-Ervas), p. 393, NATO ASI Series E: Applied Sciences, Vol. 173, Kluwer, Dordrecht.

Nye, J. F. (1957). *Physical properties of crystals*, Oxford University Press, Oxford.

Ohuchi, F. S. and Kohyama, M. (1991). *J. Am., Ceram. Soc.*, **74**, 1163.

Olson, G. B. (1986). In *Proceedings of the international conference on martensitic transformations*, p. 25. Japan Institute of Metals, Sendai, Japan.

Olson, G. B. (1990). In *Martensitic transformations: ICOMAT '89* (ed. B. C. Muddle), p. 89, Trans. Tech., Switzerland.

Olson, G. B. and Cohen, M. (1982). *J. Physique*, **43**, C4–75.

Omar, R. (1987). Dissertation, University of Warwick, Coventry, U.K.

Othen, P. J., Jenkins, M. L., Smith, G. D. W., and Phythian, W. J. (1991). *Phil. Mag. Lett.*, **64**, 383.

Othen, P. J., Jenkins, M. L., and Smith, G. D. W. (1994). *Phil. Mag. A*, **70**, 1.

Parrinello, M. and Rahman, A. (1981). *J. Appl. Phys.*, **52**, 7182.

Penisson, J. M., Nowicki, T., and Biscondi, M. (1988). *Phil. Mag. A*, **58**, 947.

Pollmann, J. and Pantelides, S. T. (1980). *Phys. Rev. B*, **21**, 709.

Pond, R. C., Smith, D. A., and Vitek, V. (1978). *Scr. Metall.*, **12**, 699.

Press, P. H., Flannery, B. P., Teukolsky, S. A., and Vetterling, W. T. (1989). *Numerical recipes*. Cambridge University Press, Cambridge.

Provan, J. W. and Bamiro, O. A. (1977). *Acta Metall.*, **25**, 309.

Pumphrey, P. H. (1972). *Scr. Metall.*, **6**, 107.

Ralantoson, N., Hairie, F., Hairie, A., Nouet, G., and Paumier, E., (1993). *Mat. Sci. Forum*, **126–28**, 241.

Rouviere, J. L. and Bourret, A. (1990). *J. Physique*, **51**, C1–329.

Sass, S. L. and Bristowe, P. D. (1980). In *Grain boundary structure and kinetics* (ed. R. W. Balluffi), p. 71. American Society for Metals, Metals Park, Ohio.

Schindler, R., Clemans, J. E., and Balluffi, R. W. (1979). *Phys. Stat. Sol. (a)*, **56**, 749.

Schonberger U., Andersen, O. K., and Methfessel, M. (1992). *Acta Metall. Mater.*, **40**, S1.

Schwartz, D., Vitek, V., and Sutton, A.P. (1985). *Phil. Mag. A*, **51**, 499.

Smith, D. A., Segmuller, A., and Taranko, A. R. (1987). *Mat. Res. Soc. Symp. Proc.*, **94**, 127.

Straumal, B. B., Klinger, L. M., and Shvindlerman, L. S. (1984). *Acta Metall.*, **32**, 1355.

Sun, C. P. and Balluffi, R. W. (1982). *Phil. Mag. A*, **46**, 49.

Sutton, A. P. (1982). *Phil. Mag. A*, **46**, 171.

Sutton, A. P. (1988). *Acta Metall.*, **36**, 1291.

Sutton, A. P. (1989*a*). *Phil. Mag. Lett.*, **59**, 53.

Sutton, A. P. (1989*b*). *Phil. Mag. A*, **60**, 147.

Sutton, A. P. (1989*c*). *Inst. Phys. Conf. Ser. No.* 104, 13.

Sutton, A. P. (1991*a*). *Phil. Mag. A*, **63**, 793.

Sutton, A. P. (1991*b*). *Springer Proc. Phys.*, **54**, 116.

Sutton, A. P. (1992). *Prog. Mat. Sci.*, **36**, 167.

Sutton, A. P. and Balluffi, R. W. (1987). *Acta Metall.*, **35**, 2177.

Sutton, A. P. and Balluffi, R. W. (1990). *Phil. Mag. Lett.*, **61**, 91.

Sutton, A. P. and Vitek, V. (1983*a*). *Phil. Trans. R. Soc. Lond. A*, **309**, 1.

Sutton, A. P. and Vitek, V. (1983*b*). *Phil. Trans. R. Soc. Lond. A*, **309**, 37.

Sutton, A. P. and Vitek, V. (1983*c*). *Phil. Trans. R. Soc. Lond. A*, **309**, 55.

Sutton, A. P., Balluffi, R. W., and Vitek, V. (1981). *Scr. Metall.*, **15**, 989.

Tarnow, E., Dallot, P., Bristowe, P. D., Joannopoulos, J. D., Francis G. P., and Payne, M.C. (1990). *Phys. Rev. B*, **42**, 3644.

Tasker, P. W. and Duffy, D. M. (1983). *Phil. Mag. A*, **47**, L45.

Trampert, A., Ernst, F., Flynn, C. P., Fischmeister, H. F., and Rühle, M. (1992). *Acta Metall. Mater.*, **40**, S227.

Tu, K. N., Alessandrini, E. I., Chu, W. K., Krautle, H., and Mayer, J. W. (1974). *Jap. J. Appl. Phys. Suppl.*, **2**, 669.

Tung, R. T., Gibson, J. M., and Poate, J. M. (1983*a*) *Phys. Rev. Lett.*, **50**, 429.

Tung, R. T., Gibson, J. M., and Poate, J. M. (1983*b*). *Appl. Phys. Lett.*, **42**, 888.

van den Hoek, P. J., Ravenek, W., and Baerends, E. J. (1988). *Phys. Rev. Lett.*, **60**, 1743.

Vanderbilt, D., Alerhand, O. L., Meade, R.D., and Joannopoulos, J. D. (1989). *J. Vac. Sci. Technol. B*, **7**, 1013.

Vitek, V., Sutton, A. P., Smith, D. A., and Pond, R. C. (1980). In *Grain boundary structure and kinetics* (ed. R. W. Balluffi), p. 115. American Society for Metals, Metals Park, Ohio.

Wagner, T., Kirchheim, R., and Rühle, M. (1992). *Acta Metall. Mater.*, **40**, S85.

Wang, G.J. and Vitek, V. (1986). *Acta Metall.*, **34**, 951.

Wang, G. J., Sutton, A. P., and Vitek, V. (1984). *Acta Metall.*, **32**, 1093.

Weins, M. J., Chalmers, B., Gleiter, H., and Ashby, M. F. (1969). *Scr. Metall.*, **3**, 601.

Weins, M. J., Gleiter, H., and Chalmers, B. (1970). *Scr. Metall.*, **4**, 235.

Weins, M. J., Gleiter, H., and Chalmers, B. (1971). *J. Appl. Phys.*, **42**, 2639.

Werner, J. H. and Strunk, H. P. (ed.) (1991). *Polycrystalline semiconductors II*, Springer Proc. Phys. **54**. Springer-Verlag, Berlin.

Wolf, D. (1980). *J. Physique*, **41**, C6–142.

Wolf, D. (1982). *J. Physqiue*, **43**, C6–45.

Wolf, D. (1984). *Acta Metall.*, **32**, 245.

Wolf, D. (1985). *J. Physique*, **46**, C4–197.

Wolf, D. (1990*a*). *J. Mater. Res.*, **5**, 1708.

Wolf, D. (1990*b*). *J. Appl. Phys.*, **68**, 3221.

Wolf, D. and Benedek, R. (1981). *Adv. Ceram.*, **1**, 107.

Wolf, D. and Phillpot, S. (1989). *Mater. Sci. Eng. A*, **107**, 3.

Wolf, D. and Merkle, K. L. (1992). In *Materials interfaces* (eds D. Wolf and S. Yip), p. 87. Chapman & Hall, London.

Wolf, U., Ernst, F., Muschik, T., Finnis, M. W., and Fischmeister, H. F. (1992). *Phil. Mag. A*, **66**, 991.

Zhong, Q. and Ohuchi, F. S. (1990). *J. Vac. Sci. Technol.*, **A8**, 2107.

# PART II

# THERMODYNAMICS OF INTERFACES

# 5
## Thermodynamics of interfaces

### 5.1 INTRODUCTION

In this chapter, we develop the thermodynamics of interfaces in crystalline materials. The quantities necessary to formulate the thermodynamics of these interfaces are introduced and defined, and the framework for working out relationships between the various thermodynamic quantities is presented. A number of important relationships and results are derived which will be of use in the following chapters. Typical experimental data are reviewed.

It is important to note at the onset that the thermodynamics of interfaces in crystalline materials should be regarded as part of the much larger and general field of interface thermodynamics. For example, there are many similarities between the thermodynamics of interfaces in crystalline materials (internal interfaces) and the thermodynamics of crystalline surfaces (free surfaces), which is a highly-developed field. Much of the formalism is the same for both. However, the internal interfaces are inherently more complex than the free surfaces because additional geometric thermodynamic variables must be introduced to describe the crystal misorientation and translational states associated with the internal interfaces.

The basic thermodynamics of interfaces was first developed more than 100 years ago by Gibbs (1875–78, 1957). He defined the extensive thermodynamic properties of an interface, such as internal energy, entropy, etc., as excess quantities associated with the presence of the interface in the system. These were defined by constructing a 'dividing surface' in the vicinity of the interface and then subtracting from the actual amount of a quantity present amounts characteristic of the bulk phases adjoining the dividing surface as if they were homogeneous right up to the dividing surface. More recently, Cahn (1979) has formulated an alternative, but ultimately equivalent, approach which does away with the artifice of the dividing surface. Cahn's method is appealing pedagogically and is also less cumbersome for many problems. We therefore adopt his approach in the main body of our work. However, a brief description of the Gibbs method is also included.

### 5.2 THE INTERFACE FREE ENERGY

We begin by introducing and defining the all-important quantity, $\sigma$, which will be termed the 'interface free energy (per unit interface area)' throughout this book. In order to define this quantity rigorously, consider a bicrystal consisting of two bulk phases, $\alpha$ and $\beta$, which are in contact along a planar internal interface which is growing in a container under assumed equilibrium conditions by the accretion of atoms from suitable reservoirs as shown schematically in Fig. 5.1. The entire system, including the reservoirs, is maintained at constant temperature, $T$, hydrostatic pressure, $P$, and chemical potential, $\mu_i$, of each of the components. It is assumed that the interface is incoherent, or semicoherent, and that dislocation sources/sinks are distributed throughout the bulk phases so that no problems exist in defining the chemical potentials as described in section 5.9.2.3. Now,

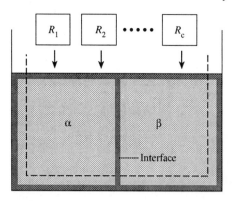

**Fig. 5.1** Formation of bicrystal consisting of $\alpha$ and $\beta$ bulk phases and interface by accretion of component atoms from reservoirs $R_1, R_2, \ldots R_C$. Width of regions influenced by the internal interface and other surfaces indicated by shaded zones. Region of primary interest is bounded by the dashed lines.

focus attention on the bicrystal system within the region bounded by the dashed surface in Fig. 5.1. All dimensions in Fig. 5.1 are large compared with the widths of any regions affected by the interfaces which are present (shown as shaded zones). For this bicrystal system we may use the usual combined form of the first and second laws of thermodynamics (Reif 1965) and write the increase in internal energy due to the accretion in the form

$$dE = T\,dS - P\,dV + \sum_{i=1}^{C} \mu_i\,dN_i + \sigma\,dA, \tag{5.1}$$

where $S$ is entropy, $V$ is volume, $N_i$ is the amount of component $i$, $C$ is the number of components, and $A$ is area of the planar interface. Equation (5.1) contains the usual bulk terms plus the added term $\sigma dA$ which is required to account for the increase in internal energy of the system which is associated with the increase in the area of the interface. From eqn (5.1), we have

$$\sigma = \left[\frac{\partial E}{\partial A}\right]_{S,V,N_i}, \tag{5.2}$$

and $\sigma$ is therefore defined as the increase in internal energy of the entire bicrystal system per unit increase in interface area at constant $S$ and $V$ of the system under closed conditions, i.e., under the condition that the $N_i$ are held constant (Cahn 1979). Further relationships for $\sigma$ may be found (Cahn 1979) by employing additional thermodynamic variables. The Gibbs free energy, $G$, the Helmholtz free energy, $F$, and the grand potential, $\Omega$, are defined by

$$G = E + PV - TS, \tag{5.3a}$$

$$F = E - TS, \tag{5.3b}$$

$$\Omega = E - TS - \sum_{i=1}^{C} \mu_i N_i. \tag{5.3c}$$

Combining each of these relations with eqn (5.1) then yields the additional three relationships:

$$dG = -S\,dT + V\,dP + \sum_{i=1}^{C} \mu_i dN_i + \sigma\,dA, \tag{5.4a}$$

$$dF = -S\,dT - P\,dV - \sum_{i=1}^{C} \mu_i dN_i + \sigma dA, \tag{5.4b}$$

$$d\Omega = -S\,dT - P\,dV - \sum_{i=1}^{C} N_i d\mu_i + \sigma dA. \tag{5.4c}$$

Therefore, $\sigma$ also corresponds to

$$\sigma = \left[\frac{\partial G}{\partial A}\right]_{T,P,N_i}, \tag{5.5a}$$

$$\sigma = \left[\frac{\partial F}{\partial A}\right]_{T,V,N_i}, \tag{5.5b}$$

$$\sigma = \left[\frac{\partial \Omega}{\partial A}\right]_{T,V,\mu_i}, \tag{5.5c}$$

Equation (5.5a) shows that $\sigma$ is equal to the increase in the Gibbs free energy of the entire system per unit increase in interface area when $T$ and $P$ are held constant and the system is again maintained closed. Equation (5.5c) shows that $\sigma$ is equal to the increase in the grand potential of the system per increase in area in an open system when $T$, $V$, and the $\mu_i$ are maintained constant. In practice the applicable equation, and the type of energy change of the system per change in interface area, which is represented by $\sigma$, will depend upon the constraints which are present. For example, the grand potential formulation will be particularly applicable for a local system of constant $V$ centred on an interface embedded in a large multicomponent bicrystal at constant $T$. Here, the local system will be open to the exhange of components with its environment, and the two large adjoining crystals will act as constant $\mu_i$ reservoirs. On the other hand, the Gibbs free energy formulation will be preferred for a closed one-component system at constant $T$ and $P$.

A further important result may be obtained by integrating eqn (5.1) by using Euler's theorem (or simply continuing the accretion process) to obtain

$$E = TS - PV + \sum_{i=1}^{C} \mu_i N_i + \sigma A, \tag{5.6}$$

or, with the help of eqn (5.3a),

$$\sigma = \frac{1}{A}\left[G - \sum_{i=1}^{C} \mu_i N_i\right]. \tag{5.7}$$

Since the quantity $\sum_{i=1}^{C} \mu_i N_i$ is the total Gibbs free energy which the homogeneous $\alpha$ and $\beta$ bulk phases would possess together if they were made up of the same amounts of the components at the same chemical potentials, we may identify $\sigma$ simply as the excess Gibbs free energy of the entire system per unit interface area due to the presence of the interface.

An interesting experiment in which the amount of grain boundary area in a one-component system can be varied under conditions of constant $T$, $P$, and $N_1$ is the diffusional creep of a polycrystalline wire (Funk *et al.* 1951) illustrated in Fig. 5.2. Here, a polycrystalline cylindrical wire with a 'bamboo' grain structure is attached to a weight. If such a specimen is held at an elevated temperature, it is known (Udin *et al.* 1949) that it will deform by diffusional creep at a constant small load (see section 12.8.2.1) and generally tend to become either longer and thinner, or, alternatively, shorter and thicker,

*Thermodynamics of interfaces*

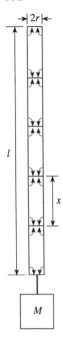

**Fig. 5.2** Cylindrical polycrystalline wire with 'bamboo' grain structure undergoing diffusional creep under the influence of a uniaxial stress produced by the mass $M$. Arrows indicate stress-motivated diffusional fluxes of atoms from the wire surface to the transverse grain boundaries.

depending upon the magnitude of the weight. This is accomplished by the diffusional transport of atoms between the lateral grain boundaries and the vertical free surface so that the wire shape changes and the areas of these interfaces change in response to the applied load at constant volume. If an elongation d$l$ occurs, and we average any anisotropies of the excess free energies associated with the free surface and the grain boundaries, the increase in the free surface free energy of the system due to the increased free surface area, d$A^S$, is $\bar{\sigma}^S$d$A^S$, where $\bar{\sigma}^S$ is the average excess free energy associated with unit area of free surface. (We note that the excess free surface free energy can be identified by an argument similar to that used to obtain eqn (5.7).). The corresponding decrease in the Gibbs free energy due to the decrease in grain boundary area is then $\bar{\sigma}^B$d$A^B$ (where $\bar{\sigma}^B$ is the average grain boundary free energy) and the total free energy change is then just $\pi r[\bar{\sigma}^S - \bar{\sigma}^B(r/x)]\cdot$d$l$, since d$A^S = \pi r$d$l$ and d$A^B = -\pi r^2$d$l/x$. On the other hand, the work done by the falling weight is d$w = mg\cdot$d$l$. The critical mass $m^*$ required to maintain the wire at constant length is then found by equating these quantities so that

$$m^*g = \pi r[\sigma^S - \sigma^B(r/x)]. \tag{5.8}$$

If the mass exceeds this value, the falling weight loses potential energy and performs work while it increases the total free energy of the wire system, while, in the reverse case, the wire system performs work by decreasing its Gibbs free energy and lifting the weight.

In order for an interface to be thermodynamically stable with respect to cleavage into two free surfaces, it is necessary that $\sigma$ is less than the free energies per unit area of the two free surfaces. This imposes an upper limit on the interfacial free energy. In the case of a grain boundary in a metal, the boundary free energy is generally less than the free energy of the free surface because the average coordination number is higher at the grain boundary. For example, the number of near-neighbour atoms around an atom in an interface is generally only slightly lower than in the perfect bulk lattice. Values of $\sigma$ vary considerably, however, depending upon the interface type, and some representative values for free surfaces and interfaces are given in Table 5.1.

**Table 5.1** Measured thermodynamic properties of some selected interfaces

| Interface | $\sigma$ (mJ m$^{-2}$) | $\Delta S$ (eqn (5.27)) (mJ m$^{-2}$ K$^{-1}$) | $\Delta V^{\#}$ (eqn (5.26)) (m$^3$ m$^{-2}$) |
|---|---|---|---|
| Au general homophase (1000 °C) | 378[a] | 0.1[a] | – |
| Ni general homophase (1060 °C) | 866[a] | 0.2[a] | – |
| Au general free surface (1000 °C) | 1400[a] | 0.43[a] | – |
| Ni general free surface (1060 °C) | 2280[a] | 0.55[a] | – |
| Au–Al$_2$O$_3$ heterphase (1000 °C) | 1725[a] | — | – |
| Ni–ThO$_2$ heterophase (1200 °C) | 2000[a] | — | – |
| Al Σ3 {551} {711} homophase | – | — | $\approx 2 \times 10^{-11}$[b] |
| Ge Σ9 {221} {221} homophase | – | — | $\approx 1 \times 10^{-11}$[c] |
| NiO Σ5 {310} {310} homophase | – | — | $\approx 4 \times 10^{-11}$[d] |

[#] Estimated from measured expansion perpendicular to the boundary plane on the assumption that the boundary core density is correspondingly decreased. (See discussion of this point in Merkle (1989).)
[a] Murr (1975, Chapter 3).)
[b] Pond and Vitek (1977).
[c] Papon *et al.* (1982).
[d] Merkle and Smith (1987).

## 5.3 ADDITIONAL INTERFACE THERMODYNAMIC QUANTITIES AND RELATIONSHIPS BETWEEN THEM

A further thermodynamic relationship may be obtained by differentiating (eqn (5.6) and subtracting the result from eqn (5.1) to obtain

$$A \, \mathrm{d}\sigma = -S \, \mathrm{d}T + V \, \mathrm{d}P - \sum_{i=1}^{C} N_i \, \mathrm{d}\mu_i. \tag{5.9}$$

We now note that all of the previous relationships have been derived for an overall system consisting of an interface and two adjoining bulk phases. In such a system, we expect the properties of the $\alpha$ and $\beta$ phases near the interface to be affected, at least to some extent, by the presence of the interface, as already illustrated in Fig. 5.1 and shown in more detail in Fig. 5.3. However, at large distances, the properties should be unaffected. The system can then be divided up into regions consisting of homogeneous regions of $\alpha$ and $\beta$ and a relatively narrow layer centred on the interface in which the phases are sensibly affected by the interface (Fig. 5.3). Now, for any homogenous region, eqn (5.9) reduces to the standard Gibbs–Duhem equation (Reif 1965),

$$\emptyset = -S \, \mathrm{d}T + V \, \mathrm{d}P - \sum_{i=1}^{C} N_i \, \mathrm{d}\mu_i. \tag{5.10}$$

and we can therefore write a set of equations for the system in the form

$$-[S]\mathrm{d}T + [V]\mathrm{d}P - \sum_{i=1}^{C} [N_i]\mathrm{d}\mu_i - \mathrm{d}\sigma = 0 \tag{5.11a}$$

$$-S^{\alpha}\mathrm{d}T + V^{\alpha}\mathrm{d}P - \sum_{i=1}^{C} N_i^{\alpha} \, \mathrm{d}\mu_i + 0 = 0 \tag{5.11b}$$

$$-S^{\beta}\mathrm{d}T + V^{\beta}\mathrm{d}P - \sum_{i=1}^{C} N_i^{\beta} \, \mathrm{d}\mu_i + 0 = 0. \tag{5.11c}$$

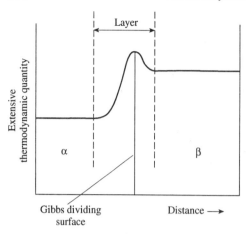

Fig. 5.3 Variation of an extensive thermodynamic quantity with distance across an interface between $\alpha$ and $\beta$ phases. The layer between
the two dashed lines includes the interface and
the regions in both phases sensibly affected by the
presence of the interface. Also shown is the
'dividing surface' of Gibbs.

The first equation refers to the layer containing the interface, and the quantities in
brackets represent the extensive thermodynamic quantities (per unit area of interface) for
the material included in this layer. The $T$s, $P$s, and $\mu$s are the same in all regions as
required for an equilibrium system. Of prime importance in this formalism is the fact
that the homogeneous $\alpha$ and $\beta$ regions contribute nothing to d$\sigma$. Therefore, the choice
of the bounds of the layer does not matter as long as the layer is thick enough to include
all inhomogeneous material affected by the interface. The individual layer quantities,
such as $[E]$ and $[S]$, clearly depend upon this choice, but the important interface quantity,
$\sigma$, is independent of it. All useful interfacial thermodynamic quantities must possess this
invariance, and we shall see in the following that this will indeed be the case.

Equations (5.11) constitute a set of $\phi + 1$ linear equations (where $\phi$ is the number of
bulk phases) which contains $3 + C$ variables (i.e., $T$, $P$, $\sigma$ and the $\mu_i$). We can therefore
eliminate any $\phi$ variables to obtain a relationship between the remaining $3 + C - \phi$
variables, of which $2 + C - \phi$ can be varied independently. We note that the number
of independent variables obtained in this way, which is generally termed the number of
'degrees of freedom', $d_F$, is simply the total number of variables minus the number of
equations; i.e.,

$$d_F = 2 + C - \phi. \tag{5.12}$$

This relation is seen to be identical to the usual phase rule obtained for bulk phases
(Gaskell 1983) when the presence of the interface is ignored, since, in that case, there
is one fewer variable but also one fewer equation. We first rewrite eqns (5.11) in a
standard format so that the first $\phi$ columns (starting on the left-hand side of the array)
are the ones containing the variables which are to be eliminated and the last one is the
one containing d$\sigma$. Using Cramer's rule for solving sets of linear equations, the desired
solution can then be obtained in the form

$$\sum_{k=1}^{3+C} (\text{Det})_k \, \mathrm{d}X_k = 0. \tag{5.13}$$

Here, $X_k$ is the $k$th variable appearing in the $k$th column counting from the left, and
$(\text{Det})_k$ is a determinant of order $(\phi + 1)$ whose first $\phi$ columns are the first $\phi$ columns
of coefficients in the standardized array, and whose last column is the column of the
coefficients associated with $X_k$. Now, $(\text{Det})_1$ through $(\text{Det})_\phi$ are identically zero (since

they each possess two columns which are identical), and, therefore, eqn (5.13) is the desired relationship between the remaining $C + 1$ variables.

We now use the above results to derive a number of important relationships involving the interface free energy. First, we solve eqn (5.13) for $d\sigma$ in the form

$$d\sigma = -\frac{1}{(\text{Det})_{C+3}} \sum_{k=1}^{C+2} (\text{Det})_k dX_k. \tag{5.14}$$

If our system containing the $\alpha$ and $\beta$ phases contains two components, then $d_F = 2$, and we can choose $T$ and $P$ as the independent variables and eliminate $\mu_1$ and $\mu_2$. Rewriting eqns (5.11) in the standard format:

$$[N_1] d\mu_1 + [N_2] d\mu_2 + [S] dT - [V] dP + d\sigma = 0 \tag{5.15a}$$

$$N_1^\alpha d\mu_1 + N_2^\alpha d\mu_2 + S^\alpha dT - V^\alpha dP + 0 = 0 \tag{5.15b}$$

$$N_1^\beta d\mu_1 + N_2^\beta d\mu_2 + S^\beta dT - V^\beta dP + 0 = 0. \tag{5.15c}$$

Then, using eqn (5.14),

$$d\sigma = -\left\{ \begin{vmatrix} [N_1] & [N_2] & [S] \\ N_1^\alpha & N_2^\alpha & S^\alpha \\ N_1^\beta & N_2^\beta & S^\beta \end{vmatrix} dT + \begin{vmatrix} [N_1] & [N_2] & -[V] \\ N_1^\alpha & N_2^\alpha & -V^\alpha \\ N_1^\beta & N_2^\beta & -V^\beta \end{vmatrix} dP \right\} \div$$

$$\begin{vmatrix} [N_1] & [N_2] & 1 \\ N_1^\alpha & N_2^\alpha & 0 \\ N_1^\beta & N_2^\beta & 0 \end{vmatrix}. \tag{5.16}$$

Therefore,

$$\left(\frac{\partial \sigma}{\partial T}\right)_P = -\frac{\begin{vmatrix} [S] & [N_1] & [N_2] \\ S^\alpha & N_1^\alpha & N_2^\alpha \\ S^\beta & N_1^\beta & N_2^\beta \end{vmatrix}}{\begin{vmatrix} N_1^\alpha & N_2^\alpha \\ N_1^\beta & N_2^\beta \end{vmatrix}}. \tag{5.17}$$

The right-hand side of eqn (5.17) has a unique physical significance which is independent of the arbitrary choice of the layer thickness. In order to show this, suppose that the layer contains enough of components 1 and 2 to make $n^\alpha$ units of $\alpha$ phase and $n^\beta$ units of $\beta$ phase. Then

$$[N_1] = n^\alpha N_1^\alpha / (N_1^\alpha + N_2^\alpha) + n^\beta N_1^\beta / (N_1^\beta + N_2^\beta) \tag{5.18}$$

$$[N_2] = n^\alpha N_2^\alpha / (N_1^\alpha + N_2^\alpha) + n^\beta N_2^\beta / (N_1^\beta + N_2^\beta). \tag{5.19}$$

Also, the excess of $S$, $\Delta S$, in the layer must be given by

$$\Delta S = [S] - n^\alpha S^\alpha / (N_1^\alpha + N_2^\alpha) - n^\beta S^\beta / (N_1^\beta + N_2^\beta). \tag{5.20}$$

Solving eqn (5.18) and (5.19) for $n^\alpha/(N_1^\alpha + N_2^\alpha)$ and $n^\beta/(N_1^\beta + N_2^\beta)$, and substituting these into eqn (5.20), we find

$$\Delta S = \frac{\begin{vmatrix} [S] & [N_1] & [N_2] \\ S^\alpha & N_1^\alpha & N_2^\alpha \\ S^\beta & N_1^\beta & N_2^\beta \end{vmatrix}}{\begin{vmatrix} N_1^\alpha & N_2^\alpha \\ N_1^\beta & N_2^\beta \end{vmatrix}} \tag{5.21}$$

which is identical to the right-hand side of eqn (5.17). Therefore,

$$\left(\frac{\partial \sigma}{\partial T}\right)_P = -\Delta S. \tag{5.22}$$

The quantity $\Delta S$ is, therefore, the excess entropy in the interface layer over what would be present in a comparison system of the $\alpha$ and $\beta$ phases containing the same amounts of components 1 and 2. This quantity is clearly independent of the choice of the layer thickness and is, therefore, a unique property of the interface, as it should be.

In similar fashion, we may use eqn (5.16) to obtain

$$\left(\frac{\partial \sigma}{\partial P}\right)_T = \frac{\begin{vmatrix} [V] & [N_1] & [N_2] \\ V^\alpha & N_1^\alpha & N_2^\alpha \\ V^\beta & N_1^\beta & N_2^\beta \end{vmatrix}}{\begin{vmatrix} N_1^\alpha & N_2^\alpha \\ N_1^\beta & N_2^\beta \end{vmatrix}} = \Delta V, \tag{5.23}$$

where $\Delta V$ is the similarly defined excess volume of the interface layer.

We emphasize that it would be incorrect to attempt to obtain an expression for a quantity such as $(\partial \sigma / \partial T)_{P, \mu_1, \mu_2}$ in the form

$$\left(\frac{\partial \sigma}{\partial T}\right)_{P, \mu_1, \mu_2} = -[S] \tag{5.24}$$

by using only eqn (5.15a). Firstly, this is an impossible variation, since $d_F = 2$ for this system, and, hence, three variables cannot be held constant at arbitrary values. In addition, the quantity $[S]$ is a layer quantity which clearly depends upon the arbitrary choice of the layer thickness. Equation (5.24) is, therefore, meaningless.

The above analysis shows that any equilibrium thermodynamic relationships involving an interface must be determined by considering the thermodynamic behaviour of the entire multiphase system which contains the interface. Since the interface and all phases present in the system are in equilibrium with each other, they may all exchange matter, energy, entropy, etc. Each bulk phase, therefore, plays a role. The equations governing this coupling are eqns (5.15), and care must be taken to ensure that all relationships are consistent with them.

For a grain boundary in a single component system $d_F = 2$, and eqns (5.11) reduce to a set of two equations. However, eqn (5.13) still holds, and eliminating $d\mu_1$, we find

$$d\sigma = -\left\{[S] - \frac{[N_1]}{N_1^\alpha} S^\alpha\right\} dT + \left\{[V] - \frac{[N_1]}{N_1^\alpha} V^\alpha\right\} dP. \tag{5.25}$$

Therefore,

$$\left(\frac{\partial \sigma}{\partial P}\right)_T = \left\{[V] - \frac{[N_1]}{N_1^\alpha}V^\alpha\right\} = \Delta V, \tag{5.26}$$

$$\left(\frac{\partial \sigma}{\partial T}\right)_P = -\left\{[S] - \frac{[N_1]}{N_1^\alpha}S^\alpha\right\} = -\Delta S \tag{5.27}$$

where the quantities $\Delta V$ and $\Delta S$ are the excess quantities in the boundary layer over what would be present in a comparison system of $\alpha$ phase containing the same number of atoms. Equations (5.26) and (5.27) are seen to be similar in form to the standard relations $(\partial G/\partial P)_T = V$ and $(\partial G/\partial T)_P = -S$ for single bulk phases (Reif 1965). Both the excess volume and excess entropy are generally positive for grain boundaries in close-packed crystals, since the atoms in the boundary core are somewhat less densely packed than in the perfect lattice and therefore vibrate in looser environments with reduced frequencies and larger amplitudes (see Section 4.3.1.10). The free energy, $\sigma$, therefore tends to decrease with increasing temperature. Typical measured values of $\Delta V$ and $\Delta S$ are listed in Table 5.1.

A great many additional thermodynamic relationships may be obtained. An important one is the Gibbs adsorption equation. Consider a grain boundary in a binary single bulk phase system containing two components where component 2 is considered the solute. Then, $d_F = 3$, and eqns (5.11) again reduce to a set of two equations. Using eqn (5.13), and eliminating $d\mu_1$, we obtain

$$d\sigma = -\left([S] - \frac{[N_1]}{N_1^\alpha}S^\alpha\right)dT + \left([V] - \frac{[N_1]}{N_1^\alpha}V^\alpha\right)dP - \left([N_2] - \frac{[N_1]}{N_1^\alpha}N_2^\alpha\right)d\mu_2, \tag{5.28}$$

and, therefore,

$$\left(\frac{\partial \sigma}{\partial \mu_2}\right)_{T,P} = -\left([N_2] - \frac{[N_1]}{N_1^\alpha}N_2^\alpha\right) = -\Delta N_2^\alpha, \tag{5.29}$$

where $\Delta N_2^\alpha$ is the 'interface excess' of component 2 (solute), i.e., the excess of component 2 in the layer over what would be present in a comparison system of alpha phase containing the same amount of component 1 (solvent). This relation expresses the well-known result (see Fig. 7.1; Hondros and Seah 1983) that a solute preferentially adsorbs ('segregates') at an interface when the free energy of the interface decreases with an increase in the solute chemical potential.

Interesting Maxwell-type relations can also be derived. Consider, for example, a grain boundary in a single bulk phase system containing three components where components 2 and 3 are considered as solutes. Then $d_F = 4$ and using eqn (5.13) and eliminating $N_1$, we obtain at constant $T$ and $P$,

$$d\sigma_{T,P} = -\left([N_2] - \frac{[N_1]}{N_1^\alpha}N_2^\alpha\right)d\mu_2 - \left([N_3] - \frac{[N_1]}{N_1^\alpha}N_3^\alpha\right)d\mu_3. \tag{5.30}$$

Therefore

$$\left(\frac{\partial \Delta N_2^\alpha}{\partial \mu_3}\right)_{T,P,\mu_2} = \left(\frac{\partial \Delta N_3^\alpha}{\partial \mu_2}\right)_{T,P,\mu_3} \tag{5.31}$$

which demonstrates the coupling which exists between the excesses of the two solute species and their chemical potentials.

As shown by Cahn (1979), the above procedure may be used for systems containing various numbers of bulk phases and components for the purpose of deriving a wide variety of thermodynamic relationships involving the interface. As might be expected, all bulk phases which are present in the system at equilibrium play a role in the formulation even when they are not contiguous to the boundary.

We have already noted that results physically identical to those obtained above can be derived by using the classical method first devised by Gibbs (1957). In Fig. 5.3 we show schematically the variation of an extensive thermodynamic quantity with distance across an interface between $\alpha$ and $\beta$ phases. The layer between the two dashed lines includes the interface and the regions in both phases which are sensibly affected by the presence of the interface. In the Gibbs approach a 'dividing surface' is first constructed in the vicinity of the interface as shown schematically in Fig. 5.3. An interface 'excess quantity' is then defined for each extensive quantity as the difference between the actual amount of that quantity in the system and the amount which would have been present if both phases were homogeneous right up to the dividing surface. Therefore,

$$\left.\begin{aligned}
E^{\mathrm{xs}} &= E - E^\alpha - E^\beta \\
S^{\mathrm{xs}} &= S - S^\alpha - S^\beta \\
V^{\mathrm{xs}} &= V - V^\alpha - V^\beta \equiv 0 \\
N_i^{\mathrm{xs}} &= N_i - N_i^\alpha - N_i^\beta
\end{aligned}\right\} \tag{5.32}$$

where, for example, $E$ is the total energy of the system, $E^\alpha$ and $E^\beta$ are the energies of the $\alpha$ and $\beta$ phases if they were homogeneous up to the dividing surface, and $E^{\mathrm{xs}}$ is the energy excess. Note especially that, since by definition $V \equiv V^\alpha + V^\beta$, $V^{\mathrm{xs}} = 0$.

Now, as previously,

$$\left.\begin{aligned}
\mathrm{d}E &= T\,\mathrm{d}S - P\,\mathrm{d}V + \sum_{i=1}^{C} \mu_i\,\mathrm{d}N_i + \sigma\,\mathrm{d}A \\
\mathrm{d}E^\alpha &= T\,\mathrm{d}S^\alpha - P\,\mathrm{d}V^\alpha + \sum_{i=1}^{C} \mu_i\,\mathrm{d}N_i^\alpha \\
\mathrm{d}E^\beta &= T\,\mathrm{d}S^\beta - P\,\mathrm{d}V^\beta + \sum_{i=1}^{C} \mu_i\,\mathrm{d}N_i^\beta
\end{aligned}\right\} . \tag{5.33}$$

Therefore, combining eqns (5.32) and (5.33), we have

$$\mathrm{d}E^{\mathrm{xs}} = T\,\mathrm{d}S^{\mathrm{xs}} + \sum_{i=1}^{C} \mu_i\,\mathrm{d}N_i^{\mathrm{xs}} + \sigma\,\mathrm{d}A, \tag{5.34}$$

which may be integrated to produce

$$E^{\mathrm{xs}} = TS^{\mathrm{xs}} + \sum_{i=1}^{C} \mu_i N_i^{\mathrm{xs}} + \sigma A. \tag{5.35}$$

Finally, dividing by $A$, we obtain

$$\sigma = E^{\mathrm{xs}} - TS^{\mathrm{xs}} - \sum_{i=1}^{C} \mu_i \Gamma_i, \tag{5.36}$$

where $\Gamma_i \equiv N_i^{xs}/A$, and $E^{xs}$ and $S^{xs}$ are excess quantities per unit interface area. Obviously, the individual values of these excess quantities depend upon the arbitrary choice of the position of the dividing surface. However, it can be shown that the magnitudes of all physically significant quantities (e.g., the value of $\sigma$) obtained using this formalism are independent of the location of the dividing surface. This situation is analogous to the situation in our previous 'layer' formalism where all physically significant results were independent of the choice of the layer thickness as long as it was thick enough to include all material affected by the interface. Both approaches will yield the same results for physically significant interface quantities as long as they are used properly. The choice of which one to use is merely a matter of convenience.

## 5.4 INTRODUCTION OF THE INTERFACE STRESS AND STRAIN VARIABLES

Up to this point, we have taken the interface free energy, $\sigma$, to be a function of the usual thermodynamic variables T, P, and $\mu_1, \mu_2, \ldots \mu_C$ and have developed a method for obtaining $d\sigma$ as a function of derivatives of these variables through the use of eqn (5.14). However, interfaces in solids can be subjected to non-hydrostatic stresses and corresponding strains, and their structures and properties can be altered by the stretching associated with such straining. This behaviour is in contrast to that of interfaces in liquids whose structures and properties are unaffected by such straining since atoms in liquids are highly mobile and can freely enter or leave the interface in order to allow it to accommodate rapidly to any imposed shape changes. For interfaces in solids, therefore, a more complete thermodynamic description of the interface free energy requires the introduction of interface strains as additional thermodynamic variables.

In order to carry out this task, it is first necessary to introduce the concept of the 'interface stress' as an excess thermodynamic quantity which is generally associated with any interface (Shuttleworth 1950, Herring 1951*b*, Mullins 1962) and which is conjugate to interface strain (Cahn 1979, 1980). In the case of a crystalline interface, the atoms in the 'surfaces' of each crystal which face each other across the interface can be expected to have a different coordination and bonding from corresponding atoms deep within the bulk. They will, therefore, tend to have different atomic volumes and configurations. However, since they are coherent with their respective bulk crystals, this tendency will be resisted, and internal stresses will be generated throughout the bicrystal leading to a so-called 'interface stress', as shown below.

As a simple preliminary example, consider the free surface of a crystal. In the case of metals, there is extensive evidence (Needs and Godfrey, 1987) that the bonding between atoms in the surface layer is stronger than in the interior, since there are fewer bonds in which the valence electrons reside. Hence, the ions in the surface layer would like to adopt generally closer spacings than those in the interior. If the surface layer does not reconstruct and if no defects (e.g. dislocations) are introduced near the surface, the atomic structure of the surface layer is the same as the bulk (i.e. it is completely coherent), and internal stresses parallel to the interface are then required in order to force this coherency. The situation is shown schematically in Fig. 5.4(a) in the cross section of a crystal slab of dimensions $L \times L \times l$. A distribution of internal tensile stress, $\tau_{11}$, parallel to the surface is shown (see figure) which is highly tensile in the thin surface layer and then becomes weakly compressive throughout the bulk. Mechanical equilibrium in the $x$ direction requires that the integral along $z$ of the tensile stresses in the thin surface layers be equal to the integral of the compressive bulk stresses. If $L \gg l$, the compressive

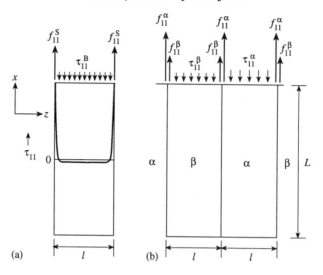

**Fig. 5.4**  (a) Cross section of slab of thickness $l$, height $L$ and thickness $L$ (into paper). Distribution of internal $\tau_{11}$ stress field component across the slab thickness due to the existence of the surface stress, $f_{11}$, is shown. Also shown are the forces (stresses) which must be applied to the slab surface in order to balance the internal stresses. (b) Same as (a) except that specimen now consists of alternating layers of $\alpha$ and $\beta$ phases, and only the surface forces (stresses) which must be applied to the specimen surface in order to balance the internal stresses due to the interface stresses $f_{11}^{\alpha}$ and $f_{11}^{\beta}$ are shown.

bulk stress will be essentially constant at the value $\tau_{11}^{B}$ across the cross-section, and the relationship between the total tensile force, $f_{11}^{S}$, present in each surface layer (per unit distance along $y$) and $\tau_{11}^{B}$ is therefore

$$f_{11}^{S} = -\tau_{11}^{B} l/2. \tag{5.37}$$

We note that we have assumed above that the properties of the two surfaces are identical which will only be true for centrosymmetric crystals. When this is not the case, a different set of forces would be required at each surface. A similar analysis of the internal stresses, $\tau_{22}^{B}$, exerted by either surface layer along $y$ in Fig. 5.4(a) would yield $f_{22}^{S}$, the total force in each surface layer acting along $y$ (per unit distance along $x$). Finally, excess internal shear stresses, $\tau_{12}^{B}$, could be present in the bulk because of the presence of the surfaces (depending upon the orientation of the $x$ and $y$ axes relative to the crystal axes), and these would be balanced by corresponding shear forces in the surface layers denoted by $f_{12}^{S}$. We note that $f_{33}^{S} = f_{13}^{S} = f_{23}^{S} = 0$, since there are no constraints on the surface layers which would generate forces in these directions. By rotating the crystal coordinate system around $z$, it can be shown (e.g., Mullins 1962) that is $f_{ij}^{S}$ is a 2D symmetric second rank tensor. The quantity $f_{ij}^{B}$ is universally termed the 'surface stress', which is unfortunate since it actually does not have the dimensions of a stress.

An important relation between the surface stress, $f_{ij}^{S}$, and the surface free energy, $\sigma^{S}$, may now be obtained by applying a small homogeneous variational strain, $\delta\varepsilon_{ij}$, to the slab in Fig. 5.4(a). The change in free energy of the slab in this variation is

$$\delta G = \sum_{ij} \left[ \frac{\partial (2\sigma^{S} L^{2})}{\partial \varepsilon_{ij}} + lL^{2} \frac{\partial w^{B}}{\partial \varepsilon_{ij}} \right] \delta\varepsilon_{ij}, \tag{5.38}$$

where $2\sigma^S L^2$ is the total surface free energy, and $w^B$ is the strain energy density due to the weak $\tau_{ij}^B$ bulk stresses. Now (Sommerfeld 1950),

$$\frac{\partial w^B}{\partial \varepsilon_{ij}} = \tau_{ij}^B, \tag{5.39}$$

and a generalization of eqn (5.37) yields

$$\tau_{ij}^B = -2f_{ij}^S/l. \tag{5.40}$$

At equilibrium, $\delta G = 0$ in eqn (5.38), and since the $\delta \varepsilon_{ij}$ can be varied independently, we may combine eqns (5.38), (5.39), and (5.40) to obtain

$$f_{ij}^S A = \sigma \frac{\partial A}{\partial \varepsilon_{ij}} + A \frac{\partial \sigma^S}{\partial \varepsilon_{ij}}. \tag{5.41}$$

The first term on the right-hand side is proportional to the increase in area due to the strain. As might be expected, the proportionality factor is $\sigma^S$, i.e., the increase in free energy per creation of unit area of surface under conditions where the physical nature of the surface remains unchanged. The second term is proportional to the change in $\sigma^S$ caused by the strain, and the proportionality factor is simply the surface area. Since $\partial A/\partial \varepsilon_{ij} = A$ when $i = j$ and zero otherwise, eqn (5.41) can be written

$$f_{ij}^S = \delta_{ij} \sigma^S + \frac{\partial \sigma^S}{\partial \varepsilon_{ij}}, \tag{5.42}$$

where $\delta_{ij} = 1$ when $i = j$ and zero otherwise. Again we note that eqn (5.42) only holds for components with $ij = 11$, 22, and 12.

The relationship between the surface stress and surface free energy can be expressed in simpler form (Cahn 1979, Cahn and Larché 1982) if we measure the surface area using a crystal coordinate system (i.e. a Lagrangian system) to measure surface area rather than the actual physical (laboratory) coordinates which were employed above. In the Lagrangian system the area of a given patch of surface containing a fixed number of atoms possesses a constant area, regardless of whether or not it is strained, since area is always measured in a fixed reference state of the crystal which may be taken as the state of zero strain. In the following, all surface quantities measured using Lagrangian coordinates will be denoted by a subscript L, while no subscript will be used, as previously, for quantities measured using physical coordinates. Using this convention, the relation between the areas of a given patch of surface measured in the two coordinate systems is

$$A = A_L(1 + \varepsilon_{ii}), \tag{5.43}$$

where $\varepsilon_{ii}$ is the trace of the strain tensor in the plane of the surface, i.e. $\varepsilon_{ii} = \varepsilon_{11} + \varepsilon_{22}$. Since the free energy of the given patch of surface must be the same in both systems

$$\sigma^S A = \sigma_L^S A_L, \tag{5.44}$$

and, therefore,

$$\sigma_L^S = (1 + \varepsilon_{ii})\sigma^S. \tag{5.45}$$

By combining eqns (5.42) and (5.45) we then obtain the relatively simple result

$$f_{ij}^S = \frac{\partial \sigma_L^S}{\partial \varepsilon_{ij}}. \tag{5.46}$$

Equation (5.46) provides another perspective regarding the nature of $f_{ij}^S$. It shows directly that it is equal to the work which must be done in order to strain a given patch of surface corresponding to unit area in the Lagrangian system.

In general, $f_{ij}^S$ may be positive or negative in different materials and also have a relatively large absolute magnitude which may be comparable to $\sigma_L^S$ (Herring 1951b, Mullins 1962). Note that $\sigma$ must always be positive for otherwise the bulk phase would not be stable.

Our previous analysis shows that the components of the $f_{ij}^S$ tensor can be regarded as the forces (per unit length) acting on the surface perimeter in the plane of the surface which maintain any coherency which exists between the surface layer and the underlying bulk. The determination of surface stress, therefore, has many of the features of the determination of the stresses (forces) which are generated when thin, misfitting layers are epitaxially deposited on thick substrates (Frank and Van der Merwe, 1949: also Section 2.10.6). In the example we have chosen, we have assumed complete coherency between the surface layer and the bulk. At least some of this coherency could conceivably be eliminated by the introduction of dislocations in the surface layer (Herring 1951b) or by various forms of reconstruction (Herring 1951b, Needs and Godfrey 1987). In such cases, the surface stresses could be markedly reduced. However, it must be realized that no principle regarding the achievement of equilibrium by minimization of the surface stress exists.

Having this background, we can now turn to the more complicated problem of analysing the 'interface stress' associated with a planar interface in a bicrystal. For purposes of analysis consider the interfaces in the specimen consisting of alternate thin slabs of $\alpha$ and $\beta$ phases illustrated in Fig. 5.4(b) where we again assume centrosymmetric crystals for simplicity so that the interfaces are identical. In this more complicated situation weak bulk stress, which differ in each phase, will generally be present because of the presence of the interface. We designate these by $\tau_{ij}^\alpha$ and $\tau_{ij}^\beta$ respectively. Again, these will be in balance with opposing forces ('interface stresses') lying in the interface layer as illustrated in Fig. 5.4(b). By employing the same argument which led to eqns (5.37) and (5.40) we then have

$$\tau_{ij}^\alpha = -2f_{ij}^\alpha/l; \quad \tau_{ij}^\beta = -2f_{ij}^\beta/l. \tag{5.47}$$

In order to explain the presence of these forces and stresses the interface may be thought of (Brooks 1952) as a double sheet, with the side facing the $\alpha$ crystal exerting a set of interface forces on that crystal while the side facing the $\beta$ crystal exerts a different set of forces on that crystal. If we now apply a homogeneous perturbing strain, $\delta\varepsilon_{ij}$, to the entire specimen in Fig. 5.4(b), and use Lagrangian coordinates with the $\alpha$ crystal acting as the reference crystal, (we have for the change in the free energy of the specimen

$$\delta G = N \sum_{ij} \left[ 2L^2 \frac{\partial\sigma_L}{\partial\varepsilon_{ij}} + lL^2 \left( \frac{\partial w^\alpha}{\partial\varepsilon_{ij}} + \frac{\partial w^\beta}{\partial\varepsilon_{ij}} \right) \right] \delta\varepsilon_{ij}, \tag{5.48}$$

where $N$ is the number of $\alpha$ slab pairs. Using eqns (5.39) and (5.47) and the same procedure which led to eqn (5.41), we obtain

$$f_{ij} \equiv f_{ij}^\alpha + f_{ij}^\beta = \frac{\partial\sigma_L}{\partial\varepsilon_{ij}}, \tag{5.49}$$

which may be compared to eqn (5.46). In the interface case it may be seen that an overall interface stress, $f_{ij}$, exists which is the sum of $f_{ij}^\alpha$ and $f_{ij}^\beta$ and which is equal to the work required to strain a given patch of interface corresponding to unit area in the Lagrangian coordinate system.

However, Cahn and Larché (1982) have pointed out that the $\varepsilon_{ij}$ strain applied above to both crystals at the interface is not the most general strain which may be applied to many interfaces. For a coherent interface, maintenance of the coherency at the interface at the interface requires that the strain increments applied to both crystals be the same, as indeed was the case in our previous analysis. However, for an incoherent interface this requirement may be relaxed, since an incoherent interface is unable to sustain shear stresses parallel to the interface (Section 12.8.1.1). In order to obtain more generality we may follow Cahn and Larché (1982) and define a set of strains in the plane of the interface where the strains in the $\alpha$ and $\beta$ crystals may be different and are related by

$$\varepsilon_{ij}^{\beta} = \varepsilon_{ij}^{\alpha} - \varepsilon_{ij}^{\beta,c} - \varepsilon_{ij}^{\beta,s}. \tag{5.50}$$

Here, $\varepsilon_{ij}^{\beta,s}$ is an arbitrary strain in the $\beta$ crystal, while $\varepsilon_{ij}^{\beta,c}$ is the fixed strain in the $\beta$ crystal relative to that in the $\alpha$ crystal which is just equal to that required to make the interface coherent in the absence of any $\varepsilon_{ij}^{\beta,s}$ strain. Therefore, for the straining of a coherent interface, $\delta\varepsilon_{ij}^{\beta,s} = \delta\varepsilon_{ij}^{\beta,c} = 0$, and $\delta\varepsilon_{ij}^{\beta} = \delta\varepsilon_{ij}^{\alpha}$. On the other hand, for the general straining of an incoherent interface, $\delta\varepsilon_{ij}^{\beta} = \delta\varepsilon_{ij}^{\alpha} - \delta\varepsilon_{ij}^{\beta,s}$. We may therefore describe the general straining of an interface in terms of the two independent variables $\varepsilon_{ij}^{\alpha}$ and $\varepsilon_{ij}^{\beta,s}$. The interface free energy can then be expressed as a function of these independent variables, i.e., $\sigma_{L} = \sigma_{L}(\varepsilon_{ij}^{\alpha}, \varepsilon_{ij}^{\beta,s})$, and, therefore,

$$d\sigma_{L} = \sum_{ij} [(\partial\sigma_{L}/\partial\varepsilon_{ij}^{\alpha}) \, d\varepsilon_{ij}^{\alpha} + (\partial\sigma_{L}/\partial\varepsilon_{ij}^{\beta,s}) \, d\varepsilon_{ij}^{\beta,s}]. \tag{5.51}$$

Here, $\partial\sigma_{L}/\partial\varepsilon_{ij}^{\alpha}$, which is the change in $\sigma_{L}$ due to $\varepsilon_{ij}^{\alpha}$ straining in the absence $\varepsilon_{ij}^{\beta,s}$ straining, corresponds to equal strain increments in the $\alpha$ and $\beta$ crystals, and, therefore, according to eqn (5.49),

$$f_{ij} = \partial\sigma_{L}/\partial\varepsilon_{ij}^{\alpha}. \tag{5.52}$$

The second coefficient, $\partial\sigma_{L}/\partial\varepsilon_{ij}^{\beta,s}$, represents the change in a $\sigma_{L}$ due to straining the $\beta$ crystal at constant $\varepsilon_{ij}^{\alpha}$. In this process the interface area must be held fixed, and, in addition, no free surface can be generated anywhere due to any change in shape of the $\beta$ crystal. This can be accomplished by adjusting the $x$ and $y$ dimensions of the $\beta$ crystal to eliminate any overhang while suitably altering its thickness (Cahn and Larché 1982). In the straining process the structure of the interface will, of course, be changed. Setting

$$g_{ij} \equiv \partial\sigma_{L}/\partial\varepsilon_{ij}^{\beta,s}, \tag{5.53}$$

we therefore have

$$d\sigma_{L} = \sum_{ij} [f_{ij} \, d\varepsilon_{ij}^{\alpha} + g_{ij} \, d\varepsilon_{ij}^{\beta,s}], \tag{5.54}$$

where the first term represents the contribution to $d\sigma_{L}$ due to straining the interface at constant structure while the second term represents the contribution by a strain which changes the structure. The quantity $g_{ij}$ may be regarded as an effective interface stress which acts on the $\beta$ crystal during the latter process.

Equation (5.51) holds in Lagrangian coordinates. A corresponding equation for $d\sigma$ in physical (laboratory) coordinates is readily found. For this purpose it is convenient to first express $\sigma$ as a function of the two independent strain variables $\varepsilon_{ij}^{\alpha}$ and $\varepsilon_{ij}^{\beta}$, i.e. $\sigma = \sigma(\varepsilon_{ij}^{\alpha}, \varepsilon_{ij}^{\beta})$, so that

$$d\sigma = \sum_{ij} [(\partial\sigma/\partial\varepsilon_{ij}^\alpha)\, d\varepsilon_{ij}^\alpha + (\partial\sigma/\partial\varepsilon_{ij}^\beta)\, d\varepsilon_{ij}^\beta]. \qquad (5.55)$$

Using eqn (5.50) and $\sigma_L = (1 + \varepsilon_{ii}^\alpha)\sigma$, eqn (5.55) becomes

$$d\sigma = \sum_{ij} [(\delta_{ij}\sigma + \partial\sigma/\partial\varepsilon_{ij}^\alpha + \partial\sigma/\partial\varepsilon_{ij}^\beta)\, d\varepsilon_{ij}^\alpha - (\delta\sigma/\partial\varepsilon_{ij}^\beta)\, d\varepsilon_{ij}^{\beta,s}]. \qquad (5.56)$$

Comparing eqn (5.56) with (5.54), we see that

$$f_{ij} = \delta_{ij}\sigma + \partial\sigma/\partial\varepsilon_{ij}^\alpha + \partial\sigma/\varepsilon_{ij}^\beta \qquad (5.57)$$

$$g_{ij} = -\partial\sigma/\partial\varepsilon_{ij}^\beta. \qquad (5.58)$$

These relations, which hold in physical coordinates, may be compared with eqns (5.52) and (5.53) which hold in Lagrangian coordinates. We note that eqn 5.57) has been obtained previously by Brooks (1952). Finally, again using $\sigma_L = (1 + \varepsilon_{ii}^\alpha)\sigma$, we may rearrange the above relationship to obtain the expression for $d\sigma$ in physical coordinates

$$d\sigma = \sum_{ij} [(f_{ij} - \delta_{ij}\sigma)\, d\varepsilon_{ij}^\alpha + g_{ij}\, d\varepsilon_{ij}^{\beta,s}], \qquad (5.59)$$

where $f_{ij}$ and $g_{ij}$ are given by eqns (5.57) and (5.58).

Having these results, we may now return to the problem of introducing the strain variables into the general formulation for $d\sigma$. The contribution to $d\sigma$ represented eqn (5.59) can be simply added (Cahn 1979) to the value of $d\sigma$ given by eqn (5.14), which holds for this same system in the absence of the strain variables to obtain the total variation, i.e.

$$d\sigma = -\frac{1}{(\mathrm{Det})_{C+3}} \sum_{k=1}^{C+2} (\mathrm{Det})_k \, dX_k + \sum_{ij} [(f_{ij} - \delta_{ij}\sigma)\, d\varepsilon_{ij}^\alpha + g_{ij}\, d\varepsilon_{ij}^{\beta,s}]. \qquad (5.60)$$

The variables appearing on the right-hand side of eqn (5.60) (i.e. the $X_k$, $\varepsilon_{ij}^\alpha$, and $\varepsilon_{ij}^{\beta,s}$) are, therefore, independent. We emphasize that our previous work indicates that the only non-zero terms appearing in the second sum in eqn (5.60) will be the $ij = 11$, 22, and 12 terms corresponding to surface stresses and strains parallel to the interface.

A number of additional thermodynamic relations of interest which involve the strains can now be derived (Cahn, 1979) with the aid of eqn (5.60). A simple Maxwell-type relation is, for example,

$$\frac{\partial (f_{ij} - \delta_{ij}\sigma)}{\partial\varepsilon_{kl}^\alpha} = \frac{\partial (f_{kl} - \delta_{kl}\sigma)}{\partial\varepsilon_{ij}^\alpha}. \qquad (5.61)$$

For a grain boundary in a single bulk phase system containing two components, where component 2 is considered the solute, $d\mu_1$ may be eliminated, and eqn (5.60) takes the form

$$dS = -\left([S] - \frac{[N_1]}{N_1^\alpha} S^\alpha\right) dT + \left([V] - \frac{[N_1]}{N_1^\alpha} V^\alpha\right) dP - \left([N_2] - \frac{[N_1]}{N_1^\alpha} N_2^\alpha\right) d\mu_2$$

$$+ \sum_{ij} [(f_{ij} - \delta_{ij}\sigma)\, d\varepsilon_{ij}^\alpha + g_{ij}\, d\varepsilon_{ij}^{\beta,s}]. \qquad (5.62)$$

Therefore, for example,

$$\frac{\partial (f_{ij} - \delta_{ij}\sigma)}{\partial \mu_2} = - \frac{\partial \left( [N_2] - \dfrac{[N_1]}{N_1^\alpha} N_2^\alpha \right)}{\partial \varepsilon_{ij}^\alpha} \tag{5.63}$$

and,

$$\frac{\partial (f_{ij} - \delta_{ij}\sigma)}{\partial T} = - \frac{\partial \left( [S] - \dfrac{[N_1]}{N_1^\alpha} S^\alpha \right)}{\partial \varepsilon_{ij}^\alpha} . \tag{5.64}$$

The last two relations show that variations of the quantity $(f_{ij} - \delta_{ij}\sigma)$ with respect to the chemical potential of the solute or with respect to the temperature are related, in turn, to variations of the solute adsorption with respect to strain and variations of the interface entropy with respect to strain.

Unfortunately, no measurements of interface stresses have yet been made for internal interfaces to the authors' knowledge. However, calculations of the interface stress at $\Sigma 5 \langle 100 \rangle$ twist boundaries and $\Sigma 5$ (310) symmetric $\langle 100 \rangle$ tilt boundaries in NaCl have been made by Chen and Kalonji (1989, 1992). In the former case the interface stress changed sign as the temperature was raised: in the latter case the anisotropy of the interface stress decreased with increasing temperature. On the other hand, measurements of the surface stress for free surfaces have been performed (Needs and Godfrey 1987) by, for example, measuring the lattice parameters of very small particles in which the internal strain throughout the bulk due to the surface stress is measurable. In addition, a number of calculated results are available (Broughton and Gilmer 1983, Needs and Godfrey 1987).

In principle, interface stress should be capable of producing significant physical effects. An interesting example (Cahn and Larché 1982) is that of a small spherical particle of $\alpha$ phase of radius $R$ embedded in an infinite matrix of $\beta$ phase such as might occur during a precipitation process. If the volume of the $\alpha$ particle is larger than the volume available for it in the $\beta$ matrix, the $\alpha$ particle will be in a state of hydrostatic compression and, in turn, will exert an outward radial stress on the surrounding $\beta$ matrix. If we initially ignore any effects due to the interface stress of the $\alpha/\beta$ interface, the solution of this elastic problem is identical to the solution of the elastic 'ball-in-a-hole' problem described in section 7.3.2.1. It is shown there that the hydrostatic compressive stress in the $\alpha$ particle is

$$\tau_{rr} = 3K^\alpha \varepsilon (a_3 - 1) = - \frac{12\mu^\beta K^\alpha \varepsilon}{(3K^\alpha + 4\mu^\beta)} \tag{5.65}$$

where $\mu^\beta$ is the shear modulus of the matrix, $K^\alpha$ is the bulk modulus of the particle, and $\varepsilon$ is the misfit strain between the particle and matrix. We can now treat this problem more exactly by taking into account the added effect of the interface stress, $f_{ij}$, associated with the $\alpha$-$\beta$ interface. If the interface stress is positive, it will contribute a compressive stress on the $\alpha$ particle as illustrated in Fig. 5.5. Here, a patch of curved interface of dimensions $dl \times dl$ is shown which is subjected to a tangential interface force of magnitude $f dl$ along each edge $dl$ of its perimeter. (For simplicity, we have assumed that $f_{ij}$ is isotropic and equal to $f$ in all directions.) By analysing the force diagram in the side view shown in Fig. 5.5(b), it is seen that the resultant inward force exerted by the surface stress is $F = 2 \cdot 2f dl \cdot \sin \theta/2 = 2f(dl)^2/R$. The corresponding inward pressure is, therefore, $\Delta \tau_{rr} = F/(dl)^2 = 2f/R$. The effect of this interface stress will be to produce an

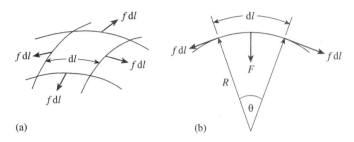

**Fig. 5.5** (a) Diagram illustrating forces exerted on patch of spherically curved interface by an isotropic surface stress. (b) Side view of (a) showing resultant inward force on patch, $F$, due to the surface stress.

inward displacement, $u_o$, of the $\alpha$-$\beta$ interface and to increase the uniform hydrostatic pressure on the particle. According to eqn (7.5), the displacements in the matrix and particle will then be of the form

$$u(r) = u_o R^2/r^2 \text{ (for the matrix)} \tag{5.66}$$

$$u(r) = (u_o/R)r \text{ (for the particle).} \tag{5.67}$$

Since $\tau_{rr} = 2\mu(\partial u/\partial r)$, the change in stress in the matrix is $-4\mu^\beta u_o R^2/r^3$ and that in the particle $3K^\alpha u_o/r$. The difference between these at the interface must equal the stress contributed by the interface stress, i.e.

$$\frac{3K^\alpha u_o}{R} - \left(\frac{-4\mu^\beta u_o}{R}\right) = -\frac{2f}{R} \tag{5.68}$$

Therefore, $u_o = -2f/(3K^\alpha + 4\mu^\beta)$, and the change in stress in the particle is just $3K^\alpha u_o/R = [-2f/R]\cdot[3K^\alpha/(3K^\alpha + 4\mu^\beta)]$.

The final stress (hydrostatic pressure) in the particle is then

$$\tau_{rr} = -\left\{\frac{12\mu^\beta K^\alpha \varepsilon}{(3K^\alpha + 4\mu^\beta)} + \frac{2f}{R} \cdot \frac{3K^\alpha}{(3K^\alpha + 4\mu^\beta)}\right\} \tag{5.69}$$

where the second term is the added increment due to the interface stress. Further discussion of the role of interface stress in the energetics of precipitation has been given by Cahn and Larché (1982).

Another effect due to interface stress can occur during the faceting of an originally flat interface. Here, differences in the interface stresses associated with adjacent facets can produce lines of force at their junctions and thereby influence the length scale of the faceting as discussed in Section 4.3.1.9.

## 5.5 INTRODUCTION OF THE GEOMETRIC THERMODYNAMIC VARIABLES

The thermodynamic properties of an interface in a crystalline material will depend upon its detailed atomic structure which, in turn, will depend upon the macroscopic variables (i.e. macroscopic degrees of freedom) which are required to specify its geometry as discussed in Section 1.4.2. These geometrical variables therefore emerge as *bona fide* thermodynamic variables which must be included in any thermodynamic treatment of crystalline interfaces. As discussed in Section 1.4.2, five such variables must be specified: also, they may be chosen in a variety of ways.

For present purposes we shall choose as our variables the two independent direction cosines required to specify the rotation axis $\hat{\rho}$, the rotation angle, $\theta$, and the two independent direction cosines required to specify $\hat{n}$. Adding these variables to our previous variables which included $T$, $P$ and the $\mu_i$ (eqn (5.11a)) and the $\varepsilon_{ij}$ [Section 5.4], we have in general:

$$\sigma = \sigma(T, P, \mu_i, \varepsilon_{ij}^\alpha, \varepsilon_{ij}^{\beta, s}, \rho_1, \rho_2, \theta, n_1, n_2). \tag{5.70}$$

However, the variables on the right-hand side of the above relationship are not all independent because of the physical coupling between the interface and the adjoining phase/phases. Taking this into account, as we did previously in writing eqn (5.60), we may therefore write the total variation of $\sigma$ in the form

$$d\sigma = -\frac{1}{(\text{Det})_{C+3}} \sum_{k=1}^{C+2} (\text{Det})_k \, dX_k + \sum_{ij} [(f_{ij} - \delta_{ij}\sigma) d\varepsilon_{ij}^\alpha + g_{ij} \, d\varepsilon_{ij}^{\beta, s}] + \frac{\partial\sigma}{\partial\rho_1} \, d\rho_1$$

$$+ \frac{\partial\sigma}{\partial\rho_2} \, d\rho_2 + \frac{\partial\sigma}{\partial\theta} \, d\theta + \frac{\partial\sigma}{\partial n_1} \, dn_1 + \frac{\partial\sigma}{\partial n_2} \, dn_2, \tag{5.71}$$

where the first term is the value of $d\sigma$ given by eqn (5.14) which holds for the system in the absence of both the strain and the geometrical variables. We emphasize that all variables appearing on the right-hand side of eqn (5.71) are now independent.

A relatively large number of these variables is involved, and eqn (5.71) is obviously quite cumbersome. However, the dependence of $\sigma$ on the various geometric variables can be revealed by moving the system along paths (trajectories) in the multivariable 'phase space' along which all variables except one are held constant. A number of important cases where this is achieved are described in the following.

## 5.6 DEPENDENCE OF σ ON THE INTERFACE INCLINATION

In many situations, interfaces migrate locally in order to change their inclination and reduce their energy while all other variables (including the crystal misorientation) remain fixed. The only geometric variables which can then vary are $n_1$ and $n_2$ (i.e. the two independent direction cosines $\hat{n}$), and the change in $\sigma$ which accompanies such a variation is represented by the last two terms on the right-hand side of eqn (5.71).

### 5.6.1 The Wulff plot

A convenient way to represent the dependence of interface free energy on inclination under the above conditions is to employ a polar plot representing $\sigma(\hat{n})$, i.e. a Wulff plot, as illustrated schematically in 2D in Fig. 5.6(a). Here, the energy for each inclination is represented by a vector in the direction of $\hat{n}$ and of magnitude proportional to $\sigma$, and the tails of all vectors are gathered at the origin. As discussed below, the surface of such a plot is expected in many situations to exhibit cusps at the inclinations of interfaces which possess relatively low energies, and these are indicated schematically in Fig. 5.6(a). (We note that in a strict mathematical sense, these singularities should not be termed 'cusps'. 'Pointed minima' would perhaps be preferred. However, we shall follow common usage and continue to refer to them as 'cusps'.) In three dimensions, the puckered surface of such a plot would somewhat resemble that of a raspberry as suggested by Frank (1962). Cusps may be expected quite generally in any situation where inclinations of particularly low energy exist and when the temperature is low enough so that the interfacial free energy

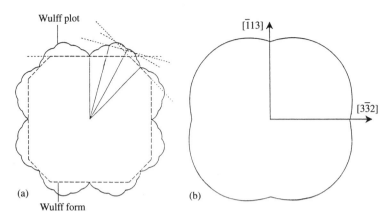

**Fig. 5.6** (a) Cross section of a 3D Wulff plot (solid cusped line) showing how the free energy of an interface of fixed crystal misorientation might vary with its inclination as specified by $\hat{n}$. To construct this plot, the boundary energy for each inclination is represented by a radius vector in the direction of $\hat{n}$ and of length proportional to $\sigma(\hat{n})$. Also shown is the Wulff form (dashed line) corresponding to the shape of a crystallite embedded in a matrix which minimizes the total free energy of the crystallite/matrix interface. This form corresponds to the inner envelope of the dashed lines drawn perpendicular to each Wulff plot radius vector. (From Herring 1951*a*) (b) Experimentally determined section of a Wulff plot for a Σ11 grain boundary in Cu. (From Omar and Mykura (1988).)

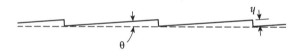

**Fig. 5.7** Vicinal interface consisting of steps separating patches of a singular interface which is nearby in inclination.

is dominated by the interface energy. In such cases, interfaces whose inclinations vary slightly from those of particularly low energy will adopt a vicinal stepped structure which preserves as much of the structure of the nearby low energy singular interface as possible (Fig. 5.7). In a simple 2D model which neglects any step interactions, the energy as a function of $\Delta\theta$, the angular deviation from the singular inclination, is then simply

$$\sigma(\Delta\theta) = \sigma(O) + (E^{\mathrm{S}}/h)\Delta\theta, \qquad (5.72)$$

where $E^{\mathrm{S}}$ is the step energy (per unit step length) and $h$ is the step height. Equation (5.72) indicates the existence of a cusp in the $\sigma(\hat{n})$ surface as $\theta$ passes through the $\Delta\theta = 0$ inclination. In a more refined model, it may be assumed that the steps repel each other. This is likely for steps in internal interfaces since their cores consist of bad material which generally causes an expansion (Lomer 1957), and they will, therefore, act as lines of positive dilation which will repel each other elastically. Also, in many cases, they may possess dislocation character as well as step character (see Section 1.7.1), and this will also produce a repulsion since the Burgers vectors will all be identical. In such a case, as shown by Landau (1965), a cusp is expected at every nearby inclination at which it is geometrically possible to distribute the steps with uniform periodicities. Since this occurs at essentially all inclinations, the result is a $\sigma(\hat{n})$ surface which might be termed 'everywhere continuous but nowhere differentiable'. Of course, the cusps due to long wavelength periodicities would be extremely weak. The result is then a $\sigma(\hat{n})$ surface containing a considerable number of cusps of widely varying depths, as illustrated in Fig. 5.6(a). As the temperature is increased, and the interface entropy becomes increasingly important causing the interface to disorder and roughen, the cusps will tend to become shallower and many of the shallower cusps will progressively 'wash out' and disappear when the corresponding singular interfaces undergo roughening transitions, and the step free energies approach zero (see further discussion and references regarding interface roughening transitions in Section 6.3.1.1. An experimentally determined section of a Wulff plot for a Σ11 grain boundary in copper at a relatively low temperature is shown in Fig. 5.6(b). In this case, only two major types of cusps were clearly resolved.

### 5.6.2 Equilibrium shape (Wulff form) of embedded second-phase particle

The Wulff plot is useful as an aid in investigating the equilibrium shape of a small second-phase particle, or crystallite, embedded in a large matrix. Such a particle will tend to adopt a shape which minimizes the total free energy of its anisotropic interface. However, the particle may be mismatched with the matrix, and misfit strains may be generated throughout the system which will be a function of the particle shape (Eshelby 1957). If the morphology which minimizes the interfacial energy differs from the morphology which minimizes the misfit strain energy, the equilibrium particle shape will be a complex function of both the interfacial energy and strain energy. However, the ratio of the strain energy to the interfacial energy generally increases with the particle size (Johnson and Cahn 1984), and interfacial energy considerations will usually be dominant when the particle is small. We therefore consider only the case of particles which are sufficiently small so that any misfit strain energy can be neglected.

Our first step is to find the shape of minimum interface energy based simply on a knowledge of the anisotropy of the macroscopic surface energy, i.e., on the form of the Wulff plot. This involves adjusting the shape to minimize the interface energy integral, $\iint \sigma(\hat{n}) \, dA$, while maintaining constant volume. Complications due to the additional energy of the line defects present along possible facet intersections and other defects which may be introduced to reduce the overall energy (Pond and Dibley 1989) will be discussed later. The problem is identical to the classic problem (Herring 1951*a*, 1953, Frank 1962, Mullins, 1962, Landau and Lifshitz, 1958) of finding the equilibrium shape of a small free crystallite of fixed volume maintained in a vacuum. The shape may be found directly from the Wulff plot by a simple geometrical construction (Herring 1951*a*, 1953). First, draw radius vectors to the surface of the Wulff plot (i.e. $r_1, r_2 \ldots$ in Fig. 5.6(a)); then draw planes perpendicular to these radii at their points of intersection (i.e. $P_1, P_2 \ldots$). The interior envelope of this family of planes is then a convex form whose shape is that of the equilibrated particle. This unique form is commonly termed the Wulff form. When the Wulff plot possesses relatively deep cusps (i.e. relatively low energy singular inclinations), the interior envelope will tend to consist of sections lying perpendicular to the radius vectors going to the cusps, and the Wulff form will tend to be faceted at the expense of increased area as in Fig. 5.6(a). However, depending upon the detailed form of the Wulff plot, the Wulf from can contain both facets and smoothly curved sections along with a variety of edges and corners as discussed, for example, by Herring (1951*a*, 1953). An observed example of the shape of a Cu crystallite embedded in a NiO matrix possessing well defined facets is shown in Fig. 5.8.

The above construction can be justified (Landau and Lifshitz 1958) by showing that

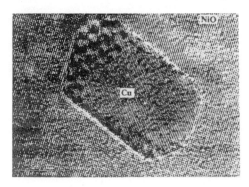

**Fig. 5.8** Experimentally observed faceted shape of Cu crystallite embedded in a NiO matrix. (From Merkle and Shao (1988).)

it produces a shape consistent with a minimization of the total particle interface energy at constant particle volume. If the form of the interface is represented by

$$z = z(x, y), \tag{5.73}$$

$\hat{n}$ is given by (Hildebrand 1962)

$$\hat{n} = \nabla g / |\nabla g|, \tag{5.74}$$

where $g(x, y, z) \equiv z - z(x, y)$. Therefore,

$$\hat{n} = \left[ -\left(\frac{\partial z}{\partial x}\right) i - \left(\frac{\partial z}{\partial y}\right) j + k \right] \cdot \left[ \left(\frac{\partial z}{\partial x}\right)^2 + \left(\frac{\partial z}{\partial y}\right)^2 + 1 \right]^{-\frac{1}{2}}. \tag{5.75}$$

An element of surface area, $dA$, is given by

$$dA = \left[ \left(\frac{\partial z}{\partial x}\right)^2 + \left(\frac{\partial z}{\partial y}\right)^2 + 1 \right]^{\frac{1}{2}} dx \, dy. \tag{5.76}$$

The total interface energy which must be minimized is

$$\int \sigma(\hat{n}) dA = \iint \sigma(\partial z/\partial x, \, \partial z/\partial y) \cdot \left[ \left(\frac{\partial z}{\partial x}\right)^2 + \left(\frac{\partial z}{\partial y}\right)^2 + 1 \right]^{\frac{1}{2}} dx \, dy, \tag{5.77}$$

and the constant volume constraint condition is

$$\iint z \, dx \, dy = \text{constant}. \tag{5.78}$$

Multiplying eqn (5.78) by a constant Lagrangian multiplier, $-2/\lambda$, and adding it to eqn (5.77), we obtain the quantity

$$\iint \left\{ \sigma(\partial z/\partial x, \, \partial z/\partial y) \cdot \left[ \left(\frac{\partial z}{\partial x}\right)^2 + \left(\frac{\partial z}{\partial y}\right)^2 + 1 \right]^{\frac{1}{2}} - 2z/\lambda \right\} dx \, dy. \tag{5.79}$$

The function, $z = z(x, y)$, which minimizes the above integral, then yields the desired form of the interface of minimum energy. This is a classic variational problem, and we therefore seek a solution of the Euler equation (Margenau and Murphy 1956),

**Fig. 5.9** Junction dislocation at the intersection of two faceted interfaces possessing different translation vectors, $t^{(1)}$ and $t^{(2)}$. The Burgers vector of the junction dislocation is then $b = t^{(1)} - t^{(2)}$.

$$\frac{\partial I}{\partial z} - \frac{\partial}{\partial x}\left[\frac{\partial I}{\partial\left(\frac{\partial z}{\partial y}\right)}\right] - \frac{\partial}{\partial y}\left[\frac{\partial I}{\partial\left(\frac{\partial z}{\partial y}\right)}\right] = 0 \qquad (5.80)$$

where $I$ is the integrand in eqn (5.79). It may be verified that a solution is

$$z = x\left(\frac{\partial z}{\partial x}\right) + y\left(\frac{\partial z}{\partial y}\right) + \sigma\lambda\left[1 + \left(\frac{\partial z}{\partial x}\right)^2 + \left(\frac{\partial z}{\partial y}\right)^2\right]^{\frac{1}{2}}. \qquad (5.81)$$

Now, a radius vector to any point on the equilibrium interface is

$$r = xi + yj + zk. \qquad (5.82)$$

The unit vector normal to the interface at that point is given by eqn (5.75), and therefore, using eqns (5.82), (5.81), and (5.75),

$$r \cdot \hat{n} = \lambda\sigma, \qquad (5.83)$$

which is just the condition that $r$ lies on the inner envelope of the Wulff construction. Alternative methods of dealing with the particle shape problem treated above have been given elsewhere (Frank 1962, Mullins 1962, Hoffman and Cahn 1972, Cahn and Hoffman 1974). Frank employs, in a particularly elegant fashion, a reciprocal Wulff plot. On the other hand, Cahn and Hoffman employ the capillarity vector, $\xi$ (Section 5.6.4), and construct a $\xi$-plot which, as shown below, turns out to have the same form as the Wulff form.

As already mentioned, the above analysis neglects any effects due to the extra energy contributed by the line defects lying along possible facet junctions. Each of these line defects may, or may not, exhibit dislocation character depending upon the translational states adopted by the intersecting facets. If each facet is considered as an individual interface with its own characteristic translation vector, $t^{(i)}$, as in Fig. 5.9, then each line of intersection between facets has the possibility of being a 'junction dislocation' with a Burgers vector given by

$$b = t^{(1)} - t^{(2)}. \qquad (5.84)$$

Such junction dislocations would possess both core energy and elastic strain energy. Direct observations of the elastic displacement fields around such line junctions possessing dislocation character have been made, for example, by Pond and Vitek (1977). Differences in the interface stresses associated with adjoining facets will also introduce lines of force at the junctions as described in Section 4.3.1.9, and these lines of force will introduce additional elastic energy. A more exact analysis of the equilibrium particle shape would involve taking these additional energies into account. No treatments of this

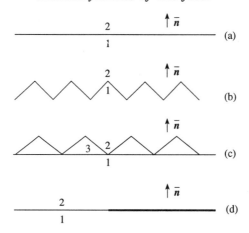

**Fig. 5.10** (a) Flat, single-phase interface between crystals 1 and 2. (b) Transformation of interface in (a) into a faceted 'hill and valley' structure containing two new interface phases (facets). (c) Transformation of interface in (a) by dissociation into a structure containing three new interface phases and occluded volumes of a new crystal misorientation 3. (d) Transformation of interface in (a) into interface composed of two congruent phases (indicated by the thin and thick lines).

more complicated problem have yet appeared. However, an analysis of the roles of these additional energies in the faceting of an initially flat interface is given in Section 4.3.1.9.

The Wulff form for a perfect embedded crystal particle, as determined above, should have point symmetry consistent with the space group of the dichromatic complex which has been described in Section 1.5.2 (Kalonji and Cahn 1982, Pond and Dibley 1990). However, as discussed in Section 1.8, an embedded particle may develop a defect structure which will allow it to change its symmetry and morphology and thereby possibly decrease the overall energy of the system by decreasing the interfacial energy even though the extra defects tend to increase the energy. The symmetry of the dichromatic complex is generally lower than that of the matrix crystal, and Pond and Dibley show that, by introducing suitable defects, particle morphologies consistent with the higher symmetry of the matrix can be obtained. Such morphologies are referred to as 'fully compensated' by the introduction of the defects. Intermediate cases, which are not fully compensated, can also be obtained. The necessary defects include, for example, twins, faults, disinclinations, and interface junction dislocations. As previously, a full treatment of the particle shape would require taking this further possible complication into account.

### 5.6.3 Faceting of initially flat interface

Another phenomenon of great interest related to the Wulff plot is the tendency of a large, initially flat, interface to assume a faceted 'hill and valley' structure (Fig. 5.10(a,b)) in order to reduce its overall free energy. This will tend to occur whenever the facets (singular interfaces) are of sufficiently low free energy (per unit area) to more than compensate for the increase in interface area which occurs. The development of such a structure allows the interface to equilibrate locally while maintaining both its average inclination and crystal misorientation. This situation would be typical of a large patch of interface in a system where the length scales are too large to allow global equilibration. Experimentally observed examples are seen in Fig. 5.11(a,b,e).

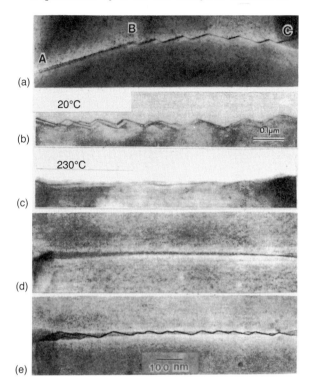

**Fig. 5.11** (a) Grain boundary in Cu showing a single-phase flat region at the inclination existing between A and B and a two-phase faceted structure at the average inclination existing between B and C. (From Ference and Balluffi (1989).) (b) and (c) Effect of an increase in temperature on eliminating the faceting of a Σ3 asymmetric ⟨111⟩ tilt grain boundary in Al. Same boundary region shown in each case. (From Hsieh and Balluffi (1989).) (d) and (e) Effect of segregated Bi on inducing the faceting of a grain boundary in Cu. Same boundary region shown in each case. Boundary in (d) is free of any Bi: boundary in (e) contains segregated Bi. (From Ference and Balluffi (1988).)

This phenomenon may be analysed by, again, initially ignoring any energy contributed by the facet junctions. Following Cabrera (1963), we again represent the form of an interface by eqn (5.73) and then define a new function, $\sigma_p(\hat{n})$, related to the Wulff plot function, $\sigma(\hat{n})$, which is the free energy of the interface of inclination, $\hat{n}$, per unit projected area on the $z = 0$ plane, and is therefore given by

$$\sigma_P(\hat{n}) = \sigma_p(\partial z/\partial x, \partial z/\partial y) = \sigma(\partial z/\partial x, \partial z/\partial y) \cdot \left[1 + \left(\frac{\partial z}{\partial x}\right)^2 + \left(\frac{\partial z}{\partial y}\right)^2\right]^{\frac{1}{2}}. \tag{5.85}$$

A plot of the $\sigma_P$ surface as a function of the inclination variables defining the interface normal, $\partial z/\partial x$ and $\partial z/\partial y$, might be expected to look like Fig. 5.12, where cusps corresponding to singular inclinations are present at the inclinations $C_1$, $C_2$, and $C_3$. Note that in this plot, the cusps protrude 'outwards' since $\sigma_P$ is relatively small in these regions. Cabrera (1963) has shown that the stability of any interface with respect to the possible formation of a faceted structure can be conveniently tested with this plot using a common tangent plane construction. In this construction, all planes tangent to the surface of the plot which do not intersect the surface are constructed at all possible points

*Thermodynamics of interfaces*

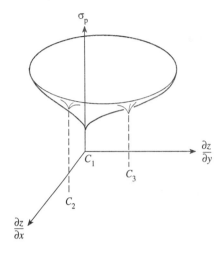

**Fig. 5.12** Possible 3D plot of $\sigma_P$ versus the inclination of the interface as defined by $\partial z/\partial x$ and $\partial z/\partial y$, where the form of the interface is given by $z = z(x, y)$ (see eqn (5.85)). $\sigma_P$ is a function of the inclination of the interface (i.e. $\sigma_P = \sigma_P(\hat{n})$) and is defined as the interface free energy per unit projected area on the $z = 0$ plane.

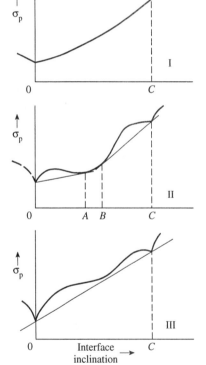

**Fig. 5.13** Three possible 2D plots of $\sigma_P$ versus interface inclination. The common tangent constructions shown predict that for case II all interfaces with average inclinations between 0 and A and B and C will be faceted, while interfaces between A and B will remain flat. Similarly, for case III all interfaces between 0 and C will be faceted.

on the surface, as illustrated schematically in 2D in Fig. 5.13. Then, the only inclinations (facets) which can appear in any equilibrium faceted interface structure must lie in these tangent planes. Several cases present themselves as illustrated schematically in Fig. 5.13. In case I, all inclinations between 0 and C are contained in the tangent planes, and, therefore, all interfaces with inclinations between 0 and $C$ will remain flat and stable. In case II, inclinations between 0 and $A$ are not contained in the tangent planes. Instead, the tangent plane for these inclinations is the common tangent to the points on the surface at 0 and $A$. All interfaces with inclinations between 0 and A can then reduce their energies by breaking up into stable structures consisting of facets with inclinations 0 and $A$.

Similarly, all inclinations between $A$ and $B$ will remain flat and stable, and all stable inclinations between $B$ and $C$ will be faceted into $B$ and $C$ inclinations. In case III, all inclinations between 0 and $C$ will be faceted into stable 0 and $C$ inclinations. This construction is readily generalized to 3D where the tangent planes can now be tangent to three points on the surface of the plot corresponding to a 3D faceted interface possessing three sets of facets in equilibrium with each other. Further analysis (Herring 1951a) also shows that if the initially flat surface is not part of the Wulf form, its free energy can always be reduced by the development of a faceted structure. On the other hand, if the initially flat surface is part of the Wulff form, no reduction of free energy is possible by faceting. The common tangency construction described above in 3D is seen to be exactly analogous to the well known common tangency construction for testing the stability of bulk solid solutions against a breakup into phases of different composition when their bulk free energy surfaces are known as functions of composition (Pelton 1983). The two inclination variables, $\partial z/\partial x$ and $\partial z/\partial y$, correspond directly to the two compositional variables required to specify the composition in the three-component solid solution. Simple lever-type rules hold (Cabrera 1963) for finding the fractional areas of the different facets projected on the average interface plane which are similar to the lever rules (Pelton 1983) that yield the amounts of the different phases which form when a solid solution is unstable.

As in the case of the embedded particle problem in the previous section, a more exact analysis of the faceting problem would include any energy contributed by the facet junctions which may act as both dislocations and lines of force. An analysis of this energy is given in Section 4.3.1.9. As discussed there, the junction energy may control the length scale of the faceting.

### 5.6.4   The capillarity vector, $\xi$

Having described the Wulff plot and some of its uses, we now introduce the capillarity vector, $\xi$, which is closely related to it and is very useful in describing the properties of interfaces whose energies vary with inclination. This vector, first introduced by Hoffman and Cahn (1972) and Cahn and Hoffman (1974), is defined as the gradient of the scalar field, $r \cdot \sigma(\hat{n})$:

$$\xi \equiv \nabla[r \cdot \sigma(\hat{n})]. \tag{5.86}$$

The value of the scalar field at any point a distance $r$ from the origin in a direction $\hat{n}$ is simply the value of $\sigma(\hat{n})$ for the interface with unit normal vector $\hat{n}$ multiplied by $r$. Equation (5.86) shows directly that $\xi$ is a function only of $\hat{n}$; i.e., $\xi = \xi(\hat{n})$. Using the identity

$$\left.\begin{array}{l} d(r\sigma) = \nabla(r\sigma) \cdot dr, \\ r\,d\sigma + \sigma\,dr = \xi \cdot d(r\hat{n}) = r\xi \cdot d\hat{n} + \xi \cdot \hat{n}\,dr \end{array}\right\}. \tag{5.87}$$

Therefore,

$$d\sigma = \xi \cdot d\hat{n} \tag{5.88}$$

$$\sigma = \xi \cdot \hat{n}. \tag{5.89}$$

Differentiating eqn (5.89) and subtracting eqn (5.88), we also have

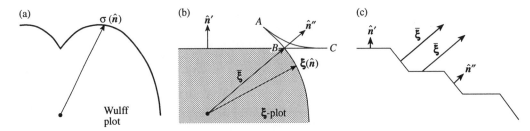

**Fig. 5.14** (a) 2D section of a Wulff plot for an interface (e.g. see Fig. 5.6(a)). (b) Section of the $\xi$ plot derived from the section of the Wulff plot shown in (a). Note that the form of the $\xi$ plot is the same as that of the Wulff form which would be obtained from the same Wulff plot (e.g. see Fig. 5.6(a)). The 'ear' at ABC represents unstable (or metastable) inclinations lying between $\hat{n}'$ and $\hat{n}''$. (c) Diagram showing the equality of the capillarity vectors (i.e. $\xi$ vectors) for adjacent facets contained in an equilibrated faceted interface.

$$\hat{n}\cdot d\xi = 0. \tag{5.90}$$

These equations represent the fundamental properties of $\xi(\hat{n})$.

The vector quantity $\xi$ can be conveniently displayed in 3D by again employing a polar plot (i.e. $\xi$-plot). Knowing the Wulff plot, $\sigma(\hat{n})$, the rules for the construction of such a plot are embodied in eqns (5.89) and (5.90). Equation (5.89) states that the projection of the vector $\xi$ on $\hat{n}$ is equal to $\sigma(\hat{n})$. Equation (5.90) states that the inclination of the surface of the $\xi$ plot at the point $\xi(\hat{n})$ is normal to $\hat{n}$. (Note, of course, that $\xi(\hat{n})$ is generally not parallel to $\hat{n}$.) A simple 2D example of a section of a $\xi$-plot derived from a corresponding section of a Wulff plot by means of these rules is illustrated in Fig. 5.14. Shown hatched in Fig. 5.14(b) is the Wulff form obtained from the Wulff plot in Fig. 5.14(a). It may be seen in Fig. 5.14(b) that the $\xi$-plot is identical to the Wulff Form with the exception of the 'ear' at $ABC$ on the $\xi$-plot. This ear corresponds to interface inclinations between $\hat{n}'$ and $\hat{n}''$ which are missing on the equilibrium Wulff form and, hence, are unstable (or metastable) because they are not inclinations whose planes are part of the inner envelope which produces the Wulff form. If this non-equilibrium portion is discarded, we have the simple result that the $\xi$-plot is geometrically similar to the Wulff form. This is a direct consequence of the close similarity between the rules for obtaining the Wulff form and the $\xi$-plot. For example, eqn (5.89) is quite analogous to eqn (5.83).

The $\xi(\hat{n})$ vector generally possesses a component normal to the interface,

$$\xi^{n} = \sigma\hat{n}, \tag{5.91}$$

and tangential to the interface

$$\xi^{t} = \xi - \sigma\hat{n}, \tag{5.92}$$

(see Fig. 5.15) and is highly useful for formulating the capillary forces which tend to either shrink or rotate the interface. A tangential shrinkage force is always present, since the interface represents excess energy and, therefore, always tends to shrink, while a rotational force (torque) exists whenever the interface energy varies with inclination and can be reduced by a rotation of the interface.

This can be put on a quantitative basis by considering a small patch of interface of area $A$ represented by the vector $A = \hat{n}A$ (Fig. 5.16). Using eqn (5.89), the work required to create this area is

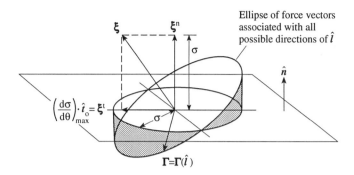

**Fig. 5.15** Diagram showing $\xi = \xi(\hat{n})$ at a patch of interface of inclination $\hat{n}$. Also shown are the normal and tangential components of $\xi$, i.e. $\xi^n$ and $\xi^t$, respectively. The capillary force $\Gamma = \Gamma(\hat{l})$, exerted by the interface on a unit vector, $\hat{l}$, lying in the interface plane is given as a function of the direction of $\hat{l}$, by the ellipse indicated. When $\hat{l}$ is perpendicular to $\xi^t$, $|\Gamma(\hat{l})|$ is a maximum: when $\hat{l}$ is parallel to $\xi^t$, $|\Gamma(\hat{l})|$ is a minimum.

$$\sigma A = \xi \cdot A. \tag{5.93}$$

Making an arbitrary change in the area, or inclination, of the patch then requires the work

$$d(\sigma A) = \xi \cdot dA + A \cdot d\xi \tag{5.94}$$

But, according to eqn (5.90), $A \cdot d\xi = 0$, and, therefore,

$$d(\sigma A) = \xi \cdot dA \tag{5.95}$$

Now, delineate the area by the vectors $L_1$ and $L_2$ so that $A = L_1 \times L_2$, as in Fig. 5.16, and also move $L_1$ parallel to itself by the distance $dS$ in order to change the area of $A$ and also rotate it. The increase in $A$ is then $dA = -L_1 \times dS$, and

$$d(\sigma A) = \xi \cdot dA = -\xi \cdot (L_1 \times dS) = -(L_1 \times \xi) \cdot dS. \tag{5.96}$$

We see that the right-hand side of eqn (5.96) has the form of a force, $L_1 \times \xi$, times a distance, $dS$, and that the work is positive when the area is increased. We therefore define $\Gamma$ as the effective capillary force exerted by the interface on a unit vector, $\hat{l}$, parallel to $L_1$ (which is just the negative of the above force) so that

$$\Gamma = \xi \times \hat{l}. \tag{5.97}$$

Equation (5.97) clearly depends upon the convention that the patch of interface bounded by $\hat{l}$ extends in the direction of $\hat{n} \times \hat{l}$. The tangential component of this force, which always tends to shrink the element, is then

$$\Gamma^t = \xi^n \times \hat{l} \tag{5.98}$$

which is perpendicular to $\hat{l}$ in the direction of the interface and is always of magnitude $\sigma$ regardless of the direction of $\hat{l}$ because, as indicated by eqn (5.91), $|\xi^n| = \sigma$. The normal component, which tends to rotate the element, and is therefore often called a 'torque', is

$$\Gamma^n = \xi^t \times \hat{l} \tag{5.99}$$

and has a magnitude which depends upon $|\xi^t|$ and the orientation of $\hat{l}$.

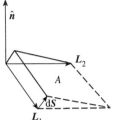

Fig. 5.16 Small patch of interface area defined by $L_1$ and $L_2$. Area $= A = \hat{n}A$. Area perturbed by the displacement of $L_1$ by $dS$.

A more physical interpretation of $|\xi^t|$ can be obtained by considering a small rotation of the interface surface element by the angle $d\theta$ so that $|d\hat{n}| = d\theta$. Therefore, using eqn (5.88),

$$\xi \cdot (d\hat{n}/d\theta) = d\sigma/d\theta. \qquad (5.100)$$

The unit tangent vector in the direction of $d\hat{n}$ is then $t = d\hat{n}/d\theta$, and, therefore, $\xi \cdot t = d\sigma/d\theta$. When $t$ lies parallel to $\xi^t$ (i.e. when $t = t_o$) $d\sigma/d\theta$ will assume its maximum value, i.e. $(d\sigma/d\theta)_{max}$, and therefore

$$\xi^t = (d\sigma/d\theta)_{max} \cdot t_o. \qquad (5.101)$$

The magnitude of $\xi^t$ is therefore equal to the maximum rate of change of $\sigma$ with respect to rotation of the surface element by the angle $\theta$.

Since $\Gamma$ is always perpendicular to $\xi$, the locus of all $\Gamma$ vectors generated by turning $\hat{l}$ through 360° in a surface is an ellipse whose plane is perpendicular to $\xi$ and whose projection normal to the surface is a circle of radius $\sigma$, as illustrated in Fig. 5.15. When $\hat{l}$ is parallel to $\xi^t$, the torque is zero; when $\hat{l}$ is perpendicular to $\xi^t$, it is a maximum.

### 5.6.5 Capillary pressure associated with smoothly curved interface

When an interface is curved, the tangential and normal capillary forces described above will generally exert a net normal force on each patch of the interface. This net force then corresponds to an effective capillary pressure which tends to move the patch of interface normal to itself. An expression for this pressure is readily found with the use of the $\Gamma$ vector.

Consider the differential patch of curved interface area ABCD, shown in Fig. 5.17. The two principal radii of curvature (Weatherburn 1927) are $R_1$ and $R_2$, $AB = DC = dl_2 = R_2 d\theta_2$, and $AD = BC = dl_1 = R_1 d\theta_1$, and the arc lengths $dl_1$ and $dl_2$ are orthogonal. We first find the forces, $t_1$ and $t_2$ (Fig. 5.17), exerted on the edge $AB$ by the remainder of the interface and the corresponding forces, $t_3$ and $t_4$, exerted on the edge $DC$. From our previous work, $t_1$ and $t_3$ can be identified as tangential forces of magnitudes $|t_1| = \sigma R_2 d\theta_2$ and

$$|t_3| = (\sigma + \partial\sigma/\partial\theta_1)R_2 d\theta_2.$$

Also, $t_2$ and $t_4$ and are normal forces of magnitudes

$$|t_2| = (\partial\sigma/\partial\theta)_{max} \cdot \cos\psi \cdot R_2 d\theta_2$$

$$|t_4| = \{(\partial\sigma/\partial\theta)_{max} \cdot \cos\psi + \partial[(\partial\sigma/\partial\theta)_{max} \cdot \cos\psi]/\partial\theta_1 \cdot d\theta_1\} \cdot R_2 d\theta_2.$$

The net force directed towards the centre of curvature (i.e. along $-\hat{n}_o$) is then

$$|T| = \{|t_1| + |t_3|\} \sin(d\theta_1/2) + \{|t_4| - |t_2|\} \cos(d\theta_1/2). \qquad (5.102)$$

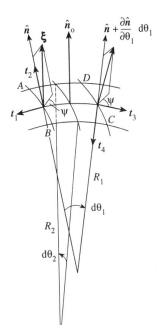

**Fig. 5.17** Diagram used to determine capillary pressure, $\Delta P$, associated with a smoothly curved interface. Differential patch of interface corresponds to $ABCD$. The two principal radii of curvature are $R_1$ and $R_2$. $\xi$ vectors at midpoints of $AB$ and $CD$ shown. Tangential and normal forces exerted on the $AB$ and $CD$ edges are indicated by $t_1$, $t_2$ and $t_3$, $t_4$ respectively.

From an examination of the corresponding Wulff plot, it can be verified that $(\partial\sigma/\partial\theta)_{max} \cdot \cos\psi = \partial\sigma/\partial\theta_1$, since these derivatives follow a simple cosine type projection law (Hoffman and Cahn 1972). Since the area of the patch is $dl_1 \cdot dl_2$, we can put all of the above results into eqn (5.102) to obtain the force per unit area due to the principal curvature, $(1/R_1)$, in the form

$$(\sigma + \partial^2\sigma/\partial\theta_1^2)/R_1. \qquad (5.103)$$

Repeating the analysis for the force per unit area due to the principal curvature $(1/R_2)$, we finally obtain the expression for the total capillary force per unit area (i.e. capillary pressure, $\Delta P$):

$$\Delta P = \sigma(1/R_1 + 1/R_2) + (1/R_1)\partial^2\sigma/\partial\theta_1^2 + (1/R_2)\partial^2\sigma/\partial\theta_2^2. \qquad (5.104)$$

The quantity $(1/R_1 + 1/R_2)$ is known as the 'mean curvature' and can be conveniently obtained from the standard relation (Weatherburn, 1927)

$$\nabla \cdot \hat{n} = (1/R_1 + 1/R_2). \qquad (5.105)$$

Equation (5.104) shows that the local capillary pressure depends upon the mean curvature and also the rates of variation of the torque forces with rotation of the interface. A curved interface will always tend to move in response to this pressure, and this result will be useful in our discussion of the motion of curved interfaces in Section 9.4.2.

### 5.6.6 Equilibrium lattice solubility at a smoothly curved heterophase interface

The equilibrium solubility of a solute species in the lattice at a heterophase interface is generally dependent upon the curvature of the interface. This phenomenon is of central importance in a number of capillary phenomena including, for example, the coarsening of a distribution of second phase particles (Section 10.4.2.1). Consider the case of a curved interface between $\alpha$ and $\beta$ phases in a binary system composed of A and B atoms. We assume an incoherent or semicoherent interface which can act as a source or sink for atoms (Section 5.9.2.1), and distributions of lattice dislocations throughout the bulk phases which can also act as sources/sinks. The $\beta$ phase is B-rich, and we wish to determine the equilibrium solubility of B in the $\alpha$ phase as a function of the interface curvature. This is most readily accomplished by finding the equilibrium solubility at the curved interface relative to the corresponding solubility at a flat interface.

In step 1 we transfer $\delta N$ atoms, composed of $\delta N_A$ A atoms and $\delta N_B$ B atoms, from the $\alpha$ phase to the $\beta$ phase across the flat interface under equilibrium conditions. The $\delta N_A$ and $\delta N_B$ numbers are proportioned so that

$$\delta N_A/\delta N = X_A^\beta; \quad \delta N_B/\delta N = X_B^\beta, \tag{5.106}$$

where the $X_i^\beta$ are the atom fractions in the $\beta$ phase. In step 2 we transfer the same numbers of atoms from the $\alpha$ phase to the $\beta$ phase across the curved interface under equilibrium conditions. If the curved interface is convex outwards into the $\alpha$ phase, the change in free energy of the $\alpha$ phase in step 2 is greater than in step 1 because additional work must be done against the effective capillary pressure, $\Delta P$, at the curved interface (given by eqn (5.104)) when the $\delta N$ atoms are inserted in step 2. This work is just $\Delta P \Omega^\beta \delta N$, where $\Omega^\beta$ = atomic volume in the $\beta$ phase. Using the standard relation $\delta G = \mu_A^\alpha \delta N_A + \mu_B^\alpha \delta N_B$, we therefore have

$$\mu_A^\alpha(R)\,\delta N_A + \mu_B^\alpha(R)\,\delta N_B - \Delta P \Omega^\beta\,\delta N = \mu_A^\alpha(\infty)\,\delta N_A + \mu_B^\alpha(\infty)\,\delta N_B, \tag{5.107}$$

where $(R)$ and $(\infty)$ indicate quantities at the curved and flat interfaces respectively. The chemical potentials can be written in the usual form $\mu_i = \mu_i^\circ + kT\ln\gamma_i X_i$, and using this and eqn (5.106), eqn (5.107) takes the form

$$X_A^\beta \ln\left[\frac{\gamma_A^\alpha(R)X_A^\alpha(R)}{\gamma_A^\alpha(\infty)X_A^\alpha(\infty)}\right] + X_B^\beta \ln\left[\frac{\gamma_B^\alpha(R)X_B^\alpha(R)}{\gamma_B^\alpha(\infty)X_B^\alpha(\infty)}\right] = \frac{\Delta P \Omega^\beta}{kT}. \tag{5.108}$$

However, all quantities associated with the curved interface differ only slightly from those associated with the flat interface, and we may therefore expand eqn (5.108) to first order and rewrite it in the form

$$X_A^\beta \frac{\Delta X_A^\alpha}{X_A^\alpha(\infty)}\left[1 + \frac{\partial\ln\gamma_A^\alpha}{\partial\ln X_A^\alpha}\right] + X_B^\alpha \frac{\Delta X_B^\alpha}{X_B^\alpha(\infty)}\left[1 + \frac{\partial\ln\gamma_B^\alpha}{\partial\ln X_B^\alpha}\right] = \frac{\Delta P \Omega^\beta}{kT}, \tag{5.109}$$

where $\Delta X_i^\alpha = X_i^\alpha(R) - X_i^\alpha(\infty)$. Using the Gibbs–Dubem equation, it may be shown that

$$\frac{\partial\ln\gamma_A^\alpha}{\partial\ln X_A^\alpha} = \frac{\partial\ln\gamma_B^\alpha}{\partial\ln X_B^\alpha}, \tag{5.110}$$

and, since $X_A^\beta + X_B^\beta = 1$, eqn (5.109) may be written finally as

$$\frac{X_B^{\alpha,\,eq}(R)}{X_B^{\alpha,\,eq}(\infty)} = 1 + \left[\frac{1 - X_B^{\alpha,\,eq}(\infty)}{X_B^\beta - X_B^{\alpha,\,eq}(\infty)}\right]\frac{\Omega^\beta}{kT\phi}\left[\sigma\left(\frac{1}{R_1} + \frac{1}{R_2}\right) + \frac{1}{R_1}(\partial^2\sigma/\partial\theta_1^2) + \frac{1}{R_2}(\partial^2\sigma/\partial\theta_2^2)\right], \tag{5.111}$$

where we have used eqn (5.104) to substitite for $\Delta P$, $\phi \equiv (1 + \partial\ln\gamma_B^\alpha/\partial\ln X_B^\alpha)$, and the eq superscript has been added to emphasize that we are concerned with equilibrium quantities. Equation (5.111) is generally known as the Gibbs–Thompson equation and shows that the equilibrium solubility increases with increased convex curvature.

### 5.6.7 Equilibrium solubility at embedded second-phase particle

Following Johnson (1965), we now find the equilibrium solubility of a $\beta$ phase particle embedded in an $\alpha$ matrix in a binary system composed of A and B atoms. In this equilibrium situation the particle possesses its equilibrium shape corresponding to the Wulff plot construction described in Section 5.6.2. Using the same approach as in Section 5.6.6, we find the equilibrium solubility at the particle interface relative to the corresponding solubility at a flat interface by transferring $\delta N$ atoms from the $\alpha$ phase to the $\beta$ phase across a flat interface in step 1 and across the particle–matrix interface in step 2 without allowing any change in the equilibrium shape of the particle. In this case the change in free energy of the $\alpha$ phase in step 2 must be greater than in step 1 because of the additional work, $\delta G^S$, which must be done to increase the interfacial area of the particle as its volume is increased. The equation relating the energy changes of the $\alpha$ phase in these two steps is therefore of the same form as our previous eqn (5.107), i.e.

$$\mu_A^\alpha(P)\,\delta N_A + \mu_B^\alpha(P)\,\delta N_B - \delta G^S = \mu_A^\alpha(\infty)\,\delta N_A + \mu_B^\alpha(\infty)\,\delta N_B, \tag{5.112}$$

where $(P)$ indicates a quantity at the particle. We must now calculate $\delta G^S$. This can be accomplished by employing the scaling parameter

$$\lambda = (r \cdot \hat{n})/\sigma(\hat{n}), \tag{5.113}$$

which appeared earlier in eqn (5.83), and is proportional to the size (linear dimensions) of the particle. Using this parameter, the volume of the particle can be written in the form $V = K\lambda^3$ where $K$ is a constant 'shape factor'. The change in volume in step 2 is $\delta V = \Omega^\beta\,\delta N = 3K\lambda^2\delta\lambda$, and the change in $\lambda$ is then

$$\delta\lambda = \frac{\Omega^\beta}{3K\lambda^2}\,\delta N. \tag{5.114}$$

Now, the total surface energy of the particle is

$$G^S = \int \sigma(\hat{n})\mathrm{d}A = \frac{1}{\lambda}\int r \cdot \hat{n}\,\mathrm{d}A, \tag{5.115}$$

and, therefore, with the help of the divergence theorem,

$$G^S = \frac{1}{\lambda}\int \nabla \cdot r\,\mathrm{d}V = \frac{3V}{\lambda} = 3K\lambda^2, \tag{5.116}$$

since $\nabla \cdot r = 3$. Finally, with the help of eqns (5.116) and (5.114),

$$\delta G^S = \frac{\delta G^S}{\delta\lambda}\,\delta\lambda = \frac{2\Omega^\beta}{\lambda}\,\delta N. \tag{5.117}$$

Putting eqn (5.117) into eqn (5.112), and following the same procedure as that used to develop eqn (5.111) from eqn (5.107), we obtain

$$\frac{X_B^{\alpha,\,eq}(P)}{X_B^{\alpha,\,eq}(\infty)} = 1 + \left[\frac{1 - X_B^{\alpha,\,eq}(\infty)}{X_B^\beta - X_B^{\alpha,\,eq}(\infty)}\right]\frac{2\Omega^\beta}{kT\phi\lambda}. \tag{5.118}$$

The parameter $\lambda$ is readily found from knowledge of the particle size and the Wulff plot. Since $\lambda$ increases linearly with particle size, the solubility is seen to increase with decreasing particle size, as might be expected. In the simple case where the surface energy is isotropic, and the equilibrium particle is then a sphere of radius $R$, $\lambda = R/\sigma$, and eqn (5.118) reduces to

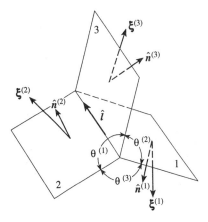

**Fig. 5.18** Diagram of triple junction where three interfaces, indicated by 1,2, and 3, meet along $\hat{l}$. The unit normal and $\xi$ vector are shown for each interface along with the dihedral angles, $\theta^{(i)}$, between pairs of adjacent interfaces.

$$\frac{X_B^{\alpha,\,eq}(P)}{X_B^{\alpha,\,eq}(\infty)} = 1 + \left[\frac{1 - X_B^{\alpha,\,eq}(\infty)}{X_B^\beta - X_B^{\alpha,\,eq}(\infty)}\right]\frac{2\Omega^\beta\sigma}{kT\phi R}. \tag{5.119}$$

We note that eqn (5.111) for the smoothly curved interface also reduces to this relatively simple form when the interface energy is isotropic, and the interface is therefore spherical. These solubility relationships will be of use in later chapters when we consider topics such as particle coarsening and interface stability.

### 5.6.8   Equilibrium interface configurations at interface junction lines

Another problem which is easily treated with the aid of the $\Gamma$ vector is that of finding the equilibrium interface configuration at interface line junctions. A common example is the case of a triple junction where three planar interfaces (1, 2, and 3) separating three crystals meet along a line as in Fig. 5.18.

We adopt a convention that the positive direction of each interface, $\hat{n}^i$, points in a clockwise direction when looking along positive $\hat{l}$ which, in turn, lies along the junction. The force exerted by the $i$th interface on unit length of the junction is then just $\Gamma^i = \xi^i \times \hat{l}$. (Note that our convention that the patch of interface bounded by $\hat{l}$ is in the direction of $\hat{n} \times \hat{l}$ is satisfied for each interface in this arrangement.)

The work required to produce a virtual parallel displacement of the line, $\delta S$, is

$$\delta W = [\Gamma^{(1)} + \Gamma^{(2)} + \Gamma^{(3)}] \cdot \delta S. \tag{5.120}$$

Since at equilibrium, $\delta W = 0$, we have the relatively simple result (Herring 1951b, Hoffman and Cahn 1972)

$$\sum_{i=1}^{3} \Gamma^i = \sum_{i=1}^{3} [\Gamma^{t,\,i} + \Gamma^{n,\,i}] = 0. \tag{5.121}$$

Since all of the vectors in eqn (5.121) are perpendicular to $\hat{l}$, this condition can be represented by a vector force diagram on a plane perpendicular to $\hat{l}$. The diagram becomes particularly simple when the torque terms can be neglected. Then, only tangential forces are present, and the solution of the vector force diagram yields

$$\frac{\sigma^{(1)}}{\sin\theta^{(1)}} = \frac{\sigma^{(2)}}{\sin\theta^{(2)}} = \frac{\sigma^{(3)}}{\sin\theta^{(3)}}. \tag{5.122}$$

By using the $\boldsymbol{\xi}$ vector it may be verified (Cahn and Hoffman 1974) that the capillary force exerted on a junction line between adjacent facets which are present on an interface which contains these facets as part of its equilibrium form is identically zero, as might be expected. To show this, consider again the $\boldsymbol{\xi}$-plot in Fig. 5.14(b). As we have already discussed in Section 5.6.4, all inclinations corresponding to the 'ear' are unstable. However, Cahn and Hoffman (1974) show that the $\boldsymbol{\xi}$ vector which touches the Wulff form at the intersection of two (or more) discontinuous patches on the Wulff Form (i.e. the $\boldsymbol{\xi} = \bar{\boldsymbol{\xi}}$ vector in Fig. 5.14(b)) is the $\boldsymbol{\xi}$ vector which must be the $\boldsymbol{\xi}$ vector of these patches directly adjacent to their intersection (i.e. the $\hat{n}'$ and $\hat{n}''$ interfaces at their junction). The $\boldsymbol{\xi}$ vectors which represent these two interfaces at the junction are therefore identical, and it is then easily seen that $\Sigma_i \, \boldsymbol{\Gamma}^i = 0$ at the junction, where $\boldsymbol{\Gamma}^i$ is the force exerted by the $i$th patch. Note that in order to obtain this result it was necessary to follow our convention (see above) regarding the sign of $\boldsymbol{\Gamma}^i$. Similarly, for an interface which is decomposed into $\hat{n}'$ and $\hat{n}''$ facets as illustrated in Fig. 5.14(c), the $\boldsymbol{\xi}$ vectors for all facets are equal to $\bar{\boldsymbol{\xi}}$, and the net force on each junction is again zero.

### 5.6.9   Further thermodynamic relationships involving changes in interface inclination

We now return to the general eqn (5.71) and show how further thermodynamic relationships involving changes in interface inclination can be obtained. First, we put the last two terms (right-hand side) in more explicit form using eqn (5.88). Since $d\hat{n}$ is parallel to the interface,

$$d\sigma = \xi_1^t \, dn_1 + \xi_2^t \, dn_2 + \xi_3^t \, dn_3. \tag{5.123}$$

However, since $n_1^2 + n_2^2 + n_3^2 = 1$, we may eliminate $n_3$ to obtain

$$\frac{\partial \sigma}{\partial n_1} \, dn_1 + \frac{\partial \sigma}{\partial n_2} \, dn_2 = \left[ \xi_1^t - \frac{n_1 \xi_3^t}{(1 - n_1^2 - n_2^2)^{\frac{1}{2}}} \right] dn_1 + \left[ \xi_2^t - \frac{n_2 \xi_3^t}{(1 - n_1^2 - n_2^2)^{\frac{1}{2}}} \right] dn_2. \tag{5.124}$$

A large number of thermodynamic relationships can now be obtained for an interface which is free to change its inclination while the misorientation remains fixed. For example, for such an interface in a two-phase system with two components which is unstrained, eqn (5.71) becomes

$$d\sigma = -\frac{1}{(\text{Det})_{C+3}} \sum_{k=1}^{C+2} (\text{Det})_k \, dX_k + \sum_{i=1}^{2} \left[ \xi_i^t - \frac{n_i \xi_3^t}{(1 - n_1^2 - n_2^2)^{\frac{1}{2}}} \right] dn_i. \tag{5.125}$$

Using our standard procedure to evaluate eqn (5.125) while eliminating $dT$ and $dP$ (see Section 5.3), we obtain

$$d\sigma = \sum_{i=1}^{2} \left[ \xi_i^t - \frac{n_i \xi_3^t}{(1 - n_1^2 - n_2^2)^{\frac{1}{2}}} \right] dn_i - \left\{ \begin{vmatrix} [N_i] & [V] & [S] \\ N_i^\alpha & V^\alpha & S^\alpha \\ N_i^\beta & V^\beta & S^\beta \end{vmatrix} \div \begin{vmatrix} V^\alpha & S^\alpha \\ V^\beta & S^\beta \end{vmatrix} \right\} d\mu_i. \tag{5.126}$$

Equation (5.126) then predicts a Maxwell-type relation, such as

$$\frac{\partial \left[ \xi_1^t - \dfrac{n_1 \xi_3^t}{(1 - n_1^2 - n_2^2)^{1/2}} \right]}{\partial \mu_1} = - \frac{\left\{ \begin{vmatrix} [N_1] & [V] & [S] \\ N_1^\alpha & V^\alpha & S^\alpha \\ N_1^\beta & V^\beta & S^\beta \end{vmatrix} \div \begin{vmatrix} V^\alpha & S^\alpha \\ V^\beta & S^\beta \end{vmatrix} \right\}}{\partial n_1} \tag{5.127}$$

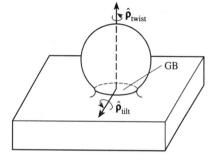

**Fig. 5.19** 'Rotating ball' experiment. A single crystal sphere is sintered to a single crystal plate of a different crystal orientation. A grain boundary is produced in the neck region, and the ball generally rotates in order to reduce the energy of the grain boundary by changing its misorientation.

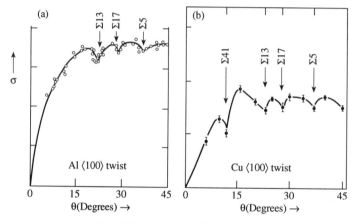

**Fig. 5.20** Measured variation of $\sigma$ with twist angle, $\theta$, for $\langle 100 \rangle$ twist boundaries in (a) Al (Otsuki and Mizuno 1986) and (b) Cu (Mori *et al.* 1988). Resolvable low energy cusps are found at a number of short-period (low-$\Sigma$) boundaries.

which relates changes in adsorption with interface inclination to changes in the components of the capillarity vector (which control the interface torque) with chemical potential. Relationships of this type have been developed and applied for free surfaces (Shewmon and Robertson, 1962).

## 5.7 DEPENDENCE OF $\sigma$ ON THE CRYSTAL MISORIENTATION

In a number of experiments (Chan and Balluffi 1985, 1986, Gao *et al.* 1989), small single-crystal balls were sintered to large single-crystal plates at different misorientations, and interfaces which ran parallel to the plate were formed in the neck regions as illustrated in Fig. 5.19. The misorientations were either of the twist or tilt type depending upon whether the rotation axis was perpendicular or parallel to the interface. Upon subsequent thermal annealing, the balls spontaneously rotated around their twist or tilt axes in order to reduce the interface energy. The boundary inclination remained parallel to the plate in this process, and the only geometric variable which changed was then the crystal misorientation angle, $\theta$. The change in $\sigma$ which occurs when the system follows this path (trajectory) then corresponds to the $(\partial \sigma / \partial \theta)\,d\theta$ term (right-hand side) of eqn (5.71). In a number of cases the rotating balls became trapped at specific rotation angles, indicating the presence of minima on the $\sigma = \sigma(\theta)$ function at these angles.

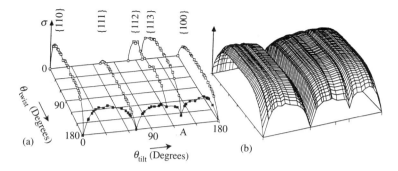

**Fig. 5.21** (a) Calculated energies of grain boundaries produced by various twist and/or tilt rotations. The five curves running parallel to the $\theta_{\text{twist}}$ axis represent the energies of {110}, {111}, {112}, {113}, and {100} pure twist boundaries as functions of $\theta_{\text{twist}}$. However, when $\theta_{\text{twist}} =$ 180°, these boundaries become identical to symmetric ⟨110⟩ tilt boundaries possessing the $\theta_{\text{tilt}}$ misorientation angles shown. Boundaries with 0° < $\theta_{\text{twist}}$ < 180° may therefore be regarded as either pure twist boundaries or as mixed boundaries containing a $\theta_{\text{tilt}}$ component plus a negative $\theta_{\text{twist}}$ component. (b) 3D energy surface produced by interpolation of the results in (a). (From Wolf (1990*b*).)

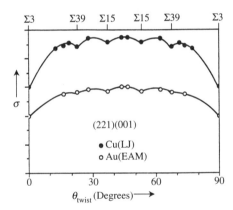

**Fig. 5.22** Calculated energies of grain boundaries in Cu and Al, with (221) in one crystal parallel to (100) in the other, as a function of $\theta_{\text{twist}}$ about the boundary normal. LJ and EAM indicate Lennard–Jones and embedded atom method interatomic potentials used respectively. Small cusps appear at relatively short period (low-Σ) boundaries. (From Wolf (1990*a*).)

Measurements of σ as a function of the rotation angle $\theta_{\text{twist}}$ (rotation angle with $\hat{\rho} = \hat{n}$) have been made (Otsuki and Mizuno 1986, Mori *et al.* 1988) for a variety of pure twist grain boundaries, as shown in Fig. 5.20. The large cusps at the extremes of these curves and the smaller intermediate cusps at the short period misorientations can be understood on the basis of primary and secondary dislocation models (Section 2.10.3). Numerous calculations have also been made of σ for pure twist grain boundaries as a function of $\theta_{\text{twist}}$. Some results (Wolf 1990*b*) are shown in Fig. 5.21(a).

As pointed out in Section 1.4.2, the five independent geometrical variables can also be selected as the two direction cosines necessary to specify $\hat{n}^{\text{b}}$ in a coordinate system in the black crystal, the two to specify $\hat{n}^{\text{w}}$ in a coordinate system in the white crystal, and the angle $\theta_{\text{twist}}$ which specifies the rotation of the black crystal relative to the white crystal about $\hat{n}^{\text{w}}$ (or $\hat{n}^{\text{b}}$). Calculations have been made (Wolf 1990*a*) of σ over trajectories where $\theta_{\text{twist}}$ was varied while holding $\hat{n}^{\text{b}}$ and $\hat{n}^{\text{w}}$ fixed as in Fig. 5.22.

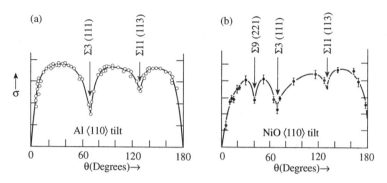

**Fig. 5.23** Measured variation of $\sigma$ with tilt angle, $\theta$, for $\langle 110 \rangle$ symmetric tilt grain boundaries in (a) Al (Otsuki and Mizuno 1986) and (b) NiO (Dhalenne *et al.* 1983). Resolvable low-energy cusps are found at several short period (low-$\Sigma$) boundaries.

## 5.8 DEPENDENCE OF $\sigma$ ON SIMULTANEOUS VARIATIONS OF THE INTERFACE INCLINATION AND CRYSTAL MISORIENTATION

Trajectories involving simultaneous variations of the interface inclination and crystal misorientation generally involve simultaneous variations of a number of the variables that we have employed above. However, special trajectories exist in such situations where the changes in boundary geometry can be expressed conveniently in terms of a single geometric variable. For example, we may fix the mean boundary plane (Section 1.4.3.1) and the rotation axis and create a series of symmetric (or asymmetric) tilt boundaries by varying the single geometrical parameter, $\theta_{\text{tilt}}$, i.e. the total angle of rotation or 'tilt angle'. Measurements of $\sigma$ as a function of $\theta_{\text{tilt}}$ for several series of symmetric tilt boundaries are shown in Fig. 5.23. Again, the large energy cusps at the extremes and the smaller intermediate cusps can be understood on the basis of primary and secondary dislocation models (Section 2.10.3). Numerous calculations of $\sigma$ as a function of $\theta_{\text{tilt}}$ have also been performed as seen, for example, in Fig. 5.21(a), where the curve along the $\theta_{\text{twist}} = 180°$ axis turns out to represent $\sigma$ as a function of $\theta_{\text{tilt}}$ for symmetric tilt boundaries having the mean boundary plane {110} (i.e., $m = \langle 110 \rangle$ and $\hat{\rho} = 2^{-\frac{1}{2}}\langle 110 \rangle$. We remark that the 3D diagram in Fig. 5.21(b) may be regarded as a display of the energy of the particular class of boundaries produced by first generating a symmetric tilt boundary by a $\theta_{\text{tilt}}$ rotation around $\hat{\rho} = 2^{-\frac{1}{2}}\langle 110 \rangle$ (while holding $m = \langle 110 \rangle$) and then applying a negative $\theta_{\text{twist}}$ rotation around the boundary normal, $\hat{n}$. As we have already seen, all of the boundaries on the trajectories parallel to $\theta_{\text{twist}}$ may be classified as pure twist boundaries with twist angle $\theta_{\text{twist}}$. The ambiguity that occurs in describing these boundaries as either pure twist, mixed, or pure tilt (when they lie on the $\theta_{\text{tilt}}$ axis) is typical of the ambiguities (already mentioned) which can arise in assigning geometric variables to a boundary. For example, the boundary at $A$ on the $\theta_{\text{tilt}}$ axis may be regarded as either a symmetric tilt boundary produced by a $\theta_{\text{tilt}} = 129.52°$ rotation around $\hat{\rho} = 2^{-\frac{1}{2}}\langle 110 \rangle$ (with $m = \langle 110 \rangle$), or a pure {113} twist boundary produced by a $\theta_{\text{twist}} = 180°$ rotation around $\langle 113 \rangle$. The approximate energy surface obtained by an interpolation of the results in Fig. 5.21(a) is displayed in Fig. 5.21(b).

## 5.9 CHEMICAL POTENTIALS AND DIFFUSION POTENTIALS, $M_i$, IN NON-UNIFORM SYSTEMS CONTAINING INTERFACES

### 5.9.1 Introduction

As discussed in Chapter 10, interfaces are often capable of acting as sources/sinks for diffusional fluxes of atoms. In non-uniform systems containing differences in stress, strain, and composition, thermodynamic forces are often present which tend to transport atoms from/to any interfaces which are present. In such cases the interfaces will tend to respond by acting as sources/sinks for the diffusing atoms. Examples include the coarsening of a distribution of small embedded second-phase particles (Section 10.4.2.1) where solute atoms generally diffuse to the interfaces enclosing the larger particles and away from the interfaces enclosing the smaller particles under a driving force associated with the decrease in interfacial energy which accompanies the overall process. Another example is the diffusional creep of polycrystalline material (Section 12.8.2.1), where atoms diffuse between interfaces which have different inclinations in the applied non-hydrostatic stress field. Here, the shape of the specimen changes as the diffusional mass transport occurs, and the driving force for this transport is associated with the work done by the applied stresses as the specimen changes shape.

The rate at which these processes can occur depends upon the ability of the interfaces to act as sources/sinks. When the source/sink action is easy and rapid, the sources/sinks will generally be able to maintain the concentrations of the diffusing species in their immediate vicinities close to those corresponding to local equilibrium. The transport kinetics then becomes 'diffusion-controlled' (see Section 10.2.1). Otherwise, when the source/sink action is relatively slow, the concentrations may deviate considerably from the local equilibrium values. The kinetics then becomes 'interface-controlled', and the overall rate of transport to/from the sources/sinks is reduced (Section 10.2.1).

In order to be able to analyse such situations we now develop a general formalism for describing the diffusional transport which will occur in such systems. This is accomplished by following Larché and Cahn (1973, 1978a,b, 1985) and first analysing such a system when it is in thermochemical equilibrium. This requires the introduction of an important new thermodynamic quantity, the 'diffusion potential', $M_i$, for a component which must be constant throughout the system when equilibrium prevails. When the system is out of equilibrium, gradients of the $M_i$ exist, and diffusion currents then develop which may be taken as linearly proportional to the gradients of the $M_i$. The gradient of $M_i$ is therefore a measure of the tendency of the component $i$ to redistribute throughout the system.

### 5.9.2 Analysis of system at equilibrium: introduction of the diffusion potential, $M_i$

We begin by analysing the equilibrium state of a relatively simple bicrystal system composed of an $\alpha$ phase and a $\beta$ phase separated by an interface A. Following Larché and Cahn (1973, 1978a,b, 1985), each phase may be non-hydrostatically stressed in an inhomogeneous manner and be composed of both substitutional and interstitial components which can be distributed non-uniformly. The interface lies entirely within the system, and each phase can possibly gain (lose) atoms at the interface in a manner depending upon the coherent or incoherent nature of the interface, as will be discussed below. No variations are allowed on the outer periphery of the system, and, for simplicity, any effects due to the excess thermodynamic properties of the interface are excluded. In essence, the interface is taken to be merely a mathematically defined surface where either abutting phase may possibly grow or shrink. A more complete analysis taking excess interface properties into account has been carried out recently by Leo and Serkerka (1989). However, the results are considerably more complicated and beyond the scope of the present book: the reader is referred to their work for a detailed discussion of these effects.

Our goal is to investigate the conditions which prevail when this non-uniform system is in elastomechanical equilibrium. Under the above conditions, the local internal energy per unit volume, $e$, of such a system can be expressed as a function of the entropy per unit volume, $s$, the strain (or deformation) and the compositions of the various components, which in the present section we represent by the symbols $c_i$ (moles per unit volume). Together these form a complete set of state variables. The total number of components is again $C$. Since the system is isolated, the condition of equilibrium is the usual one of minimum total energy subject to the constraint of constant total entropy and mass of all components. The problem of finding the conditions of equilibrium is, therefore, reduced to a classic variational problem in which Lagrange multipliers are introduced in order to deal with the various constraints (Margenau and Murphy 1956). In order to treat such a system in a rigorous manner, it is necessary to introduce a book-keeping system to keep track of the substitutional and interstitial species. This can be accomplished by using a 'network' model for each crystal in which the substitutional sites in each crystal structure serve as the nodes of a network. Substitutional species can, therefore, occupy various sites on the nodes of the network, whereas interstitial species (generally present at relatively low concentrations) can occupy the various interstices. An important conservation restriction applies to the network; all substitutional network sites must be occupied by the substitutional species. If vacancies are present, they must be considered as a *bona fide* substitutional species. In addition, network sites cannot be removed from or added to the network of either crystal except possibly at the interface which then acts as a source/sink for network sites, or at lattice dislocations distributed throughout the bulk which can climb and act as sources/sinks (Section 10.3.2.1).

The deformation in each crystal is defined relative to a uniform reference configuration of the network which may be taken to correspond to the ideal crystal in a state of zero strain. The actual strained configuration, $r$, is then related to the corresponding reference configuration, $r'$, by

$$r = r(r') = r' + u(r') \tag{5.128}$$

where $u(r')$ is the displacement from the reference configuration. (In the following, primed quantities will be referred to the reference configuration, whereas non-primed quantities will be referred to the actual configuration.) The deformation is described by the tensor, $F_{ij}$:

$$F_{ij} = \frac{\partial r_i}{\partial r_j'} = \delta_{ij} + \frac{\partial u_i}{\partial r_j'}, \tag{5.129}$$

so that an infinitesimal vector $dr$ in the actual configuration is related to its corresponding vector $dr'$ in the reference configuration by

$$dr = F_{ij} \cdot dr'. \tag{5.130}$$

Since the system is generally inhomogeneously strained, the various density quantities (per unit volume in the actual configuration) contain strain effects. Therefore, it is more convenient to work with quantities referred to the uniform reference configuration, and in this framework we may finally write

$$e' = e'(F_{ij}, s', c_1', c_2' \ldots c_C').$$ (5.131)

A variation of the total energy of this system may now be written in the form

$$\delta E = \int \left[ \sum_{ij} \left( \frac{\partial e'^\alpha}{\partial F_{ij}} \right) \delta F_{ij} + \left( \frac{\partial e'^\alpha}{\partial s'} \right) \delta s' + \sum_{i=1}^{C} \frac{\partial e'^\alpha}{\partial c_i'} \delta c_i' \right] dV'$$

$$+ \int \left[ \sum_{ij} \left( \frac{\partial e'^\beta}{\partial F_{ij}} \right) \delta F_{ij} + \left( \frac{\partial e'^\beta}{\partial s'} \right) \delta s' + \sum_{i=1}^{C} \frac{\partial e'^\beta}{\partial c_i'} \delta c_i' \right] dV'$$ (5.132)

$$+ \int e'^\alpha \cdot \delta y'^\alpha \cdot dA' + \int e'^\beta \cdot \delta y'^\beta \cdot dA'.$$

The first two integrals are taken over the volumes of the $\alpha$ and $\beta$ phases, whereas the last two are taken over the area, $A'$, of the interface. These latter two integrals represent the change in total energy due to a variation corresponding to the infinitesimal displacement of the interface which adds local volumes $\delta y'^\alpha \, dA'$ to the $\alpha$ phase and $\delta y'^\beta \, dA'$ to the $\beta$ phase. The distances $\delta y'^\alpha$ and $\delta y'^\beta$ are measured along the normal vectors to the interface $\hat{n}'^\alpha$ and $\hat{n}'^\beta$, where $\hat{n}'^\alpha$ faces towards the $\beta$ phase and $\hat{n}'^\beta$ towards the $\alpha$ phase. Consider now the constraints on this variation. Conservation of the total entropy and mass of each component over the entire system now requires that

$$\delta S = \int \left( \delta s'^\alpha + \delta s'^\beta \right) dV' + \int \left( s'^\alpha \delta y'^\alpha + s'^\beta \delta y'^\beta \right) dA' = 0$$ (5.133)

and

$$\int \left( \delta c_i'^\alpha + \delta c_i'^\beta \right) dV' + \int \left( c_i'^\alpha \delta y'^\alpha + c_i'^\beta \delta y'^\beta \right) dA' = 0 \qquad \text{(for } i = 1, 2, \ldots C).$$ (5.134)

If $L$ components are interstitial components and $C$-$L$ are substitutional, there is a network constraint on the substitutional densities for each phase of the form

$$\sum_{i=L+1}^{C} c_i' = c_o' = \text{constant}.$$ (5.135)

Also, adding all equations of the type of eqn (5.134) for the substitutional species (i.e. for $i = L + 1$ to $C$) and using eqn (5.135), we obtain

$$\int \left( c_o'^\alpha \delta y'^\alpha + c_o'^\beta \delta y'^\beta \right) dA' = 0.$$ (5.136)

which may be used in place of eqn (5.134) for $i = C$. Finally, we must introduce any further constraints associated with the displacements at the interface. These will be quite different, depending upon whether the interface is incoherent or coherent and will lead to quite different equilibrium conditions at the interface.

### 5.9.2.1 Incoherent interface

When the interface is incoherent (and general with respect to all degrees of freedom), there is no lattice (network) matching requirement across the interface, and a layer of material may therefore be added/removed to/from either the $\alpha$ or $\beta$ phase independently in the variation (see Section 9.2.4). Using the previous notation, the local vector displacements of the two phases in the reference configuration are then $\delta y'^{\alpha} \hat{n}'^{\alpha}$ and $\delta y'^{\beta} \hat{n}'^{\beta}$, respectively. Using eqn (5.130), these local displacements in the actual configuration are $F_{ij}^{\alpha} \cdot \hat{n}'^{\alpha} \delta y'^{\alpha}$ and $F_{ij}^{\beta} \cdot \hat{n}'^{\beta} \delta y'^{\beta}$, respectively. In order to find the overall displacements in the actual configuration, $\delta s^{\alpha}$ and $\delta s^{\beta}$, we must add in the displacements $\delta u^{\alpha}$ and $\delta u^{\beta}$ due to the long-range accumulation of strain in each phase. Therefore, finally,

$$\left. \begin{aligned} \delta s^{\alpha} &= \delta u^{\alpha} + F_{ij}^{\alpha} \cdot \hat{n}'^{\alpha} \delta y'^{\alpha} \\ \delta s^{\beta} &= \delta u^{\beta} + F_{ij}^{\beta} \cdot \hat{n}'^{\beta} \delta y'^{\beta} \end{aligned} \right\}. \tag{5.137}$$

For a general incoherent interface, the $\gamma$-surface is flat (see Section 4.3.1), and there is no resistance to interface sliding (see Section 12.8.1). The two phases may therefore freely slide past each other in directions parallel to the interface. Since the two phases must remain in contact, the only constraint on $\delta s^{\alpha}$ and $\delta s^{\beta}$ is, therefore,

$$\delta s^{\alpha} \cdot \hat{n}^{\alpha} = \delta s^{\beta} \cdot \hat{n}^{\alpha} \tag{5.138}$$

where $\hat{n}^{\alpha} = -\hat{n}^{\beta}$ is normal to the interface in the actual configuration.

In order to solve the variational problem for the two bulk phases, we now set $\delta E = 0$ and introduce the constant Lagrange multipliers $\theta$, $M_1, M_2 \ldots M_{C-1}, M_C$ which are associated with the constraints represented by eqn (5.133), eqns (5.134) for $i = 1, 2 \ldots$ $(C - 1)$, and eqn (5.136) respectively. The details of the solution are outlined by Larché and Cahn (1973, 1978b), and the following results are obtained throughout the volumes of the two bulk phases at equilibrium:

$$\theta = \partial e'/\partial s' = \text{constant} \tag{5.139}$$

$$\text{div}' \left( \partial e'/\partial F \right) = 0 \tag{5.140}$$

$$M_i = \begin{cases} \partial e'/\partial c_i' = \text{constant (for } i = 1, 2, \ldots L) & \tag{5.141a} \\ \partial e'/\partial c_i' - \partial e'/\partial c_C' = \text{constant (for } i = [L+1] \ldots [C-1]). & \tag{5.141b} \end{cases}$$

Equation (5.139) identifies $\theta$ with the temperature which, as expected, must be constant everywhere in the system at equilibrium. Equation (5.140) is recognized as the well-known equation of mechanical equilibrium (Sokolnikoff 1956) which, of course, must be satisfied everywhere. Equation (5.141a) shows that the quantity $M_i = \partial e'/\partial c_i'$ must be constant throughout the bulk phases for each interstitial component. The quantity $\partial e'/\partial c_i'$ is identified as proportional to the chemical potential, $\mu_i$, of the interstitial component $i$ (i.e. $\partial e'/\partial c_i' = N_o \mu_i$ as discussed further below), and we therefore conclude that the chemical potential of each interstitial component must be constant everywhere. In contrast to this, eqn (5.141b) indicates that it is only the difference quantity $M_i = (\partial e'/\partial c_i' - \partial e'/\partial c_C')$ for each of the $(C - L - 1)$ substitutional components corresponding to $i = [L + 1] \ldots [C - 1]$ which must be constant throughout the two bulk phases at equilibrium. Also, no general condition specifying the Lagrange multiplier, $M_C$, for the $C$th species is found, and it therefore remains unidentified throughout the bulk phases.

These results may be understood intuitively on the basis of the physical restrictions imposed on the substitutional components by the network. Substitutional components cannot be added independently at interior points in the bulk networks. If a substitutional component is added in the interior, then network site conservation requires that another be removed. Therefore, individual quantities such as $\partial e'/\partial c_i'$ have no physical meaning for the substitutional components within the bulk phases. On the other hand, the difference quantity $M_i = (\partial e'/\partial c_i' - \partial e'/\partial c_C')$ has a clear physical meaning; it is the energy change which occurs when component $i$ is replaced by component $C$. At equilibrium, this quantity might be expected to be a constant for just $(C - L - 1)$ of the substitutional components, as was, indeed, found above, since only a total of $(C - L - 1)$ quantities can be identified with physical processes in which one of the $(C - L)$ substitutional components present in the system is replaced by another on a site. In contrast to this, there are no restrictions on the addition of interstitial components to the network at interior interstitial sites, and the quantity $M_i = \partial e'/\partial c_i'$ has a clear physical meaning which allows it to be identified as proportional to the chemical potential of the interstitial component $i$ throughout the bulk of each phase, i.e $M_i = N_o\mu_i$.

The situation at the incoherent interface differs from that throughout the bulk phases, however, since, as we have seen, special constraints apply there which differ from those for the bulk. In general, the network constraints are less restrictive since network sites can be independently added (subtracted) locally at either phase at such an interface, and it is only necessary to maintain the two phases in contact with no limitations being put on possible sliding (eqn (5.138)). Larché and Cahn (1978b, 1985) determined the equilibrium conditions which apply at the interface from further analysis of the variational problem and found that

$$\tau_{ij}^\alpha n_j^\alpha = w^\alpha n_i^\alpha; \quad \tau_{ij}^\beta n_j^\beta = w^\beta n_i^\beta \tag{5.142}$$

$$w^\alpha = w^\beta, \tag{5.143}$$

where $\tau_{ij}$ is the stress tensor, and the quantity $w$ corresponds to the grand potential, i.e.

$$w = e - Ts - \sum_{i=1}^{C-1} M_i c_i - M_C c_o. \tag{5.144}$$

As might be expected at equilibrium, eqns (5.142) and (5.143) require that all shear stresses parallel to the interface must be zero and that the stress in each crystal normal to the interface must be equal to one another. The constant multiplier $M_C$ now appears in the results, and furthermore, it can be associated with the chemical potential of component $C$, i.e. $M_C = N_o\mu_C$. A chemical potential can therefore be found for each substitutional component at the interface, since the above set of equations is independent of which component is chosen as the $C$th component. Solving eqns (5.142) for $w^\alpha$ and $w^\beta$, we have

$$w^\alpha = \tau_{ij}^\alpha n_i^\alpha n_j^\alpha; \quad w^\beta = \tau_{ij}^\beta n_i^\beta n_j^\beta, \tag{5.145}$$

and using these results in eqn (5.144) with $M_C = N_o\mu_C$, the chemical potentials are given by

$$\mu_C = \left[e^\alpha - Ts^\alpha - \sum_{i=1}^{C-1} M_i c_i^\alpha - \tau_{ij}^\alpha n_i^\alpha n_j^\alpha\right]/c_o N_o$$

$$= \left[e^\beta - Ts^\beta - \sum_{i=1}^{C-1} M_i c_i^\beta - \tau_{ij}^\beta n_i^\beta n_j^\beta\right]/c_o N_o. \tag{5.146}$$

This result may be attributed to the fact that the quantities $\partial e'/\partial c_i'$ have clear physical meanings at the interface, since the substitutional components can be independently added or removed there. Each of these quantities can, therefore, be identified with a chemical potential, i.e $\partial e'/\partial c_i' = N_o\mu_i$. Therefore, $M_i = N_o(\mu_i - \mu_C)$ at the interface.

At equilibrium, all of the $M_i$ 's and $\mu_i$'s are required to be constant along the incoherent interface.

If readily climbing dislocations are distributed throughout the bulk phases a generally similar analysis would show that the chemical potential of each substitutional component is also defined at each dislocation, since, again, substitutional components can be independently added, or removed, at such defects. In such a case the chemical potential of each component would be established at defined values in their vicinities throughout the bulk.

### 5.9.2.2  *Coherent interface*

When the interface is coherent, lattice (network) matching must be preserved across the interface. The network constraints are therefore considerably more restrictive and resemble those within the bulk of the network in the absence of dislocations. Sites must be conserved (i.e. $\hat{n}'^\alpha \delta y'^\alpha = \hat{n}'^\beta \delta y'^\beta$) and substitutional components cannot be independently added to either phase at a coherent interface in contrast to the situation at an incoherent interface (see Section 10.3.3.1). Also, there can be no interface sliding (see Section 12.8.1.1) and, therefore,

$$\delta s^\alpha = \delta s^\beta \tag{5.147}$$

in the actual configuration. The solution of the variational problem with these constraints (Larché and Cahn 1978b, 1985) produces results for the bulk phases which are identical to those obtained previously for the bulk phases adjoining an incoherent interface. At the interface, however, it is no longer possible to define individual chemical potentials for the substitutional components, and it is only the difference quantity $M_i = (\partial e'/\partial c_i' - \partial e'/\partial c_C')$ which must be constant for all but the $C$th substitutional component.

### 5.9.2.3  *Summary*

We may summarize the previous results as follows:

(1) A non-uniform network solid (i.e. a crystalline solid) containing an interface can reach an equilibrium state under non-hydrostatic stress. The conditions at the interface are different and separable from those in the interiors of the abutting crystals, and this has a direct influence on the chemical potentials which can be defined throughout the system.

(2) Chemical potentials, $\mu_i = M_i/N_o$ for individual interstitial components can be defined everywhere in the system, and at equilibrium must be constant everywhere.

(3) Chemical potentials for individual substitutional components are undefined within the bulk phases unless dislocation sources/sinks are distributed throughout the bulk. In the absence of dislocations, only the difference quantity $M_i = (\partial e'/\partial c_i' - \partial e'/\partial c_C')$, i.e. the diffusion potential corresponding to the energy change produced by replacing component $C$ by component $i$, can be defined. When lattice dislocations are distributed throughout the bulk phases, the chemical potentials of the individual substitutional components can be defined in their vicinities. At equilibrium, $M_i$ must be constant everywhere for each substitutional component.

(4) Chemical potentials for all individual substitutional components can be defined at incoherent interfaces but not at coherent interfaces. Under equilibrium conditions at an incoherent interface, these chemical potentials, $\mu_i = (1/N_o)\partial e'/\partial c_i'$, as well as the difference quantities, $M_i = (\partial e'/\partial c_i' - \partial e'/\partial c_C') = N_o(\mu_i - \mu_C)$, must be constant everywhere on the interface.

(5) At a coherent interface the situation is similar to that in the bulk in the absence of dislocations, and only the difference quantities, $M_i$, can be defined for the substitutional components.

Larché and Cahn have termed the $M_i$ difference quantities 'diffusion potentials', since when the system is out of equilibrium, they will not be equal everywhere, and diffusion fluxes of the components will then be generated in an effort to establish equilibrium. In fact, we can assume phenomenologically (Cahn and Larché 1983) that the diffusion fluxes will be linearly proportional to the gradients of the diffusion potentials. This allows a formulation of the diffusional transport as shown below.

### 5.9.3 Diffusional transport in non-equilibrium systems

We assume phenomenologically (Cahn and Larché 1983) that the diffusion fluxes which occur in non-equilibrium systems are proportional to the gradients of the diffusion potentials, $M_i$, introduced in the previous section. The simplest possible case is one in which a single mobile interstitial component is present. The interstitial flux (measured relative to the network as a coordinate system) is then given by

$$J_1 = -B_1 \cdot \nabla M_1. \tag{5.148}$$

In this case, the diffusion potential, $M_1$, is, as stated in Section 5.9.2.3, proportional to a chemical potential, and the flux is then simply proportional to the gradient of $M_i$. The proportionality factor, i.e. the $B$ quantity, is termed a 'mobility'.

For a system consisting exclusively of $C$ substitutional components, we may write the general set of linear equations

$$J_j = -\sum_{i=1}^{C} B_{ji} \cdot \nabla M_i, \tag{5.149}$$

where, as always, $M_i = (\partial e'/\partial c_i' - \partial e'/\partial c_C')$. In addition, the relation

$$\sum_{j=1}^{C} J_j = 0, \tag{5.150}$$

must be satisfied, since the network restraint requires that every jump of a substitutional component be compensated by an equal and opposite jump of a substitutional component.

For the relatively simple case of a two-component system consisting of a single atom type (component A) and vacancies (component V), the vacancy may be taken as the $C$th component, and eqn (5.149) then assumes the form

$$\left.\begin{aligned} J_A &= -B_{AA} \cdot \nabla M_A - B_{AV} \cdot \nabla M_V \\ J_v &= -B_{va} \cdot \nabla M_1 - B_{vv} \cdot \nabla M_v \end{aligned}\right\}. \tag{5.151}$$

However, since $M_V = 0$, and $J_A + J_V = 0$,

$$J_A = -J_V = -B_{AA} \cdot \nabla M_A. \tag{5.152}$$

The quantities $B_{AA}$ and $M_A$ are generally functions of the local composition and stress, and, as discussed further below, diffusion problems concerned with the redistribution of the atoms and vacancies throughout the system can then be solved through the use of eqn (5.152) and any relevant boundary conditions on the $M_A$ which hold at the interfaces and dislocations present which can act as sources/sinks.

It is important to note that all problems concerned with the distribution or redistribution of substitutional components (under either equilibrium or non-equilibrium conditions) under the influence of non-hydrostatic stresses can be solved through the use of the diffusion potentials, $M_i$. Larché and Cahn (1985) discuss the thermochemical quantities which are needed and show how a large range of problems may be solved. They emphasize that it is generally unnecessary to be able to identify the individual chemical potentials in the interior of the network. The inability to define chemical potentials within the network is, therefore, of no consequence.

When interfaces are present which act as highly efficient sources/sinks, local equilibrium will be maintained at the interfaces, and the $M_i$ in the direct vicinities of the interfaces will assume local equilibrium values which will be functions of local stress and composition. If differences in these $M_i$ exist throughout the system, diffusion will occur, and the transport problem in the absence of significant densities of lattice dislocation sources/sinks will then be a boundary value diffusion problem in which the values of $M_i$ along the interfaces are fixed by the local equilibrium conditions. On the other hand, when the interfaces act as relatively poor sources/sinks, detailed models must be constructed in order to account for the source/sink action as discussed in Chapter 10.

# REFERENCES

Brooks, H. (1952). In *Metal interfaces*, p. 20. American Society for Metals, Metals Park, Ohio.
Broughton, J. Q. and Gilmer, G. H. (1983). *Acta Metall.*, **31**, 845.
Cabrera, N. (1963). In *Symposium on properties of surfaces*, ASTM Tech. Publ. No. 340, p. 24. American Society for Testing and Materials, Philadelphia, Pennysylvania.
Cahn, J. W. (1979). In *Interfacial segregation* (eds W. C. Johnson and J. M. Blakely) p. 3. American Society for Metals, Metals Park, Ohio.
Cahn, J. W. (1980). *Acta Metall.*, **28**, 1333.
Cahn, J. W. and Hoffman, D. W. (1974). *Acta Metall.*, **22**, 1205.
Cahn, J. W. and Larché, F. (1982). *Acta Metall.*, **30**, 51.
Cahn, J. W. and Larché, F. C. (1983). *Scripta Metall.*, **17**, 927.
Chan, S. W. and Balluffi, R. W. (1985). *Acta Metall.*, **33**, 1113.
Chan, S. W. and Balluffi, R. W. (1986). *Acta Metall.*, **34**, 2191.
Chen, L.-Q. and Kalonji, G. (1989). *Phil. Mag. A*, **60**, 525.
Chen, L.-Q. and Kalonji, G. (1992). *Phil. Mag. A*, **66**, 11.
Dhalenne, G., Dechamps, M., and Revcolevschi, A. (1983). In *Character of grain boundaries*, Advances in Ceramics Vol. 6, p. 139. American Ceramics Society, Columbus, Ohio.
Eshelby, J. D. (1957). *Proc. Roy. Soc. (London)*, A, **241**, 376.
Ference, T. G. and Balluffi, R. W. (1988). *Scripta Metall.*, **22**, 1929.
Ference, T. G. and Balluffi, R. W. (1989). Unpublished research.
Frank, F. C. (1962). In *Metal surfaces*, p. 1. American Society for Metals, Metals Park, Ohio.
Frank, F. C. and van der Merwe, J. H. (1949). *Proc. R. Soc. London*, **198**, 216.
Funk, E. R., Udin, H., and Wulff, J. (1951). *Trans. AIME*, **191**, 1206.
Gao, Y., Dregia, S. A., and Shewmon, P. G. (1989). *Acta Metall.*, **37**, 1627.
Gaskell, D. R. (1983). In *Physical metallurgy* (eds R. W. Cahn and P. Haasen). North-Holland, Amsterdam.
Gibbs, J. W. (1875–78). *Trans. Conn. Acad.*, **III**, 108–248, 343–524.
Gibbs, J. W. (1957). *Collected works*, Vol. 1. Yale University Press, New Haven, Connecticut.
Herring, C. (1951a). *Phys. Rev.*, **82**, 87.
Herring, C. (1951b). In *The physics of powder metallurgy* (ed. W. E. Kingston) p. 143. McGraw-Hill, New York.

Herring, C. (1953). In *Structure and properties of solid surfaces* (eds R. Gomer and C. S. Smith) p. 5. University of Chicago Press, Chicago, Illinois.

Hildebrand, F. B. (1962). *Advanced calculus for applications*, p. 287. Prentice-Hall, Englewood Cliffs, New Jersey.

Hoffman, D. W. and Cahn, J. W. (1972). *Surf. Sci.*, **31**, 368.

Hondros, E. D. and Seah, M. P. (1983). In *Physical metallurgy* (eds R. W. Cahn and P. Haasen) p. 855. North-Holland, Amsterdam.

Hsieh, T. E. and Balluffi, R. W. (1989). *Acta Metall.*, **37**, 2133.

Johnson, C. A. (1965). *Surf. Sci.*, **3**, 429.

Johnson, W. C. and Cahn, J. W. (1984). *Acta Metall.*, **32**, 1925.

Kalonji, G. and Cahn, J. W. (1982). *J. de Physique*, **43**, Colloq. C6, Supple. au no 12, C6-25.

Landau, L. D. (1965). Collected papers of L. D. Landau (ed. D. Ter Haar) p. 540. Gordon and Breach, New York.

Landau, L. D. and Lifshitz, E. M. (1958). *Statistical physics*, p. 460. Pergamon Press, London.

Larché, F. C. and Cahn, J. W. (1973). *Acta Metall.*, **21**, 1051.

Larché, F. C. and Cahn, J. W. (1978a). *Acta Metall.*, **26**, 53.

Larché, F. C. and Cahn, J. W. (1978b). *Acta Metall.*, **26**, 1579.

Larché, F. C. and Cahn, J. W. (1985). *Acta Metall.*, **33**, 331.

Leo, P. H. and Sekerka, R. F. (1989). *Acta Metall.*, **37**, 3119.

Lomer, W. M. (1957). *Phil. Mag.*, **2**, 1053.

Margenau, H. and Murphy, G. M. (1956). The mathematics of physics and chemistry, p. 198. Van Nostrand, New York.

Merkle, K. L. (1989). *Scripta Metall.*, **23**, 1487.

Merkle, K. L. and Shao, B. (1988). *Mats. Res. Soc. Symp. Proc.*, **122**, 69.

Merkle, K. L. and Smith, D. J. (1987). *Phys. Rev. Lett.*, **59**, 2887.

Mori, T., Miura, H., Tokita, T., Haji, J., and Kato, M. (1988). *Phil. Mag. Letters*, **58**, 11.

Mullins, W. W. (1962). In *Metal surfaces*, p. 17. American Society for Metals, Metals Park, Ohio.

Murr, L. E. (1975). *Interfacial phenomena in metals and alloys*. Addison-Wesley, London.

Needs, R. J. and Godfrey, M. J. (1987). *Physica Scripta*, **T19**, 391.

Omar, R. and Mykura, H. (1988). *Mats. Res. Soc. Symp. Proc.*, **122**, 61.

Otsuki, A. and Mizuno, M. (1986). In *Grain boundary structure and related phenomena*, Proc. of JIMIS-4, Suppl. to *Trans. Japan Inst. of Mets.*, **27**, 789.

Papon, A. M., Petit, M., Silvestre, G., and Bacmann, J. J. (1982). In *Grain boundaries in semiconductors* (eds H. J. Leamy, G. E. Pike, and C. H. Seager), Materials Research Society Symposium, Vol 5, p. 27. North-Holland, New York.

Pelton, A. D. (1983). In *Physical metallurgy* (eds R. W. Cahn and P. Haasen) p. 327. North-Holland, Amsterdam.

Pond, R. C. and Dibley, P. E. (1990). *J. Physique*, **51**, C1-25.

Pond, R. C. and Vitek, V. (1977). *Proc. R. Soc. London B*, **357**, 453.

Read, W. T. and Shockley, W. (1950). *Phys. Rev.*, **78**, 275.

Reif, F. (1965). *Fundamentals of statistical and thermal physics*. McGraw-Hill, New York.

Shewmon, P. G. and Robertson, W. M. (1963). In *Metal surfaces*, p. 67. American Society for Metals, Metals Park, Ohio.

Shuttleworth, R. (1950). *Proc. Phys. Soc. A*, **63**, 444.

Sokolnikoff, I. S. (1956). *Mathematical theory of elasticity*. McGraw-Hill, New York.

Sommerfeld, A. (1950). *Mechanics of deformable bodies*. Academic Press, New York.

Udin, H., Shaler, A. J., and Wulff, J. (1949). *Trans. AIME*, **185**, 186.

Weatherburn, C. (1927). *Differential geometry of three dimensions*. Cambridge University Press, Cambridge.

Wolf, D. (1990a). *Acta Metall. Mater.*, **38**, 791.

Wolf, D. (1990b). *J. Mater. Res.*, **5**, 1708.

# 6

# Interface phases and phase transitions

## 6.1 INTRODUCTION

In this chapter, we develop the concept that interfaces can be regarded generally as two-dimensional structures composed of one or more interface phases, and that a rich variety of transitions between these interface phases is possible. We have already seen in Section 5.6.3 that an initially flat interface with unit normal $\hat{n}$ can, under certain circumstances, achieve a state of minimum free energy (per unit area projected along $\hat{n}$) by breaking up into a faceted structure composed of one or more different types of flat patches, each possessing a distinct structure. As described in the next section, multipartite interface structures consisting of patches of different structure can arise in other situations when they lead to minimum interface free energy (per unit projected area). Furthermore, transitions between the structures, or different types of structures, which are present can occur when the thermodynamic variables controlling the system are changed. We have also seen that any flat region of interface of uniform structure can be described thermodynamically by a unique relation of the type represented by eqn (5.9) which may be regarded as an 'interface Gibbs–Duhem' type relation. In view of this, we are justified in treating such a distinct interface region as an individual interface phase in direct analogy with the usual treatment of bulk phases. We also recall (Section 5.6.3) that there is an exact analogy between the breakup of an interface into individual facets and the breakup of a bulk solid solution into individual coexisting bulk phases. A faceted interface, therefore, may be regarded as a multiphase structure consisting of individual coexisting phases (facets). A question about the concept of an interface phase naturally arises when an interface region is continuously curved since the structure then changes continuously. However, as shown below, such a region can be regarded as a single phase as long as the curvature is continuous and no discontinuities of any type are present. Such a situation is analogous to a single phase solid solution of continuously varying composition.

It should be pointed out that the subject of distinct interface phases and their phase transitions is well-developed for free crystal surfaces (Blakely and Thapliyal 1979; Blakely 1986, 1988), and that many similarities exist between these phenomena at free surfaces and interfaces in crystalline materials. Both free surfaces and interfaces can possess uniform structures of single phase or multipartite structures consisting of patches of different phase under appropriate conditions. Various transitions between structures of these types can occur when the thermodynamic variables controlling the overall system are changed. The study of such phase transitions at interfaces in crystalline materials has lagged behind comparable studies at free surfaces for a variety of reasons. Firstly, interfaces have a greater potential for complexity since the number of the thermodynamic variables, which, as we have seen, include the geometrical variables, is greater. Also, interfaces in crystalline materials are 'buried', and, hence, are less amenable to the wide range of powerful electron-optical probes which have been successfully applied to free surfaces (Woodruff and Delchar 1988, Walls 1989). In addition, ideal free surfaces are more easily prepared, and surface cleaning and deposition techniques using ultra-high vacuum are available which can be used to produce surfaces which are either ideally clean or possess well controlled concentrations of various chemical components. It is only fairly recently, therefore, that the study of phases and phase transitions at interfaces has emerged as an established field.

In the following we describe some of the main features of a variety of interface phase transitions. However, because of the huge number of transitions which are possible in the great variety of different types of interfaces which exist (see Table 4.1, p. 00), complete coverage of the field is impossible. Reviews containing additional information include: Hart (1972), Cahn (1982), Rottman (1988a, b), Pontikis (1988), Rabkin *et al.* (1991), and Rottman (1992).

## 6.2 INTERFACE PHASE EQUILIBRIA

We proceed by developing further the results in Chapter 5 in order to represent a general system containing a possible multiphase interface by a set of equations consisting of one 'interface Gibbs–Duhem' type equation, e.g. eqn (5.11a), for each interface phase present and one Gibbs–Duhem equation, e.g. eqn (5.11b), for each bulk volume phase present. We assume that capillarity effects due to any curvature can be neglected and that each phase is uniform and free of strain. However, we want to include the effects of the geometrical variables, and they must be added to the interface equations in the same way that they were added earlier in Section 5.5 in order to obtain eqn (5.71). We therefore have the set of equations

$$
\left.
\begin{aligned}
&\frac{\partial \sigma'}{\partial n_1'}\mathrm{d}n_1' + \frac{\partial \sigma'}{\partial n_2'}\mathrm{d}n_2' + \frac{\partial \sigma'}{\partial \theta'}\mathrm{d}\theta' + \frac{\partial \sigma'}{\partial \rho'_1}\mathrm{d}\rho_1' + \frac{\partial \sigma'}{\partial \rho_2'}\mathrm{d}\rho_2' \\
&\qquad\qquad - \mathrm{d}\sigma' - [S']\,\mathrm{d}T + [V']\,\mathrm{d}P - \sum_{i=1}^{C} [N_i']\,\mathrm{d}\mu_i = 0 \\[6pt]
&\frac{\partial \sigma''}{\partial n_1''}\mathrm{d}n_1'' + \frac{\partial \sigma''}{\partial n_2''}\mathrm{d}n_2'' + \frac{\partial \sigma''}{\partial \theta''}\mathrm{d}\sigma'' + \frac{\partial \sigma''}{\partial \rho_1''}\mathrm{d}\rho_1'' + \frac{\partial \sigma''}{\partial \rho_2''}\mathrm{d}\rho_2'' \\
&\qquad\qquad - \mathrm{d}\sigma'' - [S'']\,\mathrm{d}T + [V'']\,\mathrm{d}P - \sum_{i=1}^{C} [N_i'']\,\mathrm{d}\mu_i = 0 \\[6pt]
&\vdots \qquad\qquad\qquad\qquad\qquad\qquad \vdots \\[6pt]
&0 + 0 + 0 + 0 + 0 + 0 - S^{\alpha}\,\mathrm{d}T + V^{\alpha}\,\mathrm{d}P - \sum_{i=1}^{C} N_i^{\alpha}\,\mathrm{d}\mu_i = 0 \\[6pt]
&0 + 0 + 0 + 0 + 0 + 0 - S^{\beta}\,\mathrm{d}T + V^{\beta}\,\mathrm{d}P - \sum_{i=1}^{C} N_i^{\beta}\,\mathrm{d}\mu_i = 0
\end{aligned}
\right\} . \tag{6.1}
$$

Note that in eqns (6.1) we have assumed equilibrium between all coexisting interface and bulk phases and have set $T$, $P$, and the $\mu_i$'s equal throughout all parts of the system, as we did, for example, in writing eqns (5.11). We shall continue with this assumption throughout the present chapter. Cases in actual systems where this may not be true because of inadequate time for equilibration are discussed in Section 6.3.2. Equations (6.1) allow us to define a multidimensional phase space whose coordinates are the independent variables (i.e. 'state' variables) which must be specified to fix the state of the entire system. Phase diagrams and phase rules for the coexisting bulk and interface phases may then be established.

For the simplest case of a single phase interface in a single bulk volume phase, the above equations reduce to only two equations containing $C + 8$ variables, i.e. $T, P$, the $\mu_i$'s, $n_1, n_2$, $\theta, \rho_1$, $\rho_2$, and $\sigma$. The number of independent variables, $d_F$, i.e. the number of variables minus the number of equations, is then $d_F = C + 6$, which we might choose, for example, as $T$, the $\mu_i$'s, $n_1, n_2, \theta, \rho_1$, and $\rho_2$. If these variables are specified, all remaining variables are therefore automatically fixed. These $C + 6$ variables are therefore the state variables of the system, and we may visualize (Cahn 1982) a $C + 6$-dimensional phase space for the system in which each point represents a different state of the system. If we fix $T$, the $\mu_i$'s, and the misorientation (i.e. $\theta, \rho_1$, and $\rho_2$), the two inclination variables $n_1$ and $n_2$ remain independent. This means that a single-phase interface can be arbitrarily curved when all of the other state variables are fixed. This result is consistent with regarding a continuously curved interface as a single phase.

We can now proceed to more complex cases where the interface lying in a single bulk volume phase may consist of more than one interface phase, such as illustrated schematically in Fig. 5.10 (p. 00). The cases shown include the faceted case in Fig. 10(b) (already introduced in Chapter 5), the dissociated case in Fig. 10(c) (which is discussed below in Section 6.3.1.3), and also the situation in Fig. 10(d) where patches of two different interface phases are present lying parallel to the interface. In such cases, the detailed equilibrium structure of the interface, including the local misorientation and inclination of each interface phase, will correspond to a minimum of the overall free energy of the interface per unit area projected along its average normal, i.e. $\bar{n}$. This is expected to be a unique function of $T, P$, the $\mu_i$'s, and the geometrical parameters necessary to describe the interface as a whole, i.e. its average inclination (specified here by $\bar{n}_1$ and $\bar{n}_2$) and the overall misorientation across it (specified by $\theta, \rho_1$, and $\rho_2$). For example, for all of the boundaries illustrated in Fig. 5.10, $\bar{n}_1$ and $\bar{n}_2$ correspond to the inclination of the horizontal plane while $\theta, \rho_1$, and $\rho_2$ describe the misorientation between crystals 1 and 2. The interface free energy, $\bar{\sigma}$, is therefore a function of $T, P$, the $\mu_i$'s, $\bar{n}_1, \bar{n}_2, \theta, \rho_1$, and $\rho_2$, and combining this relation involving $C + 8$ variables with the Gibbs–Duhem equation for the bulk volume phase, we again find a phase space for the system which involves $C + 6$ independent variables. In some regions of the phase space, the boundary may consist of one interface phase; in another region, several interface phases may be present.

Consider some specific examples: when the patches of the interface phases are parallel to the interface plane as in the case illustrated in Fig. 10(d), the misorientations and inclinations of the interface phases are identical. In addition, interface equilibrium requires that the free energies of the interface phases be the same; i.e. $\sigma' = \sigma'' = \bar{\sigma}$. For faceted interfaces, the $\sigma_P$ energy surface (see Fig. 5.12) is specified whenever $T, P$, the $\mu_i$'s and $\sigma, \rho_1$, and $\rho_2$ are specified. If $\bar{n}_1$ and $\bar{n}_2$ are now specified, all of the facets (phases) are uniquely determined by the common tangent plane construction. In addition, the proportions of the different facet areas projected on the average boundary plane are known, and hence $\bar{\sigma}$ is known.

Phase rules and phase diagrams may be obtained for the coexisting interface phases which are present in different regions of the $C + 6$ dimensional phase space just described (Cahn 1982). If the radius of curvature of the interface of average inclination is large compared with the extent of any of the interface patches (phases) which are present, eqns (6.1) should still hold locally. If $\phi^I$ interface phases coexist, we then have $\phi^I + 1$ equations with each interface equation containing $C + 8$ variables. However, as we have already pointed out, each of these sets of $C + 8$ variables is coupled to the $C + 8$ variables describing the overall interface (i.e $\bar{\sigma}$, $T$, $P$, $\mu_i$, $\bar{n}_1$, $\bar{n}_2$, $\theta$, $\rho_1$, $\rho_2$) by the requirement of the minimization of the overall interface free energy (per unit projected area along $\bar{n}$). Therefore, the number of variables which is actually independent is (Cahn 1982)

$$d_F^I = (C + 8) - (\phi^I + 1) = C + 7 - \phi^I. \tag{6.2}$$

Equation (6.2) is, therefore, a local interface phase rule which establishes the number of variables which can be varied independently in the presence of $\phi^I$ interface phases in a local region of the interface. Phase diagrams mapping out the regions of interface phase coexistence can be constructed which, of course, must be consistent with this interface phase rule. The examination of the complete $C + 6$ dimensional hyperspace is extremely complicated, however. It can be simplified somewhat by taking advantage of the fact that coexisting phases have the same T and $\mu_i$'s. Therefore, any tie lines on the phase diagram connecting phases in equilibrium with each other must lie on sections of constant $T$ and $\mu_i$ in the hyperspace. However, these sections would still possess the five dimensions associated with the geometric variables. The hyperspace can then be visualized as an evolution of these 5D sections as $T$ and the $\mu_i$'s are continuously varied.

The situation becomes particularly simple when all coexisting phases have the same misorientation as in the case of faceted interfaces. Further sectioning is then possible, since each 5D section can be regarded as the evolution of 2D sections (with $\bar{n}_1$ and $\bar{n}_2$ as coordinates) as the misorientation ($\theta$, $\rho_1$, and $\rho_2$) is continuously varied (Cahn 1982). The phase rule for such a section is

$$d_F^I = (C + 7 - \phi^I) - (C + 1 + 3) = 3 - \phi^I \tag{6.3}$$

which is exactly analogous to a constant $T$, $P$ section of a bulk phase diagram for a system with three components (Pelton 1983) where the number of degrees of freedom for the volume phases, $d_F^V$, is related to the number of volume phases, $\phi^V$, by

$$d_F^V = 3 - \phi^V. \tag{6.4}$$

Equation (6.4) is obtained by realizing that the equilibration of $\phi^V$ volume phases involves $\phi^V$ equations of the type given by eqn (5.10) involving $C + 2$ bulk variables, i.e. $T, P$, and the $\mu_i$'s. The phase rule for the bulk phases is then the well-known relation $d_F^V = C + 2 - \phi^V$ (see eqn (5.12)) which reduces to eqn (6.4) when $C = 3$ and $T$ and $P$ are constant. In the 2D section, with $\bar{n}_1$ and $\bar{n}_2$ as coordinates (see Fig. 6.1), one-phase fields have two degrees of freedom ($d_F^I = 2$), two-phase regions have one degree of freedom ($d_F^I = 1$) and are bounded by curves which anchor tie-lines connecting coexisting phases with different specified inclinations, and three-phase regions have zero degrees of freedom ($d_F^I = 0$) and are delineated by a tie-line triangle. Within this three-phase triangle, the boundary consists of three different sets of facets whose fixed inclinations are given by the corners of the triangle.

The above results demonstrate again the analogy between phase equilibria for faceted interfaces and bulk volume solutions where a common tangent construction dictates the equilibrium facet inclinations in the former case and the equilibrium compositions in the latter case. When coexisting phases with different misorientations are present, such as in dissociated interfaces, the construction of phase diagrams becomes much more complicated, since simple diagrams of low dimensionality cannot be constructed (Cahn 1982). We also note that the previous discussion assumed that all coexisting interface phases were different. Special complications arise when they are related by symmetry, as discussed by Cahn (1982).

When more than one bulk phase is present, the situation becomes even further complicated, since we then have $C + 8$ variables and $\phi^{\mathrm{I}} + \phi^{\mathrm{V}}$ equations, where $\phi^{\mathrm{V}}$ is again the number of volume phases. The local phase rule is then

$$d_{\mathrm{F}}^{\mathrm{I}} = C + 8 - (\phi^{\mathrm{I}} + \phi^{\mathrm{V}}). \tag{6.5}$$

## 6.3  INTERFACE PHASE TRANSITIONS

Interface phase transitions may occur along any trajectory in the multidimensional phase space that we have described in the previous section. These would be signaled by the appearance of singularities in the overall free energy of the interface, $\bar{\sigma}$, and phase diagrams based on the loci of these singularities could be constructed. In view of the high dimensionality of this space, the search for these transformations and the construction of these phase diagrams is a daunting task. To date, relatively little systematic work in this direction has been pursued, and only a tiny fraction of the phase space has been explored.

Despite this situation, a variety of interface transitions has been either observed experimentally or identified theoretically. These have occurred along trajectories involving changes in temperature, composition (i.e chemical potential), interface inclination and interface misorientation. They can conveniently be classified into two major groups, i.e. 'non-congruent' and 'congruent' (Cahn 1982). Non-congruent transitions are those in which the interface phases with different inclinations are involved. They are amongst the most easily and commonly observed and include faceting and dissociation. On the other hand, in congruent transitions, the interface remains flat (or smoothly curved to a large radius) while the atomic structure of the core changes locally as illustrated shematically in Fig. 5.10(d). The restructuring of the core may take many forms including, for example, the formation of new structural units, melting and various types of ordering. At the present time, there does not seem to be any simple and useful scheme for classifying these transitions further, and we therefore discuss them within this framework. In view of the many similarities between the thermodynamics of phase equilibria at internal interfaces and free surfaces, we shall occasionally turn briefly to the extensive literature dealing with phases and phase transitions at free surfaces.

### 6.3.1  Non-congruent phase transitions

#### 6.3.1.1  *Faceting of initially flat interfaces*

The basic thermodynamics of faceting has been discussed already in Section 5.6.3. Figure 5.11(a) shows a faceting transition brought about by a change in the average inclination of the interface plane, $\bar{n}$, while all other variables are held constant. At the inclination of AB, the interface consists of one phase. However, when the average inclination changes to BC, the interface adopts a two-phase structure. This transition corresponds to moving from the one-phase field of Fig. 6.1 to the two-phase field.

Faceting transitions can also be induced by heating and cooling. Figure 5.11(b) shows a two-phase faceted interface at a relatively low temperature. Upon heating, the faceted structure eventually disappears and is replaced by a slightly undulating one-phase roughened interface (Fig. 5.11(c)). Upon cooling, the transition is reversed, and the two-phase faceted interface is recovered (Hsieh and Balluffi 1989b). The transition on heating may be called a roughening/de-faceting transition, since the initially flat singular facets roughen during the heating and are eventually replaced by a single roughened phase. The interface roughening transition in this case is undoubtedly very similar to the well-known

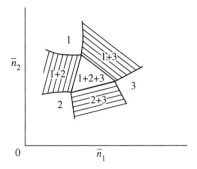

**Fig. 6.1** Phase diagram for grain boundary faceting. Diagram corresponds to a 2D section of the grain boundary phase space with $T, \theta, \rho_1, \rho_2$, and the $\mu_i$ held constant so that $d_F^I = 3 - \phi^I$ (eqn 6.3)). $\bar{n}_1$ and $\bar{n}_2$ measure the average inclination of the faceted boundary. In the central three-phase triangle, $d_F^I = 0$, and all facets are fixed. In the two-phase fields, $d_F^I = 1$, and the pairs of facets present are specified by the ends of the tie lines. In the one-phase fields, $d_F^I = 2$, and the flat boundary can assume any inclination within each field.

roughening transition for free surfaces which has been analysed extensively (Leamy *et al.* 1975, Weeks 1980). In this transition, the initially singular interface develops a defect structure during heating which consists of surface steps and various types of point defects. A similar roughening has occurred in the internal interface shown in Fig. 5.11(c), since it is a Σ3 interface which can easily form small pure steps in the short-period Σ3 CSL. The entropy of this disorder drives the process progressively and eventually the structure of the interface becomes completely uncorrelated and disordered (roughened) and the step energy goes to zero.

Reversible faceting transitions can also be induced by changes in composition (Ference and Balluffi 1988) while all other variables are held constant. Figures 5.11(d,e) show an example of a grain boundary in Cu for which equilibrium faceting was induced by introducing Bi solute atoms from the vapor phase. Bi is known to segregate strongly to grain boundaries in Cu (see Fig. 7.3, p. 00) and, evidently, the Bi atoms altered the Wulff plot and corresponding $\sigma_p$ interface energy plot (Fig. 5.12, p. 00) sufficiently to produce the necessary faceting condition. When the Bi was subsequently removed via the vapor phase, the boundary in Fig. 5.11(e) de-faceted and regained its original appearance as seen in Fig. 5.11(d).

### 6.3.1.2 *Faceting of embedded particle interfaces*

Non-congruent transitions which are closely related to the faceting transitions described above are expected to occur on the interfaces of small embedded particles when the temperature, or composition, is changed. A calculated example for the analogous case of the free surface of a small single crystal particle is shown in Fig. 6.2. Here, the Wulff plot was calculated as a function of temperature (Rottman and Wortis 1984). The corresponding Wulff form (or, alternatively, the $\xi$-plot) was then determined. At low temperature, the interface consists of various facets (interface phases) intersecting one another at edges and corners. Upon increasing the temperature, smoothly curved patches develop until, eventually, the interface becomes continuously curved and converts into a single phase. When two co-existing phases meet along a sharp edge, the inclination may be discontinuous (as in Fig. 6.2(a,b)), and the structures of the two co-existing phases are, therefore, discontinuous. In such a case, the form of the corresponding $\xi$-plot for

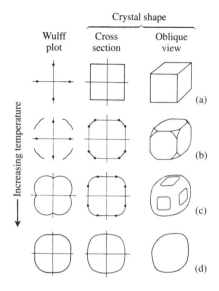

Increasing temperature

Crystal shape

Wulff plot | Cross section | Oblique view

(a)
(b)
(c)
(d)

**Fig. 6.2** Calculated Wulff plot and corresponding equilibrium shape (i.e. Wulff form) of a single crystal particle as a function of temperature. (From Rottman and Wortis (1984).)

all nearby inclinations of the interface might resemble that shown in Fig. 5.14. As we have pointed out in Section 5.6.4, the $\xi$-plot possesses an 'ear' (Fig. 5.14(b)) corresponding to the inclinations between $n'$ and $n''$ which are missing on the equilibrium form (shown cross-hatched). All inclinations associated with the 'ear' are unstable (or metastable) because they are not inclinations whose planes are part of the inner envelope which produces the Wulff form. The two coexisting phases separated by this discontinuity differ by a first-order phase transition (Cahn 1982). On the other hand, when a smoothly curved phase joins a flat facet smoothly as in Fig. 6.2(c), the 'ear' is absent, and the transition is second order.

Transitions on the interface can also occur when the composition is changed. For example, the addition of oxygen has been shown (Sundquist 1964) to change significantly the Wulff form and corresponding Wulff plot for the free surfaces of small particles of a number of fcc metals. Similar effects may be expected for the internal interfaces of small embedded single crystal particles. We note that another type of non-congruent transition could occur with small embedded particles if two different $\xi$-plots corresponding to different interface structures exist and intersect as shown in Fig. 6.3. The equilibrium form would then correspond to the inner envelope as illustrated, and the two coexisting phases would differ by a first-order transition (Cahn 1982).

### 6.3.1.3 *Interface dissociation*

Interface dissociation is illustrated schematically in Fig. 6.4. In Fig. 6.4(b) the single interface between crystals 1 and 2 shown in Fig. 6.4(a) has split (dissociated) to produce a new intermediate crystal 3 and two new interfaces. Such a structure will be stable whenever $\sigma^{1,2} > (\sigma^{1,3} + \sigma^{2,3})$. The more complex type of dissociation illustrated in Fig. 6.4(d) may occur if one of the boundaries shown in Fig. 6.4(b) tends to facet in order to reduce the energy still further. The final structure shown in Fig. 6.4(d) may then be visualized as the result of the series of steps illustrated in Figs 6.4(a) through 6.4(d). The energetics of the faceting of the 2,3 interface can be analysed in terms of the $\sigma_p$ function used previously, and, again, the energy contribution of any junction dislocations would have to be added in a more exact analysis.

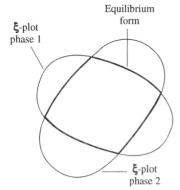

ξ-plot
phase 1

Equilibrium
form

ξ-plot
phase 2

**Fig. 6.3** Intersection of two different ξ plots (or, alternatively, Wulff forms) for an embedded single crystal particle.

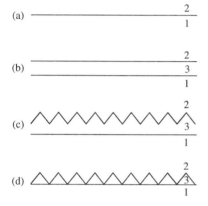

(a)

(b)

(c)

(d)

**Fig. 6.4** Interface dissociation. (b) Single interface shown in (a) has split to produce a new crystal layer 3 and two new interfaces. (c) Interface 2,3 in (b) has now faceted. (d) Interfaces 1,3 and 2,3 in (c) now abut one another.

An example of a dissociated interface in gold is shown in Fig. 6.5(b), and a dissociation transition induced by a change in boundary inclination is illustrated in Fig. 6.5(a). Here, the smoothly curved single phase boundary between A and B changes inclination rather abruptly at B and dissociates into a three-phase interface between B and C. The energetics of this particular transition can be understood qualitatively since, of the three interfaces which are present in the dissociated structure, the one whose inclination is essentially parallel to the average boundary inclination is a Σ3(111) twin boundary which is known (Goodhew *et al.* 1978) to be of exceedingly low energy. Hence, when the boundary assumes that average inclination, the dissociation becomes energetically feasible.

Another type of dissociation is illustrated in Fig. 6.6. Here, a new grain forms at a triple junction formed initially by the three crystals 1, 2 and 3. Such a structure will be stable whenever

$$l^{1,4}\sigma^{1,4} + l^{2,4}\sigma^{2,4} + l^{3,4}\sigma^{3,4} < l^{1,4}\sigma^{1,2} + l^{3,4}\sigma^{2,3}. \tag{6.6}$$

Dissociations of this type have long been thought to be the origin of Σ(111) annealing twin boundaries in f.c.c. metals (Cahn 1983). In this case, the 2,4 interface is the Σ3(111) twin boundary which is of sufficiently low energy to satisfy occasionally the inequality in eqn (6.6).

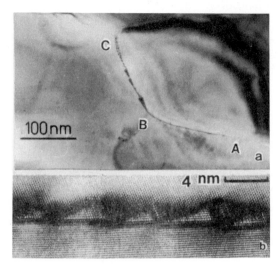

Fig. 6.5 (a) Dissociation of grain boundary in Au into three phases along the length BC, while the boundary along the length AB exists as a single phase. (From Goodhew *et al.* (1978).) (b) Dissociation of boundary in Au into three boundary phases in the manner illustrated schematically in Fig. 6.4(d). (From Krakow and Smith (1986).)

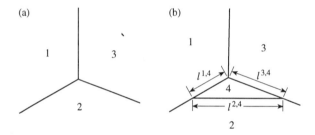

Fig. 6.6 Dissociation of boundaries at a triple junction to form a new occluded crystal (i.e. crystal 4) and three new boundary segments.

### 6.3.2 Congruent phase transitions

As pointed out by Cahn (1982), congruent phase changes will not generally appear in sections of the phase space corresponding to constant $T, \mu_i$, and misorientation where $d_F^I = 3 - \phi^I$ corresponds to the inclination degrees of freedom. As mentioned above, equilibrium between two congruent phases in an interface requires that $\sigma'(\bar{n}) = \sigma''(\bar{n})$. This is equivalent to demanding that $\sigma_P'(\bar{n}) = \sigma_P''(\bar{n})$, where $\sigma_P$ is the projected interface free energy function introduced previously in Section 5.6.3 (see Figs 5.12 and 5.13). This condition could be satisfied if the $\sigma_P$ curves for the two phases intersected as illustrated in Fig. 6.7. However, the two phases would not be able to coexist congruently in equilibrium at the intersection point (where $\sigma_P'(\bar{n}) = \sigma_P''(\bar{n})$), since according to the common tangent construction (see Fig. 5.13), the interface would facet into facets with differing inclinations. Therefore, congruent equilibrium requires that the two $\sigma_P$ curves just touch each other without crossing. This will be a rare event, and, generally, other state variables will have to be adjusted in order to achieve congruent equilibrium. As Cahn (1982) states, 'the search for congruent transformations has all the aspects of a search for a needle in a $C + 6$-dimensional haystack'.

A generalized Clausius–Clapeyron type relationship describing the condition for the maintenance of equilibrium between two congruent interface phases is readily derived.

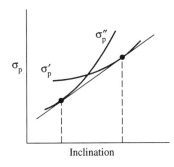

**Fig. 6.7** Intersection of $\sigma_P$ curves for two different interface phases produced by changes in their inclinations. $\sigma_P$ is the interface free energy per unit area projected on the $z = 0$ plane (see Section 5.6.3).

For two congruent interface phases (indicated here by single and double primes) and two adjoining bulk phases, we may use the methods of Section 5.3 (see eqn 5.14)) to combine eqns (6.1) into the following two relationships:

$$d\sigma' = \frac{1}{(\text{Det}')_{C+3}} \sum_{k=1}^{C+2} (\text{Det}')_k \, dX_k + \frac{\partial\sigma'}{\partial\rho_1} d\rho_1 + \frac{\partial\sigma'}{\partial\rho_2} d\rho_2 + \frac{\partial\sigma'}{\partial\theta} d\theta + \frac{\partial\sigma'}{\partial n_1} dn_1 + \frac{\partial\sigma'}{\partial n_2} dn_2;$$

$$(6.7)$$

$$d\sigma'' = \frac{1}{(\text{Det}'')_{C+3}} \sum_{k=1}^{C+2} (\text{Det}'')_k \, dX_k + \frac{\partial\sigma''}{\partial\rho_1} d\rho_1 + \frac{\partial\sigma''}{\partial\rho_2} d\rho_2 + \frac{\partial\sigma''}{\partial\theta} d\theta + \frac{\partial\sigma''}{\partial n_1} dn_1 + \frac{\partial\sigma''}{\partial n_2} dn_2.$$

$$(6.8)$$

At equilibrium we must have $d\sigma' = d\sigma''$, and, therefore,

$$\left[ \frac{1}{(\text{Det}')_{C+3}} \sum_{k=1}^{C+2} (\text{Det}')_k - \frac{1}{(\text{Det}'')_{C+3}} \sum_{k=1}^{C+2} (\text{Det}'')_k \right] dX_k + \frac{\partial(\sigma' - \sigma'')}{\partial\rho_1} d\rho_1$$

$$+ \frac{\partial(\sigma' - \sigma'')}{\partial\rho_2} d\rho_2 + \frac{\partial(\sigma' - \sigma'')}{\partial\theta} d\theta + \frac{\partial(\sigma' - \sigma'')}{\partial n_1} dn_1 + \frac{\partial(\sigma' - \sigma'')}{\partial n_2} dn_2 = 0. \quad (6.9)$$

The locus of the congruent two phase equilibrium is therefore a surface in the multi-dimensional space defined by $\rho_1, \rho_2, \theta, n_1, n_2$, and the $X_k$'s which survive in the above equation when the determinants corresponding to $(\text{Det}')_k$ and $(\text{Det}'')_k$ are evaluated.

### 6.3.2.1 *Various transitions induced by changes in temperature, composition, or crystal misorientation*

We now describe a variety of congruent transitions in which the interface core structure changes locally in response to changes in temperature, composition, or crystal misorientation. Not included are transitions involving wetting and melting which are discussed separately below.

Consider, first, the effect of increasing the temperature. On very general grounds, we may expect interfaces to disorder in various ways during heating to produce structures of higher entropy. This may include the formation of various types of point defects and line defects, such as steps. We have already seen experimental evidence for a non-congruent roughening phase transition on heating a faceted boundary. Presumably, congruent roughening transitions should also occur in large-angle singular or vicinal boundaries, but these have not yet been demonstrated in a clear-cut manner. Rottman (1986*a*, *b*, *c*) has carried out approximate calculations which indicate the possible roughening of small-angle boundaries.

A number of calculations for large-angle grain boundaries has shown evidence for at least some disordering in the absence of an abrupt phase transition, except at temperatures essentially at the bulk melting point, $T_m$, where complete interface melting occurs very suddenly as we shall describe later on. In one molecular dynamics study (Ciccotti *et al.* 1983) of a symmetric $\Sigma 5 \langle 100 \rangle (310)$ tilt boundary, the structure factor of the boundary core region was calculated as a function of increasing temperature. This quantity is defined by

$$|F_k| = | \sum_{i=1}^{N} \exp{(ik \cdot r_i)}|, \qquad (6.10)$$

where $k$ is a reciprocal lattice vector of the periodic boundary structure which exists at $T = 0$, and the sum is taken over $N$ atoms which are situated at the positions $r_i$ in the boundary core. For example, for a perfect crystal lattice (of basis unity) containing $N$ atoms without thermal vibrations $|F_k| = N$, while for a disordered liquid $|F_k| \simeq 0$. The quantity $|F_k|$ is, therefore, a useful parameter for indicating the degree of long range disorder. A monotonic decrease in $|F_k|$ with increasing temperature was found by Ciccotti *et al.*, which was considerably more rapid than the corresponding decrease in $|F_k|$ for the bulk lattice. However, the boundary structure factor did not extrapolate to zero until the bulk melting temperature was reached, and this may be taken as evidence that the boundary developed a defect structure at elevated temperatures while remaining 'crystalline' (i.e. while retaining its long-range order). In other work (Kikuchi and Cahn 1980), a 2D lattice gas model for a symmetric $\Sigma 5$ boundary showed early signs of the onset of disordering at $0.5\,T_m$. This disorder (as measured by the boundary density profile) progressed steadily with increasing temperature and culminated finally with complete boundary melting and wetting occurring rather abruptly at $T_m$. Additional molecular dynamics studies (Broughton and Gilmer 1986, Lutsko *et al.* 1989) have also failed to reveal evidence for unusually rapid losses of long-range order on heating except essentially at the bulk melting point where boundary melting occurs.

Another form of temperature-induced order–disorder may occur in boundaries of the type which possess structural multiplicity (Section 4.3.1.8) and therefore consist of mixtures of different types of structural units possessing only slightly differing energies. The structural units exhibiting multiplicity may be expected to have at least some interactions with one another, and this leads to possible order–disorder transitions as the units tend to minimize the boundary energy by forming ordered arrangements. This phenomenon has been analysed for the first time by Vitek *et al.* (1985) for a boundary containing one type of unit exhibiting multiplicity in the form of two variants using a regular solution model. For a range of structural unit interaction parameters, the analysis yields an order–disorder transition with a transition temperature either below or above the bulk melting temperature, $T_m$. Calculated results for a $\Sigma 5 \{210\}\{210\}$ boundary in Cu indicate a transition temperature above $T_m$; so, for this particular boundary, a gradual disordering of the units would be expected on heating. A more extended analysis of boundary transitions of this general type has been given by Rottman (1989).

Experimentally, no evidence for a sudden disordering transition (including melting) has been found during *in situ* electron microscope observations of the secondary grain boundary dislocation structure of a variety of boundaries in Al during heating from $0.32\,T_m$ to $0.96\,T_m$ (Hsieh and Balluffi 1989*a*). In these experiments, which involved $\Sigma 5$, $\Sigma 13$, $\Sigma 17$, and plane matching boundaries, the secondary grain boundary dislocations remained localized and apparently unaffected by the heating. Since these dislocations

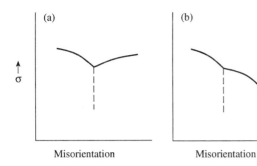

**Fig. 6.8** 2D sections of grain boundary free energy versus boundary misorientation surface for 'rotating balls-on-a-plate' experiments. (a) Trapping cusp present. (b) Non-trapping cusp present.

undoubtedly would have delocalized if extensive disordering had occurred, the results indicate the absence of such disordering. We may also cite 'rotating balls-on-a-plate' experiments which are related to those already described in Section 5.7 (see Fig. 5.19), since results obtained with these experiments have on occasion been taken (Gleiter 1982, Shvindlerman and Straumal 1985) as evidence for temperature-induced boundary disordering phase transitions. In this work, large numbers of single crystal balls were initially distributed on single crystal plates at random misorientations. Upon subsequent annealing, the balls rotated into a considerable number of discrete misorientations corresponding to trapping cusps on the $\sigma = \sigma(\rho, \theta)$ energy surface (with $\hat{n}_{\text{plate}}$ held fixed) which were associated with boundaries of relatively low energy. As the temperature was raised, the apparent number of clearly resolved misorientations (i.e. cusps) at which the balls accumulated appeared to decrease somewhat. Also, the particularly strong accumulations observed at a few special misorientations increased. This has been interpreted as due to changes in the $\sigma = \sigma(\rho, \theta)$ surface in which many of the minor trapping cusps essentially disappeared at elevated temperatures while a number of major cusps remained. This loss of cusps presumably enlarged the capture radii of the surviving cusps and therefore increased the accumulations there. It has been suggested (Gleiter 1982, Shvindlerman and Straumal 1985) that the apparent disappearance of cusps in this way could conceivably have been due to temperature-induced phase transitions in which relatively ordered boundaries corresponding to cusps at low temperatures were replaced by less ordered boundaries with higher entropies at higher temperatures. However, it is important to point out that this interpretation may not necessarily be correct. It is plausible that the entropies of many of the relatively low energy boundaries corresponding to cusp misorientations were lower than those of the nearby more general boundaries so that their free energies decreased less rapidly with increasing temperature. Many of the cusps may, therefore, have simply become shallower, and many may have been converted from trapping to non-trapping types, as illustrated schematically in 2D in Fig. 6.8. In any case, a relatively large number of significantly trapping cusps were retained to temperatures as high as $0.98\ T_{\text{m}}$, and this result indicates that many boundaries retained significant degrees of long-range correlation and order to those temperatures.

In a number of cases evidence for sudden boundary structural transitions on heating has been found. We have already discussed in Section 2.12.1 the sudden change in reference structure and secondary grain boundary dislocation structure with temperature which can occur in a non-cubic system such as Zn due to the temperature dependence of the $c/a$ ratio (see also Fig. 2.36). Several calculations have also revealed sudden structural phase transitions upon heating in certain boundaries. Guillopé (1986) found a sudden transition from a distinct low temperature ordered core structure to a distinct high-temperature structure in a Σ5 symmetric ⟨100⟩ (210) tilt boundary in a pure f.c.c.

lattice. Calculations by Najafabadi *et al.* (1990) and Chen and Kalonji (1992) have revealed transitions between different boundary structures induced by heating. Also, Maunier and Pontikis (1990) calculated by molecular dynamics a structural transformation in a $\Sigma 5$ symmetric $\langle 100 \rangle$ (210) boundary in $CaF_2$ which occurred during heating near $T = 0.2\,T_m$.

We may conclude from the above that, in general, interfaces can undergo both roughening and other types of phase transitions during heating. However, many (and, perhaps, most) interfaces can apparently be heated to essentially the bulk melting temperature without the appearance of phase transitions. During this process, they undoubtedly develop thermally induced disorder in the form of point and line defect structures but they retain significant degrees of long-range order and correlation and do not undergo a *bona fide* phase transition until they reach essentially $T_m$ where they melt quite suddenly as discussed further below.

Transitions between distinct structures have also been obtained by introducing solute atoms at the interface. In general, when solute atoms are present in a bulk system, they may tend to concentrate preferentially, i.e. 'segregate' at the interface to produce an excess concentration, $\Delta N_2^\alpha$, as already described thermodynamically in Chapter 5 by eqn (5.29). This important phenomenon is taken up separately in Chapter 7. In the present discussion, we restrict ourselves to a limited number of special cases where the presence of solute atoms in the interface core produces a distinct new structure via a phase transition.

Molecular statics calculations (Hashimoto *et al.* 1982) have shown that the addition of approximately a monolayer of P atoms to the core of an Fe grain boundary causes a reconstruction to a unique structure in which the local atomic environment around each P atom becomes similar to that in bulk $Fe_3P$ (see also Section 7.3.3). Similarly, calculations (Vitek and Wang 1984) show that the addition of Bi to a Cu grain boundary produces a restructuring to a distinctly new atomistic structure.

Experimentally, the equilibrium segregation of Nb to stacking faults in Co has been studied (Herschitz *et al.* 1985, Herschitz and Seidman 1988), and evidence has been obtained by electron diffraction and atom probe-field ion microscopy for a phase transition involving the formation of two coexisting interface phases, one of which is a 2D Nb-rich ordered phase of composition $Co_2Nb$. In other work, the addition of either Au (Sickafus and Sass 1987), Sb (Lin and Sass 1988*a*), or S (Lin and Sass 1988*b*) solute atoms to [001] twist boundaries in Fe caused the formation of patches of a new coexisting phase in which the square screw dislocation network which was initially present in the pure Fe was replaced by a new network which possessed a larger Burgers vector and spacing and was rotated by 45°. There was some evidence that the new interface phase was more solute-rich and, therefore, could have been stabilized by a stronger solute atom-dislocation interaction due to the larger Burgers vector of its dislocations. Additional work (Michael *et al.* 1988) has shown that Au segregates strongly to the cores of the larger Burgers vector dislocations in a manner consistent with this idea.

Luzzi (1991), Luzzi *et al.* (1991), and Yan *et al.* (1993) have shown that Bi atoms segregated to $\Sigma 3$ (111) twin boundaries in Cu can produce a new boundary phase consisting of a hexagonal layer of Bi atoms as described in Section 7.2 and illustrated in Fig. 7.8. Also, the form of the Fowler–Guggenheim isotherm for the segregation of Se to grain boundaries in BCC Fe in Fig. 7.11 suggests the strong clustering of Se and possibly the coexistence of two boundary phases as discussed in Section 7.4.3.

Several of the above experiments indicate the possibility of the formation and equilibration of multiphase congruent structures at interfaces in the presence of solute atoms.

For a two-component system containing one bulk phase and $\phi^I$ congruent interface phases all in equilibrium with each other, eqn (6.2) reduces to

$$d_F^I = 3 - \phi^I \tag{6.11}$$

for the interface phase rule for an interface of fixed misorientation and inclination at constant pressure. This phase rule is analogous to the usual isobaric rule for bulk phases in a two-component bulk system, and similar phase equilibria could, therefore, presumably exist. However, the situation would be altered if solute atom equilibrium was not achieved between the interface and bulk phases as may, indeed, have been the case, for example, in the work of Sickafus and Sass (1987). Various degrees of equilibration can be visualized. One extreme of interest is the case where the interface acts as a closed system and only a local equilibrium in the interface between the interface phases is achieved. This can occur under certain conditions because of the relatively high rate of diffusion of solute atoms along interfaces compared to bulk diffusion rates as described in Chapter 8. Such a situation is analogous to the formation and equilibration of 2D interface phases which occurs when a fixed number of foreign atoms is deposited on the free surface of a substrate crystal and allowed to equilibrate to form surface phases in a so-called 'closed system' (Blakely 1988).

Finally, we point out that the change in structure which occurs when a boundary goes through a cusp on the generalized $\sigma = \sigma(\rho, \theta)$ surface should be regarded as a congruent phase transition induced by a change in crystal misorientation. Consider, for example, a simple symmetric tilt boundary described by the structural unit model (Section 4.3.1). At the cusp, the boundary is a delimiting one consisting of all A units. As the boundary approaches the cusp from one side, it consists of a mixture of A-type majority units and B-type minority units. Directly on the other side, it consists of a mixture of A-type majority units and C-type minority units which, of course, differ from the B-units. The discontinuity in energy which occurs at the cusp signifies a phase transition.

### 6.3.2.2 *Interface wetting by a solid phase*

The 'wetting' of an interface by a solid phase may occur under certain conditions when it becomes energetically favorable to insert a thin layer of a new phase at the interface. This condition can often be met when bulk phases become unstable as in the vicinity of critical points as discussed by Cahn (1977). Consider, for example, the phase diagram in Fig. 6.9(a) containing the critical point at the temperature $T_c$. Within the solid solubility gap the microstructure will generally contain $\alpha/\alpha$, $\beta/\beta$, and $\alpha/\beta$ interfaces. As $T$ is raised towards $T_c$, the free energy of the $\alpha/\beta$ interface will tend to decrease rapidly (Cahn 1977), since above $T_c$ the $\alpha$ and $\beta$ phases become indistinguishable. Also, the free energies of the bulk $\alpha$ and $\beta$ phases will tend to converge. At temperatures slightly below $T_c$ it may then become energetically feasible to form a layer of $\beta$ phase at a $\alpha/\alpha$ grain boundary in such a manner that the $\alpha/\alpha$ boundary is replaced by a $\beta/\beta$ boundary and two low-energy $\alpha/\beta$ interfaces, or, more simply, by two $\alpha/\beta$ interfaces. Experimental evidence for a wetting transition of the former type has been presented by Rabkin *et al.* (1990, 1991) in the Sn–In system at $\gamma/\gamma'$ grain boundaries. In this system a bulk $\gamma' \rightarrow \beta$ phase transition occurs at a critical point upon heating which causes a sharp discontinuity in the bulk diffusivity of In. Rather similar discontinuities were found in the grain boundary diffusivity parameter $p = \delta D^B$ (see Section 8.2.2) as bulk $\gamma'$ specimens initially containing $\gamma'/\gamma'$ grain boundaries were heated up to the bulk $\gamma' \rightarrow \beta$ phase transition temperature. These discontinuities occurred about $20°$ below the bulk transition

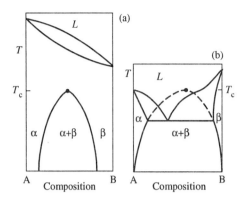

**Fig. 6.9** Binary phase diagrams. (a) Solid solution solubility gap exists with a critical point at $T_c$. (b) Eutectic system with a metastable solubility gap and critical point at $T_c$. (From Straumal et al. (1992).)

temperature indicating the possible onset of a wetting transition in the boundary corresponding to the formation of a wetting layer consisting of the $\beta$ phase.

In systems capable of developing long-range order, two-phase regions of coexistence between an ordered and a disordered phase may exist. Calculations by Kikuchi and Cahn (1979) show that the disordered phase will then coat (wet) antiphase boundaries in the ordered phase near a critical point.

### 6.3.2.3  *Interface wetting by a liquid phase in alloy systems*

Interfaces between solid phases may also be wet by liquid phases under appropriate conditions in alloy systems. Consider, for example, the eutectic phase diagram in Fig. 6.9(b). Such a system contains a metastable critical point shown at the temperature $T_c$ on the dashed metastable phase boundary. According to the same reasoning employed in the previous subsection, the free energy of the $L/\beta$ interfaces present may be expected to decrease relative to the energy of the $\beta/\beta$ grain boundaries as the temperature is increased towards $T_c$. At some point, therefore, a wetting transition will occur in which a $\beta/\beta$ grain boundary is replaced by a slab of liquid phase bounded by two $L/\beta$ interfaces. Many wetting transitions of this general type have been observed in the literature (see reviews in Rabkin *et al.* (1991) and Straumal *et al.* (1992)). Very recently, Straumal *et al.* (1992) have studied the wetting of grain bondaries in symmetric ⟨110⟩ tilt grain boundaries in Cu by a Cu(In) melt. A wetting transition was observed at 960 °C for a $\theta = 77$ ° boundary and at 930 °C for a $\theta = 141$ ° boundary. The energy of the 77 ° boundary is about 40 per cent lower than that of the 141 ° boundary, and the higher temperature of the wetting transition of the 77 ° boundary can therefore be understood on this basis.

### 6.3.2.4  *Grain boundary melting in a one-component system*

Grain boundaries generally possess excess volume and larger atomic vibrational amplitudes than the bulk (see Section 4.3.1.10) and, therefore, it has long been speculated that grain boundaries in one-component systems may undergo a melting transition at temperatures significantly below the bulk melting transition. Such melting has often been analysed as a wetting transition in which a liquid phase, possessing physical properties which approach those of the true bulk liquid phase, wets the boundary as the bulk melting point is approached. The energetics can be discussed in terms of a simple macroscopic model (Bolling 1968, Kristensen and Cotterill 1977) in which the free energy change associated with the transition is expressed as

$$W = \left( \sigma^{(1)} + \sigma^{(2)} + \Delta G + \eta \right) - \sigma^{(0)} \tag{6.12}$$

where $\sigma^{(0)}$ = energy of original interface, $\sigma^{(1)}$ and $\sigma^{(2)}$ are the energies of the two new interfaces associated with the new layer phase, $\Delta G$ = bulk free energy increase involved in the formation of the new layer phase, and $\eta$ is a correction term which takes into account effects due to non-uniformities of the layer and interactions of the adjoining bulk phases across the layer. Such a transition, therefore, becomes more likely if $(\sigma^{(1)} + \sigma^{(2)}) < \sigma^{(0)}$, and the $\Delta G$ increase is relatively small.

Crude analyses based on the use of eqn (6.12) (Bolling 1968, Kristensen and Cotterill 1977) have indicated that a liquid-like layer of thickness $\simeq 2$ nm could conceivably form in grain boundaries at temperatures within 1–2 per cent of $T_{\mathrm{m}}$. The atomistic calculations which we have already cited in the discussion of boundary disordering on heating also indicate that melting to a liquid-like layer approximating bulk liquid can only be expected at temperatures very close to $T_{\mathrm{m}}$. In addition, the lattice gas model results (Kikuchi and Cahn 1980) indicate a rather abrupt onset of melting near $T_{\mathrm{m}}$ where the boundary entropy diverges as $- \ln (T_{\mathrm{m}} - T)$. Also, the boundary thickness in this model increases rapidly in this regime. That this thickening must accompany final true melting at $T_{\mathrm{m}}$ is to be expected on simple physical grounds, since the liquid boundary material can only attain true bulk-like liquid properties when its thickness is large enough to essentially eliminate the ordering influence of the adjoining crystals. Atomistic calculations have shown that this ordering effect is significant. For example, molecular dynamics calculations of the structure of the liquid/solid interface using a Lennard–Jones interatomic potential (Ladd and Woodcock 1978) indicate that the interface is significantly wide and extends over six, or more, atomic layers.

Experimentally, direct observations have been made (Hsieh and Balluffi 1989a) of the degree of melting (wetting) of boundaries in a partially melted Al thin film specimen containing a pool of bulk liquid and a network of boundaries which either intersected the liquid/solid interface or lay directly adjacent to it. In this arrangement, the boundaries were maintained either essentially at the bulk melting temperature, $T_{\mathrm{m}}$, or very close to it and were observed *in situ* by hot-stage transmission electron microscopy. No evidence for significant wetting by a phase resembling the liquid was found for temperatures $\leqslant 0.999 \, T_{\mathrm{m}}$. However, observable wetting was seen at temperatures estimated to be within $< 1°$K of $T_{\mathrm{m}}$. Also, this wetting was observed only at large-angle boundaries and not at small-angle boundaries composed of arrays of discrete lattice dislocations.

Additional evidence is available in the literature which appears to be consistent with the above picture. Grain boundary internal friction measurements on Al (Boulanger 1954) indicate the onset of complete grain boundary melting only at temperatures within about $0.5 \, °$C, or less, of $T_{\mathrm{m}}$. In addition, grain boundary diffusivities are found to become essentially equal to liquid diffusivities at $T_{\mathrm{m}}$ on a reduced Arrhenius plot (see Fig. 8.5 of Section 8.2.3). However, the activation energy for grain boundary diffusion is higher than for diffusion in the liquid, and extrapolated values of the diffusivity in the bulk liquid at temperatures below $T_{\mathrm{m}}$ are, therefore, higher than measured grain boundary diffusivities. This behaviour, also found in atomistic modelling work (Ciccotti *et al.* 1983), suggests that the grain boundary structure below $T_{\mathrm{m}}$ is less disordered than the bulk liquid and that it is only at $T_{\mathrm{m}}$ that the structures become essentially similar.

We may finally conclude from the above theoretical and experimental work that boundaries generally disorder to varying degrees upon heating but the phenomenon which may be termed melting does not occur until the temperature essentially reaches $T_{\mathrm{m}}$. The onset of melting very near $T_{\mathrm{m}}$ is relatively sudden and is characterized by a rapid increase of the boundary width and complete wetting of the adjacent grains.

An interesting congruent melting transition induced by a change in boundary misorientation at an elevated temperature almost equal to $T_m$ has been found (Glicksman and Vold 1972) for symmetric $\langle 110 \rangle$ tilt boundaries in Bi. *In situ* hot stage transmission electron microscopy was used to measure the degree of wetting of these boundaries as a function of tilt angle, $\theta$, as they were maintained very near $T_m$ in the region directly adjacent to the solid/liquid interface in partially melted thin film specimens. As the tilt angle increased from $\theta = 0°$, the boundary abruptly became completely wet at $\theta \simeq 15°$ indicating the onset of a first-order phase transition. This result appears to be consistent with the observation described above (Hsieh and Balluffi 1989a) that large-angle boundaries in Al held essentially at $T_m$ were wet by the liquid, whereas small-angle boundaries were not. The result may be understood generally on the basis of eqn (6.12), since $\sigma^{(0)}$ is expected to increase steadily as $\theta$ is increased in the range $0° < \theta \lesssim 20°$, while the temperature is held constant very near $T_m$. At some critical point, it seems possible that significant wetting could then occur.

## REFERENCES

Blakely, J. M. (1986). In *Encyclopedia of materials science and engineering* (ed. M. B. Bever) Vol. 7, p. 4962. Pergamon Press, Oxford.

Blakely, J. M. (1988). *J. de Physique*, **49**, Colloq. C5, supple. au no 10, C5–351.

Blakely, J. M. and Thapliyal, H. V. (1979). In *Interfacial segregation* (eds W. C. Johnson and J. M. Blakely) p. 137. American Society for Metals, Metals Park, Ohio.

Bolling, G. F. (1968). *Acta Metall.*, **16**, 1147.

Boulanger, C. (1954). *Rev. Metall.*, **51**, 210.

Broughton, J. Q. and Gilmer, G. H. (1986). *Phys. Rev. Lett.*, **56**, 2629.

Cahn, J. W. (1977). *J. Chem. Phys.*, **66**, 3667.

Cahn, J. W. (1982). *J. de Physique*, **43**, Colloq. C6, supple au n° 12, C6–199.

Cahn, R. W. (1983). In *Physical metallurgy* (eds R. W. Cahn and P. Haasen) p. 1656. North-Holland, Amsterdam.

Chen, L.-Q. and Kalonji, G. (1992). *Phil. Mag. A*, **66**, 11.

Ciccotti, G., Guillopé, M., and Pontikis, V. (1983). *Phys. Rev. B*, **27**, 5576.

Ference, T. G. and Balluffi, R. W. (1988). *Scripta Metall.*, **22**, 1929.

Gleiter, H. (1982). *Mats. Sci. Eng.*, **52**, 91.

Glicksman, M. E. and Vold, C. L. (1972). *Surf. Sci.*, **31**, 50.

Goodhew, P. J., Tan, T. Y., and Balluffi, R. W. (1978). *Acta Metall.*, **26**, 557.

Guillopé, M. (1986). *J. de Physique*, **47**, 1347.

Hart, E. W. (1972). In *The nature and behavior of grain boundaries* (ed. H. Hu) p. 155. Plenum Press, New York.

Hashimoto, M., Ishida, Y., Yamamoto, R., Doyoma, M., and Fujiwara, T. (1982). *Scripta Metall.*, **16**, 267.

Herschitz, R. and Seidman, D. N. (1988). *J. de Physique*, **49**, Colloq. C5, supple. au n° 10, C5–469.

Herschitz, R., Seidman, D. N., and Brokman, A. (1985). *J. de Physique*, **46**, Colloq. C4, supple au n° 4, C4–451.

Hsieh, T. E. and Balluffi, R. W. (1989a). *Acta Metall.*, **37**, 1637.

Hsieh, T. E. and Balluffi, R. W. (1989b). *Acta Metall.*, **37**, 2133.

Kikuchi, R. and Cahn, J. W. (1979). *Acta Metall.*, **27**, 1337.

Kikuchi, R. and Cahn, J. W. (1980). *Phys. Rev.*, B **21**, 1893.

Krakow, W. and Smith, D. A. (1986). In *Grain boundary structure and related phenomena*, Suppl. to *Trans. Japan Inst. Mets.*, **27** (ed. Y. Ishida) p. 277. Japan Institute of Metals, Sendai.

Kristensen, J. K. and Cotterill, R. M. J. (1977). *Phil. Mag.*, **36**, 437.

Ladd, A. J. C. and Woodcock, L. V. (1978). *J. Phys. C*, **11**, 3565.

Leamy, H. J., Gilmer, G. H., and Jackson, K. A. (1975). In *Surface physics of materials* (ed. J. M. Blakely) p. 121. Academic Press, New York.

Lin, C. H. and Sass, S. L. (1988*a*). *Scripta Metall.*, 22, 735.

Lin, C. H. and Sass, S. L. (1988*b*). *Scripta Metall.*, 22, 1569.

Lutsko, J. F., Wolf, D., Phillpot, S. R., and Yip, S. (1989). *Phys Rev. B*, 40, 2841.

Luzzi, D. E. (1991). *Phil. Mag. Letters*, 63, 281.

Luzzi, D. E., Yan, M., Sob, M., and Vitek, V. (1991). *Phys. Rev. Letters*, 67, 1894.

Maunier, C. and Pontikis, V. (1990). *J. de Physique*, 51, Colloque Cl, Supple. au n° 1, Cl-245.

Michael, J. R., Lin, C. H., and Sass, S. L. (1988). *Scripta Metall.*, 22, 1121.

Najafabadi, R., Srolovitz, D. J., and Lesar, R. (1990). *J. Mater. Res.*, 5, 2663.

Pelton, A. D. (1983). In *Physical metallurgy* (eds R. W. Cahn and P. Haasen) p. 327. North-Holland, Amsterdam.

Phillpot, S. R., Lutsko, J. F., Wolf, D., and Yip, S. (1989). *Phys Rev. B*, 40, 2831.

Pontikis, V. (1988). *J. de Physique*, 49, Colloque C5, Supple. au n° 10, C5-327.

Rabkin, E. I., Shvindlerman, L. S., and Straumal, B. B. (1990). *J. Less Common Mets*, 158, 23.

Rabkin, E. I., Shvindlerman, L. S., and Straumal, B. B. (1991). *Int. J. Mod. Phys.*, B, 5, 2989.

Rottman, C. (1986*a*). *Phys. Rev. Letters*, 57, 735.

Rottman, C. (1986*b*). *Mats. Sci. Eng.*, 81, 553.

Rottman, C. (1986*c*). *Acta Metall.*, 34, 2465.

Rottman, C. (1988*a*). *J. de Physique*, Colloque C5, Supple. au n° 10, C5-313.

Rottman, C. (1988*b*). *Mats. Res. Soc. Symp. Proc.*, 122, 151.

Rottman, C. (1989). *Scripta Metall.*, 23, 1037.

Rottman, C. (1992). *Mat. Res. Soc. Symp. Proc.*, 238, 191.

Rottman, C. and Wortis, M. (1984). *Phys. Rpts.*, 103, 59.

Shvindlerman, L. S. and Straumal, B. B. (1985). *Acta Metall.*, 33, 1735.

Sickafus, K. E. and Sass, S. L. (1987). *Acta Metall.*, 35, 69 (1987).

Straumal, B., Muschik, T., Gust, W., and Predel, B. (1992). *Acta Metall. Mater.*, 40, 939.

Sundquist, B. E. (1964). *Acta Metall.*, 12, 67.

Vitek, V. and Wang, G.-J. (1984). *Surf. Sci.*, 144, 110.

Vitek, V., Minonishi, Y and Wang, G. J. (1985). *J. de Physique*, 46, Colloq. C4, supple au n° 4, C4-171.

Walls, J. M. (ed.) 1989. *Methods of surface analysis*. Cambridge University Press, Cambridge.

Weeks, J. D. (1980). In *Ordering in strongly fluctuating condensed matter systems* (ed. T. Riste) p. 293. Plenum Press, New York.

Woodruff, D. P. and Delchar, T. A. (1988). *Modern techniques in surface science*. Cambridge University Press, Cambridge.

# 7

# Segregation of solute atoms to interfaces

## 7.1 INTRODUCTION

The different solute atoms which are present in multicomponent systems containing interfaces often adsorb preferentially at the interfaces, thereby building up interface 'excess' concentrations. This adsorption has already been described on a macroscopic thermodynamic basis in Chapter 5 by the Gibbs adsorption isotherm, i.e. eqn (5.29), which predicts that preferential segregation of a component to an interface tends to occur at constant $T$ and $P$ whenever an increase in its chemical potential decreases the interface free energy. The phenomenon may also be understood at the atomic level by realizing that interfaces generally contain a large variety of sites for atoms which have environments which differ from those of lattice sites and which are therefore often capable of preferentially attracting solute atoms.

This general phenomenon has been termed interface 'adsorption', or, more commonly, 'segregation', and we shall therefore use the latter term. Interface segregation is of great importance since it generally affects the structure and chemistry of interfaces, and, hence, influences many of their important physical properties. This phenomenon, and its ramifications, has therefore been the subject of a great deal of research over the years (Guttmann 1977, Guttman and McLean 1977, Hondros and Seah 1977a,b, Balluffi 1979, Hondros and Seah 1983, Hofmann 1985, 1987, Militzer and Wieting 1987, Yan et al. 1983a; Grovenor 1985, Briant 1990, Seidman 1991, Cabane and Cabane 1991, Briant 1991).

In the following we discuss physical models which account for the basic tendency of solute atoms to segregate to interfaces. Statistical mechanical models are then introduced in order to describe the degree of segregation which may be achieved as a function of bulk solute atom concentration and temperature. Both experimental results and calculated results are cited. By far the most quantitative information has been obtained for metallic systems, and most attention is therefore focused on these. Less extensive information is available for ionic systems, where interesting ionic 'charge cloud' effects are associated with interface segregation, and for covalent solids, where the local electronic structure has been found to play a role.

It is important to point out that solute atom depletion rather than enrichment (i.e. segregation) can occur in certain cases. For example, in binary Pt/Au alloys Monte Carlo calculations (Udler and Seidman 1992b) indicate that Au atoms which are in dilute solution in Pt tend to segregate at grain boundaries, whereas Pt atoms which are in dilute solution in Au tend to be depleted. Such depletion effects, however, are of relatively minor interest in the present context and will not be taken up in the following.

## 7.2 OVERVIEW OF SOME OF THE MAIN FEATURES OF INTERFACE SEGREGATION IN METALS

Since interface segregation has been more extensively studied in metals than in other solid types, we begin with an overview of some of the main features of segregation in metals. The degree of solute atom segregation has often been measured along isotherms where

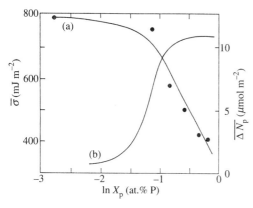

**Fig. 7.1** (a) Experimentally determined decrease in the average energy of grain boundaries in polycrystalline b.c.c. Fe containing P, $\bar{\sigma}$, caused by an increase in the bulk P concentration, $X_P$. (b) Corresponding increase in the average interface excess of $P$, $\overline{\Delta N}_P$. (From Hondros and Seah (1983).)

interface excesses, or other convenient measures of the extent of segregation, are determined as a function of lattice solute concentration at constant temperature (and usually constant pressure). The Gibbs adsorption isotherm, already mentioned above, provides a basis for such measurements. For a dilute solute in a binary system obeying Henry's Law (Gaskell 1983),

$$\mu_2 = \mu_2^{\circ} + kT \ln KX_2, \tag{7.1}$$

where $K$ = Henry's law constant and $X_2$ is the atom fraction of solute. Substituting eqn (7.1) into eqn (5.29) representing the Gibbs isotherm, we obtain

$$\Delta N_2^{\alpha} = -\frac{1}{kT} \left( \frac{\partial \sigma}{\partial \ln X_2} \right)_{T,P}. \tag{7.2}$$

Measurements of $(\partial \sigma / \partial \ln X_2)_{T,P}$ as a function of $X_2$ then yields values of the interface concentration of the solute, $\Delta N_2^{\alpha}$, as a function of $X_2$ directly. The results of measurements of this type for grain boundaries in polycrystalline b.c.c. Fe containing P solute atoms are shown in Fig. 7.1 where both the measured decrease in average $\sigma$ with increasing $X_P$ and the corresponding average values of $\Delta N_P^{\alpha}$ obtained from eqn (7.2) are included. The average grain boundary solute excess at constant temperature is seen to increase with increasing lattice solute concentration (chemical potential) and then saturate at a value corresponding to about 1/3 of a monolayer of P when the bulk concentration reaches a value near 0.1 at.% P. This result shows that a high degree of solute segregation (relative to the bulk concentration) can be reached for certain solutes. This saturation behaviour is not universal, however. Measurements of the segregation isotherms for Sn solute atoms at grain boundaries in polycrystalline b.c.c. Fe, obtained by use of the Gibbs adsorption isotherm and also from direct Auger electron spectroscopy measurements (Fig. 7.2), show no signs of saturation, and the interface excess at 550 °C is seen to reach the considerably higher average level of $\sim 1.5$ monolayers. It is also seen that the boundary segregation increases markedly as the temperature is lowered (at constant bulk composition), and also that the segregation to average grain boundaries is markedly lower than the segregation to average free surfaces. This latter result is found quite generally, and may be expected intuitively because internal interfaces are much 'weaker' planar defects in the lattice than free surfaces.

In general, the average strength of the segregation at grain boundaries in metal polycrystals is found to vary greatly from alloy system to alloy system. A rough guide for rationalizing these wide variations can be established by comparing the strength of the

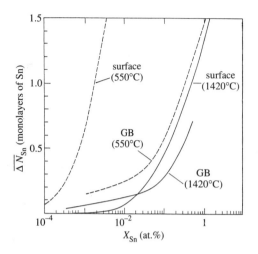

**Fig. 7.2** Experimentally determined increase in the average interface excess, $\overline{\Delta N}_{Sn}$, at grain boundaries in polycrystalline b.c.c. Fe containing Sn caused by an increase in the bulk Sn concentration, $X_{Sn}$. Data for the corresponding increase in the interface excess at free surfaces are also shown. (From Hondros and Seah (1983).)

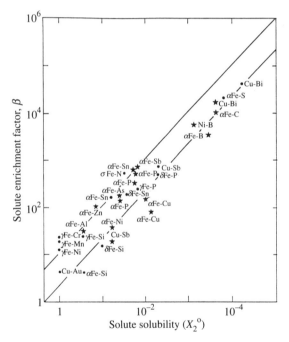

**Fig. 7.3** Collection of experimentally determined grain boundary segregation data for various systems showing the increase in solute enrichment factor, $\beta$ (defined by eqn (7.3)), which is associated with a decrease in the solubility of the solute. (From Hondros and Seah (1983).)

boundary segregation of a solute with its solubility in the bulk lattice. On simple physical grounds, one might expect to find an inverse relationship between the strength of segregation and the bulk solubility, since, in a general way, the same factors which dictate a low solubility should dictate a strong tendency towards segregation. For example, large solute atoms, with small bulk solubilities, should be able to find special highly attractive boundary sites with loose open atomic environments. That this is indeed the case is shown in Fig. 7.3, where the degree of segregation, as measured by the 'solute enrichment factor', $\beta$, is plotted as a function of the bulk lattice solubility of the solute for boundaries in polycrystalline specimens in a wide variety of binary systems. The solute enrichment factor is a convenient dimensionless parameter defined (Hondros and Seah 1977*a*) by

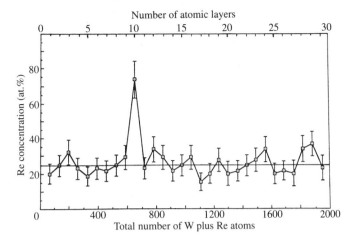

**Fig. 7.4** Experimentally determined concentration of Re atoms in atomic planes lying parallel to a segregated grain boundary in a W–Re alloy. The boundary is centered at plane number 10, and the interface excess of Re is essentially concentrated in this atomic plane. Data obtained by atom probe-field ion microscopy. (From Seidman *et al.* (1990).)

$$\beta = \frac{\Delta N_2}{\Delta N_2^\circ \cdot X_2} = -\frac{1}{\Delta N_2^\circ} \cdot \frac{1}{kT} \left(\frac{\partial \sigma}{\partial X_2}\right)_{T,P}, \tag{7.3}$$

where $\Delta N_2^\circ$ is the boundary excess corresponding to a monolayer of solute at the interface, and $\partial \sigma / \partial X_2$ is evaluated at the dilute limit where $X_2 \rightarrow 0$. The available results span about four orders of magnitude and show a high degree of correlation. Further data consistent with this trend are presented by Briant (1990) and are discussed below with reference to Fig. 7.9.

Numerous experiments using field ion microscopy and Auger electron microscopy (Balluffi 1979) and also calculations have shown that the great majority of the segregated solute atoms in sharp large-angle boundaries in metallic systems are generally located within one or two atomic distances of the boundary midplane indicating that the core sites are generally the strongest segregation sites. Some recent experimental and calculated results are shown in Figs 7.4 and 7.5. In Fig. 7.4 the experimentally determined solute distribution is shown to be highly localized in the core region. In Fig. 7.5 the main portion of the calculated distribution is again highly localized in the core but a relatively small oscillatory 'tail' is seen to extend a short distance into the crystal on the left side of the asymmetric boundary. This interesting result is reminiscent of the oscillating segregation profiles observed at crystalline surfaces (Seidman 1991). The general result that the segregation is mainly to the boundary core may be anticipated on quite general grounds, since it is only in the bad material of the core that we expect to find atom sites which differ considerably from those in the lattice. Outside the core the lattice sites are merely elastically strained, and binding energies which are relatively small are therefore expected. Since for large-angle boundaries the width of the region which is significantly strained elastically is relatively narrow, relatively little segregation may be expected outside of the core region.

The degree of segregation to grain boundaries in a given alloy system will obviously depend upon the detailed boundary structure, and therefore will vary from boundary to

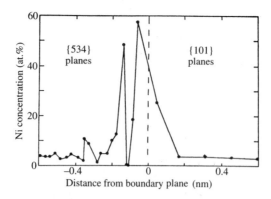

**Fig. 7.5** Calculated concentration of Ni atoms in atomic planes lying parallel to a segregated grain boundary in a Pt–Ni alloy. The boundary has a $\Sigma5$ misorientation and lies parallel to {101} and {534} lattice planes in the adjoining lattices. The calculations employed the Monte Carlo method and embedded atom interatomic potentials (see Sections 4.3.1.6 and 3.7, respectively). (From Seidman *et al.* (1990).)

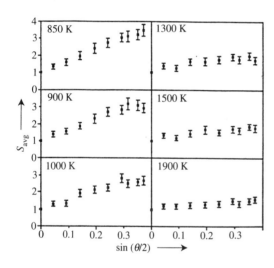

**Fig. 7.6** The average segregation enhancement factor, $S_{avg}$, vs. $\sin(\theta/2)$ for temperatures $T = 850, 900, 1000, 1300, 1500, 1900\,K$ for (001) twist boundaries in a Pt–1 at.%Au alloy computed by the Monte Carlo method. (From Seki *et al.* (1991a).)

boundary. Small-angle boundaries generally exhibit smaller amounts of segregation than large-angle boundaries under similar thermodynamic conditions. This is readily understood, since as pointed out above, strong solute segregation should occur only in, or very near, the bad core material. Small-angle boundaries contain smaller amounts of bad core material than large-angle boundaries, since the bad material is restricted to the cores of the discrete primary dislocations which comprise the boundaries and is not distributed throughout the boundary planes as in large-angle boundaries. The degree of segregation should therefore increase as the misorientation angle increases and the dislocation density increases. This behaviour is clearly evident in many experimental results reported in the literature, e.g. Watanabe (1985) and also in a number of calculated results as shown, for example, in Fig. 7.6. The atomistic calculations leading to the results in Fig. 7.6 showed directly that the positions of the solute atoms were closely correlated with the cores of the primary dislocations. The increase in the degree of segregation with misorientation can therefore be attributed to the increase in primary dislocation density with misorientation. For a detailed discussion of these calculations see Section 7.5. (Also, see Figs 7.18 and 7.19.) In other relevant work, direct experimental evidence for the preferential segregation of Au to the cores of primary screw dislocations in small-angle [001] twist boundaries in Fe/Au alloys has been obtained by Michael *et al.* (1988) through the use of scanning transmission microscopy.

For large-angle boundaries, variations up to $\pm 30$ per cent from the average may be expected (Briant 1983) in polycrystalline specimens, and even this may be exceeded in particular cases. There is evidence (see review by Balluffi 1979; also Watanabe 1985) that the level of segregation is often lower at singular large-angle boundaries of relatively low energy and small free volume than at general boundaries of higher energy. Presumably, the former boundaries contain fewer sites which are attractive to solute atoms. For example, the degree of segregation at the $\Sigma 3$ {111} twin boundary is generally found to be exceedingly small, whereas it is larger at the $\Sigma = 3$ {112} twin boundary. Ogura *et al.* (1978) found large variations in the degree of P segregation to boundaries in polycrystalline f.c.c. Fe containing Ni and Cr (see Fig. 7.7). In many cases lower segregation was found at boundaries of relatively low energy. Negligible segregation was found at {111} twin boundaries, whereas some segregation was found at {112} twin boundaries.

However, the result that segregation at f.c.c. {111} twin boundaries is negligible is not univerally true and depends upon the particular combination of solute and solvent. In the Cu–Bi system it is known experimentally that Bi segregation may induce faceting of grain boundaries in Cu (Donald and Brown 1979, Menyhard *et al.* 1989; Blum *et al.* 1990) and that it is reversible (Ference and Balluffi 1988) as already discussed in Section 6.3.1.1. Segregation of Bi is not observed to occur at pre-existing $\Sigma 3$ {111} twin boundaries (Luzzi 1991*a,b*; Yan *et al.* 1993). However, in the same specimens, strong segregation is observed to occur to $\Sigma 3$ {111} twin boundaries which are formed by the transformation (during annealing) of other boundaries already containing segregated Bi, especially curved boundaries with $\Sigma 3$ misorientations. In these segregated twin boundaries both atomic resolution electron microscopy and computer simulation using Finnis–Sinclair type interatomic potentials (Yan *et al.* 1993) show that there is a single atomic layer of Bi at the boundary, as shown in Fig. 7.8. Each Bi atom in the layer occupies an area equal to three times that of a Cu atom in a {111} layer. This is the densest packing of Bi atoms that is commensurate with a {111} layer and which avoids large Bi–Bi overlap. The Cu layer immediately below the Bi layer contains a vacant site below each Bi atom; thus a third of the Cu sites in this layer are vacant. These vacanct sites allow a decrease in the boundary expansion and a corresponding increases in the atom packing density. Furthermore, the spacing in the segregated hexagonal Bi array is close to the spacing which Bi adopts in the hexagonal basal plane of pure Bi. We note that this special form of ordered segregate may be regarded as a grain boundary phase as discussed in Section 6.3.2.1. Evidently, no segregation is possible at pre-existing $\Sigma 3$ {111} twin boundaries since there is a large barrier to the nucleation and growth of such a phase under such conditions. However, the boundary phase is able to form along a path corresponding to the transformation during annealing of other boundaries already containing segregated Bi.

There is extensive experimental evidence that the degree of segregation to grain boundaries tends to increase as the average interplanar spacing of the lattice planes running parallel to the boundary in each adjoining lattice [i.e. $\bar{d} = \frac{1}{2}\{d_1(hkl) + d_2(hkl)\}$] decreases, i.e. as the boundary becomes less singular. Such boundaries will generally tend to be of relatively high energy and large free volume, and therefore may be expected to be prone to segregation. Suzuki *et al.* (1981) found that the segregation of P to boundaries

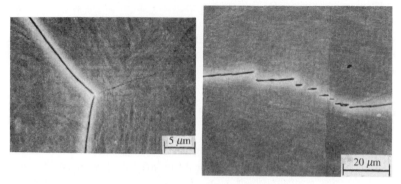

**Fig. 7.7** Electron micrographs showing differences in the degree of segregation of P to grain boundaries in a polycrystalline f.c.c. Fe–3.5%Ni–1.7%Cr–0.3%C steel doped with 0.06% P. The specimen was attacked by an aqueous solution which preferentially etched boundaries containing segregated P. Therefore, the dark appearing boundaries (or boundary segments) contain relatively large amounts of segregated P. (From Ogura *et al.* (1978).)

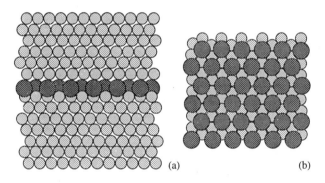

**Fig. 7.8** Calculated relaxed structure of the (111) symmetrical tilt boundary (twin) in Cu (grey) with segregated Bi atoms (black). (a) Side view looking along [1$\bar{1}$0]. (b) Plan view looking on the boundary plane. (From Yan *et al.* (1993).)

in Fe increased as $\bar{d}$ decreased. Ogura *et al.* (1987) and Bouchet and Priester (1987) found a similar trend in other segregating systems. Also, calculations by Larere *et al.* (1988) showed a higher degree of segregation of S and P to symmetric Σ11 tilt boundaries in Ni parallel to (332) rather than (113). More recently, Hofmann *et al.* (1992) and Hofmann and Lejcek (1991) have systematically measured the degree of segregation of Si, P, and C at a number of symmetric and asymmetric [001] tilt boundaries in b.c.c. Fe containing 3.5 at.% Si. Some of their results, including those for both symmetric and asymmetric boundaries in the Σ5 CSL, are shown in Fig. 7.9. Here, the absolute value of the energy of segregation for each solute is plotted versus the angle Ψ, which measures the inclination of the boundary in the Σ5 CSL away from its symmetric position when {013}$_1$ is parallel {013}$_2$. All segregation energies were negative, and therefore low values of the absolute segregation energy correspond to low degrees of segregation. All of the minima in the diagram correspond to relatively large values of $\bar{d}$. Furthermore, the degree of the segregation of the Si, P and C increases as the maximum solubility of these solutes decreases in agreement with the trend seen earlier in Fig. 7.3. In additional work,

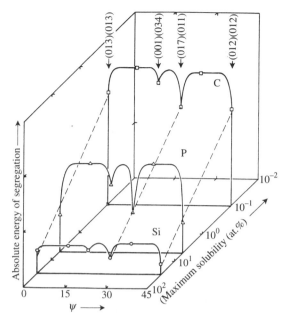

**Fig. 7.9** 3D diagram indicating the experimentally determined degree of segregation of Si, P, and C to $\Sigma 5$ [001] tilt boundaries in a b.c.c. Fe–3.5at.%Si alloy doped with P and C. The energy of segregation was negative in all cases, so that an increase in the absolute value of the energy of segregation in the diagram indicates an increase in the degree of segregation. The angle $\psi$ measures the inclination of the boundary in the $\Sigma 5$ CSL away from its symmetric position when $\{013\}_1$ is parallel to $\{013\}_2$. All of the minima in the diagram correspond to boundaries for which the average spacing of the lattice planes parallel to the boundary, $\bar{d}$, is relatively large. Also, the degree of segregation for Si, P, and C follow the same trend with solute solubility shown earlier in Fig. 7.3. (From Hofmann and Lejcek (1991).)

Hofmann *et al.* (1992) constructed a similar diagram for [001] symmetric tilt boundaries (with $\Psi$ replaced by the tilt angle, $\theta$) and again found minima in the degree of segregation at the boundaries with relatively large $\bar{d}$ spacings.

For large-angle twist boundaries $\bar{d}$ remains constant as the twist angle is varied. According to the simple $\bar{d}$ criterion, the degree of segregation should therefore remain essentially independent of twist angle. However, the results of Seki, *et al.* (1991*a,b*) and Udler and Seidman (1992*b*) for twist boundaries indicate instead a modest average increase in segregation with twist angle in the large-angle regime (at least at low temperatures as seen, for example, in Fig. 7.6). Also, in the case of the [110] twist boundary series in Fig. 7.10 (Seidman 1991), a strong localized decrease in the segregation is seen at the most singular shortest period boundary in the series, i.e. the $\Sigma 3$ boundary.

All of the above results taken together indicate that there are no simple rules which are capable of relating the degree of segregation to the geometric parameters (thermodynamic variables) of a boundary. Segregation is often relatively low at singular boundaries of large $\bar{d}$ spacing or short period, but this trend is not universal. Clearly, the nature of the solvent and solute atoms and their bonding characteristics are important variables which must be considered.

In certain metallic systems, segregation isotherms show regions where the segregation to average boundaries increases very rapidly, such as at A in Fig. 7.11. As shown below

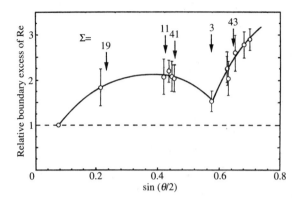

**Fig. 7.10** Measured boundary segregation enhancement factor of Re atoms in segregated near-[110] twist boundaries in a W–25 at.%Re alloy at 1913 K as a function of twist angle. The enhancement factor is the solute concentration in the two planes in the boundary core divided by the bulk concentration. A marked minimum at the short-period $\Sigma 3$ boundary is observed. Data obtained by atom probe-field ion microscopy. (From Seidman (1991).)

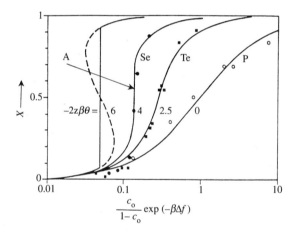

**Fig. 7.11** Experimental data showing Fowler–Guggenheim and McLean type grain boundary segregation behaviour as described in Sections 7.4.2 and 7.4.3. The various curves are drawn according to the Fowler–Guggenheim isotherm (eqn (7.69)) which reduces to the McLean isotherm [eqn (7.32)] when $\theta$ goes to zero. Here, $X$ = grain boundary site occupancy, $c_o$ = lattice site occupancy, $z$ = local coordination number for the solute atoms in the boundary, $\theta$ = solute atom interaction energy parameter (attractive when negative), $\Delta f$ = free energy of segregation independent of any solute atom interaction, and $\beta = 1/kT$. When $2z\beta\theta$ reaches the critical value $-4$, phase separation should just occur due to solute atom attraction. When $2z\beta\theta = -6$, the dashed curve indicates a metastable condition where two coexisting phases of differing composition should be present in the boundary. Data are for P in b.c.c. Fe (Erhart and Grabke 1981), and Te and Se in b.c.c. Fe (Pichard *et al.* 1975). (From Hofmann (1987).)

in Section 7.4.3, this is an indication of Fowler–Guggenheim behaviour presaging possible phase separation in the boundary segregate brought about by solute–solute attractive interactions in the boundary. Interactions of solute atoms with themselves as well as with the boundary are therefore often of prime importance in dictating the form of the segregation.

Additional segregation phenomena of interest appear in grain boundaries in multi-component metallic systems containing two, or more, solutes. Besides the wider range of solute atom–solute atom and solute atom–host atom interactions which are possible (Briant 1991), there is also possible segregation 'site competition'. In this phenomenon, different solute atoms may compete for the same boundary segregation sites. Under certain conditions, when the different solute atoms have sufficiently different binding energies to the available boundary sites, it is possible for one segregating component to force out the other and essentially displace it in the segregate (Briant 1988, 1990, 1991). An example of this is shown in Fig. 7.12.

In view of the range of phenomena described above, the types of segregation which occur at various interfaces under different conditions may be expected to cover an extremely wide spectrum. At one end of this spectrum is the simple dilute limit where a single component at a low bulk concentration segregates weakly to a dilute concentration of special sites in the host interface without sensibly changing its atomic structure. At the other end are complex cases where one or more strongly interacting components, often at higher bulk concentrations, segregate strongly to produce new multilayer structures at the interface. As we have already seen, this type of segregation may produce distinctly new structures and hence possibly induce interface phase transitions (Chapter 6).

Still a further factor which must be considered is whether or not the segregate is fully equilibrated with respect to both itself and with the adjoining bulk system. Important forms of non-equilibrium segregation may occur in many situations including, for example, solute precipitation, quenching and annealing, the application of stress and exposure to radiation damage. The results, of course, depend upon the specimen history. These complex phenomena are considered elsewhere (e.g. Hondros and Seah 1977a, 1983, Yan *et al.* 1983c) and will not be taken up here.

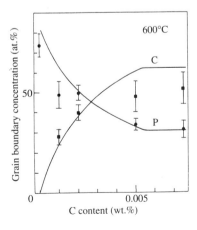

**Fig. 7.12** Data for the segregation of P and C to grain boundaries in b.c.c. Fe illustrating site competition. C competes with P for sites in the boundaries, and when the C content (activity) in the bulk is increased, the degree of C segregation in the boundaries increases while that of the P decreases. (From Briant (1990).)

## 7.3  PHYSICAL MODELS FOR THE INTERACTION BETWEEN SOLUTE ATOMS AND INTERFACES

### 7.3.1  Introduction

The spatial extent of the interaction between a solute atom and an interface varies considerably with the nature of the interaction. It may occur over large distances as in an elastic interaction or an electrostatic interaction in an insulator, or it may be short-range as in an electronic interaction. A solute atom may be attracted to an interface over a large distance, but once it arrives at the interface it may lower the energy of the system considerably by finding a particularly favourable bonding configuration, which is clearly a short-range electronic interaction. On the other hand, short-range attraction of solute to an interface will deplete the solute concentration immediately adjacent to the interface, and this may eventually establish a long-range concentration gradient as the 'denuded zone' extends into the adjoining crystals.

For a long-range interaction the atomic structure of the interface is not so significant as the fact that the interface is an abrupt discontinuity in the elastic or dielectric properties of the bicrystal, or a narrow region where these properties are different from the adjoining crystals. It is then possible to use continuum theories of elasticity or electrostatics to obtain reasonable descriptions of these interactions. Elastic interactions are usually divided into 'size effects' and 'elastic modulus effects', depending on whether the solute atom is regarded as a misfitting inclusion or an elastic inhomogeneity respectively. These interactions are described in Section 7.3.2, and it will be seen that they incorporate certain assumptions about the degree of elastic relaxation of the solute atom at the interface. However, the qualitative insight such models provide is useful, particularly the aymptotic form of the interaction at long range. There are other forms of elastic interaction, such as image interactions, and these are mentioned briefly at the end of the section. Electrostatic interactions may arise in any non-metal, for example from image interactions created by the change in the dielectric constant across a heterophase interface (see Section 3.11), or the formation of charged interfaces in insulators with compensating space charge layers as described in Section 7.6. Although these interactions may drive solute over large distances towards an interface, the solute will generally alter the structure of the interface once it arrives, and often in a way that cannot be described accurately by elasticity theory. To describe the relaxation in such cases it is necessary to use an atomistic model.

Whether segregation takes place at all is a trade-off between the free energy of the solute in the bulk and in the interface, at a given set of chemical potentials, temperature and pressure. For example, in the previous section it was noted that segregation of solute is more likely to occur in those alloys where the equilibrium solubility of the solute in the bulk is limited. Thus, we may say that segregation occurs either as a result of solute being attracted to the unique environment of the interface or as a result of it being rejected by the adjoining crystals.

The factors controlling the solubility of solute in the bulk of metals are systematized in the famous Hume–Rothery rules of alloy formation. The most important of these involves the difference in atomic sizes of the solute and host atoms. When the atomic diameters differ by more than about 15 per cent the solubility in the bulk is small. The strain associated with a solute atom with a large size misfit in the bulk may be relieved, at least partially, at an interface. This is the basis of the 'size effect' interaction. Next in importance is the electrochemical factor, which is the difference in electronegativities of the elements. A substantial difference in electronegativity between the solute and host promotes the precipitation of compounds from solution, and therefore limits solid solubility. As discussed by Pettifor (1987), the role of electronegativity in metals is quite different from its role in insulators. In contrast to an insulating compound, the difference in electronegativity between the constituents of a metallic alloy does not lead to a partially ionic bond because each atom is perfectly screened and remains charge neutral at all times. Instead, the difference in electronegativity leads to the formation of a common, partially occupied band of greater width than the bands of either constituent. It is the greater band width, and hence lower energy of bonding states, that leads to the stability of the compound. As the limit of solid solubility is approached, precipitation of the new compound may occur heterogeneously at interfaces, the precursor of which may be described as non-equilibrium segregation of solute to interfaces. The third Hume–Rothery rule concerns the electron–atom ratio (e/a). When the two factors described above are small, particular values of e/a in certain ranges of composition correlate well with the occurrence of compounds with particular crystal structures. Similarly, the competition between rival crystal structures across the transition metal series is governed by the e/a ratio associated with filling the d-band. Each rival structure has its own density of electronic states, and the structure with the lowest sum of occupied d-states varies systematically in the 4d and 5d series as the e/a ratio varies across the series. In general, we can expect the local densities of states in an interface and the adjoining crystals to differ because the atomic environments differ. Depending on the Fermi energy (which is determined by the e/a ratio) a lower electronic energy may be achieved if the solute is in the interface rather than in solution in the bulk.

Once the solute arrives at the interface, atomic and electronic relaxation occur to lower the energy of the system. These modes of relaxation are frequently separated in models of the interaction, which usually treat either the atomic or the electronic relaxation, but in reality they are fully coupled. At the most fundamental level all interactions may be described as electronic in origin, but considerable insight may be obtained through the simpler models we have already mentioned. For example, the notion of an atomic size misfit stems ultimately from the boundary condition that is placed on the radial part the wave function of the valence electrons of the solute atom (Mott 1962). But, as shown in Section 7.3.2, by making simplifying assumptions about the relaxation of the misfit of the solute atom in the interface the elastic model may be solved analytically to yield approximate segregation energies and the functional form of the long-range interaction between the solute atom and the interface.

The principal shortcoming of the elastic interaction models is the rather arbitrary assumption that is made about the extent of the elastic relaxation in the interface. In addition, new ordered structures may be formed in the interface, which cannot be readily accounted for by elasticity theory. For these reasons several atomistic studies have been performed to study the relaxation at the interface and to calculate segregation energies to various sites at interfaces. Some of these studies are described in more detail in Section 7.3.3. In such a study explicit functional forms for the atomic interaction potentials are used to obtain the relaxed structure at 0 K. While these studies give a better description of the relaxation and segregation energy than elastic interaction models they are very limited in their usefulness as a description of equilibrium segregation as a function of temperature. Atomistic models which take into account the influence of a finite temperature are described in Section 7.5, after the statistical mechanics of segregation has been introduced in Section 7.4.

Two mechanisms of electronic relaxation may be distinguished: weak and strong. In a weak interaction perturbation theory may be used to calculate the segregation energy. An example of this is dopant segregation to grain boundaries in semiconductors, and is described in Section 7.3.4. The solute occupies a site in the interface which is very similar to a site in the bulk, and the atomic relaxation in the interface is negligible. For this reason the weak interaction may be described as purely electronic in character. Solute segregates to the interface in this case because the interface is slightly more responsive ('softer') electronically than the bulk, allowing a greater degree of electronic relaxation for the whole system. By contrast, a strong interaction results in the formation of new bonds in the interface, and, in general, the concomitant atomic relaxation is significant. An example of this is the formation of an ordered compound in the interface, with the solute atoms occupying specific sites that are usually not found in the bulk. For a strong interaction the large structural rearrangements that take place in the interface, and/or the new bonds that are formed, lead to a new density of electronic states in the interface that cannot be treated by perturbation theory. The existence of an electronic interaction, either weak or strong, is dependent critically on the electron–atom ratio, the electronegativity difference between the solute and host, and the angular character of their valence orbitals. These three factors have not been taken into account in virtually all atomistic simulations of segregation using interatomic potentials.

### 7.3.2 Elastic interaction models

The total elastic interaction of solute atoms with interfaces is generally complex and involves a number of physically distinguishable features which are of varying importance depending upon the particular system under consideration. A range of models therefore exists in which the different models emphasize various aspects of the total elastic interaction. In the following we describe three of the more widely considered models in some detail and mention several others only briefly.

#### 7.3.2.1 *Size accommodation model*

This model assumes that the main interaction energy is simply the decrease in elastic strain energy which occurs when a misfitting solute atom, which is either too large or too small in the lattice, leaves the lattice and finds an interface site in which it enjoys a better fit. A 'ball-in-a-hole' model (Christian 1975) is employed in which the solute atom is represented by an elastic sphere fitted into a spherical hole in an elastic matrix continuum. The stress-free volume of the ball generally differs from that of the hole in order to mimic the difference in 'size' between the solute atom and the site in which it sits. The total elastic energy stored in the strained ball and matrix is then a measure of the relative energy of the solute atom at that site, and the difference between the total elastic energy when the solute atom occupies a boundary site and a bulk lattice site is a measure of the binding energy to the boundary site.

In order to evaluate this energy, we require the solution of the ball-in-a-hole elasticity problem. Assuming isotropic elasticity, the desired solution for the displacement field around the ball, $u(r)$, which is spherically symmetric and which satisfies the standard Navier equation (Sokolnikoff 1956),

$$(\lambda + 2\mu) \nabla (\nabla \cdot u) + \mu \nabla \times (\nabla \times u) = 0, \tag{7.4}$$

is of the general form

$$u(r) = a_1 r + a_2 r^{-2}. \tag{7.5}$$

If $r_0$ is the initial radius of the stress-free hole and $r_0(1 + \varepsilon)$ the initial radius of the stress-free ball, we suppose that the final relaxed radii after insertion will be $r_0(1 + a_3\varepsilon)$, where $a_3 < 1$. Assuming initially an infinite matrix, the boundary conditions are then $u(r_0) = a_3 r_0 \varepsilon$ and $u(\infty) = 0$ for the matrix and $u(r_0) = (a_3 - 1)r_0\varepsilon$ and $u(0) = 0$ for the ball. Equation (7.5) then becomes

$$u(r) = a_3 r_0^3 (\varepsilon/r^2) \quad \text{(for the matrix)}; \tag{7.6}$$

$$u(r) = (a_3 - 1)\varepsilon r \quad \text{(for the ball)}. \tag{7.7}$$

The constant $a_3$ is found by matching the normal radial stress, $\tau_{rr}$, across the ball-matrix interface. In general, $\tau_{rr} \equiv 2\mu(\partial u/\partial r)$, and, therefore, at the interface in the matrix, $\tau_{rr} = -4\mu^M a_3 \varepsilon$, where $\mu^M$ is the matrix shear modulus. On the other hand, the stress in the ball is purely hydrostatic, and therefore at the interface in the ball $\tau_{rr} = K^B(\Delta V/V) = 3K^B\varepsilon(a_3 - 1)$, where $K^B$ is the bulk modulus of the ball. Setting these stresses equal, we then find $a_3 = 3K^B/(3K^B + 4\mu^M)$. Our solution, so far, is for an infinite matrix. In order to obtain a solution for the ball at the centre of a finite matrix sphere of radius $R$ with a stress-free surface, we first remove all matrix material beyond $r = R$ and apply radial forces at the surface in order to maintain the radial stress there at the initial value $\tau_{rr}(R) = -4\mu^M a_3 \varepsilon(r_0/R)^3$. We then apply an equal and opposite set of surface forces to cancel these forces and produce a stress-free surface. This operation produces a homogeneous hydrostatic 'image' stress throughout the matrix given by $\tau_{rr}' = 4\mu^M a_3 \varepsilon(r_0/R)^3$.

We can now find the total elastic strain energy by integrating the elastic strain energy density, $w$, over both the ball and matrix volumes. For the present spherical coordinates

$$w = \tfrac{1}{2}\left[\tau_{rr}\varepsilon_{rr} + \tau_{\theta\theta}\varepsilon_{\theta\theta} + \tau_{\phi\phi}\varepsilon_{\phi\phi}\right]. \tag{7.8}$$

The various stresses and strains in eqn (7.8) for the ball and matrix are readily derived from our previous results by employing the standard equations of elasticity relating stress, strain and displacement (Sokolnikoff 1956), and the total strain energy obtained by integrating over their volumes is finally

$$W = \frac{6\pi K^B r_0^3 \varepsilon^2}{(1 + 3K^B/4\mu^M)}. \tag{7.9}$$

The binding energy of a solute atom to a boundary core site, or energy of segregation, $\Delta e_{\text{seg}}$, is taken as the energy of the system with the solute atom in the boundary site minus the energy when the solute atom is in the bulk lattice. The maximum possible misfit binding energies in this model would then occur at special boundary core sites where the solute atom is fully relaxed elastically. In such a case, the energy of segregation would be just $\Delta e_{\text{seg}} = -W$, where $W$ is the elastic energy associated with the solute atom in a bulk site given by eqn (7.9). Estimates of this quantity for different solute–matrix systems (McLean 1957) yield values of the order of minus several tenths of an electron volt which is in the range measured experimentally from adsorption isotherms (see below). On the other hand, if the misfit is only partially relaxed, the segregation energy will be reduced and can be obtained from eqn (7.9) by suitably adjusting $\varepsilon$.

### 7.3.2.2 *Hydrostatic pressure (P$\Delta$V) and elastic inhomogeneity models*

The first of these models applies for a misfitting solute atom when the boundary generates a hydrostatic pressure in its vicinity: the second applies when the solute atom can be regarded as either an elastically 'soft', or elastically 'hard' region in the elastic continuum (i.e. an elastic inhomogeneity) and is therefore either attracted to, or repelled from, regions of high strain energy density associated with the stress field of the boundary.

General expressions for both of these interactions can be obtained by employing a procedure due to Eshelby (1977). Our first task is to obtain an expression for the interaction energy of a misfitting solute atom with a general stress field in a bulk elastic material. Following Eshelby, we accomplish this by first imagining a massive block which is perfectly rigid at all times and contains a cavity in it which is almost a perfect unit cube. The cavity deviates from a unit cube by being slightly too small in volume and possessing a shape which is slightly parallelpipedal. We now take a stress-free unit cube of the bulk elastic material which is the focus of our interest and force it into the rigid cavity by first squeezing it down to the required volume using hydrostatic pressure and then shearing it (at constant volume) so that it exactly fits. This will produce a general stress field in the inserted elastic material having an elastic strain energy density, $w$, which can be taken as the sum of the strain energy density due to the dilation, $w_d$, and that due to the shear straining, $w_s$, i.e.

$$w = w_d + w_s. \tag{7.10}$$

Writing the dilation of the inserted cube in the form $e = \Delta V_c / V_c$, where $\Delta V_c$ is its change in volume, and $V_c = 1$, we have $-P = Ke$, where $K$ is its bulk elastic modulus and $P$ is the hydrostatic pressure to which it is subjected ($P$ is taken positive for compression). Therefore, the dilatational strain energy density in the inserted cube can be written in the form

$$w_d = -\tfrac{1}{2}P(\Delta V_c / V_c) = Ke^2/2. \tag{7.11}$$

The shear strain energy density will always be proportional to the shear modulus, and for present purposes it can be written in the similar form

$$w_s = \mu s^2/2, \tag{7.12}$$

where s is a function of the various shear strains present. (Note that the surrounding rigid block remains unchanged during this process and all of the following operations: it simply provides a means of applying a general stress to the inserted cube.)

We next imagine two processes in which $N$ solute atoms (represented by isotropic centres of dilation) are introduced into the elastically strained cube in a random distribution. In the first (process 1) we start with our initial stress-free unit cube and insert the solute atoms before it is forced into the cavity. Ignoring any interactions between the solute atoms, this requires an amount of work equal to $NE$, where $E$ is the work per solute atom. We next force the cube into the rigid cavity. In this step the work required will be $w + \delta w$. The additional $\delta w$ term occurs because the work required to insert the cube differs from w because of the presence of the newly added solute atoms. Using eqns (7.10), (7.11), and (7.12),

$$\delta w = \delta(w_d + w_s) = \tfrac{1}{2}[2Ke\delta e + e^2 \delta K] + \tfrac{1}{2}[2\mu s\delta s + s^2 \delta \mu]. \tag{7.13}$$

Letting $X$ = atom fraction of solute atoms,

$$\frac{\delta w}{\delta X} = Ke\frac{\delta e}{\delta X} + \frac{1}{2}e^2\frac{\delta K}{\delta X} + \frac{1}{2}s^2\frac{\delta \mu}{\delta X}, \tag{7.14}$$

since $\delta s/\delta X = 0$ because there is no shape change in shear when the isotropic centres of dilation are randomly inserted (Balluffi and Simmons 1960). The total work required in Process 1 is then

$$w(1) = NE + w + \delta w = NE + w + X(\delta w/\delta X)$$

$$= NE + w + XKe\frac{\delta e}{\delta X} + \frac{1}{2}Xe^2\frac{\delta K}{\delta X} + \frac{1}{2}s^2X\frac{\delta \mu}{\delta X}. \tag{7.15}$$

In the second process we first force the unit cube into the rigid cavity in the absence of any solute atoms and then add the solute atoms. The work required here may be written in the form

$$w(2) = w + NE + NE_{int}, \tag{7.16}$$

where $E_{int}$ is the interaction energy of each solute atom with the stress field present in the elastically strained cube. Since $w(1)$ must equal $w(2)$, we then have the result

$$\cdot E_{int} = \frac{XKe}{N}\frac{\delta e}{\delta X} + \frac{Xe^2}{2N}\frac{\delta K}{\delta X} + \frac{Xs^2}{2N}\frac{\delta \mu}{\delta X}. \tag{7.17}$$

Using the relations $X/N = \Omega$ and $-P = Ke$ and eqns (7.11) and (7.12), eqn (7.17) may be further written as

$$E_{int} = -P\Omega\frac{\delta e}{\delta X} + \Omega w_d\frac{1}{K}\frac{\delta K}{\delta X} + \Omega w_s\frac{1}{\mu}\frac{\delta \mu}{\delta X}. \tag{7.18}$$

Now, if $\Delta V$ = volume change due to the insertion of a solute atom, $\delta e = -\Delta V \delta N$. Note that $\delta e$ is negative when $\Delta V$ is positive. Therefore, $\delta e/\delta X = -\Delta V/\Omega$, and we finally obtain the expression

$$E_{int} = P\Delta V + \left[\frac{\Omega}{K}\frac{\delta K}{\delta X}w_d + \frac{\Omega}{\mu}\frac{\delta \mu}{\delta X}w_s\right]. \tag{7.19}$$

Equation (7.19) shows that the interaction energy consists of two terms. The first, $P\Delta V$, is the interaction of a solute atom (centre of dilation) with the hydrostatic component, $P$, of the stress field; the second, bracketed, is the so-called elastic inhomogeneity interaction. This is present because whenever a solute atom acts as either a 'hard' or 'soft' spot, i.e. as an elastic heterogeneity which increases or decreases an elastic modulus, an interaction energy arises which is proportional to the appropriate local strain energy density, i.e. either $w_d$ or $w_s$. For example, if a solute increases $K$ (i.e. $\delta K/\delta X$ is positive) and therefore acts as a hard spot, a positive interaction will occur, since $w_d$ is always positive. Such a 'hard' solute atom will therefore be repelled from a region of high dilatational strain energy density as might be expected intuitively.

Having these results, the segregation energy of a solute atom in the vicinity of an interface, $e_{seg}$, can be obtained by associating $P$, and $w_d$ and $w_s$, with the stress field of the interface and setting $e_{seg} = E_{int}$. If the stress field of the boundary possesses a hydrostatic component, the $P\Delta V$ interaction can be determined by using our previous ball-in-a-hole model results to calculate $\Delta V$. For the solute atom ('ball') in an infinite matrix, $\Delta V^\infty = 4\pi r_0^2 \cdot u(r_0) = 4\pi a_3 r_0^3 \varepsilon$. The hydrostatic image stress then produces an additional volume change given by $\Delta V' = 4\pi R^3/3K^M \cdot \tau'_{rr} = 16\pi\mu^M a_3 r_0^3\varepsilon/3 \cdot K^M$. Since $\Delta V = \Delta V^\infty + \Delta V'$, the total $P\Delta V$ segregation energy is then

$$\Delta e_{seg} = P\Delta V = 4\pi r_0^3\varepsilon\frac{K^B(3K^M + 4\mu^M)}{K^M(3K^B + 4\mu^M)}P. \tag{7.20}$$

A comparison of eqns (7.9) and (7.20) shows that $\Delta e_{seg}$, given by eqn (7.20), is often expected to be smaller in magnitude than that given by eqn (7.9), since high hydrostatic pressures, $P$, near the shear modulus in magnitude, would be required to make them comparable. The elastic inhomogeneity interaction can be estimated using information gleaned from measurements of the effect of the solute atoms on the elastic constants.

These types of interactions have been widely considered in the literature for a variety of point defects at, or near, different boundaries. For example, the binding energies of a number of isovalent solute atoms of different sizes to a ⟨110⟩ {211} tilt boundary in NiO have been calculated atomistically using the energy minimization method (Section 4.3.1.6) by Duffy and Tasker (1984). Here, electrostatic effects should be minimal, and it was found that the binding energy increased with increasing $\Delta V$ in a manner consistent with the $P\Delta V$ interaction. Recent attempts to evaluate $P\Delta V$ and elastic inhomogeneity interactions for solute atoms segregating to small-angle twist and tilt boundaries, which illustrate the strengths and weaknesses of the approach, have been made by Udler and Seidman (1992a).

### 7.3.2.3  *Further elastic models*

Further elastic models may be required to describe additional situations, e.g. Cochart *et al.* (1955), Bacon (1972), Guyot and Simon (1975), Djafari-Rouhani *et al.* (1980), and Deymier *et al.* (1989). Included are cases where: (i) the solute atoms act as anisotropic centres of dilation in the lattice, as, for example, interstitial C atoms in b.c.c. Fe (Cochart *et al.* 1955). In such cases the solute atoms can be represented elastically by sets of double forces (Hirth and Lothe 1982) which are unequal; (ii) an interaction occurs because the effective elastic constants of the boundary region differ from those of the matrix (Deymier *et al.* 1989). If the boundary is generally elastically 'softer' than the adjoining crystals, as it is expected to be in many cases, the strain energy associated with a solute atom will tend to be decreased when it enters the boundary causing an attractive inter-action. The elastic interaction energy then varies as $d^{-4}$, where $d$ is the separation of the point defect from the grain boundary; (iii) non-linear elastic interactions are present which may be significant because of unusually large strains in the boundary core; and (iv) an interaction occurs due to differences in the elastic constants of the crystals adjoining the boundary (Bacon 1972). Here, for grain boundaries the effective elastic constants in the adjoining crystals will differ because of the crystal misorientation and the elastic anisotropy, and an image force will be established which will either attract or repel the solute, depending upon its location in the vicinity of the boundary. For heterophase interfaces, the elastic constants will differ across the interface irrespective of any anisotropy, and an additional image force will exist. As was first shown by Bacon (1972), the elastic image interaction energy varies as $d^{-3}$, and may be repulsive or attractive. A point defect is repelled by an elastically harder medium on the other side of the interface and attracted by a softer medium. We note that image forces of these types also exist for dislocations near an interface and that these are described in Section 12.3.

### 7.3.3  **Atomistic models at 0 K**

In order to obtain a more realistic description of the atomic relaxation around impurity atoms at interfaces than can be obtained with a simple elastic model, it is necessary to consider atomic interactions explicitly. Several atomistic simulations have been performed in which the internal energy, or enthalpy, is minimized at absolute zero. In most of these studies atomic interactions have been described by non-directional interatomic potentials such as pair potentials or embedded atom or Finnis–Sinclair type potentials. In such cases the relaxation in the boundary is governed by packing considerations, namely how densely the atoms in the boundary may be packed given that the solute and host atoms have different sizes. Here we mention the main results of these simulations and illustrate them with a few examples.

One of the first simulations of impurity segregation was carried out by Sutton and Vitek (1982) for Bi and Ag segregating substitutionally to sites at grain boundaries in Cu and Au. Atomic interactions were represented by pair potentials augmented with a local volume dependent contribution. Atomic relaxation around the impurity atoms, both in the bulk and at the grain boundaries, was shown to be crucial, because without relaxation one cannot even be confident of the sign of the energy of segregation. The extent of the relaxation normal to the boundary plane increases with the misfit between the sizes of the solute and host atoms, and in general it extends as far as the relaxation of the clean host boundary, i.e. of the order of the boundary period. The energy of segregation was found to vary markedly from one boundary site to another. Even in cases where there is very limited solid solubility in the bulk, such as Bi in Cu, there may be sites where segregation is unfavourable. Using the interatomic potentials it is possible to define a local stress tensor at each atomic site (see eqn (4.84)). At a grain boundary there is a range of sites varying between hydrostatic tension and compression, and a bulk site is in a state of zero hydrostatic stress. In general, a site of hydrostatic tension is characterized by a larger local volume than that of a bulk crystal site. A large misfitting atom, such as Bi in Cu, is not attracted to a site of hydrostatic compression because the neighbouring host Cu atoms are closer than they are at a site in the bulk, and this leads to a greater short-range repulsion. Thus, Bi atoms seek sites of hydrostatic tension in the boundary plane. This amounts to a refinement of the size effect in elastic models where it is recognized now that there is a range of possible sites for the segregating impurity, some of which may be entirely unsuitable. This is the concept of site selectivity, and it is likely to be valid at all interfaces. It is the reason why the level of segregation may vary from one boundary to another, because the ranges of available sites differ in general.

The segregation of P and B in $\alpha$-Fe (b.c.c. Fe) has also been modelled. Using pair potentials, Hashimoto *et al.* (1984) considered the segregation of P and B to the $\Sigma 5$ (310) and $\Sigma 9$ (114) symmetrical tilt boundaries in Fe. Phosphorus occupies a substitutional site in $\alpha$-Fe, whereas B dissolves interstitially. The smaller size of the B atom, compared with the P atom, was reflected in the pair potentials between the host and each impurity. The clean (310) and (114) boundaries comprise capped trigonal prisms and pentagonal bipyramids, as shown in Fig. 7.13(a), (b). A monolayer of impurity atoms was introduced substitutionally or interstitially for P and B respectively into the boundaries and relaxed at 0 K. The B atoms occupied interstitial sites at the centres of the capped trigonal prisms and pentagonal bipyramids. The relaxation induced by the B atoms was small. But the P atoms induced a complete restructuring of both the (310) and (114) boundaries, in which the P atoms no longer occupy substitutional sites but interstitial sites at the centres of newly formed capped trigonal prisms. This is illustrated in Fig. 7.13(c), (d). At the same time the energies of the boundaries were lowered considerably. It is interesting to note that the structure of the crystalline phase $Fe_3P$ consists of P atoms at the centres of trigonal prisms of Fe atoms. The final sites of the P atoms are thus very similar to those of P in $Fe_3P$, and to those of B atoms in the grain boundaries. At lower levels of segregation of P the boundaries did not undergo the complete restructuring induced by a monolayer.

In view of the fact that pair potentials were used to describe the atomic interactions the restructuring of the boundary was driven only by packing considerations, since the

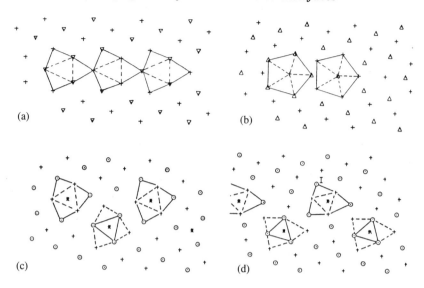

**Fig. 7.13** (a) The relaxed structure of the Σ5 (310) symmetrical tilt boundary in α-Fe, seen in projection along the [001] tilt axis. Note the capped trigonal prisms. (b) The relaxed structure of the Σ9 (114) symmetrical tilt boundary in α-Fe, seen in projection along the [1$\bar{1}$0] tilt axis. Note the pentagonal bipyramids. (c) The relaxed structure of the (310) symmetrical tilt boundary in α-Fe with segregated P atoms at sites denoted by stars. Note that the P atoms occupy interstitial sites at the centres of capped trigonal prisms, which are not at the same location in (a). (d) The relaxed structure of the (114) symmetrical tilt boundary in α-Fe with segregated P atoms at sites denoted by stars. Note that the P atoms occupy interstitial sites at the centres of capped trigonal prisms which are not present in (b). (From Hashimoto *et al.* (1984).)

formation of chemical bonds cannot be described by pair potentials. There are many 'interstitial compounds', such as TiC and WC, in which the crystal structure adopted is dependent on the ratio of the sizes of the constituent atoms (see Hume-Rothery *et al.* 1969). However, even though packing considerations may determine the crystal structure there is little doubt that directional bonds are established, and that these are the source of the exceptional hardnesses and high melting points of these materials. Such directional bonding between d-orbitals of the Fe and p-orbitals of the P has been revealed directly in recent simulations, based on density functional theory, of P segregation to a (111) twin in α-Fe [Tang *et al.* (1993)]. But the important feature of the result obtained by Hashimoto *et al.* is that, while packing considerations forbid P occupying an interstitial site in bulk α-Fe, sufficient free volume at the grain boundaries in α-Fe can be created by restructuring, and that this restructuring lowers the energy of the system considerably. The net result is that the P atoms transform the grain boundary into an environment similar to that of Fe$_3$P.

Another case of interest is the formation of the ordered segregate (grain boundary phase) at a Σ3 {111} twin boundary in the Cu–Bi system already discussed in Section 7.2 and illustrated in Fig. 7.8. Here, the Bi layer is highly commensurate with the adjacent Cu layers and is accommodated by a layer of vacant sites in the Cu which allows the packing density to be increased. Furthermore, the atomic spacing in the hexagonal Bi layer is close to the spacing which Bi adopts in the hexagonal basal plane of pure Bi. Since the atomic interactions used in the calculations were described by Finnis–Sinclair potentials (Yan *et al.* 1988), packing considerations were again of paramount importance in the determination of the final structure.

### 7.3.4  Electronic interaction models

There are certain cases where it is possible to assign the energy of segregation almost entirely to an electronic origin. This is usually rare because in general the interaction between a solute atom and an interface involves ionic relaxation, as well as the formation of chemical bonds which may be quite different in character from those found in the bulk. In Section 7.3.1 we classified the former as a weak electronic interaction and the latter as a strong interaction. To model strong interactions, a formulation of atomic interactions must be used that is capable of describing bond formation and ionic relaxation. This is only just beginning to be possible, and an example is given later in this section.

In a weak interaction the difference in the energy of ionic relaxation around the impurity in the bulk and in the interface is negligible. The energy of segregation is then dominated by the electronic contribution to the total energy of the system. A good example of this is the segregation of dopant atoms to tetrahedral sites at grain boundaries in Si and Ge. There is experimental evidence that, under equilibrium conditions, n-type dopants, such as P and As, segregate to grain boundaries in Si and Ge, whereas p-type dopants such as B do not (for a review see Maurice (1991)). An explanation for this phenomenon that is based entirely on an electronic mechanism was proposed by Arias and Joannopoulos (1992). They observed that in the clean $\Sigma 5$ (310) symmetrical tilt boundary in Ge there is a shallow, unoccupied interfacial electronic state about 10 meV below the conduction band edge. This state mixes with the shallow donor state associated with As in the bulk to yield a new state approximately 0.1 eV below the conduction band edge. The result is an energy of segregation of approximately 0.1 eV, which arises almost entirely from the energy lowering of the highest occupied state of the dopant atom. Since 0.1 eV is four times larger than $kT$ at room temperature, the n-type dopant is not expected to be ionized at room temperature, and therefore the dopant segregation should increase the resistivity of the material through the reduction in the number of carriers. (However, this mechanism is not sufficient to account for the rapid increase in resistivity with segregation observed experimentally, which is more consistent with deeper states being introduced into the gap and pinning the Fermi energy (Grovenor 1985).)

To understand the mechanism more quantitatively consider the change in the energy of occupied electronic states in the bicrystal following the segregation. Let $\Delta N(E)$ denote the change in the total density of electronic states with the dopant atom at the interface compared with the dopant atom in the bulk. Assuming there is no change in the Fermi energy of the system (i.e. the Fermi energy is assumed to be pinned by the relatively large number of dopant atoms in the bulk), the change in the energy of the occupied states is given by

$$\Delta E = \int^{E_F} E \Delta N(E) \, dE. \tag{7.21}$$

Since $\Delta N(E)$ is small we may use second-order perturbation theory to obtain

$$\Delta E^{(2)} = -\sum_m^{\text{occ}} \sum_n^{\text{unocc}} \frac{|\Delta H_{mn}|^2}{E_n - E_m} \tag{7.22}$$

where $\Delta H$ is the change of the electronic Hamiltonian accompanying the exchange of a host atom in the grain boundary with a substitutional dopant atom in the bulk. The index m refers to an occupied and n to an unoccupied eigenstate of the bicrystal *before* the exchange takes place. Note that $\Delta E^{(2)}$ is negative, because each term in the double sum is positive. The most important contribution arises when the denominator is smallest, which occurs when m refers to the highest occupied state and n refers to the lowest unoccupied state. The highest occupied state is the donor state on the n-type donor in the bulk, which is just below the conduction band. The lowest unoccupied state is the interface state also just below the conduction band. The matrix element $\Delta H_{mn}$ is maximized when the donor impurity is in the interface because the interface state is localized at the interface. Thus, the interaction between the bulk donor state and the shallow interface state leads to a favourable energy of segregation.

Let us now suppose the interface were associated with a localized state just above the valence band. Would an interaction between this state and that of a bulk dopant acceptor state lead to a favourable energy of segregation? From eqn (7.22) we see that the answer depends on whether the clean interface state is occupied or unoccupied. If it is unoccupied then there is no energy gain from mixing with the acceptor state of the dopant because both states would then be unoccupied. But if it is occupied then there is an energy lowering of the system if the dopant segregates to the interface, and the acceptor state will probably drop into the valence band. The experimental observation that acceptor dopant atoms do not segregate to grain boundaries suggests that either there are no shallow interface states just above the valence band, or, if they exist, that they are unoccupied when the boundary is clean. In the example of the $\Sigma 5$ (310) symmetrical tilt boundary considered by Arias and Joannopoulos (1992) the clean boundary is not associated with any shallow states above the valence band, and therefore they find that the segregation of Ga to the boundary is not favourable. There is also an unoccupied donor state associated with Ga, but the interaction between this state and the unoccupied interface state just below the conduction band does not affect the total energy in eqn (7.22) because both states are unoccupied.

Unfortunately, the theoretical prediction (Arias and Joannopoulos 1993) that the segregation energy of As to grain boundaries in Ge should be about 0.1 eV is inconsistent with experimental measurements of a segregation energy of As to grain boundaries in Si of about 0.65 eV (Grovenor *et al.* 1984). One possible reason for this discrepancy is that segregation takes place preferentially to defects in the interface such as dislocations and steps.

Heggie *et al.* (1990) identified a strong interaction between P atoms and the core of a 90° partial dislocation is Si, involving the breaking of reconstructed bonds in the core, a concomitant release of strain energy, and the creation of a particularly favourable environment for the P atom. When we are considering the donor atoms of Group V we should keep in mind that their preferred coordination is three-fold. This follows because of the four $sp^3$ hybrids that may be formed from the valence s and p states, one of them must be doubly occupied since there are five valence electrons. The two electrons that do not participate in bonding form a 'lone-pair'. Heggie *et al.* found that if a pair of P atoms replaces the pair of Si atoms at either end of the reconstructed bond in the core of the 90° partial dislocation in Si then the reconstructed bond is broken. In this way both P atoms become three-fold coordinated, and the strain in the system arising form the presence of the reconstructed bond is reduced. The lone pairs on each P atom are in the valence band and the gap is cleared of electronic states. The energy lowering of the system for each pair of segregating P atoms was estimated to be 2.3 eV. This is consistent with the experimental observations, referred to in Heggie *et al.*, of strong pinning of dislocations in Si by dopant atoms.

In metals eqn (7.22) indicates the electronic energy of the system is lowered by segregation if the interface has a high local density of states at the Fermi energy and if the solute atom couples states just above and below the Fermi energy in the interface. This is closely related to the 'frontier orbital' approach to chemisorption on metal surfaces (Hoffmann 1988), which has been discussed in the context of grain boundary segregation by Dal Pino *et al.* (1993).

## 7.4 STATISTICAL MECHANICAL MODELS OF SEGREGATION

### 7.4.1 Introduction

Having discussed the physical basis for the interactions of solute atoms with interfaces, we are now interested in statistical mechanical models which predict the extent of the segregation as a function of solute concentration and temperature. Simple models have been developed differing in the assumptions they make about interactions between segregated solute atoms and whether the number of sites for segregation is fixed. In this section we give a general discussion of the problem, which is illustrated in the next section by two particular models of atomic interactions.

Consider a bicrystal containing A and B atoms. The interface may be of the homophase type, e.g. a grain boundary separating misoriented crystals of an A/B alloy or an anti-phase boundary separating two ordered A/B alloy crystals, or of the heterophase type, e.g. an interface separating crystals of different phase. Whatever the interface type, it is assumed that all atomic sites are occupied either by A or B atoms, which limits the treatment to substitutional alloys. It is not assumed that the positions of the atomic sites are fixed. The problem at hand is to determine the equilibrium distribution of A and B atoms in the vicinity of the interface, at a given temperature and pressure. The problem is still not fully defined until we specify whether the numbers of A and B atoms are fixed or whether the chemical potentials of A and B atoms are fixed. We assume the latter, since the crystals on either side of the interface act as large reservoirs of A and B atoms, at fixed chemical potentials $\mu_A$ and $\mu_B$ which can exchange atoms with the interfacial region.

When we speak of the distribution of A and B atoms in the interface what do we mean? Even at equilibrium there are fluctuations in the atomic structure and solute distribution. Thus, if we took two snapshots of the interface at different instants we might see different positions of solute atoms. But if we averaged the position and occupancy of each site over a large period of time we would expect, at equilibrium, to obtain convergent values. They are convergent in the sense that if we increase the period of time over which the averaging is done we do not change the average values. Thus the problem at hand addresses the time-averaged atomic structure and solute distribution within the interface. Let $p_i$ denote the occupancy of site $i$ at a given instant in time. We define $p_i$ as follows:

$$p_i = 1 \text{ if site } i \text{ is occupied by a B atom}$$

$$p_i = 0 \text{ if site } i \text{ is occupied by an A atom} \tag{7.23}$$

Let the time average of $p_i$ be $\langle p_i \rangle = c_i$. We call $c_i$ the occupancy of site $i$, with the understanding that it means the average occupancy of the site at equilibrium. The site occupancy is a number lying between 0 and 1. For example, if $c_i$ is 0.73 then the probability of site $i$ being occupied by a B atom is 73 per cent. Let $R_i$ be the position of site $i$ at any given instant and let the time average of $R_i$ be $\langle R_i \rangle = r_i$. In statistical mechanics the time-averaged quantities, at equilibrium, are the expectation values for those quantities computed in the appropriate ensemble. We are working in the grand canonical ensemble since the numbers of A and B atoms are not fixed. We assume that there are no vacant sites, although this assumption may be relaxed by regarding the vacant sites as the third component of a ternary alloy. In this ensemble the grand potential is minimized at equilibrium. We shall write down an expression for the grand potential in terms of the sets $\{c_i\}$ and $\{r_i\}$ and demand that it is minimized with respect to these variables. This will involve a number of approximations, but we aim to avoid specific assumptions about the nature of atomic interactions until Section 7.4.3.

We express the grand potential, $\Omega$, as follows:

$$\Omega = F(\{c_i\}, \{r_i\}) - TS_c - \mu_A N_A - \mu_B N_B. \tag{7.24}$$

$F$ is the Helmholtz free energy of the ensemble excluding the configurational entropy:

$$F(\{c_i\}, \{r_i\}) = E(\{c_i\}, \{r_i\}) - TS_v(\{c_i\}, \{r_i\}) \tag{7.25}$$

where $E$ is the internal energy and $S_v$ is the vibrational entropy. In eqn (7.24) for $\Omega$ the configurational entropy is denoted by $S_c$. The equilibrium state of the segregated system may, in general, be found by minimizing $\Omega$. This is carried out in the following after invoking various approximations.

### 7.4.2 Regular solution model

In the Bragg–Williams approximation we assume that the configurational entropy, $S_c$, is that of an ideal (i.e. non-interacting) mixture of A and B atoms. This, of course, is an approximation because the interactions introduce correlations between the occupancies of the sites which alter the configurational entropy. This approximation for $S_c$ is also made in the regular solution model. In this approximation $S_c$ is given by:

$$S_c = -k \sum_i c_i \ln c_i + (1 - c_i) \ln(1 - c_i), \tag{7.26}$$

where $k$ is Boltzmann's constant. $N_A$ and $N_B$ are the numbers of $A$ and $B$ atoms:

$$N_A = \sum_i (1 - c_i) \tag{7.27a}$$

$$N_B = \sum_i c_i \tag{7.27b}$$

Inserting eqns (7.26–7.27) into eqn (7.24) for $\Omega$ and minimizing with respect to $c_k$ we obtain:

$$\frac{\partial F}{\partial c_k} + kT \ln \frac{c_k}{1 - c_k} = \mu_B - \mu_A. \tag{7.28}$$

At equilibrium, therefore, the local chemical potential difference, $\mu_B^k - \mu_A^k$, which equals the left-hand side of eqn (7.28), is the same at all sites and equal to $\mu_B - \mu_A$. The equilibrium condition involves the difference in chemical potentials $\mu_B - \mu_A$ because, in a substitutional alloy, equilibrium is attained by exchanging atoms between sites. (Remember that we are ignoring vacant sites.) In an interstitial alloy the equilibrium condition would involve absolute chemical potentials because the number of occupied sites is not conserved. The variation of the site occupancy from site to site is a result of the variation of $\partial F / \partial c_k$ from site to site. We can make this more explicit by rewriting eqn (7.28) as follows:

$$c_k = \frac{1}{1 + \exp\left[\beta(\partial F/\partial c_k - (\mu_B - \mu_A))\right]}, \tag{7.29}$$

where $\beta = 1/kT$. The function on the right of eqn (7.29) is a Fermi–Dirac distribution with $\mu_B - \mu_A$ taking the role of the Fermi energy. Fermi–Dirac statistics arises from the obvious restriction that a site can be occupied only by one atom. Thus if $\partial F/\partial c_k \gg (\mu_B - \mu_A)$ then $c_k \to 0$ the site is occupied by an A atom almost all the time, but if $\partial F/\partial c_k \ll (\mu_B - \mu_A)$ then $c_k \to 1$ (the site is occupied by a B atom almost all the time). We note that $\mu_B$ and $\mu_A$ themselves vary with temperature, pressure, and the compositions of the adjoining crystals.

Since eqn (7.28) must also hold for a site in either crystal, far from the interface, we have

$$\frac{\partial F}{\partial c_k} + kT\ln\frac{c_k}{1 - c_k} = \mu_B - \mu_A = \frac{\partial F}{\partial c_o} + kT\ln\frac{c_o}{1 - c_o}, \tag{7.30}$$

where $c_o$ is the average occupancy of a site in the bulk far from the interface. For an ordered alloy site $o$ can be any one of the non-equivalent sites within the unit cell, while for a disordered alloy we can choose any site and $c_o$ is simply the concentration of B in the disordered alloy. Rearranging eqn (7.30) we obtain:

$$\frac{c_k}{1 - c_k} = \frac{c_o}{1 - c_o}\exp\left[-\beta\left[\frac{\partial F}{\partial c_k} - \frac{\partial F}{\partial c_o}\right]\right]. \tag{7.31}$$

If we assume that (i) all the sites in the interface are equivalent, but different from those of the bulk, or, perhaps more realistically, that only certain sites are favourable for segregation and share the same environment; and (ii) that $\partial F/\partial c_k$ does not vary with occupancy of sites, then we obtain the McLean isotherm (McLean 1957):

$$X/(1 - X) \simeq X_o/(1 - X_o)\exp(-\beta\Delta f), \tag{7.32}$$

where $\Delta f$ is $\partial F/\partial c$ evaluated at an interface site at which the occupancy is $X$ minus $\partial F/\partial c$ evaluated at a bulk crystal site at which the concentration of B atoms is $X_o$. The quantity $\Delta f$ may be identified as the 'free energy of segregation' which we shall denote by $\Delta f_{seg}$, i.e. $\Delta f = \Delta f_{seg}$. The assumptions of the McLean isotherm are most likely to be valid for a dilute alloy where the level of segregation is insufficient for solute–solute interactions to be important.

The generalization to ternary alloys is straightforward. For a substitutional ternary ABC alloy the grand potential is expressed as follows:

$$\Omega = (\{c_i^A\}, \{c_i^B\}, \{r_i\}) + kT\sum_i c_i^A\ln c_i^A + c_i^B\ln c_i^B + (1 - c_i^A - c_i^B)\ln(1 - c_i^A - c_i^B)$$

$$- \mu_A\sum_i c_i^A - \mu_B\sum_i c_i^B - \mu_C\sum_i(1 - c_i^A - c_i^B), \tag{7.33}$$

where we have used $c_i^C = 1 - c_i^A - c_i^B$. Again this may be minimized with respect to all $\{c_i^A\}$, $\{c_i^B\}$, and $\{r_i\}$. Minimization with respect to $c_i^A$ and $c_i^B$ yields the following equations:

$$\frac{\partial F}{\partial c_i^A} + kT\ln\frac{c_i^A}{1 - c_i^A - c_i^B} = \mu_A - \mu_C$$

$$\frac{\partial F}{\partial c_i^B} + kT\ln\frac{c_i^B}{1 - c_i^A - c_i^B} = \mu_B - \mu_C. \tag{7.34}$$

Since the right-hand sides of these equations are site independent the equations must hold also when site $i$ is a bulk crystal site $o$. If the bulk concentrations of A and B are $c_o^A$ and $c_o^B$ then we obtain

$$\frac{c_i^A}{1 - c_i^A - c_i^B} = \frac{c_o^A}{1 - c_o^A - c_o^B} \exp\left[-\beta\left[\frac{\partial F}{\partial c_i^A} - \frac{\partial F}{\partial c_o^A}\right]\right]$$

and

$$\frac{c_i^B}{1 - c_i^A - c_i^B} = \frac{c_o^B}{1 - c_o^A - c_o^B} \exp\left[-\beta\left[\frac{\partial F}{\partial c_i^B} - \frac{\partial F}{\partial c_o^B}\right]\right]. \qquad (7.35)$$

These equations are quite general in that they do not rely on any assumptions about the absence of solute interactions in the interface. Nor do they assume that the atom positions are fixed because the grand potential must be simultaneously minimized with respect to the atom positions $\{r_i\}$, which in turn affect $\partial F/\partial c_i^A$ and $\partial F/\partial c_i^B$. In a ternary alloy site competition (see Section 7.2) may occur between segregating A and B solute atoms. This is described by the coupled eqns (7.35), which show that a strongly segregating solute can expel a weakly segregating solute even when it is present at relatively low bulk concentration. An example of this general type of behaviour has already been given in Fig. 7.12 for a case involving solute atoms which are normally interstitial in the lattice. Here, the strongly segregating C atoms replace the P atoms in the boundary as the C concentration in the lattice increases.

### 7.4.3  Mean field models

In this section we illustrate the ideas of the previous section with a particular model of segregation in a bicrystal consisting of A and B atoms in which atomic interactions are described by pair potentials. Although the assumption of pair potentials is perhaps rather simplistic, in general, the model does give considerable insight into the effect of interactions between segregating atoms. In addition the model illustrates the coupling between the degree of segregation and the local atomic structure at the interface. We include also a brief discussion of the extension of the analysis to $N$-body interactions of the Finnis–Sinclair type.

Let the A and B atoms of the bicrystal of Section 7.4.1 interact via pair potentials. We assume $\varepsilon^{AA}(r)$, $\varepsilon^{BB}(r)$, and $\varepsilon^{AB}(r)$ potentials describe the energy of A–A, B–B, and A–B atomic interactions, separated by the distance $r$. We denote the energy of interaction between an A atom at site $i$ and a B atom at site $j$ by $\varepsilon_{ij}^{AB} = \varepsilon_{ji}^{AB}$, with the understanding that $\varepsilon_{ij}^{AB}$ is a function of $|R_i - R_j|$. Similar symbols are used for A–A and B–B interaction. The Hamiltonian of the system is given by

$$\mathcal{H} = \frac{1}{2}\sum_{\substack{i,j \\ i \neq j}}\left[p_i p_j \varepsilon_{ij}^{BB} + \left[p_i(1 - p_j) + p_j(1 - p_i)\right]\varepsilon_{ij}^{AB} + (1 - p_i)(1 - p_j)\varepsilon_{ij}^{AA}\right]$$

$$- \mu_A \sum_i (1 - p_i) - \mu_B \sum_i p_i, \qquad (7.36)$$

where the $p_i$'s are defined by eqn (7.23). It is convenient to rewrite this as follows:

$$\mathcal{H} = \sum_{\substack{i,j \\ i \neq j}} p_i p_j \theta_{ij} + \sum_i (p_i \alpha_i + d_i^A), \qquad (7.37)$$

where

$$\theta_{ij} = \tfrac{1}{2} \left[ \varepsilon_{ij}^{BB} + \varepsilon_{ij}^{AA} - 2\varepsilon_{ij}^{AB} \right], \tag{7.38}$$

$$\alpha_i = \sum_{j \neq i} \left( \varepsilon_{ij}^{AB} - \varepsilon_{ij}^{AA} \right) - \left( \mu_B - \mu_A \right), \tag{7.39}$$

$$d_i^A = \left( \frac{1}{2} \sum_{j \neq i} \varepsilon_{ij}^{AA} \right) - \mu_A. \tag{7.40}$$

We shall now obtain expressions for the grand potential at a given set of atom positions, although the atom positions are allowed to relax to minimize the grand potential. The connection between the grand potential and the Hamiltonian is the grand partition function, defined as follows:

$$Z = \mathrm{Tr}\, e^{-\beta \mathcal{H}} \tag{7.41}$$

Here $\beta = 1/kT$ and $e^{-\beta \mathcal{H}}$ is an operator meaning

$$e^{-\beta \mathcal{H}} = 1 - \beta \mathcal{H} + (\beta \mathcal{H})^2/2! - (\beta \mathcal{H})^3/3! + \ldots \tag{7.42}$$

and Tr denotes trace of the operator. Taking the trace of an operator means summing the expectation values of the operator over all states of the system. In this case, since atom positions are (temporarily) fixed, each state is determined by the set of integers $p_1, p_2, p_3, \ldots$, each of which can be zero or one. The grand potential is obtained from the grand partition function as follows:

$$\Omega = -kT \ln Z. \tag{7.43}$$

We can obtain an *exact* formula for the expectation values $\langle p_k \rangle = c_k$ by evaluating the trace in a particular way due to Callen (1963) (see Balcerzak (1991) for a review of such identities). The Hamiltonian $\mathcal{H}$ is separated into a part $\mathcal{H}_k$ that contains all those parts of $\mathcal{H}$ which depend on site $k$ and the remainder of $\mathcal{H}$ is called $\mathcal{H}'$. Thus

$$\mathcal{H} = \mathcal{H}_k + \mathcal{H}', \tag{7.44}$$

where

$$\mathcal{H}_k = p_k \left( \alpha_k + 2 \sum_{j \neq k} p_j \theta_{jk} \right) + d_k^A = p_k \gamma_k + d_k^A \tag{7.45}$$

with $\gamma_k$ being the 'local field' at site $k$:

$$\gamma_k = \alpha_k + 2 \sum_{j \neq k} p_j \theta_{jk} \tag{7.46}$$

and

$$\mathcal{H}' = \sum_{\substack{j \neq k \\ i \neq j}} \sum_{i \neq k} p_i p_j \theta_{ij} + \sum_{i \neq k} (p_i \alpha_i + d_i^A). \tag{7.47}$$

The expectation value of $p_k$ is defined by

$$\langle p_k \rangle = \frac{\mathrm{Tr}\, p_k e^{-\beta \mathcal{H}}}{Z}. \tag{7.48}$$

The trick is to split the trace in the numerator into a trace over index $k$, denoted by Tr($k$), and a trace over all other site indices, Tr($'$). The trace over all sites in eqn (7.48) is thus Tr($'$)Tr($k$). We obtain

LP

$$\langle p_k \rangle = \frac{\text{Tr}(') \text{Tr}(k) p_k e^{-\beta(\mathcal{H} + \mathcal{H}')}}{Z}$$

$$= \frac{1}{Z} \text{Tr}(') e^{-\beta\mathcal{H}'} [\text{Tr}(k) e^{-\beta\mathcal{H}_k}] \left( \frac{\text{Tr}(k) p_k e^{-\beta\mathcal{H}_k}}{\text{Tr}(k) e^{-\beta\mathcal{H}_k}} \right) \tag{7.49}$$

The term in round brackets may be evaluated since it involves traces over only site $k$, each of which consists of only two terms corresponding to $p_k = 0$ or $1$:

$$\frac{\text{Tr}(k) p_k e^{-\beta\mathcal{H}_k}}{\text{Tr}(k) e^{-\beta\mathcal{H}_k}} = \frac{1}{1 + e^{\beta\gamma_k}}, \tag{7.50}$$

and therefore

$$\langle p_k \rangle = \frac{1}{Z} \text{Tr} \, e^{-\beta\mathcal{H}} \frac{1}{1 + e^{\beta\gamma_k}} = \left\langle \frac{1}{1 + e^{\beta\gamma_k}} \right\rangle. \tag{7.51}$$

This formula is exact. All the standard approximations such as mean field theory may be applied to it. The formula is perhaps deceptively simple. The local field $\gamma_k$, eqn (7.46), depends on the site occupancy operators $p_j$ of sites neighbouring site $k$. In fact this simple formula captures all the correlations between site occupancies in a succint form. To make them explicit we rewrite (following Balcerzak (1991)) the formula as follows:

$$\langle p_k \rangle = \left\langle \int_{-\infty}^{+\infty} d\omega \frac{1}{1 + e^{\beta\omega}} \delta(\omega - \gamma_k) \right\rangle \tag{7.52}$$

where $\delta(\omega - \gamma_k)$ is the Dirac delta function. Using the complex exponential representation of the delta function we obtain:

$$\langle p_k \rangle = \frac{1}{2\pi} \int_{-\infty}^{+\infty} d\omega \frac{1}{1 + e^{\beta\omega}} \int_{-\infty}^{+\infty} dt \, e^{-i\omega t} \langle e^{i\gamma_k t} \rangle. \tag{7.53}$$

We see that $c_k$, which equals $\langle p_k \rangle$, depends on the Fourier transform of $\langle e^{i\gamma_k t} \rangle$, which depends on all one centre, two-centre, three-centre, ... correlations among the sites neighbouring site $k$:

$$\langle e^{i\gamma_k t} \rangle = e^{i\alpha_k t} \left\langle \exp\left( 2i \sum_{j \neq k} \theta_{jk} p_j t \right) \right\rangle = e^{i\alpha_k t} \left\langle 1 + 2i \sum_{j \neq k} \theta_{jk} p_j t + \left[ 2i \sum_{j \neq k} \theta_{jk} p_j t \right]^2 \bigg/ 2! + \ldots \right\rangle$$

$$= e^{i\alpha_k t} \left[ 1 + 2it \sum_{j \neq k} \theta_{jk} \langle p_j \rangle + \frac{(2it)^2}{2!} \sum_{j \neq k} \sum_{m \neq k} \theta_{jk} \theta_{mk} \langle p_j p_m \rangle + \ldots \right]. \tag{7.54}$$

In the mean field approximation, and in the autocorrelation approximation discussed below, correlations in the occupancies of different sites are ignored. We can then write

$$\left\langle \exp\left( 2i \sum_{j \neq k} \theta_{jk} p_j t \right) \right\rangle = \prod_{j \neq k} \langle \exp(2i\theta_{jk} p_j t) \rangle. \tag{7.55}$$

Furthermore, in the mean field approximation it is assumed that

$$\langle p_j^n \rangle \simeq \langle p_j \rangle^n, \tag{7.56}$$

which implies that

$$\langle \exp(2i\theta_{jk} p_j t) \rangle = \exp(2i\theta_{jk} \langle p_j \rangle t) \tag{7.57}$$

and hence

$$\langle e^{i\gamma_{kl}} \rangle = e^{i\alpha_{kl}} \exp \left[ 2i \sum_{j \neq k} \theta_{jk} \langle p_j \rangle t \right]. \tag{7.58}$$

Inserting this into the integral expression for $\langle p_k \rangle$, eqn (7.53), we finally obtain

$$c_k = \langle p_k \rangle = \frac{1}{1 + e^{\beta \gamma_k^{MFA}}} \tag{7.59}$$

where

$$\gamma_k^{MFA} = \alpha_k + 2 \sum_{j \neq k} c_j \theta_{jk}. \tag{7.60}$$

The mean field Hamiltonian is then

$$\mathcal{H}^{MFA} = \sum_i p_i \left( \alpha_i + 2 \sum_{j \neq i} \theta_{ij} c_j \right) + d_i^A, \tag{7.61}$$

where the term in brackets is the local field $\gamma_i^{MFA}$. Comparing this with the exact Hamiltonian, eqn (7.37), we see that $2p_i p_j \theta_{ij}$ is approximated by $2(p_i c_j + p_j c_i)\theta_{ij}$. On average, therefore, $\theta_{ij}$ interactions are double counted in the mean field approximation. Each atom interacts with neighbouring sites at which the occupancy is not $p_j$ but the expectation value of $p_j$, namely $c_j$. A more physically transparent form of the local field is obtained by using eqns (7.38) and (7.39) for $\theta_{ij}$ and $\alpha_i$:

$$\gamma_i^{MFA} = \sum_{j \neq i} [\varepsilon_{ij}^{BB} c_j + \varepsilon_{ij}^{AB}(1 - c_j) - \varepsilon_{ij}^{AA}(1 - c_j) - \varepsilon_{ij}^{AB} c_j] - (\mu_B - \mu_A). \tag{7.62}$$

The summation term is the energy of replacing an A atom at site $i$ with a B atom. The local field is this energy difference relative to the chemical potential difference between a B atom and an A atom. It is noted that $\partial F / \partial c_k$ of eqn (7.28) is equivalent to the energy of replacing an A atom at site $k$ with a B atom in our pair potential model.

The partition function in the mean field approximation is given by

$$Z^{MFA} = \text{Tr } e^{-\beta \mathcal{H}^{MFA}} = \text{Tr} \prod_i \exp \left( -\beta (p_i \gamma_i^{MFA} + d_i^A) \right)$$

$$= \prod_i e^{-\beta d_i^A}(1 + e^{-\beta \gamma_i^{MFA}}) \tag{7.63}$$

from which we obtain the grand potential as follows:

$$\Omega^{MFA} = -kT \ln Z^{MFA} = \sum_i d_i^A - kT \ln \left( 1 + e^{-\beta \gamma_i^{MFA}} \right) \tag{7.64}$$

We may obtain the following physically transparent form for the grand potential $\Omega^{MFA}$ by using standard identities for the thermodynamic state variables in terms of the partition function:

$$\Omega^{MFA} = \frac{1}{2} \sum_{\substack{i,j \\ i \neq j}} \left[ c_i c_j \varepsilon_{ij}^{BB} + [c_i(1 - c_j) + c_j(1 - c_i)] \varepsilon_{ij}^{AB} + (1 - c_i)(1 - c_j) \varepsilon_{ij}^{AA} \right]$$

$$+ kT \sum_i c_i \ln c_i + (1 - c_i) \ln (1 - c_i)$$

$$- \mu_A \sum_i (1 - c_i) - \mu_B \sum_i c_i. \tag{7.65}$$

The first term is the internal energy, which is obtained by replacing the site occupancy operators by their expectation values in the terms of the Hamiltonian, eqn (7.36), involving the potentials. In our pair potential model this 'internal energy' is equivalent to the Helmholtz free energy $F(\{c_i\}, \{r_i\})$ in eqn (7.25) and, therefore, it is also intended to account for the contribution to the free energy of the system from the vibrational entropy. The second term is the contribution from the configurational entropy. The final term is the contribution from the chemical potentials of the A and B atoms. Thus, in the mean field approximation each A or B atom is replaced by a hybrid atom, which varies in the degree of its A-ness or B-ness with the local atomic environment in a self-consistent manner.

Equation (7.65) is a suitable form of the grand potential for variational calculations. It is easily shown that minimization of the grand potential with respect to $c_k$ leads again to eqn (7.59). But in eqn (7.65) the grand potential is also a function of all the atomic coordinates in the system, through the pair potentials. We should, therefore, minimize the grand potential with respect to all $4N$ variables, where $N$ is the total number of sites. In this way the atomic structure of the interface changes as the degree of segregation changes, for example because the chemical potential difference $\mu_B - \mu_A$ is changed. We may think of $-\partial\Omega/\partial c_i$ as a generalized force, conjugate to the site occupancy $c_i$.

It is also possible to obtain an isotherm relating the occupancy at site $k$ to the bulk solute concentration as a function of temperature:

$$\frac{c_k}{1 - c_k} = \frac{c_o}{1 - c_o} \exp - \beta(\gamma_k^{MFA} - \gamma_o^{MFA}) \tag{7.66}$$

where $c_o$ denotes the site occupancy of a bulk crystal site, at which the local field is $\gamma_o^{MFA}$. The energy $\gamma_k^{MFA} - \gamma_o^{MFA}$ is the energy of exchanging an $A$ atom at site $k$ in the interface with a B atom at a bulk crystal site. Thus eqn (7.66) is exactly what we would predict from the law of mass action. Substituting eqn (7.60) for the local field we obtain

$$\gamma_k^{MFA} - \gamma_o^{MFA} = \left(\alpha_k - \alpha_o - 2c_o \sum_{j \neq o} \theta_{oj}\right) + \left[2 \sum_{j \neq k} \theta_{kj} c_j\right]. \tag{7.67}$$

### 7.4.3.1 *McLean isotherm*

If we assume that only certain well separated sites are favourable for segregation, each of which has the same local field $\gamma_k^{MFA}$, we can ignore the term in square brackets in eqn (7.67). We then obtain the McLean isotherm (given by eqn (7.32)) in which the free energy of segregation, given by $\Delta f_{seg} = \gamma_k^{MFA} - \gamma_o^{MFA}$, is independent of the site $k$ and the degree of segregation. It has been used with apparent success to describe the average segregation behaviour to large-angle boundaries in polycrystalline samples of a number of binary systems, e.g. S in Cu (Moya and Moya-Gontier 1975), and P in Fe (Erhart and Grabke 1981). Note that S in Cu is a substitutional impurity but P in Fe is an interstitial impurity. For P in Fe, $\Delta e_{seg} = -34.3 \text{ kJ mol}^{-1}$, and $\Delta s_{seg} = 0.0215 \text{ kJ mol}^{-1}\text{K}^{-1}$, where $\Delta f_{seg} = \Delta e_{seg} - T\Delta s_{seg}$. Saturation levels are generally less than a monolayer, as should be the case.

Many systems do not follow simple McLean type isotherms, and, as we have already mentioned in Section 7.2, exhibit more complex behaviour involving, for example, various solvent–solvent, solvent–solute, and solute-solute interactions and the buildup of multilayer segregates. The assumptions of the McLean isotherm are rather severe, especially when the level of segregation is high and the interaction terms $\theta_{kj} c_j$ in eqn (7.67) can no longer be ignored.

### 7.4.3.2 *Fowler–Guggenheim isotherm*

In the next simplest level of approximation we again assume that segregation occurs only at certain equivalent interfacial sites, and that the local coordination number is $z$. We then obtain from eqn (7.67)

$$\gamma_k^{\text{MFA}} - \gamma_0^{\text{MFA}} = \Delta f + 2z\theta X = \Delta f_{\text{seg}} \tag{7.68}$$

where $\Delta f$ is the value of $\gamma_k^{\text{MFA}} - \gamma_0^{\text{MFA}}$ appearing in the McLean isotherm, i.e. in the absence of solute–solute interactions, $X$ is the occupancy of the sites, and $\Delta f_{\text{seg}}$ is the free energy of segregation. We have assumed nearest neighbour interactions only in eqn (7.68). Substituting eqn (7.68) into the isotherm equation, eqn (7.66), we obtain a Fowler-Guggenheim type equation (Fowler and Guggenheim 1939):

$$\frac{X}{1 - X} = \frac{c_0}{1 - c_0} e^{-\beta(\Delta f + 2z\theta X)}, \tag{7.69}$$

which predicts separation of the boundary phase into coexisting solute-rich and solute-lean phases if the solute atoms sufficiently attract one another, and if the temperature is sufficiently low (i.e. if the parameter $2z\beta\theta$ is sufficiently large and negative).

The forms of the isotherms predicted by this model often fit experimental data for various systems, as shown, for example, in Fig. 7.11. When the parameter $2z\beta\theta$ reaches the critical value $-4$, phase separation should just occur: the dashed section of the $-6$ curve is a metastable region where two coexisting phases of differing composition should be present in the boundary under equilibrium conditions. The data for P in Fe exhibit ideal $2z\beta\theta = 0$ behaviour, while, on the other hand, the data for Se in Fe suggests strong Se atom clustering and even, perhaps, the formation of two coexisting boundary phases as discussed, for example, by Militzer and Weitung (1987).

### 7.4.3.3 *Multiple segregation site models*

For both the McLean and Fowler-Guggenheim isotherms the simplifying assumption is made that there is only one type of site to which segregation occurs. However, some authors argue that the energy, $\Delta f$, which appears in the exponential represents an average taken over many sites to which segregation occurs. They then relate this average energy to the average level of segregation measured experimentally. This cannot be justified mathematically using eqn (7.66). The quantities that may be averaged are $c_k/(1 - c_k)$ and $\exp - \beta(\gamma_k^{\text{MFA}} - \gamma_0^{\text{MFA}})$, neither of which is directly related to the average of $c_k$ and average of $\gamma_k^{\text{MFA}}$.

A more useful approach to deal with the possibility of multiple segregation sites is to introduce the concept of a density of sites for the local field $\gamma_k^{\text{MFA}}$ (White and Coghlan 1977, Mütschele and Kirchheim 1987). Using eqn (7.62) for the local field we can express $\gamma_i^{\text{MFA}}$ as follows:

$$\gamma_i^{\text{MFA}} = S_i - (\mu_{\text{B}} - \mu_{\text{A}}) \tag{7.70}$$

where $S_i$ is the summation term in eqn (7.62) and represents the energy of replacing an A atom at site $i$ with a B atom. The average occupancy at site $i$ is given by eqn (7.59) and may be rewritten as follows:

$$c_i = \int_{-\infty}^{+\infty} dS \frac{\delta(S - S_i)}{1 + \exp \beta(S - (\mu_B - \mu_A))} \tag{7.71}$$

where $\delta(S - S_i)$ is a Dirac $\delta$-function. Summing the $c_i$'s over all interface sites we obtain the following expression for the total number of B atoms in the interface:

$$n = \int_{-\infty}^{-\infty} dS \frac{N(S)}{1 + \exp \beta(S - (\mu_B - \mu_A))}, \tag{7.72}$$

where $N(S)$ is the interfacial density of sites for B atoms:

$$N(S) = \sum_i \delta(S - S_i) \tag{7.73}$$

and the sum is taken over all interface sites. The conceptual advantage gained by introducing the density of sites is that a clear separation is achieved between the types of interfacial site available and the statistics of their occupancies. There is an obvious analogy with the total number of electrons in a system being determined by the electronic density of states and Fermi–Dirac statistics. Just as in the electron case the density of sites is a characteristic feature of an interface, which allows us to make clear distinctions between the propensity for segregation to different interfaces. For example, in Fig. 7.14(a) we have sketched the density of sites that we would expect for a short period twin boundary. Because there are few non-equivalent sites in the boundary the density of sites is 'spiky', each spike corresponding to one of the non-equivalent sites. The expected density of states for a macroscopically curved interface is sketched in Fig. 17.14(b). Now the interface contains a wider range of sites and this is reflected by the smoother and broader density of sites. For both cases the occupancy of sites with an energy $S$ is determined by the proximity of $S$ to $\mu_B - \mu_A$ and the temperature.

We note that the density of sites for a particular interface is not a fixed quantity, independent of the level of segregation. As the level of segregation increases, solute–solute interactions occur and the energies $S_i$ change (see eqn (7.62)). The energies $S_i$ also change because the atomic structure of the interface relaxes and hence the pair potentials

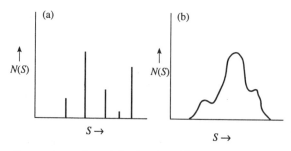

**Fig. 7.14** (a) Schematic representation of the density of boundary segregation sites, $N(S)$, expected for a short-period twin boundary. $N(S) dS$ is the number of sites possessing segregation energies between $S$ and $S + dS$. $N(S)$ for this short-period boundary is 'spiky', since only a few non-equivalent sites will exist. (b) Density of sites for a curved or general boundary. In this case $N(S)$ is smoothed out, since a wide range of sites, closely spaced in energy, will exist.

change. Thus the density of sites is a characteristic feature of an interface at a given level of segregation. Similarly, in the electron case the density of states does not change with occupation of states only in the independent electron approximation.

Mütschele and Kirchheim (1987) used a simple version of this multiple segregation site model to interpret data for the segregation of H to grain boundaries in Pd. In this system, however, the relatively small H atoms occupy interstitial sites everywhere, and the previous model, which holds for substitutional systems, must be modified accordingly. This is readily accomplished by letting the B atoms be the interstitial atoms and the A atoms the host substitutional atoms. An interstitial site between the A atoms is then either filled or empty and has an average occupancy, $c_i$. Equations (7.24), (7.25), (7.26), and (7.27b) still hold in this situation, and following the same procedure used previously to obtain eqn (7.28), we obtain

$$\frac{\partial F}{\partial c_k} + kT \ln \frac{c_k}{1 - c_k} = \mu_B. \tag{7.74}$$

Equation (7.74) differs from eqn (7.28) only on the right-hand side where $\mu_B$ appears rather than $\mu_B - \mu_A$. This result is readily understood since an interstitial B atom can be simply inserted at an empty $k$-type interstitial site, and it is not necessary to first remove an A atom as in our previous substitutional system. Rearranging eqn (7.74), we therefore have the result

$$c_k = \frac{1}{1 + \exp[\beta(\partial F/\partial c_k - \mu_B)]}, \tag{7.75}$$

which may be compared with eqn (7.29). Adopting a simple Fowler–Guggenheim approach (see Section 7.4.3), our previous results allow us to write the occupancy at lattice interstitial sites as

$$c_0 = \frac{1}{1 + \exp[\beta(S_o - \mu_B)]}, \tag{7.76}$$

and the occupancy of boundary interstitial sites in the approximate form

$$c = \frac{1}{1 + \exp[\beta(S + 2z\theta\bar{X} - \mu_B)]}. \tag{7.77}$$

Here, $S_o$ and $S$ are appropriate energies corresponding to the lattice and boundary sites respectively in the absence of solute interactions, and $2z\theta\bar{X}$ approximates the interaction energy. Here, $\bar{X} = n/N$, where $n$ is the number of occupied sites in the boundary, and $N$ is the number of boundary sites. Mütschele and Kirchheim (1987) made the simple assumption that the density of sites in the boundary had a Gaussian distribution centred around an average energy, $\bar{S}$. The total density of sites, $N_{tot}(S)$, may then be written as the sum of the distribution in the lattice, $N_0(S)$, and the distribution in the boundaries, $N(S)$, in the form

$$N_{tot}(S) = N_o(S) + N(S) = N_o \cdot \delta(S - S_o) + N \cdot (\sigma\sqrt{\pi})^{-1} \cdot \exp(-(S - \bar{S})^2/\sigma^2), \tag{7.78}$$

where $N_o$ = number of lattice sites, and $\sigma$ = width of the Gaussian. By multiplying each term in eqn (7.78) by the occupancy, obtained from either eqn (7.76) or (7.77), and integrating over all energies, the total number of occupied sites in the polycrystal, i.e.

$(n_o + n)$, where $n_o$ is the number in the lattice and $n$ is the number in the boundaries, which is equal to the total number of solute atoms, $n_s$, is then

$$n_s = n_o + n = \frac{N_o}{1 + \exp[\beta(S_o - \mu_B)]} + \frac{N}{\sigma\sqrt{\pi}} \int_{-\infty}^{\infty} \frac{\exp[-(S - \bar{S})^2/\sigma^2]\, dS}{1 + \exp[\beta(S + 2z\theta\bar{X} - \mu_B)]}. \tag{7.79}$$

It was found (Mütschele and Kirchheim 1987) that the measured hydrogen concentration in the boundaries versus the lattice could be explained only if the segregated hydrogen atoms attracted one another in the boundaries (i.e. $2z\theta$ was chosen negative). Good fits of the model to the data were obtained by setting $2z\theta = -30\,\text{kJ mol}^{-1}$, $\sigma = 15\,\text{kJ mol}^{-1}$, $S_o = 3.9\,\text{kJ mol}^{-1}$ (for equilibrium with 1 atm of $H_2$), $\bar{S} = 9.2\,\text{kJ mol}^{-1}$, and $N/N_o = 0.3/0.7$.

Finally, in this section we discuss the extension of the mean field theory to the case of $N$-body atomic interactions. Up to now we have assumed pair-wise atomic interactions, and we have noted that this assumption is often inadequate. In the following we consider a Finnis–Sinclair model of atomic interactions in an AB alloy (Rafii-Tabar and Sutton 1991). The Hamiltonian becomes

$$\mathcal{H} = \frac{1}{2} \sum_{\substack{i,j \\ i \ne j}} (p_i p_j V_{ij}^{BB} + [p_i(1 - p_j) + p_j(1 - p_i)]V_{ij}^{AB} + (1 - p_i)(1 - p_j)V_{ij}^{AA})$$

$$- \sum_i [(1 - p_i)b_A + p_i b_B]\sqrt{\rho_i} - \mu_A \sum_i (1 - p_i) - \mu_B \sum_i p_i, \tag{7.80}$$

where

$$\rho_i = \sum_{j \ne i} p_i p_j \varphi_{ij}^{BB} + [p_i(1 - p_j) + p_j(1 - p_i)]\varphi_{ij}^{AB} + (1 - p_i)(1 - p_j)\varphi_{ij}^{AA}. \tag{7.81}$$

The pair potentials $V^{AA}$, $V^{BB}$, and $V^{AB}$ describe short-range repulsive interactions between atoms while cohesion is provided by the terms in $\sqrt{\rho_i}$. Each $\rho_i$ is expressed as a sum of pair potentials $\varphi^{AA}$, $\varphi^{BB}$, and $\varphi^{AB}$. $b_A$ and $b_B$ are constants. In the mean field approximation we replace each $p_i$ operator by its expectation value $c_i$ to arrive at the following expression for the grand potential:

$$\Omega = \frac{1}{2} \sum_{\substack{i,j \\ i \ne j}} (c_i c_j V_{ij}^{BB} + [c_i(1 - c_j) + c_j(1 - c_i)]V_{ij}^{AB} + (1 - c_i)(1 - c_j)V_{ij}^{AA})$$

$$- \sum_i [(1 - c_i)b_A + c_i b_B]\sqrt{\rho_i}$$

$$+ kT \sum_i c_i \ln c_i + (1 - c_i)\ln(1 - c_i)$$

$$- \mu_A \sum_i (1 - c_i) - \mu_B \sum_i c_i, \tag{7.82}$$

where

$$\rho_i = \sum_{j \ne i} c_i c_j \varphi_{ij}^{BB} + [c_i(1 - c_j) + c_j(1 - c_i)]\varphi_{ij}^{AB} + (1 - c_i)(1 - c_j)\varphi_{ij}^{AA}. \tag{7.83}$$

Proceeding as before we require that $\Omega$ is minimized with respect to all the $\{c_j\}$ and the atom positions $\{r_j\}$, in terms of which the pair potentials $V$ and $\phi$ are defined. Each $c_i$ is again found to be a Fermi–Dirac distribution, eqn (7.59), but the the local field $\gamma_i^{MFA}$ is now given by:

$$\gamma_i^{MFA} = \sum_{j \neq i} [V_{ij}^{BB}c_j + V_{ij}^{AB}(1 - c_j) - V_{ij}^{AA}(1 - c_j) - V_{ij}^{AB}c_j]$$

$$-\frac{1}{2} \sum_{j \neq i} \left( [b_A(1 - c_i) + b_Bc_i]\frac{1}{\sqrt{\rho_i}} + [b_A(1 - c_j) + b_Bc_j]\frac{1}{\sqrt{\rho_j}} \right) \times$$

$$(\varphi_{ij}^{BB}c_j + \varphi_{ij}^{AB}(1 - c_j) - \varphi_{ij}^{AA}(1 - c_j) - \varphi_{ij}^{AB}c_j)$$

$$- (\mu_B - \mu_A) \tag{7.84}$$

Two points are worth making about this local field. The first is that, unlike the simple pair potential model of eqn (7.62), the local field at site $i$ is not simply the energy of replacing an A by a B atom at that site. The non-linear $N$-body terms in $\{\sqrt{\rho_j}\}$ no longer permit this simple interpretation. In addition the local field at site $i$ is determined not only by the neighbours of site $i$ (defined as those which interact directly with site $i$ through the pair potentials $\varphi$ and $V$) but also by the neighbours of the neighbours of site $i$. Secondly, we can define effective pair potentials to define the local field so that it has the appearance of eqn (7.62). However, these effective pair potentials are themselves dependent on the site occupancies. Thus we can express the local field as

$$\gamma_i^{MFA} = \sum_{j \neq i} [\tilde{\varepsilon}_{ij}^{BB}c_j + \tilde{\varepsilon}_{ij}^{AB}(1 - c_j) - \tilde{\varepsilon}_{ij}^{AA}(1 - c_j) - \tilde{\varepsilon}_{ij}^{AB}c_j] - (\mu_B - \mu_A) \tag{7.85}$$

where

$$\tilde{\varepsilon}_{ij}^{AA} = V_{ij}^{AA} - \frac{1}{2} \left( [b_A(1 - c_i) + b_Bc_i]\frac{1}{\sqrt{\rho_i}} + [b_A(1 - c_j) + b_Bc_j]\frac{1}{\sqrt{\rho_j}} \right) \varphi_{ij}^{AA}, \tag{7.86}$$

and similar expressions apply for $\tilde{\varepsilon}_{ij}^{BB}$ and $\tilde{\varepsilon}_{ij}^{AB}$. Thus, we can think of the effect of introducing $N$-body interactions as making the effective pair interactions $\tilde{\varepsilon}^{AA}(r)$, $\tilde{\varepsilon}^{BB}(r)$, and $\tilde{\varepsilon}^{AB}(r)$ vary with the local chemical composition. Having defined the effective pair interactions in this way we can now recover the interpretation of the local field as the energy of replacing an A atom at site $i$ by a B atom. However, it must be stressed that this replacement energy is defined in terms of effective pair interactions which are themselves dependent on the local atomic environment.

## 7.4.4 Beyond mean field models

In the mean field theory described in the previous section we ignored correlations between the occupancies on different sites. Perhaps more significantly, in eqn (7.56), we ignored autocorrelations, i.e. we assumed that $\langle p_j^n \rangle = \langle p_j \rangle^n$. Using the pair potential model described in the previous section it is possible to make a significant improvement on the mean field approximation, which is known as the autocorrelation approximation (AA) (Balcerzak 1991). In this approximation we continue to ignore correlations between occupancies at different sites but we take account of correlations on the same site, i.e. we ensure that $\langle p_j^n \rangle = \langle p_j \rangle$, rather than $\langle p_j^n \rangle = \langle p_j \rangle^n$ of the mean field approximation. Experience in magnetic systems (Balcerzak *et al.* 1990), indicates that these self-

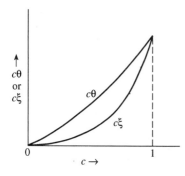

**Fig. 7.15** Schematic representation of $c\theta$ and $c\xi$ as a function of $c$ according to eqn (7.88). $c$ is the site occupation probability and $\theta$ is a function of atomic interaction parameters (see eqn (7.36)).

correlations give the largest correction to the mean field approximation of all the correlations that exist in the system.

Let us reconsider the grand partition function, and $e^{-\beta \Im C}$ in the pair potential model, eqn (7.36). It is helpful to rewrite $\exp(-\beta \theta_{ij} p_i p_j)$ as follows:

$$\exp(-\beta \theta_{ij} p_i p_j) = 1 - \beta \theta_{ij} p_i p_j + (\beta \theta_{ij} p_i p_j)^2/2 - \dots$$

$$= 1 + p_i[-\beta \theta_{ij} p_j + (\beta \theta_{ij} p_j)^2/2 - \dots]$$

$$= 1 + p_i[\exp(-\beta \theta_{ij} p_j) - 1] \tag{7.87}$$

where we have used $p_i^n = p_i$. Inserting this expression for $\exp(-\beta \theta_{ij} p_i p_j)$ into $e^{-\beta \Im C}$ we obtain:

$$e^{-\beta \Im C} = \prod_n \exp(-\beta(p_n \alpha_n + c_n^A)) \prod_i \prod_{j \neq i} [1 - p_i + p_i \exp(-\beta \theta_{ij} p_j)]. \tag{7.88}$$

This equation is still exact. In the mean-field approximation the term $\exp(-\beta \theta_{ij} p_j)$ is replaced by $\exp(-\beta \theta_{ij} \langle p_j \rangle) = \exp(-\beta \theta_{ij} c_j)$. In the autocorrelation approximation it is replaced by

$$\langle \exp(-\beta \theta_{ij} p_j) \rangle = \langle 1 + p_j(\exp(-\beta \theta_{ij}) - 1) \rangle$$

$$= 1 + c_j(\exp(-\beta \theta_{ij}) - 1)$$

$$= \exp - (\beta \xi_{ij} c_j) \tag{7.89}$$

where $\xi_{ij}$ is defined by

$$\xi_{ij} = -\frac{kT}{c_j} \ln[1 + c_j(\exp(-\beta \theta_{ij}) - 1)]. \tag{7.90}$$

We can think of the $\xi_{ij}$ as effective interaction parameters which result from the improved treatment of the statistical averaging. Note that as $c_j \to 0$ then $\xi_{ij} \to kT(1 - e_{ij}^{-\beta\theta})$, which tends to $\theta_{ij}$ as $T \to 0$. Also, as $c_j \to 1$ then $\xi_{ij} \to \theta_{ij}$. But for $0 < c_j < 1$ the effective pair potential $\xi_{ij}$ is less than $\theta_{ij}$, as shown in Fig. 7.15.

The Hamiltonian in the autocorrelation approximation is

$$\Im C^{AA} = \sum_i p_i \gamma_i^{AA} + d_i^A, \tag{7.91}$$

where the local field, $\gamma_i^{AA}$, is given by

$$\gamma_i^{AA} = \alpha_i + 2 \sum_{j \neq i} \xi_{ij} c_j. \tag{7.92}$$

The self-consistency condition is also of the same form as in the mean field approximation:

$$c_i = \frac{1}{1 + e^{\beta \gamma_i^{AA}}}. \tag{7.93}$$

Similarly, the grand potential $\Omega^{AA}$ is the same as $\Omega^{MFA}$, eqn (7.64), except that the interaction parameters $\theta_{ij}$ are replaced by $\xi_{ij}$. We conclude that the only difference between the mean field and autocorrelation approximations is the change in definition of the local field, with the interaction parameters $\theta_{ij}$ being replaced by effective interaction parameters $\xi_{ij}$, which are temperature and concentration dependent. The forms of the various isotherms remain the same.

The previous models are obviously oversimplified and can only be expected to provide the framework for a semi-quantitative understanding. A particular shortcoming of both the mean field and autocorrelation approximations is their treatment of relaxation around solute atoms with large misfits. To demonstrate this let us consider an extreme example. Consider a single f.c.c. crystal of A atoms containing vacant sites, in which atomic interactions are modelled by pair potentials $\varepsilon_{ij}^{AA}$. In this case the 'alloy' is between A atoms and vacancies. Rather than treating each site as being occupied ($p_i = 0$) or vacant ($p_i = 1$) the mean field approximation treats all sites in the crystal as having the same vacancy occupancy $c$, which is determined by the following self-consistency condition:

$$c = 1/ \left[ 1 + \exp \left( \beta \left( \mu_A - (1 - c) \sum_{j \neq i} \varepsilon_{ij}^{AA} \right) \right) \right]. \tag{7.94}$$

(Although we are discussing the mean field approximation here our remarks apply equally to the autocorrelation approximation.) We have assumed that the vacancies are in thermal equilibrium so that their chemical potential is zero. Since we expect the vacancy concentration, $c$, to be small we may replace $(1 - c)$ in the exponential by 1 and thus obtain a solution for $c$. The failure of the mean field approximation in this case lies in the inadequacy of its description of the relaxation around each vacancy. In reality there is relaxation around each vacant site in the crystal which affects the formation energy to a significant degree. This relaxation renders sites that are occupied by atoms non-equivalent. The only form of relaxation that appears in the mean field theory is that the lattice parameter of the f.c.c. crystal is altered very slightly, and all sites remain equivalent. Thus the mean field treatment essentially neglects the relaxation energy of each vacancy. Similarly the relaxation energy in any dilute AB alloy is underestimated in the mean field approximation. In such cases the approximation will be most successful where the relaxation energy is small.

### 7.4.5 Some additional models

Finally, we briefly cite a few relatively special additional models, employing various approximations, which have appeared in the literature. A simple model which attempts to describe multilayer segregation without placing limits on the number of segregation sites in the boundary is the so-called BET model (Hondros and Seah 1977*a,b*) derived originally for multilayer gas adsorption on free surfaces. In this highly approximate model, it is assumed that solutes adsorbed in the first layer have a free energy of binding

$g_1^b$ whereas those in successive layers have a different binding energy, $g_L^b$. The adsorbed atoms have no lateral interaction in the layers and each layer does not have to be completed before the next is started.

Further models involving other features have been invoked (Hondros and Seah 1977*a,b*, Guttmann 1977, Guttman and McLean 1977) in order to interpret different types of complex segregation, particularly in multicomponent systems. Models featuring various repulsive and attractive nearest-neighbour solute interaction energies have been employed in efforts to understand the general features of segregation in multi-component alloy systems (such as steels) where the judicious use of alloying elements can often influence the segregation and alleviate the potentially deleterious effects of certain segregated components.

## 7.5  ATOMISTIC MODELS AT A FINITE TEMPERATURE

In Section 7.3.3 we reviewed briefly atomistic simulations of segregation at 0 K. In this section we shall show how these simulations have been extended to a finite temperature. As the temperature of the system is increased, entropic contributions to the free energy become increasingly important. The aim of an atomistic simulation is to take both the vibrational and configurational entropy into account to model the degree of segregation and the possible occurrence of phase changes at the interface as a function of temperature.

The bulk composition of the alloy is determined by the chemical potentials of the constituents, temperature and pressure. As described in Section 7.4 the exchange of matter between the interface and the adjoining crystals is simulated by the use of the grand canonical statistical ensemble. Thus during a simulation the numbers of atoms of different species may change, but the chemical potential of each species is kept constant. In the following we shall discuss two techniques that have been applied to segregation to grain boundaries at elevated temperatures in binary substitutional metallic alloys. In both cases embedded atom potentials (see Section 3.7) have been used to describe atomic interactions in alloys of f.c.c. metals. We shall begin by describing the Monte Carlo method and then go on to a much quicker method, described in Section 7.3.4, based on direct minimization of an approximate free energy functional. We will then discuss some results that have been obtained with the two methods.

In the Monte Carlo method (Foiles 1990) we generate a series of atomic configurations each with a probability appropriate to the grand canonical distribution. Let the alloy be in some initial configuration, labelled $i$, in which there are $N_A$ A atoms and $N_B$ B atoms at positions $r_1, r_2, r_3, \ldots r_N$, where $N = N_A + N_B$. The probability, $P_i$, of this configuration occurring in the grand canonical ensemble is proportional to (Kubo 1990)

$$\left[\frac{V}{\Lambda_A^3}\right]^{N_A} \left[\frac{V}{\Lambda_B^3}\right]^{N_B} \exp - \{(E_i - N_A\mu_A - N_B\mu_B)/kT\}, \tag{7.95}$$

where $V$ is the volume of the system, $\Lambda_A$ is the de Broglie wavelength of atom $A$, equal to $(h^2/2\pi m_A kT)^{\frac{1}{2}}$, and $m_A$ is the mass of atom A. The internal energy of the configuration, which is determined by the interatomic potentials and the atomic positions, is denoted by $E_i$ and $\mu_A$ and $\mu_B$ are the chemical potentials of atoms A and B.

To model the vibrational entropy of the system, and to allow individual atomic relaxation to take place, we generate a new configuration, labelled $j$, in which a randomly chosen atom is displaced by a small amount ($\leqslant 0.1$ Å). Let the internal energy of this new configuration be $E_j$. Then the probability of the new configuration occurring in the grand canonical ensemble, relative to the old configuration $i$, is given by

$$P_j/P_i = \exp - \{(E_j - E_i)/kT\} \tag{7.96}$$

where the chemical potentials and de Broglie wavelengths have been cancelled because the numbers of A and B atoms are the same in the two configurations. If $P_j/P_i$ is greater than unity then the new configuration $j$ is accepted. Otherwise the new configuration is retained with a probability given by $P_j/P_i$. This is achieved by comparing $P_j/P_i$ with a random number generated by the computer between 0 and 1. If $P_j/P_i$ is greater than the random number then configuration $j$ is accepted. Otherwise the system is returned to configuration $i$. The accepted configuration contributes to the statistical averages that are accumulated during the simulation. Clearly the higher the temperature the greater the chance that the new configuration will be accepted. For the next atomic displacement the initial configuration is the accepted configuration from the previous step. The decision-making process we have just described is called the Metropolis algorithm. It is repeated many times (typically between 100 and 1000) for each atom in the system.

Composition variations are modelled in a Monte Carlo simulation by 'transmuting' the atom at a site from an A atom to a B atom or vice versa. From eqn (7.95) the probability of accepting a step in which an A atom is converted into a B atom is given by

$$P_{A \to B} = \frac{\Lambda_A^3}{\Lambda_B^3} \exp - \{(\Delta E + \mu_A - \mu_B)/kT\} \tag{7.97}$$

where $\Delta E$ is the concomitant change in internal energy. There is a similar expression for transmuting a B atom into an A atom. The de Broglie wavelengths may be incorporated into the chemical potentials by working with the excess chemical potentials defined by

$$\tilde{\mu}_A = \mu_A - \mu_A^{id} \tag{7.98}$$

where $\mu_A^{id}$ is the chemical potential of A for an ideal gas of A and B atoms of the same composition and volume. Then we may write

$$P_{A \to B} = \exp - \{(\Delta E + \tilde{\mu}_A - \tilde{\mu}_B)/kT\}. \tag{7.99}$$

A transmutation step is attempted each time an atom is moved, and the combination of a transmutation attempt and an atomic movement attempt is called a microstep.

The simulation of equilibrium grain boundary segregation proceeds as follows. We carry out first a simulation of the bulk AB alloy at a given temperature, excess chemical potential difference $\tilde{\mu}_A - \tilde{\mu}_B$, and applied constant pressure $P$. The purpose of this simulation is to find out what composition of the alloy this generates and the equilibrium lattice parameter (for a cubic crystal structure). During this simulation the volume of the computational cell is allowed to fluctuate; in addition, individual atom displacements and atomic transmutations occur. The volume of the cell is changed by allowing uniform contractions or dilations. From eqn (7.95) the probability of accepting configuration $j$ in which the volume of the computational cell is $V_j$, relative to the probability of configuration $i$ in which the volume is $V_i$, is given by

$$\frac{P_j}{P_i} = \left[\frac{V_j}{V_i}\right]^N \exp - \{(E_j - E_i + P(V_j - V_i))/kT\} \tag{7.100}$$

where $P$ is the desired pressure in the system (which might be zero). An attempt to change the volume is made after approximately all the atoms in the computational cell have undergone one microstep. If a particular composition of the bulk alloy is desired then this is achieved by adjusting the excess chemical potential difference.

Having established the composition and excess chemical potential difference for the bulk, the Monte Carlo procedure is applied to a bicrystal with the same chemical potential difference (Foiles 1989, Seki *et al.* 1991*a*). Periodic boundary conditions are applied in all three directions, which means that there are two grain boundaries of equal and opposite misorientations in the computational cell (see Section 4.3.1.6). It is therefore important to establish that the interactions between the grain boundaries do not introduce errors. This may be ascertained by ensuring that between the grain boundaries the crystal reaches the composition found in the bulk crystal simulation. In addition, each grain boundary will have an equilibrium translation state parallel to the boundary plane and an equilibrium expansion normal to the boundary plane. If these are to be attained then the computational cell must be allowed to undergo changes of shape and volume, such that the average taken over many Monte Carlo steps of all six components of the total stress tensor inside the computational cell equal zero. Changes of the shape of the computational cell have apparently not been allowed in Monte Carlo simulations to date, although uniform volume changes of the whole computational cell have been included for bicrystals using eqn (7.100) (Seki *et al.* 1991*a*). Instead, the existence of alternative structures with different rigid body translations parallel to the interface has been explored by incorporating different rigid body translations into the initial configurations, and then allowing only uniform changes of the volume of the computational cell.

The equilibration of a Monte Carlo simulation is of paramount importance, because it is only when equilibrium has been achieved that the statistical averages are meaningful. It is checked by calculating some physical property such as the internal energy, or the volume of the cell, or the number of A atoms in the cell at each microstep, and examining where the averages for these properties becomes constant. Seki *et al.* found that between 400 and 2000 microsteps per atom were required to achieve equilibrium. Thus, in a cell containing 5000 atoms at least $2 \times 10^6$ microsteps are needed just for equilibration. Monte Carlo simulations are therefore very expensive computationally. In general we expect the convergence to be slower for A and B atoms that differ more in atomic size. That is because the probability of a transmutation step being accepted decreases as the difference in size increases because the transmutation takes place, at least initially, in the absence of any relaxation around the atom. Thus if a smaller atom is transmuted into a larger atom then the larger atom finds itself in a relatively compressed site and $\Delta E$ is therefore large and positive. In that case the transmutation attempt is unlikely to be successful, and many attempts may be needed to attain the equilibrium distribution.

The direct minimization of an approximate free energy functional methods use quasi-harmonic theory for the vibrational free energy (see Section 3.9) and the mean field approximation and the Bragg–Williams approximation (see Section 7.4.3) for the configurational free energy. This was first done by Najafabadi *et al.* (1991), who used the Einstein model in the classical limit of eqn (3.214) for the vibrational degrees of freedom, and eqns (7.24)–(7.26) for the compositional degrees of freedom, with atomic interactions described by embedded atom potentials. The approach is to minimize the grand potential of eqn ( 7.24) directly with respect to the site occupancies, all atomic positions, the rigid body translation parallel to the interface and the expansion normal to the interface. This is done using an energy minimization algorithm such as conjugate gradients, and the boundary conditions shown in Fig. 4.23. The structure, composition and lattice parameter of the perfect crystals of region II (see Fig. 4.23), are determined by the temperature, pressure, and chemical potential difference between A and B atoms that are being used for the simulation of the interface. The method is much faster than a Monte Carlo simulation. If it were exact then the site occupancies and the atomic positions which minimized the free energy functional would equal the average occupancies and atomic positions from a Monte Carlo simulation.

Only one test has been carried out to check whether this is indeed the case. Najafabadi *et al.* (1991) have compared segregation profiles at three (001) twist boundaries (10.4°, 22.6°, and 36.9°) in the Cu–Ni system with those obtained by Monte Carlo simulations by Foiles (1989). Three bulk alloy compositions were selected: 10, 50, and 90 per cent Cu. The results are shown in Fig. 7.16, where it is seen that overall good agreement is obtained. The average error in the concentration per layer is 30 per cent, with the worst error 50 per cent and the best 7 per cent. In general the error increases with decreasing bulk copper concentration and decreasing misorientation. However, even in the worst case the shape of the segregation profile is reasonably well described. It should be noted that the difference in the lattice parameters of Cu and Ni is only 2.6 per cent. The agreement between the methods is not expected to be as good at larger size differences because the mean field approximation is better when the relaxation displacements are smaller (see Section 7.4.4).

A thorough study of the importance of the vibrational entropy in grain boundary segregation, and the Einstein approximation (eqn (3.214)) to it, has been made by Rittner *et al.* (1993). These authors calculated the free energy of segregation in the dilute limit for all 30 binary combinations of Ag, Au, Cu, Ni, Pd, and Pt to a monovacancy, a (111) free surface, and 10.4°, 36.9° (001) twist boundaries and the $\Sigma 5$ (310) symmetrical tilt boundary. Embedded atom potentials were used to describe atomic interactions. The free energy of segregation was calculated by the 'overlapping distribution Monte Carlo' method (ODMC) which, in principle, is exact. At 1000 K it was found that for ten of the alloy combinations 25 per cent or more of the segregation energy was attributed to the change in the vibrational free energy. Thus, the change of the vibrational free energy on segregation cannot be neglected.

Comparison between the ODMC and direct free energy minimization method of

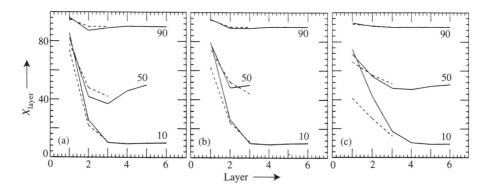

**Fig. 7.16**  Concentration of Cu atoms, $X_{\text{layer}}$ (at.%), averaged over (002) layers parallel to (a) $\Sigma 5$ $\theta = 36.9°$, (b) $\Sigma 13$ $\theta = 22.6°$ and (c) $\Sigma 61$ $\theta = 10.4°$ [001] twist boundaries in three Cu–Ni alloys with bulk Cu concentrations, $X_{\text{bulk}}$, of 10, 50, and 90 at.%, plotted against the number of (002) layers from the geometrical boundary midplane. The solid lines were obtained by direct minimization of an approximate free energy functional by Najafabadi *et al.* (1991). The dashed lines were obtained by Monte Carlo simulations by Foiles (1989) using the same atomic interactions. (From Najafabadi *et al.* (1991).)

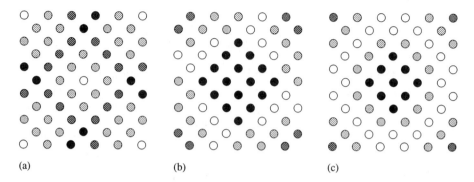

**Fig. 7.17** Distribution of Cu atoms at a $\Sigma = 61$ $\theta = 10.4\,°$ [001] twist boundary in $Ni_{0.9}Cu_{0.1}$ at 800 K. In (a), (b), and (c) the circles show the unrelaxed atomic positions in layers 1, 2, and 3 respectively, where layer 1 is adjacent to the geometrical boundary midplane. The darker the circle the higher the probability of the site being occupied by a Cu atom. (From Najafabadi *et al.* (1991).)

Najafabadi *et al.* (1991) for free energies of segregation in the dilute limit to grain boundaries allowed a test of the Einstein approximation for the vibrational free energy. It was found that good agreement is obtained for alloys where the size difference is small (such as Cu–Ni) but the errors increase as the size difference increases. This is expected because the relaxation displacements increase as the size difference increases and therefore the atomic interactions become increasingly anharmonic. Under these conditions the quasiharmonic approximation will eventually break down because higher derivatives of the potential energy than are taken into account in quasiharmonic theory will become important.

Insufficient comparative studies of the Monte Carlo and direct free energy minimization methods have been carried out to reach any definite conclusions about the accuracy of the latter. However, it appears that trends in segregation behaviour are reasonably well described by the free energy minimization methods. One obvious reason for the dearth of such comparative work is the huge computational effort involved in Monte Carlo simulations. Indeed, this is one of the main reasons why free energy minimization methods were developed. In addition, only the free energy minimization methods enable contact to be made analytically with segregation isotherms (as described in Section 7.4.3), which are used extensively to represent experimental data.

Figure 7.17 shows the site occupancies in the first three (002) planes of a 10.4° (001) twist boundary in $Ni_{0.9}Cu_{0.1}$ at 800 K, obtained by direct minimization of an approximate free energy (Najafabadi *et al.* 1991). In this small-angle twist boundary there is an orthogonal grid of $\frac{1}{2}\langle 110\rangle$ screw dislocations, between which there are patches of relatively unstrained crystal. In the first layer adjacent to the geometrical boundary plane it is seen that Cu atoms segregate to the screw dislocations and tend to avoid the unstrained crystal between them. This pattern is reversed in the second and third layers. These patterns are almost identical to those obtained by Monte Carlo simulations (Foiles 1989), indicating that, although this is the boundary and alloy composition for which the worst quantitative agreement between the absolute magnitudes of the segregation in the two methods was obtained, the qualitative pictures of the segregation are virtually the same.

Wang *et al.* (1993) applied the same direct free energy minimization method to a range of (001) twist boundaries in $Ni_{0.95}Cu_{0.05}$ alloys at temperatures between 600 and 1100 K.

They found that Cu always segregates to the boundaries. The copper atoms always selected sites that were associated with hydrostatic tension, where the slightly larger size of the Cu atom could be better accommodated. As a result the variations of the hydrostatic stress from site to site in the interfaces were reduced after segregation had taken place. On the other hand, the atomic structure of the boundaries did not change significantly after segregation had taken place, with the original structural units remaining clearly identifiable. This is not surprising in view of the small difference in atomic sizes between Cu and Ni. The local deviatoric stresses changed much less after segregation because they are sensitive to the local symmetry of the atomic environment, which remained almost the same. These results indicate that the driving force for segregation in the model Ni–Cu alloys is primarily the size effect and packing of atoms. However, the use of atomic interactions that depend only on distances between atoms, such as embedded atom potentials, precludes other possible driving forces that may or may not exist in real Ni–Cu alloys, such as the formation of directional bonds.

Seki *et al.* (1991*a*) have carried out Monte Carlo simulations of segregation of Au in a Pt–1at. %Au alloy at 850, 900, 1000, 1300, 1500, and 1900 K to (001) twist boundaries with misorientations 5.0, 10.4, 16.3, 22.6, 28.1, 33.9, 41.1, and 43.6°. At all these temperatures the model bulk alloy is a single phase solid solution. The distance between successive boundaries in each periodically repeated computational cell is 10 (002) planes, or 1.96 nm. Embedded atom potentials were used to describe atomic interactions in the Pt–Au alloy.

Figure 7.18 shows the Au concentration in the (002) planes of each computational cell at 850 K. It is seen that Au segregation takes place to all the boundaries studied, and primarily to the (002) plane on either side of the geometrical boundary plane. The width of the segregation profile decreases as $\theta$ increases to 33.9°, and is then roughly constant. The effect of increasing $T$ is to broaden the profiles slightly. The solute enrichment at the grain boundaries is also seen to increase as $\theta$ increases, and for a given $\theta$ it was found that the degree of segregation decreased exponentially with $T$. More quantitatively, Seki *et al.* define the average solute enrichment factor, $S_{avg}$, as the Au concentration in the two planes to either side of the geometrical interface divided by the bulk concentration. As shown in Fig. 7.6, they found that $S_{avg}$ increased linearly with $\sin(\theta/2)$, which is inversely proportional to the primary $\frac{1}{2}\langle 110 \rangle$ dislocation spacing in the boundaries, up to $\theta \simeq 35°$. These results demonstrate that the segregation saturates at the $\Sigma = 5$ (001) twist boundary where the boundary may be described as a contiguous slab of $\frac{1}{2}\langle 110 \rangle$ dislocation cores, with no perfect crystal structural units in between them (see Section 4.3.1.8). At misorientations greater than 36.9° ($\Sigma = 5$) the boundary comprises two kinds of structural unit of the $\Sigma = 5$ boundary (36.9° and 53.1°) separated by $\frac{1}{10}\langle 310 \rangle$ secondary screw dislocations of the $\Sigma = 5$ DSC lattice. The results displayed in Fig. 7.6 indicate that the degree of segregation is affected only by the density of perfect crystal structural units, and when the misorientation is such that they are no longer present the degree of segregation saturates.

Seki *et al.* found that the McLean isotherm, eqn (7.32), describes the temperature dependence of the average solute enrichment factor reasonably well. Since the solute concentration in the bulk and at the interface is of the order of a few per cent, the McLean isotherm may be expressed approximately as $S_{avg} = c_{gb}/c_o = \exp(-\Delta f_{seg}/kT)$. Here, $c_{gb}$ is the solute concentration in the grain boundary, $c_o$ the bulk concentration, and $\Delta f_{seg}$ is the free energy of segregation which may be expressed as $\Delta f_{seg} = \Delta e_{seg} - T\Delta s_{seg}$. Figure 7.19 demonstrates that $S_{avg}$ does indeed display Arrhenius behaviour. By least squares fitting of straight lines to the data points in Fig. 7.19 it is found that the energy

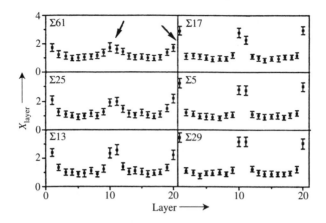

**Fig. 7.18** The concentration of Au, $X_{\text{layer}}$ (at.%), in the (002) layers parallel to $\theta = 10.4°$ ($\Sigma 61$), 16.3° ($\Sigma 25$), 22.6° ($\Sigma 13$), 28.1° ($\Sigma 17$), 36.9° ($\Sigma 5$) and 43.6° ($\Sigma 29$) (001) twist boundaries in a Pt-1at.% Au alloy, as computed by the Monte Carlo method. In each computational cell there are two interfaces (arrows) because of the periodic boundary conditions. The concentrations are averaged over $(3–5) \times 10^6$ microsteps, and the error bars correspond to 4 standard deviations. (From Seki *et al.* (1991a).)

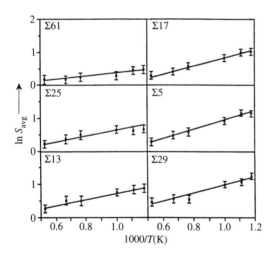

**Fig. 7.19** Arrhenius plots of the average segregation enhancement factor, $S_{\text{avg}}$, for $\theta = 10.4°$ ($\Sigma 61$), 16.3° ($\Sigma 25$), 22.6° ($\Sigma 13$), 28.1° ($\Sigma 17$), 36.9° ($\Sigma 5$), and 43.6° ($\Sigma 29$) (001) twist boundaries in a Pt-1at.% Au alloy, as computed by the Monte Carlo method. (From Seki *et al.* (1991a).)

of segregation, $\Delta e_{\text{seg}}$, is always negative, which is consistent with segregation being favourable. Table 7.1 shows the fitted enthalpies and entropies of segregation to the nine (001) twist boundaries.

The uncertainties in the Table 7.1 entries represent $\pm 1$ standard deviation. The energies of segregation increase in magnitude with $\theta$ up to about 30° and then saturate at about $-0.1$ eV per atom. The negative entropies of segregation arise from the change of the vibrational entropy of the system.

The McLean isotherm assumes that all boundary sites are equally suitable for segregation. But in the Monte Carlo simulations of Seki *et al.* it was found that segregation occurred primarily to the cores of $\frac{1}{2}\langle 110 \rangle$ screw dislocations. Therefore the increasing magnitude of $\Delta e_{\text{seg}}$ in Table 7.1 with increasing $\theta$, up to about 35°, does not necessarily

**Table 7.1** Energies, $\Delta e_{seg}$, and entropies, $\Delta s_{seg}$, of Au segregation for (001) twist boundaries in a Pt-1at.%Au alloy as a function of misorientation, $\theta$. (From Seki *et al.* (1991*a*).

| $\theta(°)$ | $\Delta e_{seg}$ (eV atom$^{-1}$) | $\Delta s_{seg} \times k^{-1}$ |
|---|---|---|
| 5 | $-0.011 \pm 0.009$ | $0.16 \pm 0.09$ |
| 10.4 | $-0.040 \pm 0.009$ | $-0.12 \pm 0.10$ |
| 16.3 | $-0.057 \pm 0.009$ | $-0.07 \pm 0.10$ |
| 22.6 | $-0.079 \pm 0.009$ | $-0.18 \pm 0.10$ |
| 28.1 | $-0.102 \pm 0.008$ | $-0.35 \pm 0.08$ |
| 33.9 | $-0.098 \pm 0.007$ | $-0.20 \pm 0.08$ |
| 36.9 | $-0.121 \pm 0.009$ | $-0.46 \pm 0.09$ |
| 41.1 | $-0.102 \pm 0.007$ | $-0.21 \pm 0.08$ |
| 43.6 | $-0.107 \pm 0.009$ | $-0.26 \pm 0.09$ |

imply an increasing magnitude of $\Delta e_{seg}$ to the same type of boundary site. The increasing solute enrichment factor, $S_{avg}$, could simply be a consequence of the increased density of suitable sites as $\theta$ increases, until saturation is reached at about 35°. In a companion paper Seki *et al.* (1991*b*) demonstrated that all the results of Table 7.1 could be explained by assuming that segregation took place only to the cores of $\frac{1}{2}\langle 110 \rangle$ screw dislocations of a fixed width of $1.6|b|$, and that the segregation energy was the same at all sites in those cores. The enthalpy of segregation to one of these dislocation core sites was then $-0.095 \pm 0.01$ eV, and the entropy of segregation was $-0.49 \pm 0.10 \, k$.

Taking the Monte Carlo and direct free energy minimization simulations of segregation in f.c.c. metallic alloys together we see that the size effect is the dominant driving force for segregation in these alloy systems. We see also that the effect of raising the temperature is simply to reduce the degree of segregation, as predicted by the simple isotherms described in Section 7.4.3. But the results of these simulations should not be taken as representative for all alloy systems. All the simulations described in this section have used embedded atom potentials, in which no directional bonding is included. Qualitatively different results might be found in alloy systems in which the differences in atomic sizes are insignificant, but where segregation and directional bonding may lead to the formation of unique environments at interfaces.

## 7.6 INTERFACE SEGREGATION IN IONIC SOLIDS

In considering the segregation of solutes to interfaces in ionic solids, an entirely new factor must be considered, i.e. the fact that the interface tends to be electrically charged and that a region of compensating space charge tends to form in the regions directly adjacent to the interface. This region of space charge influences the distribution of any solute atoms that are present which carry an excess charge (e.g. aliovalent species such as divalent solutes in an alkali halide) and, therefore, must be considered in any description of the segregation of such species.

We begin by showing that space charge regions develop at interfaces which act as sources/sinks (see Chapter 10) for charged point defects in ideally pure ionic solids whenever the formation energies of the differently charged point defects which are created under usual thermal equilibrium conditions are different (as is generally the case). We then introduce charged solute atoms and solve for their equilibrium distribution in the presence of the new distribution of space charge which ensues. A wide variety of situations can develop in different systems depending upon the types of charged point defects which are created under thermal equilibrium conditions (i.e. anion and cation vacancies and interstitials) and the types of charged aliovalent solute atoms which are present. However, the basic principles underlying the solution of this problem for all systems are essentially the same. We therefore follow Kliewer and Koehler (1965) and present a prototypical analysis for a simple NaCl-type system in which doubly charged cation solute atoms (e.g. $Ca^{++}$) may be present, and the thermal equilibrium defect population consists of anion and cation vacancies (i.e. Schottky type disorder). In this case, the $Ca^{++}$ ions and anion vacancies each carry an effective unit positive charge, $e$, while the cation vacancies carry unit negative charge.

Consider, first, such a crystal without solute solute atoms bounded by two planar interfaces at $x = 0$ and $x = 2L$. The electrostatic potential, $\phi(x)$, due to the distribution of charged vacancies created at the interfaces in the process of establishing thermodynamic equilibrium must satisfy Poisson's equation (Purcell 1965) for the distribution of charge under the boundary conditions

$$\phi = 0 \qquad (x = 0, 2L) \tag{7.101}$$

$$\frac{d\phi}{dx} = 0 \qquad (x = L). \tag{7.102}$$

Neglecting any preferred association between the Schottky defects, the free energy of the region $0 \leqslant x \leqslant L$, after the creation of the vacancies, can then be written (Kliewer and Koehler 1965) as

$$G = \int_0^L G_o + n^+(x)G^+ + n^-(x)G^- + [\rho(x) \cdot \phi(x)/2 - TS_c(x)] \, dx. \tag{7.103}$$

Here, $G_o$ is the free energy prior to the introduction of the defects, $n^+(x)$ and $n^-(x)$ are the densities of cation and anion vacancies, respectively, $G^+$ and $G^-$ are the respective vacancy formation free energies at the interfaces where $\phi = 0$, $\rho(x)$ is the charge distribution, the quantity $\rho(x) \cdot \phi(x)/2$ is the electrostatic energy (Purcell 1965), and $S_c(x)$ is the usual configurational entropy gained from randomly mixing the two different species on the available sites. By writing out $S_c$ (see, for example, eqn (7.26)) and minimizing $G$ with respect to the independent quantities $n^+$ and $n^-$, we obtain

$$n^+(x) = n \cdot \exp[-\{G^+ - e\phi(x)\}/kT] \tag{7.104}$$

$$n^-(x) = n \cdot \exp[-\{G^- + e\phi(x)\}/kT] \tag{7.105}$$

where $n$ is the density of anion or cation sites. Equations (7.104) and (7.105) have the expected Boltzmann form for equilibrium defect concentrations where, for example, the total energy to form a cation vacancy (carrying a positive effective charge $e$) in a region of positive potential is just $[G^+ - e\phi(x)]$. The potential, $\phi(x)$, can now be obtained by solving Poisson's equation

$$\nabla^2\phi(x) = -4\pi\rho(x)/\varepsilon = 4\pi e\{n^+(x) - n^-(x)\}/\varepsilon, \tag{7.106}$$

where $\varepsilon$ is the dielectric constant, and $\rho(x)$ is the charge distribution. The solution, subject to the boundary conditions, eqns (7.101) and (7.102), is rather tedious to obtain (Kliewer and Koehler 1965) but leads in a straightforward way to the results sketched schematically in Fig. 7.20 for the case $G^- > G^+$ which holds, for example, for NaCl. As might be expected, the lower formation energy for cation vacancies causes the formation of more cation vacancies than anion vacancies. The extra cations are deposited in the

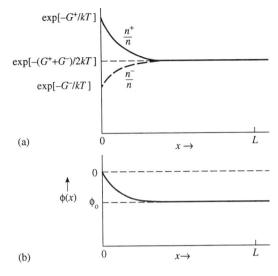

**Fig. 7.20** (a) Concentration of cation vacancies, $n^+/n$, and anion vacancies, $n^-/n$, as a function of distance, $x$, from a grain boundary located at $x = 0$ in a pure NaCl-type ionic solid. (b) Corresponding electrostatic potential in solid, $\phi$, as a function of $x$. (From Kliewer and Koehler (1965).)

interface and form a sheet of excess positive charge. This is compensated by a region of excess negative charge adjacent to the interface containing an excess of cation vacancies and a deficit of anion vacancies (Fig. 7.20(a)). These parallel layers of opposite charge constitute a dipole layer. The resulting potential (Fig. 7.20(b)) is zero at the interface, as required, and becomes negative at essentially the constant value, $\phi(L)$, in the bulk where the cation and anion vacancy concentrations become equal (Fig. 7.20(a)). This bulk condition, i.e. $n^+(L) = n^-(L)$ may have been anticipated directly, since it corresponds to the condition of electrical neutrality which is achieved by establishing a potential in the bulk which suppresses the cation vacancy concentration and enhances the anion vacancy concentration in order to make them equal. The necessary value of $\phi(L)$ is easily found by setting $n^+$ and $n^-$ in eqns (7.104) and (7.105) equal to each other and solving for $\phi$. The result is

$$\phi(L) = (1/2e) [G^+ - G^-],\qquad(7.107)$$

which is negative, since $G^- > G^+$. Therefore, in the bulk,

$$n^+(L) = n^-(L) = n \cdot \exp\left[-(G^+ + G^-)/2kT\right]\qquad(7.108)$$

as indicated in Fig. 7.20(a). Finally, the thickness of the space charge region depends on the dielectric constant and is given approximately by $\kappa^{-1}$ where

$$\kappa^2 = \frac{8\pi n e^2}{\varepsilon kT} \exp\left[-(G^+ + G^-)/2kT\right].\qquad(7.109)$$

If $Ca^{++}$ solute ions are now added to the bulk system at a density, $n_s$, the problem becomes more complicated, since three types of charged point defects are now present, i.e. anion and cation vacancies and isolated $Ca^{++}$ ions. However, the problem can again be solved (Kliewer and Koehler 1965) by using the same basic procedure employed previously. The results in the bulk may again be anticipated very simply by requiring electrical neutrality there, i.e.

$$n^-(L) + n_s(L) - n^+(L) = 0.\qquad(7.110)$$

For dilute systems, eqns (7.104) and (7.105) still hold, and therefore,

$$n^-(L) \cdot n^+(L) = n^2 \cdot \exp\left[-(G^+ + G^-)/kT\right]. \tag{7.111}$$

Using eqn (7.110), we then have

$$[n^+(L) - n_s(L)]n^+(L) = n^2 \cdot \exp\left[-(G^+ + G^-)/kT\right]. \tag{7.112}$$

Now, according to eqn (7.108), the bulk cation vacancy concentration in the pure material [denoted by $n_o^+(L)$] would be given by $n \cdot \exp[-(G^+ + G^-)/2kT]$. Therefore,

$$[n^+(L) - n_s(L)]\, n^+(L) = [n_o^+(L)]^2. \tag{7.113}$$

Equation (7.113) may be solved for $n^+(L)$ to obtain

$$n^+(L) = [n_s(L)/2] \cdot \{1 + \sqrt{1 + 4[n_o^+(L)/n_s(L)]^2}\}. \tag{7.114}$$

Equation (7.114) takes two limiting forms. When $n_o^+(L) \gg n_s(L)$, we have $n^+(L) = n_o^+(L)$, and the equilibrium cation vacancy concentration is essentially unaffected by the presence of the aliovalent impurity. This condition tends to be satisfied at high temperatures (where the thermal population in the pure material, $n_o^+(L)$, is large) and also in high purity material (where $n_s(L)$ is small). At the other extreme, when $n_o^+(L) \ll n_s(L)$, we have $n^+(L) = n_s(L)$. In this situation, the equilibrium thermal cation vacancy population is negligible, and cation vacancies are then generated in the impure material in order to maintain electrical neutrality. This condition will tend to be satisfied at low temperatures or in low purity material. The former behaviour is termed 'intrinsic', since the impurity atoms play essentially no role, while the latter is termed 'extrinsic', since the equilibrium defect structure is controlled by the presence of the impurity atoms.

A material of fixed purity (fixed aliovalent solute concentration) will generally tend to undergo a transition from intrinsic to extrinsic behaviour as the temperature is decreased. A special temperature, $T_o$, will generally be encountered in this transition, where the condition of electrical neutrality in the bulk is satisfied without the need of generating the potential $\phi(L)$. This will occur when the difference between the thermal populations of cation and anion vacancies in the absence of any potential becomes exactly equal to the solute concentration. This temperature, $T_o$, is known as the 'isoelectric temperature' and can be easily found by setting $\phi = 0$ in eqns (7.104) and (7.105) and then combining them with the electrical neutrality condition, eqn (7.110), to obtain

$$\exp[-G^+/kT_o] - \exp[-G^-/kT_o] = n_s/n. \tag{7.115}$$

The bulk potential and space charge, therefore, disappear at $T_o$. The situations are different above and below $T_o$, and some results from the complete solution obtained by Kliewer and Koehler are sketched schematically in Fig. 7.21. Above $T_o$, the interface is positively charged, as it would be in the pure material, and there is an adjacent region of negative excess space charge where the $Ca^{++}$ ion concentration, $n_s$, is reduced. Below $T_o$, the situation is reversed, and the $Ca^{++}$ ions are segregated preferentially in the positive space charge region adjacent to the interface which is negatively charged. Detailed calculations for the NaCl system using the above model (modified somewhat (Kliewer 1965) to allow relatively high solute concentrations) indicate that the accumulation of solute atoms in the space charge region at $T < T_o$ can be strong and, therefore, must be considered in any analysis of segregation.

Unfortunately, the above analysis, complex as it is, is incomplete, since it neglects any local interactions, such as, for example, elastic misfit and electrostatic interactions

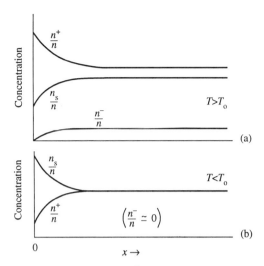

**Fig. 7.21** Concentration of cation vacancies, $n^+/n$, anion vacancies, $n^-/n$, and doubly charged cation solute atoms, $n_s$, as a function of distance, $x$, from a grain boundary located at $x = 0$ in a NaCl-type ionic solid containing doubly charged cation solute atoms. (a) Situation at temperatures above the isoelectric temperature, $T_0$. (b) Situation below $T_0$. (From Kliewer and Koehler (1965).)

between the charged species which are present and the core of the interface. Any self-consistent analysis must take these interactions and their accompanying free energy changes into account in order to obtain the final distribution of the various species in the presence of the coupled space charge region which is formed. Yan *et al.* (1983*b*) obtained numerical solutions for this coupled problem for a grain boundary in KCl containing divalent cation solutes by assuming a simple elastic misfit interaction between the segregating solutes and the boundary of the type represented by eqn (7.9). The results of a large number of calculations show that the solute segregation is significantly influenced by both the local core interactions and the presence of the space charge layer as might be expected. The width of the space charge layer ranged from ten to hundreds of angstroms.

In other work, Duffy and Tasker (1985) have carried out calculations for the situation at [100] and [110] symmetric tilt boundaries in pure NiO, where the important defects were cation vacancies and holes assumed to be localized on the cations in the form of $Ni^{3+}$ ions. Detailed atomistic energy minimization calculations (Section 4.3.1.6) were first made of the interactions (binding energies) of the cation vacancies and holes with the boundaries (in the absence of any space charge), and the binding energies of the cation vacancies were found to be greater. Self-consistent calculations of the defect distributions and space charge were then made which revealed a negatively charged boundary screened by a space charge region with an enhanced hole concentration and a diminished cation vacancy concentration. For the (211) symmetric tilt boundary at 1000 K, the screening length was about 2 nm. The segregation of various isovalent and aliovalent solute atoms to these boundaries was considered next. Solute binding energies (again in the absence of any space charge) were calculated atomistically, and rather strongly attractive sites of one type or another were found in all of the boundaries indicating that all of the solutes studied would have a tendency to segregate. These binding energies generally increased with both the size and charge differences associated with the solute atoms (see Section 7.3). Binding energies larger than 1 eV were found in some cases. Several simple estimates were made of the degree of segregation of isovalent solutes (which, of course, would not interact with the space charge). For the (211) symmetric tilt boundary at 1000 K with 100 ppm fraction of solute present in the bulk, the boundary concentration of Mg was found to be $3 \times 10^{-4}$, while, on the other hand, Sr atoms achieved full coverage and occupied essentially all of the favourable segregation sites. Further calculated results for boundaries in NiO are reported elsewhere (Duffy and Tasker 1984).

A number of experimental studies of the space charges occurring at interfaces in ionic systems has been carried out (see Ikeda and Chiang (1993) and Ikeda *et al.* (1993) and the references in these papers). In the latest and most detailed quantitative work, Ikeda and Chiang (1993) and Ikeda *et al.* (1993) studied space charge phenomena in poly-crystalline $TiO_2$ containing the aliovalent solute cations $Al^{3+}$ and $Ga^{3+}$ (acceptors) and $Nb^{5+}$ (donors). These selected cations are of closely the same ionic size as the host $Ti^{4+}$ cations, and, therefore, they are not expected to exhibit strong local segregation to the grain boundary cores due to the size effect. The experimental situation is therefore simplified to the extent that the distribution of the solute atoms should be governed primarily by electrostatic interactions. The accumulation (or depletion) of solute atoms throughout the space charge region was determined directly by analytical electron micro-scopy under different conditions, and the results closely fit the space charge model described earlier. The space charge potential was negative in undoped and acceptor-doped $TiO_2$ and positive at high donor concentrations as predicted: good agreement was found with respect to the isoelectric point and the composition and temperature dependence of the potential. Also, variations in the space charges at different boundaries were found, indicating clear differences in the formation energies of the point defects which were involved.

We may conclude that interfacial segregation in ionic solids is a highly complex phenomenon involving the interplay of a variety of interactions including, for example, both misfit and electrostatic interactions. Efforts to analyse this phenomenon to date have utilized various approximations and have revealed its main features. In general, the segregation tends to be relatively strong compared to that in other types of solids, particularly for aliovalent solutes having large charge differences. The segregation tends to be ubiquitous in common polycrystalline ceramics, since these materials generally contain appreciable concentrations of various solute atoms (Kingery 1974, Rühle 1982, Clarke 1985). In fact, many polycrystalline ceramic materials contain sufficiently high concentrations of impurities so that distinct phases, including layers of amorphous phases (Clarke 1985), often form at the boundaries. We shall not attempt to describe these complex phenomena.

## REFERENCES

Arias, T. A. and Joannopoulos, J. D. (1992). *Phys. Rev. Letters*, **69**, 3330.
Bacon, D. J. (1972). *Phys. Stat. Sol.* (*b*), **50**, 607.
Balcerzak, T. (1991). *J. Magn. Magn. Mat.*, **97**, 152.
Balcerzak, T., Mielnicki, J., Wiatrowski, G., and Urbaniak-Kucharczyk, A. (1990). *J. Phys. Condens. Matter*, **2**, 3955.
Balluffi, R. W. (1979). In *Interfacial segregation* (eds W. C. Johnson and J. M. Blakely) p. 193. American Society for Metals, Metals Park, Ohio.
Balluffi, R. W. and Simmons, R. O. (1960). *J. Appl. Phys.*, **31**, 2284.

Blum, B., Menyhard, M., Luzzi, D. E. and McMahon, Jr., C. J. (1990). *Scripta Metall. Mater.*, **24**, 2169.

Bouchet, D. and Priester, L. (1987). *Scripta Metall.*, **21**, 475.

Briant, C. L. (1983). *Acta Metall.*, **31**, 257.

Briant, C. L. (1988). *Acta Metall.*, **36**, 1805.

Briant, C. L. (1990). *Mats. Res. Soc. Bulletin*, **15**(10), 26.

Briant, C. L. (1991). In *Structure and property relationships for interfaces* (eds J. L. Walter, A. H. King and K. Tangri) p. 43. ASM International, Metals Park, Ohio.

Cabane, J. and Cabane, F. (1991). In *Interface segregation and related processes in materials* (ed. J. Nowotny) p. 1. Trans Tech. Publications, Zürich.

Callen, H. B. (1963). *Phys. Letters*, **4**, 161.

Christian, J. W. (1975). *The theory of transformations in metals and alloys*. Pergamon Press, Oxford.

Clarke, D. R. (1985). *J. de Physique*, **46**, Colloq. C-4, C4-51.

Cochart, A. W., Schoeck, G., and Wiedersich, H. (1955). *Acta Metall.*, **3**, 533.

Dal Pino, A., Galvan, M., Arias, T. A., and Joannopoulos, J. D. (1993). *J. Chem. Phys.*, **98**, 1606.

Deymier, P., Janot, L., Li, J., and Dobrzynski, L. (1989). *Phys. Rev. B*, **39**, 1512.

Djafari-Rouhani, B., Dobrzynski, L., Maradudin, A. A., and Wallis, R. F. (1980). *Surf. Sci.*, **91**, 618.

Donald, A. M. and Brown, L. M. (1979). *Acta Metall.*, **27**, 59.

Duffy, D. M. and Tasker, P. W. (1984). *Phil. Mag. A*, **50**, 155.

Duffy, D. M. and Tasker, P. W. (1985). *J. de Physique*, **46**, Colloq. C-4, C4-185.

Erhart, H. and Grabke, H. J. (1981). *Met. Sci.*, **15**, 401.

Eshelby, J. D. (1977). In *Vacancies '76* (eds R. E. Smallman and J. E. Harris) p. 3. The Metals Society, London.

Ference, T. G. and Balluffi, R. W. (1988). *Scripta Metall.*, **22**, 1929.

Foiles, S. M. (1989). *Phys. Rev. B* **40**, 11502.

Foiles, S. M. (1990). In *Surface segregation phenomena* (eds P. A. Dowben and A. Miller) p. 79. CRC Press, Boca Raton, Florida.

Fowler, R. H. and Guggenheim, E. A. (1939). *Statistical thermodynamics*. Cambridge University Press, Cambridge.

Gaskell, D. R. (1983). In *Physical metallurgy* (eds R. W. Cahn and P. Haasen). North-Holland, Amsterdam.

Grovenor, C. R. M. (1985). *J. Phys. C: Solid State Phys.*, **18**, 4079.

Grovenor, C. R. M., Batson, P. E., Smith, D. A., and Wong, C. (1984). *Phil. Mag. A*, **50**, 409.

Guttmann, M. (1977). *Metall. Trans.*, **8A**, 1383.

Guttmann, M. and McLean, D. (1977). In *Interfacial segregation* (eds W. C. Johnson and J. M. Blakely) p. 261. American Society for Metals, Metals Park, Ohio.

Guyot, P. and Simon, J.-P. (1975). *J. de Physique*, **36**, Colloq, C-4, C4-141.

Hashimoto, M., Ishida, Y., Yamamoto, R., and Doyama, M. (1984). *Acta Metall.*, **32**, 1.

Heggie, M. I., Jones, R., and Umerski, A. (1991). *Phil. Mag. A*, **63**, 571.

Hirth, J. P. and Lothe, J. (1982). *Theory of dislocations*, Wiley, New York.

Hoffman, R. (1988). *Solids and surfaces*. VCH, New York.

Hofmann, S. (1985). In *Scanning electron microscopy III*, p. 1071. SEM, Inc. AMF O'Hare, Chicago, Illinois.

Hofmann, S. (1987). *J. de Chimie Physique*, **84**, 141.

Hofmann, S. and Lejcek, P. (1991). *Scripta Metall. Mater.*, **25**, 2259.

Hofmann, S., Lejcek, P., and Adamek, J. (1992). *Surf. Interface Analysis*, **19**, 601.

Hondros, E. D. and Seah, M. P. (1977a). *International. Mets. Revs.*, No. 222, 262.

Hondros, E. D. and Seah, M. P. (1977b). *Metall. Trans.*, **8A**, 1363.

Hondros, E. D. and Seah, M. P. (1983). In *Physical metallurgy* (eds R. W. Cahn and P. Haasen) p. 855. North-Holland, Amsterdam.

Hume-Rothery, W., Smallman, R., and Haworth, C. W. (1969). *The structure of metals and alloys.* Institute of Metals, London.

Ikeda, J. A. S. and Chiang, Y.-M. (1993). *J. Am Ceram. Soc.*, **76**, 2437.

Ikeda, J. A. S., Chiang, Y.-M., Garratt-Reed, A. J., and Vander Sande, J. B. (1993). *J. Amer. Ceram. Soc.*, **76**, 2447.

Kingery, W. D. (1974). *J. Amer. Ceram. Soc.*, **57**, 1; 74.

Kliewer, K. L. (1965). *Phys. Rev.*, **140**, A1241.

Kliewer, K. L. and Koehler, J. S. (1965). *Phys. Rev.*, **140**, A1226.

Kubo, R. (1990). *Statistical mechanics.* North-Holland. Amsterdam.

Larere, A. Guillope, M. and Masuda-Jindo, K. I. (1988). *J. de Physique*, **49**, Colloque C5, supple. au n° 10, C-447.

Luzzi, D. E. (1991*a*). *Ultramicroscopy*, **37**, 180.

Luzzi, D. E. (1991). *Phil. Mag. Letters*, **63**, 281.

Maurice, J. L. (1991). In *Polycrystalline semiconductors II* (eds J. H. Werner and H. P. Strunk) p. 166, Springer-Verlag, Berlin.

McLean, D. (1957). *Grain boundaries in metals.* Clarendon Press, Oxford.

Menyhard, M., Blum, B., and McMahon, Jr. C. J. (1989). *Acta Metall.*, **37**, 549.

Michael, J. R., Lin, C.-H., and Sass, S. L. (1988). *Scripta Metall.*, **22**, 1121.

Militzer, M. and Wieting, J. (1987). *Acta Metall.*, **35**, 2765.

Mott, N. F. (1962). *Rep. Prog. Phys.*, **25**, 218.

Moya, F. and Moya-Gontier, G. E., 1975. *J. de Physique*, **36**, Colloq. C4, supple. au n° 10, C4-157.

Mütschele, T. and Kirchheim, R. (1987). *Scripta Metall.*, **21**, 135.

Najafabadi, R., Wang, H. Y., Srolovitz, D. J., and LeSar, R. (1991). *Acta Metall. Mater.*, **39**, 3071.

Ogura, T., McMahon, C. J., Feng, H. C., and Vitek, V. (1978). *Acta Metall.*, **26**, 1317.

Ogura, T., Watanabe, T., Karashima, S., and Masumoto, T. (1987). *Acta Metall.*, **35**, 1807.

Pettifor, D. G. (1987). *Sol. Stat. Phys.*, **40**, 43.

Pichard, C., Guttmann, M., Rieu, J., and Goux, C. (1975). *J. de Physique*, **36**, Colloq. C4, Supple. au n° 10, C4-151.

Purcell, E. M. (1965). *Electricity and magnetism.* McGraw-Hill, New York.

Rafii-Tabar, H. and Sutton, A. P. (1991). *Phil. Mag. Letters*, **63**, 217.

Rittner, J. D., Foiles, S. M., and Seidman, D. N. (1993). Private communication.

Rühle, M. (1982). *J. de Physique*, **43**, Colloq. C-6, C6-115.

Seidman, D. N. (1991). *Mat. Sci. and Eng.*, **A137**, 57.

Seidman, D. N., Hu, J. G., Kuo, S.-M., Krakauer, B. W., Oh, Y., and Seki, A. (1990). *Colloque de Physique, Colloque Cl*, Supple. au n° 1, *J de Physique*, **51**, Cl-47.

Seki, A., Seidman, D. N., Oh, Y., and Foiles, S. M. (1991*a*). *Acta Metall., Mater.*, **39**, 3167.

Seki, A., Seidman, D. N., Oh, Y., and Foiles, S. M. (1991*b*). *Acta Metall. Mater.*, **39**, 3179.

Sokolnikoff, I. S. (1956). *Mathematical theory of elasticity.* McGraw-Hill, New York.

Sutton, A. P. and Vitek, V. (1982). *Acta Metall.*, **30**, 2011.

Suzuki, S., Abiko, K., and Kimura, H. (1981). *Scripta Metall.*, **15**, 1139.

Tang, S., Freeman, A. J., and Olson, G. B. (1993). *Phys. Rev. B*, **47**, 2441.

Udler, D. and Seidman, D. N. (1992*a*). *Scripta Metall. Mater.*, **26**, 449, 803.

Udler, D. and Seidman, D. N. (1992*b*). *Phys. Stat. Sol. (b)*, **172**, 267.

Watanabe, T. (1985). *J. de Physique*, **46**, Colloque C4, supple. au n° 4, C4-555.

White, C. L. and Coghlan, W. A. (1977). *Met. Trans.*, **8A**, 1403.

Yan, M. F., Cannon, R. M., and Bowen, H. K. (1983*a*). In *Character of grain boundaries: advances in ceramics*, Vol. 6 (eds M. F. Yan and A. H. Heuer) p. 255. American Ceramic Society, Columbus, Ohio.

Yan, M. F., Cannon, R. M., and Bowen, H. K. (1983*b*). *J. Appl. Phys.*, **54**, 764.

Yan, M. F., Cannon, R. M. and Bowen, H. K. (1983*c*). *J. Appl. Phys.*, **54**, 779.

Yan, M., Sob, M., Luzzi, D. E., Vitek, V., Ackland, G. J., Methfessel, M., and Rodriguez, C. O. (1993). *Phys. Rev. B*, **47**, 5571.

# PART III

# INTERFACIAL KINETICS

# 8

## Diffusion at interfaces

### 8.1 INTRODUCTION

The rate of atomic diffusion at interfaces generally differs from the corresponding rate (or rates) in the adjoining crystal lattices, and this phenomenon plays an important role in many processes which involve the diffusional transport of atoms in polycrystalline materials. In considering diffusion at interfaces, two basically different processes must be distinguished, i.e. diffusion along the interface (i.e. parallel to the interface plane) and diffusion transverse to the interface (i.e. directly across the interface between the two adjoining crystals).

In this chapter, we shall be concerned with the role of interfaces in promoting (or hindering) the diffusional transport of atoms over relatively long distances corresponding to many atomic jump distances. Interfaces can often promote such diffusion by means of relatively fast diffusion along their cores. The rate at which atoms can diffuse transversely across interfaces is therefore usually of little interest, since this type of diffusion cannot lead to long-range transport. However, it can become of critical importance in certain special situations in polycrystals (see Section 8.4 below) where the overall diffusion distance of interest through the material is larger than the grain size, and transverse diffusion across the intervening interfaces is sufficiently slow in comparison with the rate of lattice diffusion so that the interfaces act as effective barriers to the overall diffusion.

In metallic, ionic, and covalent solids, atoms which are normally substitutional in the crystal lattice generally diffuse considerably more rapidly along the cores of interfaces than in the crystal lattice. This phenomenon has, therefore, become known as 'short-circuit' diffusion in an analogy with the short-circuiting which occurs along low resistance paths in electrical circuits. This behaviour is less marked for small atoms which are interstitial in the crystal lattice and therefore diffuse in the lattice at relatively rapid rates by jumping easily between interstitial sites. Under certain conditions, it has been found that the diffusion rates of such atoms along boundaries can actually be somewhat slower than in the crystal lattice. In the vast majority of these systems it does not appear that transverse boundary diffusion rates could be slow enough to cause the boundaries to act as significant barriers to long-range diffusional transport through polycrystalline materials with usual grain sizes. However, in special types of polycrystalline solids, e.g. certain ionic materials known as 'fast ion conductors', this appears to be the case. In these materials, the diffusion of certain species in single crystals is highly anisotropic and also extraordinarily rapid. However, in polycrystals it is found that the grain boundaries can apparently hinder rapid diffusion across them and that they therefore act almost as 'open circuits' for diffusion through the material (see Section 8.4).

Diffusion at interfaces in different systems can therefore exhibit a wide range of behaviour depending upon the types of diffusing species and the types of crystalline solids which adjoin the interfaces. The full range of possible crystal lattice versus interface diffusion behaviour has still not been fully investigated. We shall therefore focus our discussion on phenomena which have been investigated to at least some degree.

## 8.2 FAST DIFFUSION ALONG INTERFACES OF SPECIES WHICH ARE SUBSTITUTIONAL IN THE CRYSTAL LATTICE

### 8.2.1 Slab model and regimes of diffusion behaviour

All available evidence (see below) indicates that the relatively fast diffusion of substitutional atoms along interfaces takes place in the relatively narrow core region. This may be expected on general grounds since this region contains bad material, and it is in this highly disturbed structure that we may expect a high atomic mobility (see Section 8.2.6 below). Since the structure of this region varies with distance from the midplane, the local boundary diffusivity will be a function of distance away from the midplane. However, the width of the region sensibly affected by the boundary is relatively thin and it is usually very small compared with the distance over which diffusion occurs. For the purpose of describing the overall diffusion, it is therefore permissible to ignore all of these local complications and to replace the boundary by a thin, homogeneous slab of effective thickness, $\delta$, possessing an effective averaged diffusivity for diffusion along the boundary, $D^B$, which is larger than the corresponding crystal lattice diffusivity, $D^L$.

Having designated interfaces as thin slabs of relatively high diffusivity, we may identify a wide range of diffusion situations which can occur in materials containing interfaces. Consider, for example, self-diffusion in a polycrystal containing a network of grain boundaries. In the temperature range where atoms are mobile in both the crystal lattice and the grain boundaries, some, or all, of the migrating atoms may spend portions of their time diffusing both in the crystal lattice and more rapidly in the grain boundaries. A variety of situations may then exist, depending upon the geometrical arrangement of the boundaries, the relative magnitudes of the diffusion rates in the boundaries and the crystal lattice, and the time available for the diffusion. On the other hand, when the temperature is sufficiently low, the crystal lattice diffusion rates become negligible, and any atomic mobility becomes restricted entirely to the grain boundary network. If the boundary network remains stationary, mass transport is then entirely limited to the thin grain boundary regions.

However, the situation may be dramatically altered if the boundaries move during the diffusion. There are many situations in which this is the case, as, for example, during grain growth and recrystallization (Section 9.1.2). When the boundaries move, atoms in the lattice which are normally immobile may be overrun by a boundary, be given a chance to diffuse in the boundary, and then be deposited back in the lattice in the wake of the moving boundary. Atoms may also enter the boundary from a source and then be deposited in the lattice behind the moving boundary, or leave the boundary for a sink after having been collected by the moving boundary. Any classification of the regimes of diffusional transport behaviour which may exist must, therefore, take into account whether the boundaries are either stationary or moving.

First, take the case of stationary boundaries. If the total diffusion time is $t$, the 'diffusion distance' (i.e., root mean square displacement) of atoms in the lattice is given approximately by $(D^L t)^{\frac{1}{2}}$ (Shewmon 1989). If this distance is appreciably greater than the spacing between boundaries, $s$, each diffusing atom (on average) will have diffused along several boundaries as well as in the lattice in between. The overall diffusion in the polycrystal can then be described by an effective diffusivity, $\bar{D}$, averaged over the total diffusion which occurred both in the lattice and along the boundaries. An expression for $\bar{D}$ is readily found (Hart, 1957) by realizing that the fraction of the total diffusion time spent by a migrating atom in boundaries is just the fraction of all available atomic sites

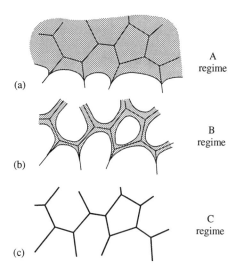

**Fig. 8.1** The three regimes for grain boundary and lattice diffusion in a polycrystal with a stationary grain boundary network according to Harrison (1961). In the A regime the diffusion length in the lattice is large compared with the grain size. In the B regime significant lattice diffusion occurs, but the diffusion length in the lattice is smaller than the grain size. In the C regime the diffusion length in the lattice is negligible, and diffusion occurs only along the grain boundary network.

which is located in the boundary slabs. For an equiaxed boundary network, this fraction is $\eta \simeq 3\delta/s$. The mean squared displacement along the boundaries is then $D^B \eta t$, while the corresponding displacement in the lattice is $D^L(1 - \eta)t$. The overall mean squared displacement, $\bar{D}t$, is then the sum of these, i.e.

$$\bar{D}t = D^L (1 - \eta)t + D^B \eta t. \tag{8.1}$$

Therefore, since $\eta \ll 1$,

$$\bar{D} \simeq D^L + (3/s)D^B\delta \ldots (D^L t > s^2). \tag{8.2}$$

The distribution of the vast majority of the diffusing species will then be very similar to that expected for bulk diffusion in a homogeneous material having an effective diffusion coefficient $\bar{D}$ as illustrated in Fig. 8.1(a).

At the other extreme, i.e. short time or low temperature, $D^L t < \lambda^2$ (where $\lambda$ is the interatomic distance), but time and temperature are sufficient for significant boundary diffusion to occur, so that $D^B t > \lambda^2$. Under this condition, diffusion occurs only along the boundary network, and there is negligible penetration anywhere into the crystal lattice as illustrated in Fig. 8.1(c).

In between these extremes is the case where $\lambda^2 < D^L t < s^2$, with significant penetration occurring from the grain boundaries into the adjacent lattices, but with insufficient time available for lattice diffusion to cover distances in the lattice as large as the spacings between different segments of the grain boundary network (see Fig. 8.1(b)). In the first situation, leading to eqn (8.2), each diffusing atom visits several boundaries, and we therefore term this the 'multiple' boundary regime. In the latter two situations, this is not the case, and the diffusion field in the lattice adjoining each

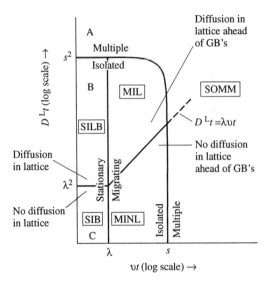

**Fig. 8.2** The five regimes for grain boundary and lattice diffusion in a polycrystal with either moving or stationary grain boundaries according to Cahn and Balluffi (1979). SILB ≡ *s*tationary boundaries, *i*solated boundary diffusion, *l*attice and *b*oundary diffusion occur; SIB ≡ *s*tationary boundaries, *i*solated boundary diffusion, only *b*oundary diffusion occurs; MIL ≡ *m*oving boundaries, *i*solated boundary diffusion, *l*attice diffusion occurs ahead of the boundaries; MINL ≡ *m*oving boundaries, *i*solated boundary diffusion, *no* *l*attice diffusion occurs ahead of the boundaries; SOMM ≡ *s*tationary *or* *m*oving boundaries, *m*ultiple boundary diffusion. Harrison's A, B, and C regimes appear on left side in stationary areas. $\lambda$ = interatomic distance. $s$ = spacing between grain boundary segments. $v$ = boundary velocity.

boundary segment is isolated, i.e. no overlap occurs. We therefore term these latter situations as 'isolated' boundary diffusion regimes.

Consider now the general case where the grain boundaries may move. Figure 8.2 (Cahn and Balluffi 1979) depicts schematically the five regimes of behaviour to be expected over the full range of possible situations. The axes are taken to be $D^L t$ and $vt$, where $v$ is the boundary velocity, and ln scales have been used in order to show the details near the origin, since $s/\lambda$ is usually $\geqslant 10^3$. When $vt < \lambda$, the boundaries may be regarded as essentially stationary, and the three regimes discussed above, therefore, appear at the extreme left of the figure. (We note that these regimes are also indicated by A, B, and C to correspond to the original nomenclature of Harrison (1961) who first identified them.)

When the distance $[(D^L t)^{\frac{1}{2}} + vt] > s$, the multiple boundary diffusion regime is obtained everywhere. It is emphasized that eqn (8.2) then holds even when $D^L$ is effectively zero as long as each atom is visited by several moving boundaries. The proof for this result is identical to that already given, since it did not depend on whether the diffusing atoms visit the boundaries or vice versa. On the other hand, when $[(D^L t)^{\frac{1}{2}} + vt] < s$, the isolated boundary diffusion regime holds everywhere. However, this regime is subdivided further according to whether the boundary is stationary, $vt < \lambda$, or moving, $vt > \lambda$. Also, as shown in Fig. 8.2, the Harrison classification for the presence, or absence, of lattice diffusion further subdivides this regime. However, the isolated, moving boundary regime divides according to a new criterion, depending upon whether or not lattice diffusion is rapid enough to produce measurable composition changes ahead of

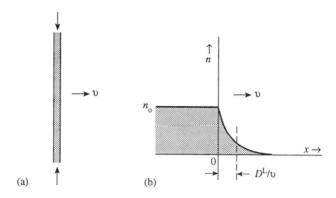

(a)                                       (b)

**Fig. 8.3** (a) A boundary core, which is moving with a velocity $v$, but, at the same time, is receiving a flux of atoms via rapid boundary diffusion. (b) Steady state distribution of diffusant maintained in the vicinity of the moving boundary in (a) when significant lattice diffusion also occurs. (See eqn (8.4).)

the moving boundary. This criterion may be obtained by considering the quasi-steady state diffusion field in the lattice ahead of a boundary moving with velocity, $v$, whose core slab is being fed laterally by rapid boundary diffusion as illustrated in Fig. 8.3(a). Putting the boundary at the origin of a coordinate system moving with respect to the lattice at the velocity, $v$, as in Fig. 8.3(b), the flux in the lattice in the region ahead of the boundary is $J = [ - D^L \cdot (\mathrm{d}n/\mathrm{d}x) - vn]$, where $n$ is the bulk concentration. The steady state condition requires that

$$- \operatorname{div} J = - \frac{\mathrm{d}}{\mathrm{d}x} [ - D^L \frac{\mathrm{d}n}{\mathrm{d}x} - vn] = 0, \tag{8.3}$$

and therefore

$$D^L \frac{\mathrm{d}n}{\mathrm{d}x} + vn = A = \text{constant}. \tag{8.4}$$

Since $\mathrm{d}n/\mathrm{d}x \to 0$, and $n \to 0$ as $x \to \infty$, $A = 0$, and eqn (8.4) may be integrated to obtain

$$n = n_\mathrm{o} \cdot \exp[ - v/D^L) \cdot x], \tag{8.5}$$

where $n_\mathrm{o}$ is the steady state concentration maintained directly adjacent to the boundary. The approximate width of the diffusion zone maintained ahead of the boundary, Fig. 8.3(b), is then measured by $D^L/v$ and is negligible when $D^L/v < \lambda$. The boundary between the MIL and MINL regimes in Fig. 8.2, therefore, lies along the curve $D^L t = \lambda \cdot vt$ as indicated. The isolated boundary region bounded by $D^L t = \lambda \cdot vt$ and $vt = \lambda$ is therefore a transition region between the above behaviour and the classical B-type diffusion associated with stationary boundaries. As the boundary velocity increases, the difference between the composition profiles ahead and behind the moving boundary becomes more extreme until finally diffusional changes are entirely confined to the trailing side.

It should be stressed that the positions of the boundaries separating the various regimes in Fig. 8.2 are only approximate and that they do not represent sharp lines of demarcation. The figure is particularly instructive for visualizing the sequence of diffusion regimes which occurs as the diffusion time increases. This sequence will generally follow a trajectory which starts at the origin ($t = 0$) and moves out into the ($D^L t, vt$) plane. It is readily seen that a given system may pass through several regimes as diffusion proceeds and that all polycrystalline systems will eventually enter the multiple boundary regime, i.e. SOMM, if they are diffused for a long enough time.

### 8.2.2 Mathematical analysis of the diffusant distribution in the type A, B, and C regimes

As discussed above, diffusion in the type A regime resembles the diffusion which would occur in a homogeneous bulk material in which the diffusant has an effective bulk diffusion coefficient, $\bar{D}$, given by eqn (8.2). The problem of analysing the distribution of diffusant under these conditions has been extensively treated elsewhere (Carslaw and Jaeger 1959, Crank 1957) and therefore will not be considered in any detail here.

The more complicated problem of analysing diffusion in the type B regime has been solved for a variety of initial and boundary conditions. The results generally yield expressions containing the combined interface diffusion parameter $p = \delta D^B$, i.e. the product of the boundary diffusivity and its thickness, and they therefore provide a basis for the experimental determination of this important quantity.

In the classic analysis of type B tracer self-diffusion by Fisher (1951), a useful approximate solution of the problem illustrated in Fig. 8.4(a) for a stationary boundary was obtained. Here, the tracer concentration at the free surface is maintained constant, and rapid diffusion occurs along the boundary slab in the $y$ direction, while at the same time, tracer atoms leak out of the boundary into the adjoining lattices transversely in the $x$ direction at a relatively slower rate. The diffusion fields in the boundary slab and the adjoining bulk crystals are obviously coupled in this problem, and the equation for diffusion in the boundary may then be written

$$\frac{\partial n^B}{\partial t_1} = \frac{\partial^2 n^B}{\partial y_1^2} + 2 \left( \frac{\partial n^L}{\partial x_1} \right)_{x_1 = 0}, \tag{8.6}$$

where the densities of diffusant (atoms per unit volume) in the boundary and lattice are designated by $n^B$ and $n^L$ respectively. The first term (RHS) is the usual divergence of the diffusion flux along the boundary, the second term is the rate of loss due to transverse leakage out of the boundary, and we have made use of the dimensionless variables $x_1 = x/\delta$, $y_1 = (y/\delta) \cdot (D^L/D^B)^{\frac{1}{2}}$ and $t_1 = tD^L/\delta^2$. Over a considerable range of conditions (see below), the boundary rapidly becomes 'saturated', and the overall diffusion field in the system at the reduced time $t_1$ is almost the same as that which would have been obtained if the instantaneous concentration distribution in the boundary slab at the time $t_1$ had been maintained fixed at that distribution since $t_1 = 0$. In addition, the effects of the relatively small gradients along $y$ in the lattice may be ignored. The concentration in the lattice in any plane of fixed $y$ is then given as a function of $x$ by the standard error function (i.e. erf) solution for diffusion into a half-space along $x$ from a surface of fixed concentration (Carslaw and Jaeger 1959). Therefore,

$$n^L (x_1, y_1, t_1) = n^B (y_1, t_1) [1 - \text{erf} (x_1/2t_1^{\frac{1}{2}})]. \tag{8.7}$$

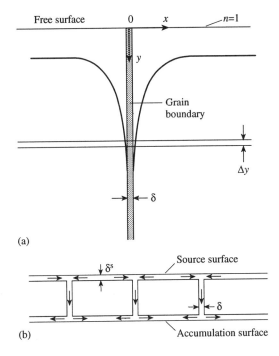

(a)

(b)

**Fig. 8.4** (a) Type B diffusion profile in vicinity of stationary grain boundary lying perpendicular to free surface. Concentration of diffusant maintained at $n = 1$ at the surface. (b) Surface accumulation method for measuring grain boundary diffusion. Atoms diffuse from the 'source' surface to the 'accumulation' surface through the transverse grain boundaries. Since $D^B < D^S$ (see Fig. 8.5), the measured rate of accumulation on the accumulation surface is controlled by $D^B$ which can then be determined.

Furthermore, it may be assumed that the rapid 'saturation' of the boundary corresponds to a quasi-steady state situation in which the dependence of the boundary concentration on $y_1$ at time $t_1$ can be obtained from eqns (8.6) and (8.7) by setting $\partial n^B / \partial t_1 = 0$ in eqn (8.6) so that

$$0 = \frac{\partial^2 n^B (y_1, t_1)}{\partial y_1^2} - 2 \cdot n^B (y_1, t_1) \cdot \left[ \frac{\partial}{\partial x_1} \mathrm{erf}(x_1 / 2 t_1^{\frac{1}{2}}) \right]_{x_1 = 0}. \tag{8.8}$$

Since $\mathrm{d\,erf}(z)/\mathrm{d}z = 2 \cdot \pi^{-\frac{1}{2}} \cdot \exp(-z^2)$, eqn (8.8) is readily solved in the form

$$n^B (y_1, t_1) = \exp[-(4/\pi t_1)^{\frac{1}{4}} \cdot y_1], \tag{8.9}$$

which satisfies the surface condition $n^B(0, t_1) = 1$. Therefore, finally,

$$n^L (x_1, y_1, t_1) = \exp[-(4/\pi t_1)^{\frac{1}{4}} y_1] [1 - \mathrm{erf}(x_1 / 2 t_1^{\frac{1}{2}})]. \tag{8.10}$$

Equation (8.10) has been widely used in the experimental determination of $D^B \delta$. If the diffusion specimen shown in Fig. 8.4(a) at time $t$ is sectioned parallel to $y$ in slices $\Delta y$ thick, and the integrated average concentration, $\bar{n}$, in each slice is measured as a function of $y$, it may be seen from the form of eqn (8.10) that the resulting curve of $\ln \bar{n}$ versus $t$ should have a slope given by

$$\mathrm{d} \ln \bar{n} / \mathrm{d}y = - [ (4 D^L / \pi t)^{\frac{1}{4}} \cdot (\delta D^B)^{-\frac{1}{2}} ]. \tag{8.11}$$

The quantities $\delta$ and $D^B$ appear in the form of the combined boundary diffusion parameter $p = \delta D^B$ which can then be calculated directly if $D^L$ is known. More exact analyses of this diffusion problem, and related ones in the B-type regime, have been extensively reviewed by Kaur and Gust (1989). The results show that the approximate Fisher analysis given above is usually acceptably accurate when the dimensionless quantity $\beta \equiv \delta D^B/[2(D^L)^{\frac{1}{2}}t^{\frac{1}{2}}] \geqslant 10$. Otherwise, more accurate (but complicated) analyses must be used (Kaur and Gust 1989).

When the diffusing species are dilute solute atoms, the previous analysis must be modified in order to take into account possible segregation at the boundary. This can be done in an approximate manner by making the reasonable assumption that the solute concentrations in the boundary slab and in the directly adjacent lattice are maintained in local equilibrium with each other during the diffusion process. If the further simplifying assumption is made that the equilibrium segregation ratio, i.e. $k = n^B/n^L$ (where $n^B$ is the concentration in the slab and $n^L$ is the concentration in the lattice directly adjacent to it), is constant everywhere along the boundary slab (i.e. independent of concentration, as is closely the case for the McLean isotherm, eqn (7.32), for dilute segregation), eqn (8.7) takes the simple form

$$n^L(x_1, y_1, t_1) = [n^B(y_1, t_1)/k] \cdot [1 - \mathrm{erf}(x_1/2t_1^{\frac{1}{2}})]. \tag{8.12}$$

By carrying through the entire analysis again, it is readily confirmed that the combined diffusion parameter, $p = k\delta D^B$, appears in the final eqn (8.11) rather than the parameter $k = \delta D^B$ as previously. This parameter can be determined experimentally, as before, but it now contains information about both the boundary segregation and diffusivity. Unfortunately, information about the boundary diffusivity parameter itself, i.e. $\delta D^B$, can then be extracted only if information about $k$ is available from independent segregation measurements. For concentrated systems, $k$ will exhibit concentration dependent behaviour (see Chapter 7), and then a more complicated analysis must be employed for an exact treatment.

In the Type C diffusion regime, where lattice diffusion is frozen out, and diffusion is therefore entirely restricted to the high diffusivity interfaces which are available, it is difficult to measure actual diffusion concentration profiles in the very thin interfaces. Accumulation methods have, therefore, been employed where the diffusing atoms migrate along the interfaces to regions where they accumulate to an extent where they can be readily detected. An example is given in Fig. 8.4(b), where the cross section of a thin slab containing a columnar grain boundary structure is shown. If the diffusing atoms are initially deposited on the source surface and allowed to diffuse, they will then diffuse through the slab along the transverse grain boundaries and spread out on the accumulation surface. Since the diffusion along the free surfaces is generally more rapid than along the grain boundaries (see Fig. 8.5), the rate-limiting step is the diffusion along the grain boundaries, and measurements of the accumulation rate, therefore, yield information about the grain boundary diffusion rates. Analyses of this type of diffusion have been given by Hwang and Balluffi (1979) and Ma and Balluffi (1993).

### 8.2.3 Experimental observations

We now summarize briefly some of the main results of a large number of measurements of rapid short-circuiting diffusion of substitutional species along interfaces in metallic, ionic, and covalent materials. In many cases, the data are rather unreliable and influenced to varying degrees by unknown factors such as segregation, precipitation, etc.

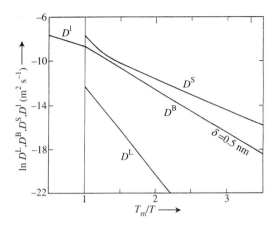

**Fig. 8.5** Arrhenius plot of various self-diffusivities in f.c.c. metals employing a reduced reciprocal temperature, $T_m/T$, where $T_m$ = melting point. $D^S$ = free surface diffusivity, and $D^l$ = diffusivity in the liquid. (From Kaur and Gust (1989)).

Discrimination must therefore be exercised in evaluating the results. Detailed data and reviews are available in Gleiter and Chalmers (1972), Gjostein (1973), Kingery (1974), Philibert (1975), Martin and Perraillon (1975, 1980), Cabane-Brouty and Bernardini (1982), Atkinson (1984, 1985, 1988a, 1988b, 1989), Monty and Atkinson (1989), Kaur and Gust (1989), and Kaur *et al.* (1989).

### 8.2.3.1 *Some major results for diffusion along interfaces*

*Metals.* The data for diffusion along grain boundaries in metals are considerably more extensive and reliable than those for ionic and covalent materials. Averaged results for a variety of f.c.c. metals for self-diffusivities in the crystal lattice, $D^L$, along general grain boundaries, $D^B$, on the free surface, $D^S$, and in the liquid above the melting temperature, $D^l$, are compared in Fig. 8.5. The $D^B$ values were obtained by measuring the parameter $\delta D^B$ and assuming $\delta = 0.5$ nm. As discussed below, this is the value which is expected to correspond closely to the actual thickness of the interface core region over which essentially all of the rapid diffusion occurs. An Arrhenius plot using a reduced reciprocal temperature scale, i.e. $T_m/T$, is employed in an effort to normalize the data for the different metals. The average of the normalized grain boundary diffusion data is represented approximately by a straight line corresponding to

$$D^B = D_o^B \exp\left(-Q^B/kT\right), \tag{8.13}$$

where, according to the usual theory of thermally activated processes, $Q^B$ is an effective activation energy and $D_o^B$ is an effective pre-exponential factor. It is noted, however, that the straight line representation, which implies a constant $D_o^B$ and a constant $Q^B$, may not be the most realistic representation of the averaged data. As discussed below, boundary diffusion is a complex process involving many different thermally activated jumping processes in the boundary core, and, therefore, the overall effective activation energy in theory should not be strictly constant. A more realistic representation may therefore be a curved line as, for example, has been used to represent the surface diffusivity data, $D^S$, in Fig. 8.5. The majority of the grain boundary diffusion data used to construct Fig. 8.5 were obtained at intermediate temperatures in the B-type kinetics

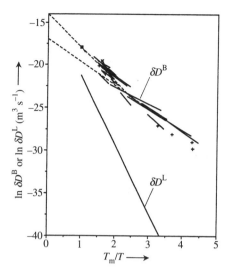

**Fig. 8.6** Arrhenius plot of data collected for grain boundary diffusion in Ag, Au, and Ag/Au alloys employing reduced reciprocal temperature, $T_m/T$, where $T_m$ = melting point. Corresponding curve for $\delta D^L$ also shown. (From Ma *et al.* (1993).)

regime. There is some evidence that averaged grain boundary diffusivities measured at lower temperatures in the C-type kinetics regime have lower values of $D_o^B$ and $Q^B$ and that the reduced Arrhenius plot should therefore be significantly curved when examined over a large temperature range. Data collected for boundary diffusion in Ag, Au, and Ag/Au alloys (Ma *et al.* 1993), shown in Fig. 8.6, may be consistent with this behaviour.

The results in Fig. 8.5 display a diffusivity spectrum in which the diffusivity along the free surface, $D^S$, is larger than that along grain boundaries, which, in turn, is larger than that through the bulk, i.e.

$$D^S > D^B > D^L. \tag{8.14}$$

Such a result might be expected intuitively, since the atomic environment for jumping becomes progressively less free in the sequence, free surface → grain boundary → bulk crystal. In a manner consistent with this, the activation energies for these diffusion processes follow the reverse behaviour,

$$Q^S < Q^B < Q^L. \tag{8.15}$$

The ratio $D^B/D^L$ is approximately $10^7$–$10^8$ at half the bulk melting temperature, and $Q^B$ is smaller than $Q^L$ by a factor $\simeq 0.6$. Also, for the f.c.c. metals, $D^B$ is seen to approach the diffusivity in the liquid at $T_m$ where it has the value $\simeq 10^{-9}\,\mathrm{m^2\,s^{-1}}$. As we have already pointed out in Section 6.3.2, the observation that values of the diffusivity in the liquid extrapolated to below $T_m$ are larger than corresponding grain boundary diffusivities is an indication that the grain boundary cores are not liquid-like at elevated temperatures approaching $T_m$.

In binary substitutional alloy systems there is evidence (Balluffi 1984, Balluffi and Cahn 1981) that the two species generally diffuse in boundaries at different rates and therefore that net transport phenonena should occur in boundaries which are similar to those which occur in the Kirkendall effect in bulk crystals (see Section 10.4.2). These results indicate that the boundary diffusion must generally occur by some type of defect mechanism as discussed further in Section 8.2.4. Studies of a number of special solute atoms which are 'fast' diffusers in the lattice show that they are 'super fast' diffusers in grain boundaries (Bernardini *et al.* 1989, Dyment *et al.* 1991). 'Fast' diffusers in the lattice are generally solute atoms which are normally substitutional in the lattice but are somewhat undersize and possess atomic radii which are up to about 0.85 that of the host atoms. They therefore spend a fraction of their time diffusing in the lattice very rapidly by a mechanism which most probably involves interstitial configurations (Shewmon 1989). Because of this, their overall diffusion rates are orders of magnitude greater than their host atoms. Bernardini *et al.* (1989) found that Ag, Au, and Ni solutes in Pb, which diffuse $\simeq 10^6$ to $10^8$ faster than Pb in the lattice, diffuse along grain boundaries in Pb at rates which are $\simeq 10^5$ to $10^6$ faster than Pb! It is therefore likely that they also diffuse along grain boundaries via some type of interstitial type process as described below in Section 8.2.5. Dyment *et al.* (1991) have found rather similar behavior for the 'fast' solute diffusers Fe, Co, and Ni in $\alpha$-Zr. On the other hand, the diffusion of Ag in $\alpha$-Zr, both in the lattice and along grain boundaries, is 'normal' in the sense that it resembles the lattice and grain boundary self-diffusion behaviour of Zr in $\alpha$-Zr (Vieregge and Herzig 1989). This is consistent with the fact that Ag is considerably less undersize in $\alpha$-Zr than Fe, Co, and Ni. We may therefore conclude that the boundary diffusion of solute atoms which are essentially substitutional in the lattice may exhibit a range of behaviour and occur by different mechanisms (as discussed in Section 8.2.7) depending on factors which include the important one of their size relative to the host.

In many other experiments the addition of solute atoms has been found (e.g. Gleiter and Chalmers 1972, d'Heurle and Gangulee 1972, Gupta *et al.* 1978, Cabane-Brouty and Bernardini 1982, Atkinson 1985, Gas *et al.* 1989, Bernardini 1989, and Kaur and Gust 1989) to have a wide variety of effects on the grain boundary diffusivities of both solvent and solute species including combinations of enhancing or retarding effects which also can be strong or weak. Such effects may be expected quite generally since, as discussed in Chapter 7, solute atoms tend to segregate to boundaries where they can occupy various favourable sites and also alter the core structure. As discussed below, boundary diffusion generally occurs mainly by the jumping of atoms amongst boundary sites in the core between which the jumps are relatively easy. Any elimination or alterations of these sites (including their environments) by segregated solute atoms will therefore affect the rate of boundary diffusion. Solute segregation and boundary diffusion are therefore inextricably coupled. For example, if solute atoms strongly bind to core sites which are normally responsible for fast diffusion, or produce new structures in a manner which slows down the rate of atom jumping, the core may become effectively 'clogged', thus causing a strong retardation of the boundary diffusion rate. This often appears to be the case when solute atoms of low solubility segregate strongly to the boundary. An example is shown in Fig. 8.7, where the segregation of Sn to grain boundaries in Fe at concentrations corresponding to about a monolayer produces a large decrease in the boundary diffusivities of both Sn (the solute) and Fe (the solvent). In this work, the combined parameter $p = k\delta D^B$ was measured, and values of $D^B$ were then extracted using independently obtained information about the segregation ratio, $k$. However, in some cases, the segregation of a strongly segregating solute has little apparent effect. An example is found in the segregation of Ni to boundaries in Ag. In this system, Ni atoms are known (Cabane-Brouty and Bernardini 1982) to segregate strongly. However, the boundary diffusivities of the Ag and Ni are found to be essentially independent of the amount of Ni segregation (Gas *et al.* 1989). Evidently, in this case, the Ni segregates strongly to boundary sites in

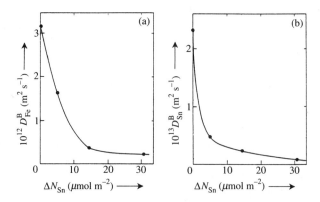

**Fig. 8.7**  The reduction in the grain boundary diffusivities of (a) Fe solvent atoms (i.e. $D_{Fe}^B$) and (b) Sn solute atoms (i.e. $D_{Sn}^B$) in Fe–Sn alloys caused by an increase in the boundary segregation of Sn. (From Bernardini *et al.* (1982) and Kaur and Gust (1989).)

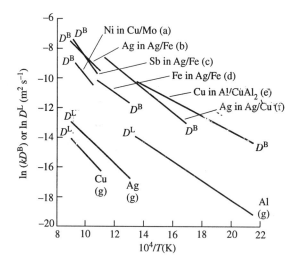

**Fig. 8.8**  Arrhenius plot of collected data showing fast diffusion along various heterophase boundaries in metal systems: (a) Ni in Cu/Mo interface; (b) Ag in Ag/Fe interface; (c) Sb in Ag/Fe interface; (d) Fe in Ag/Fe interface; (e) Cu in Al/CuAl$_2$ interface. (Data obtained from figures on pp. 476, 91, 97, 95, and 186, respectively, of Kaur *et al.* (1989).) (f) Ag in Ag/Cu interface from Sommer (1992); (g) lattice self-diffusion in Cu, Ag and Al (From Peterson (1978).)

a manner which does not markedly affect the jumping rate in the core. The bulk solubility of Ni in Ag is very small and this could, therefore, be due conceivably to some type of clustering of the Ni. Unfortunately, not enough is known at present about the atomic details of both the segregation and the diffusional jumping in these boundaries to allow a reliable explanation. Further discussion is given by Gas *et al.* (1989) and Bernardini (1989).

Relatively rapid diffusion along heterophase boundaries has also been generally observed in metallic systems. Some results are given in Fig. 8.8, where the diffusivities of a number of species along various heterophase boundaries are displayed. In each case, the boundary diffusivity is compared with the faster of the two lattice self-diffusivities in the system and is seen to be larger by four to five orders of magnitude. This fast heterophase boundary diffusion behaviour is, therefore, generally similar to the fast grain boundary diffusion behaviour displayed in Fig. 8.5. Further support for the similarity between fast grain boundary and heterophase boundary diffusion has been found recently by Dyment *et al.* (1991).

*Ionic materials (including materials which are partially covalent).* Numerous experiments have shown that fast diffusional transport of substitutional anions and cations occurs along general grain boundaries in these materials under many conditions. However, as emphasized by Atkinson (1984), the results are often conflicting and confusing. The problem is considerably more complex than in the case of the metals, and the detailed interpretation of the results is, therefore, often problematical. Complicating factors include the following:

(1) at least two different types of diffusing species (i.e. cations and anions) are involved;
(2) many materials are non-stoichiometric and hence of variable composition;
(3) electrical interactions and space charge effects (Section 7.6) are generally expected;
(4) impurity segregation is usually strong and is ubiquitous in most laboratory specimens.

With respect to this, it should be pointed out that many segregated solute atoms may be expected to exert exceptionally strong blocking effects on boundary diffusion in ionic solids because of the strong electrostatic forces which are present. For example, in NiO the boundary diffusion of Ni (which, as shown in Section 8.2.7, occurs by a vacancy mechanism) is strongly reduced by $Cr^{+++}$ doping (Atkinson 1988*a*, 1988*b*). In other work, calculations by Duffy and Tasker (1986) show that Ni-vacancy exchange near a $Ce^{+++}$ ion in a NiO boundary is essentially blocked by the attractive interaction between the impurity ion and the vacancy. It appears generally that ions of higher charge can block diffusion by occupying sites normally required for fast diffusion or else occupying neighbouring sites where they bind vacancies and hinder their motion (Atkinson 1988*a*).

In order to discuss the available grain boundary diffusion results for these materials, it is convenient (following Atkinson 1984) to group them as follows: (a) metal oxides which tend to be non-stoichiometric such as NiO, $Cr_2O_3$, and $Cu_2O$; (b) fluorite structure oxides (e.g. $CaF_2$, $ThO_2$), (c) refractory oxides (e.g. MgO and $Al_2O_3$), and (d) alkali halides.

The materials in (a) tend to be more non-stoichiometric than the others and can, therefore, possess relatively large concentrations of 'defects' induced by the non-stoichiometry and therefore exhibit intrinsic behaviour which is readily interpretable. The most extensive results have been obtained for the transition-metal oxide NiO (Atkinson 1984, 1985, 1988*a, b*, 1989, Monty and Atkinson, 1989). In this material, the cations are more mobile than the anions in the bulk, and the deviation from stoichiometry on the cation lattice, i.e. $Ni_{1-x}O$ is large enough at $T = 0.5 T_m$ so that the concentration of the intrinsic defects in the non-stoichiometric material is not sensibly affected by impurity concentrations at the ppm level. Intrinsic bulk diffusion behaviour can, therefore, be obtained to temperatures as low as $T = 0.35 T_m$. Some results for grain boundary and lattice diffusion are shown in Fig. 8.9. For pure ionic solids, the diffusion measurements generally measure the parameter $k\delta D^B$, where $k$ is not necessarily unity, since, as we have seen in Section 7.6, the boundary may be charged non-stoichiometrically. Here, $\delta$ has been set equal to 1 nm, and values of $kD^B$ are reported with the understanding that $k$ is probably of order unity. It may be seen that the Ni cations diffuse much faster than

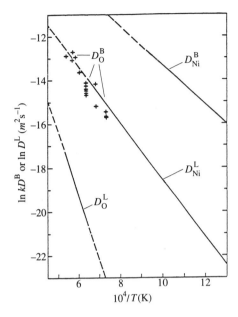

**Fig. 8.9** Arrhenius plot of the grain boundary diffusivities of Ni and O (i.e. $D_{Ni}^{B}$ and $D_{O}^{B}$) and lattice diffusivities of Ni and O (i.e. $D_{Ni}^{L}$ and $D_{O}^{L}$) in NiO. (From Atkinson (1989).)

the O anions in both the lattice and the boundaries. In addition, for both species, the boundary diffusivities are larger than the lattice diffusivities by a factor $10^5$–$10^6$. Also, for the Ni, $Q^B/Q^L \simeq 0.7$, and $D^B$ extrapolates to $\simeq 10^{-9} \, \mathrm{m^2 s^{-1}}$ at $T_m$. The fast boundary diffusion behaviour of the cations is, therefore, remarkably similar to that found previously in the metallic systems. In considering these results it must be noted, however, that differing results, due undoubtedly to impurity effects, have been obtained using 'high purity' NiO polycrystals obtained by oxidation, and lower purity bicrystals and sintered polycrystals (Atkinson and Taylor 1981, Atkinson *et al.* 1986, Barbier *et al.* 1988, Moya *et al.* 1990). However, on the basis of all of this work it appears most likely that the results given in Fig. 8.9 are representative of intrinsic NiO.

In the case of $Cu_2O$, which is also a cation-deficient oxide (i.e. $Cu_{2-x}O$), it is found (Monty and Atkinson 1989) that the grain boundary diffusivity of O is orders of magnitude larger than the bulk diffusivity of O, with $Q^B/Q^L \simeq 0.7$. Somewhat similar results have been obtained for the carbide, $UC_{1+x}$, which is a good electronic conductor and where, for at least $UC_{1+x}$, intrinsic behaviour can be achieved. The results (Routbort and Matzke 1975, Atkinson 1984) show that, for the U atoms, $D^B/D^L$ is in the range $10^4$–$10^6$ and $Q^B/Q^L \simeq 0.6$.

The fluorite structure oxides can be either good electronic or ionic conductors, depending upon the doping and stoichiometry. The fast grain boundary diffusion of cations in $ThO_2$ (possibly under extrinsic conditions) and stabilized $ZrO_2$ has been found. Of interest is the result (Atkinson 1984) that the data extrapolate to $kD^B \simeq 10^{-9} \, \mathrm{m^2 s^{-1}}$ at $T_m$ (when $\delta = 1 \mathrm{nm}$) as was the case for the metals (Fig. 8.5) and NiO (Fig. 8.9).

Data showing fast transport along grain boundaries in the refractory oxides (especially MgO (Monty and Atkinson 1989)), and the alkali halides have all appeared in the literature, but the results are difficult to interpret (Atkinson 1984, 1988*a*, Atkinson and Monty 1989). The alkali halides are stoichiometric ionic solids, and the concentrations

**Table 8.1** Data comparing grain boundary and lattice diffusion parameters for different solutes in Si. All $kD^B$ values are for average boundaries in polycrystals.

| Solute | T(K) | $kD^B(m^2\,s^{-1})$ | $kD^B/D^L$ |
|---|---|---|---|
| P[a] | 1250 | $1.4 \times 10^{-13}$ | $\simeq 9 \times 10^4$ |
| P[b] | 1250 | $1.3 \times 10^{-14}$ | $\simeq 8 \times 10^3$ |
| As[c] | 1250 | $5.0 \times 10^{-15}$ | $\simeq 2 \times 10^4$ |
| Sb[a] | 1250 | $1.2 \times 10^{-13}$ | $\simeq 5 \times 10^5$ |
| Sb[d] | 1250 | $2.9 \times 10^{-14}$ | $\simeq 1 \times 10^5$ |

[a] Liotard *et al.* (1982).
[b] Spit and Bakker (1986).
[c] Campbell *et al.* (1975).
[d] Spit *et al.* (1985).

of intrinsic point defects are very small ($\simeq 10^{-9} - 10^{-8}$ at $0.5\,T_m$ for NaCl). Any diffusional transport near this temperature will, therefore, be dominated by impurities (i.e. extrinsic behaviour will pertain). In addition, there is extensive evidence that grain boundary segregation is strong in these materials and that boundaries commonly contain second-phase precipitates. The situation is generally similar in the refractory oxides, which are highly stoichiometric and possess extremely small ionic and electronic conductivities. Again, the intrinsic defect concentrations are small and extrinsic behaviour is expected. The possible influence of second phases at boundaries in MgO has been discussed by Roshko (1987).

*Covalent materials.* Extensive evidence exists for fast grain boundary diffusion in covalent materials (Kauer *et al.* 1989). Grain boundary self-diffusion results for Ge and Si are not available, but some results for the fast boundary diffusion of various solutes in Si are presented in Table 8.1. Values of $kD^B/D^L$ in the range $10^4$–$10^6$ are evident. Data for additional systems are given by Kauer *et al.* (1989).

### 8.2.3.2 *Effects of interface structure*

As might be expected, the rate of fast boundary diffusion is found to depend directly upon the detailed atomistic structure of the boundary (Gleiter and Chalmers 1972, Peterson 1980, Balluffi 1984, Atkinson 1985). This is particularly evident when comparing grain boundary diffusion rates along small-angle versus large-angle grain boundaries. As discussed in Section 2.1, a small-angle grain boundary consists of an array of discrete primary dislocations embedded in the crystal lattice. The cores of the dislocations consist of bad material along which fast diffusion may generally occur. On the other hand, the surrounding material is essentially perfect crystalline material (slightly strained, of course) and, hence, possesses a much lower diffusion rate characteristic of the crystal lattice. From a diffusional standpoint, a small-angle boundary may, therefore, be described as a network of fast diffusion 'pipes' embedded in a relatively low diffusivity medium. In contrast to this, a large-angle boundary is a considerably more continuous slab of bad high-diffusivity material.

The way in which grain boundary self-diffusion along the tilt axis of a series of symmetric [001] tilt boundaries might be expected to vary systematically with misorientation over the entire small-angle to large-angle regime can be obtained (Balluffi and

Brokman 1983) through use of the structural unit model. This model has already been described in Section 4.3.1, where it was pointed out that it should be applicable to any grain boundary property that depends directly on the core structure. For [001] tilt boundaries in the f.c.c. structure, we may select the short period $\Sigma 1(100)$, $\Sigma 5(310)$, $\Sigma 5(210)$, and $\Sigma 1(110)$ boundaries as the delimiting boundaries of the series and designate the units which make up these boundaries as A, B, C, and D, respectively. All boundaries between the $\Sigma 1(100)$ and $\Sigma 5(310)$ boundaries are, therefore, made up of mixtures of A and B units, as illustrated schematically in Fig. 8.10. The minority B units in all boundaries up to the $\theta_3$ boundary are primary dislocations with $b = [100]$. For diffusion parallel to the tilt axis, the boundary diffusion parameter $p$ for any boundary may be written as

$$p = D_{\text{eff}}^{\text{B}} \cdot \delta_{\text{eff}} \tag{8.16}$$

where $D_{\text{eff}}^{\text{B}}$ and $\delta_{\text{eff}}$ are effective quantities which must be determined by suitably averaging the contributions to the diffusion along the tilt axis made by the various structural units present. For the boundaries in Fig. 8.10, we may therefore write:

$$\left. \begin{aligned} p_1 &= [\lambda_1]^{-1} \cdot [D_\alpha^{\text{B}} A_\alpha] \\ p_2 &= [\lambda_2]^{-1} \cdot [D_\alpha^{\text{B}} A_\alpha + 2(D_\gamma^{\text{B}} A_\gamma + D_\delta^{\text{B}} A_\delta)] \\ p_3 &= [\lambda_3]^{-1} \cdot [D_\gamma^{\text{B}} A_\gamma + D_\delta^{\text{B}} A_\delta] \\ p_4 &= [\lambda_4]^{-1} \cdot [D_\gamma^{\text{B}} A_\gamma + D_\beta^{\text{B}} A_\beta + D_\delta^{\text{B}} A_\delta] \\ p_5 &= [\lambda_5]^{-1} \cdot [D_\beta^{\text{B}} A_\beta] \end{aligned} \right\} \tag{8.17}$$

Here, the numerical subscripts refer to the boundary designations in Fig. 8.10, and the Greek letter subscripts refer to the different types of boundary patches corresponding to the regions lying between the horizontal dashed lines drawn in Fig. 8.10. The quantity $\lambda_i$, shown in Fig. 8.10, is the distance in the $i$th boundary measured normal to the tilt axis which contains the patches contributing to the $p_i$ in eqns (8.17). $D_j^{\text{B}}$ is the diffusivity of a $j$-type patch, along the tilt axis, and $A_j$ is its cross-sectional area. We emphasize that it is necessary to employ four different types of patches in order to describe properly the contributions of all regions of all boundaries. For example, the use of two types of patches corresponding to the original A and B units is insufficient, since it does not take into account differences between the local regions where different structural units are joined together. It is readily seen that $p_2$ can be expressed as a linear combination of $p_1$ and $p_3$, and $p_4$ as a linear combination of $p_3$ and $p_5$, and that eqns (8.17) can therefore be written as:

$$\left. \begin{aligned} p_1 &= p_1 \\ p_2 &= [\lambda_2]^{-1} \cdot [\lambda_1 p_1 + 2\lambda_3 p_3] \\ p_3 &= p_3 \\ p_4 &= [\lambda_4]^{-1} \cdot [\lambda_3 P_3 + \lambda_5 p_5] \\ p_5 &= p_5 \end{aligned} \right\} \tag{8.18}$$

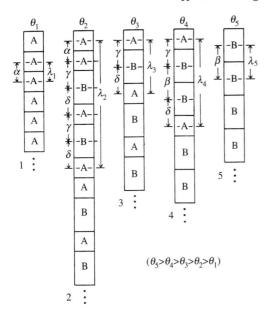

**Fig. 8.10** Schematic representation of idealized structural unit model for a series of symmetric tilt boundaries as viewed along the tilt axis. Boundaries 1 and 5 are delimiting boundaries made up of all A units and all B units respectively. Boundaries 2, 3, and 4 are of intermediate misorientations and are composed of various mixtures of the A and B units.

It was shown in Section 4.3.1 that the rule of structural continuity disallows the juxtaposition of B units in the misorientation range between the 1 and 3 boundaries and the juxtaposition of A units in the range between the 3 and 5 boundaries. As a result, $p$ for all boundaries in the misorientation range between the 1 and 3 boundaries, such as the $p_2$ boundary, can be expressed as a linear combination of $p_1$ and $p_3$. Similarly, $p$ for all boundaries in the misorientation range between the 3 and 5 boundaries can be expressed as a linear combination of $p_3$ and $p_5$. When the known values of $\lambda_i$ are substituted into eqns (8–18) and the results are plotted versus $\theta$, it is found that the variation of $p$ with $\theta$ is essentially linear in the two ranges. Our general result, therefore, is that $p$ in the misorientation range between two delimiting boundaries should follow linear behaviour for all practical purposes in each of the two regions separated by the misorientation corresponding to the boundary made up of equal numbers of the structural units belonging to the two delimiting boundaries. We note that a similar result is obtained for the variation of the core energy with misorientation in Section 4.3.1.

Available diffusion data for [001] and [110] symmetric tilt boundaries, whose structures have been found in Section 4.3.1 to be describable by the structural unit model, exhibit a dependence on $\theta$ which can be interpreted in terms of the above model as demonstrated in Figs 8.11 and 8.12. In each case, the dashed lines represent conceivable curves which are reasonably consistent with both the idealized model and the available data. In considering the results in Figs 8.11 and 8.12, it must be emphasized that the idealized linear model invoked here is based on the assumption that the basic structural units of the delimiting boundaries bounding each successive range of angular misorientation remain undistorted over each range. In actuality, as discussed in Section 4.3.1, the units will always be systematically distorted to at least some extent, and this will cause $p$ to vary non-linearly with $\theta$ over each range. The accuracy of the model can always be

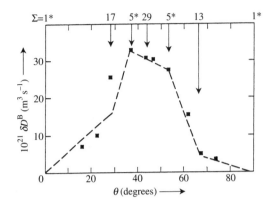

**Fig. 8.11** Fit of idealized structural unit model to data of Sommer *et al.* (1989) for boundary self-diffusion along the tilt axes of ⟨001⟩ symmetric tilt boundaries in Ag as a function of tilt angle, $\theta$. The delimiting boundaries chosen are indicated by asterisks, and, from left to right are, $\Sigma1^*(100)$, $\Sigma5^*(310)$, $\Sigma5^*(210)$, and $\Sigma1^*(110)$.

improved by employing additional delimiting boundaries, thereby decreasing the width of the various ranges and the corresponding distortions of the structural units. The net result will be generally smoother, less jagged, $p$ versus $\theta$ curves, which, of course, will fit the data more and more accurately as the number of delimiting boundaries is increased. As may be seen in Figs 8.11 and 8.12, reasonably good fits of the idealized model to the often sparse available data in the present cases can be obtained by using a fairly small number of delimiting boundaries. In the [110] series, both the $\Sigma3(111)$ and $\Sigma11(113)$ delimiting boundaries are singular boundaries with unusually low energies (Fig. 5.23) and are known (Balluffi 1984) to possess relatively tightly packed core structures. The values of $p$ for these boundaries, as well as for the $\Sigma1$ crystal lattice delimiting 'boundaries', are therefore relatively very small, as seen in Fig. 8.12. In the [001] series, the $\Sigma5$ (310) and $\Sigma5$ (210) delimiting boundaries are considerably less tightly packed and ordered than the two $\Sigma1$ crystal lattice delimiting 'boundaries', and the units of the two $\Sigma5$ delimiting boundaries therefore act as fast-diffusing units across the series.

We note that the structural unit model at small angles is equivalent to the well-known 'dislocation pipe' model (Turnbull and Hoffman 1954) where the primary dislocations which comprise the boundary are visualized as fast diffusion 'pipes' embedded in the slow-diffusing crystal lattice, and $D^B$ therefore increases linearly with the density of dislocations. In the structural unit model the cores of the dislocation pipes are, of course, just fast-diffusing minority structural units. It is important to realize that the linear behaviour predicted near $\Sigma = 1$ for the simple model may break down when the deviation from $\Sigma = 1$ becomes sufficiently small. Under this condition, the dislocation spacing becomes correspondingly large, and the strengths of any interactions between the dislocations in the array are correspondingly reduced. They then may become free enough to relax and dissociate into partial dislocations and patches of stacking faults in various complex ways not predicted by the simple dislocation core/structural unit model in order to reduce the total energy. Detailed calculations by Li and Chalmers (1963) show explicitly how elastic interactions between the dislocations in an array tend to oppose such dissociations and cause the degree of dissociation to decrease as the dislocation spacing decreases. Also relevant to this point is the result, obtained in Section 2.11.2, that in the Peierls–Nabarro model for a tilt boundary the dislocation core width decreases as the dislocation spacing decreases (see eqn (2.190)). As an example, according to the ideal simple

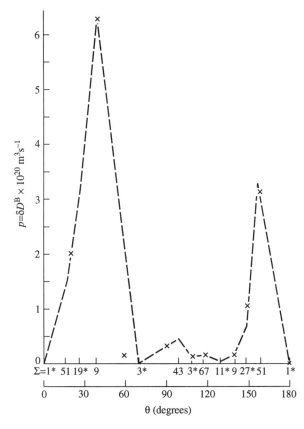

**Fig. 8.12** Fit of idealized structural unit model to data of Herbeuval and Biscondi (1974) for the diffusion of Zn along the tilt axes of [110] symmetric tilt boundaries in Al as a function of tilt angle, $\theta$. The delimiting boundaries chosen are indicated by asterisks, and, from left to right, are $\Sigma 1^*(110)$, $\Sigma 19^*(331)$, $\Sigma 3^*(111)$, $\Sigma 11^*(113)$, $\Sigma 27^*(115)$, and $\Sigma 1^*(001)$.

structural unit model, the dislocations (structural units) making up the [001] tilt boundaries in the small-angle regime near $\theta = 0°$ are straight parallel lattice dislocations running along [001] with $b = [100]$. However, Darby and Balluffi (1977) have found in Au that when $\theta \leqslant 10°$, these dislocations adopt a serrated geometry in which individual segments lie along $\langle 110 \rangle$ directions. This presumably allows them to reduce their energy by dissociating on {111} planes into partial dislocations bounded by ribbons of stacking faults, which, of course, drastically alters the core structure and causes the linear model to break down. A large variety of dissociations of this general type is possible in different types of boundaries and may be expected whenever suitable low energy faults and partial dislocations are available (Hirth and Lothe 1982).

A considerable number of measurements (Balluffi 1970, 1984, Gjostein 1973) indicates that the fast pipe diffusivity along lattice dislocation cores which are undissociated is of the same order of magnitude as diffusivities along the cores of large-angle boundaries. This is readily understood on the basis of the above model, since the structural units which make up the undissociated dislocation cores also appear in the large-angle boundary cores. Diffusivities along dissociated lattice dislocations are appreciably lower, however (Balluffi 1970, 1984, Gjostein, 1973). This may be attributed, in an intuitive way, to the increased relaxation accompanying the dissociation which produces a core structure which is less open and 'loose'.

The general picture which emerges is therefore one in which $p$ increases approximately linearly with increasing misorientation in the small-angle regime (except possibly at very small misorientations) and then reaches a plateau at large misorientations which may contain local minima if highly singular boundaries are present in the series as in Fig. 8.12. We note that the very deep minima at large misorientations in Fig. 8.12 must be regarded as unusual due to the exceptional singularity of the $\Sigma 3(111)$ and $\Sigma 11(113)$ boundaries.

The diffusional behaviour of a series of general boundaries not describable by the structural unit/grain boundary dislocation model should exhibit the same overall features (minus any strong local minima associated with highly singular boundaries, of course). The structures of such a series will make a transition from an array of discrete primary dislocation pipes at small misorientations to a continuous slab of bad material at large misorientations corresponding to boundaries with quite similar diffusion characteristics.

The boundary diffusivity may also be expected to vary with direction in any given boundary. In general, the diffusivity is a second-rank tensor (Nye 1957) and, therefore, in a 2D coordinate system lying in the boundary plane, we may express this in the form

$$\mathbf{D}^B = \begin{pmatrix} D_{11}^B & D_{12}^B \\ D_{21}^B & D_{22}^B \end{pmatrix}. \tag{8.19}$$

Measurements have been made of boundary diffusion both parallel and perpendicular to the tilt axis in [001] symmetric tilt boundaries in both metallic and ionic systems. Results are shown in Fig. 8.13 where the diffusion is seen to be highly anisotropic with $D^B(\text{para}) > D^B(\text{perp})$ at all tilt angles. Also, the ratio, $D^B(\text{para})/D^B(\text{perp})$, increases as $\theta$ decreases. This general behaviour is readily understood on the basis of the dislocation or structural unit model. At small angles, rapid diffusion occurs along the isolated dislocation pipes which lie parallel to the tilt axis, while, on the other hand, diffusion perpendicular to the tilt axis is essentially 'open-circuited' by the relatively very slow diffusion across the patches of good crystal lattice between the dislocations. As $\theta$ increases, and the dislocation spacing decreases, the good crystal lattice patches eventually disappear, and the diffusion then becomes considerably more isotropic. However, some anisotropy still remains as might be expected because of the intrinsically anisotropic atomic structure of large-angle tilt boundaries.

### 8.2.4  Mechanisms for fast grain boundary diffusion

Considerable progress has been made in determining the detailed atomic mechanisms by which substitutional atoms diffuse in the crystal lattice. In essentially all cases it has been found that the diffusion occurs via interchanges of the diffusing atom with vacancy or interstitial point defects. On the other hand, the problem of determining the mechanism (or mechanisms) for the fast diffusion of substitutional atoms in grain boundaries has proven to be much more difficult and is still not well resolved. The reasons for this state of affairs are clear: grain boundary structures are generally complex and irregular, and an infinite number of different types exists. Different mechanisms may therefore be dominant in different boundaries. In addition, the thin nature of the boundary core region precludes many of the types of critical experiments which have been used success-fully to study diffusion mechanisms in bulk crystal lattices. Nevertheless, a significant effort has been made to elucidate boundary diffusion mechanisms.

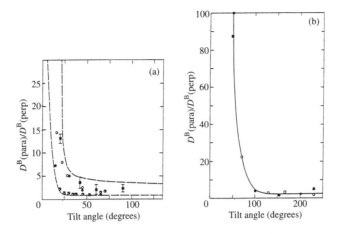

**Fig. 8.13** Ratio of boundary diffusivity parallel to tilt axis, $D^B$(para), and perpendicular to tilt axis, $D^B$(perp), for $\langle 001 \rangle$ symmetric tilt boundaries as a function of tilt angle, $\theta$. (a) Values for metal systems. Open circles, Ag[110] tracer in Ag; filled circles, Al[28] tracer in Al; filled squares, Zn[63] tracer in Al; half-filled squares, Ag[110] tracer in Cu. (b) Values for ionic systems. Open circles, Cr[51] tracer in Al$_2$O$_3$; filled circles, Cr[51] tracer in MgO; filled squares, Cr[51] tracer in MgAl$_2$O$_4$; open squares, Cr[51]tracer in MgO plus 0.45 at % Cr. (From Stubicon (1985).)

As shown below, grain boundaries are generally expected to be capable of supporting various types of localized vacancy and interstitial point defects in their cores in thermal equilibrium. It is therefore possible that substitutional atoms diffuse along boundaries by exchanging places with these defects in much the same manner as they diffuse in the crystal lattice via defect exchanges. However, before discussing these mechanisms in any detail it is necessary to establish the nature of point defects in grain boundaries.

### 8.2.4.1 *Equilibrium point defects in the grain boundary core*

In order to treat point defects in the boundary core it is instructive to consider first the simpler problem of point defects in the crystal lattice. It is well known (Agullo-Lopez *et al.* 1988) that vacancy and interstitial point defects can be produced in crystal lattices which are generally localized in structure and increase the internal energy of the crystal as well as alter its vibrational entropy. Since all lattice sites are identical, the Gibbs free energy of a crystal containing $N_A^L$ atoms and a dilute concentration of $N_V^L$ vacancies, can then be written in the form

$$G^L = N_A^L G_A^\circ + N_V^L G_V^{L,f} + kT \cdot \left\{ N_A^L \ln \left[ \frac{N_A^L}{N_A^L + N_V^L} \right] + N_V^L \ln \left[ \frac{N_V^L}{N_A^L + N_V^L} \right] \right\}, \quad (8.20)$$

where $G_V^{L,f}$ = free energy of formation of a vacancy at a lattice site, $G_A^\circ$ = free energy per atom of the crystal before the introduction of the vacancies, and the last term is the ideal free energy of mixing of the atoms and vacancies. The free energy of formation is defined as the change in free energy (exclusive of the mixing contribution) which occurs when an atom is removed from an internal lattice site and incorporated back into the crystal at a surface kink site on its free surface. At equilibrium, the free energy will be at a minimum with respect to the number of vacancies which are introduced, i.e. the condition $\partial G^L/\partial N_V^L = 0$ will hold. Applying this condition, we therefore obtain

$$\frac{N_V^{L,eq}}{N_V^{L,eq} + N_A^L} = c_V^{L,eq} = \exp(-G_V^{L,f}/kT) = \exp(S_V^{L,f}/k)\cdot\exp(-E_V^{L,f}/kT). \quad (8.21)$$

Here, $N_V^{L,eq}$ is the number of vacancies present in thermal equilibrium, and $c_V^{L,eq}$ is the probability that a crystal lattice site is occupied by a vacancy. Also, use has been made of the relation $G_V^{L,f} = E_V^{L,f} - TS_V^{L,f}$, where $E_V^{L,f}$ and $S_V^{L,f}$ are the internal energy and vibrational entropy of formation of a vacancy respectively, and the small difference between the enthalpy of formation and $E_V^{L,f}$ has been neglected.

In a similar way the probability that a particular type of interstitial site is occupied by an interstitial can be derived in the form

$$c_I^{L,eq} = \gamma_I^L\cdot\exp[-G_I^{L,f}/kT] = \gamma_I^L\cdot\exp[S_I^{L,f}/k]\cdot\exp[-E_I^{L,f}/kT], \quad (8.22)$$

where $G_I^{L,f}$ is the change in free energy (exclusive of the mixing entropy contribution) which occurs when an atom is removed from a kink site on the free surface of the crystal and inserted as an extra atom at the interstitial site in the lattice, and $\gamma_I^L$ is a purely geometrical factor which is determined by the number of energetically equivalent, but geometrically distinguishable, configurations which the interstitial atom can adopt at the site.

The introduction of either type of defect causes an increase in the internal energy of the system which, in turn, increases the free energy. However, there is also an increase in the entropy (due to the combined mixing entropy and change in vibrational entropy) which decreases the free energy. These opposing effects lead to a minimum in the total free energy at a finite defect concentration corresponding to the equilibrium concentration.

Consider now the introduction of point defects into the core of a grain boundary. The situation is now obviously more complicated than in the lattice, since the structure is irregular and possesses a variety of different types of sites. First of all, we are faced with the natural question of whether it is legitimate to assume that point defects can exist in the cores of boundaries as *bona fide* localized point defects with defined properties. It has not been possible up to now to investigate this problem experimentally because of the extremely small number of atoms which lie in grain boundary cores, and we must therefore appeal to the results of computer simulation. In this work, boundary structures of minimum internal energy have been calculated first. Vacancies, or interstitials, have then been created by either removing or inserting an atom at a selected site in the core, and then allowing the entire structure to relax. The relaxed boundary structures have then been examined to determine the forms of the resulting defects. Complications then arise, since the structure of each boundary is different, and, in addition, each boundary contains a variety of different sites in which the defect may be located. The structure (and energy) of a given type of defect (i.e. vacancy or interstitial) in each of these sites will therefore be different, and a very large number of configurations is therefore possible. Nevertheless, all calculations have produced relaxed structures which can be regarded as *bona fide* point defects in the sense that the missing or added atom increases the energy of the system and is localized in a relatively small volume about which the surrounding boundary structure is relatively undisturbed. Some examples (Brokman *et al.* 1981) are given in Figs 8.14–8.16. The displacements around the defects are generally small (i.e. small compared to interatomic distances) at distances larger than a few atomic

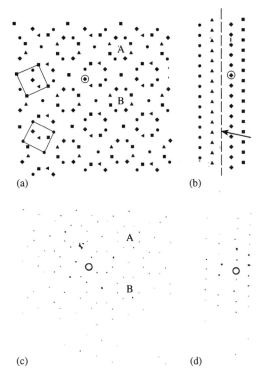

(a)  (b)

(c)  (d)

**Fig. 8.14** Atom displacement field around a vacancy in Σ5⟨001⟩ twist boundary in b.c.c. Fe calculated by molecular statics using pair potential model. (a) and (b) show plan and edge-on views of relaxed boundary structure before insertion of vacancy at encircled atom. Arrow indicates boundary midplane. (c) and (d) show corresponding views of atomic relaxations around the vacancy. Each atom displacement is represented by a vector projected on the plane of the paper. A and B introduced for reference purposes. (From Brokman *et al.* (1981).)

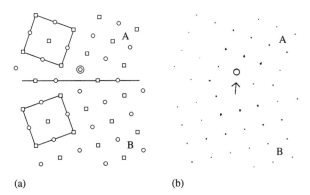

(a)  (b)

**Fig. 8.15** Atom displacement field around a vacancy in Σ5⟨001⟩(310) symmetric tilt boundary in f.c.c. calculated by molecular statics using pair potential model. (a) Edge-on view along tilt axis of relaxed boundary structure before insertion of vacancy at encircled atom. (b) Atomic relaxations around the vacancy. Each atom displacement is represented by a vector projected on the plane of the paper. A and B introduced for reference purposes. (From Brokman *et al.* (1981).)

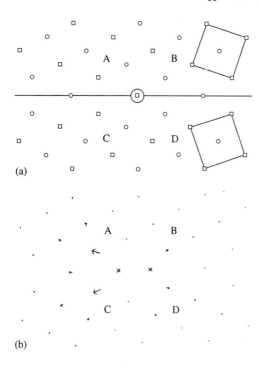

(a)

(b)

**Fig. 8.16** Atom displacement field around an interstitial in $\Sigma5\langle001\rangle(310)$ symmetric tilt boundary in b.c.c. Fe calculated by molecular statics using pair potential model. (a) Edge-on view along tilt axis of relaxed boundary structure before insertion of interstitial at encircled site. (b) Atomic relaxations around the interstitial. Each atom displacement is represented by a vector projected on the plane of the paper. The interstitial was inserted by first slightly displacing the atom already occupying the site. A, B, C, and D introduced for reference purposes. (From Balluffi (1984).)

distances from the centre of the defect. In some cases, relatively large displacements occur very near the centre of the defect. For example, in Fig. 8.15, the vacancy is essentially symmetrically 'split' as a result of the large inward displacement of one atom. In Fig. 8.16, the atomic displacements in the first shell of atoms around the inserted interstitial are relatively large as might be expected on the basis of the strong repulsion which exists between ion cores. As expected, the various displacements are generally larger and more widely distributed than the corresponding displacements around the same type of defect in the crystal lattice. Even though the point defects in the boundary may be relatively highly relaxed and extended in this way, they may be identified with a missing or added atom at specific sites in the core. It is important to emphasize that features of the unique minimum energy structure of the core such as unusually open or crowded regions cannot be regarded as defects, since they are part of the intrinsic minimum energy structure.

The formation energies of vacancy and interstitial point defects at different sites in boundaries have also been calculated in many cases; some characteristic results are listed in Table 8.2 for the $\Sigma5$ [001] (310) symmetric tilt boundary shown in Fig. 8.17. Note that in order to distinguish between sites we have introduced the index $h$, so that, for example, $E_V^{h,f}$ is the energy of formation of a vacancy at a boundary site of type $h$. As expected, the energies differ considerably from site to site. The energies for the vacancies at the boundary core sites A through D are considerably lower than for the interstitials. Also, the energies at these core sites are lower than at crystal lattice sites. However, the energy of the interstitial at the I site is relatively low and is comparable to the vacancy formation energies. This may be attributed to the relatively large open space (interstice) in the core structure and the absence of strong ion core repulsion at the I site. At a finite temperature, we may, therefore, visualize the boundary as containing an equilibrium population of

**Table 8.2** Calculated formation energies for vacancy and interstitial point defects ($E_V^{h,f}$ and $E_I^{h,f}$, respectively) in the crystal lattice and at various sites (indicated by the index $h$) in the $\Sigma5$ [001] (310) symmetric tilt boundary in b.c.c. Fe shown in Fig. 8.17. Also listed is the number of vacancy jumps, $N_V^h$, into each site during a calculated vacancy jumping sequence at 1300 K involving 195 jumps. (From Balluffi *et al.* (1981), Balluffi (1984).)

| $h$ | $E_V^{h,f}$ (kJ mol$^{-1}$) | $E_I^{h,f}$ (kJ mol$^{-1}$) | $N_V^h$ |
|---|---|---|---|
| $L$ (lattice site) | 130 | 458 | 0 |
| A | 128 | 247 | 3 |
| B | 90 | 247 | 126 |
| C | 121 | 319 | 20 |
| D | 113 | 224 | 32 |
| E | – | – | 7 |
| F | – | – | 6 |
| G | – | – | 1 |
| I | – | 103 | – |

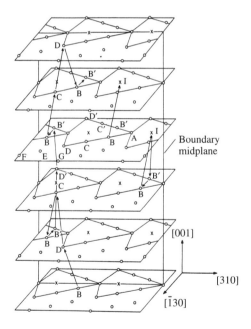

**Fig. 8.17** Atom jumping events in $\Sigma5\langle001\rangle$ (310) symmetric tilt boundary in b.c.c. Fe calculated by molecular dynamics using pair potential model. On the left is a jumping trajectory of a vacancy initially inserted at B on bottom plane. The vacancy jumps preferentially in the boundary parallel to the tilt axis amongst sites A, B, C, and D rather than into the sites E, F, and G which are further from the boundary midplane. The arrow at the centre shows an atom at a B site jumping into an interstitial I site, thus creating a boundary interstitial and boundary vacancy. The arrows on the right show interchanges of atoms between sites B and I, B and B, and I and B'. The ratios of the scales used in the drawing are [$\bar{1}30$] : [310] : [001] = 1 : 1 : 5. (From Balluffi *et al.* (1981).)

point defects with the sites of low formation energy being more heavily populated on average than the sites of higher energy. At equilibrium, any given defect will jump around between the various sites by means of thermal activation with a certain average jump frequency but will spend most of its time in low energy sites. As it jumps and visits the different sites, its configuration and energy will change correspondingly. The defect is, of course, free to leave the boundary core and enter the lattice, but such excursions will be rare if an appreciable number of sites exist in the boundary which are lower in energy than those in the lattice.

For a grain boundary a simple generalization of the statistical mechanical argument leading to eqns (8.21) and (8.22) yields

$$c_V^{h,eq} = \exp(S_V^{h,f}/k) \cdot \exp(-E_V^{h,f}/kT), \tag{8.23}$$

for the probability that a type-h site in the minimum energy boundary structure is occupied by a vacancy at equilibrium. Similarly,

$$c_I^{h,eq} = \exp(S_I^{h,f}/k) \cdot \exp(-E_I^{h,f}/kT), \tag{8.24}$$

for the probability that a type-h interstial site is occupied. Here, the geometrical factor, $\gamma_I$, which appeared in eqn (8.22) is set equal to unity, because of the generally low symmetry of interstitial sites in grain boundaries.

### 8.2.4.2 'Ring', vacancy, interstitialcy, and interstitial mechanisms

Conceivable elementary mechanisms for the diffusional jumping of an atom in a boundary (either with, or without, the participation of point defects) are illustrated schematically in Figs 8.18–8.21. These include the rotational 'ring' mechanism (Fig. 8.18), the vacancy mechanism (Fig. 8.19), the interstitialcy mechanism (Fig. 8.20), and the interstitial mechanism (Fig. 8.21).

Figure 8.18(a) shows a boundary structure of minimum energy containing a marked (shaded) atom. A simple mechanism by which this atom can execute a jump in the core without the assistance of a point defect is by a direct interchange with a near-neighbour atom by the rotational 'ring' mechanism illustrated in Fig. 8.18(a,b). In the exchange

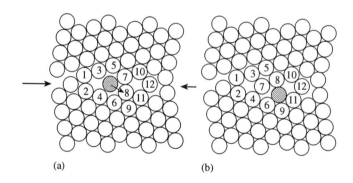

(a)                                    (b)

**Fig. 8.18** Schematic representation of rotational 'ring' mechanism for grain boundary diffusion. In the sequence (a) → (b), the marked (shaded) atom replaces atom 8, 8 replaces 7, and 7 replaces the marked atom.

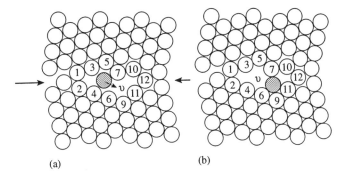

**Fig. 8.19** Schematic representation of vacancy exchange mechanism for grain boundary diffusion. In the sequence (a) → (b), the marked (shaded) atom exchanges places with a neighbouring vacancy.

shown, the marked atom, along with atoms 7 and 8, rotate simultaneously in an anti-clockwise direction so that the marked atom replaces 8, 8 replaces 7, and 7 replaces the marked atom.

Exactly the same jump of the marked atom can be accomplished by a vacancy exchange mechanism as illustrated in Fig. 8.19(a,b). Here, a vacancy appears as a near-neighbour of the marked atom, and the jump occurs by the exchange of the marked atom with the vacancy.

Exactly the same jump of the marked atom can also occur by a mechanism involving interstitial point defects as illustrated in Fig. 8.20. Figure 8.20(a) shows the same minimum energy structure as in Fig. 8.18(a) except that three possible low energy sites for interstitial defects are shown by the three × symbols. If an interstitial defect, corresponding to atom 13, is introduced at the first × site, we have the configuration in Fig. 8.20(b). The interstitial atom (i.e. atom 13) can then replace the marked atom by a replacement sequence in which the marked atom is forced into an interstitial position at the second × site, and atom 13 replaces the marked atom as in Fig. 8.20(c). Next, a second replacement sequence occurs in which the marked atom replaces atom 8 which in turn is forced into the third interstitial position as in Fig. 8.20(d). This type of transfer is generally termed an 'interstitialcy' mechanism and has the advantage that large amounts of ion core repulsion can often be avoided as the atoms squeeze by one another in the replacement processes.

Finally, the marked atom can also migrate purely as an interstitial as illustrated in Fig. 8.21. Again we start with the minimum energy structure (Fig. 8.21(a)). If the marked atom is introduced at the first interstitial site as in Fig. 8.21(b), it can then simply jump directly into the second interstitial site as in Fig. 8.21(c). This type of transfer is generally termed an 'interstitial' mechanism, since it involves the jumping of an atom from one interstitial site to another while it remains interstitial. This mechanism is usually expected for relatively small solute atoms which normally occupy interstitial sites in the lattice (see Section 8.3) and, also, possibly, for less undersize solute atoms which normally occupy substitutional sites in the lattice but might be small enough to occupy and jump between interstitial sites in the less dense boundary core. As discussed later, it could conceivably make a contribution to self-diffusion in the core if suitable paths of interstitial sites were present in the structure. We note that the ring exchange mechanism differs fundamentally

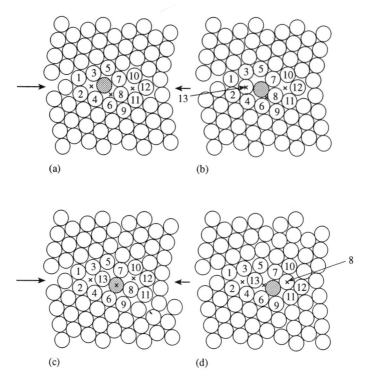

**Fig. 8.20** Schematic representation of interstitialcy mechanism for grain boundary diffusion. (Interstitial sites in the boundary core are indicated by ×'s.) In the sequence (a) → (b), an interstitial defect, corresponding to atom 13, occupies an interstitial site adjacent to the marked (shaded) atom. In the sequence (b) → (c), the interstitial atom 13 replaces the marked atom and forces it into a second interstitial site. In the sequence (c) → (d), the interstitial marked atom replaces atom 8 and forces it into a third interstitial site.

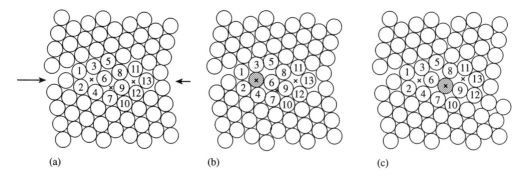

**Fig. 8.21** Schematic representation of interstitial mechanism for grain boundary diffusion. (Interstitial sites in the boundary are indicated by ×'s.) In the sequence (a) → (b), the marked (shaded) atom occupies an interstitial site. In the sequence (b) → (c), the marked atom squeezes between atoms 4, 6, and 7 and occupies a second interstitial site.

from all of the other mechanisms, since it does not allow the diffusional transport of a net number of atoms along the boundary. Every forward jump of a given atom is exactly balanced by the net backwards exchange jumping of other atoms. On the other hand, the vacancy, interstitialcy, and interstitial defect mechanisms allow a net flux of atoms. In the vacancy exchange mechanism, a net flux of atoms occurs which is exactly equal to the corresponding opposite flux of vacancies. In the interstitialcy and interstitial mechanisms a net flux of atoms occurs without the participation of an opposite flux of defects.

Finally, we note that it is possible for more complex higher order forms of these various mechanisms to occur. For example, processes involving multiple jumps occurring over longer distances could be significant, particularly at elevated temperatures where considerable thermal activation is available and the boundary becomes increasingly disordered (Ciccotti *et al.* 1983, Kwok *et al.* 1984, Plimpton and Wolf 1990).

### 8.2.5 Models for grain boundary self-diffusivities via the different mechanisms

Experimental evidence, discussed below in Section 8.2.7, rules out the exchange mechanism as a major contributor in the vast majority of cases, and we are, therefore, left with the vacancy, interstitialcy, and interstitial mechanisms. We now follow Ma and Balluffi (1994) and derive general expressions for the grain boundary self-diffusivity associated with elementary forms of these mechanisms.

Consider a marked (radioactive tracer) atom which diffuses in the core by executing a sequence of jumps by one of the above mechanisms. In general, these jumps will be of different types as the diffusing atom moves about in the core and visits various sites having different environments. If the marked atom makes $n$ jumps in the sequence in the time $\tau$, its total displacement in the $z$ direction is

$$Z = l_1 + l_2 + l_3 + \ldots l_n, \tag{8.25}$$

where $l_i$ is the component of the $i$th jump along $z$. The square of the displacement is then

$$Z^2 = \sum_{i=1}^{n} l_i^2 + 2 \sum_{i=1}^{n-1} \sum_{j=i+1}^{n} l_i l_j. \tag{8.26}$$

The diffusivity of the marked atom in the $z$ direction is given by the standard Einstein relation

$$D^B = \lim_{\tau \to \infty} \frac{\langle Z^2 \rangle}{2\tau}, \tag{8.27}$$

where the symbol $\langle \ \rangle$ represents the operation of averaging over a large number of jumping sequences. Combining eqns (8.26) and (8.27), we then have

$$D^B = \lim_{\tau \to \infty} \left( \left\langle \sum_{i=1}^{n} l_i^2 \right\rangle + 2 \left\langle \sum_{i=1}^{n-1} \sum_{j=i+1}^{n} l_i l_j \right\rangle \right) / 2\tau. \tag{8.28}$$

Since $n \to \infty$ as $\tau \to \infty$, and since the summation and averaging operations are interchangeable,

$$D^B = \lim_{\tau \to \infty} \left( \sum_{i=1}^{\infty} \langle l_i^2 \rangle + 2 \sum_{i=1}^{\infty} \left\langle \sum_{m=1}^{\infty} l_i l_{i+m} \right\rangle \right) / 2\tau. \tag{8.29}$$

We now assume that the grain boundary is periodic, and that the atom is therefore restricted to making $R$ different types of jumps (which is the number geometrically possible within the unit cell of the periodic boundary). If $p$ and $q$ represent two jumps of the same type,

$$\langle l_p^2 \rangle = \langle l_q^2 \rangle = l_p^2;$$

$$\left\langle \sum_{m=1}^{\infty} l_p l_{p+m} \right\rangle = \left\langle \sum_{m=1}^{\infty} l_q l_{q+m} \right\rangle,$$

and therefore

$$D^B = \lim_{\tau \to \infty} \sum_{\alpha=1}^{R} N_\alpha \left[ l_\alpha^2 + 2 \left\langle \sum_{m=1}^{\infty} l_\alpha l_{\alpha+m} \right\rangle \right] / 2\tau, \tag{8.30}$$

where $N_\alpha$ is the number of type-$\alpha$ jumps in the sequence. Now, let $\dot{r}_\alpha$ be the type-$\alpha$ jump rate, i.e.

$$\dot{r}_\alpha = \lim_{\tau \to \infty} \frac{N_\alpha}{\tau}. \tag{8.31}$$

Then eqn (8.31) can be rewritten as

$$D^B = \frac{1}{2} \sum_{\alpha=1}^{R} \dot{r}_\alpha l_\alpha^2 \left[ 1 + 2 \left\langle \sum_{m=1}^{\infty} l_\alpha l_{\alpha+m} \right\rangle / l_\alpha^2 \right]. \tag{8.32}$$

The first summation (inside the brackets) is over a jumping sequence of the marked atom following a type-$\alpha$ jump. The second summation is over all of the different types of jumps in the sequence. We may define the quantity in brackets by

$$f_\alpha^B \equiv 1 + 2 \left\langle \sum_{m=1}^{\infty} l_\alpha l_{\alpha+m} \right\rangle / l_\alpha^2, \tag{8.33}$$

so that

$$D^B = \sum_{\alpha=1}^{R} D_\alpha^B, \tag{8.34}$$

where

$$D_\alpha^B = \tfrac{1}{2} \dot{r}_\alpha l_\alpha^2 f_\alpha^B. \tag{8.35}$$

Equation (8.34) shows that the overall boundary diffusivity of the marked atom, $D^B$, has the form of a sum of 'partial diffusivities', $D_\alpha^B$, each of which is associated with a different type-$\alpha$ jump and is given by eqn (8.35). Associated with each partial diffusivity is the quantity, $f_\alpha^B$, the 'partial correlation factor' for the type-$\alpha$ jump, which is a measure of the degree of correlation which exists between the type-$\alpha$ jump and all successive jumps which follow a type-$\alpha$ jump. If the successive jumps were uncorrelated, $f_\alpha^B$ would be unity, since the quantity

$$\left\langle \sum_{m=1}^{\infty} l_\alpha l_{\alpha+m} \right\rangle$$

in eqn (8.33) would average out to zero because of the random nature of the jumping sequence. However, as discussed below, successive jumps during grain boundary self-diffusion are generally correlated, and correlation factors therefore emerge as important quantities in determining the rates of self-diffusion.

In an alternative formulation we may write

$$D^B = f^B \cdot \frac{1}{2} \sum_{\alpha=1}^{R} \dot{r}_\alpha l_\alpha^2,$$  (8.36)

where the quantity $f^B$ is given by

$$f^B = \left[ \sum_{\alpha=1}^{R} \dot{r}_\alpha l_\alpha^2 f_\alpha^B \right] \left[ \sum_{\alpha=1}^{R} \dot{r}_\alpha l_\alpha^2 \right]^{-1}.$$  (8.37)

The quantity $f^B$ is known as the 'total correlation factor' and would also be equal to unity in the absence of correlation.

The physical basis for correlation in boundary diffusion is easily recognized. In general, all of the significant jumping will be confined to the narrow core region. In addition, the core structure is highly non-uniform, and the fast diffusion of a marked atom will therefore tend to be confined to certain preferred paths consisting of sequences of relatively easy and more difficult jumps. Correlation then arises from two different sources. The first is the fact that the marked atom can diffuse over long distances only by executing a series of jumps of differing difficulty. For example, in the interstitial mechanism, if the marked atom, which is diffusing as an interstitial atom, makes a successful jump, and if all of the next jumps which are geometrically possible require a higher activation energy than the backwards reverse jump, then the next jump will most probably be the backwards reverse jump. The successive jumps which occur will then be correlated. We note that this correlation does not exist for the diffusion of an interstitial atom in the crystal lattice, as, for example, when interstitial C atoms diffuse in b.c.c. Fe, since all interstitial jumps in the uniform b.c.c. lattice are identical and have the same probability.

The second source of correlation, which is present only for the vacancy and interstitialcy mechanisms, is the fact that the point defects which are responsible for the jumping of the marked atom induce a correlation between the successive jumps of the marked atom, since the positions of these defects are not randomized with respect to the position of the marked atom after each successive jump. For example, in the vacancy mechanism the vacancy responsible for the jump of a marked atom inevitably ends up a near-neighbour of the marked atom after a jump of the marked atom. In its next jump the marked atom therefore has a higher probability than a random one of simply executing a reverse jump back into the neighbouring vacancy which erases its previous jump. We note that this source of correlation also exists for the diffusion of a marked atom in the crystal lattice by a vacancy or interstitialcy mechanism. However, it will be more important in a boundary because of the confinement of the diffusion to the boundary core and the higher probability of a reverse jump.

Equation (8.32) may now be further developed. By representing the frequency of a type-$\alpha$ jump of the marked atom (from a type-h site to a type-d site) by $\Gamma_M^{\alpha(h \to d)}$, and the probability that the marked atom occupies a type-h site by $c_M^h$, the type-$\alpha$ jump rate is then

$$\dot{r}_\alpha = N_\alpha/\tau = \Gamma_M^{\alpha(h \to d)} c_M^h.$$  (8.38)

From eqn (8.32), the diffusivity of the marked atom is therefore of the form

$$D^B = \sum_{\alpha=1}^{R} D^B_\alpha = \frac{1}{2} \sum_{\alpha=1}^{R} \Gamma_M^{\alpha(h \to d)} c_M^h l_\alpha^2 f_\alpha^B. \tag{8.39}$$

More explicit expressions for $D^B$ can now be written (Ma and Balluffi 1994) for the vacancy, interstitialcy, and interstitial mechanisms respectively.

### 8.2.5.1 *Vacancy mechanism*

In this case all jumping takes place on grain boundary 'lattice' sites (i.e. sites which are occupied by atoms when the boundary is in its minimum energy configuration). The jump of the marked atom from boundary lattice site $h^\lambda$ to $d^\lambda$ can be considered as a two-step process: first, a vacancy occupies site $d^\lambda$, then the vacancy jumps into site $h^\lambda$. Thus,

$$\Gamma_M^{\alpha(h^\lambda \to d^\lambda)} = c_V^{d^\lambda} \cdot \Gamma_V^{-\alpha(d^\lambda \to h^\lambda)}, \tag{8.40}$$

where $c_V^{d^\lambda}$ is the occupation probability of a vacancy at site $d^\lambda$, and $\Gamma_V^{-\alpha(d^\lambda \to h^\lambda)}$ is the jumping frequency of the vacancy from site $d^\lambda$ to site $h^\lambda$. When local equilibrium prevails, detailed balance requires

$$c_V^{d^\lambda} \cdot \Gamma_V^{-\alpha(d^\lambda \to h^\lambda)} = c_V^{h^\lambda} \cdot \Gamma_V^{\alpha(h^\lambda \to d^\lambda)}. \tag{8.41}$$

Therefore

$$\Gamma_M^{\alpha(h^\lambda \to d^\lambda)} = c_V^{h^\lambda} \cdot \Gamma_V^{\alpha(h^\lambda \to d^\lambda)}. \tag{8.42}$$

Now, at the dilute vacancy concentration limit, and with local equilibrium prevailing, the probability that the marked atom is in any grain boundary lattice site is the same. Therefore, $c_M^{h^\lambda}$ is simply the fraction of type-$h^\lambda$ sites in the boundary lattice, and

$$c_M^{h^\lambda} = N^{h^\lambda}/N, \tag{8.43}$$

where $N^{h^\lambda}$ is the number of type-$h^\lambda$ sites in a grain boundary unit cell, and $N$ is the corresponding total number of grain boundary lattice sites. Putting these results into eqn (8.39), the diffusivity of the marked atom is then

$$D^B = \sum_{\alpha=1}^{R} D^B_\alpha = \frac{1}{2N} \sum_{\alpha=1}^{R} c_V^{h^\lambda} \cdot \Gamma_V^{\alpha(h^\lambda \to d^\lambda)} l_\alpha^2 f_\alpha^B N^{h^\lambda}. \tag{8.44}$$

### 8.2.5.2 *Interstitialcy mechanism*

Here, the type-$\alpha$ jumps can be conveniently divided into two separate categories:

1. In the first, the marked atom is initially at a grain boundary lattice site, $h^\lambda$, and jumps to a grain boundary interstitial site, $d^i$, as in Fig. 8.20(b) and (c). Clearly, an interstitial at any of the near-neighbour interstitial sites, $s^i$, except at $d^i$, has a probability to cause such a jump. Therefore,

$$\Gamma_M^{\alpha(h^\lambda \to d^i)} = \sum^{s^i \neq d^i} c_I^{s^i} \Gamma_{Icy}^{\alpha(s^i \to h^\lambda, h^\lambda \to d^i)}, \tag{8.45}$$

where $c_I^{s^i}$ is the occupation probability of an interstitial at $s^i$, and $\Gamma_{Icy}^{\alpha(s^i \to h^\lambda, h^\lambda \to d^i)}$ is the jumping frequency of the marked atom from $h^\lambda$ to $d^i$ caused by an interstitial jumping from $s^i$ to $h^\lambda$. The summation is taken over all of the $s^i$ near-neighbour interstitial sites except $d^i$. Also, the probability of finding the marked atom at a type-$h^\lambda$ A site is again given by eqn (8.43).

2. In the second, the marked atom is initially at an interstitial site, $h^i$, and jumps to a boundary lattice site, $d^\lambda$ (as in Fig. 8.20(c) and (d)), causing the atom initially at the site $d^\lambda$ to jump to a near-neighbour interstitial site $s^i$. Therefore,

$$\Gamma_M^{\alpha(h^i \to d^\lambda)} = \sum_{}^{s^i \neq h^i} \Gamma_{Icy}^{\alpha(h^i \to d^\lambda, d^\lambda \to s^i)}, \tag{8.46}$$

where $\Gamma_{Icy}^{\alpha(h^i \to d^\lambda, d^\lambda \to s^i)}$ is the jumping frequency of the marked atom from $h^i$ to $d^\lambda$. Since the interstitial concentration is dilute, the probability that the marked atom is in a type-$h^i$ site is

$$c_M^{h^i} = c_I^{h^i} N^{h^i}/N, \tag{8.47}$$

where $c_I^{h^i}$ is the occupation probability of an interstitial at a type-$h^i$ site, and $N^{h^i}$ is the number of type-$h^i$ interstitial sites in a boundary unit cell.

Combining the above jumps, we therefore have

$$D^B = \sum_{\alpha 1 = 1}^{R1} D_{\alpha 1}^B + \sum_{\alpha 2 = 1}^{R2} D_{\alpha 2}^B = \frac{1}{2N} \left[ \sum_{\alpha 1 = 1}^{R1} \sum_{}^{s^i \neq d^i} c_I^{s^i} \Gamma_{Icy}^{\alpha 1(s^i \to h^\lambda, h^\lambda \to d^i)} l_{\alpha 1}^2 f_{\alpha 1}^B N^{h^\lambda} + \cdots \right.$$

$$\left. \cdots \sum_{\alpha 2 = 1}^{R2} \sum_{}^{s^i \neq h^i} c_I^{h^i} \Gamma_{Icy}^{\alpha 2(h^i \to d^\lambda, d^\lambda \to s^i)} l_{\alpha 2}^2 f_{\alpha 2}^B N^{h^i} \right]. \tag{8.48}$$

### 8.2.5.3 *Interstitial mechanism*

If the marked atom jumps as an interstitial, then,

$$\Gamma_M^{\alpha(h^i \to d^i)} = \Gamma_I^{\alpha(h^i \to d^i)}, \tag{8.49}$$

where all the sites are now interstitial sites, and $\Gamma_I^{\alpha(h^i \to d^i)}$ is the type-$\alpha$ interstitial jump frequency. Also, $c_M^{h^i}$ should have the same form as for the interstitialcy mechanism (see eqn (8.47)). Substituting these results in eqn (8.39) we then have

$$D^B = \frac{1}{2N} \sum_{\alpha = 1}^{R} c_I^{h^i} \Gamma_I^{\alpha(h^i \to d^i)} l_\alpha^2 N^{h^i} f_\alpha^B. \tag{8.50}$$

### 8.2.6 *General characteristics of the models for boundary self-diffusion*

The diffusivities derived above on the basis of the three different mechanisms (eqns (8.44), (8.48), and (8.50)) are all sums of partial diffusivities, $D_\alpha^B$, each of which is associated with a characteristic type-$\alpha$ jump of the marked atom in the boundary. The magnitude of each partial diffusivity is proportional to the jump distance squared, the rate of jumping (given by the product of $c$ and $\Gamma$), and the partial correlation factor of its characteristic jump. The terms comprising the rate of jumping can be developed more explicitly by substituting either eqn (8.21) or eqn (8.22) for the occupation probability, $c$, of the relevant point defect at the h-type or s-type site and writing the jumping frequency, $\Gamma_\alpha$, in the form

$$\Gamma_\alpha = \nu_{\alpha o} \exp(-G_\alpha^m/kT) = \nu_{\alpha o} \exp(S_\alpha^m/k) \cdot \exp(-E_\alpha^m/kT). \tag{8.51}$$

Equation (8.51) follows from the fact that the energy of the system during a type-$\alpha$ jump will vary with the position of the marked atom (or, alternatively, with the position of the point defect which is involved) in the general asymmetric manner illustrated in

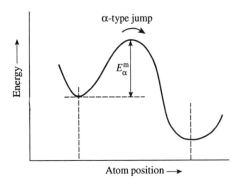

**Fig. 8.22** Schematic representation of the variation of the energy of the system with the position of an atom executing a jump in a grain boundary. $E_\alpha^m$ is the activation energy for the jump.

Fig. 8.22. According to standard rate theory, we then have eqn (8.51) with $\nu_{\alpha o}$ = atomic vibrational frequency, $S_\alpha^m$ = entropy of migration, and $E_\alpha^m$ = activation energy for the jump. The contribution of each type of jump to the overall diffusivity may then be expressed in the general form

$$D_\alpha^B = \gamma_\alpha f_\alpha^B l_\alpha^2 \nu_{\alpha o} \cdot \exp\left[-(G_\alpha^m + G_\alpha^f)/kT\right] = \gamma_\alpha f_\alpha^B l_\alpha^2 \nu_{\alpha o} \cdot \exp\left[(S_\alpha^m + S_\alpha^f)/k\right]$$

$$\cdot \exp\left[-(E_\alpha^m + E_\alpha^f)/kT\right], \tag{8.52}$$

i.e. the product of a purely geometrical factor, $\gamma_\alpha$; a partial correlation factor, $f_\alpha^B$; a jump distance squared, $l_\alpha^2$; an atomic vibrational frequency, $\nu_{\alpha o}$; an entropy factor containing the entropy of migration, $S_\alpha^m$, and entropy of formation, $S_\alpha^f$, of the relevant point defect; and a Boltzmann type factor containing the total activation energy for the jump, i.e. $(E_\alpha^m + E_\alpha^f)$.

The partial correlation factor, $f_\alpha^B$, is a highly complicated quantity which is sensitive to essentially all aspects of the boundary structure and energetics and plays an essential role in determining the magnitude of each $D_\alpha^B$ (and hence the contribution of each jump to the overall diffusivity). An analytical expression for $f_\alpha^B$ is therefore difficult to obtain. However, it can be evaluated conveniently by a computer simulation method (Ma and Balluffi 1994) when the boundary structure and the defect formation and migration energies and entropies are known at all sites. In this method the marked atom is first allowed to make a successful type-$\alpha$ jump, and the subsequent jumping sequence is then calculated with a computer using a probabilistic model. $f_\alpha^B$ is then evaluated using eqn (8.33), and the process is repeated until a reliable average value is obtained.

In an effort to establish some of the characteristic behaviour of $f_\alpha^B$ (and also $f^B$), Ma and Balluffi (1994) calculated the magnitudes of these quantities at different temperatures for jumps in a number of idealized boundaries having adjustable diffusional characteristics. For the vacancy and interstitialcy mechanisms, it was found that the total correlation factor tends to decrease as the diffusion of the marked atom is increasingly confined to certain paths in the boundary core. This may be attributed to an increased number of encounters between the marked atom and the defect responsible for its diffusion. These result in the cancellation of marked atom jumps and a reduced mean displacement and diffusivity of the marked atom. If the jumping tends to be confined to easy paths consisting of single strings of sites, the partial correlation factors of the rapid jumps along these paths is relatively small, and the contributions of these rapid jumps to $D^B$ are reduced accordingly. Furthermore, the partial correlation factors of these easy jumps are often temperature dependent in a manner which can be described approximately by a Boltzmann factor containing an effective activation energy, i.e.

$$f_\alpha^B(T) \simeq f_{\alpha o}^B \cdot \exp[-e_\alpha^{eff}/kT]. \tag{8.53}$$

The effective activation energy of the contributions to $D^B$ of the easy jumps along the easy path is then increased by $e_\alpha^{eff}$ and therefore tends to be closer to the higher activation energies for the jumps along the next easiest path in the boundary structure which is coupled to the easy path.

For all mechanisms, the total correlation factor also decreases as the differences between relatively easy and more difficult jumps, which are intermixed along confined paths in the boundary core, increase. This occurs because the partial correlation factors of the easy jumps decrease as the result of the cancellations of many of these jumps by easy reverse jumps. The contributions of the easy and more difficult jumps to $D^B$ then become comparable and have closely the same effective activation energy which corresponds approximately to that of the more difficult jumps.

The overall effective activation energy for $D^B$ is therefore a complex quantity determined by the differing effective activation energies, of the $D_\alpha^B$'s which contribute significantly to $D^B$. We therefore cannot expect $D^B$ to follow a strictly Arrhenius form with a constant activation energy as assumed earlier in Fig. 8.5. and eqn (8.13). However, the contributions of the fastest jumps in the core region will often be reduced by correlation effects as described above, and the effective activation energies of the major contributors to $D^B$ will therefore tend be fairly similar. Hence, $D^B$ should exhibit a nearly constant effective activation energy over limited ranges of temperature as assumed in Fig. 8.5. Over larger ranges of temperature, deviations may be expected as the jumps in the mix of jumps which contribute significantly to $D^B$ change systematically. Some indication of such a deviation over a relatively wide temperature range (in the form of a curved rather than a straight-lined Arrhenius plot) has already been presented in Fig. 8.6.

The above results also lead to insight into the grain boundary 'thickness' parameter, $\delta$, which has not been well defined physically. All of the above expressions for $D^B$ are seen to be proportional to $1/N$. The value of $N$ obviously depends upon the number of sites which are chosen to lie in the unit cell of the boundary. To a good approximation

$$N = A\delta/\bar{\Omega}, \tag{8.54}$$

where $A$ = area of unit cell in the boundary plane and $\bar{\Omega}$ = average atomic volume in the core. Since $A$ is fixed by the boundary geometry, $N$ is then closely proportional to $\delta$. We can therefore visualize what happens as we increase $N$ (and therefore $\delta$). In general, all of the sites which support jumps which are rapid compared to crystal lattice jumps will lie in the narrow core region. As we move outwards, the sites quickly assume the relatively slow diffusional properties of the lattice sites. Once the cell is chosen to be wide enough to include all sites which support jumps which are rapid compared to lattice jumps, there will be no further significant increase in the total jump rate in the boundary, and the summation over $R$ in all of the above expressions for $D^B$ will become constant for all practical purposes. When this condition holds, the diffusivity, $D^B$, will become proportional to $\delta^{-1}$, i.e. the reciprocal of the boundary width which is chosen.

In almost all boundary diffusion measurements it is the combined parameter $\delta D^B$ which is measured rather than $D^B$ by itself (Kaur and Gust 1989). According to the above discussion, the expressions for $D^B$ derived in the present chapter will yield expressions for the parameter $\delta D^B$ which are independent of the choice of $\delta$ as long as

$N$ (and therefore $\delta$) is chosen large enough to include all sites at which significantly fast jumping occurs compared to the jumping rate in the lattice.

We may now ask: what is the most physically satisfying way to describe $\delta$ and $D^B$ separately? Clearly, this should be done by choosing $N$ (and therefore $\delta$) to just include all sites at which signicantly fast jumping occurs. Choosing $N$ (and $\delta$) larger than this will produce artificially large $\delta$'s and correspondingly small $D^B$'s.

Finally, we note that the expressions derived above in Section 8.2.5 hold only for periodic boundaries. However, the values of $D^B$ for non-periodic boundaries could be obtained with very little approximation by representing them by periodic boundaries of long period which are nearby in the grain boundary phase space and are either of the exact coincidence or forced near-coincidence type. Of course, $R$ would then become relatively large.

### 8.2.7  On the question of the mechanism (or mechanisms) of fast grain boundary diffusion

The lattice diffusion of normally substitutional atoms in a wide variety of solid types has been found to occur by a variety of defect mechanisms. In f.c.c., b.c.c., and h.c.p. metals the diffusion occurs predominantly by the vacancy mechanism (Shewmon 1989). However, a number of somewhat undersize solute atoms may diffuse by an interstitial process and behave as 'fast' diffusers as already mentioned in Section 8.2.3. In covalently bonded Ge, self-diffusion apparently occurs by a vacancy mechanism, while in Si it occurs primarily via vacancies at low temperatures and interstials at elevated temperatures (Tan and Gösele 1985). In ionic materials the lattice diffusion of Ni in NiO occurs by a vacancy mechanism, whereas in $Cu_2O$ the diffusion of O occurs via an interstitial process (Atkinson 1988b). In the alkali halides, Schottky defects predominate, and the diffusion of the anions and cations occurs by a vacancy mechanism. On the other hand, in AgBr, Frenkel defects predominate, and the smaller Ag cations diffuse by an interstitialcy mechanism (Shewmon 1989). No simple and general conclusions regarding mechanisms can therefore be reached regarding diffusion in the lattice.

As might be expected, the problem of establishing the mechanism (or mechanisms) of boundary diffusion is considerably more difficult. So far, experimental information about the dominant mechanism (or mechanisms) of boundary diffusion in different types of boundaries in different systems is very limited. As already mentioned, this can be attributed in large part to the narrow width of the boundary core and the relatively small number of atoms which reside there. The unwanted segregation of impurities has also caused problems in many cases. We must therefore appeal to rather indirect experimental observations and also the results of atomistic calculations.

#### 8.2.7.1  *Metals*

We consider boundary self-diffusion first. There is extensive experimental evidence (e.g. Balluffi 1984) that a net diffusional flux of atoms can be induced along general boundaries in metals under numerous conditions. This, therefore, tends to rule out the ring exchange mechanism as an important process. For example, a net transport of atoms along boundaries occurs during Coble-type diffusional creep of polycrystalline material (see Section 12.8.2) when atoms diffuse along the boundary network from boundaries under normal compression to boundaries under normal tension. Net boundary transport is also observed during the sintering of voids (Section 10.2.3) which are connected to boundaries, and atoms diffuse along the boundaries in order to fill the voids. In still other work, net boundary transport is observed during electromigration and thermomigration experiments. It is easily realized that all of the previously considered defect mechanisms, i.e. the vacancy, interstitialcy, and interstitial mechanisms) are not inconsistent with these experimental observations.

Attempts have been made to gain information about the dominant mechanism from values of activation volumes obtained from measurements of the effects of hydrostatic pressure on the boundary diffusivity. In order to apply this method an explicit expression linking the pressure derivative of the diffusivity with the activation volume must be obtained. This can be accomplished by using eqns (8.34) and (8.52) to write $D^B$ in the general form

$$D^B = \gamma f^B l^2 \nu_o \exp(-G^B/kT), \tag{8.55}$$

where it is understood that all quantities are suitably adjusted so that they represent to the greatest extent possible the total average effect produced by all jumps in the boundary core which contribute to $D^B$. Differentiating $D^B$ with respect to the hydrostatic pressure, $P$, and using the standard relation $(\partial G^B/\partial P)_T = V^B$, then yields

$$\left[\frac{\partial \ln D^B}{\partial P}\right]_T = -\frac{1}{kT} \cdot V^B + \left[\frac{\partial \ln (\gamma f^B l^2 \nu_o)}{\partial P}\right]_T, \tag{8.56}$$

where $V^B =$ is a corresponding effective activation volume for boundary diffusion involving the various volumes of formation and migration of the relevant point defects in a complicated manner. The second term on the right-hand side of eqn (8.56) is expected to be relatively small and can therefore be neglected. Equation (8.56) is therefore the desired relationship. Measurements of $V^B$ for Ag self-diffusion in general boundaries and Zn diffusion in general boundaries in Al (Kedves and Erdelyi 1989) yield values which are in reasonable agreement with measured activation volumes for lattice self-diffusion by a vacancy mechanism. Zinc is a substitutional solute in Al, and neglecting complications due to solute atom effects on boundary diffusion in the Zn/Al experiments (which are probably small), these results therefore appear to be at least consistent with a vacancy mechanism for boundary diffusion. There is reason to believe that the effective activation volume for the interstitialcy mechanism may be smaller than for the vacancy, since significant free volume should be available in the boundary core to accommodate the formation of interstitial atoms, and the volume expansion required to accommodate interstitialcy replacements during migration should be modest. It is possible therefore that these results favour the vacancy or, possibly, even the interstitial mechanisms. However, none of the relevant parameters is known with sufficient precision to settle the question, and their significance therefore remains in doubt.

In other experiments, Robinson and Peterson (1972) determined the so-called 'isotope mass effect' for self-diffusion in general boundaries in Ag by measuring the boundary diffusivities of isotopes possessing two different masses. These results indicate (see Balluffi 1984) that: (i) the effective boundary correlation factor, $f^B$, has a magnitude $\geqslant 0.46$, and also that: (ii) the fraction of the total kinetic energy of the saddle point configuration which is carried by a jumping atom in a thermally activated atomic jump in the boundary is $\geqslant 0.46$. The first result implies that the grain boundary diffusion is relatively uncorrelated, i.e. it occurs in the core region by means of a network of alternate jump paths which allows the atom, and the defect responsible for its jump, to decorrelate rather effectively after a successful jump. The second result implies that only a relatively small number of atoms is strongly involved in a typical jump, and hence the jumping process is fairly localized in the boundary. Unfortunately, none of the defect mechanisms under consideration appear to be definitely inconsistent with these results.

Dranova and Mikhaylovskiy (1984) used field ion microscopy to study the rate at which interstitial W atoms diffused along grain boundaries in W and found that they diffused more slowly than in the lattice. This was attributed to the presence in the boundaries of many sites at which the interstitials experienced binding energies as large as $\simeq 0.2\,eV$. This interesting result indicates that dilute interstitial point defects may often diffuse more slowly in boundaries than in the lattice because of trapping in the non-uniform boundary core structure as argued many years ago by Friedel (1964) for the analogous case of the diffusion of interstitials along the bad material in dislocation cores. However, no experimental information is available regarding the formation energies of interstitials in grain boundary cores, and this result therefore cannot be used to rule out interstitial type mechanisms for boundary self-diffusion.

The above experimental results therefore fail to establish clearly any unique mechanism (or mechanisms). We must therefore turn to calculations. Several tests for the mechanism of boundary self-diffusion have been made by computer simulation using the methods of both molecular statics and dynamics. In the case of the $\Sigma = 5\,[001]$ symmetric tilt boundary in b.c.c. Fe (see Fig. 8.17 and Table 8.2) the properties of the point defects involved in the different mechanisms were either calculated or estimated and then compared with each other in order to test for mechanisms (Balluffi *et al.* 1981, Kwok *et al.* 1981, 1984, Balluffi 1984). The calculated formation energies for vacancies at Sites A through D (Table 8.2) are seen to be considerably lower than those for the interstitials. However, the formation energy of the interstitial at Site I is considerable lower than at the other sites and is comparable to the vacancy formation energies. This may be attributed to the relatively large open space in the core structure at this site. Using this information, the total equilibrium concentrations of vacancies and interstitials in the boundary may be calculated as functions of temperature by means of eqns (8.23) and (8.24) using reasonable estimates for the formation entropies. The total defect concentrations obtained in this way yield closely straight lines on Arrhenius plots corresponding to an effective energy of formation $E_V^f = 97\,kJ\,mol^{-1}$ and an only slightly larger value $E_I^f = 102\,kJ\,mol^{-1}$.

In order to investigate defect migration in the boundary, vacancies were created in the boundary, and their subsequent jumping sequences in the boundary were calculated by molecular dynamics at several temperatures. The following results were obtained:

1. Essentially all jumps were confined to the core and occurred between A, B, C, and D sites in Fig. 8.17. A typical trajectory is shown on the left of Fig. 8.17. The number of times a vacancy jumped into a given type of site during a sequence at 1300 K is listed in Table 8.2. As might be expected, there was a clear correlation in which the lower the formation energy the more frequent the rate of visitation.

2. The vacancy migration took place predominantly along the tilt axis rather than perpendicular to it. This would produce a correspondingly large anisotropy in the self-diffusion in at least qualitative agreement with the experimental results in Fig. 8.13.

3. The vacancy jump frequency, $\nu_V$, (averaged over all jumps) was found to obey closely the Arrhenius form

$$\nu_V = A \cdot \exp\left(-E_V^m / kT\right), \tag{8.57}$$

with $A = 4.9 \times 10^{13}\,s^{-1}$ and $E_V^m = 49\,kJ\,mol^{-1}$. The constant pre-exponential factor should be given approximately by $A = z \cdot \nu_0 \cdot \exp(S_V^m / k)$ where $z = $ number of near-neighbours in the core, and $\nu_o$ = atomic vibrational frequency. Estimating $z \simeq 8$, $\nu_o = \nu_D/2$ (where $\nu_D$ = Debye frequency = $7.1 \times 10^{12}\ s^{-1}$) and $\exp(S_V^m/k) \simeq 2$, we predict $A \simeq 5.7 \times 10^{13}\ s^{-1}$ in satisfactory agreement with the value found above.

4. Atoms on B sites occasionally jumped into interstitial I sites by a process illustrated in the centre of Fig. 8.17 which is essentially the thermally activated formation of a Frenkel pair. Inspection of Table 8.2 that this type of Frenkel pair must have a considerably lower formation energy than any other possible Frenkel pair, and this is evidently the reason for its occurrence. The vacancy formed in this way often diffused away leaving the interstitial behind at I. However, the interstitial at I remained completely immobile and could only be eliminated by mutual annihilation with a neighbouring vacancy. The immobility of the interstitial at I is readily understood on the basis of the results in Table 8.2 where it is seen that the formation energies of the interstitial at the other interstitial sites are larger by at least 121 kJ mol$^{-1}$. The activation energy for the interstitial to migrate must therefore be larger than 121 kJ mol$^{-1}$, and this process was therefore not observed.

The above results indicate that the vacancy mechanism should be predominant in this particular boundary. Neglecting complications due to any temperature dependence of the effective correlation factor, the calculated effective activation energy for boundary diffusion, i.e. $Q^B = (E_V^f + E_V^m)$, is equal to $(97 + 49) = 146\ kJ\ mol^{-1}$ for the vacancy mechanism and is markedly lower than $Q^B = (E_I^f + E_I^m) = (103 + 121) = 224\ kJ\ mol^{-1}$ for the interstitial-related mechanisms. The remaining terms constituting the boundary diffusivity are not greatly different for diffusion by the vacancy and interstitialcy-related mechanisms, and the result that $Q^B$ is much lower for the vacancy mechanism therefore indicates that it should be pedominant. Also, the measured activation energy for lattice self-diffusion is $Q^L = 240\ kJ\ mol^{-1}$ (Peterson 1978), and therefore $Q^B/Q^L = 146/240 = 0.61$, which is in the expected range.

In additional work Pontikis (1982) has performed calculations on a $\Sigma5$ tilt boundary in the f.c.c. structure with results which are generally similar to those described above. Also, calculations by Kwok and Ho (1984) indicate fast boundary diffusion via vacancies in a f.c.c. $\Sigma5$ tilt boundary. More recently, Nomura *et al.* (1991) and Nomura and Adams (1992) have calculated vacancy formation and migration energies in [001] twist boundaries in Cu using molecular statics and found values of $Q^B = (E_V^f + E_V^m)$ which were less than half that of $Q^L$, indicating that fast boundary diffusion could occur via vacancies in these boundaries.

However, other calculations indicate that interstitial-related mechanisms may also be of importance, at least in certain cases. Ma *et al.* (1993) have calculated boundary diffusivities along the tilt axes of several [001] tilt boundaries in Ag, again using molecular statics. Diffusivities by the vacancy, interstitialcy, and interstitial mechanisms (including correlation factors) were calculated through use of the expressions developed in Section 8.2.5. Of considerable interest is the result that fast diffusion was found for all three mechanisms and that in several cases interstitial-related mechanisms were *faster* than the vacancy mechanism. It was therefore concluded that interstitial-related mechanisms most probably make important (and perhaps dominant) contributions to the diffusion along the tilt axes in these boundaries. In another study, Huang *et al.* (1989) have concluded from molecular dynamics calculations that the interstitialcy mechanism can be nearly as important as the vacancy mechanism for fast pipe diffusion along dissociated lattice edge dislocations in Cu.

At this point we may therefore conclude that the vacancy mechanism is probably dominant for self-diffusion in many boundaries. However, interstitial-related mechanisms may also be important (and even dominant) under many circumstances in the wide range of boundary structures which exist. Also, as suggested by the work of Kwok *et al.* (1984), more complex higher order defect mechanisms involving multiple jumps over longer distances as well as direct interchanges may contribute, particularly at elevated temperatures where increased thermal activation is available, and increased boundary disordering may be expected (Ciccotti *et al.* 1983, Plimpton and Wolf 1990). Unfortunately, no simple generalizations seem possible.

In considering the mechanism of the grain boundary diffusion of solute atoms, which are normally substitutional in the lattice, there is experimental evidence (see Section 8.2.3.1) that the two substitutional species in binary alloys diffuse at different rates in boundaries as they do in the Kirkendall effect in the lattice (see Section 10.4.2.2). In this phenomenon the two different species diffuse at unequal rates by exchanging places with point defects at different rates, and a net flux of atoms ensues. These experimental results are clearly not inconsistent with any of the defect mechanisms which we have considered above for boundary diffusion. As discussed in Section 8.2.3.1, solute atoms, which are somewhat undersize and are 'fast' diffusers in the lattice via interstitial defects, are 'superfast' diffusers in grain boundaries. It therefore seems likely that these atoms diffuse by some type of interstitial type mechanism in the boundaries as well. On the other hand, solute atoms which are less undersize (such as Ag in $\alpha$-Zr) diffuse more 'normally' in a manner similar to self-diffusion. In such cases the mechanism in the lattice, as well as in grain boundaries, is probably (but not certainly) a vacancy mechanism.

### 8.2.7.2 *Ionic materials*

In NiO, the relatively rapid diffusion of the Ni cations in the lattice (Fig. 8.9) occurs by a vacancy mechanism (Atkinson 1985) evidently via a mixture of singly and doubly charged vacancies. Evidence for this is provided by the increase in the lattice diffusivity of Ni which occurs when the partial pressure of oxygen, $p_{O_2}$, in equilibrium with NiO is increased as seen in Fig. 8.23(a). This may be understood by analysing the equilibrium between $O_2$ and the defect structure of non-stoichiometric NiO (Kofstad 1972). First, consider the reaction in which an oxygen atom is transferred from the gas phase to the NiO to produce a neutral cation vacancy, $V_M^X$, i.e.

$$\tfrac{1}{2} O_2 = V_M^X \tag{8.58}$$

Each neutral vacancy may be regarded as being associated with two holes, and these may be 'excited' by transferring electrons from existing $Ni^{++}$ ions to the cation vacancy, thereby producing a localized hole, $h$, at each $Ni^{+++}$ ion which is formed and a negatively charged cation vacancy. If one hole is excited, a singly charged cation vacancy, $V_M'$, is produced, i.e.

$$V_M^X = V_M' + h. \tag{8.59}$$

If two are excited, a doubly charged vacancy, $V_M''$, is produced, i.e.

$$V_M^X = V_M'' + 2h. \tag{8.60}$$

If these reactions are in equilibrium, we may take them in turn and write

$$[V_M^X]/p_{O_2}^{\frac{1}{2}} = K_1, \tag{8.61}$$

$$[V_M'] \cdot [h]/[V_M^X] = K_2, \tag{8.62}$$

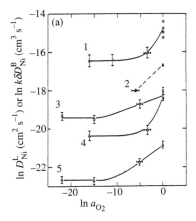 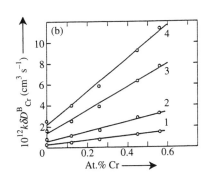

**Fig. 8.23** (a) The influence of the O activity, $a_{O_2}$, on the lattice, small-angle and large-angle boundary diffusion of Ni in NiO. Curve 1: $D_{Ni}^L$ at 690 °C. Curve 2: $k\delta D_{Ni}^B$ for large-angle boundaries at 700 °C. Curve 3: $D_{Ni}^L$ at 522 °C. Curve 4: $k\delta D_{Ni}^B$ for small-angle boundaries at 690 °C. Curve 5: $k\delta D_{Ni}^B$ for large-angle boundaries at 522 °C. (From Atkinson (1985).) (b) The diffusion parameter $k\delta D_{Cr}^B$ for the diffusion of Cr along the tilt axis of a $\langle 001 \rangle$ tilt boundary ($\theta = 15\,°$) in MgO as a function of the bulk concentration of $Cr^{+++}$. Curves 1, 2, 3, and 4 correspond to 1200, 1295, 1383, and 1444 °C respectively. (From Stubicon (1985).)

and

$$[V_M''] \cdot [h]^2 / [V_M^X] = K_3, \tag{8.63}$$

where the bracket indicates a concentration, and the $K_i$ are equilibrium constants. Also, electrical neutrality requires that

$$[h] = [V_M'] + 2[V_M'']. \tag{8.64}$$

These equations may now be solved simultaneously to show how $[V_M']$ and $[V_M'']$ increase with increasing $p_{O_2}$. Two simple limiting case are particularly interesting. If $[V_M']$ is negligible, then $[V_M'] \propto p_{O_2}^{\frac{1}{4}}$, and if $[V_M']$ is negligible $[V_M''] \propto p_{O_2}^{\frac{1}{6}}$. The lattice diffusivity of Ni increases with $p_{O_2}$ about as $p_{O_2}^{\frac{1}{5}}$ (Atkinson 1985) indicating, therefore, that the diffusion occurs by means of a mixture of singly and doubly charged cation vacancies.

The results presented in Fig. 8.23(a) show that the boundary diffusivity of Ni increases with $p_{O_2}$ in about the same manner as the lattice diffusivity. This result constitutes strong evidence that the boundary diffusion occurs by the same basic vacancy mechanism.

Several other experiments along similar lines indicate a strong correlation between lattice and grain boundary diffusion. As discussed by Atkinson (1988*b*), measurements of the diffusion of O in $Cu_2O$ (Perinet 1987) show that both $D^L$ and $\delta D^B$ vary with $O_2$ pressure closely as $p_{O_2}^{\frac{1}{2}}$. In this system, this result indicates that both the lattice and grain boundary diffusion of O involve an uncharged O interstitial defect. When the lattice diffusion of Cr in $Cr_2O_3$ is studied as a function of $p_{O_2}$, it is found (Atkinson 1985) that it goes through a minimum as $p_{O_2}$ increases. The diffusion of Cr in small-angle boundaries follows parallel behaviour (Atkinson 1985), and this may well be due to diffusion via interstitials at low pressures and diffusion via vacancies at high pressures in both the lattice and in the boundaries. Finally, both the lattice and boundary diffusivities of U in UC increase (Routbort and Matzke 1975) as the material becomes non-stoichiometric at the composition $UC_{1-x}$.

*Diffusion at interfaces*

**Table 8.3** Measured and calculated activation energies for lattice and grain boundary diffusion of Ni and O in NiO. (From Atkinson (1988b).) Calculated results obtained by Duffy and Tasker (1986).

| | Measured | | Calculated | |
|---|---|---|---|---|
| | $Q^L$(eV) | $Q^B$(eV) | $Q^L$(eV) | $Q^B$(eV) |
| Ni diffusion (via vacancies) | 2.56 | 1.78 | 2.9 | 1.6–2.2 |
| O diffusion (via interstitials) | 5.6 | 2.5 | 5.3 | 2.5 |

The above results suggest that, in many cases, grain boundary diffusion in ionic materials occurs by the same basic defect mechanism as lattice diffusion. The observed faster diffusion in the boundary is then presumably due to higher defect concentrations and/or higher defect jump rates than in the lattice. A question then arises regarding the possible width of this region which, conceivably, could be wider than it is in metals because of possible space charge effects (Section 7.6). However, experiments carried out in the B and C Harrison regimes (Fig. 8.2) indicate (Atkinson 1985) that the width, $\delta$, of the region of high boundary diffusivity for Ni in NiO and Cr in $Cr_2O_3$ is $\simeq 1$ nm in agreement with the value found previously for metals. This result is also consistent with detailed theoretical estimates (Yan *et al.* 1977) which show that diffusion distributed over the space charge region, exclusive of the boundary core, can only make a modest contribution at most to the observed fast boundary diffusion.

Atomistic calculations provide further information. Calculations of the formation energies of cation and anion vacancies in NiO (Duffy and Tasker 1984, 1985) indicate that they are substantially lower at a variety of sites in the cores of different tilt boundaries than in the lattice. Further calculations (Duffy and Tasker 1986) of the activation energies for lattice and grain boundary diffusion yield the results shown in Table 8.3.

Relatively good agreement is obtained between experimentally measured values and calculated values corresponding to Ni cation diffusion via the vacancy mechanism and O anion diffusion via the interstitialcy mechanism in both the lattice and grain boundaries. These results are also seen to be consistent with our previous conclusion that the Ni cations diffuse via vacancies.

The above results all support a general picture in which the fast grain boundary diffusion in many ionic materials occurs by a defect mechanism similar to that in the lattice. Furthermore (as in the metals), the fast diffusion is restricted to a narrow core region ($\delta \simeq 1$ nm) where the defects are more numerous (and probably more mobile) than in the lattice.

### 8.2.7.3 *Covalent materials*

Considerably less information is available regarding the mechanism of fast grain boundary diffusion in covalent materials. For Ge and Si, the bulk lattices are relatively open compared with the close-packed metals, i.e. there is more space between the relatively 'hard' ion cores. The formation energies of vacancies and interstitials are therefore expected to be more equal than in the close-packed metals where the interstitial formation energy is very large because of the large amount of hard core repulsion which must be overcome when an interstitial is forced into an interstice. As a consequence, lattice self-diffusion in Si occurs primarily via interstitial defects at elevated temperatures, as we have already mentioned. In addition, the lattice diffusion of solute atoms is dominated by either vacancy or interstitial migration in various cases. We may therefore expect similar complications for fast grain boundary diffusion. However, at present, there is no significant information which can be used to clarify the situation.

## 8.3 DIFFUSION ALONG INTERFACES OF SOLUTE SPECIES WHICH ARE INTERSTITIAL IN THE CRYSTAL LATTICE

Solute atoms which are small compared to the host atoms will generally occupy interstitial sites in the lattice and diffuse in the lattice by jumping between these interstitial sites by the interstitial mechanism. This type of lattice diffusion is generally fast compared with, for example, lattice self-diffusion because of the small size of the jumping atoms which allows them to squeeze their way between the interstices rather easily. In addition, no lattice point defects are required, and unoccupied interstitial sites are usually available adjacent to the jumping atoms at the interstitial concentrations which are normally present.

Such small solute atoms may also be expected to occupy interstitial sites in boundaries and to diffuse between them by the interstitial mechanism already illustrated in Fig. 8.21. To date, only a relatively small number of measurements have been made of the diffusivities of interstitial species both in the lattice and in interfaces. Also, in many cases the measurements have been relatively crude and approximate.

A wide range of results has been obtained for the interstitial grain boundary diffusion of the very small atom H. There is extensive evidence in several systems that its diffusion rate along grain boundaries is concentration dependent and that, at best, its diffusion rate in boundaries is not very rapid relative to its rate in the lattice.

Tsuru and Latanision (1982) found $D^B/D^L \simeq 60$ for the diffusion of H in the grain boundaries of polycrystalline Ni at room temperature, while Ladna and Birnbaum (1987) found $D^B/D^L \simeq 8$–$17$ for a $\Sigma 9 [110]$ symmetric tilt grain boundary but no evidence for enhancement (i.e. $D^B/D^L > 1$) for a $\Sigma 11 [110]$ symmetric tilt boundary. More recently, Harris and Latanision (1991) have reported values of $D^B/D^L \simeq 40$, and Palumbor *et al.* (1991) have found values near $\simeq 23$. In additional work, Mütschele and Kirchheim (1987) studied the diffusion of H in grain boundaries in extremely fine-grained nanocrystalline Pd as a function of concentration and found that it was strongly concentration dependent. At low concentration it was actually slower than the corresponding diffusion in the lattice by a factor as large as $\simeq 7$, while at higher concentrations it became faster by a factor $\simeq 4$, as seen in Fig. 8.24. Also, in general agreement with these results, Yao and Cahoon (1991) reported that the diffusion of H along grain boundaries in Ni is concentration dependent and that it is essentially stopped at low concentrations in the boundary.

The slow rate of diffusion of H along these grain boundaries has been attributed to the strong trapping of the relatively small numbers of H atoms present at various sites in the boundary cores (Mütschele and Kirchheim 1987, Yao and Cahoon 1991). (At high concentrations, these relatively deep traps are filled, allowing the additional H interstitials present to diffuse more freely.)

We now follow Kirchheim (1982) and Mütschule and Kirchheim (1987) and develop a simplified multiple segregation site boundary diffusion model (Section 7.4.3.3) containing a spectrum of interstitial sites which is capable of explaining the observed results along these lines. We assume a boundary containing a range of different types of interstitial sites and define $n_i$ as the number (per unit volume) of interstitials occupying

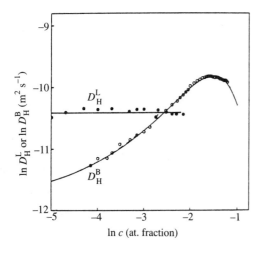

**Fig. 8.24** Lattice diffusivity of H, $D_H^L$, and grain boundary diffusivity of H, $D_H^B$, in Pd as a function of H concentration. The concentration $c$ in each case is the total average H concentration in the system. (From Mütschele and Kirchheim (1987).)

sites of type $i$. Diffusion then occurs by the jumping of the interstitials between the different available sites. We may therefore write the diffusion flux approximately as

$$J = \gamma \sum_i \sum_k \left\{ \left[ n_i \Gamma_{ik} l_{ik} \right]_x - \left[ n_i \Gamma_{ik} l_{ik} \right]_{x+dx} \right\}, \qquad (8.65)$$

where $\Gamma_{ik}$ = jump frequency of an interstitial from site $i$ to site $k$, $l_{ik}$ = jump distance between site $i$ and site $k$, and $\gamma$ is a generalized geometrical factor which takes into account the fact that all jumps are not in the forward direction. Consequently, $\gamma < 1$. Simplifying eqn (8.65) by using an effective averaged jump distance, $l$, and by employing the usual Taylor expansion around $x$ (with d$x = l$), we obtain

$$J = - [\gamma l^2] \cdot \sum_i \sum_k \frac{\partial}{\partial x} (n_i \Gamma_{ik}) . \qquad (8.66)$$

The jump frequency may again be expressed in the usual Arrhenius form

$$\Gamma_{ik} = \nu_{io} \cdot (1 - c_k) \cdot \exp(-G_{ik}^m / kT) \qquad (8.67)$$

where $\nu_{io}$ = atomic vibrational frequency, $G_{ik}^m$ = free energy of migration between sites $i$ and $k$, and $(1 - c_k)$ = probability that the $k$ site is unoccupied. (Note that we do not set $(1 - c_k) \simeq 1$, since we will be interested in diffusion at high interstitial concentrations). Neglecting differences between the atomic vibrational frequencies, $\nu_{io} = \nu_o =$ constant, and

$$J = - \gamma l^2 \nu_o \cdot \sum_i \sum_k \exp(-G_{ik}^m / kT) \cdot \frac{\partial}{\partial x} [n_i (1 - c_k)] . \qquad (8.68)$$

Since local equilibrium should prevail throughout the diffusion zone, the quantities $n_i$ and $c_k$ can be evaluated by use of the Fermi–Dirac type occupation probabilities developed in Section 7.4.3.3 [see eqn (7.77)]. As a convenient approximation, we assume that the different types of sites are randomly distributed, therefore neglecting the correlations between particular types of sites which are undoubtedly present even in general boundaries. Introducing the operator $\partial()/\partial x = [\partial()/\partial n_b] \cdot \partial n_b / \partial x$, and replacing the sums by integrals, we then obtain

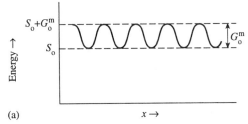

(a)

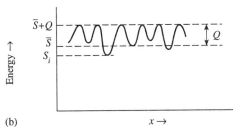

(b)

**Fig. 8.25** (a) Plot of energy versus distance for interstitial atom diffusing between identical interstitial sites. (b) Same as (a) except that the sites along the diffusion path now vary. The form of the energy trace corresponds to the approximation of Kirchheim (1982).

$$J = -\frac{\partial n}{\partial x} \cdot \frac{\gamma l^2 \nu_o}{N} \int\!\!\int_{-\infty}^{+\infty} \exp(-G_{ik}^m/kT) \cdot \frac{\partial}{\partial n} \frac{n(S_i)}{(1+y_i)} \cdot \frac{y_k n(S_k)}{(1+y_k)} \cdot dS_i \, dS_k. \tag{8.69}$$

Here, $n$ = total number of interstitials (per unit volume), $y_i = \exp(S_i + 2z\theta n/N - \mu_B)/kT$ (from eqn 7.77)), $n(S_i)$ is the density of sites, i.e. $n(S_i)\,dS_i$ = number of boundary sites of type $i$ (per unit volume) with energy between $S_i$ and $S_i + dS_i$ as in the multiple segregation site model of Section 7.4.3.3, and $N$ = number of boundary sites (per unit volume). The effective boundary diffusivity, defined by $J \equiv -D^B(\partial n/\partial x)$, is then

$$D^B = \frac{\gamma l^2 \nu_o}{N} \int\!\!\int_{-\infty}^{+\infty} \exp(-G_{ik}^m/kT) \cdot \frac{\partial}{\partial n} \frac{n(S_i)}{(1+y_i)} \cdot \frac{y_k n(S_k)}{(1+y_k)} \cdot dS_i \, dS_k. \tag{8.70}$$

In order to evaluate $D^B$ in eqn (8.70), the quantities $G_{ik}^m$ and $n(S_i)$ must be known. The simplest case is when only one type of site is available and, in addition, the interstitial concentration is very small. The energy trace for a diffusing interstitial will then appear as shown schematically in Fig. 8.25(a), and therefore $G_{ik}^m = G_o^m$, and $n(S) = N \cdot \delta(S - S_o)$, where $\delta$ is a delta function. Therefore, $y_i = y_k \simeq N/n$, and by evaluating eqn (8.70), we obtain the simple result

$$D^B \simeq \gamma l^2 \nu_o \cdot \exp(-G_o^m/kT). \tag{8.71}$$

Equation (8.71) is of the same form as eqn (8.52) and, as expected, consists of a geometrical factor, the jump distance squared, an atomic vibrational frequency, and a Boltzmann factor which in this case involves only the free energy of migration between the identical interstitial sites. Note that the correlation factor in this case is equal to unity.

Equation (8.70) may also be evaluated more generally at higher interstitial concentrations in the presence of a spectrum of site energies and migration activation energies by adopting the scheme illustrated in Fig. 8.25(b) where an energy trace for a typical migrating interstitial is shown. In this scheme, the energy of the interstitial in every saddle point is the same, but the site energies and also the activation energies for migration,

$$G_{ik}^m = \bar{S} + Q - S_i, \tag{8.72}$$

vary from site to site. This approximation is obviously artificial, but it has the advantage (Kirchheim 1982) that large numbers of jumps between two adjacent sites of relatively low energy separated by a low energy barrier are generally avoided. As pointed out by Kirchheim, such jumps would not contribute significantly to long-range diffusion, and hence should not be counted. (We note that this is due to the small partial correlation factors that they would possess as discussed in Section 8.2.5.) Putting eqn (8.72) into eqn (8.70) and rearranging, we obtain

$$D^B = \bar{D}^B \frac{\partial}{\partial n} \exp\left(\frac{\mu_B - \bar{S} - 2z\theta n/N}{kT}\right) \cdot \frac{1}{N} \int\!\!\!\int_{-\infty}^{+\infty} \frac{y_i n(S_i)}{1 + y_i} \frac{y_k n(S_k)}{1 + y_k} dS_i \cdot dS_k, \quad (8.73)$$

where $\bar{D}^B = \gamma l^2 \nu_0 \exp(-Q/kT)$ Reference to eqn (8.71) shows that $\bar{D}^B$ is just the hypothetical diffusivity which would exist in the simple case of a dilute boundary containing only sites of constant energy $\bar{S}$. The double integral has the value $(N - n)^2$ and, therefore, finally

$$D^B = \bar{D}^B \frac{\partial}{\partial \bar{X}} (1 - \bar{X})^2 \exp\left[\frac{\mu_B - (\bar{S} + 2z\theta\bar{X})}{kT}\right] \quad (8.74)$$

where $\bar{X} = n/N$.

In a numerical calculation of $D^B$ using eqn (8.74), Mütschele and Kirchheim (1987) obtained a good fit with the concentration dependence of their measured boundary diffusivities of H in nanocrystalline Pd by setting $2z\theta = -30 \text{ kJ mol}^{-1}$ and $\sigma = 15 \text{ kJ mol}^{-1}$ as seen in Fig. 8.24. Note that these values, and those of the other parameters required, are identical to those found by Mütschele and Kirchheim (1987) in Chapter 7 to produce a good fit between the measured H concentration in the boundaries and in the lattice using eqn (7.79) (see discussion following eqn (7.79)).

Very recently, Arantes *et al.* (1993) have studied H diffusion along grain boundaries in nanocrystalline Ni and obtained results generally similar to the above results for Pd. Parallel studies of H diffusion along boundaries in coarser grained specimens indicated possibly slower boundary diffusion than in the nanocrystalline material. However, this result may have been due to a lack of measurement sensitivity (Kirchheim, private communication, 1993).

In apparent contrast to the above results, relatively rapid diffusion of H along grain boundaries in Si has been found. Calder *et al.* (1973) find $D^B/D^L \simeq 10^8$ for $185° > T > 35°C$ while Ginley and Hellmer (1985) find $D^B/D^L \simeq 2 \cdot 10^4$ for $T = 350°C$. Unfortunately, no studes have been made of $D^B$ as a function of H concentration. Also, no modelling has been done to investigate whether differences in the bonding (i.e. metallic versus covalent) could be a dominating factor.

There is considerable evidence that the larger interstitial atom, C, is a relatively fast diffuser along grain boundaries in a number of metals. Lesage and Huntz (1976) report $D^B/D^L \simeq 6.10^6$ for C in Mo; Bokshtein (1985) reports that $D^B/D^L \simeq 10^4$ for C in b.c.c. Fe and $D^B/D^L \simeq 10^6$ for C in Zr. These results could conceivably be due to generally weaker trapping effects for the larger C atoms than for the smaller H atoms discussed previously.

Interstitial C and N atoms have also been found to diffuse rapidly along heterophase boundaries in metallic systems. Dirks and Meijering (1972) found $D^B/D^L \simeq 10^7$ for the diffusion of N along $\alpha/\gamma$ interfaces (where $\alpha = $ b.c.c. Fe–Ni phase, and $\gamma = $ Fe–Ni phase); also, C showed about the same behaviour. In addition, Dirks and Meijering (1975)

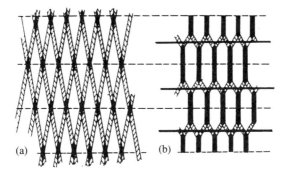

**Fig. 8.26** (a) Schematic view of unrelaxed pure twist boundary in a layered structure such as Na-$\beta$ alumina. The boundary is perpendicular to the layers in each crystal and is viewed at normal incidence. Dark areas represent 'non-blocking' sites for the fast diffusion across the boundary of ions diffusing between the layers. Shaded areas represent 'blocking' sites. (b) Same boundary as in (a) after relaxation into highly localized screw dislocations (indicated by the horizontal lines). (From DeJonghe (1979).)

found fast diffusion of N along a $\gamma'/\gamma''$ interface, where $\gamma'$ and $\gamma''$ are two different f.c.c. Fe–Ni phases. Further examples are cited by Bokshtein (1985).

## 8.4  SLOW DIFFUSION ACROSS INTERFACES IN FAST ION CONDUCTORS

Single crystals of the fast ion conductor Na $\beta$-alumina possess a layered structure consisting of layers of four close-packed planes of O and Al ions in a spinel-like structure (i.e. spinel blocks) and intervening layers consisting of loosely packed planes occupied by Al–O–Al pillars, which support the spinel blocks (May and Hooper 1978, DeJonghe 1979). Freely migrating Na ions are also present. The crystal structures may be either hexagonal or rhombohedral, depending upon the stacking sequence of the close-packed layers. The Na ions diffuse at very high rates along the planes between the closely packed spinel blocks and not at all perpendicular to them. (The anisotropy ratio is approximately $10^{15}$ at 300 °C!) The unidirectional diffusion rate of the Na ions through polycrystalline specimens is found (May and Hooper 1978; DeJonghe 1979) to be markedly reduced by the presence of the grain boundaries which act as barriers to the normally rapid lattice diffusion. It is easily seen that, in this highly anisotropic material, the high diffusivity planes in the crystals adjoining any grain boundary will generally be out of registry at the grain boundary over much of the grain boundary area and that any rapid diffusion of the Na ions across the boundary will therefore be reduced. This is illustrated schematically in Fig. 8.26(a) for an unrelaxed pure twist boundary where it is clear that the boundary will constitute an effective local diffusion barrier. The strength of the barrier could presumably be reduced by a relaxation which would produce an array of localized screw dislocations as illustrated in Fig. 8.26(b). However, it is doubtful that such a full relaxation would occur (particularly at large twist angles) since the Burgers vector of the screw dislocations would correspond to the spinel block spacing, i.e. 11.3 Å and hence be very large. A detailed discussion of a variety of boundary types and the barriers they present to the diffusion is given by DeJonghe (1979).

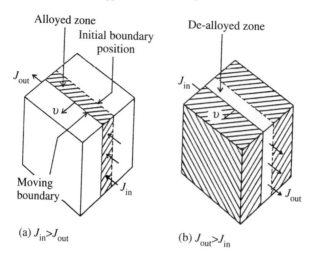

**Fig. 8.27** Schematic diagram of DIGM in a section of a grain boundary. (a) Solute atoms diffuse into a grain boundary in an initially pure host material. The diffusion flux, $J$, causes the boundary to move with a velocity, $v$, leaving an alloyed zone behind in its wake. (b) Opposite case where solute atoms are diffused out of a grain boundary in an initially alloyed material, and the moving boundary leaves a de-alloyed zone behind.

## 8.5 DIFFUSION-INDUCED GRAIN BOUNDARY MOTION (DIGM)

As a final topic we briefly consider the relatively complex phenomenon of diffusion-induced grain boundary motion (or, more commonly, 'DIGM'), in which the chemical diffusion of solute atoms along a grain boundary induces transverse motion of the boundary. The basic phenomenon is illustrated schematically in Fig. 8.27 for a binary alloy system. In Fig. 8.27(a) solute atoms are diffused into a grain boundary in an initially pure host material, and the boundary is induced to move transversely with a velocity, $v$, leaving an alloyed zone behind in its wake. The opposite case is illustrated in Fig. 8.27(b) where solute atoms are diffused out of a grain boundary in an initially alloyed material, and the moving boundary leaves behind a de-alloyed zone in its wake. Examples of the alloying of Cu by Zn by means of DIGM are shown in Fig. 8.28. Here, the Zn was diffused into a Cu bicrystal from the vapour phase, and the boundary is observed edge-on. In Fig. 8.28(a) the boundary is a $\Sigma 19$ symmetric tilt boundary, and short segments of the originally straight boundary have moved into either crystal 1 or crystal 2 in the form of bulges of varying shapes to produce alloyed zones. For the symmetric boundary the movement of the bulges into either crystal occurred with about equal probability. For the asymmetric tilt boundary shown in Fig. 8.28(b) DIGM again occurred along a number of boundary segments, but the boundary always moved into crystal 2. The forms of the alloyed zones in Fig. 8.28 are typical in the sense that DIGM appears to nucleate at various places along a boundary and advances into either one, or both, of the adjoining crystals in the form of bulges. Upon further diffusion the bulges generally expand further, and, as in the case of Fig. 8.28(a), may impinge and advance along an irregular front. Another result of interest is the observation that the extent of DIGM depends upon boundary structure. For example, in the case of a series of symmetric [011] tilt boundaries in the Cu–Zn system (King 1991) the short period $\Sigma 5(310)$ and $\Sigma 5(210)$ boundaries exhibited the least DIGM. Also, in a series of [110] asymmetric

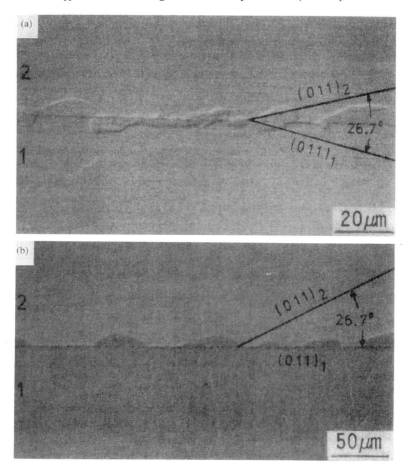

**Fig. 8.28** Alloyed zones formed by DIGM by diffusing Zn from the vapour phase into Cu bicrystals containing ⟨110⟩ tilt boundaries which are viewed edge-on. (a) DIGM at symmetric Σ19 boundary. (b) DIGM at-asymmetric Σ19 boundary with (011) in crystal 1 parallel to the boundary. (From Schmelzle *et al.* (1992).)

tilt boundaries with ($\bar{1}$10) in one crystal parallel to the boundary (King 1991) the Σ3 boundary exhibited negligible DIGM in comparison with the other boundaries in the series.

In general, DIGM is most clearly evident at relatively low temperatures where $D^B \gg D^L$, and significant alloying, or de-alloying, of the bulk material is achieved in the absence of significant lattice diffusion. The deposition of solute atoms in the bulk material is achieved by means of the relatively fast diffusion of solute atoms along the boundary and their deposition in the lattice as the boundary moves through the lattice. This process has already been discussed in Section 8.2.1 (see Fig. 8.2). However, in the case of DIGM, the boundary motion is coupled directly to the diffusion in the boundary and is not independent of it, as might be the case if the boundary motion is induced, for example, by capillary forces (see Chapter 9).

DIGM has been observed under a wide range of conditions in a broad range of materials including metallic, covalent, and ionic systems. Comprehensive reviews of the available observations and the state of our knowledge of the phenomenon up to the period

1987–88 have been given by King (1987) and Handwerker (1988). It is clear from this, and subsequent work, that DIGM is ubiquitous in many systems and may contribute to diffusional alloying (or de-alloying) at relatively low temperatures where usual lattice diffusion is essentially frozen out. In addition, it may cause considerable difficulty in the reliable measurement of chemical diffusion rates along grain boundaries, particularly in the type C diffusion regime where lattice diffusion is relatively insignificant (see Section 8.2.1 and Fig. 8.2). As such, it stands as an important phenomenon associated with chemical grain boundary diffusion.

A basic question which has been asked about DIGM since it was first recognized is: why does it occur at all? Clearly, the overall motivation is the decrease in free energy which occurs as a result of the alloying, or de-alloying, which is achieved (Hillert and Purdy 1978, Kajihara and Gust 1991). However, a detailed model for the way in which the diffusion causes the boundary to experience a force so that at least a portion of the available free energy decrease is dissipated in the accompanying boundary motion has been difficult to establish and is still a matter of discussion.

Some of the models which have been suggested include the following (which we merely list very briefly): (1) the climb along the boundary of grain boundary dislocations possessing step character which is driven by a diffusional grain boundary Kirkendall effect (Balluffi and Cahn 1981, Smith and King 1981); (2) a difference between the coherency strain energy in the two crystals adjoining the boundary due to the inward diffusion of misfitting solute atoms coming from the boundary core (Hillert 1983, Handwerker 1988); (3) solute-atom-induced variations of the grain boundary energy (Louat *et al.* 1985); (4) solute-atom-induced structural transformations within the boundary (Kasen 1986); (5) a difference between the stored energy of plastic strain in the two crystals adjoining the boundary due to a diffusional lattice Kirkendall effect associated with the inward diffusion of solute atoms coming from the boundary core (Handwerker 1988); (6) a diffusional flux of vacancies across the boundary (Fournelle 1993); and (7) the gradient energy stored in the region of steep chemical gradient which exists when the crystals adjoining the boundary are penetrated by diffusion from the boundary core as in (2) and (5) above (Rabkin *et al.* (1993), Rabkin (1994)). Still further suggested models are described by King (1987) and Handwerker (1988). No single one of these models, at least in a simple form, has been able to explain all of the many complex features of DIGM which have been observed in different systems. It is therefore possible that one, or the other, or more than one in tandem, may play a significant role in specific cases. For example, in model (1) listed above, the boundary moves by means of the motion along it of grain boundary dislocations possessing step character. (This general type of boundary movement is discussed extensively in Chapter 9.) Obviously, this mechanism could not be operative in general boundaries which are incapable of supporting localized grain boundary dislocations. On the other hand, it is conceivable that it could play at least some role in boundaries containing localized dislocations with step character.

Despite this complicated situation, considerable experimental evidence has accumulated that the coherency strain energy model ((2) above) is dominant in a number of cases. In this model it is visualized that solute atoms initially diffuse into the boundary. They then begin to penetrate the two adjoining crystals. Since there is generally a size difference between the solute and solvent atoms, coherency stresses are generated in the thin diffused layers on the adjoining crystals if they remain coherent with the bulk. If the boundary is asymmetric, as is usually the case, the elastic strain energies in these layers will differ across the boundary because of elastic anisotropy. They will also differ if coherency is lost in one crystal by the generation of dislocations. In such cases the boundary will be urged to move towards the crystal with the higher strain energy in order to consume it. Once begun, this process can continue in a self-sustaining manner. Rhee *et al.* (1987) have tested this model in the Mo–Ni–(Co–Sn) system where the coherency strain energy can be systematically varied and even set equal to zero by control of the chemical composition. A zero rate of DIGM was found at zero coherency strain in a manner consistent with the basic model. In other work, King and coworkers (Chen and King 1988, Chen *et al.* 1990, King and Dixit 1990, King 1991) have studied DIGM in several series of [100] and [110] tilt boundaries in the Cu–Zn system. For asymmetric [100] tilt boundaries (with {100} in one crystal always parallel to the boundary) and also for asymmetric [110] tilt boundaries (with {110} in one crystal always parallel to the boundary), the direction of DIGM was always in the direction predicted by the coherency strain model. Interestingly, this included a reversal in the DIGM direction with increasing tilt angle in the [110] tilt boundary series. DIGM was also observed in symmetric [100] tilt boundaries, a phenonenon which would not be predicted by the coherency strain model operating by itself. In this case it appears that the initial symmetry must have been broken locally by some other DIGM mechanism. Once the boundary became asymmetric, the coherency strain model could then have become controlling.

It is important to realize that the velocity of DIGM can be limited by either the magnitude of the driving force or the mobility of the boundary, or both in a mixed situation. The magnitude of the driving force, in turn, is dependent upon the diffusion along the boundary. King (1991) has cited examples for certain tilt boundaries in the Cu–Zn system where DIGM was limited by the lack of a driving force because of slow grain boundary diffusion and, conversely, where DIGM was apparently limited by a low boundary mobility.

A particularly convincing demonstration of the relationship between DIGM and boundary diffusion in the Cu–Zn system has been provided by Schmelzle *et al.* (1992). Here, the Zn was diffused from the vapour phase into [110] symmetric tilt boundaries in initially pure Cu bicrystals. The boundaries, which were transverse to the surfaces exposed to the Zn, underwent DIGM, and measurements of the boundary displacement and the Zn concentration in the zone alloyed with Zn were made as a function of the distance, $z$, measured along each boundary from its intersection with the surface. It was found that over the regime studied the boundary velocity, $v$, was essentially constant and independent of $z$. In addition, the Zn concentration in the alloyed zone decreased exponentially with increasing $z$. The results were therefore consistent with a simple quasi-steady-state model in which the gain in Zn of any element of the boundary by boundary diffusion is just balanced by the loss of Zn due to the boundary motion. Under these conditions the governing equation is simply

$$k\delta D^{B} \cdot \partial^2 n/\partial z^2 = vn, \qquad (8.75)$$

and therefore

$$n(z) = n_{o} \exp(-z/L), \qquad (8.76)$$

where $L$ is a characteristic length given by

$$L = (k\delta D^{B}/v)^{\frac{1}{2}}, \qquad (8.77)$$

and $n_{o} = n(0)$. Using eqn (8.76) and (8.77), the parameter $k\delta D^{B}$ was determined from the data, and is shown in Fig. 8.29 as a function of the tilt angle, $\theta$. The shape of the

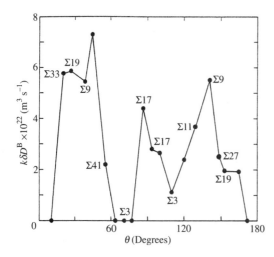

**Fig. 8.29** Values of the diffusion parameter $k\delta D^B$ versus tilt angle, $\theta$, obtained from DIGM data through use of eqn (8.76) and (8.77) for various [110] symmetric tilt boundaries in the Cu–Zn system. Low $\Sigma$ boundaries are indicated. Note the similarity of these results with those in Fig. 8.12. (From Schmelzle *et al.* (1992).)

$k\delta D^B$ versus $\theta$ curve is seen to be quite similar to that previously shown for Zn diffusion along [110] symmetric tilt boundaries in the Al–Zn system (Fig. 8.12). Also, the magnitudes of $k\delta D^B$ agree substantially with those obtained previously by more conventional methods by Klotsman *et al.* (1969) and Hässner (1977). In considering these results it must be stressed that they offer no explanation of why the boundary moved with a constant velocity in response to the boundary diffusion.

We conclude this short discussion of DIGM by noting that many aspects of the phenomenon are still not completely understood or successfully modelled. As a further example, Kuo and Fournelle (1991) have observed the DIGM of curved boundaries in the Al–Cu system and concluded that the coherency strain energy which could be developed in this system was not sufficient to drive the boundaries against their observed curvatures. Unanswered questions therefore remain.

## REFERENCES

Agullo-Lopez, F., Catlow, C. R. A., and Townsend, P. D. (1988). *Point defects in materials*. Academic Press, London.

Arantes, D. R., Huang, X. Y., Marte, C., and Kirchheim, R. (1993). *Acta Metall. Mater.*, **41**, 3215.

Atkinson, A. (1984). *Solid State Ionics*, **12**, 309.

Atkinson, A. (1985). *J. Physique*, **46**, Colloq. C4, Supple. au no. 4, C4–379.

Atkinson, A. (1988*a*). *Solid State Ionics*, **28–30**, 1377.

Atkinson, A. (1988*b*). *Mat. Res. Soc. Symp. Proc.*, **122**, 183.

Atkinson, A. (1989). In *Surfaces and interfaces of ceramic materials* (eds. L.-C. Dufour, C. Monty, and G. Petot-Ervas) p. 273. Kluwer, Dordrecht.

Atkinson, A. and Taylor, R. I. (1981). *Phil. Mag. A*, **43**, 979.

Atkinson, A., Moon, D. P., Smart, D. W., and Taylor, R. I. (1986). *J. Mats. Sci.*, **21**, 1747.

Balluffi, R. W. (1970). *Phys. Stat. Sol.*, **42**, 11.

Balluffi, R. W. (1984). In *Diffusion in crystalline solids* (eds G. E. Murch and A. S. Nowick) p. 320. Academic Press, New York.

Balluffi, R. W. and Brokman, A. (1983). *Scripta Metall.*, **17**, 1027.

Balluffi, R. W. and Cahn, J. W. (1981). *Acta Metall.*, **29**, 493.

Balluffi, R. W., Kwok, T., Bristowe, P. D., Brokman, A., Ho, P. S., and Yip, S. (1981). *Scripta Metall.*, **15**, 951.

Barbier, F., Monty, C., and Dechamps, M. (1988). *Phil. Mag. A*, **58**, 475.

Bernardini, J. (1989). *Defect and Diff. Forum*, **66–69**, 667.

Bernardini, J., Gas, P., Hondros, E. D., and Seah, M. P. (1982). *Proc. Roy. Soc. A*, **379**, 159.

Bernardini, J., Bennis, S., and Moya, G. (1989). *Defect and Diff. Forum*, **66–69**, 805.

Bokshtein, S. Z. (1985). *Diffusion and structure of metals*. Oxonian Press, New Dehli.

Brokman, A., Bristowe, P. D., and Balluffi, R. W. (1981). *J. Appl. Phys.*, **52**, 6116.

Cabane-Brouty, F. and Bernardini. J. (1982). *J. de Physique*, **43**, Colloq. C6, Supple. au no. 12, C6-163.

Cahn, J. W. and Balluffi, R. W. (1979). *Scripta Metall.*, **13**, 499.

Calder, R. D., Elleman, T. S., and Verghese, K. (1973). *J. Nucl. Mats.*, **46**, 46.

Campbell, D. R., Tu, K. N., and Schwenter, R. O. (1975). *Thin Solid Films*, **25**, 213.

Carslaw, H. S. and Jaeger, J. C. (1959). *Conduction of heat in solids*, 2nd edn. Oxford University Press, Oxford.

Chen, F.-S. and King, A. H. (1988). *Acta Metall.*, **36**, 2827.

Chen, F.-S., Dixit, G., Aldykiewicz, A. J., and King, A. H. (1990). *Met. Trans.*, **21A**, 2363.

Ciccotti, G., Guillope, M., and Pontikis, V. (1983). *Phys. Rev. B*, **B27**, 5576.

Crank, J. (1957). *The mathematics of diffusion*. Oxford University Press, Oxford.

Darby, T. P. and Balluffi, R. W. (1977). *Phil. Mag.*, **36**, 53.

DeJonghe, L. C. (1979). *J. Mats. Sci.*, **14**, 33.

d'Heurle, F. M. and Gangulee, A. (1972). In *The nature and behavior of grain boundaries*, p. 339. Plenum Press, New York.

Dirks, A. G. and Meijering, J. L. (1972). *Acta Metall.*, **20**, 1101.

Dirks, A. G. and Meijering, J. L. (1975). *Acta Metall.*, **23**, 217.

Dranova, Z. I. and Mikhaylovskiy, I. M. (1984). *Phys. Met. Metall.*, **57**, 121.

Duffy, D. M. and Tasker, P. W. (1984). *Phil. Mag. A.*, **50**, 143.

Duffy, D. M. and Tasker, P. W. (1985). *J. de Physique*, **46**, Colloq. C4, Supple. au no 4, C4-185.

Duffy, D. M. and Tasker, P. W. (1986). *Phil. Mag. A*, **54**, 759.

Dyment, F., Iribarren, M. J., Vieregge, K., and Herzig, C. (1991). *Phil. Mag. A*, **63**, 959.

Fisher, J. C. (1951). *J. Appl. Phys.*, **22**, 74.

Fournelle, R. A. (1993). *Mats. Sci. Forum*, **126–128**, 383.

Friedel, J. (1964). Dislocations, p. 292. Addison-Wesley, London.

Gas, P., Poize, S., Bernardini, J., and Cabane, F. (1989). *Acta Metall.*, **37**, 17.

Ginley, D. S. and Hellmer, R. P. (1985). *J. Appl. Phys.*, **58**, 871.

Gjostein, N. A. (1973). In *Diffusion* (ed. H. I. Aaronson) p. 241, American Society for Metals, Metals Park, Ohio.

Gleiter, H. and Chalmers, B. (1972). *High angle grain boundaries*, p. 94. Pergamon Press, New York.

Gupta, D., Campbell D. R., and Ho, P. S. (1978). In *Thin films—interdiffusion and reactions*, p. 161, Wiley, New York.

Handwerker, C. A. (1988). In *Diffusion phenomena in thin films and microelectronic materials* (eds. D. Gupta and P. S. Ho) p. 245. Noyes, Park Ridge, New Jersey.

Harris, T. M. and Latanision, R. M. (1991). *Met. Trans. A*, **22**, 351.

Harrison, L. G. (1961). *Trans. Faraday Soc.*, **57**, 1191.

Hart, E. W. (1957). *Acta Metall.*, **5**, 597.

Hässner, A. (1977). *Wiss. Z. Techn. Hochsch. Kark-Marx-Stadt*, **19**, 619.

Herbeuval, I. and Biscondi, M. (1974). *Canad. Metall. Quart.*, **13**, 171.

Hillert, M. (1983). *Scripta Metall.*, **17**, 237.

Hillert, M. and Purdy, G. R. (1978). *Acta Metall.*, **26**, 333.

Hirth, J. P. and Lothe, J. (1982). *Theory of dislocations*. Wiley, New York.

Huang, J., Meyer, M., and Pontikis, V. (1989). *Phys. Rev. Lett.*, **63**, 628.

Hwang, J. C. M. and Balluffi, R. W. (1979). *J. Appl. Phys.*, **50**, 1339; 1349.

Kajihara, M. and Gust, W. (1991). *Acta Metall. Mater.*, **39**, 2565.

Kasen, M. B. (1986). *Phil Mag. A*, **54**, L31.

Kaur, I. and Gust, W. (1989). *Fundamentals of grain and interphase boundary diffusion.* Ziegler Press, Stuttgart.

Kaur, I., Gust, W., and Kozma, L. (1989). *Handbook of grain and interphase boundary diffusion data*, Vols. 1 and 2, Ziegler, Stuttgart.

Kedves, F. J. and Erdelyi, G. (1989). *Defect and Diff. Forum*, **66–69**, 175.

King, A. H. (1987). *Int. Mets. Revs.*, **32**, 173.

King, A. H. (1991). *Mat. Res. Symp.*, **229**, 343.

King, A. H. and Dixit, G. (1990). *Colloque de Physique*, Colloque C1, Supple. au n°1, **51**, C1-545.

Kingery, W. D. (1974). *J. Amer. Ceram. Soc.*, **57**, 1; 74.

Kirchheim, R. (1982). *Acta Metall.*, **30**, 1069.

Klotsman, S. M., Rabovskiy, Y. A., Talinskiy, V. K., and Timofeyev, A. N. (1969). *Phys. Metals Metallogr.*, **28**, 66.

Kofstad, P. (1972). *Nonstoichiometry, diffusion, and electrical conductivity in binary metal oxides.* Wiley, New York.

Kuo, M. and Fournelle, R. A. (1991). *Acta Metall. Mater.*, **39**, 2835.

Kwok, T. and Ho, P. S. (1984). *Surf. Sci.*, **144**, 44.

Kwok, T., Ho, P. S., and Yip, S. (1984). *Phys. Rev. B*, **29**, 5354; 5363.

Kwok, T., Ho, P. S., Yip, S., Balluffi, R. W., Bristowe, P. D., and Brokman, A. (1981). *Phys. Rev. Lett.*, **47**, 1148.

Ladna, B. and Birnbaum, H. K. (1987). *Acta Metall.*, **35**, 2537.

Lesage, B. and Huntz, A. M. (1976). *Mem. Sci. Rev. Metall.*, **73**, 19.

Li, J. C. M. and Chalmers, B. (1963). *Acta Metall.*, **11**, 243.

Liotard, J. J., Biberian, R. and Cabane, J. (1982). *J. Physique*, **43**, Colloque C1, Supple. au no. 10, C1-213.

Louat, N., Pande, C. S. and Rath, B. B. (1985). *Phil Mag. A*, **51**, L73.

Ma, Q. and Balluffi, R. W. (1993). *Acta Metall Mater.*, **41**, 133.

Ma, Q. and Balluffi, R. W. (1994). *Acta Metall. Mater.*, **42**, 1.

Ma, Q., Liu, C. L., Adams, J. B., and Balluffi, R. W. (1993). *Acta Metall. Mater.*, **41**, 143.

Martin, G. and Perraillon, B. (1975). *J. de Physique*, **36**, Colloq. C4, Supple. au no. 10, C4-165.

Martin, G. and Perraillon, B. (1980). In *Grain boundary structure and kinetics* (ed. R. W. Balluffi). American Society for Metals, Metals Park, Ohio.

May, G. J. and Hooper, A. (1978). *J. Mats. Sci.*, **13**, 1480.

Monty, C. and Atkinson, A. (1989). *Cryst. Latt. Def. and Amorph. Mat.*, **18**, 97.

Moya, E. G., Deyme, G., and Moya, F. (1990). *Scripta Metall. Mater.*, **24**, 2447.

Mütschele, T. and Kirchheim, R. (1987). *Scripta*, **21**, 135.

Nomura, M. and Adams, J. B. (1992). *J. Mater. Res.*, **7**, 3202.

Nomura, M., Lee, S.-Y., and Adams, J. B. (1991). *J. Mater. Res.*, **6**, 1.

Nye, J. F (1957). *Physical properties of crystals.* Oxford University Press, Oxford.

Palumbo, G., Doyle, D. M., El-Sherik, A. M., Erb, U., and Aust, K. T. (1991). *Scripta Metall. Mater.*, **25**, 679.

Perinet, F. (1987). Thesé, Université de Paris-Sud, Orsay.

Peterson, N. L. (1978). *J. Nucl. Mats.*, **69, 70**, 3.

Peterson, N. L. (1980). In *Grain boundary structure and kinetics* (ed. R. W. Balluffi) p. 209. American Society for Metals, Metals Park, Ohio.

Philibert, J. (1975). *J. Physique*, **36**, Collloq. C4, Supple. au no. 10, C4-411.

Plimpton, S. J. and Wolf, E. D. (1990). *Phys. Rev. B*, **41**, 2712.

Pontikis, V. (1982). *J. Physique*, **43**, Colloq. C6, Supple. au no. 12, C6-65.

Rabkin, E. I. (1994). *Scripta Metall. Mater.*, **30**, 1443.

Rabkin, E. I., Shvindlerman, L. S., and Gust, W. (1993). *Interface Sci.*, **1**, 131.

Rhee, W.-H., Song, Y.-D., and Yoon, D. N. (1987). *Acta Metall.*, **35**, 57.

Robinson, J. T. and Peterson, N. L. (1972). *Surf. Sci.*, **31**, 586.

Roshko, A. (1987). Ph.D. Thesis, MIT.

Routbort, J. L. and Matzke, H. J. (1975). *J. Amer. Ceram. Soc.*, **58**, 81.

Schmelzle, R., Giakupian, B., Muschik, T., Gust. W., and Fournelle, R. A. (1992). *Acta Metall. Mater.*, **40**, 997.

Shewmon, P. (1989). *Diffusion in solids*, 2nd edn. Minerals, Metals and Materials Society, Warrendale, Pennsylvania.

Smith, D. A. and King, A. H. (1981). *Phil Mag. A*, **44**, 333.

Sommer, J. (1992). Ph. D. Thesis, Westfälischen Wilhelms-Universität Münster.

Sommer, J., Herzig, C., Mayer, S., and Gust, W. (1989). *Defect and Diff, Forum*, **66–69**, 843.

Spit, F. H. M. and Bakker, H. (1986). *Phys. Stat. Solidi*, **97**, 135.

Spit, F. H. M., Albers, H., Lubbes, A., Rijke, Q. J. A., Ruijven, L. J. V., Westerveld, J. P. A., Bakker, H., and Radelaar, S. (1985). *Phys. Stat. Solidi.*, **89**, 105.

Stubicon, V. S. (1985). In *Transport in nonstoichiometric compounds* (eds G. Simkovich and V. S. Stubicon) p. 345. Plenum Press, New York.

Tan, T. Y. and Gösele, U. (1985). *Appl. Phys. A*, **37**, 1.

Tsuru, T. and Latanision, R. M. (1982). *Scripta Metall.*, **16**, 575.

Turnbull, D. and Hoffman, R. E. (1954). *Acta Metall.*, **2**, 419.

Vieregge, K. and Herzig, C. (1989). *Defect and Diff. Forum*, **66–69**, 811.

Yan, M. F., Cannon, R. M., Bowen, H. K., and Coble, R. L. (1977). *J. Amer. Ceram. Soc.*, **60**, 120.

Yao, J. and Cahoon, J. R. (1991). *Acta Metall. Mater.*, **39**, 119.

# Conservative motion of interfaces

## 9.1 INTRODUCTION

We begin by first introducing the concept of interface motion and then distinguishing between 'conservative' motion (which is the subject of this chapter) and 'non-conservative' motion (which is taken up in Chapter 10). We then introduce a number of important general aspects of conservative motion and go on to discuss the basic mechanisms by which the various types of interfaces listed in Table 4.1 move in this fashion: models are considered, and related experimental observations are cited. In many cases, reliable experimental observations are sparse and incomplete, and definitive models are not available. In such cases, we present plausible models which appear to be consistent with experimental observations. Other important aspects of conservative interface motion are taken up as well, including the equations of motion which apply in certain simple situations and the impediments to motion provided by solute atom drag and interface pinning at embedded particles and free surfaces.

### 9.1.1 'Conservative' versus 'non-conservative' motion of interfaces

In general, interface motion is motion of an interface relative to the crystal lattices which adjoin it. In some cases the two crystals adjoining a moving interface will remain stationary with respect to each other. In such a case the motion of the interface relative to each adjoining crystal will be the same: in other cases the two adjoining crystals will move relative to one another, and the motion of the interface relative to each crystal will be different. A full description of the interface motion therefore requires a description of the motion relative to each crystal. This could be obtained experimentally, for example, by measuring the distances between the average midplane of the moving interface and inert fiduciary markers buried nearby in each adjoining crystal. Such markers would be similar to those used to measure (daSilva and Mehl 1951) the Kirkendall effect during chemical diffusion in crystal lattices (see Section 10.4.2.2).

'Conservative' interface motion is defined as the motion of an interface which occurs in the absence of a diffusion flux of any component in the system to/from the interface. In view of the conservation of each type of atom, this type of motion can only be sustained under quasi-steady-state conditions when the two crystals adjoining the inter-face are of the same composition (measured in terms of the atom fraction of each component). One adjoining crystal then simply grows at the expense of the other. Since there is no compositional difference across the boundary, there is no need for the long-range diffusion of different types of atoms to/from the moving interface to sustain the process: atoms will simply be transferred locally by one means or another across the interface. In certain types of boundaries it may be necessary to transport atoms from one region of the boundary to another by short-range diffusion in order to allow this process to occur. However, there will be no net transfer to/from the boundary as a whole, and the overall boundary motion will therefore be conservative.

Since lattice sites in the vicinity of the interface are conserved, the positions of the two crystals adjoining the interface will remain fixed during this type of motion except in the case of heterophase interfaces where differences between the atomic volumes of the two adjoining phases will generally produce a small dilational effect. The motion of the interface relative to each crystal will therefore be the same except when such dilational effects are present.

Conservative interface motion may be contrasted with 'non-conservative' interface motion which occurs when the interface movement is coupled to long-range diffusional fluxes of one, or more, of the components in the system to/from the interface. In this situation each flux is either generated or terminated at the interface. The interface therefore acts as a source/sink for that component, and the interface motion is non-conservative with respect to that component. An example of non-conservative motion is the case where there is a composition difference across a moving heterophase interface during a phase change, and fluxes of certain types of atoms to/from the interface are required to sustain the process. Another example is when a grain boundary in a pure material creates/annihilates a flux of point defects, and a corresponding flux of host atoms to/from the boundary is therefore required. In all cases the interface acts as a net source/sink for atoms fluxes of one type or another, and interface motion, which is coupled to these fluxes, occurs. This type of motion is taken up in detail in Chapter 10.

### 9.1.2 Driving pressures for conservative motion

The conservative motion of an interface will generally occur in response to a driving pressure which will exist whenever the conservative motion of the interface causes the total free energy of the system to decrease. This pressure is given by

$$p = -\frac{\delta G}{\delta x},\tag{9.1}$$

where $\delta G$ is the change in free energy of the system when unit area of interface moves a distance $\delta x$ normal to itself.

Driving pressures may originate from a wide variety of sources. During a phase transformation, a heterophase interface between an unstable parent phase and a more stable product phase may be induced to move towards the parent phase by a driving pressure provided by the decrease in bulk free energy which occurs as the interface advances (see Table 9.1). In other circumstances, an interface may be subjected to a pressure by applied stresses whenever the motion of the interface causes the macroscopic shape of the specimen to change. In this case work is done by the applied stresses during the shape change (see Table 9.1).

During recrystallization, the interface between a recrystallized grain and the plastically deformed matrix moves under an average driving force supplied by the stored energy of plastic deformation. This phenomenon (Himmel 1963, Haessner 1978, Cahn 1983) often occurs in material which is first plastically deformed at a sufficiently low temperature so that the excess lattice defects (point defect aggregates and dislocations which are generated during the plastic deformation) are retained. Upon subsequent thermal annealing, new defect-free grains nucleate throughout the material and then grow at the expense of the deformed matrix until the material is converted entirely into an aggregate of new relatively defect-free grains. The average driving pressure for this growth process originates in the stored energy of plastic deformation which becomes available when relatively large areas of the interfaces separating the recrystallized grains from the deformed matrix advance (see Table 9.1). We note that the defect state produced by

**Table 9.1** Approximate common pressures, $p$, exerted on interfaces

| Type of interface | Source of pressure | $p$ (Pa) | $g_m = p\Omega$ (eV)* |
|---|---|---|---|
| Heterophase interface between non-equilibrium phases during phase transformation | Bulk free energy difference between phases, per unit volume, $\Delta G$. $$p = -\frac{\delta G}{\delta x} = -\Delta G$$ | $10^7 - 10^9$ | $10^{-3} - 10^{-1}$ |
| Interface whose movement causes a macroscopic shape change | Applied stress, $\tau$. $$p = \tau\theta$$ (see eqn (9.2)) | $10^5 - 10^8$ | $10^{-5} - 10^{-2}$ |
| Grain boundary between recrystallized grain and deformed matrix | Average stored energy of plastic deformation, per unit volume, $\Delta G_d$. $$\bar{p} = -\frac{\delta G}{\delta x} = -\Delta \bar{G}_d$$ | $10^6 - 10^7$ | $10^{-4} - 10^{-3}$ |
| Interface in stressed polycrystal | Difference in elastic strain energy density across interface caused by elastic incompatibility, $\Delta G_s$. $$p = -\Delta G_s$$ | $10^4 - 10^5$ | $10^{-6} - 10^{-5}$ |
| Curved interface | Boundary curvature. $$p = \Delta P \simeq \sigma\left(R_1^{-1} + R_2^{-1}\right)$$ (see eqn (5.104)) | $10^3 - 10^5$ | $10^{-7} - 10^{-5}$ |

* $g_m = p\Omega$ = energy available from system to transfer an atom across the interface during its motion.

plastic deformation is inhomogeneous and that local driving pressures can therefore exist which deviate considerably from the average.

When a polycrystal is subjected to applied stress, differences in stress across the interfaces are generally produced because of elastic incompatibility as described in Section 12.2. This leads to differences in the elastic strain energy density across the interfaces, $\Delta G_s$, and therefore to pressures given by $p = -\Delta G_s$. An upper limit for $\Delta G_s$ is approximately $E\varepsilon_c^2/2$, where $E$ is the elastic modulus and $\varepsilon_c$ is the critical elastic strain required for plastic deformation. Taking $\varepsilon_c \simeq 10^{-3}$, we then obtain the estimates listed in Table 9.1.

A curved interface will always tend to move in response to the capillary pressure arising from its curvature. As already pointed out in Section 5.6.5, a curved interface is subjected to a capillary pressure which operates against the curvature and tends to straighten it out and reduce its area (see Table 9.1). This occurs, for example, during 'grain growth' when a fine-grained polycrystalline aggregate is annealed at an elevated temperature causing the average grain size to increase (Rhines and Craig 1974, Atkinson 1988, Mullins and Vinals 1989). The overall driving force in this process is the decrease in total grain boundary area (and associated free energy) which occurs as the average grain size increases. More locally, capillary pressures due to boundary curvature exist at various boundaries present in the aggregate, and these cause them to move in such a manner that, on average, larger grains grow at the expense of smaller grains, and the average grain size increases.

The approximate common pressures listed in Table 9.1 are seen to extend over a wide range covering about six orders of magnitude. The decrease in the energy of the system which is achieved when an atom is transferred across an interface during its motion in the presence of the pressure $p$ is simply $g_m = p\Omega$, and values of this quantity are also listed. In general, a significant rate of motion will only be achieved if the available energy is large enough to drive the mechanism responsible for the motion at a significant rate.

### 9.1.3 Basic mechanisms: correlated versus uncorrelated processes

The transfer of atoms across the interface which occurs during conservative motion can occur by a variety of basic mechanisms depending upon the interface type (Table 4.1). These are summarized in Table 9.2 and described below in Sections 9.2 and 9.3. As may be seen, these mechanisms range from the motion of interfacial dislocations in semi-coherent interfaces (by both glide and climb), to uncorrelated atom shuffling at incoherent interfaces, to atom jumping by self-diffusion as in the case of the motion of diffuse antiphase boundaries. In all cases atoms move around in the interfacial core region in a way which causes one adjoining lattice to grow and the other to shrink.

In some of these mechanisms the atom transfers occur in a highly correlated manner, as, for example, during the glissile motion of interfacial dislocations (Table 9.2). In such cases each atom in the shrinking crystal is moved to a predetermined position in the growing crystal as it is overrun by the moving dislocation and shuffled across the interface. In addition, when the dislocation moves, all atoms in the bicrystal change their positions to at least some degree in a pre-ordained manner which is determined by the changes in the long-range displacement field of the dislocation which occur as it moves. Since the position of the interface is correlated with the position of the dislocation, the positions of all atoms in the bicrystal are correlated in a precise manner with the position of the interface. We may then expect a possible change in the macroscopic shape of the bicrystal depending upon the numbers and types of glissile dislocations which are present. This highly organized type of interface motion is often termed 'military' (Christian 1975) to distinguish it from the disorganized 'civilian' type of motion which occurs, for example, when an incoherent interface moves by random (uncorrelated) shuffling of atoms from the shrinking crystal to the growing crystal (Table 9.2).

In addition, a number of the mechanisms listed in Table 9.2 require widely different degrees of thermal activation. As discussed below, the combination of all these factors causes the motion of different types of interfaces by these mechanisms to have quite different physical characteristics.

### 9.1.4 Impediments to interface motion

It is important to note that interface motion can be profoundly influenced by various impediments. Solute atoms, either in the form of controlled alloying additions or uncontrolled impurities, will always be present in solution to at least some degree. When the temperature is high enough so that they can execute thermally activated diffusional jumps at a significant rate, they may be able to accumulate at a moving interface and exert a strong impeding effect, i.e. 'solute drag', on the interface motion. In addition, interfaces can be pinned at embedded second phase particles and at grooves where they

**Table 9.2** Mechanisms for the conservative motion of interfaces

| | Mechanism | Examples |
|---|---|---|
| | Glissile motion of interfacial dislocations | Small-angle grain boundaries (Section 9.2.1) <br> Singular and vicinal large-angle grain <br> boundaries (Section 9.2.1) <br> Martensitic interfaces (Section 9.2.1) |
| | Glide and climb of interfacial dislocations | Small-angle grain boundaries (Section 9.2.2) <br> Large-angle singular or vicinal CSL and <br> near CSL grain boundaries (Section 9.2.2) <br> Singular or vicinal massive transformation <br> interfaces (Section 9.2.2) |
| Sharp interfaces | Shuffling motion of pure steps | Large-angle singular or vicinal CSL grain <br> boundaries (Section 9.2.3) |
| | Uncorrelated atom shuffling | General grain boundaries (Section 9.2.4) <br> General heterophase interfaces <br> (Section 9.2.4) |
| | Uncorrelated diffusional transport | General grain boundaries (Section 9.2.4) <br> General heterophase interfaces <br> (Section 9.2.4) |
| Diffuse interfaces | Propagation of non-linear elastic wave <br> or <br> Glissile motion of coherency dislocations | Displacive transformation interfaces <br> (Section 9.3.1) |
| | Self-diffusion | Antiphase domain boundaries <br> (Section 9.3.2) |

intersect free surfaces and therefore have their motion severely impeded. These effects are taken up in Sections 9.2.5 and 9.5.

## 9.2  MECHANISMS AND MODELS FOR SHARP INTERFACES

### 9.2.1  Glissile motion of interfacial dislocations

#### 9.2.1.1  *Small-angle grain boundaries*

A small-angle boundary may be regarded as a semicoherent interface consisting of coherent patches separated by dislocations which destroy the continuity of the reference lattice (i.e. the crystal lattice). Under certain conditions such boundaries can be induced to move in purely glissile fashion, i.e. they can move entirely by means of the simultaneous forward glissile motion of their dislocations. The simplest example is that of a symmetric tilt boundary of the type already illustrated in Figs 2.1 and 2.5 which consists of a simple stack of anticoherency primary dislocations. If such a boundary is present in a bicrystal subjected to a shear stress such as shown in Fig. 9.1(a), use of the Peach–Koehler equation (Hirth and Lothe 1982) shows that each primary dislocation is subjected to a force (per unit length), $|F| = \tau b$, urging it to glide forward on its glide

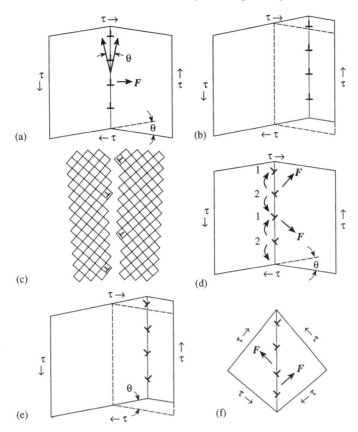

**Fig. 9.1** Responses of small-angle symmetric tilt boundaries to applied shear stresses, $\tau$. (a) and (b) Boundary consists of one type of edge dislocation, and the shear stress exerts a glide force, $F$, on each dislocation producing an overall pressure on the boundary to move forward. The boundary then advances in glissile fashion causing a specimen shape change. (c) and (d) Boundary now consists of two sets of edge dislocations, and the applied stress exerts a climb force, $F$, on each dislocation as seen in (d). (e) Shape change which occurs when the boundary shown in (d) moves forward by the simultaneous forward motion of each of its edge dislocations by combined glide and climb. The diffusion currents between the two types of dislocations associated with the climb are shown in (d). (f) Same boundary as in (d) subjected to applied shear stress rotated by 45°. Here, the stress exerts glide forces on the edge dislocations in different directions tending to split the boundary, and no overall pressure is therefore exerted on the boundary.

plane. The total force on the boundary is then the sum of the forces on the dislocations, and, since the dislocation spacing is $d \simeq b/\theta$, the overall pressure (force per unit area) is $p = \tau b/d \simeq \tau\theta$. In response to this pressure the boundary will therefore move forward by means of the simultaneous glissile motion of all its dislocations without any change in their local distribution as illustrated in Fig. 9.1(b). The uniform motion of the boundary in this fashion causes the bicrystal to change its shape. Such motion corresponds simply to translating the boundary by dislocation glide without changing its structure and putting it in a new position in the bicrystal. Since the crystal misorientation associated with the boundary (a simple tilt in the present case) must accompany the boundary, it is evident that a macroscopic shape change must accompany this displacement of the boundary. This change in shape is easily confirmed by constructing the bicrystal from its initial reference structure so that it first occupies its position in Fig. 9.1(a) and then its position in Fig. 9.1(b). The shape change may be generally attributed to the correlated atom displacements throughout the bicrystal which are associated with the glissile motion of the dislocations. Because of this shape change, the shear stress, $\tau$, exerts a force on the boundary of a magnitude which can be calculated by noting that when the boundary moves a distance $\delta x$ to the right, the change in shape (i.e. shear) of the bicrystal allows the applied stress to do the work $\delta x \cdot \theta \tau$ (per unit boundary area). Using eqn (9.1), the pressure is then

$$p = -\delta G/\delta x = \tau\theta, \tag{9.2}$$

which, as might be expected, is exactly equal to the pressure calculated above from the sum of the forces on the individual dislocations. This result can be readily understood by realizing that the force exerted on an individual dislocation by an applied stress is a result of the change in crystal shape which occurs when the dislocation moves (Hirth and Lothe 1982). Since the change in bicrystal shape due to the motion of the boundary is just the sum of all the shape changes due to the motion of all the individual dislocations, the total force must be just the sum of the forces on the individual dislocations.

We note that these considerations involve only the primary (anticoherency) disloca- tions, and not the additional coherency dislocations for reasons which should be evident from our discussion of the Frank–Bilby equation in Section 2.3. There it was pointed out that the lattice transformation, **T**, is related to either the anticoherency dislocation content or the coherency dislocation content of the boundary by the Frank–Bilby equation. In the present case **T** is a simple rotation around the tilt axis, and we have invoked the relationship between **T** and the anticoherency dislocation content. Finally, it should be remarked that when the boundary as a whole moves by this type of simultaneous glissile dislocation motion, it may be conveniently regarded as a glissile surface dislocation as discussed in Section 2.3.

Experimental verification of this type of motion has been obtained by Li *et al.* (1953), Bainbridge *et al.* (1954), Fukutomi and Horiuchi (1981), and Horiuchi *et al.* (1987), who demonstrated (see Fig. 9.2) that small-angle tilt boundaries of this type in Zn and Al bicrystals could be forced to move in glissile fashion by the application of a shear stress exactly as illustrated in Fig. 9.1(a,b). Even though the major basic mechanism for boundary motion in these experiments was clearly the forward glissile motion of the edge dislocations, the kinetics was found to be a complicated function of temperature and stress. At low temperatures, the boundaries in Zn could be forced to move by an appropriate stress which had to be continuously increased to induce continued motion. At elevated temperatures, steady-state motion at constant stress was achieved for both Zn and Al: the motion was thermally activated with an activation energy close to that for lattice self-diffusion. Evidently, the experimental conditions were not ideal, and the boundaries encountered obstacles as they moved, such as other bulk lattice dislocations, or else, in some cases, they contained relatively small numbers of additional intrinsic dislocations of other types which had to climb in order to move along appropriately (see Section 9.2.2). These impediments were evidently overcome by sufficiently high applied stress or self-diffusional relaxation processes.

Consider next the more complex case of a symmetric tilt boundary containing two sets of edge dislocations in a bicrystal again subjected to an applied shear stress as illustrated in Fig. 9.1(c,d). The applied stress now exerts a pure climb force $|F| = \tau b$, on each type of dislocation as indicated. Since the dislocation spacing is $d = b/\sqrt{2}\theta$, the overall pressure on the boundary is $p = |F| \cdot \cos 45°/d = \tau\theta$. Upon application of the stress, we

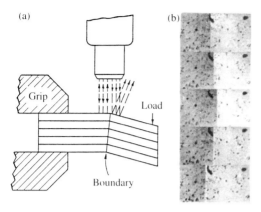

**Fig. 9.2** Experimental observation of glissile movement of small-angle tilt boundary in Zn of the type illustrated in Fig. 9.1(a, b). (a) Experimental arrangment. (b) Observed reversible glissile movement of boundary caused by reversing the load shown in (a). (From Li *et al.* (1953).)

would expect each dislocation to begin to climb in the direction of the applied force, $F$. However, mutual forces between the dislocations in the array will tend to keep them at a regular spacing corresponding to their equilibrium structure of minimum energy, and the final result will be the movement of the entire regularly spaced array in the direction normal to the boundary without any change in its structure. This type of motion corresponds simply to translating the boundary in the forward direction. Each pair of 1 and 2 dislocations is then equivalent to a single edge dislocation with Burgers vector $b = b_1 + b_2$: the situation is therefore similar to that in Fig. 9.1(a,b), and the resulting shape change will then be as illustrated in Fig. 9.1(e). According to eqn (9.2), a pressure will therefore be exerted on the boundary given by $p = \tau\theta$, consistent with our previous calculation. Movement of the boundary in this fashion requires each dislocation to move normal to the boundary plane simultaneously by combined glide and climb. A detailed analysis of the kinetics of this motion is given below in Section 9.2.2. Note that a distribution of coherency dislocations will also be present in the interface. However, they can always be easily moved around conservatively, and they need not be considered explicitly in the present discussion.

The applied stress in Fig. 9.1(d) is seen to exert no forces on the individual dislocations urging them to glide forward on their intersecting glide planes. No pressure on the boundary therefore exists for its purely glissile forward motion as a whole by this mechanism. This can be confirmed by considering the macroscopic shape change which would be produced if the dislocations did move forward simultaneously in this way. A straightforward, but tedious, analysis of the detailed intersecting shear displacements produced throughout the crystal in this process (Shockley 1952) shows that the overall resulting shape change is exactly zero, as required. It may also be seen that a different applied shear stress of the type shown in Fig. 9.1(f) exerts pure glide forces, $|F| = \tau b$, on the two sets of dislocations. However, they are in different directions which will tend to cause the boundary to split. This splitting will be resisted by attractive forces between the two sets of dislocations which tend to keep them together to maintain the equilibrium boundary structure, and there is obviously no net pressure tending to move the boundary as a whole in any direction. Further aspects of these various types of boundary motion have been discussed by Read (1953).

Despite the fact that the tilt boundary in Fig. 9.1(d) was not urged to move forward by the glissile motion of its edge dislocations on their intersecting slip planes by applied shear stresses, it can be urged to do so by capillary forces as illustrated in Fig. 9.3(a,b,c).

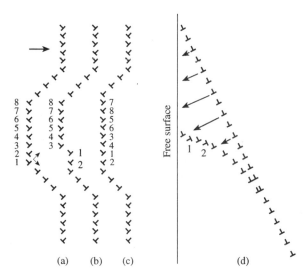

(a)    (b)    (c)                          (d)

Fig. 9.3 Glissile motion of tilt boundaries under capillary forces. (a–c)
Bowed-out section straightens out by the
glissile motion of its edge dislocations.
The dislocation motion occurs in a
sequential manner which produces steps
in the bowed-out section (see (b)) which
propagate along the boundary. (From
Scholz and Bauer (1990).) (d) Glissile
motion of boundary intersecting a free
surface.

Here, the boundary contains a bowed-out section, and a capillary force is present urging
it to straighten out by a pure glissile process, as illustrated in Fig. 9.3(b,c). In Fig. 9.3(b),
dislocations 1 and 2 have moved by glide to the positions shown and have thereby
produced a boundary step at the corner of the bow-out. This step can then propagate
upwards by the similar glide motion of dislocations 3–8 to produce the configuration
shown in Fig. 9.3(c). Repetition of this inhomogeneous process will progressively reduce
the overall boundary energy and eventually eliminate the bow-out. Scholz and Bauer
(1990) have directly observed the elimination of bow-outs by such boundary step
propagation and conclude that the boundary motion most likely occurs by the glissile
mechanism just described. However, during this motion, the dislocations must essentially
pass either through each other, or very near each other, thereby producing configurations
of relatively high energy which will act as barriers to the continued glissile motion. It
appears that these barriers can be surmounted locally by thermal activation with an
activation energy of the order of that for lattice self-diffusion (Scholz and Bauer 1990).
We therefore have a situation where the boundary motion is fundamentally glissile but
must be assisted at critical junctures by thermal activation. We note that the boundary
would encounter strong resistance to any straightening process in which its dislocations
move in the direction of the large arrow shown in Fig. 9.3(a) by combined glide and climb.
Motion of this type will quickly produce a highly non-equilibrium boundary structure,
and back-stresses will develop which will oppose the motion.

Another interesting situation is shown in Fig. 9.3(d) where a symmetric tilt boundary
intersects a free surface. Here, a capillary force is present urging the boundary to round
off and reduce its total energy by the glissile motion of existing and newly formed
dislocations as illustrated. In this process, pre-existing dislocations in the boundary
structure glide toward the free surface and are progressively annihilated while smaller
numbers of new dislocations (such as those at 1 and 2) are generated progressively at the
free surface and glide away from the surface in the boundary configuration. The
boundary structure changes as it migrates, and new dislocations must be introduced in
order to avoid the formation of a highly non-equilibrium boundary structure which would
generate long range back-stresses opposing the continued glissile motion of the boundary.

The generation of new dislocations therefore becomes an additional obstacle to boundary motion.

The above examples of glissile boundary motion under conditions where the boundary structure remains fixed and also where it changes with time have been particularly simple and have all involved arrays of parallel dislocations. Numerous other examples of varying degrees of complexity may be imagined. The most complex cases will involve more general boundaries composed of networks of several different types of dislocation segments joined at nodes (see Section 2.8). The glide planes of the different segments present in such boundaries will generally not lie on a common zone with its zone axis out of the boundary plane, and any forward glissile motion will therefore inevitably involve significant disruptions of the equilibrium boundary structure as the different segments glide on their non-parallel glide planes. Even though it is topologically possible for forward motion to occur in this glissile manner, very large forces would be required, and such motion will therefore not generally be possible. In certain extreme cases the the boundary may be irreversibly torn apart in the presence of large forces (Hackney and Lillo 1990). We shall not attempt any detailed discussion of these complex cases. The more general motion of small-angle boundaries by the combined glide and climb of their dislocations is taken up below in Section 9.2.2.1.

### 9.2.1.2 *Large-angle grain boundaries*

It is also possible, in a strictly topological sense, for a large-angle grain boundary to move forward in glissile fashion by the simultaneous glide motion of its primary dislocations. However, it will not generally be possible to induce this type of motion at the pressures available except possibly in very special situations where: (1) the dislocations are all of the same type and can simultaneously glide on parallel slip planes without hindering one another as in the case of symmetrical tilt boundaries of the type in Fig. 9.1(a); and (2) the large shape changes which accompany the boundary motion can be readily accommodated. This was apparently the case in the experiments of Fukutomi and Horiuchi (1981) and Horiuchi *et al.* (1987) already cited above in Section 9.2.1.1, where it was found that in some cases symmetric tilt boundaries in the configuration of Fig. 9.1(a) responded to an applied shear stress by the simultaneous forward glide of their edge dislocations at tilt angles as large as $\simeq 26°$. However, at still higher tilt angles (and smaller dislocation spacings) the response of the boundaries to the applied shear stress changed to the extent that localized shearing occurred at the boundaries (i.e. boundary sliding occurred) in the absence of any detectible forward motion. In these cases the forward glissile motion of the edge dislocations became prohibitively difficult, and grain boundary sliding of the type described in Section 12.8.1 for general large-angle boundaries (which occurs by quite different mechanisms involving viscous shearing and diffusional accommodation processes in the boundary) became predominant. This result is understandable, since in a simple model the pressure required to force the forward simultaneous glide motion of the edge dislocations in the boundary of Fig. 9.1(a) should increase with an increase in the number of dislocations per unit boundary area which must be moved (i.e the tilt angle). On the other hand, as the misorientation increases, and the boundary enters the large-angle regime where the dislocations are close enough so that the core becomes a continuous slab of bad material, the rate of boundary sliding (as described in Section 12.8.1) becomes significant. At some point the mode of the bicrystal response should therefore switch as was indeed observed.

In more general situations the primary dislocations in large-angle boundaries will be very closely spaced, their simultaneous glissile motion will require glide on intersecting slip planes leading to repulsive interactions (as was the case for the boundary in Fig. 9.1(d)), and there will be bulk restraints on any shape changes. There will therefore be a high resistance to glissile boundary motion via the simultaneous glissile motion of the primary boundary dislocations.

The problem of determining the pressure required to move a large-angle boundary forward by the simultaneous glissile motion of its primary dislocations may also be approached in an approximate manner by again regarding the sharp large-angle boundary (with its closely spaced overlapping dislocation cores) as a narrow slab of bad material. The forward motion of this narrow slab via atomic displacements corresponding to the simultaneous glide motion of all of its dislocations will require significant changes in its core structure as the atoms shuffle and change their positions. This, in turn, will produce significant changes in the overall core energy. For example, if the boundary is in its position of minimum energy in the bicrystal, and a forward movement of $\Delta x = 0.05$ nm requires only a nominal 5 per cent increase in its core energy, the required pressure would be $p \simeq 0.05\sigma/\Delta x = 5 \times 10^8$ Pa cm$^{-2}$ which is large on the scale of the pressures listed in Table 9.1.

This situation for sharp interfaces may be contrasted to that for diffuse interfaces (Table 4.1) which is physically quite different. For diffuse interfaces the core straddles an appreciable number of atomic distances and does not contain any bad material. For displacive transformation interfaces (see Section 4.2.3) the structure of one adjoining lattice grades to that of the other in a quasi-continuous fashion over a relatively large distance. Such an interface can move easily by means of a large number of relatively small atomic displacements distributed throughout the wide core region. Some of these displacements will increase the energy of the boundary while others will decrease it, and the net effect will average essentially to zero. An approximate analysis of the variations in overall core energy which accompany such motion has been performed (Cahn 1960), and the results show that they tend to decrease as the interface core width increases as might be expected. Such motion will therefore be very easy, and diffuse interfaces will therefore move forward easily in a uniform manner. We remark on the similarity between this phenomenon and the lattice resistance to the homogenous glide motion of lattice dislocations where the energy barrier to the motion (i.e. the Peierls energy) decreases as the dislocation cores become wider, i.e., more 'diffuse' (see Section 2.11.2.2 and Hirth and Lothe (1982)).

Sharp large-angle grain boundaries will therefore not generally be able to move by the simultaneous forward glissile motion of their primary dislocations at usual pressures (Table 9.1), and they will therefore adopt other mechanisms requiring less pressure. An alternative, and easier, glissile mechanism is possible for singular or vicinal boundaries which are capable of supporting localized secondary dislocations having step character which are glissile in the boundary plane. The lateral sweeping movement of such dislocations across such a semi-coherent boundary will therefore cause the boundary to move forward inhomogeneously. In this process the line defects provide special locations in the interface where network sites can be transferred across the interface with relatively small increases in energy as they glide across the otherwise unperturbed interface. When the line defect is glissile in the interface, the site transfer occurs in a correlated military fashion at the line defect by local atom shuffling, making the boundary as a whole glissile.

One of the best-known examples of this type of motion is that exhibited by the large-angle singular $\Sigma 3\{111\}$ twin boundary in the f.c.c. system (Read 1953, Christian 1975, Hirth and Lothe 1982), which was discussed in Section 2.2 and illustrated in Fig. 2.12(a). This interface can support the localized extrinsic secondary dislocation shown in

Fig. 2.12(b) which is glissile in the boundary (since $h^w = h^b$) and possesses a step height $h = (h^b + h^w)/2 = h^w = h^b = \frac{1}{3}|\langle 111 \rangle|$ and a Burgers vector $b = \frac{1}{6}\langle 112 \rangle$ parallel to the interface. The motion of this dislocation across the semicoherent interface therefore advances the interface by $h$ and displaces (shears) the upper crystal with respect to the lower one by $b$. If such dislocations glide across the interface successively at a rate $\dot{n}$, the interface will move forward at the average rate

$$v = \dot{n}h, \tag{9.3}$$

and the bicrystal will change shape macroscopically as a result of the successive shears. This type of motion can be induced by applying the shear stress $\tau$ (Fig. 2.12(b)) which urges each dislocation to glide and shear the bicrystal in the direction of the applied stress. If the interface moves a distance $\delta x$ in response to the applied stress, the work done (per unit boundary area) during the shape change is $(\delta x/h)b\tau$. The pressure exerted on the boundary by the stress is then

$$p = -\delta G/\delta x = b\tau/h. \tag{9.4}$$

Alternatively, the stress exerts a force, $\tau b$, per unit length on each dislocation. If $n$ dislocations traverse unit distance along the boundary, the work done by the applied stress is $nb\tau$. This is exactly equal to the corresponding work per unit area done by the force on the boundary, since the distance moved by the boundary is $nh$, and the work is then $nh \cdot p = nh \cdot (b\tau/h) = nb\tau$.

This type of glissile interface motion is known to occur during mechanical twinning when twinned regions bounded by these interfaces grow within a matrix crystal under the influence of applied shear stresses as illustrated schematically in Fig. 2.7. Such twinning is a form of plastic deformation, since the material changes shape permanently in response to applied forces. In fact, we may regard the process as a form of shearing, or boundary sliding (as defined in Chapter 12) which is coupled directly to the boundary motion. The twinned crystals generally grow in thin lenticular form in order to reduce the elastic strain energy which would otherwise be produced throughout the material by the large shear associated with their growth. If the twinned region intersects the surface as in Fig. 9.4, surface relief effects will be observed at the intersections of the interfaces with the surface.

The large numbers of dislocations which must be generated on successive {111} planes in order to allow the process to continue, can be obtained in a number of ways. In general, they will not be formed by the homogeneous nucleation of small dislocation loops under the influence of the average applied shear stress, since the stress required is too large as may be deduced from eqn (10.53). (See also Hirth and Lothe 1982.) Therefore, some type of heterogeneous process is required. A likely candidate is a pole mechanism (Sleeswyk 1974, Hirth and Lothe 1982) such as, for example, illustrated in Fig. 9.5. Figure 9.5(a) shows an initial configuration consisting of a triple junction of lattice dislocations in the parent crystal. The components of the Burgers vectors of dislocations OA and OB along the normal of the {111} planes are each equal to the {111} interplanar spacing, and the crystal, therefore, is actually a helical ramp as illustrated. Under the applied stress, the dislocation OC dissociates into two partial dislocations with Burgers vectors of the form $\frac{1}{6}\langle 112 \rangle$. As seen in Fig. 9.5(b), one partial then glides down the helical ramp and wraps itself around OB, while the other glides upwards and wraps itself around OA in the opposite sense. The continued expansion of successive loops on successive {111} planes (Fig. 9.5(c)) then converts the region between O' and O" into a twinned crystal of the general form shown in Fig. 2.7, and the dislocation O'O" left in the twin is a lattice

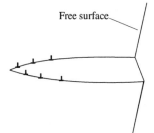

Free surface

**Fig. 9.4** Surface relief effects at intersection of a lenticular mechanical twin with a free surface.

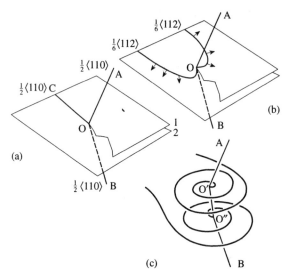

$\frac{1}{6}\langle 112 \rangle$

$\frac{1}{6}\langle 112 \rangle$

$\frac{1}{2}\langle 110 \rangle$

$\frac{1}{2}\langle 110 \rangle$ C

A

O

$\frac{1}{2}$

B

(b)

A

O

(a)

$\frac{1}{2}\langle 110 \rangle$ B

A

O'

O''

(c)

B

**Fig. 9.5** Pole mechanism for the glissile formation of a mechanical twin in an f.c.c. crystal. (a) Initial configuration consisting of triple junction of three lattice dislocations with Burgers vectors of the type $\frac{1}{2}\langle 110 \rangle$. Dislocations OA and OB possess screw components normal to the lattice planes 1 and 2 which convert the lattice into a helical ramp as illustrated. (b) Dislocation OC splits into two partial dislocations which wrap themselves around the pole dislocations OA and OB in opposite directions. (c) Formation of lenticular twin between O' and O'' by the continual wrapping motion and expansion of the partial dislocations. (From Sleeswyk (1974).)

dislocation with Burgers vector of the type $\frac{1}{2}\langle 110 \rangle$. Additional possible heterogeneous dislocation formation mechanisms have been discussed by Hirth and Lothe (1982). These include additional varieties of pole mechanisms and also formation due to high stress concentrations at surfaces, dislocation pile-ups, and grain boundaries. (The nucleation of dislocation loops in regions of high stress concentration is discussed in Chapter 10: see, e.g., eqn (10.53).)

The velocity of the glissile interface will be controlled by the magnitude of the applied stress and the mobility of the dislocations. The interface velocities which are achieved can be very large, approaching the speed of sound. The dynamics of the moving dislocations will be characteristic of usual glissile dislocation dynamics (Hirth and Lothe 1982) where acceleration occurs in a manner controlled by the inertial mass of the line defect, and damping forces produced by interactions of the line defect with phonons and electrons.

Boundary motion due primarily to the glissile motion of extrinsic secondary interfacial dislocations possessing similar step character along the boundary under the influence of stress has also been induced in other singular or vicinal high-angle boundaries; i.e. in $\Sigma 9$, 11, and 17 boundaries in Al and in near-coincidence $\Sigma 15$ and $\Sigma 29$ boundaries in Zn (Fukutomi and Kamijo 1985, Horiuchi *et al.* 1987). In all cases, the dislocations in the semicoherent boundaries possessed Burgers vectors parallel to the interface and step heights $h = h^w = h^b$, and their motion was induced by an applied shear stress. The resulting coupled boundary displacement, $M$, and shear displacement parallel to the boundary, $S$ (i.e., boundary sliding as defined in Chapter 12), were measured directly from the distortion of fiduciary grids as illustrated in Fig. 9.6, and their ratio, $M/S$, was found to agree with the ratio, $h/b$, for the dislocations as required geometrically. All of the boundaries studied were large-angle symmetrical tilt boundaries, and in no cases did these boundaries move forward by the simultaneous glissile movement of their primary edge dislocations in response to the shear stresses which were applied. This result is consistent with our expectation (discussed above) that higher shear stresses (and corresponding pressures on the boundaries) would be required to force the forward simultaneous motion of the primary dislocations. Clearly, the response of a boundary to applied shear stress via the lateral motion of extrinsic secondary dislocations as described above may be regarded as either boundary motion in response to the applied force, or, alternatively, as boundary shearing (or sliding, as defined in Chapter 12). Further discussion of boundary sliding, and possible sources of the interfacial dislocations which are required, is given in Section 12.8.1.

Even though the above results were clearly due primarily to the glissile motion of dislocations, the process only occurred at elevated temperatures at the stresses employed, and, in the case of the Zn experiments, appeared to be thermally activated with an activation energy near that expected for grain boundary diffusion. The exact role played by the thermal activation is not obvious: clearly, it must have been associated with either the production of the dislocations or the process of moving them along the boundary.

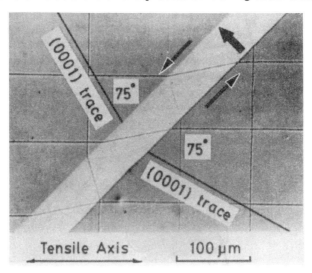

**Fig. 9.6** Coupled boundary displacement and shearing observed for a near-$\Sigma 15$ symmetric $[10\bar{1}0]$ tilt boundary in Zn subjected to a tensile stress along the axis shown. The boundary plane is $(1\bar{2}11)$. The boundary displacement and shearing are revealed by the distortion of the initially uniform fiduciary grid of scribed lines and are indicated by the arrows. (From Horuichi *et al.* (1987).)

None of the boundaries studied was ideally flat, and, therefore, a small amount of dislocation climb may have been required to allow their motion along the boundaries (see Section 12.8.1.5). Also, the boundaries were not ideal exact coincidence boundaries, and arrays of intrinsic secondary dislocations were therefore unfortunately present. These may have provided obstacles to the dislocation motion which could be overcome only by thermally activated relaxation processes.

### 9.2.1.3 *Heterophase interfaces*

Sharp heterophase interfaces (Table 4.1) meeting certain geometrical requirements may also be able to move forward in purely glissile fashion by the glissile motion of discrete interfacial dislocations. Again, as previously, network sites are then transferred across the semicoherent interface locally in correlated military fashion by atom shuffling at the gliding dislocations, and a specimen shape change occurs. Such interfaces play an important role in displacive phase transformations when, for example, a martensite phase (Cohen, *et al.* 1979, Christian 1982, Olson and Cohen 1986, Olson and Owen 1992) grows at the expense of a parent phase under the influence of either a decrease in chemical free energy or a mechanical driving force.

The geometry and glissile nature of sharp displacive transformation (martensitic) interfaces may be understood by considering a few illustrative examples. Consider first the particularly simple 2D example illustrated in Fig. 9.7. Here, the martensite and parent structures are identical along the $z$ axis, and the martensite phase ($\beta$ phase) can be generated from the parent $\alpha$ phase by a homogeneous lattice transformation, $\mathbf{T}$, consisting of a simple shear parallel to $x$, of the form

$$\begin{vmatrix} 1 & -s & 0 \\ 0 & 1 & 0 \\ 0 & 0 & 1 \end{vmatrix} \tag{9.5}$$

followed by a dilation parallel to $x$, of the form

$$\begin{vmatrix} (1-\epsilon) & 0 & 0 \\ 0 & 1 & 0 \\ 0 & 0 & 1 \end{vmatrix} \tag{9.6}$$

so that

$$\mathbf{T} = \begin{vmatrix} (1-\epsilon) & -s(1-\epsilon) & 0 \\ 0 & 1 & 0 \\ 0 & 0 & 1 \end{vmatrix} \tag{9.7}$$

This particular transformation has a plane associated with it (Fig. 9.7(b)) which is 'invariant', i.e. it is both undistorted and unrotated (Christian 1982). Use of eqn (9.7) shows that it lies at the angle

$$\theta = \tan^{-1}\left[\frac{\varepsilon}{-s(1-\varepsilon)}\right] \tag{9.8}$$

with respect to the $(x, z)$ plane. This invariant plane will generally be irrational in both crystals and, since there is perfect macroscopic matching across it, we may expect the system to select it as the habit plane of the transformation (as shown in Fig. 9.7(b)) in order to avoid any long-range strain energy or diffusion when it moves. Since the

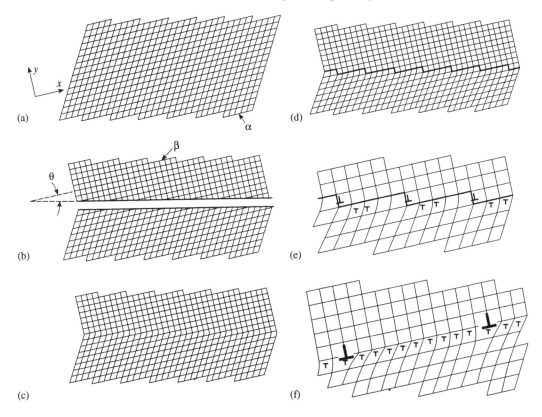

**Fig. 9.7** Schematic 2D example of sharp displacive transformation interface. (a) Initial $\alpha$ phase crystal. (b) Homogeneous transformation of upper half of crystal to $\beta$ phase. Invariant plane is present along the gap. (c) Coherent configuration obtained by joining $\alpha$ and $\beta$ crystals along the invariant plane. (d) Relaxation of interface into set of facets parallel to dense lattice planes and associated steps. (e) Sets of glissile coherency dislocations associated with relaxed stepped interface structure in (d). (From Dahmen (1987).) (f) New semicoherent interface produced by the rotation of the interface in (e) by the glissile loss of its coherency dislocation/steps and the introduction of anticoherency lattice dislocations.

transformation is homogeneous, and since there is exact macroscopic matching, the two lattices in their unrelaxed state will be coherent across the interface as shown more clearly in Fig. 9.7(c). 'Corresponding' lattice planes and lattice directions (i.e. lattice planes and directions which are related by the transformation, T) are, of course, continuous through the interface but change inclination in a manner prescribed by the transformation. It is evident that the uniform forward motion of such an unrelaxed interface would produce a macroscopic shape change of the bicrystal given by T.

The interface in Fig. 9.7(c) will tend to remain sharp upon relaxation as long as the system is relatively far from any condition of mechanical instability. (When this is not the case, the interface is expected to become diffuse upon relaxation as discussed in Section 4.2.3.) When such an interface is almost parallel to relatively dense crystal lattice planes it will tend to relax by decomposing into a set of facets running parallel to these planes (Fig. 9.7(d)) and separated by discrete and localized coherency dislocations possessing step character as illustrated in Fig. 9.7(e). The interface will remain completely coherent during this relaxation but will develop the distribution of localized coherency dislocations illustrated in detail in Fig. 9.7(e). Two sets will develop with Burgers vector strengths which cancel (and long-range stress fields which cancel), since the interface is an invariant plane of the transformation, and the total strength of the coherency dislocations must therefore be zero. The coherency dislocations which are localized at the steps between the facets have both dislocation and step character with $h = h^w = h^b$, and are similar in many respects to the coherency dislocations with step character illustrated previously in Fig. 2.8. The dislocations in Fig. 9.7(e) are readily glissile, and the interface as a whole can then move forward inhomogeneously in a glissile conservative manner if the interface dislocations, with their step character, move laterally to the left without appreciably changing their spacing. In this process the interface will remain closely parallel to the invariant plane, and each moving dislocation will transfer atoms (by shuffling) from sites in the parent crystal to corresponding prescribed sites in the martensite crystal while preserving the lattice correspondence across the interface. The movement of the interface by this means will therefore produce a macroscopic shape change corresponding to **T**. This type of interface motion is seen to be similar to the motion of the vicinal grain boundary illustrated previously in Fig. 2.7. In that case, the corresponding singular boundary was an invariant plane of the lattice transformation relating the two lattices (i.e. a simple twinning shear). The atomic positions and lattice planes across the boundary were again related by the lattice correspondence, and a shape change was produced by the boundary motion corresponding to the lattice transformation.

We note that it would be topologically possible for the relaxed interface illustrated in Figs. 9.7(d,e) to move forward uniformly by means of the collective 'conservative climb' of the coherency dislocations distributed along the interface. However, this is not expected for this sharp interface, since it would be difficult energetically for the reasons discussed in the previous section.

If the coherency dislocations associated with the steps in Fig. 9.7(e) change their spacing during the motion the interface plane will rotate away from the invariant plane, and long-range stresses will develop in order to maintain lattice coherency. In the extreme case where they all move off, the interface would rotate into the inclination shown in Fig. 9.7(f). The long-range stresses which would then tend to develop could then be cancelled by introducing the new array of anticoherency crystal lattice edge dislocations (stress annihilators) shown, and the interface would then be semicoherent and free of long-range stress. The total dislocation content would consist of the new anticoherency dislocations plus a set of coherency dislocations (see Fig. 9.7(e)), and the total Burgers vector strength would sum to zero. Also, the interface would no longer be glissile: forward motion would require the anticoherency crystal lattice dislocation array to climb.

An experimental observation of an essentially completely coherent interface capable of moving via the lateral motion of coherency dislocations possessing step character is shown in Fig. 9.8. Similarly, Hayzelden *et al.* (1991) have resolved glissile coherency dislocations in a coherent f.c.c./h.c.p. martensite interface in a Co-alloy.

More generally, in a displacive (martensitic) phase transformation the lattice transformation does not leave either an undistorted or an invariant plane in contrast to the special examples just discussed. By 'undistorted', we mean that the dimensions of the plane are unchanged, but that it may be rotated in space. In order for a lattice transformation to leave an invariant plane (which is both undistorted and unrotated) it can be shown (Christian 1975) that three conditions must be satisfied: (i) one principal strain must be zero; (ii) one principal strain must be positive; and (iii) one principal strain must

**Fig. 9.8** Observation of glissile coherency dislocations with step character in martensite interface between orthorhombic (o) and monoclinic (m) phases in $ZrO_2$. Fringes and facets parallel to the parallel (001) planes of both lattices. Blurring at dislocation/steps due to their motion during the photographic exposure time. (From Chiao and Chen (1990).)

be negative. The problem of producing a glissile interface between the parent and martensite phases therefore becomes more complex. The full 3D crystallography of martensitic transformations has been fully described elsewhere (Bullough and Bilby 1956, Christian 1965, Wayman 1964 and 1983, Olson and Cohen 1986). It may be shown that, if a lattice transformation, **T**, exists which produces an undistorted line, it is then possible to produce a plane which is undistorted macroscopically. This may be done by shearing the martensite by a further shear, **S**, which shears the material but leaves the crystal lattice unchanged as, for example, during shearing by the motion of perfect lattice dislocations. The undistorted plane created by the two successive deformations, **T** and **S** can then be finally converted to an invariant plane by means of a rigid body rotation, **R**, which will rotate it so that macroscopically it exactly matches the parent phase in its original position which is then the habit plane of the transformation. The lattice invariant shear, **S**, can be accomplished by passing a set of glissile crystal lattice dislocations through the martensite. The end result is the semicoherent interface illustrated schematically in Fig. 9.9 which will generally be irrational in both adjoining phases. The interface is a generalized 3D version of the 2D coherent interface previously illustrated in Fig. 9.7(d, e) containing, however, an added array of superimposed anticoherency crystal lattice dislocations which are just those dislocations required to shear the martensite in order to produce the invariant habit plane. The total Burgers vector strength of the anticoherency crystal dislocations (stress annihilators) and coherency dislocations (stress generators) is again zero. The crystal lattice dislocations lie along the intersections of their slip planes with the interface, and the interface as a whole is glissile, since it can advance by means of the glissile lateral motion of the well-spaced coherency dislocations with step character coupled with the forward glissile motion of the crystal lattice anticoherency dislocations on their inclined slip planes. Clearly, the military motion of the interface will cause a macroscopic shape change, **M**, given by

$$\mathbf{M} = \mathbf{RST}. \tag{9.9}$$

Since the interface habit plane is invariant, the macroscopic deformation, **M** can be decomposed into a shear parallel to the interface combined with a dilation normal to the interface which is necessary to account for the volume change of the transformation. Even though in writing eqn (9.9) we have visualized the creation of the martensite phase by a successive lattice transformation, **T**, a lattice invariant shear, **S**, and a rotation, **R**, it is evident that these operations can be taken in any order, and that whatever happens at the interface as it advances must always be regarded as the result of all of these operations occurring simultaneously.

We note that the structure of the interface illustrated in Fig. 9.9 can perhaps be most easily visualized as the result of first performing the lattice transformation **T** and then forcibly joining the martensite crystal directly to the parent phase to produce a coherent interface of the type illustrated in Fig. 9.7(c,d). This will generally require a large amount of long-range elastic strain, since no undistorted plane is available at this point. However, if the martensite is now sheared by passing the anticoherency crystal lattice dislocations through it so that they end up in the interface and produce a macroscopically invariant plane there, we will relieve all of the long-range elastic strain and obtain the final invariant plane strain interface shown in Fig. 9.9. An analogous sequence of steps is shown in Fig. 2.5 for a simple tilt grain boundary.

As in the case of mechanical twins, the martensitic regions which are produced within the parent phase will usually adopt thin lenticular macroscopic forms which run parallel to the invariant plane (habit plane) in order to reduce the large strain energy which would otherwise be associated with the shear and dilation associated with the shape change, and the transformed regions will again appear as illustrated in Fig. 2.7. If the martensitic regions intersect a free surface, surface relief effects will occur similar to those in Fig. 9.4. As already mentioned, a force can be exerted on the interface by a suitable applied stress because of the shape change associated with the interface motion. The work done per unit area (when the interface advances a distance $\delta x$) by an applied tensile stress normal to the interface, $\tau_n$, in the dilation, $\varepsilon$, associated with the invariant plane strain deformation is $\varepsilon \cdot \delta x \cdot \tau_n$. The corresponding work done in the accompanying shear, $s$, by an applied shear stress, $\tau_s$, parallel to the shear direction is $s \cdot \delta x \cdot \tau_s$. The total pressure due to a general stress with the components $\tau_n$ and $\tau_s$ is therefore

$$p = -\delta G/\delta x = \varepsilon\tau_n + s\tau_s. \tag{9.10}$$

This mechanical pressure augments the pressure due to the chemical driving force in the transformation. In thermoelastic martensite these two contributions to the total driving force are roughly in equilibrium. The martensitic phase may then be induced to grow, or shrink, by application of stress or by varying the temperature. In this way thermoelastic alloys display the 'shape memory effect' (Reed-Hill and Abbaschian 1992).

Again, as in the case of mechanical twinning, the glissile motion of the interface can be very rapid, in some instances approaching the speed of sound and can be generally described within the framework of glissile dislocation dynamics. Also, the required dislocations can be obtained from pole-type and other mechanisms as during mechanical twinning. Detailed discussion of many additional points of interest, along with examples, is given by Christian (1982), Olson and Cohen (1986), and Olson and Owen (1992).

### 9.2.2 Glide and climb of interfacial dislocations

#### 9.2.2.1 *Small-angle grain boundaries*

The purely glissile motion of small-angle boundaries via the simultaneous glide motion of their component primary dislocations has already been described in Section 9.2.1.1. More generally, the motion of small-angle boundaries will involve the simultaneous glide and climb of their dislocations. A particularly simple example is provided by the boundary illustrated in Fig. 9.1(c,d). As pointed out in Section 9.2.1.1, an applied shear stress of the type shown in Fig. 9.1(d) exerts a force on this boundary urging it to move

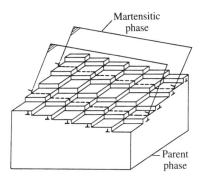

**Fig. 9.9** Schematic view of glissile semicoherent martensite interface in 3D. Interface contains arrays of glissile coherency dislocations with step character (as in Fig. 9.7) and anticoherency crystal lattice dislocations. (From Dahmen (1987).)

to the right by a process involving the motion of each primary anticoherency dislocation in a direction normal to the boundary plane by means of combined glide and climb. Successive dislocations in the array must then execute alternating positive and negative climb, and this can be achieved by establishing local self-diffusion currents of atoms between adjacent dislocations which then act as line sources/sinks for these currents, as illustrated in Fig. 9.1(d), while allowing the overall boundary motion to be conservative. The motion of this semicoherent boundary in this fashion will then cause the specimen shape change illustrated in Fig. 9.1(e). The two basic underlying processes which transport atoms from the shrinking crystal to the growing crystal (shown separately in Fig. 9.1(c)) are the local shuffles which occur as the dislocations glide and the diffusional transport which occurs when they climb.

A discussion of the atomistic details of the climb of primary dislocations is given in Section 10.3.2.1. Assuming a vacancy diffusion mechanism, two limiting kinetic situations can be visualized. At one limit the creation and destruction of the vacancies at the dislocation cores necessary to sustain the diffusional transport required for the climb is relatively easy and rapid, and the overall rate of motion is then controlled by the rate at which atoms can diffuse between the dislocations (i.e. the rate is 'diffusion-controlled'). At the other limit the situation is reversed, and the rate is controlled by the rate at which the necessary vacancies can be created/destroyed at the dislocation cores (i.e. the rate is 'source/sink-controlled'). In general, at the diffusion-controlled limit the diffusion potentials of the diffusing species at the dislocations will be maintained essentially at their local equilibrium values, and the dislocations will, therefore, act as 'ideal' line sources/sinks for the atoms. The resulting rates of mass transport and corresponding dislocation motion will then be the maximum possible. (An extensive discussion of these limiting forms of kinetics, along with an analysis of the way in which the driving free energy is dissipated in each case, is given in Sections 10.2.1 and 10.2.2.)

The diffusion-controlled rate can be easily obtained in approximate form, since this merely involves solving a relatively simple boundary value diffusion problem. Since the edge dislocations constitute line sources/sinks where network sites can be independently added/removed, we can specify the equilibrium diffusion potentials in their direct vicinities (see Section 5.9.2) and use the diffusion potential formalism developed in Section 5.9.3 to solve for the resulting diffusion flux of atoms through the lattice. For a unary system, in which diffusion occurs via vacancies, the flux of atoms is given by eqn (5.152) and has the form

$$J_A = - B_{AA}^L \nabla M_A,$$ 

(9.11)

if we take the vacancies to be the $C$th component. Since the vacancy concentration is exceedingly small, the free energy of a local region in the lattice can be written in the same form as eqn (8.20), i.e.

$$G = N_A G_A^o + N_V G_V^f + kT \left[ N_A \cdot \ln\left(\frac{N_A}{N_A + N_V}\right) + N_V \cdot \ln\left(\frac{N_V}{N_A + N_V}\right) \right] \tag{9.12}$$

where the last term is the usual ideal free energy of mixing. The quantities $N_A$ and $N_V$ are the numbers of atoms and vacancies, respectively, $G_A^o$ is the free energy per atom, and $G_V^f$ is the vacancy free energy of formation. With these assumptions, we may employ eqn (5.141b) and write

$$M_A \equiv \left[\frac{\partial e'}{\partial \rho_A'}\right]_{s',F,\rho_V'} - \left[\frac{\partial e'}{\partial \rho_V'}\right]_{s',F,\rho_A'} = N_o \left( \left[\frac{\partial G}{\partial N_A}\right]_{T,N_V} - \left[\frac{\partial G}{\partial N_V}\right]_{T,N_A} \right) \tag{9.13}$$

and, therefore, by differentiating eqn (9.12),

$$M_A = N_o [G_A^o - G_v^f - kT\ln(N_V/N_A)]. \tag{9.14}$$

Putting eqn (9.14) into eqn (9.11), we therefore have

$$J_A = (B_{AA}^L N_o kT / X_V) \cdot \nabla X_V. \tag{9.15}$$

However, the vacancy diffusion coefficient, $D_V^L$, is defined by

$$J_V = - (D_V^L/\Omega) \cdot \nabla X_V, \tag{9.16}$$

and, since $J_A + J_V = 0$, we find

$$B_{AA}^L = \frac{D_V^L X_V}{N_o \Omega kT}. \tag{9.17}$$

The coefficient for self-diffusion by a vacancy mechanism is given by $D^L = f^L D_V^L X_V^{eq}$ (Shewmon 1989), where $f^L$ is the lattice diffusion correlation coefficient, and $X_V^{eq}$ is the equilibrium vacancy concentration in the stress-free lattice. Since $X_V \simeq X_V^{eq}$, we may, therefore, write eqn (9.11) for the flux of atoms in the lattice as

$$J_A = - \frac{D^L}{N_o \Omega kT f^L} \cdot \nabla M_A. \tag{9.18}$$

The value of $M_A$ at a dislocation source/sink may be obtained using eqn (9.12) with the understanding that $G$ is now the free energy of the entire system and that the atoms and vacancies are inserted at the climbing dislocation. If an atom is inserted at a dislocation of Type 1 (Fig. 9.1(d)), a unit length of dislocation will move forward a distance $\sqrt{2}\Omega b$. The applied force per unit length in that direction is $\tau b/\sqrt{2}$, and therefore, $\partial G/\partial N_A = G_A^o - (\tau b/\sqrt{2})(\sqrt{2}\Omega/b) = G_A^o - \tau\Omega$. If a vacancy is produced there, and, if the vacancy concentration is in local equilibrium, $\partial G/\partial N_V = 0$. Therefore, $M_A^{(1)} = N_o(G_A^o - \tau\Omega)$. For a type 2 dislocation, $M_A^{(2)} = N_o(G_A^o + \tau\Omega)$. Since. $M_A^{(2)} > M_A^{(1)}$, atoms will flow from type 2 to type 1 dislocations as expected. We can estimate the flux by substituting the average gradient, $\overline{\nabla M_A} \simeq 2N_o\tau\Omega/d$, into eqn (9.18). Then, since the average area through which the flux flows is $\bar{A} \simeq d/2$, the total current, $I_A$, entering each type 1 dislocation (per unit length) is

$$I_A = 2\bar{A}J_A = \frac{2D^L\tau}{kTf^L}, \tag{9.19}$$

and the diffusion-controlled boundary velocity is then

$$v \simeq I_A \cdot \sqrt{2}\Omega/b = \frac{2\sqrt{2}D^L\Omega}{kTbf^L} \cdot \tau. \tag{9.20}$$

Since $d = b/\sqrt{2}\theta$ and $p = \tau\theta$, eqn (9.20) can be rewritten in the form

$$v \simeq \left[\frac{4D^L\Omega d}{kTf^Lb^2}\right] \cdot \theta\tau = M \cdot p. \tag{9.21}$$

The velocity can therefore be expressed as the product of a pressure, $p$, and a mobility, $M = [4D^L\Omega d/kTf^Lb^2]$ which is a function of the boundary structure and is proportional to $D^L$. Expressing $D^L$ in the standard Arrhenius form $D^L = D_o^L \cdot \exp(-Q^L/kT)$, the activation energy for boundary motion is seen to be that of lattice self-diffusion, $Q^L$, as anticipated for a lattice diffusion controlled process.

Finally, we note that we have assumed above that all of the lattice dislocations were uniformly spaced and moved forward in unison. However, as pointed out in Section 10.3.2.1, this will not be possible for an array of climbing dislocations, and perturbations in their spacings are inevitable. These will generally cause small increases in energy which may conceivably produce threshold effects at small driving forces (see detailed discussion of this effect in Section 10.3.2.1).

When the kinetics is controlled by the rate of vacancy creation/destruction at the dislocation cores, the boundary velocity will be reduced, and a model for the detailed processes occurring at the cores is required. This is taken up in Section 10.3.2 along with discussion of the effectiveness of primary dislocations as line sources/sinks for diffusional fluxes of atoms.

Innumerable other situations exist for the motion of small-angle boundaries, particularly when the boundary possesses a complex structure consisting of networks of anticoherency mixed dislocation segments joined at nodes. The boundary motion will then generally occur by combined dislocation glide and climb, and it will again tend to be climb-limited. However, it is important to point out that when networks with nodes are present, it is possible that at least a part of the diffusional transport between the various segments required to support their climb will be contributed by fast short-circuit diffusion along the dislocation cores (Chapter 8). The mix of lattice diffusion and dislocation core diffusion which results will then depend upon both the geometry of the dislocation array and the relative rates of lattice and dislocation core diffusion, and would be highly complex to analyse. Activation energies for boundary motion ranging between those of lattice diffusion and dislocation short-circuit diffusion could then result if the dislocations act as perfect sources/sinks.

### 9.2.2.2 *Large-angle grain boundaries*

Topologically, large-angle grain boundaries should be able to move conservatively via the simultaneous glide and climb of their primary dislocations in generally the same way as small-angle boundaries. However, singular or vicinal large-angle boundaries, which are capable of supporting localized secondary dislocations, tend to preserve their reference structures which consist of closely spaced primary dislocations arranged in regular fashion and are of relatively low energy. Such boundaries will therefore be particularly resistant to motion via the simultaneous glide and climb of their primary dislocations, since this process would inevitably cause significant perturbations in the regular arrangement of these closely spaced dislocations and, therefore, significant perturbations in the boundary energy. These boundaries should therefore be able to move more easily as semicoherent interfaces via the lateral movement of secondary dislocations possessing step character by combined glide and climb. Such a process is generally similar to the motion of large-angle boundaries via the lateral glissile motion of line defects as described in Section 9.2.1.2. However, in the present case a climb component is also required. If a variety of line defects of this type is present whose non-conservative lateral motion can cause the white and black crystals to grow or shrink, it is possible for the boundary as a whole to move conservatively. In this process atoms are transported between the different climbing line defects (acting as line sources/sinks) in a manner which is conservative for the boundary as a whole while, at the same time, their collective lateral movements cause one crystal to grow and the other to shrink. In essence, the sweeping line defects provide special sites where atoms can be easily transported across the interface by shuffles as they glide and by diffusional transport as they climb, while the remainder of the interface remains unperturbed. Several examples are discussed in the following.

*Singular or vicinal CSL boundaries.*   We focus here on the important case of singular or vicinal CSL boundaries where the required line defects are extrinsic and may be present either initially or be produced during the·interface motion. To start with, it is easily demonstrated that line defects possessing dislocation and step character cannot be homogeneously nucleated on these boundaries in the form of dislocation loops under the influence of the pressures listed in Table 9.1 for grain boundaries. The energy to form a loop nucleus of radius $R$ possessing both step and dislocation character may be approximated by

$$\Delta G^L = \frac{\mu b^2 R}{2(1 - \nu)} \left[ \ln \left( \frac{4R}{r_o} \right) - 1 \right] + 2\pi R h \sigma^S - \pi R^2 h p. \qquad (9.22)$$

Here, the first term is the elastic energy contributed by the dislocation character of the loop (Hirth and Lothe 1982) where $r_o$ is the usual cutoff radius which is of the order of an interatomic distance. The second term is the core energy of the step where $\sigma^S$ is the energy, per unit area of the step and will be approximately equal to the interfacial free energy, $\sigma$. The last term is the energy gained when the interface advances the distance $h$ over the area of the loop. The radius of the critical nucleus, $R^*$, is determined from the condition $\partial \Delta G^L / \partial R = 0$, and the critical energy for nucleation, $\Delta G^{L*}$, can then be found by substituting $R^*$ into eqn (9.22) (Christian 1975). The nucleation rate is proportional to $\exp(-\Delta G^{L*}/kT)$ (Christian 1975), and significant nucleation rates generally require that $\Delta G^{L*} < 50\,kT$ (Cahn 1960). Using the values of $p$ in Table 9.1 for grain boundaries, and reasonable values of $h$, $\sigma^S$, $r_o$, and $b$, it is readily shown that $\Delta G^{L*}$ is too large to allow a significant nucleation rate, i.e. $\Delta G^{L*} > 50\,kT$.

However, such line defects can be present as the result of the impingement of lattice dislocations on the boundary. The impingement of a lattice dislocation on a singular boundary in the $\Sigma 5$ CSL is illustrated in Figs 9.10(a) and 9.11(a). The overall configuration is shown in Fig. 9.10(a), while the detailed boundary structure viewed edge-on along CB is shown in Fig. 9.11(a). In Fig. 9.11(a) the lattice dislocation has entered the boundary along the lattice glide plane shown dashed. The boundary is represented throughout in a rigid unrelaxed configuration within the framework of the $\Sigma 5$ DSC lattice for purposes of clarity: in this representation, the introduction of the lattice dislocation in Fig. 9.11(a) has opened up a gap of width $h^w = \hat{n} \cdot b = \hat{n} \cdot s^w$. (In a relaxed configuration the gap would be eliminated, the pairs of sites which are encircled would

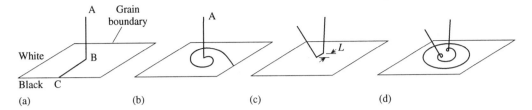

**Fig. 9.10** (a) Impinged length of lattice dislocation, CB, on grain boundary. (b) Spiralling of impinged dislocation around lattice dislocation (acting as a pole) by climb motion in the boundary. (c) Lattice dislocation impinged on boundary along length $L$. (d) Dislocation configuration in (c) acting as a Bardeen–Herring source by the generation of successive dislocation loops in the grain boundary via their climb motion.

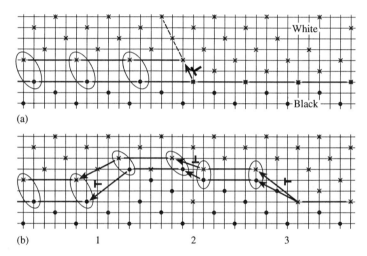

**Fig. 9.11** (a) View of impinged lattice dislocation segment CB shown in Fig. 9.10(a) when the grain boundary is a [001] symmetric $\Sigma 5$ tilt boundary. Configuration viewed along CB in Fig. 9.10(a) within the framework of the $\Sigma 5$ DSC lattice in an unrelaxed configuration. In this unrelaxed state a gap is present along the boundary on the left side of the dislocation. In the relaxed state the gap would be eliminated by pinching the encircled atoms together and discarding all overlapping atoms. (b) Dissociation of lattice dislocation in (a) into three DCS lattice dislocations possessing various dislocation and step characters. Again, the configuration is shown unrelaxed.

be pinched together, and all overlapping atoms would be discarded.) The result is a boundary dislocation with $s^w = b$, $s^b = 0$, $h^w = b \cdot \hat{n}$, $h^b = 0$. The lateral sweeping motion of this dislocation by combined glide and climb would cause the white crystal to either grow or shrink depending upon its direction of motion. If similar line defects were produced on the 'internal surface' of the black crystal, the interface could then move relative to both crystals as a result of the coordinated sweeping motion of both dislocations. One dislocation would act as a local source while the other would act as a sink causing one crystal to shrink and the other to grow. The directions of motion of the line defects would be determined by the direction of the pressure acting on the boundary, and the overall process for the boundary would be conservative.

However, such an impinged lattice dislocation will usually tend to dissociate in the boundary into DSC dislocations with smaller Burgers vectors causing a decrease in the energy of the system as discussed in detail in Section 12.4.3.1. One such conceivable dissociation into three DSC lattice dislocations possessing various dislocation and step character is illustrated in Fig. 9.11(b), again in an unrelaxed configuration. The total Burgers vector and step heights on each crystal are conserved in such a dissociation as discussed in Section 12.4.3.1 (see eqns (12.33) and (12.34)). Therefore,

$$\left.\begin{array}{l} b = b_1 + b_2 + b_3, \\ h^w = s^w \cdot \hat{n} = s_1^w \cdot \hat{n} + s_2^w \cdot \hat{n} + s_3^w \cdot \hat{n}, \\ h^b = s^b \cdot \hat{n} = 0 = s_1^b \cdot \hat{n} + s_2^b \cdot \hat{n} + s_3^b \cdot \hat{n}. \end{array}\right\} \tag{9.23}$$

If we approximate the elastic energy (per unit length) by $\mu b^2$ (corresponding to the well-known constant line tension approximation for a dislocation (Read 1953)) and the step core energy (per unit length) by $\sigma^S h$ (as previously), it is easily verified that the energy will be substantially reduced in this case by the dissociation.

These product dislocations will also be able to glide and climb in the boundary, and hence promote boundary motion. It should be noted, however, that in general, the line defects will not be able to glide/climb in the boundary via the periodic motion along them of well-defined elementary jog/kinks. In order for this to be possible the area swept out by the moving defect during each fluctuation producing its motion would have to be equal to (or an integer multiple of) the area of the primitive 2D unit cell defining its 2D periodic structure. In general, the area swept out will be smaller than this, and the advance of the line defect will therefore, be more irregular. For example, the area swept out when a single network site is created/destroyed must be $A = \Omega/(b \cdot \hat{n})$. For dislocation 3 in Fig. 9.11(b), $\Omega = a^3/4$, $|b| = a/\sqrt{10}$, and, therefore, $A = a^2\sqrt{10}/4$. However, the area of its 2D unit cell is $a^2\sqrt{10}/2$. Therefore, two network sites must be created/destroyed in order to sweep out one unit cell area.

The connected lattice dislocation segment AB has converted the white crystal into a helical ramp of pitch $b \cdot \hat{n} = h^w$, and it is therefore topologically possible for any product dislocation whose step height on the white crystal is equal to the pitch of the ramp to climb and wrap itself up in the form of a spiral similar to that illustrated in Fig. 9.10(b). Such a process would promote boundary motion, of course, since it would increase the line defect density. In the present example, dislocation 1 in Fig. 9.11(b) satisfies this requirement. Such a climbing dislocation would have to overcome any restraining line tension force and also any repulsive forces offered by the remaining dislocations as it passes by them during each revolution. These restraints could therefore produce threshold effects for boundary motion by this mechanism. An estimate of the minimum radius of curvature which can be achieved in the absence of repulsive forces can be obtained by representing the restoring line tension force (per unit length) by $(\mu b^2 + h\sigma^S)/R$ and the forward driving force (supplied by $p$) by $g_m h/\Omega = ph$. Equating these forces,

$$R_{\min} \simeq \frac{\mu b^2 + h\sigma^S}{ph}. \tag{9.24}$$

The impinged lattice dislocation may also appear as illustrated in Figs 9.10(c,d), in which case it could act as a Bardeen–Herring type source of dislocation loops (Bardeen and Herring 1951) only if $L \geqslant 2R_{\min}$. This follows from the fact that in order to form each loop, the dislocation segment $L$ (Fig. 9.10(c)) must be forced to bulge out between the pole dislocations by glide and climb in the form of a semicircle of radius $R_c = L/2$. This is the critical step, because $L/2$ is the minimum radius required at any stage of the process. We note that there is a great similarity between the present problem of producing line defects with step character on special interfaces and that of producing steps on the free surfaces of crystals during crystal growth or shrinkage. In the latter case it is well known (Frank 1949, Burton *et al.* 1951, Hirth and Pound 1963) that spiral steps are produced on the free surfaces of growing/shrinking crystals as a result of their intersections with lattice dislocations possessing screw character, and that these steps often control the kinetics of growth/shrinkage.

For dislocation 1 we find $R_{min} \simeq 2 \times 10^{-7}$ m, when $p = 10^7$ Pa and $2 \times 10^{-3}$ m when $p = 10^3$ Pa (see values of $p$ in Table 9.1), when $\mu = 4 \times 10^{10}$ Pa, $a = 4 \times 10^{-10}$ m, and $\sigma^S = 400$ mJ/m$^{-2}$. Therefore, at small driving pressures it will only be possible to activate sources with large effective values of $R_{min}$ or $L$. This will require large boundary areas and generally will not produce significantly high dislocation densities. However, the situation may often be alleviated locally at heterogeneities in the system. If applied stresses were present at some stage of the specimen history, the production of useful dislocations could be aided by stress concentrations. For example, in the source configuration illustrated in Fig. 9.10(c,d), a concentrated shear stress, $\tau_c$, would produce a force on the segment $L$ of magnitude $\simeq \tau_c b$. According to our previous discussion, sources with lengths as small as $L \simeq 2(\mu b^2 + h\sigma^S)/b\tau_c$ could then become active. In the case of grain boundary motion during recrystallization, local values of $p$ larger than the $\bar{p}$ values listed in Table 9.1 could occur leading to reduced values of $R_{min}$ or $L$.

The singular CSL boundary discussed above (Fig. 9.11) was of short period. On the basis of the discussion in Section 2.11.1, the products of the dissociation of the lattice dislocation should therefore be preserved as localized interfacial dislocations as was indeed assumed to be the case. However, for a singular CSL boundary of long period, with a low index plane in the white crystal parallel to a high index plane in the black crystal, as in Fig. 9.12, we would expect any product dislocations from the dissociation of a lattice dislocation to be delocalized parallel to the boundary. The dissociation products of an impinged lattice dislocation from the white crystal (Fig. 9.12(a)) could then minimize their total energy by forming a localized line defect corresponding essentially to a pure step of height $h = h^w = d^w(hkl) = h^b = m^b d^b(hkl)$ and a group of $m^b$ localized interface edge dislocations with small Burgers vectors of magnitude $d^b(hkl)$ dispersed along the interface as in Fig. 9.12(b). ($m^b = 5$ in the particular example shown.) Note that the energy driving the reaction is the decrease of the elastic energy of the dislocations which is achieved. The pure step would then be able to move conservatively in the interface by means of shuffling as described in Section 9.2.3, and it would be topologically possible for it to rotate and expand in the boundary using the impinging lattice dislocation as a pole. In many cases this would be a relatively easy process because of the relatively small line tension of the step given approximately by $h\sigma^S$.

*Singular or vicinal near-CSL boundaries.* For these boundaries one, or both, of the adjoining lattices will have low index planes parallel to the boundary in the reference structure (Section 4.3.1.5): also, since there is no exact CSL matching, it will not be possible to form pure steps as in the exact CSL case described above. However, a variety of localized line defects possessing both step and dislocation character may again exist depending upon the degree to which the spacings of the planes parallel to the boundary match and the length of the period of the grain boundary in the reference bicrystal. Under

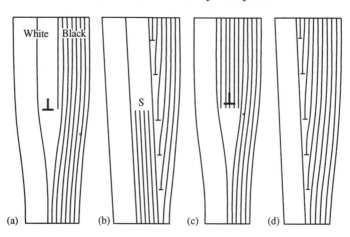

**Fig. 9.12** (a) Impingement of lattice dislocation in white crystal on a singular CSL grain boundary of long period. Low index plane in white crystal parallel to grain boundary and to high index plane in black crystal with $d^w(hkl) = 5d^b(hkl)$. (b) Dissociation of impinged lattice dislocation into a pure step of height $d^w(hkl)$ at S and five grain boundary edge dislocations, each with relatively small Burgers vectors, $b = d^b(hkl)$. (c) Impingement of lattice dislocation in black crystal on singular grain boundary where low index plane of white crystal is again parallel to the boundary and to high index plane of black crystal. (d) Dissociation of impinged lattice dislocation into grain boundary edge dislocations with relatively small burgers vectors, $b = d^b(hkl)$.

near-CSL conditions, the thickness of $m^w$ planes in the white lattice parallel to the boundary will be almost equal to the thickness of $m^b$ planes in the black lattice, and a line defect which is almost a pure step can be formed having a height $h \simeq m^b d^b(hkl) \simeq m^w d^w(hkl)$. In most cases the period of the grain boundary in the reference bicrystal will be long, and the strength of any Burgers vector component parallel to the boundary of the line defect will therefore be delocalized (see Section 4.3.1.5). The effective localized Burgers vector component of the line defect will therefore be perpendicular to the boundary and of strength $b = [m^w d^w(hkl) - m^b d^b(hkl)]$. When $d^w$ is considerably larger than $d^b$, an integral number of $d^b$ spacings will almost match one $d^w$ spacing, and a line defect can be produced with $b$ small in comparison with $h$. Such a line defect would be relatively efficient in transferring atoms across the interface, since its lateral motion would cause the addition/removal of a layer of material of thickness $h^w$ to/from the white crystal, the addition/removal of a layer of thickness $h^b$ to/from the black crystal, and require only a relatively small diffusional current of atoms to/from the line defect to make up the difference.

Line defects of the above type could again be produced at impinging lattice dislocations as illustrated schematically in Fig. 9.12. However, in this case the line defect at $S$ could not be a pure step as in the previous CSL case but would instead possess the small dislocation component just described. Such a line defect could again use the impinging lattice dislocation as a pole, since its step height on the white crystal is exactly equal to the pitch of the helical ramp in the white crystal produced by the impinging lattice dislocation. Of course, the driving pressure for boundary motion would again have to be sufficient to drive the line defect against its line tension and other forces tending to restrict its motion. Experimentally observed examples of spiral-like extrinsic dislocation configurations which resemble those under discussion are presented below in Section 9.2.6.2.

*Construction of a comprehensive model.* The construction of a comprehensive model for the overall rate of boundary motion in any of the above cases would inevitably be highly complex. Each line defect would experience a force urging it to move in a direction which would cause the boundary to advance in the direction of the pressure, $p$, acting on the boundary. As already mentioned, an ensemble of defects would generally be required, some acting as atom sources and some as atom sinks, in order to make the overall process conservative. The necessary diffusional transport of atoms between them could occur both along the boundary and through the two adjoining lattices. In addition, if no change in the macroscopic shape of the specimen were to occur (as would often be the case), the moving dislocations would have to be of the extrinsic type (as assumed above), and their Burgers would have to sum to zero. This is necessary, because the condition of no shape change requires that the net Burgers vector strength of all the dislocations passing any point in the interface must sum to zero. Since this cannot be satisfied by any intrinsic array, it can only be satisfied by employing a mix of cancelling extrinsic dislocations. The model would therefore have to quantify: (i) the types of line defects which participate; (ii) the geometrical arrangements of these defects in the interface; and (iii) the rate at which these line defects glide and climb in the interface taking into account any possible impediments to their motion which could slow the rate of boundary motion and even cause threshold effects.

In some cases where the dominant line defects are almost perfect steps, it is conceivable that the rate of motion could be controlled by the local rate of shuffling at the cores rather than by the rate of diffusional transport, thereby greatly simplifying the model. However, in all cases many of the critical factors are difficult to quantify. The most serious problem is the lack of any way to predict the density (and arrangement) of the necessary line defects in the boundary. As discussed previously, this will generally involve a variety of factors unrelated to the intrinsic properties of the boundary itself and will often depend upon the boundary history. In many cases it is clear that only low densities may be present causing low mobilities. Efforts to construct models for the motion on the basis of various simplifying assumptions have been made by Gleiter (1969) and Smith *et al.* (1980). In view of the complexity of the problem, and the difficulties in achieving a realistic description, we shall not attempt to develop them further.

### 9.2.2.3 *Heterophase interfaces*

There are numerous situations in so-called 'massive' polymorphic transformations (Massalski 1970, Barrett and Massalski 1980, Hornbogen 1983) where an interface separating two phases of the same composition but different crystal structure moves conservatively in a non-glissile manner by means of thermally activated atomic jumping processes. The observed interfaces can be either smoothly curved or faceted (Kittl *et al.* 1967), and, in some cases, surface relief is observed (Menon *et al.* 1988). The latter two observations indicate that, in many cases, the interfaces are singular (or vicinal) and move by means of the lateral motion of line defects possessing localized dislocation and step character. Plichta *et al.* (1984) and Menon *et al.* (1988) have suggested that these types of interfaces may actually be dominant in many massive transformations.

Models for the thermally activated motion of such interfaces will closely resemble those discussed above for singular (or vicinal) large-angle grain boundaries. It must be be recalled, however, that exact CSL's will not exist for any heterophase interfaces, and that the near-CSL model will therefore apply. The sweeping motion of all line defects possessing step character will therefore be non-conservative to at least some extent. The rate at which atoms of species $i$ must diffuse to/from a unit length of line defect when it moves along the interface with a velocity $v$ will be be given by

$$\dot{N}_i = vX_i[h^b/\Omega^b - h^w/\Omega^w], \qquad (9.25)$$

where $X_i$ is the atom fraction of $i$ (which must be the same in both adjoining phases) and $\Omega^b$ and $\Omega^w$ are the atomic volumes in the adjoining phases. It may be seen that $\dot{N}_i$ will not be zero, in general, since $h^b \neq h^w$, and $\Omega^b \neq \Omega^w$. In addition, it is important to recognize that the rate at which such line defects will be able to move will depend upon the properties of *both* of the different phases adjoining the interface. For example, if one phase has a much higher cohesive energy than the other, the rates of the cooperative shuffling and diffusional transport necessary to sustain the line defect motion would tend to be controlled by that phase.

Suitable extrinsic line defects could again be produced by the impingement of lattice dislocations: it also seems possible that, in contrast to the situation for grain boundaries, they could be homogeneously nucleated in the form of loops during the interface motion in some cases because of the relatively high driving pressure which could be present due to the free energy of the transformation (Table 9.1). Use of eqn (9.22) shows that this is conceivable if $b$ is relatively small, and $p$ is near $10^8$ Pa.

### 9.2.3  Shuffling motion of pure steps

Singular or vicinal large-angle grain boundaries with exact CSL's which are capable of supporting localized pure steps ($h^w = h^b$, $b = 0$) can move within the framework of the CSL by the lateral motion of such steps. Here again, line defects are present which provide sites at which atoms can be easily transported across the boundary. When step heights of the order of atomic dimensions exist, the steps can move in a particularly simple manner by means of thermally activated shuffles. Since the line defects possess no dislocation character, the process is conservative at each step and no changes in specimen shape occur. Recently, a molecular dynamics computer simulation model (Jhan and Bristowe 1990a,b) has directly demonstrated such boundary motion in $\Sigma = 5$, 13, 17, and 29 CSLs. Bow-outs, consisting of stepped terraces, were constructed in otherwise flat [001] twist boundaries running parallel to the (002) planes of the white and black crystals in the CSL as illustrated in Fig. 9.13(a,b). Upon heating, the bow-outs, which, of course, were subjected to an inward capillary pressure, were eliminated by boundary motion which occurred by the progressive inward motion of the steps as illustrated in Fig. 9.13(b). The steps were found to migrate by means of the thermally activated shuffles of small groups of atoms at the steps which transferred atoms from one crystal to the other as seen in Fig. 9.13(c–f). Experimental evidence for this shuffling mechanism is discussed below in Section 9.2.6 (see, e.g., Fig. 9.14). Additional molecular dynamics studies (Majid and Bristowe 1987) have also shown that small 'pillbox' steps can form by thermally activated fluctuations on initially flat $\Sigma = 5$ [001] twist boundaries as illustrated schematically in Fig. 9.15. In this work, the atoms in the first plane of the black crystal facing the boundary in a unit cell of the CSL spontaneously shuffled to produce the displacements indicated by the vectors. These conservative shuffles caused the boundary to migrate temporarily into the black crystal by a distance equal to the step height, $a/2$, over the area of the CSL cell.

When the CSL is sufficiently dense in the plane parallel to the boundary, long lengths of boundary steps can assume ordered structures possessing well-defined kinks of relatively small dimensions as shown schematically in Fig. 9.16(a). The shortest lengths of

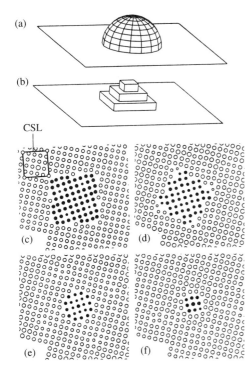

**Fig. 9.13** (a) Bow-out in otherwise flat grain boundary. (b) Bow-out in (a) resolved into atomic ledges. (c–f) Progressive removal of an atomic ledge (and the bow-out) in a Σ17 [001] twist boundary by the shuffling of atoms from crystal 1 (filled circles) to crystal 2 (open circles). Boundary observed along [001]. (From Jhan and Bristowe (1990b).)

the kink segments will correspond to the shortest available vectors of the CSL in the boundary plane. These steps can then migrate across the boundary by means of the motion of these kinks along their lengths just as kinks migrate along steps on free surfaces during crystal growth (Hirth and Pound 1963). The kink motion will only be periodic when small (elementary) kinks are involved, and the shuffling of a relatively small number of atoms suffices to move the elementary kink from one periodic position to the next. When this is not the case, the kink will have to change its configuration as it moves, and it will then move along by a sequence consisting of a variety of different types of shuffles.

The boundary velocity, $v$, can be readily expressed in terms of the average rate of thermally activated shuffling at the steps. If $n$ is the average number of atoms transferred in a shuffle, and $N$ is the total number of sites on steps (kinks) per unit boundary area at which shuffles can occur, then

$$v = n\Omega \cdot N\nu, \tag{9.26}$$

where $\nu$ is the average rate of at which shuffles occur. The energy change which occurs during an average shuffle in configuration space is shown in Fig. 9.17(a). Here, $G^s$ is the free energy barrier to the shuffle, and $\Delta g$ is the decrease in energy which is achieved when the boundary moves via the shuffle in the direction of the applied pressure. Therefore, according to standard rate theory, the rate of forward shuffles is given by

$$N \cdot \nu(+) = N \cdot \nu_0 \exp[-(G^s - \Delta g/2)/kT],$$

while the reverse rate will be

$$N \cdot \nu(-) = N \cdot \nu_0 \exp[-(G^s + \Delta g/2)/kT].$$

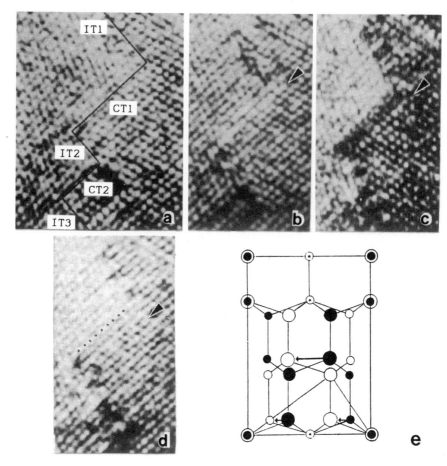

**Fig. 9.14** Experimental observation of the motion of a $\Sigma 3$ tilt boundary in Si by step motion via atomic shuffling within the framework of the $\Sigma 3$ CSL. (a) Initial stepped boundary structure. The facets consist of orthogonal $\{111\}$ $\{111\rangle$ and $\{112\}$ $\{112\}$ boundary segments labeled IT1 and CT1, respectively. (b) and (c) Changes in imaging contrast immediately preceding movement of facet labelled CT1 from initial position at arrow to final position indicated by row of dots in (d). In this motion the facet has moved forward by three $\{111\}$ interplanar spacings. (e) Atomic shuffles within framework of $\Sigma 3$ CSL unit cell capable of explaining the step motion. Shuffle displacement vectors are of the type $\frac{1}{6}\langle 112 \rangle$. (From Ichinose and Ishida (1990).)

The frequency, $\nu_o$, is of the order of the lattice Debye frequency. Therefore,

$$N\nu = N[\nu(+) - \nu(-)] = N\nu_o[\exp[-(G^s - \Delta g/2)/kT] - \exp[-(G^s + \Delta g/2)/kT]].$$
(9.27)

Since $\Delta g = pn\Omega$ and $\Delta g \ll kT$ under all expected conditions, we may expand eqn (9.27) to first order in $\Delta g/2kT$, set $G^s = E^s - TS^s$, and write eqn (9.26) in the form

$$v = \left[ \frac{Nn^2\Omega^2\nu_o \exp(S^s/k)}{kT} \cdot \exp(-E^s/kT) \right] \cdot p$$

$$= [M_o^B \cdot \exp(-Q^B/kT)] \cdot p = M^B \cdot p.$$
(9.28)

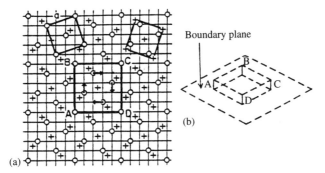

**Fig. 9.15** Formation of 'pillbox' shaped step by shuffling in initially flat [001] Σ5 twist boundary calculated by molecular dynamics using empirical interatomic potential for Cu. (a) Boundary viewed along [001]. Crosses indicated first (002) plane of crystal 1 below boundary midplane: circles indicate first (002) plane of crystal 2 above interface. Arrows indicate the shuffles which produced the pillbox shaped protuberance of crystal 1 into crystal 2 (of height a/2) illustrated in (b). (From Babcock and Balluffi (1989).)

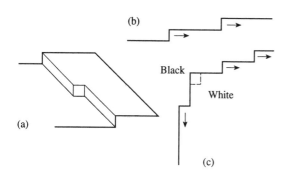

**Fig. 9.16** (a) Kink on a boundary step; (b) train of steps on a boundary; (c) heterogeneous nucleation of steps at a boundary corner.

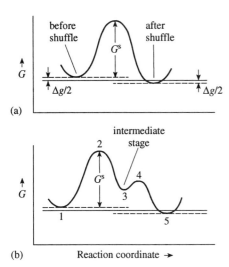

**Fig. 9.17** (a) Free energy of the system, $G$, as a function of the generalized reaction coordinate during an atomic shuffle: $G^s$ is the energy barrier to a shuffle; $\Delta g$ is the decrease in free energy of the system when the boundary moves via a shuffle. (b) Free energy of the system, $G$, as a function of reaction coordinate during a two-stage shuffle when a metastable energy minimum occurs at an intermediate stage of the overall shuffle.

Here, $Q^B = E^s$ = activation energy for the boundary motion and $M_o^B$ is the pre-exponential mobility factor.

It is also possible that vacancies present in the boundary core may play an important role in facilitating the shuffling process by 'loosening up' the local core structure at the kinks. If this is the case, the general forms of eqns (9.26) and (9.27) still remain valid. However, the quantity $N$ (i.e. the number of kink sites where successful shuffles can occur) must now be multiplied by the probability, $\Phi$, of finding a vacancy in a suitable nearby core site. We write this probability in the simple form

$$\Phi = z \cdot X_V^{eq} = z \exp(-G^f/kT) = [z \exp(S^f/k)] \cdot \exp(-E^f/kT), \qquad (9.29)$$

where $z$ is the number of suitable nearby sites, and it is assumed that all other quantities are suitably averaged over these sites. Therefore, using this result and eqns (9.26) and (9.27), the corresponding equations for the vacancy-assisted boundary velocity may be written as

$$v = \left[ \frac{zNn^2\Omega^2 v_o \exp[(S^s + S^f)/k]}{kT} \cdot \exp[-(E^s + E^f)/kT] \right] \cdot p.$$

$$= [M_o^B \cdot \exp(-Q^b/kT)] \cdot p = M^B \cdot p. \qquad (9.30)$$

The activation energy for boundary motion under these conditions is then $Q^B = E^s + E^f$. Again, as in the case of eqn (9.21) for the diffusion climb motion of a small-angle boundary, the velocity in both of the above situations takes the form of a simple product of a driving pressure, $p$, and a thermally activated mobility, $M^B$.

The above mobilities are directly proportional to $N$, the density of sites at which the shuffles can occur. In general, this important quantity will depend upon the density of steps available and also upon the density of favourable sites on these steps. The situation is therefore highly complicated, and $N$ may vary widely causing low mobilities in many situations. As discussed in Section 6.3.1.1, the thermodynamics of an interface which is stepped within the framework of a dense CSL should be very similar to that of a stepped free surface: both types of interfaces can therefore undergo roughening transitions during which a high density of pure steps develops, and the step energy approaches zero. Therefore, at temperatures above its roughening temperature (if one exists below the melting point) the interface will possess a high density of steps and a correspondingly high mobility. However, below its roughening temperature the interface will tend to be smooth with a greatly reduced mobility. In this situation steps may be present for intrinsic geometrical reasons as illustrated in Fig. 9.16(b), but they will tend to be eliminated during the interface motion. Such steps can therefore provide only a limited amount of movement.

In order to allow more extensive interface motion new steps must be generated. It is readily shown that these cannot be nucleated homogeneously on a smooth interface by means of thermal fluctuations in the presence of the pressures available (Table 9.1). At temperatures below the roughening transition a simple classical model can be employed (Burton *et al.* 1951, Cahn 1960) where the nucleus consists of a pillbox-shaped step of radius $R$ and height $h$ sitting on an otherwise flat interface. The calculation is then similar to that carried out previously for the homogeneous nucleation of a dislocation loop employing eqn (9.22). The energy required to form the pillbox step, $\Delta G^S$ is then the same as that given for $\Delta G^L$ by eqn (9.22) but without the elastic energy term contributed by the dislocation character of the loop. Therefore,

$$\Delta G^S = 2\pi R h \sigma^S - \pi R^2 h p. \qquad (9.31)$$

The critical nucleus radius, found from the condition $\partial \Delta G^S / \partial R = 0$ (Christian 1975), is then

$$R^{S*} = \sigma^S / p, \qquad (9.32)$$

and the corresponding critical energy for nucleation is

$$\Delta G^{S*} = \pi h (\sigma^S)^2 / p. \qquad (9.33)$$

Using values of $p$ from Table 9.1, and reasonable values of $h$ and $\sigma^S$, it is easily shown that $\Delta G^{S*}$ will be too large under all expected conditions to allow the spontaneous nucleation of any significant number of steps.

Any new steps must therefore be nucleated heterogeneously. Possible sites for such nucleation may be grain corners such as illustrated in Fig. 9.16(c) which would allow the black crystal to grow at the expense of the white crystal (but not vice versa). Any steps formed in this way would inevitably possess curvature as they migrated, and this would produce a backwards restraining force on their motion because of their line tension (energy per unit length). The effects of such curvature on the step migration kinetics have been ignored in developing eqns (9.28) and (9.30), but they are readily incorporated. The step line tension is $h\sigma^S$, and the backwards force (per unit length) due to it is $h\sigma^S/R$. The forward force due to $p$ is $ph$, and the net forward force is then $(ph - h\sigma^S/R)$. The energy $\Delta g$ in eqn (9.27), previously taken equal to $pn\Omega$, must therefore be replaced simply by $(pn\Omega - \sigma^S n\Omega/R)$, and $p$ in eqns (9.28) and (9.30) must be replaced by $(p - \sigma^S/R)$. Migration is therefore impossible when $R < \sigma^S/p$. (We note that this critical value corresponds to the critical radius for nucleation given by eqn (9.32).) It may therefore be concluded generally, that, in view of the above difficulties, interface mobility due to the motion of pure steps at temperatures below any roughening temperature may be extremely small under many circumstances.

### 9.2.4 Uncorrelated atom shuffling and/or diffusional transport

The situation at general large-angle grain boundaries and heterophase interfaces which are general with respect to all degrees of freedom is quite different than in singular or vicinal interfaces, since here the closely spaced primary dislocations are arranged irregularly, and the interface, which is incoherent, cannot support sufficiently localized and physically significant line defects (i.e. secondary dislocations or steps) to facilitate the transfer of atoms across the interface. Such interfaces are generally curved to at least some extent and are everywhere parallel to high index planes of the two adjoining lattices over distances longer than the scale of any local facetting (see Section 4.3.2). They pass through a wide variety of local regions in their dichromatic patterns, all possessing different atomic environments, and their core structures, therefore, vary along the interface. Consequently, they are incapable of supporting localized line defects which are of any physical significance. When such an inhomogeneous interface is moved normal to itself through the dichromatic pattern, the structure at any fixed point in the interface will therefore vary with time. However, as the interface moves, various local regions with characteristic structures will continuously appear and disappear while the overall energy remains essentially constant.

In contrast to the situation for singular or vicinal interfaces, there will be many places in the irregular interface structure where it can be perturbed relatively easily. It is therefore likely that atoms can be transported across the boundary by either of the two basic mechanisms illustrated in Figs 9.18 and 9.19 for the case of grain boundaries. Atoms

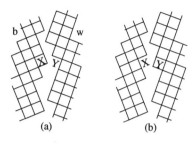

**Fig. 9.18** Transfer (a) → (b) of single atom from black crystal to white crystal across the boundary via a localized shuffle. Boundary is general with respect to all degrees of freedom.

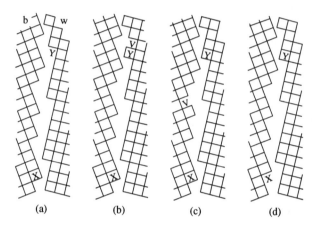

**Fig. 9.19** Transfer of single atom from black crystal to white crystal across boundary via a diffusional transport process. Boundary is general with respect to all degrees of freedom. (a) Initial configuration; (b) formation of vacancy in white crystal next to Y; (c) diffusion of vacancy away from Y and across the boundary; (d) further diffusion of vacancy to the point where it displaces the atom in the black crystal at X and is annihilated.

in the boundary can always be assigned to either the black crystal or the white crystal as in the unrelaxed configuration shown. In the first mechanism (Fig. 9.18), network sites are simply transferred across the interface at various places by localized atom shuffling. For the reasons given above, these transfers in a general interface should be uncorrelated, and we therefore term this mechanism 'uncorrelated atom shuffling'. In Fig. 9.18 the shuffle involves only one atom. However, in general, more than one atom may be involved as was the case in our previous treatment of correlated shuffling at interfacial steps in Section 9.2.3 (see eqn (9.28)) where it was assumed that, on average, $n$ atoms were transferred per shuffle, where $n$ is typically between 1 and 10.

In the second mechanism (Fig. 9.19), atoms (network sites) are transferred across the interface via the diffusion of point defects which are created and destroyed at different places in the interface in a manner which is again uncorrelated. In the example shown, which is again for a grain boundary, a vacancy is created in the white crystal at a location next to Y, abutting the interface, thereby creating a new network (lattice) site in that crystal (Fig. 9.19(a,b)). This vacancy then diffuses along the interface [Fig. 9.19(c)] and eventually arrives at a location in the black crystal abutting the interface, X, where it replaces an atom and is annihilated, thereby annihilating a network site in the black crystal (Fig. 9.19(d)). The net result is the transfer of a lattice site from the black crystal to the white crystal while conserving the total number of lattice sites. In this process the interface in the region of the white crystal gaining the site acts as a local vacancy source, and the interface in the region of the black crystal losing the site acts as a vacancy sink. We therefore term this mechanism 'uncorrelated diffusional transport', since relatively long-range defect (atom) diffusion is involved. For interfaces where an interstitial point defect related diffusion mechanism is dominant (see Chapter 8), an interstitial point defect can be created by detaching an atom from the black crystal at the interface, thereby annihilating a network site in that crystal. It can then diffuse to another location in the interface where it can join the white crystal, thereby creating a network site in that crystal. Again, the result is the transfer of a site from the black crystal to the white crystal.

Each of these mechanisms should be able to operate inhomogeneously throughout the interface by utilizing various sites which are momentarily favourable. Each successful fluctuation would, of course, perturb the structure locally, but further rearrangements (relaxations) would then occur to reduce any cumulative effects caused by these perturbations and restore the equilibrium structure to the maximum degree possible. The boundary would then move forward inhomogeneously in this way while keeping its structure close to its equilibrium structure. Examples are the sites indicated the Xs and Ys in the initial configurations shown schematically in Figs 9.18(a) and 9.19(a) for the two mechanisms. We note that, strictly speaking, we should regard the Y sites in the final configurations, i.e. Figs 9.18(b) and 9.19(d), as bound interstitials in the white crystal and the X sites as bound vacancies in the black crystal, since it is not established that the added atoms at the Y sites in the white crystals are fully incorporated in the white crystals and that the vacancies at the X sites in the black crystals are fully destroyed vacancies. However, it must be realized that the boundary is not fully equilibrated everywhere at any stage of its motion, and that these defects are presumably at least strongly bound at these locations. Further readjustments of the boundary structure may be necessary to make these transfers permanent. Many events of these types will occur during the motion, and statistically, many of them will lead to the permanent addition of network sites to the white crystal and corresponding loss of sites at the black crystal as the boundary structure undergoes continuous change at the local level during its motion. We may characterize this situation by saying that there is a multitude of local minimum energy structures which the boundary may access.

Boundary motion by these uncorrelated atom transfer mechanisms will be of the 'civilian' type in contrast to the 'military' type which is produced by the highly correlated transfers which accompany the sweeping lateral movement of secondary dislocations or steps. Because of the uncorrelated nature of these civilian type transfers, no macroscopic shape change will be produced.

The above discussion has focused only on grain boundaries. Clearly, however, similar mechanisms will hold for general heterophase interfaces.

### 9.2.4.1 *Uncorrelated atom shuffling*

We consider first a simple model for the motion of a general grain boundary by the uncorrelated shuffling mechanism. An energy diagram for a simple two-stage shuffle process is illustrated in reaction coordinate space in Fig. 9.17(b). We imagine that a successful shuffle involves an initial shuffle from the initial state at 1 to an intermediate metastable state at 3. This is then followed by a further relaxational shuffle to the final state at 5 which is relatively easy, i.e. the barrier $(G_4 - G_3)$ is relatively small.

In the steady state, there are $N$ sites in states 1 and 5 and $N_3$ sites in state 3. Therefore,

$$N \cdot \nu_o \exp[-(G_2 - G_1)/kT] - N_3 \cdot \nu_o \exp[-(G_2 - G_3)/kT]$$
$$= N_3 \cdot \nu_o \exp[-(G_4 - G_3)/kT] - N \cdot \nu_o \exp[-(G_4 - G_5))kT] = 0.$$

Since $(G_2 - G_1) > (G_4 - G_5)$, and $(G_2 - G_3) > (G_4 - G_3)$,

$$N_3 \simeq N \cdot \exp[-(G_3 - G_5)/kT],$$

and the net forward rate is then

$$Nv \simeq N\nu_o \exp[-(G_2 - G_1)/kT] - N_3 \nu_o \exp[-(G_2 - G_3)/kT]$$
$$\simeq N\nu_o [\exp[-(G_2 - G_1)/kT] - \exp[-(G_2 - G_5)/kT]]. \qquad (9.34)$$

This result is identical to that obtained previously, i.e. eqn (9.27), for the simpler reaction path illustrated in Fig. 9.17(a) and indicates that the kinetics will be essentially the same in both cases as long as no large barriers exist for the complete relaxation of the initial shuffle. The boundary velocity is therefore again given by an equation of the form of either eqn (9.28) or (9.30), depending upon whether or not the shuffles are aided by the presence of nearby vacancies in the core. However, the quantity $N$ is now an intrinsic geometric property of the boundary corresponding to the average density of sites where successful shuffles may occur. The mobility is therefore independent of $p$, and the velocity is linearly proportional to $p$, so that we may write

$$v = M^B \cdot p = M_o^B \exp(-Q^B/kT) \cdot p, \qquad (9.35)$$

where $M^B$ is given by eqn (9.28) or (9.30) with $N \neq N(p)$. It is anticipated that the intrinsic mobilities of general grain boundaries by the above shuffle mechanism will tend to be higher than that of special boundaries because the disordered character of general boundaries will provide a relatively high density of intrinsic sites where the shuffling is relatively easy.

Finally, we note that a generally similar model may be readily developed for the motion of a general heterophase interface by the same basic mechanisms.

### 9.2.4.2 *Uncorrelated diffusional transport*

Consider next a simple model for the motion of a general grain boundary by the uncorrelated diffusional transport mechanism. Unfortunately, little is known specifically about the structural details of the creation/destruction of point defects (network sites) at general boundaries. However, a rudimentary model for the boundary motion can be constructed by taking the boundary to be a thin slab of bad material and simply assuming that at any instant there are $Z$ sites distributed in the slab (per unit area) where point defects can be thermally generated as, for example, at the site next to the $Y$ site in the white crystal in Fig. 9.19. We may assume that half of these sites are associated with each abutting crystal. Defects are then generated in the boundary at the total rate

$$\dot{N}(+) = 2\{Z/2\}\nu \exp[(S^f + S^m)/k] \exp[-(E^f + E^m)/kT]\}, \qquad (9.36)$$

where $\nu$ is an atomic vibrational frequency of the order of the Debye lattice frequency, and $E^f$ and $E^m$ and $S^f$ and $S^m$ are suitably averaged energies and entropies of defect formation and migration in the boundary slab. Also, they are destroyed at a total rate which is proportional to their concentration, $X$, and the rate at which they jump, which, in turn is proportional to $\exp(S^m/k) \cdot \exp(-E^m/kT)$. Therefore,

$$\dot{N}(-) = AX \exp(S^m/kT) \exp(-E^m/kT), \qquad (9.37)$$

where $A$ is a constant. At equilibrium, $\dot{N}^{eq}(+) = \dot{N}^{eq}(-)$, and $X = X^{eq} = \exp(S^f/k) \cdot \exp(-E^f/kT)$. Therefore, $A = Zv$.

We now put the boundary into motion by applying a pressure, $p$, and allowing a steady state to develop. The energy of point defect formation at the $Z/2$ sites associated with one side of the interface is now increased by $p\Omega/2$ and decreased by $p\Omega/2$ at the $Z/2$ sites associated with the other side. This causes the total rate of defect production to be decreased on one side and increased on the other. However, the total rate of defect production, given by

$$\dot{N}(+) = (Z/2)v\exp[(S^f + S^m)/k] \cdot \exp[-(E^f + E^m)/kT] \cdot$$
$$[\exp(p\Omega/2kT) + \exp(-p\Omega/2kT]$$
$$\simeq 2\{(Z/2)v\exp[(S^f + S^m)/k] \cdot \exp[-(E^f + E^m)/kT], \qquad (9.38)$$

remains essentially unchanged. (We note that eqn (9.38) was obtained by expanding the exponentials to first order in $p\Omega/2kT$, since, as usual, $p\Omega/2 \ll kT$.) Since $\dot{N}(+) = \dot{N}(-)$ in the quasi-steady state, and $\dot{N}(-)$ is simply proportional to $X$, the destruction rate at each side, and also the point defect concentration, remain unchanged. However, since the production rate is increased on one side and reduced on the other (eqn (9.38)), atoms are transferred across the boundary at a steady rate corresponding to

$$\Delta\dot{N} = \{v\exp[(S^f + S^m)/k]\exp[-(E^f + E^m)/kT]\}(Z\Omega/4kT)p. \qquad (9.39)$$

According to the results in Chapter 8 (i.e. eqn (8.52)), the self-diffusivity in the boundary is essentially the product of a combined geometrical factor of order unity, a correlation factor, a jump distance squared, an atomic vibrational frequency, an entropy factor, and a Boltzmann factor which involves the sum of the formation and migration energies of the migrating point defects which produce the self-diffusion. We may therefore replace the bracketed term in eqn (9.39) approximately by

$$v\exp[(S^f + S^m)/k] \cdot \exp[-(E^f + E^m)/kT] \simeq D^B/l^2 f^B, \qquad (9.40)$$

where $D^B$ is the boundary self-diffusivity, $l$ is the jump distance, and $f^B$ is the correlation factor. Therefore, since $v = \Delta\dot{N}\Omega$, we finally have, by combining eqns (9.39) and (9.40),

$$v \simeq \left[\frac{Z\Omega^2 D^B}{4l^2 f^B kT}\right]p = M^B p = M_o^B \exp(-Q^B/kT) \cdot p, \qquad (9.41)$$

where $M^B$ is independent of $p$. The boundary velocity by the diffusional transport mechanism, i.e. eqn (9.41), is therefore of the same general form as for the shuffle mechanism, i.e. eqn (9.35). However the mobilities are quite different. Little is known about the critical parameters $N$ and $Z$ in these models, and it is therefore difficult to reach any conclusion, theoretically, about the relative importance of the shuffle mechanism versus the diffusional transport mechanism for general grain boundaries. However, the two models are tested against available experimental data below in Section 9.2.6.1 where it is found that the atom shuffling mechanism is apparently favored, at least in the case of general grain boundaries in Al.

Finally, we note that a generally similar model could be developed for the motion of a general heterophase interface by the same basic mechanism. However, it would be more complicated because of the asymmetry associated with the different structures and properties of the two phases adjoining the interface.

### 9.2.5  Solute atom drag

Up to now, we have described boundary motion in ideally pure materials and have ignored possible effects due to the interaction of the boundary with solute atoms which may be present in solution in the system, either by design, or inadvertently. When the temperature is high enough to allow thermally activated solute atom jumping, there is extensive evidence (see Section 9.2.6) that solute atoms which interact with the boundary, as discussed in Chapter 7, can often be forced to diffuse along with the boundary under the influence of the interaction forces. Under many circumstances, the rate at which they can diffuse along with the boundary is slow relative to the rate at which the boundary could otherwise move, and they therefore exert a backwards 'drag force' on the boundary which tends to impede its motion. In many cases, this drag force dominates the kinetics and therefore must be taken into account.

Models for this effect, for general grain boundaries, have been developed by a number of workers, e.g. Cahn (1962), Lücke and Stüwe (1971), Hillert and Sundman (1976), and Westengen and Ryum (1978). However, the problem is highly complex, and critical parameters are often unknown. Therefore, none of the models can be regarded as being definitive. The early model of Cahn (1962), which involves a number of practical simplifications and is essentially equivalent to that of Lücke and Stüwe, contains many of the essential physical aspects of the problem, and we shall therefore describe it in some detail.

The goal is to calculate the steady state motion of a general boundary in the presence of diffusing solute atoms with which it interacts. It is assumed initially that the solute binding energy to all of the available sites in the boundary and adjoining lattice, $g^b(x)$, varies with distance from the boundary core, $x$ in the simplified manner illustrated in Fig. 9.20. The solute atoms will then be diffusing in the force field provided by the interaction. For a system consisting of host atoms (species A), solute atoms (species B), and vacancies, eqn (5.149) for the diffusion of the solute atoms takes the form

$$J_B = -B_{BB}\nabla M_B - B_{BV}\nabla M_V, \qquad (9.42)$$

if the $A$ species is taken as the $C$th component. For a dilute solution, and neglecting solute–solute and solute–vacancy interactions, we then follow the same procedure used to derive eqn (9.14) to obtain

$$M_B = N_o[G_B^o + G_B^b(x) - G_A^o + kT\ln X_B]; \qquad (9.43)$$

$$M_v = N_o[G_v^f - G_A^o + kT\ln X_v]. \qquad (9.44)$$

Furthermore, assuming that the vacancies are maintained at equilibrium, $M_v = -N_o G_A^o$. Substituting these results into eqn (9.42), and dropping the B subscript, the diffusional flux of solute atoms is then

$$J = -N_o B\left[\frac{\partial g^b}{\partial x} + \frac{kT}{X}\cdot\frac{\partial X}{\partial x}\right]. \qquad (9.45)$$

In the lattice region, $\partial g^b/\partial x = 0$, and $J = -D\cdot\partial n/\partial x$, where n is the solute concentration (atoms per unit volume). Therefore, $B = Dn/N_o kT$. The flux in eqn (9.45) is measured with respect to a coordinate system fixed to the material in which the diffusion is occurring. In order to obtain a steady state solution, we therefore switch to a coordinate system fixed to the moving boundary in which the flux is given by

$$j = J - vn. \qquad (9.46)$$

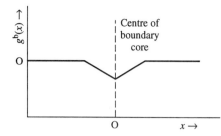

Fig. 9.20 Variation of solute atom binding energy to the grain boundary, $g^b(x)$, with distance from the centre of the boundary core, $x$, assumed in the solute atom drag model of Cahn (1962).

The diffusion equation in this coordinate system in the steady state is then

$$\frac{\partial n}{\partial t} = -\operatorname{div} j = \frac{\partial}{\partial x}\left[D\frac{\partial n}{\partial x} + \frac{Dn}{kT}\cdot\frac{\partial g^b}{\partial x} + vn\right] = 0, \tag{9.47}$$

where $D$ and $g^b$ are functions of position relative to the boundary, i.e. $D = D(x)$ and $g^b = g^b(x)$. As may be verified by direct substitution, a solution of eqn (9.47) may be written in the form

$$n(x) = n^L\cdot v\cdot\exp\left[-\frac{g^b(x)}{kT} - v\int_{x_0}^{x}\frac{\mathrm{d}\eta}{D(\eta)}\right]\cdot\int_{-\infty}^{x}\exp\left[-\frac{g^b(\xi)}{kT} + v\int_{x_0}^{\xi}\frac{\mathrm{d}\eta}{D(\eta)}\right]\cdot\frac{\mathrm{d}\xi}{D(\xi)}, \tag{9.48}$$

where $n^L$ is the lattice concentration far from the boundary. The drag force exerted on the boundary by the total distribution of solute atoms may now be found by summing the force exerted by each solute atom over the distribution, $n(x)$. The force per atom is given by $\partial g^b/\partial x$, as may be verified by realizing that the change in energy of the system due to a virtual displacement of the boundary relative to a fixed solute atom is $\mathrm{d}w = -(\partial g^b/\partial x)\cdot\mathrm{d}x$. Using eqn (9.1), the total drag force (per unit area) on the boundary is then

$$p_d = \int_{-\infty}^{\infty} n\frac{\partial g^b}{\partial x}\,\mathrm{d}x. \tag{9.49}$$

In order to find the velocity, $v(p, n^L, T)$, which is achieved upon the application of an applied pressure, $p$, in the presence of the bulk solute atom concentration, $n^L$, which produces the drag pressure, $p_d = p_d(v, n^L, T)$, it is assumed as an approximation that the applied pressure must be equal to the drag pressure plus an intrinsic pressure, $p_i$, which is equal to the pressure which would move the same boundary with the velocity $v$ in the absence of solute atoms. We note that this assumption may be unrealistic in some situations, since it ignores any changes in boundary structure due to the solute atoms and other possible effects. Therefore,

$$p(v, n^L, T) = p_i(v, T) + p_d(v, n^L, T). \tag{9.50}$$

We note that eqn (9.50) may be readily interpreted in terms of the energy dissipated at the interface. If we multiply it through by $v$, we obtain

$$pv = p_iv + p_dv \tag{9.51}$$

which expresses the fact that the total rate of energy dissipated at the interface, $pv$, is the sum of that contributed by the intrinsic migration process, $p_iv$, and that contributed by the drag, $p_d v$. Also, we can use the basic relation $v = Mp$, defining the mobility, to write eqn (9.50) in the equivalent form

$$\frac{1}{M} = \frac{1}{M_i} + \frac{1}{M_d},$$  (9.52)

where $M_d$ is defined as a 'solute atom drag' mobility. Unfortunately, eqn (9.50) is an implicit relation for $v(p, n^L, T)$, and an exact closed form solution cannot be obtained. However, approximate solutions for the high and low velocity extremes may be obtained, and the general intermediate behaviour of the model may be worked out by interpolation (see Cahn (1962) for details). Some of the main results, which are quite complicated, since $v = v(p, n^L, T)$ are illustrated in Fig. 9.21.

The concentration profile of the segregated atoms at the boundary is shown in 9.21(a) for different boundary velocities. When $v = 0$, the distribution is symmetric, as expected, and there is no drag force. However, as $v$ increases, the distribution becomes increasingly non-symmetric, and in addition, increased numbers of segregated solute atoms are lost. A drag force is therefore present because of the non-symmetric form of the distribution. The magnitude of the drag force, of course, depends in a complex manner upon both the numbers of solute atoms present and the degree to which they are distributed non-symmetrically. The final boundary velocity (at constant $T$) as a function of the driving force for different solute concentrations, $n^L$, is shown in Fig. 9.21(b). For the pure material, the boundary movement is intrinsic, and $v$ is proportional to $p$ as, for example, in eqn (9.35). When $n^L > 0$, the boundary velocity is always reduced. At sufficiently small driving forces, we again have $v$ proportional to $p$. However, in this regime, solute atoms are strongly segregated, the drag force is dominant, and the boundary velocity is controlled by the forward rate of solute diffusion (i.e. we are completely in the extrinsic regime). However, as $p$ (and therefore $v$) increases, solute atoms are progressively desorbed allowing the boundary to move faster. Eventually, at sufficiently high driving forces (and large $v$'s), solute atoms can no longer keep up with the boundary and are essentially desorbed, and the boundary velocity approaches intrinsic behaviour. For sufficiently high solute concentrations, a region of instability (dashed) is seen to occur between the extrinsic and intrinsic regimes as the solute atoms desorb. This is accompanied by a 'breakaway' phenomenon when the boundary suddenly breaks free of interacting solute atoms. In Fig. 9.21(c), it is seen that intrinsic behaviour tends to be achieved at elevated temperature (because of the thermal desorption of the solute atoms) and extrinsic behaviour takes over at low temperatures in a manner which is exacerbated by increasing solute concentrations. The observed activation energy (obtained from the slope of the ln v versus $1/T$ curve) in the intrinsic regime corresponds, for example, to $Q^B$ in eqn (9.35), whereas in the extrinsic regime it corresponds to the activation energy required for the solute atom diffusion, which in Fig. 9.21(c) is taken to be larger than $Q^B$ as is generally expected. Finally, in Fig. 9.21(d) it is seen that intrinsic behaviour can be obtained over a range of solute concentration as long as the driving force is sufficiently high. This range decreases with decreasing $p$ until eventually it becomes negligible. In general, therefore, the impeding effects due to solute segregation tend to become increasingly important as the solute concentration increases, and the driving force and temperature decrease as might be expected intuitively.

Qualitatively similar solute atom drag effects may be expected for both small-angle and special large-angle grain boundaries. In the former case, the solute atoms will interact in a similar manner with the cores of the lattice dislocations comprising the boundary. In the latter case, preferential drag interactions with the sweeping boundary line defects

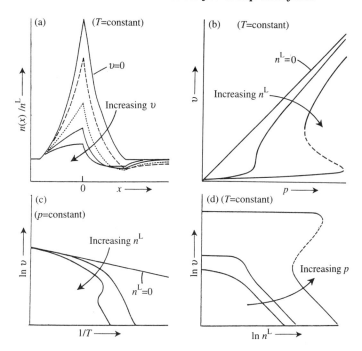

**Fig. 9.21** Behaviour of solute atom drag model for grain boundaries according to model of Cahn (1962). (a) Concentration profile, $n(x)/n^L$, of solute atoms segregated at grain boundary as a function of increasing boundary velocity, $v$. $n(x)$ = solute atom concentration, $n^L$ = solute concentraion in bulk lattice. (b) Boundary velocity as a function of pressure on the boundary, $p$, as influenced by increasing $n^L$. (c) $\ln v$ as a function of $1/T$ as influenced by increasing $n^L$. (d) $\ln v$ as a function of $\ln n^L$ as influenced by increasing $p$. ((a) and (b) from Cahn (1962). (c) and (d) from Simpson *et al.* (1976).)

which produce the boundary motion will be involved. Also, similar behavior should be found for various types of heterophase interfaces.

### 9.2.6 Experimental observations of non-glissile (thermally activated) grain boundary motion in metals

We now review relevant experimental observations of non-glissile (thermally activated) motion and discuss them in terms of the models described in the previous sections. By far the most studies have been made on grain boundaries in metal systems, and we therefore focus on these observations.

#### 9.2.6.1 *General large-angle grain boundaries*

We begin with grain boundaries which are general with respect to all degrees of freedom, since they have received the most attention. As might be expected, direct electron microscope observations of such incoherent boundaries when they are moving (e.g. Babcock and Balluffi 1989) have failed to reveal the basic mechanisms of migration. This, of course, can be attributed to the absence of any localized line defects and the small number of atoms which are involved in the proposed mechanisms at any instant. We are, therefore, restricted to an analysis of measured boundary mobilities.

An immense literature exists in which boundary velocity has been measured in a wide range of situations involving differing driving forces, temperatures, and solute atom (impurity) concentrations (e.g. Gleiter and Chalmers 1972, Simpson and Aust 1972, Lücke, *et al.* 1972, Higgins *et al.* 1974, Simpson *et al.* 1976, Haessner and Hofmann 1978, Smith *et al.* 1980, Bauer 1982, Bauer and Lanxner 1986, Bauer 1989). Unfortunately, it appears likely that in a majority of cases, the observed kinetics were strongly influenced by solute atom drag effects. Solute atoms, often unidentified and at very small concentrations, were frequently present at sufficient strength to influence the kinetics drastically. For example, Aust and Rutter (1962) found that general boundaries in Al moved up to three orders of magnitude faster in 12-pass zone refined Al than 4-pass Al under otherwise similar conditions. Also, as seen in Fig. 9.22, the mobilities of general boundaries were two orders of magnitude lower in 99.9992 at.% Al than in the higher purity 99.999 95 at.% Al in some cases. The fact that such large effects can be produced by such small impurity concentrations makes the interpretation of the measurements unusually treacherous. A major question in many cases is whether the observed boundary motion was 'intrinsic', i.e. free of solute atom drag effects and therefore describable in terms of our previous models for such motion, or else 'extrinsic', and therefore dominated by solute drag effects.

The simplest case is intrinsic motion. As discussed in Section 9.2.5, this type of motion should be favoured by high purity, high driving force, and elevated temperature. Also,

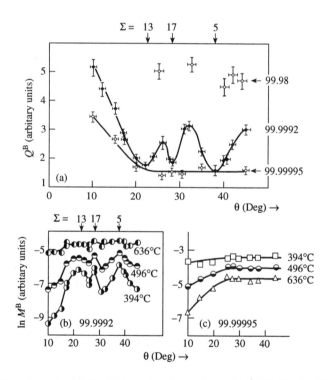

**Fig. 9.22** Mobility data for ⟨001⟩ tilt boundary motion in Al of 99.98, 99.9992, and 99.999 95 purity. (a) Mobility activation energy, $Q^B$, as a function of tilt angle, $\theta$. Tilt angles corresponding to $\Sigma = 13, 17$, and 5 indicated by arrows. (b) ln $M^B$ as a function of $\theta$ for 99.9992 purity $M^B$ = boundary mobility). (c) Same as (b) except for 99.999 95 purity. (From Fridman *et al.* (1975).)

since our models suggest that general boundaries should have higher intrinsic mobilities than singular or vicinal boundaries, and since solute atom drag tends to slow down boundary motion, the intrinsic mobilities of general boundaries should be the highest of any large-angle boundaries. Unfortunately, intrinsic motion appears to have been achieved in only relatively few experiments. By far the most mobility data, over the widest temperature range, have been obtained for the metal Al, and in Fig. 9.23 we have plotted selected data (solid lines) which appear to correspond to intrinsic motion. In order to obtain these data boundary velocities were measured under relatively large known driving forces provided by the stored energy of plastic deformation or 'reversed capillarity' using the 'Sun–Bauer' technique (Sun and Bauer 1970a), and values of $M$ were determined. The data are characteristic of the average of the fastest rates at which the boundaries were found to move in the different investigations. In some of the experiments, some boundaries were observed to move significantly more slowly than others depending upon boundary type, temperature, and specimen purity. These boundaries were evidently either general boundaries whose motion was hindered by solute drag or singular (or vicinal) boundaries whose intrinsic motion was slower than that of general boundaries as discussed below. The data for these boundaries were therefore ignored. The data (Fig. 9.23) show less than an order of magnitude scatter and can be fit reasonably well by the mobility relation $M^B = M_o^B \exp(-Q^B/kT)$ (see eqns (9.35) and (9.41)) with $M_o^B = 1.4 \times 10^{-3}\,\mathrm{m^4\,J^{-1}\,s^{-1}}$ and $Q^B = 55\,\mathrm{kJ/mole}$. We note that the AR and FKS data (dashed lines), which are also shown, were obtained using driving forces which were lower than those used in the other experiments, and they were evidently influenced by solute atom drag effects which reduced the mobilities.

We now attempt to understand these results in terms of the uncorrelated shuffling models which led to eqn (9.35). The observed energy $Q^B = 55\,\mathrm{kJ/mole}$ is in the range conceivable for boundary shuffles with or without the assistance of nearby vacancies. Also, setting $M_o^B$ in eqns (9.28) and (9.30) equal to the observed value $1.4 \times 10^{-3}\,\mathrm{m^4\,J^{-1}\,s^{-1}}$, and adopting the reasonable values $\nu_o = 3.9 \times 10^{12}\,\mathrm{s^{-1}}$ (i.e. half

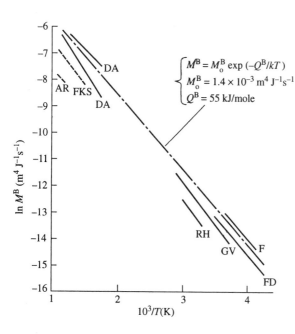

$$\begin{cases} M^B = M_o^B \exp(-Q^B/kT) \\ M_o^B = 1.4 \times 10^{-3}\,\mathrm{m^4\,J^{-1}s^{-1}} \\ Q^B = 55\,\mathrm{kJ/mole} \end{cases}$$

**Fig. 9.23** Measured $\ln M^B$ versus $1/T$ for grain boundaries in high-purity Al ($M^B$ = boundary mobility). AR = Aust and Rutter (1962); FKS = Fridman *et al.* (1975); DA = Demianczuk and Aust (1975); RH = Rath and Hu (1966): GV = Gordon and Vandermeer (1962); FD = Frois and Dimitrov (1962): F = Fromageau (1969).

the Debye frequency), $n = 4$, $S^s = 4k$, and $T = 400K$, we find from eqn (9.28) that $N = 8.2 \times 10^{18} \, m^{-2}$ and from eqn (9.30), $Nz \exp [S^f/k] = 8.2 = 10^{18} \, m^{-2}$. Since the total density of atomic sites in the boundary core is $\simeq 1.7 \times 10^{19} \, m^{-2}$, these values are in the range which might be expected for either the non-vacancy-assisted or vacancy-assisted models, i.e. the number of sites available for fluctuations is relatively large and is a substantial fraction of the total number of boundary sites. We may conclude, therefore, that the basic shuffle model, in which several atoms may be involved per shuffle and in which the shuffles may occur at a large fraction of the sites along the boundary core, appears to be not inconsistent with the results in Fig. 9.23. Unfortunately, it is impossible to distinguish clearly between the vacancy assisted and non-vacancy assisted models. However, there is a mass of evidence in the literature which indicates that the presence of vacancies tends to increase rates of boundary motion (Gleiter and Chalmers 1972, Simpson *et al.* 1976, Haessner and Hofmann 1978, Atwater and Thompson 1988). This indicates that the mobility may be enhanced by vacancies, but, of course, it does not necessarily mean that boundary motion under local equilibrium conditions is always vacancy-assisted.

Consider next the applicability of the diffusional transport model which led to eqn (9.41). Here, the observed activation energy for the mobility, $Q^B = 55 \, kJ/mole$, should be equal to the activation energy for boundary self-diffusion. According to Gust *et al.* (1985), this latter quantity should be $\simeq 69 \, kJ/mole$ which is somewhat higher than observed. However, more importantly, the observed value of $M^B$ at 465 K is $10^{-9} \, m^4 \, J^{-1} \, s^{-1}$. According to Fig. 8.5, the value of the boundary diffusivity should be $D^B = 10^{-13} \, m^2 \, s^{-1}$, and taking $l = 2.86 \times 10^{-10} \, m$, we then find from eqn (9.41) that $Z$ must have the value $Z = 7.6 \times 10^{22} \, m^{-2}$. This value is about four orders of magnitude too large to be physically reasonable, and we therefore conclude that the diffusional transport mechanism is probably too slow to account for the observed mobilities of general boundaries in Al. At this point the shuffle mechanism is therefore preferred.

### 9.2.6.2  *Singular (or vicinal) large-angle grain boundaries*

Even though the lateral motion of dislocations possessing step character which might be responsible for the thermally activated large-scale motion of singular or vicinal boundaries (as described in Section 9.2.2.2) should often be detectable by electron microscopy, very few observations relevant to this mechanism have been made. In Fig. 9.24 we show examples of extrinsic grain boundary dislocations with spiral-like configurations which have been observed in a few instances in the literature and which resemble some of the model configurations discussed in Section 9.2.2.2. Unfortunately, these structures have not been well characterized and therefore cannot be discussed quantitatively. Gleiter (1969) and also Rae and coworkers (Smith *et al.* 1980, Rae and Smith 1980, Rae 1981, 1982) have observed grain boundary dislocations moving in boundaries during their motion by means of *in situ* electron microscopy. However, no quantitative measurements were made relating the total amount of dislocation activity to the amount of boundary movement. In further work, Babcock and Balluffi (1989) observed the motion of vicinal $\Sigma 5$ boundaries in Au which contained arrays of DSC dislocations which moved along with the moving boundaries. In this case measurements proved that the motion of the dislocations, with their associated steps, could not possibly have accounted for the boundary motion. We may therefore conclude that there is relatively little direct evidence for the importance of the dislocation step model for large scale boundary motion.

On the other hand, boundary motion via the lateral motion of pure steps has been

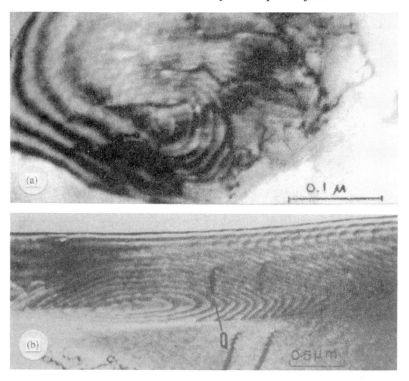

**Fig. 9.24** Observations of extrinsic grain boundary dislocations in the form of spirals. (Compare with, for example, Fig. 9.10.) (a) Al–0.39wt.% Cu alloy. (From Gleiter (1969).) (b) Pure Al. (From Dingley and Pond (1979).)

observed directly in a few cases. Figure 9.25 shows the motion of a $\Sigma 3$ tilt boundary by the occasional lateral motion of steps (supersteps). The boundary is almost parallel to the $\Sigma 3$ {111} twin inclination and contains large steps consisting of facets apparently parallel to the orthogonal $\Sigma 3$ {112} twin inclination. The step motion must have occurred by atom shuffling within the framework of the $\Sigma 3$ CSL. A direct observation of shuffling within the $\Sigma 3$ CSL has been obtained by electron microscope lattice imaging (Ichinose and Ishida 1990) as seen in Fig. 9.14(a–d). A diagram of the most likely detailed shuffle displacements is given in Fig. 9.14(e). Further indirect evidence for the motion of vicinal boundaries within the framework of dense CSLs by shuffling at steps is the work of Babcock and Balluffi (1989) already cited above. In this case the observed motion must have been due to the shuffling motion of pure steps in the $\Sigma 5$ CSL.

Further more indirect information must be gleaned from measurements of boundary mobility. According to our models, the intrinsic mobilities of singular or vicinal boundaries should generally be lower than those of general boundaries because of the need for appropriate boundary line defects in the former cases. However, demonstrating this through the use of experimental data is complicated by possible solute drag effects. Since solute segregation is generally expected to be lower at many singular boundaries than general boundaries (Chapter 7), there may be situations in which singular boundaries move faster than general boundaries under extrinsic conditions even though they would have moved more slowly under intrinsic conditions. Nevertheless, there is extensive

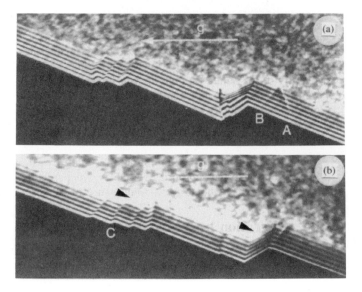

**Fig. 9.25** Movement of Σ3 tilt boundary via the lateral motion of pure steps. Step motion indicated by arrows in (b). Boundary plane is almost parallel to the Σ3 {111} twin boundary inclination, and the steps consist of facets apparently parallel to the orthogonal {112} twin inclination. (From Smith *et al.* (1980).)

experimental evidence (described below) that singular (or vicinal) boundaries often move more slowly than general boundaries under otherwise similar conditions. Since any solute atom drag effects were probably not stronger than at the general boundaries, we may take this as evidence that the intrinsic mobilities of the singular (or vicinal) boundaries were indeed lower than those of general boundaries. (We note that several cases where singular (or vicinal) boundaries moved more rapidly than general boundaries are described in the following. However, as discussed there, this most probably may be attributed to reduced solute atom drag at the former boundaries.)

The most extreme case of a singular boundary possessing a relatively low mobility is the Σ3 {111} twin in the f.c.c. structure which is usually found to have an essentially negligible mobility (e.g. Masteller and Bauer 1979). This is consistent with the need for line defects corresponding to either pure steps, as in Figs 9.14 and 9.25, or secondary dislocations with step character. Evidence for increased mobility due to an increase in the density of boundary dislocations with step character comes from the work of Howell *et al.* (1978). Here, it was found that Σ3 {111} twin boundaries containing essentially no line defects were practically immobile in the annealed state, but became mobile once they had picked up boundary dislocations from the impingement of lattice dislocations (see Section 12.4) during creep plastic deformation.

Masteller and Bauer (1978, 1979) have measured relatively low mobilities for several singular ⟨110⟩ tilt boundaries in Al with misorientations near the Σ = 3, 9, and 11 misorientations. The mobilities were orders of magnitude lower than those of ⟨100⟩ tilt boundaries of comparable purity. Also, the configurations of the moving boundaries which were observed in the experiments were non-planar, and they tended to develop non-uniform curvature and facets during their motion. These results indicate that much of the observed motion must have occurred by a step mechanism. We note that there are considerably more high density planes parallel to ⟨110⟩ than ⟨100⟩ in the f.c..c.

structure, and, therefore, a larger number of singular boundaries may be generally expected in a series of ⟨110⟩ tilt boundaries than ⟨100⟩ tilt boundaries as was apparently the case.

In other experiments (Gastaldi and Jourdan 1981, Jourdan and Gastaldi 1981, Bauer *et al.* 1988; Bauer 1989, Gastaldi *et al.* 1990), the growth of newly nucleated island crystallites into deformed matrix material has been observed during the recrystallization of Al. The island crystallites became progressively more faceted as they grew (Fig. 9.26(a)–(f)) with the facets tending to form parallel to dense lattice planes of the growing crystallites and not those of the matrix. General boundaries (with high index planes parallel to the boundary plane) had the highest mobilities and were eliminated by growth selection which favoured the survival of the less mobile singular boundaries. (In this process, the more mobile inclinations of the overall boundary delineating a crystallite eliminated themselves by moving forward rapidly while leaving behind the less mobile inclinations.) Measurements of the velocities of facets (i.e. singular boundaries) parallel to {111}, {100}, and {110} planes of the growing crystallite) are shown in Fig. 9.26(g). The velocities decrease in the order {110} → {100} → {111} which is the same as the order in which the interplanar spacing increases. The mobility therefore decreased as the boundary became more singular. As might be expected, the apparent activation energies for the motion of the singular facets were considerably higher than those measured for the corresponding motion of general boundaries in Al. The result that relatively slow-moving facets only formed when low index planes of the growing crystallites were parallel to the boundary plane can be understood if the facet mobilities were controlled by the motion of almost pure steps produced by the impingement of lattice dislocations. As may be seen from Fig. 9.12(a,b), a lattice dislocation impinging from an adjoining crystal with a low index plane parallel to the boundary can produce an almost pure boundary step. However, as seen in Fig. 9.12(c,d), this is not so when the crystal has a high index plane parallel to the boundary. Since the matrix had a much higher lattice dislocation density than the island crystallites, only boundaries parallel to low index planes of the crystallites (rather than the matrix) had a low step density and a correspondingly low mobility.

A considerable number of additional experiments, for example those cited by Gleiter and Chalmers (1972), show that the boundary mobility depends upon the boundary inclination and the indices of the crystal planes parallel to the boundary plane. Many of these indicate that low mobilities result when low index planes, particularly {111} planes, are parallel to the boundary in agreement with the previous results. A particularly clear example in Al is shown in Fig. 9.27. Here a new crystal was nucleated in a deformed single crystal matrix somewhere near A and grew anisotropically into the deformed matrix in the form of a long narrow finger. The immobile straight boundary segment along BC bounding this grain is a ⟨111⟩ $\theta = 45°$ twist boundary, i.e. a singular boundary parallel to low index {111} planes in both adjoining crystals. On the other hand, the highly mobile segment along DC is a general boundary.

### 9.2.6.3 *Solute atom drag effects*

Many observations have been made of the effects of solute atom drag on boundary motion. Most results appear to be broadly consistent with the model described in Section 9.2.5, but in many cases significant discrepancies are found (e.g. Lucke *et al.* 1972, Higgins *et al.* 1974, Simpson *et al.* 1976). This is not too surprising, since the drag model is clearly highly simplified and cannot be expected to hold in all situations. For example, possibly important phenomena such as solute–solute interactions (Hillert and Sundman 1976) and changes in the structure of the moving boundaries due to the presence of the

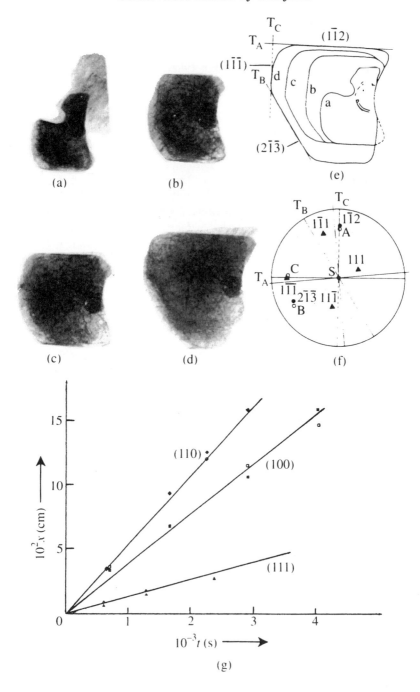

**Fig. 9.26** (a)–(d) Observed growth of freshly nucleated island crystallite in a plastically deformed matrix during the recrystallization of Al. The crystallite growth is anisotropic, and the crystallite becomes progressively more faceted along inclinations corresponding to low index planes of the crystallite as it grows (see pole figure construction in (f)). (From Bauer *et al.* (1988).) (g) Displacement, $x$, versus time for the motion of different low-index crystallite facets during growth. Facet indices refer to the crystallite. (From Gastaldi *et al.* (1990).)

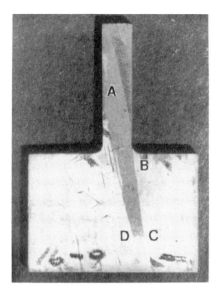

**Fig. 9.27** Al specimen in which a new crystal, nucleated somewhere near A, grew anisotropically into a deformed matrix in the form of a long narrow finger. The almost immobile straight boundary segment along BC bounding this finger grain is a singular $\langle 111 \rangle$ $\theta = 45°$ twist boundary. The highly mobile short segment along DC is a general boundary. (From Lücke (1974)).

segregated solute atoms (Simpson *et al.* 1976) have been ignored entirely. We therefore discuss a few selected results.

The results collected in Fig. 9.28(b,c,d) show drag effects measured as a function of driving pressure, temperature, and solute concentration which appear to be broadly consistent in many respects with the corresponding model behaviour predicted in Figs. 9.21(b,c,d). Evidence for a transition from intrinsic boundary behavior to extrinsic boundary behaviour is seen in each case. However, some discrepancies are present. For example, the curves in Fig. 9.28(b) do not show the predicted tendency to approach the intrinsic asymptote as $p$ is increased. In Fig. 9.29 we show data which indicate a distinct double desorption of solute atoms which occurred as the temperature was increased. Apparently, in this case, two markedly different interaction energies were involved. Further discussion of these complex phenomena may be found in the above references and in the reviews cited earlier.

The results in Fig. 9.22 provide evidence for solute atom drag effects which were stronger at general boundaries than singular boundaries because of generally stronger segregation at the general boundaries (see Chapter 7). Figure 9.22(a) shows that the effective-activation energy for boundary motion of Al of intermediate purity (99.9992 Al) was found to be lower for singular low-$\Sigma$ boundaries than general high-$\Sigma$ boundaries. This generally caused the corresponding mobilities to be higher (Fig. 9.22(b)). Other measurements (Aust and Rutter 1959, Rutter and Aust 1965 (shown in Fig. 9.28(a)) have shown similar effects. However, when the material was of higher purity, all of the large-angle boundaries moved at rates which were considerably higher and which were almost the same, irrespective of their $\Sigma$ values [Fig. 9.22(c)]. It may also be seen in Fig. 9.22(b) for the 99.9992 Al material that the effect of boundary misorientation decreased as the temperature was increased. This was apparently due to a decrease in the preferential segregation to the general high-$\Sigma$ boundaries. Similar behaviour is seen in Fig. 9.28(a) for boundaries in Pb where the effect of boundary misorientation ($\Sigma$ value) was also reduced by an increase in temperature.

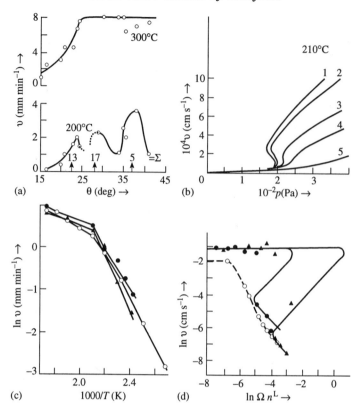

**Fig. 9.28** (a) Measured velocity, $v$, of near- $\langle\,001\,\rangle$ tilt boundaries in Pb as a function of tilt angle, $\theta$, at 200 and 300 °C. Values of $\theta$ corresponding to $\Sigma = 13, 17$, and 5 indicated by arrows along abscissa. (From Rutter and Aust (1965).) (b) Measured $v$ versus pressure, $p$, for general boundaries in Pb containing increasing concentrations of Sn solute atoms. 1 = double zone refined; 2 = 100 ppm; 3 = 500 ppm; 4 = 1000 ppm; 5 = 3000 ppm. (Simpson *et al.* (1976), (1971).) (c) Measured $\ln v$ versus $1/T$ for general boundaries in zone refined Pb. (From Simpson *et al.* (1976).) (d) Measured $\ln v$ versus $\ln \Omega n^L$ for Al containing increasing concentrations of solute atoms ($n^L$ = solute concentration in bulk alloy). Triangles denote Mg additions (data of Frois and Dimitrov (1962)); filled circles denote Cu additions (data of Frois and Dimitrov (1962)); open circles denote Cu additions (data of Cordon and Vandermeer (1962)). (From Lücke *et al.* (1972).)

### 9.2.6.4  *Small-angle grain boundaries*

The thermally activated motion of non-glissile small-angle grain boundaries has usually been found to be slower, and, certainly, never faster, than general large-angle grain boundaries under the same driving force in material of reasonably high purity (Rutter and Aust 1965 (see Fig. 9.28(a)), Sun and Bauer 1970b, Viswanathan and Bauer 1973, Fridman *et al.* 1975 (see Fig. 9.22(b,c)), Bauer 1982). Also, the process has appeared to be thermally activated with an activation energy which is often close to that of lattice self-diffusion, or somewhat higher. This suggests that the rate of motion was often determined in some way by dislocation climb limited by lattice self-diffusion as, for example, in the model leading to eqn (9.21). However, this mechanism was evidently not rate-controlling in all cases. In some instances, obstacles to the glide of dislocations or obstacles to changes in boundary structure during the boundary motion may have been

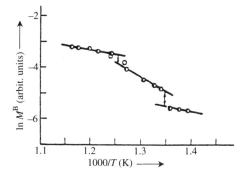

**Fig. 9.29** Measurements of $\ln M^B$ versus $1/T$ for a near-$\Sigma 19$ $\theta = 46.50\,°$ $\langle\,111\,\rangle$ tilt boundary in Al containing dilute Fe solute atoms ($m^B$ = boundary mobility). Two successive and clearly defined detachments of segregated solute atoms due to increasing temperature are evident. (From Molodov *et al.* (1981).)

important as discussed in Section 9.2.1.1. Also, solute drag may have been a factor. Activation energies higher than that of lattice self-diffusion may then have been required. In any event, it is clear that processes limiting the climb and/or glide motion of individual dislocations were involved which required relatively high activation energies. In contrast to this, as already discussed, the intrinsic motion of general large-angle boundaries appears to occur by the relatively easy transfer of atoms across the boundary core by a thermally activated shuffling process having an activation energy considerably lower than that of lattice self-diffusion and closer to that of grain boundary short-circuit diffusion. When strong drag effects by slow-diffusing solute atoms are not a major factor, the motion of general large-angle boundaries is therefore faster than that of small-angle boundaries.

We may conclude that the available experimental observations discussed above can be understood qualitatively on the basis of the models which we have presented. It should also be mentioned that it will be seen below in Chapter 10 that sharp interfaces act as net sources/sinks for diffusional fluxes of atoms and engage in non-conservative motion via essentially the same basic atomic mechanisms that we have invoked for their conservative motion in this chapter. As discussed in Chapter 10, there is considerable experimental support for models based on these mechanisms, and this may be taken as further support for the models for conservative motion described in this chapter. Nevertheless, further work will be required to establish them definitively, and many details remain to be worked out.

## 9.3 MECHANISMS AND MODELS FOR DIFFUSE INTERFACES

The structures and energies of diffuse interfaces in systems near mechanical or chemical instabilities have been analysed in Section 4.2. Since these interfaces straddle appreciable numbers of atomic planes, it was found that they could be analysed using continuum methods. Continuum methods can therefore also be used to model their motion, as shown below.

### 9.3.1  Propagation of non-linear elastic wave (or, alternatively, coherency dislocations)

The structures of diffuse interfaces between lattices related by a displacive type transformation have been described in Section 4.2.3 (see Fig. 4.5(a)). In their static configurations, these interfaces are slabs of material in which the core atoms are displaced in a progressively increasing manner so that one adjoining lattice is eventually continuously transformed into the other across the width of the core. Such an interface may be expected to move forward uniformly by the simultaneous shuffling of the atoms throughout its core in a correlated manner so that one crystal continuously transforms into the other as the interface advances. In contrast to the situation for sharp interfaces containing high densities of anticoherency dislocations, this type of motion should be relatively easy, as already discussed in Section 9.2.1.2. Motion of this type has an obvious wave-like character, and the overall process may therefore be described as the motion of a solitary non-linear elastic wave (Barsch and Krumhansl 1984). As shown by Barsch and Krumbansl (1984), the equation of motion of such a wave can be obtained in terms of parameters which are related to elastic constants and phonon dispersion data. An expression for the velocity can then be obtained by making various approximations. Details are given by Barsch and Krumhansl (1984).

It should be pointed out that the motion of the same interface (Fig. 4.5(a)) could be treated in an alternative fashion within the framework of a dynamic dislocation model. The interface lies along an invariant plane of the transformation relating the two adjoining lattices, and it may therefore be regarded as composed of the three sets of coherency edge dislocations shown in Fig. 4.5(b) whose Burgers vector strength sums to zero. In this figure we have drawn the interface in the hypothetical configuration that it would adopt if it were sharp. The dislocations are, therefore, highly localized. However, the interface is actually diffuse (Fig. 4.5(a)), and the coherency dislocations must therefore be partially delocalized as indicated schematically in Fig. 4.5(c). Such an interface can therefore be regarded as a slab of material containing a distribution of partially delocalized coherency dislocations, and its motion could therefore be modelled within the framework of dislocation dynamics (Hirth and Lothe 1982). The results of such an analysis should lead, of course, to exactly the same results as those obtained by using the solitary elastic displacement wave approach. Further discussion of the equivalence of soliton and dislocation models is given by Barsch and Krumhansl (1991).

### 9.3.2  Self-diffusion

The continuum approach may also be used (Allen and Cahn 1979) to analyse the motion of diffuse antiphase boundaries arising from non-equilibrium curvature. For a curved antiphase boundary the free energy of the system will not be at a minimum with respect to possible variations in the local long range order parameter, $\eta$. The parameter, $\eta$, is not a conserved quantity, and the central assumption of the Allen and Cahn analysis is that the degree of local order, $\eta$, in the moving interface core will change with time at a rate which is proportional to the rate at which the local energy (per atom) in the non-equilibrium (and non-uniform) system, $g$, changes with the degree of local order, i.e.

$$\frac{\partial \eta}{\partial t} = -\alpha \frac{\delta g}{\delta \eta},\tag{9.53}$$

where $\alpha$ is a phenomenologically based 'kinetic coefficient'. The basic mechanism for the motion is self-diffusion which allows the different types of atoms to jump from one sublattice to the other in order to change the degree of local order. In most cases this will occur by means of vacancy exchange.

The total free energy of the system, $G$, is given by eqn (4.29). If a small variation of order from the equilibrium state is made throughout the system, given by $\delta\eta(x)$, a variational calculation (Margenau and Murphy 1943) shows that

$$\delta G = 1/\Omega \int_{-\infty}^{\infty} \left[ \frac{\partial I}{\partial \eta} - \frac{d}{dx} \frac{\partial I}{\partial(d\eta/dx)} \right] \delta\eta(x)\, dx, \tag{9.54}$$

where,

$$I = \Delta g(\eta) + K(d\eta/dx)^2. \tag{9.55}$$

Therefore,

$$\delta g = \left( \frac{\partial I}{\partial \eta} - \frac{d}{dx} \frac{\partial I}{\partial(d\eta/dx)} \right) \delta\eta(x) \tag{9.56}$$

and using eqns (9.53), (9.55), and (9.56), we obtain the basic relationship for $\eta = \eta(x, y, z, t)$ anywhere in the interface,

$$\frac{\partial \eta}{\partial t} = -\alpha \frac{\partial \Delta g}{\partial \eta} + 2K\alpha \nabla^2\eta. \tag{9.57}$$

(Note that in writing eqn (9.57) we have replaced $\partial\eta^2/\partial x^2$ with its 3D counterpart, $\nabla^2\eta$.) A simple expression for the velocity of the interface can be obtained directly from eqn (9.57) for the common case where the principal radii of the curved interface are large compared with its thickness. For this purpose we switch to a curvilinear coordinate system $(q_1, q_2, q_3)$ in which the $q_1$ coordinate curve is normal to the iso-$\eta$ surfaces of the interface. Using standard curvilinear coordinate relationships (Hildebrand 1949), $\nabla^2\eta$ then takes the form

$$\nabla^2\eta = \frac{1}{h_2 h_3} \frac{\partial}{\partial s} \left( h_2 h_3 \frac{\partial \eta}{\partial s} \right), \tag{9.58}$$

where $h_2$, $h_3$ are the usual scale factors for measuring distances along the $q_2$ and $q_3$ coordinate curves, and $s$ is distance measured along $q_1$. If $\hat{s}$ is a unit vector along the interface normal, we also have (Hildebrand 1949)

$$\nabla \cdot \hat{s} = \frac{1}{h_2 h_3} \left[ \frac{\partial}{\partial s} (h_2 h_3) \right]. \tag{9.59}$$

Therefore, by combining eqns (9.58) and (9.59),

$$\nabla^2 = \frac{\partial^2\eta}{\partial s^2} + \frac{\partial \eta}{\partial s} \nabla \cdot \hat{s}. \tag{9.60}$$

However, according to eqn (5.105),

$$\nabla \cdot \hat{s} = -(1/R_1 + 1/R_2), \tag{9.61}$$

when $\hat{s}$ is directed towards the concave side of the interface, and $R_1$ and $R_2$ are the principal radii of curvature. Therefore, combining eqns (9.61), (9.60), and (9.57),

$$\frac{\partial \eta}{\partial t} = -\alpha \frac{\partial \Delta g}{\partial \eta} + 2K\alpha \left[ \frac{\partial^2 \eta}{\partial s^2} - \frac{\partial \eta}{\partial s} \left( \frac{1}{R_1} + \frac{1}{R_2} \right) \right]. \tag{9.62}$$

For a relatively gently curved interface, the $\eta$ profile along $s$ should not vary appreciably from the profile along $x$ through a flat interface at equilibrium, for which according to eqn (9.57)

$$\frac{\delta \Delta g}{\partial \eta} - 2K \, \mathrm{d}^2\eta/\mathrm{d}x^2 = 0. \tag{9.63}$$

Therefore, combining eqns (9.62) and (9.63),

$$\frac{\partial \eta}{\partial t} = -2K\alpha \frac{\partial \eta}{\partial s} \left( \frac{1}{R_1} + \frac{1}{R_2} \right). \tag{9.64}$$

The velocity, $v$, of the APB along $s$ corresponds to $(\partial s/\partial t)_\eta$, and, since

$$\left[ \frac{\partial s}{\partial t} \right]_\eta = - \left[ \frac{\partial \eta}{\partial t} \right]_s \bigg/ \left[ \frac{\partial \eta}{\partial s} \right]_t, \tag{9.65}$$

we obtain, finally, the simple result

$$v = 2K\alpha \left( \frac{1}{R_1} + \frac{1}{R_2} \right). \tag{9.66}$$

Equation (9.66) predicts that the curved antiphase boundary will spontaneouly migrate in order to reduce its curvature at a velocity proportional to the combined parameter $2K\alpha$, which has the dimensions of a diffusion coefficient. The result that the velocity is proportional to the kinetic coefficient, $\alpha$, might be expected, since this parameter controls the rate at which the required ordering or disordering takes place throughout the interface as it migrates.

A remarkable aspect of this result is that the interfacial energy, $\sigma$, does not appear in eqn (9.66), and the velocity is not described by the product of a mobility and a driving pressure which, for a curved interface, would be proportional to $\sigma$. Instead, the boundary velocity is related to the rate at which ordering occurs in response to the driving energy for ordering in the non-equilibrium core structure, and $\sigma$ is not directly involved. In a further calculation Allen and Cahn (1979) verified that the total rate of energy dissipation due to the changes in order which occur throughout the antiphase boundary as it migrates is exactly equal to the rate at which interfacial energy is dissipated due to decreasing interfacial area. Therefore, no energy conservation principle is violated.

The applicability of eqn (9.66) was experimentally tested by Allen and Cahn (1979) by measuring the rate of antiphase domain coarsening in non-stoichiometric Fe–Al alloys as a function of temperature near $T_c$. Results were obtained which were consistent with eqn (9.66) and inconsistent with a velocity proportional to $\sigma$. We note that $\sigma$ is expected to vary widely near $T_c$, and that the measurement of boundary velocity as a function of temperature was therefore a conclusive way to distinguish between velocities which were either proportional to $\sigma$, or not. We may draw the general conclusion from this work that expressions for interface velocity may take various forms which are not necessarily proportional to the driving pressure.

At lower temperatures eqn (9.66) loses its applicability for non-stoichiometric Fe–Al alloys, since significant variations in composition develop at the antiphase boundary due to interface segregation, and strong solute-atom drag effects then occur which tend to reduce the velocity. Analyses and experimental results regarding this effect are given by Krzanowski and Allen (1986).

## 9.4 EQUATIONS OF INTERFACE MOTION

In this section we consider the important problem of predicting the shape (position) of a conservatively moving interface as a function of time. Since conservative interface motion is not coupled to long range diffusion fields in the adjoining crystals, the problem is simplified to the extent that the interface motion depends only upon local conditions at the interface. Unfortunately, the motion process is still highly complex as should be evident from our previous discussions of mechanisms and models. However, as seen previously, under certain conditions we may assume that the interface motion can be described by a relatively simple rate equation. An example is the case where the velocity is linearly proportional to the existing pressure through a constant mobility as in eqn (9.35). Under these conditions, as we shall see, the motion of the boundary can often be predicted as a function of time.

A large number of situations is possible. We begin by analysing two important cases, i.e. one in which the local velocity is a function only of the local interface inclination so that $v = v(\hat{n})$ (Section 9.4.1) and the other in which the velocity is proportional to the pressure via a constant mobility, and the pressure corresponds to capillary pressure (Section 9.4.2). More general situations are then considered briefly in Section 9.4.3.

### 9.4.1 Motion when $v = v(\hat{n})$

We follow Frank (1958) and first demonstrate that interface motion for which the interface velocity, $v$, depends upon the interface inclination, $\hat{n}$, (i.e. $v = v(\hat{n})$) can be obtained when the interface moves via the lateral movement of line defects with step character. We then analyse some of the general characteristics of this type of motion which, of course, can also occur under other conditions when boundaries move by other mechanisms.

Consider the 2D motion of the interface illustrated in Fig. 9.30. The interface in 3D is a ruled surface parallel to $z$ and contains a distribution of line defects on it which possess step character and run parallel to $z$. The form of the interface in the $(x, y)$ plane at a given instant is indicated in the figure. In general, the form will vary with time, so that $y = y(x, t)$. The function $y = y(x, t)$ is then a surface in the $(y, x, t)$ space of Fig. 9.30 representing the time evolution of the shape of the interface. We define $k = k(x, t)$ as the step density (i.e. the number of steps per unit length along $x$) and $q = q(x, t)$ as the step flux (i.e. the number of steps passing a fixed point along $x$ per unit time). The velocity of the interface parallel to $y$ is then

$$v_y = (\partial y/\partial t)_x = hq, \tag{9.67}$$

and the interface inclination is

$$(\partial y/\partial x)_t = -hk. \tag{9.68}$$

Also, step conservation along $x$ must hold, and, therefore,

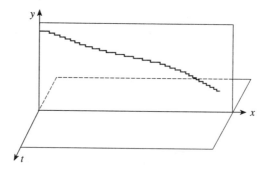

**Fig. 9.30** Representation in $(x, y, t)$ space of an interface containing a continuous distribution of steps which in 3D $(x, y, z)$ space is a ruled surface running parallel to $z$. The form of the interface in the $(x, y)$ plane at a given time is shown.

$$(\partial k/\partial t)_x = -\,\mathrm{div}\,q = -(\partial q/\partial x)_t. \tag{9.69}$$

We now assume quite generally that the local step velocity depends upon the proximity of other steps, i.e. the local step density. Since the local step flux will be the product of the local step density and step velocity, we then have

$$q = q(k). \tag{9.70}$$

(Note that this includes the special case where the velocities of all steps are equal.) Using eqn (9.70) and $dq = (dq/dk)dk$,

$$\left[\frac{\partial q}{\partial x}\right]_t = \frac{dq}{dk}\left[\frac{\partial k}{\partial x}\right]_t. \tag{9.71}$$

Combining eqns (9.71) and (9.69), we then have

$$\frac{dq}{dk}\left[\frac{\partial k}{\partial x}\right]_t + \left[\frac{\partial k}{\partial t}\right]_x = 0. \tag{9.72}$$

Since $k = k(x, t)$,

$$\frac{dk}{dt} = \left[\frac{\partial k}{\partial x}\right]_t \cdot \frac{dx}{dt} + \left[\frac{\partial k}{\partial t}\right]_x, \tag{9.73}$$

and combining eqns (9.73) and (9.72),

$$\frac{dk}{dt} = \left[\frac{\partial k}{\partial x}\right]_t \cdot \left[\frac{dx}{dt} - \frac{dq}{dk}\right]. \tag{9.74}$$

Now follow the motion of a point on the $y = y(x, t)$ interface where the inclination of the interface remains constant. From eqns (9.68) and (9.70) this will be a point of constant $k$ and $q$. From eqn (9.74), the motion of this point on the moving interface will project on the $(x, t)$ plane of Fig. 9.30 as a straight line of slope corresponding to $dx/dt = dq/dk = c(k)$.

Consider next the projection of this same point on the $(x, y)$ plane of Fig. 9.30. Here,

$$\frac{dy}{dx} = \left[\frac{\partial y}{\partial x}\right]_t + \left[\frac{\partial y}{\partial t}\right]_x \cdot \frac{dt}{dx}. \tag{9.75}$$

Therefore, using eqns (9.67) and (9.68) and $dx/dt = c$,

$$\frac{dy}{dx} = -kh + hq/c = -h(k - q/c). \tag{9.76}$$

Since $c$, $k$, and $q$ are all constant, the projection on the $(x, y)$ plane is also a straight line. We therefore conclude that points in the interface where the inclination is constant travel along straight line trajectories in the $(x, y)$ plane. Such lines are called 'characteristics'.

Further useful results can be obtained by switching to a vector formulation. The unit normal to the interface, $\hat{n}$, may be expressed in terms of our previous quantities in the form

$$\hat{n} = (1 + h^2k^2)[hk\hat{i} + \hat{j}], \tag{9.77}$$

where $\hat{i}$ and $\hat{j}$ are unit vectors along $x$ and $y$. The rate of advance of the interface normal to itself, $v$, is then

$$v = \left[\frac{\partial y}{\partial t}\right]_x \hat{j} \cdot \hat{n} = hq\hat{j} \cdot \hat{n} = hq(1 + h^2k^2)^{-\frac{1}{2}} = v(\hat{n}). \tag{9.78}$$

Note that we obtain the above result that $v = v(\hat{n})$, since $k$ is related to $\hat{n}$ by eqn (9.77), and $q = q(k)$. Having this result, we can now construct a 'slowness' vector, $s$, which is parallel to $\hat{n}$ but of magnitude equal to the reciprocal of $v(\hat{n})$, i.e.

$$s = s(\hat{n}) = \hat{n}/v(\hat{n}) = (hq)^{-1}[hk\hat{i} + \hat{j}]. \tag{9.79}$$

A polar plot of $s(\hat{n})$, i.e. a 'slowness' polar plot, can now be constructed. The tangent to this plot at the point $s(\hat{n})$ has the direction

$$t = ds/dk = -[c/hq^2] \cdot [h(k - q/c)\hat{i} + \hat{j}]. \tag{9.80}$$

Also, according to eqn (9.76), the straight line characteristic has the constant slope $dy/dx = -h(k - q/c)$ in the $(x, y)$ plane, and it therefore lies parallel to the vector

$$c = \hat{i} + (dy/dx)\hat{j} = \hat{i} - h(k - q/c)\hat{j}. \tag{9.81}$$

Therefore, since $c$ is perpendicular to $t$, we obtain the useful and elegant result (Frank 1958) that the characteristic for a point in the interface of constant inclination, $\hat{n}$, is parallel to the normal to the slowness polar plot at $s(\hat{n})$. If $\hat{n}'$ is a unit vector along the characteristic, i.e. along $c$, the velocity with which this point travels along its characteristic is then equal to $v(\hat{n})/(\hat{n}' \cdot \hat{n})$.

It may be verified (Frank 1958, Cahn *et al.* 1991, Taylor *et al.* 1992) that the above vector construction should hold in any situation where the interface velocity depends in a unique manner on its inclination, i.e. $v = v(\hat{n})$, and does not depend, for example, upon the existence of a step mechanism. In addition, it may be generalized to 3D.

In Fig. 9.31 we show calculated growth forms for a crystal enclosed by an interface for which $v = v(\hat{n})$ is a known function. The polar diagram of $v(\hat{n})$ is shown along with corresponding growth forms as a function of increasing $t$. The straight line characteristics are shown, and it may be seen directly that segments of constant inclination travel along these trajectories.

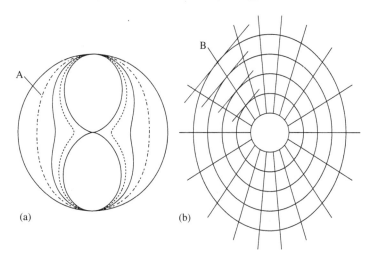

**Fig. 9.31** (a) Possible $v(\hat{n})$ polar plots for an interface whose velocity, $v$, is a function of the unit normal $\hat{n}$. (b) Growth forms at increasing times of an initially circular interface of relatively small radius whose $v(\hat{n})$ plot is indicated by A in (a). Various growth characteristics are shown, and the tangent lines constructed along the characteristic denoted by B show how a point in the interface of constant inclination travels along a characteristic. (From Cahn *et al.* (1991).)

### 9.4.2 Motion of curved interfaces under capillary pressure

We now consider the problem of the evolving shape of an interface when its local velocity is linearly proportional to the pressure according to $v = Mp$, with $M$ constant (e.g., eqn (9.35)), and the pressure, $p$, is due to capillarity, i.e. $p = \Delta P$. The capillarity induced pressure, $\Delta P$ (eqn (5.104)), generally includes both curvature and torque contributions, but the torque contribution is usually relatively small and will be neglected here in a first approximation. The capillary pressure at any point along the interface is then due only to curvature and is directed towards the centre of curvature with a magnitude given by $\Delta P = \sigma K$, where $K$ is the curvature. Under these conditions each point will respond by moving towards the local centre of curvature with a velocity proportional to the curvature. We proceed by finding the general governing equation of motion for this situation in the practically important 2D case where the interface lies in a flat sheet specimen and is everywhere a ruled surface perpendicular to the sheet surfaces. Any additional forces which might act on the interface in this geometry due, for example, to differences in the surface or bulk free energies of the two crystals adjoining the interface are neglected.

To obtain the governing equation, we follow Mullins (1956) and employ polar coordinates $(r, \theta)$ as in Fig. 9.32(a). Two successive positions of the interface, $r = r(\theta, t)$, are shown at $t$ and $t + \Delta t$. It may be seen that $\Delta r$ (at constant $\theta$) is related to $\Delta t$ by $\Delta r = -v\Delta t/\sin \psi = -M\sigma K\Delta t/\sin \psi$. Also, for plane curves, we have the standard relations, $K = \partial \beta/\partial s$ (where s is the arc length) and $\sin \psi = r(\partial \theta/\partial s)$. Therefore,

$$\lim_{\Delta t \to 0} \frac{\Delta r}{\Delta t} = \frac{\partial r}{\partial t} = -\frac{M\sigma K}{\sin \psi} = -\frac{M\sigma}{r}\frac{\partial \beta}{\partial \theta}. \tag{9.82}$$

This non-linear partial differential equation cannot be solved analytically in the general case. However, special solutions may be mapped out by making use of the separation

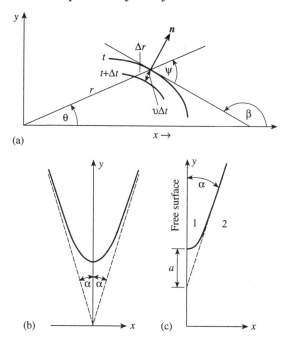

(a)

(b)

(c)

**Fig. 9.32** (a) Construction showing successive positions of a section of interface at $t$ and $t + \Delta t$ when it moves with local velocity $v$ along $\hat{n}$. (b) Curved interface evolved from initial dashed wedge shape under capillary pressure, $\Delta P$, when $v = M \cdot \Delta P$ with $M = $ constant. (c) Shape predicted from the result in (b) for an initially wedge-shaped grain (dashed) impinging on a free surface.

of variables method which assumes a product solution of the form $r(\theta, t) = R(\theta) \cdot T(t)$. Solutions of this special form will be shape preserving, i.e. they will exhibit invariance under magnification, since all dimensions will simply scale linearly with the magnitude of the time dependent function $T(t)$. Substitution of this assumed solution into eqn (9.82) yields

$$T \frac{\mathrm{d}T}{\mathrm{d}t} = -\frac{M\sigma}{R^2} \frac{\partial \beta}{\partial \theta} = c, \tag{9.83}$$

where $c$ is a constant. The time dependent function is readily obtained by integration, and is given by

$$T(t) = [T^2(O) + 2ct]^{\frac{1}{2}}. \tag{9.84}$$

The determination of the spatial function, $R(\theta)$, which must satisfy

$$\frac{\partial \beta}{\partial \theta} = -\frac{c}{M\sigma} R^2, \tag{9.85}$$

or, equivalently,

$$\beta_2 - \beta_1 = -\frac{c}{M\sigma} \int_{\theta_1}^{\theta_2} R^2 \,\mathrm{d}\theta, \tag{9.86}$$

is more problematic. It may be seen immediately from eqns (9.85) and (9.86) that interface shapes will generally expand away from the origin when $c > 0$ and shrink towards the origin when $c < 0$. As discussed by Mullins (1956), a number of shapes may be deduced directly from these equations. When $c = 0$, $\partial\beta/\partial\theta = 0$, and we then have the trivial case of a straight line, which, according to eqn (9.82), is stationary as might be expected. When $c < 0$, and has the value $c = -M\sigma/R^2$, $\partial\beta/\partial\theta = 1$, and we have a circle which will shrink in the form of a circle at an ever increasing rate which can be readily obtained from eqn (9.82). When $c > 0$, we have hyperbola-like curves having asymptotes which pass through the origin and are symmetrically disposed as shown in Fig. 9.32(b). As pointed out by Mullins (1956), this shape is consistent with eqn (9.86), which states that the tangent must turn through an angle, $(\beta_2 - \beta_1)$, which is proportional to the area subtended by the corresponding radius vectors. This interface shape would be expected to develop during the annealing of an interface in a bicrystal sheet specimen containing an initially wedge-shaped grain impinging on a free surface as illustrated in Fig. 9.32(c). The distance a in Fig. 9.32(c) should vary with time in a manner proportional to the function $T(t)$. Since $c > 0$, it should therefore increase parabolically with time according to $a = A \cdot \sqrt{t}$ where $A = \sqrt{2c}$. This behaviour has been verified experimentally by Sun and Bauer (1970$a$,$b$) who also showed that $A = [2M\sigma \cdot f(\alpha)]^{\frac{1}{2}}$, where $f(\alpha)$ is a numerically determined function of the angle $\alpha$. Experiments with such specimens can therefore be carried out (Sun and Bauer 1970$a$,$b$) to measure the boundary mobility $M$ under advantageous conditions where the capillary pressure on the boundary is initially high because of the large curvature which exists at short times.

The shapes of interfaces moving under capillary pressures have also been studied by other investigators (Antonov *et al.* 1972, Aristov *et al.* 1973) using variational methods and a variety of approximations. Further aspects of this type of interface motion are discussed and analysed in this work.

### 9.4.3   More general conservative motion

The analysis of more general types of conservative motion has been reviewed recently by Taylor *et al.* (1992). More specifically, they examine so-called 'geometric' models in which the local velocity normal to the interface of a point in the interface can depend only on its position, *r*, the local shape of the interface, and the values that field variables, such as temperature or chemical composition, take at the point. Any dependence on variables away from the interface, such as, for example, the parameters required to describe diffusion fields which are coupled to the movement of the interface as in non-conservative interface motion (Chapter 10), is not allowed. The velocity therefore depends only upon local conditions at the interface, a situation characteristic of conservative interface motion.

The conservative interface motion in a large number of geometric models can be described (Taylor *et al.* 1992) by the linearized relationship

$$v = M(r, t, \hat{n})\,[p(r, t) + \Delta P(r, t)], \tag{9.87}$$

where $M$ is the interface mobility which may be a function of *r*, *t*, and $\hat{n}$, $\Delta P$ is the capillary pressure, and *p* is the pressure due to all other sources (see Table 9.1). This rather general equation has of course, already appeared in various special forms throughout this chapter. Taylor *et al.* (1992) discuss a variety of methods for finding the interface motion when eqn (9.87) holds in certain forms. Details are given in their paper.

## 9.5  IMPEDIMENTS TO INTERFACE MOTION DUE TO PINNING

The motion of interfaces may be impeded by a variety of means. We have already described the 'drag' force which is exerted on a moving interface by diffusing solute atoms which interact with it (Section 9.2.5). Other important impediments include the pinning of an interface by embedded second-phase particles which are dispersed throughout the bulk and the pinning of an interface at a free surface which can occur at grooves which develop at the intersection of the interface with the surface. These impediments can often exert a decisive influence on the development of interfacial structure in polycrystalline materials. For example, dispersions of fine particle are often added in order to inhibit grain growth at elevated temperatures. Pinning at free surface grooves can be of critical importance in inhibiting the motion of interfaces in polycrystalline thin films.

### 9.5.1  Pinning effects due to embedded particles

A variety of situations may occur when a dispersion of embedded particles tends to pin an interface depending upon the particle sizes, the number of particles per unit volume, the temperature, and other possible factors. When the particles are immobile, and the pressure applied to the interface is low, the particles may be able to pin the interface and hold it stationary. At higher pressures the particles may no longer be able to pin the interface, and it will then be able to move directly through the dispersion. At elevated temperatures there is also the possibility of thermally activated unpinning if the temperature is high enough, or the particles small enough. Alternatively, the interface may remain pinned at the particles but may be able to move forward together with the particles if the particles are able to move forward by means of diffusive mass transport.

#### 9.5.1.1  *Pinning at stationary particles at low temperatures*

In order to analyse the impeding effect of a distribution of stationary dispersed embedded particles at low temperatures in the absence of any thermally activated unpinning, we first consider the effect produced by a single stationary particle. For simplicity, we assume a spherical particle interacting with an interface in the matrix which is subjected to an applied pressure urging it to move forward, as illustrated in Fig. 9.33. The interface is constrained at the particle because of interfacial energy considerations, as discussed below, and it therefore tends to bulge out around the particle in the dimpled configuration shown. If we assume local equilibrium along the junction line at $J$, and neglect any interface torque terms, the angle $\alpha$ is fixed by the relation

$$\sigma^{p,b} + \sigma^{w,b} \cos \alpha = \sigma^{p,w} \tag{9.88}$$

which guarantees that the net capillary force tangent to the particle at $J$ is zero. The net force exerted by the interface on the particle in the forward $y$ direction (and therefore the opposite of the force exerted by the particle on the interface) is then

$$F = 2\pi R \cos \phi \cdot \sigma^{w,b} \cos (\alpha - \phi). \tag{9.89}$$

This force reaches a maximum, $F_{max}$, when $\partial F / \partial \phi = 0$, corresponding to $\phi = \alpha/2$, and, therefore,

$$F_{max} = \pi R \sigma^{w,b} (1 + \cos \alpha). \tag{9.90}$$

The maximum impeding force (i.e. the pinning force) which the particle can exert on the interface is therefore a function of the particle size, and the three interfacial energies which are involved. When $\sigma^{p,w} \simeq \sigma^{p,b}$, $\cos \alpha \simeq 0$, $\phi \simeq 45°$, and

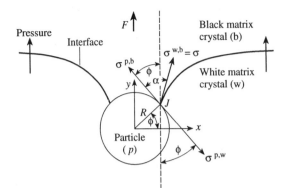

**Fig. 9.33** Spherical particle undergoing a force, $F$, exerted by a pinned interface, which in turn, is subjected to a pressure.

$$F_{\max} \simeq \pi R\sigma, \tag{9.91}$$

where we have set $\sigma$ equal to $\sigma^{w,b}$ the energy of the matrix interface. The maximum force is then dependent only upon $R$ and $\sigma$. The effect of changing the shape of the particle has been considered by Ryum *et al.* (1983), who represented the particle by an ellipsoid of varying eccentricity. Significant effects were found, depending upon the eccentricity and the inclination of the boundary with respect to the ellipsoid axes. Further aspects of the interaction between a single particle and interface have been discussed by Ashby *et al.* (1969), Ashby (1980), Nes *et al.* (1985), Wörner and Cabo (1987), and Wörner and Hazzledine (1992).

Consider next an interface which is attempting to move through a dispersion of stationary randomly distributed particles in the bulk under an applied pressure. The quantity of interest is the critical pressure which must be applied to the interface to drive it directly through the dispersion. The basic problem is to find the average number of particles which is attached to the interface at any instant (per unit area) taking into account the flexibility of the interface. Knowledge of that number and the impeding force supplied by each particle allows a calculation of the total impeding force (per unit area) which must then be equal to the critical applied pressure. This is a difficult problem which has not yet been solved exactly and which has been tackled by employing a variety of approximate models involving somewhat differing assumptions (see reviews by Wörner and Cabo (1987) and Wörner and Hazzledine (1992)). For our purposes we shall present the relatively simple analysis of Nes *et al.* (1985), which yields a critical pressure which agrees within an order of magnitude with the results of all of the other main investigations.

The interface will tend to bulge out between the particles as a result of the applied pressure as illustrated schematically in Fig. 9.34(a). For simplicity, we assume that all particles are of the same radius, we have a homophase interface, and $\cos\alpha \simeq 0$. Also, since the boundary can interact with each particle with maximum force by making only small local readjustments in position, we assume that this is the case. From eqn (9.91), the maximum impeding pressure which the particles can produce is then

$$p_i = \frac{F_{\max}}{A} = \frac{\pi R\sigma}{A}, \tag{9.92}$$

where $A$ is the average boundary area per particle. The determination of the critical parameter, $A$, is difficult, but it can be estimated fairly simply by realizing that when

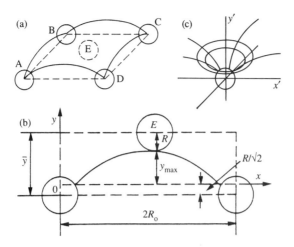

**Fig. 9.34** (a) Interface bulging out from pinning points established by particles A, B, C, and D. (b) Situation where the interface in (a) has bulged out sufficiently to make contact with particle E. (c) Doubly curved surface formed by bulging interface in vicinity of a particle. (From Nes *et al.* (1985).)

the applied pressure, $p_a$, is only infinitesimally larger than $p_i$, a steady state situation will exist in which, on average, the boundary advances by losing contact with particles at the same rate that it makes contact with new particles. In order to make each new contact, it must be able to bulge out between particles a sufficient distance to reach a new particle while, on average, losing contact with a trailing particle. A typical contact event is illustrated schematically in Figs. 9.34(a,b) where E is the new particle, and A, for example, may become the trailing particle. Contact with E requires that the boundary bulges out under the influence of the applied pressure a sufficient distance to encounter it as illustrated in Fig. 9.34(b). As may be seen in Fig. 9.34(b), the boundary will contact a randomly distributed new particle if the particle lies in the volume given approximately by $\bar{y}\pi R_o^2$. The condition for contact is therefore

$$\bar{y}\pi R_o^2 = \frac{4\pi R^3}{3f} = \bar{y}A, \tag{9.93}$$

where $f$ is the fraction of total volume occupied by the particles, and

$$\bar{y} = y_{max} + \frac{R}{\sqrt{2}} + R. \tag{9.94}$$

It now remains to calculate $y_{max}$, and this requires a determination of the shape of the bulged out boundary. Neglecting any torque terms, the boundary is forced outwards by the applied pressure, equal to $p_a$, and inwards by its capillary pressure. Using eqn (5.104), the equilibrium condition is then given by

$$p_a = -\sigma\left(\frac{1}{\rho_1} + \frac{1}{\rho_2}\right) \tag{9.95}$$

where $\rho_1$ and $\rho_2$ are the two principal radii of curvature which are present as illustrated in Fig. 9.34(c). Assuming cylindrical symmetry, these radii are given by

$$\rho_1 = \frac{[1 + (dy/dx)^2]^{\frac{1}{2}}}{d^2y/dx^2} \qquad (9.96)$$

$$\rho_2 = \frac{x[1 + (dy/dx)^2]^{\frac{1}{2}}}{dy/dx}. \qquad (9.97)$$

Assuming $dy/dx \ll 1$, the equation for the shape is then

$$\frac{d^2y}{dx^2} + \frac{1}{x}\frac{dy}{dx} = -p_a/\sigma \qquad (9.98)$$

which has the solution

$$y = -\frac{p_a}{4\sigma}x^2 + c_1 \ln x + c_2, \qquad (9.99)$$

where $c_1$ and $c_2$ are constants. Setting $A = \pi R_o^2$ in eqn (9.92), and $p_a = p_i$, we find

$$p_a = R\sigma/R_o^2. \qquad (9.100)$$

Using this result and applying the boundary conditions: (i) $dy/dx = 0$ at $x = R_o$; and (ii) $x = R/\sqrt{2}$ at $y = 0$, to eqn (9.99), we find

$$y = -\frac{R}{4R_o^2}\left[x^2 - 2R_o^2 \ln x + 2R_o^2 \ln\frac{R}{\sqrt{2}} - \frac{R^2}{2}\right]. \qquad (9.101)$$

This result is not fully consistent with our assumption that $dy/dx \ll 1$ everywhere and the assumption that the particle is exerting the maximum possible restraining force on the boundary. However, the discrepancies are relatively small (Nes *et al.* 1985), and eqn (9.101) may be taken as a reasonable approximation. Therefore, since $y_{max} = y(R_o)$, and taking $R \ll R_o$, we have

$$y_{max} = \frac{R}{2}\left(\ln\left[\frac{\sqrt{2}R_o}{R}\right] - \frac{1}{2}\right). \qquad (9.102)$$

By combining eqns (9.93), (9.94), and (9.102), we then have

$$\frac{4}{3f}\left[\frac{R}{R_o}\right] + \frac{1}{2}\ln\frac{R}{R_o} = 1.63 = 0. \qquad (9.103)$$

This implicit relation may be solved graphically with the result $R/R_o = 1.15f^{0.46}$, and putting this relationship into eqn (9.92) we obtain, finally, the expression for the maximum impeding pressure exerted by the particles

$$p_i = \frac{1.3\sigma f^{0.92}}{R} \simeq \frac{1.3\sigma f}{R}. \qquad (9.104)$$

This pressure is of the same order of magnitude as that obtained in all of the other main investigations of this problem including the original one of Zener in 1948 (Smith 1948, Wörner and Cabo 1987, Wörner and Hazzledine 1992). It is noted that, according to eqn (9.104), $p_i$ depends only on $\sigma$, $f$, and $R$ and is independent of any macroscopic curvature of the pinned boundary which may be present, as, for example, during grain growth: several models (see above references) lead to a $p_i$ which is more complicated and has a weak dependence upon possible macroscopic boundary curvature.

### 9.5.1.2 *Thermally activated unpinning*

We now consider the possibility of thermally activated unpinning when $F < F_{\text{max}}$. If the force which the particle exerts on the interface just before the fluctuation is $F$, the force which it would exert on the interface during all stages of the unpinning is, from eqns (9.89) and (9.91)

$$F' = 2\pi R\sigma \cos\phi \sin\phi = F_{\text{max}} \sin 2\phi = 2F_{\text{max}}(\eta/R) \cdot [1 - (\eta/R)^2]^{\frac{1}{2}}, \quad (9.105)$$

which is shown in Fig. 9.35 as a function of the interface displacement, $\eta$, which is in the direction of the $y$ axis and is equal to zero at $y = 0$ in the coordinate system shown in Fig. 9.33. The energy which must be supplied by the thermal activation (i.e. the activation energy) is then the shaded area in Fig. 9.35 which is given by

$$\Delta G^* = \int_F^{F_{\text{max}}} \Delta\eta\,(F')\,\mathrm{d}F' \quad (9.106)$$

The quantity $\Delta\eta(F') = \eta_2(F') - \eta_1(F')$ may be found by realizing that eqn (9.105) is satisfied by two values of $\phi$ for a given value of $F'$, i.e. $\Phi_1$ and $\phi_2 = 90 - \phi_1$. Since $\eta/R = \sin\phi$, $\Delta\eta = \eta_2 - \eta_1 = R(\cos\phi_1 - \sin\phi_1)$. Therefore, $\Delta\eta^2 = R^2\,(1 - \sin 2\phi_1)$, and, using eqn (9.105), $\Delta\eta(F') = R(1 - F'/F_{\text{max}})^{\frac{1}{2}}$. Putting this result into eqn (9.106) and integrating, we finally obtain

$$\Delta G^* = \frac{2F_{\text{max}}R}{3}\left[1 - \frac{F}{F_{\text{max}}}\right]^{\frac{3}{2}}. \quad (9.107)$$

The activation energy therefore is seen to decrease with decreasing particle size and increasing applied force, $F$, as might be expected intuitively. The rate of unpinning would then be of the usual form $\nu \cdot \exp(-\Delta G^*/kT)$, where $\nu$ is an attempt frequency.

A kinetic model for the forced quasi-steady-state motion of an interface through a dispersion of particles in the presence of thermally activated unpinning could then be constructed using concepts similar to those employed in our previous model for interface motion in the absence of thermally activated unpinning. Further aspects of this phenomenon and its role in grain growth are discussed by Gore *et al.* (1989).

### 9.5.1.3 *Diffusive motion of pinned particles along with the interface*

Finally, we consider the case where both the pinned particles and the interface move forward together by a diffusive process. For simplicity, we assume that the particle and matrix phases are both pure single component phases with negligible mutual solubility and equal atomic volumes. In order for the particle to move relative to the matrix, matrix atoms must diffuse from matrix sites at the interface in front of the particle to matrix sites at the interface at the rear of the particle. Various parts of the interface must therefore act as local sources or sinks for the diffusing atoms as illustrated in Fig. 9.36, and the diffusion can take place either through the matrix (i.e. $J_{\text{m}}^{\text{m}}$) or along the particle/matrix interface (i.e. $J_{\text{m}}^{\text{p,m}}$). However, in certain cases this process may not be sufficient to allow the forward motion of the particle because of constraints imposed by the surrounding matrix. For example, for a non-diffusing particle to move forward relative to the matrix, it would be necessary for shear to occur between the particle and matrix at the particle diameter shown dashed in Fig. 9.36. This problem can be overcome if an equal and opposite diffusion flux of particle atoms occurs through the particle and along the particle/matrix interface as illustrated by $J_{\text{p}}^{\text{p}}$ and $J_{\text{p}}^{\text{p,m}}$ respectively in Fig. 9.36. This one-to-one coupling is apparently often necessary during particle dragging (Ashby

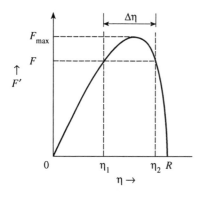

**Fig. 9.35** Pinning force, $F'$, exerted by a spherical particle of radius $R$ on an interface as a function of the displacement of the interface, $\eta$, which is in the direction of the $y$ axis and is equal to zero at $y = 0$ in the coordinate system shown in Fig. 9.33.

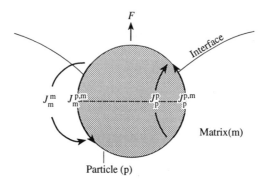

**Fig. 9.36** Diagram of a particle undergoing diffusive motion under the influence of a force, $F$, exerted by a pinned interface. Diffusion currents of matrix atoms, $J_m^m$ and $J_m^{p,m}$, travel towards the rear of the particle and correspond to diffusion through the matrix and along the particle/matrix interface, respectively. Diffusion currents of particle atoms, $J_p^p$ and $J_p^{p,m}$, travel towards the front and correspond to diffusion through the particle and along the particle/matrix interface, respectively.

1980), and we shall therefore follow Ashby (1980) and assume that this is the case.

An approximate expression for the particle mobility, $M$, defined by

$$v = M \cdot F \tag{9.108}$$

(where $v$ is the particle velocity due to the drag force, $F$), which is achieved under the above circumstances is readily obtained. In order to move the particle forward by the distance $\delta x$, $\delta N = \pi R^2 \delta x / \Omega$ atoms must be transported from one hemisphere to the other. The required current, $I$, of either matrix or particle atoms is therefore related to $v$ by

$$v = \frac{\delta x}{\delta t} = \frac{\Omega}{\pi R^2} \frac{\delta N}{\delta t} = \frac{\Omega}{\pi R^2} \cdot I. \tag{9.109}$$

The rate at which work is done is $Fv$, and this, of course, will equal the rate at which energy is dissipated by the various atom transport processes. If we assume that the interface acts as an ideal source/sink, the particle motion is 'diffusion-controlled' (see Section 10.2.1), and all of the energy is dissipated (see Section 10.2.2) by the diffusive processes illustrated in Fig. 9.36.

Consider first the transport of matrix atoms. For the $J_m^m$ flux, the average diffusion distance $\simeq 4R$ and the average gradient in diffusion potential is therefore $\overline{\nabla M}_m = \overline{\Delta M}_m / 4R$, where $\overline{\Delta M}_m$ is the average difference between the diffusion potential in front of and behind the particle. Therefore, using eqn (9.18),

$$J_m^m \simeq \frac{-D_m^m}{N_o \Omega k T f_m^m} \cdot \frac{\overline{\Delta M}_m}{4R}. \tag{9.110}$$

The average area through which diffusion occurs is $\simeq 3\pi R^2$, and therefore the diffusion current is

$$I_m^m \simeq \frac{D_m^m}{N_o \Omega k T f_m^m} \frac{\overline{\Delta M}_m}{4R} \cdot 3\pi R^2. \tag{9.111}$$

For the $J_m^{p,m}$ flux. average diffusion distance is $\simeq 2R$, and the average area is $\simeq 2\pi R \cdot \delta$ where $\delta$ is the effective thickness of the interface for diffusion (see Section 8.2.1). Therefore,

$$I_m^{p,m} \simeq \frac{D_m^{p,m}}{N_o \Omega k T f_m^{p,m}} \cdot \frac{\overline{\Delta M}_m}{2R} \cdot 2\pi R \cdot \delta. \tag{9.112}$$

Similar expressions for the transport of the particle atoms may be obtained in the forms

$$I_p^p \simeq \frac{D_p^p}{N_o \Omega k T f_p^p} \frac{\overline{\Delta M}_p}{2R} \cdot \pi R^2 \tag{9.113}$$

$$I_p^{p,m} \simeq \frac{D_p^{p,m}}{N_o \Omega k T f_p^{p,m}} \cdot \frac{\overline{\Delta M}_p}{2R} \cdot 2\pi R \cdot \delta. \tag{9.114}$$

The rate of energy dissipation associated with each diffusion current is just the product of the current and the average decrease in the diffusion potential which drives it. Therefore,

$$vF = \frac{v^2}{M} = \frac{1}{N_o} [(I_m^m + I_m^{p,m})\overline{\Delta M}_m + (I_p^p + I_p^{p,m})\overline{\Delta M}_p]. \tag{9.115}$$

However,

$$(I_m^m + I_m^{p,m}) = (I_p^p + I_p^{p,m}) = \frac{\pi R^2}{\Omega} \cdot v, \tag{9.116}$$

and, using eqns (9.111), (9.112), (9.113), (9.114), and (9.116) in eqn (9.115), we finally find

$$\frac{1}{M} = \frac{1}{[M_m^m + M_m^{p,m}]} + \frac{1}{[M_p^p + M_p^{p,m}]}, \tag{9.117}$$

where

$$M_m^m = \frac{3\Omega D_m^m}{4\pi k T f_m^m R^3}, \tag{9.118}$$

$$M_m^{p,m} = \frac{\Omega D_m^{p,m} \delta}{\pi k T f_m^{p,m} R_4}, \tag{9.119}$$

$$M_p^p = \frac{\Omega D_p^p}{2\pi k T f_p^p R^3},$$ (9.120)

$$M_p^{p,m} = \frac{\Omega D_p^{p,m}\delta}{\pi k T f_p^{p,m} R^4}.$$ (9.121)

The above quantities may be regarded as the 'mobilities' associated with the different mechanisms which are responsible for the overall mobility of the particle, $M$. These have different dependences upon temperature and particle size, and different mechanisms may therefore control the particle mobility under different conditions. For example, at low temperatures and small particle size, we might expect the mobilities corresponding to interface diffusion to become relatively large (i.e. $M_m^{p,m} \gg M_m^m$ and $M_p^{p,m} \gg M_p^p$), and eqn (9.117) would then have the form

$$M = \frac{M_m^{p,m} M_p^{p,m}}{(M_m^{p,m} + M_p^{p,m})}.$$ (9.122)

If $M_p^{p,m} \ll M_m^{p,m}$, $M \simeq M_p^{p,m}$, and the overall particle motion would be controlled by the rate at which particle atoms could diffuse around the particle along the interface.

The situation may be further complicated by the diffusion of matrix atoms through the particle or particle atoms through the matrix if some solubility exists. In addition, the interface may not act as a perfect source/sink, and the kinetics could become partially or essentially completely 'interface controlled' as discussed in Section 10.2.1. At least some energy would then be dissipated locally in the source/sink action at the interface rather than entirely in the long-range diffusion process as assumed above. As shown by Ashby (1980), these processes contribute additional mobility terms to the general expression for the particle mobility, eqn (9.117). In further work, Ashby (1980) showed how particle mobility diagrams may be constructed that show which mechanisms may be expected to dominate in different regions of 'temperature–particle size space'.

### 9.5.2   Pinning at free surface grooves

Any stationary internal interface which intersects a free surface will tend to establish an equilibrium configuration at the intersection and form a groove in the surface as illustrated in Fig. 9.37(a). Such a groove constitutes a position of minimum energy for the interface, and it will therefore tend to pin the interface there against any subsequent motion. When there is no pressure acting on the interface, as in Fig. 9.37(a), the condition of local equilibrium at the triple interface junction is given by eqn (5.121) which, we note, holds equally well when the interfaces involved are either internal interfaces or free surfaces. If we take the interface to be a homophase boundary, and use the results in Section 5.6.8 while ignoring any torque terms, the equilibrium angle, $\psi$, in Fig. 9.37(a) is given by

$$\sin \psi = \sigma^{b,w}/2\sigma^S.$$ (9.123)

The groove may form by mass transport due to fast diffusion along the surface (see Section 8.2.3.1), evaporation–condensation or diffusion through the lattice. The kinetics of groove formation by these three mechanisms has been analysed by Mullins (1957, 1960). The results indicate that surface diffusion will usually dominate except for cases where the material has a relatively high vapour pressure. When surface diffusion is

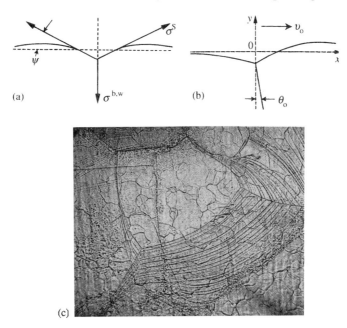

**Fig. 9.37** (a) Groove formed at intersection of a grain boundary with a free surface when the grain boundary is free of pressure. Local equilibrium pertains at the triple junction. (b) Steady state asymmetric groove shape obtained when boundary is subjected to a pressure, and the entire configuration moves with a velocity, $v$. The vertical scales in both (a) and (b) are exaggerated by a factor $\simeq 3$. (c) View of series of surface grooves on surface of 0.127 cm thick Cu sheet left behind by grain boundaries which presumably became sequentially pinned and unpinned during grain growth. (From Mullins (1958).)

dominant, the shape of the groove remains constant with increasing time, and all dimensions increase in scale as $t^{\frac{1}{3}}$.

A variety of situations with respect to the formation of a groove exists under dynamic conditions when the boundary is subjected to a pressure urging it to move in a direction parallel to the surface. If the pressure is relatively high, and the boundary velocity is correspondingly large, there is insufficient time for any significant grooving to occur and any effect due to the surface is then negligible. On the other band, at lower pressures and velocities, significant grooving can occur, and the groove then tends to act as an impediment to the motion. Under appropriate conditions, the groove can pin the boundary at the surface, or move along with it in a steady state configuration, or periodically pin and release the boundary. The dynamical problem has been analysed by Mullins (1958) and Aristov *et al.* (1978).

In order to establish some of the main features of the dynamic behaviour we begin by deriving the basic differential equation governing the form of the groove. Following Mullins (1958), we employ the coordinate system shown in Fig. 9.37(b). The $x$ position of the origin is fixed at the root of the groove which is moving relative to the bicrystal with the velocity $v(t)$; the $y$ position is fixed at the height of the average bicrystal surface. The bicrystal surface is a ruled surface lying parallel to the root of the groove of the form $y = y(x)$. In this moving coordinate system, the number of atoms passing a point $x$ (per unit distance along the root of the groove) can be written as

$$I_A = - \frac{D^S \delta}{N_o \Omega k T f^S} \nabla M_A - \frac{v(t)}{\Omega} y. \tag{9.124}$$

Here, we have formulated the fast surface diffusion transport using the same 'thin slab' model that we used in Section 8.2.1 to describe fast grain boundary diffusion: $D^S$ is the surface diffusivity, $\delta$ is the thickness of the surface slab, and $f^S$ is the surface diffusion correlation coefficient. We have also assumed that the flux in the surface slab can be written in the same form as eqn (9.18). The diffusion potential in the surface slab, $M_A$, may be evaluated by employing the same thermodynamic formalism used to find $M_A$ at an incoherent interface in Section 5.9.2.1, since substitutional atoms can be independently added or removed at a general surface in the same way as at an incoherent interface. According to eqn (5.104), the curved surface is subjected to a capillary pressure given by $\Delta P = \sigma^S/R$, where $1/R$ is the local curvature. In Chapter 12 it is shown that the results in Section 5.9.2.1 yield a value for $M_A$ given by eqn (12.65), i.e $M_A = N_o \Omega (f - \tau_n)$, for an incoherent interface subjected to a normal traction $\tau_n$ in a system containing A type atoms and point defects in local equilibrium with the interface. (Here, $f$ is a bulk free energy which may be taken as constant.) Using eqns (5.75), (5.104), and (5.105) we then find $\tau_n = -\Delta P = -\sigma^S/R = \sigma^S \partial^2 y/\partial x^2$, when $\partial y/\partial x \ll 1$. Putting this result into the above expression for $M_A$, we obtain

$$M_A = N_o \Omega f - N_o \Omega \sigma^S \cdot \partial^2 y/\partial x^2. \tag{9.125}$$

Substituting eqn (9.125) into eqn (9.124), and dropping the subscript A, then yields

$$I = \frac{D^S \delta \sigma^S}{k T f^S} \cdot \frac{\partial^3 y}{\partial x^3} - \frac{v(t)}{\Omega} \cdot y. \tag{9.126}$$

Since $\partial y/\partial t = -\Omega(\partial I/\partial x)$, we then have

$$By'''' - v(t)y' + \frac{\partial y}{\partial t} = 0, \tag{9.127}$$

where $B = \Omega D^S \delta \sigma^S/k T f^S$, and the primes indicate partial differentiation with respect to $x$. The travelling groove consists of two branches, i.e. $y_-$ for $x \leqslant 0$, and $y_+$ for $x \geqslant 0$. The boundary conditions are then

$$y_-(0) = y_+(0), \tag{9.128}$$

$$y_+'(0) - y_-'(0) = \sigma^{b,w}/\sigma^S \equiv M, \tag{9.129}$$

and

$$C \cdot \{\Omega \sigma^S [y_+''(0) - y_-''(0)] + \Delta \mu^o\} = y_+'''(0) = y_-'''(0). \tag{9.130}$$

Equation (9.128) guarantees continuity at the root, while, as shown below, eqn (9.129) guarantees local equilibrium of the capillary forces at the root. We anticipate that the boundary will rotate by the angle $\theta$ in the dynamical situation, and since both $\psi$ and $\theta$ will be small over the range of interest, it is readily seen that

$$y_+'(0) \simeq \theta + \psi$$
$$y_-'(0) \simeq \theta - \psi. \tag{9.131}$$

Since, from eqn (9.123), $\psi \simeq \sigma^{b,w}/2\sigma^S$, we therefore obtain eqn (9.129). The right-hand side of eqn (9.130) expresses the fact that the diffusional currents, which are proportional to $y'''$ according to eqn (9.126), must be equal in and out of the root. The left-hand side expresses the fact that this current must also be equal to the current of atoms which is transferred across the boundary core in order to produce its velocity $v(t)$. According to the development leading up to eqn (9.35), for example, this current is proportional to the difference in free energy for atoms across the boundary. The difference in free energy per atom is made up of the difference in chemical potential due to the surface curvature obtained from eqn (9.125), and the quantity $\Delta\mu^{\circ}$ which is taken to be the difference due to all other sources (e.g. strain energy due to cold work, etc.). The quantity $C$ is then a constant proportional to the boundary mobility. Mullins (1958) argues that under most conditions where grooving will be significant, the quantity $\Delta\mu^{\circ}$ which tends to drive the boundary forward is small enough so that it can be neglected in eqn (9.130). Also, $C$ will be large enough so that we may set

$$y''_+(0) \simeq y''_-(0). \tag{9.132}$$

The steady-state solution can now be found by solving eqn (9.127) (with $v = v_{\mathrm{o}} = $ constant, and $\partial y/\partial t = 0$) subject to the conditions (9.128), (9.129), (9.130), and (9.132). The solution is

$$y_+(x) = -\frac{2M}{3\alpha} \cdot \cos\left(\frac{\sqrt{3}}{2}\alpha x + \frac{\pi}{3}\right) \cdot \exp\left(-\frac{\alpha x}{2}\right)$$

$$y_-(x) = -\frac{M}{3\alpha} \cdot \exp(\alpha x), \tag{9.133}$$

where $\alpha^3 = v_{\mathrm{o}}/B$. Since $x$ and $y$ are multiplied by $\alpha$ everywhere, all steady state grooves have the same characteristic shape which is shown in Fig. 9.37(b). Using eqns (9.131) and (9.133), the steady state value of $\theta$ is found to be

$$\theta_{\mathrm{o}} = \frac{M}{6} = \frac{\sigma^b}{6\sigma^s} \tag{9.134}$$

which is, of course, independent of groove size or velocity. Using the relationship $v_{\mathrm{o}} = B\alpha^3$ and eqn (9.133), the steady state velocity can be written in the form

$$v_{\mathrm{o}} = B\left(\frac{M}{3d}\right)^3, \tag{9.135}$$

where $d$ is the depth of the groove, i.e. $d = -y_-(0)$. The velocity therefore decreases rapidly with increasing groove size.

In further work, Mullins (1958) used eqn (9.127) to analyse the time-dependent behaviour of grooves for cases where $\theta < \theta_{\mathrm{o}}$ and showed that they continuously increase in size and therefore decelerate as might be expected intuitively. In fact, their motion rather quickly stops for all practical purposes and the boundary then becomes pinned at the groove which is essentially stationary.

The result that boundaries will become pinned at surface grooves when the driving pressure on the boundary is small, or moderate (i.e. $\Delta\mu^{\circ}$ in eqn (9.130) can be neglected) and $\theta < \theta_{\mathrm{o}}$ has interesting consequences. In general, the value of $\theta$ which will be established at a groove will be a complex function of the groove mobility, the boundary mobility, the magnitude of the pressure exerted on the boundary, and any condition imposed on the boundary at its other end away from the surface. At small pressures, the boundary may bow out slightly in the $x$ direction under conditions where $\theta$ does not exceed $\theta_{\mathrm{o}}$. Such a boundary will then remain pinned at the essentially immobile groove. However, at higher pressures, the boundary may be forced to bow out further so that

$\theta > \theta_0$. Under these conditions, the boundary will become unpinned. If the unpinned end then moves forward considerably more rapidly than the section farther from the surface the boundary will then straighten considerably slowing down the unpinned end. The unpinned end can then generate another groove with $\theta < \theta_0$ and again become unpinned. However, the interior section will then bow out again and the entire cycle may repeat. At still higher pressures, where $\Delta\mu^\circ$ in eqn (9.130) becomes dominant, the present analysis breaks down, the entire boundary moves much more rapidly, and any effects due to grooving are expected to be negligible. There is extensive experimental evidence for the cyclic pinning and unpinning of boundaries at surfaces in the presence of low or moderate driving pressures such as the one provided by capillary forces during the migration of curved boundaries. Figure 9.37(c) shows the surface grooves left behind by boundaries which evidently became periodically pinned and unpinned during grain growth. Additional experimental evidence for this behaviour has been given by Aristov *et al.* (1978) along with further theoretical analysis of the grooving phenomenon.

## REFERENCES

Allen, S. M. and Cahn, J. W. (1979). *Acta Metall.*, **27**, 1085.

Antonov, A. V., Kopezkii, C. V., Shvindlerman, L. S., and Mukovskii, Y. M. (1972). *Phys. Stat. Sol. (a)*, **9**, 45.

Aristov, V. Y., Fridman, Y. M., and Shvindlerman, L. S. (1973). *Fiz. Metall. Metalloved.*, **35**, 859.

Aristov, V. Y., Fradkov, V. Y., and Shvindlerman, L. S. (1978). *Phys. Met. Metall.*, **45**, 83.

Ashby, M. F. (1980). In *Recrystallization and grain growth of multi-phase and particle containing materials* (eds N. Hansen, A. R. Jones, and T. Leffers) p. 325. Riso National Lab., Denmark.

Ashby, M. F., Harper, J., and Lewis, J. (1969). *Trans. A.I.M.E.*, **245**, 413.

Atkinson, H. V. (1988). *Acta Metall.*, **36**, 469.

Atwater, H. A. and Thompson, C. V. (1988). *Appl. Phys. Lett.*, **53**, 2155.

Aust, K. T. (1969). In *Interfaces conference, Melbourne* (ed. R. C. Gifkins) p. 307. Butterworths, London.

Aust, K. T. and Rutter, J. W. (1959). *Trans. A.I.M.E.*, **215**, 119.

Aust, K. T. and Rutter, J. W. (1962). In *Ultra-high-purity metals*, p. 115. American Society for Metals, Metals Park, Ohio.

Babcock, S. E. and Balluffi, R. W. (1989). *Acta Metall.*, **37**, 2357; 2367.

Bainbridge, D. W., Li, C. H., and Edwards, E. H. (1954). *Acta Metall.*, **2**, 322.

Bardeen, J. and Herring, C. (1951). In *Atom movements* (ed. J. H. Holloman) p. 87. American Society for Metals, Cleveland, Ohio.

Barrett, C. S. and Massalski, T. B. (1980). *Structure of metals*. Pergamon Press, Oxford.

Barsch, G. R. and Krumhansl, J. A. (1984). *Phys. Rev. Lett.*, **53**, 1069.

Barsch, G. R. and Krumhansl, J. A. (1991). In *Martensite* (eds G. B. Olson and W. S. Owen) p. 125. ASM International, Metals Park, Ohio.

Bauer, C. L. (1982). *J. de Physique*, **43**, Colloq. C6, supple. au n° 12, C6-187.

Bauer, C. L. (1989). *Defects and Diff. Forum*, **66–69**, 749.

Bauer, C. L. and Lanxner, M. (1986). In *Grain boundary structure and related phenomena* (ed. Y. Ishida) p. 411. Supple. to *Trans. Jap. Inst. Mets.* **27**.

Bauer, C. L., Gastaldi, J., Jourdan, C., and Grange, G. (1988). In *Mat. Res. Soc. Symp. Proc.*, Vol. 122 (eds M. H. Yoo, W. A. T. Clark, and C. L. Briant) p. 199. Materials Research Society, Pittsburgh, Pennsylvania.

Bullough, R. and Bilby, B. A. (1956). *Proc. Phys. Soc.*, **69B**, 1276.

Burton, W. K., Cabrera, N., and Frank, F. C. (1951) *Phil. Trans. Roy. Soc. (London)*, **243**, 299.

Cahn, J. W. (1960). *Acta Metall.*, **8**, 554.

Cahn, J. W. (1962). *Acta Metall.*, **10**, 789.

Cahn, R. W. (1983). In *Physical metallurgy* (eds R. W. Cahn and P. Haasen) p. 1595. North-Holland, Amsterdam.

Cahn, J. W., Taylor, J. E., and Handwerker, C. A. (1991). In *Sir Charles Frank, an 80th birthday tribute* (eds R. G. Chambers, J. E. Enderby, A. Keller, A. R. Lang, and J. W. Steeds) p. 88. Hilger, New York.

Chiao, Y.-H. and Chen, I.-W. (1990). *Acta Metall. Mater.*, **38**, 1163.

Christian, J. W. (1965). *The theory of transformations in metals and alloys*. Pergamon Press, Oxford.

Christian, J. W. (1975). *Transformations in metals and alloys*. Part I, 2nd edn. Pergamon Press, Oxford.

Christian, J. W. (1982). *Met. Trans.*, **13A**, 509.

Cohen, M., Olson, G. B., and Clapp, P. C. (1979). In *Proceedings of the third international conference on martensitic transformations*, ICOMAT 79, p. 1.

Dahmen, U. (1987). *Scripta Metall.*, **21**, 1029.

daSilva, L. C. C. and Mehl, R. F. (1951). *Trans. A.I.M.E.*, **191**, 155.

Demianczuk, D. W. and Aust, K. T. (1975). *Acta Metall.*, **23**, 1149.

Dingley, D. J. and Pond, R. C. (1979). *Acta Metall.*, **27**, 677.

Drolet, J. P. and Galibois, A. (1971). *Met. Trans.*, **2**, 53.

Frank, F. C. (1949). *Disc. Faraday Soc.*, No. 5, **48**, 67.

Frank, F. C. (1958). In *Growth and perfection of crystals* (eds R. H. Doremus, B. W. Roberts, and D. Turnbull) p. 411. Wiley, New York.

Fridman, E. M., Kopezky, C. V., and Shvindlerman, L. S. (1975). *Z. Metallk.*, **66**, 533.

Frois, C. and Dimitrov, O. (1962). *Mem. Sci. Rev. Metall.*, **59**, 643.

Fromageau, R. (1969). *Mem. Sci. Rev. Metall.*, **66**, 287.

Fukutomi, H. and Horiuchi, R. (1981). *Trans. Jap. Inst. Mets.*, **22**, 633.

Fukutomi, H. and Kamijo, T. (1985). *Scripta Metall.*, **19**, 195.

Gastaldi, J. and Jourdan, C. (1981). *J. Crystal Growth*, **52**, 949.

Gastaldi, J., Jourdan, C., and Grange, G. (1990). *J. de Physique*, **51**, Colloque Cl, Supple. au n° 1, Cl–405.

Gleiter, H. (1969). *Acta Metall.*, **17**, 565; 853.

Gleiter, H. and Chalmers, B. (1972). *High-angle grain boundaries*. Pergamon Press, New York.

Gordon, P. and Vandermeer, R. A. (1962). *Trans. A.I.M.E.*, **224**, 917.

Gore, M. J., Grujicic, M., Olson, G. B., and Cohen, M. (1989). *Acta Metall.*, **37**, 2849.

Gust, W., Mayer, S., Bögel, A., and Predel, B. (1985). *J. de Physique*, **46**, Colloq. C4, supple. au n° 4, C4–537.

Hackney, S. A. and Lillo, T. (1990). *Scripta Metall.* **24**, 1653.

Haessner, F. (ed.) (1978). *Recrystallization of metallic materials*. Riederer Verlag, Stuttgart.

Haessner, F. and Hofmann, S. (1978). In *Recrystallization of metallic materials* (ed. F. Haessner) p. 63. Riederer Verlag, Stuttgart.

Hayzelden, C., Chattopadhyay, K., Barry, J. C., and Cantor, B. (1991). *Phil. Mag.*, **A63**, 461.

Higgins, G. T., Grey, E. A., and Gordon, P. (1974). *Canad. Metall. Quart.*, **13**, 309.

Hildebrand, F. B. (1949). *Advanced calculus for engineers*. Prentice-Hall, New York.

Hillert, M. and Sundman, B. (1976). *Acta Metall.*, **24**, 731.

Himmel, L. (ed.) (1963). *Recovery and recrystallization of metals*. Interscience, New York.

Hirth, J. P. and Lothe, J. (1982). *Theory of dislocations*. Wiley, New York.

Hirth, J. P. and Pound, G. M. (1963). *Condensation and evaporation*. Macmillan, New York.

Horiuchi, R., Fukutomi, H., and Takahashi, T. (1987). In *Fundamentals of diffusion bonding* (ed. Y. Ishida) p. 347. Elsevier, Amsterdam.

Hornbogen, E. (1983). In *Physical metallurgy* (eds. R. W. Cahn and P. Haasen) p. 1075. North-Holland, Amsterdam.

Howell, P. R., Nilsson, J. O., and Dunlop, G. L. (1978). *Phil. Mag. A*, **38**, 39.

Ichinose, H. and Ishida, Y. (1990). *J. de Physique*, **51**, Colloq. Cl, supple. au n° 1, Cl–185.

Jhan, R. J. and Bristowe, P. D. (1990a). In *Atomic scale calculations of structure in materials*,

*Mats. Res. Soc. Symp. Proc.* **193** (eds M. S. Daw and M. A. Schlüter) p. 189. Materials Research Society, Pittsburgh, Pennsylvania.

Jhan, R. J. and Bristowe, P. D. (1990*b*). *Scripta Metall.* **24**, 1313.

Jourdan, C. and Gastaldi, J. (1981). *J. Crystal Growth*, **54**, 361.

Kittl, J. E., Serebrinsky, H., and Gomez, M. P. (1967). *Acta Metall.*, **15**, 1703.

Krzanowski, J. E. and Allen, S. M. (1986). *Acta Metall.*, **34**, 1035, 1045.

Li, C. H., Edwards, E. H., Washburn, J., and Parker, E. R. (1953). *Acta Metall.*, **1**, 223.

Lücke, K. (1974). *Canad. Metall. Quart.*, **13**, 261.

Lücke, K. and Stüwe, H. P. (1971). *Acta Metall.*, **19**, 1087.

Lücke, K., Rixen, R., and Rosenbaum, F. W. (1972). In *The nature and behavior of grain boundaries* (ed. H. Hu) p. 245. Plenum Press, New York.

Majid, I. and Bristowe, P. D. (1987). *Scripta Metall.*, **21**, 1153.

Margenau, H. and Murphy, G. M. (1943). *The mathematics of physics and chemistry.* Van Nostrand, New York.

Massalski, T. B. (1970). In *Phase transformations* (ed. H. I. Aaronson) p. 433. American Society for Metals, Metals Park, Ohio.

Masteller, M. S. and Bauer, C. L. (1978). *Phil. Mag. A*, **38**, 697.

Masteller, M. S. and Bauer, C. L. (1979). *Acta Metall.*, **27**, 483.

Menon, E. S. K., Plichta, M. R., and Aaronson, H. I. (1988). *Acta Metall.*, **36**, 321.

Molodov, D. A., Kopetskii, Ch.V., and Shvindlerman, L. S. (1981). *Sov. Phys. Solid State*, **23**, 1718.

Mullins, W. W. (1956). *J. Appl. Phys.*, **27**, 900.

Mullins, W. W. (1957). *J. Appl. Phys.*, **28**, 333.

Mullins, W. W. (1958). *Acta Metall.*, **6**, 414.

Mullins, W. W. (1960). *Trans. A.I.M.E.*, **218**, 354.

Mullins, W. W. and Vinals, J. (1989). *Acta Metall.*, **37**, 991.

Nes, E., Ryum, N. and Hunderi, O. (1985). *Acta Metall.*, **33**, 11.

Olson, G. B. and Cohen, M. (1986). In *Dislocations in solids*, Vol. 7 (ed. F. R. N. Nabarro) p. 295. Elsevier, Oxford.

Olson, G. B. and Owen, W. S. (eds.) (1992). *Martensite.* ASM International, Metals Park, Ohio.

Plichta, M. R., Clark, W. A. T., and Aaronson, H. I. (1984). *Met. Trans. A.*, **15A**, 427.

Rae, C. M. F. (1981). *Phil. Mag. A*, **44**, 1395.

Rae, C. M. F. (1982). Ph. D. Thesis, Oxford University.

Rae, C. M. F. and Smith, D. A. (1980). *Phil. Mag. A*, **41**, 477.

Rath, B. B. and Hu, H. (1966). *Trans. A.I.M.E.* **236**, 1193.

Read, W. T. (1953). *Dislocations in crystals.* McGraw-Hill, New York.

Reed-Hill, R. E. and Abbaschian, R. (1992). *Physical metallurgy principles*, 3rd edn., PWS-Kent, Boston, Massachusetts.

Rhines, F. N. and Craig, K. R. (1974). *Met. Trans.*, **5**, 413.

Rutter, J. W. and Aust, K. T. (1965). *Acta Metall.*, **13**, 181.

Ryum, N., Hunderi, O. and Nes, E. (1983). *Scripta Metall.*, **17**, 1281.

Scholz, R. and Bauer, C. L. (1990). *J. de Physique*, **51**, Colloque Cl, supple. au n° 1, Cl–623.

Shewmon, P. (1989). *Diffusion in solids*, 2nd edn. Minerals, Metals and Materials Society, Warrendale, Pennsylvania.

Shockley, W. (1952). In *l'Etat solide*, report of the 9th international solvay conference p. 431. Stoops, Brussels.

Simpson, C. J. and Aust, K. T. (1972). *Surf. Sci.*, **31**, 479.

Simpson, C. J., Winegard, W. C., and Aust, K. T. (1976). In *Grain boundary structure and properties* (eds G. A. Chadwick and D. A. Smith) p. 201. Academic Press, New York.

Sleeswyk, A. W. (1974). *Phil. Mag.*, **29**, 407.

Smith, C. S. (1948). *Trans. A.I.M.E.*, **175**, 15.

Smith, D. A., Rae, C. M. F., and Grovenor, C. R. M. (1980). In *Grain boundary structure and kinetics* (ed. R. W. Balluffi) p. 337. American Society for Metals, Metals Park, Ohio.

Sun, R. C. and Bauer, C. L. (1970*a*). *Acta Metall.*, **18**, 635.

Sun, R. C. and Bauer, C. L. (1970*b*). *Acta Metall.*, **18**, 639.

Taylor, J. E., Cahn, J. W., and Handwerker, C. A. (1992). *Acta Metall. Mater.*, **40**, 1443.

Viswanathan, R. and Bauer, C. L. (1973). *Acta Metall.*, **21**, 1099.

Wayman, C. M. (1964). *Introduction to the crystallography of martensitic transformations.* Macmillan, New York.

Wayman, C. M. (1983). In *Physical metallurgy* (eds R. W. Cahn and P. Haasen) p. 1031. North-Holland, Amsterdam.

Westengen, H. and Ryum, N. (1978). *Phil. Mag. A*, **38**, 279.

Wörner, C. H. and Cabo, A. (1987). *Acta Metall.*, **35**, 2801.

Wörner, C. H. and Hazzledine, P. M. (1992). *J. Minerals, Mets., and Mats. Soc.*, **44** (9), 16.

# Non-conservative motion of interfaces: interfaces as sources/sinks for diffusional fluxes of atoms

## 10.1 INTRODUCTION

The concepts of conservative and non-conservative interface motion have been introduced in Section 9.1.1, and conservative motion has been treated in Chapter 9. We now take up non-conservative motion.

As already pointed out in Section 9.1.1, non-conservative motion of an interface occurs when the interface acts as a source/sink for diffusion fluxes of one, or more, components in the system, and the interface motion is coupled to those fluxes. In this process, atoms are added to, or removed from, one/both of the lattices abutting the interface: this, in turn, causes one/both of the lattices adjoining the interface to grow/shrink and the interface to move non-conservatively with respect to one/both of the lattices.

There is a host of situations in materials in which interfaces act as sources/sinks and engage in non-conservative motion. For example, in a pure material undergoing diffusional creep, sintering, or point defect annealing (see Section 10.3.3), fluxes of the lattice atoms diffuse via point defects from/to grain boundaries acting as sources/sinks for those atoms. The creation/destruction of these fluxes requires the creation/destruction of lattice sites (network sites) at the boundaries which, in turn, causes the boundaries to move relative to their adjoining crystals. In multicomponent multiphase systems various phases of differing compositions may grow, or shrink, as a result of the diffusion of the different components between them. In such cases, the heterophase interfaces between the phases act as sources/sinks for the fluxes of the different types of atoms which are involved. However, here the fluxes may, in general, be created/destroyed by two basic mechanisms, i. e. the creation/destruction of lattice sites and the shuffling of atoms across the interface from/to the phase supporting the flux to/from the adjoining phase. Again, the source/sink action causes the interface to move relative to the two adjoining crystals.

The rate at which any of these processes can proceed will clearly depend upon the ease with which atoms can be added to, or removed from, one/both of the abutting lattices at the interface. As already seen in Chapter 9, the ability of an interface to move conservatively also depends upon the ease with which atoms can be removed from one abutting lattice and added to the other as they are transferred across the interface. There are therefore close connections between a number of the mechanisms involved in conservative interface motion (see Table 9.2) and those involved in non-conservative motion: these will become apparent as we proceed.

In the following sections we first discuss a number of important general aspects of interfaces as sources/sinks for fluxes of atoms and then go on to describe in detail the manner in which relevant interfaces listed in Table 4.1 act as sources/sinks and engage in non-conservative motion in a wide range of situations: models are considered, and related experimental observations are cited. As might be expected, the results are sensitive to the structure of the interface and to the amount of energy available to drive the source/sink process. Again, as in the case of conservative interface motion, definitive and well proven models are not available in all cases: we then present plausible models which appear consistent with available observations.

## 10.2 GENERAL ASPECTS OF INTERFACES AS SOURCES/SINKS

### 10.2.1 'Diffusion-controlled', 'interface-controlled', and 'mixed' kinetics

In all cases of interest, we are concerned with a non-equilibrium situation in which a flux of atoms diffuses over some distance to/from a region of the lattice directly adjacent to an interface under the influence of a gradient in diffusion potential (Section 5.9.3). A critical issue is then the ability of the interface to maintain equilibrium in this localized region by acting as a source/sink for this flux. This will generally depend both upon how rapidly the flux of atoms can be created/destroyed at the interface and upon how rapidly the atoms can diffuse from/to the local interface region. However, two comparatively simple limiting situations may be identified. In the first the interface source/sink operates rapidly enough so that equilibrium is essentially maintained in the local interface region, i.e., the local diffusion potential is maintained essentially at the value required for local equilibrium as described in Section 5.9.2. Under these conditions, the quasi-steady rate at which atoms leave/enter the interface depends only upon the rate at which they are able to diffuse from/to the interface. If several such interfaces are distributed throughout the system in regions where they are able to maintain different local equilibrium conditions, the rate at which atoms diffuse between them then corresponds to the solution of the diffusion boundary value problem defined by the local equilibrium conditions maintained at the sources/sinks. The rate of transport will then be the maximum possible and is only limited by the rate at which atoms can diffuse between the interface sources and sinks. When the sources/sinks operate this rapidly, we regard them as 'ideal', and the resulting kinetics is termed 'diffusion-controlled'. However, the sources/sinks may operate more slowly so that they are not able to maintain local equilibrium in their vicinities. The rate of transport is then reduced. In the limit where the rate of emission/absorption of atoms becomes much slower than the rate at which the atoms can diffuse between the sources/sinks, the transport becomes entirely determined by the rate at which the atoms can be emitted/absorbed at the interface sources/sinks, i.e. the kinetics becomes 'interface-controlled'. In intermediate situations we term the kinetics 'mixed'. It must be emphasized, however, that the above limiting forms of behaviour can never be achieved exactly, and that the kinetics in any real situation will always be 'mixed' to at least some small degree. Despite this, we shall use the terms diffusion-controlled and interface-controlled to describe kinetics where the deviations from these limiting cases are not considered significant.

In order to demonstrate the above behaviour quantitatively consider a binary system consisting of A and B atoms in which spherical B-rich $\beta$ phase particles precipitate in an A-rich $\alpha$ matrix phase in which there is a dilute concentration of supersaturated interstitial B atoms. The B atoms diffuse to the particles (which are distributed throughout the volume) where they join the $\beta$ phase lattice at the $\alpha/\beta$ interface. If we neglect any effects of volume constraints on the particle growth (see, for example, Li and Blakely (1966) and Li and Oriani (1968), and assume that the $\alpha/\beta$ interfaces act as ideal sinks, we can follow Ham (1958) and treat this problem as a straightforward diffusion boundary value problem and solve for the resulting rate of precipitation. Each $\alpha/\beta$ interface will then maintain the concentration of B atoms in the $\alpha$ phase, $n^\alpha$, in its direct vicinity at its local equilibrium value, $n^{\alpha,eq}$. We shall take this concentration to be

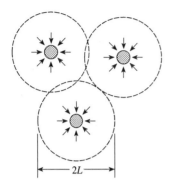

**Fig. 10.1** Diffusion field around spherical precipitate particles which are receiving fluxes of solute-atoms from the surrounding matrix. Dashed circles represent spherical Wigner–Seitz diffusion cells of radii $L$ centred on the particles.

constant and neglect its small variation with particle radius given by eqn (5.119). (We note, however, that this variation cannot be neglected when we consider particle coarsening below in Section 10.4.2.) An approximately spherically symmetric diffusion field will then be established locally around each particle as illustrated in Fig. 10.1. As a reasonable approximation, each particle can then be regarded as being at the centre of a spherical Wigner–Seitz diffusion cell of average radius, $L$, on whose surface the boundary condition $\partial n^{\alpha}/\partial r = 0$ applies. Since the cells must fill all of the available space, $L$ is given by the condition $4\pi L^3/3 = 1/N$, where $N$ is the number of $\beta$-particles per unit volume. We first find a solution for the diffusion in the cell for the case where the particle is assumed (artificially) to remain fixed in size at the radius $R$ during the precipitation and then go on to discuss the more realistic case where it grows (in response to the incoming diffusion flux (i.e. its interface is allowed to move non-conservatively). From eqn (5.148), the flux equation for the B atoms is

$$J^{\alpha} = -B^{\alpha}\nabla M^{\alpha} = -B^{\alpha}\nabla\mu^{\alpha}. \tag{10.1}$$

Assuming Henry's Law (Gaskell 1983) for the interstitials, $\mu^{\alpha} = \mu^{\alpha o} + kT\ln K^{\alpha}n^{\alpha}$, where $K^{\alpha} = $ constant, we obtain

$$J^{\alpha} = -(B^{\alpha}kT/n^{\alpha})\nabla n^{\alpha} = -D^{\alpha}\nabla n^{\alpha}, \tag{10.2}$$

where $D^{\alpha}$ is the interstitial diffusivity in the $\alpha$ lattice. Since $D^{\alpha}$, may be taken as independent of concentration, the diffusion equation is then

$$\partial n^{\alpha}/\partial t = D^{\alpha}\nabla^{\alpha}n^{\alpha}. \tag{10.3}$$

The initial condition is

$$n^{\alpha}(r,O) = n^{\alpha}_{o}, \tag{10.4}$$

and the boundary conditions are

$$n^{\alpha}(R,t) = n^{\alpha,\mathrm{eq}}; \tag{10.5}$$

$$[\partial n^{\alpha}(r,t)/\partial r]_{r=L} = 0. \tag{10.6}$$

As shown by Ham (1958), the standard separation of variables method then yields a series solution

$$n^{\alpha} = n^{\alpha,\mathrm{eq}} + \sum_{n=0}^{\infty} a_n \exp(-\lambda_n^2 D^{\alpha}t)\sin\left[\lambda_n(r-R)\right](1/r), \tag{10.7}$$

which satisfies eqns (10.5) and (10.6) if the eigenvalues, $\lambda_n$ are roots of

$$\tan\left[\lambda_n(L-R)\right] = \lambda_n L. \tag{10.8}$$

Also, eqn (10.4) is satisfied if

$$a_n = \frac{2(n_o^\alpha - n^{\alpha,\mathrm{eq}})R(\lambda_n^2 L^2 + 1)}{\lambda_n[\lambda_n L^2(L-R) - R]}. \tag{10.9}$$

The diffusion current into the sink carried by the $n$th eigenfunction is then

$$I_n = 4\pi R^2 D^\alpha (\partial n^\alpha / \partial r)_{r=R}$$

$$= 4\pi D^\alpha (n_o^\alpha - n^{\alpha,\mathrm{eq}})R\frac{2R(\lambda_n^2 L^2 + 1)}{[\lambda_n^2 L^2(L-R) - R]}\exp(-\lambda_n^2 D^\alpha t). \tag{10.10}$$

The total diffusion current is therefore a sum of eigenfunctions of the form of eqn (10.10). Each eigenfunction decays exponentially with time with a characteristic relaxation time, $\tau_n$, given by $\tau_n = 1/\lambda_n^2 D^\alpha$. When $R \ll L$, Ham (1958) showed that all of the higher-order terms die out very rapidly compared with the lowest order term which is given to a good approximation by

$$I = 4\pi D^\alpha (n_o^\alpha - n^{\alpha,\mathrm{eq}})R\exp(-t/\tau_o), \tag{10.11}$$

where $\tau_o = L^3/3D^\alpha R$. If $\bar{n}^\alpha$ = average concentration in the cell,

$$\frac{d\bar{n}^\alpha}{dt} = -\frac{I}{(4/3)\pi L^3} = -\frac{(n_o^\alpha - n^{\alpha,\mathrm{eq}})}{\tau_o}\exp(-t/\tau_o). \tag{10.12}$$

Integrating eqn (10.12), we obtain $(\bar{n}^\alpha - n^{\alpha,\mathrm{eq}}) = (n_o^\alpha - n^{\alpha,\mathrm{eq}})\exp(-t/\tau_o)$, which, when combined with eqn (10.11), yields the remarkably simple result

$$I = 4\pi D^\alpha R(\bar{n}^\alpha - n^{\alpha,\mathrm{eq}}). \tag{10.13}$$

This analysis shows that the annealing current to an ideal spherical sink quickly settles down to a quasi-steady rate given by eqn (10.13). The early transients, corresponding to the rapid decay of the higher-order eigenfunctions, are relatively unimportant and can be neglected. Because of this behaviour, we can also use the above solution to describe to a good approximation the instantaneous diffusion current of atoms to a particle which is growing in size continuously as the annealing progresses. Since any transients are always small, the instantaneous current will always correspond closely to the quasi-steady-state value given by eqn (10.13). The rate of growth of the particle (i.e. the rate of non-conservative motion of the particle/matrix interface) is directly coupled to this current and therefore can be obtained readily by using eqn (10.13).

   It is useful to point out that the above result (i.e. eqn (10.13)) can be obtained by an even simpler approximation in which the instantaneous rate of annealing is calculated by assuming a true steady state in the Wigner–Seitz cell under the boundary conditions $n^\alpha(R) = n^{\alpha,\mathrm{eq}}$ and $n^\alpha(L) = \bar{n}^\alpha$. Equations (10.3) then becomes simply

$$\nabla^2 n^\alpha = 0 \tag{10.14}$$

which has the solution $n^\alpha = a_1 + a_2/r$, where $a_1$ and $a_2$ are constants. Fitting this solution to the boundary conditions, we obtain

$$n^\alpha - n^{\alpha,\mathrm{eq}} = (\bar{n}^\alpha - n^{\alpha,\mathrm{eq}})\frac{LR}{(L-R)}\left(\frac{1}{R} - \frac{1}{r}\right) \simeq (\bar{n}^\alpha - n^{\alpha,\mathrm{eq}})\left(1 - \frac{R}{r}\right), \qquad (10.15)$$

when $R \ll L$. Therefore, $I = 4\pi r^2 D^\alpha \cdot \partial n^\alpha/\partial r = 4\pi D^\alpha R(\bar{n}^\alpha - n^{\alpha,\mathrm{eq}})$, which is identical to eqn (10.13)! The boundary condition on the Wigner–Seitz cell boundary assumed here is fictitious, of course, but the concentration far from the sink is always close to $\bar{n}^\alpha$, and the approximation is therefore reasonable physically.

Consider now a spherical $\alpha/\beta$ interface sink which is incapable of maintaining the local solute concentration in its vicinity at the equilibrium value corresponding to $n^{\alpha,\mathrm{eq}}$. If, instead, the concentration there becomes supersaturated at the value $n^{\alpha,I}$, the quasi-steady rate of annealing (eqn (10.13)) will be $I = 4\pi D^\alpha R(\bar{n}^\alpha - n^{\alpha,I})$. We may then adopt a simple model for the sink action by assuming that the rate at which B atoms in solution transfer to the particle is simply proportional to the rate at which they impinge on the particle. This will be proportional to $n^{\alpha,I}$, and the total rate of transfer will then be $(4\pi R^2 \kappa n^{\alpha,I} - 4\pi R^2 \kappa n^{\alpha,\mathrm{eq}})$ where $\kappa$ is a rate constant. The first term is the rate of impingement from solution, while the second term is the rate of emission from the particle, which we take as equal to the rate of impingement which would occur under equilibrium conditions when detailed balance prevails. The total rate of transfer must equal $I$, and therefore

$$I = 4\pi D^\alpha R\left(\bar{n}^\alpha - n^{\alpha,I}\right) = 4\pi R^2 \kappa n^{\alpha,I} - 4\pi R^2 \kappa n^{\alpha,\mathrm{eq}}. \qquad (10.16)$$

Solving eqn (10.16) for $n^{\alpha,I}$, we then obtain the annealing rate

$$I = 4\pi D^\alpha R\left(\bar{n}^\alpha - n^{\alpha,\mathrm{eq}}\right)\eta, \qquad (10.17)$$

where

$$\eta = \frac{1}{[1 + D^\alpha/R\kappa]}. \qquad (10.18)$$

When the sink annihilation capability is high, i.e. when $\kappa$ is large so that $k \gg D^\alpha/R$, $\eta \simeq 1$, $n^{\alpha,I} \simeq n^{\alpha,\mathrm{eq}}$, and kinetics is 'diffusion-controlled' with the sink acting as an ideal sink. At the other extreme, when $\kappa$ is small, so that $\kappa \ll D^\alpha/R$, $\eta \ll 1$, and $n^{\alpha,I} \simeq \bar{n}^\alpha$, the solute concentration at the sink is determined by local kinetic factors, and the kinetics is 'interface-controlled'. Otherwise, the kinetics is 'mixed'. This simple example suggests the use of $\eta$ as a 'source/sink efficiency' defined as

$$\eta \equiv \frac{\text{flux of atoms created/destroyed at actual source/sink}}{\text{flux of atoms created/destroyed at corresponding 'ideal' source/sink}}, \qquad (10.19)$$

so that an ideal source/sink operates with an efficiency of unity.

In general, we can expect $\eta$ to exhibit a range of behaviour depending upon the type of interface which is involved and the magnitude of the energy which is available to drive the source/sink action. Since the creation/destruction of a flux of atoms at an interface is generally dependent upon a number of different physical processes working together, $\eta$ is generally an inherently complicated quantity involving a number of parameters. Under certain conditions, however, these processes conspire to produce essentially ideal diffusion-controlled conditions with $\eta \simeq 1$, and the details of the physical processes at the interface core may then be conveniently ignored in an overall description of the kinetics.

## 10.2.2 Dissipation of energy during source/sink action

When diffusion-controlled conditions prevail, the sources/sinks operate easily, and relatively little energy dissipation occurs there i.e. essentially all of the overall energy which is being dissipated in the non-equilibrium system is dissipated in the long range diffusion field. Conversely, when the process becomes interface-controlled, essentially all of the available energy is dissipated in the source/sink process directly at the interface. The partition of the dissipated energy in this fashion can be demonstrated quantitatively in the case of the precipitation system which we have just analysed. It is readily shown that the total free energy which is dissipated in the overall precipitation, $\Delta g$, may be represented graphically as shown in Fig. 10.2. Here, we plot the free energy per atom, $g$, versus composition, $X$, (atom fraction of B) curves for the bulk $\alpha$ and $\beta$ phases. The compositions of the equilibrium $\alpha$ and $\beta$ phases, i.e. $X^{\alpha,\mathrm{eq}}$ and $X^{\beta,\mathrm{eq}}$, correspond, as usual, to the points of common tangency. We are interested in the energy change which occurs when a small amount of $\beta$ phase, whose composition we may take to be equal to $X^{\beta,\mathrm{eq}}$, is formed from the $\alpha$ phase of average composition $\bar{X}^{\alpha}$. If we start with an amount of $\alpha$ phase of composition $\bar{X}^{\alpha}$ containing $(N + N^{\beta})$ atoms at time $t_1$ and form a small amount of $\beta$ phase containing $N^{\beta}$ atoms at a slightly later time $t_2$, conservation of B atoms requires

$$[N + N^{\beta}]\bar{X}^{\alpha}(t_1) = N^{\beta}X^{\beta,\mathrm{eq}} + N\bar{X}^{\alpha}(t). \qquad (10.20)$$

The corresponding change in free energy is

$$\Delta G = [g^{\alpha}(t_2)N + g^{\beta}N^{\beta}] - g^{\alpha}(t_1)[N + N^{\beta}], \qquad (10.21)$$

and therefore, combining eqns (10.20) and (10.21), we have

$$\Delta g = \Delta G/N^{\beta} = \left\{ g^{\beta} - \left[ g^{\alpha} + (X^{\beta,\mathrm{eq}} - \bar{X}^{\alpha})\left( \frac{\partial g}{\partial X^{\alpha}} \right)_{\bar{X}^{\alpha}} \right] \right\}, \qquad (10.22)$$

since $N^{\beta} \ll N$ and $[g^{\alpha}(t_2) - g^{\alpha}(t_1)]/[\bar{X}^{\alpha}(t_2) - \bar{X}^{\alpha}(t_1)] = \partial g/\partial X^{\alpha}$ at $X^{\alpha} = \bar{X}^{\alpha}$. From eqn (10.22), it may be seen that, $\Delta g$, the energy which must be dissipated, per atom of precipitate formed, is given by the tangent construction shown in Fig. 10.2. Now, if the sink efficiency, $\eta$, during the precipitation of the $\beta$ phase approaches zero, and the concentration is then essentially uniform in the $\alpha$ phase at the level $\bar{X}^{\alpha}$ (as in Fig. 10.3(a)), no significant chemical gradients will be present and essentially no dissipation will occur in the diffusion field. Essentially all of the available energy, $\Delta g$, will then be dissipated at the interface (if we neglect the energy required to increase the $\alpha/\beta$ interfacial area). In an intermediate mixed case where $0 < \eta < 1$, the concentration maintained at the sink interface, $X^{\alpha,\mathrm{I}}$, will have an intermediate value $X^{\alpha,\mathrm{eq}} < X^{\alpha,\mathrm{I}} < \bar{X}^{\alpha}$ (as in Fig. 10.3(b)), and the physical situation at the sink will be the same as if the concentration were uniform everywhere in the $\alpha$ phase. The energy dissipated at the interface, $\Delta g^{\mathrm{I}}$, will then be given by the same tangent construction, but the tangent will have to be taken at $X^{\alpha,\mathrm{I}}$, rather than $\bar{X}^{\alpha}$. The energy dissipated in the diffusion field, $\Delta g^{\mathrm{d}}$, will then be $\Delta g^{\mathrm{d}} = \Delta g - \Delta g^{\mathrm{I}}$ as shown. Finally, at the other limit when $\eta \to 1$, $X^{\alpha,\mathrm{I}} \to X^{\alpha,\mathrm{eq}}$ (as in Fig. 10.3(c)), and the diffusion gradients will be the steepest possible. The energy dissipated at the interface will then be zero ($\Delta g^{\mathrm{I}} = 0$), and all of the energy will be dissipated in the diffusion field.

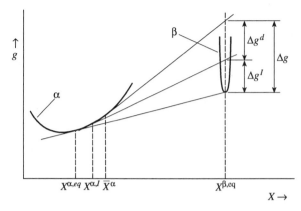

**Fig. 10.2**   Free energy per atom, $g$, of binary $\alpha$ and $\beta$ phases as a function of their composition
($X$ = atom fraction of B). $X^{\alpha,\,eq}$ and $X^{\beta,\,eq}$ are the compositions of the $\alpha$ and $\beta$ phases which
would be in equilibrium with each other in alloys with overall compositions ranging between $X^{\alpha,\,eq}$
and $X^{\beta,\,eq}$. $\Delta g$ is the free energy (per atom of $\beta$ phase) which must be dissipated if a small amount
of $\beta$ phase, of composition $X^{\beta,\,eq}$, is precipitated from a large amount of supersaturated $\alpha$ phase
of composition $\bar{X}^{\alpha}$. $\Delta g^{I}$ represents the portion of the energy, $\Delta g$, which is dissipated at the $\alpha/\beta$
interface if the $\alpha/\beta$ interface acts as an imperfect sink ($0 < \eta < 1$) so that the composition at the
interface is maintained at $X^{\alpha,\,I}$ during the precipitation. $\Delta g^{d}$ is then the portion which is dissipated
in the diffusion field.

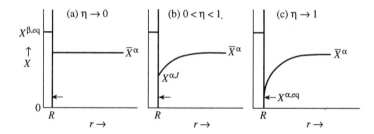

**Fig. 10.3**   Concentration profiles of B atoms, $X(r)$, in the vicinity of B-rich $\beta$ phase precipitate
particle growing by diffusion in an $\alpha$ matrix as illustrated in Fig. 10.1. (a) Profile when $\alpha/\beta$ interface
is: (a) a poor sink for B atoms with $\eta \to 0$; (b) a moderately effective sink with $0 < \eta < 1$; (c) an
essentially ideal sink with $\eta \to 1$.

### 10.2.3   The maximum energy available to drive the source/sink action

In Chapter 9 we found that the pressure, $p$, exerted on a conservatively moving interface,
and the corresponding free energy available from the system to transfer an atom across
the interface during its motion, $g_m$, were physically important quantities. In general, all
of the available energy, $g_m$, was dissipated at the interface, and this quantity had to be
large enough to force the various mechanisms for motion to operate effectively. In a
similar way, it will be seen in the present chapter that the maximum energy available from
the system to drive the source/sink action of an interface is of critical importance. In
source/sink situations this available energy, which we designate by $g_s$ (per atom
emitted/absorbed), is generally dissipated both at the source/sink and in the surrounding
diffusion field as has been demonstrated above in Section 10.2.2. When $g_s$ is relatively

**Table 10.1** Approximate common values of $g_s$, the maximum energy available to drive the source/sink action of interfaces (per atom emitted/absorbed)

| Type of interface | Source of $g_s$ | $g_s$ (eV) |
|---|---|---|
| Homophase interface | Annihilation of supersaturated points defects during quenching or radiation damage (see eqn (10.23)) | $10^{-1}$–$10^0$ |
| Heterophase interface between precipitate and matrix | Growth of precipitates (see eqn (10.22)) | $10^{-2}$–$10^{-1}$ |
| Homophase interface | Shrinkage of nearby dislocation loops during annealing (see eqn (10.25)) | $10^{-2}$–$10^{-1}$ |
| Homophase interface | Shrinkage of nearby voids during sintering (see eqn (10.26)) | $10^{-6}$–$10^{-4}$ |
| Homophase interface | Nearby stressed interface during diffusional creep (see eqn (10.27)) | $10^{-6}$–$10^{-4}$ |

large compared with the energy required to force the source/sink mechanisms to operate effectively, the source/sink will operate with a relatively high efficiency. Otherwise, lower efficiencies will result.

The magnitude of $g_s$ will vary widely in different source/sink situations as seen in Table 10.1. In the precipitation process considered above, $g_s$ is given by eqn (10.22) and can have values in the range $10^{-2}$–$10^{-1}$ eV. During quenching and irradiation, supersaturated point defects (i.e. vacancies and interstitials) will be present and will tend to anneal out at interfaces acting as atom sources for the vacancies and atom sinks for the interstitials. The available energy, $g_s$, is then equal to the chemical potentials of the point defects. For vacancies in a unary system eqn (9.12) holds, and since the vacancy chemical potential is given by $\mu_V = \partial G/\partial N_V$, we have $\mu_V = g_V^f + kT \ln X_V$, where $X_V$ is the atom fraction. However, $X_V^{eq} = \exp(-g_V^f/kT)$, and, therefore, $\mu_V = kT \ln(X_V/X_V^{eq})$. A similar development holds for interstitials, and therefore,

$$g_s = \mu = g^f + kT \ln X = kT \ln(X/X^{eq}). \tag{10.23}$$

It is easily verified that values of $g_s$ in the range $10^{-1}$–$10^0$ eV can then be present.

During quenching, or irradiation, the supersaturated point defects often precipitate in the form of dislocation loops (Cotterill *et al.* 1965, Seeger *et al.* 1970). Upon further annealing, these loops may anneal out by gaining/losing atoms via lattice diffusion from/to nearby boundaries acting as atom sources/sinks. In this case $g_s$ arises from the decrease in the self-energy of the loop as it shrinks. Consider, for example, a circular prismatic loop of radius $R$ and Burgers vector $b$ containing $\pi R^2 b/\Omega$ precipitated point defects. The decrease in $R$ per atom exchanged with the interface is then $\Omega/2\pi bR$. The self energy of the loop is given (Hirth and Lothe 1982) by

$$W = \frac{\mu b^2 R}{2(1-\nu)} \left[ \ln\left(\frac{8R}{r_o}\right) - 1 \right], \tag{10.24}$$

where $r_o$ is the usual cutoff radius (which is of the order of an interatomic distance), and, therefore,

$$g_s = [\mu b \Omega / 4\pi (1 - \nu)] \ln(8R/r_o) \cdot (1(R). \tag{10.25}$$

Values of $g_s$ in the range $10^{-2}$–$10^{-1}$ eV are then present when $R$ is in the range $10^2$–$10^3$ Å.

During sintering, voids are eliminated by the diffusional transport of atoms to the voids from nearby grain boundaries acting as atom sources (Alexander and Balluffi 1957). Here, $g_s$ arises from the decrease in the surface energy of the void as it gains atoms and shrinks. For a spherical void of radius $\rho$, the change in surface area $\delta A$ due to a change in volume $\delta V$ is $\delta A = (2/\rho)\delta V$. The decrease in area per atom gained is then $\delta A = 2\Omega/\rho$, and the decrease in energy is then

$$g_s = 2\sigma^S \Omega / \rho, \tag{10.26}$$

where $\sigma^S$ = surface free energy per unit area. Values of $g_s$ in the range $10^{-6}$–$10^{-4}$ eV are then present when $\rho$ is in the range $10^{-2}$–$10^{-4}$ cm.

During the diffusional creep of polycrystals under uniaxial tension (Section 12.8.2.1), atoms are transferred from longitudinal boundaries acting as sources to transverse boundaries acting as sinks under the influence of the applied stress, $\tau$. Here, $g_s$ arises from the work done by the applied stress as the atoms are transferred, and the polycrystal elongates. Therefore,

$$g_s = \tau \Omega. \tag{10.27}$$

Common values of $g_s$ are then in the range $10^{-6}$–$10^{-4}$ eV.

The values of the energy $g_s$ listed in Table 10.1 are seen to extend over a wide range corresponding to about six orders of magnitude. Note that in many cases these energies are considerably larger than the energies, $g_m$, available to drive the conservative motion of interfaces listed in Table 9.1.

## 10.3  GRAIN BOUNDARIES AS SOURCES/SINKS FOR FLUXES OF ATOMS

### 10.3.1  Introduction

When grain boundaries act as sources/sinks, the compositions of the two adjoining crystals in the immediate vicinity of the boundary are the same. Under these circumstances the boundaries can act as sources/sinks for fluxes of atoms and engage in non-conservative motion coupled to these fluxes only by creating/destroying network sites. Since there is no composition difference across the boundary, fluxes of the different types of atoms in the system cannot be created/destroyed by the shuffling of atoms across the interface from/to the phase supporting the flux to/from the adjoining phase.

### 10.3.2  Small-angle grain boundaries

#### 10.3.2.1  *Models*

For small-angle boundaries the only viable mechanism available for the creation/ destruction of network sites is the climb of the primary dislocations in the boundary having Burgers vector components normal to the boundary plane. In this mechanism (Balluffi 1969, Balluffi and Granato 1979, Hirth and Lothe 1982) point defects are created/destroyed at jogs on the climbing dislocations by means of thermally activated fluctuations. These sites are unique places where these events can occur relatively easily; for example, they cannot possibly occur in the surrounding crystal. The continuous creation/destruction of point defects in this manner allows fluxes of point defects (or, alternatively, fluxes of atoms) to diffuse through the lattice to/from the individual dislocations causing the boundary to act as an overall net source/sink.

Consider, first, as the simplest possible case, a pure symmetric tilt boundary consisting of a planar array of a single set of parallel edge dislocations in a crystal composed only of A-type atoms and supersaturated vacancies as illustrated in Fig. 10.4(a). The crystal is free of applied stress, the equilibrium vacancy concentration in the stress-free bulk is $n_V^{eq}$, and self-diffusion occurs by a vacancy mechanism. Supersaturated vacancies are present at the concentration $\bar{n}_V$ in the bulk (at a large distance, $L$, from the boundary) and can diffuse to the boundary where they can be destroyed at jogs on the dislocations (Fig. 10.4(a)) causing the dislocations to climb and the boundary as a whole to act as a source of atoms which replace the incoming vacancies. The unrelaxed structure of a boundary of this type with its plane parallel to $(14\ \bar{1}\ 0)$ planes of the abutting lattices is shown in Fig. 2.1(a). It may be seen that the continuous climb of its edge dislocations will cause the continuous removal of $(14\ \bar{1}\ 0)$ planes from the surfaces of both abutting crystals. This, in turn, will cause the two crystals to move towards each other and the interface to move towards each crystal.

The overall ability of the boundary to act as a planar atom source then depends upon a variety of parameters such as the density of its dislocations, the extent of the vacancy diffusion field around it, and the ability of the individual dislocations to destroy the incoming vacancies. We now follow Balluffi (1969) and Hirth and Lothe (1982) and develop an approximate model for this idealized boundary which shows explicitly how the source efficiency, $\eta$, depends upon these parameters. The destruction of vacancies at the dislocation lines is itself a complex process which involves the jumping of vacancies into the core, and the diffusion of these vacancies along the core to jogs where they are

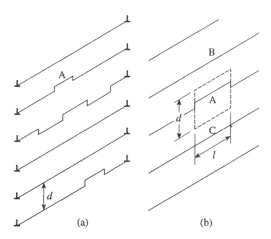

**Fig. 10.4** (a) Small-angle symmetric tilt boundary, composed of parallel edge dislocations, acting as a sink for supersaturated vacancies. Excess vacancies are destroyed at jogs on the dislocation causing them to climb. (b) Schematic diagram showing the equivalence of a double jog (at A) and a weak dislocation loop (dashed) possessing Burgers vector $b^l = bh^J/d$, where $b$ = edge dislocation Burgers vector and $h^J$ = jog height.

replaced by atoms and destroyed. We assume initially (Balluffi 1969) that vacancy destruction (and creation) at the jogs is sufficiently rapid so that the vacancy concentration in the direct vicinity of the jogs is maintained in local equilibrium at the value $n_V^{eq}$ and that all the dislocations climb together in unison at the same quasi-steady rate. The overall kinetics is then dictated by the rate at which the vacancies can diffuse to the jogs.

From eqn (9.15) the flux of vacancies in the lattice is given by

$$J_V = -J_A = -[B_{AA}^L N_o kT/X_V]\nabla X_V = -D_V^L \nabla n_V, \tag{10.28}$$

where $D_V^L$ is the vacancy diffusivity in the lattice which can be taken as concentration independent because of the low density of vacancies. The quasi-steady-state rate of annealing of the excess vacancies may now be obtained by using some of our previous results. According to the discussion of fast diffusion along grain boundaries and dislocations in Section 8.2.3.2, we expect an average vacancy to experience a binding energy to the dislocation cores and also to diffuse along the cores more rapidly than in the lattice. As a necessary preliminary we inquire first into the mean diffusion distance that an attached vacancy would diffuse along a dislocation core between the time it jumped on and the time it jumped off in the hypothetical situation where no jogs were present. Following the developments in Sections 8.2.5 and 8.2.6, the diffusivity of an attached vacancy along the dislocation may be written approximately as

$$D_V^D = b^2 \nu \cdot \exp[-(G_V^m - \Delta G_V^m)/kT], \tag{10.29}$$

where $b$ is the average interatomic distance, $\nu$ is an atomic vibrational frequency, $G_V^m$ is the vacancy migration energy in the lattice, $G_V^m - \Delta G_V^m$ is the migration energy of an attached vacancy in the core, and correlation effects have been neglected. The mean time of stay of a vacancy in the core is then

$$\tau = \nu^{-1} \cdot \exp[(G_V^m + \Delta G_V^f)/kT], \tag{10.30}$$

where $\Delta G_V^f$ is the binding energy of the attached vacancy to the core. The mean migration distance along the core during this time is therefore

$$\bar{Z} = (D_V^D \tau)^{\frac{1}{2}}, \tag{10.31}$$

or, combining eqns (10.31), (10.30), and (10.29),

$$\bar{Z} = b \cdot \exp[(\Delta G_V^m + \Delta G_V^f)/2kT]. \tag{10.32}$$

Since the average attached vacancy diffuses a distance $\bar{Z}$ along the core before jumping off, each jog is capable of maintaining the concentration of attached vacancies along the core on either side of it in approximate equilibrium over a distance of about $\bar{Z}$. Each jog, aided by the rapid diffusion along the core, therefore tends to set up a quasi-steady-state diffusion field around it. The jog, in combination with the two adjoining segments of fast diffusivity core, therefore, constitutes essentially an ellipsoidal vacancy sink (Fig. 10.5) of semiaxes $\simeq b$ and $\simeq \bar{Z}$ on whose surface the vacancy concentration is maintained at the equilibrium concentration $n_V^{eq}$. Even though the jog is moving due to the absorption of vacancies we may assume, on the basis of our previous discussion of quasi-steady-state sink annealing kinetics, that it is stationary and solve for the quasi-steady diffusion current into the ellipsoidal sink surface. This may be done by using a convenient result, pointed out by Flynn (1964), that the quasi-steady-state current to a sink of any shape in a Wigner–Seitz cell in a situation where $\partial n/\partial t = D^L \nabla^2 n$ is given by the relation

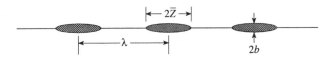

**Fig. 10.5**  Jogs arrayed along a dislocation line at a spacing $\lambda$. In order to analyse the diffusion of excess vacancies to the jogs, each jog is represented by an ellipsoidal sink of semi-axes $b$ and $\bar{Z}$.

$$I = 4\pi D^L \Delta n \cdot r_{eff} = 4\pi D^L \Delta n \cdot C \qquad (10.33)$$

where $\Delta n$ is the difference between the concentrations maintained at the sink surface and the cell boundary and $r_{eff}$ is the effective sink radius which is equal to $C$, the electrostatic capacitance of a conductor of the same shape as the sink surrounded by a conducting shell of the same geometry as the Wigner–Seitz cell. (We note that the capacitance of a sphere inside a much larger sphere is closely equal to its radius, and therefore eqn (10.33) yields our previous eqn (10.13) directly.) This result is a consequence of the close agreement between quasi-steady-state and true steady state solutions and the similarity between the concentration fields $c(x, y, z)$, and electrostatic potential fields $V(x, y, z)$, which are obtained by solving Laplace's equation in steady state diffusion ($\nabla^2 n = 0$) and electrostatic ($\nabla^2 V = 0$) problems, respectively. The effective radius (capacitance) of the ellipsoidal sink in a large space is $\simeq \bar{Z}/\ln(2\bar{Z}/b)$ when $\bar{Z} > b$ (Morse and Feshbach 1953), and, according to eqn (10.33), the steady state current which would enter the sink if it were isolated would then be

$$I = 4\pi D_V^L (\bar{n}_V - n_V^{eq}) \cdot [\bar{Z}/\ln(2\bar{Z}/b)]. \qquad (10.34)$$

Also, the corresponding diffusion field around the sink, at distances $r > \bar{Z}$, would be given by

$$n_V - n_V^{eq} = (\bar{n}_V - n_V^{eq})\{1 - r_{eff}/r\} = (\bar{n}_V - n_V^{eq})\{1 - (\bar{Z}/\ln[2\bar{Z}/b])/r\}. \qquad (10.35)$$

However, the jog sinks are not isolated but instead are strung out along the dislocations at the spacing $\lambda$ (Fig. 10.5). It is therefore necessary to take into account the overlap between the diffusion fields set up by the individual jogs. This is a complicated problem in general, but simple approximate solutions can be obtained for some particular situations. The simplest case occurs when $\bar{Z} > \lambda/2$, and the jogs are close enough together so that their overlapping diffusion fields maintain the vacancy concentration at equilibrium everywhere along each dislocation (i.e. each dislocation acts as an ideal line sink). In this situation each dislocation will tend to establish a cylindrical diffusion field around itself out to a radial distance which is about half the spacing between dislocations. We can therefore construct a cylindrical Wigner–Seitz cell around each dislocation of radius $d/2$ and find the steady-state diffusion current, $I^D$, entering each dislocation if the vacancy concentration on the surface of the cell is $n_V^s$ and the concentration on the surface of the dislocation core (taken to be a cylinder of radius b) is $n_V^{eq}$. The capacitance of a cylinder of radius $b$ inside a cylinder of radius $d/2$ is just $C = 1/2\ln(d/2b)$, and using eqn (10.33) we therefore obtain

$$I^D = 2\pi D_V^L (n_V^s - n_V^{eq})/\ln(d/2b). \qquad (10.36)$$

The total current entering unit area of boundary is then $I = I^D/d$. On the other hand, the current diffusing to unit area of boundary as a whole in the perpendicular direction from the large distance $L$ is approximately

$$I = 2D_V^L(\bar{n}_V - n_V^s)/L. \tag{10.37}$$

Since these two currents must be equal, we can solve for $n_V^s$ to obtain

$$I = \frac{2D_V^L(\bar{n}_V - \bar{n}_V^{eq})}{L[1 + (d/\pi L) \cdot \ln(d/2b)]}. \tag{10.38}$$

If the boundary would act ideally as an ideal planar sink, the current would be given by $2D_V^L(\bar{n}_V - n_V^{eq})/L$, and, using eqn (10.19), the sink efficiency is therefore

$$\eta = \frac{I}{2D_V^L(\bar{n}_V - n_V^{eq})/L} = \frac{1}{[1 + (d/\pi L) \cdot \ln(d/2b)]}. \tag{10.39}$$

Equation (10.39) shows that a small-angle boundary consisting of an array of dislocations acting as ideal vacancy line sinks will act as an ideal planar sink whenever the average diffusion distance to the boundary, $L$, is sufficiently large compared with the distance between the dislocations, $d$, so that $[(d/\pi L) \cdot \ln(d/2b)] \ll 1$ (as will be the case in many actual situations).

However, the array can still act as an ideal planar sink even when $2\bar{Z} < \lambda$, and the individual dislocations are not ideal line sinks. The overlapping diffusion fields established by the individual jog sinks must then add up so that the contributions of all the jogs to the overall diffusion field produce the appropriate equilibrium vacancy concentration in the vicinity of each jog sink and also the concentration $\bar{n}_v$ at the large distance $L$. When $\lambda > d > 2\bar{Z}$, this can be accomplished in an approximate manner by fictitiously reducing the effective radius of each jog sink by the factor $A$. The condition on $A$ is then

$$\frac{A \cdot r_{eff}}{r_{eff}} + A \cdot r_{eff} \cdot \sum_i \frac{1}{r_i} = 1 \tag{10.40}$$

where the first term is the contribution of the jog under consideration, and the second term is the sum of the contributions of all the other jogs located at distances $r_i$ in the boundary plane. The planar sink density is $(\lambda d)^{-1}$, and replacing the sum by an integral with a cutoff at $r_i = L$ (since sinks at distances greater than about $L$ will not contribute), we have

$$A = \frac{1}{\{1 + [\bar{Z}/\ln(2\bar{Z}/b)] \cdot [2\pi L/\lambda d]\}}. \tag{10.41}$$

Therefore, using eqn (10.33), the total current entering unit area of boundary is $I = 4\pi D_V^L(\bar{n}_V - n_V^{eq}) \cdot [A\bar{Z}/\ln(2\bar{Z}/b)] \cdot (1/\lambda d)$, and the current which would enter if it were an ideal planar sink is again $2D_V^L(\bar{n}_V - n_V^{eq})/L$. The efficiency is then

$$\eta = \frac{I}{2D_V^L(\bar{n}_V - n_V^{eq})/L} = \frac{1}{[1 + (1/\pi) \cdot (\lambda/2\bar{Z}) \cdot (d/L) \cdot \ln(2\bar{Z}/b)]}. \tag{10.42}$$

Again there will be many situations in which the values of the various parameters which are involved will conspire to produce ideal sink behaviour, i.e. $\eta \simeq 1$.

The above model is still incomplete, however, since it does not provide any basis for establishing the jog spacing, $\lambda$. During the quasi-steady-state climb, positive and negative jogs will continuously migrate along the dislocations and eventually destroy one another by mutual annihilation (Fig. 10.4(a)). New jogs must therefore be continuously formed. This can occur by the nucleation of jog pairs, such as at A in Fig. 10.4(a), via the aggregation of supersaturated vacancies on the dislocation lines (Thomson and Balluffi 1962,

Balluffi and Thomson 1962, Balluffi 1969, Hirth and Lothe 1982) at a rate which will depend upon the jog energy, $W_j$, and the degree of vacancy supersaturation. For example, an approximate analysis of the steady state nucleation rate when the dislocations are not acting as ideal line sinks yields a relation for the steady state jog spacing (Balluffi and Thomson 1962, Balluffi 1969) of the form

$$\lambda \simeq \lambda^{eq}/(S+1)^{\Lambda/2}, \tag{10.43}$$

where $S \equiv [\bar{n}_V/n_V^{eq}] - 1$ is the vacancy supersaturation, $\Lambda b$ is the minimum spacing between positive and negative jogs in a fully formed jog pair, and $\lambda^{eq}$ is the equilibrium jog spacing given by $\lambda^{eq} = b \exp(W_j/kT)$. The spacing will therefore be near its equilibrium value at small $S$ and become smaller as the supersaturation increases as might be expected. The jog energy will vary considerably from material to material depending upon the degree to which the primary dislocations are dissociated into partial dislocations separated by ribbons of stacking fault (Hirth and Lothe 1982). In materials with high stacking fault energies, the dislocations will be barely dissociated causing the jogs to have relatively low energies. On the other hand, the dislocations in low stacking fault energy materials will be more dissociated and will, therefore, have higher jog energies (Hirth and Lothe 1982). Also, the parameter $\Lambda$ will tend to increase as the dissociation increases.

A wide variety of situations is therefore possible. In materials with low jog energies a jog spacing close to the equilibrium value may be sufficient to produce essentially ideal sink behaviour. In that case $\lambda \leqslant \lambda^{eq}$, and the vacancy concentration will be maintained close to equilibrium in the vicinity of the boundary. In materials with high jog energies this will not be possible. However, if the supersaturation is very large, a steady state situation may develop in which $\lambda$ is sufficiently reduced (i.e., the dislocations become sufficiently 'joggy') so that the boundary operates as a highly efficient sink. It is therefore possible for boundaries in high jog energy materials to act as nearly ideal sinks if the vacancy supersaturation is sufficiently high. We note that this is an explicit example of the importance of the parameter $g_s$, i.e. the maximum energy available to drive the source/sink action. According to eqn (10.23), $g_s = kT \ln(S+1)$, and when this is sufficiently large the fraction of it which must be dissipated at the boundary in order to drive the sink action can be relatively small, resulting in relatively ideal sink action.

So far, we have assumed that the dislocations in the array are uniformly spaced and that during the sink action they are all able to climb in an unimpeded fashion while maintaining their spacing uniform. However, this is unrealistic, and a number of inhibiting factors of varying importance will generally be present. As emphasized, for example, by Gleiter (1979) and King and Smith (1980), the inhomogeneous nature of the dislocation climb process makes it impossible for any array to climb in a perfectly uniform fashion. In order to demonstrate this explicitly consider an initially uniform array as in Fig. (10.4(b)), where one of the dislocations develops a double jog at A and climbs by a unit jog height as indicated. The accompanying removal of atoms will cause the two crystals to move together over the area $l \times d$ by a distance $bh^J/d$, where $h^J =$ unit jog height. This produces a boundary defect which is equivalent to the insertion of a dislocation loop with a small effective Burgers vector $b^{eff} = bh^J/d$ corresponding to the dashed rectangle in Fig. 10.4(b). We recognize this dislocation as a relatively weak secondary dislocation, i.e. a perturbation in the spacing of the primary dislocations as discussed in Section 2.10.3. The insertion of such a loop will clearly cause an increase in the overall energy. The continued climb of the array would proceed with the least further increase in energy if the double jog at A continued to expand, or if double jogs formed sympathetically on neighbouring primary dislocations at B or C. The entire process therefore corresponds to the nucleation and growth of a secondary dislocation loop. In reality, however, the primary dislocations in small-angle boundaries will generally be non-uniformly spaced for intrinsic geometrical reasons, and they will therefore already contain arrays of secondary dislocations corresponding to the perturbations in the spacings of the primaries (Section 2.10.3). They may also contain extrinsic secondary dislocations due to the impingement of lattice dislocations as discussed in Section 9.2.2. These secondary dislocations could then climb preferentially as in the case of the nucleated secondary dislocations just described.

It must also be recalled that the continuous destruction of vacancies via dislocation climb will cause the two abutting crystals to move towards each other. If volume constraints exist which oppose this macroscopic shape change, bulk stresses will be established which will exert forces on the climbing dislocations opposing their continued climb motion. In addition, the climbing array will not be of infinite extent, and various restrictive conditions could exist around its periphery. For example, if the array comprises one boundary in a polycrystalline 'subgrain' structure, appropriate numbers of dislocations would have to be injected or removed at its junctions with other boundaries in order to maintain a steady rate of climb.

The above complications will increase the difficulty of the sink action and increase the amount of energy dissipated in the boundary, thereby decreasing the sink efficiency. At high values of the driving energy, $g_s$, these phenomena will tend to be unimportant, and all of the primary dislocations will be driven to climb simultaneously as assumed earlier. However, at lower values of $g_s$ the sink action could become restricted to the climb of secondary dislocations: at sufficiently low $g_s$ values it could become impossible to drive the restraining processes, and threshold effects could then occur.

The above climb model can be extended to include boundaries consisting of various arrays of primary dislocations including networks of mixed dislocation segments joined together at nodes. Such networks would also have to undergo significant perturbations in their geometry in order to climb. A comprehensive analysis, including the details of this phenomenon as well as others such as the possible heterogeneous nucleation of jogs at dislocation nodes and the effects of the possible dissociation of dislocations into partial dislocation configurations, would be extremely complex, and hence will not be pursued any further here. Discussion of some of these topics for individual dislocations is available elsewhere (see Balluffi 1969).

A further potential complication arises from the fact that we have neglected any effects of the local stress field of the boundary on the diffusion rate of the vacancies to the boundary. In general, there will be an energy of interaction between the vacancies and the stress field, as described in Section 7.3.2.2. and they will then experience a 'drift force' proportional to the gradient of the interaction energy. This drift force will then bias the vacancy jumps in the direction of the force and cause an overall drift of the diffusing defects in the direction of the force. This phenomenon can be analysed quantitatively (Ham 1959) by adding an appropriate drift term (proportional to the drift force) to the equation for the diffusion flux. In quasi-steady-state situations when the range of the interaction (i.e. the distance over which the interaction energy falls to below $kT$) is smaller than $\simeq d/2$, the effect of the interaction can be approximated rather well by simply changing the effective radius (or capacitance) of the dislocations and retaining the present formalism (Ham 1959, Brailsford and Bullough 1981). It is readily seen from our previous analysis that this can have, at most, only a small effect on the boundary sink efficiency. Since the stress field cannot extend appreciably beyond $\simeq d/2$, the range of interaction cannot extend beyond this distance as well, and any effects of the stress field will therefore be relatively small.

Still a further complicating phenomenon may occur when solute atoms are present in the system. When a current of excess point defects diffuses towards a boundary sink, the diffusing defects will generally interact with the solute atoms and induce a coupled flux of solute atoms either towards or, even possibly, away from the boundary (Anthony 1975, Okamoto and Rehn 1979, Wiedersich *et al.* 1979, Balluffi and King 1983). The coupling of these fluxes is complex and depends upon the details of the interactions between the solute atoms and the defects and the various ways in which they jump in each other's presence. When solute atoms are 'dragged' into the sinks by such a mechanism, the defects are destroyed, and the solute atoms then remain deposited at, or near, the sink. This form of non-equilibrium segregation can be strong and, under certain circumstances, a sufficient solute concentration can be built up to induce even the formation of non-equilibrium second phase segregates which would then affect virtually all of the atomic processes associated with the sink action of the boundary as discussed in detail by Balluffi and King (1983).

### 10.3.2.2 *Experimental observations*

We now review experimental observations of the source/sink behaviour of small-angle grain boundaries in metallic systems for which a significant number of quantitative results is available. In doing this, it is also relevant to include observations of the closely related source/sink behaviour of relatively isolated lattice dislocations. Detailed reviews have been given elsewhere (Seidman and Balluffi 1967, Balluffi 1969, Balluffi 1975, Balluffi and Granato 1979, Siegel *et al.* 1980, Balluffi 1980, Balluffi and King 1983, Balluffi 1984). The information comes from a variety of sources involving vacancy quenching, dislocation loop annealing, and sintering experiments. Of primary interest is the determination of the source/sink efficiency, $\eta$, and its dependence upon $g_s$. The basic procedure is to calculate the maximum possible rate of source/sink action assuming ideal source/sink behaviour and to compare the results with measured values.

The results for relatively isolated dislocations indicate that climbing lattice dislocations distributed throughout bulk crystals of the relatively low stacking fault energy metal Au act as highly efficient ($\eta \simeq 1$) vacancy sinks in downquenched and upquenched specimens when $g_s$ is large ($\simeq 0.7 - 0.3$ eV). However, there is apparently some drop-off in $\eta$ when $g_s$ becomes lower, i.e. $< 0.2$ eV. Sintering experiments with the relatively low stacking fault energy metal Cu with $g_s$ at the low value $g_s \leqslant 10^{-5}$ eV indicate a possibly low value of $\eta$ (i.e. $\eta \leqslant 0.05$). Dislocation loops in high stacking fault energy metals (e.g. Al) shrink (or grow) by climb with a high source/sink efficiency ($\eta \simeq 1$) when $g_s$ is moderately large (0.01–0.1 eV). On the other hand, loops in low stacking fault energy metals shrink (and also probably grow) with lower efficiencies. The general picture which emerges is that all dislocations, including relatively non-dissociated dislocations in high stacking fault energy metals and dissociated dislocations in lower stacking fault energy metals, operate as highly efficient sources/sinks when $g_s$ is large. When $g_s$ decreases, lower efficiencies are found for the lower stacking fault metals which eventually may become very small. The efficiencies for the higher stacking fault metals appear to fall off less rapidly. The results may be generally understood on the basis of the tendency of the dislocations to become more joggy as $g_s$ increases and the greater difficulty in forming jogs on dissociated dislocations than non-dissociated ones.

Results for small-angle boundaries (Siegel *et al.* 1980), have shown that boundaries with misorientations as low as 2° operate with high efficiencies as sinks for supersaturated vacancies in Au (with $g_s \simeq 0.7$) and also Al (with $g_s$ somewhat lower than 0.1 eV). An example of a small-angle boundary in Au which acted as an efficient sink for highly supersaturated vacancies in a down-quenched specimen (Siegel *et al.* 1980) is shown in Fig. 10.6. Here, the supersaturated vacancies in the bulk precipitated and collapsed in the form of small stacking fault tetrahedra. However, the sink action of the boundary reduced the vacancy concentration in its vicinity and produced the precipitate-denuded zone seen in Fig. 10.6(a). The number of vacancies destroyed at the boundary was determined by measuring the vacancies missing from the precipitates in the denuded zone (Fig. 10.6(b)). On the other hand, Dollar and Gleiter (1985) found that small-angle boundaries in Au with misorientations in the range 2–11° acted as poor point defect sinks at values of $g_s$ estimated near 0.1 eV. These results appear to be generally consistent with those found above for relatively isolated lattice dislocations as far as the effects of the driving energy, the stacking fault and jog energies, and possible restraints are concerned.

We also mention the experiments of Chan and Balluffi (1986), in which the climb of edge dislocations in small-angle tilt boundaries in Au acting as sources/sinks was observed directly by electron microscopy. Here, the dislocations climbed with a high efficiency with $g_s$ near $10^{-2}$ eV. However, all of the climbing dislocation segments were very short (less than $\simeq 10^{-5}$ cm), and they all intersected the free surface of the specimen: the climb process may, therefore, have been considerably easier than in comparable bulk situations because of the relatively easy nucleation of jogs at the free surface intersections where the degree of dislocation dissociation would have been lower than in the bulk.

### 10.3.3　Large-angle grain boundaries

#### 10.3.3.1　*Models for singular or vicinal grain boundaries*

As already pointed out in Section 9.2.2.2, singular or vicinal large-angle boundaries which are capable of supporting localized secondary dislocations tend to preserve their reference structures which are of relatively low energy and consist of regularly and closely spaced primary dislocations. Such boundaries will therefore be resistant to the simultaneous climb of their primary dislocations, since this process would perturb significantly these reference structures. These boundaries will therefore act as sources/sinks more readily

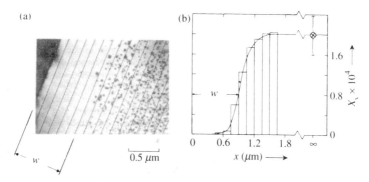

**Fig. 10.6**　(a) Zone denuded of vacancy precipitation in the vicinity of a small-angle grain boundary which acted as a sink for supersaturated vacancies in down-quenched Au. (b) Measured concentration of precipitated vacancies as a function of distance from the boundary, $x$. (From Siegel *et al.* (1980).)

via the climb (and possibly accompanying glide) of secondary dislocations in the boundary plane. An extensive discussion of the topology of secondary dislocations in both singular CSL and vicinal near-CSL boundaries has already been given in Section 9.2.2.2 and elsewhere in the present book and therefore will not be repeated here.

A possible exception to the above rule is the particular case where the lattice planes parallel to the boundary are of low index on one side and high index on the other, and the high index planes are far (with respect to inclination) from any low index planes in the same crystal. According to the discussion in Section 9.2.2.2, the Burgers vector strengths of any boundary dislocations with $b$ normal to the boundary will then be very weak (see Fig. 9.12). The situation on the high index side of such a highly asymmetric boundary will resemble somewhat the situation in a general boundary as described below in Section 10.3.3.2. It is then possible (see below) that point defects could be created/destroyed at various favorable places without sensibly affecting the low index planes on the other side. In such a case the source/sink action would not necessarily be restricted to localized climbing secondary dislocations.

When the source/sink action is restricted to localized climbing secondary dislocations, the situation will resemble in many respects that in small-angle boundaries (Section 10.3.2). Important differences will exist, however: for the large-angle boundaries, the diffusion of the lattice point defects to/from the climbing dislocations can occur not only by diffusion through the adjoining lattices (as in the small-angle case) but also by possible rapid short-circuit diffusion along the boundaries in which they are embedded which then act as 'defect collector plates'. The model then resembles models (Hirth and Pound 1963) for single crystal growth/shrinkage by condensation/evaporation where adatoms diffuse on the crystal free surface and are added/removed at surface steps (i.e. line defects) which sweep across the surface. Also, the secondary dislocations with their smaller Burgers vectors will generally not be able to climb via the periodic motion along them of elementary jog/kinks as already discussed in Section 9.2.2.2.

The source/sink efficiency may be modelled by means of the same basic approach used above for small-angle boundaries if proper account is taken of the additional diffusion path to the climbing dislocations. We shall only attempt a qualitative discussion of this highly complex problem, again for the case where supersaturated vacancies are annihilated at the boundary. A number of situations may be visualized, depending upon the geometry of the dislocations and the relative magnitudes of the diffusion rates through the lattice, along the boundary, and along the secondary dislocation cores. Consider, first, a single dislocation segment in the boundary. If, for example, the jog/kink density is high and pipe diffusion along the segment by a vacancy mechanism is considerably faster than along the boundary, the segment will tend to maintain the vacancy concentration at equilibrium everywhere along a strip in the boundary plane centred on the segment and of width $\simeq 2\bar{Z}_b$ where $\bar{Z}_b$ is now a characteristic vacancy diffusion length in the boundary similar to the characteristic diffusion length, $\bar{Z}$, along the dislocation discussed earlier (see eqn (10.32)). The segment will then act as an ideal 'strip sink', and the overall boundary sink efficiency will be determined by the capacitance of an array of such strips corresponding to the arrangement of dislocation segments throughout the boundary. On the other hand, if the pipe diffusion rate along the segments is not appreciably faster than the diffusion rate along the singular boundary, diffusion along the segments becomes unimportant, and each jog/kink on the segment will tend to maintain the vacancies in equilibrium over a patch in the boundary centred on it of a size and shape dependent upon the magnitude and anisotropy of $\bar{Z}_b$. The overall boundary sink efficiency will then be determined by the capacitance of these patches.

The climbing secondary dislocations may be present in many different arrangements depending upon the prevailing conditions and the previous history of the system. Any intrinsic array which is present will always tend to maintain itself in the form of a uniform network of minimum energy, but, as discussed earlier in our treatment of small-angle boundaries (Section 10.3.2.1), such a network must inevitably be disrupted to at least some extent during the climb process, causing an increase in the energy of the system. (In this respect we note that perturbations in the spacings of secondary dislocations may be regarded as even higher order secondary dislocations.) If the density of intrinsic dislocations is low (or zero as in an initially singular boundary), a sufficient density of extrinsic dislocations may be produced to dominate the kinetics. This may occur as a result of the impingement of lattice dislocations as already described in Section 9.2.2.2. It was pointed out there that dislocations impinged on a boundary subjected to a pressure, $p$, could use their connected lattice dislocation segments as poles and wrap themselves up into spirals or produce loops (Fig. 9.10), if the pressure was sufficiently high. The situation is generally similar in source/sink situations except that the energy available to drive this process is often considerably higher (compare Tables 9.1 and 10.1). Spirals or loops of smaller radius are then possible. An estimate of the maximum dislocation curvature which can be achieved can be obtained by the same procedure used previously to obtain eqn (9.24) or eqn (9.32). The restoring force (per unit length) due to the dislocation line tension is again $(\mu b^2 + h\sigma^S)/R$, while the forward driving force supplied by $g_s$ is $g_s(b \cdot \hat{n})/\Omega$. Equating these forces, we obtain

$$R_{min} \simeq \frac{(\mu b^2 + h\sigma^S)\Omega}{g_s(b \cdot \hat{n})}. \tag{10.44}$$

For example, for dislocation 3 in Fig. 9.11(b), $R_{min} \simeq 6 \times 10^{-8}$ m when $g_s = 10^{-2}$ eV (Table 10.1), $\mu = 4 \times 10^{10}$ Pa, $a = 4 \times 10^{-10}$ m, and $\sigma^S = 4 \times 10^2$ mJ m$^{-2}$.

If applied stress is present, the production of useful dislocations may also be aided by stress concentrations. Various types of dislocation sources, such as those illustrated in Fig. 9.10, could become active in regions high stress concentration as discussed in Section 9.2.2.2.

When very large values of $g_s$ exist (Table 10.1), it is possible that extrinsic dislocations could be homogeneously nucleated on boundaries in the form of loops. For example, during particle irradiation or quenching, loops could be nucleated by the aggregation and collapse of excess point defects. In this case the energy to form a nucleus of radius R can be approximated by

$$\Delta G = \frac{\mu b^2 R}{2(1 - \nu)} \left[ \ln\left(\frac{4R}{r_o}\right) - 1 \right] + 2\pi R h\sigma^S - \pi R^2 (b \cdot \hat{n}) g_s/\Omega, \tag{10.45}$$

where the first two terms represent the self-energy of the loop (as previously in eqn (9.22)), and the last term is the energy supplied by $g_s$. For example, for dislocation 3 in Fig. 9.11(b) the maximum in $\Delta G$ as a function of $R$ corresponds to a critical free energy for nucleation $\Delta G^* = 2.6$ eV when $g_s = 0.3$ eV (see Table 10.1), $a = 4 \times 10^{-10}$ m, $\mu = 4 \times 10^{10}$ Pa and $\sigma^S = 4 \times 10^2$ mJ m$^{-2}$. This value of $\Delta G^*$ is in the range $\Delta G^* < 50kT$ generally required for significant nucleation rates. We note, however, that the radius of the critical nucleus in the above example is only about $8 \times 10^{-10}$ m (i.e. a few atomic spacings). The above model which, of course, is formulated in terms of macroscopic quantities, must therefore be regarded as only a rough approximation.

The source/sink efficiency can therefore vary widely depending upon the density of secondary dislocations which is available and the sizes of the collector strips or patches. It will, of course, tend to be high whenever the strips or patches are closely spaced and the long range diffusion distance to the boundary in the lattice, $L$, is relatively large. A major stumbling block in the construction of quantitative models is, as always, determining the density of line defects responsible for the source/sink action. When $g_s$ becomes relatively small, constraints on the climb motion of the dislocations, which are similar to those discussed for the primary dislocations in small-angle boundaries in Section 10.3.2.1, may again play a role.

### 10.3.3.2 *Models for general grain boundaries*

In order to model the source/sink action of grain boundaries which are general with respect to all degrees of freedom, we may adopt basic elements of the uncorrelated diffusional transport model developed previously in Section 9.2.4.2 to explain their intrinsic conservative motion. In this model the boundary is incoherent and is taken as a slab of bad material containing a density of favorable sites distributed along it where point defects can be thermally generated/destroyed in an uncorrelated fashion. This model is consistent with the fact that general grain boundaries of the present type, with their closely and irregularly spaced primary dislocations, are unable to support localized and well established dislocations. Again, taking the density of favourable sites to be $Z$, the rate of point defect generation will be given by eqn (9.36). They will then be destroyed at a rate given by eqn (9.37) with $A = Z\nu$. We consider the case where vacancies are the dominant point defects, and the boundary is acting (as in our other examples) as a net sink for excess vacancies which are diffusing to it through the lattice from a relatively long distance, $L$, where the concentration is maintained at the value $X^L$. Therefore, in the quasi-steady state in the slab,

$$Z\nu \exp[(S^f + S^m)/k] \cdot \exp[-(E^f + E^m)/kT] + 2D_V^L(X^L - X)/\Omega L$$
$$= Z\nu X \exp(S^m/k) \cdot \exp(-E^m/kT). \tag{10.46}$$

The second term is the flux of vacancies arriving at the boundary slab via long-range diffusion in the lattice ($D_V^L$ = vacancy diffusivity in the lattice). Using $X^{eq} = \exp(S^f/k) \cdot \exp(-E^f/kT)$, eqn (10.46) may be rearranged in the form

$$\frac{X - X^{eq}}{X^L - X} = \frac{2D_V^L}{\Omega L Z\nu \exp(S^m/k) \cdot \exp(-E^m/kT)}. \tag{10.47}$$

The boundary sink efficiency, according to eqn (10.19), is defined as

$$\eta = \frac{X^L - X}{X^L - X^{eq}} = \frac{1}{1 + \left[\dfrac{X - X^{eq}}{X^L - X}\right]}. \tag{10.48}$$

Therefore,

$$\eta = \frac{1}{\left[1 + \dfrac{2D_V^L}{\Omega L Z\nu \exp(S^m/k) \cdot \exp(-E^m/kT)}\right]}. \tag{10.49}$$

However, inspection of eqn (9.40) shows that $\nu \cdot \exp(S^m/k) \cdot \exp(-E^m/kT)$ should be approximately equal to $D_V^B/l^2$, where $D_V^B$ is the vacancy diffusivity in the slab, and $l$ is the vacancy jump distance. Therefore, finally

$$\eta = \cfrac{1}{\left[1 + \cfrac{2D_V^L l^2}{\Omega L Z D_V^B}\right]}.$$  (10.50)

Since $D_V^L/D_V^B \ll 1$, $l/L \ll 1$, and $\Omega \simeq l^3$, it is easily seen from eqn (10.50) that high efficiencies (i.e. $\eta \simeq 1$) will be obtained for very modest densities of favourable sites, $Z$. We therefore expect high efficiencies under usual circumstances.

### 10.3.3.3  *Experimental observations*

The interfacial dislocation climb model for singular or vicinal grain boundaries has been directly confirmed in a number of cases by direct electron microscopy. Chan and Balluffi (1986) observed the climb of intrinsic edge dislocations in different regions of vicinal large-angle tilt boundaries in Au as they acted as either sources/sinks. In these experiments the climb caused many of the intrinsic edge dislocations to climb completely out of the boundaries and thus change their crystal misorientations. Komem *et al.* (1972) observed the climb of extrinsic dislocations in boundaries vicinal to $\Sigma 5$ {001} twist boundaries in Au as they acted as sinks for excess self-interstitial point defects produced by ion irradiation (Fig. 10.7). The Burgers vectors of these dislocations were almost perpendicular to the boundary. The distance climbed by the dislocations was too small to produce Bardeen–Herring type source spirals of the type illustrated in Fig. 9.10. These boundaries also contained square networks of intrinsic screw DSC dislocations which were unable to climb because of their pure screw character. Babcock and Balluffi (1989) observed the climb/glide motion of secondary dislocations in vicinal $\Sigma 5$ boundaries in Au acting as sources for stress-motivated fluxes of atoms. Direct measurements verified the expected coupling between the motion of the dislocations and: (i) the translation produced between the two adjoining crystals corresponding to the sum of the Burgers vectors of the moving

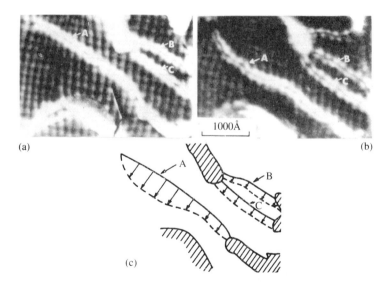

**Fig. 10.7**  Observed climb of extrinsic dislocations in vicinal $\Sigma 5$ (001) twist boundary in Au as it acted as a sink for excess self-interstitial atoms produced by Au ion irradiation. Background consists of a square network of intrinsic DSC screw dislocations. (a) Before irradiation. (b) After irradiation. (c) Schematic illustration of the climb motion. (From Komem *et al.* (1972).)

dislocations; and (ii) the non-conservative motion of the boundary relative to each adjoining crystal (due to the addition/removal of layers of thickness and $|h^w|$ and $|h^b|$ to/from the abutting crystals. In some cases new dislocations were observed to form at the intersections of the boundaries with the free surface in the thin film specimens, presumably in regions of high stress concentrations. King and Smith (1980) observed $\Sigma 3$ {111} twin boundaries in Cu as they acted as sinks for self-interstitial point defects produced by electron irradiation. Here, the supersaturated interstitials continuously nucleated extrinsic prismatic dislocation loops on the boundaries which then grew by absorbing further interstitials until they annihilated one another by mutual impingement (Fig. 10.8). The similar nucleation and growth of dislocation loops on $\Sigma 3$ {111} twins via the aggregation of supersaturated point defects has been observed in quenched and irradiated Al and irradiated Ni (Yamakawa and Shimomura 1991).

As might be expected, no details of the mechanism by which general grain boundaries act as sources/sinks have yet been revealed by any direct observations.

Semi-quantitative evaluations of the magnitudes of the source/sink efficiencies of a number of singular, vicinal and general large-angle grain boundaries have been reviewed by Balluffi (1980) and are summarized in Table 10.2 along with additional results. The information was gleaned from a number of different types of experiments involving a wide range of values of the maximum driving energy, $g_s$. Briefly, general boundaries appeared to act as efficient sources/sinks over a wide range of $g_s$ values extending to as low as $\simeq 10^{-5}$ eV. (By 'efficient', we mean that $\eta$ was of order unity.) However, apparent threshold effects were often observed to appear in the range $10^{-4}$–$10^{-5}$ eV. On the other hand, vicinal boundaries frequently exhibited generally lower efficiencies and evidently became essentially inoperative at low values of $g_s$ where general boundaries continued to operate. The highly singular $\Sigma 3$ {111} twin boundary was operative to same extent in Au when $g_s$ had very large values near 1 eV and was inoperative in Al when $g_s < 10^{-2}$ eV.

These results appear to be at least qualitatively consistent with the models discussed earlier. Evidently, the creation/destruction of defects at general boundaries is an easy, process which can occur at numerous places in the boundary, i.e. $Z$ in eqn (10.50) is large enough to maintain $\eta$ near unity, and no large barriers exist for the creation/destruction of the defects as we have assumed. On the other hand, for many singular or vicinal boundaries the creation/destruction is restricted to sites at interfacial dislocations. The density of these sites is often relatively small, particularly at small values of $g_s$ where the driving force is insufficient to nucleate large numbers of jog/kinks or significantly

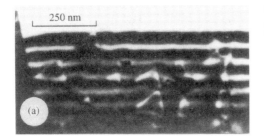
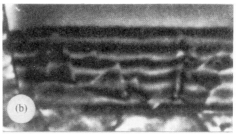

**Fig. 10.8** Nucleation and growth of extrinsic prismatic dislocation loops on a $\Sigma 3$ (111) twin boundary in Cu by the aggregation of excess self-interstitial atoms produced by electron irradiation. (From King and Smith (1980).)

**Table 10.2**  Summary of selected experiments involving large-angle grain boundaries as atom sources/sinks under different maximum driving energies, $g_s$

| Experiment | Estimated $g_s$ (eV) (see Table 10.1) | Result |
|---|---|---|
| Vacancy precipitate denudation at boundaries in quenched Au[a] | $\simeq 1$ | General boundaries efficient[*] vacancy sinks; $\Sigma 3$ {111} singular twin boundary operative but poorer sink |
| Dislocation loops annealing to boundary sinks in irradiated Au[b] | $\simeq 10^{-1}$ | General boundaries efficient sources/sinks; vicinal $\Sigma 13$ boundary poor source/sink |
| Vacancy precipitate denudation at boundaries in cooled Al[c] | $\gtrsim 10^{-2}$ | General boundaries efficient vacancy sinks; $\Sigma 3$ {111} singular twin boundary essentially inoperative |
| Diffusional creep of Cu[d] | $10^{-5}$ | General boundaries efficient vacancy sources |
| Diffusional creep of Cu[d] | $2 \times 10^{-5}$ | Vicinal boundaries (i.e. near-$\Sigma 3,5,7,9,11,15$ boundaries) essentially inoperative as vacancy sources |
| Diffusional creep of pure metals[e] | $> 10^{-4}$ | General boundaries efficient sources/sinks. Threshold effects found in range $10^{-5} < g_s < 10^{-4}$ eV |
| Sintering of Cu[f] | $10^{-5}$ | General boundaries efficient vacancy sinks |

[*] By 'efficient' we mean that $\eta$ was of order unity.
[a] Siegel *et al.* (1980).
[b] Dollar and Gleiter (1985).
[c] Basu and Elbaum (1965).
[d] Jaeger and Gleiter (1978). The boundaries were not examined directly. However, the boundaries close to the low $\Sigma$ boundaries were presumably vicinal.
[e] Burton (1977), Arzt *et al.* (1983).
[f] Alexander and Balluffi (1957).

increase the dislocation line length. Also, various constraints on the climb of the dislocations may be present. In addition the ability of the boundaries to deliver/carry away point defects to/from the dislocations should increase with increasing point defect diffusivity and binding energy to the boundary. Both of these latter properties will often decrease as the boundaries become more singular. All of these effects can then conspire to produce relatively low efficiencies for singular or vicinal boundaries, particularly at low $g_s$ values.

The apparent threshold effects observed at low values of $g_s$ can be explained in a number of different ways. For general boundaries a very small value of $g_s$ may be required to force the boundary structure into various slightly non-equilibrium configurations as it emits/absorbs point defects: also, bulk volume constraints may exist which will resist the necessary expansion/contraction normal to the boundary. For many singular or vicinal boundaries a critical value of $g_s$ may be required to: (i) copiously nucleate jog/kinks; (ii) produce significant dislocation line length; or (iii) overcome the various restraints on dislocation climb discussed earlier.

## 10.4  SHARP HETEROPHASE INTERFACES AS SOURCES/SINKS FOR FLUXES OF ATOMS

In the previous section we saw that homophase interfaces (grain boundaries) were able to act as sources/sinks for fluxes of atoms only by means of the creation/destruction of network sites for reasons given in Section 10.3.1. Heterophase interfaces may also utilize this mechanism, but when the two adjoining phases differ in composition, they may also act as sources/sinks for particular types of atoms by shuffling atoms across the interface.

For example, at an interface between an $\alpha$ phase and $\beta$ phase in a binary system where the $\beta$ phase is richer in B atoms than the $\alpha$ phase, B atoms may arrive in the $\alpha$ phase at the interface by long-range diffusion in the $\alpha$ phase and then shuffle across the interface, often with additional A-type atoms as well, and become incorporated in the $\beta$ phase. This process would be conservative for network sites, but not for the flux of B atoms in the $\alpha$ phase for which the interface would act as a sink. Such a shuffling process can, in fact, be regarded as a form of chemical reaction at the interface and could be described within the framework of usual chemical reaction kinetics (e.g. Laidler 1965) with reactants (i.e. A and B atoms in the $\alpha$ phase at the interface before the shuffle) and products (i.e. the A and B atoms in the $\beta$ phase after the shuffle). A chemical reaction rate constent could then be introduced to describe the reaction rate, and the kinetics could be cast in this form, if desired.

### 10.4.1  Models

Detailed mechanisms for the source/sink action and non-conservative interface motion of singular, vicinal and general heterophase interfaces will be similar in many respects to a number of the basic processes already described in Chapter 9 and the present chapter. However, the problem is complicated by the fact that the abutting crystals are different phases with generally different compositions, structures, and interatomic bonding. This basic asymmetry can generally cause the behaviour of the phases on each side of the interface to be different, and a range of asymmetric source/sink behaviour may be expected depending upon the properties of the two adjoining phases and the types of interactions which exist between them across the interface. A large variety of situations can exist which we shall comment on only briefly.

#### 10.4.1.1  *Singular or vicinal heterophase interfaces*

For most interfaces of this type, the required creation/destruction of network sites and shuffling will be restricted to discrete localized interfacial dislocations for reasons which are similar to those discussed previously for grain boundaries. In order to illustrate some of the phenomena possible, consider the line defects in the vicinal interface illustrated in Fig. 10.9 where a near-CSL corresponding to $\Sigma^\alpha = 3/\Sigma^\beta = 4$ exists. Since an exact CSL does not exist, all line defects possess at least some dislocation character. In Fig. 10.9 the defects at A and B possess significant dislocation and step character, whereas the defect at C is almost a pure step. The movement of any of these line defects along the interface will require both the creation/destruction of network sites and the shuffling of atoms of the different components across the interface. If the line defect moves along the interface with velocity $v$, so that the $\beta$ phase crystal grows with respect to the $\alpha$ phase

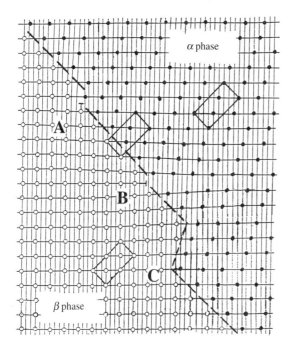

**Fig. 10.9** Various line defects in vicinal heterophase interface between $\alpha$ and $\beta$ phases for which a near-CSL corresponding to $\Sigma^\alpha = 3/\Sigma^\beta = 4$ exists. A differently inclined interface between these same phases has already been illustrated in Fig. 2.10. Again, as in Fig. 2.10, the interface is shown within a common DSC lattice. (From Balluffi *et al.* (1982).)

crystal, the rate at which atoms of species $i$ must join the $\beta$ phase crystal (per unit time per unit line length), $\dot{N}_i^\beta$, and the corresponding rate at which they must leave the $\alpha$ phase crystal, $\dot{N}_i^\alpha$, are given by

$$\dot{N}_i^\beta = vh^\beta X_i^\beta/\Omega^\beta; \qquad \dot{N}_i^\alpha = vh^\alpha X_i^\alpha/\Omega^\alpha, \tag{10.51}$$

where $\dot{X}_i^\beta$ and $\dot{X}_i^\alpha$ and are the atom fractions of $i$, and $\Omega^\beta$ and $\Omega^\alpha$ are the atomic volumes in the two phases. If $\dot{N}_i^\beta > \dot{N}_i^\alpha$, the rate at which $i$ type atoms must be shuffled across the interface is then $\dot{N}_i^\alpha$, and the total rate at which network sites must be created is

$$\dot{N} = \sum_i (\dot{N}_i^\beta - \dot{N}_i^\alpha) = v(h^\beta/\Omega^\beta - h^\alpha/\Omega^\alpha). \tag{10.52}$$

The currents of species required to sustain the movement of the various line defects which are present will be associated with a complex diffusion field involving both atoms and defects which extends throughout the boundary and the two abutting crystals. As the line defects move, network sites will be shuffled from one phase to the other, and additional network sites will be created/destroyed as described by eqns (10.51) and (10.52). These local processes will be coupled to the diffusion field, and the overall source/sink action and motion of the interface will depend upon how rapidly the entire coupled process is able to proceed.

The line defects at A and C in Fig. 10.9 can be active only with the participation of both phases. However, this is not necessary for the line defect at B which can move by creating/destroying network sites in the $\beta$ phase crystal without any participation of the $\alpha$ phase crystal. The kinetics at these different types of line defects can then be very different, depending upon the properties of the two phases and the nature of the bonding in the interface core. For example, if the $\alpha$ phase in Fig. 10.9 has a much higher cohesive energy and melting point than the $\beta$ phase, then the line defect at B could remain active as a source/sink for atoms in the $\beta$ crystal under conditions where kinetic processes in the $\alpha$ phase and at the line defects at both A and C were frozen out. The interface would therefore exhibit very asymmetric behaviour.

In the cases discussed so far, the dislocation motion has been restricted to the interface. However, for certain interfaces this is not necessary. A common heterophase interface is the vicinal $\Sigma 1$ interface illustrated in Fig. 2.2, which contains an intrinsic grid of edge dislocations with lattice Burgers vectors which are parallel to the interface plane. Such an interface can create/destroy network sites if the dislocations climb out of the interface plane. This substantially increases the energy of the interface, of course, and will therefore require a large value of $g_s$ (Table 10.1). As pointed out by Pieraggi *et al.* (1990), under large driving forces the dislocations which have climbed out of the interface in this fashion may often be able to dissociate into other lattice dislocations which can then glide back into the interface under the influence of the attractive force which would exist. Repetition of this process would then allow the interface to act as a continuous source/sink.

Many other situations are, of course, possible, including cases where the lattice planes parallel to the interface of one of the abutting phases are of very high index. The creation/destruction of network sites in that side of the interface would then resemble that in a general interface as discussed below.

The problem of determining the overall source/sink efficiency for these models is generally similar to the corresponding problem for grain boundaries discussed earlier. Again, the rates of diffusion of atoms to/from the migrating line defects can often be aided by rapid short-circuit diffusion along the interface. The efficiency will again depend upon the spacing of the line defects, the extent of the diffusion field adjoining the interface, and the ease with which the diffusion currents can be accommodated at the line defects at various types of jog/kinks. Depending upon the circumstances, the necessary line defects can be intrinsic dislocation arrays or extrinsic dislocations generated by the mechanisms described in Chapter 9 and also above. For heterophase interfaces very large stress concentrations can be achieved, particularly during the precipitation and growth of second phases (see Section 10.4.2.1), and the heterogeneous nucleation of dislocation loops in these highly stressed regions is then a further possibility. The energy to form such a loop may be approximated by

$$\Delta G = \frac{\mu b^2 R}{2(1 - \nu)} \left[ \ln\left(\frac{4R}{r_o}\right) - 1 \right] + 2\pi R h \sigma^S - \pi R^2 \tau b, \tag{10.53}$$

where the first two terms represent the self-energy of the loop (as previously in eqn (10.45), and the last is the work done by the stress, $\tau$, as the loop is formed with the help of the constant force $\tau b$. Equation (10.53) yields a critical free energy of nucleation, $\Delta G^* = 2.3\,\text{eV}$, when $\tau = 2 \times 10^9\,\text{Pa}$, $\mu = 4 \times 10^{10}\,\text{Pa}$, $b = 10^{-10}\,\text{m}$, $h = 2 \times 10^{-10}\,\text{m}$, and $\sigma^S = 4 \times 10^2\,\text{mJ}\,\text{m}^{-2}$. Significant nucleation rates would then be possible.

Again, as for homophase interfaces, many of the parameters necessary for a quantitative source/sink model are difficult to establish, particularly the line defect density. Obviously, a wide range of efficiencies and possibly highly asymmetric behaviour can be expected.

### 10.4.1.2  *General heterophase interfaces*

As in the case of general grain boundaries, no localized line defects of any physical significance will be present in general heterophase interfaces. The required creation of network sites and transfer of sites across the interface will then occur in uncorrelated fashion by the same types of processes distributed throughout the interface described previously (Sections 9.2.4 and 10.3.3.2) for general grain boundaries. When these processes are easy for the atoms of both of the two different phases which abut the interface, we may expect relatively high source/sink efficiencies. However the rates of these processes can vary considerably depending upon the bonding and the cohesive energies of the two phases and the material in the interface core.

## 10.4.2  Experimental observations

An immense literature exists in which heterophase boundaries act as sources/sinks and engage in non-conservative motion. As anticipated above, a wide range of source/sink efficiency is found. In the following we discuss a number of representative cases where experimental results are available.

### 10.4.2.1  *Growth, coarsening, shape-equilibration, and shrinkage of small precipitate particles*

An important phenomenon is the precipitation of solute atoms in the form of stable (or metastable) second-phase particles from a supersaturated solid solution (Kelly and Nicholson 1963, Russell and Aaronson 1978). The precipitation is generally a complex process which passes through several stages. In the first stage, particles are nucleated and then grow in the presence of a relatively high solute atom supersaturation. As the supersaturation falls, the nucleation rate decreases. Eventually, no new particles form, and the total volume of the precipitated particles becomes essentially constant. However, larger particles in the distribution of particles which has formed tend to grow at the expense of the smaller particles by the diffusional transport of solute atoms through the matrix, and the number of particles begins to decrease, i.e. particle 'coarsening' takes place. At the same time, any particles which may have been forced to grow into non-equilibrium shapes for kinetic reasons during the growth will tend to relax to their equilibrium shapes given by the Wulff construction (Section 5.6.2), i.e. 'shape equilibration' also takes place. Two main regimes can therefore be distinguished. In the first, particle nucleation and growth occurs under a driving force provided by the solute atom supersaturation. In the second, the supersaturation is much reduced, and the particles then tend to undergo coarsening and shape equilibration under a driving force provided by the excess interfacial energy which is present in the system. These regimes are not exclusive, of course, and will generally overlap to varying degrees. In contrast to this, if an existing distribution of particles is suddenly heated to an elevated temperature so that the matrix becomes highly subsaturated, the particles will all tend to shrink and possibly assume new non-equilibrium shapes.

The kinetics of each of the above processes depend directly upon the abilities of the particle/matrix interfaces to act as sources/sinks for the necessary diffusional fluxes of solute atoms.

*Particle growth.*  In one important type of precipitation system particular misorientations between the matrix and precipitate exist at which low index lattice planes in the matrix and precipitate are parallel and nearly match in structure. Singular coherent interfaces parallel to these planes, which are of relatively low energy, can therefore be produced by the application of small strains. During precipitation the particles therefore assume platelet forms with their broad faces corresponding to these interfaces in order to minimize their total interfacial energies as shown in Fig. 10.10(a). During their growth, the platelets may be driven into highly non-equilibrium shapes having much larger diameter to thickness ratios (see Fig. 10.10(a)) than those corresponding to their equilibrium shapes of minimum energy if the singular interfaces at the broad faces act as less efficient solute atom sinks than the platelet edges.

Especially extensive studies of platelet growth have been made of Ag-rich $Ag_2Al$ hexagonal phase platelets growing in solid solutions of f.c.c. Al supersaturated with Ag (Aaronson *et al.* 1970, Ferrante and Doherty 1979, Howe *et al.* 1986, 1987, Howe 1988, Rajab and Doherty 1989, Doherty and Rajab 1989, Howe and Prabhu 1990, Prabhu and Howe 1990, Aiken and Plichta 1990) and $CuAl_2$ tetragonal phase platelets growing in solid solutions of Al supersaturated with Cu (Aaronson *et al.* 1970, Weatherly 1971, Merle and Fouquet 1981, Merle and Merlin 1981, Bouazra and Reynaud 1984). In these systems, close-packed {0001} basal planes of the $Ag_2Al$ are parallel to commensurate close-packed {111} planes of the Al as seen in Fig. 10.10(a), and {100} planes of the $CuAl_2$ are parallel to commensurate {100} planes of the Al. An example of a $CuAl_2$ platelet (viewed at normal incidence to its broad face) is shown in Fig. 10.10(b). Many electron microscope observations have shown that the singular (or vicinal) interfaces at the broad faces operate as solute sinks via the lateral motion of line defects (coherency dislocations) possessing both dislocation and step character as described in Section 10.4.1.1. A train of such defects is evident on the broad face of the platelet in Fig. 10.10(b). High-resolution electron microscopy studies of similar line defects on $Ag_2Al$ platelets (Howe 1988, Howe and Prabhu 1990, Prabhu and Howe 1990) show that they move via the lateral motion of kinks as shown in Fig. 10.11. The observed line defects generally possess a wide variety of step heights, and kinks of different sizes exist in a number of crystallographically equivalent forms. The most elementary line defect in the $Ag_2Al$ interfaces has already been illustrated in Fig. 1.20: its Burgers vector is

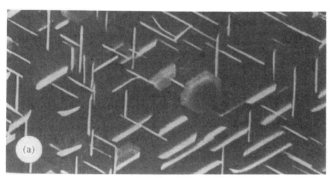
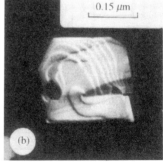

**Fig. 10.10** (a) Platelet $Ag_2Al$ precipitates formed in a Al–Ag alloy. The platelets lie parallel to {111} of the matrix which is almost pure Al. (From Rajab and Doherty (1989).) (b) Platelet $CuAl_2$ precipitate formed in a Al–Cu alloy. The platelet lies parallel to {100} of the matrix which is almost pure Cu. A train of line defects possessing both step and dislocation character is visible on the broad face. (From Weatherly (1971).)

parallel to the boundary and its step height corresponds to two close-packed interplanar spacings. Experimentally, it is observed (Howe *et al.* 1986) that line defects with step heights corresponding to two, four, and six interplanar spacings appear: this can be understood by realizing (Aiken and Plichta 1990) that line defects with these step heights will tend to form successively in trains in order to avoid high energy stacking sequences of the parallel close-packed planes. When the step height corresponds to a multiple of six interplanar spacings, the line defect is a pure step. Otherwise, it possesses a Burgers vector which is parallel to the interface. When any of these defects sweeps across the interface, network sites are conserved, and $\{111\}$ planes of the matrix are simply converted into $\{0001\}$ planes of the silver-rich $Ag_2Al$ phase. This process requires the diffusion of Ag atoms to the defects through the Al matrix and may be regarded in a sense as a dislocation glide (shuffling) process coupled to the long range diffusional transport. It is noted that in this special case $\dot{N}$ in eqn (10.52) is zero, and, therefore, if no other line defects are operative, no point defects can be created/destroyed at the interface. Under these conditions the interface cannot operate as an ideal source/sink for all species, since, even though it can absorb/emit solute atoms, it is unable to create/destroy vacancies and therefore maintain the vacancies in the lattice in local equilibrium. We may then expect a rather curious constrained diffusion situation to develop in the lattice in which the line defects act as sources/sinks for the solute atoms while the vacancies become 'backed up' in a non-equilibrium distribution which will suppress the normal Kirkendall effect in the lattice (see Section 10.4.2.2) and cause both the flux of vacancies and the sum of the fluxes of solute and matrix atoms to the interface to be zero, i.e. $J_A + J_B + J_V = 0$, $J_V = 0$, and $J_A + J_B = 0$.

In general, the overall solute atom sink efficiency at the broad faces will be controlled by the line defect density and the rate at which the solute atoms are able to diffuse to the line defects and become incorporated in the platelets. There is clear experimental evidence that the broad faces are often incapable of acting as highly efficient sinks particularly during the early stages of growth because of a lack of line defects. Measurements in both the Ag/Al system (Ferrante and Doherty 1979; Rajab and Doherty 1989; Doherty and Rajab 1989, Aiken and Plichta 1990) and the Cu/Al system (Merle and Fouquet 1981, Merle and Merlin 1981) show that the platelets quickly developed ratios, $\rho$, of diameter, $d$, to thickness, $t$, which far exceeded the ratio corresponding to their equilibrium shape, which according to the Wulff construction (Section 5.6.2) is given by

$$\rho^{eq} = (d/t)^{eq} = \sigma^e/\sigma^f, \tag{10.54}$$

where $\sigma^e$ and $\sigma^f$ are the energies (per unit area) of the platelet edge (assumed isotropic) and face, respectively. This is evidence that the broad faces acted as less efficient solute sinks than the platelet edges. If the platelet interface acted as an ideal sink everywhere during growth, we would expect each platelet to have maintained its equilibrium shape, barring unlikely growth instabilities of the type discussed in Section 10.6. The solute concentration in the matrix in local equilibrium with each platelet would then have been equal everywhere around the interface at the level given by eqn (5.118), and the platelet diameter and thickness would have grown at appropriate rates in the solute diffusion field (Ham 1959, Horvay and Cahn 1961). Detailed observations in the Ag/Al system (Rajab and Doherty 1989, Doherty and Rajab 1989) revealed that the platelets, which were growing on various $\{111\}$ planes of the matrix, eventually intersected. At this point the sink efficiency of the broad faces increased, and the platelets grew everywhere under essentially diffusion-controlled conditions while maintaining their large diameter to

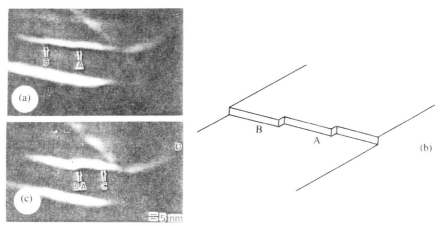

**Fig. 10.11** (a) Electron micrograph of two kinks (at A and B) on a line defect possessing both step and dislocation character which is sweeping across the broad face of a growing $Ag_2Al$ platelet precipitate in a Al–Ag alloy. (b) Schematic drawing of the two kinks. (c) *In situ* observation of the merging of the two kinks observed in (a) after 6 s of further platelet growth. (From Howe (1988).)

thickness ratios constant. Apparently, the broad faces acted as poor sinks in the early stages because of a lack of line defects. However, when the platelets intersected, a high step density became available due to easy heterogeneous step nucleation at the intersections, and the sink efficiency then approached unity. Doherty and Rajab (1989) were unable to find a plausible model for the heterogeneous nucleation, however, and the detailed mechanism therefore remains unknown.

Considerable further evidence for the relatively low sink efficiency of the broad faces of growing platelet precipitates due to a lack of growth steps has been found in other studies. Details may be found in the above references. (We note that they have also been found to act as poor sources/sinks during platelet coarsening and shape equilibration when the supersaturation (i.e. $g_s$) fell to relatively low levels as described below.)

It is important to note that the chemically induced lateral motion of the dislocations possessing step character on the broad faces has the potential to produce a macroscopic specimen shape change. When a succession of such dislocations sweeps across a face, the matrix will be displaced relative to the particle by the sum of the Burgers vectors of the dislocations, and this will tend to produce a shape change of the region surrounding the particle. This can be mitigated, however if a number of symmetry related variants of the Burgers vector is available, as is usually the case. Then, a mix of dislocations can be employed whose Burgers vectors sum to zero. Also, since the diffusion which feeds the particle growth occurs over relatively long distances, any shape changes produced by the sweeping dislocations can, in principle, be cancelled by an appropriate redistribution of matter by diffusion (Christian 1975). Despite these mitigating possibilities, evidence for shape changes associated with the diffusional growth of platelet particles has often been found in the form of surface relief in regions where the growing particles intersected free surfaces (Liu and Aaronson 1970). The effect is similar to the surface relief produced by the sweeping motion of gliding interface dislocations during the motion of glissile interfaces (see Section 9.2.1.2 and Fig. 9.4).

In another class of precipitation systems the crystal structure of the precipitate nearly matches that of the matrix in 3D (rather than 2D), and it can therefore generate a $\Sigma 1$ CSL with the matrix with the help of a small strain. The particles in these systems then assume an equiaxed form (approximately spherical during the earlier stages of growth) which is coherent with the matrix within the framework of the 3D $\Sigma 1$ CSL. Examples (Ardell 1988) are $Ni_3Al$ particles in Al supersaturaterad with Ni, Co-rich particles in Cu supersaturated with Co, and $Al_3Li$ particles in Al superaturated with Li. Since the particles are almost spherical, their interfaces must possess high densities of coherency dislocations which are essentially steps (with very small dislocation strengths) in the $\Sigma 1$ CSLs, and they should therefore operate as highly efficient sinks for incoming solute atoms. This appears to be confirmed by a variety of nucleation, growth, and coarsening measurements (Servi and Turnbull 1966, Ardell 1988, Xiao and Haasen 1991).

*Particle coarsening and shape equilibration.* We consider first the coarsening of a distribution of equiaxed particles whose equilibrium shape is equiaxed and where any simultaneous shape equilibration is therefore not an important factor. Even in this relatively simple situation the coarsening turns out to be a complex phenomenon which is very difficult to treat realistically. In fact, no completely satisfactory model has yet been developed. In view of this, we begin by describing the classic 'mean field' model of Wagner (1961) and Lifshitz and Slyozov (1961), which has played a central role in the development of the subject. This model involves a number of rather drastic simplifications and is only partially successful in explaining the available data. Nevertheless, it embodies several essential elements of the coarsening phenomenon and serves as a useful starting point. For brevity, we present a somewhat simplified version developed by Greenwood (1969) which produces results which are practically identical to those obtained by the fuller theory.

Consider a binary system made up of A and B atoms containing a distribution of B-rich $\beta$ phase particles in an A-rich $\alpha$ phase matrix. The A and B atoms are substitutional everywhere, and we shall assume that the particles are spherical with isotropic interfacial energies, any constraints on volume changes can be ignored, and the particle/matrix interfaces act as ideal sources/sinks. If we also assume, for simplicity, that the particles are essentially pure B, the $\alpha$ phase is dilute in B, and the activity coefficients of A and B are independent of concentration in the matrix, eqn (5.119) shows that the concentration of B in the $\alpha$ matrix in equilibrium with a $\beta$ phase particle of radius $r$, $n^{\alpha, eq}(r)$, is larger than the corresponding concentration in equilibrium with a particle of infinite radius, $n^{\alpha, eq}(\infty)$, by an amount given by

$$\frac{n^{\alpha, eq}(r)}{n^{\alpha, eq}(\infty)} = \, \simeq 1 + \frac{2\sigma\Omega}{kTr}. \tag{10.55}$$

The concentration of B in the $\alpha$ matrix in the direct vicinity of small particles will therefore be larger than at large particles. Consequently, on average, B atoms will tend to diffuse through the matrix from the smaller particles, acting as sources, to the larger particles, acting as sinks, causing: (i) the smaller particles to shrink; (ii) the larger particles to grow; (iii) the total number of particles to decrease; and (iv) the average particle size to increase. The particles of various sizes will be distributed throughout the matrix more, or less, randomly, and the problem of solving for the diffusional transport between them in any rigorous manner is obviously difficult. If the volume fraction of the particles is small (i.e. the spacing between particle centres is on average at least several particle diameters), our previous analysis of precipitate growth in Section 10.2.1 indicates that each particle will tend to establish a quasi-steady-state diffusion field around it which is highly localized in the sense that the regions of steep gradient only extend outwards from each particle a distance of the order of the particle radius (see eqn (10.15)). In the present situation we may therefore visualize an average solute concentration, $\bar{n}^\alpha$, which exists throughout the system in all regions farther from the particles than their radii as illustrated schematically in Fig. 10.12. This 'mean field' concentration is expected to be lower than that in equilibrium with the small particles (so that they will shrink) and higher than that in equilibrium with the larger particles (so that they will grow) as shown by the flux arrows in Fig. 10.12. According to Section 10.2.1 (eqn (10.13)) the diffusion current gained by a particle of size $r$ is then

$$I = 4\pi D^\alpha r[\bar{n}^\alpha - n^{\alpha,\,\text{eq}}(r)], \tag{10.56}$$

if we make the simplifying assumption that the current can be described in terms of an effective composition-independent diffusivity. The rate of particle growth is then

$$\frac{dr}{dt} = \frac{\Omega I}{4\pi r^2} = \frac{\Omega D^\alpha}{r}[\bar{n}^\alpha - n^{\alpha,\,\text{eq}}(r)]. \tag{10.57}$$

During the coarsening (after precipitation is complete) the total number of solute atoms contained in the particles will be constant, and, therefore,

$$\sum_p 4\pi r^2 \frac{dr}{dt} = 0, \tag{10.58}$$

where the sum is over all the particles. Combining eqn (10.55), (10.57), and (10.58), we obtain

$$[\bar{n}^\alpha - n^{\alpha,\,\text{eq}}(\infty)] \cdot \sum_p r = \frac{2N\sigma\Omega\, n^{\alpha,\,\text{eq}}(\infty)}{kT}, \tag{10.59}$$

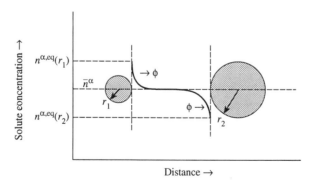

Distance →

**Fig. 10.12** Schematic diagram of solute diffusion field around large and small B-rich $\beta$ phase particles during coarsening of a distribution of $\beta$ phase particles according to the diffusion-controlled 'mean field' model. $\bar{n}^\alpha$ = mean solute concentration in the $\alpha$ phase matrix. The solute concentration in equilibrium with the small particle, $n^{\alpha,\,\text{eq}}(r_1)$ is higher than $\bar{n}^\alpha$: the solute concentration in equilibrium with the large particle, $n^{\alpha,\,\text{eq}}(r_2)$, is lower than $\bar{n}^\alpha$. The small particle therefore shrinks, while the large particle grows as indicated by the diffusion fluxes, $\phi$.

where N is the total number of particles. The average particle radius is $\bar{r} = \sum_p r/N$, and therefore

$$\bar{n}^\alpha - n^{\alpha,\text{eq}}(r) = \frac{2\sigma\Omega\, n^{\alpha,\text{eq}}(\infty)}{kT}[1/\bar{r} - 1/r]. \tag{10.60}$$

The average concentration is therefore the concentration in equilibrium with the particle of average size, as might have been expected. Combining eqns (10.60) and (10.57), we then find

$$\frac{dr}{dt} = \frac{2\sigma D^\alpha \Omega^2 n^{\alpha,\text{eq}}(\infty)}{kTr}[1/\bar{r} - 1/r], \tag{10.61}$$

which has the form plotted in Fig. 10.13. Particles with radii larger than $\bar{r}$ therefore grow, and particles with radii smaller than $\bar{r}$ shrink. The shrinkage rate of small particles increases very rapidly as $r$ decreases (i.e. approximately as $r^{-2}$), while the growth rate of large particles varies more slowly and has a maximum at $r = 2\bar{r}$. These results suggest that the distribution of particle sizes, denoted by $N(r, t)$, which is coarsening as $\bar{r}$ increases with time, should tend to reach a steady state with respect to the reduced variable $r/\bar{r}$. Any very small particles (i.e. particles with $r/\bar{r} \ll 1$) should tend to disappear, since they shrink at very high rates. Any large particles (i.e. particles with $r/\bar{r} \gtrsim 2$) would continue to grow, but they would grow at a rate which is lower than that for somewhat smaller particles. Hence, they would tend to be overtaken and disappear from the distribution. The quasi-steady-state distribution should therefore contain essentially no very small particles or large particles and should be peaked somewhere in between.

In order to be more quantitative we must solve the differential equation describing the time evolution of the distribution. This equation may be obtained by visualizing the particles as inhabiting a 'particle size' space described by $r$. The growing/shrinking particles constitute a flux of particles in this space given by $\phi = (dr/dt)\cdot N(r, t)$. Therefore, conservation of the total number of solute atoms leads to the continuity equation

$$\frac{\partial N(r, t)}{\partial t} = -\operatorname{div}\phi = -\frac{\partial}{\partial r}[(dr/dt)\cdot N(r, t)]. \tag{10.62}$$

Using the reduced variable, $r/\bar{r}$, Wagner (1961) assumed a solution of this equation of the form

$$N(r, t) = g(t)\cdot(r/\bar{r})^2\cdot h(r/\bar{r}), \tag{10.63}$$

where $h(r/\bar{r}) \to 1$ when $r \to 0$, and showed that this is indeed a solution which evolves into a steady state at long times becoming independent of both time and the form of the initial distribution. The limiting steady state distribution has the form shown in Fig. 10.14, which is of the form expected intuitively. Furthermore, in the quasi-steady state it can be shown (Wagner 1961) that the mean particle size increases with time according to

$$\bar{r}(t)^3 = \bar{r}(0)^3 + \frac{8\sigma D^\alpha \Omega^2\, n^{\alpha,\text{eq}}(\infty)}{9kT}\cdot t = A + Bt. \tag{10.64}$$

We note that essentially the same result can be deduced very simply by making the approximation that $d\bar{r}/dt$ will be the same as the maximum value of $dr/dt$, which, from eqn (10.61), occurs for $r = 2\bar{r}$ and is therefore given by

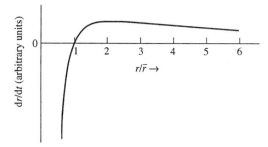

**Fig. 10.13** Rate of growth of particle of radius $r$ during coarsening of a distribution of particles versus $r/\bar{r}$ according to the diffusion-controlled 'mean field' model. $\bar{r}$ = mean particle radius. Particles with $r < \bar{r}$ shrink, while particles with $r > \bar{r}$ grow.

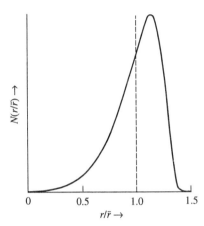

**Fig. 10.14** Limiting steady-state size distribution of particles, $N(r/\bar{r})$, which is achieved during coarsening according to the diffusion-controlled 'mean field' model. $r$ = particle radius: $\bar{r}$ = mean particle radius.

$$\frac{\mathrm{d}\bar{r}}{\mathrm{d}t} \simeq \left(\frac{\mathrm{d}r}{\mathrm{d}t}\right)_{\max} = \frac{\sigma D^\alpha \Omega^2 n^{\alpha,\,\mathrm{eq}}(\infty)}{2kT} \cdot (\bar{r})^{-2}. \tag{10.65}$$

Integration of eqn (10.65) then yields

$$\bar{r}(t)^3 \simeq \bar{r}(0)^3 + \frac{3\sigma D^\alpha \Omega^2 n^{\alpha,\,\mathrm{eq}}(\infty)}{2kT} \cdot t, \tag{10.66}$$

which compares favourably with eqn (10.64).

An analysis along similar lines can be made for coarsening under assumed interface-controlled conditions. If it is again assumed, as in developing eqn (10.16), that the rate of solute atom transfer across the particle/matrix interfaces is controlled by a rate constant, $\kappa$, Greenwood (1969) shows that

$$\bar{r}(t)^2 = \bar{r}(0)^2 + \frac{64\kappa\sigma\Omega\, n^{\alpha,\,\mathrm{eq}}(\infty)}{81kT} \cdot t. \tag{10.67}$$

Note that $D^\alpha$ does not appear in eqn (10.67) and has been replaced by the rate constant $\kappa$. Note also the $\bar{r}(t)^2$ dependence rather than the $\bar{r}(t)^3$ dependence for the diffusion-controlled kinetics. In further results, it is found that a considerably wider steady state size distribution than that illustrated in Fig. 10.14 for diffusion-controlled kinetics is obtained.

The systems which might be expected to follow the above simple diffusion-controlled coarsening model most closely are those relatively ideal systems such as the Ni/Al, Co/Cu, and Li/Al systems (described above) where the particles are approximately spherical and possess coherent interfaces with a high step density which should be efficient sources/sinks for solute atoms via shuffling. Many features of the observed coarsening in systems of this type are indeed found to be consistent with the model: these include substantial agreement with the form of eqn (10.64), i.e $\bar{r}(t)^3 = A + Bt$, approximate agreement of the constant B with calculated values based on values of the interfacial energy, $\sigma$, and the effective solute atom diffusivity estimated from available data, and a number of features of the evolution of the size distribution (Greenwood 1969, Ardell 1969, Martin and Doherty 1976, Wendt and Haasen 1983, Xiao and Haasen 1991, Ardell 1988). However, significant deviations are often found (Ardell 1988, Brown 1985, 1989, 1991). This is not too surprising because of the many approximations made in the simple model. Particularly troublesome are the assumption of well separated particles, the neglect of possible effects due to volume constraints at the growing and shrinking particles, and the basic problem of finding the proper statistical way to describe the diffusional transport between the 'large' and 'small' particles in the distribution. Many efforts have been made to improve the analysis over the years (Tsumuraya and Miyata 1983, Voorhees and Glicksman 1984, Ardell 1988, Brown 1985, 1989, 1991, DeHoff 1991). For example, DeHoff drops the mean field approximation and develops a model based on the concept of 'commumicating neighbours'. However, it must be stated that at this point no completely satisfactory model has yet been developed. Despite these problems, it appears, on balance, that coarsening in the relatively ideal systems cited above is largely diffusion-controlled, as anticipated.

The problem is even more complex for distributions of particles which have grown into non-equilibrium shapes with large diameter to thickness ratios. Here, simultaneous coarsening and shape equilibration may be expected, and, in addition, the broad faces may be poor sources/sinks. Any platelet with a diameter-to-thickness ratio which is larger than the equilibrium ratio, $\rho^{eq}$, given by eqn (10.54), will tend to thicken by transporting atoms from its edge to its faces. It will tend to establish a solute concentration in the matrix in local equilibrium with its edge, $n^{\alpha,\,eq}(e)$, which is larger than the corresponding concentration, $n^{\alpha,\,eq}(f)$, in local equilibrium with its faces. Diffusion will then occur from the edge regions to the face regions through the matrix and also possibly along the interface. In order to find these equilibrium concentrations (Ferrante and Doherty 1979) we use the same basic procedure used previously to derive eqn (5.111) in Section 5.6.6. We take the platelet to be an ideal disc of diameter, $d$, and thickness, $t$, and first transfer $\delta N$ atoms from the matrix to the platelet at its edge while keeping its thickness constant. The $\delta N$ atoms are composed of $\delta N_A$ A atoms and $\delta N_B$ B atoms with $\delta N_B/\delta N = X_B^\beta$ the atom fraction of B in the $\beta$ phase particle. The increase in interfacial energy of the platelet is then

$$\delta G = \pi(\sigma^e t + \sigma^f d)\delta d, \tag{10.68}$$

where $\delta d$ is the increase in diameter. However, $\delta d = (2\Omega^\beta/\pi t d)\delta N$, where $\Omega^\beta$ is the volume per atom in the $\beta$ phase, and therefore

$$\delta G = 2\Omega^\beta(\sigma^e/d + \sigma^f/t)\delta N. \tag{10.69}$$

We next imagine a similar transfer to a $\beta$ phase particle of infinite radius from a region of the $\alpha$ matrix in equilibrium with such a particle. The change in interfacial energy in this transfer is vanishingly small, and therefore the free energy change in the first transfer is greater than in the second by $\delta G$ given by eqn (10.69). The chemical potentials of A and B in the matrix in equilibrium with the platelet edge, i.e. $\mu_A^{\alpha,\,eq}(e)$ and $\mu_B^{\alpha,\,eq}(e)$, and with the infinite particle, i.e. $\mu_A^{\alpha,\,eq}(\infty)$ and $\mu_B^{\alpha,\,eq}(\infty)$, must then be related to $\delta G$ by

$$[\delta N_A \mu_A^{\alpha,\,eq}(e) + \delta N_B \mu_B^{\alpha,\,eq}(e)] - [\delta N_A \mu^{\alpha,\,eq}(\infty) + \delta N_B \mu^{\alpha,\,eq}(\infty)] =$$

$$\delta G = 2\Omega^\beta (\sigma^e/d + \sigma^f/t)\delta N. \tag{10.70}$$

Equation (10.70) is of the same general form as eqn (5.107), as might have been expected, and following the same basic procedure as that used to derive eqn (5.111), we obtain

$$\frac{n_B^{\alpha,\,eq}(e)}{n_B^{\alpha,\,eq}(\infty)} = \frac{X_B^{\alpha,\,eq}(e)}{X_B^{\alpha,\,eq}(\infty)} = 1 + \left[\frac{1 - X_B^{\alpha,\,eq}(\infty)}{X_B^\beta - X_B^{\alpha,\,eq}(\infty)}\right]\left[\frac{2\Omega^\beta\sigma^e}{kT\phi d}\right]\left[1 + \frac{\rho}{\rho^{eq}}\right]. \tag{10.71}$$

A similar exercise, in which the $\delta N$ atoms are transported to the platelet faces (while keeping $d$ constant), yields $\delta G = (4\Omega^\beta\sigma^e/d)\delta N$, and therefore

$$\frac{n_B^{\alpha,\,eq}(f)}{n_B^{\alpha,\,eq}(\infty)} = \frac{X_B^{\alpha,\,eq}(f)}{X_B^{\alpha,\,eq}(\infty)} = 1 + \left[\frac{1 - X_B^{\alpha,\,eq}(\infty)}{X_B^\beta - X_B^{\alpha,\,eq}(\infty)}\right]\frac{4\Omega^\beta\sigma^e}{kT\phi d}. \tag{10.72}$$

We note that when the platelet has its equilibrium shape, $\rho = \rho^{eq}$, and $n_B^{\alpha,\,eq}(e) = n_B^{\alpha,\,eq}(f)$, and the solute concentration in the matrix in local equilibrium with the platelet is constant everywhere around the platelet as required. In addition, this concentration agrees with that predicted by eqn (5.119). Analyses of the rate at which an isolated platelet would undergo shape equilibration under the influence of the above concentration differences have been given by Shiflet *et al.* (1977), for volume diffusion-controlled kinetics, and by Merle and Doherty (1982), for interface short-circuit diffusion-controlled kinetics. Experiments in which both coarsening and shape equilibration occurred indicate that the broad faces again generally acted as relatively poor sinks. Rajab and Doherty (1989) and Doherty Rajab (1989) studied the coarsening of the platelet particles in the Ag/Al system which had grown into highly non-equilibrium shapes with large $\rho = d/t$ ratios. It was found that these non-equilibrium shapes were maintained for long times of coarsening, and that the platelet thickening was strongly inhibited. The experiments also indicated that the platelet edges of different platelets were able to come into equilibrium with each other more readily than were the edge and faces of an individual platelet. These results indicate that the faces acted as poor sinks due to a lack of growth steps. Even though platelet intersections were present (see above), any heterogeneous nucleation of line defects at the intersections was evidently too slow to produce ideal sink action in the presence of the relatively low values of $g_s$ which existed during the late stages of coarsening. In further work Aiken and Plichta (1990) found that the platelet thickening rate during coarsening was inversely proportional to the spacing between steps, again indicating interface control at the broad faces. In the Cu/Al system Merle and Fouquet (1981) and Merle and Merlin (1981) found that platelet thickening during coarsening was interface-controlled and related to the growth step spacing on the broad faces. In agreement with our previous discussion of models, these results, taken together, indicate that singular (or vicinal) interfaces such as the broad faces of platelet precipitates can exhibit a wide range of sink efficiencies depending upon the density of growth steps. The growth step density, in turn, often depends upon such factors as the availability of heterogeneous nucleation sites and the magnitude of the driving energy, $g_s$.

*Particle shrinkage.* The dissolution of particles has been less studied than their growth. In general, the kinetics of shrinkage may be expected to differ from that of growth (Aaron and Kettler 1971, Sagoe-Crentsil and Brown 1984, 1991). For example, the edges of platelet particles may serve as preferred sites for the nucleation of dislocations with strong step character causing dissolution in a manner similar to the ledge nucleation illustrated in Fig. 9.16(c). Attempts to study the question of diffusion-controlled versus interface-controlled kinetics during the shrinkage of relatively large well-annealed platelet precipitates have been made by investigating the diffusion fields in the vicinities of the platelets directly by means of electron-probe microanalysis. In the $Al/Ag_2Al$ system some evidence was obtained (Abbott and Haworth 1973) that the shrinkage of the $Ag_2Al$ platelets was interface-controlled, whereas in the $Al/CuAl_2$ system the shrinkage was apparently diffusion-controlled (Hall and Haworth 1970).

However, Sagoe-Crentsil and Brown (1984) showed that studies of the diffusion fields must be complemented with kinetic measurements of the shrinkage in order to obtain conclusive results. Detailed studies of the shrinkage of $Ag_2Al$ platelets (Sagoe-Crentsil and Brown 1984) along these lines showed that the dissolution rate at the broad faces was parabolic with time while the rate at the edges was linear. This caused the axial ratio of the platelets to decrease during the dissolution. Furthermore, the dissolution rate at the broad faces was at least an order of magnitude slower than that expected for volume diffusion control. In later work, Sagoe-Crentsil and Brown (1991) found that different interface dislocation/step structures may be present during platelet dissolution and platelet growth. Clearly, the overall kinetics of these processes are complex and are generally different due to differences in the types and densities of the interfacial dislocation/steps which are generated under the different kinetic conditions.

As another example of interface-controlled particle shrinkage we cite studies (Nolfi *et al.* 1970*a,b*) of the dissolution of $Fe_3C$ (cementite) precipates in b.c.c. Fe (ferrite). Here, the dissolution rate was found to be interface-controlled rather than controlled by the rate of diffusion of interstitial C away from the C-rich precipitates. The result was attributed to the probable presence of vicinal or singular interfaces enclosing the particles which possessed low densities of suitable line defects. We note that this result is in substantial agreement with coarsening experiments carried out with $Fe_3C$ particles in b.c.c. Fe (Heckel and DeGregorio 1965), where interface-controlled kinetics were also found.

### 10.4.2.2  *Growth of phases in the form of flat parallel layers*

Another phenomenon of great interest is the growth of phases in the form of flat parallel layers. In many cases flat free surfaces are exposed to new chemical environments causing layers of new phases to form (as during oxidation), or two bulk phases are joined along a flat interface and then interact to form layers of intermediate phases. The rate of growth (or shrinkage) of these layered phases then depends in general upon the rate of diffusion of the different components to/from the interfaces which separate them and the abilities of the interfaces to act as sources/sinks for the fluxes, i.e. we again may have either diffusion-controlled, interface-controlled, or mixed kinetics. Innumerable situations can then develop depending upon the types of phases and interfaces involved.

Many observations have been made of the kinetics of layer growth in systems containing phases in which the diffusional transport to/from the interfaces involves substitutional species which diffuse via a vacancy mechanism. In order to evaluate the experimentally observed performance of the interfaces in these systems as sources/sinks we must first analyse the anticipated behaviour of such a system under ideal conditions where the interfaces act as ideal sources/sinks for all species. The overall kinetics is then diffusion-controlled, and the problem reduces to finding the solution of a boundary value diffusion problem.

We start with the relatively simple two-phase system illustrated in Fig. 10.15 consisting of $\alpha$ and $\beta$ phases, each composed of A and B atoms and vacancies: here, the two phases, of initially uniform compositions $n_B^\alpha(-\infty)$ and $n_B^\beta(\infty)$, were joined at $x = 0$ and then annealed isothermally. The bulk $\alpha$ phase is supersaturated with respect to B, while the $\beta$ phase is subsaturated: the interface, acting as an ideal source/sink, has quickly established local equilibrium at the interface. Continued diffusion has produced a long-range diffusion field in both phases and movement of the interface. Also, in general, there will be a tendency for a Kirkendall diffusion effect to exist in each phase in a system of this type. In this phenomenon (Shewmon 1989) the A and B atoms diffuse at different rates with respect to the network and generate a vacancy flux corresponding to a flow of matter in the diffusion zone in each phase. As we shall see below, maintenance of such a bulk Kirkendall effect generally requires an interface, acting as an ideal source/sink, to act as a source/sink for network sites.

For simplicity, we shall assume that the atomic volumes of all species are constant and equal, and therefore that no volume changes occur during the diffusion, i.e. the distance $L$ (Fig. 10.15) remains constant. (We note that volume changes do not introduce any new phenomena of present interest but merely complicate the analysis (e.g. Balluffi (1960)).) The coordinate system shown in Fig. 10.15, with its origin at the position of the original interface, is then fixed with respect to the ends of the specimen and will be referred to as the 'laboratory' coordinate system. The diffusion-controlled velocity of the interface (i.e. the rate of growth of the $\alpha$ phase) relative to this coordinate system will be represented by $v^I$.

Our first step is to establish the necessary diffusion equations for the different species in each phase. According to eqn (5.149), the equations for the fluxes in the $\alpha$ phase measured relative to the network as a local coordinate system are

$$\left.\begin{aligned} J_A^\alpha &= -B_{AA}^\alpha \nabla M_A^\alpha - B_{AB}^\alpha \nabla M_B^\alpha \\ J_B^\alpha &= -B_{BA}^\alpha \nabla M_A^\alpha - B_{BB}^\alpha \nabla M_B^\alpha \\ J_V^\alpha &= -B_{VA}^\alpha \nabla M_A^\alpha - B_{VB}^\alpha \nabla M_B^\alpha \end{aligned}\right\}, \tag{10.73}$$

if the vacancies are taken as the $C$th component. We may assume, as is usually the case (Shewmon 1989), that a sufficient density of sources/sinks are present throughout the bulk in the form of lattice dislocations to maintain the vacancies essentially at their local equilibrium concentrations and provide sources and sinks for the fluxes of the A and B atoms. Therefore, $\mu_V^\alpha = 0$, and from eqn (5.141b), $\nabla M_A^\alpha = N_o \nabla \mu_A^\alpha$ and $\nabla M_B^\alpha = N_o \nabla \mu_B^\alpha$. However, according to the Gibbs–Duhem equation,

$$n_A^\alpha \nabla \mu_A^\alpha + n_B^\alpha \nabla \mu_B^\alpha + n_V^\alpha \nabla \mu_V^\alpha = n_A^\alpha \nabla \mu_A^\alpha + n_B^\alpha \nabla \mu_B^\alpha = 0. \tag{10.74}$$

Therefore, substituting these results into eqn (10.73),

$$\left.\begin{aligned} J_A^\alpha &= -N_o [B_{AA}^\alpha - B_{AB}^\alpha (n_A^\alpha / n_B^\alpha)] \nabla \mu_A^\alpha \\ J_B^\alpha &= -N_o [B_{BB}^\alpha - B_{BA}^\alpha (n_B^\alpha / n_A^\alpha)] \nabla \mu_B^\alpha \end{aligned}\right\}. \tag{10.75}$$

However, as usual, $\mu_i^\alpha = \mu_i^{\alpha\circ} + kT \ln(\gamma_i^\alpha n_i^\alpha \Omega)$. Therefore, for one-dimensional diffusion along $x$

$$\left.\begin{aligned} J_A^\alpha &= -N_o kT [B_{AA}^\alpha / n_A^\alpha - B_{AB}^\alpha / n_B^\alpha] \,[1 + \mathrm{d}\ln \gamma_A^\alpha / \mathrm{d}\ln n_A^\alpha] \,\mathrm{d}n_A^\alpha / \mathrm{d}x \\ J_B^\alpha &= -N_o kT [B_{BB}^\alpha / n_B^\alpha - B_{BA}^\alpha / n_A^\alpha] \,[1 + \mathrm{d}\ln \gamma_B^\alpha / \mathrm{d}\ln n_B^\alpha] \,\mathrm{d}n_B^\alpha / \mathrm{d}x \end{aligned}\right\}. \tag{10.76}$$

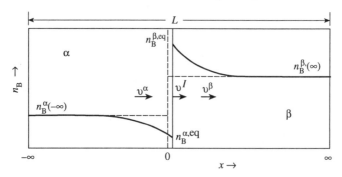

**Fig. 10.15** Concentration profiles during interdiffusion between infinitely thick layers of $\alpha$ and $\beta$ phases. The two phases, of initially uniform compositions $n_B^\alpha(-\infty)$ and $n_B^\beta(\infty)$ were initially joined along $x = 0$. The sharp $\alpha/\beta$ interface acts as an ideal source/sink and maintains the compositions of the $\alpha$ and $\beta$ phases in its vicinity in equilibrium with each other at the values $n_B^{\alpha,\,\text{eq}}$ and $n_B^{\beta,\,\text{eq}}$, respectively. The bulk $\alpha$ phase is supersaturated with respect to B, while the bulk $\beta$ phase is subsaturated with respect to B. The lattices of both phases are flowing with local velocities $v^\alpha$ and $v^\beta$ due to the Kirkendall effect, and the $\alpha/\beta$ interface is moving with a velocity $v^I$.

We now assign 'intrinsic' diffusivities (Shewmon 1989) to the A and B atoms defined by

$$\left.\begin{aligned}J_A^\alpha &= -D_A^\alpha(\mathrm{d}n_A^\alpha/\mathrm{d}x)\\ J_B^\alpha &= -D_B^\alpha(\mathrm{d}n_B^\alpha/\mathrm{d}x)\end{aligned}\right\}. \tag{10.77}$$

Therefore,

$$\left.\begin{aligned}D_A^\alpha &= N_\mathrm{o}kT[B_{AA}^\alpha/n_A^\alpha - B_{AB}^\alpha/n_B^\alpha]\,[1 + \mathrm{d}\ln\gamma_A^\alpha/\mathrm{d}\ln n_A^\alpha]\\ D_B^\alpha &= N_\mathrm{o}kT[B_{BB}^\alpha/n_B^\alpha - B_{BA}^\alpha/n_A^\alpha]\,[1 + \mathrm{d}\ln\gamma_B^\alpha/\mathrm{d}\ln n_B^\alpha]\end{aligned}\right\}. \tag{10.78}$$

Since, according to eqn (5.150), $J_A^\alpha + J_B^\alpha + J_V^\alpha = 0$, and since, in general, $D_A^\alpha \neq D_B^\alpha$, we must have a flux of vacancies relative to the local network given by

$$J_V^\alpha = -(J_A^\alpha + J_A^\alpha) = D_A^\alpha(\mathrm{d}n_A^\alpha/\mathrm{d}x) + D_B^\alpha(\mathrm{d}n_B^\alpha/\mathrm{d}x) = (D_B^\alpha - D_A^\alpha)\,(\mathrm{d}n_B^\alpha/\mathrm{d}x). \tag{10.79}$$

Since this flux is ultimately destroyed in the bulk ahead of the local network causing a contraction, the local network must be moving forward (i.e. flowing) with a velocity, $v^\alpha$, relative to the laboratory coordinate system given by

$$v^\alpha = \Omega J_V^\alpha = \Omega(D_B^\alpha - D_A^\alpha)\,(\mathrm{d}n_B^\alpha/\mathrm{d}x); \tag{10.80}$$

i.e. a Kirkendall effect must be present.

The flux of *B* relative to the laboratory coordinate system is then

$$j_B^\alpha = J_B^\alpha + v^\alpha n_B^\alpha = \Omega(n_A^\alpha J_B^\alpha - n_B^\alpha J_A^\alpha). \tag{10.81}$$

Therefore, in this coordinate system

$$\partial n_B^\alpha/\partial t = -\operatorname{div} j_B^\alpha = \frac{\partial}{\partial x}[\,(\Omega n_A^\alpha D_B^\alpha + \Omega n_B^\alpha D_A^\alpha)\,(\partial n_B^\alpha/\partial x)\,]. \tag{10.82}$$

Finally, we set

$$D^\alpha = \Omega n_A^\alpha D_B^\alpha + \Omega n_B^\alpha D_A^\alpha, \tag{10.83}$$

and write eqn (10.82) as

$$\partial n_B^\alpha / \partial t = \frac{\partial}{\partial x} [D^\alpha (\partial n_B^\alpha / \partial x)]. \tag{10.84}$$

The quantity $D^\alpha$ is known as the 'chemical' diffusivity (Shewmon 1989) and describes the diffusion in the laboratory coordinate system in a manner which takes full account of the local mass flow which accompanies the diffusion. We emphasize that $D^\alpha$ is a function of composition and of the intrinsic diffusivities which describe the diffusion locally in the flowing network. An exactly parallel set of relationships can be derived for the diffusion in the $\beta$ phase.

We now show that the interphase boundary will advance 'parabolically' with time (i.e. its displacement will be proportional to $t^{\frac{1}{2}}$) as long as the diffusion has not reached the far ends of the specimen. At the present stage of our analysis, $n_B^\alpha$, $n_B^\beta$, and all other dependent quantities are functions of the two independent variables $x$ and $t$. However, it is well known (Boltzmann 1894) that this boundary value problem can be converted into one which involves only the single independent variable $\lambda \equiv x/t^{\frac{1}{2}}$. Consider first the diffusion in the $\alpha$ phase. We assume $n_B^\alpha = n_B^\alpha (x, t) = n_B^\alpha (\lambda)$. The intrinsic diffusivities are functions of composition, e.g. $D_B^\alpha = D_B^\alpha (n_B^\alpha)$. Therefore, since $n_B^\alpha = n_B^\alpha (\lambda)$, $D_B^\alpha = D_B^\alpha (\lambda)$, and, therefore, according to eqn (10.83), $D^\alpha = D^\alpha (\lambda)$. Applying the change of variable $\lambda = x/t^{\frac{1}{2}}$ to eqn (10.84) then yields

$$- (\lambda/2) \, dn_B^\alpha (\lambda)/d\lambda = \frac{d}{d\lambda} [D^\alpha (\lambda) \cdot \{ dn_B^\alpha (\lambda)/d\lambda \}], \tag{10.85}$$

which is consistent with our assumption that $n_B^\alpha = n_B^\alpha (\lambda)$. A similar result is obtained for the $\beta$ phase. It can also be shown that all initial and boundary conditions can be expressed solely in terms of $\lambda$, thus completing the demonstration. (Note that this will not be the case if diffusion reaches the specimen ends.) The composition at the interface in the $\alpha$ phase is maintained constant at its equilibrium value, and therefore $\lambda$ at the interface must also be constant at a value designated by $\lambda^I$. Since $\lambda^I = x^I/t^{\frac{1}{2}}$, we then have the parabolic result $x^I = \lambda^I t^{\frac{1}{2}}$ directly.

An expression for the interface velocity can be obtained from the condition of flux continuity at the interface, i.e.

$$v^I n_B^{\beta, \mathrm{eq}} - v^I n_B^{\alpha, \mathrm{eq}} + (j_B^\alpha)_I - (j_B^\beta)_I = 0. \tag{10.86}$$

Use of equations such as eqns (10.81) and (10.77) for both phases in eqn (10.86) then yields

$$v^I = [1/(n_B^{\beta, \mathrm{eq}} - n_B^{\alpha, \mathrm{eq}})] [D_B^\alpha (\partial n_B^\alpha / \partial x) - v^\alpha n_B^{\alpha, \mathrm{eq}} - D_B^\beta (\partial n_B^\beta / \partial x) + v^\beta n_B^{\beta, \mathrm{eq}}]_I. \tag{10.87}$$

Note that a transformation of the right hand side of the above equation using $\lambda = x/t^{\frac{1}{2}}$ gives the result $v^I = \mathrm{constant} \cdot t^{-\frac{1}{2}}$: then, since $v^I = dx^I/dt$, we again find parabolic motion of the interface.

Having these results we can now examine the source/sink action which must take place at the interface. There are three different velocities there, i.e. $v^I$, $v^{\alpha, I}$, and $v^{\beta, I}$. The $\alpha$ phase network is moving into the interface with a relative velocity $(v^{\alpha, I} - v^I)$, while the $\beta$ phase network is moving out of the interface with a relative velocity $(v^{\beta, I} - v^I)$. Therefore, new network sites (i.e. vacancies) must be generated at the interface at the rate

$$\dot{N} = (1/\Omega) [ (v^{\beta, I} - v^I) - (v^{\alpha, I} - v^I)] = (1/\Omega) [v^{\beta, I} - v^{\alpha, I}] = J_V^\beta - J_V^\alpha]_I$$
$$= [(D_B^\beta - D_A^\beta) (\partial n_B^\beta / \partial x) - (D_B^\alpha - D_A^\alpha) (\partial n_B^\alpha / \partial x)]_I, \tag{10.88}$$

where use has been made of equations of the type (10.80).

To summarize, in order for the interface to act as an ideal source/sink for all species, as assumed, it must generate network sites, i.e. vacancies, at the rate $\dot{N}$ given by eqn (10.88). Of course, at the same time network sites must be transported across the interface by shuffling, since, in general, $v^I$ differs from both $v^{\alpha,I}$ and $v^{\beta,I}$. Further details of these phenomena have been discussed by van Loò *et al.* (1990) and Pieraggi *et al.* (1990). In addition, the interface will move parabolically with time as long as the appropriate boundary conditions at the extremes of the specimen are maintained. We note that in the example chosen the boundary conditions at the extremes corresponded to two large half-spaces which were never penetrated by the diffusion. Another common boundary condition is one in which the surface of the layered specimen is maintained at a constant concentration while the interior is a large half-space which again is not penetrated by the diffusion. Under these conditions the boundary value problem can again be transformed by the $\lambda = x/t^{\frac{1}{2}}$ change of variable, and parabolic behavior results. Examples include oxidation systems and other vapour/solid diffusion situations where a volatile component is diffused into the face of a solid from the vapour.

The above analysis can be readily extended to describe the ideal behaviour of a more complex multiphase system whose phase diagram contains a series of intermediate phases. If two phases, separated by one or more intermediate phases, are initially joined and then isothermally diffused, the system will form layers of all of the intermediate phases in successive order across the diffusion zone. All interfaces will act as ideal sources/sinks, local equilibrium will be maintained at each interface, and the situation at each interface will be similar to that described above for the two-phase system. Again, it can be shown by means of the $\lambda = x/t^{\frac{1}{2}}$ transformation that all interfaces in the system will move parabolically, and all intermediate phases will grow, as long as the proper boundary conditions are maintained at the extremes of the terminal phases.

Of course, actual systems may not behave in this ideal manner, since the formation of the intermediate phases requires their nucleation in the diffusion zone, and this may not occur for kinetic reasons (Allen and Sargent 1986). However, if an intermediate phase is missing for this reason we can still have ideal source/sink action at the remaining interfaces and parabolic interface motion. In this case the two phases facing each other across the missing phase will achieve metastable equilibrium with each other at the interface between them, and this metastable interface, along with the others, will then execute parabolic motion. In essence, the system will 'forget' that one of its equilibrium phases is missing.

The behaviour of a more general system which drops the *a priori* assumption of ideal interface sources/sinks and allows for possible mixed or interface-controlled kinetics can be investigated by again employing the simple model used previously in Section 10.2.1 (see eqn (10.16)) to treat the non-ideal sink behaviour of precipitate particles. Here, the rate of transport of a component across the interface is taken to be proportional (via a rate constant $\kappa$) to its concentration in the adjoining lattice. Using this model, it has been shown (Guy and Oikawa 1969, Tu 1985, 1986) that all multiphase systems will act non-ideally (i.e non-parabolically) when the time is sufficiently short. Then, at longer times, ideal source/sink action, accompanied by parabolic growth, tends to be achieved. This general result is easily understood: at short times the layers are exeedingly thin, and very large chemical gradients and diffusion fluxes must therefore be sustained in order to achieve ideal behaviour. This requires a rate of source/sink action which is larger than can be achieved. At longer times as the layer thicknesses increase, the required gradients and fluxes decrease tending to make closely ideal behaviour possible eventually. The point at which any system approximates ideal behaviour then depends upon the magnitudes of the controlling parameters. Systems with relatively large rate constants may exhibit ideal behaviour during all times of practical observation, while those with small rate constants may never approach ideal behaviour.

In certain multiphase systems some of the interfaces between intermediate phases which might be expected on the basis of assumed diffusion-controlled behaviour may act as relatively poor sources/sinks. In such cases some of the expected intermediate phases may be missing for kinetic reasons as explained in detail by Tu (1986) and Tu *et al.* (1992). Briefly, phases which are growing under interface-control tend to thicken at a linear rate, whereas those growing under diffusion-control grow at a rate which decreases with increasing time. At short times (or at small layer thicknesses) the phases which would grow under interface-controlled conditions cannot compete with the phases growing under diffusion-controlled conditions and hence cannot appear. However, at longer times when the rate of growth of the diffusion-controlled phases decreases sufficiently, the missing phases may compete successfully, and they will then appear.

The growth of layered phases has been studied over wide ranges of times and temperatures and layer thicknesses which have varied between $\simeq 10^{-6}$ cm and $\simeq 10^{-1}$ cm. In particular, the observation of exeedingly thin multilayered diffusion zones formed at relatively low temperatures in thin-film systems has been greatly stimulated by their importance in microelectronic devices (Tu and Mayer 1978, Tu 1985, 1986, Tu *et al.* 1992). As might be expected, a wide range of interfacial source/sink behaviour has been observed. The most definitive experiments have included both: (i) measurements of the rates of layer growth; and (ii) measurements of the compositions in the direct vicinities of the interfaces which provide direct tests for the establishment of local equilibrium. In experiments of this type, carried out at relatively elevated temperatures and involving relatively large layer thicknesses, essentially ideal interfacial source/sink behavior has been observed at numerous interfaces, e.g. the $\alpha/\gamma$ interface in the Fe/C system (Purdy and Kirkaldy 1963); interfaces in the Ni/Al system (Janssen and Rieck 1967, Janssen 1973); interfaces in the Ti/Ni system (Bastin and Rieck 1974); and interfaces in the Ag/Zn system (Hurley and Dayananda 1970, Williams *et al.* 1981). An example of ideal parabolic layer growth in the Ti/Ni system is shown in Fig. 10.16. However, non-ideal behaviour has also been found. For example, Castleman and Seigle (1958) observed a transition from non-parabolic to parabolic layer growth kinetics in the Ni/Al system as the thickness increased which is similar to that predicted by the simple model for mixed kinetics cited above: Eifert *et al.* (1968) found direct evidence for non-equilibrium concentrations at the $\alpha/\beta$ interface in the Cu/Al system.

In many of these experiments exhibiting essentially ideal behaviour a strong Kirkendall effect accompanied the layer growth, and we may therefore conclude that the various interfaces which were present must have acted as highly efficient sources/sinks for point defects as well as sites where atoms could be efficiently transferred from one crystal to the other by shuffling. Unfortunately, essentially no information is available regarding the structures of the various interfaces. However, in all cases polycrystalline materials were used, and the interfaces were often produced by solid state bonding techniques. It therefore seems likely that many of the interfaces were of the general type.

Many experiments have also been carried out at generally lower temperatures using thin film specimens where layer thicknesses were as small as $\simeq 10^{-6}$ cm (or even less). A range of kinetics spanning the spectrum from interface-controlled to diffusion-controlled has been found (Balluffi and Blakely 1975, Poate *et al.* 1978, Tu 1985, 1986, Tu *et al.* 1992). These experiments have often been complicated by: (i) short-circuit diffusion along

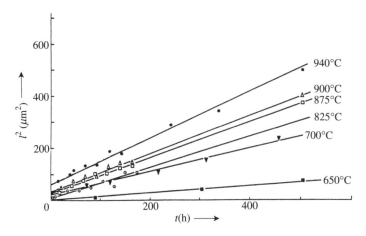

**Fig. 10.16** Plot of the square of the layer thickness, $l$, versus time for the growth of intermediate $Ti_2Ni$ layers in initially $Ti/TiNi_3$ layered diffusion couples. (From Bastin and Rieck (1974).)

interfaces transverse to the layers (ubiquitous in polycrystalline specimens); (ii) difficulties in nucleating various intermediate phases expected from the equilibrium phase diagram; and (iii) experimental problems in measuring local compositions at the closely spaced interfaces. In the extensively investigated metal/Si systems it has been found that the interfaces between refractory metals (e.g. W, Mo) and their disilicides are generally poor sources/sinks causing interface-controlled kinetics, whereas interfaces between noble metals (Pt, Pd) and their silicides are efficient sources/sinks causing diffusion-controlled kinetics. This can be understood (Tu and Mayer 1978, Tu 1986) by realizing that refractory metal atoms are essentially insoluble in Si at relatively low temperatures and are of no assistance in the breaking of Si–Si bonds and the release of Si atoms at the interface which is necessary for any source/sink action. On the other hand, the noble metal atoms can dissolve in Si and diffuse there by a probable dissociative type interstitial mechanism. Evidently, they are therefore able to penetrate the Si at the interface causing bond weakening (due to charge transfer) and the release of Si atoms. Again, unfortunately, little reliable information is available regarding the detailed structures of the various interfaces. However, many of the specimens consisted of layers which were fabricated initially by vapour deposition techniques or else were nucleated from vapour-deposited layers. It is therefore likely that preferred textures were often present and that many of the interfaces may have been singular or vicinal.

### 10.4.2.3 *Annealing of supersaturated vacancies*

Little information about the ability of heterogeneous interfaces to act as sinks for point defects is generally available from experiments in which the annealing of excess defects was observed in a more or less direct fashion. An exception is the observation of the annealing of supersaturated vacancies in Al to various $Al/Al_2O_3$ interfaces. Here, sheets of Al, possessing a layer of $Al_2O_3$ on the surface were equilibrated at an elevated temperature and then cooled (Doherty and Davis 1959, Basu and Elbaum 1965, Anthony 1970). During the cooling the thermal population of vacancies in the Al became supersaturated and tended to anneal out. In regions near the surface the excess vacancies annealed out at grain boundaries intersecting the surface and also at pits (voids) in the Al which nucleated at the $Al/Al_2O_3$ interface as illustrated in Fig. 10.17. In no cases

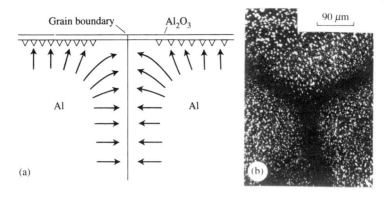

**Fig. 10.17** (a) Diffusion fluxes of supersaturated vacancies near the intersection of a grain boundary with an oxidized surface of Al during cooling from an elevated temperature. A thin film of $Al_2O_3$ is present on the surface. The grain boundary acts as an efficient sink for the supersaturated vacancies which are present, whereas the $Al/Al_2O_3$ interface does not. Instead, vacancies are annihilated at pits in the Al which nucleate and grow at the $Al/Al_2O_3$ interface. The grain boundary sink action produces a zone denuded of pits centred on the intersection of the grain boundary with the $Al_2O_3$ film. (b) Denuded zone at three intersecting grain boundaries observed at normal incidence to the surface. (From Basu and Elbaum (1965).)

were any vacancies annihilated at the $Al/Al_2O_3$ interface. The vacancy annihilation at the transverse grain boundaries produced zones in the $Al/Al_2O_3$ interface which were denuded of pits in regions where the grain boundaries intersected the $Al/Al_2O_3$ interface as may be seen in Fig. 10.17. The results demonstrate directly that the grain boundaries in the Al acted as much more efficient sinks for the vacancies than the $Al/Al_2O_3$ interfaces which remained essentially inoperative at all stages. Lower limits for the values of $g_s$ at which the $Al/Al_2O_3$ interface remained inoperative can be estimated from the vacancy supersaturations which were present when the pits nucleated and intervened by draining away the excess vacancies. Singular $Al/Al_2O_3$ interfaces, where the $Al_2O_3$ was epitaxial to (111) planes of the Al lying parallel to the interface, remained inoperative during cooling through the point where the pits nucleated at a vacancy undercooling as large as $\simeq 108$ K relative to an original equilibration temperature of $\simeq 888$ K (Anthony 1970). From eqn (10.23) the value of $g_s$ which was available at this juncture is given by $g_s = kT \ln (X_V/X_V^{eq})$. Therefore, since $X_V^{eq} = \exp(S_V^f/k) \cdot \exp(-E_V^f/kT)$ and $E_V^f = 0.67$ eV (Balluffi 1978), $g_s = E_V^f(108/888) = 0.08$ eV. At more general $Al/Al_2O_3$ interfaces, the pits nucleated at smaller undercoolings which were in the range 4–10 K (Doherty and Davis 1959, Basu and Elbaum 1965, Anthony 1970). Therefore, the lower limits on $g_s$ which can be set for these interfaces are considerably smaller and are of order $3 \times 10^{-3}$ eV.

We may therefore conclude that values of $g_s$ greater than $\simeq 3 \times 10^{-3}$ eV and $\simeq 0.08$ eV are required to force general and singular $Al/Al_2O_3$ interfaces respectively to act as sinks for excess vacancies in the Al. Furthermore, all $Al/Al_2O_3$ interfaces acted as less efficient vacancy sinks than the co-existing grain boundaries that were present in the Al. We may speculate that this result is due to the strong bonding on the $Al_2O_3$ side of the $Al/Al_2O_3$ interfaces.

In further experiments Derby and Qin (1992) found that general grain boundaries in Ni were more efficient sinks for vacancies diffusing from void surface sources than heterophase Ni/ZrO$_2$ interfaces. This observation appears to be consistent with the above results for Al/Al$_2$O$_3$ interfaces. In addition, Harris (1978) has suggested that ceramic/metal interfaces may often be relatively poor sinks when the ceramic has strong bonding and a high melting temperature compared with the metal.

### 10.4.2.4  *Diffusional accommodation of boundary sliding at second phase particles*

As discussed in Section 12.8.1, the two crystals adjoining an approximately flat general grain boundary can be readily sheared with respect to one another across the boundary by an applied shear stress at elevated temperature, i.e. grain boundary sliding can be induced as illustrated in Fig. 12.25. However, if a distribution of hard second phase particles is introduced in the boundary, the particles will act as 'pegs' which block the sliding. When such particles are present, and the shear stress is first applied, opposing elastic stresses will quickly build up at the matrix/particle interfaces until their components in the direction of shear balance the applied stress. In this situation the particles therefore bear the full applied shear load. The stress at a cuboidal second phase particle at this stage is then as shown in Fig. 10.18(a). In constructing this figure it has been assumed that the particle/matrix interface is of the general type: therefore, all existing stresses must be normal to the interface, since no shear stresses can be sustained at any of the interfaces present. However, the stresses (normal tractions) shown in Fig. 10.18(a) now provide a diffusion potential which will tend to redistribute material around the particle by diffusion.

The above situation is similar in many respects to the situation in Section 12.8.1.3 (see Fig. 12.27(c)) where the grain boundary sliding of two interlocked crystals is accommodated by diffusion. If the $\alpha$ matrix is composed only of A type atoms and vacancies, and if the heterophase particle/matrix interface acts as an ideal source/sink for the matrix atoms with $\mu_V = 0$, this potential, according to eqn (12.65), is given by $M_A = N_o\Omega(f^\alpha - \tau_n)$, where $\tau_n$ is the local traction normal to the interface and $f^\alpha$ is a bulk free energy. Matrix atoms will then leave regions of the interface in compression and enter regions in tension. During the early stages of this process the distribution of normal stresses will change as atoms are added or removed in different parts of the interface. However, in a relatively short time a quasi-steady-state situation will be reached in which no further changes in $\tau_n$ occur. The normal stresses will then be distributed so that the diffusion potential and the flux which enters or leaves each part of the interface is regulated in a manner which allows the upper crystal to slide compatibly with respect to the lower crystal at a constant rate. A steady-state rate of sliding will therefore be established which is diffusion-controlled as long as the particle/matrix interfaces act as ideal sources/sinks. The diffusion transport can occur by both volume diffusion through the matrix and rapid diffusion along the heterophase interface as illustrated in Fig. 10.18(b) where the expected forms of the diffusion fields around a spheroidal particle are shown schematically.

In order to determine the rate of diffusion accommodated sliding we follow the same basic method which is used in Section 12.8.1.3 to solve the diffusional accommodation problem illustrated in Fig. 12.27(c). The boundary value diffusion problem illustrated in Fig. 10.18(b) is first solved. Then the fluxes which accommodate the sliding are determined. For the lattice transport, the flux is given by eqn (9.18), i.e. $J_A = -(D^L/N_o\Omega kTf^L)\nabla M_A$, and in view of the steady state, we must therefore find a solution of the Laplacian $\nabla^2 M_A = 0$ (since $-\text{div } J_A = (D^L/N_o\Omega kTf^L)\nabla^2 M_A = 0$) subject to the boundary conditions on $M_A$ resulting from the steady-state distribution of normal stress. Raj and Ashby (1971) have found an approximate solution to this problem

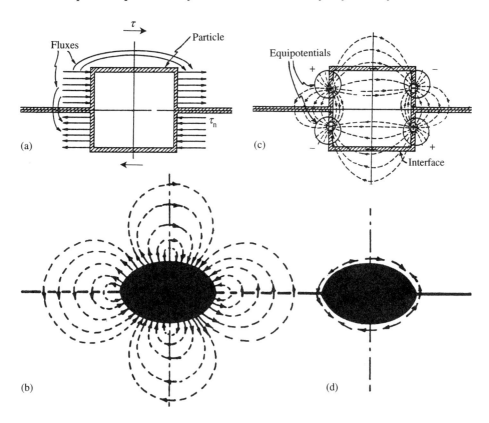

**Fig. 10.18** (a) Hard second-phase cuboidal particle in grain boundary opposing shearing of the boundary subjected to the shear stress, $\tau$. The faces of the cube are subjected to the normal stresses, $\tau_n$, which generate the lattice diffusion fluxes shown. (b) Lattice diffusion fluxes around spheroidal particle subjected to the same conditions as the cuboidal particle in (a). (c) Electrostatic analogue of the diffusion situation in (a) in the quasi-steady state. Positive and negative electric charges are distributed as shown generating an electric field which is similar to the expected diffusion field in (b). (d) Corresponding diffusion fluxes along interface arising from the shear stress, $\tau$. (From Raj and Ashby (1971, 1972).)

by taking advantage of the analogy between solutions of steady state diffusion problems and electrostatics problems pointed out previously in Section 10.3.2.1. Here, the diffusion flux field is analogous to the electric field: for simplicity, Raj and Ashby (1971) took the particle to be cuboidal and approximated the form of the expected flux field around it (which would be generally similar to that shown in Fig. 10.18(b)) by the electric field produced by the electrical charges shown in Fig. 10.18(c). The solution of the electrostatics problem was then used to find the solution of the diffusion problem, thus allowing a calculation of the relevant diffusion fluxes and the corresponding sliding rate due to the lattice diffusion (see Raj and Ashby (1971) for details). The additional sliding rate due to diffusional transport along the interface (illustrated in Fig. 10.18(d)) was determined in a generally similar way, and an approximate expression for the total sliding rate was finally obtained in the form

$$\dot{S} \simeq \frac{1.6\tau_a \Omega \lambda^2 D^L}{kTs^3} \left[ 1 + \frac{5\delta D^B}{sD^L} \right], \tag{10.89}$$

where $\tau_a$ is the applied shear stress, $\lambda$ is the spacing of particles in the grain boundary, $s$ is the particle size, $\delta$ is the thickness of the particle/matrix interface, and $D^B$ is the diffusivity in the particle/matrix boundary. When the particle/matrix boundary acts as an ideal source/sink, we therefore expect the sliding rate to be proportional to the applied stress and the matrix diffusivity at elevated temperatures where the contribution of the boundary transport term is relatively small, i.e. where $5\delta D^B/sD^L \ll 1$. However, at lower temperatures where $5\delta D^B/sD^L \gg 1$, the boundary transport dominates, and the sliding rate will then be proportional to the boundary diffusivity.

Raj and Ashby (1972) performed experiments in which $Al_2O_3$ particles were introduced into grain boundaries in Ag and their effects on subsequent grain boundary sliding at relatively small applied stresses were studied. The sliding in the Ag without particles was generally anisotropic in individual boundaries, and its magnitude varied from boundary to boundary (see Section 12.8.1.5). However, when the particles were introduced, the sliding became remarkably similar to that predicted by the ideal source/sink model described above. The sliding became isotropic and essentially identical at practically all boundaries. Also, the rate varied linearly with applied shear stress and was thermally activated with an activation energy at high temperatures near that of lattice diffusion and at lower temperatures near that expected for interface diffusion. In addition, the absolute magnitudes of the observed sliding rates were in reasonably good agreement with those calculated for the model on the basis of available diffusivity data and measured particle sizes and spacings.

One significant deviation from the ideal source/sink model was found, however. A relatively small threshold stress for sliding corresponding to $\tau_a = 7 \times 10^{-6}\mu$ was detected ($\mu$ = shear modulus). Since the stress supported at each particle was approximately $\tau = \tau_a(\lambda/s)^2$, this threshold stress corresponds to a threshold value of the energy $g_s = \tau\Omega$ approximately equal to $2 \times 10^{-4}$ eV. The vast majority of at least one of the two $Ag/Al_2O_3$ interfaces associated with each particle were most likely general interfaces, and we may therefore conclude that these interfaces acted as reasonably good sources/sinks when the above threshold stress was exceeded. We note that this threshold is significantly lower than the limits established for the general $Al/Al_2O_3$ interfaces discussed above in Section 10.4.2.3. It is higher, however, than some of those indicated for general large-angle homophase boundaries in Table 10.2.

## 10.5  DIFFUSE HETEROPHASE INTERFACES AS SOURCES/SINKS FOR SOLUTE ATOMS

A diffuse heterophase interface in a miscibility gap system of the type whose equilibrium configuration was discussed in Section 4.2.1 (see Fig. 4.2(a)) will tend to act as a source/sink under the conditions shown in Fig. 10.19. Here, the overall situation is similar in a number of respects to that for the sharp interface illustrated in Fig. 10.15 and analysed in Section 10.4.2.2. The system is again made up of A and B atoms and vacancies, and the bulk $\alpha$ phase is again supersaturated with respect to B, while the $\beta$ phase is subsaturated. As previously, the interface will attempt to establish local equilibrium in its vicinity, and, in general, this will establish a long range diffusion field in the adjoining phases and cause the interface to move with a velocity, $v^I$, relative to the ends of the specimen. During the motion, one half-space will grow at the expense of the other, and a divergence of the diffusion flux of solute atoms will occur in the relatively narrow interface region as the interface acts as a source/sink for the solute atoms. It is evident that two length scales are present. The first, $\xi$, is relatively short and is of the order of the width of the diffuse interface, while the second, $L$, is very much longer, and is of the order of the long range diffusion distance in the bulk crystal lattice. The details of the concentration profile at the diffuse interface over the length, $\xi$, are shown in Fig. 10.19(b). In cases where the interface is successful in maintaining essential equilibrium in its vicinity the region within $\xi$ will appear as in Fig. 4.2(a) corresponding to the interface at equilibrium.

In the present case we consider the case illustrated in Fig. 4.2(b), where anticoherency edge dislocations are present, and assume that these sources/sinks, along with other lattice dislocations distributed throughout the bulk, are capable of maintaining the vacancy concentration close to equilibrium everywhere during the diffusion (i.e. $\mu_V \simeq 0$). However, we do not assume *a priori* that the solute concentration is maintained in local equilibrium at the interface as we did in our previous analysis of the sharp interface illustrated in Fig. 10.15: instead, we allow for a possible non-equilibrium situation where the solute concentrations in the lattice in the direct vicinity of the interface deviate from equilibrium by the increments $\delta n_B^\alpha = n_B^\alpha - n_B^{\alpha,\,eq}$ and $\delta n_B^\beta = n_B^\beta - n_B^{\beta,\,eq}$ as shown in Fig. 10.19(a). We then investigate the conditions under which local equilibrium can be achieved so that the diffuse interface can act as an ideal source/sink for the solute atoms as well as vacancies.

Since the bicrystal is a relatively gradually varying continuum, we can again use continuum methods and analyse the diffusion throughout the entire specimen by employing a single diffusion equation as described by Langer and Sekerka (1975). Extremely steep chemical gradients exist in the diffuse boundary region, and the effects of these must be taken into account. This can be accomplished (Langer and Sekerka 1975) by including

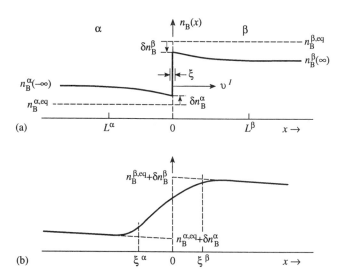

**Fig. 10.19** (a) Interdiffusion at a diffuse heterophase interface between binary $\alpha$ and $\beta$ phases. $n_B^{\alpha,\,eq}$ and $n_B^{\beta,\,eq}$ are the concentrations of the bulk $\alpha$ and $\beta$ phases which would be in equilibrium with each other at the diffusion temperature. The bulk $\alpha$ phase, at the composition $n_B^\alpha(-\infty)$, is supersaturated with $B$, while the bulk $\beta$ phase, at the composition $n_B^\beta(\infty)$, is subsaturated with $B$. The concentrations of $B$ in the $\alpha$ and $\beta$ phases at the interface deviate from equilibrium by the amounts $\delta n_B^\alpha$ and $\delta n_B^\beta$, respectively. (b) Detailed view of concentration profile, $n_B(x)$, in the diffuse interface region lying within the relatively narrow $\xi$ region indicated in (a). (From Langer and Sekerka (1975).)

the extra energy associated with the gradient in the construction of the diffusion potential as described by Cahn and Hilliard (1958) and Cahn (1960). In the same spirit in which eqn (9.53) was written for the rate of ordering in a diffuse antiphase boundary, we take the rate of solute atom transport to be proportional to the gradient of the rate at which the local free energy (per atom) in the non-equilibrium (and non-uniform) system changes with the local concentration, i.e.

$$j_B = -N_o B \frac{\partial}{\partial x} \left( \frac{\delta g}{\delta X_B} \right), \tag{10.90}$$

where $B$ is a phenomenologically based mobility. The total free energy of the system is given by eqn (4.11) with an additional term added to the integrand to account for the energy contributed by the misfit dislocations. We may assume the variation of lattice parameter with composition to be linear (Vegard's law), and under this condition the local density of misfit dislocations will be proportional to the gradient in composition. The local energy (per atom) contributed by the dislocations, $g_d$, will be a function of their local density, and therefore also a function of the local gradient in composition, so that $g_d = g_d(dX_B/dx)$.

If a small compositional variation from the equilibrium state is made throughout the system given by $\delta X_B(x)$, a variational calculation similar to that used to derive eqn (9.54) shows that

$$\delta g = \left[ \frac{\partial I}{\partial X_B} - \frac{d}{dx} \frac{\partial I}{\partial (dX_B/dx)} \right] \delta X_B, \tag{10.91}$$

with $I = [g'(X_B) + g_d + K(dX_B/dx)^2]$. Therefore,

$$j_B = -N_o B \frac{\partial}{\partial x} [ (\partial g'/\partial X_B) - \frac{\partial}{\partial x} \left[ \frac{\partial g_d}{\partial (dX_B/dx)} \right] - 2K(\partial^2 X_B/\partial x^2) ]. \tag{10.92}$$

It is readily shown that eqn (10.92) is consistent with eqn (10.81) when the extra energy associated with the gradient is neglected and $\mu_V = 0$ everywhere. Since from Fig. 4.1(a), $\partial g'/\partial X_B = \mu_B - \mu_A$, eqn (10.92) reduces to

$$j_B = -N_o B(1 + n_B/n_A)\nabla\mu_B \tag{10.93}$$

with the help of eqn (10.74). On the other hand, eqn (10.81), with the help of eqns (10.75), and (10.74), may be written in the form

$$j_B = -N_o \Omega(B_B n_A + B_A n_B^2/n_A)\nabla\mu_B, \tag{10.94}$$

if $B_B \equiv [B_{BB} - B_{BA} n_B/n_A]$ and $B_A \equiv [B_{AA} - B_{AB} n_A/n_B]$. In both eqns (10.93) and (10.94) the flux of B atoms in the laboratory coordinate system is seen to be proportional to a single compositional dependent mobility and the gradient of the chemical potential of B.

The final diffusion equation for B atoms is then

$$\partial n_B/\partial t = -\text{div} j_B = \frac{\partial}{\partial x} \left\{ N_o B \frac{\partial}{\partial x} \left[ (\partial g'/\partial n_B) - \frac{\partial}{\partial x} \left[ \frac{\partial g_d}{\partial (dX_B/dx)} \right] - 2K(\partial^2 n_B/\partial x^2) \right] \right\}. \tag{10.95}$$

An analytical solution of eqn (10.95) for the present problem cannot be found. However, Langer and Sekerka (1975) examined the behaviour of the system by assuming that the interface profile does not deviate greatly from the equilibrium one and neglecting the $g_d$ term (or, alternatively, any elastic energy which may possibly have been present due to lattice parameter variations as discussed in Section 4.2.1). Solutions were found by piecing together approximate solutions valid in the interfacial region with corresponding solutions valid far from the interface. In order to assess the effect of a possibly low mobility across the diffuse interface, a mobility of the plausible form

$$B(x) = \frac{B^\circ}{1 + \rho \operatorname{sech}^2(x/2\xi)}, \tag{10.96}$$

(where $\rho$ and $B^\circ$ are constants) was assumed. The quantity $\rho$ is therefore a measure of the possible decrease in B at the centre of the diffuse interface. Such slow diffusion would correspond to a resistance to the transport of B atoms in the interfacial region and would reduce the effectiveness of the interface as a source/sink for the solute atoms. Expressions for the corresponding deviations from local equilibrium at the interface (i.e. $\delta n_B^\beta$ and $\delta n_B^\alpha$) caused by this resistance were then obtained and were found to be proportional to $\rho$. They also contain a term proportional to the flux through the interface. This latter term may be expected because of the greater difficulty in maintaining local solute atom equilibrium as the solute atom flux increases. Significant deviations from local equilibrium at the interface were predicted only when $\rho$ is very large, i.e. at least a few per cent of the ratio $L/\xi$.

Inclusion of the $g_d$ term should not alter the qualitative aspects of these results, and we may therefore conclude that a diffuse interface capable of maintaining $\mu_V \simeq 0$ should act as a closely ideal source/sink for solute atoms whenever there is no unusually large decrease in the mobility in the interfacial region. Quantitative details, along with further insights are given by Langer and Sekerka (1975).

When the interface is incapable of acting as a source/sink for vacancies, as in a completely coherent case, the analysis becomes much more complex because of the presence of non-equilibrium vacancy concentrations in the interfacial region. Under these conditions approximate local solute atom equilibrium could still be achieved, but the overall diffusion kinetics would be different. Further analysis is required to deal with this case.

## 10.6 ON THE QUESTION OF INTERFACE STABILITY DURING SOURCE/SINK ACTION

We have previously found that interfaces advancing in a diffusion field may assume non-equilibrium shapes for kinetic reasons. In Section 10.4.2.1 it was pointed out that growing platelet precipitates may develop highly non-equilibrium diameter-to-thickness ratios if the broad faces perform less efficiently as solute atom sinks than the platelet edges. However, interfaces advancing in a diffusion field can become unstable and develop non-equilibrium morphologies for kinetic reasons even when the interface acts everywhere as an ideal source/sink. The origin of this type of instability is illustrated schematically in Fig. 10.20(a) and may be called the 'flux focusing' effect. If a small perturbation in the form of a hump appears on an otherwise relatively smooth interface, it will tend to collect more flux from the diffusion field than neighbouring areas which are relatively recessed and therefore shielded. Under certain conditions (to be explored below) such a protuberance will continue to grow. Such protuberances can form initially on an interface by local fluctuations of one type or another, and if they are capable of growing as described

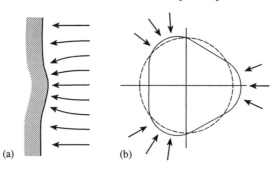

**Fig. 10.20** (a) Protuberance developing at an unstable interface due to the 'flux focusing' effect in the adjacent diffusion field. (b) Cross section of a spherical particle perturbed by a growing $l = 3$ harmonic. (From Mullins and Sekerka (1963).)

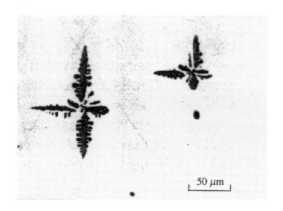

**Fig. 10.21** $\gamma$ phase particles precipitated in dendritic form in a Cu/Zn alloy from the $\beta$ phase supersaturated with Zn. (From Bainbridge and Doherty (1969).)

above, the interface will be unstable and develop a highly non-equilibrium morphology. The situation is similar in many respects to the well-known interfacial instability which occurs during solidification where the advancing liquid/solid interface becomes unstable in a combined temperature/composition field and develops a dendritic form (Flemings 1974). In fact, as illustrated in Fig. 10.21, it is possible for a crystalline interface advancing in a diffusion field to develop a dendritic morphology which is strikingly similar to those developed in solidification systems.

Up to this point we have ignored this type of possible instability. For example, we assumed in Section 10.2.1 that the interfaces enclosing growing spherical precipitates remained spherical and in Section 10.4.2.2 that the interfaces between growing layered phases remained flat. It is therefore necessary to investigate this phenomenon, identify the factors which may play a role, and judge its significance.

It is useful to begin with the classic analysis of Mullins and Sekerka (1963) for the kinetic stability of an interface enclosing a precipitate particle growing in a diffusion field as described in Section 10.2.1. This analysis takes into account the following two important factors which influence stability and, which, furthermore, oppose one another: (i) 'flux focusing' as described above, and (ii) 'capillary smoothing' of interfaces. This latter effect arises from the tendency of any protuberances on an interface to be smoothed off and eliminated because of the reduction in interfacial energy which is then achieved. The analysis assumes diffusion-controlled growth everywhere at the interface and tests the stability of the particle shape during growth to small perturbations in shape. If a perturbation grows, the interface is judged to be unstable with respect to that perturbation and presumably would then degenerate into a highly complex non-spherical shape. On the other hand, if the perturbation decays, the interface is judged to be stable.

We assume an initially spherical B-rich $\beta$ phase particle of radius $R$ and fixed composition, $n^\beta$, growing under quasi-steady-state diffusion-controlled conditions in an $\alpha$ phase matrix as described in Section 10.2.1. However, all atoms are now assumed to be substitutional. Working in spherical coordinates $(r,\theta,\phi)$, we then perturb its shape to

$$r = R + \delta_1 Y_1(\theta), \tag{10.97}$$

where $Y_1(\theta)$ is an associated Legendre function (Hildebrand 1949) given by

$$\left.\begin{array}{l} Y_1(\theta) = \cos\theta \\ Y_2(\theta) = (3\cos 2\theta + 1)/4 \\ Y_3(\theta) = (5\cos 3\theta + 3\cos\theta)/8 \ldots \end{array}\right\} \tag{10.98}$$

and $\delta_1$ is the amplitude of the perturbation, which is small compared with $R$. After a short diffusion transient, quasi-steady-state diffusion-controlled conditions will be re-established, and the concentration of the solute B in the matrix in the direct vicinity of the particle will again be maintained in local equilibrium with the particle. According to eqn (5.119), this concentration $n^{\alpha,\,eq}(\kappa)$, will vary with the local curvature of the interface, $\kappa$, as

$$\frac{n^{\alpha,\,eq}(\kappa)}{n^{\alpha,\,eq}(\infty)} = 1 + \Gamma\kappa, \tag{10.99}$$

where $\Gamma = \Omega^\beta \sigma [1 - X^{\alpha,\,eq}(\infty)]/kT\phi[X^\beta - X^{\alpha,\,eq}(\infty)]$, if the interfacial energy, $\sigma$, is taken to be isotropic. The curvature of the perturbed particle is given by $\kappa = \nabla \cdot \hat{n}$ (see eqn (5.105)), while $\hat{n}$ is given by $\hat{n} = \nabla g/|\nabla g|$ (see eqn (5.74)). In the present spherical coordinate problem $g(r,\theta) = r - R - \delta_1 Y_1(\theta)$ (Hildebrand 1949), and, since $\delta_1$ is small compared to $R$, $\kappa \simeq \nabla^2 g$. Therefore,

$$\kappa = 2/r - \delta_1 \Lambda \cdot Y_1(\theta), \tag{10.100}$$

where $\Lambda$ is the angular part of the Laplacian operator, i.e. $(1/\sin\theta)\,(\partial[\sin\theta \cdot \partial/\partial\theta]/\partial\theta)$. Using eqn (10.97) in eqn (10.100), and expanding to first order,

$$\kappa = 2/R(1 - \delta_1 Y_1/R) - \delta_1 \Lambda Y_1/R^2. \tag{10.101}$$

Since $\Lambda Y_1 = -l(l+l)Y_1$, eqn (10.99) becomes

$$n^{\alpha,\,eq}(r,\theta)/n^{\alpha,\,eq}(\infty) = 1 + 2\Gamma/R + \Gamma\delta_1 Y_1(l+2)(l-1)/R^2. \tag{10.102}$$

In order to investigate whether the perturbation grows or shrinks we must now solve the quasi-steady-state diffusion equation, $\nabla^2 n^\alpha = 0$, subject to the boundary conditions $n^\alpha \to n^{\alpha,\,eq}(r,\theta)$ at the particle, and $n^\alpha \to \bar{n}^\alpha$ as $r \to \infty$. An appropriate solution is

$$n^\alpha(r,\theta) = \bar{n}^\alpha + A/r + B\delta_1 Y_1(\theta)/r^{l+1}, \tag{10.103}$$

where

$$A = R[n^{\alpha,\,eq}(\infty) - \bar{n}^\alpha] + 2\Gamma n^{\alpha,\,eq}(\infty);$$

$$B = R^l[n^{\alpha,\,eq}(\infty) - \bar{n}^\alpha] + R^{l-1}\Gamma \cdot (l+1)l \cdot n^{\alpha,\,eq}(\infty).$$

Since the perturbation is small, we have, from eqn (10.97),

$$v = dr/dt = dR/dt + Y_1(\theta)\, d\delta_1/dt. \tag{10.104}$$

Also, mass balance at the moving interface yields the relation

$$v = D^\alpha [n^\beta - n^{\alpha,\,eq}(R)]^{-1} \cdot (\partial n^\alpha/\partial r)_{\text{Int}}, \tag{10.105}$$

where $n^{\alpha,\,eq}(R)$ is the concentration which would be in equilibrium with a sphere of radius $R$, i.e. $n^{\alpha,\,eq}(R) \equiv n^{\alpha,\,eq}(\infty)[1 + 2\Gamma/R]$. By setting eqn (10.104) equal to eqn (10.105), and using eqn (10.103), $d\delta_1/dt \equiv \dot\delta_1$ can then be expressed in the form

$$\dot\delta_1/\delta_1 = \frac{D^\alpha(l-1)}{[n^\beta - n^{\alpha,\,eq}(R)]}\left[G - \frac{\Gamma(l+1)(l+2)n^{\alpha,\,eq}(\infty)}{R^2}\right], \tag{10.106}$$

where $G \equiv [\bar n^\alpha - n^{\alpha,\,eq}(R)]/R$ is the gradient which would be present if the particle were a non-perturbed sphere of radius $R$. Note that this expression for $G$ can be obtained directly from our previous eqn (10.15).) The average rate of growth of the slightly perturbed sphere is, to a good approximation,

$$\frac{dR}{dt} = \dot R = v = \frac{D^\alpha}{[n^\beta - n^{\alpha,\,eq}(R)]}\left[\frac{\partial n^\alpha}{\partial r}\right]_{\text{Int}} = \frac{D^\alpha G}{[n^\beta - n^{\alpha,\,eq}(R)]} = \frac{D^\alpha[\bar n^\alpha - n^{\alpha,\,eq}(R)]}{R[n^\beta - n^{\alpha,\,eq}(R)]}. \tag{10.107}$$

Therefore, combining eqn (10.106) an (10.107),

$$\frac{(\dot\delta_1/\delta_1)}{(\dot R/R)} = \frac{(l-1)R}{[\bar n^\alpha - n^{\alpha,\,eq}(R)]}\left[G - \frac{\Gamma(l+1)(1+2)n^{\alpha,\,eq}(\infty)}{R^2}\right]. \tag{10.108}$$

Equation (10.108) can be put into a convenient form by introducing the radii $R^*$ and $R_1^c$. The radius $R^*$ is the radius of the sphere which would be in equilibrium with the solute concentration at long distances, $\bar n^\alpha$, so that $R^* = 2\Gamma n^{\alpha,\,eq}(\infty)/[\bar n^\alpha - n^{\alpha,\,eq}(\infty)]$. The radius $R_1^c$ is the radius of the sphere for which the $l$th harmonic will just grow: it can be obtained in the form $R_1^c = [1 + (l+1)(l+2)/2] \cdot R^*$ from eqn (10.106) by setting $\dot\delta_1 = 0$, and using the previous expressions for $G$ and $R^*$ Introducing these quantities into eqn (10.108) then yields

$$\frac{(\dot\delta_1/\delta_1)}{(\dot R/R)} = (l-1)\frac{[1 - R_1^c/R]}{[1 - R^*/R]}. \tag{10.109}$$

When $(\dot\delta_1/\delta_1)/(\dot R/R) > 1$, the particle shape will become increasingly perturbed with time, and the interface will be unstable. Equation (10.108) is seen to consist of two terms. The first, proportional to the gradient $G$, is positive and therefore tends to produce instability. It may be identified with the 'flux focusing' effect, since steep gradients at the particle will tend to produce large fluxes and thereby increase the effect. The second term is negative and therefore tends to promote stability. It is proportional to the surface energy and may be identified with the 'capillary smoothing' effect. It may be seen from eqn (10.109) that $(\dot\delta_1/\delta_1)/(\dot R/R)$ for the $1 = 2$ harmonic will be less than unity for all $R > R_2^c$, since $R_2^c > R^*$. However, for the $l = 3$ harmonic, $R^* = R_3^c/11$, and $(\dot\delta_3/\delta_3)/(\dot R/R) > 1$ when $R > 21R^*$. The particle will therefore begin to become increasingly perturbed (unstable) when it reaches this size. A cross-section of a particle, perturbed by the $l = 3$, harmonic is illustrated in Fig. 10.20(b). Further discussion of instability behaviour along with an analysis of the stability of flat interfaces is given by Mullins and Sekerka (1963).

Despite its complexity, the above stability analysis is still incomplete and is therefore only of limited applicability, since it has assumed diffusion-controlled kinetics, and, furthermore, has neglected a number of factors which can be of importance. Additional factors which must be considered include:

(1) lattice diffusion through the particle: significant lattice diffusion through the particle would make an additional contribution to the 'capillary smoothing' of the interface, and, hence increase stability;

(2) short-circuit diffusion along the interface: transport along this path would also increase stability;

(3) possible interface-controlled or mixed kinetics: under these conditions the gradient at the particle, $G$, would be lowered, thus reducing the flux focusing and increasing stability;

(4) anisotropic interface energy: it would generally be more difficult energetically to generate protuberances on anisotropic interfaces than isotropic interfaces because of the existence of additional torque terms (Section 5.6.4), thus promoting stability.

These additional phenomena, all of which tend to promote stability, have been discussed elsewhere (Shewmon 1965, Nichols and Mullins 1965, Sekerka 1968, Doherty 1983).

Experimentally, as discussed by Doherty (1983), instability is actually observed only rarely at crystalline interfaces. This state of affairs may be compared with liquid/solid interfaces during solidification where instabilities leading to dendritic morphologies are ubiquitous. Evidently, the stabilizing phenomena mentioned above (which are largely absent at liquid/solid interfaces), and the values of the physical parameters which control stability conspire to produce stability under most conditions. Isolated examples include the one shown in Fig. 10.21 and others found, for example, by Malcolm and Purdy (1967) and Ricks *et al.* (1983). We note that a deliberate attempt to promote instability at Ag-rich precipitates in Cu failed (see Doherty 1983). We may therefore conclude that the instability of crystalline interfaces can usually (but not always) be discounted.

# REFERENCES

Aaron, H. B. and Kottler, G. R. (1971). *Met. Trans.*, **2**, 393.
Aaronson, H. I., Laird, C., and Kinsman, K. R. (1970). In *Phase transformations* (ed. H. I. Aaronson) p. 313. American Society for Metals, Metals Park, Ohio.
Abbott, K. and Haworth, C. W. (1973). *Acta Metal.*, **21**, 951.
Aiken, R. M. and Plichta, M. R. (1990). *Acta Metall. Mater.*, **38**, 77.
Alexander, B. H. and Balluffi, R. W. (1957). *Acta Metall.*, **5**, 666.
Allen, C. W. and Sargent, G. A. (1986). *Mat. Res. Soc. Symp. Proc.*, **54**, 97.
Anthony, T. R. (1970). *Acta Metall.*, **18**, 471.
Anthony, T. R. (1975). In *Diffusion in solids* (eds A. S. Nowick and J. J. Burton) p. 353. Academic Press, New York.
Ardell, A. J. (1969). In *The mechanism of phase transitions in crystalline solids*, p. 111. Institute of Metals, London.
Ardell, A. J. (1988). In *Phase transformations '87* (ed. G. W. Lorimer) p. 485. Institute of Metals, London.
Arzt, E., Ashby, M. F., and Verrall, R. A. (1983). *Acta Metall.*, **31**, 1977.
Babcock, S. E. and Balluffi, R. W. (1989). *Acta Metall.*, **37**, 2357; 2367.

Bainbridge, B. G. and Doherty, R. D. (1969). In *Quantitative relation beween properties and microstructure* (eds D. G. Brandon and A. Rosen) p. 427. Israel University Press, Jerusalem.

Balluffi, R. W. (1960). *Acta Metall.*, **8**, 871.

Balluffi, R. W. (1969). *Phys. Stat. Sol.*, **31**, 443.

Balluffi, R. W. (1975). *Proceedings of the conference on fundamental aspects of radiation damage in metals*, p. 852. National Technical Information Service, US Department of Commerce, Springfield, Virginia.

Balluffi, R. W. (1978). *J. Nucl. Mats.*, **69–70**, 240.

Balluffi, R. W. (1980). In *Grain boundary structure and kinetics* (ed. R. W. Balluffi) p. 297. American Society for Metals, Metals Park, OH.

Balluffi, R. W. (1984). In *Diffusion in crystalline solids* (ed. G. E. Murch and A. S. Nowick) p. 320. Academic Press, New York.

Balluffi, R. W. and Blakely, J. M. (1975). *Thin Solid Films*, **25**, 363.

Balluffi, R. W. and Granato, A. V. (1979). In *Dislocations in solids*, Vol. 4 (ed. F. R. N. Nabarro) p. 1. North-Holland, Amsterdam.

Balluffi, R. W. and King, A. H. (1983). In *Phase transformations during irradiation* (ed. F. V. Nolfi) p. 147. Applied Science Publications, Essex.

Balluffi, R. W. and Thomson, R. M. (1962). *J. Appl. Phys.*, **33**, 817.

Balluffi, R. W., Brokman, A., and King, A. H. (1982). *Acta Metall.*, **30**, 1453.

Bastin, G. F. and Rieck, G. D. (1974). *Met. Trans.*, **5**, 1817; 1827.

Basu, B. K. and Elbaum, C. (1965). *Acta Metall.*, **13**, 1117.

Boltzmann, L. (1894). *Ann. Physik*, **53**, 959.

Bouazra, Y. and Reynaud, F. (1984). *Acta Metall.*, **32**, 529.

Brailsford, A. D. and Bullough, R. (1981). *Phil. Trans. Roy. Soc.* (*London*) A, **302**, 87.

Brown, L. C. (1984). *Met. Trans.*, **15A**, 449.

Brown, L. C. (1985). *Acta Metall.*, **33**, 1391.

Brown, L. C. (1989). *Acta Metall.*, **37**, 71.

Brown, L. C. (1991). *Scripta Metall.*, **25**, 261.

Burton, B. (1977). *Diffusion creep of polycrystalline materials*. Trans Tech, Aedermannsdorf, Switzerland.

Cahn, J. W. (1961). *Acta Metall.*, **9**, 795.

Cahn, J. W. and Hilliard, J. E. (1958). *J. Chem. Phys.*, **28**, 258.

Castleman, L. S. and Seigle, L. L. (1958). *Trans. A.I.M.E.*, **212**, 569.

Chan, S.-W. and Balluffi, R. W. (1986). *Acta Metall.*, **34**, 2191.

Christian, J. W. (1975). *Transformations in metals and alloys*, Part 1, 2nd edn. Pergamon Press. Oxford.

Cotterill, R. M. J., Doyama, M., Jackson, J. J., and Mishii, M. (eds) (1965). *Lattice defects in quenched metals*. Academic Press, New York.

DeHoff, R. T. (1991). *Acta Metall. Mater.*, **39**, 2349.

Derby, B. and Qin, C.-D. (1992). *Acta Metall. Mater.*, **40**, S53.

Doherty, R. D. (1983). In *Physical metallurgy* (eds R. W. Cahn and P. Haasen) p. 933. North-Holland, Amsterdam.

Doherty, P. E. and Davis, R. S. (1959). *Acta Metall.*, **7**, 118.

Doherty, R. D. and Rajab, K. E. (1989). *Acta Metall.*, **37**, 2723.

Dollar, M. and Gleiter, H. (1985). *Scripta Metall.*, **19**, 481.

Eifert, J. R., Chatfield, D. A., Powell, G. W., and Spretnak, J. W. (1968). *Trans. A.I.M.E.*, **242**, 66.

Ferrante, M. and Doherty, R. D. (1979). *Acta Metall.*, **27**, 1603.

Flemings, M. C. (1974). *Solidification processing*. McGraw-Hill New York.

Flynn, C. P. (1964). *Phys. Rev.*, **133**, A587.

Gaskill, D. R. (1983). In *Physical metallurgy*, 3rd edn. (eds R. W. Cahn and P. Haasen) p. 271. North-Holland, Amsterdam.

Gleiter, H. (1979). *Acta Metall.*, **27**, 187.

Greenwood, G. W. (1969). In *The mechanism of phase transformations in crystalline solids*, p. 103. Institute of Metals, London.

Guy, A. G. and Oikawa, H. (1969). *Trans. A.I.M.E.*, **245**, 2293.

Hall. M. G. and Haworth, C. W. (1970). *Acta Metall.*, **18**, 331.

Ham, F. S. (1958). *J. Phys. Chem. Solids*, **6**, 335.

Ham, F. S. (1959). *J. Appl. Phys.*, **30**, 915.

Harris, J. E. (1978). *Acta Metall.*, **26**, 1033.

Heckel, R. W. and DeGregorio, R. L. (1965). *Trans. A.I.M.E.*, **233**, 2001.

Hildebrand, F. B. (1949). *Advanced calculus for engineers*. Prentice-Hall, New York.

Hillert, M. and Sundman, B. (1976). *Acta Metall.*, **24**, 731.

Hirth, J. P. and Lothe, J. (1982). *Theory of dislocations*, 2nd edn. Wiley, New York.

Hirth, J. P. and Pound, C. M. (1963). *Condensation and evaporation*. Macmillan, New York.

Horvay, G. and Cahn, J. W. (1961). *Acta Metall.*, **9**, 695.

Howe, J. M. (1988). In *Phase transformations '87* (ed. G. W. Lorimer) p. 637. Institute of metals, London.

Howe, J. M. and Prabhu, N. (1990). *Acta Metall. Mater.*, **38**, 881.

Howe, J. M., Aaronson, H. I., and Gronsky, R. (1986). *Acta Metall.*, **33**, 639, 649.

Howe, J. M., Dahmen, U., and Gronsky, R. (1987). *Phil. Mag. A*, **56**, 31.

Hurley, A. L. and Dayananda, M. A. (1970). *Met. Trans.*, **1**, 139.

Jaeger, W. and Gleiter, H. (1978). *Scripta Metall.*, **12**, 675.

Janssen, M. M. P. (1973). *Met. Trans.*, **4**, 1623.

Janssen, M. M. P. and Rieck, G. D. (1967). *Trans. A.I.M.E.*, **239**, 1372.

Kelly, A. and Nicholson, R. B. (1963). *Progress in Materials Science*, Vol. 10. Pergamon Press, Oxford.

King, A. H. and Smith, D. A. (1980). *Metal Sci.*, **14**, 57.

Komem, Y., Petroff, P., and Balluffi, R. W. (1972). *Phil. Mag.*, **26**, 239.

Laidler, K. J. (1965). *Chemical kinetics*. McGraw-Hill, New York.

Langer, J. S. and Sekerka, R. F. (1975). *Acta Metall.*, **23**, 1225.

Li, C.-Y. and Blakely, J. M. (1966). *Acta Metall.*, **14**, 1397.

Li, C.-Y. and Oriani, R. A. (1968). In *Oxide dispersion strengthening* (eds G. S. Ansell, T. D. Cooper, and F. V. Lenel) p. 431. Gordon and Breach, New York.

Lifshitz, I. M. and Slyozov, V. V. (1961). *J. Phys. Chem. Solids*, **19**, 35.

Liu, Y. C. and Aaronson, H. I. (1970). *Acta Metall.*, **18**, 845.

Malcomb, J. A. and Purdy, G. R. (1967). *Trans. A.I.M.E.*, **239**, 1391.

Martin, J. W. and Doherty, R. D. (1976). *Stability of microstructure in metallic systems*. Cambridge University Press, Cambridge.

Merle, P. and Doherty, R. D. (1982). *Scripta Metall.*, **16**, 357.

Merle, P. and Fouquet, F. (1981). *Acta Metall.*, **29**, 1919.

Merle, P. and Merlin, J. (1981). *Acta Metall.*, **29**, 1929.

Morse, P. M. and Feshbach, H. (1953). *Methods of theoretical physics*, p. 1308. McGraw-Hill, New York.

Mullins, W. W. and Sekerka, R. F. (1963). *J. Appl. Phys.*, **34**, 323.

Nichols, F. A. and Mullins, W. W. (1965). *Trans. A.I.M.E.*, **233**, 1840.

Nolfi, F. V., Shewmon, P. C., and Foster, J. S. (1970). *Met. Trans.*, **1**, 798; 2291.

Okamoto, P. R. and Rehn, L. E. (1979). *J. Nucl. Mats.*, **83**, 2.

Pieraggi, B., Rapp, R. A., van Loo, F. J. J., and Hirth, J. P. (1990). *Acta Metall. Mater.*, **38**, 1781.

Poate, J. M., Tu, K. N., and Mayer, J. W. (eds.) (1978). *Thin film–interdiffusion and reactions*. Wiley, New York.

Prabhu, N. and Howe, J. M. (1990). *Acta Metall. Mater.*, **38**, 889.

Purdy, G. R. and Kirkaldy, J. S. (1963). *Trans. A.I.M.E.*, **227**, 1255.

Raj, R. and Ashby, M. F. (1971). *Met. Trans.*, **2**, 1113.

Raj, R. and Ashby, M. F. (1972). *Met. Trans.*, **3**, 1937.

Rajab, K. E. and Doherty, R. D. (1989). *Acta Metall.*, **37**, 2709.

Ricks, R. A., Porter, A. J., and Ecob, R. C. (1983). *Acta Metall.*, **31**, 43.

Russell, K. C. and Aaronson H. I. (eds.) (1978). *Precipitation processes in solids*. A.I.M.E., New York.

Sagoe-Crentsil, K. K. and Brown, L. C. (1984). *Met Trans.*, **15A**, 1969.

Sagoe-Crentsil, K. K. and Brown, L. C. (1991). *Phil. Mag. A*, **63**, 447.

Seeger, A., Schumacher, D., Schilling, W., and Diehl, J. (eds) (1970). *Vacancies and interstitials in metals*. North-Holland, Amsterdam.

Seidman, D. N. and Balluffi, R. W. (1967). In *Lattice defects and their interactions* (ed. R. R. Hasiguti) p. 911. Gordon and Breach, New York.

Sekerka, R. F. (1968). *J. Crystal Growth*, **4**, 71.

Servi, I. S. and Turnbull, D. (1966). *Acta Metall.*, **14**, 161.

Shewmon, P. G. (1965). *Trans. A.I.M.E.*, **233**, 736.

Shewmon, P. (1989). *Diffusion in solids*, 2nd edn. Minerals, Metals and Materials Society, Warrendale, Pennysylvania.

Shiflet, G. J., Aaronson, H. I., and Courtney, T. H. (1977). *Scripta Metall.*, **11**, 677.

Siegel, R. W., Chang, S. M., and Balluffi, R. W. (1980). *Acta Metall.*, **28**, 249.

Thomson, R. M. and Balluffi, R. W. (1962). *J. Appl. Phys.*, **33**, 803.

Tsumuraya, K. and Miyata, Y. (1983). *Acta Metall.*, **31**, 437.

Tu, K. N. (1985). *Ann. Rev. Mater. Sci.*, **15**, 147.

Tu, K. N. (1986). In *Advances in electronic materials* (eds B. Wessels and G. Y. Chen) p. 147. American Society for Metals, Metals Park, Ohio.

Tu, K. N. and Mayer, J. W. (1978). In *Thin films—interdiffusion and reactions* (eds J. M. Poate, K. N. Tu, and J. W. Mayer) p. 359. Wiley, New York.

Tu, K. N., Mayer, J. W., and Feldman, L. C. (1992). *Electronic thin film science*. Macmillan, New York.

van Loo, F. J. J., Pieraggi, B., and Rapp, R. A. (1990). *Acta Metall. Mater.*, **38**, 1769.

Voorhees, P. W. and Glicksman, M. E. (1984). *Acta Metall.*, **32**, 2001, 2013.

Wagner, C. (1961). *Z. Electrochem.*, **65**, 581.

Weatherly, G. C. (1971). *Acta Metall.*, **19**, 181.

Wendt, H. and Haasen, P. (1983). *Acta Metall.*, **31**, 1649.

Wiedersich, H., Okamoto, P. R., and Lam, N. Q. (1979). *J. Nucl. Mats.*, **83**, 98.

Williams, D. S., Rapp, R. A., and Hirth, J. P. (1981). *Met. Trans.*, **12A**, 639.

Xiao, S. Q. and Haasen, P. (1991). *Acta Metall. Mater.*, **39**, 651.

Yamakawa, K. and Shimomura, Y. (1991). *Scripts Metall. Mater.*, **25**, 2423.

# PART IV

## INTERFACIAL PROPERTIES

# 11

# Electronic properties of interfaces

## 11.1 INTRODUCTION

The electronic properties of interfaces are among the most important for technological exploitation and development. The basic building blocks of semiconductor devices almost invariably involve interfaces between semiconductors, metals, or insulators. In metals grain boundaries are a source of electrical resistance and grain boundaries in superconductors bring with them both beneficial and detrimental effects. In this chapter our aim is to provide a basic understanding of the underlying physics and chemistry controlling the electronic properties of interfaces. We suggest that those readers more interested in the utilization of interfaces in semiconductor devices consult Sze (1981).

One of the most pervasive concepts in the electronic properties of interfaces is 'bending' of the valence and conduction bands. Band bending can occur at all interfaces except those between two metals. In the remainder of this section we shall describe the pheno-menon for the simplest possible case, namely that of a free surface of a semiconductor or insulator. The modifications of the analysis required for an interface with another material are minor.

Consider a semi-infinite semiconductor as shown in Fig. 11.1(a). Let the semiconductor be doped uniformly with a density of $N_d$ donor atoms per unit volume. The majority carriers are therefore electrons and the material is n-type. The analysis for p-type material is very similar, except the direction of bending of the bands is reversed. It is assumed that there is no surface segregation of the dopant, so that the density of dopant is constant throughout the semi-infinite space occupied by the semiconductor. Let the valence band maximum and conduction band minimum be called $E_v$ and $E_c$ respectively, and let $E_F$ denote the Fermi energy. Let the total density of states in the semi-infinite crystal be $N_s(E)$. Let the total density of states for the same number of atoms in an infinite crystal environment be $N_b(E)$. The value of $E_F$ in the bulk is defined by the condition that

$$\int_{-\infty}^{+\infty} n_b(E) f(E)\, dE = \nu \qquad (11.1)$$

where $\nu$ is the average number of electrons per atom in the bulk, $n_b(E)$ is the average density of states per atom in the bulk, and $f(E) = 1/(1 + \exp(E - E_F)/kT)$ is the Fermi–Dirac distribution function at temperature $T$. Since the material is doped n-type $E_F$ will be just below $E_c$ in the bulk. The excess density of states at the surface is defined by $\delta N(E) = N_s(E) - N_b(E)$. If

$$\int_{-\infty}^{+\infty} \delta N(E) f(E)\, dE = 0 \qquad (11.2)$$

then there is no excess charge at the surface and the energy bands are flat as shown in Fig. 11.1(a). But suppose the integral in eqn (11.2) is positive. This would arise if there are states in the energy gap localized at the surface because, as explained in Section 11.2.4, they take weight from both the valence and conduction bands. Since $E_F$ is close to $E_c$

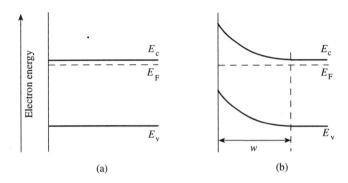

(a)                                                          (b)

**Fig. 11.1** The lowermost energy, $E_c$, of the conduction band, and the uppermost energy, $E_v$, of the valence band as a function of position in the vicinity of a free surface of an n-type semiconductor. In (a) and (b) the free surface is on the left and $E_F$ is the Fermi energy. The absence of any excess charge at the surface in (a) leads to 'flat' energy bands. In (b) localized surface states have trapped an excess number of electrons, which have been taken from n-type dopant atoms in a region of width $w$ below the surface, thereby depleting this region of free carriers. The electrostatic potential that is developed in the depletion region 'bends' the bands.

before any band bending takes place, integrating over the states in the gap not only replenishes the weight lost from the valence band but also includes some weight from the conduction band. Hence the integral is likely to be positive in n-type material if there are states in the energy gap at the surface. In that case there is an excess of electronic charge at the surface, which must have come from ionized donor atoms in a region of width $w$ beneath the surface. In this region, which is called a depletion layer, there is a positive space charge $N_d e$ per unit volume, where $e$ is the magnitude of the electronic charge. If the excess electronic charge at the surface is $Q_s$ then overall charge neutrality requires

$$Q_s = - N_d e w \tag{11.3}$$

The positive space charge of the depletion layer sets up an electrostatic potential which 'bends' the bands as shown in Fig. 11.1(b). The effect of this band bending is to reduce the excess surface charge $Q_s$. As we shall see shortly the reduction in $Q_s$ is so great as to eliminate it almost completely! But although the surface becomes almost charge neutral after band-bending takes place the very small amount of residual excess charge, $Q_s$, is crucial, because without it there would be no depletion layer and hence no band bending at all.

Let $V(x)$ be the electrostatic potential at $x$ from the surface, due to the space charge of the depletion layer. The problem has been reduced to 1D by assuming that the surface states are distributed uniformly within the surface. The potential, $V(x)$, and electric field, $- dV/dx$, are set to zero at the edge of the depletion layer, $x = w$. Poisson's equation then determines the variation of the potential and electric field within the depletion layer:

$$\frac{d^2 V}{dx^2} = - \frac{N_d e}{\varepsilon \varepsilon_o} \quad \text{for } 0 \leqslant x \leqslant w, \tag{11.4}$$

where $\varepsilon$ is the dielectric constant of the semiconductor and $\varepsilon_o$ is the permittivity of free space. Integrating this equation we find that the electric field, $\Xi = - dV/dx$, varies linearly with $x$ inside the depletion layer:

$$\Xi = -\frac{N_d e}{\varepsilon\varepsilon_o}(w - x), \tag{11.5}$$

and the potential varies quadratically:

$$V = -\frac{N_d e}{2\varepsilon\varepsilon_o}(w - x)^2 \tag{11.6}$$

The additional potential energy for an electron in the depletion layer is $-eV$, which is positive. That is why the bands bend up in energy in Fig. 11.1(b), which depicts the energies of electrons. The maximum upward shift of the bands is reached at the surface $(x = 0)$ where

$$V_{max} = -\frac{N_d e w^2}{2\varepsilon\varepsilon_o}$$

$$= -\frac{Q_s^2}{2\varepsilon\varepsilon_o N_d e} \tag{11.7}$$

where we have used eqn (11.3). The energy $-eV_{max}$ (which is positive) must be supplied in order for an electron to pass through the depletion layer and reach the surface. At an interface $V_{max}$ is a potential barrier to the passage of an electron across the interface. Using eqn (11.3) again we may also obtain the following useful expression for w:

$$w = \sqrt{\frac{2\varepsilon\varepsilon_o |V_{max}|}{N_d e}}. \tag{11.8}$$

The depletion layer is always larger than typical interatomic separations by several orders of magnitude. For example, with $\varepsilon = 10$, $V_{max} = 0.5$ V, $\varepsilon_o = 8.85 \times 10^{-12}$ F m$^{-1}$, $N_d = 10^{22}$ m$^{-3}$, and $e = 1.6 \times 10^{-19}$ C we obtain $w = 2.4 \times 10^{-7}$ m. In the following sections it is important to keep in mind that the length scale relating to the formation of bonds at interfaces is much smaller than that relating to the formation of depletion layers.

The excess surface charge, $Q_s$, is given by eqn (11.3), and for the above values of $N_d$ and $w$ we obtain $Q_s = 2.4 \times 10^{15}$ electrons m$^{-2}$. For a Si(111) surface there are approximately $0.8 \times 10^{19}$ atoms m$^{-2}$. Therefore $Q_s = 3 \times 10^{-4}$ electrons per surface atom. It is clear, therefore, that the surface is almost charge neutral once band bending has occurred.

Finally, in this section we introduce the concept of 'pinning' of the Fermi energy by surface states. Suppose, for simplicity, that there is just one energy, $E_s$, in the gap at which there are surface states:

$$N_s(E) = Z\delta(E - E_s) \tag{11.9}$$

where there are $Z$ such centres in the surface per unit area. The Fermi level is said to be pinned at $E_s$ when these surface states are only partially occupied. If further electrons are transferred to the surface they occupy states at $E_s$ that were formerly unoccupied: the Fermi energy remains at $E_s$. On the other hand, if electronic charge is removed from the surface, some of the states at $E_s$ that were formerly occupied become unoccupied and again the Fermi energy remains at $E_s$. The condition for pinning of the Fermi level is that the flow of electronic charge, to or from the surface, involves changing the occupancy of only those surface states at the energy where $E_F$ is pinned.

We can estimate the minimum value of $Z$ required to pin the Fermi energy at $T = 0\,\text{K}$ as follows. (The effect of a finite temperature is negligible in this order of magnitude estimate.) If we assume that each surface state can absorb one electron then the maximum electronic charge that can be stored at the surface per unit area is $-Ze$. If the Fermi energy coincides with $E_s$ at the surface after band bending takes place, then $eV_{max} = E_F - E_s$, where $E_F$ is the Fermi energy before band bending takes place. For $E_s$ in the middle of the gap $eV_{max}$ is about half the band gap energy. The amount of charge in the depletion layer is $N_d ew$, which, using eqn (11.8), is $\sqrt{2\varepsilon\varepsilon_o N_d eV_{max}}$. Overall charge neutrality requires:

$$Ze = \sqrt{2\varepsilon\varepsilon_o N_d (E_F - E_s)}. \tag{11.10}$$

Using the same values for these parameters as before, together with $E_F - E_s = 0.5\,\text{eV}$, we obtain $Z = 2.4 \times 10^{15}\,\text{m}^{-2}$, which is the same as $Q_s$. Therefore only about 1 in 1000 atoms at the surface, or more, has to be associated with a surface state in order to pin the Fermi energy. The same result applies to interfaces as well.

## 11.2  METAL–SEMICONDUCTOR INTERFACES

### 11.2.1  Introduction

The vast majority of metal–semiconductor interfaces act as diodes since the electric current they pass depends exponentially on the forward bias. This rectifying interface is known as a Schottky contact, after Schottky (1938) who first identified the origin of the rectification as the existence of a depletion layer in the semiconductor. The current–voltage relation at some metal–semiconductor interfaces satisfies Ohm's law and they are consequently known as ohmic contacts. An ohmic contact has a sufficiently low resistance for the current to be limited by the resistance of the bulk semiconductor rather than by properties of the contact itself.

In a sense the electrical properties of metal–semiconductor interfaces should be the most simple to understand of all interfaces involving semiconductors because only one depletion layer is involved. As we shall see in Section 11.5, the band bending at grain boundaries in semiconductors may be described as two 'back-to-back' Schottky barriers. Similarly, in Section 11.3.1 it will be seen that the band bending at semiconductor heterojunctions is also closely related to that of two Schottky barriers. Yet despite the relative simplicity of Schottky barriers there is still no predictive theory, although some important physics has been uncovered. As will shortly become apparent this is largely because the experimental data that have accumulated over the past 50 years consist almost entirely of average barrier heights for polycrystalline specimens of varying degrees of contamination. It is only much more recently that experiments on single, structurally characterized interfaces in high-purity materials have been carried out, and they form the basis of much of our current understanding; they also present the greatest challenges to theory.

Our emphasis is on understanding the factors that control the Schottky barrier height. Readers more interested in transport properties are guided to the excellent books by Sze (1981) and Rhoderick and Williams (1988).

### 11.2.2  The Schottky model

The reason for the rectifying behaviour of a Schottky contact is a potential barrier on the semiconductor side of the contact. The Schottky model for the formation of the

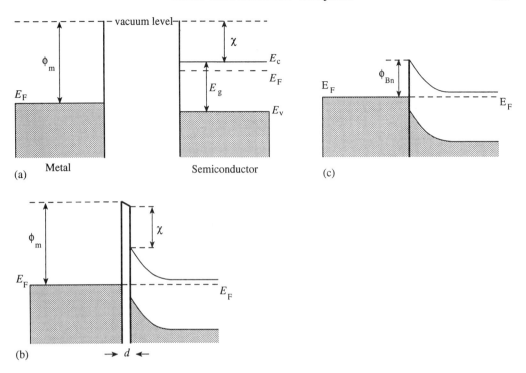

**Fig. 11.2** To illustrate the alignment of the energy bands during the formation of a Schottky barrier between a metal and an n-type semiconductor. (a) The metal and semiconductor are well separated. The vacuum level is the common zero of energy. The shaded areas indicate occupied electronic states. The work function in the metal is $\phi_m$, and the electron affinity in the semiconductor is $\chi$. In (b) the separation is reduced to $d$, which is sufficiently small for some electrons to tunnel across the gap, thereby creating a depletion layer in the semiconductor, with attendant band bending. In (c) intimate contact is achieved and the potential barrier, $\phi_{Bn}$, is equal to $\phi_m - \chi$.

barrier for an n-type semiconductor may be understood with reference to Fig. 11.2. In Fig. 11.2(a) we see the electronic energy bands in the metal and semiconductor prior to any interaction between them. The common zero of energy is the vacuum level. The metal comprises a band that is filled to the Fermi energy, $E_F$, and the work function of the metal is denoted by $\phi_m$. The work function is the energy required to remove an electron from the bulk of the metal to a point just outside the metal. It has two contributions: a bulk contribution and a surface contribution due to a dipole at the surface. Different surfaces of a metal are associated with different atomic arrangements and hence different dipoles. Therefore, there is generally a significant variation in the work function of a metal, depending on the orientation of the surface from which the electron exits the metal. The electron affinity in the semiconductor is denoted by $\chi$. It is defined as the energy difference between the vacuum level and the bottom of the conduction band. It is assumed that there are no states on the semiconductor surface so that there is no band bending within the semiconductor prior to bonding to the metal. It is also assumed that both the metal and semiconductor are clean.

The Fermi energy is the electrochemical potential of electrons in the metal and semi-conductor. As the separation, $d$, between the metal and semiconductor surfaces is decreased (Fig. 11.2(b)) an electric field is set up, equal to the difference in Fermi energies divided by $d$. As $d$ decreases electrons tunnel across the gap to equalize the difference in Fermi energies. The donors in a depletion layer adjacent to the surface of the semiconductor are ionized as electrons leave the semiconductor for the metal. The depletion layer thickens until eventually (Fig. 11.2(c)), when intimate contact is achieved, the system is in equilibrium with the same electrochemical potential throughout. Assuming the density of dopant atoms is constant, the potential in the depletion layer varies quadratically with distance from the interface, as shown in eqn (11.6). The electronic charge that is transferred to the metal resides on the metal surface. It is the dipole that is set up by the excess electronic charge on the metal surface and the positive space charge of the depletion layer that equalizes the Fermi energies in the metal and semiconductor.

In the depletion layer there is a steep gradient in the concentration of electron carriers. Immediately adjacent to the metal the concentration of electron carriers is very low owing to the large difference, $\phi_{Bn}$, between the Fermi energy and the bottom of the conduction band. The difference $\phi_{Bn}$ is called the potential barrier, or simply the Schottky barrier. Since $\phi_{Bn}$ is much greater than $kT$ the concentration of free carriers there is $N_d \exp - (\phi_{Bn}/kT)$, where $N_d$ is the concentration of donor atoms. At the other side of the depletion layer the concentration of free carriers is $N_d$. This very large concentration gradient sets up a diffusional drift of free carriers towards the metal. At equilibrium the diffusional flux is balanced by the flux away from the metal due to the electrostatic repulsion from the negative sheet of charge on the surface of the metal, which is incompletely screened within the depletion layer. In the semiconductor outside the depletion layer the concentration of free carriers is everywhere $N_d$ and the electric field due to the dipole at the interface is zero. Therefore, the whole system is in equilibrium.

Relative to the top of the valence band at the interface we may express the potential barrier, $\phi_{Bn}$, as follows:

$$\phi_{Bn} = E_g - E_F, \tag{11.11}$$

where $E_g$ is the energy of the gap. From Figs 11.2 (a) and (b) we see that we can also express $\phi_{Bn}$ as follows:

$$\phi_{Bn} = \phi_m - \chi. \tag{11.12}$$

A similar analysis for a p-type semiconductor yields a potential barrier for holes given by

$$\phi_{Bp} = E_F$$
$$= E_g - (\phi_m - \chi) \tag{11.13}$$

where the Fermi energy is again measured with respect to the top of the valence band at the interface. From eqns (11.12) and (11.13) we see that

$$\phi_{Bn} + \phi_{Bp} = E_g. \tag{11.14}$$

In an ohmic contact to an n-type semiconductor either the work function of the semiconductor is greater than that of the metal, in which case there is no barrier at all, or, more commonly, the level of doping in the semiconductor is so high that the depletion layer is sufficiently narrow for electrons to tunnel through the potential barrier.

It is clear that in the Schottky model the metallic work function and the electron affinity of the semiconductor (or at least their difference) are assumed to remain unchanged on forming the contact. It is also assumed that intimate contact between the metal and semiconductor is achieved, and that there are no intervening layers, such as reaction layers or oxide layers.

A key feature of the model is the expectation that the barrier height is proportional to the metal work function, as seen in eqn (11.12). Experimentally it is found that the barrier height is only weakly dependent on the metallic work function, and for some semiconductors it is almost independent of $\phi_m$. In Fig. 11.3 we see that measured Schottky barriers on Si and GaAs are independent of the metal to within about 0.1 eV. But the Schottky barrier for more ionic semiconductors, such as ZnS and ZnSe, increases as the electronegativity of the metal increases. In general a metal with a large electronegativity tends to have a large work function, but there is no precise relationship between them. This is partly because the concept of electronegativity itself is not very well defined for a solid (it is really an atomic property), and partly because the work function varies with the orientation of the surface normal. The variation of the work function from one surface to another of the same metal can be of the order of 1 eV. The assumption in the Schottky model that the work functions of the metal and semiconductor remain unchanged after the interface is formed seem even more implausible when one considers the interfacial dipoles that are likely to be formed as a result of the localized charge transfer accompanying the formation of bonds between the metal and semiconductor.

With these limitations in the Schottky model it is perhaps no surprise that it fails to agree with experimental observations in many cases. The fact that the observed barrier heights are often independent of the metal work function, especially for the small band gap semiconductors, suggests that the interior of the semiconductor is *screened* from what is happening at the interface. The Fermi energy is somehow pinned at some value in the energy gap at the interface. Various models have been proposed to explain this, all of which are modifications in one way or another of the original idea due to Bardeen (1947).

### 11.2.3 The Bardeen model

Bardeen noted that electronic states at the interface, which were capable of absorbing or supplying charge, could explain the apparent insensitivity of the barrier height to the work function (or electronegativity) of the metal. Any charge redistribution across the

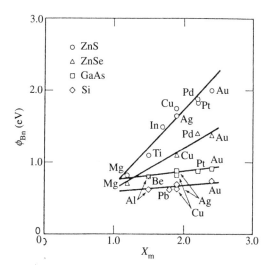

**Fig. 11.3** Experimental values of the Schottky barrier heights for four semiconductors in contact with various metals. $X_m$ is the electronegativity of the metal in the Pauling–Gordy scale. (From Louie *et al.* (1977).)

interface would be effected by adjusting the occupancy of these interfacial states at the Fermi energy. Provided the density of interfacial states were sufficiently high the charge redistribution would take place without altering the Fermi energy. In this way the Fermi energy is pinned and the same barrier height is obtained for any metal. The numerous modifications of this basic idea differ in the origin ascribed to the localized states. It may appear odd at first sight that localized interfacial states may appear side by side with, and at the same energies as, delocalized states in the metal. A thin insulating layer at the interface, originating from oxide films on the metal and semiconductor surfaces, would enable localized states at the semiconductor–oxide interface to be separated spatially from the metal. But we shall see in the next section that there are states which are localized in the semiconductor, normal to the interface, even when direct contact is established between it and the metal.

The nature, distribution and origin of the interfacial states has been a subject of considerable debate. It seems that the states must, in some sense, be a property of the semiconductor because of the insensitivity of the barrier height to the electronegativity of the metal. This is particularly true of the narrow band gap semiconductors, such as Si, Ge, and GaAs. But for semiconductors with wider band gaps the barrier height is dependent on the metal, as seen in Fig. 11.3. One suggestion for the origin of interface states is that they are the same states found at the interface between the semiconductor and the thin oxide surface layer prior to bonding to the metal. But this would not explain the origin of states at interfaces formed between clean metal and semiconductor surfaces under UHV conditions. From both UHV experiments on clean surfaces and accurate total energy calculations (e.g. Zangwill 1988) we know that, sometimes, clean non-defective surfaces relax and reconstruct in such a way as to eliminate states in the gap. Furthermore, when two clean surfaces are bonded together to form an interface the change in bonding is so great there is no reason to believe that the electronic structure of the free surface is at all related to that of the interface.

### 11.2.4 Metal-induced gap states (MIGS)

Heine (1965) pointed out that Bloch states of the metal with energies coinciding with the gap of the semiconductor decay exponentially into the semiconductor. They become Bloch states of the semiconductor with complex wave vector components normal to the interface. These 'metal-induced gap states' (MIGS) have a significant amplitude only a few layers from the interface, and at those layers they give rise to a continuum of states in the gap. This is seen clearly in Fig. 11.4 where we show Louie and Cohen's (1976) calculation of the charge density normal to an Al–Si (111) interface in the energy range of the gap. The charge density has been averaged parallel to the interface. It was calculated self-consistently in the local density approximation by treating the metal as jellium of the same electron density as Al. It is seen that by the third (111) double layer of the Si the charge density in the energy range of the gap has decayed to almost zero. This is also seen in the layer projected local densities of states, shown in Fig. 11.5, where the six regions are identified in Fig. 11.6. In the third double layer from the interface, region VI, we see that the gap (between 0 and 1.2 eV) is almost clear of states. But as we approach the interface, regions V and IV, there is an increasing, continuous density of MIGS throughout the gap. This phenomenon is called metallization of the semiconductor.

It is helpful to compare a metal-induced gap state with the decay of a Bloch state at the surface of a metal. A Bloch state of the metal decays exponentially into the vacuum

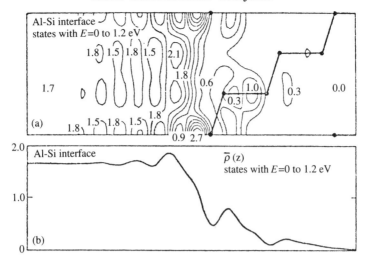

**Fig. 11.4** (a) Electronic charge density contours for interfacial states in the Si energy gap at an Al–Si (111) interface. The direction normal to the page is [1–10] and the interface runs from top to bottom in the middle of the panel. The Al is modelled by jellium. The Si atoms are represented by dots. (b) The integrated charge density of states in the Si energy gap normal to the interface. Notice that the states decay within a few Si bond lengths in the Si. These states in the energy gap are MIGS. (From Louie and Cohen (1976).)

at a free surface because the component of the wave vector normal to the surface is complex in the vacuum. At the surface the Bloch function inside the metal is matched to a linear superposition of eigenstates of the vacuum, which are plane waves parallel to the surface with an amplitude that decays exponentially normal to the surface. At a metal–semiconductor interface a Bloch state of the metal with an energy coinciding with the gap of the semiconductor cannot propagate in the semiconductor, and decays exponentially. In the semiconductor it becomes a state with a complex wave vector component normal to the interface. This metal-induced gap state is a linear combination of Bloch states of the bulk semiconductor with amplitudes that decay exponentially normal to the interface.

There is a sum rule which states that at any site the integrated local density of states, including occupied and unoccupied states, is a constant. The constant is just the zeroth moment of the local density of states. The weight associated with the valence and conduction bands at each site is reduced by an amount which equals the weight associated with metal induced gap states. States in the gap are made up of linear combinations of bulk valence and conduction band states. They are composed primarily of those band states that are closest in energy. Thus states towards the bottom of the gap tend to be derived from valence band states, and those towards the top from conduction band states. Overall charge neutrality of the semiconductor therefore requires that MIGS up to some energy, $\phi_0$, in the gap are occupied. The energy $\phi_0$ is called the *charge neutrality level* because if MIGS higher (lower) than $\phi_0$ are occupied then there is an excess (deficit) of electronic charge in them, and the semiconductor as a whole is charged. The charge neutrality level corresponds to the energy at which MIGS change over from being derived mainly from the valence band to being derived mainly from the conduction band. For

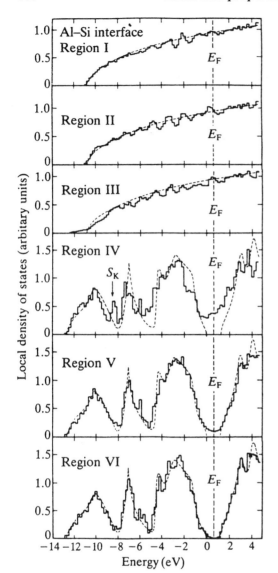

Local density of states (arbitary units)

**Fig. 11.5** Local densities of states for six regions parallel to the Al-Si(111) interface. The regions are identified in Fig. 11.6. Each local density of states is averaged parallel to the interface. The broken line shows the corresponding bulk densities of states in Al (jellium) and Si. Notice the MIGS in regions IV and V, and that they have almost completely decayed in region VI. (From Louie and Cohen (1976).)

many semiconductor band structures this is near the middle of the indirect gap (Tersoff 1985).

In the absence of MIGS overall charge neutrality requires that the charge, $Q_m$, transferred to the metal surface, plus the space charge, $Q_d$, of the depletion layer equal zero. This charge transfer is a result of the difference in electronegativities of the metal and semiconductor, and it may also be expected to vary with the atomic structure and chemistry of the interface.

With a continuum of MIGS at the interface the overall charge balance becomes:

$$Q_{MIGS} + Q_d + Q_{is} = 0, \qquad (11.15)$$

where $Q_{MIGS}$ is the charge excess or deficit in occupied MIGS. $Q_{is}$ is the electronic charge in occupied interface states. The Fermi energy at the interface gravitates towards the

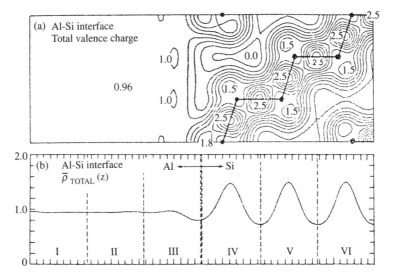

**Fig. 11.6**  (a) Electronic charge density contours of all the occupied states viewed in a [1–10] direction, normal to an Al–Si(111) interface. See caption to Fig. 11.4. (b) Integrated charge density, over all occupied states normal to the interface. Notice the Friedel oscillations in the charge density on the Al side, and the net transfer of charge from Al to Si. (From Louie and Cohen (1976).)

charge neutrality level as the following thought experiment shows. Consider a metal–semiconductor interface for which the one-electron eigenstates of the bicrystal have been determined self-consistently. We now imagine an infinitesimally thin barrier along the interface which prevents electrons from flowing between the metal and semiconductor. We occupy the eigenstates of the bicrystal in the metal side to a level $E_F$ such that the metal is charge neutral. (The error in $E_F$ due to the charge located in the tails of MIGS which have spilt over into the semiconductor is negligible). Similarly we occupy eigenstates of the bicrystal in the semiconductor side to a level $\phi_0$ such that the semiconductor is charge neutral. It is at this stage that the depletion layer in the semiconductor is required in order to effect the band bending which ensures that only those MIGS with an energy up to $\phi_0$ are occupied. Now suppose that $E_F$ of the metal lies above $\phi_0$ of the semiconductor. There is an unscreened dipole across the barrier equal to $(E_F - \phi_0)$. On removing the barrier electrons flow from the metal to the semiconductor in order to screen the dipole, and $E_F$ approaches $\phi_0$. A similar argument applies if the Fermi energy of the metal is below the charge neutrality point of the semiconductor. In the semiconductor $\phi_0$ plays a role analogous to the Fermi energy in the metal.

The resultant screened dipole, $D$, may be written as

$$D = (E_F - \phi_0)/S \tag{11.16}$$

where $S$ is typically about 0.1 for small band gap semiconductors, $D$ is around 1 eV and $E_F - \phi_0$ is about 0.1 eV (Flores *et al.* 1989).

In a zeroth-order approximation the charge neutrality level is metal independent, if it is determined only by the band structure of the bulk semiconductor for complex wave vectors. In this approximation the charge neutrality level is sometimes called 'universal' or 'intrinsic'. In practice, however, the charge neutrality level is metal dependent and it also depends on the atomic structure of the interface; it is then called 'extrinsic' (Flores *et al.* 1989). For example, at a clean surface the charge neutrality level coincides with the centre of gravity of states in the gap. Different surfaces of the same semiconductor in general have different densities of states in the energy gap, and therefore the charge neutrality level changes from one surface to another. In the case of metal–semiconductor interfaces $\phi_o$ is determined by the density of states induced by the metal in the semiconductor at all energies. Changes to the local density of states in the valence band are just as important as changes in the gap because one has to integrate over both to find the charge neutrality level. In general, we would expect this induced density of states to be dependent on the type of metal and on the atomic structure and bonding at the interface. The charge neutrality level will be the 'intrinsic' value only for induced densities of states in the semiconductor that are broad and featureless.

As we saw in Section 11.1 the minimum density of interfacial states required to pin the Fermi energy is remarkably small, i.e. about one state per electron volt for every 1000 atoms in the plane of the interface. With higher densities, charge flow between the metal and semiconductor involves only the interfacial states at or near the Fermi energy. The depletion layer is then constant in size and the barrier is given by

$$\phi_{Bn} = E_g - \phi_o. \tag{11.17}$$

Here we have ignored the small difference between $E_F$ and $\phi_o$ at the interface. Thus, the amount of charge in the depletion layer, and the potential barrier, are independent of the work function of the metal, provided $\phi_o$ is the intrinsic value. This is the Bardeen limit. In this limit (the zeroth order approximation described above) details of the atomic structure and bonding at the interface are unimportant. In effect, MIGS screen the interior of the semiconductor from the atomic scale interfacial dipoles (Tersoff 1984). The very low density of interfacial states required to pin the Fermi energy translates, in the Thomas–Fermi approximation, into a density of only one interfacial state/eV for every 50 atoms to give a screening length of 3 Å. The Bardeen limit is in contrast to the Schottky limit where there are no interfacial states of any kind to pin the Fermi energy. In practice metal–semiconductor interfaces fall between these two extremes.

Although MIGS are always present their decay length into the semiconductor decreases rapidly as the band gap increases (Tejedor *et al.* 1977, Louie *et al.* 1977). The less they penetrate the semiconductor the less they will be involved in charge transfer between atoms of the semiconductor. In other words, the less effective they will be at screening the interior of the semiconductor from local contributions to the interfacial dipole (Tersoff 1984). We have already noted that the distribution of MIGS through the gap is expected to depend, in general, on the structure of the interface. Similarly, the extent to which MIGS penetrate a given semiconductor is expected to depend also on the interface normal since it alters the matching of the wave functions of MIGS at the interface. Indeed, if the interface is faceted or stepped we would expect the local position of the Fermi energy, relative to the band gap edges, to vary along the interface. These variations would reflect the varying induced densities of interfacial states as well as the varying effectiveness of MIGS at screening local dipoles at the interface.

### 11.2.5 The defect model

Defects at interfaces have also been proposed as the origin of interface states that pin the Fermi level (Spicer *et al.* 1980). Dislocations, steps, vacancies, anti-site defects etc., at or near the interface in the semiconductor have all been invoked as the source of interfacial states. If the defects are isolated from each other they may be expected to be associated with discrete energies in the gap. Occupation of these states is unlikely, therefore, to pin the Fermi energy at $\phi_0$ but at some other energy which does not lead to charge neutrality in each layer. The effectiveness of these states at pinning the Fermi energy decreases dramatically as the defect centre comes within the decay length of MIGS. That is because the continuum density of states of MIGS is usually much higher than that of isolated defects, and the Fermi energy is again pinned near the charge neutrality level, $\phi_0$. In effect, MIGS screen the local dipoles that occupation of defect states causes. Provided the decay length of MIGS is more than 1 or 2 Å, defects at the interface plane, such as misfit dislocations and steps, will not be effective pinning centres, and only bulk-like ('native') defects further in the semiconductor, such as vacancies and anti-site defects, will be effective. The implication of this conclusion is that the barrier height is not expected to be sensitive to the detailed atomic structure of the interface, but is controlled by Fermi level pinning due to defects some distance from the interface. As we shall see below, this disagrees with experimental observations and self-consistent calculations. Although defect states may be very effective at pinning the Fermi energy at free surfaces their effectiveness at metal-semiconductor interfaces is severely reduced through screening by MIGS.

## 11.2.6 The development of the Schottky barrier as a function of metal coverage

There has been a considerable amount of experimental work on following the development of a Schottky barrier by vapour depositing increasing amounts of metal on a semiconductor substrate. The metal coverage ranges from sub-monolayer to several monolayers. Mönch (1990) has reviewed these experiments which are all based on surface sensitive techniques, such as photoelectron spectroscopy or scanning tunnelling microscopy (STM). Cao *et al.* (1987) observed the initial stages of the Schottky barrier formation at GaAs (110) surfaces as a function of metal coverage at low temperatures. GaAs (110) surfaces produced by cleavage under UHV conditions are free of gap states, and there is no band bending. Cao *et al.* achieved metal coverages as low as $10^{-3}$ monolayers, and their experimental data for In, Al, Ag, and Au deposited at 83 K on (110) surfaces of p-type GaAs are shown in Fig. 11.7. For coverages up to about 0.3 monolayers it is found that the measured positions of the Fermi energy can be fitted by assuming that each metal atom is associated with a single donor level of 0.87, 0.76, 0.68, and 0.49 eV above the valence band for In, Al, Ag, and Au respectively. These fits are shown by the solid lines. Above 0.3 monolayers it is seen that the measured Fermi energies start to converge to the same value, which is consistent with the measured barrier height at bulk metal GaAs contacts.

The change of behaviour at about 0.3 monolayers coverage suggests a change of mechanism, and a plausible explanation is that islands of metal atoms start to form. As noted by Tersoff (1984) isolated metal atoms chemisorbed on the semiconductor surface behave like isolated defects. The wave functions of adatoms with energy levels in the band gap decay into the semiconductor. The asymmetric distribution of electronic charge of each adatom induces a dipole at the interface, which affects the position of the Fermi energy at the surface of the semiconductor. If the adatom state lies above (below) the charge neutrality level, $\phi_0$, then it has acceptor (donor) character because it is positively (negatively) charged. But, it is only when a continuum of MIGS forms, through the formation of islands of metal atoms and the concomitant introduction of metallic screening, that the Fermi level is pinned at an energy consistent with the Schottky barrier measured for bulk metal–semiconductor contacts.

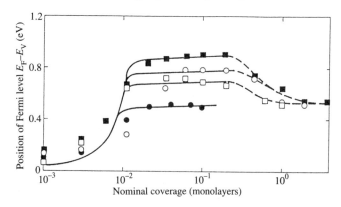

**Fig. 11.7** Measured positions of the Fermi energy, $E_F$, above the valence band maximum, $E_v$, as a function of the nominal amount of metal coverage on clean p-GaAs (110) surfaces. Experimental data (from Cao *et al.* 1987) is shown for In (black squares), Al (open circles), Ag (open squares), and Au (black circles). The curves are calculated for surface donor states at 0.87, 0.76, 0.68, and 0.49 eV respectively above the top of the valence band. Notice that at more than a monolayer coverage the position of the Fermi energy converges rapidly to a single value. (From Mönch (1990).)

An STM experiment was carried out by First *et al.* (1989) to make real space observations of the electronic states associated with single Fe clusters grown by molecular beam epitaxy (MBE) on a GaAs (110). The Fe atoms diffused on the surface and formed isolated clusters, varying in size from 100 to $1500\,\text{Å}^3$, corresponding to 9–127 atoms, assuming they adopt a b.c.c. structure. After placing the STM tip above a cluster, the tunnelling current was measured as a function of voltage. When the tip was placed above a patch of the GaAs (110) surface that was remote from any Fe clusters the differential conductance, $dI/dV$, was zero throughout the bias range corresponding to the gap of GaAs. This is in agreement with the absence of any surface states on clean GaAs (110). The conductance was positive at all voltages for tunnelling into a large cluster ($1150\,\text{Å}^3$), indicating that the cluster was metallic. However, for smaller clusters ($150–400\,\text{Å}^3$) semiconducting behaviour was found, where the differential conductance was zero for a finite range around zero bias. For these small clusters, gaps of 0.1 to 0.5 eV were observed. This is a quantum size effect where the spacing between energy levels in the cluster decreases with increasing number of atoms. First *et al.* estimated that clusters containing more than about 35 atoms were metallic. Therefore, pinning of the Fermi level by clusters smaller than about 35 Fe atoms corresponds to the defect model of the origin of interface states, while pinning by MIGS occurs with clusters larger than about 35 Fe atoms.

One of the most remarkable observations by First *et al.* (1989) concerned the energy dependence of the decay length of MIGS in the GaAs surface surrounding the larger metallic clusters. This was done by measuring the variation of the differential conductance at different voltages in the GaAs band gap. The differential conductance is proportional to the local density of states. Spatial variations of $dI/dV$ at a voltage $V$ reflect the spatial variations of the local density of gap states at an energy, eV, relative to the Fermi energy. MIGS at 0.15 eV (0.25 eV) below the conduction band minimum were observed to decay over about $17\,\text{Å}$ ($11\,\text{Å}$). The smallest decay length measured was about $3\,\text{Å}$ for states in the middle of the band gap. The results are shown in Fig. 11.8

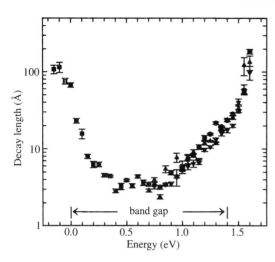

**Fig. 11.8** The decay length of MIGS surrounding Fe clusters on GaAs (110) surfaces as a function of the energy of the MIGS. The valence band maximum is at the zero of energy. Data from p-type material are represented by the circles and triangles, and from n-type by squares. (From First *et al.* (1989).)

where the energy zero has been shifted to the top of the valence band. These observations provide direct evidence of MIGS in the semiconductor and the energy dependence of their spatial decay.

### 11.2.7 Schottky barriers on Si

Barrier heights are determined from current–voltage (*I–V*), capacitance–voltage (*C–V*), and photoresponse–wavelength characteristics. Barrier heights obtained with *I/V* and photoelectric methods tend to be substantially smaller than those measured by *C/V* techniques (Rhoderick and Williams 1988). The most studied Schottky barriers are those to Si, owing to the dominance of Si in device technology. It is found that chemical reactions and intermixing often take place when intimate contacts are made between Si and metals. This complicates the interpretation of experimental measurements considerably, and is one reason for a lack of reproducibility even in experiments carried out under UHV conditions. These reactions, which result in the formation of silicides, often take place at room temperature and sometimes even at liquid nitrogen temperatures. Contacts to surfaces that have been oxidized or chemically etched are usually even less reproducible owing to the variety of surface preparations and heat treatments ('ageing') that is given to the specimens.

Figure 11.9 shows measured barrier heights for intimate contacts between vapour-deposited metal films and cleaved Si (111) surfaces as a function of the metal work function. The barrier heights fall between 0.4 and 0.85 eV. Metals with large work functions, such as Au and Pt, tend to be associated with large barriers, while those with low work function, such as Mg and Ca, tend to be associated with small barriers. Note the considerable scatter in the data for a given metal. The results of Thanailakis and Rasul (1976) are more reliable because they were obtained entirely under UHV conditions. In most cases the metal film is polycrystalline with a grain size and texture that depend on the deposition conditions. Sometimes the metal overlayer is in an epitaxial relationship with the substrate, and then the presence of elastic coherency strains in the overlayer, or anticoherency misfit dislocations at the interface, may be expected to have an influence on the measured barrier height, because elastic strains alter the band structure of both the metal and semiconductor. But in general little attention has been paid to the atomic

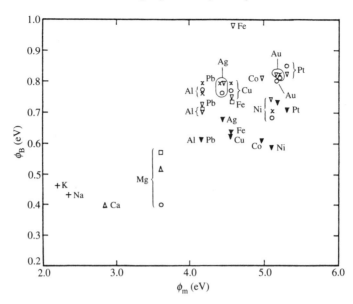

**Fig. 11.9** Measured barrier heights for intimate contacts between vapour deposited metal films and cleaved Si (111) surfaces as a function of the metal work function. (From Rhoderick and Williams (1988).)

structure of metal–semiconductor interfaces, and therefore there have been very few opportunities for quantitative comparisons with models and simulations.

The work of Tung (1984) was, therefore, a major advance because it established a direct dependence of the Schottky barrier height on the atomic structure of the interface. In Section 4.4.4 we discussed the structures of the type A and B interfaces between $NiSi_2$ and Si(111). In the type A orientation the $NiSi_2$ grows in a parallel orientation to the Si(111) substrate. In the type B orientation the $NiSi_2$ is rotated by 180° about the common [111] normal. As discussed in Section 4.4.4 experimental and theoretical studies indicate that the structures of type A and B interfaces are as shown in Figs. 4.58(a) and 4.59(a) respectively. The first and second neighbours of atoms at both interfaces are the same. Differences between them appear only at third neighbours. Tung (1984) found that the n-type Schottky barriers, $\phi_{Bn}$ (see eqn (11.11)), in the type A and B interfaces are 0.65 and 0.79 eV respectively.

Das *et al.* (1989) carried out local density functional calculations of the type A and B interfaces. The Schottky barrier is determined by the interfacial dipole that aligns the Fermi energies in the Si and $NiSi_2$ phases. The absolute values of the calculated barriers differed from experiment by about 0.4 eV, which is due to a non-local correction to the local density approximation (see the discussion at the end of Section 3.2.1). However, the calculated difference in the barrier heights was 0.14 eV, in exact agreement with experiment, and the non-local error arising from the local density approximation cancelled in the difference. In both interfaces Das *et al.* found a partially occupied interfacial state which arose because Ni atoms at the interface are 7-fold rather than 8-fold coordinated by Si atoms. The difference between the A and B Schottky barriers was ascribed to a more complete screening of this state in the type A interface than in the type B.

Schottky barriers of epitaxial interfaces between $NiSi_2$ and Si(100) have been corre-lated with transmission electron microscopy observations of their structure by Tung *et al.* (1991). The $NiSi_2$ grows in a parallel orientation on the Si substrate, but at low temperatures the interface is faceted on {111} planes. The facets are typically 50–100 Å wide and over 500 Å long. The {111} facets have the type A structure described above. Tung *et al.* found that the n-type Schottky barrier of this faceted (100) interface was 0.65 eV, the same as the uniform type A (111) interface. By varying the growth conditions and subsequent heat treatment they were also able to grow interfaces which were not faceted onto {111} planes but remained parallel to (100). The n-type Schottky barrier measured for this flat (100) interface was 0.40 eV. This is 0.25 eV lower than the Schottky barrier of (111) type A and almost 0.40 eV lower than that of (111) type B.

It is interesting to note that in the faceted (100) interface $\frac{1}{4}\langle 111 \rangle$ dislocations may exist at {111} facet junctions (see Section 1.6.2). However, it appears that these dislocations have no influence on the Schottky barrier since it is the same as that of the uniform type A interface. Also, in the flat (100) interface Tung *et al.* observed the $\alpha$ and $\beta$ complex variants (see Section 1.6.2) separated by $\frac{1}{4}\langle 111 \rangle$ dislocations. These disloca-tions are associated with partial steps of $\frac{1}{4}[100]$ height.

Tung (1993) has reviewed Schottky barrier measurements at other epitaxial interfaces and given further evidence of the dependence of the Schottky barrier on interfacial structure.

### 11.2.8 Discussion of models for Schottky barriers

In the Bardeen limit the Fermi level at the interface is pinned by interfacial states at an energy that is universal in the sense that it is independent of the metal. In terms of the MIGS model this universal level is the 'intrinsic' charge neutrality level of the semi-conductor. In the Schottky limit there is no pinning and the Fermi level is dependent on the metal.

In practice it appears that neither of these limits holds. While calculations indicate that the Fermi level is pinned at the charge neutrality level by the high density of MIGS, Tung's experiments and self-consistent calculations indicate that the charge neutrality point is not universal but varies markedly with the structure and composition of the interface. This was also the view reached by van Schilfgaarde and Newman (1991) on the basis of a series of density functional calculations of Schottky barriers on GaAs (110). Thus, in each metal–semiconductor interface the Fermi level is pinned by MIGS, but the pinning level varies from case to case depending on the structure of the interface. In the terminology of Flores *et al.* (1989) the Fermi level is pinned at the 'extrinsic' charge neutrality level, which varies throughout some range centred on the 'intrinsic' charge neutrality level. Unfortunately, the extrinsic charge neutrality level cannot be predicted without a full self-consistent calculation, in which the atomic structure is relaxed subject to whatever constraints apply in reality (such as forced elastic coherence in thin epitaxial films).

In our view a better understanding of the Schottky barrier will emerge only from further experiments and simulations carried out on fully characterized interfaces.

### 11.2.9 Inhomogeneous Schottky barriers

The experiments and self-consistent calculations on characterized epitaxial interfaces described in Section 11.2.7 have revealed a strong dependence of the Schottky barrier height on the atomic structure of the interface. It follows that at an inhomogeneous interface, such as one containing facets or regions of different composition, the local Schottky barrier will vary on the scale of the inhomogeneities. Similarly, in a polycrystalline contact the Schottky barrier will vary, in general, from grain to grain.

The length scales of these structural or compositional variations within a grain may range from angstroms upwards. There may be rapid variations of the electrostatic potential on an atomic scale, but they are not important for transport because electrons can tunnel through them. On the other hand, if the length scale is greater than the thickness of the depletion layer then transport through each region may be treated independently. Therefore, the significant length scales for lateral variations of the band bending, from the point of view of generating new physics, are greater than tunnelling distances and smaller than the depletion layer thickness. Depending on the level of doping, they will often be less than the width of the depletion layer. Conventional band bending diagrams, such as Fig. 11.2, are misleading because they ignore these variations of the band bending along the interface. A more realistic visualization is given in Fig. 11.10. The consequences of these spatial variations in the local Schottky barrier height for transport properties, such as their non-idealities in current–voltage characteristics (Werner and Güttler 1991, Sullivan *et al.* 1992) have only just begun to be considered. One of the interesting points that has emerged is the phenomenon of 'pinch-off' (Tung *et al.* 1991). If lateral variations of the Schottky barrier are less than the width of the depletion layer then patches of low Schottky barrier height may not be accessible to carriers if they are surrounded by regions of high barrier height. The low barrier patches are 'pinched off' by surrounding patches of high barrier. In that case the current flowing through a low barrier patch does not depend on the height of the barrier at the interface, but on the height of the saddle point in front of it. It is only when patches are very large, or the doping level is very high, that the electrostatic fields governing transport through different patches may be treated independently.

## 11.3  SEMICONDUCTOR HETEROJUNCTIONS

### 11.3.1  Introduction

A heterojunction is a bicrystal formed between two different semiconductors. Each semiconductor may be doped p or n type as in a p–n (homo)junction. In contrast to an ordinary p–n (homo)junction, however, there are different band gaps, dielectric constants and electron affinities on either side of the interface. Heterojunctions have found widespread applications in microelectronics, particularly in lasers, photodetectors, and solar cells (Sze 1981, Grovenor 1989). Applications have diversified significantly since the advent of MBE and other vapour phase epitaxial growth techniques which have enabled superlattices of semiconductors to be grown with atomic level precision (Eskai 1986). The ability to grow interfaces one atomic layer at a time enables devices to be engineered at the atomic scale. In particular, the semiconductor band gap can be graded by varying the composition of the deposited layers, and atomic-scale variations in the composition of interfaces can be introduced to engineer interfacial properties (see Margaritondo 1988 and Niles and Margaritondo 1992 for reviews).

Anderson (1962b) noted that the changes in the band gap and electron affinities introduce discontinuities, $\Delta E_v$ and $\Delta E_c$, in the edges of valence and conduction bands respectively, as illustrated in Fig. 11.11. These discontinuities are sometimes called *band offsets*. In Fig.11.11(a) we see the energy bands in the two semiconductors before contact

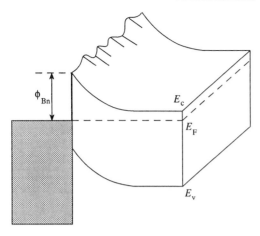

**Fig. 11.10** Two-dimensional band diagram of an inhomogeneous Schottky contact. Notice that the conduction and valence band edges vary with position in the interface, and that the barrier height varies with position. (After Werner and Güttler (1991).)

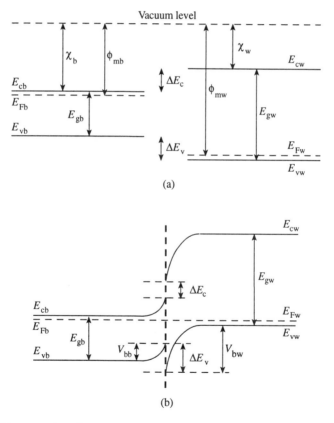

**Fig. 11.11** To illustrate the alignment of the energy bands during the formation of a heterojunction between a black and white semiconductor. In (a) the semiconductors are well separated and the vacuum level is the common zero of energy. Intimate contact in (b) leads to charge transfer and band bending in each semiconductor. The conduction and valence band offsets are $\Delta E_c$ and $\Delta E_v$.

has been made. It has been assumed in Fig. 11.11(a) that there are no surface states so
that the energy bands are flat. The common zero of energy is the vacuum level. The left
and right hand semiconductors are doped n and p type respectively. There are discontinuities in the conduction band minima and valence band maxima given by

$$\left.\begin{array}{l} \Delta E_c = \chi_b - \chi_w \\ \Delta E_v = \Delta E_g - \Delta E_c \end{array}\right\} \tag{11.18}$$

where $\Delta E_g = E_{gb} - E_{gw}$ is the discontinuity in the band gaps. The left and right semiconductors have been designated black and white respectively. Equation (11.18) is known
as the 'electron affinity rule'. When the interface is formed it is assumed that it is
atomically sharp and that no intermixing takes place. This is often not far from the truth
in modern MBE grown samples. The Fermi energies align by charge transfer which bends
the bands in both semiconductors, as shown in Fig. 11.11(b). The total 'built-in' voltage,
$V_{bi}$, is the difference in work functions, $\phi_{mb} - \phi_{mw}$, which is equal to the sum of the
partial built-in voltages, $V_{bb}$ and $V_{bw}$. The partial built in voltages arise from the band
bending in each half of the junction. Because of the band offsets at the interface the
barriers to electrons and holes will differ, and so the current through a heterojunction
will usually be carried by either electrons or holes. The magnitudes of the band offsets
$\Delta E_c$ and $\Delta E_v$ are then central to the transport properties of the heterojunction. Similarly, the band offsets $\Delta E_c$ and $\Delta E_v$ between successive semiconductor phases in compositional superlattices determine the one-dimensional periodic confining potential for
electrons and holes (Eskai 1986). As shown in Fig. 11.12 the carriers become trapped in
one-dimensional 'quantum wells', the depths of which are governed by the band offsets
$\Delta E_c$ and $\Delta E_v$. The energy levels of the carriers are quantized by the small controllable
width of the wells. The quantization of the energy levels has led to novel optical devices,
and transistors based on resonant tunnelling (see Margaritondo 1988).

   A significant focus of research on semiconductor heterojunctions has been concerned
with the determination of the band offsets $\Delta E_c$ and $\Delta E_v$. This is also the focus of our
discussion below. Not only are these band offsets of central importance in devices, but
their determination represents one of the most fundamental problems in the science of
interfaces. As in the case of Schottky barriers, a great deal of the early work on semiconductor heterojunctions was blighted by contamination and lack of structural characterization. However, a considerable body of reliable data has now been collected and will
be used in the next section to assess models for the band offsets.

### 11.3.2   The band offsets

In deriving eqn (11.18) we have assumed that the band offsets $\Delta E_c$ and $\Delta E_v$ are determined by bulk properties of the two semiconductors concerned. The structure of
the interface itself was not considered to play any role. This is very reminiscent of the
Schottky model for metal–semiconductor interfaces, eqn (11.12). Our account of
the Schottky barrier in Section 11.2 emphasized the role of the interface in affecting the
Schottky barrier. In this section we shall consider various models that have been proposed
for the offsets $\Delta E_c$ and $\Delta E_v$, which consider the role of the interface to varying degrees.
At the outset it should be said that the issue is only how $\Delta E_g$ is divided between $\Delta E_c$ and
$\Delta E_v$, since $\Delta E_c + \Delta E_v = \Delta E_g$.

   The electron affinity rule was shown (Niles and Margaritondo 1986) by synchrotron
radiation photoemission experiments to fail for the Ge–ZnSe heterojunction. In contrast

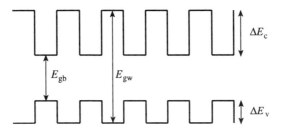

**Fig. 11.12** The conduction and valence band edges in a quantum well superlattice structure of two semiconductors designated as black and white. The band offsets $\Delta E_c$ and $\Delta E_v$ lead to 2D spatial confinement of electrons and holes respectively.

to many earlier measurements the surfaces and interfaces in this experiment were well characterized. It may be concluded, therefore, that the electron affinity rule does not apply to all semiconductor heterojunctions.

Another simple rule, called the common anion rule, was based on the observation that in a polar semiconductor valence band states derive more of their weight from the anion s and p states than from those of the cation. This suggests that for an interface between two semiconductors sharing the same anion, and the same crystal structures with very small lattice mismatch, the valence band offset should be very small. This 'rule' was initially given experimental support, but later experiments showed that it failed (Margaritondo 1988). Wei and Zunger (1987) argued that the reason for the failure was that *cation* d-orbitals also contribute significantly to the valence band maximum.

A general discussion of the theoretical difficulties in predicting band offsets has been given by Tersoff (1987), and we shall now follow this work closely.

In general, whenever two materials are brought into contact there will be some charge transfer resulting from the difference in Fermi energies. The charge transfer creates a dipole at the interface which aligns the Fermi energies. The effect of the dipole is to effect a rigid, uniform shift in the electrostatic potential on either side of the interface. The electrostatic potential is a linear functional of the charge density. Therefore, the electron affinity rule would be valid only if the charge density at the interface is a linear superposition of the charge densities of the two free surfaces prior to contact. To be more precise let $\rho(b)$ and $\rho(w)$ denote the charge densities of the free surfaces of the black and white crystals respectively. Let $\rho(I)$ denote the charge density of the final interface between them. Then the electron affinity rule would be correct if $\rho(I) = \rho(b) + \rho(w)$. The difference, $\delta\rho = \rho(I) - \rho(b) - \rho(w)$, averages to zero owing to overall charge neutrality, but in general there is a dipole associated with $\delta\rho$. This is the interfacial dipole. The difference between the measured $\Delta E_c$ and the value predicted by the electron affinity rule is the potential shift associated with the dipole $\delta\rho$.

As stressed by Tersoff the interfacial dipole $\delta\rho$ is not uniquely defined because it is defined as the difference between two charge densities. It is defined relative to a reference interface with a charge density $\rho(b) + \rho(w)$. Since $\rho(b)$ and $\rho(w)$ refer to charge densities of free surfaces, which may be relaxed or unrelaxed, they are quite arbitrary. Therefore, the calculated dipole may be large or small depending on the choice of reference surfaces.

When two metals come into contact the metal with the greater work function attracts electrons from the other metal. Here we are using the Fermi energy of each metal as a reference level. The work function is simply the position of the vacuum level relative to the Fermi level. The vacuum level is a suitable common zero of energy for all isolated pieces of metal and enables us to position their Fermi energies on the same absolute energy scale. The charge flow stops when the relative shift in potential due to the dipole aligns the Fermi levels in the two metals. The charge density of the dipole is located within an atomic layer on either side of the interface because the screening length in metals is about a bond length. The work function is dependent on the orientation of the surface of the metal at which it is measured owing to the presence of a dipole at the surface. Therefore the work function of a metal is not unique, and the magnitude of the interfacial dipole, for a given pair of metals, depends on the orientations of the two surfaces prior to contact. But whatever the initial difference in the work functions of the metals the Fermi level is constant throughout the bicrystal once contact has been made.

In Section 11.2.4 we introduced the notion of a charge neutrality level for a semiconductor and pointed out that at an interface it plays a role analogous to that of the Fermi level in a metal. If the neutrality levels of two semiconductors are equal then no charge transfer will take place between them when they contact. On the other hand, if they differ then electronic charge will flow from the semiconductor with the higher charge neutrality level to the lower. The charge neutrality level is thus a suitable reference level for a semiconductor. Once the position of the valence band maximum is known relative to the charge neutrality level in each bulk semiconductor then $\Delta E_v$ can be calculated straightforwardly. Provided the charge neutrality level is a property of the bulk semiconductor then we would be able to calculate the band offsets between any pair of semiconductors.

In Section 11.2.4 we also distinguished between intrinsic and extrinsic charge neutrality levels. For the present we shall consider only the intrinsic charge neutrality level, which is determined by the bulk band structure of the semiconductor concerned. We shall return to whether extrinsic neutrality levels should be considered later.

Following Flores and Tejedor (1979) and Tersoff (1987) we shall now apply linear response theory to calculate the interface dipole. In the absence of any dipole then $\Delta E_v$ is equal to $\Delta E_v^{\circ}$, where

$$\Delta E_v^{\circ} = E_v^{\circ}(b) - E_v^{\circ}(w) \tag{11.19}$$

and $E_v^{\circ}(b)$ is the position of the valence band maximum in the black crystal relative to the vacuum level, and a similar meaning for $E_v^{\circ}(w)$. Let $E_n^{\circ}(b)$ and $E_n^{\circ}(w)$ denote the charge neutrality levels of the isolated semiconductors relative to the vacuum level. If $E_n^{\circ}(b)$ and $E_n^{\circ}(w)$ coincide then there is no charge transfer and $\Delta E_v = \Delta E_v^{\circ}$. But if $E_n^{\circ}(b)$ and $E_n^{\circ}(w)$ differ then a dipole will be produced through charge transfer which moves $E_n^{\circ}(b)$ and $E_n^{\circ}(w)$ to new (screened) positions $E_n(b)$ and $E_n(w)$. The shift in potential energy, $\Delta V$, due to the dipole is given by

$$\Delta V = -\alpha [E_n(b) - E_n(w)]. \tag{11.20}$$

The susceptibility, $\alpha$, is dimensionless. For a metal $\alpha \to \infty$, and for an insulator $\alpha \to 0$. For a semiconductor it is finite. Below we will see that it is related to the static, long wavelength dielectric constant.

The screened mismatch in the neutrality levels is therefore

$$\left. \begin{aligned} \Delta E_n = E_n(b) - E_n(w) &= [E_n^{\circ}(b) - E_n^{\circ}(w)] + \Delta V \\ &= [E_n^{\circ}(b) - E_n^{\circ}(w)] - \alpha \Delta E_n \end{aligned} \right\} \tag{11.21}$$

for which the self-consistent solution is

$$\Delta E_n = (1 + \alpha)^{-1} \Delta E_n^{\circ} \tag{11.22}$$

where $\Delta E_n^o = E_n^o(b) - E_n^o(w)$. Thus, the initial mismatch in the charge neutrality levels is reduced by a factor of $(1 + \alpha)$ by the interfacial dipole. Similarly,

$$\Delta V = -\frac{\alpha}{1 + \alpha} \Delta E_n^o. \tag{11.23}$$

In the metallic limit, where $\alpha \to \infty$, we have $\Delta V = -\Delta E_n^o$ and $\Delta E_n = 0$: the dipole at the interface exactly cancels the difference in the neutrality levels. In the insulating, non-polarizable limit, where $\alpha \to 0$, we have $\Delta V = 0$ and $\Delta E_n = \Delta E_n^o$.

The valence band offset is given by

$$\Delta E_v = \Delta E_v^o + \Delta V. \tag{11.24}$$

Thus, the electron affinity rule, for which $\Delta E_v = \Delta E_v^o$, corresponds to the insulating limit for the dielectric response of the interface ($\Delta V = 0$). The charge neutrality rule, for which $\Delta E_n = 0$, corresponds to the metallic limit. In that limit

$$\Delta E_v = \Delta E_v^o - \Delta E_n^o. \tag{11.25}$$

If we denote the charge neutrality level relative to the valence band maximum by $E_n^v$ then $\Delta E_v$, in the metallic limit, becomes:

$$\Delta E_v = -E_n^v(b) + E_n^v(w) = -\Delta E_n^v. \tag{11.26}$$

A semiconductor lies between these two extremes. Tersoff (1987) related $\alpha$ to the static, long-wavelength bulk dielectric constant, $\varepsilon$, by the following thought experiment. Consider an infinite single crystal of a semiconductor. Imagine a plane which divides the crystal into two halves. On one side of this plane we make the material more electro-negative, i.e. we decrease $E_v^o$ and $E_n^o$ on that side. We can do this in two ways. Either we introduce a downward rigid shift in potential energy to all atoms on that side or we insert a dipole at the plane separating the two crystal halves.

If we shifted the potential, and hence $E_v^o$ and $E_n^o$, throughout the crystal half by an amount $\Delta V_x$, the initial band discontinuity, $\Delta E_v^o = \Delta V_x$, at the 'interface' is screened and becomes

$$\Delta E_v = \frac{1}{1 + \alpha} \Delta V_x, \tag{11.27}$$

where we have used eqns (11.23) and (11.24).

Now suppose that the shift in potential was effected by introducing a dipole sheet, $\Delta V_x$, along a plane in a perfect crystal of a semiconductor. The screened dipole in the crystal is given by

$$\Delta V = \Delta V_x / \varepsilon \tag{11.28}$$

where $\varepsilon$ is the static long-wavelength dielectric constant. The screened valence band edge discontinuity, $\Delta E_v$, is then simply $\Delta V$. Therefore,

$$\alpha + 1 = \varepsilon. \tag{11.29}$$

For typical elemental and III–V semiconductors $\varepsilon \approx 10$. Inserting $\alpha = 9$ in eqn (11.22) we find that 90 per cent of the initial difference in the charge neutrality levels is screened out. Therefore, in these semiconductors the screening is much closer to the metallic limit, where the charge neutrality rule holds, than to the insulating limit, where the electron affinity rule holds. Lining up the charge neutrality levels of the semiconductors, as in eqn (11.26), is the best zeroth order approximation. The error is the difference between eqn (11.24) and eqn (11.25):

$$\text{error} = \frac{\Delta E_n^o}{1 + \alpha} = -\frac{1 + \alpha}{\alpha} \Delta V \frac{1}{1 + \alpha} = -\frac{\Delta V}{\alpha} = \frac{\Delta E_v^o - \Delta E_v}{\alpha} \tag{11.30}$$

where we have used eqn (11.23) and eqn (11.24). The difference $\Delta E_v - \Delta E_v^o$ is the error in the electron affinity rule, which is typically less than 0.5 eV. Therefore the error in the neutrality line-up is about 0.05 eV.

In Section 11.2.4 it was argued that at a metal–semiconductor interface there is a continuum of MIGS. If these states are occupied up to the charge neutrality level then the semiconductor is charge neutral. In this section we have again used the concept of a charge neutrality level and we have implicitly assumed that there are continua of states in the gaps of both semiconductors to enable an arbitrary dipole to be developed at the interface. Where do these gap states come from at a semiconductor heterojunction? To answer this question we again follow Tersoff (1987) and return to our hypothetical homojunction, shown in Fig. 11.13, consisting of a single crystal in which the potential in one half has been shifted rigidly relative to the other. We shall consider the quantum mechanical origin of the screening of the imposed interfacial dipole, $\Delta V_x$.

The bands on the left of Fig. 11.13 are shifted down in energy by $\Delta V_x$, but otherwise the band structures are identical on either side of the interface. The Fermi energy lies above the valence band on the right side and below the conduction band on the left side. An electron in a state at the top of the valence band on the right cannot propagate into the left because there are no eigenstates in the left side with that energy. Therefore, the wave function for this state decays exponentially into the left side. This tunnelling of valence band states from the right into the left hand sides results in some electronic charge being transferred from right to left: a dipole. However, this is not the whole story because empty conduction band states tunnel from the left to the right sides where the conduction band in the left overlaps the gap on the right. Although no charge is carried by these empty states they affect the charge distribution at the interface owing to the sum rule on the local densities of states. This is a subtle point which we now explain.

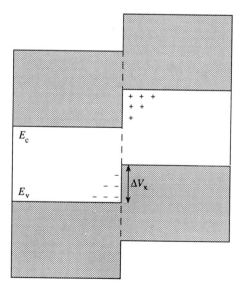

**Fig. 11.13** Band diagram of a single semiconductor showing a discontinuity $\Delta V_x$ in the valence and conduction band edges. This discontinuity could be induced either by imposing an external step potential or by introducing a dipole sheet at the interface. The '+' and '−' symbols indicate the net local charge due to the presence of gap states. (After Tersoff (1987).)

The semiconductor is neutral when the valence band states are full and the conduction band states empty. A gap state just above the valence maximum takes most of its weight from the valence band but also a little from the conduction band. If that state is filled then not only is the component, which originated from the valence band, of the state occupied, but also a fraction of a conduction band state is also effectively occupied. Thus, a filled gap state leads to a net negative charge. If the state lies higher in the energy gap it has more conduction band character and its occupation leads to a greater net negative charge.

On the other hand, an unoccupied state near the top of the gap leads to a small net positive charge (because a small part of a valence band state is not occupied) while an unoccupied state near the bottom of the gap leads to a larger net positive charge. It follows that the unoccupied conduction band states that tunnel into the gap on the right of Fig. 11.13 are associated with a net positive charge. The occupied valence band states that tunnel into the gap on the left are associated with a net negative charge. This dipole of induced charge screens the shift in potential across the interface. Thus, we have a microscopic picture of the dielectric response of the medium to the shift in potential $\Delta V_x$.

At a heterojunction valence and conduction band states of either semiconductor tunnel into the band gap of the other semiconductor where they overlap in energy. By varying the occupation of these continua of interface-induced states the charge neutrality levels of the semiconductors become aligned.

We can calculate an intrinsic charge neutrality level for each semiconductor, as described in Section 11.2.4. This energy is a property of the bulk for each semiconductor, and it is calculated from the bulk band structure. At this energy, states in the gap change over from being derived more from the valence to the conduction band. The charge neutrality level is the reference level for each semiconductor in the same way as the Fermi energy is the reference level for a metal. The charge neutrality line-up of the bands at a semiconductor heterojunction consists of rigidly shifting the bands on either side of the interface until the charge neutrality levels are equal.

In Table 11.1 we show the calculated intrinsic charge neutrality levels of a number of semiconductors from Tersoff (1987). In each case they are given relative to the valence band maximum, $E_n^v$.

Lining up the charge neutrality levels gives a valence band offset, $\Delta E_v$, given by eqn (11.26). Table 11.2 compares valence band offsets calculated in this way with experiment. It is seen that on the whole the values of $\Delta E_v$ predicted by the charge neutrality theory agree with experiment to within about 0.2 eV, although there are cases where the disagreement is larger. However, the accuracy is not sufficient for quantitative device design where an accuracy of the order of $kT$ is necessary, i.e. 0.026 eV at room temperature (Kroemer 1983). Nevertheless, the data listed in Table 11.2 can be used to guide the selection of promising candidates for specific device applications. But, aside from its moderate success at predicting band offsets, perhaps the most important aspect of the charge neutrality level theory is that it treats semiconductor heterojunctions and metal–semiconductor interfaces on the same footing. This is a significant conceptual achievement.

Local density functional calculations have been carried out for a number of lattice matched semiconductor heterojunctions by Van de Walle and Martin (1987) and by Massidda *et al.* (1987). Both groups used thp local density approximation which underestimates band gaps by about 50 per cent. Since it is not clear how much of this error may be attributed to the valence and conduction band edges separately, one does not

**Table 11.1** Calculated intrinsic charge neutrality levels. $E_n^v$ (relative to the valence band maximum), for various semiconductors (Tersoff 1987)

| Semiconductor | $E_n^v$ (eV) |
| --- | --- |
| Si | 0.36 |
| Ge | 0.18 |
| AlP | 1.27 |
| GaP | 0.81 |
| InP | 0.76 |
| AlAs | 1.05 |
| GaAs | 0.50 |
| InAs | 0.50 |
| AlSb | 0.45 |
| GaSb | 0.07 |
| InSb | 0.01 |
| ZnSe | 1.70 |
| MnTe | 1.6 |
| ZnTe | 0.84 |
| CdTe | 0.85 |
| HgTe | 0.34 |

know the magnitude of the error in the calculated $\Delta E_v$. For the GaAs/AlAs (110) interface Van de Walle and Martin found that the screened interfacial dipole was confined to the first (220) plane of each crystal on either side of the interface. This dipole gave rise to a small shift in the average electrostatic potential of either crystal of 0.035 eV. To determine $\Delta E_v$ two further calculations are required, one for each bulk crystal, in which the position of the valence band maximum is found relative to the average electrostatic potential. In this way Van de Walle and Martin obtained $\Delta E_v = 0.37$ eV, compared with an experimental value of 0.55 eV (Batey and Wright 1986). Van de Walle and Martin further calculated that $\Delta E_v = 0.37$ and 0.39 eV at (100) and (111) GaAs/AlAs interfaces, indicating that $\Delta E_v$ is not sensitive to interface orientation, for lattice matched heterojunctions, in agreement with experiment (Wang *et al.* 1985). Indeed, for (001), (110), and (111) lattice-matched GaAs/AlAs heterojunctions Baldereschi *et al.* (1988) calculated the same band offsets to within 0.01 eV. Moreover, they found that the band offsets were not affected by intermixing across the GaAs/AlAs interface. The results of self-consistent interface calculations obtained by Van de Walle and Martin are included in the last column in Table 11.2 for comparison with experiment and the calculated charge neutrality line-up values. It is seen that the agreement with experiment is on the whole no better than Tersoff's values.

Up to now we have assumed that the band offsets are determined by the alignment of the intrinsic charge neutrality levels of the bulk semiconductors. If this were true then the band offsets would not be affected by the structure and composition of the interface. An elegant linear response theory was given by Baroni *et al.* (1989) which revealed clearly the circumstances under which the band offsets at lattice-matched heterojunctions may be expected to depend on the structure of the interface. In summary, the band offsets

**Table 11.2** Experimental and theoretical valence band offsets

| Heterojunction | $\Delta E_n^v$ (expt.) | $\Delta E_n^v$ (theor.)[d] | $\Delta E_n^v$ (theor.)[e] |
|---|---|---|---|
| Si–Ge | 0.17[a] | 0.18 | |
| AlAs–Ge | 0.95[a] | 0.87 | 1.05[c] |
| GaAs–Ge | 0.35[a] | 0.32 | 0.63[c] |
| InAs–Ge | 0.33[a] | 0.32 | |
| GaSb–Ge | 0.20[a] | −0.11 | |
| GaP–Ge | 0.80[a] | 0.63 | |
| InP–Ge | 0.64[a] | 0.58 | |
| GaAs–Ge | 0.35[a] | 0.32 | |
| ZnSe–Ge | 1.52[b] | 1.52 | 2.17[c] |
| GaAs–Si | 0.05[a] | 0.14 | |
| InAs–Si | 0.15[a] | 0.14 | |
| GaSb–Si | 0.05[a] | −0.29 | |
| GaP–Si | 0.80[a] | 0.45 | 0.61[c] |
| InP–Si | 0.57[a] | 0.40 | |
| AlAs–GaAs | 0.50[b] | 0.55 | 0.37[c] |
| InAs–GaSb | 0.51[b] | 0.43 | 0.38[c] |
| GaAs–InAs | 0.17[b] | 0.00 | |
| ZnSe–GaAs | 0.96[b] | 1.20 | 1.59[c] |
| AlSb–GaSb | 0.45[c] | 0.38 | 0.38[c] |

[a] Margaritondo (1985).
[b] quoted in Tersoff (1987).
[c] van de Walle and Martin (1987).
[d] obtained from eqn (11.26) using calculated $E_n^v$ values listed in Table 11.1.
[e] obtained from self-consistent interface calculations.

at isovalent junctions, such as GaAs/AlAs, are not expected to vary with interface features. By contrast, at heterovalent interfaces, such as Ge/GaAs, the band offsets are expected to depend on interfacial structure and composition. The essence of the approach is to define a reference *single* crystal from which the two real crystals on either side of the interface are obtained by small perturbations. A suitable choice of reference is one in which the ionic pseudopotentials for the cation and anion sublattices are arithmetic averages of the pseudopotentials for the cations and anions respectively in the two crystals. There is obviously no band offset in the reference crystal because it is a single crystal. Bare perturbations $+\Delta V_{ionic}$ and $-\Delta V_{ionic}$ ionic are then applied to the reference crystal to generate the two ionic real crystals on either side of the interface. After these perturbations have been screened the valence band offset, measured far from the interface, may be expressed as $\Delta E + \Delta v$ (compare Van de Walle and Martin 1987). The first term, $\Delta E$, is determined by the bulk properties of the two adjoining crystals and is equal to the difference in the valence band maxima measured relative to the average electrostatic potential of each bulk semiconductor. The second term, $\Delta v$, is equal to the difference in the average electrostatic potentials on either side of the interface. $\Delta v$ is determined by the local dipole that may be established at the interface, and, in principle, it depends on the structure and composition of the interface. Therefore, it is only the second term that may vary with the details of the interface. $\Delta v$ is entirely electrostatic in nature, and it is determined, through Poisson's equation, by the change in the screened charge density induced by the perturbation $\pm \Delta V_{ionic}$ on either side of the interface.

Consider the case of GaAs/AlAs lattice matched heterojunctions. The reference crystal is $\langle Ga_{0.5}Al_{0.5} \rangle As$. This is a fictitious crystal in which the anion sublattice is occupied by As atoms and the cation sublattice is occupied by a hybrid atom whose ionic pseudopotential is an arithmetic average of the Ga and Al ionic pseudopotentials. Baroni *et al.* (1989) showed that the linear response of the reference crystal to replacing hybrid atoms $\langle Ga_{0.5}Al_{0.5} \rangle$ on one side of the interface by Ga atoms, and replacing hybrid atoms by Al atoms on the other side of the interface, is independent of the structure and compositional width of the interface, assuming each atom continues to occupy a site of tetrahedral symmetry. There is an electrostatic contribution to the band offsets in this case but it is independent of the interfacial structure. The key point to note is that the net charge induced by replacing a $\langle Ga_{0.5}Al_{0.5} \rangle$ hybrid atom by a Ga or Al atom is zero. Therefore, there are no point charges induced by the perturbation in going from the reference to the real crystals.

As a counter example consider the heterovalent case of Ge/GaAs interfaces. The reference crystal is a hypothetical zincblende structure $\langle Ge_{0.5}Ga_{0.5} \rangle/\langle Ge_{0.5}As_{0.5} \rangle$, in which the cation has valence charge 3.5 and the anion has valence charge 4.5. In this case replacing the hybrid cations on either side of the interface by Ge or Ga atoms and the hybrid anions by Ge or As atoms results, in linear response theory, in semi-infinite lattices of screened point charges. Each screened point charge has a magnitude of $1/(2\varepsilon)$, where $\varepsilon$ is the bulk dielectric constant of the *reference* crystal. These lattices of point charges give an electrostatic contribution to the band offsets which clearly depends on the detailed structure of the interface. The actual structure attained during MBE growth depends on growth parameters as well as thermodynamic stability. For example, growing GaAs on Ge or growing Ge on GaAs may be expected to produce quite different structures for the same substrate normal. Thus a valence band offset of $+0.17\,eV$ was measured for GaAs grown on Ge(001), and $-0.54\,eV$ for Ge grown on GaAs(001) (Biasiol *et al.* 1992). These measurements were found to be highly reproducible provided the same growth conditions were maintained.

Nicolini *et al.* (1994) used a monochromatic X-ray photoemission spectrometer to measure the valence band offsets at ZnSe–GaAs (001) interfaces grown by MBE. Zn and Se were co-deposited at varying pressure ratios onto GaAs (001) substrates. This produced interfaces with varying compositions, and the valence band offset was found to vary from 1.20 eV for Zn-rich to 0.58 eV for Se-rich interfaces. Nicolini *et al.* also carried out self-consistent local density functional calculations and found that the experimental band offsets were consistent with electrically neutral interfaces of varying composition.

The fact that the band offsets are dependent on the structure and composition of heterovalent interfaces suggests that it may be possible to 'tune' band offsets to desired values by introducing an atomic layer of a third material at the interface. In this way it may be possible to engineer semiconductor heterojunctions to have required properties. For example, Niles *et al.* (1985) obtained, changes in the valence band offset of 0.1–0.3 eV by inserting layers of 0.5–2 Å thickness of Al at interfaces between CdS and Si or Ge. Band offsets at the isovalent GaAs/AlAs (001) interface have been modified with Si and Ge intralayers (Sorba *et al.* 1991).

## 11.4  GRAIN BOUNDARIES IN METALS

Grain boundaries are a source of electrical resistance in metals, where both the change in crystal orientation and the short range disorder at the interface introduce scattering of electrons. Until recently the only experiments that had been conducted were on polycrystalline specimens for which only average grain boundary resistivities could be measured. In addition about half of the specimens were in the form of a thin film where surface scattering is an important contribution to the total resistivity. Experimental results on single grain boundaries were obtained at 4.2 K in zone-refined Al and Al–0.0050 at %Mg alloy by Nakamichi (1990). These measurements reveal a dependence of the grain boundary resistivity on the tilt and twist misorientations, and on the boundary plane.

The specific grain boundary resistivity, $\rho_{gb}$, is defined as the excess resistivity per unit area of boundary perpendicular to the current flow in a unit volume of material. Lormand and Chevreton (1981) compiled a list of measured average values of $\rho_{gb}$ for metals. Typical values vary from $10^{-16}$ to $10^{-14}\,\Omega m^2$. This paper also reviews the theoretical attempts that have been made to calculate the specific grain boundary resistivity.

The experimental difficulty in measuring $\rho_{gb}$ for a single grain boundary may be appreciated by noting that the contribution to the resistance of a centimetre cube of metal containing an area, A, of 1 cm$^2$ of grain boundary normal to the current flow is $\rho_{gb}/A = 10^{-16}/10^{-4}\,\Omega = 10^{-12}\,\Omega$. This should be compared with the contribution to the resistance arising from an impurity content of 1 part per million, for which the resistivity is $10^{-12}\,\Omega m$, equal to $10^{-12} \times 10^{-2}/10^{-4}\,\Omega = 10^{-10}\,\Omega$. We see that the grain boundary contribution is only 1 per cent of such an impurity contribution. This explains the need for ultra-high-purity specimens. A similar comparison with the phonon contribution to the resistivity indicates that the temperature should be as low as possible.

The specimen geometry used in Nakamichi's measurements is shown in Fig. 11.14. By decreasing the distance between the leads 2 and 3 the grain boundary contribution to the resistance was amplified. All leads were also made of zone-refined Al and were spot-welded onto the specimen. The macroscopic degrees of freedom of each boundary were determined by X-ray techniques. The resistivity was measured with a SQUID (super-conducting quantum interference device) at 4.2 K. Resistances between leads 1 and 2, 2 and 3, and 3 and 4 were measured to determine the resistances of crystal 1, the boundary region and crystal 2 respectively. Using the distances between the leads and the cross-sectional area of the specimen, these resistances were converted to resistivities, $\rho_{1-2}$, $\rho_{2-3}$, and $\rho_{3-4}$ from which the specific grain boundary resistivity was computed as follows:

$$\rho_{gb}A/V = \rho_{2-3} - (\rho_{1-2} + \rho_{3-4})/2, \tag{11.31}$$

where $A$ is the cross-sectional area of the specimen and $V$ is the volume between leads 2 and 3. The factor $A/V$ is the inverse of the distance between leads 2 and 3. The subtraction on the right hand side of this equation removes contributions to $\rho_{2-3}$ from scattering sources in the bulk such as residual impurities, the specimen surfaces and phonons.

Figure 11.15 shows the measured specific grain boundary resistivities in zone refined

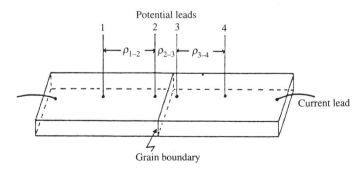

**Fig. 11.14** The specimen geometry used by Nakamichi (1990) to measure the electrical resistivity of a single grain boundary.

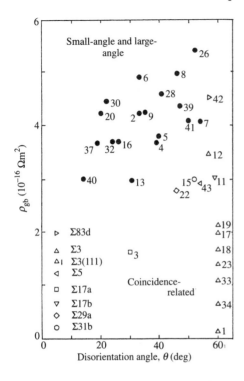

**Fig. 11.15** Measured specific grain boundary resistivities in zone-refined Al as a function of the disorientation angle, $\theta$ (Nakamichi 1990). The open symbols denote boundaries with, or close to, coincidence site lattice orientations and the filled symbols denote other disorientations. The numbers beside the symbols identify the specimens.

Al as a function of the disorientation angle, $\theta$. The average value of $\rho_{gb}$ is $3.26 \times 10^{-16}\,\Omega\mathrm{m}^2$. Specimen numbers 1, 34, 33, 23, 18, 17, and 19 are all in the $\Sigma = 3$ coincidence system. Specimen number 1 is a (111) twin for which $\rho_{gb}$ is $0.15 \times 10^{-16}\,\Omega\mathrm{m}^2$, while specimen number 19 has $\rho_{gb} = 2.1 \times 10^{-16}\,\Omega\mathrm{m}^2$. It is clear that the boundary resistivity is not only a function of misorientation, but also of the boundary plane. In general, non-coincidence-related boundaries have a higher resistivity than coincidence-related boundaries.

In Fig. 11.16 we show the specific grain boundary resistivity for Al–0.0050 at.% Ag alloy (filled symbols) plotted against disorientation, $\theta$. The resistivity is up to 10 times greater in the alloy for general boundaries and up to 100 times greater in the alloy for the $\Sigma = 3$ boundaries. In contrast to the case for zone-refined Al, general boundaries in the alloy are seen, on the whole, to have lower resistivity than $\Sigma = 3$ boundaries. Also, in contrast to the case for zone-refined Al, the resistivity displays a minimum as a function of disorientation at about 40°. It is particularly interesting that the specific boundary resistivities are higher for the $\Sigma = 3$ boundaries in the alloy. It is not clear whether this implies a solute excess or deficit at these boundaries compared with other boundaries in the system.

An exact, general, single particle formulation, of the electronic conductance for elastic scattering was given by Todorov *et al.* (1993). This formulation has been applied to the conductance of an intrinsic (111) stacking fault in an orthonormal tight-binding, s-band f.c.c. metal as a function of the fault vector, $t$. It was found that as the fault vector varied continuously throughout the c.n.i.d. the conductance per atom varied smoothly between 0.65 and 0.81, in units of $2e^2h^{-1}$ ($7.748 \times 10^{-5}\,\Omega^{-1}$), with the maximum at the perfect crystal and ideal stacking fault translations. The minimum coincided with the fault configuration where one (111) layer was directly above another. To calculate the maxi-

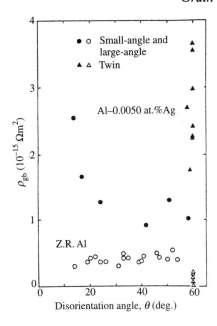

Fig. 11.16 The specific grain boundary resistivity Al–0.0050 at.% Ag alloy (filled symbols) plotted against disorientation, $\theta$ (Nakamichi 1990). The data for zone-refined Al are also shown by open symbols for comparison. The circles represent general boundaries and the triangles represent boundaries from the $\Sigma = 3$ coincidence system.

mum specific interfacial resistance for a stacking fault from these figures we first calculate the maximum excess resistance per atom: $(1/0.65 - 1/0.81) \times (7.748 \times 10^{-5})^{-1} = 3900\,\Omega$ per atom. We *multiply* this resistance by the area of interface associated with one atom, i.e. $a^2\sqrt{3}/8$. For Al we take $a = 4$ Å, and hence we obtain $\rho = 1.35 \times 10^{-16}\Omega m^2$. This figure compares reasonably well with those seen in Fig. 11.15 for grain boundaries in pure Al.

The formulation of Todorov *et al.* (1993) makes explicit the factors controlling the conductance for an interface, in the absence of inelastic scattering. For simplicity consider a twist grain boundary. The generalization to other homophase and heterophase interfaces is straightforward. Let the conductance per unit area of the bicrystal be $g_{gb}$, and let the conductance per unit area of the same plane in the black and white perfect crystals be $g_b = g_w = g$. Then the specific resistance of the grain boundary is given by

$$\rho_{gb} = \frac{1}{g_{gb}} - \frac{1}{2}\left(\frac{1}{g_b} + \frac{1}{g_w}\right)$$

(11.32)

$$= \frac{1}{g_{gb}} - \frac{1}{g}$$

which may be compared with eqn (11.31). To calculate $g_{gb}$ we introduce two cuts parallel to the boundary as shown in Fig. 11.17. The cuts must be sufficiently far from the interface that they lie in perfect crystal. In making these cuts we isolate electronically the material between the cuts from the remainder of the bicrystal. The bicrystal now consists of an isolated slab containing the 'bad' material of the grain boundary between two semi-infinite perfect crystals. We may define density operators $n_b^o(E)$ and $n_w^o(E)$ for the black and white semi-infinite crystals using standard expressions involving their Green functions, namely

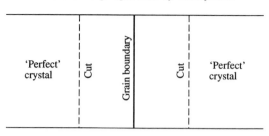

**Fig. 11.17**  To illustrate the cuts that are involved in deriving eqn (11.35) to separate the grain boundary 'bad' material from the 'perfect' semi-infinite adjoining crystals.

$$n_b^o(E) = -\operatorname{Im} G_b^o(E + i0)/\pi \left.\begin{array}{c} \\ \\ \end{array}\right\}.$$
$$n_w^o(E) = -\operatorname{Im} G_w^o(E + i0)/\pi$$

$$(11.33)$$

Let $V_{gb}$ denote the change in the Hamiltonian when the slab is recoupled to the semi-infinite crystals. The $t$-matrix for the bicrystal is then defined by

$$t_{gb} = V_{gb} + V_{gb}G^o t_{gb} \tag{11.34}$$

where $G^o$ is the Green's function for the *decoupled* system comprising the two semi-infinite perfect crystals and the slab of bad material. Then the conductance for the bicrystal is given by

$$g_{gb} = \frac{2e^2}{h} 4\pi^2 \operatorname{Tr} \left[ n_b^o(E_F) t_{gb}^\dagger(E_F) n_w^o(E_F) t_{gb}(E_F) \right] \tag{11.35}$$

which is valid in the limits of zero temperature and zero voltage.

The conductance of the perfect crystal is found by repeating the above procedure for a single crystal. The operator representing the recoupling of the two semi-infinite pieces of the same crystal and the slab, which now also comprises perfect crystal material, is denoted by $V_b$ (or $V_w$). In this way we can define a t-matrix for the perfect crystal, $t_b$ (or $t_w$):

$$t_b = V_b + V_b G^o t_b. \tag{11.36}$$

Here, $G^o$ is the Green function for the decoupled perfect crystal comprising two semi-infinite crystals and the slab of perfect crystal material. The conductance in the perfect crystal is obtained from the same formula for $g_{gb}$, eqn (11.35), but with $t_{gb}$ replaced by $t_b$ (or $t_w$).

In this formulation it is seen that the effect of the grain boundary is to alter the $t$-matrix in the slab of 'bad' material. The density operators that enter the conductance formulae for each perfect crystal and the bicrystal are the same. The $t$-matrix is an operator that is spatially localized normal to the interface and therefore the influence of the grain boundary is short-ranged. The semi-infinite nature of the adjoining crystals is implicit in the Green functions $G_b^o$ and $G_w^o$. The $t$-matrix encapsulates all the multiple scattering events that take place when the semi-infinite crystals are recoupled through the slab in a succinct and exact manner. This formulation has not yet been applied extensively to interfaces, but one illustrative example of its usefulness may be found in Todorov and Sutton (1993).

## 11.5 GRAIN BOUNDARIES IN SEMICONDUCTORS

Polycrystalline semiconductors are used extensively in modern device technology, such as thin-film interconnects, resistors, diodes, transistors, capacitors, solar cells, and varistors. They are usually formed by vapour deposition, which often results in fine-grained polycrystalline material. Grain boundaries in these materials are sometimes regarded as detrimental to the performance of the device, e.g. in limiting the efficiency of solar cells, but some devices exploit the electrical properties of grain boundaries to good effect, e.g. varistors and capacitors.

The electrical properties of grain boundaries in semiconductors stem from the fact that they trap an excess number of majority carriers—either electrons or holes depending on whether the material is doped n-type or p-type respectively. We shall consider n-type material here. The treatment for p-type material follows the same lines. In order for boundaries to trap electrons in n-type material there must be localized states, such that the integral

$$S_{gb} = \int_{-\infty}^{+\infty} \delta N(E) f(E) \, dE \tag{11.37}$$

is positive, where

$$\delta N(E) = N_{gb}(E) - N_b(E) \tag{11.38}$$

is the excess density of states in the bicrystal. Here, $N_{gb}(E)$ is the total density of states in the bicrystal and $N_b(E)$ is the total density of states for the same number of atoms in the bulk. The Fermi–Dirac distribution is represented by $f(E)$ as before. The integral, $S_{gb}$, is the excess number of occupied states in the bicrystal. Assuming that each state contains one electron when it is neutral, as in a dangling bond state, then each state absorbs one more electron when it is filled. The excess boundary charge is then

$$Q_{gb} = -eS_{gb}, \tag{11.39}$$

where $e$ is the magnitude of the electronic charge. It is frequently written that for $S_{gb}$ to be positive there must be states in the energy gap at the boundary. But this is not necessarily true because $S_{gb}$ is positive if there are no states in the gap but, at the boundary, some weight is transferred from the conduction band to the valence band. For example, this could be effected by a local resonance in the valence band. Such a transfer of weight would satisfy the sum rule on the local densities of states, because the increase in weight in the valence band would be compensated by a decrease in weight in the conduction band.

Overall charge neutrality requires compensating depletion layers on either side of the boundary. The charge neutrality condition is that

$$S_{gb} = 2N_d w. \tag{11.40}$$

In Fig. 11.18 we show the band bending at a boundary in n-type material. Solving Poisson's equation for this geometry, rather than the free surface geometry of Section 11.1, we find that the maximum band bending, which is called the potential barrier, $V_B$, is given by

$$V_B = \frac{Q_{gb}^2}{8\varepsilon\varepsilon_o N_d e}. \tag{11.41}$$

The corresponding width, $w$, of each depletion layer is given by

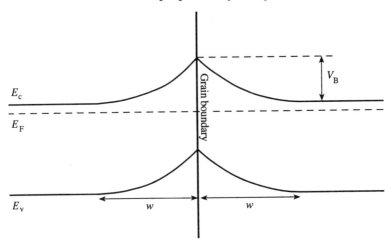

**Fig. 11.18**  The band bending at a grain boundary in an n-type semiconductor. The negative charge on the grain boundary is balanced by positive space charges in the adjoining depletion layers of width $w$. This gives rise to a bending of the bands on both sides of the grain boundary, and a potential barrier $V_B$.

$$w = S_{gb}/2N_d = \sqrt{\frac{2\varepsilon\varepsilon_o V_B}{eN_d}}.$$                    (11.42)

In deriving these two equations we have assumed that the grain size, $D$, is larger than $2w$. In that case the excess boundary charge, $Q_{gb}$, saturates. Increasing the dopant concentration, $N_D$, increases the screening of the boundary charge which reduces $w$ and $V_B$. But in fine-grained material, where $2w < D$, the boundary charge does not saturate. For example, for $S_{gb} = 10^{16}\,\mathrm{m^{-2}}$ and $N_d = 10^{22}\,\mathrm{m^{-3}}$ the minimum grain size required to achieve saturation of the boundary states is $1\,\mu\mathrm{m}$. In that case the appropriate boundary conditions for Poisson's equation are that the potential and electric field are zero at $x = D/2$ and we obtain:

$$V_B = \frac{eN_d D^2}{8\varepsilon\varepsilon_o}$$                    (11.43)

and

$$S_{gb} = N_d D.$$                    (11.44)

Thus, the potential barrier and the trapped grain boundary charge increase linearly with the dopant density, $N_d$. Combining eqns (11.41) and (11.43) the variation of the potential barrier with dopant density for a fixed grain size is shown in Fig. 11.19 (Seto 1975).

A number of the electrical properties of grain boundaries in semiconductors may now be understood in the light of the above analysis.

The depletion layers compensating the trapped charge at the grain boundary endow the grain boundary with a capacitance. For the geometry depicted in Fig. 11.18 the capacitance, $C$, per unit area of grain boundary is given by

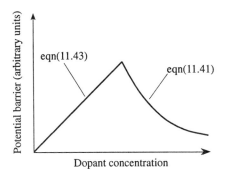

**Fig. 11.19** The variation of the potential barrier with dopant density for a fixed grain size. (After Seto (1975).)

$$C = \frac{\varepsilon \varepsilon_o}{2w} \tag{11.45}$$

and can be very large if the depletion layer is narrow and if $\varepsilon$ is high.

When a voltage is applied to the grain boundary it responds in the same way as two back-to-back Schottky barriers, one of which is forward biased and the other is reversed. The current–voltage ($I$–$V$) relation is highly non-linear when electrical breakdown occurs and the current, $I$, is proportional to the voltage, $V$, to some power $\alpha$:

$$\alpha = d \ln I / d \ln V. \tag{11.46}$$

This voltage-dependent resistance is the basis of the varistor effect, which is used for protection against voltage surges. Values of $\alpha$ between 15 and 200 have been observed. The origin of the effect is the voltage dependence of the charge trapped at the grain boundary, which controls the barrier height and hence the current flow across the boundary. In commercial varistors a second phase is added when the material is sintered to compaction. The second phase is located along the grain boundaries and it increases the charge that is trapped, e.g. $Bi_2O_3$ is added to polycrystalline ZnO varistors.

A non-equilibrium distribution of majority and minority carriers is produced (i.e. electron–hole pairs) in a polycrystalline semiconductor when electromagnetic radiation of a sufficiently high frequency, or an electron beam of sufficiently high kinetic energy, is incident on it. Unlike majority carriers the minority carriers are attracted electrostatically to the sheets of charge trapped at the grain boundaries. Once the minority carriers drift into the depletion layer they are accelerated to the grain boundary where they meet majority carriers and recombine. The recombination of minority and majority carriers at the grain boundary reduces the majority carrier charge trapped at the boundary and the band bending is reduced correspondingly. Further majority carriers are now attracted to the vacant states at the grain boundary. An equilibrium is established in which the fluxes of majority and minority carriers to the boundary are equal. This equilibrium determines the rate at which recombination of the carriers proceeds. A detailed analysis (Seager 1981, 1982) shows that this rate is proportional to $\exp(V_B/kT)$. The use of polycrystalline semiconductors for solar cells is limited by recombination, and a variety of 'passivating' techniques is used to reduce the potential barriers and hence reduce the rate of recombination.

Experimental techniques for measuring barrier heights, and the density of states and other properties of localized electronic states at boundaries have been reviewed by Grovenor (1985) and, more quantitatively, by Broniatowski (1985). Here we shall outline them very briefly.

Perhaps the most direct measurement of the barrier height is from the $I-V$ relation, in the limit of zero-bias. However, this technique does not reveal the presence of unoccupied electronic states at the grain boundary.

By applying voltage pulses we can populate states that are unoccupied at equilibrium. (However, as we shall see at the end of this section, there is an upper limit to the number of states per unit area that may be occupied.) By monitoring the transient in the boundary capacitance, or the rate at which carriers are emitted from the boundary, after the pulse, we can determine the energy distribution of trapping states and their capture cross-section. This is the basis of capacitance transient and deep level transient spectroscopy (DLTS) techniques.

Another popular technique is admittance spectroscopy wherein a small alternating bias, $V_{ac}e^{i\omega t}$, is superposed on an applied direct current bias, $V_{dc}$. The resulting current also has direct and alternating components: $j = j_{dc} + j_{ac}e^{i\omega t}$. The complex admittance is defined by $j_{ac}/V_{ac}$ and it depends on the frequency $\omega$, the bias, $V_{dc}$, and the temperature. The imaginary part of the admittance contains information about the grain boundary trapping states, such as the characteristic time required to fill and empty states (Greuter and Blatter 1990).

Recombination of carriers at grain boundaries is observed directly by EBIC (electron-beam-induced current) techniques. Here one generates electron–hole pairs by illuminating the specimen with an electron beam which is rastered across the surface of the specimen in a scanning electron microscope. The carriers are separated and collected by a p–n junction or Schottky barrier and the short-circuit current is measured as a function of position of the electron beam. Carriers that are generated near grain boundaries, and other defects, may become trapped and recombine at these defect centres. In that case the current flowing through the specimen is reduced. The variation of the current flow as a function of position of the rastered electron beam on the surface is translated into an image, in which centres of recombination appear dark. The spatial resolution of the technique is about 1 $\mu$m.

Figure 11.20 shows an EBIC image of a hexagonal network of Shockley partial dislocations in a (111) twin boundary in Si obtained by Ast *et al.* (1982). The dislocations appear as dark lines in the image. It is very interesting to observe that the twin boundary plane itself is not a recombination centre but the dislocations certainly are. Similarly, Dianteill and Rocher (1982) found no EBIC contrast at a (221) twin boundary in Si. Ruterana *et al.* (1982) observed a strong inclination dependence of EBIC contrast at embedded grains in Si. These and other observations indicate that not all grain boundaries are centres of recombination, and that defects within a boundary such as dislocations and precipitates may be effective recombination centres (see Maurice 1991 for a review).

The microscopic origin of the formation of the potential barrier, and other electrical effects at grain boundaries, is the difference in the local densities of states at grain boundaries compared with the bulk. There has been a long-standing debate about the origin of this difference, that is, whether it is an intrinsic effect arising from the structural disorder at the grain boundary, or whether it is an extrinsic effect resulting from impurity segregation. As mentioned in Section 4.3.3 there is a growing consensus that the origin is extrinsic rather than intrinsic, although the matter is not fully resolved.

Experimental evidence for an intrinsic origin was provided by Petermann and Haasen (1989) who, using a variety of techniques, detected a density of trapping states of $10^{16}\,eV^{-1}\,m^{-2}$ at the (710) twin in very high purity (float-zone) Si, in which the dopant density was approximately $10^{19}\,m^{-3}$. These states were present before any 'activating anneal'. Lenahan and Schubert (1983) used electron spin resonance (ESR) to detect spin-

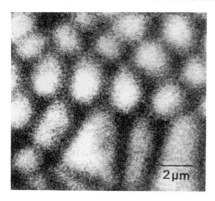

**Fig. 11.20** An EBIC image of a hexagonal network of Shockley partial dislocations in a (111) twin boundary in Si. From (Ast *et al.* (1982).)

dependent trapping of majority carriers at grain boundaries in polycrystalline Si specimens. The measured density of such states is again about $10^{16}\,\mathrm{eV^{-1}\,m^{-2}}$. Only those states containing one electron contribute to an ESR signal since empty or filled states have no spin. The origin of the states is not clear. They could be dangling bond states associated with the intrinsic structure of the boundary, or they could be associated with impurities that have segregated to the boundary.

Indirect evidence of an extrinsic origin for the states comes from the observations that an 'activating anneal' is often required to introduce potential barriers at grain boundaries or to make them effective recombination centres (see Sutton 1991 and references therein). During the anneal impurity segregation takes place. Transition metal impurities (Cu, Ni, Fe, Co) are particularly effective. Metal-rich precipitates appear in the boundaries in colonies, each colony containing 10–100 small precipitates, each of which is 20–60 nm in size. There is a one-to-one correspondence between those boundaries on which precipitation of the metal-rich particles occurs and those showing EBIC contrast. Moreover, the higher the density of precipitates, the greater the EBIC contrast (Bary and Nouet 1989).

Two features of transition metal impurities in Si, such as Cu and Ni, make them more likely than other impurities to precipitate on grain boundaries. The first is that their solubility is high at high temperatures, and decreases rapidly at lower temperatures. For example, the equilibrium concentration of Cu in Si at 1000°C is $3 \times 10^{23}\,\mathrm{m^{-3}}$ and about $10^{20}\,\mathrm{m^{-3}}$ at room temperature. Secondly their diffusivity is high, e.g. the diffusivity of Cu in Si at 1000°C is $10^{-8}\,\mathrm{m^2\,s^{-1}}$. The Cu content of as-grown Si may not be sufficiently high to lead to precipitation, but Cu is often introduced during processing, e.g. during cutting, etching, annealing, and even handling with metallic tweezers. Taken together these facts explain why (i) a 750–900°C anneal is required to activate the material, and (ii) why the cooling rate has to be sufficiently fast. (i) is explained by the requirement of getting enough Cu into solid solution. (ii) ensures that a supersaturation of Cu exists, which can be relieved in the time available only by heterogeneous nucleation on the grain boundaries.

A particularly interesting experimental study of the effect of copper-rich precipitation on the recombination activity of a (710) twin bicrystal in Czochralski-grown Si, has been performed by Broniatowski (1989, 1993). Both n-type and p-type bicrystals were studied. No potential barriers were detected in the as-grown bicrystals. Copper was chemically plated onto the specimens, which were then subjected to a variety of heat treatments. Colonies of Cu-rich precipitates were observed directly by transmission electron microscopy. The area density of precipitates was estimated to be about $10^{14}\,\mathrm{m^{-2}}$. Microanalysis of the precipitates revealed that they are copper-rich silicides, with a disordered b.c.c. crystal structure. Boundaries containing these Cu rich precipitates showed strong EBIC contrast, and potential barriers, and a single peak in DLTS spectra, both in n-type and p-type specimens. Broniatowski (1989) proposed a model that fits the experimental data, in which the precipitates are assumed to be metallic and to form Schottky contacts to the adjoining Si crystals. In this way the standard model (Fig. 11.18) of band bending at a grain boundary as that of two back-to-back Schottky barriers acquires a literal meaning. It was estimated that each precipitate was associated with about 10 trapped carriers. The origin of the boundary states is presumably MIGS (see Section 11.2.4) that penetrate the Si crystals from the metallic silicide particles.

Taking these observations together it is clear that grain boundaries in semiconductors do not behave as homogeneous sheets of stored charge or recombination activity. Structural defects such as dislocations, facet junctions and steps, as well as second phases in the boundary plane have been observed to act as centres of recombination activity. We conclude that grain boundaries display inhomogeneous electrical properties. Experimental evidence of the existence of spatially varying electrostatic potentials at grain boundaries has been provided by admittance spectroscopy measurements (Werner 1985). This may be compared with spatially inhomogeneous Schottky barriers discussed in Section 11.2.9.

The experimentally observed densities of occupied states at grain boundaries in semiconductors are usually less than $10^{17} \, \text{m}^{-2}$ and sometimes as low as $10^{14} \, \text{m}^{-2}$ in material where the doping density is particularly low. These values are typical of other semiconductor interfaces as well, such as heterojunctions and Schottky barriers. On the other hand the planar density of atoms is of the order of $10^{19} \, \text{m}^{-2}$. Therefore, the maximum planar density of occupied states corresponds to only 1 per cent of the available atomic sites in the boundary plane, and it can be as low as 0.001 per cent. How do we interpret this observation? One possibility is simply that the planar density of available states is between 0.001 per cent and 1 per cent of the available atomic sites at the interface. In that case, at least a significant fraction of the available states is occupied, and it is quite feasible that the states that are unoccupied at equilibrium can be filled by a voltage pulse during a DLTS experiment for example. A second possibility is that the area density of available states is much higher, perhaps approaching 100 per cent of the atomic sites, but the energy associated with occupying more than 1 per cent of them is prohibitive. In that case the vast majority of the unoccupied interfacial states could not be detected by voltage pulse techniques which attempt to fill them.

In the following we estimate the electrostatic energy associated with the interfacial charge and the depletion layers. We will show that the energy varies as the cube of $Q_{\text{gb}}$ and inversely with $N_d$ and $\varepsilon$. Our conclusion is that the electrostatic energy of occupying more than $10^{17}$ states $\text{m}^{-2}$ is prohibitively high, unless the dopant density is much higher. It follows that if the density of unoccupied states at the interface is higher than this it cannot be detected by techniques that rely on filling these states.

To keep the calculation simple we assume that the grain boundary is a homogeneous sheet of charge, and that the charge density in the depletion layers is also uniform. To estimate the electrostatic energy it is convenient to assign a width, $2l$, to the grain boundary, where $l \ll w$. The charge density per unit volume in the grain boundary is then $-N_d e(w-l)/l$, and the total grain boundary charge, per unit area, $Q_{\text{gb}}$, is $-2N_d(w-l)e$. Figure 11.21 shows the charge density and potential. Solving Poisson's equation we find that, for $l \leqslant x \leqslant w$, the potential is given by

$$V(x) = -\frac{N_d e}{2\varepsilon\varepsilon_o}(w - x)^2. \tag{11.47}$$

For $0 \leqslant x \leqslant l$ the solution to Poisson's equation is

$$V(x) = -\frac{N_d e(w - l)(wl - x^2)}{2\varepsilon\varepsilon_o l}. \tag{11.48}$$

These solutions are shown in Fig. 11.21(b). It is seen that the only effect of regarding the boundary width as being finite is to make the potential smooth at $x = 0$. The error incurred in taking the limit $l \to 0$ is therefore negligible. We shall find it convenient to take this limit below in order to simplify the final expressions.

The electrostatic energy, per unit area of boundary, associated with the two depletion layers is given by:

$$E_1 = 2 \times \frac{1}{2} \int_l^w N_d e(-)\frac{N_d e}{2\varepsilon\varepsilon_o}(w - x)^2 \, dx$$

In the limit of $l \to 0$ we obtain

$$E_1 = -\frac{(N_d e)^2 w^3}{6\varepsilon\varepsilon_o}. \tag{11.49}$$

The electrostatic energy, per unit area of boundary, associated with the charged boundary is given by

$$E_2 = 2 \times \frac{1}{2} \int_0^l \frac{N_d e(w - l)}{1} \frac{N_d e(w - l)}{2\varepsilon\varepsilon_o l}(wl - x^2) \, dx. \tag{11.50}$$

In the limit of $l \to 0$ we obtain

$$E_2 = \frac{(N_d e)^2 w^3}{2\varepsilon\varepsilon_o}. \tag{11.51}$$

Adding $E_1$ and $E_2$ to get the total electrostatic energy per unit area, $E_{es}$, we find

$$E_{es} = \frac{|Q_{gb}|^3}{24N_d e\varepsilon\varepsilon_o} \tag{11.52}$$

where we have used $|Q_{gb}| = 2N_d we$.

For $N_d = 10^{22}\,\mathrm{m}^{-3}$, $\varepsilon = 10$ and $|Q_{gb}| = 10^{16}\,\mathrm{e\,m}^{-2}$ we obtain $E_{es} = 1.2\,\mathrm{mJ\,m}^{-2}$, whereas for $Q_{gb} = 10^{17}\,\mathrm{e\,m}^{-2}$ we obtain $E_{es} = 1.2\,\mathrm{J\,m}^{-2}$. Since a typical grain boundary energy is about $1\,\mathrm{J\,m}^{-2}$ it is unreasonable for $Q_{gb}$ to be as high as $10^{17}\,\mathrm{e\,m}^{-2}$ unless the dopant concentration, $N_d$, is much higher. An increase in the dopant concentration by 3 orders of magnitude will compensate an increase in the boundary charge by one order of magnitude.

## 11.6 GRAIN BOUNDARIES IN HIGH TEMPERATURE SUPERCONDUCTORS

In 1986 Bednorz and Müller (Bednorz and Müller 1986) reported the discovery of superconductivity in $La_{2-x}Ba_xCuO_{4-y}$ at a critical temperature, $T_c$, of 35 K. Following that

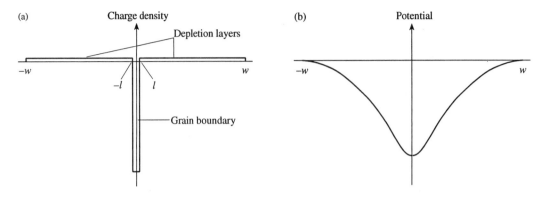

**Fig. 11.21** (a) The charge density as a function of position at and near a grain boundary in an n-type semiconductor. The grain boundary has a width $2l$, and the negative charge stored within it is balanced by the positive charge in the much larger depletion layers of width $w$-$l$. (b) The electrostatic potential associated with the charge density shown in (a). Notice that the only effect of including the large negative charge density at the grain boundary is to make the potential smooth in the grain boundary.

initial discovery a spate of other oxides was found for which $T_c$ increased dramatically. One of the most studied is $YBa_2Cu_3O_{7-x}$ which has $T_c = 92$ K, above the temperature of liquid nitrogen. This oxide has an orthorhombic crystal structure, which is typical of a class of perovskite-based copper oxide structures. The lattice parameters and $T_c$ depend on the oxygen content. The superconductivity is ascribed to electron hopping along Cu–O–Cu chains within Cu–O planes of the structure, and it is strongly dependent on the oxygen content. For an oxygen content of 6.9 the critical temperature is 92 K, and falls to 60 K as the oxygen content is decreased to 6.75.

From the point of view of applications the single most important parameter of super-conductors is the critical current density, $J_c$, at which the material becomes normal. (In the following 'normal' will be taken to mean non-superconducting.) The value of $J_c$ limits the potential applications of the material for magnetic coils, power transmission, electronic circuits, and many other applications. Critical current densities of at least $10^5$–$10^6$ A cm$^{-2}$ will be required for many potential commercial applications of these materials to be realized. But polycrystalline samples of the material often have critical current densities much lower than these values.

The high $T_c$ oxide materials are type II superconductors. This means that above a critical magnetic field $B_{c1}$, the material is penetrated by a magnetic field in the form of magnetic flux lines. However, the material continues to be superconducting as the magnetic field is increased above $B_{c1}$ until a second critical field, $B_{c2}$, is reached when it becomes normal. Inside each flux line the material is normal, and there are currents circulating around it whose magnetic field cancels the field inside the flux line. Each magnetic flux line experiences a lateral Lorentz force due to the supercurrent flowing through the specimen. If there were nothing to oppose this force the flux lines would move out of the specimen and the current flowing above $B_{c1}$ would be zero. But in a type II superconductor this is not observed. Instead the flux lines are pinned in position by defects within the material, and grain boundaries may contribute to this pinning. For example, in $Nb_3Sn$, which is a conventional superconductor, flux pinning is effected by grain boundaries, and the pinning strength is inversely proportional to the grain diameter

(Dew-Hughes 1987). The nature of the pinning is simply that when the flux line is located at the defect, which is usually a region of normal material, two normal regions have been replaced by one. Since the superconducting state has a lower free energy density than the normal state, at temperatures below $T_c$, motion of the flux line away from the pinning centre will raise the free energy of the system.

Anderson (1962a) proposed that flux lines may undergo thermally activated motion between pinning centres under the influence of the Lorentz force acting on them. This process, which is called flux creep, is dissipative and if it occurs the material acquires a resistance. It is, therefore, very desirable to prevent flux creep by introducing sufficiently large pinning forces for the flux lines. However, it is clear that, for a given pinning force, flux creep will increase exponentially with temperature since it is a thermally activated process. This is one of the most significant factors limiting the usefulness of high $T_c$ materials in practice (Dew-Hughes 1988).

Another consequence of the higher value of $T_c$ in the oxide superconductors is that the coherence length, $\xi$, is shorter (Rosenberg 1990). Superconductivity arises from the pairing of electronic states. The coherence length is a measure of the distance over which the state of one electron is correlated with that of another, and it defines the minimum thickness of a transition layer between normal and superconducting regions of a material. In $YBa_2Cu_3O_{7-\delta}$ the coherence length is highly anisotropic: parallel to the Cu–O planes it is about 1.5 nm and normal to Cu–O planes it is about 0.3 nm. These distances are very significant because, as we shall see, they are comparable with the structural and compositional widths of grain boundaries in the material (Babcock 1992). It is possible, therefore, that grain boundaries may act as barriers to supercurrents.

The first direct experimental evidence that grain boundaries in $YBa_2Cu_3O_{7-\delta}$ act as 'weak links' was obtained by Chaudhari *et al.* (1988). A weak link is a region of weakened superconductivity. These authors carried out measurements of $J_c$ for individual grain boundaries in bicrystals grown epitaxially on $SrTiO_3$ bicrystals. They found that $J_c$ was indeed less in a bicrystal than in a single crystal, proving that the grain boundary is a barrier to supercurrent flow. They also concluded that the grain boundaries were not homogeneous, with some parts of the boundary being a more effective barrier than others. Dimos *et al.* (1990) extended this work and found that $J_c$ in the bicrystal was less than 10 per cent of $J_c$ in a single crystal except for small angle grain boundaries, with misorientation less than 11°. In their specimens $J_c$ in the single crystal was $1.6 \times 10^7 \,A\,cm^{-2}$. Figure 11.22 shows the measured misorientation dependence (up to 45°) of $J_c$ at 5 K for bicrystals containing [001] tilt, [100] tilt, and (100) twist boundaries, normalized to the corresponding bulk value of $J_c$. It is seen that at a misorientation of about 11° the normalized value of $J_c$ in the bicrystal drops by almost 2 orders of magnitude.

Why is the critical current density lower in the bicrystal than in a single crystal? If the grain boundary has a width, $w$, several times greater than the coherence length, $\xi$, then states in the superconducting regions on either side of the boundary are decoupled, and the boundary region behaves like a slab of normal, resistive material sandwiched between the superconductors (see Fig. 11.23(a)). But if $w$ is comparable with $\xi$ then Cooper pairs may tunnel through the grain boundary, which then forms a weak link and behaves like a Josephson junction (see Fig. 11.23(b)). It is also possible that when a grain boundary facets, or contains a sufficiently low density of dislocations, that parts of the boundary will be resistive and other parts will act as weak links, as shown in Fig. 11.23(c).

One of the key questions is what determines the relevant width, $w$? It could be the extent to which the structural disorder at the grain boundary penetrates the adjoining

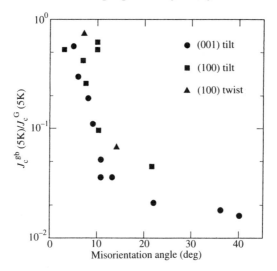

**Fig. 11.22** Measured misorientation dependence (up to 45°) of $J_c$ at 5 K for bicrystals containing [001] tilt, [100] tilt, and (100) twist boundaries in $YBa_2Cu_3O_{7-\delta}$, normalized to the corresponding bulk value of $J_c$. (From Dimos *et al.* (1990).)

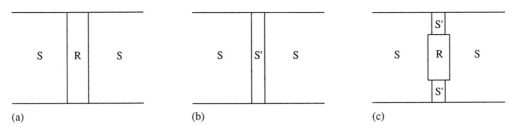

(a)                                    (b)                                    (c)

**Fig. 11.23** Schematic illustration of three types of grain boundary in a superconductor (S). (a) The grain boundary has a width several times thicker than the coherence length so that the grain boundary region is resistive (R). (b) The grain boundary has a width comparable to the coherence length so that the grain boundary remains superconducting (S') but it acts as a Josephson junction weak link. (c) The grain boundary has isolated regions that are resistive and other regions that are superconducting weak links.

crystals. It could also be the width of the region in which the composition of the material is different from that of the bulk, either due to segregation of impurities or due to local changes in stoichiometry. As discussed in Section 7.4, the structural and compositional widths are not independent.

Transmission electron microscope observations of arrays of grain boundary dislocations at large angle boundaries in $YBa_2Cu_3O_{7-\delta}$ have been reported by Babcock and Larbalestier (1990). The spacings of the dislocations were on the scale of a few coherence lengths. Babcock and Larbalestier (1989) used energy-dispersive X-ray microanalysis to observe local compositional variations at grain boundaries in $YBa_2Cu_3O_{7-\delta}$ in a high resolution, scanning transmission electron microscope. They found considerable compositional variations both parallel and perpendicular to the boundary plane. The compositional variations parallel to the boundary plane were regular and oscillatory with a wavelength of about 25 nm, suggesting that they may be related to periodic structural features in the boundary. It was found that whenever the local Cu concentration increased the local oxygen concentration decreased, and vice versa. At some boundaries, but not all, it was found that the local copper content was higher and the oxygen content was lower than in the bulk. Kroeger *et al.* (1988) came to a similar conclusion by applying Auger spectroscopy to fracture surfaces of sintered material. They found a Cu rich layer at grain boundaries, about 1.5–5.0 nm thick, which was deficient in oxygen. In this layer the oxygen content was 15–20 per cent lower than in the bulk. This could be very significant because $T_c$ in bulk $YBa_2Cu_3O_{7-\delta}$ is strongly dependent on the oxygen content.

A particularly interesting attempt to rationalize the rapid decrease in $J_c$ in Fig. 11.22 at a misorientation of about 11° in terms of the structure and composition of small angle boundaries was made by Gao *et al.* (1991*a*). These authors observed dislocations in small-angle grain boundaries in $YBa_2Cu_3O_{7-\delta}$ grown by metallo-organic chemical vapour deposition (MOCVD) by using high resolution electron microscopy and image simulation. They found discrete crystal lattice dislocations, with a Burgers vector $\mathbf{b} = a[100]$ (0.39 nm) and a core width of about 1 nm. Image simulation led to the conclusion that the cores were rich in Cu, indicating that they may be composed of CuO. They postulated that, owing to the local change of composition and the large local strains, the dislocation cores were not superconducting material. It was presumed that the supercurrent flowed across these boundaries in the relatively unstrained, bulk-like material between the cores. As the misorientation increases the cores approach each other. The boundary becomes a slab of contiguous, dislocation cores at a misorientation of $|\mathbf{b}|/D$, where $D$ is twice the core radius (2 nm), which is 11°! At this misorientation the whole boundary becomes a weak link, in agreement with Fig. 11.22.

If all large-angle grain boundaries were weak links or resistive layers there would be little hope that these materials could ever be used for high current density applications. Indeed, Fig. 11.22 is not very encouraging in this regard. However, a set of 90° large angle grain boundaries in $YBa_2Cu_3O_{7-\delta}$ has been found which show no weak link behaviour and for which $J_c$ is as high as in a single crystal (Chan *et al.* 1990, Babcock *et al.* 1990, Eom *et al.* 1991, 1992). It is encouraging that these boundaries occurred naturally in polycrystalline specimens, whereas those shown in Fig. 11.22 were manufactured. In Fig. 11.24 we show a schematic representation of three of these 90° boundaries. Fig. 11.24(a) depicts a [100] or [010] 90° twist boundary in which the Cu–O 'c' planes are misoriented by 90°. Provided there is little structural and/or compositional disorder at the boundary this boundary should be an excellent conductor for the supercurrent because each Cu–O plane on one side contacts many Cu–O planes on the other. The [100] or [010] 90° asymmetric tilt boundary shown in Fig. 11.24(b) is not expected to carry a supercurrent across it very easily because the Cu–O planes on one side are parallel to the boundary plane. In view of the very short coherence length normal to the Cu–O planes (3 Å) conduction in that direction is almost negligible. On the other hand the symmetric [100] or [010] 90° tilt boundary facet shown in Fig. 24(c) is expected to allow a supercurrent to flow across it because the Cu–O planes are inclined to the boundary plane on both sides of the boundary.

Gao *et al.* (1991b) observed 90° [100] tilt boundaries by high-resolution electron micro-scopy in $YBa_2Cu_3O_{7-\delta}$ films grown by MOCVD. The boundaries were atomically sharp, contained no amorphous or second phases and were faceted along asymmetric (010)/(001) facets (Fig. 11.24(b)) and symmetric (013) facets (Fig. 11.24(c)). An example is shown in Fig. 11.25. The asymmetric facets are presumably non-superconducting, but the symmetric facets would presumably allow a supercurrent to flow.

In conclusion, it appears that the local structure and chemical composition of grain boundaries in these high $T_c$ oxide materials determines whether or not they act as weak

(a)

Type A

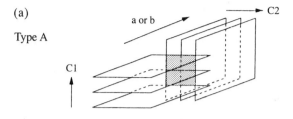

(b)

Type B

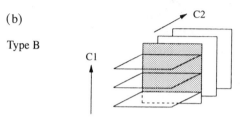

(c)

Type C

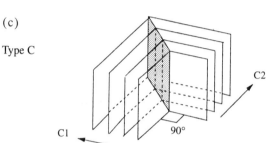

**Fig. 11.24** Schematic illustrations of three types of 90° boundaries (shaded) in $YBa_2Cu_3O_{7-\delta}$. (a) depicts a [100] or [010] 90° twist boundary in which the Cu–O 'c' planes are misoriented by 90°. (b) shows a [100] or [010] 90° asymmetric tilt boundary in which the Cu–O 'c' planes are parallel to the boundary in grain 2. (c) shows a symmetric [100] or [010] 90° tilt boundary, in which the Cu–O 'c' planes are inclined to the boundary on both sides. (From Eom *et al.* (1992).)

links. By 'local' we mean on the length scale of the coherence length, which is of the order of a few nanometres. In those samples where the grain boundaries are not weak links the critical current density is limited by the degree to which magnetic flux lines within the material are pinned. Flux pinning is effected by microstructural features such as defects and second phase particles. The challenge for the future is to develop processing routes which bring about effective flux pinning, but which do not turn the grain boundaries into weak links.

## REFERENCES

Anderson, P. W. (1962a). *Phys. Rev. Lett.*, **9**, 309.
Anderson, R. L. (1962b). *Solid-State Electronics*, **5**, 341.
Ast, D. G., Cunningham, B., and Strunk, H. (1982). *Mat. Res. Soc. Symp. Proc.*, **5**, 167.
Babcock, S. E. (1992). *MRS Bulletin*, **17**(8), 20.
Babcock, S. E. and Larbalestier, D. C. (1989). *Appl. Phys. Lett.*, **55**, 393.
Babcock, S. E. and Larbalestier, D. C. (1990). *J. Mater. Res.*, **5**, 919.
Babcock, S. E., Cai, X. Y., Kaiser, D. L., and Larbalestier, D. C. (1990). *Nature*, **347**, 167.
Baldereschi, A., Baroni, S., and Resta R. (1988). *Phys. Rev. Lett.*, **61**, 734.
Bardeen, J. (1947). *Phys. Rev.*, **71**, 717.
Baroni, S., Resta, R., Baldereschi, A., and Peressi, M. (1989). *Spectroscopy of semiconductor microstructures* (eds G. Fasol, A. Fasolino, and P. Lugli). Plenum, New York.
Bary, A. and Nouet, G. (1989). *J. Physique*, **51**, C1–423.
Batey, J. and Wright, S. L. (1986). *J. Appl. Phys.*, **59**, 200.

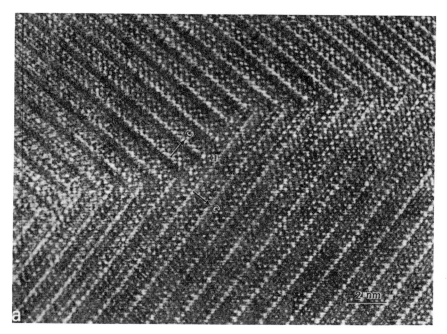

**Fig. 11.25** HREM image of a 90° tilt boundary in $YBa_2Cu_3O_{7-\delta}$ showing two symmetric facets, where the 'c' planes are equally inclined to the boundary plane (shown schematically in Fig. 11.24(c)) separated by an asymmetric facet labelled 'H' (shown schematically in Fig. 11.24(b)). (From Gao *et al.* (1991b).)

Bednorz, J. G. and Müller, K. A. (1986). *Z. Physik B*, **64**, 189.

Biasiol, G., Sorba, L., Bratina, G., Nicolini, R., Franciosi, A., Peressi, M., Baroni, S., Resta, R., and Baldereschi, A. (1992). *Phys. Rev. Lett.*, **69**, 1283.

Broniatowski, A. (1985). In *Polycrystalline semiconductors* (ed. G. Harbeke)., Springer Series in Solid-State Sciences, Vol. 57, p. 95. Springer-Verlag; Berlin.

Broniatowski, A. (1989). *Phys. Rev. Lett.*, **62**, 3074.

Broniatowski, A. (1993). *Mat. Sci. Forum*, **126–128**, 721.

Cao, R., Miyano, K., Kendelewicz, T., Lindau, I., and Spicer, W. E. (1987). *J. Vac. Sci. Technol. B*, **5**, 998.

Chan, S. W., Hwang, D. M., Ramesh, R., Sampere, S. M., Nazar, L., Gerhardt, R., and Pruna, P. (1990). In *High T_c Superconducting Thin Films: Processing, Characterization and Applications* (ed. R Stockbaur), American Institute of Physics Proceedings, No. 200, p. 172. American Institute of Physics, New York.

Chaudhari, P., Mannhart, J., Dimos, D., Tsuei, C. C., Chi, J., Oprysko, M. M., and Schevermann, M. (1988). *Phys. Rev. Lett.*, **60**, 1653.

Das, G. P., Blochl, P., Andersen, O. K., Christensen, N. E., and Gunnarsson, O. (1989). *Phys. Rev. Lett.*, **63**, 1168.

Dew-Hughes, D. (1987). *Phil. Mag. B*, **55**, 459.

Dew-Hughes, D. (1988). *Cryogenics*, **28**, 674.

Dianteill, C. and Rocher, A. (1982). *J. Physique*, **43**, C1–75.

Dimos, D., Chaudhari, P., and Mannhart, J. (1990). *Phys. Rev. B.*, **41**, 4038.

Eom, C. B., Marshall, A. F., Suzuki, Y., Boyer, B., Pease, R. F. W., and Geballe, T. H. (1991). *Nature*, **353**, 544.

Eom, C. B., Marshall, A. F., Suzuki, Y., Geballe, T. H., Boyer, B., Pease, R. F. W., van Dover, R. B., and Phillips, J. M. (1992). *Phys. Rev. B.*, **46**, 11902.

Eskai, L. (1986). *IEEE Journal of quantum electronics*, **QE–22**, 1611.

First, P. N., Stroscio, J. A., Dragoset, R. A., Pierce, D. T., and Celotta, R. J. (1989). *Phys. Rev. Lett.* **63**, 1416.

Flores, F. and Tejedor, C. (1979). *J. Phys. C: Solid state physics*, **12**, 731.

Flores, F., Munoz, A., and Duran, J. C. (1989). *Appl. Surf. Sci.*, **41/42**, 144.

Gao, Y., Merkle, K. L., Bai, G., Chang, H. L. M., and Lam, D. J. (1991*a*). *Physica C*, **174**, 1.

Gao, Y., Bai, G., Lam, D. J., and Merkle, K. L. (1991*b*). *Physica C*, **173**, 487.

Greuter, F. and Blatter, G. (1990). *Semicond. Sci. Technol.*, **5**, 111.

Grovenor, C. R. M. (1985). *J. Phys. C: Solid State Phys.*, **18**, 4079.

Grovenor, C. R. M. (1989). *Microelectronic materials.* Hilger, Bristol.

Heine, V. (1965). *Phy. Rev.* **138A**, 1689.

Kroeger, D. M., Choudhury, A., Brynestad, J., Williams, R. K., Padgett, R. A., and Coghlan, W. A. (1988). *J. Appl. Phys.*, **64**, 331.

Kroemer, H. (1983). *Surf. Sci.*, **132**, 543.

Lenahan, P. M. and Schubert, W. K. (1983). *Phys. Rev. B*, **30**, 1544.

Lormand, G. and Chevreton, M. (1981). *Phil. Mag. B*, **44**, 389.

Louie, S. G. and Cohen, M. L. (1976). *Phys. Rev. B*, **13**, 2461.

Louie, S. G., Chelikowsky, J., and Cohen, M. L. (1977). *Phys. Rev. B*, **15**, 2154.

Margaritondo, G. (1985). *Phys. Rev. B*, **31**, 2526.

Margaritondo, G. (1988). *Electronic structure of semiconductor heterojunctions*, Perspectives in condensed matter physics, Vol. 1. Kluwer; Dordrecht.

Massida, S., Min, B. I., and Freeman, A. J. (1987). *Phys. Rev. B*, **35**, 9871.

Maurice, J. L. (1991). *Springer Proc. Phys.*, **54**, 166.

Mönch, W. (1990). *Rep. Prog. Phys.*, **53**, 221.

Nakamichi, I. (1990). *J. Sci. Hiroshima Univ.*, Ser. A, **54**, 49.

Nicolini, R., Vanzetti, L., Mula, G., Bratina, G., Sorba, L., Franciosi, A., Peressi, M., Baroni, S., Resta, R., Baldereschi, A., Angelo, J. E., and Gerberich, W. W. (1994). *Phys. Rev. Lett.*, **72**, 294.

Niles, D. W. and Margaritondo, G. (1986). *Phys. Rev. B*, **34**, 2923.

Niles, D. W. and Margaritondo, G. (1992). *Materials Interfaces* (eds D. Wolf and S. Yip), chapter 22. Chapman and Hall, London.

Niles, D. W., Margaritondo, G., Perfetti, P., Quaresima, C. and Capozi, M. (1985). *Appl. Phys. Lett.*, **47**, 1092.

Petermann, G. and Haasen, P. (1989). *Springer Proc. Phys.*, **35**, 332.

Rhoderick, E. H. and Williams, R. H. (1988). *Metal-semiconductor contacts.* Oxford University Press, Oxford.

Rosenberg, H. M. (1990). *The solid state*, 3rd edn. Oxford University Press, Oxford, p. 254.

Ruterana, P., Bary, A. and Nouet, G. (1982). *J. Physique*, **43**, Cl-27.

Schottky, W. (1938). *Naturwissenschaften*, **26**, 843.

Seager, C. H. (1981). *J. Appl. Phys.*, **52**, 3960.

Seager, C. H. (1982). *Appl. Phys. Lett.*, **40**, 471.

Seto, J. Y. W. (1975). **46**, 5247.

Spicer, W. E., Lindau, I., Skeath, P., Su, C. U., and Chye, P. W. (1980). *Phys. Rev. Lett.*, **44**, 420.

Sorba, L., Bratina, A., Antonini, J. F., Walker, J. F., Mikovic, M., Ceccone, G., and Franciosi, A. (1991). *Phys. Rev. B*, **43**, 2450.

Sullivan, J. P., Tung, R. T., Schrey, F., and Graham, W. R. (1992). *J. Vac. Sci. Technol. A*, **10**, 1959.

Sutton, A. P. (1991). *Springer Proc. Phys.*, **54**, 116.

Sze, S. M. (1981). *Physics of semiconductor devices.* Wiley, New York.

Tejedor, C., Flores, F., and Louis, E. (1977). *J. Phys. C: Solid State Physics*, **10**, 2163.

Tersoff, J. (1984). *Phys. Rev. Lett.*, **52**, 465.

Tersoff, J. (1985). *Phys. Rev. B*, **32**, 6968.

Tersoff, J. (1987). In *Heterojunction band discontinuities: physics and device applications* (eds F. Capasso and G. Margaritondo). North Holland, Amsterdam, p. 3.

Thanailakis, A. and Rasul, A. (1976). *J. Phys. C: Solid State Phys.*, **9**, 337.

Todorov, T. N. and Sutton, A. P. (1993). *Phys. Rev. Lett.*, **70**, 2138.

Todorov, T. N., Briggs, G. A. D. and Sutton, A. P. (1993). *J. Phys.: Condens. Matter*, **5**, 2389.

Tung, R. T. (1984). *Phys. Rev. Lett.*, **52**, 461.

Tung, R. T. (1993). *J. Vac. Sci. Technol. B*, **11**, 1546.

Tung, R. T., Levi, A. F. J., Sullivan, J. P., and Schrey, F. (1991). *Phys. Rev. Lett.*, **66**, 72.

Van de Walle, C. and Martin, R. M. (1987). *Phys. Rev. B*, **35**, 8154.

van Schilfgaarde, M. and Newman, N. (1991). *Phys. Rev. Lett.*, **67**, 2746.

Wang, W. I., Kuan, T. S., Mendez, E. E., and Eskai, L. (1985). *Phys. Rev. B*, **31**, 6890.

Wei, S. H. and Zunger, A. (1987). *Phys. Rev. Lett.*, **59**, 144.

Werner, J. (1985). In *Polycrvstalline semiconductors* (ed. G. Harbeke), Springer Series in Solid-State Sciences, Vol. 57, p. 76. Springer-Verlag, Berlin.

Werner, J. H. and Güttler, H. H. (1991). *J. Appl. Phys.*, **69**, 1522.

Zangwill, A. (1988). *Physics at surfaces*. Cambridge University Press, Cambridge.

# Mechanical properties of interfaces

## 12.1 INTRODUCTION

Interfaces exert profound effects on the mechanical properties of bicrystals and poly-crystals. These effects appear in a wide variety of forms and stem from a wide range of sources. In this chapter we attempt to describe as many of these as possible within the space available. In so far as is possible, relatively elementary phenomena are described first, and then more complex processes, often involving combinations of these more elementary phenomena, are dealt with later. Temperature plays a decisive role in deter-mining many of the mechanical properties of interfaces, and in a number of cases properties at 'low' temperatures and at 'high' temperatures are described in separate sections.

## 12.2 COMPATIBILITY STRESSES IN BICRYSTALS AND POLYCRYSTALS

When a bicrystal is subjected to either an overall average applied elastic stress, a plastic strain, or a uniform change in temperature, the individual responses of the two adjoining crystals may differ in a manner which tends to produce a dilatational mismatch along the interface. If compatibility is to be retained along the interface, an additional set of stresses must then be generated in order to conserve this compatibility. Of course, these 'compatibility stresses' may be relieved if interface sliding is possible, as described in Section 12.8.1. In many cases, e.g. at low temperatures, this may require a considerable length of time, and quasi-static stress distributions may be achieved which vary slowly as the compatibility stresses are progressively relieved by time-dependent interface sliding. In the present discussion we shall not consider these complex time-dependent cases but will consider only the limiting case where no sliding is allowed, and the compatibility stresses are therefore at a maximum. As discussed later, these compatibility stresses may play important roles in the mechanical behaviour of the material ranging from influencing the behaviour of lattice dislocations near the interfaces to promoting intergranular frac-ture. It is noted that compatibility stresses may also be generated if certain grains in a polycrystal undergo a phase transformation which changes their individual shapes and sizes. However, we shall not pursue this subject in any detail.

### 12.2.1 Compatibility stresses caused by applied elastic stress

In single phase bicrystals or polycrystals, subjected to an overall applied stress, the tendency of differently oriented crystals to undergo different strains in any given direction will be due to their elastic anisotropy. In the case of multiphase polycrystals, the tendency will arise because the elastic constants of the different phases will generally be different as well as anisotropic.

Two simple examples are illustrated in Fig. 12.1. In the first (Fig. 12.1(a, b, c)), a bicrystal composed of a black and a white crystal is compressed along the 3 axis between

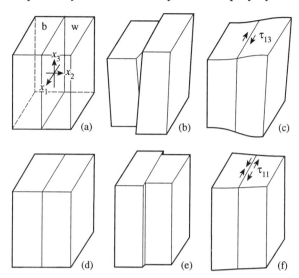

**Fig. 12.1** (a) Bicrystal composed of black and white crystals which are of the same phase but are elastically anisotropic. (b) Black and white crystals undergoing opposite strains due to compression along the 3 axis in the absence of any compatibility. (c) Restoration of compatibility between crystals in (b) by the introduction of compatibility stress $\tau_{13}$. (d) Bicrystal composed of black and white crystals which are of different phase but are elastically isotropic. (e) Black and white crystals undergoing different $\varepsilon_{11}$ strains due to stressing along the 3 axis in the absence of compatibility. (f) Restoration of compatibility between crystals in (e) by introduction of compatibility stress $\tau_{11}$.

a pair of platens, thereby generating a compressive stress $-\tau_{33}$. If the crystals are anisotropic, the general linear relationship between strain and stress can be written (Hirth and Lothe 1982) in the form

$$
\begin{bmatrix}
\varepsilon_{11} \\
\varepsilon_{22} \\
\varepsilon_{33} \\
2\varepsilon_{23} \\
2\varepsilon_{13} \\
2\varepsilon_{12}
\end{bmatrix}
=
\begin{bmatrix}
S_{11} & S_{12} & S_{13} & S_{14} & S_{15} & S_{16} \\
S_{12} & S_{22} & S_{23} & S_{24} & S_{25} & S_{26} \\
S_{13} & S_{23} & S_{33} & S_{34} & S_{35} & S_{36} \\
S_{14} & S_{24} & S_{34} & S_{44} & S_{45} & S_{46} \\
S_{15} & S_{25} & S_{35} & S_{45} & S_{55} & S_{56} \\
S_{16} & S_{26} & S_{36} & S_{46} & S_{56} & S_{66}
\end{bmatrix}
\begin{bmatrix}
\tau_{11} \\
\tau_{22} \\
\tau_{33} \\
\tau_{23} \\
\tau_{13} \\
\tau_{12}
\end{bmatrix}
\tag{12.1}
$$

where the $s_{ij}$ are the usual compliances. The stress $-\tau_{33}$ will therefore generally induce a $\varepsilon_{13}$ shear strain. If the black and white crystals are of the same phase and are oriented so that $\varepsilon_{13}^b = -\varepsilon_{13}^w$, and were not joined at the interface, they would shear in opposite directions and assume the shapes shown in Fig. 12.1(b). However, since they actually are joined, the final configuration will appear as in Fig. 12.1(c) where the crystals are forced into compatibility at the interface by $\tau_{13}$ shear stresses as shown schematically. These stresses are most intense at the interface and decay with distance away from the interface: they are compatibility stresses caused by the elastic anisotropy.

In the second example (Fig. 12.1(d, e, f,)) we show a bicrystal composed of two crystals which are isotropic, but of different phase, which is stretched elastically in tension along the $x_3$ axis. The tensile stress, $\tau_{33}$, will produce a strain, $\varepsilon_{11}$, in each crystal. If the black and white crystals were not joined at the interface, the $\varepsilon_{11}$ strains in each crystal would be different because of their different elastic constants as shown in Fig. 12.1(e). However, since they actually are joined, the final configuration will appear as in Fig. 12.1(f) where the crystals are forced into compatibility at the interface by means of $\tau_{11}$ stresses. Again, these stresses are most intense at the interface: they are compatibility stresses caused by the different elastic properties of the black and white crystals.

An approximate general analysis for the compatibility stresses in the direct vicinity of an interface in a bulk bicrystal specimen at a distance relatively far from any free surfaces may be obtained in a comparatively simple manner following Gemperlova *et al.* (1989). We begin with a finite bicrystal (as in Fig. 12.1(a)) and assume the most general case where the two crystals are now of different phase and also elastically anisotropic. The surfaces are then loaded with applied forces, $F_j$, so that the applied stresses produced at the surfaces by these forces, $\tau_{ij}^A$, take on constant values. The relationship between these forces and stresses is then

$$F_j = \sum_i \hat{n}_i \tau_{ij}^A, \tag{12.2}$$

where $\hat{n}$ is the surface normal. Equation (12.2), with each $\tau_{ij}^A$ established at a constant value, therefore holds. We note that this surface loading would produce constant stresses throughout the body corresponding to each $\tau_{ij}^A$ if the body were homogeneous. We shall call these hypothetical uniform stresses the 'applied stressess'. In actuality, additional stresses, i.e. compatibility stresses designated by $\tau_{ij}^C$, will appear because of the inhomogeneity associated with the bicrystalline nature of the body. The compatibility stresses can then be regarded as the differences between the actual total stresses in the body, $\tau_{ij}^T$, and the applied stresses, i.e.

$$\tau_{ij}^C = \tau_{ij}^T - \tau_{ij}^A. \tag{12.3}$$

The total stresses, $\tau_{ij}^T$, must fulfil the equilibrium boundary conditions

$$F_j = \sum_i \hat{n}_i \tau_{ij}^A = \sum_i \hat{n}_i \tau_{ij}^T \tag{12.4}$$

at the surface, and

$$\tau_{i2}^{T,b} = \tau_{i2}^{T,w} \tag{12.5}$$

at the interface. (Note the use of the superscripts b and w to indicate quantities in the black and white crystals when necessary.) In addition, the total strains produced by the total stresses, i.e. $\varepsilon_{ij}^T$, must satisfy the following compatibility relations at the interface:

$$\varepsilon_{11}^{T,b} = \varepsilon_{11}^{T,w}; \quad \varepsilon_{33}^{T,b} = \varepsilon_{33}^{T,w}; \quad \varepsilon_{13}^{T,b} = \varepsilon_{13}^{T,w}. \tag{12.6}$$

The solution of this elastic problem is highly complicated and will generally yield non-uniform $\tau_{ij}^C$ and $\tau_{ij}^T$ stresses which depend upon the shape of the bicrystal. However, the compatibility stresses which are generated in the direct vicinity of the interface at locations relatively far from the external surfaces can be determined approximately in a relatively simple manner by allowing the bicrystal to expand into an infinite body consisting of two crystal halfspaces. In this situation the compatibility stresses in each half space must assume constant values as discussed (for example) by Gemperlova *et al.* (1989). In finite bodies they will vary throughout the body as mentioned above and could presumably be obtained by using the constant values obtained for the infinite body and adding suitable image terms to account for the presence of the surface. These image terms would make only relatively small contributions at large distances from the surface, and, thus the solution for the infinite body may be used as an approximation for the solution for the finite body in the vicinity of the interface at locations far from the surface.

For the infinite bicrystal containing uniform stresses in each halfspace and subjected to the applied stress, $\tau_{ij}^{A}$, mechanical equilibrium requires

$$\left.\begin{array}{ll}\tau_{i2}^{T,b} = \tau_{i2}^{T,w} = \tau_{i2}^{A}; & \tau_{11}^{T,b} = \tau_{11}^{T,w} = 2\tau_{11}^{A} \\ \tau_{33}^{T,b} = \tau_{33}^{T,w} = 2\tau_{33}^{A}; & \tau_{13}^{T,b} = \tau_{13}^{T,w} = 2\tau_{13}^{A}\end{array}\right\}. \qquad (12.7)$$

Therefore, using eqn (12.3),

$$\left.\begin{array}{lll} \tau_{i2}^{C,b} = \tau_{i2}^{C,w} = 0 \\ \tau_{11}^{C,b} = -\tau_{11}^{C,w}; & \tau_{33}^{C,b} = -\tau_{33}^{C,w}; & \tau_{13}^{C,b} = -\tau_{13}^{C,w} \end{array}\right\}. \qquad (12.8)$$

Therefore, no compatibility shear stresses parallel to the interface or normal stresses perpendicular to the interface exist, and the remaining three compatibility stresses reverse sign at the interface.

We may now relate compatibility stresses and strains. Let

$$\Delta\varepsilon_{ij}^{C} \equiv \varepsilon_{ij}^{C,b} - \varepsilon_{ij}^{C,w}, \qquad (12.9)$$

where $\varepsilon_{ij}^{C}$ is the strain due to the compatibility stress, $\tau_{ij}^{C}$. Using eqns (12.1) and (12.8)

$$\left.\begin{array}{l} \Delta\varepsilon_{11}^{C} = (s_{11}^{b} + s_{11}^{w})\tau_{11}^{C,b} + (s_{13}^{b} + s_{13}^{w})\tau_{33}^{C,b} + (s_{15}^{b} + s_{15}^{w})\tau_{13}^{C,b} \\ \Delta\varepsilon_{33}^{C} = (s_{13}^{b} + s_{13}^{w})\tau_{11}^{C,b} + (s_{33}^{b} + s_{33}^{w})\tau_{33}^{C,b} + (s_{35}^{b} + s_{35}^{w})\tau_{13}^{C,b} \\ 2\Delta\varepsilon_{13}^{C} = (s_{15}^{b} + s_{15}^{w})\tau_{11}^{C,b} + (s_{35}^{b} + s_{35}^{w})\tau_{33}^{C,b} + (s_{55}^{b} + s_{55}^{w})\tau_{13}^{C,b} \end{array}\right\}. \qquad (12.10)$$

A further relation for $\Delta\varepsilon_{ij}^{C}$ may be obtained from the requirement that

$$\varepsilon_{ij}^{T} = \varepsilon_{ij}^{A} + \varepsilon_{ij}^{C}. \qquad (12.11)$$

Using eqns (12.6), (12.9), and (12.11), we then obtain

$$\Delta\varepsilon_{ij}^{C} = \varepsilon_{ij}^{A,w} - \varepsilon_{ij}^{A,b}. \qquad (12.12)$$

We now have all the relationships necessary to calculate the compatibility stresses for a given loading of the bicrystal. Such a loading establishes the applied stresses, $\tau_{ij}^{A}$, through eqn (12.2). The strains $\varepsilon_{ij}^{A,w}$ and $\varepsilon_{ij}^{A,b}$ can then be calculated using eqn (12.1). This allows the calculation of the $\Delta\varepsilon_{ij}^{C}$ by means of eqn (12.12), and, finally, the calculation of the $\tau_{ij}^{C,b}$ by means of eqn (12.10).

Gemperlova *et al.* (1989) present a number of detailed results for various cases. For an isotropic material

$$\left.\begin{array}{ll} s_{11} = s_{33} = 1/E; & s_{15} = s_{35} = 0 \\ s_{13} = -\nu/E; & s_{55} = 2(1 + \nu)/E = 1/\mu \end{array}\right\}. \qquad (12.13)$$

where $E$ is Young's modulus, $\nu$ is Poisson's ratio, and $\mu$ is the shear modulus. If the black and white crystals are isotropic, but have different elastic constants, we then have

$$\left.\begin{array}{l} \tau_{11}^{C,b} = \dfrac{E^{b}E^{w}[(E^{b} + E^{w})\Delta\varepsilon_{11}^{C} + (\nu^{b}E^{w} + \nu^{w}E^{b})\Delta\varepsilon_{33}^{C}]}{[E^{b}(1 + \nu^{w}) + E^{w}(1 + \nu^{b})][E^{b}(1 - \nu^{w}) + E^{w}(1 - \nu^{b})]} \\[4mm] \tau_{33}^{C,b} = \dfrac{E^{b}E^{w}[(\nu^{b}E^{w} + \nu^{w}E^{b})\Delta\varepsilon_{11}^{C} + (E^{b} + E^{w})\Delta\varepsilon_{33}^{C}]}{[E^{b}(1 + \nu^{w}) + E^{w}(1 + \nu^{b})][E^{b}(1 - \nu^{w}) + E^{w}(1 - \nu^{b})]} \\[4mm] \tau_{13}^{C,b} = \dfrac{2\mu^{b}\mu^{w}}{(\mu^{b} + \mu^{w})}\Delta\varepsilon_{13}^{C}. \end{array}\right\} \qquad (12.14)$$

When the elastic constants of the two crystals are the same, $\Delta \varepsilon_{ij}^{C} = 0$ from eqn (12.12), and eqn (12.14) predicts $\tau_{ij}^{C,b} = 0$, as expected. It is readily seen from eqns (12.12) and (12.14) that compatibility stresses which are of the same order as the applied stresses can be reached depending upon the differences between the elastic constants of the two crystals.

When both crystals are anisotropic the results become considerably more complicated. Gemperlova *et al.* (1989) present a straightforward, but lengthly, formalism, based on the previous results, for calculating the compatibility stresses in such cases which we shall not reproduce here. They then calculate a number of results for homophase interfaces in Fe/Si and Cu/Zn alloys and show that the elastic anisotropy can produce compatibility stresses which are often of the order of the applied stress depending upon the geometrical arrangement.

As already pointed out, the above results are only applicable for finite bicrystals at positions near the interface and relatively far from any surface. For example, they should predict the correct compatibility stresses at the interface near the origin in the center of the finite bicrystal in Fig. 12.1(f) which is subjected to a tensile stress along $x_3$ and where we expect the compatibility stress $\tau_{11}^{C}$ to have a large magnitude which reverses sign across the interface. For the finite bicrystal, however, $\tau_{11}^{C}$ must eventually fall to zero on the free surface corresponding to $x_1 = $ constant in order to satisfy the boundary condition there. This latter result may be attributed to a surface image stress, and its calculation would be considerably more complicated than our simple infinite bicrystal calculation.

A number of detailed calculations have been made of compatibility stresses in various finite bicrystals under various loadings, taking account of the specimen shape and surface boundary conditions, and in some cases employing approximations (Chou and Hirth 1970, Meyers and Ashworth 1982, Hashimoto and Margolin 1983, Kitagawa *et al.* 1986). Generally, it has been found that the compatibility stresses are most intense at the interface and are distributed in complicated ways depending upon the loading direction and specimen geometry. Celinski and Kurzydkowski (1982) and Kurzydkowski *et al.* (1980) have calculated compatibility stresses by the finite element method in Cu bicrystals and tricrystals respectively and found total stresses in certain regions at the interfaces which exceeded the applied stress by factors as large as about 3 (see discussion by Varin *et al.* (1987). In still further work, Tvergaard and Hutchinson (1988) and Ghahremani *et al.* (1990) have considered compatibility stresses in polycrystals in 2D at junctions where grains meet along a line and also in 3D at vertices where grains meet at a point. In the latter case the vertices were approximated by representing one crystal as a cone which is embedded into a surrounding matrix. Stress concentrations at the junctions and vertices were found and studied as a function of mismatches due to elastic anisotropy and differences in the elastic constants of the grains. The results are generally complicated, and details may be found in the above references.

Compatibility stresses can cause important physical effects. They can cause interface decohesion (fracture), which is a known failure mode for axially strained composites (Kelly and Davies 1965, Hirth 1972), and they can also contribute to cracking in brittle polycrystalline materials as discussed in Section 12.9.2. They can also exert forces on lattice dislocations near interfaces. For example, Chou and Hirth (1970) analyse the motion of lattice screw dislocations in the compatibility stress field of an interface between elastically 'hard' and 'soft' crystals and show how the dislocations will be forced to converge in the harder crystal at the interface in a manner which could promote fracture. In addition, as discussed in Section 12.5.2, compatibility stresses can aid in the generation of lattice dislocations in the vicinities of interfaces and therefore play an important role in the plastic deformation of bicrystals and polycrystals.

## 12.2.2 Compatibility stresses caused by plastic straining

In the previous section we discussed the compatibility stresses which are generally produced when polycrystals are subjected to applied forces (stresses), and all of the induced strain is purely elastic. We now consider the further compatibility stresses which may be generated when the stresses exceed the flow stress and reach sufficiently high levels to induce plastic straining by means of the glide of lattice dislocations. In general, the individual crystals in a polycrystal will then tend to undergo different plastic strains, since they will generally be subjected to different local stresses, and, in addition, their potential slip systems will be oriented differently in the local stress fields which are present. When different plastic strains occur in adjacent crystals and no boundary sliding is allowed, mismatches will tend to develop, and compatibility stresses must then be generated.

As demonstrated below, the compatibility stress due exclusively to any plastic deformation (in the absence of boundary sliding) may be readily calculated through use of the results obtained in Section 12.2.1 when the local plastic deformation in each adjoining crystal is known. The total compatibility stress present at a boundary in a polycrystal subjected to applied forces will then be the sum of the compatibility stress due to elastic incompatibility as described in Section 12.2.1 and the additional compatibility stress due to the plastic deformation.

The difference between the plastic strains in the two crystals adjoining a boundary may be expressed in terms of the amounts of shear contributed by the active slip systems in the two crystals. Each slip system is described in terms of the unit vector parallel to the slip direction, $\hat{d}$, and the unit vector normal to the slip plane, $\hat{p}$. The strain produced by a shear, $\gamma$, on this system then has the form

$$\varepsilon_{ij}^{P\prime} = \gamma \begin{bmatrix} 0 & \frac{1}{2} & 0 \\ \frac{1}{2} & 0 & 0 \\ 0 & 0 & 0 \end{bmatrix}, \tag{12.15}$$

in a Cartesian coordinate system (indicated by the primed superscript) where the 1 axis is parallel to $\hat{d}$, and the 2 axis is parallel to $\hat{p}$. However, if we employ a coordinate system in which the components of $\hat{d}$ and $\hat{p}$ are $(d_1, d_2, d_3)$ and $(p_1, p_2, p_3)$ respectively, the strain field given by eqn (12.15) takes the form

$$\varepsilon_{ij}^{P} = \gamma \begin{bmatrix} d_1 p_1 & \frac{1}{2}[d_1 p_2 + d_2 p_1] & \frac{1}{2}[d_1 p_3 + d_3 p_1] \\ \frac{1}{2}[d_1 p_2 + d_2 p_1] & d_2 p_2 & \frac{1}{2}[d_2 p_3 + d_3 p_2] \\ \frac{1}{2}[d_1 p_3 + d_3 p_1] & \frac{1}{2}[d_2 p_3 + d_3 p_2] & d_3 p_3 \end{bmatrix}. \tag{12.16}$$

The total plastic strain may then be found by referring all of the strains contributed by the various slip systems to a common coordinate system and summing them to obtain

$$\varepsilon_{ij}^{P,T} = \frac{1}{2} \sum_{k} (d_i^k p_j^k + d_j^k p_i^k) \gamma^k, \tag{12.17}$$

where k indicates the $k$th slip system.

When the total plastic strain, $\varepsilon_{ij}^{P,T}$, in the two crystals adjoining a boundary differ, the strain mismatch must be compensated by an elastic compatibility strain and a corresponding compatibility stress. Using results derived previously in Section 12.2.1, expressions for the compatibility stresses are readily obtained for the simple case where elastic anisotropy can be ignored, and the black and white crystals are of the same phase. Employing eqns (2.10) and (12.13), these stresses are then

$$\tau_{11} = \left(\Delta\varepsilon_{11}^{P,T} + \nu\Delta\varepsilon_{33}^{P,T}\right)\left(\mu/[1-\nu]\right)$$

$$\tau_{33} = \left(\nu\Delta\varepsilon_{11}^{P,T} + \Delta\varepsilon_{33}^{P,T}\right)\left(\mu/[1-\nu]\right) \tag{12.18}$$

$$\tau_{13} = \Delta\varepsilon_{13}^{P,T}\,\mu,$$

where $\Delta\varepsilon_{ij}^{P,T}$ is the difference between the plastic strains, $\varepsilon_{ij}^{P,T}$, in the two crystals (Gemperlova *et al.* 1989).

Of course, the above result is still incomplete, since we now require values of the $\Delta\varepsilon_{ij}^{P,T}$ strains. These could be determined in a self-consistent manner in particular cases from knowledge of the magnitudes of the stresses present, the orientations of the various slip systems available, and the critical shear stresses required to activate slip on the various systems.

### 12.2.3   Compatibility stresses caused by heating/cooling

Compatibility stresses will also be generated whenever a polycrystal is heated or cooled and the thermal expansion coefficients of the individual grains are different due to thermal expansion anisotropy or differences in phase (or both). In such cases adjacent grains will attempt to change dimensions and develop mismatches by amounts controlled by the parameter $\Delta\alpha\cdot\Delta T$, where $\Delta\alpha$ is the difference between the thermal expansion coefficients in the appropriate directions, and $\Delta T$ is the temperature change. One of the convenient techniques used to analyse the required compatibility stresses when no boundary sliding occurs is the procedure of Eshelby (Eshelby 1957, Evans 1978) in which the grains are separated and allowed to undergo unconstrained thermal expansion/contraction. Surface forces are then applied to restore them to their original shapes and they are then rejoined. Finally, interface tractions are imposed to establish appropriate stress continuity throughout the system.

Analyses of the compatibility stresses due to thermal expansion mismatch have been made by a number of investigators: Evans (1978), Clarke (1980), Davidge (1981), Fu and Evans (1985), Tvergaard and Hutchinson (1988), and Ghahremani *et al.* (1990). These studies have ranged from 2D cases where the grains have been assumed to have anisotropic thermal expansion coefficients but to be elastically isotropic, to 3D studies where both thermal expansion coefficients and elastic constants were anisotropic. In recent work (Ghahremani *et al.* (1990), 2D studies at junctions (where grains meet along lines) and 3D vertices (where grains meet at a point) for cases involving both thermal expansion mismatch and elastic constant mismatch were performed. Again, as described in Section 12.2.1, the vertices were approximated by representing one crystal as a cone embedded into a surrounding matrix. The stress concentrations found in these studies are relevant to the nucleation of cracks, particularly in brittle polycrystalline materials as discussed in Section 12.9.2.

## 12.3 ELASTIC INTERACTIONS BETWEEN DISLOCATIONS AND INTERFACES

Lattice dislocations will generally experience several types of forces when they are in the vicinity of interfaces. Firstly, such a dislocation will feel the intrinsic stress field, $\tau_{ij}^I$, of the interface itself which has already been described in Sections 2.10.2, 2.10.4, and 2.10.5. Secondly, if an applied stress is present, the dislocation will feel an additional stress equal to the sum of the applied stress, $\tau_{ij}^A$, and any compatibility stress, $\tau_{ij}^C$, as just described above in Section 12.2. The force on the dislocation, $F$, due to the sum of these stresses, will then be given by the Peach–Koehler equation (Hirth and Lothe 1982) in the form

$$F = (b \cdot [\tau^I + \tau^A + \tau^C]) \times \hat{\xi}, \tag{12.19}$$

where $b$ is the Burgers vector and $\hat{\xi}$ is the unit vector tangent to the dislocation line. Of course, if any additional sources of stress were present, such as extrinsic interface dislocations or nearby lattice dislocations, their stresses would have to be added in. However, still a further source of force will generally be present due to the different elastic properties of the crystals which abut the interface. If the dislocation is displaced with respect to the interface, the fraction of its total elastic stress field which lies in each crystal will vary. Since the elastic properties of the two crystals generally differ, the total elastic self-energy of the dislocation will vary with its position. It will therefore experience a force. As discussed below, this force can generally be described in terms of the forces produced by fictitious 'image' dislocations, and we therefore term this force the 'image force', $F^{IM}$, produced by the interface.

A simple example is one in which a right-handed lattice screw dislocation lies parallel to an interface between two elastically isotropic crystals of different phase which possess different shear moduli, $\mu^b$ and $\mu^w$, as illustrated in Fig. 12.2(a). As we now demonstrate, an image force will be present which can be found conveniently by introducing two right-handed screw image dislocations. The real dislocation, A, with Burgers vector $b$ is at $x = a$: the first image dislocation, B, with Burgers vector $\beta b$ is also at $x = a$: the second, C, with Burgers vector $\gamma b$ is at $x = -a$. The stress field in the black crystal is due to dislocations A and C, while that in the white crystal is due to B. Therefore, the elastic displacements are given by (Hirth and Lothe 1982),

$$\left. \begin{array}{l} u_z^b(x, y) = (b/2\pi) \cdot \tan^{-1}[y/(x-a)] + (b\gamma/2\pi) \cdot \tan^{-1}[y/(x+a)] \\ u_z^w(x, y) = (\beta b/2\pi) \cdot \tan^{-1}[y/(x-a)] \end{array} \right\}. \tag{12.20}$$

Since we must have $u_z^b(0, y) = u_z^w(0, y)$ at the interface,

$$\beta = 1 - \gamma. \tag{12.21}$$

In addition, we must have $\tau_{xz}^b(0, y) = \tau_{xz}^w(0, y)$ in order to satisfy mechanical equilibrium. Using the standard relations $\tau_{xz} = 2\mu\varepsilon_{xz}$ and $\varepsilon_{xz} = (\partial u_z/\partial x + \partial u_x/\partial z)/2$, we then find

$$\left. \begin{array}{l} \tau_{xz}^b = -\dfrac{\mu^b by}{2\pi[(x-a)^2 + y^2]} - \dfrac{\gamma\mu^b by}{2\pi[(x+a)^2 + y^2]} \\[3mm] \tau_{xz}^w = -\dfrac{\mu^w \beta by}{2\pi[(x-a)^2 + y^2]} \end{array} \right\}. \tag{12.22}$$

The condition of mechanical equilibrium therefore requires

$$\mu^b(1 + \gamma) = \mu^w \beta, \tag{12.23}$$

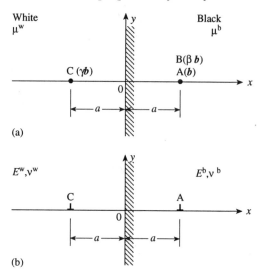

**Fig. 12.2**  (a) Lattice screw dislocation, $A$, with Burgers vector $b$, lying along $z$ at $(a, 0)$ in the black crystal of a bicrystal. The interface is at $x = 0$, and the black and white crystals are of different phase but isotropic. Image screw dislocations B and C, parallel to $z$. are present at $(a, 0)$ and $(-a, 0)$ and possess Burgers vectors $\beta b$ and $\gamma b$, respectively. (b) Lattice edge dislocation, $A$, lying along $z$ at $(a, 0)$ in the black crystal of a bicrystal. The black and white crystals are again of different phase but elastically isotropic. Image edge dislocation, $C$, parallel to $z$, is present at $(-a, 0)$.

and by combining eqns (12.21) and (12.23) we obtain

$$\gamma = \frac{\mu^{\text{w}} - \mu^{\text{b}}}{\mu^{\text{w}} + \mu^{\text{b}}}; \quad \beta = \frac{2\mu^{\text{b}}}{\mu^{\text{w}} + \mu^{\text{b}}}. \tag{12.24}$$

Application of the Peach–Koehler equation shows that dislocation A will experience an image force directed along $x$ given by $F^{\text{IM}} = b\tau_{yz}^{\text{IM}}\hat{\imath}$, where $\tau_{yz}^{\text{IM}}$ is the stress in the black crystal produced by the image dislocation C. This stress is simply $\tau_{yz}^{\text{IM}} = \mu^{\text{b}}b\gamma/4\pi a$, and therefore with the help of eqn (12.24), we finally obtain

$$F^{\text{IM}} = \frac{\mu^{\text{b}}(\mu^{\text{w}} - \mu^{\text{b}})b^2}{(\mu^{\text{w}} + \mu^{\text{b}})4\pi a}\hat{\imath}. \tag{12.25}$$

When the white crystal is 'harder' than the black crystal in shear (i.e. when $\mu^{\text{w}} > \mu^{\text{b}}$), the dislocation is repelled from the interface, and vice versa when $\mu^{\text{b}} > \mu^{\text{w}}$. This is in agreement with the intuitive prediction that the dislocation will strive to have as little of its strain field in the hard crystal, and as much in the soft crystal, as possible. We note that in the limit when $\mu^{\text{w}} = 0$, $F^{\text{IM}} = -(\mu^{\text{b}}b^2/4\pi a)\hat{\imath}$, and we recover the well known expression (Hirth and Lothe 1982) for the image force tending to pull a screw dislocation out of a single halfcrystal at a free surface. The image force given by eqn (12.25) can be significant. For example, if the difference between $\mu^{\text{b}}$ and $\mu^{\text{w}}$ is 10 per cent, as is easily possible, and the dislocation is located at a distance $10^2 b$ from the interface, $|F^{\text{IM}}| \simeq 4 \times 10^{-5}\mu b$.

The corresponding problem for an edge dislocation is more complicated, since a more complex stress field is involved. However, Head (1953) obtained a solution for the edge dislocation shown at A in Fig. 12.2(b) where the two crystals are isotropic with different elastic constants $(E^w, \nu^w)$ and $(E^b, \nu^b)$ respectively. The image force on the dislocation at A for glide (i.e., along $x$) can be represented by the effect of an image edge dislocation located at C having a Burgers vector which depends upon the elastic constants in both crystals in a complicated manner. However, when $\nu^b = \nu^w$ and $E^w > E^b$, the results show directly that the image dislocation has the same sign as the real dislocation and that the real dislocation is again repelled from the interface. Again, the dislocation will attempt to reduce the extent of its strain field in the 'harder' crystal. Further behaviour may be deduced from the detailed results of Head (1953).

More general solutions which also take into account the anisotropy of the two abutting crystals have also been obtained. Anisotropy should be included, of course, since it will affect the degree to which the misoriented crystal facing the dislocation across the interface may appear elastically 'harder' or 'softer' to the dislocation than its host crystal. Gemperlova and Saxl (1968) and Barnett and Lothe (1974) solved the problem for dislocations parallel to the interface, while Belov *et al.* (1983) solved it for a dislocation impinging upon an interface at an angle and leaving it on the other side at a different angle (i.e. for a dislocation 'refracted' at the interface). Additional results and discussion are given by Bonnet (1987). The results for the refracted dislocation are complex and can only be evaluated numerically: we therefore refer the reader to the original paper for details. The results for parallel dislocations are simpler (Barnett and Lothe 1974) and can be described in terms of a relatively simple image force theorem. This theorem states that a straight dislocation parallel to the interface at a distance $h$ experiences a force normal to the interface given by

$$F^{\mathrm{IM}} = (E^\infty - E^{\mathrm{b,w}})/h = \Delta E/h, \tag{12.26}$$

where $E^\infty$ is the 'pre-logarithmic' energy factor of the same dislocation in an infinite crystal which is elastically identical to the crystal in which it resides, and $E^{\mathrm{b,w}}$ is the pre-logarithmic energy factor of the same dislocation located at the interface of the bicrystal. (We note that the elastic energy of a dislocation generally consists of the product of a logarithmic term, containing the dimensions of the body in which it lies along with a cutoff core radius, and a pre-logarithmic term (Hirth and Lothe 1982).) Furthermore, the energies $E^\infty$ and $E^{\mathrm{b,w}}$ depend only on the crystallographic direction of the dislocation line (i.e., the tangent vector, $\xi$) and its Burgers vector relative to the black and white crystals. When these directions are fixed, the image force is the same for all interfaces whose zonal axis is parallel to the dislocation line as long as the distance $h$ is held constant. The energy $E^\infty$ can be calculated relatively easily using standard results for dislocations in anisotropic media (Hirth and Lothe 1982), while $E^{\mathrm{b,w}}$ can be calculated by simple numerical integration using a formalism developed by Barnett and Lothe (1974).

It is easily seen that the above image force can be significant under many conditions. For example, $E^\infty$ will be of order $\mu b^2/4\pi$. When $E^{\mathrm{b,w}} \simeq 0.9 E^\infty$, as it may well be, $F^{\mathrm{IM}} = 0.1 \,\mu b^2/4\pi h$. Then, when $h = 10^2 b$, $F^{\mathrm{IM}} \simeq 10 \times {}^{-4} \mu b$. This force is of the same magnitude as the image force found previously for a screw dislocation near an interface separating two half-crystals with different shear moduli. However, $F^{\mathrm{IM}}$ will vary considerably depending upon the geometry of the situation. Despite the fact that $F^{\mathrm{IM}}$ is independent of interface rotation around $\xi$ the interaction of the dislocation with the interface is highly complex. Five degrees of freedom are associated with $\Delta E$ in eqn (12.26) for a dislocation with a fixed Burgers vector. These may be chosen, for example, as two direction cosines to establish $\xi$, and three parameters to describe the misorientation between the black and white crystals as discussed in Section 1.4.2. Khalfallah *et al.* (1990) have carried out a large number of calculations for the interaction between a dislocation and a homophase interface in b.c.c. Fe in order to reveal the range of possibilities. Various interfaces with misorientation axes $\langle 100 \rangle$, $\langle 110 \rangle$, and $\langle 111 \rangle$ were considered, and the dislocation (with $b$ fixed at $\frac{1}{2} \langle 111 \rangle$) was aligned along about 3000 different directions, $\xi$. Large variations in the interaction were found including cases where the image force was either zero, attractive, or repulsive. The original paper may be consulted for details.

So far, we have considered image forces on lattice dislocations which arise because of various differences between the relevant elastic constants of the white and black crystals. However, we can identify still a further image force which will be present to at least some extent whenever the thin core region of the interface possesses effective elastic constants which differ from those of the adjoining black and white bulk crystals. In the case of a large-angle grain boundary we expect the effective elastic constants of the bad material in the boundary core to be somewhat lower than the corresponding constants of the adjoining crystals. An image force will then exist which will urge a lattice dislocation towards the boundary core region, since such a displacement will increasingly relax its stress field and reduce its elastic energy. The extent to which the stress field and the elastic energy are reduced for a screw dislocation located in the centre of such a core has been estimated by Lim (1987*a*) who simply took the core region to be a thin homogeneuos elastic slab possessing a reduced shear modulus. The stress field of the dislocation lying in this composite material was then calculated through the use of an infinite array of images, and the elastic energy was evaluated numerically. The calculations indicate that only a modest effect can be expected. For example, if the boundary core slab is two lattice parameters thick, and the shear modulus is lower by 20 per cent, as is conceivable, the elastic energy of the dislocation is reduced by about 4 per cent. This relatively modest result may be attributed to the fact that even though the reduction of modulus in the core is significant, the boundary core region is relatively very thin.

## 12.4  INTERFACES AS SINKS, OR TRAPS, FOR LATTICE DISLOCATIONS

### 12.4.1  Introduction

When bicrystals (or polycrystals) are plastically deformed by slip, many of the lattice dislocations which are generated in the component crystals and glide through them impinge on the interfaces which are present. Under many conditions these dislocations can enter the interfaces and interact with them in a manner which lowers the total energy of the system. A wide range of interactions is possible in different types of interfaces. These include, for example, the rearrangement of the dislocation structure of the interface upon the introduction of the lattice dislocation, or the dissociation of the lattice dislocation into interface dislocations of different types and the subsequent interaction of these dislocations with the boundary structure. In the ideal limit, an impinged lattice dislocation will become fully incorporated into the interface as a result of these interactions, thereby generating a new intrinsic interface structure which differs from the initial structure because of the change in the total Burgers vector content. In this final intrinsic state all long-range stresses associated with the original impinged dislocation will have been reduced to zero, and the interface will have achieved a new structure corresponding to a minimization of its energy with respect to the addition of the impinged dislocation. As discussed in Section 2.2, the Frank–Bilby equation will then be satisfied, and the impinged dislocation will have made the transition from 'extrinsic' to 'intrinsic' status.

In many real situations this final ideal state will not be reached if the dislocation is limited in its ability to interact. For example, the Burgers vector of the impinged dislocation will not, in general, be parallel to the boundary. Therefore, either the impinged dislocation, or at least a portion of its dissociation products, will have to climb in the boundary in order to become fully incorporated. At low temperatures this process will be essentially frozen out. In such a case the impinged dislocation can become only partially incorporated and it will therefore retain much of its extrinsic status. When the degree of incorporation is sufficient to bind the impinged dislocation permanently, we may regard the interface as a dislocation sink. On the other hand, under certain conditions an impinged lattice dislocation which has not been significantly incorporated may be removed (stripped) from the interface by forces which are present, and in such a case we must regard the interface as merely a trap. An example of this is shown in Fig. 12.3, where a segment of a dislocation which is trapped in a boundary (and is attached to a segment of lattice dislocation) is removed from the boundary.

As discussed later in this chapter, the extent to which interfaces can act as sinks for lattice dislocations and incorporate them into their intrinsic structures will have a strong effect on the mechanical properties of materials. The detailed processes which occur will depend directly upon the initial structure of the interface, the magnitudes of the forces

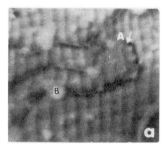

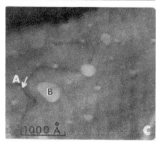

**Fig. 12.3** *In situ* electron microscope sequence showing removal of trapped extrinsic dislocation segment from boundary by an applied stress. (a) Initial configuration showing extrinsic segment in boundary passing through bubble (at **B**) and attached to lattice dislocation on the right which intersects the specimen surface at **A**. (b) Partial removal of extrinsic boundary dislocation by motion of lattice dislocation segment to the left under an applied stress. (c) Further removal of the boundary dislocation. Vicinal Σ5 ⟨001⟩ twist boundary in Au. (From Balluffi *et al.* (1972).)

present, and the degree of thermal activation available. Innumerable situations are
therefore possible. We shall therefore be content in the following to discuss a number
of representative examples involving different types of interfaces.


### 12.4.2  Small-angle grain boundaries

The sink action of small-angle boundaries can be described completely in terms of the
interactions which occur between the incoming lattice dislocations and the primary
dislocations comprising the initial boundary. A particularly simple and instructive
example is the case of lattice edge dislocations impinging upon a tilt boundary as
illustrated in Fig. 12.4. We consider the ideal limiting case where the system maintains
a state of minimum energy as the dislocations leave the lattice and become fully
incorporated in the boundary. In this process the dislocations remaining in the lattice
continuously adjust their positions (by glide and climb) so that they remain uniformly
distributed, and the dislocations in the boundary continuously adjust their positions (by
climb) so that they accommodate the added dislocations and maintain a boundary
structure of mimimum energy. As discussed by Nye (1953), when the lattice dislocations
remain uniformly distributed, the lattice is uniformly bent, and no long-range stresses
are present. We note that a situation related to this has already been discussed in Section
2.2 (see Fig. 2.5(c)) where a distribution of lattice dislocations (stress annihilators) is
present which cancels the long-range stress which would otherwise be present because of
lattice bending. Also, of course, no long-range stresses are generated by the boundary
as long as the added dislocations are fully incorporated and the Frank-Bilby equation
is satisfied. As the added dislocations become incorporated, the total Burgers vector
content of the boundary increases, and the average dislocation spacing, $\bar{d}$, therefore
decreases. This will cause a continuous increase in the tilt angle, $\theta$, which is easily
calculated. Letting $N_d^B$ be the number of dislocations in the boundary in distance $s$ along
$y$, we have $\bar{d} = s/N_d^B$, and since $\bar{d} = b/\theta$, the increase in $\theta$ per dislocation added in the
distance s is

$$d\theta/dN_d^B = b/s. \tag{12.27}$$

We are interested next in the energy which is available to drive the above process. As
already pointed out in Chapter 2, the energy per unit length of each dislocation in the
boundary, $E_d^B$ is given approximately by eqn (2.141), since the stress field of each dislocation extends over a distance corresponding approximately to the dislocation spacing,
$\bar{d}$ in the boundary. The energy of a patch of boundary extending $s$ along $y$ and unity
along $z$ is then

$$E^B = N_d^B E_d^B = N_d^B \left[ \frac{\mu b^2}{4\pi(1-\nu)} \ln(s/N_d^B r_0) + E_c \right]. \tag{12.28}$$

Therefore,

$$\partial E^B/\partial N_d^B = \frac{\mu b^2}{4\pi(1-\nu)} \left[ \ln(\bar{d}/r_0) - 1 \right] + E_c. \tag{12.29}$$

Similarly, the stress field of each lattice dislocation extends over a distance corresponding
to the distance $R$ in Fig. 12.4 which is approximately the dislocation spacing in the lattice
and is given to an acceptable degree of approximation by $\pi R^2 N_d^L = A$, where $N_d^L$ is the
number of lattice dislocations threading area $A$. The energy contributed by each lattice

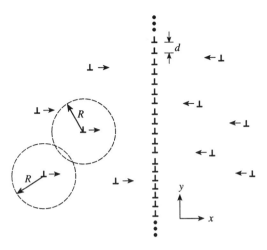

Fig. 12.4  Lattice edge dislocations impinging upon a small-angle tilt boundary. $R$ is approximately the lattice dislocation spacing.

dislocation, $E_d^L$, is then of the same form as that corresponding to eqn (12.28), and, therefore, the energy of the dislocations threading area $A$ (per unit dislocation length) is

$$E^L = N_d^L E_d^L = N_d^L \left[ \frac{\mu b^2}{8\pi(1-\nu)} \ln(A/\pi N_d^L r_o^2) + E_c \right]. \tag{12.30}$$

Using this result,

$$\partial E^L / \partial N_d^L = \frac{\mu b^2}{4\pi(1-\nu)} \left[ \ln(R/r_o) - \tfrac{1}{2} \right] + \cdot E_c. \tag{12.31}$$

The change in total energy which occurs when a dislocation leaves the lattice and joins the boundary is then

$$\Delta E = [\partial E^B / \partial N_d^B - \partial E^L / \partial N_d^L] = -\frac{\mu b^2}{4\pi(1-\nu)} \left[ \ln(R/\bar{d}) + \tfrac{1}{2} \right]. \tag{12.32}$$

The energy change is seen to depend directly upon the square of the Burgers vector and more weakly, via the logarithmic term, on the geometrical parameters $R$ and $\bar{d}$. Since $R \gg \bar{d}$ (see Fig. 12.4), $\Delta E$ will be strongly negative: the energy will therefore decrease significantly, causing the incorporated dislocation to be tightly bound and the boundary to act as an effective sink. Of course, if the Burgers vector of the lattice dislocations and grain boundary dislocations were of opposite sign, mutual annihilation would occur, and the energy decrease would be even larger.

We may conclude on the basis of this simple example that any small-angle boundary should be capable of acting as a sink for lattice dislocations if the incoming dislocations are able to incorporate themselves sufficiently into an intrinsic boundary structure. At the very least, this incorporation will reduce the total elastic energy of the system and in many cases will also decrease the total dislocation line length. In many cases, as in the above example, the incorporation process will require dislocation climb and therefore thermal activation.

The above example is highly relevant to the well known process of polygonization (Cahn 1949, Nye 1953, Young 1958) which occurs when an initial distribution of lattice dislocations in a plastically bent crystal becomes incorporated during subsequent annealing into small-angle tilt boundaries. An example of this phenomenon is shown in

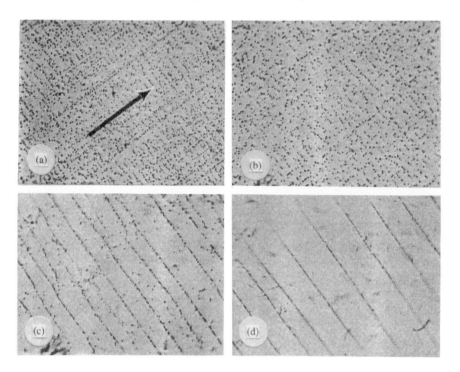

**Fig. 12.5** Formation of small-angle tilt boundaries by polygonization in a plastically bent single crystal of Cu. Arrow is parallel to single operating slip planes. Dislocations are visible via etch pits. (a) Dislocation distribution in early stage after annealing at 500 °C (b) More advanced stage after annealing at 700 °C. (c) After annealing at 900 °C, the dislocations are now mostly arranged in walls (small-angle tilt boundaries) running perpendicular to the original slip planes. (d) After annealing at 1000 °C, the tilt boundary spacing has coarsened. (From Young (1958).)

Fig. 12.5. The distribution of lattice dislocations present in Cu in an early stage of poly-gonization after plastic bending (via single slip on the slip plane indicated) and annealing at 500 °C is shown in Fig. 12.5(a). Upon more extensive annealing at successively higher temperatures, the dislocations progressively rearrange themselves into small-angle tilt boundaries running perpendicular to the slip plane as seen in Figs. 12.5(b, c). Once well defined tilt boundary segments form, the structure coarsens (Fig. 12.5(d)) by the coalescence of the segments at 'Y' junctions of the type visible in Fig. 12.5(c) (Gilman 1955, Young 1958). It is easily verified, through use of eqn (12.28), that the total energy of the system is reduced by this coalescence. Even though the polygonization does not occur by the addition of single lattice dislocations to existing boundaries as visualized in the ideal example analysed previously, the decrease in total energy (per dislocation) which is achieved by the process is given approximately by eqn (12.32). In essence, the phenomenon may be regarded as a dislocation 'precipitation' process with the small-angle boundaries acting as the dislocation sinks.

Innumerable situations exist where portions of lattice dislocations impinged on boundaries are embedded in the boundary structure. In such cases the boundary can be regarded as either a sink or a trap for the embedded segments which remain connected to the lattice dislocations and are extrinsic to the boundary structure. Examples are seen in Fig. 12.6 for several different types of lattice dislocations impinged on a small-angle

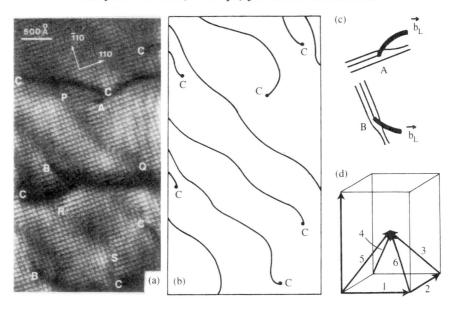

**Fig. 12.6** (a) Extrinsic grain boundary dislocations in a vicinal $\Sigma 1$ [001] twist boundary ($\theta = 2.5°$) in Au. Intrinsic background structure consists of square network of primary screw dislocations possessing Burgers vectors $b_1 = \frac{1}{2}[110]$ and $b_2 = \frac{1}{2}[1\overline{1}0]$, shown as vectors 1 and 2 in (d). Lattice dislocations impinge on the boundary at points A, or C, where they are connected with extrinsic grain boundary dislocations embedded in the network. Lattice dislocations which enter at A or B have the same Burgers vector as one or the other of the intrinsic primary screw dislocations in the boundary and produce configurations of the types illustrated in (c). The lattice dislocations which enter the boundary at points C have the Burgers vector $b_4$ corresponding to the vector 4 in (d). These produce the extrinsic embedded dislocations illustrated in (b), and as seen along PQ and RS in (a), the intrinsic grid is offset across these dislocations. (From Sun and Balluffi (1982).)

[001] twist boundary in MgO whose intrinsic structure would otherwise consist of a uniform square network of screw dislocations. The structures of these configurations can be discussed within the framework of the $\Sigma = 1$ DSC lattice for this interface (Fig. 12.6(d)) which, of course, is identical to the crystal lattice. The network screw dislocations possess Burgers vectors $b_1$ and $b_2$, and lattice dislocations with these same Burgers vectors enter the boundary at points indicated by A and the B's respectively. (Note that $b_i$ corresponds to the $i$th vector in Fig. 12.6(d).) The screw dislocations initially present in the network have adjusted their positions in order to accommodate the embedded segments to the maximum extent possible. In fact, as illustrated in Fig. 12.6(c), these inserted segments appear essentially as defects in the network which resemble edge dislocations in 2D. In this accommodation process the stress fields of the embedded dislocations are considerably reduced, since the added Burgers vector strength has become widely distributed in the boundary as a result of the relaxation of the network. The elastic energy associated with the embedded segments is therefore reduced, and they are therefore bound to the interface.

   The lattice dislocations which impinge at the C points have the Burgers vector $b^L = b_4$ which possesses a component $a/2$ normal to the boundary. The embedding of these dislocations is therefore more complex as illustrated in Fig. 12.7 for such a dislocation lying at 45 ° with respect to the network. We may imagine that the final structure (shown

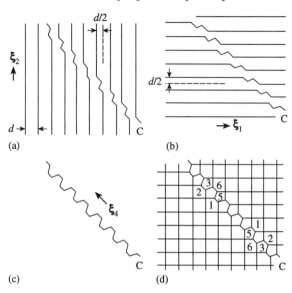

**Fig. 12.7** Series of imaginary steps by which lattice dislocations of the type which enter the boundary of Fig. 12.6 at points C become embedded in the boundary. (a) First, shear and serrate the $b_2$ screw dislocations along the path of the anticipated impinged lattice dislocation. (b) Repeat this process for the $b_1$ screw dislocations. (c) Next, deform the $b_4$ lattice dislocation (before impingment) as shown. (d) Finally, lay the $b_4$ dislocation in the boundary. After all overlapping dislocation segments have interacted, the final structure is obtained. The Burgers vector of each dislocation segment, $b_i$, is identified by the number shown which corresponds to $i$. These vectors are shown in Fig. 12.6(d) and are identified by their $i$'s. The $\xi_i$'s are the dislocation tangent vectors.

in Fig. 12.7(d)) occurred by a series of steps. First, the $b_2$ dislocations in the network are sheared and serrated as illustrated in Fig. 12.7(a). Next, the $b_1$ dislocations are sheared and serrated as shown in Fig. 12.7(b). Then, the $b_4$ lattice dislocation is laid in the boundary and serrated as shown in Fig. 12.7(c). Finally, the overlapping segments produced in the overall structure by these operations react to create new $b_3, b_5$, and $b_6$ segments as follows:

$$b_4 - b_1 = b_3; \quad b_2 + b_4 = b_5; \quad b_2 + b_4 - b_1 = b_6.$$

This produces the final structure shown in Fig. 12.7(d). The shearing of the $b_1$ and $b_2$ network dislocations shown in Figs. 12.7(a,b) constitutes a relaxation of the network which serves to accommodate the embedded lattice dislocation and reduce the energy of the system. This may be seen by realizing that the effective Burgers vector of the localized embedded dislocation in Fig. 12.7(d) is given by

$$b_{\mathrm{eff}} = b_4 + b_2/2 - b_1/2 = \tfrac{1}{2}[001],$$

which is smaller than the original $b_4$ Burgers vector of the lattice dislocation. The shearing of the network in Fig. 12.7(a) contributes the $b_2/2$ component to $b_{\mathrm{eff}}$ in the above relation, while the shearing in Fig. 12.7(b) contributes $-b_1/2$. As a result of this shearing (network relaxation), the component of the original $b_4$ Burgers vector parallel to the boundary plane is widely dispersed in the boundary, and the embedded dislocation is tightly bound to the interface because of the reduction in energy caused by the substantial decrease in the square of its effective Burgers vector.

### 12.4.3  Large-angle grain boundaries and heterophase boundaries

#### 12.4.3.1  *Singular boundaries*

When a lattice dislocation impinges upon a singular large-angle grain boundary which possesses an exact CSL misorientation, it will always be topologically possible for it to dissociate into an integral number of localized DSC dislocations. This follows from the fact that the total Burgers vector must be conserved (which must hold for any dislocation reaction in a continuum) and from the fact that all lattice vectors are also DSC lattice vectors. The Burgers vector dissociation reaction may therefore be expressed entirely within the framework of the DSC lattice, i.e.

$$b^{L} = N_1 b_1 + N_2 b_2 + N_3 b_3, \tag{12.33}$$

where all vectors are vectors of the DSC lattice, and the $N_i$ are integers. In addition, the total step height must be conserved (King and Smith 1980). This is easily demonstrated by realizing that

$$h^{L,w} = \sum_i N_i h_i^w; \quad h^{L,b} = \sum_i N_i h_i^b, \tag{12.34}$$

where $h^{L,w}$ and $h^{L,b}$ are the step heights associated with impinged lattice dislocation on the white and black crystals respectively (before dissociation), and the $h_i^w$ and $h_i^b$ are the step heights associated with the product dislocations on the white and black crystals respectively. These relations hold since the heights of both the black and white crystals obviously remain unchanged on either side of the dissociating dislocation and its reaction products. Using eqn (1.83) we then obtain the conservation relationship

$$h^{L} = \tfrac{1}{2}(h^{L,w} + h^{L,b}) = \sum_i N_i \tfrac{1}{2}(h_i^w + h_i^b) = \sum_i N_i h_i. \tag{12.35}$$

The formation of a variety of different steps is possible, topologically. Unfortunately, there are no simple rules based on geometry for finding the minimum energy structure which presumably would be preferred.

A specific example of a possible dissociation of a lattice dislocation which demonstrates the above conservation rules is given in Section 9.2.2.2 (see Fig. 9.11). The energy change upon dissociation will be equal to the total elastic and core energy of the product dislocations minus the elastic and core energy of the original lattice dislocation. The core energies must include, of course, the contributions of any steps. If the product dislocations become well dispersed in the boundary, their total elastic energy will be proportional to the sum of the squares of their Burgers vectors. The change in elastic energy is then approximately

$$\Delta E^{el} \simeq A \cdot [ (b^{L})^2 - (N_1 b_1^2 + N_2 b_2^2 + N_3 b_3^2) ], \tag{12.36}$$

where $A$ is a constant. (We note that this approximation ignores additional factors which are generally less important such as interactions of the dislocations with other dislocations in the boundary, interactions with the boundary itself as discussed in Section 12.3, and other more global factors such as those discussed below in Section 12.4.4.) The contribution of the steps can be approximated in a rough way by simply taking each step as an extra thin strip of boundary area. A unit length of step in the boundary then makes an energy contribution $h\sigma$. This approximation has already been used in developing eqns (9.22) and (9.31). The energy $\Delta E^{el}$ is generally large and negative because of the relatively small $b_i$'s of the product dislocations, and, hence, a strong driving force for dissociation is present which is usually sufficient to overcome possible opposing contributions from step energies and the other sources mentioned above. However, in some cases step energies may be more significant and therefore play a decisive role in determining the form of the possible dissociation. It is also possible that the dissociation of an impinged dislocation may produce partial rather than perfect boundary dislocations (Pond 1977). In such cases the dissociation cannot be analysed within the framework of the DSC lattice. However, both Burgers vector and step height conservation still apply. Many situations may therefore exist.

The kinetics of dissociation will generally depend upon the types of product dislocations formed and the temperature. Product dislocations which are glissile in the boundary will be able to glide away quickly in the absence of obstacles even at relatively low temperatures. If climb is required, the dissociation will be limited to temperatures high enough to allow the necessary diffusional transport.

The operation of singular grain boundaries as sinks for lattice dislocations via dissociation is a common phenomenon which has been widely observed (e.g. Schober and Balluffi (1971), Pond and Smith (1977), Darby, *et al.* (1978), Clark and Smith (1979), Dingley and Pond (1979), Liu and Balluffi (1984), Putaux and Thibault-Desseaux (1990), Lee *et al.* (1990a,b)). We now present a few selected examples.

An unusually detailed study of symmetric [011] tilt boundaries in Si and Ge as sinks for lattice dislocations has been carried out by Bacman *et al.* (1982), Elkajbaji and Thibault-Desseaux (1988), Thibault-Desseaux *et al.* (1988), Putaux and Thibault-Desseaux (1990), and Thibault *et al.* (1991). Bacmann *et al.* (1982) injected lattice dislocations into an initially $\Sigma 9$ symmetric tilt boundary ($\theta = 38.9\,°$) in a Ge bicrystal at 490°C by plastically straining it in tension as illustrated schematically in Fig. 12.8(a, b). The deformation occurred by essentially single primary slip on symmetrically disposed (111) slip planes as shown, causing lattice dislocations to accumulate in the boundary along the [011] tilt axis where they eventually became incorporated into the intrinsic structure of the boundary. Inspection of Fig. 12.8(a, b) reveals that the Burgers vector strength added to the boundary in this fashion should cause the tilt angle to decrease with increasing tensile strain, $\varepsilon$. The relationship between $\theta$ and $\varepsilon$ is readily found. The strain produced by 2d$N$ dislocations per unit distance along the unit vector $\hat{t}$ is d$\varepsilon = (b^{L,b} \cdot \hat{t})$d$N$, while the added Burgers vector strength normal to the boundary plane per dislocation is $2b^{L,b} \cdot \hat{n}$. The change in tilt angle is then d$\theta = -2(b^{L,b} \cdot \hat{n})$d$N$, and therefore

$$\Delta\theta = \int d\theta = -\int \frac{2(b^{L,b} \cdot \hat{n})}{(b^{L,b} \cdot t)}\, d\varepsilon, \tag{12.37}$$

which is analogous to eqn (12.27) in many respects. This relationship was confirmed experimentally by Bacmann *et al.* (1982) over the relatively wide range $20\,° \leqslant \theta \leqslant 38.9\,°$.

Studies of the detailed mechanisms by which the lattice dislocations injected into such grain boundaries during plastic deformation in both tension and compression become incorporated into the boundaries and alter their structures have been carried out by Elkajbaji and Thibault-Desseaux (1988), Thibault-Desseaux *et al.* (1988), Putaux and Thibault-Desseaux (1990), and Thibault *et al.* (1991). In general, the impinged dislocations first dissociate in the boundary: their dissociation products then rearrange in the boundary by glide and climb and interact in various ways. At sufficiently high temperatures the boundary 'digests' the added dislocations almost completely into a new intrinsic structure. At lower temperatures the digestion process is less complete and a residue inherited from the added dislocations remains in the form of extrinsic dislocations.

Consider, for example, the $\Sigma 9$ grain boundary in compression as in Fig. 12.8(c, d) rather than tension as in Figs. 12.8(a, b). In such a case the Burgers vectors of the incoming lattice dislocations are just reversed, and $\theta$ therefore increases with increasing strain. The observed impingement and dissociation of a $b^{L,w}$ dislocation is shown in Fig. 12.9. The $b^{L,w}$ dislocation in the lattice is normally dissociated into $b_1^w$ and $b_4^w$ partial dislocations connected by a ribbon of stacking fault (Fig. 12.9(a)). Upon impingement, the leading $b_4^w$ partial, which reaches the boundary first, dissociates into $b_2^w$ and $b_3^w$ dislocations, and the $b_3^w$ dislocation, which is glissile in the boundary, quickly glides away (Fig. 12.9(b)). Finally, the $b_1^w$ dislocation enters the boundary while the $b_2^w$ dislocation moves away by climb, and the $b_3^w$ dislocation moves farther away by glide (Fig. 12.9(c)). The overall dissociation is therefore

$$b^{L,w} = b_1^w + b_2^w + b_3^w.$$

All of the Burgers vectors of these dislocations are vectors of the $\Sigma 9$ DSC lattice as shown at the lower right of Fig. 12.9. The dissociation of a $b^{L,b}$ dislocation would occur in an analogous way according to $b^{L,b} = b_1^b + b_2^b + b_3^b$ (see vectors in Fig. 12.9). However, $b_1^w = b_2^b = b_2$, $b_3^b = -b_3^b$, and $b_1^w + b_1^b = b_2$, since the components of $b_1^w$ and $b_1^b$ along the tilt axis cancel because of symmetry. Therefore, it is possible for all of the product dislocations to interact to form three $b_2$ dislocations according to

$$b^{L,w} + b^{L,b} = b_1^w + b_2^w + b_3^w + b_1^b + b_2^b + b_3^b = 3b_2.$$

Further $b_2$ dislocations could be produced by this process, and the end result would be an intrinsic array of $b_2$ pure edge dislocations in the boundary which is superimposed on the original $\Sigma 9$ structure and causes a decrease in the original tilt angle. In

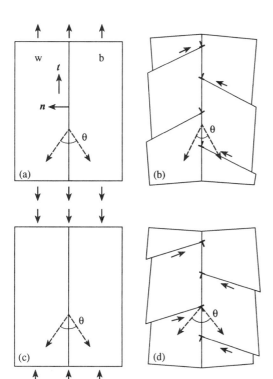

**Fig. 12.8** Symmetric tilt boundaries consisting of white and black crystals undergoing plastic deformation in tension or compression. Tilt axis normal to paper. (a) Initial bicrystal subjected to tension. (b) Geometry after tensile deformation. Dislocations added to boundary in directions of arrows produce a decrease in the tilt angle, $\theta$. (c) Same initial bicrystal as in (a) but subjected to compression. (d) Geometry after compressive deformation. Added dislocations now produce an increase in $\theta$.

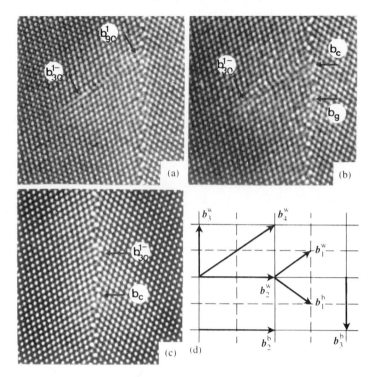

**Fig. 12.9** The impingement of a dissociated lattice dislocation on a symmetric ⟨110⟩ Σ9 tilt boundary in Si. (a) Dissociated dislocation (arrows) approaches boundary from white crystal on left on {111} slip plane. (b) Dissociation of leading lattice partial dislocation into two grain boundary dislocations (arrows). (c) Entrance of trailing lattice partial dislocation into boundary (top arrow). (d) DSC lattice of boundary including Burgers vectors of the boundary dislocations of interest (see text). All vectors are in the plane of the paper except $b_1^w$ and $b_1^b$ which have equal and opposite components normal to paper of magnitude $a/2\sqrt{2}$. (From Elkajbaji and Thibault-Desseaux (1988).)

the development of this structure the Burgers vector components normal to the boundary of the impinged lattice dislocations have been conserved, while the components parallel to the boundary have been eliminated. Structures of this type have been found over appreciable areas of the boundary when the deformation is performed at high enough temperatures to allow the dissociation products to interact sufficiently (Thibault-Desseaux *et al.* (1988)). At lower temperatures the interactions are less complete, and a variety of extrinsic dissociation and reaction products remain in the boundary (Thibault-Desseaux *et al.* (1988), Elkajbaji and Thibault-Desseaux (1988)). At large compressive strains, large increases in the tilt angle are observed and heterogeneous non-equilibrium boundary structures are obtained. Also, various types of grain boundary dislocation structures are obtained which can be interpreted in terms of structural multiplicity within the framework of the structural unit model for [011] symmetric tilt boundaries (Putaux and Thibault-Desseaux (1990)). Generally similar phenomena have been found with the Σ19 ($\theta = 26.5°$) [011] tilt boundary (Thibault *et al.* (1991)). Further observations of the dissociation of lattice dislocations in boundaries by *in situ* methods have been presented by Clark *et al.* (1989) and Lee *et al.* (1990).

Many other observations have also been made of configurations where portions of lattice dislocations impinged on grain boundaries are dissociated in the boundary. Figure 12.10 shows an impinged lattice dislocation which has dissociated into five boundary dislocations with Burgers vectors belonging to the $\Sigma 29$ DSC lattice. Figure 12.11(a) shows an extrinsic grain boundary dislocation segment embedded in a vicinal $\Sigma 5$ [001] twist boundary which can be derived from an impinged lattice dislocation in a manner quite similar to the way in which the extrinsic segment in the vicinal $\Sigma 1$ (small-angle) [001] twist boundary illustrated in Fig. 12.7 was derived. The $\Sigma 5$ DSC lattice is shown in Fig. 12.11(c), and the initial vicinal $\Sigma 5$ boundary structure contains a square network of intrinsic $b_1$ and $b_2$ screw DSC dislocations as may be seen on the right side of Fig. 12.11(b). In the first step of the process we deposit an extrinsic lattice dislocation with $b^L = (\tfrac{1}{2})$ [011] along CD in Fig. 12.11(b). Then, let it dissociate according to $b^L = b_2 + b_4$ and let the $b_2$ screw dislocation move far away and become accommodated in the existing screw dislocation network. Next, shear the $b_1$ dislocations along CD by $d/2$ and let the resulting overlapping segments along CD react according to $b_1 + b_4 = b_3$. Finally, if the $b_2$ screws are moved apart at DC to a spacing of $1.5d$, the final configuration shown in Fig. 12.11(b) is obtained. Again, as in Fig. 12.7, the Burgers vector component of the original lattice dislocation parallel to the boundary is widely dispersed in the boundary, and the effective Burgers vector of the remaining localized line defect is $b_{\text{eff}} = b_4 + b_1/2 - b_2/2 = \tfrac{1}{2}$ [001].

We note that we have chosen to analyse the above lattice dislocation dissociation, and also the one in Fig.12.7 (which both occur in vicinal low $\Sigma$ grain boundaries), in terms of reactions in the DSC lattices of the nearby vicinal $\Sigma$ boundary followed by rearrangements of the dislocations in the existing intrinsic dislocation arrays. We could equally well have regarded both boundaries as being exact high-$\Sigma$ boundaries with DSC lattices consisting of a $b_3 = \tfrac{1}{2}$ [001] primitive vector and exceedingly small $b_1$ and $b_2$ vectors in the boundary plane. The dissociations would then have been simple dissociations in these DSC lattices producing the $b_3 = b_{\text{eff}} = \tfrac{1}{2}$ [001] dislocations and $b_1$ and $b_2$ dislocations (with exceedingly small Burgers vectors) widely dispersed in the boundary. The final results of both analyses are identical, of course, since the dispersed $b_1$ and $b_2$ dislocations in the latter analysis correspond exactly to the small perturbations in the

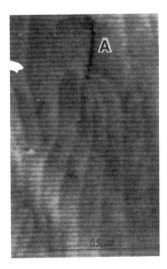

**Fig. 12.10** Dissociation of lattice dislocation (at A) impinged upon a near-$\Sigma 29$ grain boundary in stainless steel into five boundary dislocations possessing Burgers vectors belonging to the $\Sigma 29$ DSC lattice. (From Bollmann *et al.* (1972).)

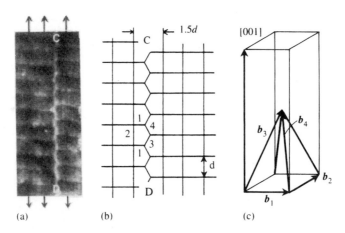

**Fig. 12.11** (a) Extrinsic grain boundary dislocation (along CD) in a vicinal Σ5 [001] twist boundary in MgO derived from an impinged lattice dislocation. (b) Detailed diagram of dislocation arrangement. The Burgers vector of each dislocation segment, $b_i$, is identified by the number shown which corresponds to $i$. These vectors are shown in (c) within the framework of the Σ5 DSC lattice. $b_1 = \frac{1}{10}[3\bar{1}0]$; $b_2 = \frac{1}{10}[130]$. (From Sun and Balluffi (1982).)

spacings of the existing intrinsic dislocation array which occur when these dislocations are rearranged in the former analysis.

It is readily seen that the main features of the dissociation will be dominated by the degree of eccentricity of the boundary DSC lattice. If one, or two, of the primitive vectors of the DSC lattice are exceedingly small, the components of the impinged lattice dislocation having Burgers vector components in the directions of these vectors will tend to dissociate into a distribution of product dislocations corresponding for all practical purposes to an infinite number of dislocations possessing infinitesimal Burgers vectors. The resulting structure will therefore consist of product dislocations, some of which are widely spread out in distributions of essentially the type just described. A number of such cases has been observed experimentally by Darby *et al.* (1978). When the dissociation requires climb of the product dislocations, the rate of dissociation will tend to be limited by diffusional transport in the boundary. A treatment of the dissociation kinetics under these conditions is given below in Section 12.4.3.2.

So far, we have restricted our discussion mainly to singular grain boundaries possessing exact CSL's. However, as we have seen in earlier chapters, a variety of other types of singular large-angle grain boundaries and heterophase boundaries exist which possess near-CSL DSC lattices with at least one primitive vector of significant magnitude. These boundaries can therefore also act as sinks (or traps) for impinged lattice dislocations in a similar manner.

### 12.4.3.2  *General boundaries*

Large-angle grain boundaries and heterophase boundaries, which are general with respect to all degrees of freedom, are not able to support any localized interfacial dislocations of physical significance. A lattice dislocation impinged on such a boundary will therefore tend to become widely dissociated into a distribution of boundary dislocations which, for all practical purposes, may be regarded as an infinite number of dislocations possessing infinitesimal Burgers vectors. The Burgers vector of an impinged dislocation will generally lie at an oblique angle to the boundary, and its component in the boundary plane can then be readily dissociated by glide. However, the dissociation of its component normal to the boundary will require climb. An experimentally observed example is shown in Fig. 12.12 where several lattice dislocations have entered a grain boundary along parallel slip planes. The dislocations in Fig. 12.12(a) have probably already lost any Burgers vector components parallel to the boundary by glissile processes. However, at the least, their components normal to the boundary are still intact and localized. Upon heating, these components progressively dissociate, presumably by climb, as indicated by the disappearance of the diffraction contrast produced in the electron microscope by their strain fields. We may conclude that general boundaries will always act as highly effective sinks whenever the temperature is high enough to provide the thermal activation necessary for the required climb. It is worth noting that this same phenomenon occurs during recrystallization when a recrystallized grain grows at the expense of a plastically deformed and dislocated polycrystalline matrix (Section 9.1.2). In this case the advancing interface sweeps up lattice dislocations which then dissociate, interact and are largely digested by the interface (Balluffi 1976).

An expression for the rate of dissociation of an impinged edge dislocation in a general grain boundary is easily obtained if we assume that no large barriers to the dissociation exist and that the rate is determined by diffusion-controlled climb. The dislocation lies along $z$ and its Burgers vector is along $y$ as illustrated in Fig. 12.13(a), the partially dissociated dislocation is represented by a continuous distribution of edge dislocations with infinitesimal Burgers vectors represented by $\rho(x, t)$, so that $\rho(x, t)$ is the Burgers vector lying between $x$ and $x + \mathrm{d}x$ at time $t$, and $\int \rho(x, t)\mathrm{d}x = b$. (Note that this

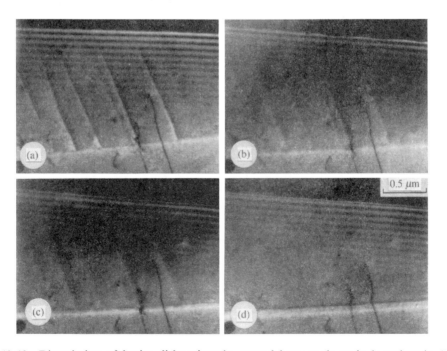

**Fig. 12.12** Dissociation of lattice dislocations in general large-angle grain boundary in Al–Mg alloy as a result of annealing. (a) Configuration at room temperature. (b), (c), and (d) after 30 s heating at 120, 135 and 145 °C respectively. (From Pumphrey and Gleiter (1974).)

distribution function is the same as that used previously in the Peierls–Nabarro model in Section 2.11.2). The infinitesimal dislocations repel each other so that all dislocations at $x > 0$ experience a force towards $+x$, and all at $x < 0$ a force towards $-x$. The dislocations will then respond by climbing in these directions thus widening the distribution. The required diffusional transport occurs via diffusion along the boundary from the dislocations at $x < 0$ to those at $x > 0$ as illustrated in Fig. 12.13. A reasonably accurate expression for the rate of dissociation can be obtained (Johannesson and Tholen 1972) in a simple manner by coalescing the total Burgers vector strength at $x > 0$ and at $x < 0$ into two localized 'half-dislocations' as illustrated in Fig. 12.13(b), and simply calculating the rate at which they climb away from each other assuming that they act as perfect line sources and sinks. The problem can also be solved more exactly by allowing the distribution to remain continuous (Lojkowski and Grabski 1981). However, this method involves an integral equation and is more complicated. Since both approaches yield essentially the same results (differing only by relatively inconsequential differences in constant factors) we shall present the simpler model. We assume that the phenomenological equation for the diffusional flux of atoms in the grain boundary slab can be written in the same form as eqn (9.18), i.e.

$$J_A^B = - \frac{D^B}{N_o \Omega k T f^B} \nabla M_A, \tag{12.38}$$

where all quantities now refer to the boundary slab. The force experienced by each half-dislocation (per unit length) is of magnitude $F = \mu b / 8\pi (1 - \nu) w$. When each climbs by $dw/2$, the number of atoms transferred is $bdw/4\Omega$. The work done by the two forces is then $Fdw$, and the difference between $M_A$ at the two half-dislocations is $\Delta M_A = 4 N_o F \Omega / b$. Therefore, $\nabla M_A = 4 N_o F \Omega / bw$, and from eqn (12.38)

$$J_A^B = - \frac{D^B \mu b}{k T f^B 2\pi (1 - \nu) w^2}. \tag{12.39}$$

Now,

$$\frac{dw}{dt} = \frac{4\Omega}{b} \delta J_A^B = \frac{2\Omega \delta D^B \mu}{k T f^B \pi (1 - \nu)} \cdot \frac{1}{w^2}, \tag{12.40}$$

where $\delta$ is the grain boundary thickness. Therefore, upon integration,

$$w^3 = \frac{6\Omega \delta D^B \mu}{k T f^B \pi (1 - \nu)} \cdot t. \tag{12.41}$$

Note that the result is independent of the magnitude of $b$. Therefore, the rate of dissociation of the climb component of an impinged dislocation will be independent of the angle between $b$ and $\hat{n}$. Also, of course, eqn (12.40) may be used to calculate the rate of dissociation of a lattice dislocation into two localized boundary dislocations if desired.

Many observations of the kinetics of dissociation of impinged lattice dislocations in general grain boundaries in the above fashion have been made using the electron microscope. The strain field diffraction contrast generally disappears when $w$ exceeds a few extinction distances, and observations of the time required for disappearance may then be used to estimate values of $D^B$ through use of eqn (12.41). The results (Lojkowski and Grabski 1981, Grabski 1985, Swiatnicki *et al.* 1986, Lojkowski *et al.* 1989) yield values of $D^B$ at relatively low temperatures which are not inconsistent with values expected on the basis of other measurements.

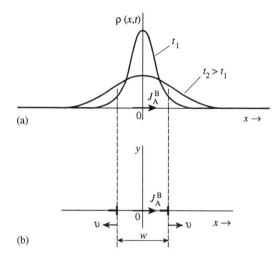

**Fig. 12.13** Model for the dissociation of an impinged edge dislocation in a general grain boundary by climb. (a) the dissociating dislocation is represented by a continuous distribution of dislocations, $\rho(x, t)$, which broadens with increasing time by dislocation climb controlled by the rate of boundary diffusion. (b) Approximate representation of the dissociation process by replacing continuous dislocation distribution by two localized 'half-dislocations'.

### 12.4.4 On the global equilibration of impinged lattice dislocations

The dislocations which are initially impinged on an interface during plastic deformation are generally distributed non-uniformly because of the inhomogeneous nature of the slip process in the adjoining crystals where the gliding lattice dislocations are restricted to discrete bands of slip originating at various dislocation sources. Full boundary equilibration then requires that the non-uniformly distributed impinged dislocations dissociate, interact and rearrange themselves over relatively long distances in order to produce a boundary structure which is free of extrinsic dislocations and in full compliance with the Frank–Bilby equation. This will not occur to any degree at low temperatures, and will be accomplished to an increasing degree as the temperature is raised. However, the climb distances are frequently sufficiently large, and the rates of the required interactions sufficiently slow, so that approximate equilibrium can be approached only at the very highest temperatures. We have already seen examples of this in the work of Thibault-Desseaux *et al.* (Section 12.4.3.1) where residues of extrinsic product dislocations were found in single grain boundaries in plastically deformed bicrystals at intermediate temperatures, and full equilibration was only approached at the highest temperatures.

The difficulty in achieving global equilibration is especially severe in the case of polycrystals. Here, different boundaries may experience different degrees of impingement and also react with the impinged dislocations in different ways. Each boundary is then just one element in an array of interconnected and interacting boundaries as illustrated schematically in Fig. 12.14. Since the incorporation of extrinsic dislocations in a single boundary generally requires a change in the crystal misorientation at that boundary, this process cannot occur independently at each boundary because of constraints on rotations of the grains which adjoin it which are imposed by neighbouring grains. The equilibration processes in the various boundaries are therefore coupled. In the equilibration process, dislocations which are initially impinged on one boundary may enter other boundaries and redistribute themselves in the array (while conserving their total Burgers vector strength) as illustrated schematically in Fig. 12.14. This will generally tend to change the crystal misorientation at each boundary and also produce a certain amount of sliding which must be accommodated at the triple junctions and adjacent boundaries in the array.

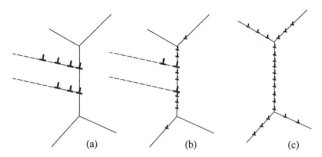

**Fig. 12.14**  Schematic view of the dissociation of impinged lattice dislocations in a polycrystalline grain boundary network. (a) Lattice dislocations impinge on a single network segment. (b) Lattice dislocations partially dissociated with dissociation products concentrated mainly in original network segment. (c) Dissociation products now spread out more widely in adjacent segments.

At the same time, boundary migration and changes in boundary inclination may also be expected as the system of coupled boundaries seeks equilibrium. Full global equilibration then requires interactions over relatively long distances of the order of at least the grain size, and will usually not be achieved, even at high temperatures.

On the basis of the previous discussion it may be seen that the ability of interfaces to act as sinks for lattice dislocations and 'digest' them to produce equilibrated structures should exert a strong influence on the overall mechanical properties of polycrystalline materials. At low temperatures where the impinged lattice dislocations are unable to climb, substantial components will remain in place in the boundaries and oppose the entry of additional lattice dislocations on their slip planes thereby producing pileups of these dislocations in the adjoining crystals (see Fig. 12.14(a)). The boundaries will therefore be resistant to the entry of large numbers of lattice dislocations. (Of course, such situations can be relieved to some extent if sufficient stress is developed to cause the transmission of slip through the boundaries, or backwards into the original grains, as discussed below in Section 12.6.) On the other hand, at elevated temperatures where impinged dislocations can dissociate and redistribute throughout the boundary network, any such pileups can be largely relaxed, and large numbers of lattice dislocations can enter the boundaries as illustrated in Fig. 12.14. In a sense, each boundary then acts as a 'collector plate' capable of absorbing and redistributing large numbers of lattice dislocations. However, even at elevated temperatures we can often expect this process to be incomplete, and significant densities of extrinsic boundary dislocations and long range stresses due to incompatibilities between the grains generated by the dislocation sink action of the various boundaries will persist (Grabski and Korski 1970, Lojkowski *et al.* 1990, Lojkowski 1991). There is some experimental evidence (see reviews by Grabski (1985), Valiev *et al.* Kaibyshev (1986), and Valiev and Gertsman (1990)) that these residual non-equilibrium effects can significantly affect the properties of the interfaces in polycrystals and also the overall mechanical properties of polycrystals in complex ways which we shall not attempt to discuss here.

## 12.5  INTERFACES AS SOURCES OF BOTH INTERFACIAL AND LATTICE DISLOCATIONS

When a system containing an interface is subjected to applied stress, the interface may act as a source of either interfacial dislocations or lattice dislocations. In the former case, interfacial dislocations are generated in special interfaces which are capable of supporting localized interfacial dislocations, and these dislocations remain in the interface as localized dislocations. In the latter case, lattice dislocations are generated at interfaces and move away from the interfaces into the adjoining crystals. As discussed later, these phenomena may play significant roles in determining the overall mechanical behaviour of the system.

### 12.5.1   Interfaces as sources of interfacial dislocations

We have already discussed the generation of interfacial dislocations in different types of singular or vicinal interfaces under a variety of conditions in Chapters 9 and 10. These included generation at sources of various types such as glissile pole sources (Section 9.2.1.2, and Fig. 9.5), Bardeen–Herring type glide/climb sources (Section 9.2.2.2, and Figs 9.10 and 9.24), and also generation by the nucleation of individual loops under large driving forces (Sections 9.2.2.3, 10.3.3.1, 10.4.1.1, and Fig. 10.8).

The generation of interfacial dislocations under applied stress at low temperatures where climb is frozen out can occur by purely glissile processes at sources of the general types shown in Fig. 9.10 if: (1) their Burgers vector is parallel to the interface; and (2) the interface is perfectly flat. These two conditions guarantee that the product dislocations can be glissile in the boundary. Such a source would then be classified as an interfacial Frank–Read source. However, these conditions are quite restrictive and will therefore be met only rarely. They are most likely to be met at faceted interfaces associated with low-$\Sigma$ CSL's and corresponding DSC lattices possessing vectors (i.e. Burgers vectors) parallel to the interface. Sources of this type have have been proposed, for example, by Gleiter *et al.* (1968) and Baillin *et al.* (1990). Even though the glissile motion of trains of dislocations along grain boundaries with the above characteristics has been frequently observed (e.g. Mori and Tangri (1979), no direct and definitive experimental observations of such boundary sources have yet been made. We shall see in Section 12.5.2 (Fig. 12.17) that trains of glissile boundary dislocations can be injected into boundaries via the impingement of piled up lattice dislocations (see also Fig. 12.9). Therefore, observations of trains of glissile dislocations in boundaries, by themselves, cannot be taken as clear proof for the existence of boundary sources. The relative importance of such sources is therefore largely undetermined at present.

The probability of obtaining active boundary sources is greatly enhanced when the temperature is high enough so that the product dislocations can climb as well as glide in the boundary. A probable example of such a source operating an elevated temperature under almost glissile conditions in a slightly non-planar grain boundary is shown in Fig. 12.15. Here (Kegg *et al.* 1973), the boundary was subjected to an applied shear stress and boundary sliding (Section 12.8.1) took place. The elevated temperature was required in order to allow the emitted dislocations to climb sufficiently to negotiate the slightly uneven boundary. It seems likely that similar sources were operative to at least some extent in the grain boundary sliding experiments of Fukutomi and Kamijo (1985) and Horiuchi *et al.* (1987) discussed earlier in Section 9.2.1.2.

Even though such sources must exist, their relative importance as sources of boundary dislocations in stressed materials is not well established. As will be discussed below in Section 12.8.1.1, the sliding of singular boundaries under applied shear stresses must occur by the glide/climb of boundary dislocations along the boundary. The experimental results (Section 12.8.1.5) indicate that boundary sliding rates are markedly increased as a result of the impingement of lattice dislocations. Presumably, the increased sliding rate

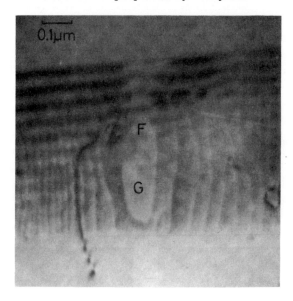

**Fig. 12.15**  Grain boundary dislocation source observed in symmetrical ⟨011⟩ boundary in Al under applied stress at 360 °C. Boundary is of the plane matching type with {110} planes nearly matching across the interface. Boundary dislocation bowed out between pinning points at F and G. Train of emitted dislocations confined to boundary plane. (From Kegg *et al.* (1973).)

could result either from the movement along the boundary of an increased number of boundary dislocations picked up directly from impinged lattice dislocations, or, conceivably, from an increased number of boundary dislocations generated at an increased density of pole-type sources produced by impinged dislocations.

### 12.5.2  Interfaces as sources of lattice dislocations

There is extensive experimental evidence (cited below) that both special and general interfaces can act as effective sources of lattice dislocations under applied stress during plastic deformation. In this process lattice dislocations are generated at the interface and are then injected into the adjoining crystals. Two distinct situations may be distinguished as illustrated schematically in Fig. 12.16. In the first (Fig. 12.16(a)), the source lies directly in the interface and half-loops expand from the interface plane into one adjoining crystal, while the remaining loop segments (shown dashed) are accommodated in the boundary in one way or another. In the second, loops are generated either in the perfect lattice, or at lattice dislocation sources, in a region of crystal 1 or 2 very close to the boundary. Half-loops again expand into an adjoining crystal as illustrated in Fig. 12.16(b), while the remaining loop segments are accommodated in the boundary. If this process occurs preferentially in the boundary region in response to stress concentrations generated because of incompatibility stresses (Section 12.2) or nearby dislocation pileups at the boundary (see Section 12.6), we may regard the boundary as the effective source of the lattice dislocations.

Unfortunately, little reliable information is available from experiments about the details of how dislocations are produced at boundaries. No direct observations of the actual generation events are available, and we must therefore rely on less conclusive

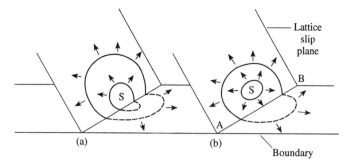

**Fig. 12.16** Two possible types of sources of lattice dislocations at grain boundaries. (a) Source, S, lies directly in the interface: one set of half-loops expands into the adjoining crystal, while the other is accommodated in the boundary. (b) Loops are now generated initially in the lattice very close to the boundary.

information. For example, efforts to obtain information by electron microscopy have been largely inconclusive. Here, the generated dislocations are either observed in thin film sections *after* their formation in bulk specimens, or the generation events are observed in thin film sections during *in situ* straining under conditions which may not be representative of bulk conditions (Murr 1981, Valiev *et al.* 1986, Varin *et al.* 1987). In addition, the generation often occurs rapidly in regions where details are difficult to resolve. Obviously, the problem in distinguishing between the two mechanisms illustrated in Fig. 12.16 would be difficult under any circumstances.

### 12.5.2.1 *Singular interfaces*

An example of the generation of lattice dislocations at a highly stressed region of a singular grain boundary is shown in Fig. 12.17. Here, lattice dislocations have impinged on a $\Sigma 3$ {111} twin boundary from crystal 1 on two different slip systems, A and B. Interactions between these dislocations in the boundary and high stresses produced by pile-ups of further impinging dislocations then generate new lattice dislocations, which are injected into crystal 2 at D, and grain boundary dislocations which are glissile in the boundary and glide away towards the left in the direction of the arrow. However, these dislocations are held up by the step in the boundary producing a pileup of glissile grain boundary dislocations. The high stresses produced at the pile-up (Hirth and Lothe 1982) then nucleate further lattice dislocations in crystal 2 which are seen near C. We note that a shear stress of the order of $\simeq \mu/30$ is required to nucleate a complete circular dislocation loop in the perfect lattice (Hirth and Lothe 1982, Varin *et al.* 1987). Varin *et al.* (1987) have performed calculations indicating that such stresses can be reached in the near vicinity of localized grain boundary dislocations in the presence of reasonable applied stresses. However, the required stress can be reduced somewhat if the dislocations form as illustrated in Fig. 12.16 under conditions where the segments remaining in the boundary interact with the boundary in a manner which reduces their energy. This could occur via dissociation and/or interactions with other boundary dislocations. Numerous additional examples of the generation of lattice dislocations at singular boundaries in highly stressed regions may be found in the literature including cases at boundary triple junctions (Hashimoto *et al.* 1986) where high stresses are generated at elevated temperatures during grain boundary sliding.

The observed generation of lattice dislocations at a possible dislocation source lying

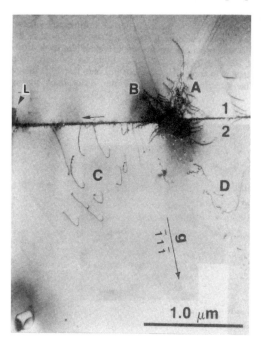

**Fig. 12.17** Generation of lattice dislocations at two highly stressed regions of a Σ3 {111} twin boundary in stainless steel. Lattice dislocations seen at D were generated by high stresses produced by piled up lattice dislocations at boundary arriving on slip systems A and B. Lattice dislocations at C were generated by high stresses produced by glissile grain boundary dislocations moving from the AB pile-up in the direction of the arrow which piled up at the boundary step indicated by the arrow at *L*. (From Lee *et al*. (1990*a*).)

in a singular Σ3 {111} twin boundary is shown schematically in Fig. 12.18(a). Here, it has been proposed (Whelan *et al*. 1957) that the lattice dislocations generated along the slip plane ABCD, and also ABEF, were produced by Frank–Read sources in the boundary of the type illustrated in Fig. 12.18(b). In this case a segment of screw dislocation with Burgers vector $\frac{1}{2}\langle 110 \rangle$ lies in the boundary. If this segment expands into one of the grains by slip as illustrated, and then cross slips as shown, successive lattice dislocation loops can be generated in each crystal adjoining the boundary in a manner consistent with the observed results. In this very special case the source is capable of producing half-loops in each adjoining crystal rather than a half-loop in one crystal and a half-loop in the boundary as in the more general case in Fig. 12.16.

A variety of other sources, some of which involve boundary dislocations or steps in different configurations, have been proposed in the literature (Gleiter *et al*. 1968, Orlov 1968, Baro *et al*. 1968/69, Mascanzoni and Buzzichelli 1970, Price and Hirth 1972). A number of these sources are quite specialized, since various restrictive conditions must be met for their operation (e.g. the existence of two lattice slip planes which intersect along a line parallel to the boundary plane as in our previous example in Fig. 12.18). They are therefore of limited general interest.

### 12.5.2.2  *General interfaces*

Extensive experimental evidence exists for the generation of lattice dislocations at general grain boundaries. In the early stages of the plastic deformation of annealed polycrystals, lattice dislocations are generated preferentially at the general boundaries which are present and are injected into the nearby adjoining crystals as shown, for example, in Fig. 12.19 (Worthington and Smith 1964, Murr and Wang 1982, Meyers and Ashworth 1982). Also, electron microscopy (Murr 1975 and 1981, Kurzydlowski *et al*. 1984) has provided evidence for the emission of trains of lattice dislocations from boundaries on

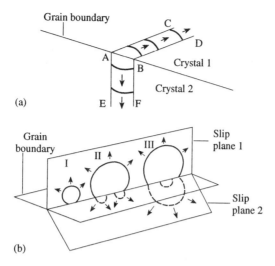

**Fig. 12.18** (a) Generation of trains of lattice dislocations in crystals 1 and 2 at Σ3 {111} twin boundary in stainless steel in configuration observed in the electron microscope by Whelan *et al.* (1957). Boundary observed edge-on. AC and BD, and AE and BF, are intersections of the slip planes in crystals 1 and 2, respectively, with the free surfaces of the thin film specimen. (b) Three successive stages, i.e. I, II and III, of model proposed by Whelan *et al.* (1957) to explain the dislocation generation observed in (a). See text for details.

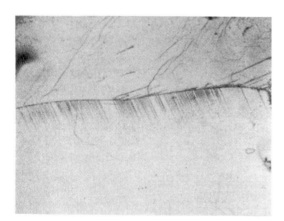

**Fig. 12.19** Preferential generation of slip (detected by etching) at grain boundary in polycrystalline Fe–Si alloy after a small plastic strain. (From Worthington and Smith (1964).)

slip systems of the adjoining crystals (Fig. 12.20). Of particular interest is the observation that lattice dislocations are copiously created at general boundaries which must, at least initially, contain essentially no detectable localized dislocations (or sources involving localized dislocations). Furthermore, the required applied stresses are in the range $\mu/3000$ to $\mu/300$, and are therefore no more than an order of magnitude lower than the stress required to nucleate dislocation loops in the crystal lattice (Varin *et al.* 1987). One source of dislocations in such cases may be nearby sources in the lattice (as in Fig. 12.16(b)) which are activated preferentially with the help of compatibility stress concentrations which, under certain circumstances, can be larger than applied stresses by factors as large as about three (Section 12.2.1). We may also suggest, tentatively, that loops can be

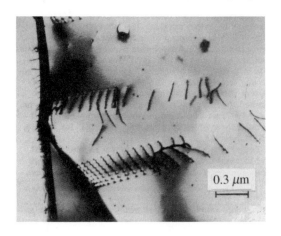

0.3 μm

**Fig. 12.20** Trains of lattice dislocations emitted from grain boundaries in polycrystalline stainless steel as a result of ≃5 per cent strain. (From Murr (1975).)

nucleated at general boundaries at high stress concentrations produced by localized grain boundary sliding. Conceivably, general boundaries may be able to shear (slide) over various relatively small local areas which are unusually flat at temperatures below those required for significant diffusional mass transport in the boundary. These localized regions then act elastically very much like small free cracks in the boundary plane and generate large local stress concentrations. For example, according to eqn (12.117a), the local shear stress stress at a distance $l$ from the edge of a mode 2 crack of length $2a$ in the plane of the crack is $\tau(l) = \tau^{\infty}(a/2l)^{\frac{1}{2}}$, where $\tau^{\infty}$ = bulk applied stress. In order to model approximately the nucleation of a lattice dislocation from the edge of such a crack we may adopt the simple dislocation emission model of Rice and Thomson (1974) and assume that it expands from the edge of the crack in the form of a semi-circular half loop on a slip plane as illustrated in Fig. 12.16(b) (with AB delineating the crack edge). The energy to form such a half-loop of radius $R$ is then approximately

$$W \simeq \frac{R(2-v)\mu b^2}{8(1-v)}\ln[(4R/r_0)-2] - 2^{-\frac{3}{2}}b\tau^{\infty}a^{\frac{1}{2}}\int_{r_0}^{R} r^{\frac{1}{2}}dr \cdot \int_{0}^{\pi}(\sin\theta)^{-\frac{1}{2}}d\theta, \quad (12.42)$$

where the first term is the self-energy of the half-loop (Hirth and Lothe 1982) and the second is the work done by the stress. The quantities $r$ and $\theta$ are polar coordinates in the slip plane, $r_0 \simeq b$ is the usual dislocation core cutoff radius, and the effective stress acting on the slip plane is assumed to be half the shear stress in the crack plane. For stainless steels, lattice dislocations are generated at grain boundaries at applied stresses near $\mu/380$ (Varin *et al.* 1987). Integrating eqn (12.42), and evaluating the results with the assumption that the bulk stress in the boundary region is twice the applied stress because of incompatibility stresses, we find spontaneous dislocation emission (i.e. $W < 0$) at crack lengths $\geqslant 0.5\,\mu$m. The establishment of stress concentrations in this fashion at general boundaries is actually similar in many respects to the establishment of stress concentrations at special boundaries by means of dislocation pileups, since the shear stress near the tip of a pile-up is of the same magnitude as the stress near the edge of a crack of similar length (Hirth and Lothe 1982). However, it is clear that more information about the ability of general boundaries to slide locally over these lengths at the stress levels which prevail must be available before the above suggestion can be accepted. Further discussion of our present knowledge of boundary sliding is presented in Section 12.8.1.

Finally, we note that localized extrinsic dislocations may be present in general boundaries in various configurations when the temperature is below the temperature required for complete dissociation (Section 12.4.3.2). Lattice dislocations could then be generated under stress by mechanisms similar to a number of those discussed above for special boundaries.

## 12.6  INTERFACES AS BARRIERS TO THE GLIDE OF LATTICE DISLOCATIONS (SLIP)

We are next interested in the role of interfaces as barriers to the propagation of plastic deformation via dislocation glide. More specifically, we imagine that slip has been initiated in crystal 1 adjoining an interface and lattice dislocations are gliding on their slip planes and impinging on the interface. Under what conditions can these impinging dislocations glide directly through the interface into crystal 2, or else pile up against the interface in various ways as the interface acts as either a partial or complete barrier?

### 12.6.1  Grain boundaries

We consider grain boundaries first. It is readily seen that the direct transmission of the incoming dislocations through grain boundaries requires special geometrical conditions which will be met only rarely. The simplest case is the very special one in which the slip planes in crystals 1 and 2 are parallel, and the slip vectors (Burgers vectors) in the two crystals are identical. Another case is the one in which the two slip planes are not parallel but intersect along a line lying in the boundary plane, and the two Burgers vectors are identical and are also parallel to this line. Under these conditions incoming lattice dislocations in screw orientation can simply cross slip from one slip system to the other as they glide through the boundary. An experimentally observed example is shown in Fig. 12.21. Here, both the incoming and outgoing dislocations are parallel to the line of intersection in the boundary as required, and the incoming dislocations are piled up against the boundary because of their resistance to cross slip (Hirth and Lothe 1982). In this case the boundary is relatively 'transparent' to the incoming dislocations but acts as a barrier because of the difficulty of the required cross slip.

The direct propagation of dislocation strength through a boundary can also occur under less restrictive conditions when the two slip planes are not parallel but meet along a line lying in the boundary, and the two Burgers vectors lie in any directions parallel to their respective slip planes. In this case the incoming dislocation with Burgers vector $b^{(1)}$ impinges on the boundary and reacts to produce a dislocation with Burgers vector $b^{(2)}$, which then glides away into crystal 2, and a second dislocation with Burgers vector $b^B$ remains in the boundary as a residual dislocation. Since the total Burgers vector must be conserved, the reaction has the form

$$b^{(1)} \rightarrow b^{(2)} + b^B. \tag{12.43}$$

If $|b^{(1)}| = |b^{(2)}|$, as will usually be the case, the energy barrier to this reaction will correspond approximately to the self-energy of the residual dislocation left behind in the boundary. However, if successive dislocations impinge on the boundary from crystal 1 and react according to eqn (12.43), a high density of residual dislocations will tend to build up in the boundary and oppose the further propagation of slip by this process. An additional barrier will then be present. Detailed studies of this type of slip propagation have been made by, for example, Jacques *et al.* (1990) and Baillin *et al.* (1990).

More generally, the above special geometrical conditions will not be met. All slip

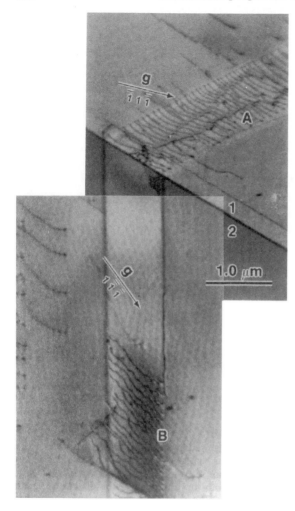

**Fig. 12.21** Composite micrograph showing direct transmission of lattice dislocations through $\Sigma 3$ {111} twin grain boundary in stainless steel. The incoming and outgoing slip systems, indicated by A and B, intersect along a common line in the boundary which is observed edge-on. All lattice dislocations have the same Burgers vector and are in screw orientation parallel to the intersection of the two slip systems. (From Lee *et al.* (1990*a*).)

systems will then be effectively terminated at the boundary, and it will be impossible for impinging lattice dislocations to propagate slip directly through the boundary. However, slip may still be propagated across the boundary, but in a less direct fashion, if piled up incoming dislocations in crystal 1 can generate sufficiently high local stresses in the vicinity of the boundary to generate dislocations on new slip systems in crystal 2 which then glide away from the boundary. An example of this type of 'indirect' slip propagation has already been shown in Fig. 12.17, where piled up dislocations incident from crystal 1 on slip systems A and B produced high enough stresses at the boundary to generate outgoing dislocations in crystal 2 on slip system D. At the same time, residual grain boundary dislocations were produced in the boundary, and these piled up to produce sufficiently high local stresses to generate outgoing dislocations in crystal 2 on slip system C as discussed in Section 12.5.2. Conservation of the total Burgers vector strength then requires

$$\sum_i \boldsymbol{b}_i^{\text{IN}} = \sum_j \boldsymbol{b}_j^{\text{OUT}} + \sum_k \boldsymbol{b}_k^{\text{B}}, \tag{12.44}$$

where the left side is the sum of the Burgers vectors of all the incoming dislocations, and the right side is the sum of the Burgers vectors of all the outgoing dislocations and those remaining behind in the boundary.

The ability of the system to generate sufficiently high local stresses to propagate slip across the boundary in the above manner will depend upon the magnitudes of the applied stresses which are driving the piled up incoming dislocations and the degree to which the system is able to relax local stress concentrations at these dislocations. As discussed in Section 12.4, this can be accomplished by spreading out the impinged dislocations in the boundary by means of glide and/or climb. The outgoing dislocations in crystal 2 may be generated at sources in the boundary or at lattice dislocation sources located within crystal 2 at small distances from the boundary (see Baillin *et al*. 1987, 1990).

Many cases of the indirect propagation of slip across boundaries have been investigated (Lim and Raj 1985*a*,*b*, Martinez-Hernandez *et al*. 1987, Baillin *et al*. 1987, Jacques *et al*. 1987, Baillin *et al*. 1990, Jacques *et al*. 1990, Shen *et al*. 1986, 1988, Clark *et al*. 1989, Lee *et al*. 1990*a*,*b*). These include interesting 'reflection' cases where incident dislocations from crystal 1 cause the emission from the boundary region of new lattice dislocations back into crystal 1.

When incident dislocations gliding on a single slip system in crystal 1 pile up at a boundary and activate outgoing dislocations in crystal 2 on a single slip system, criteria for predicting the outgoing slip system which is selected by the bicrystal (when the incoming slip system is specified) have been discussed by Clark *et al*. (1992). In general, the outgoing slip system which is chosen, at least initially, is the one which experiences the maximum resolved shear stress. However, if the operation of this system leaves behind a relatively high density of residual dislocations in the boundary it will quickly become inoperative, and another slip system which experiences significant resolved shear stress and tends to minimize the density of residual dislocations left behind in the boundary will take over. This, of course, tends to reduce the energy of the system and the back stresses acting on the incoming dislocations. The angle, $\phi$, between the traces in the boundary of the planes of the active slip systems in crystals 1 and 2 is usually not of paramount importance except in special cases such as in Fig. 12.21 where $\phi = 0$. The different ways in which boundaries can act as barriers to slip are summarized in the schematic diagrams of Fig. 12.22. In (a), the two slip systems intersect along a common line in the boundary, the Burgers vectors of all dislocations are the same and are parallel to the intersection of the slip planes in the boundary, the dislocations (in screw orientation) cross slip in the boundary, the barrier to transmission is the difficulty in cross slipping, and the transmission is direct. In (b), the traces of the slip planes are again parallel but the Burgers vectors of the dislocations are different, residual dislocations are left behind in the boundary, the barrier is essentially the energy of the residual dislocations, and the transmission is direct. In (c), the traces of the slip planes are not parallel, the Burgers vectors are different, residual dislocations are left behind, and the transmission is indirect. In (d), no transmission occurs, dislocations may enter the boundary, and the boundary acts as a complete barrier.

Finally we note that the effectiveness of grain boundaries as barriers to slip by indirect transmission should tend to increase with increasing temperature. As the temperature is raised and dislocation climb becomes possible, the system will acquire an additional degree of freedom to relax the localized stresses produced by pile-ups. This will make it more difficult to initiate outgoing slip in crystal 2.

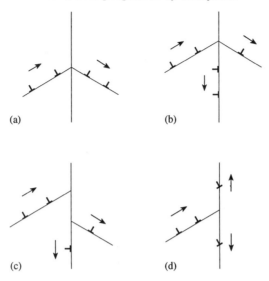

**Fig. 12.22** Different ways in which grain boundaries can act as barriers to the transmission of lattice slip. Arrows indicate directions of dislocation motion. (a) Direct transmission by the cross slip of screw dislocations belonging to slip systems which intersect along a common line in the boundary and possess identical Burgers vectors parallel to the intersection. (b) Direct transmission involving slip systems which intersect along a common line in the boundary but possess different Burgers vectors. Residual dislocations deposited in the boundary. (c) Indirect transmission involving slip systems which do not intersect along a common line in the boundary and possess different Burgers vectors. Residual dislocations deposited in the boundary. (d) Case of no transmission when dislocations belonging to a slip system in crystal 1 are deposited in the boundary without the activation of any slip in crystal 2.

### 12.6.2  Heterophase interfaces

Heterophase interfaces should act as barriers to slip in many of the same ways as grain boundaries. However, some important differences should be noted. In general, the slip vectors of the two adjoining phases will always differ, and, therefore, it will always be necessary to leave a residual dislocation component behind in the interface in any transmission event. In addition, there may be large differences between the slip systems and critical stresses required to activate slip in the two adjoining phases. For example, if one phase is very soft compared with the other, the interface will act as a strong barrier to dislocations impinging from the soft side irrespective of the detailed geometry of the slip systems adjoining the interface.

## 12.7  EFFECTS OF INTERFACES ON THE PLASTIC DEFORMATION OF BICRYSTALS AND POLYCRYSTALS AT LOW TEMPERATURES

In the previous sections of this chapter we have described a number of the basic aspects of the stresses which may exist at interfaces and also the behaviour of dislocations at interfaces at both the macroscopic and microscopic levels. We now have the necessary background to discuss the relatively complex problem of the plastic deformation of bicrystals and polycrystals at relatively low temperatures where thermally activated processes such as boundary and/or lattice diffusional transport and dislocation climb are suppressed, and the deformation is produced by the glide of lattice dislocations (i.e. 'slip').

### 12.7.1  Homophase bicrystals and polycrystals

In order to understand the plastic deformation of these materials by dislocation glide it is helpful to consider the increasingly complex phenomena which occur when single crystals, and then bicrystals, and finally polycrystals are deformed by this mechanism. The deformation of single crystals is relatively simple, since no crystalline interfaces are present, and the process occurs homogeneously (at least on a macroscopic scale) throughout the material. When a single crystal is plastically strained by uniaxial tension, slip will generally begin on the slip system experiencing the maximum resolved shear stress. This causes the crystal axes to rotate relative to the stress axis, and eventually the resolved shear stress on another slip system with a different orientation relative to the stress axis will become large enough to activate slip on that system initiating so-called multiple slip (Reid 1973, Havner 1992). The rate of strain hardening (i.e. the slope of the flow stress-plastic strain curve) is relatively small in the single slip ('easy-glide') regime and increases sharply when multiple slip occurs due to the interactions and interferences which occur between the dislocations gliding on the different slip systems.

Consider next the straining of the bicrystal in Fig. 12.1(a) by uniaxial tension along $x_3$. When the stress is first applied, elastic incompatibility will produce stress concentrations in the region near the boundary as described in Section 12.2. If the boundary is a general one, we may then expect it to act as a source of lattice dislocations (Section 12.5.2) which will expand into the region near the boundary and produce small 'micro-plastic' strains there which will at least partially relieve the compatibility stresses (see Section 12.2 and Fig. 12.19). In a later stage of the deformation at larger plastic strains, plastic straining will be induced everywhere in the bicrystal. However, since the two crystals of the bicrystal have different crystal orientations with respect to the stress axis and maintain compatibility across the grain boundary, this deformation will be inhomogeneous on a macroscopic scale. The inhomogeneous nature of these larger plastic strains may be easily recognized by ignoring any elastic strains (which are negligible compared to the plastic strains at this point) and imagining a process in which the two undeformed crystals of the bicrystal are first separated along the boundary and are then deformed individually by the applied tensile stress. In this operation each crystal will be deformed homogeneously, but differently, since different slip systems will generally be activated in each crystal because of their different orientations. Finally, the two crystals are rejoined along the interface in a manner which restores the original compatibility at the interface. This can be accomplished by plastically deforming the crystals locally in the regions near the interface so that they can be joined together compatibly. The result is therefore a bicrystal which is inhomogeneously deformed. (We note that we could have restored the original compatibility by deforming each crystal homogeneously rather than inhomogeneously. However, in general, this would have required more energy.) The inhomogeneous deformation will generally require the introduction of a distribution of excess lattice dislocations in the region near the interface as described by Ashby (1970). These dislocations are required geometrically and have therefore been termed 'geometrically necessary' (Ashby 1970).

The slip in the region near the boundary will clearly be more complicated than elsewhere in the bicrystal in this regime of strain, and we now inquire into the minimum

number of slip systems which must be active there. In order to maintain compatibility
there we must have

$$\varepsilon_{11}^{A} = \varepsilon_{11}^{B}; \quad \varepsilon_{33}^{A} = \varepsilon_{33}^{B}; \quad \varepsilon_{13}^{A} = \varepsilon_{13}^{B}, \tag{12.45}$$

where all of the strains are now plastic strains in the boundary region rather than elastic
strains as in eqn (12.6). (Note that elastic strains have again been neglected.) We now pose
the question, if one slip system is active in crystal B, how many must be active in crystal
A in order to maintain compatibility? According to eqn (12.17), the contribution to $\varepsilon_{ij}$
of a shear of magnitude $\gamma^{k}$ on slip system k may be expressed in the form

$$\varepsilon_{ij}^{k} = g_{ij}^{k} \cdot \gamma^{k}, \tag{12.46}$$

where $g_{ij}^{k}$ is a purely geometrical factor. Using eqns (12.45) and (12.46), we therefore
have the set of four linear equations

$$\left.\begin{aligned}
\sum_{k} g_{11}^{k,A} \cdot \gamma^{k,A} &= g_{11}^{1,B} \cdot \gamma^{1,B} \\[2mm]
\sum_{k} g_{33}^{k,A} \cdot \gamma^{k,A} &= g_{33}^{1,B} \cdot \gamma^{1,B} \\[2mm]
\sum_{k} g_{13}^{k,A} \cdot \gamma^{k,A} &= g_{13}^{1,B} \cdot \gamma^{1,B} \\[2mm]
\text{constant} &= g_{33}^{1,B} \cdot \gamma^{1,B}
\end{aligned}\right\}. \tag{12.47}$$

in the variables $\gamma^{k,A}$ and $\gamma^{1,B}$, where the last specifies the amount of strain imposed
along the tensile axis. It may be seen that three slip systems are then required in crystal
A, since we will then have a set of four linear equations which can be solved consistently
for the four unknowns $\gamma^{1,A}$, $\gamma^{2,A}$, $\gamma^{3,A}$, and $\gamma^{1,B}$. If we assume next that two slip
systems are active in crystal B, a similar exercise shows that two slip systems must be
active in crystal A. These results show that a total of at least four slip systems must be
involved and that multiple slip is required in at least one crystal (Livingston and Chalmers
1957).

   In further work Hauser and Chalmers (1961) attempted to simulate more closely the
deformation in a polycrystal by studying the deformation of specimens in which crystal
B was totally surrounded by either crystal A or a more general polycrystalline layer (see
Fig. 12.23). In these cases maintenance of compatibility requires even larger numbers of
slip systems. For example, when one crystal is entirely surrounded by another, a total
of six slip systems may be required in certain regions (Hauser and Chalmers 1961).

   The general picture which emerges at this point is one in which the different mechanical
responses of the crystals adjoining interfaces and the requirements of compatibility cause
the plastic deformation of systems containing interfaces to be inhomogeneous. Higher
degrees of multiple slip will generally be required in the regions near the interfaces, and,
in addition, concentrations of excess 'geometrically necessary' lattice dislocations will tend
to accumulate there. However, so far we have considered only the macroscopic aspects
of the overall plastic deformation. Microscopic effects must also be included. As discussed in Section 12.6, boundaries will generally act as barriers to the transmission of
slip causing incoming dislocations to pile up against them. In addition, dislocations may
enter boundaries and accumulate there. These local pile-ups and accumulations will
produce complex local stress fields around them which, in turn, can activate further slip

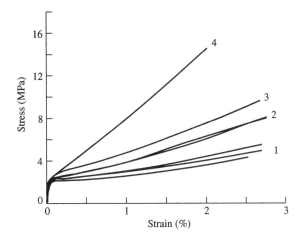

**Fig. 12.23** Stress–strain curves for Ag specimens pulled in tension. Curves 1: single crystals pulled along ⟨210⟩. Curves 2: 'compatible' bicrystals in which crystals A and B were pulled along a common ⟨210⟩ direction, and the slip systems in the two crystals possessed mirror symmetry across the boundary plane. Curve 3: 'totally surrounded' bicrystal in which the A and B crystals were oriented as for Curves 2 but the A crystal totally surrounded the B crystal axially. Curve 4: polycrystal. (From Hauser and Chalmers (1961).)

systems in the boundary region. The slip induced in this manner will generally extend over distances of the order of the spacings between the pileups (slip bands) and will therefore further increase the complexity of slip and the dislocation density accumulated in the boundary region. Further aspects of the complex elastic and plastic deformation which can occur in different systems containing boundaries have been described, for example, by Hook and Hirth (1967), Margolin and Stanescu (1975), Rey and Zaoui (1980, 1982), Paidar *et al.* (1986), Rey *et al.* (1986), and Sittner and Paidar (1989).

As might be anticipated, bicrystals and polycrystals generally tend to be more resistant to plastic flow than corresponding single crystals, particularly at small to moderate strains. This may be attributed to the greater multiplicity of slip required in the boundary regions and the higher density of lattice dislocations which accumulate there and act as impediments to further dislocation motion. Put simply, the boundary regions are in a more advanced state of strain hardening. Some characteristic results comparing corresponding single crystals and bicrystals and also polycrystals are shown in Fig. 12.23.

At large strains, however, the strengthening effects of grain boundaries tend to fade away. In the large strain regime all regions of the specimen have generally been extensively strained in multiple slip, and large densities of residual dislocations, often collected into walls (Hansen 1992), have accumulated everywhere. Differences between the crystal interiors and regions near the boundaries are then reduced.

Another phenomenon of interest is the increase in the yield stress, i.e. the stress required to produce significant plastic deformation ('macro-yielding'), which is achieved when the grain size of a polycrystal composed of an array of equiaxed crystals (grains) is reduced and the density of interfaces is increased. This effect can be usually represented reasonably well by the so-called Hall–Petch relation

$$\tau_Y = \tau_Y^\circ + kd^{-1/m}, \dots \dots (1 \leqslant m \leqslant 3) \tag{12.48}$$

where $\tau_Y$ is the yield stress, $\tau_Y^0$ and $k$ are constants, and $d$ is the grain size (Hirth 1972, Hirth and Lothe 1982, Meyers and Ashworth 1982, Meyers and Chawla 1984, Shen *et al.* 1988). Many models have been proposed to explain this relationship (see reviews in the above references). Due to the obvious complexity of the problem, these models vary considerably and place different emphases on different features of the overall deformation process. For present purposes we shall describe briefly the model of Meyers and Ashworth (1982), which incorporates many of the basic phenomena which we have already discussed.

The model is illustrated schematically in Fig. 12.24. Compatibility stresses are first generated in the boundary regions under purely elastic conditions (Fig. 12.24(a)), and these are then attenuated by the plastic strain produced by the glide of dislocations which are created there and accumulate as shown in Fig. 12.24(b,c). At this stage the material is essentially a composite made up of a continuous network corresponding to a 'grain boundary film' enclosing discontinuous 'islands' of single crystal bulk material as indicated by the shaded and unshaded regions in Fig. 12.24(d). Furthermore, the flow stress of the boundary network region is higher than that of the bulk crystal islands because of its higher dislocation density. As the applied load is increased further, the boundary network continues to act as a strong framework which inhibits plastic flow in the crystal islands. Eventually, however, plastic flow in the boundary region results in increased stress and eventual plastic flow in the crystal islands which initiates macro-yielding (Fig. 12.24e). Finally, the dislocation densities in the bulk and boundary regions become equal, and any plastic incompatibility disappears (Fig. 12.24f). Meyers and Ashworth (1982) then show how the stress required to produce this macro-yielding can follow a relationship of the form of eqn (12.48).

### 12.7.2 Heterophase bicrystals and polycrystals

The plastic deformation of heterophase bicrystals and polycrystals will involve the same basic mechanisms which have been invoked above to understand the deformation of homophase materials. Many of the features of the deformation of homophase and heterophase aggregates will therefore be similar in a qualitative sense. However, important quantitative differences may exist, particularly when the different phases in a heterophase system have widely different deformation properties, e.g. hard and brittle versus soft and ductile. An immense effort has been made by the materials science and engineering community to produce heterophase polycrystalline systems, e.g. composites, with desirable mechanical properties by combining phases with widely different individual mechanical properties in different geometries. For example, systems containing hard thin fibres of one phase embedded in a soft ductile matrix have been widely studied. A discussion of this work would require at least another book and will not be pursued here. Some references are: Hull (1981), Kelly and Macmillan (1986), Chawla (1987), and Mortensen (1988).

## 12.8 ROLE OF INTERFACES IN THE PLASTIC DEFORMATION OF BICRYSTALS AND POLYCRYSTALS AT HIGH TEMPERATURES

In the previous section we discussed the plastic deformation of bicrystals and polycrystals at relatively low temperatures where potentially important thermally activated processes such as diffusional transport and interface sliding were suppressed. Plastic deformation

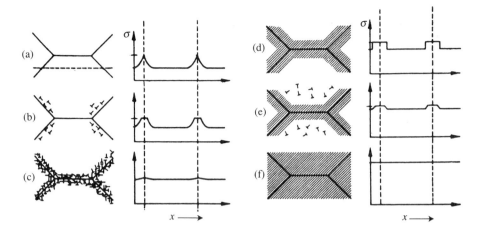

**Fig. 12.24** Stages (a)–(f) in the plastic deformation of a polycrystal according to the model of Meyers and Ashworth (1982). Schematic structures are shown at each stage along with the corresponding distribution of stress across the dashed plane shown in (a). See text for details.

was therefore restricted to dislocation glide processes. We now turn our attention to elevated temperatures where these thermally activated processes can produce further types of plastic deformation.

### 12.8.1  Interface sliding

We define interface 'sliding' in a bicrystal as any plastic shear displacement of crystal 1 with respect to crystal 2 parallel to the interface which is localized in the interface region and occurs in response to applied forces. This sliding can take a variety of forms (e.g. Valiev *et al.* (1991)) ranging from cases where the shear is highly localized, and is restricted essentially to the narrow core of the interface as in the examples shown in Fig. 12.25, to cases, such as in Fig. 9.6, where the shear is distributed more widely over a slab of material which includes the interface. As long as this region is relatively narrow compared with the thicknesses of the two adjoining bulk crystals, the two crystals act essentially as uniform blocks as one crystal shears, or 'slides', with respect to the other.

Interface sliding is an important phenomenon which occurs over a wide range of conditions at elevated temperatures, i.e. $T > 0.4T_{\mathrm{m}}$. The shearing which is associated with sliding is clearly a form of plastic deformation, and it often makes an important, and sometimes major, contribution to the overall plastic straining of bicrystals and polycrystals. In addition, it may be responsible for cavitation (Section 12.9.4), leading to catastrophic intergranular mechanical failure. Despite the apparent geometrical simplicity of interface sliding from a macroscopic point of view, we shall find that it is a complex phenomenon which may involve a variety of mechanisms depending upon deformation variables such as stress, plastic strain, and temperature, and also the characteristics of the interfaces which are involved. We therefore begin by taking up the simplest possible case, i.e. sliding at an ideally planar grain boundary in a bicrystal.

### 12.8.1.1  *Sliding at an ideally planar grain boundary*

Consider first the case where we attempt to shear a perfectly planar boundary such as illustrated in Fig. 12.26(a), simultaneously along its entire length in a reversible manner

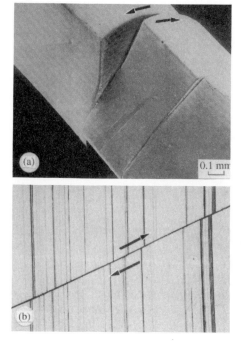

0.1 mm

Fig. 12.25  (a) Grain boundary sliding in stainless steel wire during torsional shear. (From Schneibel and Peterson (1986).) (b) Grain boundary sliding (indicated by displacements of fiduciary scratches) in Al–Mg alloy. (From Mullendore and Grant (1963).)

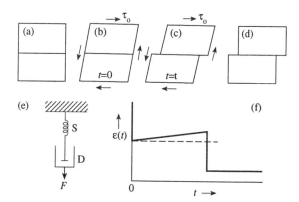

Fig. 12.26  Bicrystal acting as a Maxwell solid via viscous grain boundary sliding. (a) Unstressed bicrystal. (b) Bicrystal subjected to sudden elastic shear stress, $\tau_o$. (c) Bicrystal after grain boundary plastic shear (sliding) under constant shear stress. (d) Bicrystal after sudden removal of shear stress. (e) Mechanical analogue of bicrystal consisting of linear spring, S, and dashpot, D. (f) Shear strain versus time during steps (a)–(d).

by applying a shear stress to the bicrystal parallel to the boundary in order to shear crystal 1 with respect to crystal 2. During this operation the boundary will remain in its ground state at each stage of the translation so that the overall process will be reversible. As discussed in Sections 2.11.1 and 4.3.1.1, the shear stress required to shear the boundary under these conditions will then depend upon the shape of the $\gamma$-surface.

If the interface is periodic, the $\gamma$-surface will not be flat in general (Section 4.3.1.1), and relatively large shear stresses may then be required to shear the boundary simultaneously along its entire length in this reversible manner. Such boundaries will then prefer to shear/slide non-uniformly in a completely different manner by the passage of localized extrinsic boundary dislocations along their cores by glide and/or climb. (We note that the shearing of such a periodic boundary by the passage of dislocations is similar in many respects to the shearing of a single crystal on a lattice slip system where shearing via the passage of lattice dislocations is preferred because of the high resistance of the periodic lattice to uniform shear.) If the boundary dislocations must climb in this process, or if they possess step character, their motion along the boundary will produce boundary motion normal to itself as discussed in Chapter 9.

On the other hand, if the boundary is quasi-periodic, the $\gamma$-surface will be flat (Sections 2.11.1 and 4.3.1.1), and there will be no resistance to such ideal reversible sliding. However, in actual practice, when the boundary is sheared irreversibly at finite rates, there will be a kinetic resistance to the sliding, and in many cases the boundary may exhibit a Newtonian viscosity where the sliding rate is proportional to the applied shear stress. Since the boundary is incoherent (and unable to support localized dislocations with Burgers vector components parallel to the interface), we would expect the shearing to occur throughout the boundary by means of highly localized shear transformation events of the general type found by Deng *et al.* (1989) in their computer simulation studies of the plastic deformation of atomic glasses. In this process (see also Ke (1949, (1990)) shear transformations occur in small groups of atoms in momentarily fertile regions which percolate through the material. This process is thermally activated and causes the glassy material to possess an intrinsic Newtonian viscosity. Since the disorder of the bad material in the core of an incoherent boundary is similar in many respects to the disorder of an atomic glass, a generally similar intrinsic mechanism for shear deformation of the boundary core should hold. The activation energy for this type of shear is generally expected to be lower than that for lattice diffusion and close to that for grain boundary diffusion.

We note that boundary shear by the motion of localized boundary dislocations, as described previously, may also follow Newtonian viscous behaviour under certain conditions, particularly at small shears (and stresses) where large numbers of boundary dislocations do not accumulate and produce complex non-linear effects in the stress dependence of the sliding rate.

If an unstressed bicrystal containing an initially flat boundary capable of viscous behaviour (Fig. 12.26(a)) is suddenly subjected to a constant shear stress $\tau_0$, the bicrystal will instantaneously undergo an elastic shear strain as shown in Fig. 12.26(b). Under the constant stress the boundary will then shear plastically at a constant rate as indicated in Fig. 12.26(c). During this process the boundary is out of its ground state and is supporting a shear stress under non-equilibrium conditions. If the applied stress is then suddenly removed, the elastic strain will immediately recover, as illustrated in Fig. 12.26(d), and all deformation will cease.

From a purely mechanical point of view the bicrystal can be well represented by a linear spring connected in series to a dashpot as illustrated in Fig. 12.26(e) (Nowick and Berry 1972). Here, the linear spring represents the two elastic bulk crystals, while the dashpot (consisting of a plunger immersed in a viscous liquid) represents the viscous boundary. If the system is subjected to a force $F$ (and we neglect inertial effects) the linear spring will elongate a distance $\Delta x^S$ corresponding to

$$\Delta x^S = a^S F, \tag{12.49}$$

where $a^S$ is the spring constant. In addition, the dashpot will elongate at a rate proportional to $F$, i.e.

$$d(\Delta x^D)/dt = a^D F, \tag{12.50}$$

where the constant $a^D$ is related to its effective viscosity. We may then define a shear strain which is proportional to the total elongation, i.e.

$$\varepsilon = a^\varepsilon (\Delta x^S + \Delta x^D) = a^\varepsilon \Delta x, \tag{12.51}$$

where $a^\varepsilon$ is a constant, and a shear stress proportional to $F$, i.e.

$$\tau = (1/a^\tau)F, \tag{12.52}$$

where $a^\tau$ is a constant. Combining these relationships we then obtain the equation of motion

$$\frac{d\varepsilon}{dt} = (a^\varepsilon a^S a^\tau)\frac{d\tau}{dt} + (a^D a^\tau a^\varepsilon)\tau, \tag{12.53}$$

which contains two independent constants (bracketed) corresponding to the two mechanical elements in the system.

Consider now the behaviour of this system. When we suddenly impose the constant shear stress, $\tau_0$, there will be an immediate elastic strain given by $\varepsilon(0) = (a^\varepsilon a^S a^\tau) \cdot \tau_0$. This result is obtained using eqns (12.49), (12.52), and (12.51) with $\Delta x^D = 0$, since the viscous element has had no time to respond. In the subsequent constant stress regime $\tau = \tau_0$, and eqn (12.53) takes the form

$$\frac{d\varepsilon}{dt} = (a^D a^\tau a^\varepsilon) \cdot \tau_0, \tag{12.54}$$

which, upon integration, yields

$$\varepsilon(t) = (a^\varepsilon a^S a^\tau)\tau_0[1 + (a^D/a^S)t]. \tag{12.55}$$

The strain as a function of time is plotted in Fig. 12.26(f) and shows the expected linear increase due to the constant rate of boundary sliding induced by $\tau_0$. If $\tau_0$ is suddenly removed, the elastic strain will instantaneously recover as shown. Finally, under quasi-equilibrium conditions, $F$ (and therefore $\tau$) would be infinitesimally small, and the bicrystal would shear at a steady infinitesimal rate.

A material with the above characteristics is generally described as a 'Maxwell solid' (Nowick and Berry 1972). In practice, however, bicrystals with ideally flat boundaries do not exist, and there is always at least some curvature or stepping present. As we shall now see, real boundaries therefore have quite different sliding properties.

### 12.8.1.2 *Sliding at a non-planar grain boundary by means of elastic accommodation*

As mentioned above, real grain boundaries are always non-planar to at least some extent. If such a boundary (represented ideally by a sinusoid as in Fig. 12.27(a)) is subjected to an applied shear stress, it may shear (slide) by a small amount initially, but further shearing will be quickly blocked by the various asperities of the two interlocked crystals. If the applied stresses are low enough so that no plastic flow by lattice dislocation motion takes place in the two adjoining crystals, and if diffusional accommodation as described in Section 12.8.1.3 is slow enough to be negligible, the small amount of sliding which occurs will be accommodated entirely by elastic strains in the two adjoining crystals. This

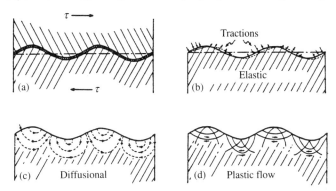

**Fig. 12.27** Mechanisms for the accommodation of sliding at a non-planar grain boundary such as in (a). (b) Elastic accommodation (showing lower crystal only). (c) Diffusional accommodation via lattice diffusion. (d) Plastic flow accommodation via the motion of lattice dislocations. (From Ashby (1972b).)

is illustrated in Fig. 12.27(b) for a general grain boundary which is everywhere parallel to z and possesses a Newtonian viscosity as discussed above in Section 12.8.1.1. Here, the boundary was subjected to the constant applied shear stress, $\tau_0$. Small plastic shears (sliding) then occurred everywhere along the boundary, since the viscous boundary was unable to sustain shear stress across it statically. This, in turn, generated a set of tractions acting on the 'internal surfaces' of the two crystals facing the boundary whose sum balanced the net shearing force exerted on the bicrystal. The result at equilibrium is an applied stress field consisting of pure tractions and no shear stresses at the boundary as illustrated in Fig. 12.27(b). In this process the upper crystal, as a whole, is displaced relative to the lower crystal by a small amount, $\Delta U$, due entirely to the sliding at the boundary and the accompanying elastic accommodation in the two adjoining crystals. Raj and Ashby (1971) have calculated $\Delta U$ in this situation for a grain boundary of the sinusoidal form $y = A \cos(2\pi x/\lambda)$ with the result that $\Delta U$ is proportional to the applied stress according to

$$\Delta U = \frac{(1 - \nu)^2 \lambda^3}{\pi^3 E A^2} \cdot \tau_0, \tag{12.56}$$

where $E$ is Young's modulus. Values of $\Delta U$ will generally be small, typically of the order of 10 nm. Since all deformation in the two crystals is purely elastic, this deformation (and also the corresponding boundary sliding) will be fully recoverable upon removal of the applied stress. As we now demonstrate, this causes the bicrystal to behave as a 'standard anelastic solid' rather than as a Maxwell solid as was the case when the grain boundary possessed a Newtonian viscosity and was ideally planar.

The mechanical analogue is shown in Fig. 12.28(a). The basic relationships for the forces and displacements associated with the two springs, S1 and S2, and the dashpot, D, are of the same form as given by eqns (12.49) and (12.50), and, therefore, for this system

$$\Delta x^{S1} = a^{S1} F^{S1}; \quad \Delta x^{S2} = a^{S2} F^{S2}, \tag{12.57 a,b}$$

$$d(\Delta x^D)/dt = a^D F^D, \tag{12.58}$$

$$\Delta x^{S1} = \Delta x^D, \tag{12.59}$$

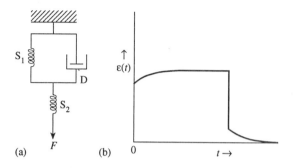

**Fig. 12.28** Bicrystal acting as a 'standard anelastic solid' via viscous boundary shearing and elastic accommodation. (a) Mechanical analogue consisting of two linear springs, $S_1$ and $S_2$, and dash-pot, D. (b) Shear strain versus time for bicrystal subjected to sudden application, and then sudden removal, of shear stress.

$$\Delta x = \Delta x^{S1} + \Delta x^{S2}, \tag{12.60}$$

$$F = F^{S2} = F^{S1} + F^{D}. \tag{12.61}$$

By using eqns (12.51) and (12.52), the above equations may be combined to obtain the equation of motion for the system in the form

$$[a^{\varepsilon}a^{\tau}(a^{S1} + a^{S2})] \cdot \tau + \left[\frac{a^{S1}a^{S2}a^{\varepsilon}a^{\tau}}{a^{D}}\right] \cdot \frac{d\tau}{dt} = \left[\frac{a^{S1}}{a^{D}}\right] \cdot \frac{d\varepsilon}{dt} + \varepsilon, \tag{12.62}$$

which contains three independent constants (bracketed) corresponding to the three mechanical elements in the system.

Consider now the mechanical behaviour of this system. If a constant shear stress, $\tau_0$, is suddenly imposed, only the S2 spring will elongate initially. According to eqns (12.57b), (12.51), and (12.52), this will produce the initial strain $\varepsilon(0) = [a^{S2}a^{\varepsilon}a^{\tau}]\tau_0$. For $t > 0$, eqn (12.62) reduces to

$$\left[\frac{a^{S1}}{a^{D}}\right] \cdot \frac{d\varepsilon}{dt} + \varepsilon = a^{\varepsilon}a^{\tau}(a^{S1} + a^{S2}) \cdot \tau_0. \tag{12.63}$$

Solving this equation, and fitting the result to the initial condition, we obtain

$$\varepsilon(t) = a^{\varepsilon}a^{\tau}a^{S2}\tau_0 + a^{\varepsilon}a^{\tau}a^{S1}\tau_0\{1 - \exp[-(a^{D}/a^{S1})t]\}. \tag{12.64}$$

This result shows that a further time dependent strain will occur which increases at an exponentially decreasing rate, with a time constant given by $a^{S1}/a^{D}$. If the stress is then suddenly removed, the elastic strain $a^{S2}a^{\varepsilon}a^{\tau}\tau_0$ is immediately recovered, and eqn (12.63) may then be solved with $\tau_0 = 0$ to show that the remaining strain will relax exponentially with the same time constant, $a^{S1}/a^{D}$. The complete curve of $\varepsilon(t)$ versus $t$ is shown in Fig. 12.28(b) and is typical of a 'standard anelastic solid' (Nowick and Berry 1972).

### 12.8.1.3 *Sliding at a non-planar grain boundary by means of diffusional accommodation*

As shown above, a small amount of boundary sliding can be accommodated elastically at non-planar boundaries exhibiting viscous behaviour. However, once full elastic accommodation is reached, further sliding is blocked. However, at elevated temperatures relatively large amounts of further sliding can be accommodated if material can be added/removed from the internal 'surfaces' of the two crystals facing the boundary by diffusion in a manner which allows the interlocked crystals to slide past one another continuously in compatible fashion as illustrated in Fig. 12.27(c). In this process appropriate regions of the boundary act as sources or sinks for diffusional fluxes of atoms (Chapter 10), and the diffusional transport occurs by diffusion through the adjoining crystals (as illustrated in Fig. 12.27(c)) and/or diffusion along the boundary. The diffusional transport is motivated, as always, by differences in the diffusional potential for atoms along the boundary, which in this case are due to differences in the normal tractions which are present.

Consider first the sliding of an incoherent general boundary. If we assume that the viscous shear mechanism in the boundary core is rapid enough so that negligible shear stresses exist across the boundary, the overall rate of sliding will be controlled by the rate of the diffusional accommodation. Under these conditions a quasi-steady-state lattice diffusion field will very quickly be established (Fig. 12.27(c)) in which appropriate numbers of atoms either enter or leave each local region of the boundary by diffusion at rates which allow one bulk crystal to translate past the other at a constant rate, $dU/dt$, without producing any gaps or overlaps. In order to achieve this steady state condition, the distribution of normal tractions along the boundary will be modified initially by transients in the diffusion currents which add or remove material locally at the boundary until the tractions at the boundary are adjusted to produce corresponding values of $M_A$ which sustain just the required quasi-steady-state diffusion field. (We note that the situation here is similar to that in Section 10.4.2.4 where the steady state diffusional accommodation of boundary sliding at blocking second phase particles was analysed.) An expression for the dependence of the diffusion potential for atoms, $M_A$, at the boundary on the local value of the normal traction, $\tau_n$, can be obtained from eqn (5.146). For a system containing atoms (species A) and vacancies, eqn (5.146) reduces to $\mu_V = (e - Ts - M_A c_A - \tau_n)/c_o N_o$, if the vacancies are chosen as the $C$th component. If we further assume that the boundary acts everywhere as a perfect source/sink, $\mu_V = 0$ at the boundary, and therefore

$$M_A = (f - \tau_n)/c_A = N_o \Omega (f - \tau_n). \tag{12.65}$$

The form of eqn (12.65) is easily underscood, since work of expansion corresponding to $-N_o \Omega \tau_n$ is required to insert a mole of atoms into the system at the interface when the normal traction is $\tau_n$.

As demonstrated by Raj and Ashby (1971), the resulting steady state diffusion problem can now be solved by solving the Laplace equation, $\nabla^2 M_A = 0$, subject to the boundary conditions on $M_A$ at the boundary. However, for present purposes, we may employ a simpler approach which yields essentially the same results. The appropriate flux equation is eqn (9.18). On average, the tractions on sections AB and BC of the boundary in Fig. 12.27(c) are $\bar{\tau}_n = \tau_o \lambda/4A$ and $-\tau_o \lambda/4A$, respectively, when the boundary has the form $y = A \cos(2\pi x/\lambda)$. Using eqn (12.65), the difference between the average diffusion potentials along the AB and BC sections is then $\Delta \bar{M}_A = N_o \Omega \tau_o \lambda/2A$. The average distance from section BC to AB is $\simeq 2\lambda/3$, and, therefore, the average gradient of the diffusion potential between the two sections is $\overline{\nabla M}_A = 3N_o \Omega \tau_o/4A$. The cross-sectional area through which diffusion occurs (on each side of the boundary) is $\simeq 2\lambda/3$, and, therefore, using eqn (9.18), the total diffusion current reaching section BC is $I = 2D^L \tau_o \lambda/ktf^L A$. The average component of the velocity $dU/dt$ normal to the boundary in section BC is then $\overline{dU_n}/dt \simeq I\Omega/\overline{BC}$, where $\overline{BC}$ is the length of the section BC. However, $U \simeq \bar{U}_n \overline{BC}/2A$, and therefore

$$\frac{\mathrm{d}U}{\mathrm{d}t} \simeq \frac{D^L \lambda \Omega}{kTf^L A^2} \cdot \tau_0. \tag{12.66}$$

This result has the same functional form as obtained by Raj and Ashby for the sinusoidal boundary and differs only by a numerical factor which is smaller than two.

An expression for the constant sliding rate due to diffusion along the boundary may be obtained by making similar approximations to obtain

$$\frac{\mathrm{d}U}{\mathrm{d}t} \simeq \frac{D^B \delta \Omega}{kTf^B A^2} \cdot \tau_0. \tag{12.67}$$

To obtain this result, the boundary was approximated in the usual way as a slab of thickness, $\delta$, and diffusivity $D^B$ (Chapter 8), and the flux equation was taken to be of the same form as eqn (9.18). Again, our result is in satisfactory agreement with that of Raj and Ashby (1971).

It may be seen that the sliding rate in both cases is linearly proportional to the applied stress, and that the non-planar boundary possesses an effective Newtonian viscosity which may be easily derived from either eqn (12.66) or (12.67). It is important to note that the sliding rate due to lattice diffusion is proportional to the boundary wavelength while the rate due to boundary diffusion is independent of it. Therefore, a short wavelength and a low temperature, where $D^B$ is much larger than $D^L$ (Chapter 8), favour the boundary diffusion mechanism.

Solutions for the rate of sliding of grain boundaries of other shapes have been given by Raj and Ashby (1971). In particular, they use the basic solution for a sinusoidal boundary and a Fourier series method to find the general solution for a boundary of arbitrary shape.

If the boundary is semicoherent and can maintain a reasonably high density of localized dislocations which can easily glide and climb along the wavy boundary, the results will be generally similar. If the sum of the Burgers vectors of the dislocations passing each point is in the direction of the shear, the diffusion fluxes required to produce the positive and negative climb of the dislocations as they traverse the wavy boundary will be the same as the diffusion fluxes calculated above for the general boundary, and we may regard the sliding as diffusionally accommodated. In fact, this latter case may be regarded as a 'quantized' version of the former.

### 12.8.1.4 *Sliding at a non-planar grain boundary by means of plastic flow accommodation in the lattice*

Finally, we point out that grain boundary sliding can also be accommodated by the plastic flow of the adjoining crystals via the motion of lattice dislocations as illustrated in Fig. 12.27(d). This, of course, will occur only at relatively high applied stresses where deformation by means of the motion of lattice dislocations is possible and will depend in a complicated manner upon the boundary as a source/sink for lattice dislocations as discussed in Sections 12.5 and 12.6. Also, the sliding rate will generally depend upon the the plastic properties of the adjoining crystals and will therefore exhibit work hardening and vary non-linearly with applied stress in contrast to the simple linear behaviour which occurs when the accommodation is purely diffusional.

Crossman and Ashby (1975) have treated this accommodation problem by assuming that the plastic deformation in the lattice occurs via dislocation motion by both glide and climb and obeys a power law creep (Weertman 1955, McClintock and Argon 1966) relation of the form

$$\dot{\varepsilon} = A\tau^n, \tag{12.68}$$

where $A$ and $n$ are constants. As described in Section 12.8.2.3, they have analysed the overall creep rate of a homophase polycrystal undergoing grain boundary sliding coupled with plastic deformation of the grains via power-law creep using finite element techniques.

### 12.8.1.5 *Experimental observations of sliding at interfaces*

Interface sliding of a variety of types has been observed experimentally over a wide range of temperature, applied stress, and geometrical conditions. Direct evidence for the sliding of singular or vicinal interfaces caused by the passage of dislocations along them has been obtained. Observations of elastically accommodated, diffusionally accommodated, and plastically accommodated sliding have also been made under appropriate conditions. In this section we briefly review a number of these observations along with several additional aspects of sliding which are of interest.

*Evidence for the passage of boundary dislocations during the sliding of singular or vicinal interfaces.* We have already described in Section 9.2.1.2 the sliding of singular CSL grain boundaries which was coupled geometrically to simultaneous boundary migration (see Fig. 9.6) and which was clearly due to the passage along the boundaries of large numbers of extrinsic boundary dislocations possessing fixed ratios of their step heights to the magnitudes of their Burgers vectors. In their experiments with Zn, Horuichi *et al.* (1987) found that the sliding rate, $\dot{S}$, was thermally activated and proceeded in a non-viscous manner at a rate approximately proportional to the square of the applied stress according to $\dot{S} = A\tau^2 \exp(-Q/kT)$, where $A$ is a constant, $\tau$ is the applied stress, and the activation energy, $Q$, was near that expected for grain boundary diffusion. We have also cited the experiments of Kegg *et al.* (1973) in Section 12.5.1, where sliding at a vicinal grain boundary was found to occur via the lateral motion of trains of grain boundary dislocations (see Fig. 12.15). In this case some of the dislocations were generated at boundary sources (see Fig. 12.15), but, according to Kegg *et al.*, most were undoubtedly produced by the impingement of lattice dislocations on the boundaries and their subsequent dissociation as discussed further below.

*Elastically accommodated grain boundary sliding.* A simple idealized form of this type of sliding has been described in Section 12.8.1.2. Here, a specimen containing a wavy boundary of the general type possessing an assumed Newtonian viscosity was subjected to a small applied shear stress insufficient to activate the motion of lattice dislocations. Also, any diffusional accommodation was negligible. The amount of sliding which was accommodated elastically was relatively small (eqn (12.56)) and was recoverable upon release of the applied stress (Fig. 12.28), and, as we have shown, caused the specimen to act as a standard anelastic solid. Such a specimen will exhibit 'internal friction', i.e. the conversion of mechanical work to heat, if it is stressed cyclically at an appropriate rate (Nowick and Berry 1972). If the bicrystal in Fig. 12.27(b) is subjected to a small oscillating shear stress with a frequency which is tuned to the reciprocal of the relaxation time for the boundary sliding by means of elastic accommodation, a maximum amount of mechanical energy will be converted to heat in each cycle in order to produce the back and forth viscous flow (sliding) which occurs at the boundary, and a peak in a plot of the 'internal friction' versus frequency will be obtained. (At lower frequencies the process will become closely reversible so that the dissipation will become very small: at high frequencies the cycles will become too short for any significant boundary sliding to occur.)

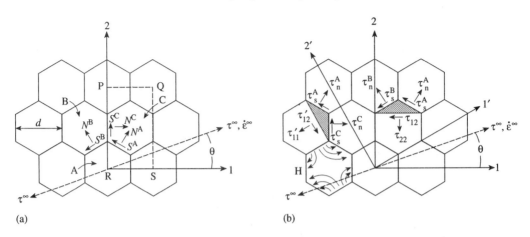

**Fig. 12.29** Idealized 2D representation of polycrystal subjected to uniaxial tension, $\tau^{\infty}$, producing the overall strain rate, $\dot{\varepsilon}^{\infty}$. Grains A, B, and C move apart along the normals to their common boundaries by the distances $N^A$, $N^B$, and $N^C$ and undergo sliding displacements of $S^A$, $S^B$, and $S^C$. (b) Shear stresses and normal stresses present at the different boundaries in (a). Coordinate system $(1', 2')$ rotated $30°$ with respect to system $(1, 2)$. (From Beeré (1978).)

Elastically accommodated sliding can therefore be investigated conveniently by means of internal friction measurements, and many such measurements have been performed using a wide variety of specimen types under different conditions (e.g. Ke 1947, 1949, 1986, 1990, Ke *et al.* 1984, Nowick and Berry 1972, Berry and Prichet 1981, Cosandey *et al.* 1991, Prieler *et al.* 1993). In considering the results of this work it must be realized that many additional forms of elastically accommodated sliding can occur besides the simple example considered so far. In general, these will depend upon the geometry of the grain boundary network structure and the structures of the individual grain boundary segments which are present. For example, in polycrystalline specimens the sliding of individual boundary segments may be blocked elastically by triple junctions at their ends (Zener 1941) as illustrated in Fig. 12.57. If the grain size is relatively large and the boundary segments are wavy on a somewhat smaller scale, effectively viscous boundary sliding may occur at the segments by means of localized diffusional accommodation as described in Section 12.8.1.3, where it is pointed out that a long boundary wave length favours diffusional accommodation by lattice diffusion rather than grain boundary diffusion. Internal friction will then occur if the relaxation time of the effectively viscous boundary sliding due to diffusional accommodation is short compared to the relaxation time of the elastic resistance to sliding at the triple junctions. On the other hand, if the grain size is orders of magnitude smaller, and the boundaries are either essentially flat, or wavy with a very short wavelength, the sliding may be either intrinsically viscous, as discussed in Section 12.8.1.1, or effectively viscous via accommodation by localized grain boundary diffusion. The grain boundary internal friction peaks in such cases will tend to occur at lower temperatures and with smaller activation energies (characteristic of intrinsic sliding or grain boundary diffusion) than in the cases of large-grained specimens where lattice diffusion will tend to be controlling. Behaviour of this type has been found by Prieler *et al.* (1993) and Bohn *et al.* (1994) in specimens of widely different grain sizes.

In many grain boundary internal friction experiments, however, the geometrical details and controlling mechanisms have not been clearly established. A further complication is the possible presence of impurity solute atoms which can alter the positions and magnitudes of the internal friction peaks in ways which are difficult to interpret (e.g. Bohn *et al.* 1994). In addition, physical processes not associated with boundary sliding can produce internal friction peaks in polycrystalline materials at elevated temperatures (Nowick and Berry 1972). It has therefore been difficult to model grain boundary internal friction in a quantitative and clearcut fashion in many experiments. Despite these difficulties, the effectively viscous sliding behaviour of elastically constrained general grain boundaries of the types which are predominant in common homophase polycrystalline materials appears to be well established by the available results.

Recently, Kato and Mori (1993) have measured grain boundary internal friction for a series of Cu bicrystals containing [001] twist boundaries having a range of twist angles, $\theta$, with the results shown in Fig. 12.30. Vicinal boundaries with misorientations near those of the singular low-$\Sigma$ short-period boundaries indicated had significantly larger activation energies for the boundary sliding causing the measured internal friction than general boundaries well away from these singular misorientations. The boundary energy cusps for these boundaries evident in Fig. 5.20 indicate that the vicinal boundaries were semi-coherent boundaries containing dislocations possessing localized Burgers vector components parallel to the boundary plane. Also, any dislocations in the general boundaries must have been delocalized in the boundary plane. The results of Kato and Mori must therefore have been due to a reduced sliding rate of the vicinal boundaries caused by the localization of dislocations in their cores. For example, the shearing rate due to localized dislocations moving along the vicinal boundaries may have been slower then the shearing rate of the general boundaries because diffusional accommodation may have been slower in the former boundaries because they acted as poorer sources/sinks for fluxes of atoms as might by expected on the basis of Table 10.2.

*Diffusionally accommodated grain boundary sliding.*   This type of sliding occurs at rates linearly proportional to the applied stress and can be observed directly in its pure form at very small stresses and high temperatures in the absence of significant lattice dislocation activity. Extensive diffusionally accommodated grain boundary sliding has been measured in wire specimens with a 'bamboo' type grain structure of the type illustrated in Fig. 12.25(a) in which single boundaries extend across the entire specimen cross-section. Raj and Ashby (1972) undoubtedly achieved this form of sliding in their studies with stressed bamboo Ag wires where the sliding often caused successive grains to be markedly offset with respect to one another. Schneibel and Petersen (1985) also observed diffusionally accommodated sliding of this tyde at low stresses in Ni bamboo wire. As we shall see in Section 12.8.2, Herring–Nabarro and also Coble type diffusional creep of homophase polycrystalline materials, which is frequently observed at small stresses and elevated temperatures, can be regarded quite correctly as a form of diffusionally accommodated grain boundary sliding. Finally, Raj and Ashby (1972) have observed diffusionally accommodated sliding in bamboo Ag wires containing second phase particles embedded in the boundaries. However, here the embedded particles, rather than any boundary waviness, effectively blocked the boundary sliding. This work is discussed in Section 10.4.2.4.

*Further aspects of grain boundary sliding.*   At higher stresses and relatively large shear displacements, evidence is often found for lattice plastic deformation in the vicinity of the boundary indicating accommodation via plastic flow. Also, there is extensive evidence that grain boundary sliding is enhanced when the adjoining crystals undergo slip, and

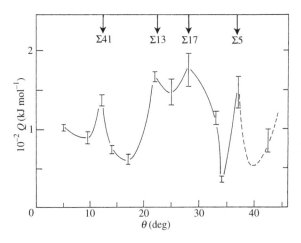

**Fig. 12.30** Activation energy for grain boundary sliding of ⟨001⟩ twist boundaries in Cu bicrystal as a function of twist angle, $\theta$. Values obtained from internal friction measurements. (From Kato and Mori (1993).)

lattice dislocations impinge on the boundary. Valiev *et al.* (1983) observed grain boundary sliding along a general boundary in a Zn bicrystal subjected to a shear stress under conditions where the basal slip planes of the two adjoining bulk crystals were oriented with respect to the applied stress so that no lattice basal slip occurred. No impingement of lattice dislocations associated with lattice basal slip therefore occurred during the sliding. They then observed the sliding of the same boundary subjected to the same shear stress under conditions where the basal planes were oriented so that extensive lattice basal slip occurred in both adjoining crystals, and large numbers of lattice dislocations therefore impinged on the boundary during the sliding. As seen in Fig. 12.31, the sliding was considerably enhanced as a result of the addition of the impinged lattice dislocations. In this case the lattice dislocations which were injected into the general boundary most probably dissociated into an infinite number of infinitesimal dislocations as discussed in Section 12.4.3.2. The dissociation products then provided large numbers of excess local environments in the boundary core where localized shear events could occur causing increased rates of sliding. In the case of vicinal boundaries we would also expect an increase in the sliding rate. Here, the impinged lattice dislocations would dissociate into localized extrinsic boundary dislocations possessing Burgers vectors belonging to the DSC lattice. In either case, the boundaries act as sinks, or traps, for impinging lattice dislocations which contribute an effective Burgers vector strength which moves along the

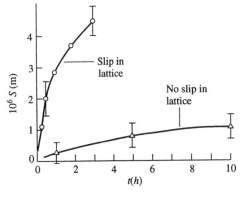

**Fig. 12.31** Amount of grain boundary sliding, $S$, versus time, for two Zn bicrystals. The two curves correspond to cases where lattice slip either occurred, or did not occur, while the boundary sliding took place. The same boundary type, subjected to the same shear stress, was studied in each case. (From Valiev *et al.* (1983).)

boundary causing sliding. Models for grain boundary sliding based on this process have, for example, been suggested by Kegg *et al.* (1973), Watanabe and Davies (1978) and Reading and Smith (1985). It is interesting to note that a certain portion of the Burgers vector strength which enters the boundary by impingement will generally be incorporated into the boundary as part of its intrinsic structure (Kokawa *et al.* 1981, Valiev and Khairullin 1990) causing corresponding changes in the boundary misorientation in a manner consistent with the Frank–Bilby equation as discussed in Section 12.4. When the boundary sliding takes place in the absence of any lattice slip, all of the processes necessary for the sliding must take place locally in the boundary region. Valiev *et al.* (1986) have termed this type of sliding 'pure sliding' in contrast to the case where impinging lattice dislocations play a significant role.

Extensive sliding is generally accompanied by 'slide hardening', i.e. a decrease in the rate of sliding at constant applied shear stress caused by a continuously increasing resistance to further sliding as seen in Figs 12.31 and 12.32. For vicinal grain boundaries, where the sliding is produced by the passage of localized grain boundary dislocations, this slide hardening must be due, at least in part, to a build-up of extrinsic dislocation density in the boundary and the formation of obstacles to their movement. When lattice slip occurs during the sliding, the intersections of lattice slip bands with the boundary may be expected to produce various obstacles. For general boundaries, devoid of localized boundary dislocations, the situation is less clear. However, the density of the distribution of infinitesimal dislocations produced by the dissociation of the impinging lattice dislocations would tend to increase during sliding (just as the density of localized dislocations in singular boundaries would), and interactions between them could then reduce the overall rate of sliding. In addition, if sliding is controlled by plastic flow accommodation in the adjacent lattices, lattice work hardening could occur in this region.

There is also extensive evidence that the rate of grain boundary sliding involving relatively large displacements is lower at singular or vicinal boundaries than at general boundaries (Lagarde and Biscondi 1974, 1975, Kokawa *et al.* 1981, Watanabe *et al.* 1984). An example is shown in Fig. 12.32. This may be due, at least in part, to the fact that impinging lattice dislocations can be more readily absorbed at general boundaries than

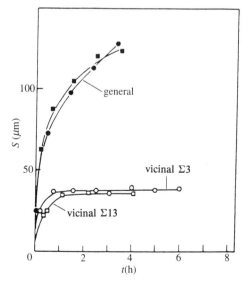

**Fig. 12.32** Amount of grain boundary sliding, $S$, versus time, for general, vicinal $\Sigma 3$, and vicinal $\Sigma 13$ grain boundaries in Al. Stress $= 1$ MPa, and $T = 800$ K. (From Kokawa *et al.* (1981).)

singular or vicinal boundaries as discussed in Section 12.4.3. It may also have been due
to the poorer source/sink action of the singular boundaries if diffusional accommodation
was important as suggested earlier in our discussion of the results of Kato and Mori (1993)
in Fig. 12.30.

*Sliding at heterophase interfaces.*   We conclude by noting briefly that heterophase
interfaces are capable of sliding by the same mechanisms already described for grain
boundary sliding, e.g. Takayuki and Izumi (1979), Eberhardt and Baudelet (1980), Suzuki
*et al.* (1982). Models for such sliding are therefore similar to those for grain boundaries,
but, of course, account must be taken of the different properties of the two adjoining
phases. For example, Hashimoto *et al.* (1991) observed sliding at b.c.c. $\alpha$-phase/f.c.c. $\gamma$-
phase interfaces in bicrystals in the Fe–Cr–Ni system. In these experiments the $\alpha$ phase
was more ductile than the $\gamma$ phase, and a narrow region of plastic deformation, associated
with plastic flow accommodation, was observed in the $\alpha$ phase directly adjacent to the
interface while no deformation was detected in the $\gamma$ phase.

## 12.8.2   Creep of polycrystals

Interfaces often play crucial roles in the creep of polycrystalline material. Creep is defined
as deformation which is able to proceed under a constant applied load and is produced
by thermally activated deformation mechanisms which are capable of continuous operation at constant stress (Evans and Wilshire 1985, Cadek 1988). For polycrystals at elevated
temperatures, these deformation mechanisms include grain boundary sliding, the diffusional transport of atoms between interfaces in a manner which changes the specimen
shape and the movement of lattice dislocations by glide and climb.

   As we shall see below, these processes are often coupled and are of varying importance
in different types of polycrystalline materials in different regimes of temperature and
stress. In order to indicate the range of creep behaviour possible for polycrystals, we
analyse below a number of situations for homophase polycrystals where each of the above
processes can become rate-limiting. In all cases we shall find that the overall behaviour
of the system is dependent upon the detailed behaviour of the boundaries which are
present.

   It is of interest to point out that the diffusionally accommodated sliding of the bicrystal
in Fig. 12.27(c), already considered above, in Section 12.8.1.3, is actually a *bona fide*
example of diffusional creep, since the bicrystal shears plastically at a constant rate under
constant applied stress as a result of the diffusional transport between the boundary
sources and sinks which are illustrated. Other forms of diffusional creep are also possible
including the case of the bamboo wire discussed in Section 5.2 (Fig. 5.2), where the wire
elongated under a constant load (crept) by means of the stress motivated diffusional
transport of atoms between the wire surface and transverse grain boundaries (in the
absence of any boundary sliding). A quantitative analysis of the creep rate of such a wire
has been given by Herring (1950) under the assumption that surface and grain boundary
energy terms could be neglected as the wire changes length (and also diameter) under the
constant creep load. However, in the following three sections we shall focus attention
on the practically more important problem of the creep deformation of polycrystalline
material when the three different deformation mechanisms mentioned above become rate-
controlling. For simplicity, we consider only the case of homophase polycrystals.

### 12.8.2.1 *Creep of homophase polycrystals controlled by diffusional transport*

For purposes of analysis we shall follow other investigators and focus on the ideal 2D homophase polycrystal composed of identical equiaxed hexagonal grains illustrated in Fig. 12.29. In this approach the actual 3D problem is considerably simplified without losing any of the basic physics. The polycrystal is subjected to an applied uniaxial tension, $\tau^\infty$ (Fig. 12.29), and we shall assume throughout this section that $\tau^\infty$ is small enough so that no lattice dislocations are activated anywhere in the system. Under this condition, the bulk of each grain remains rigid. However, under the applied stress different normal stresses (tractions) and shear stresses are established at each of the three different types of grain boundaries present (i.e., the A/B, A/C, and B/C) interfaces. (Note that the behaviour of each grain can be assumed to be identical.)

Different values of the diffusion potential will then be established at the three types of boundaries, and diffusional transport will tend to occur between these boundaries under the influence of these potentials. This transport will cause the centres of the rigid adjoining grains to move apart (or together) along the normals to their common boundaries by the distances $N^A$, $N^B$, and $N^C$, as indicated. At the same time, the shear stresses will tend to produce sliding displacements of the rigid grains at each boundary indicated by $S^A$, $S^B$, and $S^C$. These latter displacements correspond to relative displacements (parallel to the boundary) of the centres of the grains adjoining each boundary. The $N^i$ and $S^i$ displacements will then combine to produce an overall change in shape (creep deformation) of the polycrystal as measured by the deformation of the network connecting the centers of the grains. As we now demonstrate, certain relationships must exist between the above displacements, if the grains are displaced with respect to each other compatibly (i.e. without the development of any gaps or overlaps). In addition, the overall strain, $\varepsilon^\infty$, occurring along $\tau^\infty$ can be calulated directly in terms of these quantities (Beeré 1976, 1978, Pilling and Ridley 1989).

Firstly, the vertical displacement of grain C relative to grain B must be consistent with the difference between the vertical displacement of grain C with respect to grain A and that of grain B with respect to grain A. Using the (1, 2) coordinate system shown, this requires

$$3N^A - 3N^B = -\sqrt{3}\,S^A - \sqrt{3}\,S^B + 2\sqrt{3}\,S^C. \tag{12.69}$$

A similar calculation for the horizontal displacements yields

$$N^A + N^B - 2N^C = \sqrt{3}\,S^A - \sqrt{3}\,S^B. \tag{12.70}$$

A third condition is that the volume must remain constant, i.e.,

$$\varepsilon_{11} + \varepsilon_{22} = 0, \tag{12.71}$$

where $\varepsilon_{11}$ and $\varepsilon_{22}$ are the normal strains of the network connecting the centres of the grains (in the (1, 2) coordinate system) produced by the various $N$'s and $S$'s. Consider the rectangular region PQRS in Fig. 12.29(a). Since this region is representative of the entire polycrystal,

$$\varepsilon_{11} = \partial u_1/\partial x_1 = (\Delta C_1 - \Delta B_1)/d; \tag{12.72}$$

$$\varepsilon_{22} = \partial u_2/\partial x_2 = (\Delta B_2 + \Delta C_2)/\sqrt{3}d; \tag{12.73}$$

$$\varepsilon_{12} = \tfrac{1}{2}(\partial u_1/\partial x_2 + \partial u_2/\partial x_1) = (\Delta B_1 + \Delta C_1)/2\sqrt{3}d + (\Delta C_2 - \Delta B_2)/2d. \tag{12.74}$$

Here, the $\Delta B_i$'s and $\Delta C_i$'s are the components of the displacements of the centres of grains B and C relative to A and are given by

$$2\Delta B_1 = -\sqrt{3}\,S^B - N^B; \tag{12.75}$$

$$2\Delta B_2 = -S^B + \sqrt{3}\,N^B; \tag{12.76}$$

$$2\Delta C_1 = -\sqrt{3}\,S^A + N^A; \tag{12.77}$$

$$2\Delta C_2 = S^A + \sqrt{3}\,N^A. \tag{12.78}$$

Combining these relationships, we obtain

$$\varepsilon_{11} = \sqrt{3}\,(S^B - S^A)/2d + (N^A + N^B)/2d; \tag{12.79}$$

$$\varepsilon_{22} = (S^A - S^B)/2\sqrt{3}d + (N^A + N^B)/2d; \tag{12.80}$$

$$\varepsilon_{12} = (N^A - N^B)/\sqrt{3}\,d. \tag{12.81}$$

Therefore, substituting eqn (12.79) and (12.80) into (12.71),

$$3N^A + 3N^B = \sqrt{3}S^A - \sqrt{3}S^B. \tag{12.82}$$

Solving eqn (12.69), (12.70), and (12.82), we then obtain

$$\sqrt{3}N^A = S^C - S^B; \tag{12.83}$$

$$\sqrt{3}N^B = S^A - S^C; \tag{12.84}$$

$$\sqrt{3}N^C = S^B - S^A; \tag{12.85}$$

which yields

$$N^A + N^B + N^C = 0. \tag{12.86}$$

Equation (12.86) simply states that the sum of the amount of material removed/added at all boundary sources/sinks by diffusional transport must be zero, as might be expected because of the constant volume condition.

Having the above results, we may now calculate the strain rate, $\dot\varepsilon^\infty$, induced along the direction of the applied stress, $\tau^\infty$, in Fig. 12.29. Using standard tensor transformation rules, the strain, $\varepsilon^\infty$, is related to the $\varepsilon_{11}$, $\varepsilon_{22}$, and $\varepsilon_{12}$ strains in the (1, 2) coordinate system by

$$\varepsilon^\infty = \cos^2\theta\,\varepsilon_{11} + \sin^2\theta\,\varepsilon_{22} + 2\sin\theta\cos\theta\,\varepsilon_{12}. \tag{12.87}$$

Using our previous relationships, $\dot\varepsilon^\infty$ can then be expressed (Beeré 1978) either in terms of only the $\dot N^i$'s or only the $\dot S^i$'s, i.e.,

$$\dot\varepsilon^\infty = \frac{1}{d}\left[(\dot N^A + \dot N^B)(1 - 2\cos^2\theta) + \frac{(\dot N^A - \dot N^B)}{\sqrt{3}}2\sin\theta\cos\theta\right], \tag{12.88}$$

or alternatively,

$$\dot\varepsilon^\infty = \frac{1}{d}\left[\frac{(\dot S^A - \dot S^B)}{\sqrt{3}}(1 - 2\cos^2\theta) + \frac{(2\dot S^C - \dot S^A - \dot S^B)}{3}2\sin\theta\cos\theta\right]. \tag{12.89}$$

Equation (12.88) shows that the creep strain rate, $\dot\varepsilon^\infty$, can be regarded as due to diffusional transport which is accommodated by appropriate boundary sliding, while eqn (12.89) shows that it may be regarded, alternatively, as due to boundary sliding which is accommodated by diffusional transport. Raj and Ashby (1971) have pointed out earlier that diffusional creep is identical with boundary sliding accommodated by diffusional transport, and it has been widely recognized that grain boundary sliding must accompany the diffusional creep of a polycrystal if it is to occur compatibly, e.g. Lifshitz (1963).

In order to proceed further, we need to find relationships for the average tractions and shear stresses acting at the different boundaries in the presence of the uniaxial stress, $\tau^{\infty}$. These can be found by analysing the mechanical equilibrium of the two triangular shaded regions indicated in Fig. 12.29(b). The average tractions and shear stresses acting on the 'surfaces' of these regions are shown. Since the situation in each hexagon is the same, $\tau_{12}$ and $\tau_{22}$ are the stresses produced on average by $\tau$ in the (1,2) coordinate system along the dashed interface shown, while $\tau'_{11}$ and $\tau'_{12}$ are the corresponding stresses in the $(1',2')$ coordinate system produced on the other dashed interface. Simple tensor transformations show that

$$\tau_{12} = \sin\theta\cos\theta \cdot \tau^{\infty}; \tag{12.90}$$

$$\tau_{22} = \sin^2\theta \cdot \tau^{\infty}; \tag{12.91}$$

$$\tau'_{11} = \left[\tfrac{3}{4}\cos^2\theta + \frac{\sqrt{3}}{2}\sin\theta\cos\theta + \tfrac{1}{4}\sin^2\theta\right] \cdot \tau^{\infty}; \tag{12.92}$$

$$\tau'_{12} = \left[-\frac{\sqrt{3}}{4}\cos^2\theta + \tfrac{1}{2}\sin\theta\cos\theta + \frac{\sqrt{3}}{4}\sin^2\theta\right] \cdot \tau^{\infty}. \tag{12.93}$$

Equilibrium of forces in the vertical and horizontal directions on the two elements then yield the four equations:

$$-2\sqrt{3}\tau_{22} - \tau_s^B + \sqrt{3}\,\tau_n^B + \tau_s^A + \sqrt{3}\,\tau_n^A = 0; \tag{12.94}$$

$$-2\sqrt{3}\,\tau_{12} - \sqrt{3}\,\tau_s^{B} - \tau_n^B - \sqrt{3}\,\tau_s^A + \tau_n^A = 0; \tag{12.95}$$

$$-2\sqrt{3}\,\tau'_{12} + \sqrt{3}\,\tau_s^A + \tau_n^A + \sqrt{3}\,\tau_s^C - \tau_n^C = 0; \tag{12.96}$$

$$-2\sqrt{3}\,\tau'_{11} - \tau_s^A + \sqrt{3}\,\tau_n^A + \tau_s^C + \sqrt{3}\,\tau_n^C = 0. \tag{12.97}$$

By combining the above relationships it may be shown (Beeré 1976 1978) that

$$\tau_s^A + \tau_s^B + \tau_s^C = 0; \tag{12.98}$$

$$\tau_n^A + \tau_n^B + \tau_n^C = 3\tau/2. \tag{12.99}$$

The diffusional transport and the boundary shearing processes occur by independent physical mechanisms, and we now assume the limiting case where the shearing mechanism occurs much faster than the diffusional transport so that the overall creep rate is controlled by the slower rate of diffusional transport. (We note that this case is similar to our previous analysis of boundary sliding in Section 12.8.1.3 where it was assumed that the boundary shearing was rapid enough so that the rate of sliding was limited by the slower rate of diffusional accommodation.) Under these conditions the shear stresses will be essentially relaxed to zero, i.e.

$$\tau_s^A = \tau_s^B = \tau_s^C = 0. \tag{12.100}$$

The normal forces, which now support the applied stress, can then be found by combining eqn (12.100) with our previous relationships to yield

$$\tau_n^A = [\sin^2\theta + \sqrt{3}\sin\theta\cos\theta] \cdot \tau^{\infty}; \tag{12.101}$$

$$\tau_n^B = [\sin^2\theta - \sqrt{3}\sin\theta\cos\theta] \cdot \tau^{\infty}; \tag{12.102}$$

$$\tau_n^C = [3/2 - 2\sin^2\theta] \cdot \tau^{\infty}. \tag{12.103}$$

These different average tractions will establish different average diffusion potentials at the three different boundaries, which, in turn, will set up quasi-steady-state diffusion currents in each grain as shown schematically in Fig. 12.29(b) at H for the transport by lattice diffusion. As in the case of the quasi-steady-state diffusional transport which accommodated boundary sliding in Section 12.8.1.3, non-uniform distributions of tractions (and values of the diffusion potentials) will quickly be established along each type of boundary which will cause equal fluxes to enter all regions of each boundary so that the process at each boundary will be compatible. Again, as in Section 12.8.1.3, the transport can occur by lattice diffusion and by boundary diffusion, and the boundary value problem can be solved approximately by assuming ideal boundary sources/sinks and employing the same basic methods and arguments used to solve the quasi-steady-state diffusionally accommodated sliding problem described in Section 12.8.1.3. In the present case the values of $\dot{N}^A$ and $\dot{N}^B$ can be calculated from the same basic flux equations and then substituting them into the expression for $\dot{\varepsilon}^\infty$ given by eqn (12.88). The result (Beeré 1976, 1978) for lattice diffusion is

$$\dot{\varepsilon}^\infty \simeq \frac{15\,D^L\Omega}{kTf^Ld^2}T^\infty,$$  (12.104)

and for boundary diffusion

$$\dot{\varepsilon}^\infty \simeq \frac{15\pi\,D^B\delta\Omega}{kTf^Bd^3}T^\infty.$$  (12.105)

Both results are independent of the angle $\theta$. When the lattice diffusion mechanism is dominant and controlling, the creep is known as 'Herring–Nabarro creep', and when the boundary diffusion mechanism is dominant, 'Coble creep' after Nabarro (1948), Herring (1950), and Coble (1963). The expressions given by eqns (12.104) and (12.105) are of the same functional forms as those obtained by other authors for these types of creep, and their magnitudes agree to well within an order of magnitude (see Raj and Ashby 1971).

In the above analyses we calculated the instantaneous creep rate of the polycrystal in Fig. 12.29 made up of equiaxed hexagonal grains subjected to uniaxial tension. However, the hexagons will tend to become elongated along the stress axis as creep proceeds as illustrated in Fig. 12.33(a,b). It must be recalled that the grain boundaries are mobile at creep temperatures, and that they will therefore attempt to maintain a configuration where they meet at approximately equal angles (i.e. 120°) in order to satisfy the requirement of local nodal equilibrium (Section 5.6.8) as illustrated in Fig. 12.33(b). However, as creep continues, and the grains become even more elongated along the stress axis, the grain boundary network structure will eventually become metastable to a new network structure of the type illustrated by the dashed network in Fig. 12.33(b) and shown in Fig. 12.33(c). In this process 'grain switching' (Ashby and Verrall 1973, Beeré 1978, Spingarn and Nix 1978) has taken place, since, for example, the D and E grains which were neighbours in Fig. 12.33(a) become non-neighbours in Fig. 12.33(c), whereas the C and F grains, which were non-neighbours, become neighbours. In effect, the C and F grains in Fig. 12.33(a) have assumed positions between the D and E grains in order to elongate the specimen in the direction of the applied stress. We note that this switching process requires a critically large strain, and that it can recur throughout the polycrystalline structure and therefore serve to maintain relatively equiaxed grains up to very large creep strains.

The diffusional creep rates given by both eqn (12.104) and (12.105) are linearly propor-

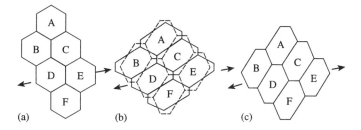

**Fig. 12.33** Schematic illustration of evolution of initially uniform hexagonal grain boundary network during diffusional creep. (a) Initial structure. (b) Grains become elongated until network becomes metastable to new more equiaxed network (dashed). (c) New network formed by grain boundary migration. (From Beeré (1978).)

tional to the applied stress. The material therefore behaves in viscous fashion (with an effective Newtonian viscosity) even though it is composed of individually rigid grains. The diffusional creep rate via lattice diffusion varies with grain size as $d^{-2}$, whereas the rate via grain boundary diffusion varies as $d^{-3}$. This often provides a convenient way to distinguish between the dominant mechanisms. In Fig. 12.34 we present a deformation 'map' (Ashby 1972a) which shows the different regions in 'stress-temperature' space in which different possible mechanisms for plastic flow are expected to dominate in a Ag polycrystal with a grain size of $d = 32\,\mu$m at a strain rate of $10^{-8}\,\mathrm{s}^{-1}$. As might be expected, diffusion creep is the dominant mechanism at low stresses and high temperatures. Furthermore, Coble creep becomes predominant at the lower temperatures because boundary diffusion tends to predominate there (Fig. 8.5).

The above results have been obtained assuming that all grain boundaries acted as ideal sources/sinks for the diffusional transport. A large body of experimental data, particularly for pure materials, is in substantial quantitative agreement with the above models (see Burton 1977, Beeré 1978; Arzt *et al.* 1983). However, small threshhold stresses, typically of the order of $10^{-5}\mu$ frequently appear (see Table 10.2). Morever, further data (Arzt *et al.* 1983), particularly for alloys and materials containing second phase particles, show more substantial disagreement. This is evidently due to the relatively poor performance of the boundaries in these materials as sources/sinks for the diffusion fluxes even at higher stresses (i.e. higher driving energies). We note that the overall source/sink action of boundaries containing second phase particles can be inhibited by poor source/sink action at the particle/matrix interfaces. Further possible causes of the inefficient performance of boundary sources/sinks have been given in Section 10.3.3.

### 12.8.2.2 *Creep of homophase polycrystals controlled by boundary sliding*

We now reverse our assumptions and investigate the limiting case where the diffusional transport is so rapid that all differences in the diffusion potentials at the different grain boundaries are essentially eliminated. The rate of boundary sliding is then the relatively slow process which controls the overall creep rate. When the diffusion potentials are equal in this manner, we must have from eqn (12.99),

$$\tau_n^A = \tau_n^B = \tau_n^C = \tau^\infty/2. \tag{12.106}$$

The shearing stresses can then be found with the use of eqns (12.90)–(12.97) with the result

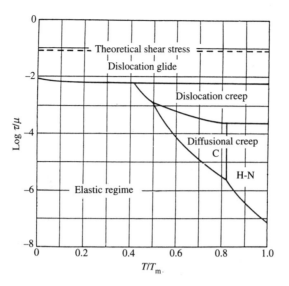

**Fig. 12.34** Deformation mechanism map for polycrystalline Ag of grain size 32 μm at a strain rate of $10^{-8}\text{s}^{-1}$. $\tau$ = applied stress and $\mu$ = shear modulus. Each mechanism is dominant within its field. The diffusional creep field is divided into two sub-fields; i.e. the Coble creep field (indicated by C) and the Herring–Nabarro creep field (H–N). (From Ashby (1972a).)

$$\tau_s^A = [-\sqrt{3}\cos^2\theta - \sin\theta\cos\theta + \sqrt{3}/2]\cdot\tau^\infty, \qquad (12.107)$$

$$\tau_s^B = [\sqrt{3}\cos^2\theta - \sin\theta\cos\theta - \sqrt{3}/2]\cdot\tau^\infty, \qquad (12.108)$$

$$\tau_s^C = 2\sin\theta\cos\theta\cdot\tau^\infty. \qquad (12.109)$$

If we now assume that the boundaries slide viscously at a rate proportional to the applied shear stress, as is conceivable on the basis of Section 12.8.1, we may use eqn (12.89) to find the instantaneous creep rate for a polycrystal composed of equiaxed hexagonal grains. If we start by making the simple assumption that the viscosity of all of the grain boundaries is the same, we have $\dot{S}^i = K\tau_s^i$, and using eqns (12.107)–(12.109), we then obtain the remarkably simple result (Beeré 1976, 1978)

$$\dot\varepsilon^\infty = (K/d)\cdot\tau^\infty. \qquad (12.110)$$

This result is, of course, independent of $\theta$, and again indicates that the polycrystal as a whole behaves as a viscous material which in this case is controlled by the rate of viscous boundary sliding.

The assumption that all of the boundaries have the same viscosity is oversimplified, however, and we now consider the more realistic situation when they differ. As we shall see, this will generally produce frictional forces which will tend to make the grains rotate, as is often observed experimentally Pilling and Ridley 1989). The physical principle is illustrated in a simple way for a polycrystal composed of square grains in Fig. 12.35. In Fig. 12.35(b) all boundaries are equally viscous, and the polycrystal is stressed as shown. In Fig. 12.35(c) the viscosity of the vertical boundaries is essentially infinite compared to that of the horizontal boundaries, and the same strain has been accomplished as in Fig. 12.35(b) by the combination of no sliding on the highly viscous vertical boundaries, sliding on the horizontal boundaries, and a rotation of the grains. This allowed the

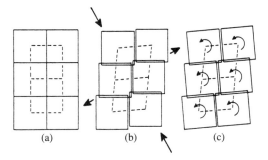

**Fig. 12.35** Illustration of grain rotations during diffusional creep in simple example assuming square grains. (a) Initial structure. (b) Situation when all boundaries are equally viscous, and grain boundary sliding without grain rotation occurs. (c) Situation when viscosity of the vertical boundary segments is essentially infinite compared to that of the horizontal segments. Sliding then occurs on the horizontal segments, and the grains rotate. (From Beeré (1978).)

achievement of the same creep strain with less sliding and the dissipation of less energy.

The general problem of finding the optimum rate of boundary rotation and the overall creep rate can be treated (Beeré 1978) by finding the minimum in the work dissipated during creep. For simplicity, we again employ the idealized polycrystal of Fig. 12.29 where there are again three different types of grains with three different types of boundaries, each, in general, having a different viscosity. Again the polycrystal is subjected to a uniaxial stress, and the behaviour of each grain is identical. $S^B$ is again the displacement of the centre of a B grain with respect to the centre of an A grain in the direction parallel to the intervening boundary. In the presence of a grain rotation rate, $\dot{\omega}$, as illustrated in Fig. 12.36, the local sliding rate at the boundary is

$$\dot{X}^B = \dot{S}^B - \dot{\omega}d. \tag{12.111}$$

Similar relations hold for the other two boundaries, and, eqn (12.89), which is still applicable, therefore has the form

$$\dot{\varepsilon}^\infty = \frac{1}{d}\left[\frac{(\dot{X}^A - \dot{X}^B)}{\sqrt{3}}(1 - 2\cos^2\theta) + \frac{(2\dot{X}^C - \dot{X}^A - \dot{X}^B)2\sin\theta\cos\theta)}{3}\right]. \tag{12.112}$$

We now assume the boundary viscous behaviour

$$\dot{X}^A = K^A \tau_s^A; \quad \dot{X}^B = K^B \tau_s^B; \quad \dot{X}^C = K^C \tau_s^C. \tag{12.113}$$

When $K^A = K^B = K^C$, there is no rotation, and we obtain our previous simple result given by eqn (12.110). In the general case the work done by the applied stress, $W$, is all dissipated in the boundary sliding, since the normal tractions are essentially zero. The rate of dissipation is then

$$\dot{W} \alpha \dot{X}^A \tau_s^A + \dot{X}^B \tau_s^B + \dot{X}^C \tau_s^C, \tag{12.114}$$

or, with the use of eqns (12.111) and (12.113),

$$\dot{W} \alpha \frac{(\dot{S}^A - \dot{\omega}d)^2}{K^A} + \frac{(\dot{S}^B - \dot{\omega}d)^2}{K^B} + \frac{(\dot{S}^C - \dot{\omega}d)^2}{K^C}. \tag{12.115}$$

The optimum rate of rotation is found by minimizing the rate of work dissipated, $\dot{W}$, at constant $\dot{S}^A$, $\dot{S}^B$, and $\dot{S}^C$. Results for grain rotation rates and overall creep rates

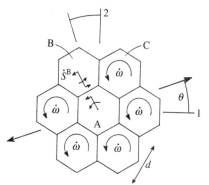

**Fig. 12.36** Grain rotations in idealized hexagonal poly-crystal during diffusional creep. A, B, and C grains same as in Fig. 12.29. Centre of grain B again moves with respect to centre of grain A with velocity $\dot{S}^B$ as shown. However, the local sliding rate at the intervening bound-ary is now different because of the grain rotations (see eqn (12.111)). (From: Beeré (1978).)

under typical conditions based on this approach have been given by Beeré (1978). As might be expected, increasing the differences between the values of $K^A$, $K^B$, and $K^C$ increases the rates of grain rotation. Grain rotations are found which are generally consistent with experimentally observed values (Beeré 1978).

Many of the phenomena discussed above occur during the superplastic deformation of fine-grained homophase polycrystals in which the material undergoes very large uniform elongations (i.e. > 200 per cent, and sometimes > 1000 per cent). Discussions are given, for example, by Beeré (1976 and 1978) and Pilling and Ridley (1989).

### 12.8.2.3 *Creep of homophase polycrystals controlled by the movement of lattice dislocations*

During the creep of the homophase polycrystals described above, the individual grains were not plastically deformed in their interiors: material was simply added/removed from their 'internal surfaces' facing the grain boundaries and coupled sliding occurred at the boundaries. However, polycrystalline materials may also creep at elevated temperatures (and higher stresses) by means of grain boundary sliding which is coupled to the deforma-tion of the grains by means of the movement of lattice dislocations within the grains. Of course, polycrystals can deform by the movement of lattice dislocations in the absence of boundary sliding, but when sliding is possible (such as at elevated temperatures) an additional deformation mechanism becomes available, and this serves to increase the overall creep rate. When this occurs, the overall process may be regarded in one sense as grain boundary sliding accommodated by plastic deformation of the lattice via the movement of lattice dislocations as already discussed in Section 12.8.1.4. Under most conditions the lattice dislocation motion will be the process which limits the overall rate of the creep deformation. As pointed out in Section 12.8.1.4, Crossman and Ashby (1975) have analysed this coupled phenomenon by assuming a power law creep relationship (see eqn (12.68)) to model the dislocation deformation in the lattice and using the finite element method to treat the overall elastic/plastic deformation problem. Since the lattice dislocation deformation process tends to be rate limiting, the macroscopic behaviour of a polycrystal undergoing simultaneous power law creep of its grains and grain boundary sliding will then also obey power law creep behaviour. However, the boundary sliding properties are important in determining the extent of the boundary sliding and the magnitude of its contribution to the overall creep. Crossman and Ashby (1975) found that the creep deformation in such situations consists of sliding at grain boundaries accommodated by bands of concentrated plastic deformation within the grains. When the boundary sliding occurs freely, the displacement and stress fields in the grains become more uniform. As the exponent, $n$, in the power law creep expression $\dot{\varepsilon} = A\tau^n$ increases from 1 to $\infty$, the plastic flow within the grains becomes increasingly concentrated into bands, and the fractional contribution of the boundary sliding shear to the total deformation increases from about $\frac{1}{6}$ to $\frac{1}{2}$. Under all conditions, active boundary sliding increases the overall creep rate. The detailed conditions under which boundaries are able to slide at sufficient rates to contribute significantly to the overall creep rate, along with many other details, are discussed by Crossman and Ashby (1975). Further aspects of the problem are analysed by Speight (1976).

### 12.8.2.4 *Further aspects of the creep of polycrystals*

*Homophase polycrystals.* In the previous sections we have described various mechanisms for the creep of homophase polycrystals using a simple idealized 2D model composed of identical grains. Various phenomena included diffusional transport, grain boundary sliding, grain rotations, 'grain switching' and power law dislocation creep within the grains. All of these phenomena have been observed experimentally under various circumstances in different regimes of temperature, applied stress, and creep strain (see previous references).

It is important to emphasize that the 2D model employing identical grains is obviously only a convenient approximation. More realistically, the problem should be modelled in 3D for polycrystals containing a distribution of grain sizes. (We note that a start in this direction has been made recently in 2D by Hazzledine and Schneibel (1993).) When grains of different size are present in 3D, the various creep processes occurring throughout the polycrystal will become heterogeneous, and several mechanisms may operate simultaneously, or sequentially, in different regions of the microstructure (Pilling and Ridley 1989). Also, additional phenomena, not included in the simple identical grain model may appear. For example, as pointed out by Raj and Lange (1985), when a distribution of grain sizes is present it will no longer be possible to maintain equilibrium configurations at all boundary junctions, while at the same time maintaining flat boundary segments, as in the idealized structures in Fig. 12.33. In such cases the boundaries will become curved as boundary migration strives to maintain local equilibrium configurations at the junctions. Raj and Lange (1985) show that this migration will occur in directions to maintain equiaxed grain structures during the deformation. However, if the strain rate is sufficiently large relative to the rate of boundary migration this may not be possible.

We may conclude this discussion by noting that the quantitative modelling of polycrystalline creep in 3D, taking full account of all of the phenomena identified above, has not yet been accomplished and clearly will be a difficult task.

*Heterophase polycrystals.* The creep of heterophase polycrystals will occur by the same basic mechanisms as those described above, but these will obviously be complicated by the different properties of the grains of different phase which are present. Some relevant references are: Gittus (1977), Suery and Baudelet (1981), Chen (1982), Pilling and Ridley (1989).

## 12.9  FRACTURE AT HOMOPHASE INTERFACES

### 12.9.1  Overview of the different types of fracture observed experimentally in homophase polycrystals

Under a variety of conditions homophase polycrystals are found to fracture preferentially along paths following their grain boundaries as shown, for example, in Fig. 12.37. Such fractures are termed 'intergranular'. When fractures occur along paths traversing the bulk grains, they are called 'transgranular'. Mixed fractures, partly intergranular and partly transgranular, also occur. Polycrystals may also flow homogeneously over macroscopic distances which are larger than the grain size. As discussed below, fracture under tensile conditions may then occur by a process in which the body necks down essentially to a point. Such fractures cannot be regarded as either transgranular or intergranular, since no fracture paths are involved.

In order to establish some perspective, we begin by presenting a broad overview of the major types of fractures (classified by mechanism) which have been observed experimentally in homophase polycrystals. This will allow us to identify and characterize the major types which involve grain boundaries; these are then taken up in more detail in later sections.

The type of fracture which a given homophase polycrystalline material undergoes is a complex matter which depends upon a host of variables including, amongst others; (1) the materials type (i.e. its type of bonding and structure); (2) the types of interfaces present; (3) the presence, or absence, of solute atoms which segregate strongly to the grain boundaries; (4) the presence, or absence, of minor inclusions (in the form of second phase particles) due to uncontrolled impurities; (5) the temperature; (6) the magnitude of the applied stress; and (7) the rate of application of the applied stress (e.g. slow loading versus shock loading).

Ashby *et al.* (1979) and Gandhi and Ashby (1979) have distinguished, in a rough way, the six types of fracture observed in polycrystalline specimens pulled in tension illustrated schematically in Fig. 12.38. They range from highly brittle (exhibiting negligible specimen elongation) to highly ductile (exhibiting considerable elongation). They also vary with temperature, which is roughly classified as either 'low' or 'high'. In several cases they may be either intergranular or transgranular. The reader should be warned, however, that the descriptors 'low' and 'high' cannot be quantified and are only useful for indicating a trend. For example, b.c.c. metals generally fracture by crack propagation only at temperatures $\leqslant 0.5T_m$, whereas materials with strong covalent bonding can fracture by crack propagation near $T_m$ (Gandhi and Ashby 1979). A brief description of each of these types of fracture (and its variants) follows.

*Propagation of cleavage cracks* (Fig. 12.38(a).   This type of fracture tends to occur at 'low' temperatures by the propagation of cleavage cracks which may be either transgranular, intergranular, or mixed. By cleavage we mean that decohesion occurs at the advancing crack tip by the breaking of atomic bonds. Ashby *et al.* (1979) and Gandhi and Ashby (1979) classify failures of this type into three subgroups (i.e. types A, B, and C cracking), depending upon the nature and extent of the dislocation activity which accompanies the cracking. As we shall see later, the propagation of cleavage cracks becomes progressively more difficult as the amount of dislocation activity (i.e. plasticity) increases. This is a result of, for example, the blunting of crack tips which may be caused by the creation and motion of dislocations and the absorption of energy which

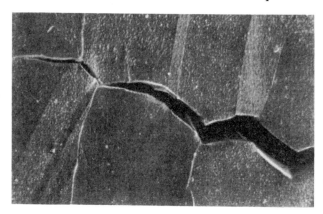

**Fig. 12.37** Intergranular crack in polycrystalline Cu–Bi alloy. Tensile axis is vertical. (From Watanabe (1989).)

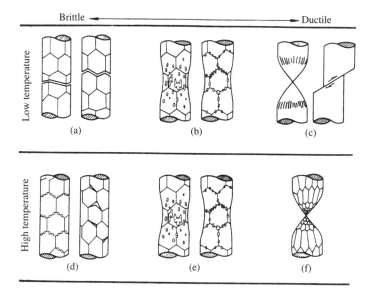

**Fig. 12.38** The simplest classification of fracture mechanisms in polycrystalline material pulled in tension. Upper row refers to 'low temperatures' (i.e. ⩽ about $0.3T_m$) where plastic flow does not depend strongly on temperature or time. Lower row refers to 'high temperatures' where materials creep. Transition from brittle to ductile behaviour from left to right. (a) Propagation of cleavage cracks (transgranular, as on left, or intergranular, as on right). (b) Growth and coalescence of cavities by plastic flow due to dislocation glide (transgranular, as on left, or intergranular, as on right). (c) Rupture by necking or shearing-off. (d) Growth and coalescence of intergranular cavities by diffusion, augmented by power-law creep and boundary sliding. (e) Growth and coalescence of cavities by power-law creep (transgranular, as on left, or intergranular, as on right). (f) Rupture enhanced by dynamic recovery or recrystallization. (From Ashby *et al.* (1979).)

accompanies plastic flow. Eventually, when the material becomes sufficiently plastic, the propagation of cleavage cracks becomes impossible.

Type A cracking occurs in materials in which the generation and motion of lattice dislocations is exceedingly difficult, and the crack propagates by a brittle cleavage process in which the material fails at the crack tip by the progressive breaking of the atomic bonds with only a very limited amount of dislocation glide activity occurring in the region of the crack tip. There is essentially no dislocation activity throughout the bulk, and cracks therefore cannot be nucleated in regions of high stress concentration produced, for example, by dislocation pile-ups (e.g. Fig. 12.17). The cracks must therefore originate at pre-existing flaws.

Type B cracking occurs when lattice dislocation activity is somewhat easier (perhaps at higher stresses or temperatures). There is then some dislocation generation and glide movement in the bulk, and cracks can be nucleated in regions of high stress at dislocation pile-ups. Also, there is more dislocation glide activity in the crack tip region.

Finally, type C cracking occurs when still further dislocation glide plasticity is possible in the bulk. These types of cracks are generally preceeded by appreciable dislocation plasticity in the bulk, and there is also a correspondingly larger amount of dislocation plasticity in the crack tip region.

*Growth and coalescence of cavities by plastic flow due to dislocation glide* (Fig. 12.38(b)).   This occurs at 'low' temperatures when there is sufficient dislocation glide plasticity to permit relatively large plastic strains without the formation of cracks. The fracture occurs by the nucleation, plastic growth and eventual coalescence of cavities. The cavities may originate at pre-existing voids (e.g. in materials produced by sintering), or at small second phase inclusions. When the densities of these flaws is higher at boundaries than in the bulk, the fracture will tend to be intergranular.

*Rupture by necking or shearing-off* (Fig. 12.38(c)).   The dislocation glide plasticity is now sufficiently high (i.e. the material is sufficiently ductile) so that both single crystals and polycrystals fail by a rupturing process in which the bulk material necks down or shears-off. In this process the material becomes plastically unstable on a macroscopic scale, and the deformation becomes localized in the neck region, or a shear band, and continues until the cross-sectional area is reduced to zero. The degree of strain localization depends upon the work hardening characteristics and strain rate sensitivity of the material (Hart 1967, Burke and Nix 1975). It is emphasized that both this type of failure and the one immediately above occur by the same basic process, i.e. plastic flow due to dislocation glide.

*Growth and coalescence of intergranular cavities by diffusion, augmented by power-law creep and boundary sliding* (Fig. 12.38(d)).   At 'high' temperatures and low stresses, creep deformation occurs, and cavities form on boundaries and grow and ultimately coalesce causing intergranular fracture. Under various conditions the cavities may be in the form of relatively equiaxed voids or 'wedge cracks' as illustrated. The cavity growth can be by diffusion alone, or by coupled mechanisms involving diffusion, power-law creep (which involves the glide and climb of dislocations), and boundary sliding.

*Growth and coalescence of cavities by power-law creep* (Fig. 12.38(e)).   This type of fracture is generally similar to fracture by the 'growth and coalescence of cavities by plastic flow due to dislocation glide' described above except that it occurs at higher temperatures where the plastic flow occurs by power-law creep rather than pure dislocation glide. The fracture may be transgranular or intergranular depending upon the conditions for the cavity nucleation.

*Rupture enhanced by dynamic recovery or recrystallization* (Fig. 12.38(f)). Again, as in 'rupture by necking' described above, the material is highly ductile and becomes unstable on a macroscopic scale and ruptures by necking down. The temperature is sufficiently high so that the formation of cavities is suppressed by recovery, or recrystallization. Recrystallization causes boundary motion and therefore eliminates boundaries as fixed sites where cavity growth and coalescence can produce intergranular fracture as described earlier. In addition, both recovery and recrystallization serve to reduce dislocation densities and local stress levels and, hence, discourage the formation and growth of cavities throughout the bulk of the material.

*Dynamic fracture.* This type of fracture (which is not illustrated in Fig. 12.38) occurs at high stresses at high rates of loading, and it involves the propagation of elastic and plastic waves. Discussion of this mode of failure is beyond the scope of the present book.

Many of the above types of fracture can occur within a single polycrystalline material under appropriate conditions. This is shown, for example, in Fig. 12.39, where a fracture map in temperature–stress space is presented for commercially pure MgO according to Gandhi and Ashby (1979). Each field on the map represents the range of conditions over which the indicated type of fracture is found to dominate. At low temperatures ($T \leqslant 0.4T_m$) the material is highly brittle and fails by transgranular Type A crack propagation at low stresses (region (a)) and either transgranular or intergranular Type B crack propagation at higher stresses (region (b)). At higher temperatures ($0.4T_m \leqslant T \leqslant 0.8T_m$) in the regime of creep plasticity, fracture occurs by: (i) type C intergranular crack propagation at the higher stresses (region (c)); (ii) transgranular growth of voids by power-law creep (region (d)); or (iii) intergranular growth of voids by diffusion, augmented by boundary sliding and power-law creep at the lower stresses (region (e)). Finally, at even higher temperatures ($T > 0.8T_m$), rupture, enhanced by recrystallization, occurs at the higher stresses (region (f)). Intergranular fracture is therefore seen to occur in MgO over a wide range of conditions ranging from crack propagation at low temperatures and high stresses to void growth and coalescence at high temperatures and low stresses.

The fracture maps for different classes of material, possessing different types of binding and structure, are generally quite different. As shown by Gandhi and Ashby (1979), for materials of commercial purity, free of solute atoms which segregate strongly to boundaries and produce strong embrittling effects (see Section 12.9.2.2), there is a steady progression in fracture behaviour as the bonding changes from metallic to ionic to covalent. Brittle crack propagation is essentially absent in the highly ductile f.c.c. metals (except under very special conditions) but appears in the b.c.c. metals and grows in importance as the bonding changes from metallic to ionic to covalent. On the other hand, the tendency to rupture decreases in the same sequence of materials. The general behaviour is consistent with an increased resistance of the lattice to dislocation activity in the sequence (Hirth and Lothe 1982).

Having this general background, we now consider in more detail the different types of intergranular fractures identified above which involve interfaces either directly or indirectly. As we have seen, these occur by diverse mechanisms and occur over a wide

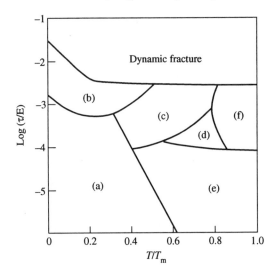

**Fig. 12.39** Fracture map for MgO under tensile stress, $\tau$, at temperature, $T$. $E$ = Young's modulus. Each fracture mechanism is dominant within its field. (a) Propagation of type A cracks (transgranular). (b) Propagation of type B cracks (transgranular or intergranular). (c) Propagation of type C cracks (intergranular). (d) Growth and coalescence of cavities by power-law creep (transgranular). (e) Growth and coalescence of cavities by diffusion, augmented by power-law creep and boundary sliding (intergranular). (f) Rupture, enhanced by recrystallization. (From Gandhi and Ashby (1979).)

range of conditions. The subject of these fractures is therefore an immense and complex field which is still not understood very well and is under active investigation. In order to keep the presentation within bounds, only major selected topics will be discussed, often only briefly. Further details may be found in the references.

## 12.9.2  Propagation of cleavage cracks

As already pointed out, the fracture of a polycrystal by cracking may be transgranular, intergranular, or mixed. However, even when the fracture is purely transgranular, it will still be influenced by the presence of the grain boundaries. We therefore organize this section by first discussing the propagation of cleavage cracks in single crystals, then along grain boundaries, and, finally, their initiation and propagation in polycrystals.

### 12.9.2.1  *Crack propagation in a single crystal*

We begin with an existing sharp crack in a single crystal which is subjected to an applied stress. Three different prototype cracks are possible (Thomson 1983, 1986) depending upon the direction of the applied forces (stresses) as illustrated in Fig. 12.40. The mode I crack is produced by a $\tau_{23}$ applied normal stress, the mode II crack by a $\tau_{12}$ applied shearing stress, and the mode III crack by a $\tau_{23}$ applied shearing stress. More complex modes, made up of combinations of these three, are, of course, possible. High stress concentrations will exist near the crack tips in each case, since less material is available in the plane containing the crack to carry the full loading force applied to the body. This is illustrated in Fig. 12.41(b) for the mode I crack in Fig. 12.41(a). The stresses in the regions near the tips have been calculated (see McClintock and Argon 1966, Meyers and

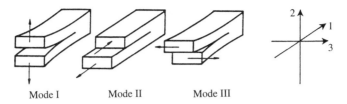

**Fig. 12.40**   The three prototype mode I, mode II, and mode III cracks. (From Thomson (1983).)

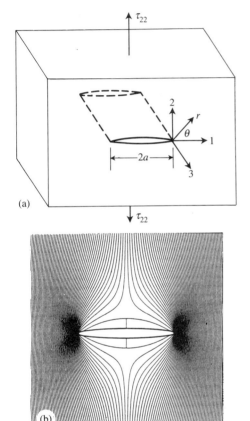

**Fig. 12.41**   (a) Mode I crack of length $2a$. (b) Representation of stress concentration in regions near crack tips. (From Thomson (1983).)

Chawla 1984, Thomson 1983, 1986). In the coordinate system shown in Fig. 12.41(a), the result for a mode I crack in 2D is

$$\tau_{ij}^{\mathrm{I}} = \frac{K^{\mathrm{I}}}{\sqrt{2\pi r}} f_{ij}^{\mathrm{I}}(\theta), \tag{12.116}$$

where $K^{\mathrm{I}}$ is known as the 'stress intensity factor' for a mode I crack, and the $f_{ij}^{\mathrm{I}}(\theta)$ are known functions of $\theta$. The stress intensity factor depends upon the applied loading system and the geometry of the crack and of the specimen. For the case where the crack is in a relatively large body subjected to a uniform $\tau_{22}$ stress at large distances from the crack, i.e. $\tau_{22}^{\infty}$, $K^{\mathrm{I}} = \tau_{22}^{\infty}\sqrt{\pi a}$. The stresses $\tau_{ij}^{\infty}$ therefore increase with the crack length as

$a^{\frac{1}{2}}$. The $\tau^{I}_{ij}$ stresses are also seen to have singularities at the crack tip origin, and fall off with distance from the tip as $r^{-\frac{1}{2}}$. It is emphasized that the $\tau^{I}_{ij}$ given by eqn (12.116) are applicable to a good approximation only near the tip region and are not complete solutions for the stresses everywhere, since, obviously, they must approach constant asymptotic values as $r \to \infty$. For example, for the case just cited, $\tau^{I}_{22}$ must approach the constant value $\tau^{\infty}_{22}$. Similar expressions hold for the stresses near mode II and mode III cracks, i.e.

$$\tau^{II}_{ij} = \frac{K^{II}}{\sqrt{2\pi r}} f^{II}_{ij}(\theta); \qquad \tau^{III}_{j} = \frac{K^{III}}{\sqrt{2\pi r}} f^{III}_{ij}(\theta). \qquad (12.117a,b)$$

When the applied stresses in large bodies containing mode II and mode III cracks are $\tau^{\infty}_{12}$ and $\tau^{\infty}_{23}$ respectively, $K^{II} = \tau^{\infty}_{12} \sqrt{\pi a}$, and $K^{III} = \tau^{\infty}_{23} \sqrt{\pi a}$.

In general, when a loaded crack, such as the one shown in Fig. 12.41, propagates forward, the energy of the system, $E$ (exclusive of the energy of the new free surface which is produced), will change. We may therefore define a force on the crack corresponding to $f = -\delta E/\delta x$, where $\delta E$ is the change in energy caused by the forward motion $\delta x$ of the crack. A detailed analysis (Thomson 1983, 1986), shows that under plane strain conditions the forces exerted on the mode I, II, and III cracks shown in Fig 12.40 (per unit length parallel to the crack tip) are

$$f^{I} = \frac{(1-\nu)\,[K^{I}]^{2}}{2\mu}; \quad f^{II} = \frac{(1-\nu)\,[K^{II}]^{2}}{2\mu}; \quad f^{III} = \frac{[K^{III}]^{2}}{2\mu}; \qquad (12.118a,b,c)$$

respectively. We note that these forces are traditionally termed 'crack energy release rates' (i.e. energy released per unit area of crack advance) in the crack mechanics literature. Equations (12.118a,b,c) are valid only for the force on a crack embedded in an ideally brittle material. When the crack is embedded in a plastic deformation zone containing dislocations the J-integral approach of Rice may be used (see discussion by Thomson 1983).

We now seek to answer the question of whether a crack will be able to propagate in a completely brittle fashion under the influence of the above force. By this, we mean that the crack tip will advance by a cleavage process involving the sequential breaking of each atomic bond across the crack tip as shown schematically in the atomic model of Fig. 12.42 for a mode I crack. Also, we specifically preclude the possibility that any lattice dislocations will be emitted from the tip in order to shield it from the applied stress as described later. A complete solution of this problem involves the recognition that the crack cannot advance if conditions do not allow the breaking of the bonds, i.e. the force applied to each bond which is stretched and eventually broken, must be capable of exceeding the maximum bond strength. At the same time, it must be realized that, as the crack expands, new free surface area is produced, and sufficient force must be applied to the crack to perform the work required to produce this new surface area. In order to find the overall condition for crack growth, consider first the condition that sufficient energy must be supplied to form the new surface area. The change in energy which occurs when the crack expands a differential distance $da$ under the force $f$ is

$$dE = 2(-f\,da + 2\sigma^{S}\,da), \qquad (12.119)$$

where the first term is the work done by the force, and the second is the surface energy term ($\sigma^{S}$ is the free surface energy per unit area). Assuming a mode I crack under a uniform load, using eqn (12.118a) for $f$, and integrating with respect to a between 0 and $a$, we obtain

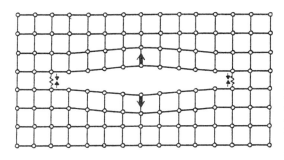

Fig. 12.42 Schematic diagram of mode I crack in simple square lattice. The crack propagates by cleavage by the sequential breaking of atomic bonds at the two tips as indicated. (From Thomson (1983).)

$$E(a) = 4\sigma^S a - \frac{\pi(1 - \nu)[\tau_{22}^\infty]^2 a^2}{2\mu},$$ (12.120)

which is often termed the 'Griffith energy function' after Griffith (1920). This function, when plotted versus $a$, is an inverted parabola (Thomson 1983) with a maximum at the value of $a$ given by

$$a = \frac{4\mu\sigma^S}{\pi(1 - \nu)[\tau_{22}^\infty]^2}.$$ (12.121)

This is also the relationship which must be satisfied when the force, $f^I$, tending to expand the crack, is exactly balanced by the tangential capillary force exerted by the two free surfaces, $2\sigma^S$, tending to contract it, i.e.

$$2\sigma^S = f^I = \frac{(1 - \nu)[K^I]^2}{2\mu} = \frac{\pi(1 - \nu)[\tau_{22}^\infty]^2 a}{2\mu}.$$ (12.122)

Accordingly, a mode I crack of length $2a$ will be in unstable equilibrium in the presence of the critical stress

$$\tau_{22}^\infty(\text{crit}) = \left[\frac{4\mu\sigma^S}{(1 - \nu)\pi a}\right]^{\frac{1}{2}},$$ (12.123)

which is generally termed the 'critical Griffith stress'. The energy absorbed per unit area of crack advance is generally termed the 'toughness' in the crack mechanics literature and denoted by $G$. In the present analysis $G = 2\sigma^S$, and the above critical Griffith condition may then be expressed in the form

$$f^I = G,$$ (12.124)

i.e. the crack energy release rate is just equal to the toughness, $G$.

If the applied stress is increased slightly above the critical Griffith stress, while holding the crack length constant, the crack length will become supercritical, and according to the Griffith energy function, the crack will then be able to expand indefinitely with a continuous decrease in the energy of the system, i.e. catastrophic crack growth will occur. Similar relationships and results may be obtained for mode II and III cracks. The surface energy, $\sigma^S$, is lower for low index lattice planes, and therefore, according to eqn (12.123), the critical stress for brittle fracture (cleavage) should be lower for fracture along low index planes.

However, the analysis is still incomplete, since we have not yet taken into account the requirement that sufficient force must be present to exceed the maximum strength of each bond which is broken sequentially at the advancing tip. A detailed analysis (Thomson 1983, 1986), shows that the energy required for this discretized process is oscillatory with each period corresponding to the stretching and eventual breaking of a single bond. Superimposing this energy onto the previous smoothly varying Griffith energy function, $E(a)$, given by eqn (12.120), we obtain the final result (i.e. a modified Griffith energy function) shown in Fig. 12.43. The phenomenon leading to the relatively small oscillatory energy barriers has been called 'lattice trapping', since it is obviously due to the discrete lattice debonding events which must occur as the crack advances. We note that these energy barriers are directly analogous to the Peierls energy barriers to the motion of lattice dislocations (Hirth and Lothe 1982) which are also a result of the discrete nature of the crystal lattice (see Peierls–Nabarro dislocation model in Section 2.11.2.2).

The modified Griffith energy function in Fig. 12.43 leads to the possibility of slow thermally activated crack growth in situations where the applied stress is insufficient to overcome the barriers, and thermal fluctuations are then necessary (Thomson 1983). Of course, rapid crack growth can occur when $\tau_{22}^{\infty}$ exceeds $\tau_{22}^{\infty}(\text{crit})$ by an amount which is sufficient to allow the overcoming of the barriers without thermal activation. The model also provides a framework for modelling chemically enhanced crack growth, since external chemical environments can alter the free surface energy, $\sigma^{S}$, and also the stretched bonds, at the crack tip (Thomson 1983, 1986).

So far, we have assumed ideally brittle behaviour in the absence of any dislocation activity in the highly stressed crack tip region. Such behaviour may actually be achieved under certain conditions in exceptionally hard brittle materials such as Si at low temperatures (see Thomson 1983). However, in many materials, sharp mode I cracks will spontaneously emit lattice dislocations from the highly stressed crack tip as indicated schematically in Fig. 12.44. These dislocations will be repelled from the cracks, and detailed elasticity calculations (Thomson 1983, 1986) show that their stress fields will reduce the stress intensity factor at the crack tip. In fact, we may now introduce a 'local stress intensity factor', $k$, which may be written in the form

$$k = K - \sum_{i} k_{i}^{D}, \tag{12.125}$$

where $K$ is the stress intensity factor due to the applied stress, and $k_{i}^{D}$ is the reduction in the stress intensity factor produced by the $i$th emitted dislocation. The emitted dislocations therefore, in essence, shield the crack tip from the applied stress, and they are therefore often called 'shielding dislocations'. It may also be seen from Fig. 12.44 that the emission of the shielding dislocations from the mode I crack serves to blunt the tip. Detailed analyses of the effects of shielding dislocations on mode I, II, and III cracks, and also mixed mode cracks, are given by Lin and Thomson (1986).

The ability of the crack to generate shielding dislocations is important in determining the energetics of crack popagation, since additional energy must be supplied by the applied stress system when the dislocations are produced. In fact, we shall see that, if the material is sufficiently ductile, and if sufficient numbers of dislocations are generated, crack propagation may become energetically impossible.

Many efforts have therefore been made to model the dislocation generation process. These have grown progressively more sophisticated over the years and include work described in the following papers and their references: Rice and Thomson (1974), Rice (1976), Mason (1979), Jokl, *et al.* (1980), Thomson (1983, 1986), Lin and Thomson (1986), Ohr (1986), Rice (1987), Chiao and Clarke (1989), Zhang *et al.* (1991), Zhou and

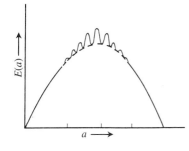

**Fig. 12.43** Modified Griffith energy function for crack, $E(a)$, as a function of $a$, where $2a$ = crack length. The modified function consists of the Griffith energy function (eqn (12.120)) augmented by relatively small oscillatory energy barriers due to lattice trapping of the crack. (From Thomson (1983).)

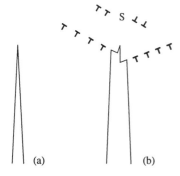

**Fig. 12.44** Generation of lattice dislocations at mode I crack. (a) Initial crack. (b) Crack after emission of dislocations from crack tip and also the generation of dislocations at a nearby source, S, in the lattice activated by the crack stress field. Dislocation emission causes blunting of the crack tip. (From Thomson (1983).)

Thomson (1991$a,b$), Rice (1992), Beltz and Wang (1992), Beltz and Rice (1992). In the first seminal model (Rice and Thomson (1974) the nucleation and expansion of a single shielding dislocation for a mode I crack was modeled in terms of: (a) its self-energy; (b) the work done by the applied stress in expanding it; and (c) the energy required for the additional surface area created at the crack tip in the form of a step as illustrated in 12.44. However, difficulty was encountered in modelling the very early stages of the dislocation formation while it was taking place over distances of the order of an atomic spacing, and rather poorly defined cut-off radii were introduced to cope with the problem. In the more recent work, an improved model for the nucleation has been developed (Rice 1992) which assumes that the dislocation emerges from a traction-free crack by a well defined sequential process along a slip plane on which a Peierls-type stress versus shear displacement relationship (see Section 2.11.2.2) is satisfied. The advantage of this approach is that it allows for the existence of a more realistic extended dislocation core during the formation of the dislocation and eliminates the need for choosing cutoff radii. Also, the model introduces a Peierls stress which may have an influence on whether a dislocation, once formed, will be able to move away from the crack tip at the required rate (Brede and Haasen 1987, Chiao and Clarke 1989). Results have also been obtained for other modes of cracks (including general mixed modes) and also the effects of other shielding dislocations in the region of the crack tip (e.g. Lin and Thomson (1986), Zhang *et al.* (1991), Zhou and Thomson (1991$b$)). In addition, the role of ledges on cracks as efficient sources of dislocations has been studied (Zhou and Thomson 1991$a$). As might be expected, in all of these models the ease of dislocation generation is strongly dependent upon geometric factors such as the crack geometry and the availability of outgoing slip systems which are oriented so that they are subjected to relatively large shear stresses. Details may be found in the above references.

Still a further complication is the possibility that additional dislocations (including both shielding and anti-shielding types) may be generated at sources in the lattice in the highly stressed region near the tip as is also illustrated in Fig. 12.44. The general result will be a cloud of dislocations in the tip region with an excess of shielding dislocations. The quasi-steady-state propagation of a crack under these conditions will now require the energy necessary for the production and motion of all dislocations which are produced and left behind in its wake as it propagates. This energy (per unit area swept by the advancing crack) and designated by $w^P$, may then be formally incorporated into the Griffith model by adding it to the surface energy term to produce a larger 'effective surface energy', or toughness, i.e.

$$\sigma_{\text{eff}} = G = 2\sigma^S + w^P. \tag{12.126}$$

This added energy will, of course, make the crack propagation more difficult. The magnitude of $w^P$ may vary widely, and in many cases it will be considerably larger than $2\sigma^S$. In addition, $w^P$ is generally expected to be an increasing function of the intrinsic surface energy, $2\sigma^S$ (Jokl *et al.* 1980, Zhang *et al.* 1991). This important point may be recognized intuitively by realizing that the critical level of stress which can be sustained at the crack tip depends upon $2\sigma^S$. This is essentially the same stress which is responsible for the generation of the dislocations whose energies contribute to $w^P$. When $\sigma^S$ is increased, this stress is increased, and the number of dislocations which is produced is therefore also increased. In the limiting case where $\sigma^S$ is sufficiently small so that the local stress intensity factor can be increased to the point where crack propagation can occur without any dislocation generation, ideally brittle crack propagation occurs, and $w^P$ is then, of course, zero (Zhang *et al.* 1991). Detailed models for $w^P$ as an increasing function of $\sigma^P$ have been developed by Jokl *et al.* (1980) and Zhang *et al.* (1991).

Finally, it is noted that the criterion for crack propagation based on the Griffith energy function, employing $\sigma_{\text{eff}}$, is strictly thermodynamic, since it is couched entirely in terms of energy considerations. However, in many cases kinetic factors, involving the motion of the crack tip and the dislocations in the vicinity of the crack tip, may be important, depending upon the applied stress level and the rate of load application as discussed, for example, by Rice (1987) and Zhou and Thomson (1991b). For example, in materials in which the Peierls stress for the lattice dislocations is high, the dislocations in the lattice may be slowed down sufficiently so that the rate at which the shielding dislocations can move away from the crack tip may become critical (Brede and Haasen 1986, Chiao and Clarke 1989). Clearly, a full description of all energetic and kinetic features of these processes (which has not yet been achieved) will be required in order to reach a fuller understanding of the complex phenomenon of crack propagation.

Despite these difficulties, a range of crack propagation modes may be identified roughly. When the stress intensity factor can reach the critical level for crack propagation without the generation of significant numbers of lattice dislocations, as in lattices in which lattice dislocation production and motion is exceedingly difficult, cracks may propagate in essentially ideally brittle cleavage fashion as in type A cracking previously defined in Section 12.9.1. When dislocation activity is relatively easier, shielding and anti-shielding dislocations are produced along with possible tip blunting during the propagation (type B cracking). During type C cracking even more dislocation activity occurs. Finally, crack propagation becomes impossible when there is so much plasticity that the local stress intensity factor is unable to reach the critical value for propagation.

Having this background, we may now turn to crack propagation along grain boundaries.

### 12.9.2.2 *Crack propagation along a grain boundary*

Crack propagation along a grain boundary may be treated within the same framework used for fracture in the lattice. However, for boundary fracture, the boundary energy, $\sigma^B$, which is consumed during the propagation of the crack contributes to the process. Therefore, the 'effective surface energy', or toughness, given previously by eqn (12.126) for lattice fracture, must be reduced by $\sigma^B$, i.e.

$$G = \sigma_{\text{eff}} = 2\sigma^S - \sigma^B + w^P. \tag{12.127}$$

A question which is now of interest is whether grain boundaries have a greater tendency to fracture by crack propagation than the lattice. This may be examined for the case of highly brittle fracture (where $w^P$ is negligible) by comparing $G^L$ for fracture in the lattice with $G^B$ for boundary fracture through the ratio

$$R = \frac{G^B}{G^L} = \frac{(2\sigma^{S,B} - \sigma^B)}{2\sigma^{S,L}} = \left[\frac{\sigma^{S,B}}{\sigma^{S,L}}\right] - \left[\frac{\sigma^B}{2\sigma^{S,L}}\right], \tag{12.128}$$

where $\sigma^{S,B}$ is the energy of the free surface produced by boundary fracture, and $\sigma^{S,L}$ is the corresponding energy produced by lattice fracture. If we compare the fracture of general grain boundaries versus lattice fracture along preferred low index fracture (cleavage) planes, $(\sigma^{S,B}/\sigma^{S,L}) > 1$, since the crystal surfaces produced by the grain boundary fracture will be of higher index than those produced by the lattice fracture. On the other hand, $0 < \sigma^B/2\sigma^{S,L} < 1$, and, therefore, whether $R > 1$, or $R < 1$, depends upon the degree of balance between the magnitudes of these ratios. For many metals at low temperatures, $(\sigma^{S,B}/\sigma^{S,L}) \simeq 1.2$. (Cottrell 1989), whereas $\sigma^B/2\sigma^{S,L} \simeq \frac{1}{6}$ (Hondros and Seah 1983). Therefore $R$ is close to unity. As discussed below in Section 12.9.2.3, polycrystals with $R$ close to unity, and are of the type which undergo brittle fracture rather than ductile fracture at low temperatures, should therefore generally fracture in a brittle transgranular mode rather than a brittle intergranular mode. This indeed seems to be the case for these metals (Gandhi and Ashby 1979, Cottrell 1989, 1990).

However, many metals which fracture at low temperatures in either a transgranular brittle mode (as above) or in a ductile mode (Ashby *et al.* 1979), are found to fracture in a intergranular mode when they contain a dilute concentration of substitutional solute atoms in the bulk of a type which segregates strongly to the grain boundaries. This may be attributed to a relatively large decrease in the quantity $(2\sigma^{S,B} - \sigma^B)$ relative to $2\sigma^{S,L}$ in eqn (12.128) caused by the segregation. Also, according to Section 12.9.2.1, this decrease in $R$ should also serve to produce a greater reduction in any possible work, $w^P$, associated with dislocation production in the case of boundary fracture than in the case of lattice fracture, and therefore further enhance the effect of the segregation on promoting intergranular fracture.

Since the substitutional solute atoms will be essentially immobile at the low temperatures, the quantity $2\sigma^{S,L}$ will hardly be affected by the addition of the dilute concentration of solute atoms, since no excess concentration of solute atoms will be present on the free surfaces produced by lattice fracture. On the other hand, high concentrations of solute atoms will be present on the surfaces produced by boundary fracture. At constant $T$ and $P$, the various interfacial energies will be functions of the surface excess of solute atoms, $\Delta N_2$, defined by eqn (5.29). Therefore, the change in $(2\sigma^{S,B} - \sigma^B)$ due to the addition of the segregated solute atoms is

$$\Delta(2\sigma^{S,B} - \sigma^B) = 2\Delta\sigma^{S,B} - \Delta\sigma^B = 2[\sigma^{S,B}(\Delta N_2/2) - \sigma^{S,B}(0)] - [\sigma^B(\Delta N_2) - \sigma^B(0)].$$
$$(12.129)$$

Here, we have assumed that half of the interface excess, $\Delta N_2$, originally segregated at the grain boundary remains on each free surface of the boundary crack. Srolovitz *et al.* (1992) have calculated all of the various quantities in eqn (12.129) for the segregation of Cu solute atoms to a $\Sigma 13$ (001) twist boundary in a 10 Cu/90 Ni (at.%) alloy at 800 K using the classical Einstein model (Section 3.9) and embedded atom potentials (Section 3.7). They obtained (in units of mJ m$^{-2}$) $2\sigma^{S,B}(\Delta N_2) = 2274$; $2\sigma^{S,B}(0) = 3120$; $\sigma^B(\Delta N_2)$ 690; and $\sigma^B(0) = 901$. Both the free surface energy and the grain boundary energy were therefore reduced by the segregation. However, $2\Delta\sigma^{S,B} = -846$ was larger in magnitude than $\Delta\sigma^B = -211$, and therefore the quantity $(2\sigma^{S,B} - \sigma^B)$ was reduced by 635 mJ m$^{-2}$. The solute atom segregation is therefore predicted to produce about a 29 per cent decrease in $R$ in eqn (12.128), which is quite significant. We note that the Cu solute atom segregation in the Cu/Ni system is not expected to be particularly strong on the basis of Fig. 7.3 because of the complete miscibility of Cu in Ni. On the other hand, less soluble solutes, such as Bi in Cu, and S, Sb, Sn, and Se in b.c.c. iron (see Fig. 7.3) and S in Ni, would be expected to produce relatively larger decreases in $R$ and therefore tend to promote intergranular fracture more strongly. This is consistent with many observations (Seah 1980, Hondros and Seah 1983, Roy *et al.* 1982, Russell and Winter 1984, McMahon 1989, Lee *et al.* 1989, Wang and Anderson 1991) which show that, indeed, these strongly segregating solute atoms produce extensive grain boundary embrittlement.

Let us now examine in more detail the widely studied effect of Bi on causing brittle intergranular fracture in the normally ductile metal Cu. A number of studies has shown that the degree of embrittlement depends upon the boundary type. Russell and Winter (1984) and Watanabe (1984) cite evidence that the very singular $\Sigma 3$ {111} twin boundary does not embrittle and that the $\Sigma 5${310} {310} boundary is considerably more resistant to brittle fracture than more general boundaries. Roy *et al.* (1982) used the ball-on-a-plate sintering technique (Herrmann *et al.* 1976, Sautter *et al.* 1977; see also Sections 5.7 and 6.3.2.1) to produce in the Cu/Bi system a wide range of different types of grain boundaries located in the neck regions (see Fig. 5.19) between a large number ($\simeq$ 10 000) of single crystal balls and a (110) single crystal plate. These boundaries were then stressed by ultrasonic means, and it was found that the particularly low energy singular boundaries in the distribution of boundary types were considerably more resistant to brittle fracture than the higher energy boundaries. Wang and Anderson (1991) carried out a detailed study of the fracture of Cu/Bi bicrystals containing $\Sigma 11$ {113} {113}, $\Sigma 9$ {221} {221}, $\Sigma 5$ {310} {310}) and more general boundaries. The results were reasonably consistent with the basic tenets of the Rice/Thomson model, in which brittle propagation of a crack by debonding competes with the generation of shielding dislocations in the crack tip region. The fracture mode depended upon both the degree to which Bi segregated to the different boundaries (and therefore reduced $G^B$) and also upon the geometry of the slip systems which were available to generate shielding dislocations. In general, the $\Sigma 11$ boundary, which is an exceptionally low energy singular boundary (see Fig. 5.23), could not be embrittled, the $\Sigma 9$ boundary, which is less singular, was less resistant, and the $\Sigma 5$ and random boundaries, which are the least singular, were most susceptible to embrittlement. As discussed in Chapter 7, the degree of segregation for these boundaries is expected to increase in about the same order: the degree of embrittlement therefore increased with the degree of segregation. In a further set of experiments it was shown that a segregated $\Sigma 9$ boundary was susceptible to brittle fracture in the [$\bar{1}$14] direction, but not in the opposite [1$\bar{1}$4] direction where a more favourably oriented slip system was available. It is evident from this work that the brittle fracture of large-angle boundaries depends in detail upon complex relationships between the boundary structure, the degree of segregation and its effect on $G^B$, and the geometry of slip in the adjoining crystals.

So far, we have discussed only the effects of segregating substitutional species. Brittle intergranular behaviour can also be induced by many other agents including fast-diffusing interstitial species such as H. Here, the diffusion may be fast enough to maintain the degree of segregation at the free surfaces generated at the crack tip near the equilibrium value required by the ambient chemical potential of the H in the overall system, in contrast to the situation for the relatively slow-diffusing substitutional species discussed above (see Hirth and Rice 1980). Also, in hydride-forming systems, the stress at the crack tip may be high enough to create a brittle hydride phase there which will promote brittle cracking (Birnbaum 1984), Hirth 1987). During intergranular 'stress corrosion cracking' other chemical species may influence the debonding process at the crack tip and so influence the cracking. Detailed discussion of all of these phenomena is beyond the scope of the present chapter. Many are discussed by Pugh (1991). Also, see Thomson (1980), Fuller and Thomson (1980), Hondros and Seah (1983), Thomson (1983), articles in Lantanision and Jones (1987), and the review of Kasul and Heldt (1989).

Finally, we note that small-angle boundaries (which always possess relatively low energies and degrees of segregation) are highly resistant to intergranular fracture (Watanabe 1984) as might be expected.

### 12.9.2.3 *Crack propagation in homophase polycrystals*

The failure of a polycrystalline body by crack propagation requires the initiation of a crack (or a number of cracks) and its (their) subsequent propagation through the material in a manner which eventually produces a single continuous pathway through the entire body. The prediction of just how such a crack will occur is a highly complex matter, since any crack propagating in a polycrystal will arrive at many junctures at which it may travel along a grain boundary, or alternatively, through a grain. The overall result may therefore be intergranular, transgranular, or mixed. Consider first the initiation of a crack.

*Crack initiation.* Grain boundaries constitute sites where cracks may form preferentially by a variety of mechanisms. For grain boundaries in materials which are ductile enough to permit at least some lattice dislocation generation and glide, cracks may be nucleated in the regions of high stress which are produced at dislocation pileups which occur at grain boundaries acting as barriers to slip (Section 12.6). The basic model (Knott 1973) is illustrated schematically in Fig. 12.45. A lattice dislocation source exists within a grain, and dislocations emanating from the source pile up at a boundary as seen in Fig. 12.45(a). The pile-up produces a stress similar to that near a mode II crack (Hirth and Lothe 1982) and may be large enough to initiate fracture as illustrated. Many variations of this basic scheme are possible, as, for example, the one illustrated in Fig. 12.46. Here, an intergranular crack has nucleated at the intersection of two lattice deformation bands with the grain boundary.

For highly brittle materials, in which there is essentially no lattice dislocation activity, the above mechanism cannot operate. However, crack nucleation may then take place at stress concentrations at boundaries caused by various flaws, such as small cavities or precipitates. These defects are frequently concentrated at boundaries in brittle materials which are often fabricated by sintering techniques and also contain small concentrations of insoluble impurities.

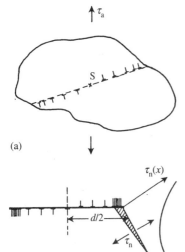

(a)

(b)

**Fig. 12.45**  Initiation of crack at pileup of lattice dislocations at grain boundary. (a) Pile-up of lattice dislocations emanating from source, S, at boundary surrounding a grain in a polycrystal. $\tau_a$ = applied stress. (b) Initiation of crack in region of high stress at end of pile-up. $\tau_n$ = normal stress which is present and acts to propagate the crack. (From Knott (1973).)

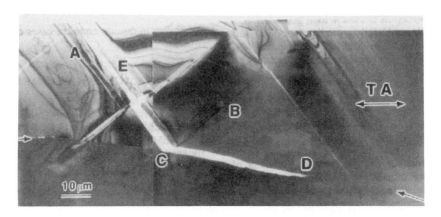

**Fig. 12.46**  Composite micrograph showing crack initiation and propagation in a Ni–S alloy. The crack was initiated at C at the intersection of the deformation bands E and B with the boundary. It then propagated along the boundary (along CD) and also into the lattice within the deformation band E. Tensile axis indicated by TA. (From Lee, *et al.* (1989).)

The initiation of cracks at boundaries in brittle materials can also occur as a result of incompatibility stresses generated by applied elastic stressing (Section 12.2.1) or heating/cooling (Section 12.2.3). As has already been pointed out, these stresses are concentrated at boundaries, and they can be especially severe in brittle materials, since no stress relief mechanisms by plastic flow are available.

*Crack propagation.*   Once formed, a crack may propagate initially along either a grain boundary or through a grain. In the former case it may reach a grain boundary junction and then propagate along another grain boundary or be deflected into a grain. In the latter case it may reach a grain boundary and continue into the next grain or be deflected into the grain boundary. The direction which the crack takes at each juncture will depend upon a host of factors including at least the initial direction of the crack, the value of $G$ for each possible direction, and the nature of the loading on the polycrystal. Relatively low values of $G^B$ for the grain boundaries, as is usually the case when strong solute atom segregation occurs, will, of course, encourage propagation along boundaries. However, at singular boundaries $G^B$ will tend to remain high, and these boundaries may often remain immune to cracking (Wanatabe 1984).

Srolovitz *et al.* (1992) have carried out a simplified computer simulation of this complex situation using a model of a polycrystal in which the atomic bonds are represented by springs which are resistant to both stretching and bending. Bond breaking occurs when the strain energy stored in a spring exceeds a critical value. Situations where $G^B$ for the boundaries is lower than $G^L$ for the grains can then be simulated in a simple way by varying the ratio $E_c^B/E_c^L$, where $E_c^B$ is the critical strain energy for breaking bonds at all of the boundaries, and $E_c^L$ is the corresponding quantity for all bonds in the lattice. Figure 12.47 shows the propagation of a crack under tensile loading when $E_c^B/E_c^L = 0.28$, corresponding to boundaries of relatively low toughness. The crack is mainly intergranular but transgranular in some sections. Of interest is the result that a segment of the crack which is intergranular is deflected into a grain at A and B where sections of the boundary network are encountered which are parallel to the tensile loading direction. At C the stress field of the crack produces a new crack at a nearby boundary segment which propagates back to join the original crack and then propagates forward in intergranular fashion. By carrying out other simulations it was shown that the fracture became essentially transgranular when $E_c^B/E_c^L > \cong 0.5$, and essentially intergranular when $E_c^B/E_c^L < \cong 0.2$. These results appear to be generally consistent with calculations (He and Hutchinson 1989*a*) for the deflection of cracks impinging on heterophase interfaces at different angles when $G^B$ for the boundary is lower than $G^L$ of the lattice (Srolovitz *et al.* 1992). These results also allow us to understand why the pure polycrystalline metals which undergo brittle fracture at low temperatures, and which should have values of $R = G^B/G^L$ close to unity according to the estimates in Section 12.9.2.2, are observed to fracture in a transcrystalline rather than intergranular mode (Gandhi and Ashby 1979, Cottrell 1989, 1990).

An interesting feature of such transgranular fractures is shown in Fig. 12.48(a) where the fracture stress is plotted versus $d^{-\frac{1}{2}}$ ($d$ = grain size). At all grain sizes, cracks are initiated at the grain boundaries by pile-ups (Fig. 12.45) very shortly after the yield stress is exceeded. When the grain size is large, these cracks can expand to long enough lengths (because of the large grain size) to produce stress intensity factors which are large enough to force them across successive grain boundaries and through successive grains to cause a transgranular fracture of the entire polycrystal. Such cracks are therefore 'initiation limited', since they can grow indefinitely once initiated. The fracture stress therefore has essentially the same dependence on the grain size as the yield stress, given earlier by eqn (12.48). However, when the grain size becomes sufficiently small, cracks which are initiated at applied stresses just above the yield stress have lengths which are restricted to the grain size, since their stress intensity factors are not large enough to force them to propagate across grain boundaries and through adjacent grains. Such a crack, termed a 'microcrack', is shown in Fig. 12.48(b). Additional applied stress, well above the yield stress, is then required to expand such cracks. In this regime the final overall fracture is therefore 'propagation limited'.

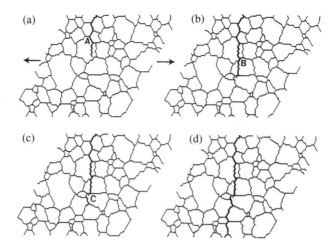

**Fig. 12.47** Propagation of brittle crack, (a)–(d), in polycrystal calculated by computer simulation using a simple atomic bond-breaking model. Specimen loaded in tension along arrows in (a). (From Srolovitz *et al.* (1992).)

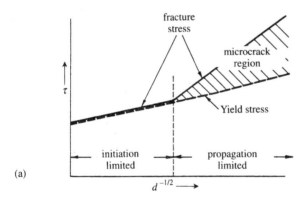

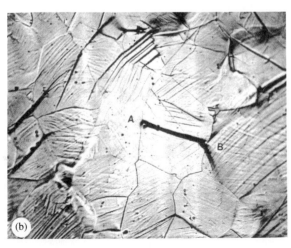

**Fig. 12.48** (a) Stress for fracture by cracking and also the yield stress as a function of $d^{-\frac{1}{2}}$ for polycrystals. $d$ = grain size. At large grain sizes, fracture by cracking is initiation-limited, whereas at amall grain sizes it is crack propagation-limited. In the latter regime microcracks, as in (b), are observed. (From Gilman (1958).) (b) Microcrack (along AB) in polycrystal-line Fe. (From Hahn *et al.* (1959).)

### 12.9.3 Growth and coalescence of cavities at grain boundaries at low temperatures by plastic flow due to dislocation glide

In this type of fracture, small cavities may be concentrated preferentially on the grain boundaries before deformation, as is often the case in polycrystalline material produced by sintering. In other cases small hard second-phase particles may be pinned and concentrated there (see Section 9.5.1). In the former case, the small pre-existing cavities grow when the material is plastically deformed until they eventually coalesce and produce localized intergranular fracture. In the latter case, cavities are nucleated at the particle/matrix interface, and these then grow and eventually cause fracture. Cavities can nucleate at the hard particles, since a hard inclusion, which is embedded in a plastic medium which strain hardens, disturbs both the elastic and plastic displacement fields generated in a deforming body (Ashby *et al.* 1979). This produces a stress concentration which builds up as the plastic strain increases until either the particle/matrix interface is separated or the inclusion itself is fractured. For relatively large particles ($\gg 1$ mm) the matrix can be treated approximately as a plastic continuum (Argon *et al.* Safoglu 1975). The maximum local stress is then 1.5 to 2.0 times the remote applied stress. However, if the particles are small, the local rate of work hardening can be much larger than the average rate (Brown and Stobbs 1971, 1976). The local stress at the particles is then a much larger multiple of the applied stress and increases faster with strain. When the local stress becomes sufficiently large a cavity is produced (Ashby *et al.* 1979, Brown and Stobbs 1976, Goods and Brown 1979).

Cavities present for either of the above reasons grow continuously during further plastic straining. An initially spherical cavity under applied tensile strain concentrates stress and elongates at a rate which is initially approximately double the overall specimen rate and therefore becomes elliptical as illustrated in Fig. 12.49 (McClintock 1968, Rice and Tracey 1969). When the cavity length eventually becomes comparable to the cavity spacing (Fig. 12.49(c)), the plasticity becomes strongly localized between the cavities, and localized shear produces necking in these regions leading to quick coalescence and complete intergranular fracture (Ashby *et al.* 1979, Thomason 1968, 1971, Brown and Embury 1973).

The final stage in this process, which produces a continuous intergranular fracture path running through the body, can be complex in a polycrystal and can take a variety of forms. Different boundary segments possessing different structures and different inclinations in the stress field may cavitate at different rates and therefore tend to undergo cavity

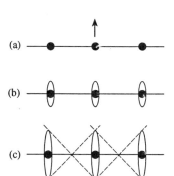

(a)

(b)

(c)

**Fig. 12.49** Schematic illustration of growth of intergranular cavities under tensile stress by plastic flow at low temperatures due to dislocation glide. (a) Initial distribution of small hard second phase particles on grain boundary. (b) Initiation and growth of cavities at the particles. (c) Cavity length now comparable to cavity spacing in the boundary.

coalescence at different times. The linking up of boundary segments which are completely fractured therefore assumes a statistical character. Furthermore, once several completely fractured boundary segments have linked up to form a relatively long 'macro-crack', the stress intensity factor in the tip region of the macro-crack will be large enough to accelerate the rate of cavitation and coalescence in boundary segments ahead of it. Such a macrocrack can then propagate in essentially ductile fashion through the entire body or else link up with other macro-cracks distributed throughout the body. On the other hand, under certain conditions the boundaries may cavitate more uniformly, and the contribution of propagating macro-cracks may be minimal. A review of further aspects of fractures caused by 'distributed damage processes', such as the formation and growth of boundary cavities, has been given by Thompson (1987).

### 12.9.4  Growth and coalescence of cavities at grain boundaries at high temperatures by diffusion, power-law creep, and boundary sliding

At high temperatures, diffusional transport, plastic deformation by power-law creep (which involves dislocation climb and is described by eqn (12.68)), and boundary sliding become possible. These new processes, often acting in tandem, can produce boundary cavitation as has already been pointed out in Fig. 12.38(d, e). Intergranular fracture resulting from such cavitation is of great technological importance and has been intensely studied. A vast number of experimental observations has been made, and many models for the complex phenomena which occur have been constructed. Analyses and reviews include: Argon *et al.* (1980), Cocks and Ashby (1982), Evans (1984), Evans and Wilshire (1985), Riedel (1987), Cadek (1988).

Experimental observations indicate that the boundary cavitation in polycrystals occurs more extensively on the boundary segments which are more nearly perpendicular to the applied tensile stress. In addition, new cavities are generally initiated continuously as earlier formed cavities grow. Also, the cavities may develop in a variety of forms ranging from almost equiaxed cavities to flat penny-shaped cavities to wedge-shaped cavities (i.e. 'wedge cracks') terminating at triple junctions as illustrated in Fig. 12.38(d). As in the case of the intergranular fracture due to cavitation described above in Section 12.9.3, final fracture in creeping polycrystals which are initially free of cavities occurs by the initiation and growth of boundary cavities and their eventual coalescence to form a continuous intergranular crack through the material. We therefore discuss these processes in the same order.

### 12.9.4.1  *Initiation of cavities*

We begin by showing that cavities cannot be nucleated homogeneously on boundaries at high temperatures by the aggregation of vacancies under the influence of the applied stresses which are typically present during creep. Consider the formation of a cavity nucleus of equilibrium shape on a boundary subjected to a remotely applied uniform traction, $\tau_n^\infty$, as illustrated in Fig. 12.50. The cavity surface is spherical with principal radii $R$, and meets the boundary at the equilibrium angle $\alpha$, which according to the results in Section 5.6.8, is given by $\alpha = \cos^{-1}(\sigma^B/2\sigma^S)$, where $\sigma^B$ and $\sigma^S$ are the boundary and surface free energies respectively. The free energy to produce this nucleus is then

$$\Delta G = - (2\pi/3)(2 - 3\cos\alpha + \cos^3\alpha)\tau_n^\infty R^3 + 4\pi(1 - \cos\alpha)\sigma^S R^2 - \pi\sin^2\alpha\cdot\sigma^B R^2.$$

$$(12.130)$$

Here, the first term (RHS) is the work done by the traction during the expansion which

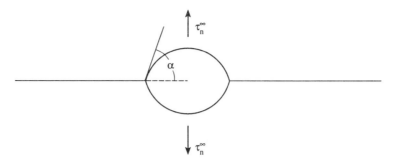

**Fig. 12.50** Cavity nucleus on grain boundary. $\tau_n^\infty$ = remotely applied tensile stress. $\alpha$ = equilibrium junction angle between cavity surface and grain boundary. Cavity surfaces are spherical.

accompanies the formation of the nucleus, the second term is the energy required to produce the new free surface, and the third term is the energy contributed by the boundary area which is eliminated. The radius of the nucleus of critical size is found in the usual way from the condition $\partial\Delta G/\partial R = 0$, and is given by

$$R^* = 2\sigma^S/\tau_n^\infty. \tag{12.131}$$

Equation (12.131) yields the critical cavity radius for which the energy of the system is stationary with respect to the transfer of atoms from the cavity surface, acting as an atom source, to the nearby boundary, acting as an atom sink. A cavity with $R > R^*$ will tend to spontaneously grow, while a cavity with $R < R^*$ will shrink (Balluffi and Seigle 1955). This result can be obtained by a somewhat more direct approach in which we recognize that when this condition holds the diffusion potentials for atoms at the cavity surface and the boundary must be equal. Assuming that the diffusion occurs by a vacancy mechanism, and that the vacancies are maintained in local equilibrium at the two interfaces, the diffusion potential for an atom at the boundary is $M_A^B = N_o\Omega(f - \tau_n)$, according to eqn (12.65). With the use of eqn (10.26) the potential at the cavity surface is

$$M_A^C = N_o\Omega\left(f - 2\sigma^S/R\right), \tag{12.132}$$

and the critical $R$ for which these potentials are equal is then given by eqn (12.131).

   Next, the critical free energy for cavity nucleation may be obtained by substituting the above value of $R^*$ into eqn (12.130). Using this, and standard nucleation rate theory (Christian 1975), it is easily demonstrated (Raj and Ashby 1975, Argon *et al.* 1980) that the homogeneous nucleation of cavities is many orders of magnitude too slow to account for the observed rates of generation of cavities on boundaries at usual applied creep stresses. The cavities must therefore be generated inhomogeneously at special regions of high local tensile stress concentration.

   A large number of models has been suggested (see references below) for the initiation of cavities in such regions. Some of these are illustrated schematically in Fig. 12.51. Included are models in which sufficiently high local stress is achieved to cause decohesion at the boundary by the rupturing of bonds and also models where cavity nuclei are formed by the aggregation of vacancies aided by the local stress. Several require boundary sliding. In Fig. 12.51(a) lattice dislocations have piled up at the boundary to produce a sufficient stress concentration to produce a crack (cavity) by decohesion much in the manner shown in Fig. 12.45(b). In Fig. 12.51(b) vacancies aggregate to produce a cavity in the region of high stress produced by a pile-up. In this case a region of high

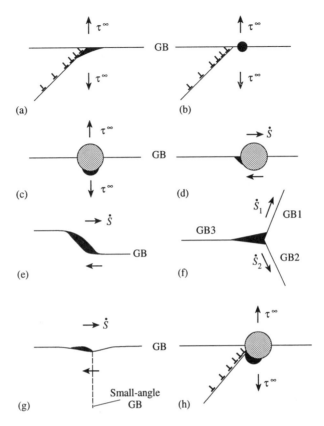

**Fig. 12.51** Various mechanisms for the initiation of cavities at grain boundaries under stress. (a) Decohesion (cracking) at high stress caused by dislocation pile-up (see Fig. 12.45). (b) Vacancy aggregation in highly stressed region caused by dislocation pile-up. (c) Decohesion or vacancy aggregation at weak particle/matrix interface under tensile stress. (d) Same as (c) except that tensile stress is caused by boundary sliding. (e) Initiation due to high tensile stress produced at boundary step by boundary sliding. (f) Same as (e) except that high stress is generated at triple junction by boundary sliding. (g) Same as (e) except that stress is produced at depression in sliding boundary at intersection with small-angle boundary. (h) Combination of mechanisms in (b) and (c).

stress is eliminated at the boundary, thereby aiding the vacancy aggregation. In an approximation, we may write the energy gained as the average strain energy density which existed originally in the region occupied by the nucleus multiplied by the volume of the nucleus. This contributes a term to $\Delta G$ in eqn (12.130) given approximately by $-(2\pi/3)(2 - 3\cos\alpha + \cos^3\alpha)R^3(\bar{\tau}_c^2/2E)$ where $\bar{\tau}_c$ is the average concentrated stress in the region, and $E =$ Young's modulus (Lim 1987, Xinggang *et al.* 1993). When this term is added, the radius of the nucleus of critical size is given by

$$R^* = \frac{2\sigma^S}{\{\tau_n^\infty + \bar{\tau}_c^2/2E\}},\qquad(12.133)$$

rather than by eqn. (12.131). Concentrated stress therefore reduces $R^*$ and consequently increases the nucleation rate.

In Fig. 12.51(c) a hard particle is present, and a cavity is produced at the particle/matrix interface under the influence of a tensile stress. A high particle/matrix interfacial energy (i.e. a mechanically weak interface) would allow possible separation by bond rupturing or the possible easy heterogeneous nucleation of a cavity on the particle/matrix interface by vacancy aggregation. Figure 12.51(d) illustrates cavity formation at a hard particle in a boundary which is sliding. As discussed previously in Section 10.4.2.4, and illustrated in Fig. 10.18(a), compressive stresses are developed on the leading faces of the particle and tensile stresses on the trailing faces. A cavity may then be nucleated at a trailing face in a manner generally similar to the cavity formation in Fig. 12.51(c). In Fig. 12.51(e) a step is present in a sliding boundary, and a cavity is produced at the step where a high tensile stress is present. In Fig. 12.51(f) tensile stresses are concentrated at a triple junction as a result of the boundary sliding shown, causing the formation of a cavity. In Fig. 12.51(g), a cavity is generated at the concentration of tensile stress produced by a depression in a sliding boundary at its intersection with a small-angle boundary. Finally, the above mechanisms may operate in tandem as illustrated, for example, in Fig. 12.51(h) where the mechanisms of Fig. 12.51(b) and (c) are operating together.

Unfortunately, it has not been possible to verify directly the detailed mechanism (or mechanisms) which dominate cavity nucleation in various common situations because of severe experimental difficulties. Many aspects of cavity nucleation are therefore still unsettled. A particularly troublesome problem is the fact that rapid mechanisms for the relaxation of stress concentrations are available at elevated creep temperatures, such as local diffusional transport and plastic flow (power law creep). Questions have therefore arisen as to whether sufficiently high stress concentrations to produce cavities can actually be raised by a number of the above mechanisms. In order to cope with this problem it has been pointed out that lattice slip and boundary sliding occur spasmodically, and it has therefore been suggested that the cavity nucleation occurs during peak periods when the stress concentrations are relatively high. In addition, impurity solute atoms may diffuse to the surfaces of nucleus embryos and stabilize them by segregating there and decreasing $\sigma^S$. Extensive discussions and analyses of various models, and the problems which have arisen in reaching a full understanding of cavity nucleation, have been given by Argon *et al.* (1980), Nix (1983), Watanabe (1983), Evans (1984), Evans and Wilshire (1985), Riedel (1987), Lim (1987b), George *et al.* (1987), Cadek (1988), Lim (1989), and Xinggang *et al.* (1993).

### 12.9.4.2 *Growth of cavities*

The growth of cavities on boundaries in stressed polycrystals is better understood than their initiation. Nevertheless, it is a complicated phenomenon whose rate can be influenced by a number of factors which include the rate of boundary diffusion, the rate of surface diffusion on the cavity surface, the rate of power-law creep in the bulk, boundary sliding, and the degree of restraint imposed by the specimen bulk on the volume expansion associated with the growth of cavities on a segment of boundary. Analyses and reviews include: Cocks and Ashby (1982), Chen (1983), Evans (1984), Evans and Wilshire (1985), Riedel (1987), and Cadek (1988).

For present purposes we start with cavity growth on a boundary segment which is transverse to the applied tensile stress, $\tau^\infty$, as illustrated in Fig. 12.52(a). Even though some sliding may occur on such a boundary if the grain size is irregular (Chen 1983), we shall neglect it for the time being.

Following Cocks and Ashby (1982), each cavity can be taken as being in the centre of a cylindrical cell of height d and diameter $2l$ = cavity spacing on the boundary Fig. 12.52(b)). Under certain conditions (to be explored later) the void growth can be controlled by either boundary diffusion, surface diffusion, or power-law creep in the

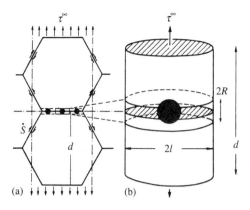

**Fig. 12.52** (a) Cavity growth on boundary segment in polycrystal transverse to remotely applied tensile stress, $\tau^\infty$. Inclined boundaries are undergoing sliding at a rate $\dot{S}$. (b) Expanded view of cylindrical cell used to analyse the cavity growth. (From Cocks and Ashby (1982).)

bulk. We therefore proceed by first analysing each of these processes as if it were completely controlling. The results will be approximate and will ignore refinements which influence the outcome by factors less than about two.

*Cavity growth controlled by boundary diffusion.* In this case atoms diffuse from the cavity surface, acting as an ideal atom source, out along the boundary, acting as an ideal atom sink. It is assumed that atoms can be redistributed rapidly enough on the cavity surface by surface diffusion to maintain the cavity at its equilibrium shape which, for all practical purpose, can be taken as a sphere of radius $R$. The diffusion potential on the cavity surface is then given by eqn (12.132) and on the boundary by eqn (12.65). The deposition of atoms along the boundary by diffusion produces a quasi-steady state in which the distribution of surface tractions along the boundary, $\tau_n(r)$, is adjusted so that an equal flux of atoms enters the boundary at every point in order to maintain proper compatibility. It is then necessary to solve the Laplace equation $\nabla^2 M_A = 0$ and employ a flux equation similar to eqn (9.18) using a procedure generally similar to that used to solve the diffusional accommodation problem outlined previously in Section 12.8.1.3. The result (Cocks and Ashby 1982) is a rate of cavity volume growth given to a good approximation by

$$\frac{\mathrm{d}V}{\mathrm{d}t} = \frac{2\pi\delta D^B\Omega\left[\tau^\infty/(1 - R^2/l^2) - 2\sigma^S/R\right]}{f^B kT(1 - R^2/l^2)\ln(l/R)}. \tag{12.134}$$

*Cavity growth controlled by surface diffusion alone.* When the rate of diffusion of atoms along the surface of the cavity, which initially possesses an almost spherical equilibrium shape, is slow enough so that it is unable to maintain this shape in the face of a relatively rapid loss of atoms away from the cavity due to grain boundary diffusion, the cavity will tend to flatten out. This flattening will continue until the cavity has a flat shape which grows primarily at its intersection with the boundary, and a quasi-steady-state diffusion flux along the cavity surface is established which controls the overall rate of cavity growth. In order to find this growth rate, we again assume ideal source/sink action everywhere and simplify the problem by employing the cavity shape illustrated in Fig. 12.53. The diffusion flux on the surface (atoms per unit length) has the form $J_A = -(\delta D^S/N_o\Omega kTf^S)\cdot\nabla M_A$ from eqn (9.124). The diffusion potential, $M_A$, at any point on the surface is given by eqn (12.132) in the form $M_A^S = N_o\Omega(f - \sigma^S[R_1^{-1} + R_2^{-1}])$, where $R_i$ are the principal radii of curvature. From eqn (12.65), the average value of $M_A$ at

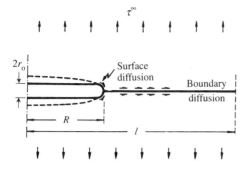

Fig. 12.53 Idealized configuration used to analyze cavity growth limited by the surface diffusion rate. Boundary subjected to remotely applied tensile stress, $\tau^\infty$. (From Cocks and Ashby (1982).)

boundary is $M_A^B = N_o\Omega(f - \tau^\infty/[1 - R^2/l^2])$. In order to have continuity in $M_A$ at the intersection of the cavity with the boundary, we must have $M_A^B = M_A^S$ there, with $R_1 = r_o$, and $R_2 = \infty$. This leads to

$$r_o = (\sigma^S/\tau^\infty)(1 - R^2/l^2). \tag{12.135}$$

Conservation of atoms requires that

$$dR/dt = \Omega J_A/r_o. \tag{12.136}$$

An approximate value of $\nabla M_A$ at the cavity edge is obtained (Cocks and Ashby 1982) by assuming that the value of $M_A$ on the sharply curved portion of the cavity surface leading into the intersection with the boundary drops from the value $N_o\Omega f$ (characteristic of a flat surface) to the value $N_o\Omega(f - \sigma^S/r_o)$ over the short curved distance $\pi r_o/2$. Therefore,

$$\nabla M_A = -(N_o\Omega\sigma^S/r_o)(\pi r_o/2)^{-1}. \tag{12.137}$$

Putting this value of $\nabla M_A$ into the equation for $J_A$, and substituting this into eqn (12.136) with $r_o$ given by eqn (12.135), we finally obtain

$$\frac{dR}{dt} = \frac{2\delta D^S\Omega}{\pi kTf^S(\sigma^S)^2(1 - R^2/l^2)^3}\cdot(\tau^\infty)^3. \tag{12.138}$$

Since eqn (12.138) gives the cavity growth rate when the rate of surface diffusion on the cavity surface is rate controlling, $D^B$ does not appear, as should be expected.

More general solutions, involving the effects of both surface and boundary diffusion, have been given by Chuang *et al.* (1979) and reviewed by Riedel (1987) and Cadek (1988). Chuang *et al.* find that cavities will grow under boundary controlled conditions while maintaining their equilibrium shape when $S < (1 + 6D^S/D^B)$ and grow under surface diffusion controlled conditions in a thin crack-like mode (as in the above analysis) when $S > (2 + 9D^S/D^B)$. Here, $S$ is the ratio of the applied stress to the stress which just equilibrates the cavity against shrinking by sintering as discussed earlier in Section 12.9.4.1. Surface diffusion controlled growth is therefore favored by high stresses, large cavity radii (since these decrease the equilibration stresses), and small values of $D^S/D^B$.

*Cavity growth controlled by power-law creep alone.* As pointed out by Cocks and Ashby (1982), when the bulk deforms by power-law creep, it is an acceptable approximation to assume that the slab of thickness $2R$ containing the cavities in Fig. 12.52 extends at a rate determined by the section stress $\tau^\infty/(1 - r^2/l^2)$, while the rest of the cylindrical element extends at a rate determined by $\tau^\infty$. This will cause the slab to dilate and the cavities to grow. Expressions for the rate of cavity growth obtained by the use of this approximation have been given by Cocks and Ashby (1982) and will not be described in any detail here.

*Cavity growth by coupled mechanisms.* Cavities on the transverse boundary in Fig. 12.52 may grow by combinations of the above mechanisms under many circumstances. When the stress is low, they tend to grow by diffusion: when high, they tend to grow by power-law creep: in between, they grow by a combination of both mechanisms. When $R^2/l^2$ is small, they tend to grow by diffusion, and when large, by power-law creep.

When cavities grow by combined boundary diffusion and power-law creep, the situation can be represented to a good approximation by the arrangement illustrated in Fig. 12.54. Two zones are now present within the cylindrical cell, i.e. a 'boundary diffusion-controlled' zone and a 'power-law creep controlled' zone. Atoms are deposited along the boundary in the diffusion zone by boundary diffusion. The insertion of these atoms causes a wedging effect which partially relaxes the local tensile stress in the diffusion zone and causes a corresponding increase in the tensile stresses acting on the power-law creep zone. In the quasi-steady state, the rate of expansion of the two zones must be equal. The general solution of this coupled problem (Cocks and Ashby 1982) yields a result in which the relative sizes of the two zones vary as the physical parameters which control the system change. In one limit the power-law creep zone shrinks to zero, and 'boundary diffusion alone' behaviour is obtained. In the other limit the diffusion zone shrinks to zero resulting in pure power-law creep control. Furthermore, it is found that the overall coupled rate of growth can be represented to a reasonable approximation by simply adding the results for the two mechanisms as if they were operating alone.

Generally similar results are obtained (Cocks and Ashby 1982) for cavity growth by coupled surface diffusion and power-law creep mechanisms. Having these results, a cavity growth map may be constructed as, for example, in Fig. 12.55, where the ranges in stress-temperature space are shown over which the various mechanisms which we have considered are rate controlling. We note that the line separating surface diffusion controlled growth from boundary diffusion controlled growth cannot be determined by simply finding the conditions under which the rates of both processes are equal as is the case for the surface diffusion versus power-law creep and boundary diffusion versus power-law creep lines. This is so because the surface diffusion and boundary diffusion must occur in series. However, when surface diffusion is controlling, the cavity will grow in a crack-like form, whereas when boundary diffusion is controlling, the cavity will grow in an approximately spherical form. Hence, by solving the coupled growth problem the resulting cavity shape can be used as a reasonable criterion to find the conditions under which either mechanism is rate controlling (Cocks and Ashby 1982).

It may be seen that cavity growth controlled by diffusion tends to occur at low stresses, while growth controlled by power-law creep occurs at high stresses in a manner generally consistent with the fracture map presented previously in Fig. 12.39. The map in Fig. 12.55 was constructed for a material with $T_m = 1356\,\mathrm{K}$, $\sigma^S = 1.72\,\mathrm{J\,m^{-2}}$, $D^B\delta = 5.12 \times 10^{-15}\exp(-105\,\mathrm{kJ\,mol^{-1}}/kT)\,\mathrm{m^3\,s}$, $^{-1}$ and $D^S\delta = 6 \times 10^{-10}\exp(-205\,\mathrm{kJ}$ $\mathrm{mol^{-1}}/kT)\,\mathrm{m^3 s^{-1}}$: $D^S$ is therefore considerably smaller than $D^B$ over much of the range of $T/T_m$, particularly at the low temperatures. The area of the surface diffusion controlled field is therefore considerably larger than it would be for a material where $D^S > D^B$ as in Fig. 8.5.

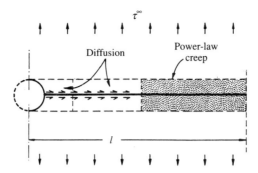

**Fig. 12.54** Details within cylindrical cell of Fig. 12.52 when the growth of the cavity on a transverse grain boundary subjected to a remotely applied tensile stress, $\tau^\infty$, occurs by coupled diffusion and power-law creep. Two zones, a diffusion zone and a power-law creep zone, are present. (From Cocks and Ashby (1982).)

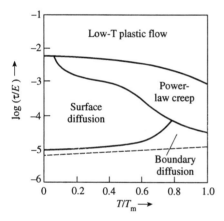

**Fig. 12.55** Cavity growth map for cavity on grain boundary of Cu with $l = 12\,\mu m$ (see Fig. 12.52), and $f_h = 0.1$, where $f_h$ is the fraction of grain boundary area occupied by cavities. $\tau$ = applied stress and $E$ = Young's modulus. Each growth mechanism is rate-controlling within its field. (From Cocks and Ashby (1982).)

*Cavity growth due to boundary sliding.* Cavity growth on boundaries which are undergoing sliding may occur when steps are present as illustrated in Fig. 12.51(e). Here, the cavity simply opens up at a rate corresponding to the sliding rate. The energetics of this process have been discussed by Evans (1984).

Also, the rate of cavity growth on a sliding boundary, which at the same time is also subjected to a normal traction, may be speeded up as discussed by Chen (1983) and illustrated in Fig. 12.56. As may be seen, the sliding tends to spread out the cavity and sharpen its edges. This stimulates the rate of surface diffusional transport. Chen finds that under certain circumstances the rate of cavity growth can be substantially enhanced by this phenomenon in regions normally controlled by both surface diffusion and boundary diffusion.

*Effects of bulk constraints on cavity growth.* So far, we have neglected the effects of possible bulk constraints on the expansion which must occur at a boundary when cavities grow on it by diffusional transport. These constraints would be essentially absent, if the cavity growth on all boundary segments were similar throughout the polycrystal and if boundary sliding occurred easily. However, more generally, different nearby boundary segments will experience different rates of cavity growth, and the cavity growth at the

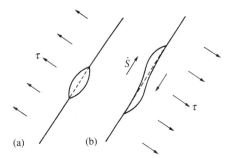

Fig. 12.56 Influence of boundary sliding, $\dot{S}$, on cavity growth on grain boundary subjected to tensile stress, $\tau$. (From Chen (1983).)

more rapidly cavitating segments will then be constrained and slowed down. The rates at which these restraints can be relieved, by, for example, power-law creep in the surrounding material, have been analysed by a number of workers. The results have been reviewed by Riedel (1987), and Cadek (1988).

### 12.9.4.3   *Coalescence of cavities and complete intergranular fracture*

Final complete fracture resulting from creep cavitation will occur by the coalescence of the cavities to produce eventually a continuous crack through the material. This process has already been described briefly in Section 12.9.3 for the case where the cavity growth and final fracture occur at low temperatures by plastic flow due to dislocation glide. The process at high temperatures, where diffusional transport and boundary sliding may now occur, and where any plastic flow occurs by power-law creep may be expected to be similar in many respects. The details may differ, but 'distributed damage' (Thompson 1987) in the form of growing cavities will again be present, and a range of fracture behaviour may again occur. Boundary segments in various regions of the specimen may undergo cavity coalescence and fracture to produce microcracks. For example, Fig. 12.57 shows a model for the formation of a wedge microcrack at a boundary segment (see Fig. 12.38(d)) by the progressive growth and coalescence of cavities in front of the advancing microcrack under the influence of the forces produced by boundary sliding at the triple point. Microcracks formed in various ways may then link up to form macro-cracks, which will then propagate preferentially by the preferential growth and coales-cence of cavities ahead of them in their highly concentrated stress fields. Eventually, these processes will produce a complete fracture path. An example of such a fracture in an intermediate state is shown in Fig. 12.58. Under other conditions the cavitation may be more uniformly distributed and the propagation and linkage of macrocracks may be less important. Detailed models are reviewed and discussed by Evans (1984), Riedel (1987), and Cadek (1988).

## 12.10   FRACTURE AT HETEROPHASE INTERFACES

Fracture at heterophase interfaces will occur by the same basic mechanisms as those discussed above for fracture at homophase interfaces (grain boundaries). However, additional complexities will be present because of the different physical properties of the two phases which adjoin the interface. A vast amount of research has been carried out on the fracture of heterophase interfaces because of its important influence on the mechanical properties of heterophase alloys, composites, coated materials and materials bonded by an intermediate layer of a second phase. Since it will be impossible to discuss

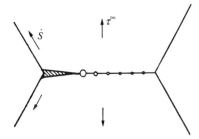

**Fig. 12.57** Formation of wedge crack by growth and coalescence of cavities under the influence of high stresses resulting from boundary sliding at a triple junction. (From Nix, *et al.* (1977).)

**Fig. 12.58** Multiple macrocracking in polycrystalline Fe–Ni–Cr alloy at 700 °C. Applied stress = 78.5 MPa. (From Söderberg (1975).)

all of this work within the confines of this section, we shall limit our discussion mainly to a number of the more important new features of fracture which appear.

Consider first interface fracture by the propagation of cleavage cracks under plane strain conditions where, as a consequence, only mode I and II crack components can be present. If the adjoining black and white crystals are isotropic with shear moduli $\mu^b$ and $\mu^w$ and Poisson ratios $\nu^b$ and $\nu^w$, the elastic field of an interface crack is governed (see Evans *et al.* 1990) by the material parameters

$$\left. \alpha = \frac{\mu^b(1-\nu^w) - \mu^w(1-\nu^b)}{\mu^b(1-\nu^w) + \mu^w(1-\nu^b)}; \quad \beta = \frac{\mu^b(1-2\nu^w) + \mu^w(1-2\nu^b)}{2[\mu^b(1-\nu^w) + \mu^w(1-\nu^b)]} \atop \varepsilon = (\pi/2)\ln[(1-\beta)/(1+\beta)] \right\} \quad (12.139a,b,c)$$

Unfortunately, when $\beta \neq 0$, the crack tip stress and displacement fields oscillate (Rice 1988). These oscillations cause contact between the crack surfaces and complicate the problem of obtaining realistic solutions. Fortunately, for many heterophase systems $\beta$ is small, and it appears that setting $\beta = 0$ can be used as an acceptable approximation for

*Mechanical properties of interfaces*

dealing with the main features of many cracks (He and Hutchinson 1989b). With this approximation, stress singularities of the form $r^{-\frac{1}{2}}$ are found at the crack tip as were also found previously for cracks in homopbase interfaces (e.g., eqn 12.116)). Also, as might be expected the condition for crack propagation is again determined by the strain energy release rate, and the toughness, $G$.

Differences between the properties of the phases adjoining the interface can have a strong influence on the trajectories of cracks at heterophase interfaces. The tendency of a propagating crack in a heterophase interface to remain in the interface or to kink out of it depends (He and Hutchinson 1989b) upon differences in the elastic properties of the two adjoining phases, the relative amounts of $K^{I}$ and $K^{II}$ loading, and the relative toughnesses of the interface and the adjoining phases. In this respect it is interesting to note that $K^{II}$ for a crack in a heterophase interface is often non-zero even when the external loading is normal to the interface because of the compatibility stresses due to the differing elastic properties of the adjoining phases. In general, a crack will tend to follow a path corresponding to the maximum crack energy release rate. However, it will also tend to follow a path of minimum toughness. He and Hutchinson (1989b) have calculated crack energy release rates for cracks which remain in the interface and which also kink out of the interface under a variety of conditions. If the toughnesses of the interface and adjoining phases are known, predictions about the possible crack path can be made. A large number of situations in which the crack is predicted to remain in the interface or else kink out of it are discussed by He and Hutchinson (1989b). Roughly speaking, the energy release rate is enhanced if the crack kinks into the more compliant phase and is diminished if it kinks into the stiffer phase. The crack will therefore have a tendency to be trapped in the interface if the compliant phase is tougher than either the interface or the stiff phase, and the stiff material is at least as tough as the interface.

Some specific results (Evans *et al.* 1990) are shown in Fig. 12.59 for crack trajectories in a metal/ceramic bicrystal for which $G^{M} \gg G^{C} > G^{B}$, where M, C, and B refer to the metal, ceramic, and boundary, respectively. The high toughness of the metal precludes any kinking into the metal, and the crack therefore either kinks into the ceramic or remains trapped in the interface. For positive values of $\psi = \tan^{-1}(K^{II}/K^{I})$, the crack kinks into the ceramic when $G^{B}/G^{C}$ exceeds the values shown, while for negative $\psi$, the crack always tends to enter the metal. But, since this is impossible, it remains at the interface, where, in the example given, it detaches ceramic 'chips' at flaws which pre-exist in the ceramic at the interface as illustrated.

The problem of determining the trajectory of a crack approaching a heterophase interface, i.e. deciding whether it will penetrate the interface or be deflected into the interface, has also been analysed by He and Hutchinson (1989a). The results depend upon the same physical parameters which determined the crack trajectory in our above discussion. In order to solve the penetration versus deflection problem, the crack energy release rates for the deflected crack $G_{\varepsilon}^{D}$, and penetrating crack, $G_{\varepsilon}^{P}$, are calculated. The impinging crack is then likely to be deflected into the interface if

$$G_{\varepsilon}^{D}/G_{\varepsilon}^{P} > G^{I}/G^{P}, \tag{12.140}$$

where $G^{P}$ is the toughness of the phase into which the crack would penetrate. Conversely, the crack will tend to penetrate the interface when the inequality is reversed. Detailed results for a variety of situations are given by He and Hutchinson (1989a).

A relatively large number of studies has been carried out of cleavage type cracking at (or near) different types of metal/ceramic interfaces. Interesting issues which are explored include, for example, decohesion of thin metal films from ceramic substrates (Hu *et al.*

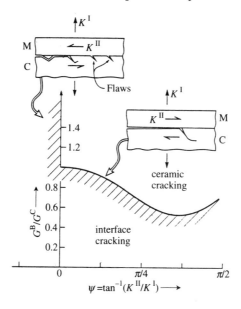

**Fig. 12.59** Crack trajectories in a metal/ceramic bicrystal for which $G^M \gg G^C > G^B$, where M, C, and B refer to the metal, ceramic, and boundary, respectively. The crack either kinks into the ceramic or remains trapped in the boundary depending upon the magnitude of $G^B/G^C$ and the relative amounts of mode I and mode II loading as measured by the angle $\psi$. $K^I$ and $K^{II}$ are the corresponding stress intensity factors. (From Evans *et al.* (1990).)

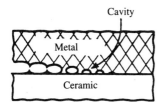

**Fig. 12.60** Fracture at metal/ceramic interface by growth and coalescence of cavities at the interface by plastic flow due to dislocation glide in the metal. (From Evans and Dalgleish (1992).)

1988, Evans *et al.* 1990); debonding at fibre/matrix interfaces in brittle ceramic matrix composites (Evans *et al.* 1960); fracture at thin metal film bonding layers between ceramics (Dalgleish *et al.* 1988, 1989, He and Evans 1991), effects of interface non-planarity on mixed mode interface fracture resistance (Evans and Hutchinson 1989); effect of plasticity on the metal side on the interface toughness (Reimanis *et al.* 1990); and emission of dislocations from cracks on the metal side (Beltz and Rice 1992, Beltz and Wang 1992).

We conclude by noting that other types of fractures, similar to those described previously at homophase interfaces, may occur at heterophase interfaces. Figure 12.60 illustrates fracture at a metal/ceramic interface by the growth and coalescence of cavities at the interface by plastic flow due to dislocation glide (Evans and Dalgleish 1992). In this case all of the plastic flow occurs in the ductile metal side. At high temperatures we also expect fracture by the growth and coalescence of cavities by diffusion, power-law creep and boundary sliding mechanisms. Again, these processes will occur asymmetrically because of differences in the properties of the two phases adjoining the interface.

# REFERENCES

Argon, A., Im, J., and Safoglu, R. (1975). *Metall. Trans.*, **6A**, 825.

Argon, A., Chen, I.-W., and Lau, C.-W. (1980). In *Creep-fracture-environment interactions* (eds R. M. Pelloux and N. S. Stoloff) p. 46. Metallurgy Society of AIME, Warrendale, Pennsylvania.

Arzt, E. Ashby, M. F., and Verrall, R. A. (1983). *Acta Metall.*, **31**, 1977.

Ashby, M. F. (1970). *Phil. Mag.*, **21**, 399.

Ashby, M. F. (1972*a*). *Acta Metall.*, **20**, 887.

Ashby, M. F. (1972*b*). *Surf. Sci.*, **31**, 498.

Ashby, M. F. and Verrall, R. A. (1973). *Acta Metall.*, **21**, 149.

Ashby, M. F., Gandhi, C., and Taplin, D. M. R. (1979). *Acta Metall.*, **27**, 699.

Bacmann, J. J., Gay, M. O., and deTournemeine, R. (1982). *Scripta Met.*, **16**, 353.

Baillin, X., Pelissier, J., Bacmann, J. J., Jacques, A., and George, A. (1987). *Phil. Mag.*, **A55**, 143.

Baillin, X., Pelissier, J., Jacques, A., and George, A. (1987). *Phil. Mag.*, **A61**, 329.

Balluffi, R. W. (1976). *Scripta Metall.*, **10**, 913.

Balluffi, R. W., Seigle, L. (1955). *Acta Metall.*, **3**, 170.

Balluffi, R. W., Komen, Y., and Schober, T. (1972). *Surf. Sci.*, **31**, 68.

Barnett, D. M. and Lothe, J. (1974). *J. Phys. F: Metal Phys.*, **4**, 1618.

Baro, G., Gleiter, H., and Hornbogen, E. (1968/69). *Mater. Sci. Eng.*, **3**, 92.

Beeré, W. (1976). *Met. Sci. J.*, **10**, 133.

Beeré, W. (1978). *Phil. Trans. Roy. Soc. London A*, **288**, 177.

Belov, A. Y., Chamrov, V. A., Indenbom, V. L., and Lothe, J. (1983). *Phys. Stat. Sol.* (*b*), **119**, 565.

Beltz, G. E. and Rice, J. R. (1992). *Acta Metall. Mater.*, **40**, S321.

Beltz, G. E. and Wang, J.-S. (1992). *Acta Metall. Mater.*, **40**, 1675.

Berry, B. S. and Prichet, W. C. (1981). *J. de Physique*, **42**, C5-1111.

Birnbaum, H. K. (1984). *J. Less Common Mets.*, **104**, 31.

Bohn, H. G., Prieler, M., Su, C. M., Trinkaus, H., and Schilling, W. (1994). To be published.

Bollmann, W., Michaut, B., and Sainfort, G. (1972). *Phys. Stat. Sol.* (*a*), **13**, 637.

Bonnet, R. (1987). In *Fundamentals of diffusion bonding* (ed. Y. Ishida) p. 329. Elsevier, Amsterdam.

Brede, M. and Haasen, P. (1987). In *Chemistry and physics of fracture* (eds. R. M. Lantanision and R. H. Jones) p. 449. Martinus Nijhoff, Dordrecht.

Brown, L. M. and Embury, J. D. (1973). In *Proceedings of the third international conference on the strength of metals and alloys*, Institute of Metals, London.

Brown, L. M. and Stobbs, W. M. (1971). *Phil. Mag.*, **23**, 1201.

Brown, L. M. and Stobbs, W. M. (1976). *Phil. Mag.*, **34**, 351.

Burke, M. A. and Nix, W. D. (1975). *Acta Metall.*, **23**, 793.

Burton, B. (1977). *Diffusion creep of polycrystalline materials*. Trans. Tech., Switzerland.

Cadek, J. (1988). *Creep in metallic materials*. Elsevier, Amsterdam.

Cahn, R. W. (1949). *J. Inst. Metals*, **76**, 121.

Celinski, Z. and Kurzydkowski, K. J. (1982). *Res. Mech.*, **5**, 89.

Chawla, K. K. (1987). *Composite materials*. Springer-Verlag, New York.

Chen, I.-W. (1982). *Acta Metall.*, **30**, 1655.

Chen, I.-W. (1983). *Metall. Trans.*, **14A**, 2289.

Chiao, Y.-H. and Clarke, D. R. (1989). *Acta Metall.*, **37**, 203.

Chou, T. W. and Hirth, J. P. (1970). *J. Composite Mats.*, **4**, 102.

Christian, J. W. (1975). *The theory of transformations in metals and alloys*, Part 1. Pergamon Press, Oxford.

Chuang, T.-J., Kagawa, K. I., Rice, J. R., and Sills, L. B. (1979). *Acta Metall.*, **27**, 265.

Clarke, D. R. (1980). *Acta Metall.*, **28**, 913.

Clark, W. A. T, and Smith, D. A. (1979). *J. Mat. Sci.*, **14**, 776.

Clark, W. A. T., Wagoner, R. H., Shen, Z. Y., Lee, T. C., Robertson, I. M., and Birnbaum, H.K.

(1992). *Scripta Metall. Mater.*, **26**, 203.

Clark, W. A. T., Wise, C. E., Shen, Z. Y., and Wagoner, R. H. (1989). *Ultramiscopy*, **30**, 76.

Coble, R. L. (1963). *J. Appl. Phys.*, **34**, 1679.

Cocks, A. C. F. and Ashby, M. F. (1982). *Prog. in Mats. Sci.*, **27**, 189.

Cosandey, F., Li, S., Cao, B., Shaller, R., and Benoit, W. (1991). *Mat. Res. Soc. Symp. Proc.*, **229**. 255.

Cottrell, A. H. (1989). *Mats. Sci. and Tech.*, **5**, 1165.

Cottrell, A. H. (1990). *Mats. Sci. and Tech.*, **6**, 121.

Crossman, F. W. and Ashby, M. F. (1975). *Acta Metall.*, **23**, 425.

Dagleish, B. J., Lee, M. C., and Evans, A. G. (1988). *Acta Metall.*, **36**, 2029.

Dagleish, B. J., Trumble, K. P., and Evans, A. G. (1989). *Acta Metall.*, **37**, 1923.

Darby, T. P., Schindler, R., and Balluffi, R. W. (1978). *Phil. Mag.*, **A37**, 245.

Davidge, R. W. (1981). *Acta Metall.*, **29**, 1695.

Deng, D., Argon, A. S., and Yip, S. (1989). *Phil. Trans. Roy. Soc. London A*, **329**, 549, 575, 595, 613.

Dingley, D. J. and Pond, R. C. (1979). *Acta Met.*, **27**, 667.

Eberhardt, A. and Baudelet, B. (1980). *Phil Mag A*, **41**, 843.

Elkajbaji, M. and Thibault-Desseaux, J. (1988). *Phil. Mag A*, **58**, 325.

Eshelby, J. D. (1957). *Proc. Roy. Soc A*, **241**, 376.

Evans, A. G. (1978). *Acta Metall.*, **26**, 1845.

Evans, H. E. (1984). *Mechanisms of creep fracture*. Elsevier, London.

Evans, A. G. and Dalgleish, B. J. (1992). *Acta Metall. Mater.*, **40**, S295.

Evans, A. G. and Hutchinson, J. W. (1989). *Acta Metall.*, **37**, 909.

Evans, R. W. and Wilshire, B. (1985). *Creep of metals and alloys*. Institute of Metals, London.

Evans, A. G., Rühle, M., Dalgleish, B. J., and Charalambides, P. G. (1990). *Mats. Sci. Eng A.*, **126**, 53.

Fu, Y. and Evans, A. G. (1985). *Acta Metall.*, **33**, 1515.

Fukutomi, H. and Kamijo, T. (1985). *Scripta Met.*, **19**, 195.

Fuller, E. R. and Thomson, R. (1980). *J. Mats. Sci.*, **15**, 1027.

Gandhi, C. and Ashby, M. F. (1979). *Acta Metall.*, **27**, 1565.

Gemperlova, J., Paidar, V. and Kroupa, F. (1989). *Czech. J. Phys B*, **39**, 427.

Gemperlova, J. and Saxl, I. (1968). *Czech. J. Phys B.*, **18**, 1085.

George, E. P., Li, P. L., and Pope, D. P. (1987). *Acta Metall.*, **35**, 2471; 2487.

Ghahremani, F., Hutchison, J. W., and Tvergaard, V. (1990). *J. Am. Ceram. Soc.*, **73**, 1548.

Gilman, J. J. (1955). *Acta Met.*, **3**, 277.

Gilman, J. J. (1958). *Trans. A.I.M.E.*, **212**, 783.

Gittus, J. H. (1977). *J. Eng. Mat. Tech.*, **99**, 244.

Gleiter, H., Hornbogen, E., and Baro, G. (1968). *Acta Metal.*, **16**, 1053.

Goods, S. H. and Brown, L. M. (1979). *Acta Metall.*, **27**, 1.

Grabski, M. W. (1985). *J. de Physique*, **46**, Colloque C4, Supple. au n°4, C4-569.

Grabski, M. W. and Korski, R. (1970). *Phil. Mag.*, **22**, 707.

Griffith, A. A. (1920). *Phil. Trans. Roy. Soc. (London) A*, **221**, 163.

Hahn, G. T., Averbach, B. L., Owen, W. S., and Cohen, M. (1959). In *Fracture* (eds B. L. Averbach, D. K. Felbeck, G. T. Hahn and D. A. Thomas) p. 91. Wiley, New York.

Hansen, N. (1992). *Scripta Metall. Mater.*, **27**, 947.

Hart, E. W. (1967). *Acta Metall.*, **15**, 351.

Hashimoto, K. and Margolin, H. (1983). *Acta Metall.*, **31**, 773.

Hashimoto, S., Fujii, T. K., Fujii, H., and Miura, S. (1986). *JIMIS 4, Grain boundary structure and related phenomena*, Supple. to *Trans. Japan Inst. Mets.*, **27**, p. 921.

Hashimoto, S., Moriwaki, F., Mimaki, T. and Miura, S. (1991). In *Superplasticity in advanced materials* (eds S. Hori, M. Tokizane and N. Furushiro) p. 23. Japan Soc. for Research on Superplasticity.

Hauser, J. J. and Chalmers, B. (1961). *Acta Metall.*, **9**. 802.

Havner, K. S. (1992). *Finite plastic deformation of crystalline solids*. Cambridge University Press, Cambridge.

Hazzledine, P. M. and Schneibel, J. H. (1993). *Acta Metall. Mater.*, **41**, 1253.

He, M. Y. and Evans, A. G. (1991). *Acta Matall. Mater.*, **39**, 1587.

He, M. Y. and Hutchinson, J. W. (1989*a*). *Int. J. Solids and Structures*, **25**, 1053.

He, M. Y. and Hutchinson, J. W. (1989*b*). *J. Appl. Mech.*, **56**, 270.

Head, A. K. (1953). *Proc. Phys. Soc.*, **66B**, 793.

Herring, C. (1950). *J. Appl. Phys.*, **21**, 437.

Herrmann, G., Baro, G., and Gleiter, H. (1976). *Acta Metall.*, **24**, 353.

Hirth, J. P. (1972). *Met. Trans.*, **3**, 3047.

Hirth, J. P. (1987). In *Chemistry and physics of fracture* (eds R. M. Lantanision and R. H. Jones) p. 538. Martinus Nijhoff, Dordrecht.

Hirth, J. P. and Lothe, J. (1982). *Theory of dislocations*. Wiley, New York.

Hirth, J. P. and Rice, J. R. (1980). *Metall. Trans.*, **11A**, 1501.

Hondros, E. D. and Seah, M. P. (1983). In *Physical metallurgy* (eds R. W. Cahn and P. Haasen) p. 855. North-Holland, Amsterdam.

Hook, R. E. and Hirth, J. P. (1967). *Acta Metall.*, **15**, 535, 1099.

Horiuchi, R., Fukutomi, H., and Takahashi, T. (1987). In *Fundamentals of diffusion bonding* (ed. Y. Ishida) p. 347. Elsevier, Amsterdam.

Hu, M. S., Thouless, M. D., and Evans, A. G. (1988). *Acta Metall.*, **36**, 1301.

Hull, D. (1981). *An introduction to composite materials*. Cambridge University Press, Cambridge.

Jacques, A., George, A., Baillin, X. and Bacmann, J. J. (1987). *Phil. Mag.*, **A55**, 165.

Jacques, A., Michaud, H.-M., Baillin, X., and George, A. (1990). *J. de Physique* **51**, Colloq. Cl, supple. au n°1, C1-531.

Johannesson, T. and Tholen, A. (1972). *Met. Sci. J.*, **6**, 189.

Jokl, M. L., Vitek, V., and McMahon, C. J. (1980). *Acta Metall.*, **28**, 1479.

Kasul, D. B. and Heldt, L. A. (1989). *Bull. of Mats. Res. Soc.*, **14**, No. 8, 37.

Kato, M. and Mori, T. (1993). *Phil. Mag. A.*, **68**, 939.

Ke, T. S. (1947). *Phys. Rev.*, **71**, 533.

Ke, T. S. (1949). *J. Appl. Phys.*, **20**, 274.

Ke, T. S. (1986). *JIMIS 4 Grain boundary structure and related phenomena*. Supple. to *Trans. Jap. Inst. Mets.*, **27**, 679.

Ke, T. S. (1990). *Scripta Metall. Mater.*, **24**, 347.

Ke, T. S., Cui, P., Yan, S. C. and Huang, Q. (1984). *Phys. Stat. Sol. (a)*, **86**, 593.

Kegg, G. R., Horton, C. A. P., and Silcock, J. M. (1973). *Phil. Mag.*, **27**, 1041.

Kelly, A. and Davies, G. J. (1965). *Met. Rev.*, **10**, 1.

Kelly, A. and Macmillan, N. H. (1986). *Strong solids*, 3rd edn. Clarendon Press, Oxford.

Khalfallah, O., Condat, M. Priester, L., and Kirchner, H. O. K. (1990). *Phil. Mag.*, *A.* **61**, 291.

King, A. H. and Smith, D. A. (1980). *Acta Cryst. A*, **36**, 335.

Kitagawa, K., Asada, H. Monzen, R., and Kichuki, M. (1986). *JIMIS 4, Grain boundary structure and related phenomena*, Supple. to *Trans. Japan Inst. Mets.*, **27**, 827.

Knott, J. F. (1973). *Fundamentals of fracture mechanics*. Butterworth, London.

Kokawa, H., Watanabe, T., and Karashima, S. (1981). *Phil. Mag.*, *A*, **44**, 1239.

Kurzydkowski, K., Celinski, Z., and Grabski, M. W. (1980). *Res. Mech.*, **1**, 283.

Kurzydlowski, K. J., Varin, R. A., and Zielinski, W. (1984). *Acta Met.*, **32**, 71.

Lagarde, P. and Biscondi, M. (1974). *Canad. Met. Quart.*, **13**, 245.

Lagarde, P. and Biscondi, M. (1975). *J. de Physique*, **36**, Colloque C4, supple. au n°10, C4-297.

Lantanision, R. M. and Jones, R. H. (eds.) (1987). *Chemistry and physics of fracture*. Martinus Nijhoff, Dordrecht.

Lee, T. C., Robertson, I. M., and Birnbaum, H. K. (1989). *Acta Metall.*, **37**, 407.

Lee, T. C., Robertson, I. M., and Birnbaum, H. K. (1990*a*). *Met. Trans.*, **21A**, 2437.

Lee, T. C., Robertson, I. M., and Birnbaum, H. K. (1990*b*). *Phil. Mag A*, **62**, 131.

Lifshitz, L. M. (1963). *Soviet Phys. JETP*, **17**, 909.

Lim, L. C. (1987*a*). *Acta Metall.*, **35**, 163.

Lim, L. C. (1987*b*). *Acta Metall.*, **35**, 1663.

Lim, L. C. (1989). *Acta Metall.*, **37**, 969.

Lim, L. C. and Raj, R. (1985*a*). *Acta Metall.*, **33**, 1577.

Lim, L. C. and Raj, R. (1985*b*). *Acta Metall.*, **33**, 2205.

Lin, I.-H. and Thomson, R. (1986). *Acta Metall.*, **34**, 187.

Liu, J. S. and Balluffi, R. W. (1984). *Mat. Res. Soc. Symp. Proc.*, **25**, 261.

Livingston, J. D. and Chalmers, B. (1957). *Acta Metall.*, **5**, 322.

Lojkowski, W. (1991). *Acta Metall. Mater.*, **39**, 1891.

Lojkowski, W. and Grabski, M. W. (1981). In *Deformation of polycrystals, mechanisms and microstructures* (eds N. Hansen, A Horswell, T. Lefers, and H. Lilholt) p. 329. Riso National Lab., Roskilde, Denmark.

Lojkowski, W., Wyrzykowski, J. W., Kwiecinski, J., Beke, D. L., and Godeny, I. (1989). *Defect and Diff. Forum*, **66-69**, 701.

Lojkowski, W., Wyzrzykowski, J., and Kwiecinski, J. (1990). *J. de Physique*, **51**, Colloque Cl, Supple. au n°l, C1-239.

Margolin, H. and Stanescu, M. S. (1975). *Acta Metall.*, **23**, 1411.

Martinez-Hernandez, M., Kirchner, H. O. K., Korner, A., George, A., and Michel, J. P. (1987). *Phil. Mag A.*, **56**, 641.

Mascanzoni, A. and Buzzichelli, G. (1970). *Phil. Mag.*, **22**, 857.

Mason, D. D. (1979). *Phil. Mag. A*, **39**, 455.

McClintock, F. A. (1968). *J. Appl. Mech.*, **35**, 363.

McClintock, F. A. and Argon, A. S. (eds) (1966). *Mechanical behavior of materials*. Addison-Wesley, Reading, Massachusetts.

McMahon, C. J. (1989). *Mats. Sci. Forum*, **46**, 61.

Meyers, M. A. and Ashworth, E. (1982). *Phil. Mag. A*, **46**, 737.

Meyers, M. A. and Chawla, K. K. (1984). *Mechanical metallurgy*. Prentice-Hall, Englewood Cliffs, New Jersey.

Mori, T. and Tangri, K. (1979). *Met. Trans.*, **10A**, 733.

Mortensen, A. (1988). In *Mechanical and physical behavior of metallic and ceramic composites* (eds S. I. Andersen, H. Lilholt, and O. B. Pedersen). Riso National Lab., Roskilde, Denmark.

Mullendore, A. W. and Grant, N. J. (1963). *Trans. A.I.M.E.*, **227**, 319.

Murr, L. E. (1975). *Met. Trans.*, **6A**, 505.

Murr, L. E. (1981). *Mats. Sci. Eng.*, **51**, 71.

Murr, L. E. and Wang, S.-H. (1982). *Res Mech.*, **4**, 237.

Nabarro, F. R. N. (1948). In *Report of a conference on the strength of solids*, p. 75. The Physical Society, London.

Nix, W. D. (1983). *Scripta Metall.*, **17**, 1.

Nix, W. D., Matlock, D. K., and Dimelfi, R. J. (1977). *Acta Metall.*, **25**, 495.

Nowick, A. S. and Berry, B. S. (1972). *Anelastic relaxation in crystalline solids*. Academic Press, New York.

Nye, J. F. (1953). *Acta Met.*, **1**, 153.

Ohr, S. M. (1986). *Scripta Metall.* **20**, 1465.

Orlov, L. G. (1968). *Soviet Phys.-Solid State*, **9**, 1836.

Paidar, V., Pal-Val, P. P., and Kadeckova, S. (1986). *Acta Metall.*, **34**, 2277.

Pilling, J. and Ridley, N. (1989). *Superplasticity in crystalline solids*. The Institute of Metals, London.

Pond, R. C. (1977). *Proc. Roy. Soc.* (*London*) *A*, **357**, 471.

Price, C. W. and Hirth, J. P. (1972). *Mats. Sci. Eng.*, **9**, 15.

Prieler, M., Bohn, H. G., Schilling, W., and Trinkaus, H. (1993). *Mat. Res. Soc. Symp. Proc.*, **308**, 305.

Pugh, S. F. (1991). *An introduction to grain boundary fracture in metals*, Institute of Metals, London.

Pumphrey, P. H. and Gleiter, H. (1974). *Phil. Mag.*, **30**, 593.

Pond, R. C. and Smith, D. A. (1977). *Phil. Mag.*, **36**, 353.

Putaux, J. L. and Thibault-Desseaux, J. (1990). *J. de Physique*, **51**, Colloque Cl, Supple. au n°l, C1-323.

Raj, R. and Ashby, M. F. (1971). *Met. Trans.*, **2**, 1113.

Raj, R. and Ashby, M. F. (1972). *Met. Trans.*, **3**, 1937.

Raj, R. and Ashby, M. F. (1975). *Acta Metall.*, **23**, 653.

Raj, R. and Lange, F. F. (1985). *Acta Metall.*, **33**, 699.

Reading, K. and Smith, D. A. (1985). *Phil. Mag A.*, **51**, 71.

Reid, C. N. (1973). *Deformation geometry for materials scientists*. Pergamon Press, Oxford.

Reimanis, I. E., Dalgleish, B. J., Brahy, M., Rühle, M., and Evans, A. G. (1990). *Acta Metall. Mater.*, **38**, 2645.

Rey, C. and Zaoui, A. (1980). *Acta Metall.*, **28**, 687.

Rey, C. and Zaoui, A. (1982). *Acta Metall.*, **30**, 523.

Rey, C., Mussot, P., and Zaoui, A. (1986). *JIMIS 4, Grain boundary structure and related phenomena*, Supple. to *Trans. Jap. Inst. Mets.*, **27**, 867.

Rice, J. R. (1976). In *Effect of hydrogen on behavior of materials* (eds A. W. Thompson and I. M. Bernstein), p. 455. Metals Society of AIME, New York.

Rice, J. R. (1987). In *Chemistry and physics of fracture* (eds R. M. Lantanision and R. H. Jones) p. 23. Martinus Nijhoff, Dordrecht.

Rice, J. R. (1988). *J. Appl. Mech.*, **55**, 98.

Rice, J. R. (1992). *J. Mech. Phys. Solids*, **40**, 239.

Rice, J. R. and Thomson, R. M. (1974). *Phil. Mag.*, **29**, 73.

Rice, J. R. and Tracey, D. M. (1969). *J. Mech. Phys. Solids*, **17**, 201.

Riedel, H. (1987). *Fracture at high temperatures*. Springer Verlag, Berlin.

Roy, A., Erb, U. and Gleiter, H. (1982). *Acta Metall.*, **30**, 1847.

Russell, J. D. and Winter, A. T. (1984). *Scripta Metall.*, **19**, 575.

Sautter, H., Baro, G., and Gleiter, H. (1977). *Acta Metall.*, **25**, 467.

Schneibel, J. H. and Petersen, G. F. (1985). *Acta Metall.*, **33**, 437.

Schneibel, J. H. and Petersen, G. F. (1986). *JIMIS 4, Grain boundary structure and related phenomena*, Supple. to *Trans. Jap. Inst. Mets.*, **27**, 859.

Schober, T. and Balluffi, R. W. (1971). *Phil. Mag.*, **24**, 165.

Seah, M. P. (1980). *Acta Metall.*, **28**, 955.

Shen, Z., Wagoner, R. H. and Clark, W. A. T. (1986). *Scripta Metall.*, **20**, 921.

Shen, Z., Wagoner, R. H. and Clark, W. A. T. (1988). *Acta Metall.* **36**, 3231.

Sittner, P. and Paidar, V. (1989). *Acta Metall.*, **37**, 1717.

Söderberg, R. (1975). *Met. Sci.*, **9**, 275.

Speight, M. V. (1976). *Acta Metall.*, **24**, 725.

Spingarn, J. R. and Nix, W. D. (1978). *Acta Metall.*, **26**, 1389.

Srolovitz, D. J., Yang, W. H., Najafabadi, R., Wang, H. Y., and LeSar, R. (1992). In *Materials interfaces* (eds D. Wolf and S. Yip) p. 691. Chapman and Hall, London.

Suery, M. and Baudelet, B. (1981). *Res. Mech.*, **2**, 163.

Sun, C. P. and Balluffi, R. W. (1982). *Phil. Mag A.*, **46**, 49, 63.

Suzuki, H., Takasugi, T., and Izumi, O. (1982). *Acta Metall.*, **30**, 1647.

Swiatnicki, W., Lojkowski, W., and Grabski, M. W. (1986). *Acta Met.*, **34**, 599.

Takayuki, T. and Izumi, O. (1979). *Acta Metall.*, **28**, 465.

Thibault, J., Putaux, J. L., Michaud, H. M., Baillin, X., Jacques, A., and George, A (1991). In *Microscopy of semiconducting materials* 1991 (eds A. G. Cullis and N. J. Long) p. 105, Inst. Phys. Conf. Ser. No. 117. Institute of Physics, Bristol.

Thibault-Desseaux, J., Putaux, J. L., Jacques, A., and Elkajbaji, M. (1988). *Mat. Res. Soc. Symp. Proc.*, **122**, 293.

Thomason, P. F. (1968). *J. Inst. Mets.*, **96**, 360.

Thomason, P. F. (1971). *Int. J. Fract. Mech.*, **7**, 409.

Thompson, A. W. (1987). In *Chemistry and physics of fracture* (eds R. M. Lantanision and R. H. Jones) p. 129. Martinus Nijhoff, Dordrecht.

Thomson, R. M. (1980). *J. Mats. Sci.*, **15**, 1014.

Thomson, R. M. (1983). In *Physical metallurgy* (eds R. W. Cahn and P. Haasen) p. 1487. North-Holland, Amsterdam.

Thomson, R. M. (1986). In *Solid state physics* (eds H. Ehrenreich and D. Turnbull) Vol. 39, p. 1. Academic Press, New York.

Tvergaard, V. and Hutchinson, J. W. (1988). *J. Am. Ceram. Soc.*, **71**, 157.

Valiev, R. Z. and Gertsman, V. Yu. (1990). *J. de Physique.*, **51**, Colloque C1, Supple au n°1, C1-679.

Valiev, R. Z. and Khairullin, V. G. (1990). *Colloque de Physique*, Colloque C1, supple. au n°1, **51**, C1-685.

Valiev, R. Z., Kaibyshev, O. A., Astanin, V. V., and Emaletdinov, A. K. (1983). *Phys. Stat Sol. (a)*, **78**, 439.

Valiev, R. Z., Gertsman, V. Yu, and Kaibyshev, O. A. (1986). *Phys. Stat. Sol. (a)*, **97**, 11.

Valiev, R. Z., Khairullin, V. G., and Sheikh-Ali, A. D. (1991). In *Structure and property relationships for interfaces* (eds J. L. Walter, A. H. King, and K. Tangri) p. 309. ASM International, Metals Park, Ohio.

Varin, R. A., Kurzydlowski, K. J. and Tangri, K. (1987). *Mat. Sci. Eng.*, **85**, 115.

Wang, J. S. and Anderson, P. M. (1991) *Acta Metall.*, **39**, 779.

Watanabe, T. (1983). *Metall. Trans.*, **14A**, 531.

Watanabe, T. (1984). *Res Mech.*, **11**, 47.

Watanabe, T. (1989). *Mats. Sci. Forum*, **46**, 25.

Watanabe, T. and Davies, P. W. (1978). *Phil. Mag A*, **37**, 649.

Watanabe, T., Kimura, S.-I., and Karashima, S. (1984). *Phil Mag A*, **49**, 845.

Weertman, J. (1955). *J. Appl. Phys.*, **26**, 1213.

Whelan, M. J., Hirsch, P. B., Horne, R. W., and Bollmann, W. (1957). *Proc. Roy. Soc A.*, **240**, 524.

Worthington, P. J. and Smith, E. (1964). *Acta Metall.*, **12**, 1277.

Xinggang, J., Jianzhong, C., and Longxiang, M. (1993). *Acta Metall. Mater.*, **41**, 539.

Young, F. W. (1958). *J. Appl. Phys.*, **29**, 760.

Zener, C. (1941). *Phys. Rev.*, **60**, 906.

Zhang, H., King, A. H. and Thomson, R. (1991). *J. Mater. Res.*, **6**, 314.

Zhou, S. J. and Thomson, R. (1991a). *J. Mater. Res.*, **6**, 639.

Zhou, S. J. and Thomson, R. (1991b). *J. Mater. Res.*, **6**, 1763.

# Index